# How to get the most from this textbook:

## Read the Book

| FEATURE | DESCRIPTION | BENEFIT | PAGE |
| --- | --- | --- | --- |
| Chapter Opener | Each chapter opener features motivating real-world applications of mathematics that are relevant and tied to specific material within the chapters. | These openers enable you to explore realistic situations using the mathematics learned from this textbook. | 169 |
| Learning Objectives | Each section begins with plainly stated, numbered objectives that are clearly labeled as they are carried throughout the exposition. | See exactly what is covered in each section and which skills you are expected to master. | 170 |
| Emphasis on Problem Solving | A six-step problem-solving method (Read, Assign a variable, Write an equation, Solve, State the answer, Check) is integrated through the text. | This method reinforces the problem-solving process for examples and exercises throughout the text. | 62 |
| Examples | Detailed examples include step-by-step solutions and extensive side comments to explain what is being done in each step. | Examples provide you with an explanation and model for solving similar problems in the exercises. | 245 |
| Function Boxes | These boxes offer a comprehensive, visual introduction to each circular and inverse circular function, and they serve as an excellent resource for reference and review. | The Function Boxes allow you to see a lot of information about one function in one place, helping you to make connections between different aspects of these key concepts. | 771 and 853 |

## Work the Problems

| FEATURE | DESCRIPTION | BENEFIT | PAGE |
| --- | --- | --- | --- |
| Now Try Exercises | Each example concludes with a reference to one or more parallel odd-numbered exercises from the corresponding exercise set. | These exercises actively engage you in the learning process, asking you to immediately apply and reinforce the skills and concepts presented in the examples. | 245 and 248 |
| Varied Exercise Sets | Exercise types include free-response, multiple choice, matching, true/false, writing, and estimation problems. | Exercises provide you with the opportunity to practice, apply, connect, and extend the concepts and skills you are learning. | 246 |
| Concept Check Exercises | These particular problems facilitate mathematical thinking and conceptual understanding. | Strategically placed, these exercises give you a chance to check your comprehension of big ideas. | 246 |

## Review

| FEATURE | DESCRIPTION | BENEFIT | PAGE |
|---|---|---|---|
| Chapter Summary | This feature includes Key Terms, New Symbols, and Quick Review of each section's content with additional examples. | The summary highlights the key concepts you need to know from the chapter. | 279 |
| Chapter Review Exercises | Each chapter provides opportunities to practice what you have learned and check your comprehension. | Use these problems to review all the concepts that you learned in the chapter and assess your knowledge. | 281 |
| Chapter Test | These tests provide you with an opportunity to simulate an in-class exam. | In this final stage of test prep, you can complete a mock test to really evaluate your understanding. | 283 |

## Optional Tools

| TOOL | DESCRIPTION | BENEFIT |
|---|---|---|
| MathXL® Tutorials on CD | This interactive tutorial CD-ROM contains algorithmically-generated practice exercises that are correlated at the objective level to the exercises in the textbook. | This software is designed to involve and support students with guided solutions for practice problems and helpful feedback for incorrect answers. Printed summaries also help students track their progress. |
| MathXL for School | This online homework, tutorial, and assessment program contains text-specific exercises that can be automatically graded to quickly and easily assess understanding. | Students can receive immediate feedback on their assignments, get assistance on problems in a variety of ways (guided solutions and step-by-step examples) and receive a personalized study plan based on quiz and test results. |

# Algebra and Trigonometry
## for College Readiness

**MARGARET L. LIAL**
American River College

**JOHN HORNSBY**
University of New Orleans

**Addison-Wesley**

Boston   Columbus   Indianapolis   New York   San Francisco   Upper Saddle River
Amsterdam   Cape Town   Dubai   London   Madrid   Milan   Munich   Paris   Montreal   Toronto
Delhi   Mexico City   Sao Paulo   Sydney   Hong Kong   Seoul   Singapore   Taipei   Tokyo

**Executive Editor:** Anne Kelly
**Associate Project Editor:** Leah Goldberg
**Editorial Assistant:** Sarah Gibbons
**Senior Managing Editor:** Karen Wernholm
**Senior Production Project Manager:** Kathleen A. Manley
**Digital Assets Manager:** Marianne Groth
**Media Producers:** Peter Silvia and Carl Cottrell
**Software Development:** Eric Gregg (MathXL) and Marty Wright (TestGen)
**Executive Marketing Manager:** Becky Anderson
**Senior Marketing Manager:** Katherine Greig
**Marketing Assistant:** Katherine Minton
**Senior Author Support/Technology Specialist:** Joe Vetere
**Senior Prepress Supervisor:** Caroline Fell
**Senior Manufacturing Manager:** Evelyn Beaton
**Senior Media Buyer:** Ginny Michaud
**Text Design, Production Coordination, Composition, Illustrations:** Nesbitt Graphics, Inc.
**Senior Designer:** Beth Paquin
**Cover Designer:** Christina Gleason
**Cover image:** *Blazing Autumn* Copyright © Lorraine Cota Manley

For permission to use copyrighted material, grateful acknowledgment is made to the copyright holders on page xvi, which is hereby made part of the copyright page.

Many of the designations used by manufacturers and sellers to distinguish their products are claimed as trademarks. Where those designations appear in this book, and Addison-Wesley was aware of a trademark claim, the designations have been printed in initial caps or all caps.

**Library of Congress Cataloging-in-Publication Data**

Lial, Margaret L.
    Algebra and trigonometry for college readiness/Margaret L. Lial, John Hornsby.
       p. cm.
    Includes index.
    ISBN 0-13-136626-2 (student ed.)
1. Algebra—Textbooks.   2. Trigonometry—Textbooks.  I. Hornsby, E. John.  II. Title.
    QA152.3.L498 2010
    512' .13—dc22                                           2009036443

Copyright © 2011 Pearson Education, Inc.

All rights reserved. No part of this publication may be reproduced, stored in a retrieval system, or transmitted, in any form or by any means, electronic, mechanical, photocopying, recording, or otherwise, without the prior written permission of the publisher. Printed in the United States of America. For information on obtaining permission for use of material in this work, please submit a written request to Pearson Education, Inc., Rights and Contracts Department, 501 Boylston Street, Suite 900, Boston, MA 02116, fax your request to 617-671-3447, or see http://www.pearsoned.com/legal/permissions.htm

2 3 4 5 6 7 8 9 10—CRK—14 13 12 11 10

www.pearsonschool.com

ISBN-13: 978-0-13-136626-8
ISBN-10:    0-13-136626-2

# Contents

| | |
|---|---|
| Preface | ix |
| Photo Credits | xvi |

## 1 Review of the Real Number System — 1

| | | |
|---|---|---|
| 1.1 | Basic Concepts | 2 |
| 1.2 | Operations on Real Numbers | 15 |
| 1.3 | Exponents, Roots, and Order of Operations | 23 |
| 1.4 | Properties of Real Numbers | 30 |
| | **Chapter 1 Summary** | 37 |
| | **Chapter 1 Review Exercises** | 39 |
| | **Chapter 1 Test** | 41 |

## 2 Linear Equations, Inequalities, and Applications — 43

| | | |
|---|---|---|
| 2.1 | Linear Equations in One Variable | 44 |
| 2.2 | Formulas | 52 |
| 2.3 | Applications of Linear Equations | 60 |
| 2.4 | Linear Inequalities in One Variable | 72 |
| 2.5 | Set Operations and Compound Inequalities | 83 |
| 2.6 | Absolute Value Equations and Inequalities | 91 |
| | **Chapter 2 Summary** | 99 |
| | **Chapter 2 Review Exercises** | 102 |
| | **Chapter 2 Test** | 105 |

## 3 Graphs, Linear Equations, and Functions — 107

| | | |
|---|---|---|
| 3.1 | The Rectangular Coordinate System | 108 |
| 3.2 | The Slope of a Line | 117 |
| 3.3 | Linear Equations in Two Variables | 130 |
| 3.4 | Linear Inequalities in Two Variables | 142 |
| 3.5 | Introduction to Functions | 147 |
| | **Chapter 3 Summary** | 161 |
| | **Chapter 3 Review Exercises** | 163 |
| | **Chapter 3 Test** | 166 |

## 4 Systems and Matrices — 169

| | | |
|---|---|---|
| 4.1 | Systems of Linear Equations in Two Variables | 170 |
| 4.2 | Systems of Linear Equations in Three Variables | 182 |

| | | |
|---|---|---:|
| 4.3 | Solving Systems of Linear Equations by Matrix Methods | 189 |
| 4.4 | Properties of Matrices | 197 |
| 4.5 | Matrix Inverses | 207 |
| 4.6 | Determinants and Cramer's Rule | 217 |
| | **Chapter 4 Summary** | 226 |
| | **Chapter 4 Review Exercises** | 230 |
| | **Chapter 4 Test** | 232 |

## 5 Exponents, Polynomials, and Polynomial Functions — 235

| | | |
|---|---|---:|
| 5.1 | Integer Exponents and Scientific Notation | 236 |
| 5.2 | Adding and Subtracting Polynomials | 249 |
| 5.3 | Polynomial Functions, Graphs, and Composition | 255 |
| 5.4 | Multiplying Polynomials | 264 |
| 5.5 | Dividing Polynomials | 272 |
| | **Chapter 5 Summary** | 279 |
| | **Chapter 5 Review Exercises** | 281 |
| | **Chapter 5 Test** | 283 |

## 6 Factoring — 285

| | | |
|---|---|---:|
| 6.1 | Greatest Common Factors; Factoring by Grouping | 286 |
| 6.2 | Factoring Trinomials | 291 |
| 6.3 | Special Factoring | 298 |
| 6.4 | A General Approach to Factoring | 303 |
| 6.5 | Solving Equations by Factoring | 306 |
| | **Chapter 6 Summary** | 314 |
| | **Chapter 6 Review Exercises** | 315 |
| | **Chapter 6 Test** | 317 |

## 7 Rational Expressions and Functions — 319

| | | |
|---|---|---:|
| 7.1 | Rational Expressions and Functions; Multiplying and Dividing | 320 |
| 7.2 | Adding and Subtracting Rational Expressions | 330 |
| 7.3 | Complex Fractions | 339 |
| 7.4 | Equations with Rational Expressions and Graphs | 345 |
| 7.5 | Applications of Rational Expressions | 351 |
| 7.6 | Variation | 361 |
| | **Chapter 7 Summary** | 371 |
| | **Chapter 7 Review Exercises** | 375 |
| | **Chapter 7 Test** | 377 |

## 8 Roots, Radicals, and Root Functions — 379

| | | |
|---|---|---:|
| 8.1 | Radical Expressions and Graphs | 380 |
| 8.2 | Rational Exponents | 386 |
| 8.3 | Simplifying Radical Expressions | 393 |
| 8.4 | Adding and Subtracting Radical Expressions | 402 |
| 8.5 | Multiplying and Dividing Radical Expressions | 406 |
| 8.6 | Solving Equations with Radicals | 413 |
| 8.7 | Complex Numbers | 420 |
| | **Chapter 8 Summary** | 427 |
| | **Chapter 8 Review Exercises** | 430 |
| | **Chapter 8 Test** | 433 |

## 9 Quadratic Equations and Inequalities — 435

- 9.1 The Square Root Property and Completing the Square — 436
- 9.2 The Quadratic Formula — 444
- 9.3 Equations Quadratic in Form — 451
- 9.4 Formulas and Further Applications — 461
- 9.5 Quadratic and Rational Inequalities — 469
- **Chapter 9 Summary** — 475
- **Chapter 9 Review Exercises** — 477
- **Chapter 9 Test** — 480

## 10 Additional Graphs of Functions and Relations — 483

- 10.1 Review of Operations and Composition — 484
- 10.2 Graphs of Quadratic Functions — 490
- 10.3 More about Parabolas and Their Applications — 500
- 10.4 Symmetry; Increasing and Decreasing Functions — 510
- 10.5 Piecewise Linear Functions — 516
- **Chapter 10 Summary** — 524
- **Chapter 10 Review Exercises** — 527
- **Chapter 10 Test** — 528

## 11 Inverse, Exponential, and Logarithmic Functions — 531

- 11.1 Inverse Functions — 532
- 11.2 Exponential Functions — 539
- 11.3 Logarithmic Functions — 547
- 11.4 Properties of Logarithms — 555
- 11.5 Common and Natural Logarithms — 562
- 11.6 Exponential and Logarithmic Equations; Further Applications — 569
- **Chapter 11 Summary** — 579
- **Chapter 11 Review Exercises** — 582
- **Chapter 11 Test** — 585

## 12 Polynomial and Rational Functions — 587

- 12.1 Synthetic Division — 588
- 12.2 Zeros of Polynomial Functions — 593
- 12.3 Graphs and Applications of Polynomial Functions — 601
- 12.4 Graphs and Applications of Rational Functions — 613
- **Chapter 12 Summary** — 626
- **Chapter 12 Review Exercises** — 629
- **Chapter 12 Test** — 631

## 13 Conic Sections and Nonlinear Systems — 633

- 13.1 The Circle and the Ellipse — 634
- 13.2 The Hyperbola and Functions Defined by Radicals — 640
- 13.3 Nonlinear Systems of Equations — 648
- **Chapter 13 Summary** — 654
- **Chapter 13 Review Exercises** — 656
- **Chapter 13 Test** — 656

## 14 Trigonometric Functions — 659

14.1 Angles — 660
14.2 Angle Relationships and Similar Triangles — 668
14.3 Trigonometric Functions — 677
14.4 Using the Definitions of the Trigonometric Functions — 683
**Chapter 14 Summary** — 692
**Chapter 14 Review Exercises** — 694
**Chapter 14 Test** — 698

## 15 Acute Angles and Right Triangles — 701

15.1 Trigonometric Functions of Acute Angles — 702
15.2 Trigonometric Functions of Non-Acute Angles — 709
15.3 Finding Trigonometric Function Values Using a Calculator — 715
15.4 Solving Right Triangles — 720
**Chapter 15 Summary** — 728
**Chapter 15 Review Exercises** — 730
**Chapter 15 Test** — 733

## 16 Radian Measure and Circular Functions — 735

16.1 Radian Measure — 736
16.2 Applications of Radian Measure — 742
16.3 The Unit Circle and Circular Functions — 749
16.4 Linear and Angular Speed — 757
**Chapter 16 Summary** — 763
**Chapter 16 Review Exercises** — 765
**Chapter 16 Test** — 767

## 17 Graphs of the Circular Functions — 769

17.1 Graphs of the Sine and Cosine Functions — 770
17.2 Translations of the Graphs of the Sine and Cosine Functions — 781
17.3 Graphs of the Tangent and Cotangent Functions — 788
17.4 Graphs of the Secant and Cosecant Functions — 796
**Chapter 17 Summary** — 803
**Chapter 17 Review Exercises** — 805
**Chapter 17 Test** — 807

## 18 Trigonometric Identities — 809

18.1 Fundamental Identities — 810
18.2 Verifying Trigonometric Identities — 815
18.3 Sum and Difference Identities for Cosine — 822
18.4 Sum and Difference Identities for Sine and Tangent — 827
18.5 Double-Angle Identities — 833
18.6 Half-Angle Identities — 839
**Chapter 18 Summary** — 845
**Chapter 18 Review Exercises** — 846
**Chapter 18 Test** — 848

## 19 Inverse Circular Functions and Trigonometric Equations  849

19.1 Inverse Circular Functions 850
19.2 Trigonometric Equations I 863
19.3 Trigonometric Equations II 870
**Chapter 19 Summary** 875
**Chapter 19 Review Exercises** 876
**Chapter 19 Test** 878

## 20 Oblique Triangles and Vectors  881

20.1 Oblique Triangles and the Law of Sines 882
20.2 The Ambiguous Case of the Law of Sines 890
20.3 The Law of Cosines 895
20.4 Vectors 903
**Chapter 20 Summary** 911
**Chapter 20 Review Exercises** 913
**Chapter 20 Test** 916

**Appendix A** Counting Theory 917
**Appendix B** Basics of Probability 927

**Glossary** G-1
**Answers to Selected Exercises** A-1
**Index** I-1

# Preface

Written with the specific needs of high school students and teachers in mind, *Algebra and Trigonometry for College Readiness* has been designed to prepare today's students for college-level mathematics courses. Research suggests that when students place into credit-bearing math courses, drop-out rates decrease, and they are more successful in the workforce and beyond. The table of contents has been specifically crafted to meet the needs of those who have completed Algebra 2 but require another year of mathematics to prepare them for credit-bearing courses in college. The text takes a traditional skill-building and problem-solving approach and includes numerous, modern applications. The many exercises in this text supply ample opportunity for students to hone their skills. Many features, outlined below, are included to facilitate the learning and teaching processes.

**MATHEMATICAL CONTENT**

This text provides ample content for a complete full-year course or two separate semester courses. All of the topics needed for success in college-level College Algebra and Precalculus courses are covered thoroughly. In addition, several topics are included that will prepare students for college courses in Finite Mathematics: basic set theory, counting theory, and probability theory; and extensive work with systems of equations and matrices. Algebra topics are covered in Chapters 1–13 and Appendices A and B, while the standard concepts of trigonometry are covered in Chapters 14–20.

The book begins with a review of the real numbers and basic algebra in Chapter 1. This material can easily be omitted if it is not needed. Chapter 2 covers both linear equations and inequalities and absolute value equations and inequalities in one variable. This chapter introduces both solution set and interval notation. These notations will be used to write the solutions of equations and inequalities throughout the text.

The function concept is a unifying theme through many chapters of the book, starting with an introduction (or review) of the function concept in Chapter 3. As they progress through the text, students will work with linear, quadratic, and higher-degree polynomial functions; the absolute value and other piecewise linear functions; the greatest integer function and other step functions; square root functions, rational functions, exponential and logarithmic functions; and trigonometric and inverse trigonometric functions. For each type of function, students learn key characteristics such as the domain and range. Graphing is emphasized so that students become proficient in both drawing and interpreting graphs. Mathematical models and other applications are presented, and they are based on all of these types of functions.

**KEY FEATURES**

The following features will help students and teachers get the most benefit from this text:

**Annotated Teacher's Edition** For easy reference, answers are provided in the margins for all of the exercises, including sample answers for the writing exercises. (See page 165.) For in-class instruction, the margins also include a Classroom Example to parallel every example in the text and frequent Teaching Tips. Answers appear directly below each Classroom Example, and solutions for these examples are available separately. (See pages 111 and 354.)

**Chapter Openers** Chapter openers feature motivating real-world applications of mathematics that are relevant to students and tied to specific material within the chapters. Topics include American spending on pets, college tuition, and video games. (See pages 1, 107, and 169—Chapters 1, 3, and 4.) Each opener also includes a list of the sections in the chapter.

**Real-Life Applications** We are always on the lookout for interesting data to use in real-life applications. As a result, we have included many examples and exercises from fields such as business, pop culture, sports, the life sciences, environmental studies, and technology that show the relevance of algebra and trigonometry to daily life. (See pages 63 and 780.) Students work with data displayed in a variety of formats, including tables and circle, line, and bar graphs. (See pages 103, 115, 137, and 533.) Students gain extensive experience in working with mathematical models.

**Figures, Photos, and Hand-Drawn Graphs**
Today's students are more visually oriented than ever. Thus, we have made a concerted effort to include mathematical figures, diagrams, tables, and graphs, including "hand-drawn" graphs, whenever possible. (See pages 110 and 144.) Many of the graphs also use a style similar to that seen by students in today's print and electronic media. Photos have been incorporated in examples and exercises to enhance applications. (See pages 124 and 141.)

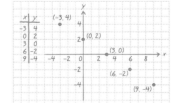

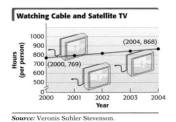

**Emphasis on Problem Solving** Introduced in Chapter 2, our six-step problem-solving method is integrated throughout the text to help students build confidence with solving problems with algebra. The six steps, *Read, Assign a variable, Write an equation, Solve, State the answer,* and *Check,* are emphasized in boldface type and repeated in examples and exercises to reinforce the problem-solving process for students. (See pages 62 and 462.)

> **Solving an Applied Problem**
>
> **Step 1** **Read** the problem, several times if necessary, until you understand what is given and what is to be found.
> **Step 2** **Assign a variable** to represent the unknown value, using diagrams or tables as needed. Write down what the variable represents. If necessary, express any other unknown values in terms of the variable.
> **Step 3** **Write an equation** using the variable expression(s).
> **Step 4** **Solve** the equation.
> **Step 5** **State the answer** to the problem. Does it seem reasonable?
> **Step 6** **Check** the answer in the words of the original problem.

**Learning Objectives** Each section begins with clearly stated, numbered objectives, and the included material is directly keyed to these objectives so that students know exactly what is covered in each section and which skills they are expected to master. (See pages 108, 130, and 170.)

> **OBJECTIVES**
> 1 Interpret a line graph.
> 2 Plot ordered pairs.
> 3 Find ordered pairs that satisfy a given equation.
> 4 Graph lines.
> 5 Find *x*- and *y*-intercepts.
> 6 Recognize equations of horizontal and vertical lines and lines passing through the origin.
> 7 Use the midpoint formula.

**Cautions** These boxes warn students about common errors and emphasize important ideas throughout the text. (See pages 109, 119, and 143.) Highlighted in yellow, the design makes them easy to spot.

> **CAUTION** The parentheses used to represent an ordered pair are also used to represent an open interval (introduced in **Section 1.1**). The context of the discussion tells whether ordered pairs or open intervals are being represented.

**Examples** Mathematical concepts introduced in the exposition are illustrated with detailed examples. Many of the skill-based examples contain multiple parts. The step-by-step solutions to the examples include extensive side comments to explain what is being done in each step. (See pages 176 and 785.) *Pointers* from the authors, highlighted in blue, provide students with important on-the-spot reminders within the examples about common pitfalls. (See pages 76, 124, and 138.) Application examples at the end of many sections guide students in the process of applying the skills learned within the section to real-life situations.

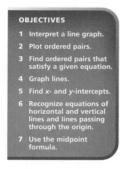

**EXAMPLE 1** Evaluating Polynomial Functions

Let $f(x) = 4x^3 - x^2 + 5$. Find each value.

(a) $f(3)$

Read this as "f of 3," not "f times 3."

$f(x) = 4x^3 - x^2 + 5$
$f(3) = 4(3)^3 - 3^2 + 5$  Substitute 3 for *x*.
$= 4(27) - 9 + 5$  Order of operations
$= 108 - 9 + 5$  Multiply.
$= 104$  Subtract; add.

(b) $f(-4) = 4(-4)^3 - (-4)^2 + 5$  Let $x = -4$.
$= 4(-64) - 16 + 5$
$= -267$

Be careful with signs.

**NOW TRY** Exercise 7.

**Now Try Exercises** To actively engage students in the learning process, each example concludes with a reference to **NOW TRY** one or more parallel, odd-numbered exercises from the corresponding exercise set. In this way, students are able to immediately apply and reinforce the concepts and skills presented in the examples. The Now Try exercises are clearly marked with yellow screens ( **29.** ) in the exercise sets so that they can be easily spotted. (See pages 204 and 205.)

7. $f(x) = -x^2 + 2x^3 - 8$

**Function Boxes** Special function boxes offer a comprehensive visual introduction to each circular and inverse circular function and serve as excellent resources for student reference and review. Each function box includes a table of values alongside traditional and calculator graphs, as well as the domain, range, and other specific information about the function. (See pages 771 and 853.)

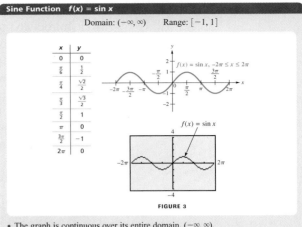

**Use of Technology** We have integrated the use of graphing calculators where appropriate, although graphing technology is not a central feature of this text. We stress that graphing calculators are an aid to understanding and that students must master the underlying mathematical concepts. Graphing calculator solutions are included for selected examples, and exercises that use graphing calculators are identified with an icon: . *This graphing calculator material is optional and can be omitted without loss of continuity.*

**Ample and Varied Exercise Sets** The text contains a wealth of exercises to provide students with opportunities to practice, apply, connect, and extend the concepts and skills they are learning. Numerous illustrations, tables, graphs, and photos help students to visualize the problems they are solving. Exercise types include writing , estimation, and graphing calculator  exercises, and a wide variety of applications, as well as the following:

- *Concept Check* exercises facilitate mathematical thinking and conceptual understanding. (See pages 126, 180, and 246.)

- **WHAT WENT WRONG?** exercises ask students to identify typical errors in solutions and work the problems correctly. (See pages 50, 80, and 246.)

- *Multiple-Choice, Matching, True or False,* and *Fill in the Blanks* exercises give students practice in answering questions presented in these formats, often found in testing situations. (See pages 125, 127, and 223.)

- *Connecting Graphs with Equations* problems provide students with opportunities to write equations for given graphs. (See pages 779 and 787.)

7. *Concept Check* The following "solution" contains a common student error.

$$8x - 2(2x - 3) = 3x + 7$$
$$8x - 4x - 6 = 3x + 7 \quad \text{Distributive property}$$
$$4x - 6 = 3x + 7 \quad \text{Combine like terms.}$$
$$x = 13 \quad \text{Subtract } 3x; \text{ add } 6.$$

**WHAT WENT WRONG?** Find the correct solution.

8. *Concept Check* When clearing parentheses in the expression
$$-5m - (2m - 4) + 5$$
on the right side of the equation in Exercise 39, the − sign before the parentheses acts like a factor representing what number? Clear parentheses and simplify this expression.

9. Explain the distinction between a conditional equation, an identity, and a contradiction.

10. *Concept Check* A student tried to solve the equation $8x = 7x$ by dividing each side by $x$, obtaining $8 = 7$. He gave the solution set as ∅. **WHAT WENT WRONG?**

*Connecting Graphs with Equations* Each function graphed is of the form $y = c + \cos x$, $y = c + \sin x, y = \cos(x - d),$ or $y = \sin(x - d)$, where $d$ is the least possible positive value. Determine the equation of the graph.

21. 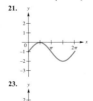  22.

23.   24.

**Ample Opportunity for Review** Each chapter concludes with a Chapter Summary featuring Key Terms, New Symbols, and a Quick Review of each section's content with additional examples. (See pages 279–281.) A comprehensive set of Chapter Review Exercises and a Chapter Test provide materials to help students retain what they have learned and prepare for classroom tests. (See pages 281–283 and 283–284.)

**Answers to Selected Exercises** To aid students in checking their work, the student edition includes answers at the back of the book to all odd-numbered exercises in the textbook and to all exercises in the Chapter Test. The *Annotated Teacher's Edition* contains answers to all exercises in the margins of the textbook.

**Glossary** A comprehensive glossary of key terms from throughout the text is included at the back of the book. (See pages G-1 to G-9.)

**TECHNOLOGY RESOURCES FOR STUDENTS**

**MathXL® for School (optional, for purchase only—access code required) www.MathXLforSchool.com** MathXL® for School is a powerful online homework, tutorial, and assessment program designed specifically for Pearson Education mathematics textbooks; however it is flexible enough to use with any math program.

With MathXL for School, students:

- Do their homework and receive immediate feedback
- Get self-paced assistance on problems in a variety of ways (guided solutions, step-by-step examples)
- Have a large number of practice problems to choose from, helping them master a topic
- Receive a personalized study plan based on quiz and test results

With MathXL for School, teachers:

- Quickly and easily create quizzes, tests, and homework assignments
- Utilize automatic grading to rapidly assess student understanding
- Receive data-driven reporting on student and group-level performance
- Prepare students for high-stakes testing
- Deliver quality instruction regardless of experience level

The new Flash-based, platform- and browser-independent MathXL Player v2 now supports Firefox® on Windows® (XP and Vista), Safari™ and Firefox® on the Macintosh, as well as Internet Explorer®. For more information, visit our Web site at www.MathXLforSchool.com, or contact your Pearson sales representative.

**MathXL® Tutorials on CD** This interactive tutorial CD-ROM provides algorithmically-generated practice exercises that are correlated at the chapter, section, and objective level to the exercises in the textbook. Every practice exercise is accompanied by an example and a guided solution designed to involve students in the solution process. The software provides helpful feedback for incorrect answers and can generate printed summaries of students' progress. A CD is available for purchase separately using ISBN-13: 978-0-13-136923-8; ISBN-10: 0-13-136923-7.

**ADDITIONAL TEACHER RESOURCES**

*The following supplements are available for qualified adopters:*

Most of the teacher supplements and resources for this text are available for download from the Instructor Resource Center (IRC). Please go to www.PearsonSchool.com/Access_Request and select "access to online instructor resources." You will be required to complete a one-time registration subject to verification before being emailed access information for download materials.

### Annotated Teacher's Edition

- Provides answers in the margins next to the corresponding problem for all exercises, including sample answers for writing exercises.
- Icons identify calculator and writing exercises.
- Margin notes also include teaching tips and extra examples for the classroom with answers.
- ISBN-13: 978-0-13-136903-0; ISBN-10: 0-13-136903-2

### Online Teacher Solutions

- Provides complete solutions for all text exercises and Classroom Examples.
- Available for download only; contact your Pearson sales representative for additional details.

### Online Testing Manual

- Provides two End of Chapter Tests, including worked problems correlating to those throughout the exercise sets, and two Test Prep Chapter Tests, geared towards preparation for standardized exams, for all chapters and appendices in the text.
- Also includes pretests, mid-term exams, and final exams.
- Available for download only; contact your Pearson sales representative for additional details.

### TestGen®

- Enables teachers to build, edit, print, and administer tests using a computerized bank of questions developed to cover all objectives of the text. TestGen is algorithmically based, allowing teachers to create multiple but equivalent versions of the same question or test with the click of a button. Teachers can also modify test bank questions or add new questions. Tests can be printed or administered online.
- Available for download only; contact your Pearson sales representative for additional details.

### PowerPoint® Slides

- Features presentations written and designed specifically for this text, including figures, examples, definitions, and key concepts.
- Available for download only; contact your Pearson sales representative for additional details.

### Classroom Example PowerPoint Slides

- Include full, worked-out solutions to all Classroom Examples from the text margins.
- Available for download only; contact your Pearson sales representative for additional details.

## ACKNOWLEDGMENTS

We have many people to thank for their invaluable contributions to this new textbook, specifically designed for third- and fourth-year mathematics students. We would first like to thank Abby Tanenbaum for the time, effort, and unsurpassable attention to detail that she put into this text. This book would not be possible without her expertise. Paul Lorczak, Steve Ouellette, and Janis Cimperman were integral to ensuring the accuracy and consistency throughout the text by reading through all page proofs, and we would also like to thank Beverly Fusfield for her work on this text, including authoring both the solutions manual and testing manual. The behind-the-scenes assistance of Terry McGinnis and Callie Daniels was also vital to the completion of this book.

Our sincere thanks go to these dedicated individuals at Pearson Education who worked long and hard to make this textbook a success: Greg Tobin, Rich Williams, Anne Kelly, Becky Anderson, Katherine Greig, Kathy Manley, Karen Wernholm, Leah Goldberg, Sarah Gibbons, Katherine Minton, Peter Silvia, Carl Cottrell, Christina Gleason, Courtney Marsh, and Andrea Sheehan. Janette Krauss and Nesbitt Graphics, Inc. provided excellent production work. A special thanks to Lucie Haskins for preparing the accurate, useful index.

We especially wish to thank the following consultants who gave input during the planning stages of this text:

  Carolyn Elder, *Stuttgart High School (AR)*
  Sherry Everding, *Cor Jesu Academy (MO)*
  Sue Jordan, *Franklin High School (TN)*
  Donna Miller, *Eagle Grove High School (IA)*
  Nathalie Smalls, *Rockdale Magnet School for Science and Technology (GA)*

And the following teachers who participated in our forum on third- and fouth-year mathematics:

  Tom Beatini, *Glen Rock High School (NJ)*
  Jeremy Beckman, *Naches Valley High School (WA)*
  Susan M. Bothman, *Ooltewah High School (TN)*
  Veronica Carlson, *Moon Valley High School (AZ)*
  Thomas M. Haver, *Winchester High School (MA)*
  Ron Millard, *Shawnee Mission South High School (KS)*

As an author team, we are committed to providing the best possible text to help teachers teach and students succeed. As we continue to work toward this goal, we would welcome any comments or suggestions you might have via e-mail to *math@pearson.com*.

Margaret L. Lial<br>
John Hornsby

# Photo Credits

**p. 1** Sara Piaseczynski; **p. 8** PhotoDisc/Getty; **p. 27** Brand X Pictures; **p. 29** Sara Piaseczynski; **p. 30** PhotoDisc/Getty; **p. 43** PhotoDisc/Getty; **p. 52** PhotoDisc/Getty; **p. 56** Terry McGinnis; **p. 58** left Getty Editorial; **p. 58** right William Manning/Corbis; **p. 63** Getty Images; **p. 69** top SuperStock; **p. 69** bottom left Getty Images; **p. 69** bottom right Digital Vision (PP); **p. 70** top Bettmann/Corbis; **p. 70** bottom PhotoDisc/Getty; **p. 71** Digital Vision (PP); **p. 79** PhotoDisc/Getty; **p. 81** Jason Reed/Reuters/Corbis; **p. 82** left PhotoDisc/Getty; **p. 82** right PhotoDisc/Getty; **p. 88** Everett Collection; **p. 90** Comstock RF; **p. 104** Photofest; **p. 107** Shutterstock; **p.117** PhotoDisc/Getty; **p. 124** Blend Images/Getty Images; **p. 135** Rubberball Productions/Getty Images; **p. 136** Stockdisc/Getty; **p. 141** Digital Vision; **p. 148** Getty Images; **p. 159** Blend Images/Getty Images; **p. 160** Corbis RF PP; **p. 167** PhotoDisc/Getty; **p. 169** Shutterstock; **p. 235, 263** Comstock RF; **p. 245** PhotoDisc/Getty; **p. 249** top Beth Anderson; **p. 249** bottom NASA/PAL; **p. 256** Digital Vision PP; **p. 258** PhotoDisc/Getty; **p. 262** PhotoDisc/Getty; **p. 283** PhotoDisc/Getty; **p. 285** Getty Images/Stone Allstock; **p. 311** AP Photo/Middletown Journal, Pat Auckerman; **p. 319, 338** PhotoDisc/Getty; **p. 352** PhotoDisc/Getty; **p. 354** B.K. Tomlin; **p. 358** PhotoDisc/Getty; **p. 359** Jim McIsaac/Getty Images; **p. 365** Digital Vision; **p. 367** Blend Images/Getty Images; **p. 369** PhotoDisc/Getty; **p. 376** PhotoDisc/Getty; **p. 378** Natalie Fobes/Corbis; **p. 379, 401** PhotoDisc/Getty; **p. 385** PhotoDisc/Getty; **p. 393** Digital Vision; **p. 418** Beth Anderson; **p. 435, 464** Corbis Royalty Free; **p. 437** PhotoDisc/Getty; **p. 443** left Corbis RF; **p. 443** right Corbis RF; **p. 453** John Madere/Corbis; **p. 478** Corbis RF; **p. 479** Digital Vision RF; **p. 483, 495** Dana Fineman/Corbis/Sygma; **p. 491** PhotoDisc/Getty; **p. 519** Getty Images; **p. 521** Beth Anderson; **p. 529** Richard Carlson www.pals.iastate.edu/carlson/; **p. 530** Beth Anderson; **p. 531, 564** Corbis RF; **p. 533** Getty Images; **p. 537** Corbis; **p. 552** PhotoDisc/Getty; **p. 563** Digital Vision; **p. 567** Photofest; **p. 568** PhotoDisc/Getty; **p. 572** PhotoDisc/Getty; **p. 574** Corbis; **p. 578** left Blend Images/Getty Images; **p. 578** right Stockdisk/Getty; **p. 582** Beth Anderson; **p. 583** PhotoDisc/Getty; **p. 586** Stockdisk/Getty; **p. 587** PhotoDisc/Getty; **p. 596** Library of Congress (PAL); **p. 609** Corbis RF; **p. 633** Michael Freeman/Corbis; **p. 640** NASA; **p. 647** PhotoDisc/Getty; **p. 659** PhotoDisc/Getty; **p. 701** Shutterstock; **p. 726** PhotoDisc/Getty; **p. 732** Shutterstock; **p. 735** Hubble Space Telescope; **p. 741** Shutterstock; **p. 758** Shutterstock; **p. 761** NASA; **p. 767** Shutterstock; **p. 769** Shutterstock; **p. 780** Corbis Royalty Free; **p. 806** Shutterstock; **p. 809** Stockbyte/Getty Images; **p. 833** iStockphoto; **p. 837** Shutterstock; **p. 839** Digital Vision; **p. 849** Michael Ochs Archives/Getty Images; **p. 860** iStockphoto; **p. 872** Shutterstock; **p. 874** Artville (Royalty Free); **p. 881** Shutterstock; **p. 889** Shutterstock; **p. 903** Digital Vision

# 1

# Review of the Real Number System

Americans are crazy about their pets. Over 71 million U.S. households owned pets in 2009. Combined, these households spent more than $35 billion pampering their animal friends. The fastest-growing segment of the pet industry is the high-end luxury area, which includes everything from gourmet pet foods, designer toys, and specialty furniture to groomers, dog walkers, boarding in posh pet hotels, and even pet therapists. (*Source:* American Pet Products Association.)

In Exercise 95 of Section 1.3, we use an *algebraic expression,* one of the topics of this chapter, to determine how much Americans have spent annually on their pets in recent years.

**1.1** Basic Concepts

**1.2** Operations on Real Numbers

**1.3** Exponents, Roots, and Order of Operations

**1.4** Properties of Real Numbers

## 1.1 Basic Concepts

**OBJECTIVES**

1. Write sets using set notation.
2. Use number lines.
3. Know the common sets of numbers.
4. Find additive inverses.
5. Use absolute value.
6. Use inequality symbols.
7. Graph sets of real numbers.

In this chapter, we review some of the basic symbols and rules of algebra.

**OBJECTIVE 1** Write sets using set notation. A **set** is a collection of objects called the **elements** or **members** of the set. In algebra, the elements of a set are usually numbers. Set braces, { }, are used to enclose the elements. For example, 2 is an element of the set {1, 2, 3}. Since we can count the number of elements in the set {1, 2, 3}, it is a **finite set**.

In our study of algebra, we refer to certain sets of numbers by name. The set

$$N = \{1, 2, 3, 4, 5, 6, \ldots\}$$ Natural (counting) numbers

is called the **natural numbers,** or the **counting numbers.** The three dots (*ellipsis points*) show that the list continues in the same pattern indefinitely. We cannot list all of the elements of the set of natural numbers, so it is an **infinite set.**

When 0 is included with the set of natural numbers, we have the set of **whole numbers,** written

$$W = \{0, 1, 2, 3, 4, 5, 6, \ldots\}.$$ Whole numbers

The set containing no elements, such as the set of whole numbers less than 0, is called the **empty set,** or **null set,** usually written $\emptyset$ or { }.

**CAUTION** Do not write $\{\emptyset\}$ for the empty set; $\{\emptyset\}$ is a set with one element: $\emptyset$. Use the notation $\emptyset$ or { } for the empty set.

To write the fact that 2 is an element of the set {1, 2, 3}, we use the symbol $\in$ (read "is an element of").

$$2 \in \{1, 2, 3\}$$

The number 2 is also an element of the set of natural numbers $N$, so we may write

$$2 \in N.$$

To show that 0 is *not* an element of set $N$, we draw a slash through the symbol $\in$.

$$0 \notin N$$

Two sets are equal if they contain exactly the same elements. For example, $\{1, 2\} = \{2, 1\}$. (Order doesn't matter.) However, $\{1, 2\} \neq \{0, 1, 2\}$ ($\neq$ means "is not equal to"), since one set contains the element 0 while the other does not.

In algebra, letters called **variables** are often used to represent numbers or to define sets of numbers. For example,

$$\{x \mid x \text{ is a natural number between 3 and 15}\}$$

(read "the set of all elements $x$ such that $x$ is a natural number between 3 and 15") defines the set

$$\{4, 5, 6, 7, \ldots, 14\}.$$

The notation $\{x \mid x \text{ is a natural number between 3 and 15}\}$ is an example of **set-builder notation.**

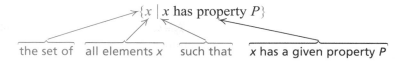

### EXAMPLE 1 Listing the Elements in Sets

List the elements in each set.

(a) $\{x \mid x \text{ is a natural number less than 4}\}$
The natural numbers less than 4 are 1, 2, and 3. This set is $\{1, 2, 3\}$.

(b) $\{y \mid y \text{ is one of the first five even natural numbers}\}$ is $\{2, 4, 6, 8, 10\}$.

(c) $\{z \mid z \text{ is a natural number greater than or equal to 7}\}$
The set of natural numbers greater than or equal to 7 is an infinite set, written with ellipsis points as

$$\{7, 8, 9, 10, \ldots\}.$$

**NOW TRY** Exercise 1.

### EXAMPLE 2 Using Set-Builder Notation to Describe Sets

Use set-builder notation to describe each set.

(a) $\{1, 3, 5, 7, 9\}$
There are often several ways to describe a set in set-builder notation. One way to describe the given set is

$$\{y \mid y \text{ is one of the first five odd natural numbers}\}.$$

(b) $\{5, 10, 15, \ldots\}$
This set can be described as $\{x \mid x \text{ is a multiple of 5 greater than 0}\}$.

**NOW TRY** Exercises 13 and 15.

**OBJECTIVE 2** **Use number lines.** A good way to get a picture of a set of numbers is to use a **number line.** To construct a number line, choose any point on a horizontal line and label it 0. Next, choose a point to the right of 0 and label it 1. The distance from 0 to 1 establishes a scale that can be used to locate more points, with positive numbers to the right of 0 and negative numbers to the left of 0. See Figure 1.

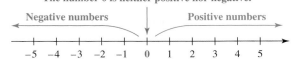

**FIGURE 1**

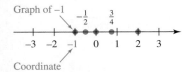

FIGURE 2

The set of numbers identified on the number line in Figure 1, including positive and negative numbers and 0, is part of the set of **integers,** written

$$I = \{\ldots, -3, -2, -1, 0, 1, 2, 3, \ldots\}. \quad \text{Integers}$$

Each number on a number line is called the **coordinate** of the point that it labels, while the point is the **graph** of the number. Figure 2 shows a number line with several points graphed on it.

The fractions $-\frac{1}{2}$ and $\frac{3}{4}$, graphed on the number line in Figure 2, are examples of *rational numbers.* A **rational number** can be expressed as the quotient of two integers, with denominator not 0. The set of all rational numbers is written

$$\left\{ \frac{p}{q} \,\middle|\, p \text{ and } q \text{ are integers, } q \neq 0 \right\}. \quad \text{Rational numbers}$$

***The set of rational numbers includes the natural numbers, whole numbers, and integers,*** since these numbers can be written as fractions. For example, $14 = \frac{14}{1}$, $-3 = \frac{-3}{1}$, and $0 = \frac{0}{1}$. A rational number written as a fraction, such as $\frac{1}{8}$ or $\frac{2}{3}$, can also be expressed as a decimal by dividing the numerator by the denominator as follows.

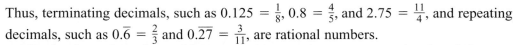

Thus, terminating decimals, such as $0.125 = \frac{1}{8}$, $0.8 = \frac{4}{5}$, and $2.75 = \frac{11}{4}$, and repeating decimals, such as $0.\overline{6} = \frac{2}{3}$ and $0.\overline{27} = \frac{3}{11}$, are rational numbers.

Decimal numbers that neither terminate nor repeat are *not* rational and thus are called **irrational numbers.** Many square roots are irrational numbers; for example, $\sqrt{2} = 1.4142135\ldots$ and $-\sqrt{7} = -2.6457513\ldots$ repeat indefinitely without pattern. (Some square roots *are* rational: $\sqrt{16} = 4$, $\sqrt{100} = 10$, and so on.) Another irrational number is $\pi$, the ratio of the distance around, or circumference of, a circle to its diameter.

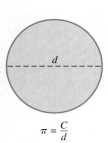

Some of the rational and irrational numbers just discussed are graphed on the number line in Figure 3 on the next page. The rational numbers together with the irrational numbers make up the set of **real numbers.** *Every point on a number line corresponds to a real number, and every real number corresponds to a point on the number line.*

**FIGURE 3**

**OBJECTIVE 3** Know the common sets of numbers.

### Sets of Numbers

**Natural numbers, or counting numbers** $\{1, 2, 3, 4, 5, 6, \ldots\}$

**Whole numbers** $\{0, 1, 2, 3, 4, 5, 6, \ldots\}$

**Integers** $\{\ldots, -3, -2, -1, 0, 1, 2, 3, \ldots\}$

**Rational numbers** $\left\{\frac{p}{q} \,\middle|\, p \text{ and } q \text{ are integers}, q \neq 0\right\}$

*Examples:* $\frac{4}{1}$ or 4, 1.3, $-\frac{9}{2}$ or $-4\frac{1}{2}$, $\frac{16}{8}$ or 2, $\sqrt{9}$ or 3, $0.\overline{6}$

**Irrational numbers** $\{x \,|\, x \text{ is a real number that is not rational}\}$

*Examples:* $\sqrt{3}, -\sqrt{2}, \pi$

**Real numbers** $\{x \,|\, x \text{ is a rational number or an irrational number}\}$*

Figure 4 shows that the set of real numbers includes both the rational and irrational numbers. Every real number is either rational or irrational. Also, notice that the integers are elements of the set of rational numbers and that the whole numbers and natural numbers are elements of the set of integers.

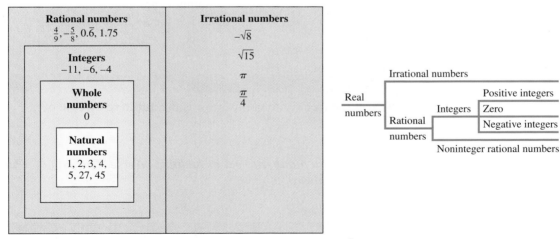

**FIGURE 4** The Real Numbers

---

*An example of a number that is not real is $\sqrt{-1}$. This number, part of the *complex number system*, is discussed in **Chapter 8**.

### EXAMPLE 3  Identifying Examples of Number Sets

Which numbers in

$$\left\{-8, -\sqrt{5}, -\frac{9}{64}, 0, 0.5, \frac{1}{3}, 1.\overline{12}, \sqrt{3}, 2, \pi\right\}$$

are elements of each set?

**(a)** Integers
$-8$, 0, and 2 are integers.

**(b)** Rational numbers
$-8$, $-\frac{9}{64}$, 0, 0.5, $\frac{1}{3}$, $1.\overline{12}$, and 2 are rational numbers.

**(c)** Irrational numbers
$-\sqrt{5}$, $\sqrt{3}$, and $\pi$ are irrational numbers.

**(d)** Real numbers
All the numbers in the given set are real numbers.

**NOW TRY** Exercise 23.

### EXAMPLE 4  Determining Relationships between Sets of Numbers

Decide whether each statement is *true* or *false*.

**(a)** All irrational numbers are real numbers.
This is true. As shown in Figure 4, the set of real numbers includes all irrational numbers.

**(b)** Every rational number is an integer.
This statement is false. Although some rational numbers are integers, other rational numbers, such as $\frac{2}{3}$ and $-\frac{1}{4}$, are not.

**NOW TRY** Exercise 25.

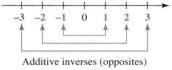

Additive inverses (opposites)

**FIGURE 5**

**OBJECTIVE 4**  Find additive inverses.  Look at the number line in Figure 5. For each positive number, there is a negative number on the opposite side of 0 that lies the same distance from 0. These pairs of numbers are called *additive inverses, opposites,* or *negatives* of each other. For example, 3 is the additive inverse of $-3$, and $-3$ is the additive inverse of 3.

#### Additive Inverse

For any real number $a$, the number $-a$ is the **additive inverse** of $a$.

*Change the sign of a number to get its additive inverse. The sum of a number and its additive inverse is always 0.*

#### Uses of the Symbol −

The symbol "$-$" is used to indicate any of the following:

1. a negative number, such as $-9$ or $-15$;
2. the additive inverse of a number, as in "$-4$ is the additive inverse of 4";
3. subtraction, as in $12 - 3$.

In the expression $-(-5)$ the symbol "−" is being used in two ways: the first − indicates the additive inverse (or opposite) of $-5$, and the second indicates a negative number, $-5$. Since the additive inverse of $-5$ is 5, then

$$-(-5) = 5.$$

This example suggests the following property.

---

**$-(-a)$**

For any real number $a$,  $-(-a) = a.$

---

| Number | Additive Inverse |
|---|---|
| 6 | −6 |
| −4 | 4 |
| $\frac{2}{3}$ | $-\frac{2}{3}$ |
| −8.7 | 8.7 |
| 0 | 0 |

Numbers written with positive or negative signs, such as $+4$, $+8$, $-9$, and $-5$, are called **signed numbers**. A positive number can be called a signed number even though the positive sign is usually left off. The table in the margin shows the additive inverses of several signed numbers. The number 0 is its own additive inverse.

**OBJECTIVE 5** Use absolute value. Geometrically, the **absolute value** of a number $a$, written $|a|$, is the distance on the number line from 0 to $a$. For example, the absolute value of 5 is the same as the absolute value of $-5$ because each number lies five units from 0. See Figure 6. That is,

$$|5| = 5 \quad \text{and} \quad |-5| = 5.$$

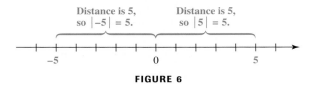

**FIGURE 6**

---

**CAUTION** *Because absolute value represents distance, and distance is never negative, the absolute value of a number is always positive or 0.*

---

The formal definition of absolute value follows.

---

**Absolute Value**

For any real number $a$, $\quad |a| = \begin{cases} a & \text{if } a \text{ is positive or } 0 \\ -a & \text{if } a \text{ is negative.} \end{cases}$

---

The second part of this definition, $|a| = -a$ if $a$ is negative, requires careful thought. If $a$ is a *negative* number, then $-a$, the additive inverse or opposite of $a$, is a positive number. Thus, $|a|$ is positive. For example, if $a = -3$, then

$$|a| = |-3| = -(-3) = 3. \quad |a| = -a \text{ if } a \text{ is negative.}$$

### EXAMPLE 5   Finding Absolute Value

Simplify by finding each absolute value.

**(a)** $|13| = 13$   **(b)** $|-2| = -(-2) = 2$   **(c)** $|0| = 0$

**(d)** $-|8|$

Evaluate the absolute value first. Then find the additive inverse.

$$-|8| = -(8) = -8$$

**(e)** $-|-8|$

Work as in part (d): $|-8| = 8$, so $-|-8| = -(8) = -8$.

**(f)** $|5| + |-2| = 5 + 2 = 7$

**(g)** $|5 - 2| = |3| = 3$

**NOW TRY** Exercises 43, 47, 49, and 53.

Absolute value is useful in applications comparing size without regard to sign.

### EXAMPLE 6   Comparing Rates of Change in Industries

The projected annual rates of change in employment (in percent) in some of the fastest-growing and in some of the most rapidly declining industries from 2002 through 2012 are shown in the table.

| Industry (2002–2012) | Annual Rate of Change (in percent) |
|---|---|
| Software publishers | 5.3 |
| Care services for the elderly | 4.5 |
| Child day-care services | 3.6 |
| Cut-and-sew apparel manufacturing | −12.2 |
| Fabric mills | −5.9 |
| Metal ore mining | −4.8 |

*Source:* U.S. Bureau of Labor Statistics.

What industry in the list is expected to see the greatest change? the least change?

We want the greatest *change,* without regard to whether the change is an increase or a decrease. Look for the number in the list with the largest absolute value. That number is found in cut-and-sew apparel manufacturing, since $|-12.2| = 12.2$. Similarly, the least change is in the child day-care services industry: $|3.6| = 3.6$.

**NOW TRY** Exercise 59.

**OBJECTIVE 6**  Use inequality symbols. The statement $4 + 2 = 6$ is an **equation**—a statement that two quantities are equal. The statement $4 \neq 6$ (read "4 is not equal to 6") is an **inequality**—a statement that two quantities are *not* equal. When two numbers are not equal, one must be less than the other. The symbol $<$ means "is less than." For example,

$$8 < 9, \quad -6 < 15, \quad -6 < -1, \quad \text{and} \quad 0 < \frac{4}{3}.$$

The symbol $>$ means "is greater than." For example,

$$12 > 5, \quad 9 > -2, \quad -4 > -6, \quad \text{and} \quad \frac{6}{5} > 0.$$

*In each case, the symbol "points" toward the smaller number.*

The number line in Figure 7 shows the graphs of the numbers 4 and 9. We know that $4 < 9$. On the graph, 4 is to the left of 9. *The smaller of two numbers is always to the left of the other on a number line.*

**FIGURE 7**

### Inequalities on a Number Line

On a number line,

$a < b$ if $a$ is to the left of $b$; $\quad a > b$ if $a$ is to the right of $b$.

We can use a number line to determine order. As shown on the number line in Figure 8, $-6$ is located to the left of 1. For this reason, $-6 < 1$. Also, $1 > -6$. From the same number line,

$$-5 < -2, \quad \text{or} \quad -2 > -5.$$

**FIGURE 8**

**CAUTION** *Be careful when ordering negative numbers.* Since $-5$ is to the left of $-2$ on the number line in Figure 8, $-5 < -2$, or $-2 > -5$. In each case, the symbol points to $-5$, the smaller number.

**NOW TRY** Exercises 65 and 73.

The following table summarizes results about positive and negative numbers in both words and symbols.

| Words | Symbols |
| --- | --- |
| Every negative number is less than 0. | If $a$ is negative, then $a < 0$. |
| Every positive number is greater than 0. | If $a$ is positive, then $a > 0$. |
| 0 is neither positive nor negative. | |

In addition to the symbols $\neq$, $<$, and $>$, the symbols $\leq$ and $\geq$ are often used.

**INEQUALITY SYMBOLS**

| Symbol | Meaning | Example |
|---|---|---|
| $\neq$ | is not equal to | $3 \neq 7$ |
| $<$ | is less than | $-4 < -1$ |
| $>$ | is greater than | $3 > -2$ |
| $\leq$ | is less than or equal to | $6 \leq 6$ |
| $\geq$ | is greater than or equal to | $-8 \geq -10$ |

| Inequality | Why It Is True |
|---|---|
| $6 \leq 8$ | $6 < 8$ |
| $-2 \leq -2$ | $-2 = -2$ |
| $-9 \geq -12$ | $-9 > -12$ |
| $-3 \geq -3$ | $-3 = -3$ |
| $6 \cdot 4 \leq 5(5)$ | $24 < 25$ |

The table in the margin shows several inequalities and why each is true. Notice the reason that $-2 \leq -2$ is true: *With the symbol* $\leq$, *if either the* $<$ *part or the* $=$ *part is true, then the inequality is true. This is also the case with the* $\geq$ *symbol.*

In the last row of the table, recall that the dot in $6 \cdot 4$ indicates the product $6 \times 4$, or 24, and $5(5)$ means $5 \times 5$, or 25. Thus, the inequality $6 \cdot 4 \leq 5(5)$ becomes $24 \leq 25$, which is true.

**NOW TRY** Exercise 95.

**OBJECTIVE 7** Graph sets of real numbers. Inequality symbols and variables are used to write sets of real numbers. For example, the set $\{x \mid x > -2\}$ consists of all the real numbers greater than $-2$. On a number line, we graph the elements of this set by drawing an arrow from $-2$ to the right. We use a parenthesis at $-2$ to indicate that $-2$ is *not* an element of the given set. See Figure 9.

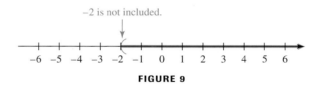

**FIGURE 9**

The set of numbers greater than $-2$ is an example of an **interval** on the number line. To write intervals, we use **interval notation.** We write the interval of all numbers greater than $-2$ as $(-2, \infty)$. The **infinity symbol** $\infty$ does not indicate a number; it shows that the interval includes all real numbers greater than $-2$. The left parenthesis indicates that $-2$ is not included. *A parenthesis is always used next to the infinity symbol.* The set of all real numbers is written in interval notation as $(-\infty, \infty)$.

**EXAMPLE 7** Graphing an Inequality Written in Interval Notation

Write $\{x \mid x < 4\}$ in interval notation and graph the interval.

The interval is written $(-\infty, 4)$. The graph is shown in Figure 10. Since the elements of the set are all real numbers *less than* 4, the graph extends to the left.

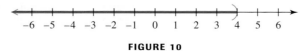

**FIGURE 10**

**NOW TRY** Exercise 101.

The set $\{x | x \leq -6\}$ includes all real numbers less than or equal to $-6$. To show that $-6$ is part of the set, a square bracket is used at $-6$, as shown in Figure 11. In interval notation, this set is written $(-\infty, -6]$.

**FIGURE 11**

**EXAMPLE 8** Graphing an Inequality Written in Interval Notation

Write $\{x | x \geq -4\}$ in interval notation and graph the interval.

This set is written in interval notation as $[-4, \infty)$. The graph is shown in Figure 12. We use a square bracket at $-4$, since $-4$ is part of the set.

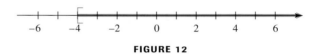

**FIGURE 12**

**NOW TRY** Exercise 103.

We sometimes graph sets of numbers that are *between* two given numbers. For example, the set $\{x | -2 < x < 4\}$ includes all real numbers between $-2$ and $4$, but *not* the numbers $-2$ and $4$ themselves. This set is written in interval notation as $(-2, 4)$. The graph has a heavy line between $-2$ and $4$, with parentheses at $-2$ and $4$. See Figure 13. The inequality $-2 < x < 4$, called a **three-part inequality,** is read "$-2$ is less than $x$ and $x$ is less than $4$," or "$x$ is between $-2$ and $4$."

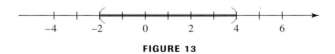

**FIGURE 13**

**EXAMPLE 9** Graphing a Three-Part Inequality

Write $\{x | 3 < x \leq 10\}$ in interval notation and graph the interval.

Use a parenthesis at $3$ and a square bracket at $10$ to get $(3, 10]$ in interval notation. The graph is shown in Figure 14. Read the inequality $3 < x \leq 10$ as "$3$ is less than $x$ and $x$ is less than or equal to $10$," or "$x$ is between $3$ and $10$, excluding $3$ and including $10$."

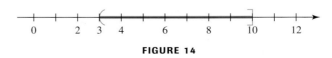

**FIGURE 14**

**NOW TRY** Exercise 109.

## 1.1 Exercises

*Write each set by listing its elements. See Example 1.*

1. $\{x \mid x \text{ is a natural number less than 6}\}$
2. $\{m \mid m \text{ is a natural number less than 9}\}$
3. $\{z \mid z \text{ is an integer greater than 4}\}$
4. $\{y \mid y \text{ is an integer greater than 8}\}$
5. $\{z \mid z \text{ is an integer less than or equal to 4}\}$
6. $\{p \mid p \text{ is an integer less than 3}\}$
7. $\{a \mid a \text{ is an even integer greater than 8}\}$
8. $\{k \mid k \text{ is an odd integer less than 1}\}$
9. $\{x \mid x \text{ is an irrational number that is also rational}\}$
10. $\{r \mid r \text{ is a number that is both positive and negative}\}$
11. $\{p \mid p \text{ is a number whose absolute value is 4}\}$
12. $\{w \mid w \text{ is a number whose absolute value is 7}\}$

*Write each set using set-builder notation. See Example 2. (More than one description is possible.)*

13. $\{2, 4, 6, 8\}$
14. $\{11, 12, 13, 14\}$
15. $\{4, 8, 12, 16, \ldots\}$
16. $\{\ldots, -6, -3, 0, 3, 6, \ldots\}$

17. *Concept Check* A student claimed that $\{x \mid x \text{ is a natural number greater than 3}\}$ and $\{y \mid y \text{ is a natural number greater than 3}\}$ actually name the same set, even though different variables are used. Was this student correct?

*Graph the elements of each set on a number line. See Objective 2.*

18. $\{-3, -1, 0, 4, 6\}$
19. $\{-4, -2, 0, 3, 5\}$
20. $\left\{-\dfrac{2}{3}, 0, \dfrac{4}{5}, \dfrac{12}{5}, \dfrac{9}{2}, 4.8\right\}$
21. $\left\{-\dfrac{6}{5}, -\dfrac{1}{4}, 0, \dfrac{5}{6}, \dfrac{13}{4}, 5.2, \dfrac{11}{2}\right\}$

*Which elements of each set are **(a)** natural numbers, **(b)** whole numbers, **(c)** integers, **(d)** rational numbers, **(e)** irrational numbers, **(f)** real numbers? See Example 3.*

22. $\left\{-8, -\sqrt{5}, -0.6, 0, \dfrac{3}{4}, \sqrt{3}, \pi, 5, \dfrac{13}{2}, 17, \dfrac{40}{2}\right\}$
23. $\left\{-9, -\sqrt{6}, -0.7, 0, \dfrac{6}{7}, \sqrt{7}, 4.\overline{6}, 8, \dfrac{21}{2}, 13, \dfrac{75}{5}\right\}$

24. *Concept Check* Give a real number that satisfies each condition.

(a) An integer between 6.75 and 7.75
(b) A rational number between $\tfrac{1}{4}$ and $\tfrac{3}{4}$
(c) A whole number that is not a natural number
(d) An integer that is not a whole number
(e) An irrational number between $\sqrt{4}$ and $\sqrt{9}$

*True or False* Decide whether each statement is true or false. If it is false, tell why. See Example 4.

25. Every integer is a whole number.
26. Every natural number is an integer.
27. Every irrational number is an integer.
28. Every integer is a rational number.
29. Every natural number is a whole number.
30. Some rational numbers are irrational.
31. Some rational numbers are whole numbers.
32. Some real numbers are integers.

**33.** The absolute value of any number is the same as the absolute value of its additive inverse.

**34.** The absolute value of any nonzero number is positive.

**35.** *Matching* Match each expression in Column I with its value in Column II. Choices in Column II may be used once, more than once, or not at all.

| I | II |
|---|---|
| (a) $-(-4)$ | A. 4 |
| (b) $|-4|$ | B. $-4$ |
| (c) $-|-4|$ | C. Both A and B |
| (d) $-|-(-4)|$ | D. Neither A nor B |

**36.** *Concept Check* For what value(s) of $x$ is $|x| = 4$ true?

*Give (a) the additive inverse and (b) the absolute value of each number. See the discussion of additive inverses and Example 5.*

**37.** 6    **38.** 8    **39.** $-12$    **40.** $-15$    **41.** $\dfrac{6}{5}$    **42.** 0.13

*Simplify by finding each absolute value. See Example 5.*

**43.** $|-8|$    **44.** $|-11|$    **45.** $\left|\dfrac{3}{2}\right|$    **46.** $\left|\dfrac{7}{4}\right|$

**47.** $-|5|$    **48.** $-|17|$    **49.** $-|-2|$    **50.** $-|-8|$

**51.** $-|4.5|$    **52.** $-|12.6|$    **53.** $|-2| + |3|$    **54.** $|-16| + |12|$

**55.** $|-9| - |-3|$    **56.** $|-10| - |-5|$

**57.** $|-1| + |-2| - |-3|$    **58.** $|-6| + |-4| - |-10|$

*Solve each problem. See Example 6.*

**59.** The table shows the percent change in population from 2000 through 2004 for selected cities in the United States.

| City | Percent Change |
|---|---|
| Las Vegas | 11.5 |
| Los Angeles | 4.1 |
| Chicago | $-1.2$ |
| Philadelphia | $-3.1$ |
| Phoenix | 7.3 |
| Detroit | $-5.4$ |

*Source:* U.S. Census Bureau.

(a) Which city had the greatest change in population? What was this change? Was it an increase or a decline?

(b) Which city had the least change in population? What was this change? Was it an increase or a decline?

**60.** The table gives the net trade balance, in millions of U.S. dollars, for selected U.S. trade partners for January 2006.

| Country | Trade Balance (in millions of dollars) |
|---|---|
| India | $-1257$ |
| China | $-17,911$ |
| Netherlands | 756 |
| France | $-85$ |
| Turkey | $-78$ |
| Australia | 925 |

*Source:* U.S. Census Bureau.

A negative balance means that imports to the U.S. exceeded exports from the U.S., while a positive balance means that exports exceeded imports.

(a) Which country had the greatest discrepancy between exports and imports? Explain.

(b) Which country had the least discrepancy between exports and imports? Explain.

*Sea level refers to the surface of the ocean. The depth of a body of water such as an ocean or sea can be expressed as a negative number, representing average depth in feet below sea level. By contrast, the altitude of a mountain can be expressed as a positive number, indicating its height in feet above sea level. The table gives selected depths and heights.*

| Body of Water | Average Depth in Feet (as a negative number) | Mountain | Altitude in Feet (as a positive number) |
| --- | --- | --- | --- |
| Pacific Ocean | −12,925 | McKinley | 20,320 |
| South China Sea | −4,802 | Point Success | 14,158 |
| Gulf of California | −2,375 | Matlalcueyetl | 14,636 |
| Caribbean Sea | −8,448 | Rainier | 14,410 |
| Indian Ocean | −12,598 | Steele | 16,644 |

*Source: World Almanac and Book of Facts.*

**61.** List the bodies of water in order, starting with the deepest and ending with the shallowest.

**62.** List the mountains in order, starting with the shortest and ending with the tallest.

**63.** *True or False* The absolute value of the depth of the Pacific Ocean is greater than the absolute value of the depth of the Indian Ocean.

**64.** *True or False* The absolute value of the depth of the Gulf of California is greater than the absolute value of the depth of the Caribbean Sea.

*Use the number line to answer* true *or* false *to each statement. See Objective 6.*

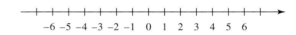

**65.** $-6 < -2$      **66.** $-4 < -3$      **67.** $-4 > -3$      **68.** $-2 > -1$
**69.** $3 > -2$      **70.** $5 > -3$      **71.** $-3 \geq -3$      **72.** $-4 \leq -4$

*Rewrite each statement with $>$ so that it uses $<$ instead; rewrite each statement with $<$ so that it uses $>$. See Objective 6.*

**73.** $6 > 2$      **74.** $4 > 1$      **75.** $-9 < 4$      **76.** $-5 < 1$
**77.** $-5 > -10$      **78.** $-8 > -12$      **79.** $0 < x$      **80.** $-2 < x$

*Use an inequality symbol to write each statement.*

**81.** 7 is greater than $y$.

**82.** −4 is less than 12.

**83.** 5 is greater than or equal to 5.

**84.** −3 is less than or equal to −3.

**85.** $3t - 4$ is less than or equal to 10.

**86.** $5x + 4$ is greater than or equal to 19.

**87.** $5x + 3$ is not equal to 0.

**88.** $6x + 7$ is not equal to −3.

**89.** $t$ is between −3 and 5.

**90.** $r$ is between −4 and 12.

**91.** $3x$ is between −3 and 4, including −3 and excluding 4.

**92.** $5y$ is between −2 and 6, excluding −2 and including 6.

*Simplify. Then tell whether the resulting statement is* true *or* false. *See Objective 6.*

**93.** $-6 < 7 + 3$      **94.** $-7 < 4 + 2$      **95.** $2 \cdot 5 \geq 4 + 6$
**96.** $8 + 7 \leq 3 \cdot 5$      **97.** $-|-3| \geq -3$      **98.** $-|-5| \leq -5$
**99.** $-8 > -|-6|$      **100.** $-9 > -|-4|$

*Write each set in interval notation and graph the interval. See Examples 7–9.*

**101.** $\{x \mid x > -1\}$      **102.** $\{x \mid x < 5\}$      **103.** $\{x \mid x \leq 6\}$

**104.** $\{x \mid x \geq -3\}$      **105.** $\{x \mid 0 < x < 3.5\}$      **106.** $\{x \mid -4 < x < 6.1\}$

**107.** $\{x \mid 2 \leq x \leq 7\}$      **108.** $\{x \mid -3 \leq x \leq -2\}$      **109.** $\{x \mid -4 < x \leq 3\}$

**110.** $\{x \mid 3 \leq x < 6\}$      **111.** $\{x \mid 0 < x \leq 3\}$      **112.** $\{x \mid -1 \leq x < 6\}$

## 1.2 Operations on Real Numbers

**OBJECTIVES**

1. Add real numbers.
2. Subtract real numbers.
3. Find the distance between two points on a number line.
4. Multiply real numbers.
5. Divide real numbers.

We review the rules for adding, subtracting, multiplying, and dividing real numbers.

**OBJECTIVE 1** Add real numbers. Recall that the answer to an addition problem is called the **sum**. The rules for adding real numbers follow.

**Adding Real Numbers**

***Same sign*** To add two numbers with the *same* sign, add their absolute values. The sum has the same sign as the given numbers.

***Different signs*** To add two numbers with *different* signs, find the absolute values of the numbers, and subtract the smaller absolute value from the larger. The sum has the same sign as the number with the larger absolute value.

**EXAMPLE 1** Adding Two Negative Real Numbers

Find each sum.

**(a)** $-12 + (-8)$

First find the absolute values.

$$|-12| = 12 \quad \text{and} \quad |-8| = 8$$

Because $-12$ and $-8$ have the *same* sign, add their absolute values.

*Both numbers are negative, so the answer will be negative.*

$-12 + (-8) = -(12 + 8)$    Add the absolute values.
$= -(20)$
$= -20$

**(b)** $-6 + (-3) = -(|-6| + |-3|)$    Add the absolute values.
$= -(6 + 3) = -9$

**(c)** $-1.2 + (-0.4) = -(1.2 + 0.4) = -1.6$

**(d)** $-\dfrac{5}{6} + \left(-\dfrac{1}{3}\right) = -\left(\dfrac{5}{6} + \dfrac{1}{3}\right)$    Add the absolute values. Both numbers are negative, so the answer will be negative.

$= -\left(\dfrac{5}{6} + \dfrac{2}{6}\right)$    The least common denominator is 6; $\dfrac{1 \cdot 2}{3 \cdot 2} = \dfrac{2}{6}$

$= -\dfrac{7}{6}$    Add numerators; keep the same denominator.

**NOW TRY** Exercise 11.

**EXAMPLE 2** Adding Real Numbers with Different Signs

Find each sum.

**(a)** $-17 + 11$

First find the absolute values.

$$|-17| = 17 \quad \text{and} \quad |11| = 11$$

Because $-17$ and $11$ have *different* signs, subtract their absolute values.

$$17 - 11 = 6$$

The number $-17$ has a larger absolute value than $11$, so the answer is negative.

$$-17 + 11 = -6 \quad \text{The sum is negative because } |-17| > |11|.$$

**(b)** $4 + (-1)$

Subtract the absolute values, 4 and 1. Because 4 has the larger absolute value, the sum must be positive.

$$4 + (-1) = 4 - 1 = 3 \quad \text{The sum is positive because } |4| > |-1|.$$

**(c)** $-9 + 17 = 17 - 9 = 8$

**(d)** $-2.3 + 5.6 = 5.6 - 2.3 = 3.3$

**(e)** $-16 + 12$

The absolute values are 16 and 12. Subtract the absolute values.

$$-16 + 12 = -(16 - 12) = -4 \quad \text{The sum is negative because } |-16| > |12|.$$

**(f)** $-\dfrac{4}{5} + \dfrac{2}{3}$

The least common denominator is 15. Write each fraction in terms of the common denominator.

$$-\frac{4}{5} = -\frac{4 \cdot 3}{5 \cdot 3} = -\frac{12}{15} \quad \text{and} \quad \frac{2}{3} = \frac{2 \cdot 5}{3 \cdot 5} = \frac{10}{15}$$

$$-\frac{4}{5} + \frac{2}{3} = -\frac{12}{15} + \frac{10}{15}$$

$$= -\left(\frac{12}{15} - \frac{10}{15}\right) \quad \text{Subtract the absolute values. } -\frac{12}{15} \text{ has the larger absolute value, so the answer will be negative.}$$

$$= -\frac{2}{15} \quad \text{Subtract numerators; keep the same denominator.}$$

**NOW TRY** Exercises 13, 15, and 17.

**OBJECTIVE 2** Subtract real numbers. Recall that the answer to a subtraction problem is called the **difference.** Compare the following two statements.

$$6 - 4 = 2$$
$$6 + (-4) = 2$$

Thus, $6 - 4 = 6 + (-4)$. To subtract 4 from 6, we add the additive inverse of 4 to 6. This example suggests the following definition of subtraction.

### Subtraction

For all real numbers $a$ and $b$,

$$a - b = a + (-b).$$

That is, to subtract $b$ from $a$, add the additive inverse (or opposite) of $b$ to $a$.

**EXAMPLE 3** Subtracting Real Numbers

Find each difference.

(a) $6 - 8 = 6 + (-8) = -2$     Change to addition. The additive inverse of 8 is $-8$.

(b) $-12 - 4 = -12 + (-4) = -16$     Change to addition. The additive inverse of 4 is $-4$.

(c) $-10 - (-7) = -10 + 7$     The additive inverse of $-7$ is 7.
$$= -3$$

(d) $-2.4 - (-8.1) = -2.4 + 8.1 = 5.7$

(e) $\dfrac{5}{6} - \left(-\dfrac{3}{8}\right) = \dfrac{5}{6} + \dfrac{3}{8}$     To subtract, add the additive inverse (opposite).

$$= \dfrac{5 \cdot 4}{6 \cdot 4} + \dfrac{3 \cdot 3}{8 \cdot 3}$$     Write each fraction with the least common denominator, 24.

$$= \dfrac{20}{24} + \dfrac{9}{24}$$

$$= \dfrac{29}{24}$$     Add numerators; keep the same denominator.

**NOW TRY** Exercises 19, 25, and 27.

When working a problem that involves both addition and subtraction, add and subtract in order from left to right. Work inside brackets or parentheses first.

**EXAMPLE 4** Adding and Subtracting Real Numbers

Perform the indicated operations.

(a) $15 - (-3) - 5 - 12$
$$= (15 + 3) - 5 - 12 \quad \text{Work from left to right.}$$
$$= 18 - 5 - 12$$
$$= 13 - 12$$
$$= 1$$

*(continued)*

**(b)** $-9 - [-8 - (-4)] + 6$
$= -9 - [-8 + 4] + 6$     Work inside brackets.
$= -9 - [-4] + 6$
$= -9 + 4 + 6$     Add the additive inverse.
$= -5 + 6$     Work from left to right.
$= 1$

**NOW TRY** Exercises 39 and 41.

**OBJECTIVE 3** Find the distance between two points on a number line. The number line in Figure 15 shows several points.

**FIGURE 15**

To find the distance between the points 4 and 7, we subtract $7 - 4 = 3$. Since distance is always positive (or 0), we must be careful to subtract in such a way that the answer is positive (or 0). Or, to avoid this problem altogether, we can find the absolute value of the difference. Then the distance between 4 and 7 is either

$$|7 - 4| = |3| = 3 \quad \text{or} \quad |4 - 7| = |-3| = 3.$$

This discussion can be summarized as follows.

### Distance

The **distance** between two points on a number line is the absolute value of the difference between their coordinates.

**EXAMPLE 5** Finding Distance between Points on the Number Line

Find the distance between each pair of points listed from Figure 15.

**(a)** 8 and $-4$

Find the absolute value of the difference of the numbers, taken in either order.

$$|8 - (-4)| = 12 \quad \text{or} \quad |-4 - 8| = 12$$

**(b)** $-4$ and $-6$

$$|-4 - (-6)| = 2 \quad \text{or} \quad |-6 - (-4)| = 2$$

**NOW TRY** Exercise 51.

**OBJECTIVE 4** Multiply real numbers. The answer to a multiplication problem is called the **product**. For example, 24 is the product of 8 and 3.

### Multiplying Real Numbers

**Same sign**    The product of two numbers with the *same* sign is positive.

**Different signs**    The product of two numbers with *different* signs is negative.

### EXAMPLE 6  Multiplying Real Numbers

Find each product.

**(a)** $-3(-9) = 27$     Same sign; product is positive.

**(b)** $-0.5(-0.4) = 0.2$

**(c)** $-\dfrac{3}{4}\left(-\dfrac{5}{6}\right) = \dfrac{15}{24}$     Multiply numerators; multiply denominators.

$\phantom{-\dfrac{3}{4}\left(-\dfrac{5}{6}\right)} = \dfrac{5 \cdot 3}{8 \cdot 3}$     Factor to write in lowest terms.

$\phantom{-\dfrac{3}{4}\left(-\dfrac{5}{6}\right)} = \dfrac{5}{8}$     Divide out the common factor, 3.

**(d)** $6(-9) = -54$     Different signs; product is negative.

**(e)** $-0.05(0.3) = -0.015$   **(f)** $-\dfrac{5}{8}\left(\dfrac{12}{13}\right) = -\dfrac{15}{26}$   **(g)** $\dfrac{2}{3}(-6) = -4$

$-6 = -\dfrac{6}{1}$

**NOW TRY** Exercises 59, 63, 65, and 71.

**OBJECTIVE 5**  **Divide real numbers.**  The result of dividing one number by another is called the **quotient**. The quotient of two real numbers $a$ and $b$ ($b \neq 0$) is the real number $q$ such that $q \cdot b = a$. That is,

$$a \div b = q \quad \text{only if} \quad q \cdot b = a.$$

For example, $36 \div 9 = 4$, since $4 \cdot 9 = 36$. Similarly, $35 \div (-5) = -7$, since $-7(-5) = 35$.

The quotient $a \div b$ can also be denoted $\dfrac{a}{b}$. Thus, $35 \div (-5)$ can be written $\dfrac{35}{-5}$. Then, $\dfrac{35}{-5} = -7$, because $-7$ answers the question, "What number multiplied by $-5$ gives the product 35?"

Now consider $\dfrac{5}{0}$. On the one hand, there is *no* number whose product with 0 gives 5. On the other hand, $\dfrac{0}{0}$ would be satisfied by *every* real number, because any number multiplied by 0 gives 0. When dividing, we always want a *unique* quotient. ***Therefore, division by 0 is undefined.***

**CAUTION** Division by 0 is undefined. However, dividing 0 by a nonzero number gives the quotient 0. For example,

$\dfrac{6}{0}$ is undefined,  but  $\dfrac{0}{6} = 0$   (because $0 \cdot 6 = 0$).

**Be careful when 0 is involved in a division problem.**

**NOW TRY** Exercises 79 and 81.

Recall that $\dfrac{a}{b} = a \cdot \dfrac{1}{b}$. ***Thus, dividing by $b$ is the same as multiplying by $\dfrac{1}{b}$.*** If $b \neq 0$, then $\dfrac{1}{b}$ is the **reciprocal**, or **multiplicative inverse,** of $b$. When multiplied, reciprocals have a product of 1. The table on the next page gives several numbers and their reciprocals. There is no reciprocal for 0 because there is no number that can be multiplied by 0 to give the product 1.

| Number | Reciprocal | | |
|---|---|---|---|
| $-\frac{2}{5}$ | $-\frac{5}{2}$ | $-\frac{2}{5}\left(-\frac{5}{2}\right) = 1$ | |
| $-6$ | $-\frac{1}{6}$ | $-6\left(-\frac{1}{6}\right) = 1$ | Reciprocals have a product of 1. |
| $\frac{7}{11}$ | $\frac{11}{7}$ | $\frac{7}{11}\left(\frac{11}{7}\right) = 1$ | |
| 0.05 | 20 | $0.05(20) = 1$ | |
| 0 | None | | |

> **CAUTION** *A number and its additive inverse have opposite signs; however, a number and its reciprocal always have the same sign.*

The preceding discussion suggests the following definition of division.

### Division

For all real numbers $a$ and $b$ (where $b \neq 0$),
$$a \div b = \frac{a}{b} = a \cdot \frac{1}{b}.$$

That is, multiply the first number by the reciprocal of the second number.

Since division is defined as multiplication by the reciprocal, the rules for signs of quotients are the same as those for signs of products.

### Dividing Real Numbers

***Same sign*** The quotient of two nonzero real numbers with the *same* sign is positive.

***Different signs*** The quotient of two nonzero real numbers with *different* signs is negative.

### EXAMPLE 7  Dividing Real Numbers

Find each quotient.

(a) $\dfrac{-12}{4} = -12 \cdot \dfrac{1}{4} = -3 \qquad \frac{a}{b} = a \cdot \frac{1}{b}$

(b) $\dfrac{6}{-3} = 6\left(-\dfrac{1}{3}\right) = -2 \qquad$ The reciprocal of $-3$ is $-\frac{1}{3}$.

(c) $\dfrac{-\frac{2}{3}}{-\frac{5}{9}} = -\dfrac{2}{3} \cdot \left(-\dfrac{9}{5}\right) = \dfrac{6}{5} \qquad$ The reciprocal of $-\frac{5}{9}$ is $-\frac{9}{5}$.

This is a *complex fraction* **(Section 7.3)**—a fraction that has a fraction in the numerator, the denominator, or both.

**(d)** $-\dfrac{9}{14} \div \dfrac{3}{7} = -\dfrac{9}{14} \cdot \dfrac{7}{3}$    Multiply by the reciprocal.

$= -\dfrac{63}{42}$    Multiply numerators; multiply denominators.

$= -\dfrac{7 \cdot 3 \cdot 3}{7 \cdot 3 \cdot 2}$    Factor.

$= -\dfrac{3}{2}$    Lowest terms

**NOW TRY** Exercises 73, 75, 83, and 85.

The rules for multiplication and division suggest the following results.

---

**Equivalent Forms of a Fraction**

The fractions $\dfrac{-x}{y}$, $\dfrac{x}{-y}$, and $-\dfrac{x}{y}$ are equivalent ($y \neq 0$).

*Example:* $\dfrac{-4}{7} = \dfrac{4}{-7} = -\dfrac{4}{7}$

The fractions $\dfrac{x}{y}$ and $\dfrac{-x}{-y}$ are equivalent ($y \neq 0$).

*Example:* $\dfrac{4}{7} = \dfrac{-4}{-7}$

---

Every fraction has three signs: the sign of the numerator, the sign of the denominator, and the sign of the fraction itself. Changing any two of these three signs does not change the value of the fraction. Changing only one sign, or changing all three, *does* change the value.

## 1.2 Exercises

**NOW TRY Exercise**

*Fill in the Blanks*  *Complete each statement and give an example.*

1. The sum of a positive number and a negative number is 0 if _____.
2. The sum of two positive numbers is a _____ number.
3. The sum of two negative numbers is a _____ number.
4. The sum of a positive number and a negative number is negative if _____.
5. The sum of a positive number and a negative number is positive if _____.
6. The difference between two positive numbers is negative if _____.
7. The difference between two negative numbers is negative if _____.
8. The product of two numbers with the same sign is _____.
9. The product of two numbers with different signs is _____.
10. The quotient formed by any nonzero number divided by 0 is _____, and the quotient formed by 0 divided by any nonzero number is _____.

*Add or subtract as indicated. See Examples 1–3.*

**11.** $-6 + (-13)$    **12.** $-8 + (-15)$    **13.** $13 + (-4)$

**14.** $19 + (-13)$    **15.** $-\dfrac{7}{3} + \dfrac{3}{4}$    **16.** $-\dfrac{5}{6} + \dfrac{3}{8}$

**17.** $-2.3 + 0.45$
**18.** $-0.238 + 4.55$
**19.** $-6 - 5$
**20.** $-8 - 13$
**21.** $8 - (-13)$
**22.** $13 - (-22)$
**23.** $-16 - (-3)$
**24.** $-21 - (-8)$
**25.** $-12.31 - (-2.13)$
**26.** $-15.88 - (-9.22)$
**27.** $\dfrac{9}{10} - \left(-\dfrac{4}{3}\right)$
**28.** $\dfrac{3}{14} - \left(-\dfrac{1}{4}\right)$
**29.** $|-8 - 6|$
**30.** $|-7 - 9|$
**31.** $-|-4 + 9|$
**32.** $-|-5 + 7|$
**33.** $-2 - |-4|$
**34.** $9 - |-13|$

*Perform the indicated operations. See Example 4.*

**35.** $-7 + 5 - 9$
**36.** $-12 + 13 - 19$
**37.** $6 - (-2) + 8$
**38.** $7 - (-3) + 12$
**39.** $-9 - 4 - (-3) + 6$
**40.** $-10 - 5 - (-12) + 8$
**41.** $-8 - (-12) - (2 - 6)$
**42.** $-3 + (-14) + (-5 + 3)$
**43.** $-0.382 + 4 - 0.6$
**44.** $3 - 2.94 - (-0.63)$
**45.** $\left(-\dfrac{5}{4} - \dfrac{2}{3}\right) + \dfrac{1}{6}$
**46.** $\left(-\dfrac{5}{8} + \dfrac{1}{4}\right) - \left(-\dfrac{1}{4}\right)$
**47.** $-\dfrac{3}{4} - \left(\dfrac{1}{2} - \dfrac{3}{8}\right)$
**48.** $\dfrac{7}{5} - \left(\dfrac{9}{10} - \dfrac{3}{2}\right)$
**49.** $|-11| - |-5| - |7| + |-2|$
**50.** $|-6| + |-3| - |4| - |-8|$

*The number line has several points labeled. Find the distance between each pair of points. See Example 5.*

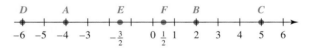

**51.** $A$ and $B$
**52.** $A$ and $C$
**53.** $D$ and $F$
**54.** $E$ and $C$

**55.** A statement that is often heard is "Two negatives give a positive." When is this true? When is it not true? Give a more precise statement that conveys this message.

**56.** Explain why the reciprocal of a nonzero number must have the same sign as the number.

*Multiply. See Example 6.*

**57.** $5(-7)$
**58.** $6(-6)$
**59.** $-8(-5)$
**60.** $-10(-4)$
**61.** $-10\left(-\dfrac{1}{5}\right)$
**62.** $-\dfrac{1}{2}(-12)$
**63.** $\dfrac{3}{4}(-16)$
**64.** $\dfrac{4}{5}(-35)$
**65.** $-\dfrac{5}{2}\left(-\dfrac{12}{25}\right)$
**66.** $-\dfrac{9}{7}\left(-\dfrac{35}{36}\right)$
**67.** $-\dfrac{3}{8}\left(-\dfrac{24}{9}\right)$
**68.** $-\dfrac{2}{11}\left(-\dfrac{99}{4}\right)$
**69.** $-2.4(-2.45)$
**70.** $-3.45(-2.14)$
**71.** $3.4(-3.14)$
**72.** $5.66(-2.1)$

*Divide where possible. See Example 7.*

**73.** $\dfrac{-14}{2}$
**74.** $\dfrac{-26}{13}$
**75.** $\dfrac{-24}{-4}$
**76.** $\dfrac{-36}{-9}$
**77.** $\dfrac{100}{-25}$
**78.** $\dfrac{300}{-60}$
**79.** $\dfrac{0}{-8}$
**80.** $\dfrac{0}{-10}$
**81.** $\dfrac{5}{0}$
**82.** $\dfrac{12}{0}$
**83.** $-\dfrac{10}{17} \div \left(-\dfrac{12}{5}\right)$
**84.** $-\dfrac{22}{23} \div \left(-\dfrac{33}{4}\right)$
**85.** $\dfrac{\dfrac{12}{13}}{-\dfrac{4}{3}}$
**86.** $\dfrac{\dfrac{5}{6}}{-\dfrac{1}{30}}$

87. $\dfrac{-27.72}{13.2}$   88. $\dfrac{-126.7}{36.2}$   89. $\dfrac{-100}{-0.01}$   90. $\dfrac{-50}{-0.05}$

*Solve each problem.*

91. The highest temperature ever recorded in Juneau, Alaska, was 90°F. The lowest temperature ever recorded there was −22°F. What is the difference between these two temperatures? (*Source: World Almanac and Book of Facts.*)

92. On August 10, 1936, a temperature of 120°F was recorded in Ponds, Arkansas. On February 13, 1905, Ozark, Arkansas, recorded a temperature of −29°F. What is the difference between these two temperatures? (*Source: World Almanac and Book of Facts.*)

93. Andrew McGinnis has $48.35 in his checking account. He uses his debit card to make purchases of $35.99 and $20.00, which overdraws his account. His bank charges his account an overdraft fee of $28.50. He then deposits his paycheck for $66.27 from his part-time job at Arby's. What is the balance in his account?

94. Kayla Koolbeck has $37.50 in her checking account. She uses her debit card to make purchases of $25.99 and $19.34, which overdraws her account. Her bank charges her account an overdraft fee of $25.00. She then deposits her paycheck for $58.66 from her part-time job at Subway. What is the balance in her account?

## 1.3 Exponents, Roots, and Order of Operations

**OBJECTIVES**

1. Use exponents.
2. Find square roots.
3. Use the order of operations.
4. Evaluate algebraic expressions for given values of variables.

Two or more numbers whose product is a third number are **factors** of that third number. For example, 2 and 6 are factors of 12, since $2 \cdot 6 = 12$. Other integer factors of 12 are 1, 3, 4, 12, −1, −2, −3, −4, −6, and −12.

**OBJECTIVE 1** Use exponents. In algebra, we use *exponents* as a way of writing products of repeated factors. For example, the product $2 \cdot 2 \cdot 2 \cdot 2 \cdot 2$ is written

$$\underbrace{2 \cdot 2 \cdot 2 \cdot 2 \cdot 2}_{\text{5 factors of 2}} = 2^5.$$

The number 5 shows that 2 is used as a factor 5 times. The number 5 is the *exponent*, and 2 is the *base*.

$$2^5 \leftarrow \text{Exponent}$$
$$\uparrow$$
$$\text{Base}$$

Read $2^5$ as "2 to the fifth power," or "2 to the fifth." Multiplying the five 2s gives

$$2^5 = 2 \cdot 2 \cdot 2 \cdot 2 \cdot 2 = 32.$$

**Exponential Expression**

If $a$ is a real number and $n$ is a natural number, then

$$a^n = \underbrace{a \cdot a \cdot a \cdot \ldots \cdot a}_{n \text{ factors of } a},$$

where $n$ is the **exponent**, $a$ is the **base**, and $a^n$ is an **exponential expression**. Exponents are also called **powers**.

### EXAMPLE 1  Using Exponential Notation

Write using exponents.

(a) $4 \cdot 4 \cdot 4$
Here, 4 is used as a factor 3 times.
$$\underbrace{4 \cdot 4 \cdot 4}_{\text{3 factors of 4}} = 4^3$$
Read $4^3$ as "4 cubed."

(b) $\dfrac{3}{5} \cdot \dfrac{3}{5} = \left(\dfrac{3}{5}\right)^2$  2 factors of $\dfrac{3}{5}$
Read $\left(\dfrac{3}{5}\right)^2$ as "$\dfrac{3}{5}$ squared."

(c) $(-6)(-6)(-6)(-6) = (-6)^4$
Read $(-6)^4$ as "$-6$ to the fourth power," or "$-6$ to the fourth."

(d) $(0.3)(0.3)(0.3)(0.3)(0.3) = (0.3)^5$

(e) $x \cdot x \cdot x \cdot x \cdot x \cdot x = x^6$

**NOW TRY** Exercises 13, 15, 17, and 19.

(a) $3 \cdot 3 = 3$ squared, or $3^2$

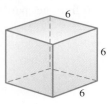

(b) $6 \cdot 6 \cdot 6 = 6$ cubed, or $6^3$

**FIGURE 16**

In parts (a) and (b) of Example 1, we used the terms *squared* and *cubed* to refer to powers of 2 and 3, respectively. The term *squared* comes from the figure of a square, which has the same measure for both length and width, as shown in Figure 16(a). Similarly, the term *cubed* comes from the figure of a cube. As shown in Figure 16(b), the length, width, and height of a cube have the same measure.

### EXAMPLE 2  Evaluating Exponential Expressions

Evaluate.

(a) $5^2 = 5 \cdot 5 = 25$   5 is used as a factor 2 times.

$5^2 = 5 \cdot 5$, **NOT** $5 \cdot 2$.

(b) $\left(\dfrac{2}{3}\right)^3 = \dfrac{2}{3} \cdot \dfrac{2}{3} \cdot \dfrac{2}{3} = \dfrac{8}{27}$   $\dfrac{2}{3}$ is used as a factor 3 times.

(c) $2^6 = 2 \cdot 2 \cdot 2 \cdot 2 \cdot 2 \cdot 2 = 64$

**NOW TRY** Exercises 21 and 27.

### EXAMPLE 3  Evaluating Exponential Expressions with Negative Signs

Evaluate.

(a) $(-3)^5 = (-3)(-3)(-3)(-3)(-3) = -243$   The base is $-3$.

(b) $(-2)^6 = (-2)(-2)(-2)(-2)(-2)(-2) = 64$   The base is $-2$.

(c) $-2^6$
There are no parentheses. The exponent 6 applies *only* to the number 2, not to $-2$.
$$-2^6 = -(2 \cdot 2 \cdot 2 \cdot 2 \cdot 2 \cdot 2) = -64 \quad \text{The base is 2.}$$

**NOW TRY** Exercises 29, 31, and 33.

Examples 3(a) and (b) suggest the following generalizations.

> The product of an *odd* number of negative factors is negative.
> The product of an *even* number of negative factors is positive.

**CAUTION** As shown in Examples 3(b) and (c), it is important to distinguish between $-a^n$ and $(-a)^n$.

$$-a^n = -1\underbrace{(a \cdot a \cdot a \cdot \ldots \cdot a)}_{n \text{ factors of } a} \quad \text{The base is } a.$$

$$(-a)^n = \underbrace{(-a)(-a) \cdot \ldots \cdot (-a)}_{n \text{ factors of } -a} \quad \text{The base is } -a.$$

*Be careful when evaluating an exponential expression with a negative sign.*

**OBJECTIVE 2** Find square roots. As we saw in Example 2(a), $5^2 = 5 \cdot 5 = 25$, so 5 squared is 25. The opposite (inverse) of squaring a number is called taking its **square root**. For example, a square root of 25 is 5. Another square root of 25 is $-5$, since $(-5)^2 = 25$. Thus, 25 has two square roots: 5 and $-5$.

We write the **positive** or **principal square root** of a number with the symbol $\sqrt{\phantom{x}}$, called a **radical sign**. For example, the positive or principal square root of 25 is written $\sqrt{25} = 5$. The **negative square root** of 25 is written $-\sqrt{25} = -5$. *Since the square of any nonzero real number is positive, the square root of a negative number, such as $\sqrt{-25}$, is not a real number.*

**EXAMPLE 4** Finding Square Roots

Find each square root that is a real number.

(a) $\sqrt{36} = 6$, since 6 is positive and $6^2 = 36$.

(b) $\sqrt{0} = 0$, since $0^2 = 0$.

(c) $\sqrt{\dfrac{9}{16}} = \dfrac{3}{4}$, since $\left(\dfrac{3}{4}\right)^2 = \dfrac{9}{16}$.

(d) $\sqrt{0.16} = 0.4$, since $(0.4)^2 = 0.16$.

(e) $\sqrt{100} = 10$, since $10^2 = 100$.

(f) $-\sqrt{100} = -10$, since the negative sign is outside the radical sign.

(g) $\sqrt{-100}$ is not a real number, because the negative sign is inside the radical sign. No *real number* squared equals $-100$.

Notice the difference among the square roots in parts (e), (f), and (g). Part (e) is the positive or principal square root of 100, part (f) is the negative square root of 100, and part (g) is the square root of $-100$, which is not a real number.

**NOW TRY** Exercises 37, 41, 43, and 47.

**CAUTION** The symbol $\sqrt{\phantom{x}}$ is used only for the *positive* square root, except that $\sqrt{0} = 0$. The symbol $-\sqrt{\phantom{x}}$ is used for the negative square root.

**OBJECTIVE 3** Use the order of operations. To simplify $5 + 2 \cdot 3$, what should we do first—add 5 and 2 or multiply 2 and 3? When an expression involves more than one operation symbol, we use the following **order of operations**.

### Order of Operations

1. Work separately above and below any **fraction bar.**
2. If **grouping symbols** such as **parentheses ( )**, **brackets [ ]**, or **absolute value bars | |** are present, start with the innermost set and work outward.
3. Evaluate all **powers, roots,** and **absolute values.**
4. **Multiply** or **divide** in order from left to right.
5. **Add** or **subtract** in order from left to right.

**EXAMPLE 5** Using the Order of Operations

Simplify.

(a) $5 + 2 \cdot 3$
$= 5 + 6$     Multiply.
$= 11$     Add.

(b) $24 \div 3 \cdot 2 + 6$

*Multiplications and divisions are done in the order in which they appear from left to right,* so divide first.

$24 \div 3 \cdot 2 + 6$
$= 8 \cdot 2 + 6$     Divide.
$= 16 + 6$     Multiply.
$= 22$     Add.

**NOW TRY** Exercises 53 and 57.

**EXAMPLE 6** Using the Order of Operations

Simplify.

(a) $10 \div 5 + 2|3 - 4|$
$= 10 \div 5 + 2|-1|$     Subtract inside the absolute value bars.
$= 10 \div 5 + 2 \cdot 1$     Take the absolute value.
$= 2 + 2$     Divide; multiply.
$= 4$     Add.

(b) $4 \cdot 3^2 + 7 - (2 + 8)$
$= 4 \cdot 3^2 + 7 - 10$     Add inside parentheses.
$= 4 \cdot 9 + 7 - 10$     Evaluate the power.
$= 36 + 7 - 10$     Multiply.
$= 43 - 10$     Add.
$= 33$     Subtract.

$3^2 = 3 \cdot 3$, NOT $3 \cdot 2$.

(c) $\frac{1}{2} \cdot 4 + (6 \div 3 - 7)$
$= \frac{1}{2} \cdot 4 + (2 - 7)$     Divide inside parentheses.
$= \frac{1}{2} \cdot 4 + (-5)$     Subtract inside parentheses.
$= 2 + (-5)$     Multiply.
$= -3$     Add.

**NOW TRY** Exercises 65 and 71.

### EXAMPLE 7  Using the Order of Operations

Simplify $\dfrac{5 + (-2^3)(2)}{6 \cdot \sqrt{9} - 9 \cdot 2}$.

Work separately above and below the fraction bar.

$$\dfrac{5 + (-2^3)(2)}{6 \cdot \sqrt{9} - 9 \cdot 2}$$

$$= \dfrac{5 + (-8)(2)}{6 \cdot 3 - 9 \cdot 2} \quad \text{Evaluate the power and the root.}$$

$$= \dfrac{5 - 16}{18 - 18} \quad \text{Multiply.}$$

$$= \dfrac{-11}{0} \quad \text{Subtract.}$$

Since division by 0 is undefined, the given expression is undefined.

**NOW TRY** Exercise 75.

**OBJECTIVE 4** Evaluate algebraic expressions for given values of variables. Any sequence of numbers, variables, operation symbols, and/or grouping symbols formed in accordance with the rules of algebra is called an **algebraic expression.**

$$6ab, \quad 5m - 9n, \quad \text{and} \quad -2(x^2 + 4y) \quad \text{Algebraic expressions}$$

Algebraic expressions have different numerical values for different values of the variables. We evaluate such expressions by *substituting* given values for the variables.

For example, if movie tickets cost $8 each, the amount in dollars you pay for $x$ tickets can be represented by the algebraic expression $8x$. We can substitute different numbers of tickets to get the costs of purchasing those tickets.

### EXAMPLE 8  Evaluating Algebraic Expressions

Evaluate each expression if $m = -4$, $n = 5$, $p = -6$, and $q = 25$.

*Use parentheses around substituted values to avoid errors.*

**(a)** $5m - 9n$

$= 5(-4) - 9(5)$    Substitute; let $m = -4$ and $n = 5$.

$= -20 - 45$    Multiply.

$= -65$    Subtract.

**(b)** $\dfrac{m + 2n}{4p}$

$= \dfrac{-4 + 2(5)}{4(-6)}$    Substitute; let $m = -4$, $n = 5$, and $p = -6$.

$= \dfrac{-4 + 10}{-24}$    Work separately above and below the fraction bar.

$= \dfrac{6}{-24} = -\dfrac{1}{4}$    Write in lowest terms; also, $\dfrac{a}{-b} = -\dfrac{a}{b}$.

**(c)** $-3m^3 - n^2(\sqrt{q})$

$= -3(-4)^3 - (5)^2(\sqrt{25})$    Substitute; let $m = -4$, $n = 5$, and $q = 25$.

$= -3(-64) - 25(5)$    Evaluate the powers and the root.

$= 192 - 125$    Multiply.

$= 67$    Subtract.

Notice the careful use of parentheses around substituted values.

**NOW TRY** Exercises 79 and 85.

## 1.3 Exercises

**NOW TRY Exercise**

*True or False* Decide whether each statement is true or false. If it is false, correct the statement so that it is true.

1. $-4^6 = (-4)^6$
2. $-4^7 = (-4)^7$
3. $\sqrt{16}$ is a positive number.
4. $3 + 5 \cdot 6 = 3 + (5 \cdot 6)$
5. $(-2)^7$ is a negative number.
6. $(-2)^8$ is a positive number.
7. The product of 8 positive factors and 8 negative factors is positive.
8. The product of 3 positive factors and 3 negative factors is positive.
9. In the exponential expression $-3^5$, $-3$ is the base.
10. $\sqrt{a}$ is positive for all positive numbers $a$.

*Concept Check* In Exercises 11 and 12, evaluate each exponential expression.

11. **(a)** $8^2$    **(b)** $-8^2$    **(c)** $(-8)^2$    **(d)** $-(-8)^2$

12. **(a)** $4^3$    **(b)** $-4^3$    **(c)** $(-4)^3$    **(d)** $-(-4)^3$

*Write each expression by using exponents. See Example 1.*

13. $10 \cdot 10 \cdot 10 \cdot 10$
14. $8 \cdot 8 \cdot 8$
15. $\dfrac{3}{4} \cdot \dfrac{3}{4} \cdot \dfrac{3}{4} \cdot \dfrac{3}{4} \cdot \dfrac{3}{4}$
16. $\dfrac{1}{2} \cdot \dfrac{1}{2}$
17. $(-9)(-9)(-9)$
18. $(-4)(-4)(-4)(-4)$
19. $z \cdot z \cdot z \cdot z \cdot z \cdot z \cdot z$
20. $a \cdot a \cdot a \cdot a \cdot a$

*Evaluate each expression. See Examples 2 and 3.*

21. $4^2$
22. $2^4$
23. $0.28^3$
24. $0.91^3$
25. $\left(\dfrac{1}{5}\right)^3$
26. $\left(\dfrac{1}{6}\right)^4$
27. $\left(\dfrac{4}{5}\right)^4$
28. $\left(\dfrac{7}{10}\right)^3$
29. $(-5)^3$
30. $(-2)^5$
31. $(-2)^8$
32. $(-3)^6$
33. $-3^6$
34. $-4^6$
35. $-8^4$
36. $-10^3$

*Find each square root. If it is not a real number, say so. See Example 4.*

37. $\sqrt{81}$
38. $\sqrt{64}$
39. $\sqrt{169}$
40. $\sqrt{225}$
41. $-\sqrt{400}$
42. $-\sqrt{900}$
43. $\sqrt{\dfrac{100}{121}}$
44. $\sqrt{\dfrac{225}{169}}$
45. $-\sqrt{0.49}$
46. $-\sqrt{0.64}$
47. $\sqrt{-36}$
48. $\sqrt{-121}$

49. *Matching* Match each square root with the appropriate value or description.

   **(a)** $\sqrt{144}$    **(b)** $\sqrt{-144}$    **(c)** $-\sqrt{144}$

   **A.** $-12$    **B.** $12$    **C.** Not a real number

50. Explain why $\sqrt{-900}$ is not a real number.

*Concept Check* In Exercises 51 and 52, a represents a positive number.

**51.** Is $-\sqrt{-a}$ positive, negative, or not a real number?

**52.** Is $-\sqrt{a}$ positive, negative, or not a real number?

*Simplify each expression. Use the order of operations. See Examples 5–7.*

**53.** $12 + 3 \cdot 4$      **54.** $15 + 5 \cdot 2$      **55.** $6 \cdot 3 - 12 \div 4$

**56.** $9 \cdot 4 - 8 \div 2$      **57.** $10 + 30 \div 2 \cdot 3$      **58.** $12 + 24 \div 3 \cdot 2$

**59.** $-3(5)^2 - (-2)(-8)$      **60.** $-9(2)^2 - (-3)(-2)$      **61.** $5 - 7 \cdot 3 - (-2)^3$

**62.** $-4 - 3 \cdot 5 + 6^2$      **63.** $-7(\sqrt{36}) - (-2)(-3)$      **64.** $-8(\sqrt{64}) - (-3)(-7)$

**65.** $6|4 - 5| - 24 \div 3$      **66.** $-4|2 - 4| + 8 \cdot 2$      **67.** $|-6 - 5|(-8) + 3^2$

**68.** $(-6 - 3)|-2 - 3| \div 9$      **69.** $6 + \frac{2}{3}(-9) - \frac{5}{8} \cdot 16$      **70.** $7 - \frac{3}{4}(-8) + 12 \cdot \frac{5}{6}$

**71.** $-14\left(-\frac{2}{7}\right) \div (2 \cdot 6 - 10)$      **72.** $-12\left(-\frac{3}{4}\right) - (6 \cdot 5 \div 3)$

**73.** $\dfrac{(-5 + \sqrt{4})(-2^2)}{-5 - 1}$      **74.** $\dfrac{(-9 + \sqrt{16})(-3^2)}{-4 - 1}$

**75.** $\dfrac{2(-5) + (-3)(-2)}{-8 + 3^2 - 1}$      **76.** $\dfrac{3(-4) + (-5)(-8)}{2^3 - 2 - 6}$

**77.** $\dfrac{5 - 3\left(\dfrac{-5 - 9}{-7}\right) - 6}{-9 - 11 + 3 \cdot 7}$      **78.** $\dfrac{-4\left(\dfrac{12 - (-8)}{3 \cdot 2 + 4}\right) - 5(-1 - 7)}{-9 - (-7) - [-5 - (-8)]}$

*Evaluate each expression if $a = -3$, $b = 64$, and $c = 6$. See Example 8.*

**79.** $3a + \sqrt{b}$      **80.** $-2a - \sqrt{b}$      **81.** $\sqrt{b} + c - a$      **82.** $\sqrt{b} - c + a$

**83.** $4a^3 + 2c$      **84.** $-3a^4 - 3c$      **85.** $\dfrac{2c + a^3}{4b + 6a}$      **86.** $\dfrac{3c + a^2}{2b - 6c}$

*Evaluate each expression if $w = 4$, $x = -\frac{3}{4}$, $y = \frac{1}{2}$, and $z = 1.25$. See Example 8.*

**87.** $wy - 8x$      **88.** $wz - 12y$      **89.** $xy + y^4$      **90.** $xy - x^2$

**91.** $-w + 2x + 3y + z$      **92.** $w - 6x + 5y - 3z$      **93.** $\dfrac{7x + 9y}{w}$      **94.** $\dfrac{7y - 5x}{2w}$

*Solve each problem.*

**95.** An approximation of the amount in billions of dollars that Americans have spent on their pets from 1998 to 2009 can be obtained by substituting a given year for $x$ in the expression

$$2.076x - 4125.$$

(*Source:* American Pet Products Association.) Approximate the amount spent in each year. Round answers to the nearest tenth.

  **(a)** 1998    **(b)** 2005    **(c)** 2009

  **(d)** How has the amount Americans have spent on their pets changed from 1998 to 2009?

**96.** An approximation of federal spending on education in billions of dollars from 2001 through 2005 can be obtained using the expression

$$y = 9.0499x - 18,071.87,$$

where $x$ represents the year. (*Source:* U.S. Department of the Treasury.)

**(a)** Use this expression to complete the table. Round answers to the nearest tenth.

| Year | Education Spending (in billions of dollars) |
|---|---|
| 2001 | 37.0 |
| 2002 | 46.0 |
| 2003 | |
| 2004 | |
| 2005 | |

**(b)** How has the amount of federal spending on education changed from 2001 to 2005?

## 1.4 Properties of Real Numbers

**OBJECTIVES**

1. Use the distributive property.
2. Use the inverse properties.
3. Use the identity properties.
4. Use the commutative and associative properties.
5. Use the multiplication property of 0.

The study of any object is simplified when we know the properties of the object. For example, a property of water is that it freezes when cooled to 0°C. Knowing this helps us to predict the behavior of water.

The study of numbers is no different. The basic properties of real numbers studied in this section reflect results that occur consistently in work with numbers, so they have been generalized to apply to expressions with variables as well.

**OBJECTIVE 1** Use the distributive property. Notice that

$$2(3 + 5) = 2 \cdot 8 = 16$$

and 
$$2 \cdot 3 + 2 \cdot 5 = 6 + 10 = 16,$$

so 
$$2(3 + 5) = 2 \cdot 3 + 2 \cdot 5.$$

This idea is illustrated by the divided rectangle in Figure 17. Similarly,

$$-4[5 + (-3)] = -4(2) = -8$$

and 
$$-4(5) + (-4)(-3) = -20 + 12 = -8,$$

so 
$$-4[5 + (-3)] = -4(5) + (-4)(-3).$$

These examples are generalized to *all* real numbers as the **distributive property of multiplication with respect to addition,** or simply the **distributive property.**

Area of left part is $2 \cdot 3 = 6$.
Area of right part is $2 \cdot 5 = 10$.
Area of total rectangle is $2(3 + 5) = 16$.

**FIGURE 17**

**Distributive Property**

For any real numbers $a$, $b$, and $c$,

$$a(b + c) = ab + ac \quad \text{and} \quad (b + c)a = ba + ca.$$

The distributive property can also be written

$$ab + ac = a(b + c) \quad \text{and} \quad ba + ca = (b + c)a.$$

and can be extended to more than two numbers as well.

$$a(b + c + d) = ab + ac + ad$$

*The distributive property provides a way to rewrite a product $a(b + c)$ as a sum $ab + ac$ or a sum as a product.*

When we rewrite $a(b + c)$ as $ab + ac$, we sometimes refer to the process as "removing" or "clearing" parentheses.

### EXAMPLE 1  Using the Distributive Property

Use the distributive property to rewrite each expression.

**(a)** $3(x + y)$
$= 3x + 3y$    Use the first form of the property to rewrite the given product as a sum.

**(b)** $-2(5 + k)$
$= -2(5) + (-2)(k)$
$= -10 - 2k$

**(c)** $4x + 8x$
$= (4 + 8)x$    Use the second form of the property to rewrite the given sum as a product.
$= 12x$

**(d)** $3r - 7r$
$= 3r + (-7r)$    Definition of subtraction
$= [3 + (-7)]r$    Distributive property
$= -4r$

**(e)** $5p + 7q$

Because there is no common number or variable here, we cannot use the distributive property to rewrite the expression.

**(f)** $6(x + 2y - 3z)$
$= 6x + 6(2y) + 6(-3z)$
$= 6x + 12y - 18z$

**NOW TRY** Exercises 11, 13, 15, and 19.

The distributive property can also be used for subtraction (Example 1(d)), so

$$a(b - c) = ab - ac.$$

**OBJECTIVE 2** Use the inverse properties. In **Section 1.1**, we saw that the *additive inverse* (or *opposite*) of a number $a$ is $-a$ and that additive inverses have a sum of 0.

$5$ and $-5$,   $-\dfrac{1}{2}$ and $\dfrac{1}{2}$,   $-34$ and $34$    Additive inverses (sum of 0)

In **Section 1.2**, we saw that the *multiplicative inverse* (or *reciprocal*) of a number $a$ is $\dfrac{1}{a}$ (where $a \neq 0$) and that multiplicative inverses have a product of 1.

$5$ and $\dfrac{1}{5}$,   $-\dfrac{1}{2}$ and $-2$,   $\dfrac{3}{4}$ and $\dfrac{4}{3}$    Multiplicative inverses (product of 1)

This discussion leads to the **inverse properties** of addition and multiplication, which can be extended to the real numbers of algebra.

> **Inverse Properties**
>
> For any real number $a$,
> $$a + (-a) = 0 \quad \text{and} \quad -a + a = 0$$
> $$a \cdot \frac{1}{a} = 1 \quad \text{and} \quad \frac{1}{a} \cdot a = 1 \quad (a \neq 0).$$

The inverse properties "undo" addition or multiplication. Think of putting on your shoes when you get up in the morning and then taking them off before you go to bed at night. These are inverse operations that undo each other.

**OBJECTIVE 3** Use the identity properties. The numbers 0 and 1 each have a special property. Zero is the only number that can be added to any number to get that number. Adding 0 to any number leaves the identity of the number unchanged. For this reason, 0 is called the **identity element for addition,** or the **additive identity.** In a similar way, multiplying any number by 1 leaves the identity of the number unchanged, so 1 is the **identity element for multiplication,** or the **multiplicative identity.** The **identity properties** summarize this discussion and extend these properties from arithmetic to algebra.

> **Identity Properties**
>
> For any real number $a$,
> $$a + 0 = 0 + a = a$$
> $$a \cdot 1 = 1 \cdot a = a.$$

The identity properties leave the identity of a real number unchanged. Think of a child wearing a costume on Halloween. The child's appearance is changed, but his or her identity is unchanged.

**EXAMPLE 2** Using the Identity Property $1 \cdot a = a$

Simplify each expression.

**(a)** $12m + m$
$= 12m + 1m$   Identity property; $m = 1 \cdot m$, or $1m$
$= (12 + 1)m$   Distributive property
$= 13m$   Add inside parentheses.

**(b)** $y + y$
$= 1y + 1y$   Identity property
$= (1 + 1)y$   Distributive property
$= 2y$   Add inside parentheses.

**(c)**  $-(m - 5n)$
$= -1(m - 5n)$   Identity property
$= -1(m) + (-1)(-5n)$   Distributive property
$= -m + 5n$   Multiply.

*Multiply each term by $-1$. Be careful with signs.*

**NOW TRY** Exercises 21 and 23.

Expressions such as $12m$ and $5n$ from Example 2 are examples of *terms*. A **term** is a number or the product of a number and one or more variables raised to powers. The numerical factor in a term is called the **numerical coefficient,** or just the **coefficient.** Some examples of terms and their coefficients are shown in the table in the margin.

Terms with exactly the same variables raised to exactly the same powers are called **like terms.** Some examples of like terms are

| Term | Numerical Coefficient |
|---|---|
| $-7y$ | $-7$ |
| $34r^3$ | $34$ |
| $-26x^5yz^4$ | $-26$ |
| $-k = -1k$ | $-1$ |
| $r = 1r$ | $1$ |
| $\frac{3x}{8} = \frac{3}{8}x$ | $\frac{3}{8}$ |
| $\frac{x}{3} = \frac{1x}{3} = \frac{1}{3}x$ | $\frac{1}{3}$ |

$$5p \text{ and } -21p \qquad -6x^2 \text{ and } 9x^2. \qquad \text{Like terms}$$

Some examples of **unlike terms** are

$$3m \text{ and } 16x \qquad 7y^3 \text{ and } -3y^2. \qquad \text{Unlike terms}$$

**OBJECTIVE 4** Use the commutative and associative properties. Simplifying expressions as in parts (a) and (b) of Example 2 is called **combining like terms.** *Only like terms may be combined.* To combine like terms in an expression such as

$$-2m + 5m + 3 - 6m + 8,$$

we need two more properties. From arithmetic, we know that

$$3 + 9 = 12 \quad \text{and} \quad 9 + 3 = 12$$
$$3 \cdot 9 = 27 \quad \text{and} \quad 9 \cdot 3 = 27.$$

The order of the numbers being added or multiplied does not matter. The same answers result. Also,

$$(5 + 7) + 2 = 12 + 2 = 14$$
$$5 + (7 + 2) = 5 + 9 = 14,$$

and

$$(5 \cdot 7) \cdot 2 = 35 \cdot 2 = 70$$
$$5(7 \cdot 2) = 5 \cdot 14 = 70.$$

The way in which the numbers being added or multiplied are grouped does not matter. The same answers result.

These arithmetic examples can be extended to algebra.

---

**Commutative and Associative Properties**

For any real numbers $a$, $b$, and $c$,

$$a + b = b + a$$
and $$ab = ba.$$ Commutative properties

Interchange the order of the two terms or factors.

Also, $$a + (b + c) = (a + b) + c$$
and $$a(bc) = (ab)c.$$ Associative properties

Shift parentheses among the three terms or factors; the order stays the same.

---

The commutative properties are used to change the *order* of the terms or factors in an expression. Think of commuting from home to work and then from work to home.

The associative properties are used to *regroup* the terms or factors of an expression. Remember, to *associate* is to be part of a group.

**EXAMPLE 3** Using the Commutative and Associative Properties

Simplify $-2m + 5m + 3 - 6m + 8$.

$$-2m + 5m + 3 - 6m + 8$$
$$= (-2m + 5m) + 3 - 6m + 8 \quad \text{Order of operations}$$
$$= (-2 + 5)m + 3 - 6m + 8 \quad \text{Distributive property}$$
$$= 3m + 3 - 6m + 8 \quad \text{Add inside parentheses.}$$

By the order of operations, the next step would be to add $3m$ and $3$, but they are unlike terms. To get $3m$ and $-6m$ together, use the associative and commutative properties. Begin by inserting parentheses and brackets according to the order of operations.

$$= [(3m + 3) - 6m] + 8$$
$$= [3m + (3 - 6m)] + 8 \quad \text{Associative property}$$
$$= [3m + (-6m + 3)] + 8 \quad \text{Commutative property}$$
$$= [(3m + [-6m]) + 3] + 8 \quad \text{Associative property}$$
$$= (-3m + 3) + 8 \quad \text{Combine like terms.}$$
$$= -3m + (3 + 8) \quad \text{Associative property}$$
$$= -3m + 11 \quad \text{Add.}$$

In practice, many of these steps are not written down, but you should realize that the commutative and associative properties are used whenever the terms in an expression are rearranged to combine like terms.

**NOW TRY** Exercise 27.

**EXAMPLE 4** Using the Properties of Real Numbers

Simplify each expression.

**(a)** $5y - 8y - 6y + 11y$
$$= (5 - 8 - 6 + 11)y \quad \text{Distributive property}$$
$$= 2y \quad \text{Combine like terms.}$$

**(b)** $3x + 4 - 5(x + 1) - 8$
$$= 3x + 4 - 5x - 5 - 8 \quad \text{Distributive property}$$

*Be careful with signs.*
$$= 3x - 5x + 4 - 5 - 8 \quad \text{Commutative property}$$
$$= -2x - 9 \quad \text{Combine like terms.}$$

**(c)** $8 - (3m + 2)$
$$= 8 - 1(3m + 2) \quad \text{Identity property}$$
$$= 8 - 3m - 2 \quad \text{Distributive property}$$
$$= 6 - 3m \quad \text{Combine like terms.}$$

**(d)** $3x(5)(y)$

$= [3x(5)]y$     Order of operations

$= [3(x \cdot 5)]y$     Associative property

$= [3(5x)]y$     Commutative property

$= [(3 \cdot 5)x]y$     Associative property

$= (15x)y$     Multiply.

$= 15(xy)$     Associative property

$= 15xy$

As previously mentioned, many of these steps are not usually written out.

**NOW TRY** Exercises 29 and 31.

**CAUTION** Be careful. The distributive property does not apply in Example 4(d), because there is no addition involved.
$$(3x)(5)(y) \neq (3x)(5) \cdot (3x)(y)$$

**OBJECTIVE 5** Use the multiplication property of 0. The additive identity property gives a special property of 0, namely, that $a + 0 = a$ for any real number $a$. The **multiplication property of 0** gives a special property of 0 that involves multiplication. The product of any real number and 0 is 0.

**Multiplication Property of 0**

For any real number $a$,
$$a \cdot 0 = 0 \quad \text{and} \quad 0 \cdot a = 0.$$

# Exercises

*Multiple Choice* Choose the correct response in Exercises 1–4.

**1.** The identity element for addition is
  **A.** $-a$    **B.** 0    **C.** 1    **D.** $\dfrac{1}{a}$.

**2.** The identity element for multiplication is
  **A.** $-a$    **B.** 0    **C.** 1    **D.** $\dfrac{1}{a}$.

**3.** The additive inverse of $a$ is
  **A.** $-a$    **B.** 0    **C.** 1    **D.** $\dfrac{1}{a}$.

**4.** The multiplicative inverse of $a$, where $a \neq 0$, is
  **A.** $-a$    **B.** 0    **C.** 1    **D.** $\dfrac{1}{a}$.

*Fill in the Blanks* Complete each statement.

**5.** The multiplication property of 0 says that the _____ of 0 and any real number is _____.

**6.** The commutative property is used to change the _____ of two terms or factors.

**7.** The associative property is used to change the _____ of three terms or factors.

**8.** Like terms are terms with the _____ variables raised to the _____ powers.

**9.** When simplifying an expression, only _____ terms can be combined.

**10.** The coefficient in the term $-8yz^2$ is _____.

*Simplify each expression. See Examples 1 and 2.*

**11.** $2(m + p)$  **12.** $3(a + b)$  **13.** $-12(x - y)$  **14.** $-10(p - q)$

**15.** $5k + 3k$  **16.** $6a + 5a$  **17.** $7r - 9r$  **18.** $4n - 6n$

**19.** $-8z + 4w$  **20.** $-12k + 3r$  **21.** $a + 7a$  **22.** $s + 9s$

**23.** $-(2d - f)$  **24.** $-(3m - n)$

*Simplify each expression. See Examples 1–4.*

**25.** $-12y + 4y + 3 + 2y$  **26.** $-5r - 9r + 8r - 5$

**27.** $-6p + 5 - 4p + 6 + 11p$  **28.** $-8x - 12 + 3x - 5x + 9$

**29.** $3(k + 2) - 5k + 6 + 3$  **30.** $5(r - 3) + 6r - 2r + 4$

**31.** $-2(m + 1) - (m - 4)$  **32.** $6(a - 5) - (a + 6)$

**33.** $0.25(8 + 4p) - 0.5(6 + 2p)$  **34.** $0.4(10 - 5x) - 0.8(5 + 10x)$

**35.** $-(2p + 5) + 3(2p + 4) - 2p$  **36.** $-(7m - 12) - 2(4m + 7) - 8m$

**37.** $2 + 3(2z - 5) - 3(4z + 6) - 8$  **38.** $-4 + 4(4k - 3) - 6(2k + 8) + 7$

*Fill in the Blanks* Complete each statement so that the indicated property is illustrated. Simplify each answer if possible.

**39.** $5x + 8x = $ _____ (distributive property)

**40.** $9y - 6y = $ _____ (distributive property)

**41.** $5(9r) = $ _____ (associative property)

**42.** $-4 + (12 + 8) = $ _____ (associative property)

**43.** $5x + 9y = $ _____ (commutative property)

**44.** $-5 \cdot 7 = $ _____ (commutative property)

**45.** $1 \cdot 7 = $ _____ (identity property)

**46.** $-12x + 0 = $ _____ (identity property)

**47.** $-\frac{1}{4}ty + \frac{1}{4}ty = $ _____ (inverse property)

**48.** $-\frac{9}{8}\left(-\frac{8}{9}\right) = $ _____ (inverse property)

**49.** $8(-4 + x) = $ _____ (distributive property)

**50.** $3(x - y + z) = $ _____ (distributive property)

**51.** $0(0.875x + 9y - 88z) = $ _____ (multiplication property of 0)

**52.** $0(35t^2 - 8t + 12) = $ _____ (multiplication property of 0)

**53.** *Concept Check* Give an "everyday" example of a commutative operation and of an operation that is not commutative.

**54.** *Concept Check* Give an "everyday" example of inverse operations.

The distributive property can be used to mentally perform calculations. For example, calculate $38 \cdot 17 + 38 \cdot 3$ as follows:

$$38 \cdot 17 + 38 \cdot 3 = 38(17 + 3) \quad \text{Distributive property}$$
$$= 38(20) \quad \text{Add inside parentheses.}$$
$$= 760. \quad \text{Multiply.}$$

*Use the distributive property to calculate each value mentally.*

**55.** $96 \cdot 19 + 4 \cdot 19$  **56.** $27 \cdot 60 + 27 \cdot 40$  **57.** $58 \cdot \frac{3}{2} - 8 \cdot \frac{3}{2}$

**58.** $8.75(15) - 8.75(5)$  **59.** $4.31(69) + 4.31(31)$  **60.** $\frac{8}{5}(17) + \frac{8}{5}(13)$

# 1 SUMMARY

## KEY TERMS

**1.1** set
elements (members)
finite set
infinite set
empty set (null set)
variable
set-builder notation
number line
coordinate
graph
additive inverse
(opposite, negative)

signed numbers
absolute value
equation
inequality
interval
interval notation
three-part inequality
**1.2** sum
difference
product
quotient

reciprocal
(multiplicative
inverse)
**1.3** factors
exponent (power)
base
exponential expression
square root
principal (positive)
square root
negative square root

algebraic expression
**1.4** identity element for
addition
identity element for
multiplication
term
coefficient (numerical
coefficient)
like terms
unlike terms
combining like terms

## NEW SYMBOLS

|  |  |
|---|---|
| $\{a, b\}$ | set containing the elements $a$ and $b$ |
| $\emptyset$ or $\{\ \}$ | empty set |
| $\in$ | is an element of (a set) |
| $\notin$ | is not an element of |
| $\neq$ | is not equal to |
| $\{x \mid x \text{ has property } P\}$ | set-builder notation |
| $\lvert x \rvert$ | absolute value of $x$ |
| $<$ | is less than |
| $\leq$ | is less than or equal to |
| $>$ | is greater than |
| $\geq$ | is greater than or equal to |
| $\infty$ | infinity |
| $-\infty$ | negative infinity |
| $(-\infty, \infty)$ | set of all real numbers |
| $(a, \infty)$ | the interval $\{x \mid x > a\}$ |
| $(-\infty, a)$ | the interval $\{x \mid x < a\}$ |
| $(a, b]$ | the interval $\{x \mid a < x \leq b\}$ |
| $a^m$ | $m$ factors of $a$ |
| $\sqrt{\ }$ | radical sign |
| $\sqrt{a}$ | positive (or principal) square root of $a$ |

## QUICK REVIEW

| CONCEPTS | EXAMPLES |
|---|---|
| **1.1 Basic Concepts** | |
| **Sets of Numbers** | |
| *Natural Numbers* $\{1, 2, 3, 4, \ldots\}$ | 10, 25, 143 |
| *Whole Numbers* $\{0, 1, 2, 3, 4, \ldots\}$ | 0, 8, 47 |
| *Integers* $\{\ldots, -2, -1, 0, 1, 2, \ldots\}$ | $-22, -7, 0, 4, 9$ |
| *Rational Numbers* $\left\{\frac{p}{q} \mid p \text{ and } q \text{ are integers}, q \neq 0\right\}$ (all terminating or repeating decimals) | $-\dfrac{2}{3}, -0.14, 0, 6, \dfrac{5}{8}, 0.33333\ldots$ |
| *Irrational Numbers* $\{x \mid x \text{ is a real number that is not rational}\}$ (all nonterminating, nonrepeating decimals) | $\pi, \sqrt{3}, -\sqrt{22}$ |
| *Real Numbers* $\{x \mid x \text{ is a rational or an irrational number}\}$ | $-3, 0.7, \pi, -\dfrac{2}{3}$ |
| **Absolute Value** $\lvert a \rvert = \begin{cases} a & \text{if } a \text{ is positive or } 0 \\ -a & \text{if } a \text{ is negative} \end{cases}$ | $\lvert 12 \rvert = 12$ <br> $\lvert -12 \rvert = 12$ |

*(continued)*

| CONCEPTS | EXAMPLES |
|---|---|

## 1.2 Operations on Real Numbers

**Addition**

*Same Sign:* Add the absolute values. The sum has the same sign as the given numbers.

$$-2 + (-7) = -(2 + 7) = -9$$

*Different Signs:* Find the absolute values of the numbers, and subtract the smaller absolute value from the larger. The sum has the same sign as the number with the larger absolute value.

$$-5 + 8 = 8 - 5 = 3$$
$$-12 + 4 = -(12 - 4) = -8$$

**Subtraction**

For all real numbers $a$ and $b$,

$$a - b = a + (-b).$$

$$-5 - (-3) = -5 + 3 = -2$$

**Multiplication and Division**

*Same Sign:* The answer is positive when multiplying or dividing two numbers with the same sign.

$$-3(-8) = 24 \qquad \frac{-15}{-5} = 3$$

*Different Signs:* The answer is negative when multiplying or dividing two numbers with different signs.

$$-7(5) = -35 \qquad \frac{-24}{12} = -2$$

**Division**

For all real numbers $a$ and $b$ (where $b \neq 0$),

$$a \div b = \frac{a}{b} = a \cdot \frac{1}{b}.$$

$$\frac{2}{3} \div \frac{5}{6} = \frac{2}{3} \cdot \frac{6}{5} = \frac{4}{5} \qquad \text{Multiply by the reciprocal.}$$

## 1.3 Exponents, Roots, and Order of Operations

The product of an even number of negative factors is positive. The product of an odd number of negative factors is negative.

$(-5)^2$ is positive: $(-5)^2 = (-5)(-5) = 25$
$(-5)^3$ is negative: $(-5)^3 = (-5)(-5)(-5) = -125$

**Order of Operations**
1. Work separately above and below any fraction bar.
2. If parentheses, brackets, or absolute value bars are present, start with the innermost set and work outward.
3. Evaluate all exponents, roots, and absolute values.
4. Multiply or divide in order from left to right.
5. Add or subtract in order from left to right.

$$\frac{12 + 3}{5 \cdot 2} = \frac{15}{10} = \frac{3}{2}$$

$$(-6)[2^2 - (3 + 4)] + 3$$
$$= (-6)[2^2 - 7] + 3$$
$$= (-6)[4 - 7] + 3$$
$$= (-6)[-3] + 3$$
$$= 18 + 3$$
$$= 21$$

## 1.4 Properties of Real Numbers

For real numbers $a$, $b$, and $c$,

**Distributive Property**

$$a(b + c) = ab + ac$$

$$12(4 + 2) = 12 \cdot 4 + 12 \cdot 2$$

**Inverse Properties**

$a + (-a) = 0 \quad$ and $\quad -a + a = 0$

$$5 + (-5) = 0 \qquad -12 + 12 = 0$$

$a \cdot \dfrac{1}{a} = 1 \quad$ and $\quad \dfrac{1}{a} \cdot a = 1$

$$5 \cdot \frac{1}{5} = 1 \qquad -\frac{1}{3}(-3) = 1$$

*(continued)*

| CONCEPTS | EXAMPLES |
|---|---|
| **Identity Properties** <br> $a + 0 = 0 + a = a$ and $a \cdot 1 = 1 \cdot a = a$ | $-32 + 0 = -32 \qquad 17.5 \cdot 1 = 17.5$ |
| **Commutative Properties** <br> $a + b = b + a$ and $ab = ba$ | $9 + (-3) = -3 + 9 \qquad 6(-4) = (-4)6$ |
| **Associative Properties** <br> $a + (b + c) = (a + b) + c$ and $a(bc) = (ab)c$ | $7 + (5 + 3) = (7 + 5) + 3 \qquad -4(6 \cdot 3) = (-4 \cdot 6)3$ |
| **Multiplication Property of 0** <br> $a \cdot 0 = 0$ and $0 \cdot a = 0$. | $4 \cdot 0 = 0 \qquad 0(-3) = 0$ |

# 1 REVIEW EXERCISES

*Graph the elements of each set on a number line.*

**1.** $\left\{-4, -1, 2, \dfrac{9}{4}, 4\right\}$

**2.** $\left\{-5, -\dfrac{11}{4}, -0.5, 0, 3, \dfrac{13}{3}\right\}$

*Find the value of each expression.*

**3.** $|-16|$

**4.** $-|-4|$

**5.** $|-8| - |-3|$

*Let* $S = \left\{-9, -\dfrac{4}{3}, -\sqrt{4}, -0.25, 0, 0.\overline{35}, \dfrac{5}{3}, \sqrt{7}, \sqrt{-9}, \dfrac{12}{3}\right\}$. *Simplify the elements of S as necessary, and then list those elements of S that belong to the specified set.*

**6.** Whole numbers

**7.** Integers

**8.** Rational numbers

**9.** Real numbers

*Write each set by listing its elements.*

**10.** $\{x \mid x \text{ is a natural number between 3 and 9}\}$

**11.** $\{y \mid y \text{ is a whole number less than 4}\}$

***True or False*** *Indicate whether each inequality is* true *or* false.

**12.** $4 \cdot 2 \leq |12 - 4|$

**13.** $2 + |-2| > 4$

**14.** $4(3 + 7) > -|40|$

*Write each set in interval notation and graph the interval.*

**15.** $\{x \mid x < -5\}$

**16.** $\{x \mid -2 < x \leq 3\}$

*Add or subtract as indicated.*

**17.** $-\dfrac{5}{8} - \left(-\dfrac{7}{3}\right)$

**18.** $-\dfrac{4}{5} - \left(-\dfrac{3}{10}\right)$

**19.** $-5 + (-11) + 20 - 7$

**20.** $-9.42 + 1.83 - 7.6 - 1.9$

21. $-15 + (-13) + (-11)$

22. $-1 - 3 - (-10) + (-7)$

23. $\dfrac{3}{4} - \left(\dfrac{1}{2} - \dfrac{9}{10}\right)$

24. $-|-12| - |-9| + (-4) - |10|$

*Multiply or divide as indicated.*

25. $-\dfrac{3}{7}\left(-\dfrac{14}{9}\right)$

26. $2(-5)(-3)(-3)$

27. $\dfrac{-2.3754}{-0.74}$

28. $\dfrac{75}{-5}$

29. *Concept Check* Which one of the following is undefined: $\dfrac{5}{7-7}$ or $\dfrac{7-7}{5}$?

*Evaluate each expression.*

30. $\left(\dfrac{3}{7}\right)^3$

31. $10^4$

32. $(-5)^3$

33. $-5^3$

*Find each square root. If it is not a real number, say so.*

34. $\sqrt{\dfrac{64}{121}}$

35. $\sqrt{400}$

36. $\sqrt{-64}$

37. $-\sqrt{0.81}$

*Simplify each expression.*

38. $-\dfrac{2}{3}[5(-2) + 8 - 4^3]$

39. $-14\left(\dfrac{3}{7}\right) + 6 \div 3$

40. $\dfrac{-5(3^2) + 9(\sqrt{4}) - 5}{6 - 5(-2)}$

*Evaluate each expression if $k = -4$, $m = 2$, and $n = 16$.*

41. $4k - 7m$

42. $-3\sqrt{n} + m + 5k$

43. $\dfrac{4m^3 - 3n}{7k^2 - 10}$

*Simplify each expression.*

44. $2q + 19q$

45. $13z - 17z$

46. $-m + 6m$

47. $5p - p$

48. $-2(k + 3)$

49. $6(r + 3)$

50. $9(2m + 3n)$

51. $-(-p + 6q) - (2p - 3q)$

52. $-3y + 6 - 5 + 4y$

53. $2a + 3 - a - 1 - a - 2$

54. $-3(4m - 2) + 2(3m - 1) - 4(3m + 1)$

*Fill in the Blanks* Complete each statement so that the indicated property is illustrated. Simplify each answer if possible.

55. $2x + 3x = $ _____ (distributive property)

56. $-4 \cdot 1 = $ _____ (identity property)

57. $2(4x) = $ _____ (associative property)

58. $-3 + 13 = $ _____ (commutative property)

59. $-3 + 3 = $ _____ (inverse property)

60. $5(x + z) = $ _____ (distributive property)

61. $0 + 7 = $ _____ (identity property)

62. $8 \cdot \dfrac{1}{8} = $ _____ (inverse property)

# 1 TEST

1. Graph $\left\{-3, 0.75, \dfrac{5}{3}, 5, 6.3\right\}$ on a number line.

Let $A = \left\{-\sqrt{6}, -1, -0.5, 0, 3, \sqrt{25}, 7.5, \dfrac{24}{2}, \sqrt{-4}\right\}$. *First simplify each element as needed, and then list the elements from A that belong to each set.*

2. Whole numbers
3. Integers
4. Rational numbers
5. Real numbers

*Write each set in interval notation and graph the interval.*

6. $\{x \mid x < -3\}$
7. $\{y \mid -4 < y \leq 2\}$

*Perform the indicated operations.*

8. $-6 + 14 + (-11) - (-3)$
9. $10 - 4 \cdot 3 + 6(-4)$
10. $7 - 4^2 + 2(6) + (-4)^2$
11. $\dfrac{10 - 24 + (-6)}{\sqrt{16}(-5)}$
12. $\dfrac{-2[3 - (-1 - 2) + 2]}{\sqrt{9}(-3) - (-2)}$
13. $\dfrac{8 \cdot 4 - 3^2 \cdot 5 - 2(-1)}{-3 \cdot 2^3 + 1}$

*The table shows the heights in feet of some selected mountains and the depths in feet (as negative numbers) of some selected ocean trenches.*

| Mountain | Height | Trench | Depth |
|---|---|---|---|
| Foraker | 17,400 | Philippine | −32,995 |
| Wilson | 14,246 | Cayman | −24,721 |
| Pikes Peak | 14,110 | Java | −23,376 |

*Source:* World Almanac and Book of Facts.

14. What is the difference between the height of Mt. Foraker and the depth of the Philippine Trench?
15. What is the difference between the height of Pikes Peak and the depth of the Java Trench?
16. How much deeper is the Cayman Trench than the Java Trench?

*Find each square root. If the number is not real, say so.*

17. $\sqrt{196}$
18. $-\sqrt{225}$
19. $\sqrt{-16}$

20. *Concept Check* For the expression $\sqrt{a}$, under what conditions will its value be **(a)** positive, **(b)** not real, **(c)** 0?

21. Evaluate $\dfrac{8k + 2m^2}{r - 2}$ if $k = -3$, $m = -3$, and $r = 25$.

22. Simplify $-3(2k - 4) + 4(3k - 5) - 2 + 4k$.

23. How does the subtraction sign affect the terms $-4r$ and 6 when $(3r + 8) - (-4r + 6)$ is simplified? What is the simplified form?

*Matching* Match each statement in Column I with the appropriate property in Column II. Answers may be used more than once.

| I | II |
|---|---|
| 24. $6 + (-6) = 0$ | A. Distributive property |
| 25. $-2 + (3 + 6) = (-2 + 3) + 6$ | B. Inverse property |
| 26. $5x + 15x = (5 + 15)x$ | C. Identity property |
| 27. $13 \cdot 0 = 0$ | D. Associative property |
| 28. $-9 + 0 = -9$ | E. Commutative property |
| 29. $4 \cdot 1 = 4$ | F. Multiplication property of 0 |
| 30. $(a + b) + c = (b + a) + c$ | |

# 2

# Linear Equations, Inequalities, and Applications

Television, first operational in the 1940s, has become the most widespread form of communication in the world. A recent study found that 106.9 million homes—98% of all U.S. households—owned at least one TV set, and average viewing time among all viewers exceeded 31 hours per week. Favorite prime-time television programs were *CSI* and *American Idol*. (*Source:* Nielsen Media Research; *Microsoft Encarta Encyclopedia*.)

In Section 2.2, we discuss the concept of *percent*—one of the most common everyday applications of mathematics—and use it in Exercises 41–44 to determine additional information about television viewing in U.S. households.

**2.1** Linear Equations in One Variable

**2.2** Formulas

**2.3** Applications of Linear Equations

**2.4** Linear Inequalities in One Variable

**2.5** Set Operations and Compound Inequalities

**2.6** Absolute Value Equations and Inequalities

# 2.1 Linear Equations in One Variable

**OBJECTIVES**

1. Decide whether a number is a solution of a linear equation.
2. Solve linear equations by using the addition and multiplication properties of equality.
3. Solve linear equations by using the distributive property.
4. Solve linear equations with fractions or decimals.
5. Identify conditional equations, contradictions, and identities.

In **Chapter 1,** we reviewed *algebraic expressions*. Examples include

$$8x + 9, \quad y - 4, \quad \text{and} \quad \frac{x^3 y^8}{z}. \quad \text{Algebraic expressions}$$

Equations and inequalities compare algebraic expressions, just as a balance scale compares the weights of two quantities. Recall from **Section 1.1** that an *equation* is a statement that two algebraic expressions are equal. It is important to be able to distinguish between algebraic expressions and equations. *An equation always contains an equals sign, while an expression does not.*

$3x - 7 = 2$ — Left side | Right side — Equation (to solve)

$3x - 7$ — Expression (to simplify or evaluate)

**NOW TRY** Exercise 5.

A *linear equation in one variable* involves only real numbers and one variable raised to the first power. Examples include

$$x + 1 = -2, \quad x - 3 = 5, \quad \text{and} \quad 2k + 5 = 10. \quad \text{Linear equations}$$

### Linear Equation in One Variable

A **linear equation in one variable** can be written in the form

$$Ax + B = C,$$

where $A$, $B$, and $C$ are real numbers, with $A \neq 0$.

A linear equation is a **first-degree equation,** since the greatest power on the variable is 1. Some equations that are not linear (that is, *nonlinear*) are

$$x^2 + 3y = 5, \quad \frac{8}{x} = -22, \quad \text{and} \quad \sqrt{x} = 6. \quad \text{Nonlinear equations}$$

**OBJECTIVE 1** Decide whether a number is a solution of a linear equation. If the variable in an equation can be replaced by a real number that makes the statement true, then that number is a **solution** of the equation. For example, 8 is a solution of the equation $x - 3 = 5$, since replacing $x$ with 8 gives a true statement. An equation is *solved* by finding its **solution set,** the set of all solutions. The solution set of the equation $x - 3 = 5$ is $\{8\}$.

**Equivalent equations** are related equations that have the same solution set. To solve an equation, we usually start with the given equation and replace it with a series of simpler equivalent equations. For example,

$$5x + 2 = 17, \quad 5x = 15, \quad \text{and} \quad x = 3 \quad \text{Equivalent equations}$$

are all equivalent, since each has the solution set $\{3\}$.

**OBJECTIVE 2** **Solve linear equations by using the addition and multiplication properties of equality.** We use two important properties to produce equivalent equations.

---

**Addition and Multiplication Properties of Equality**

*Addition Property of Equality*
For all real numbers $A$, $B$, and $C$, the equations
$$A = B \quad \text{and} \quad A + C = B + C$$
are equivalent.

That is, the same number may be added to each side of an equation without changing the solution set.

*Multiplication Property of Equality*
For all real numbers $A$ and $B$, and for $C \neq 0$, the equations
$$A = B \quad \text{and} \quad AC = BC$$
are equivalent.

That is, each side of an equation may be multiplied by the same nonzero number without changing the solution set.

---

Because subtraction and division are defined in terms of addition and multiplication, respectively, the preceding properties can be extended:

*The same number may be subtracted from each side of an equation, and each side of an equation may be divided by the same nonzero number, without changing the solution set.*

---

**EXAMPLE 1** Using the Properties of Equality to Solve a Linear Equation

Solve $4x - 2x - 5 = 4 + 6x + 3$.

The goal is to isolate $x$ on one side of the equation.

$$4x - 2x - 5 = 4 + 6x + 3$$
$$2x - 5 = 7 + 6x \quad \text{Combine like terms.}$$
$$2x - 5 + 5 = 7 + 6x + 5 \quad \text{Add 5 to each side.}$$
$$2x = 12 + 6x \quad \text{Combine like terms.}$$
$$2x - 6x = 12 + 6x - 6x \quad \text{Subtract } 6x \text{ from each side.}$$
$$-4x = 12 \quad \text{Combine like terms.}$$
$$\frac{-4x}{-4} = \frac{12}{-4} \quad \text{Divide each side by } -4.$$
$$x = -3$$

To be sure that $-3$ is the solution, check by substituting for $x$ in the *original* equation.

*Check:*

$$4x - 2x - 5 = 4 + 6x + 3 \quad \text{Original equation}$$
$$4(-3) - 2(-3) - 5 = 4 + 6(-3) + 3 \quad ? \quad \text{Let } x = -3.$$
$$-12 + 6 - 5 = 4 - 18 + 3 \quad ? \quad \text{Multiply.}$$
$$-11 = -11 \quad \text{True}$$

Use parentheses around substituted values to avoid errors.

This is *not* the solution.

The true statement indicates that $\{-3\}$ is the solution set.

**NOW TRY** Exercise 15.

**CAUTION** Notice in Example 1 that the equality symbols are aligned in a column. *Do not use more than one equality symbol in a horizontal line of work when solving an equation.*

We use the following steps to solve a linear equation in one variable.

### Solving a Linear Equation in One Variable

*Step 1* **Clear fractions.** Eliminate any fractions by multiplying each side by the least common denominator.

*Step 2* **Simplify each side separately.** Use the distributive property to clear parentheses and combine like terms as needed.

*Step 3* **Isolate the variable terms on one side.** Use the addition property to get all terms with variables on one side of the equation and all numbers on the other.

*Step 4* **Isolate the variable.** Use the multiplication property to get an equation with just the variable (with coefficient 1) on one side.

*Step 5* **Check.** Substitute the proposed solution into the original equation.

**OBJECTIVE 3** Solve linear equations by using the distributive property. In Example 1, we did not use Step 1 or the distributive property in Step 2 as given in the box. Many equations, however, will require one or both of these steps.

### EXAMPLE 2 Using the Distributive Property to Solve a Linear Equation

Solve $2(k - 5) + 3k = k + 6$.

*Step 1* Since there are no fractions in this equation, Step 1 does not apply.

*Step 2* Use the distributive property to simplify and combine terms on the left.

Be sure to distribute over *all* terms within the parentheses.

$$2(k - 5) + 3k = k + 6$$
$$2k + 2(-5) + 3k = k + 6 \quad \text{Distributive property}$$
$$2k - 10 + 3k = k + 6 \quad \text{Multiply.}$$
$$5k - 10 = k + 6 \quad \text{Combine like terms.}$$

***Step 3*** Next, use the addition property of equality.

$$5k - 10 + 10 = k + 6 + 10 \quad \text{Add 10.}$$
$$5k = k + 16 \quad \text{Combine like terms.}$$
$$5k - k = k + 16 - k \quad \text{Subtract } k.$$
$$4k = 16 \quad \text{Combine like terms.}$$

***Step 4*** Use the multiplication property of equality to isolate $k$ on the left.

$$\frac{4k}{4} = \frac{16}{4} \quad \text{Divide by 4.}$$
$$k = 4$$

***Step 5*** Check by substituting 4 for $k$ in the original equation.

$$\text{Check:} \quad 2(k - 5) + 3k = k + 6 \quad \text{Original equation}$$
$$2(4 - 5) + 3(4) = 4 + 6 \quad ? \quad \text{Let } k = 4.$$
$$2(-1) + 12 = 10 \quad ?$$
$$10 = 10 \quad \text{True}$$

*Always* check your work.

The solution checks, so $\{4\}$ is the solution set.

**NOW TRY** Exercise 19.

**OBJECTIVE 4** Solve linear equations with fractions or decimals. When fractions or decimals appear as coefficients in equations, our work can be made easier if we multiply each side of the equation by the least common denominator (LCD) of all the fractions. This is an application of the multiplication property of equality, and it produces an equivalent equation with integer coefficients.

**EXAMPLE 3** Solving a Linear Equation with Fractions

Solve $\dfrac{x + 7}{6} + \dfrac{2x - 8}{2} = -4$.

Start by eliminating the fractions. Multiply both sides by the LCD, 6.

***Step 1*** $\quad 6\left(\dfrac{x + 7}{6} + \dfrac{2x - 8}{2}\right) = 6(-4) \quad \text{The LCD is 6.}$

***Step 2*** $\quad 6\left(\dfrac{x + 7}{6}\right) + 6\left(\dfrac{2x - 8}{2}\right) = 6(-4) \quad \text{Distributive property}$

$$\dfrac{6(x + 7)}{6} + \dfrac{6(2x - 8)}{2} = -24 \quad \text{Multiply; } 6 = \tfrac{6}{1}.$$
$$x + 7 + 3(2x - 8) = -24$$
$$x + 7 + 3(2x) + 3(-8) = -24 \quad \text{Distributive property}$$
$$x + 7 + 6x - 24 = -24 \quad \text{Multiply.}$$
$$7x - 17 = -24 \quad \text{Combine like terms.}$$

***Step 3*** $\quad 7x - 17 + 17 = -24 + 17 \quad \text{Add 17.}$
$$7x = -7 \quad \text{Combine like terms.}$$

**Step 4** $$\frac{7x}{7} = \frac{-7}{7} \quad \text{Divide by 7.}$$
$$x = -1$$

**Step 5** *Check* by substituting $-1$ for $x$ in the original equation.

$$\frac{x+7}{6} + \frac{2x-8}{2} = -4$$

$$\frac{-1+7}{6} + \frac{2(-1)-8}{2} = -4 \quad ? \quad \text{Let } x = -1.$$

$$\frac{6}{6} + \frac{-10}{2} = -4 \quad ?$$

$$1 - 5 = -4 \quad ?$$

$$-4 = -4 \quad \text{True}$$

The solution checks, so the solution set is $\{-1\}$.

**NOW TRY** Exercise 59.

In **Sections 2.2 and 2.3,** we solve problems dealing with interest rates and concentrations of solutions. These problems involve percents that are converted to decimals. The equations that are used to solve such problems have decimal coefficients. We can clear these decimals by multiplying by a power of 10, such as

$$10^1 = 10, \quad 10^2 = 100, \quad \text{and so on.}$$

This allows us to obtain integer coefficients.

**EXAMPLE 4** Solving a Linear Equation with Decimals

Solve $0.06x + 0.09(15 - x) = 0.07(15)$.

Since each decimal number is given in hundredths, multiply both sides of the equation by 100. A number can be multiplied by 100 by moving the decimal point two places to the right. To multiply the second term, $0.09(15 - x)$, by 100—that is, $100(0.09)(15 - x)$—remember the associative property: To multiply three terms, first multiply any two of them. Here we multiply $100(0.09)$ first to get 9, so the product $100(0.09)(15 - x)$ becomes $9(15 - x)$.

$$0.06x + 0.09(15 - x) = 0.07(15)$$

$$0.06x + 0.09(15 - x) = 0.07(15) \quad \text{Multiply each term by 100.}$$

Move decimal points 2 places to the right.

$$6x + 9(15 - x) = 7(15)$$

$$6x + 9(15) - 9x = 105 \quad \text{Distributive property; multiply.}$$

$$-3x + 135 = 105 \quad \text{Combine like terms; multiply.}$$

$$-3x + 135 - 135 = 105 - 135 \quad \text{Subtract 135.}$$

$$-3x = -30 \quad \text{Combine like terms.}$$

$$\frac{-3x}{-3} = \frac{-30}{-3} \quad \text{Divide by } -3.$$

$$x = 10$$

*Check:* 
$$0.06x + 0.09(15 - x) = 0.07(15)$$
$$0.06(10) + 0.09(15 - 10) = 0.07(15) \quad ? \quad \text{Let } x = 10.$$
$$0.6 + 0.09(5) = 1.05 \quad ?$$
$$0.6 + 0.45 = 1.05 \quad ?$$
$$1.05 = 1.05 \quad \text{True}$$

A true statement results, so the solution set is $\{10\}$.

**NOW TRY** Exercise 63.

**OBJECTIVE 5** Identify conditional equations, contradictions, and identities. In Examples 1–4, all of the equations had solution sets containing *one* element, such as $\{10\}$ in Example 4. Some equations, however, have no solutions, while others have an infinite number of solutions. The table gives the names of these types of equations.

| Type of Linear Equation | Number of Solutions | Indication when Solving |
|---|---|---|
| **Conditional** | One | Final line is $x =$ a number. (See Example 5(a).) |
| **Identity** | Infinite; solution set {all real numbers} | Final line is true, such as $0 = 0$. (See Example 5(b).) |
| **Contradiction** | None; solution set $\emptyset$ | Final line is false, such as $-15 = -20$. (See Example 5(c).) |

**EXAMPLE 5** Recognizing Conditional Equations, Identities, and Contradictions

Solve each equation. Decide whether it is a *conditional equation,* an *identity,* or a *contradiction.*

(a) 
$$5x - 9 = 4(x - 3)$$
$$5x - 9 = 4x - 12 \quad \text{Distributive property}$$
$$5x - 9 - 4x = 4x - 12 - 4x \quad \text{Subtract } 4x.$$
$$x - 9 = -12 \quad \text{Combine like terms.}$$
$$x - 9 + 9 = -12 + 9 \quad \text{Add 9.}$$
$$x = -3$$

The solution set, $\{-3\}$, has only one element, so $5x - 9 = 4(x - 3)$ is a conditional equation.

(b) 
$$5x - 15 = 5(x - 3)$$
$$5x - 15 = 5x - 15 \quad \text{Distributive property}$$
$$5x - 15 - 5x + 15 = 5x - 15 - 5x + 15 \quad \text{Subtract } 5x; \text{ add 15.}$$
$$0 = 0 \quad \text{True}$$

The final line, $0 = 0$, indicates that the solution set is {all real numbers}, and the equation $5x - 15 = 5(x - 3)$ is an identity. (The first step yielded $5x - 15 = 5x - 15$, which is *true* for all values of $x$. We could have identified the equation as an identity at that point.)

(c) $\quad 5x - 15 = 5(x - 4)$

$\quad\quad 5x - 15 = 5x - 20 \quad$ Distributive property

$\quad 5x - 15 - 5x = 5x - 20 - 5x \quad$ Subtract 5x.

$\quad\quad\quad\quad -15 = -20 \quad$ False

Since the result, $-15 = -20$, is *false*, the equation has no solution. The solution set is $\emptyset$, so the equation $5x - 15 = 5(x - 4)$ is a contradiction.

**NOW TRY** Exercises 17, 23, and 33.

## 2.1 Exercises

**NOW TRY Exercise**

1. *Multiple Choice* Which equations are linear equations in $x$?

   **A.** $3x + x - 1 = 0$    **B.** $8 = x^2$    **C.** $6x + 2 = 9$    **D.** $\dfrac{1}{2}x - \dfrac{1}{x} = 0$

2. Which of the equations in Exercise 1 are nonlinear equations in $x$? Explain why.

3. Decide whether 6 is a solution of $3(x + 4) = 5x$ by substituting 6 for $x$. If it is not a solution, explain why.

4. Use substitution to decide whether $-2$ is a solution of $5(x + 4) - 3(x + 6) = 9(x + 1)$. If it is not a solution, explain why.

5. *Concept Check* Identify each of the following as an *expression* or an *equation*.

   (a) $3x = 6$                            (b) $3x + 6$

   (c) $5x + 6(x - 3) = 12x + 6$      (d) $5x + 6(x - 3) - (12x + 6)$

6. In Example 1, a student looked at the check and thought that $\{-11\}$ should be given as the solution set. Explain why this is not correct.

7. *Concept Check* The following "solution" contains a common student error.

   $\quad\quad 8x - 2(2x - 3) = 3x + 7$

   $\quad\quad 8x - 4x - 6 = 3x + 7 \quad$ Distributive property

   $\quad\quad\quad\quad 4x - 6 = 3x + 7 \quad$ Combine like terms.

   $\quad\quad\quad\quad\quad\quad x = 13 \quad$ Subtract 3x; add 6.

   **WHAT WENT WRONG?** Find the correct solution.

8. *Concept Check* When clearing parentheses in the expression

   $$-5m - (2m - 4) + 5$$

   on the right side of the equation in Exercise 39, the $-$ sign before the parentheses acts like a factor representing what number? Clear parentheses and simplify this expression.

9. Explain the distinction between a conditional equation, an identity, and a contradiction.

10. *Concept Check* A student tried to solve the equation $8x = 7x$ by dividing each side by $x$, obtaining $8 = 7$. He gave the solution set as $\emptyset$. **WHAT WENT WRONG?**

*Solve each equation, and check your solution. If applicable, tell whether the equation is an* identity *or a* contradiction. *See Examples 1, 2, and 5.*

11. $7x + 8 = 1$             12. $5x - 4 = 21$            13. $5x + 2 = 3x - 6$

14. $9p + 1 = 7p - 9$     15. $7x - 5x + 15 = x + 8$     16. $2x + 4 - x = 4x - 5$

17. $12w + 15w - 9 + 5 = -3w + 5 - 9$     18. $-4t + 5t - 8 + 4 = 6t - 4$

19. $3(2t - 4) = 20 - 2t$                 20. $2(3 - 2x) = x - 4$

21. $-5(x + 1) + 3x + 2 = 6x + 4$
22. $5(x + 3) + 4x - 5 = 4 - 2x$
23. $-2x + 5x - 9 = 3(x - 4) - 5$
24. $-6x + 2x - 11 = -2(2x - 3) + 4$
25. $2(x + 3) = -4(x + 1)$
26. $4(t - 9) = 8(t + 3)$
27. $3(2w + 1) - 2(w - 2) = 5$
28. $4(x - 2) + 2(x + 3) = 6$
29. $2x + 3(x - 4) = 2(x - 3)$
30. $6x - 3(5x + 2) = 4(1 - x)$
31. $6p - 4(3 - 2p) = 5(p - 4) - 10$
32. $-2k - 3(4 - 2k) = 2(k - 3) + 2$
33. $-2(t + 3) - t - 4 = -3(t + 4) + 2$
34. $4(2d + 7) = 2d + 25 + 3(2d + 1)$
35. $2[w - (2w + 4) + 3] = 2(w + 1)$
36. $4[2t - (3 - t) + 5] = -(2 + 7t)$
37. $-[2z - (5z + 2)] = 2 + (2z + 7)$
38. $-[6x - (4x + 8)] = 9 + (6x + 3)$
39. $-3m + 6 - 5(m - 1) = -5m - (2m - 4) + 5$
40. $4(k + 2) - 8k - 5 = -3k + 9 - 2(k + 6)$
41. $7[2 - (3 + 4x)] - 2x = -9 + 2(1 - 15x)$
42. $4[6 - (1 + 2x)] + 10x = 2(10 - 3x) + 8x$
43. $-[3x - (2x + 5)] = -4 - [3(2x - 4) - 3x]$
44. $2[-(x - 1) + 4] = 5 + [-(6x - 7) + 9x]$

45. *Concept Check* To solve the linear equation
$$\frac{8x}{3} - \frac{5x}{4} = -13,$$
we multiply each side by the least common denominator of all the fractions in the equation. What is this least common denominator?

46. Suppose that in solving the equation
$$\frac{1}{3}x + \frac{1}{2}x = \frac{1}{6}x,$$
we begin by multiplying each side by 12, rather than the *least* common denominator, 6. Would we get the correct solution? Explain.

47. *Concept Check* To solve a linear equation with decimals, we usually begin by multiplying by a power of 10 so that all coefficients are integers. What is the least power of 10 that will accomplish this goal in each equation?
    (a) $0.05x + 0.12(x + 5000) = 940$   (Exercise 63)
    (b) $0.006(x + 2) = 0.007x + 0.009$   (Exercise 69)

48. *Multiple Choice* The expression $0.06(10 - x)(100)$ is equivalent to which of the following?
    **A.** $0.06 - 0.06x$   **B.** $60 - 6x$   **C.** $6 - 6x$   **D.** $6 - 0.06x$

*Solve each equation, and check your solution. See Examples 3 and 4.*

49. $-\dfrac{5}{9}k = 2$
50. $\dfrac{3}{11}z = -5$
51. $\dfrac{6}{5}x = -1$
52. $-\dfrac{7}{8}r = 6$
53. $\dfrac{m}{2} + \dfrac{m}{3} = 5$
54. $\dfrac{x}{5} - \dfrac{x}{4} = 1$
55. $\dfrac{3x}{4} + \dfrac{5x}{2} = 13$
56. $\dfrac{8x}{3} - \dfrac{2x}{4} = -13$
57. $\dfrac{x - 10}{5} + \dfrac{2}{5} = -\dfrac{x}{3}$
58. $\dfrac{2r - 3}{7} + \dfrac{3}{7} = -\dfrac{r}{3}$
59. $\dfrac{3x - 1}{4} + \dfrac{x + 3}{6} = 3$
60. $\dfrac{3x + 2}{7} - \dfrac{x + 4}{5} = 2$

61. $\dfrac{4t + 1}{3} = \dfrac{t + 5}{6} + \dfrac{t - 3}{6}$

62. $\dfrac{2x + 5}{5} = \dfrac{3x + 1}{2} + \dfrac{-x + 7}{2}$

63. $0.05x + 0.12(x + 5000) = 940$

64. $0.09k + 0.13(k + 300) = 61$

65. $0.02(50) + 0.08r = 0.04(50 + r)$

66. $0.20(14{,}000) + 0.14t = 0.18(14{,}000 + t)$

67. $0.05x + 0.10(200 - x) = 0.45x$

68. $0.08x + 0.12(260 - x) = 0.48x$

69. $0.006(x + 2) = 0.007x + 0.009$

70. $0.004x + 0.006(50 - x) = 0.004(68)$

## 2.2 Formulas

**OBJECTIVES**

1. Solve a formula for a specified variable.
2. Solve applied problems by using formulas.
3. Solve percent problems.

A **mathematical model** is an equation or inequality that describes a real situation. Models for many applied problems already exist; they are called *formulas*. A **formula** is an equation in which variables are used to describe a relationship. Some formulas that we will be using are

$$d = rt, \quad I = prt, \quad \text{and} \quad P = 2L + 2W. \quad \text{Formulas}$$

**OBJECTIVE 1** Solve a formula for a specified variable. The formula $I = prt$ says that interest on a loan or investment equals principal (amount borrowed or invested), times rate (annually, in percent), times time (in years) at the given interest rate. To determine how long it will take for an investment at a stated interest rate to earn a predetermined amount of interest, it would help to first solve the formula for $t$. This process is called **solving for a specified variable** or **solving a literal equation.**

The steps used in the examples that follow are similar to those used in solving linear equations in **Section 2.1.** *When you are solving for a specified variable, the key is to treat that variable as if it were the only one; treat all other variables like numbers (constants).*

The following additional suggestions may be helpful.

### Solving for a Specified Variable

*Step 1* Transform so that all terms containing the specified variable are on one side of the equation and all terms without that variable are on the other side.

*Step 2* If necessary, use the distributive property to combine the terms with the specified variable.* The result should be the product of a sum or difference and the variable.

*Step 3* Divide both sides by the factor that is the coefficient of the specified variable.

---

*Using the distributive property to write $ab + ac$ as $a(b + c)$ is called *factoring*. See **Chapter 6.**

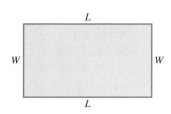

Perimeter $P$, the distance around a rectangle, is given by
$P = 2L + 2W$.

**FIGURE 1**

### EXAMPLE 1  Solving for a Specified Variable

Solve the formula $P = 2L + 2W$ for $W$.

This formula gives the relationship between the perimeter $P$ (distance around) a rectangle, the length $L$ of the rectangle, and the width $W$ of the rectangle. See Figure 1.

Solve the formula for $W$ by isolating $W$ on one side of the equals sign.

$$P = 2L + 2W$$

**Step 1**  $\quad P - 2L = 2L + 2W - 2L \quad$ Subtract $2L$ from both sides.

$\qquad\qquad P - 2L = 2W \quad$ Combine like terms.

**Step 2**  is not needed here.

**Step 3**  $\dfrac{P - 2L}{2} = \dfrac{2W}{2} \quad$ Divide both sides by 2.

$$\dfrac{P - 2L}{2} = W, \quad \text{or} \quad W = \dfrac{P - 2L}{2}$$

**NOW TRY** Exercise 3.

**CAUTION** In Step 3 of Example 1, do not simplify the fraction by dividing 2 into the term $2L$. The subtraction in the numerator must be done before the division.

$$\dfrac{P - 2L}{2} \neq P - L$$

### EXAMPLE 2  Solving a Formula Having Parentheses

The formula for the perimeter of a rectangle is sometimes written in the equivalent form $P = 2(L + W)$. Solve this form for $W$.

One way to begin is to use the distributive property on the right side of the equation to get $P = 2L + 2W$, which we would then solve as in Example 1. Another way to begin is to divide by the coefficient 2.

$$P = 2(L + W)$$

$\dfrac{P}{2} = L + W \qquad$ Divide both sides by 2.

$\dfrac{P}{2} - L = W, \quad \text{or} \quad W = \dfrac{P}{2} - L \qquad$ Subtract $L$ from both sides.

We can show that this result is equivalent to our result in Example 1 by multiplying $L$ by $\frac{2}{2}$.

$\dfrac{P}{2} - \dfrac{2}{2}(L) = W \qquad \frac{2}{2} = 1$, so $L = \frac{2}{2}(L)$.

$\dfrac{P}{2} - \dfrac{2L}{2} = W$

$\dfrac{P - 2L}{2} = W \qquad$ Subtract fractions: $\frac{a}{c} - \frac{b}{c} = \frac{a - b}{c}$.

The final line agrees with the result in Example 1.

**NOW TRY** Exercise 9.

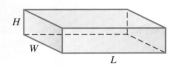

Rectangular solid
$A = 2HW + 2LW + 2LH$

**FIGURE 2**

A rectangular solid has the shape of a box but is solid. See Figure 2. The labels $H$, $W$, and $L$ represent the height, width, and length of the figure, respectively. The surface area of any solid three-dimensional figure is the total area of its surface. For a rectangular solid, the surface area $A$ is

$$A = 2HW + 2LW + 2LH.$$

### EXAMPLE 3 Using the Distributive Property to Solve for a Specified Variable

Solve the formula $A = 2HW + 2LW + 2LH$ for $L$.

To solve for the length $L$, treat $L$ as the only variable and treat all other variables as constants.

*We must isolate the L-terms.*

$$A = 2HW + 2LW + 2LH$$
$$A - 2HW = 2LW + 2LH \qquad \text{Subtract } 2HW.$$
$$A - 2HW = L(2W + 2H) \qquad \text{Distributive property}$$

*This is a key step.*
$$\frac{A - 2HW}{2W + 2H} = L, \quad \text{or} \quad L = \frac{A - 2HW}{2W + 2H} \qquad \text{Divide by } 2W + 2H.$$

**NOW TRY** Exercise 15.

**CAUTION** The most common error in working a problem like Example 3 is not using the distributive property correctly. We must write the expression so that the specified variable is a *factor;* then we can divide by its coefficient in the final step.

### OBJECTIVE 2 Solve applied problems by using formulas.

The distance formula $d = rt$ relates $d$, the distance traveled; $r$, the rate or speed; and $t$, the travel time.

### EXAMPLE 4 Finding Average Speed

Janet Branson found that usually it took her $\frac{3}{4}$ hr each day to drive a distance of 15 mi to work. What was her speed?

Find the formula for speed (rate) $r$ by solving $d = rt$ for $r$.

$$d = rt$$
$$\frac{d}{t} = \frac{rt}{t} \qquad \text{Divide by } t.$$
$$\frac{d}{t} = r, \quad \text{or} \quad r = \frac{d}{t}$$

Notice that only Step 3 was needed to solve for $r$ in this example. Now find Janet's speed by substituting the given values of $d$ and $t$ into the last formula.

$$r = \frac{d}{t}$$
$$r = \frac{15}{\frac{3}{4}} \qquad \text{Let } d = 15, t = \frac{3}{4}.$$
$$r = 15 \cdot \frac{4}{3} \qquad \text{Multiply by the reciprocal of } \frac{3}{4}.$$
$$r = 20$$

Her speed averaged 20 mph. (That is, at times she may have traveled a little faster or a little slower than 20 mph, but overall, her speed was 20 mph.)

**NOW TRY** Exercise 19.

**OBJECTIVE 3** Solve percent problems. An important everyday use of mathematics involves the concept of percent. Percent is written with the symbol %. The word **percent** means "per one hundred." One percent means "one per one hundred" or "one one-hundredth."

$$1\% = 0.01 \quad \text{or} \quad 1\% = \frac{1}{100}$$

The following formula can be used to solve a percent problem:

$$\frac{\text{partial amount}}{\text{whole amount}} = \text{percent (represented as a decimal)}.$$

For example, if a class consists of 50 students, and 32 are males, then the percent of males in the class is

$$\frac{\text{partial amount}}{\text{whole amount}} = \frac{32}{50} = 0.64, \quad \text{or} \quad 64\%.$$

**EXAMPLE 5** Solving Percent Problems

**(a)** A 50-L mixture of acid and water contains 10 L of acid. What is the percent of acid in the mixture?

The amount of the mixture is 50 L, and the part that is acid is 10 L. Let $x$ represent the percent of acid. Then the percent of acid in the mixture is

$$x = \frac{10}{50} \leftarrow \text{partial amount} \atop \leftarrow \text{whole amount}$$

$$x = 0.20, \quad \text{or} \quad 20\%.$$

**(b)** If a savings account balance of $3550 earns 6% interest in one year, how much interest is earned?

Let $x$ represent the amount of interest earned (that is, the part of the whole amount invested). Since $6\% = 6 \cdot 0.01 = 0.06$, the equation is

$$\frac{x}{3550} = 0.06 \qquad \frac{\text{partial amount}}{\text{whole amount}} = \text{percent}$$

$$x = 0.06(3550) \qquad \text{Multiply by 3550.}$$

$$x = 213.$$

The interest earned is $213.

**NOW TRY** Exercises 31 and 33.

> **EXAMPLE 6** Interpreting Percents from a Graph
>
> In 2005, people in the United States spent an estimated $35.9 billion on their pets. Use the graph in Figure 3 to determine how much of this amount was spent on pet food.
>
>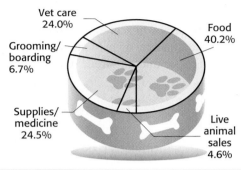
>
> *Source*: American Pet Products Manufacturers Association, Inc.
>
> **FIGURE 3**
>
> According to the graph, 40.2% was spent on food. Let $x$ represent this amount in billions of dollars.
>
> $$\frac{x}{35.9} = 0.402 \qquad \text{40.2\% = 0.402}$$
> $$x = 0.402(35.9) \qquad \text{Multiply by 35.9.}$$
> $$x = 14.4 \qquad \text{Nearest tenth}$$
>
> About $14.4 billion was spent on pet food.
>
> **NOW TRY** Exercise 37.

## 2.2 Exercises

Solve each formula for the specified variable. See Examples 1 and 2.

1. $I = prt$ for $r$   (simple interest)
2. $d = rt$ for $t$   (distance)
3. $P = 2L + 2W$ for $L$
   (perimeter of a rectangle)

4. $A = bh$ for $b$   (area of a parallelogram)

   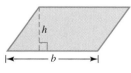

5. $V = LWH$
   (volume of a rectangular solid)
   **(a)** for $W$  **(b)** for $H$

   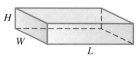

6. $P = a + b + c$
   (perimeter of a triangle)
   **(a)** for $b$  **(b)** for $c$

   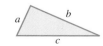

7. $C = 2\pi r$ for $r$
   (circumference of a circle)

8. $A = \dfrac{1}{2}bh$ for $h$
   (area of a triangle)

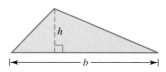

9. $A = \dfrac{1}{2}h(b + B)$ (area of a trapezoid)
   **(a)** for $h$  **(b)** for $B$

10. $S = 2\pi rh + 2\pi r^2$ for $h$
    (surface area of a right circular cylinder)

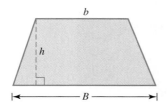

11. $F = \dfrac{9}{5}C + 32$ for $C$
    (Celsius to Fahrenheit)

12. $C = \dfrac{5}{9}(F - 32)$ for $F$
    (Fahrenheit to Celsius)

13. *Multiple Choice* When a formula is solved for a particular variable, several different equivalent forms may be possible. If we solve $A = \tfrac{1}{2}bh$ for $h$, one possible correct answer is

$$h = \dfrac{2A}{b}.$$

Which one of the following is *not* equivalent to this?

**A.** $h = 2\left(\dfrac{A}{b}\right)$  **B.** $h = 2A\left(\dfrac{1}{b}\right)$  **C.** $h = \dfrac{A}{\tfrac{1}{2}b}$  **D.** $h = \dfrac{\tfrac{1}{2}A}{b}$

14. *Concept Check* Suppose a student solved the formula $A = 2HW + 2LW + 2LH$ for $L$ as follows:

$$A = 2HW + 2LW + 2LH$$
$$A - 2LW - 2HW = 2LH$$
$$\dfrac{A - 2LW - 2HW}{2H} = L.$$

**WHAT WENT WRONG?**

*Solve each equation for the specified variable. See Example 3.*

15. $2k + ar = r - 3y$ for $r$

16. $4s + 7p = tp - 7$ for $p$

17. $w = \dfrac{3y - x}{y}$ for $y$

18. $c = \dfrac{-2t + 4}{t}$ for $t$

*Solve each problem. See Example 4.*

**19.** In 2005, Jeff Gordon won the Daytona 500 (mile) race with a speed of 135.173 mph. Find his time to the nearest thousandth. (*Source: World Almanac and Book of Facts.*)

**20.** In 2004, rain shortened the Indianapolis 500 race to 450 mi. It was won by Buddy Rice, who averaged 138.518 mph. What was his time to the nearest thousandth? (*Source:* indy500.com)

**21.** Faye Korn traveled from Kansas City to Louisville, a distance of 520 mi, in 10 hr. Find her rate in miles per hour.

**22.** The distance from Melbourne to London is 10,500 mi. If a jet averages 500 mph between the two cities, what is its travel time in hours?

**23.** As of 2009, the highest temperature ever recorded in Chicago was 40°C. Find the corresponding Fahrenheit temperature. (*Source: World Almanac and Book of Facts.*)

**24.** As of 2009, the lowest temperature ever recorded in Memphis was −13°F. Find the corresponding Celsius temperature. (*Source: World Almanac and Book of Facts.*)

**25.** The base of the Great Pyramid of Cheops is a square whose perimeter is 920 m. What is the length of each side of this square? (*Source: Atlas of Ancient Archaeology.*)

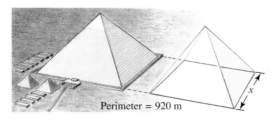

**26.** Marina City in Chicago is a complex of two residential towers that resemble corncobs. Each tower has a concrete cylindrical core with a 35-ft diameter and is 588 ft tall. Find the volume of the core of one of the towers to the nearest whole number. (*Hint:* Use the $\pi$ key on your calculator.) (*Source:* www.architechgallery.com; www.aviewoncities.com)

**27.** The circumference of a circle is $480\pi$ in. What is the radius of the circle? What is its diameter?

**28.** The radius of a circle is 2.5 in. What is the diameter of the circle? What is its circumference?

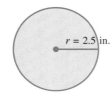

29. A sheet of standard-size copy paper measures 8.5 in. by 11 in. If a ream (500 sheets) of this paper has a volume of 187 in.$^3$, how thick is the ream?

30. Copy paper (Exercise 29) also comes in legal size, which has the same width, but is longer than standard size. If a ream of legal-size copy paper has the same thickness as the standard-size paper and a volume of 238 in.$^3$, what is the length of a sheet of legal paper?

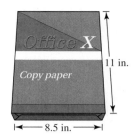

*Solve each problem. See Example 5.*

31. A mixture of alcohol and water contains a total of 36 oz of liquid. There are 9 oz of pure alcohol in the mixture. What percent of the mixture is water? What percent is alcohol?

32. A mixture of acid and water is 35% acid. If the mixture contains a total of 40 L, how many liters of pure acid are in the mixture? How many liters of pure water are in the mixture?

33. A real-estate agent earned $6300 commission on a property sale of $210,000. What is her rate of commission?

34. A certificate of deposit for 1 yr pays $221 simple interest on a principal of $3400. What is the interest rate being paid on this deposit?

*Exercises 35 and 36 deal with winning percentage in the standings of baseball teams. Winning percentage (Pct.) is commonly expressed as a decimal rounded to the nearest thousandth. To find the winning percentage of a team, divide the number of wins (W) by the total number of games played (W + L).*

35. At the start of play on May 4, 2009, the standings of the Central Division of the American League were as shown. Find the winning percentages of the following teams.

    (a) Detroit  (b) Minnesota
    (c) Chicago  (d) Cleveland

    | | W | L | Pct. |
    |---|---|---|---|
    | Kansas City | 15 | 11 | .577 |
    | Detroit | 13 | 12 | |
    | Minnesota | 13 | 13 | |
    | Chicago | 12 | 13 | |
    | Cleveland | 10 | 16 | |

36. Repeat Exercise 35 for the following standings of the Central Division of the National League.

    (a) St. Louis  (b) Milwaukee
    (c) Cincinnati  (d) Houston

    | | W | L | Pct. |
    |---|---|---|---|
    | St. Louis | 17 | 9 | |
    | Chicago | 14 | 11 | .560 |
    | Milwaukee | 14 | 12 | |
    | Cincinnati | 13 | 12 | |
    | Pittsburgh | 12 | 13 | .480 |
    | Houston | 11 | 15 | |

*An average middle-income family will spend $242,070 to raise a child born in 2004 from birth to age 17. The graph shows the percentages spent for various categories. Use the graph to answer Exercises 37–40. See Example 6.*

37. To the nearest dollar, how much will be spent to provide housing for the child?

38. To the nearest dollar, how much will be spent for health care?

39. Use your answer from Exercise 38 to find how much will be spent for transportation.

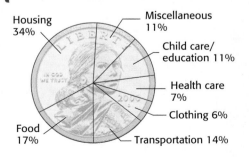

**The Cost of Parenthood**

Housing 34%
Miscellaneous 11%
Child care/education 11%
Health care 7%
Clothing 6%
Transportation 14%
Food 17%

*Source:* U.S. Department of Agriculture.

40. About $41,000 will be spent for food. To the nearest percent, what percent of the cost of raising a child from birth to age 17 is this? Does your answer agree with the percent shown in the graph?

*Television networks have been losing viewers to cable programming since 1982, as the two graphs show. Use these graphs to answer Exercises 41–44. See Example 6.*

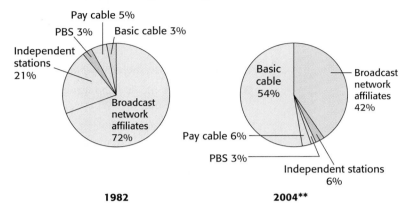

**Shifting Share of the Television Audience as More Homes Receive Cable Programming***

*Shares don't add to 100% because of viewing of multiple TV sets in some households.
**Independent stations include all superstations except TBS; broadcast affiliates include Fox; cable includes TBS.

*Source:* Nielsen Media Research.

**41.** In a typical group of 50,000 television viewers, how many would have watched basic cable in 1982?

**42.** In 1982, how many of a typical group of 110,000 viewers watched independent stations?

**43.** How many of a typical group of 35,000 viewers watched basic cable in 2004?

**44.** In a typical group of 65,000 viewers, how many watched independent stations in 2004?

## 2.3 Applications of Linear Equations

**OBJECTIVES**

1. Translate from words to mathematical expressions.
2. Write equations from given information.
3. Distinguish between expressions and equations.
4. Use the six steps in solving an applied problem.
5. Solve percent problems.
6. Solve investment problems.
7. Solve mixture problems.

**OBJECTIVE 1** **Translate from words to mathematical expressions.** Producing a mathematical model of a real situation often involves translating verbal statements into mathematical statements.

Usually, there are key words and phrases in a verbal problem that translate into mathematical expressions involving addition, subtraction, multiplication, and division. Translations of some commonly used expressions follow.

**TRANSLATING FROM WORDS TO MATHEMATICAL EXPRESSIONS**

| Verbal Expression | Mathematical Expression (where $x$ and $y$ are numbers) |
|---|---|
| **Addition** | |
| The **sum** of a number and 7 | $x + 7$ |
| 6 **more than** a number | $x + 6$ |
| 3 **plus** a number | $3 + x$ |
| 24 **added to** a number | $x + 24$ |
| A number **increased by** 5 | $x + 5$ |
| The **sum** of two numbers | $x + y$ |

| | |
|---|---|
| **Subtraction** | |
| 2 **less than** a number | $x - 2$ |
| 2 **less** a number | $2 - x$ |
| 12 **minus** a number | $12 - x$ |
| A number **decreased by** 12 | $x - 12$ |
| A number **subtracted from** 10 | $10 - x$ |
| **From** a number, **subtract** 10 | $x - 10$ |
| The **difference between** two numbers | $x - y$ |
| **Multiplication** | |
| 16 **times** a number | $16x$ |
| A number **multiplied by** 6 | $6x$ |
| $\frac{2}{3}$ **of** a number (used with fractions and percent) | $\frac{2}{3}x$ |
| $\frac{3}{4}$ **as much as** a number | $\frac{3}{4}x$ |
| **Twice** (2 times) a number | $2x$ |
| The **product** of two numbers | $xy$ |
| **Division** | |
| The **quotient** of 8 and a number | $\frac{8}{x}$ $(x \neq 0)$ |
| A number **divided by** 13 | $\frac{x}{13}$ |
| The **ratio** of two numbers or the **quotient** of two numbers | $\frac{x}{y}$ $(y \neq 0)$ |

**CAUTION** Because subtraction and division are not commutative operations, it is important to correctly translate expressions involving them. For example, "2 less than a number" is translated as $x - 2$, *not* $2 - x$. "A number subtracted from 10" is expressed as $10 - x$, *not* $x - 10$.

For division, the number *by which* we are dividing is the denominator, and the number *into which* we are dividing is the numerator. For example, "a number divided by 13" and "13 divided into $x$" both translate as $\frac{x}{13}$. Similarly, "the quotient of $x$ and $y$" is translated as $\frac{x}{y}$.

**OBJECTIVE 2** Write equations from given information. The symbol for equality, $=$, is often indicated by the word *is*. In fact, because equal mathematical expressions represent names for the same number, any words that indicate the idea of "sameness" translate to $=$.

**EXAMPLE 1** Translating Words into Equations

Translate each verbal sentence into an equation.

| Verbal Sentence | Equation |
|---|---|
| Twice a number, decreased by 3, is 42. | $2x - 3 = 42$ |
| The product of a number and 12, decreased by 7, is 105. | $12x - 7 = 105$ |
| The quotient of a number and the number plus 4 is 28. | $\frac{x}{x + 4} = 28$ |
| The quotient of a number and 4, plus the number, is 10. | $\frac{x}{4} + x = 10$ |

**NOW TRY** Exercises 7 and 17.

**OBJECTIVE 3** Distinguish between expressions and equations. An expression translates as a phrase. An equation includes the = symbol, with something on either side of it, and translates as a sentence.

**EXAMPLE 2** Distinguishing between Expressions and Equations

Decide whether each is an *expression* or an *equation*.

**(a)** $2(3 + x) - 4x + 7$

There is no equals sign, so this is an expression.

**(b)** $2(3 + x) - 4x + 7 = -1$

Because there is an equals sign with something on either side of it, this is an equation.

Note that the expression in part (a) simplifies to the expression $-2x + 13$, and the equation in part (b) has solution 7.

**NOW TRY** Exercises 21 and 23.

**OBJECTIVE 4** Use the six steps in solving an applied problem. While there is no one method that allows us to solve all types of applied problems, the following six steps are helpful.

### Solving an Applied Problem

*Step 1* **Read** the problem, several times if necessary, until you understand what is given and what is to be found.

*Step 2* **Assign a variable** to represent the unknown value, using diagrams or tables as needed. Write down what the variable represents. If necessary, express any other unknown values in terms of the variable.

*Step 3* **Write an equation** using the variable expression(s).

*Step 4* **Solve** the equation.

*Step 5* **State the answer** to the problem. Does it seem reasonable?

*Step 6* **Check** the answer in the words of the original problem.

**EXAMPLE 3** Solving a Perimeter Problem

The length of a rectangle is 1 cm more than twice the width. The perimeter of the rectangle is 110 cm. Find the length and the width of the rectangle.

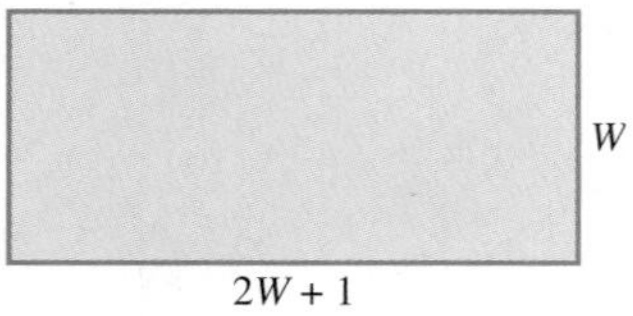

*Step 1* **Read** the problem. What must be found? The length and width of the rectangle. What is given? The length is 1 cm more than twice the width; the perimeter is 110 cm.

*Step 2* **Assign a variable.** Let $W$ = the width; then $2W + 1$ = the length. Make a sketch, as in Figure 4.

*Step 3* **Write an equation.** Use the formula for the perimeter of a rectangle.

$$P = 2L + 2W \qquad \text{Perimeter of a rectangle}$$
$$110 = 2(2W + 1) + 2W \qquad \text{Let } L = 2W + 1 \text{ and } P = 110.$$

**FIGURE 4**

$2W + 1$

$W$

*Step 4* **Solve** the equation obtained in Step 3.

$$110 = 4W + 2 + 2W \quad \text{Distributive property}$$
$$110 = 6W + 2 \quad \text{Combine like terms.}$$
$$110 - 2 = 6W + 2 - 2 \quad \text{Subtract 2.}$$
$$108 = 6W$$
$$\frac{108}{6} = \frac{6W}{6} \quad \text{Divide by 6.}$$
$$18 = W$$

*Step 5* **State the answer.** The width of the rectangle is 18 cm and the length is $2(18) + 1 = 37$ cm.

*Step 6* **Check.** The length, 37 cm, is 1 cm more than 2(18) cm (twice the width). The perimeter is $2(37) + 2(18) = 74 + 36 = 110$ cm, as required.

**NOW TRY** Exercise 29.

**EXAMPLE 4** Finding Unknown Numerical Quantities

Two outstanding major league pitchers in recent years are Randy Johnson and Johan Santana. In 2004, the two pitchers had a combined total of 555 strikeouts. Johnson had 25 more strikeouts than Santana. How many strikeouts did each pitcher have? (*Source: World Almanac and Book of Facts.*)

*Step 1* **Read** the problem. We are asked to find the number of strikeouts each pitcher had.

*Step 2* **Assign a variable** to represent the number of strikeouts for one of the men.

Let $s =$ the number of strikeouts for Johan Santana.

We must also find the number of strikeouts for Randy Johnson. Since he had 25 more strikeouts than Santana,

$s + 25 =$ the number of strikeouts for Johnson.

*Step 3* **Write an equation.** The sum of the numbers of strikeouts is 555, so

Santana's strikeouts + Johnson's strikeouts = Totals
$$s + (s + 25) = 555.$$

*Step 4* **Solve** the equation.

$$s + (s + 25) = 555$$
$$2s + 25 = 555 \quad \text{Combine like terms.}$$
$$2s + 25 - 25 = 555 - 25 \quad \text{Subtract 25.}$$
$$2s = 530$$
$$\frac{2s}{2} = \frac{530}{2} \quad \text{Divide by 2.}$$
$$s = 265$$

(Don't stop here.)

*Step 5* **State the answer.** We let $s$ represent the number of strikeouts for Santana, so Santana had 265. Then Johnson had

$$s + 25 = 265 + 25 = 290 \text{ strikeouts.}$$

*Step 6* **Check.** 290 is 25 more than 265, and $265 + 290 = 555$. The conditions of the problem are satisfied, and our answer checks.

**NOW TRY** Exercise 35.

**CAUTION** A common error in solving applied problems is forgetting to answer all the questions asked in the problem. In Example 4, we were asked for the number of strikeouts for *each* player, so there was extra work in Step 5 in order to find Johnson's number.

**OBJECTIVE 5** Solve percent problems. Recall from **Section 2.2** that percent means "per one hundred," so 5% means 0.05, 14% means 0.14, and so on.

**EXAMPLE 5** Solving a Percent Problem

In 2002, there were 301 long-distance area codes in the United States, an increase of 250% over the number when the area code plan originated in 1947. How many area codes were there in 1947? (*Source:* SBC Telephone Directory.)

*Step 1* **Read** the problem. We are given that the number of area codes increased by 250% from 1947 to 2002, and there were 301 area codes in 2002. We must find the original number of area codes.

*Step 2* **Assign a variable.**

Let $x$ = the number of area codes in 1947.

Since $250\% = 250(0.01) = 2.5$,

$2.5x$ = the number of codes added since then.

*Step 3* **Write an equation** from the given information.

the number in 1947 + the increase = 301
$$x + 2.5x = 301$$

*Step 4* **Solve** the equation.  *Note the $x$ in $2.5x$.*

$$1x + 2.5x = 301 \quad \text{Identity property}$$
$$3.5x = 301 \quad \text{Combine like terms.}$$
$$x = 86 \quad \text{Divide by 3.5.}$$

*Step 5* **State the answer.** There were 86 area codes in 1947.

*Step 6* **Check** that the increase, $301 - 86 = 215$, is 250% of 86.

$$250\% \cdot 86 = 250(0.01)(86) = 215, \text{ as required.}$$

**NOW TRY** Exercise 45.

**CAUTION** Watch for two common errors that occur in solving problems like the one in Example 5.

1. Do not try to find 250% of 301 and subtract that amount from 301. The 250% should be applied to *the amount in 1947, not the amount in 2002.*

2. Do not write the equation as

$$x + 2.5 = 301. \quad \text{Incorrect}$$

The percent must be multiplied by some amount; in this case, the amount is the number of area codes in 1947, giving $2.5x$.

**SECTION 2.3** Applications of Linear Equations **65**

**OBJECTIVE 6** **Solve investment problems.** The investment problems in this chapter deal with *simple interest.* In most real-world applications, *compound interest* (covered in a later chapter) is used.

**EXAMPLE 6** **Solving an Investment Problem**

After winning the state lottery, Mark LeBeau has $40,000 to invest. He will put part of the money in an account paying 4% interest and the remainder into stocks paying 6% interest. His accountant tells him that the total annual income from these investments should be $2040. How much should he invest at each rate?

*Step 1* **Read** the problem again. We must find the two amounts.

*Step 2* **Assign a variable.**

Let $x$ = the amount to invest at 4%;

then $40,000 - x$ = the amount to invest at 6%.

The formula for interest is $I = prt$. Here the time $t$ is 1 yr. Use a table to organize the given information.

| Principal | Rate (as a decimal) | Interest |
|---|---|---|
| $x$ | 0.04 | $0.04x$ |
| $40,000 - x$ | 0.06 | $0.06(40,000 - x)$ |
| 40,000 |  | 2040 |

Multiply principal, rate, and time (here, 1 yr) to get interest.

← Total

*Step 3* **Write an equation.** The last column of the table gives the equation.

interest at 4% + interest at 6% = total interest

$$0.04x + 0.06(40,000 - x) = 2040$$

*Step 4* **Solve** the equation. We do so without clearing decimals.

$0.04x + 0.06(40,000) - 0.06x = 2040$    Distributive property

$-0.02x + 2400 = 2040$    Combine like terms; multiply.

$-0.02x = -360$    Subtract 2400.

$x = 18,000$    Divide by $-0.02$.

*Step 5* **State the answer.** Mark should invest $18,000 of the money at 4% and $40,000 - $18,000 = $22,000 at 6%.

*Step 6* **Check** by finding the annual interest at each rate. The sum of these two amounts should total $2040.

$0.04(\$18,000) = \$720$    and    $0.06(\$22,000) = \$1320$

$\$720 + \$1320 = \$2040$, as required.

**NOW TRY** Exercise 49.

**OBJECTIVE 7** **Solve mixture problems.** Mixture problems involving rates of concentration can be solved with linear equations.

### EXAMPLE 7  Solving a Mixture Problem

A chemist must mix 8 L of a 40% acid solution with some 70% solution to get a 50% solution. How much of the 70% solution should be used?

*Step 1*  **Read** the problem. The problem asks for the amount of 70% solution to be used.

*Step 2*  **Assign a variable.** Let $x = $ the number of liters of 70% solution to be used. The information in the problem is illustrated in Figure 5.

**FIGURE 5**

Use the given information to complete a table.

| Number of Liters | Percent (as a decimal) | Liters of Pure Acid |
|---|---|---|
| 8 | 0.40 | $0.40(8) = 3.2$ |
| $x$ | 0.70 | $0.70x$ |
| $8 + x$ | 0.50 | $0.50(8 + x)$ |

Sum must equal

The numbers in the last column were found by multiplying the strengths by the numbers of liters. The number of liters of pure acid in the 40% solution plus the number of liters in the 70% solution must equal the number of liters in the 50% solution.

*Step 3*  **Write an equation.**
$$3.2 + 0.70x = 0.50(8 + x)$$

*Step 4*  **Solve.**
$$3.2 + 0.70x = 4 + 0.50x \quad \text{Distributive property}$$
$$0.20x = 0.8 \quad \text{Subtract 3.2 and 0.50}x.$$
$$x = 4 \quad \text{Divide by 0.20.}$$

*Step 5*  **State the answer.** The chemist should use 4 L of the 70% solution.

*Step 6*  **Check.** 8 L of 40% solution plus 4 L of 70% solution is
$$8(0.40) + 4(0.70) = 6 \text{ L}$$
of acid. Similarly, $8 + 4$ or 12 L of 50% solution has
$$12(0.50) = 6 \text{ L}$$
of acid in the mixture. The total amount of pure acid is 6 L both before and after mixing, so the answer checks.

**NOW TRY** Exercise 55.

### EXAMPLE 8  Solving a Mixture Problem when One Ingredient Is Pure

The octane rating of gasoline is a measure of its antiknock qualities. For a standard fuel, the octane rating is the percent of isooctane. How many liters of pure isooctane should be mixed with 200 L of 94% isooctane, referred to as 94 octane, to get a mixture that is 98% isooctane?

***Step 1*** **Read** the problem. The problem asks for the amount of pure isooctane.

***Step 2*** **Assign a variable.** Let $x$ = the number of liters of pure (100%) isooctane. Complete a table. Recall that $100\% = 100(0.01) = 1$.

| Number of Liters | Percent (as a decimal) | Liters of Pure Isooctane |
|---|---|---|
| $x$ | 1 | $x$ |
| 200 | 0.94 | 0.94(200) |
| $x + 200$ | 0.98 | $0.98(x + 200)$ |

***Step 3*** **Write an equation.** The equation comes from the last column of the table.
$$x + 0.94(200) = 0.98(x + 200)$$

***Step 4*** **Solve.**
$$x + 0.94(200) = 0.98x + 0.98(200) \quad \text{Distributive property}$$
$$x + 188 = 0.98x + 196 \quad \text{Multiply.}$$
$$0.02x = 8 \quad \text{Subtract } 0.98x \text{ and } 188.$$
$$x = 400 \quad \text{Divide by } 0.02.$$

***Step 5*** **State the answer.** 400 L of isooctane is needed.

***Step 6*** **Check** by showing that $400 + 0.94(200) = 0.98(400 + 200)$ is true.

**NOW TRY** Exercise 59.

## 2.3 Exercises

**NOW TRY Exercise**

*Concept Check* In each of the following, **(a)** translate as an expression and **(b)** translate as an equation or inequality. Use $x$ to represent the number.

1. **(a)** 12 more than a number
   **(b)** 12 is more than a number.
2. **(a)** 3 less than a number
   **(b)** 3 is less than a number.
3. **(a)** 4 less than a number
   **(b)** 4 is less than a number.
4. **(a)** 6 greater than a number
   **(b)** 6 is greater than a number.

5. *Multiple Choice* Which one of the following is *not* a valid translation of "20% of a number," where $x$ represents the number?
   **A.** $0.20x$ **B.** $0.2x$ **C.** $\dfrac{x}{5}$ **D.** $20x$

6. Explain why $13 - x$ is *not* a correct translation of "13 less than a number."

*Translate each verbal phrase into a mathematical expression. Use $x$ to represent the unknown number. See Example 1.*

7. Twice a number, decreased by 13
8. The product of 6 and a number, decreased by 12
9. 12 increased by three times a number
10. 12 more than one-half of a number
11. The product of 8 and 12 less than a number
12. The product of 9 more than a number and 6 less than the number
13. The quotient of three times a number and 7
14. The quotient of 6 and five times a nonzero number

*Use the variable $x$ for the unknown, and write an equation representing the verbal sentence. Then solve the problem. See Example 1.*

15. The sum of a number and 6 is $-31$. Find the number.
16. The sum of a number and $-4$ is 12. Find the number.

**17.** If the product of a number and $-4$ is subtracted from the number, the result is 9 more than the number. Find the number.

**18.** If the quotient of a number and 6 is added to twice the number, the result is 8 less than the number. Find the number.

**19.** When $\frac{2}{3}$ of a number is subtracted from 12, the result is 10. Find the number.

**20.** When 75% of a number is added to 6, the result is 3 more than the number. Find the number.

*Decide whether each is an* expression *or an* equation. *See Example 2.*

**21.** $5(x + 3) - 8(2x - 6)$

**22.** $-7(z + 4) + 13(z - 6)$

**23.** $5(x + 3) - 8(2x - 6) = 12$

**24.** $-7(z + 4) + 13(z - 6) = 18$

**25.** $\dfrac{r}{2} - \dfrac{r+9}{6} - 8$

**26.** $\dfrac{r}{2} - \dfrac{r+9}{6} = 8$

*Fill in the Blanks* In Exercises 27 and 28, complete the six suggested problem-solving steps to solve each problem.

**27.** Two of the leading U.S. research universities are the Massachusetts Institute of Technology (MIT) and Stanford University. In 2002, these two universities secured 230 patents on various inventions. Stanford secured 38 fewer patents than MIT. How many patents did each university secure? (*Source:* Association of University Technology Managers.)

  *Step 1* **Read** the problem carefully. We are asked to find _____.

  *Step 2* **Assign a variable.** Let $x$ = the number of patents that MIT secured. Then $x - 38$ = the number of _____.

  *Step 3* **Write an equation.** _____ + _____ = 230

  *Step 4* **Solve** the equation. $x$ = _____

  *Step 5* **State the answer.** MIT secured _____ patents, and Stanford secured _____ patents.

  *Step 6* **Check.** The number of Stanford patents was _____ fewer than the number of _____, and the total number of patents was $134 +$ _____ = _____.

**28.** *Fill in the Blanks* In a recent sample of book buyers, 70 more shopped at large-chain bookstores than at small-chain/independent bookstores. A total of 442 book buyers shopped at these two types of stores. How many buyers shopped at each type of bookstore? (*Source:* Book Industry Study Group.)

  *Step 1* **Read** the problem carefully. We are asked to find _____.

  *Step 2* **Assign a variable.** Let $x$ = the number of book buyers at large-chain bookstores. Then $x - 70$ = the number of _____.

  *Step 3* **Write an equation.** _____ + _____ = 442

  *Step 4* **Solve** the equation. $x$ = _____

  *Step 5* **State the answer.** There were _____ large-chain bookstore shoppers and _____ small-chain/independent shoppers.

  *Step 6* **Check.** The number of _____ was _____ more than the number of _____, and the total number of these shoppers was $256 +$ _____ = _____.

*Solve each problem. See Examples 3 and 4.*

**29.** The John Hancock Center in Chicago has a rectangular base. The length of the base measures 65 ft less than twice the width. The perimeter of the base is 860 ft. What are the dimensions of the base?

**30.** The John Hancock Center (Exercise 29) tapers as it rises. The top floor is rectangular and has perimeter 520 ft. The width of the top floor measures 20 ft more than one-half its length. What are the dimensions of the top floor?

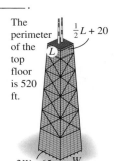

The perimeter of the top floor is 520 ft.

The perimeter of the base is 860 ft.

**31.** Grant Wood painted his most famous work, *American Gothic,* in 1930 on composition board with perimeter 108.44 in. If the painting is 5.54 in. taller than it is wide, find the dimensions of the painting. (*Source: The Gazette,* Cedar Rapids, Iowa, March 12, 2004.)

**32.** The perimeter of a certain rectangle is 16 times the width. The length is 12 cm more than the width. Find the length and width of the rectangle.

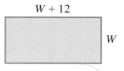

**33.** The Bermuda Triangle supposedly causes trouble for aircraft pilots. It has a perimeter of 3075 mi. The shortest side measures 75 mi less than the middle side, and the longest side measures 375 mi more than the middle side. Find the lengths of the three sides.

**34.** The Vietnam Veterans Memorial in Washington, DC, is in the shape of two sides of an isosceles triangle. If the two walls of equal length were joined by a straight line of 438 ft, the perimeter of the resulting triangle would be 931.5 ft. Find the lengths of the two walls. (*Source:* Pamphlet obtained at Vietnam Veterans Memorial.)

**35.** The two companies with top revenues in the Fortune 500 list for 2008 were Exxon Mobil and Wal-Mart. Their revenues together totaled $849 billion. Wal-Mart revenues were $37 billion less than Exxon Mobil revenues. What were the revenues of each corporation? (*Source:* www.money.cnn.com)

**36.** Two of the longest-running Broadway shows were *Cats,* which played from 1982 through 2000, and *Les Misérables,* which played from 1987 through 2005. Together, there were 14,165 performances of these two shows during their Broadway runs. There were 805 fewer performances of *Les Misérables* than of *Cats.* How many performances were there of each show? (*Source:* The League of American Theatres and Producers.)

**37.** Galileo Galilei conducted experiments involving Italy's famous Leaning Tower of Pisa to investigate the relationship between an object's speed of fall and its weight. The Leaning Tower is 804 ft shorter than the Eiffel Tower in Paris, France. The two towers have a total height of 1164 ft. How tall is each tower? (*Source: Microsoft Encarta Encyclopedia.*)

**38.** In 2003, the New York Yankees and the New York Mets had the highest payrolls in Major League Baseball. The Mets' payroll was $32.8 million less than the Yankees' payroll, and the two payrolls totaled $266.6 million. What was the payroll for each team? (*Source:* Associated Press.)

**39.** In the 2008 presidential election, Barack Obama and John McCain together received 538 electoral votes. Obama received 192 more votes than McCain. How many votes did each candidate receive? (*Source:* www.usconstitution.net.)

**40.** Ted Williams and Rogers Hornsby were two great hitters. Together, they got 5584 hits in their careers. Hornsby got 276 more hits than Williams. How many base hits did each get? (*Source:* Neft, D. S., and R. M. Cohen, *The Sports Encyclopedia: Baseball,* St. Martins Griffin; New York, 1997.)

**41.** In 2005, the number of participants in the ACT exam was 1,186,251. In 1990, a total of 817,000 took the exam. By what percent did the number increase over this period of time, to the nearest tenth of a percent? (*Source:* ACT.)

**42.** Composite scores on the ACT exam fell from 21.0 in 2001 to 20.8 in 2002. What percent decrease was the drop? (*Source:* ACT.)

**43.** In 1995, the average cost of tuition and fees at public four-year universities in the United States was $2811 for full-time students. By 2005, it had risen approximately 95%. To the nearest dollar, what was the approximate cost in 2005? (*Source:* The College Board.)

**44.** In 1995, the average cost of tuition and fees at private four-year universities in the United States was $12,216 for full-time students. By 2005, it had risen approximately 73.8%. To the nearest dollar, what was the approximate cost in 2005? (*Source:* The College Board.)

**45.** In 2005, the average cost of a traditional Thanksgiving dinner for 10, featuring turkey, stuffing, cranberries, pumpkin pie, and trimmings, was $36.78, an increase of 3.1% over the cost in 2004. What was the cost, to the nearest cent, in 2004? (*Source:* American Farm Bureau.)

**46.** Refer to Exercise 45. The first year that information on the cost of a traditional Thanksgiving dinner was collected was 1987. The 2005 cost of $36.78 was an increase of 37.5% over the cost in 1987. What was the cost, to the nearest cent, in 1987? (*Source:* American Farm Bureau.)

**47.** At the end of a day, Jeff Hornsby found that the total cash register receipts at the motel where he works amounted to $2725. This included the 9% sales tax charged. Find the amount of the tax.

**48.** Fino Roverato sold his house for $159,000. He got this amount knowing that he would have to pay a 6% commission to his agent. What amount did he have after the agent was paid?

*Solve each investment problem. See Example 6.*

**49.** Carter Fenton earned $12,000 last year by giving tennis lessons. He invested part of the money at 3% simple interest and the rest at 4%. In one year, he earned a total of $440 in interest. How much did he invest at each rate?

| Principal | Rate (as a decimal) | Interest |
|---|---|---|
| $x$ | 0.03 | |
| | 0.04 | |

**50.** Courtney Slade won $60,000 on a slot machine in Las Vegas. She invested part of the money at 2% simple interest and the rest at 3%. In one year, she earned a total of $1600 in interest. How much was invested at each rate?

| Principal | Rate (as a decimal) | Interest |
|---|---|---|
| $x$ | 0.02 | |
| | | |

**51.** Leah Goldberg won $5000 in a contest. She invested some of the money at 5% simple interest and $400 less than twice that amount at 6.5%. In one year, she earned $298 in interest. How much did she invest at each rate?

**52.** Toshiro Hashimoto invested some money at 4.5% simple interest and $1000 more than four times that amount at 6%. His total annual income for one year from interest on the two investments was $801. How much did he invest at each rate?

**53.** Vincente and Ricarda Pérez have invested $27,000 in bonds paying 7%. How much additional money should they invest in a certificate of deposit paying 4% simple interest so that the total return on the two investments will be 6%?

**54.** Rebecca Herst received a year-end bonus of $17,000 from her company and invested the money in an account paying 6.5%. How much additional money should she deposit in an account paying 5% so that the return on the two investments will be 6%?

*Solve each problem involving rates of concentration and mixtures. See Examples 7 and 8.*

**55.** Ten liters of a 4% acid solution must be mixed with a 10% solution to get a 6% solution. How many liters of the 10% solution are needed?

| Liters of Solution | Percent (as a decimal) | Liters of Pure Acid |
|---|---|---|
| 10 | 0.04 | |
| $x$ | 0.10 | |
| | 0.06 | |

**56.** How many liters of a 14% alcohol solution must be mixed with 20 L of a 50% solution to get a 30% solution?

| Liters of Solution | Percent (as a decimal) | Liters of Pure Alcohol |
|---|---|---|
| $x$ | 0.14 | |
| | 0.50 | |

**57.** In a chemistry class, 12 L of a 12% alcohol solution must be mixed with a 20% solution to get a 14% solution. How many liters of the 20% solution are needed?

**58.** How many liters of a 10% alcohol solution must be mixed with 40 L of a 50% solution to get a 40% solution?

**59.** How much pure dye must be added to 4 gal of a 25% dye solution to increase the solution to 40%? (*Hint:* Pure dye is 100% dye.)

**60.** How much water must be added to 6 gal of a 4% insecticide solution to reduce the concentration to 3%? (*Hint:* Water is 0% insecticide.)

**61.** Randall Albritton wants to mix 50 lb of nuts worth $2 per lb with some nuts worth $6 per lb to make a mixture worth $5 per lb. How many pounds of $6 nuts must he use?

| Pounds of Nuts | Cost per Pound | Total Cost |
|---|---|---|
|  |  |  |
|  |  |  |
|  |  |  |

**62.** Lee Ann Spahr wants to mix tea worth 2¢ per oz with 100 oz of tea worth 5¢ per oz to make a mixture worth 3¢ per oz. How much 2¢ tea should be used?

| Ounces of Tea | Cost per Ounce | Total Cost |
|---|---|---|
|  |  |  |
|  |  |  |
|  |  |  |

**63.** Why is it impossible to mix candy worth $4 per lb and candy worth $5 per lb to obtain a final mixture worth $6 per lb?

**64.** Write an equation based on the following problem, solve the equation, and explain why the problem has no solution:

How much 30% acid should be mixed with 15 L of 50% acid to obtain a mixture that is 60% acid?

## 2.4 Linear Inequalities in One Variable

**OBJECTIVES**

1. Solve linear inequalities by using the addition property.
2. Solve linear inequalities by using the multiplication property.
3. Solve linear inequalities with three parts.
4. Solve applied problems by using linear inequalities.

In **Section 1.1,** we used interval notation to write solution sets of inequalities, with a parenthesis to indicate that an endpoint is not included and a square bracket to indicate that an endpoint is included. We summarize the various types of intervals here.

**INTERVAL NOTATION**

| Type of Interval | Set | Interval Notation | Graph |
|---|---|---|---|
| Open interval | $\{x \mid a < x\}$ | $(a, \infty)$ | |
|  | $\{x \mid a < x < b\}$ | $(a, b)$ | |
|  | $\{x \mid x < b\}$ | $(-\infty, b)$ | |
|  | $\{x \mid x \text{ is a real number}\}$ | $(-\infty, \infty)$ | |
| Half-open interval | $\{x \mid a \leq x\}$ | $[a, \infty)$ | |
|  | $\{x \mid a < x \leq b\}$ | $(a, b]$ | |
|  | $\{x \mid a \leq x < b\}$ | $[a, b)$ | |
|  | $\{x \mid x \leq b\}$ | $(-\infty, b]$ | |
| Closed interval | $\{x \mid a \leq x \leq b\}$ | $[a, b]$ | |

An **inequality** says that two expressions are *not* equal. Solving inequalities is similar to solving equations.

> **Linear Inequality in One Variable**
>
> A **linear inequality in one variable** can be written in the form
> $$Ax + B < C,$$
> where $A$, $B$, and $C$ are real numbers, with $A \neq 0$.

(Throughout this section, we give definitions and rules only for $<$, but they are also valid for $>$, $\leq$, and $\geq$.) Examples of linear inequalities include

$$x + 5 < 2, \quad x - 3 \geq 5, \quad \text{and} \quad 2k + 5 \leq 10. \quad \text{Linear inequalities}$$

**OBJECTIVE 1** Solve linear inequalities by using the addition property. We solve an inequality by finding all numbers that make the inequality true. Usually, an inequality has an infinite number of solutions. These solutions, like solutions of equations, are found by producing a series of simpler related equivalent inequalities. **Equivalent inequalities** are inequalities with the same solution set. We use two important properties to produce equivalent inequalities.

> **Addition Property of Inequality**
>
> For all real numbers $A$, $B$, and $C$, the inequalities
> $$A < B \quad \text{and} \quad A + C < B + C$$
> are equivalent.
>
> That is, adding the same number to each side of an inequality does not change the solution set.

**EXAMPLE 1** Using the Addition Property of Inequality

Solve $x - 7 < -12$ and graph the solution set.

$$x - 7 < -12$$
$$x - 7 + 7 < -12 + 7 \quad \text{Add 7.}$$
$$x < -5$$

*Check:* Substitute $-5$ for $x$ in the *equation* $x - 7 = -12$.

$$x - 7 = -12$$
$$-5 - 7 = -12 \quad ? \quad \text{Let } x = -5.$$
$$-12 = -12 \quad \text{True}$$

This shows that $-5$ is the boundary point. Now test a number on each side of $-5$ to verify that numbers *less than* $-5$ make the inequality true. Choose $-4$ and $-6$.

$$x - 7 < -12$$

| | | |
|---|---|---|
| $-4 - 7 < -12$ ? Let $x = -4$. | $-6 - 7 < -12$ ? Let $x = -6$. |
| $-11 < -12$ False | $-13 < -12$ True |
| $-4$ is not in the solution set. | $-6$ is in the solution set. |

The check confirms that $(-\infty, -5)$, graphed in Figure 6, is the correct solution set.

**FIGURE 6**

**NOW TRY** Exercise 11.

As with equations, the addition property can be used to *subtract* the same number from each side of an inequality.

### EXAMPLE 2  Using the Addition Property of Inequality

Solve $14 + 2m \leq 3m$ and graph the solution set.

$$14 + 2m \leq 3m$$
$$14 + 2m - 2m \leq 3m - 2m \quad \text{Subtract } 2m.$$
$$14 \leq m \quad \text{Combine like terms.}$$
**Be careful.** $\quad m \geq 14 \quad \text{Rewrite.}$

The inequality $14 \leq m$ (14 is less than or equal to $m$) can also be written $m \geq 14$ ($m$ is greater than or equal to 14). Notice that in each case the inequality symbol points to the lesser number, 14.

Check:
$$14 + 2m = 3m$$
$$14 + 2(14) = 3(14) \quad ? \quad \text{Let } m = 14.$$
$$42 = 42 \quad \text{True}$$

So 14 satisfies the equality part of $\leq$. Choose 10 and 15 as test points.

$$14 + 2m < 3m$$

| $14 + 2(10) < 3(10)$ ? Let $m = 10$. | $14 + 2(15) < 3(15)$ ? Let $m = 15$. |
|---|---|
| $34 < 30$ False | $44 < 45$ True |
| 10 is not in the solution set. | 15 is in the solution set. |

The check confirms that $[14, \infty)$ is the correct solution set. See Figure 7.

**FIGURE 7**

NOW TRY Exercise 27.

**OBJECTIVE 2** Solve linear inequalities by using the multiplication property. Solving an inequality such as $3x \leq 15$ requires dividing each side by 3, using the *multiplication property of inequality*. To see how this property works, start with the true statement

$$-2 < 5.$$

Multiply each side by, say, 8.

$$-2(8) < 5(8) \quad \text{Multiply by 8.}$$
$$-16 < 40 \quad \text{True}$$

This gives a true statement. Start again with $-2 < 5$, and multiply each side by $-8$.

$$-2(-8) < 5(-8) \quad \text{Multiply by } -8.$$
$$16 < -40 \quad \text{False}$$

The result, $16 < -40$, is false. To make it true, we must change the direction of the inequality symbol to get

$$16 > -40. \quad \text{True}$$

As these examples suggest, multiplying each side of an inequality by a *negative* number requires reversing the direction of the inequality symbol. The same is true for dividing by a negative number, since division is defined in terms of multiplication.

### Multiplication Property of Inequality

For all real numbers $A$, $B$, and $C$, with $C \neq 0$,

**(a)** the inequalities

$$A < B \quad \text{and} \quad AC < BC \quad \text{are equivalent if } C > 0;$$

**(b)** the inequalities

$$A < B \quad \text{and} \quad AC > BC \quad \text{are equivalent if } C < 0.$$

That is, each side of an inequality may be multiplied (or divided) by a *positive* number without changing the direction of the inequality symbol. *Multiplying (or dividing) by a negative number requires that we reverse the inequality symbol.*

---

**EXAMPLE 3** Using the Multiplication Property of Inequality

Solve each inequality and graph the solution set.

**(a)** $5m \leq -30$

Divide each side by 5. **Since $5 > 0$, do not reverse the inequality symbol.**

$$5m \leq -30$$

$$\frac{5m}{5} \leq \frac{-30}{5} \quad \text{Divide by 5.}$$

$$m \leq -6$$

Check that the solution set is the interval $(-\infty, -6]$, graphed in Figure 8.

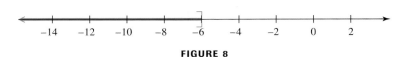

**FIGURE 8**

**(b)** $-4k \leq 32$

Divide each side by $-4$. **Since $-4 < 0$, reverse the inequality symbol.**

$$-4k \leq 32$$

$$\frac{-4k}{-4} \geq \frac{32}{-4} \quad \text{Divide by } -4\text{; reverse the symbol.}$$

$$k \geq -8$$

> Reverse the inequality symbol when dividing by a *negative* number.

Check the solution set. Figure 9 shows the graph of the solution set, $[-8, \infty)$.

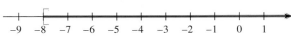

**FIGURE 9**

**NOW TRY** Exercises 15 and 19.

The steps used in solving a linear inequality are given here.

> **Solving a Linear Inequality**
>
> **Step 1** **Simplify each side separately.** Use the distributive property to clear parentheses and combine like terms as needed.
>
> **Step 2** **Isolate the variable terms on one side.** Use the addition property of inequality to get all terms with variables on one side of the inequality and all numbers on the other side.
>
> **Step 3** **Isolate the variable.** Use the multiplication property of inequality to change the inequality to the form
>
> $$x < k \quad \text{or} \quad x > k.$$

**CAUTION** *Reverse the direction of the inequality symbol when multiplying or dividing each side of an inequality by a negative number.*

### EXAMPLE 4  Solving a Linear Inequality by Using the Distributive Property

Solve $-3(x + 4) + 2 \geq 7 - x$ and graph the solution set.

**Step 1**
$$-3(x + 4) + 2 \geq 7 - x$$
$$-3x - 12 + 2 \geq 7 - x \qquad \text{Distributive property}$$
$$-3x - 10 \geq 7 - x$$

**Step 2**
$$-3x - 10 + x \geq 7 - x + x \qquad \text{Add } x.$$
$$-2x - 10 \geq 7$$
$$-2x - 10 + 10 \geq 7 + 10 \qquad \text{Add 10.}$$
$$-2x \geq 17$$

**Step 3**
$$\frac{-2x}{-2} \leq \frac{17}{-2} \qquad \text{Divide by } -2; \text{ change } \geq \text{ to } \leq.$$

*Be sure to reverse the inequality symbol.*
$$x \leq -\frac{17}{2}$$

Figure 10 shows the graph of the solution set, $\left(-\infty, -\frac{17}{2}\right]$.

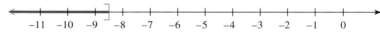

**FIGURE 10**

**NOW TRY** Exercise 31.

### EXAMPLE 5  Solving a Linear Inequality with Fractions

Solve $-\frac{2}{3}(r - 3) - \frac{1}{2} < \frac{1}{2}(5 - r)$ and graph the solution set.

To clear fractions, multiply each side by the least common denominator, 6.

$$-\frac{2}{3}(r - 3) - \frac{1}{2} < \frac{1}{2}(5 - r)$$

$$6\left[-\frac{2}{3}(r - 3) - \frac{1}{2}\right] < 6\left[\frac{1}{2}(5 - r)\right] \qquad \text{Multiply by 6.}$$

*Be careful here.*
$$6\left[-\frac{2}{3}(r - 3)\right] - 6\left(\frac{1}{2}\right) < 6\left[\frac{1}{2}(5 - r)\right] \qquad \text{Distributive property}$$

$$-4(r - 3) - 3 < 3(5 - r) \qquad \text{Multiply.}$$

**Step 1**  $\quad -4r + 12 - 3 < 15 - 3r \quad$ Distributive property
$$-4r + 9 < 15 - 3r$$
**Step 2**  $\quad -4r + 9 + 3r < 15 - 3r + 3r \quad$ Add $3r$.
$$-r + 9 < 15$$
$$-r + 9 - 9 < 15 - 9 \quad \text{Subtract 9.}$$
$$-r < 6$$
**Step 3**  $\quad -1(-r) > -1(6) \quad$ Multiply by $-1$; change $<$ to $>$.
$$r > -6$$

*Remember to reverse the inequality symbol when multiplying by a negative number.*

Check that the solution set is $(-6, \infty)$. See the graph in Figure 11.

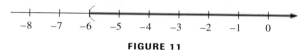

**FIGURE 11**

**NOW TRY** Exercise 37.

**OBJECTIVE 3** Solve linear inequalities with three parts. For some applications, it is necessary to work with a **three-part inequality** such as

$$3 < x + 2 < 8,$$

where $x + 2$ is *between* 3 and 8. To solve this inequality, we subtract 2 from each of the three parts of the inequality, giving

$$3 - 2 < x + 2 - 2 < 8 - 2$$
$$1 < x < 6.$$

Thus, $x$ must be between 1 and 6, so that $x + 2$ will be between 3 and 8. The solution set, $(1, 6)$, is graphed in Figure 12.

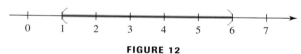

**FIGURE 12**

> **CAUTION** *In three-part inequalities, the order of the parts is important.* It would be *wrong* to write an inequality as $8 < x + 2 < 3$, since this would imply that $8 < 3$, a false statement. *In general, three-part inequalities are written so that the symbols point in the same direction and both point toward the lesser number.*

**EXAMPLE 6** Solving a Three-Part Inequality

Solve $-2 \leq -3k - 1 \leq 5$ and graph the solution set.

$$-2 \leq -3k - 1 \leq 5$$
$$-2 + 1 \leq -3k - 1 + 1 \leq 5 + 1 \quad \text{Add 1 to each part.}$$
$$-1 \leq -3k \leq 6$$
$$\frac{-1}{-3} \geq \frac{-3k}{-3} \geq \frac{6}{-3} \quad \text{Divide each part by } -3;\text{ reverse the inequality symbols.}$$
$$\frac{1}{3} \geq k \geq -2$$
$$-2 \leq k \leq \frac{1}{3} \quad \text{Rewrite in the order on the number line.}$$

Check that the solution set is $\left[-2, \frac{1}{3}\right]$, as shown in Figure 13.

**FIGURE 13**

 Exercise 53.

The following table gives examples of the types of solution sets to be expected from solving linear equations or linear inequalities.

**SOLUTION SETS OF LINEAR EQUATIONS AND INEQUALITIES**

| Equation or Inequality | Typical Solution Set | Graph of Solution Set |
|---|---|---|
| Linear equation $5x + 4 = 14$ | $\{2\}$ | |
| Linear inequality $5x + 4 < 14$ or $5x + 4 > 14$ | $(-\infty, 2)$ $(2, \infty)$ | |
| Three-part inequality $-1 \leq 5x + 4 \leq 14$ | $[-1, 2]$ | |

**OBJECTIVE 4** Solve applied problems by using linear inequalities. In addition to the familiar "is less than" and "is greater than," other expressions such as "is no more than" and "is at least" indicate inequalities, as shown in the table.

| Word Expression | Interpretation |
|---|---|
| *a* is at least *b* | $a \geq b$ |
| *a* is no less than *b* | $a \geq b$ |
| *a* is at most *b* | $a \leq b$ |
| *a* is no more than *b* | $a \leq b$ |

**EXAMPLE 7** Using a Linear Inequality to Solve a Rental Problem

A rental company charges $15 to rent a chain saw, plus $2 per hr. Tom Ruhberg can spend no more than $35 to clear some logs from his yard. What is the *maximum* amount of time he can use the rented saw?

*Step 1* **Read** the problem again.

*Step 2* **Assign a variable.** Let $h$ = the number of hours he can rent the saw.

*Step 3* **Write an inequality.** He must pay $15, plus $2h$, to rent the saw for $h$ hours, and this amount must be *no more than* $35.

$$\underbrace{15 + 2h}_{\text{Cost of renting}} \underbrace{\leq}_{\text{is no more than}} \underbrace{35}_{\text{35 dollars.}}$$

Step 4   Solve.          $2h \leq 20$      Subtract 15.
                         $h \leq 10$       Divide by 2.

Step 5   **State the answer.** He can use the saw for a maximum of 10 hr. (Of course, he may use it for less time, as indicated by the inequality $h \leq 10$.)

Step 6   **Check.** If Tom uses the saw for 10 hr, he will spend $15 + 2(10) = 35$ dollars, the maximum amount.

**NOW TRY** Exercise 67.

**EXAMPLE 8**  Finding an Average Test Score

Martha has scores of 88, 86, and 90 on her first three algebra tests. An average score of at least 90 will earn an A in the class. What possible scores on her fourth test will earn her an A average?

Let $x$ = the score on the fourth test. Her average score must be at least 90. To find the average of four numbers, add them and then divide by 4.

$$\underbrace{\frac{88 + 86 + 90 + x}{4}}_{\text{Average}} \underbrace{\geq}_{\text{is at least}} \underbrace{90}_{90.}$$

$$\frac{264 + x}{4} \geq 90 \qquad \text{Add the scores.}$$

$$264 + x \geq 360 \qquad \text{Multiply by 4.}$$

$$x \geq 96 \qquad \text{Subtract 264.}$$

She must score 96 or more on her fourth test.

Check:   $\dfrac{88 + 86 + 90 + 96}{4} = \dfrac{360}{4} = 90,$   the minimum score.

A score of 96 or more will give an average of at least 90, as required.

**NOW TRY** Exercise 65.

## 2.4 Exercises

**NOW TRY Exercise**

*Matching*  Match each inequality in Column I with the correct graph or interval in Column II.

I

1. $x \leq 3$
2. $x > 3$
3. $x < 3$
4. $x \geq 3$
5. $-3 \leq x \leq 3$
6. $-3 < x < 3$

II

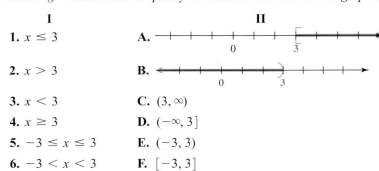

C. $(3, \infty)$
D. $(-\infty, 3]$
E. $(-3, 3)$
F. $[-3, 3]$

**7.** *Concept Check* Refer to the graph, and write an inequality or a three-part inequality for each description.

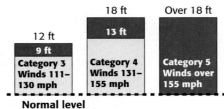

*Source:* National Oceanic and Atmospheric Administration.

**(a)** The wind speed $s$ (in miles per hour) of a Category 4 hurricane
**(b)** The wind speed $s$ (in miles per hour) of a Category 5 hurricane
**(c)** The storm surge $x$ (in feet) from a Category 3 hurricane
**(d)** The storm surge $x$ (in feet) from a Category 5 hurricane

**8.** *Concept Check* Dr. Paul Donohue writes a syndicated column in which readers question him on a variety of health topics. Reader C. J. wrote, "Many people say they can weigh more because they have a large frame. How is frame size determined?" Here is Dr. Donohue's response:

"For a man, a wrist circumference between 6.75 and 7.25 in. [inclusive] indicates a medium frame. Anything above is a large frame and anything below, a small frame."

Using $x$ to represent wrist circumference in inches, write an inequality or a three-part inequality that represents wrist circumference for a male with a

**(a)** small frame  **(b)** medium frame  **(c)** large frame.
(*Source: The Gazette,* Cedar Rapids, Iowa, October 4, 2004.)

**9.** *Concept Check* A student solved the following inequality as shown.

$$4x \geq -64$$
$$\frac{4x}{4} \leq \frac{-64}{4}$$
$$x \leq -16$$

Solution set: $(-\infty, -16]$

**WHAT WENT WRONG?** Give the correct solution set.

**10.** Explain how to determine whether to use parentheses or brackets when graphing the solution set of an inequality.

*Solve each inequality. Give the solution set in both interval and graph form. See Examples 1–5.*

**11.** $x - 4 \geq 12$  **12.** $t - 3 \geq 7$  **13.** $3k + 1 > 22$
**14.** $5z + 6 < 76$  **15.** $4x < -16$  **16.** $2m > -10$
**17.** $-\frac{3}{4}r \geq 30$  **18.** $-\frac{2}{3}x \leq 12$
**19.** $-1.3m \geq -5.2$  **20.** $-2.5y \leq -1.25$
**21.** $5t + 2 \leq -48$  **22.** $4x + 1 \leq -31$
**23.** $\frac{5z - 6}{8} < 8$  **24.** $\frac{3k - 1}{4} > 5$
**25.** $\frac{2k - 5}{-4} > 5$  **26.** $\frac{3z - 2}{-5} < 6$

**27.** $6x - 4 \geq -2x$

**28.** $-2m + 8 \leq 2m$

**29.** $m - 2(m - 4) \leq 3m$

**30.** $x + 4(2x - 1) \geq x$

**31.** $-(4 + r) + 2 - 3r < -14$

**32.** $-(9 + k) - 5 + 4k \geq 4$

**33.** $-3(z - 6) > 2z - 2$

**34.** $-2(x + 4) \leq 6x + 16$

**35.** $\dfrac{2}{3}(3k - 1) \geq \dfrac{3}{2}(2k - 3)$

**36.** $\dfrac{7}{5}(10m - 1) < \dfrac{2}{3}(6m + 5)$

**37.** $-\dfrac{1}{4}(p + 6) + \dfrac{3}{2}(2p - 5) < 10$

**38.** $\dfrac{3}{5}(k - 2) - \dfrac{1}{4}(2k - 7) \leq 3$

**39.** $3(2x - 4) - 4x < 2x + 3$

**40.** $7(4 - x) + 5x < 2(16 - x)$

**41.** $8\left(\dfrac{1}{2}x + 3\right) < 8\left(\dfrac{1}{2}x - 1\right)$

**42.** $10x + 2(x - 4) < 12x - 10$

*Solve each inequality. Give the solution set in both interval and graph form. See Example 6.*

**43.** $-4 < x - 5 < 6$

**44.** $-1 < x + 1 < 8$

**45.** $-9 \leq k + 5 \leq 15$

**46.** $-4 \leq m + 3 \leq 10$

**47.** $-6 \leq 2z + 4 \leq 16$

**48.** $-15 < 3p + 6 < -12$

**49.** $-19 \leq 3x - 5 \leq 1$

**50.** $-16 < 3t + 2 < -10$

**51.** $-1 \leq \dfrac{2x - 5}{6} \leq 5$

**52.** $-3 \leq \dfrac{3m + 1}{4} \leq 3$

**53.** $4 \leq -9x + 5 < 8$

**54.** $4 \leq -2x + 3 < 8$

*Find the unknown numbers in each description.*

**55.** Six times a number is between $-12$ and $12$.

**56.** Half a number is between $-3$ and $2$.

**57.** When 1 is added to twice a number, the result is greater than or equal to 7.

**58.** If 8 is subtracted from a number, then the result is at least 5.

**59.** One third of a number is added to 6, giving a result of at least 3.

**60.** Three times a number, minus 5, is no more than 7.

*The weather forecast by time of day for the U.S. Olympic Track and Field Trials in Sacramento, California, is shown in the figure. Use this graph to work Exercises 61–64.*

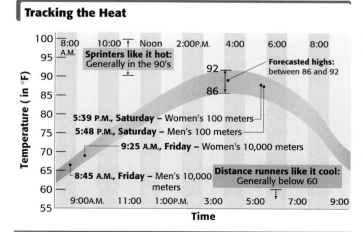

*Source:* Accuweather, Bee research.

61. Sprinters prefer Fahrenheit temperatures in the 90s. Using the upper boundary of the forecast, during what period is the temperature expected to be at least 90°F?

62. Distance runners prefer cool temperatures. During what period are temperatures predicted to be no more than 70°F? Use the lower forecast boundary.

63. What range of temperatures is predicted for the Women's 100-m event?

64. What range of temperatures is forecast for the Men's 10,000-m event?

*Solve each problem. See Examples 7 and 8.*

65. Finley Westmoreland earned scores of 90 and 82 on his first two tests in English literature. What score must he make on his third test to keep an average of 84 or greater?

66. Jack Hornsby scored 92 and 96 on his first two tests in "Methods in Teaching Mathematics." What score must he make on his third test to keep an average of 90 or greater?

67. Amber is signing up for cell phone service. She is trying to decide between Plan A, which costs $54.99 a month with a free phone included, and Plan B, which costs $49.99 a month, but would require her to buy a phone for $129. Under either plan, Amber does not expect to go over the included number of monthly minutes. After how many months would Plan B be a better deal?

68. Craig and Julie Phillips need to rent a truck to move their belongings to their new apartment. They can rent a truck of the size they need from U-Haul for $29.95 a day plus 28 cents per mile or from Budget Truck Rentals for $34.95 a day plus 25 cents per mile. After how many miles would the Budget rental be a better deal than the U-Haul one?

*A product will produce a profit only when the revenue R from selling the product exceeds the cost C of producing it. In Exercises 69 and 70, find the least whole number of units x that must be sold for the business to show a profit for the item described.*

69. Peripheral Visions, Inc., finds that the cost of producing $x$ studio-quality DVDs is $C = 20x + 100$, while the revenue produced from them is $R = 24x$ ($C$ and $R$ in dollars).

70. Speedy Delivery finds that the cost of making $x$ deliveries is $C = 3x + 2300$, while the revenue produced from them is $R = 5.50x$ ($C$ and $R$ in dollars).

71. A body mass index (BMI) between 19 and 25 is considered healthy. Use the formula

$$\text{BMI} = \frac{704 \times (\text{weight in pounds})}{(\text{height in inches})^2}$$

to find the weight range $w$, to the nearest pound, that gives a healthy BMI for each height. (*Source: Washington Post.*)

(a) 72 in.    (b) Your height in inches

**72.** To achieve the maximum benefit from exercising, the heart rate, in beats per minute, should be in the target heart rate (THR) zone. For a person aged $A$, the formula is

$$0.7(220 - A) \leq \text{THR} \leq 0.85(220 - A).$$

Find the THR to the nearest whole number for each age. (*Source:* Hockey, Robert V., *Physical Fitness: The Pathway to Healthful Living,* Times Mirror/Mosby College Publishing, 1989.)

**(a)** 35    **(b)** Your age

## 2.5 Set Operations and Compound Inequalities

**OBJECTIVES**

1. Find the intersection of two sets.
2. Solve compound inequalities with the word *and*.
3. Find the union of two sets.
4. Solve compound inequalities with the word *or*.

The table shows symptoms of an underactive thyroid and an overactive thyroid.

| Underactive Thyroid | Overactive Thyroid |
|---|---|
| Sleepiness, $s$ | Insomnia, $i$ |
| Dry hands, $d$ | Moist hands, $m$ |
| Intolerance of cold, $c$ | Intolerance of heat, $h$ |
| Goiter, $g$ | Goiter, $g$ |

*Source:* The Merck Manual of Diagnosis and Therapy, 16th Edition, Merck Research Laboratories, 1992.

Let $N$ be the set of symptoms of an underactive thyroid, and let $O$ be the set of symptoms of an overactive thyroid. Suppose we are interested in the set of symptoms that are found in *both* sets $N$ and $O$. In this section, we discuss the use of the words *and* and *or* as they relate to sets and inequalities.

**OBJECTIVE 1** Find the intersection of two sets. The intersection of two sets is defined with the word *and*.

**Intersection of Sets**

For any two sets $A$ and $B$, the **intersection** of $A$ and $B$, symbolized $A \cap B$, is defined as follows:

$$A \cap B = \{x \mid x \text{ is an element of } A \text{ and } x \text{ is an element of } B\}.$$

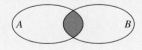

**EXAMPLE 1** Finding the Intersection of Two Sets

Let $A = \{1, 2, 3, 4\}$ and $B = \{2, 4, 6\}$. Find $A \cap B$.

The set $A \cap B$ contains those elements that belong to both $A$ and $B$: the numbers 2 and 4. Therefore,

$$A \cap B = \{1, 2, 3, 4\} \cap \{2, 4, 6\}$$
$$= \{2, 4\}.$$

**NOW TRY** Exercise 7.

A **compound inequality** consists of two inequalities linked by a connective word such as *and* or *or*. Examples of compound inequalities are

$$x + 1 \leq 9 \quad \text{and} \quad x - 2 \geq 3$$
$$2x > 4 \quad \text{or} \quad 3x - 6 < 5.$$

Compound inequalities

**OBJECTIVE 2** Solve compound inequalities with the word *and*. Use the following steps.

### Solving a Compound Inequality with *and*

**Step 1** Solve each inequality individually.

**Step 2** Since the inequalities are joined with *and*, the solution set of the compound inequality will include all numbers that satisfy both inequalities in Step 1 (the intersection of the solution sets).

### EXAMPLE 2  Solving a Compound Inequality with *and*

Solve the compound inequality

$$x + 1 \leq 9 \quad \text{and} \quad x - 2 \geq 3.$$

**Step 1** Solve each inequality individually.

$$x + 1 \leq 9 \quad \text{and} \quad x - 2 \geq 3$$
$$x + 1 - 1 \leq 9 - 1 \quad \text{and} \quad x - 2 + 2 \geq 3 + 2$$
$$x \leq 8 \quad \text{and} \quad x \geq 5$$

**Step 2** Because the inequalities are joined with the word *and*, the solution set will include all numbers that satisfy both inequalities in Step 1 at the same time. Thus, the compound inequality is true whenever $x \leq 8$ and $x \geq 5$ are both true. The top graph in Figure 14 shows $x \leq 8$, and the bottom graph shows $x \geq 5$.

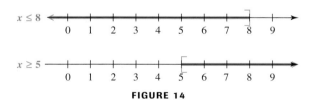

**FIGURE 14**

Find the intersection of the two graphs in Figure 14 to get the solution set of the compound inequality. Figure 15 shows that the solution set, in interval notation, is $[5, 8]$.

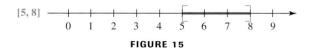

**FIGURE 15**

**NOW TRY** Exercise 25.

### EXAMPLE 3  Solving a Compound Inequality with *and*

Solve the compound inequality
$$-3x - 2 > 5 \quad \text{and} \quad 5x - 1 \leq -21.$$

**Step 1** Solve each inequality individually.

$$-3x - 2 > 5 \quad \text{and} \quad 5x - 1 \leq -21$$
$$-3x > 7 \quad \text{and} \quad 5x \leq -20$$
$$x < -\frac{7}{3} \quad \text{and} \quad x \leq -4$$

Remember to reverse the inequality symbol.

The graphs of $x < -\frac{7}{3}$ and $x \leq -4$ are shown in Figure 16.

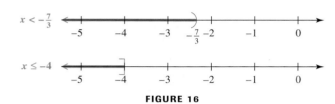

**FIGURE 16**

**Step 2** Now find all values of $x$ that satisfy both conditions; that is, find the real numbers that are less than $-\frac{7}{3}$ and also less than or equal to $-4$. As shown by the graph in Figure 17, the solution set is $(-\infty, -4]$.

**FIGURE 17**

**NOW TRY** Exercise 29.

### EXAMPLE 4  Solving a Compound Inequality with *and*

Solve $x + 2 < 5$ and $x - 10 > 2$.

First solve each inequality individually.

$$x + 2 < 5 \quad \text{and} \quad x - 10 > 2$$
$$x < 3 \quad \text{and} \quad x > 12$$

The graphs of $x < 3$ and $x > 12$ are shown in Figure 18.

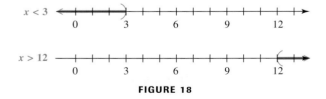

**FIGURE 18**

There is no number that is both less than 3 *and* greater than 12, so the given compound inequality has no solution. The solution set is $\emptyset$. See Figure 19.

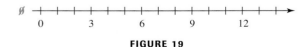

**FIGURE 19**

**NOW TRY** Exercise 23.

**OBJECTIVE 3** Find the union of two sets. The union of two sets is defined with the word *or*.

### Union of Sets

For any two sets $A$ and $B$, the **union** of $A$ and $B$, symbolized $A \cup B$, is defined as follows:

$$A \cup B = \{x \mid x \text{ is an element of } A \text{ or } x \text{ is an element of } B\}.$$

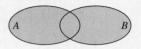

**EXAMPLE 5** Finding the Union of Two Sets

Let $A = \{1, 2, 3, 4\}$ and $B = \{2, 4, 6\}$. Find $A \cup B$.

Begin by listing all the elements of set $A$: 1, 2, 3, 4. Then list any additional elements from set $B$. In this case the elements 2 and 4 are already listed, so the only additional element is 6. Therefore,

$$A \cup B = \{1, 2, 3, 4\} \cup \{2, 4, 6\}$$
$$= \{1, 2, 3, 4, 6\}.$$

The union consists of all elements in either $A$ or $B$ (or both).

**NOW TRY** Exercise 13.

**NOTE** In Example 5, notice that although the elements 2 and 4 appeared in both sets $A$ and $B$, they are written only once in $A \cup B$.

**OBJECTIVE 4** Solve compound inequalities with the word *or*. Use the following steps.

### Solving a Compound Inequality with *or*

**Step 1** Solve each inequality individually.

**Step 2** Since the inequalities are joined with *or*, the solution set of the compound inequality includes all numbers that satisfy either one of the two inequalities in Step 1 (the union of the solution sets).

**EXAMPLE 6** Solving a Compound Inequality with *or*

Solve $6x - 4 < 2x$ or $-3x \leq -9$.

**Step 1** Solve each inequality individually.

$$6x - 4 < 2x \quad \text{or} \quad -3x \leq -9$$
$$4x < 4$$
$$x < 1 \quad \text{or} \quad x \geq 3$$

Remember to reverse the inequality symbol.

The graphs of these two inequalities are shown in Figure 20 on the next page.

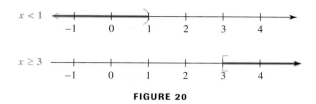

**FIGURE 20**

***Step 2*** Since the inequalities are joined with *or*, find the union of the two solution sets. The union is shown in Figure 21 and is written

$$(-\infty, 1) \cup [3, \infty).$$

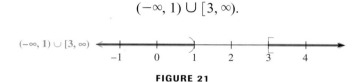

**FIGURE 21**

**NOW TRY** Exercise 41.

**CAUTION** When inequalities are used to write the solution set in Example 6, it *must* be written as

$$x < 1 \quad \text{or} \quad x \geq 3,$$

which keeps the numbers 1 and 3 in their order on the number line. Writing $3 \leq x < 1$ would imply that $3 \leq 1$, which is **FALSE**. There is no other way to write the solution set of such a union.

---

**EXAMPLE 7** Solving a Compound Inequality with *or*

Solve $-4x + 1 \geq 9$ or $5x + 3 \leq -12$.

First we solve each inequality individually.

$$-4x + 1 \geq 9 \quad \text{or} \quad 5x + 3 \leq -12$$
$$-4x \geq 8 \quad \text{or} \quad 5x \leq -15$$
$$x \leq -2 \quad \text{or} \quad x \leq -3$$

The graphs of these two inequalities are shown in Figure 22.

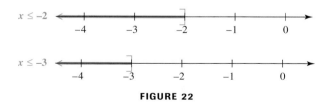

**FIGURE 22**

By taking the union, we obtain the interval $(-\infty, -2]$. See Figure 23.

**FIGURE 23**

**NOW TRY** Exercise 35.

### EXAMPLE 8  Solving a Compound Inequality with *or*

Solve $-2x + 5 \geq 11$ or $4x - 7 \geq -27$.

$$-2x + 5 \geq 11 \quad \text{or} \quad 4x - 7 \geq -27$$
$$-2x \geq 6 \quad \text{or} \quad 4x \geq -20$$
$$x \leq -3 \quad \text{or} \quad x \geq -5$$

The graphs of these two inequalities are shown in Figure 24.

**FIGURE 24**

By taking the union, we obtain every real number as a solution, since every real number satisfies at least one of the two inequalities. The set of all real numbers is written in interval notation as $(-\infty, \infty)$ and graphed as in Figure 25.

**FIGURE 25**

**NOW TRY** Exercise 45.

### EXAMPLE 9  Applying Intersection and Union

The five highest-grossing domestic films (adjusted for inflation) as of July, 2005, are listed in the table.

**FIVE ALL-TIME HIGHEST-GROSSING DOMESTIC FILMS**

| Film | Admissions | Gross Income |
| --- | --- | --- |
| Gone with the Wind | 202,044,569 | $1,293,085,000 |
| Star Wars | 178,119,595 | $1,139,965,000 |
| The Sound of Music | 142,415,376 | $911,458,000 |
| E.T. | 141,925,359 | $908,322,298 |
| The Ten Commandments | 131,000,000 | $838,400,000 |

*Source:* Exhibitor Relations Co., Inc.

List the elements of the following sets.

**(a)** The set of the top five films with admissions greater than 180,000,000 *and* gross income greater than $1,000,000,000

The only film that satisfies both conditions is *Gone with the Wind*, so the set is

$$\{Gone\ with\ the\ Wind\}.$$

**(b)** The set of the top five films with admissions less than 170,000,000 *or* gross income greater than $1,000,000,000

Here, any film that satisfies at least one of the conditions is in the set. This set includes all five films:

{*Gone with the Wind, Star Wars, The Sound of Music, E.T., The Ten Commandments*}.

**NOW TRY** Exercise 63.

## 2.5 Exercises

*True or False* Decide whether each statement is true or false. If it is false, explain why.

1. The union of the solution sets of $x + 1 = 5$, $x + 1 < 5$, and $x + 1 > 5$ is $(-\infty, \infty)$.
2. The intersection of the sets $\{x \mid x \geq 7\}$ and $\{x \mid x \leq 7\}$ is $\emptyset$.
3. The union of the sets $(-\infty, 8)$ and $(8, \infty)$ is $\{8\}$.
4. The intersection of the sets $(-\infty, 8]$ and $[8, \infty)$ is $\{8\}$.
5. The intersection of the set of rational numbers and the set of irrational numbers is $\{0\}$.
6. The union of the set of rational numbers and the set of irrational numbers is the set of real numbers.

Let $A = \{1, 2, 3, 4, 5, 6\}$, $B = \{1, 3, 5\}$, $C = \{1, 6\}$, and $D = \{4\}$. Specify each set. See Examples 1 and 5.

7. $B \cap A$
8. $A \cap B$
9. $A \cap D$
10. $B \cap C$
11. $B \cap \emptyset$
12. $A \cap \emptyset$
13. $A \cup B$
14. $B \cup D$

*Concept Check* Two sets are specified by graphs. Graph the intersection of the two sets.

15.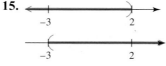
16.
17.
18.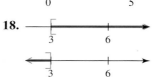

For each compound inequality, give the solution set in both interval and graph form. See Examples 2–4.

19. $x < 2$ and $x > -3$
20. $x < 5$ and $x > 0$
21. $x \leq 2$ and $x \leq 5$
22. $x \geq 3$ and $x \geq 6$
23. $x \leq 3$ and $x \geq 6$
24. $x \leq -1$ and $x \geq 3$
25. $x - 3 \leq 6$ and $x + 2 \geq 7$
26. $x + 5 \leq 11$ and $x - 3 \geq -1$
27. $-3x > 3$ and $x + 3 > 0$
28. $-3x < 3$ and $x + 2 < 6$
29. $3x - 4 \leq 8$ and $-4x + 1 \geq -15$
30. $7x + 6 \leq 48$ and $-4x \geq -24$

*Concept Check* Two sets are specified by graphs. Graph the union of the two sets.

31.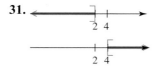
32.
33.
34.

For each compound inequality, give the solution set in both interval and graph form. See Examples 6–8.

35. $x \leq 1$ or $x \leq 8$
36. $x \geq 1$ or $x \geq 8$
37. $x \geq -2$ or $x \geq 5$
38. $x \leq -2$ or $x \leq 6$
39. $x \geq -2$ or $x \leq 4$
40. $x \geq 5$ or $x \leq 7$

**41.** $x + 2 > 7$ or $1 - x > 6$  
**42.** $x + 1 > 3$ or $x + 4 < 2$  
**43.** $x + 1 > 3$ or $-4x + 1 > 5$  
**44.** $3x < x + 12$ or $x + 1 > 10$  
**45.** $4x + 1 \geq -7$ or $-2x + 3 \geq 5$  
**46.** $3x + 2 \leq -7$ or $-2x + 1 \leq 9$  

*Concept Check* Express each set in the simplest interval form. (Hint: Graph each set and look for the intersection or union.)

**47.** $(-\infty, -1] \cap [-4, \infty)$  
**48.** $[-1, \infty) \cap (-\infty, 9]$  
**49.** $(-\infty, -6] \cap [-9, \infty)$  
**50.** $(5, 11] \cap [6, \infty)$  
**51.** $(-\infty, 3) \cup (-\infty, -2)$  
**52.** $[-9, 1] \cup (-\infty, -3)$  
**53.** $[3, 6] \cup (4, 9)$  
**54.** $[-1, 2] \cup (0, 5)$  

*For each compound inequality, decide whether* intersection *or* union *should be used. Then give the solution set in both interval and graph form. See Examples 2–4 and 6–8.*

**55.** $x < -1$ and $x > -5$  
**56.** $x > -1$ and $x < 7$  
**57.** $x < 4$ or $x < -2$  
**58.** $x < 5$ or $x < -3$  
**59.** $-3x \leq -6$ or $-3x \geq 0$  
**60.** $2x - 6 \leq -18$ and $2x \geq -18$  
**61.** $x + 1 \geq 5$ and $x - 2 \leq 10$  
**62.** $-8x \leq -24$ or $-5x \geq 15$  

*Average expenses for full-time college students at all two- and four-year institutions during a recent academic year are shown in the table.*

**COLLEGE EXPENSES (IN DOLLARS)**

| Type of Expense | Public Schools | Private Schools |
|---|---|---|
| Tuition and fees | 2928 | 16,517 |
| Board rates | 2702 | 3236 |
| Dormitory charges | 2925 | 3750 |

*Source:* U.S. National Center for Education Statistics.

*Refer to the table on college expenses above. List the elements of each set. See Example 9.*

**63.** The set of expenses that are less than $3000 for public schools *and* are greater than $5000 for private schools

**64.** The set of expenses that are less than $2800 for public schools *and* are less than $4000 for private schools

**65.** The set of expenses that are greater than $2900 for public schools *or* are greater than $5000 for private schools

**66.** The set of expenses that are greater than $4000 *or* are less than $2700

## 2.6 Absolute Value Equations and Inequalities

**OBJECTIVES**

1. Use the distance definition of absolute value.
2. Solve equations of the form $|ax + b| = k$, for $k > 0$.
3. Solve inequalities of the form $|ax + b| < k$ and of the form $|ax + b| > k$, for $k > 0$.
4. Solve absolute value equations that involve rewriting.
5. Solve equations of the form $|ax + b| = |cx + d|$.
6. Solve special cases of absolute value equations and inequalities.

In a production line, quality is controlled by randomly choosing items from the line and checking to see how selected measurements vary from the optimum measure. The differences are sometimes positive and sometimes negative, so they are expressed as absolute values. For example, a machine that fills quart milk cartons might be set to release 1 qt (32 oz), plus or minus 2 oz per carton. Then the number of ounces in each carton should satisfy the *absolute value inequality* $|x - 32| \leq 2$, where $x$ is the number of ounces.

**OBJECTIVE 1** Use the distance definition of absolute value. In **Section 1.1**, we saw that the absolute value of a number $x$, written $|x|$, represents the distance from $x$ to 0 on the number line. For example, the solutions of $|x| = 4$ are 4 and $-4$, as shown in Figure 26.

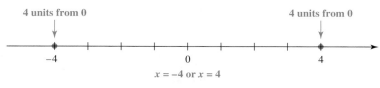

$x = -4$ or $x = 4$

**FIGURE 26**

Because absolute value represents distance from 0, it is reasonable to interpret the solutions of $|x| > 4$ to be all numbers that are *more* than four units from 0. The set $(-\infty, -4) \cup (4, \infty)$ fits this description. Figure 27 shows the graph of the solution set of $|x| > 4$. Because the graph consists of two separate intervals, the solution set is described with *or*: $x < -4$ or $x > 4$.

$x < -4$ or $x > 4$

**FIGURE 27**

The solution set of $|x| < 4$ consists of all numbers that are *less* than 4 units from 0 on the number line. Another way of thinking about this is to think of all numbers *between* $-4$ and 4. This set of numbers is given by $(-4, 4)$, as shown in Figure 28. Here, the graph shows that $-4 < x < 4$, which means $x > -4$ *and* $x < 4$.

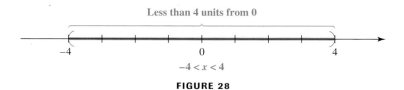

$-4 < x < 4$

**FIGURE 28**

The equation and inequalities just described are examples of **absolute value equations and inequalities.** They involve the absolute value of a variable expression and generally take the form

$$|ax + b| = k, \qquad |ax + b| > k, \qquad \text{or} \qquad |ax + b| < k,$$

where $k$ is a positive number. From Figures 26–28, we see that

$|x| = 4$ has the same solution set as $x = -4$ or $x = 4$,

$|x| > 4$ has the same solution set as $x < -4$ or $x > 4$,

$|x| < 4$ has the same solution set as $x > -4$ and $x < 4$.

Thus, we solve an absolute value equation or inequality by solving the appropriate compound equation or inequality.

### Solving Absolute Value Equations and Inequalities

Let $k$ be a positive real number and $p$ and $q$ be real numbers.

1. To solve $|ax + b| = k$, solve the compound equation

$$ax + b = k \quad \text{or} \quad ax + b = -k.$$

The solution set is usually of the form $\{p, q\}$, which includes two numbers.

2. To solve $|ax + b| > k$, solve the compound inequality

$$ax + b > k \quad \text{or} \quad ax + b < -k.$$

The solution set is of the form $(-\infty, p) \cup (q, \infty)$, which consists of two separate intervals.

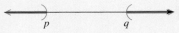

3. To solve $|ax + b| < k$, solve the three-part inequality

$$-k < ax + b < k.$$

The solution set is of the form $(p, q)$, a single interval.

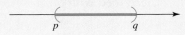

**OBJECTIVE 2** Solve equations of the form $|ax + b| = k$, for $k > 0$. *Remember that because absolute value refers to distance from the origin, an absolute value equation will have two parts.*

### EXAMPLE 1  Solving an Absolute Value Equation

Solve $|2x + 1| = 7$.

For $|2x + 1|$ to equal 7, $2x + 1$ must be 7 units from 0 on the number line. This can happen only when $2x + 1 = 7$ or $2x + 1 = -7$. This is the first case in the preceding box. Solve this compound equation as follows:

$$2x + 1 = 7 \quad \text{or} \quad 2x + 1 = -7$$
$$2x = 6 \quad \text{or} \quad 2x = -8$$
$$x = 3 \quad \text{or} \quad x = -4.$$

Check by substituting 3 and then $-4$ into the original absolute value equation to verify that the solution set is $\{-4, 3\}$. The graph is shown in Figure 29.

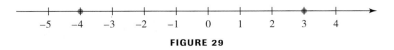

**FIGURE 29**

**NOW TRY** Exercise 11.

**OBJECTIVE 3** Solve inequalities of the form $|ax + b| < k$ and of the form $|ax + b| > k$, for $k > 0$.

**EXAMPLE 2** Solving an Absolute Value Inequality with >

Solve $|2x + 1| > 7$.
By Case 2 of the previous box, this absolute value inequality is rewritten as

$$2x + 1 > 7 \quad \text{or} \quad 2x + 1 < -7,$$

because $2x + 1$ must represent a number that is *more* than 7 units from 0 on either side of the number line. Now, solve the compound inequality.

$$\begin{aligned} 2x + 1 &> 7 \quad \text{or} \quad 2x + 1 < -7 \\ 2x &> 6 \quad \text{or} \quad 2x < -8 \\ x &> 3 \quad \text{or} \quad x < -4 \end{aligned}$$

Check these solutions. The solution set is $(-\infty, -4) \cup (3, \infty)$. See Figure 30. Notice that the graph consists of two intervals.

**FIGURE 30**

**NOW TRY** Exercise 25.

**EXAMPLE 3** Solving an Absolute Value Inequality with <

Solve $|2x + 1| < 7$.
The expression $2x + 1$ must represent a number that is less than 7 units from 0 on either side of the number line. Another way of thinking about this is to realize that $2x + 1$ must be between $-7$ and 7. As Case 3 of the previous box shows, that relationship is written as a three-part inequality.

$$\begin{aligned} -7 < 2x + 1 &< 7 \\ -8 < 2x &< 6 \quad \text{Subtract 1 from each part.} \\ -4 < x &< 3 \quad \text{Divide each part by 2.} \end{aligned}$$

Check that the solution set is $(-4, 3)$. The graph consists of the single interval shown in Figure 31.

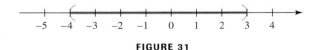

**FIGURE 31**

**NOW TRY** Exercise 39.

Look back at Figures 29, 30, and 31, with the graphs of $|2x + 1| = 7$, $|2x + 1| > 7$, and $|2x + 1| < 7$, respectively. If we find the union of the three sets, we get the set of all real numbers. This is because, for any value of $x$, $|2x + 1|$ will satisfy one and only one of the following: It is equal to 7, greater than 7, or less than 7.

> **CAUTION** When solving absolute value equations and inequalities of the types in Examples 1, 2, and 3, remember the following.
>
> 1. The methods described apply when the constant is alone on one side of the equation or inequality and is *positive*.
> 2. Absolute value equations and absolute value inequalities of the form $|ax + b| > k$ translate into "or" compound statements.
> 3. Absolute value inequalities of the form $|ax + b| < k$ translate into "and" compound statements, which may be written as three-part inequalities.
> 4. An "or" statement *cannot* be written in three parts. It would be incorrect to write $-7 > 2x + 1 > 7$ in Example 2, because this would imply that $-7 > 7$, which is *false*.

**OBJECTIVE 4** Solve absolute value equations that involve rewriting. Sometimes an absolute value equation or inequality requires some rewriting before it can be set up as a compound statement, as shown in the next example.

**EXAMPLE 4** Solving an Absolute Value Equation That Requires Rewriting

Solve $|x + 3| + 5 = 12$.

First get the absolute value alone on one side of the equals sign.

$$|x + 3| + 5 - 5 = 12 - 5 \quad \text{Subtract 5.}$$
$$|x + 3| = 7$$

Now use the method shown in Example 1 to solve $|x + 3| = 7$.

$$x + 3 = 7 \quad \text{or} \quad x + 3 = -7$$
$$x = 4 \quad \text{or} \quad x = -10$$

Check that the solution set is $\{-10, 4\}$ by substituting into the original equation.

*We write solutions in the order in which they appear on a number line.*

**NOW TRY** Exercise 65.

We use a similar method to solve an absolute value *inequality* that requires rewriting:

| | |
|---|---|
| $\|x + 3\| + 5 \geq 12$ | $\|x + 3\| + 5 \leq 12$ |
| $\|x + 3\| \geq 7$ | $\|x + 3\| \leq 7$ |
| $x + 3 \geq 7 \quad \text{or} \quad x + 3 \leq -7$ | $-7 \leq x + 3 \leq 7$ |
| $x \geq 4 \quad \text{or} \quad x \leq -10.$ | $-10 \leq x \leq 4.$ |
| Solution set: $(-\infty, -10] \cup [4, \infty)$ | Solution set: $[-10, 4]$ |

**OBJECTIVE 5** Solve equations of the form $|ax + b| = |cx + d|$. By definition, for two expressions to have the same absolute value, they must either be equal or be negatives of each other.

> **Solving $|ax + b| = |cx + d|$**
>
> To solve an absolute value equation of the form
>
> $$|ax + b| = |cx + d|,$$
>
> solve the compound equation
>
> $$ax + b = cx + d \quad \text{or} \quad ax + b = -(cx + d).$$

**EXAMPLE 5** Solving an Equation with Two Absolute Values

Solve $|z + 6| = |2z - 3|$.

This equation is satisfied either if $z + 6$ and $2z - 3$ are equal to each other or if $z + 6$ and $2z - 3$ are negatives of each other.

$$z + 6 = 2z - 3 \quad \text{or} \quad z + 6 = -(2z - 3)$$
$$z + 9 = 2z \quad \text{or} \quad z + 6 = -2z + 3$$
$$9 = z \quad \text{or} \quad 3z = -3$$
$$z = -1$$

Check that the solution set is $\{-1, 9\}$.

**NOW TRY** Exercise 71.

**OBJECTIVE 6** Solve special cases of absolute value equations and inequalities. When an absolute value equation or inequality involves a *negative constant or 0* alone on one side, use the properties of absolute value to solve the equation or inequality. Keep the following in mind.

> **Special Cases of Absolute Value**
>
> 1. The absolute value of an expression can never be negative; that is, $|a| \geq 0$ for all real numbers $a$.
> 2. The absolute value of an expression equals 0 only when the expression is equal to 0.

**EXAMPLE 6** Solving Special Cases of Absolute Value Equations

Solve each equation.

**(a)** $|5r - 3| = -4$

See Case 1 in the preceding box. *The absolute value of an expression can never be negative,* so there are no solutions for this equation. The solution set is ∅.

**(b)** $|7x - 3| = 0$

See Case 2 in the preceding box. The expression $7x - 3$ will equal 0 *only* if

$$7x - 3 = 0.$$

The solution of this equation is $\frac{3}{7}$. Thus, the solution set is $\left\{\frac{3}{7}\right\}$, with just one element. Check by substitution into the original equation.

**NOW TRY** Exercises 81 and 83.

---

### EXAMPLE 7  Solving Special Cases of Absolute Value Inequalities

Solve each inequality.

**(a)** $|x| \geq -4$

*The absolute value of a number is always greater than or equal to 0.* Thus, $|x| \geq -4$ is true for *all* real numbers. The solution set is $(-\infty, \infty)$.

**(b)**
$$|k + 6| - 3 < -5$$
$$|k + 6| < -2 \quad \text{Add 3 to each side.}$$

There is no number whose absolute value is less than $-2$, so this inequality has no solution. The solution set is $\emptyset$.

**(c)**
$$|m - 7| + 4 \leq 4$$
$$|m - 7| \leq 0 \quad \text{Subtract 4 from each side.}$$

The value of $|m - 7|$ will never be less than 0. However, $|m - 7|$ will equal 0 when $m = 7$. Therefore, the solution set is $\{7\}$.

**NOW TRY** Exercises 79, 89, and 95.

---

## 2.6 Exercises

**NOW TRY Exercise**

*Matching* Match each absolute value equation or inequality in Column I with the graph of its solution set in Column II.

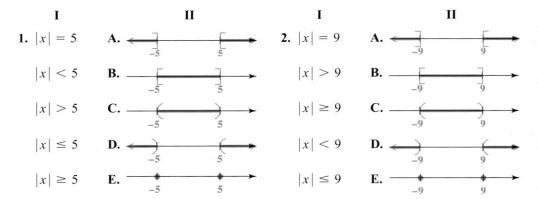

**3.** *Concept Check* How many solutions will $|ax + b| = k$ have if
  **(a)** $k = 0$   **(b)** $k > 0$   **(c)** $k < 0$?

4. Explain when to use *and* and when to use *or* if you are solving an absolute value equation or inequality of the form $|ax + b| = k$, $|ax + b| < k$, or $|ax + b| > k$, where $k$ is a positive number.

*Solve each equation. See Example 1.*

5. $|x| = 12$
6. $|k| = 14$
7. $|4x| = 20$
8. $|5x| = 30$
9. $|y - 3| = 9$
10. $|p - 5| = 13$
11. $|2x - 1| = 11$
12. $|2y + 3| = 19$
13. $|4r - 5| = 17$
14. $|5t - 1| = 21$
15. $|2y + 5| = 14$
16. $|2x - 9| = 18$
17. $\left|\frac{1}{2}x + 3\right| = 2$
18. $\left|\frac{2}{3}q - 1\right| = 5$
19. $\left|1 + \frac{3}{4}k\right| = 7$
20. $\left|2 - \frac{5}{2}m\right| = 14$

*Solve each inequality and graph the solution set. See Example 2.*

21. $|x| > 3$
22. $|y| > 5$
23. $|k| \geq 4$
24. $|r| \geq 6$
25. $|r + 5| \geq 20$
26. $|3x - 1| \geq 8$
27. $|t + 2| > 10$
28. $|4x + 1| \geq 21$
29. $|3 - x| > 5$
30. $|5 - x| > 3$
31. $|-5x + 3| \geq 12$
32. $|-2x - 4| \geq 5$

33. *Concept Check* The graph of the solution set of $|2x + 1| = 9$ is given here.

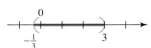

Without actually doing the algebraic work, graph the solution set of each inequality, referring to the graph shown.

(a) $|2x + 1| < 9$  (b) $|2x + 1| > 9$

34. *Concept Check* The graph of the solution set of $|3x - 4| < 5$ is given here.

Without actually doing the algebraic work, graph the solution set of the following, referring to the graph shown.

(a) $|3x - 4| = 5$  (b) $|3x - 4| > 5$

*Solve each inequality and graph the solution set. See Example 3. (Hint: Compare your answers with those in Exercises 21–32.)*

35. $|x| \leq 3$
36. $|y| \leq 5$
37. $|k| < 4$
38. $|r| < 6$
39. $|r + 5| \leq 20$
40. $|3x - 1| < 8$
41. $|t + 2| \leq 10$
42. $|4x + 1| < 21$
43. $|3 - x| \leq 5$
44. $|5 - x| \leq 3$
45. $|-5x + 3| \leq 12$
46. $|-2x - 4| \leq 5$

*Exercises 47–62 represent a sampling of the various types of absolute value equations and inequalities covered in Exercises 1–46. Decide which method of solution applies, and find the solution set. In Exercises 47–58, graph the solution set. See Examples 1–3.*

47. $|-4 + k| > 9$
48. $|-3 + t| > 8$
49. $|r + 5| > 20$
50. $|2x - 1| < 7$
51. $|7 + 2z| = 5$
52. $|9 - 3p| = 3$
53. $|3r - 1| \leq 11$
54. $|2s - 6| \leq 6$
55. $|-6x - 6| \leq 1$
56. $|-2x - 6| \leq 5$
57. $|2x - 1| \geq 7$
58. $|-4 + k| \leq 9$

**59.** $|x + 2| = 3$　　　　　　　　　　**60.** $|x + 3| = 10$
**61.** $|x - 6| = 3$　　　　　　　　　　**62.** $|x - 4| = 1$

*Solve each equation or inequality. See Example 4.*

**63.** $|x| - 1 = 4$　　**64.** $|x| + 3 = 10$　　**65.** $|x + 4| + 1 = 2$
**66.** $|x + 5| - 2 = 12$　　**67.** $|2x + 1| + 3 > 8$　　**68.** $|6x - 1| - 2 > 6$
**69.** $|x + 5| - 6 \leq -1$　　**70.** $|r - 2| - 3 \leq 4$

*Solve each equation. See Example 5.*

**71.** $|3x + 1| = |2x + 4|$　　　　　　**72.** $|7x + 12| = |x - 8|$
**73.** $\left|m - \dfrac{1}{2}\right| = \left|\dfrac{1}{2}m - 2\right|$　　　　**74.** $\left|\dfrac{2}{3}r - 2\right| = \left|\dfrac{1}{3}r + 3\right|$
**75.** $|6x| = |9x + 1|$　　　　　　　　**76.** $|13x| = |2x + 1|$
**77.** $|2p - 6| = |2p + 11|$　　　　　　**78.** $|3x - 1| = |3x + 9|$

*Solve each equation or inequality. See Examples 6 and 7.*

**79.** $|x| \geq -10$　　**80.** $|x| \geq -15$　　**81.** $|12t - 3| = -8$
**82.** $|13w + 1| = -3$　　**83.** $|4x + 1| = 0$　　**84.** $|6r - 2| = 0$
**85.** $|2q - 1| = -6$　　**86.** $|8n + 4| = -4$
**87.** $|x + 5| > -9$　　**88.** $|x + 9| > -3$
**89.** $|7x + 3| \leq 0$　　**90.** $|4x - 1| \leq 0$
**91.** $|5x - 2| = 0$　　**92.** $|4 + 7x| = 0$
**93.** $|x - 2| + 3 \geq 2$　　**94.** $|k - 4| + 5 \geq 4$
**95.** $|10z + 7| + 3 < 1$　　**96.** $|4x + 1| - 2 < -5$

**97.** The 2005 recommended daily intake (RDI) of calcium for females aged 19–50 is 1000 mg. Actual mineral needs vary from person to person. Write an absolute value inequality, with *x* representing the RDI, to express the RDI plus or minus 100 mg, and solve the inequality. (*Source:* Food and Nutrition Board, National Academy of Sciences—Institute of Medicine.)

**98.** The average clotting time of blood is 7.45 sec, with a variation of plus or minus 3.6 sec. Write this statement as an absolute value inequality with *x* representing the time, and solve the inequality.

# 2 SUMMARY

## KEY TERMS

**2.1** linear (first-degree) equation in one variable
solution
solution set
equivalent equations

conditional equation
identity
contradiction
**2.2** mathematical model
formula
percent

**2.4** inequality
linear inequality in one variable
equivalent inequalities
three-part inequality
**2.5** intersection

compound inequality
union
**2.6** absolute value equation
absolute value inequality

## NEW SYMBOLS

$1°$ one degree    $\cap$ set intersection    $\cup$ set union

## QUICK REVIEW

| CONCEPTS | EXAMPLES |
|---|---|
| **2.1 Linear Equations in One Variable** | |
| **Addition and Multiplication Properties of Equality** | Solve. |
| The same number may be added to (or subtracted from) each side of an equation to obtain an equivalent equation. Similarly, the same nonzero number may be multiplied by or divided into each side of an equation to obtain an equivalent equation. | $x - 5 = 10$ $\qquad$ $\frac{1}{2}x = 10$ <br> $x - 5 + 5 = 10 + 5$ $\qquad$ $2\left(\frac{1}{2}x\right) = 2(10)$ <br> $x = 15$ $\qquad$ $x = 20$ <br> The solution set is $\{15\}$. $\qquad$ The solution set is $\{20\}$. |
| **Solving a Linear Equation in One Variable** | Solve $4(8 - 3t) = 32 - 8(t + 2)$. |
| *Step 1* Clear fractions. | $32 - 12t = 32 - 8t - 16$ $\quad$ Distributive property |
| *Step 2* Simplify each side separately. | $32 - 12t = 16 - 8t$ |
| *Step 3* Isolate the variable terms on one side. | $32 - 12t + 12t = 16 - 8t + 12t$ $\quad$ Add 12t. |
|  | $32 = 16 + 4t$ |
|  | $32 - 16 = 16 + 4t - 16$ $\quad$ Subtract 16. |
|  | $16 = 4t$ |
| *Step 4* Isolate the variable. | $\frac{16}{4} = \frac{4t}{4}$ $\quad$ Divide by 4. |
|  | $4 = t$ |
| *Step 5* Check. | The solution set is $\{4\}$. This can be checked by substituting 4 for $t$ in the original equation. |

*(continued)*

| CONCEPTS | EXAMPLES |
|---|---|
| **2.2 Formulas** | |
| **Solving a Formula for a Specified Variable** | Solve $A = \frac{1}{2}bh$ for $h$. |
| *Step 1* Get all terms with the specified variable on one side and all terms without that variable on the other side. | $A = \frac{1}{2}bh$ |
| *Step 2* If necessary, use the distributive property to combine terms with the specified variable. | $2A = 2\left(\frac{1}{2}bh\right)$    Multiply by 2. |
| | $2A = bh$ |
| *Step 3* Divide both sides by the factor that is the coefficient of the specified variable. | $\frac{2A}{b} = h$, or $h = \frac{2A}{b}$    Divide by $b$. |
| **2.3 Applications of Linear Equations** | |
| **Solving an Applied Problem** | How many liters of 30% alcohol solution and 80% alcohol solution must be mixed to obtain 100 L of 50% alcohol solution? |
| *Step 1* Read the problem. | |
| *Step 2* Assign a variable. | Let $x$ = number of liters of 30% solution needed; then $100 - x$ = number of liters of 80% solution needed. |

| Liters of Solution | Percent (as a decimal) | Liters of Pure Alcohol |
|---|---|---|
| $x$ | 0.30 | $0.30x$ |
| $100 - x$ | 0.80 | $0.80(100 - x)$ |
| 100 | 0.50 | $0.50(100)$ |

| | |
|---|---|
| *Step 3* Write an equation. | The equation is $0.30x + 0.80(100 - x) = 0.50(100)$. |
| *Step 4* Solve the equation. | The solution of the equation is 60. Thus, 60 L of 30% solution and $100 - 60 = 40$ L of 80% solution are needed. |
| *Step 5* State the answer. | |
| *Step 6* Check. | $0.30(60) + 0.80(100 - 60) = 50$ is true. |
| **2.4 Linear Inequalities in One Variable** | |
| **Solving a Linear Inequality in One Variable** | Solve $3(x + 2) - 5x \leq 12$. |
| *Step 1* Simplify each side of the inequality by clearing parentheses and combining like terms. | $3x + 6 - 5x \leq 12$    Distributive property |
| | $-2x + 6 \leq 12$ |
| *Step 2* Use the addition property of inequality to get all terms with variables on one side and all terms without variables on the other side. | $-2x + 6 - 6 \leq 12 - 6$    Subtract 6. |
| | $-2x \leq 6$ |
| *Step 3* Use the multiplication property of inequality to write the inequality in the form $x < k$ or $x > k$. | $\frac{-2x}{-2} \geq \frac{6}{-2}$    Divide by $-2$; change $\leq$ to $\geq$. |
| | $x \geq -3$ |
| *If an inequality is multiplied or divided by a negative number, the inequality symbol must be reversed.* | The solution set $[-3, \infty)$ is graphed here. 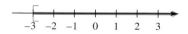 |

*(continued)*

| CONCEPTS | EXAMPLES |
|---|---|
| To solve a three-part inequality, work with all three parts at the same time. | Solve $-4 < 2x + 3 \leq 7$.<br><br>$-4 - 3 < 2x + 3 - 3 \leq 7 - 3$    Subtract 3.<br>$-7 < 2x \leq 4$<br>$\dfrac{-7}{2} < \dfrac{2x}{2} \leq \dfrac{4}{2}$    Divide by 2.<br>$-\dfrac{7}{2} < x \leq 2$<br><br>The solution set, $\left(-\dfrac{7}{2}, 2\right]$, is graphed here.<br>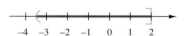 |

## 2.5 Set Operations and Compound Inequalities

| | |
|---|---|
| **Solving a Compound Inequality**<br><br>**Step 1** Solve each inequality in the compound inequality individually.<br><br>**Step 2** If the inequalities are joined with *and*, then the solution set is the intersection of the two individual solution sets.<br><br>If the inequalities are joined with *or*, then the solution set is the union of the two individual solution sets. | Solve $x + 1 > 2$ and $2x < 6$.<br><br>$x + 1 > 2$    and    $2x < 6$<br>$x > 1$    and    $x < 3$<br><br>The solution set is $(1, 3)$.<br>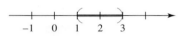<br><br>Solve $x \geq 4$ or $x \leq 0$.<br>The solution set is $(-\infty, 0] \cup [4, \infty)$.<br> |

## 2.6 Absolute Value Equations and Inequalities

| | |
|---|---|
| **Solving Absolute Value Equations and Inequalities**<br><br>Let $k$ be a positive number.<br><br>To solve $\lvert ax + b \rvert = k$, solve the compound equation<br><br>    $ax + b = k$    or    $ax + b = -k$.<br><br>To solve $\lvert ax + b \rvert > k$, solve the compound inequality<br><br>    $ax + b > k$    or    $ax + b < -k$. | Solve $\lvert x - 7 \rvert = 3$.<br><br>$x - 7 = 3$    or    $x - 7 = -3$<br>$x = 10$    or    $x = 4$<br><br>The solution set is $\{4, 10\}$.<br>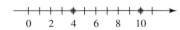<br><br>Solve $\lvert x - 7 \rvert > 3$.<br><br>$x - 7 > 3$    or    $x - 7 < -3$<br>$x > 10$    or    $x < 4$<br><br>The solution set is $(-\infty, 4) \cup (10, \infty)$.<br>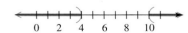 |

*(continued)*

| CONCEPTS | EXAMPLES |
|---|---|
| To solve $\|ax + b\| < k$, solve the compound inequality $$-k < ax + b < k.$$ | Solve $\|x - 7\| < 3$. $$-3 < x - 7 < 3$$ $$4 < x < 10 \quad \text{Add 7.}$$ The solution set is $(4, 10)$. |
| To solve an absolute value equation of the form $$\|ax + b\| = \|cx + d\|,$$ solve the compound equation $$ax + b = cx + d \quad \text{or} \quad ax + b = -(cx + d).$$ | Solve $\|x + 2\| = \|2x - 6\|$. $x + 2 = 2x - 6 \quad$ or $\quad x + 2 = -(2x - 6)$ $x = 8 \qquad\qquad\qquad x + 2 = -2x + 6$ $\qquad\qquad\qquad\qquad\qquad 3x = 4$ $\qquad\qquad\qquad\qquad\qquad x = \dfrac{4}{3}$ The solution set is $\left\{\dfrac{4}{3}, 8\right\}$. |

## 2 REVIEW EXERCISES

*Solve each equation.*

1. $-(8 + 3z) + 5 = 2z + 6$
2. $-\dfrac{3}{4}x = -12$
3. $\dfrac{2q + 1}{3} - \dfrac{q - 1}{4} = 0$
4. $5(2x - 3) = 6(x - 1) + 4x$

*Solve each equation. Then tell whether the equation is* conditional, *an* identity, *or a* contradiction.

5. $7r - 3(2r - 5) + 5 + 3r = 4r + 20$
6. $8p - 4p - (p - 7) + 9p + 6 = 12p - 7$
7. $-2r + 6(r - 1) + 3r - (4 - r) = -(r + 5) - 5$

*Solve each formula for the specified variable.*

8. $V = LWH$ for $L$
9. $A = \dfrac{1}{2}h(b + B)$ for $b$

*Solve each equation for x.*

10. $M = -\dfrac{1}{4}(x + 3y)$
11. $P = \dfrac{3}{4}x - 12$

12. Give the steps you would use to solve $-2x + 5 = 7$.

*Solve each problem.*

13. A rectangular solid has a volume of 180 ft³. Its length is 6 ft and its width is 5 ft. Find its height.

14. The total number of deaths from AIDS in the United States in 2003 was 17,849. In 2004, this figure had decreased to 15,798. What approximate percent decrease did this represent? (*Source:* U.S. Centers for Disease Control.)

15. Find the simple-interest rate that Francesco Castellucio is earning if his principal of $30,000 earns $7800 interest in 4 yr.

16. If the Fahrenheit temperature is 77°, what is the corresponding Celsius temperature?

*For 2005, total U.S. government spending was about $2500 billion (or $2.5 trillion). The circle graph shows how the spending was divided.*

17. About how much was spent on Social Security?

18. About how much did the U.S. government spend on education and social services in 2005?

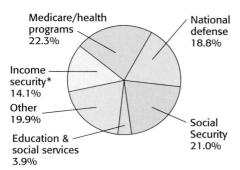

**2005 U.S. Government Spending**

- Medicare/health programs 22.3%
- National defense 18.8%
- Income security* 14.1%
- Other 19.9%
- Education & social services 3.9%
- Social Security 21.0%

*Includes pensions for government workers, unemployment compensation, food stamps, and other such programs.

*Source:* U.S. Office of Management and Budget.

*Write each phrase as a mathematical expression, using x as the variable.*

19. One-third of a number, subtracted from 9

20. The product of 4 and a number, divided by 9 more than the number

*Solve each problem.*

21. The length of a rectangle is 3 m less than twice the width. The perimeter of the rectangle is 42 m. Find the length and width of the rectangle.

22. In a triangle with two sides of equal length, the third side measures 15 in. less than the sum of the two equal sides. The perimeter of the triangle is 53 in. Find the lengths of the three sides.

23. A candy clerk has three times as many kilograms of chocolate creams as peanut clusters. The clerk has 48 kg of the two candies altogether. How many kilograms of peanut clusters does the clerk have?

24. How many liters of a 20% solution of a chemical should be mixed with 15 L of a 50% solution to get a 30% mixture?

25. How much water should be added to 30 L of a 40% acid solution to reduce it to a 30% solution?

26. Jay Jenkins invested some money at 6% and $4000 less than that amount at 4%. Find the amount invested at each rate if his total annual interest income is $840.

| Liters of Solution | Percent (as a decimal) | Liters of Pure Acid |
|---|---|---|
|  | 0.40 |  |
| $x$ |  |  |
|  | 0.30 |  |

| Principal | Rate (as a decimal) | Interest |
|---|---|---|
| $x$ | 0.06 |  |
|  | 0.04 |  |

*Solve each inequality. Express the solution set in interval form.*

**27.** $-\dfrac{2}{3}k < 6$

**28.** $-5x - 4 \geq 11$

**29.** $\dfrac{6a + 3}{-4} < -3$

**30.** $5 - (6 - 4k) \geq 2k - 7$

**31.** $8 \leq 3z - 1 < 14$

**32.** $\dfrac{5}{3}(m - 2) + \dfrac{2}{5}(m + 1) > 1$

*Solve each problem.*

**33.** The perimeter of a rectangular playground must be no greater than 120 m. One dimension of the playground must be 22 m. Find the possible lengths of the other dimension of the playground.

**34.** A group of college students wants to buy tickets to attend a performance of Monty Python's *Spamalot* at the Cadillac Palace Theatre in Chicago. The best price they can find is a group rate of $89 per ticket if 10 or more tickets are purchased at the same time. If they have $2000 available to spend on tickets and they qualify for a $50 group discount, how many tickets can they purchase?

**35.** To pass algebra, a student must have an average of at least 70 on five tests. On the first four tests, a student has scores of 75, 79, 64, and 71. What possible scores on the fifth test would guarantee the student a passing grade in the class?

**36.** While solving the inequality

$$10x + 2(x - 4) < 12x - 13,$$

a student did all the work correctly and obtained the statement $-8 < -13$. The student did not know what to do at this point, because the variable "disappeared." How would you explain to the student the interpretation of this result?

*Let $A = \{a, b, c, d\}$, $B = \{a, c, e, f\}$, and $C = \{a, e, f, g\}$. Find each set.*

**37.** $A \cap B$     **38.** $A \cap C$     **39.** $B \cup C$     **40.** $A \cup C$

*Solve each compound inequality. Give the solution set in both interval and graph form.*

**41.** $x > 6$  and  $x < 9$

**42.** $x + 4 > 12$  and  $x - 2 < 12$

**43.** $x > 5$  or  $x \leq -3$

**44.** $x \geq -2$  or  $x < 2$

**45.** $x - 4 > 6$  and  $x + 3 \leq 10$

**46.** $-5x + 1 \geq 11$  or  $3x + 5 \geq 26$

*Express each union or intersection in simplest interval form.*

**47.** $(-3, \infty) \cap (-\infty, 4)$

**48.** $(-\infty, 6) \cap (-\infty, 2)$

**49.** $(4, \infty) \cup (9, \infty)$

**50.** $(1, 2) \cup (1, \infty)$

*Solve each absolute value equation.*

**51.** $|x| = 7$

**52.** $|x + 2| = 9$

**53.** $|3k - 7| = 8$

**54.** $|z - 4| = -12$

**55.** $|2k - 7| + 4 = 11$

**56.** $|4a + 2| - 7 = -3$

**57.** $|3p + 1| = |p + 2|$

**58.** $|2m - 1| = |2m + 3|$

*Solve each absolute value inequality. Give the solution set in interval form.*

**59.** $|p| < 14$       **60.** $|-t + 6| \leq 7$

**61.** $|2p + 5| \leq 1$      **62.** $|x + 1| \geq -3$

# CHAPTER 2 TEST

*Solve each equation.*

1. $3(2x - 2) - 4(x + 6) = 3x + 8 + x$
2. $0.08x + 0.06(x + 9) = 1.24$
3. $\dfrac{x + 6}{10} + \dfrac{x - 4}{15} = \dfrac{x + 2}{6}$
4. Solve each equation. Then tell whether the equation is a *conditional equation*, an *identity*, or a *contradiction*.
   - (a) $3x - (2 - x) + 4x + 2 = 8x + 3$
   - (b) $\dfrac{x}{3} + 7 = \dfrac{5x}{6} - 2 - \dfrac{x}{2} + 9$
   - (c) $-4(2x - 6) = 5x + 24 - 7x$
5. Solve $-16t^2 + vt - S = 0$ for $v$.
6. Solve $ar + 2 = 3r - 6t$ for $r$.

*Solve each problem.*

7. The 2005 Indianapolis 500 (mile) race was won by Dan Wheldon, who averaged 157.603 mph. What was Wheldon's time to the nearest thousandth of an hour? (*Source: World Almanac and Book of Facts.*)
8. A certificate of deposit pays $2281.25 in simple interest for 1 yr on a principal of $36,500. What is the rate of interest?
9. In 2005, there were 37,142 offices, stations, and branches of the U.S. Postal Service, of which 27,385 were actually classified as post offices. What percent, to the nearest tenth, were classified as post offices? (*Source:* U.S. Postal Service.)
10. Tyler McGinnis invested some money at 3% simple interest and some at 5% simple interest. The total amount of his investments was $28,000, and the interest he earned during the first year was $1240. How much did he invest at each rate?

*Solve each inequality. Give the solution set in both interval and graph form.*

11. $4 - 6(x + 3) \leq -2 - 3(x + 6) + 3x$
12. $-\dfrac{4}{7}x > -16$
13. $-6 \leq \dfrac{4}{3}x - 2 \leq 2$
14. *Multiple Choice* Which one of the following inequalities is equivalent to $x < -3$?
    - **A.** $-3x < 9$    **B.** $-3x > -9$    **C.** $-3x > 9$    **D.** $-3x < -9$

*Solve each problem.*

15. A student must have an average of at least 80 on the four tests in a course to get a B. The student had scores of 83, 76, and 79 on the first three tests. What minimum score on the fourth test would guarantee the student a B in the course?
16. A product will break even or produce a profit only if the revenue $R$ (in dollars) from selling the product is at least equal to the cost $C$ (in dollars) of producing it. Suppose that the cost to produce $x$ units of carpet is $C = 50x + 5000$, while the revenue is $R = 60x$. For what values of $x$ is $R$ at least equal to $C$?

17. Let $A = \{1, 2, 5, 7\}$ and $B = \{1, 5, 9, 12\}$. Find each of the following sets.
    (a) $A \cap B$    (b) $A \cup B$

*Solve each compound or absolute value inequality.*

18. $3k \geq 6$  and  $k - 4 < 5$
19. $-4x \leq -24$  or  $4x - 2 < 10$
20. $|4x + 3| \leq 7$
21. $|5 - 6x| > 12$
22. $|7 - x| \leq -1$
23. $|-3x + 4| - 4 < -1$

*Solve each absolute value equation.*

24. $|3k - 2| + 1 = 8$
25. $|3 - 5x| = |2x + 8|$
26. *Concept Check*  If $k < 0$, what is the solution set of
    (a) $|8x - 5| < k$    (b) $|8x - 5| > k$    (c) $|8x - 5| = k$?

# Graphs, Linear Equations, and Functions

In 1998, the average undergraduate tuition and fees at four-year public colleges was $3247. This rose to $4098 in 2002, $5126 in 2004, and $6484 in 2008. (*Source:* The College Board.) We can estimate what average undergraduate tuition and fees will be in 2012 by determining the *average rate of change*. In this example, tuition and fees rose at about $324 per year from 1998 to 2008. If these costs continue to rise at the same rate, the tuition and fees will be about $7780 in 2012.

The *average rate of change* is one of the fundamental concepts in mathematics. It allows you to see the relationship between two quantities that are changing and then to make conjectures and interpretations based on the given data.

In this chapter, we will see how average rate of change relates to the *slope of a line*. We will use the notion of average rate of change to interpret data and make predictions in Exercises 83–88 of Section 3.2.

**3.1** The Rectangular Coordinate System

**3.2** The Slope of a Line

**3.3** Linear Equations in Two Variables

**3.4** Linear Inequalities in Two Variables

**3.5** Introduction to Functions

# 3.1 The Rectangular Coordinate System

**OBJECTIVES**

1. Interpret a line graph.
2. Plot ordered pairs.
3. Find ordered pairs that satisfy a given equation.
4. Graph lines.
5. Find $x$- and $y$-intercepts.
6. Recognize equations of horizontal and vertical lines and lines passing through the origin.
7. Use the midpoint formula.

**OBJECTIVE 1** Interpret a line graph. The line graph in Figure 1 shows personal spending (in billions of dollars) on medical care in the United States from 1997 through 2003. About how much was spent on medical care in 2002? (We will answer this question shortly.)

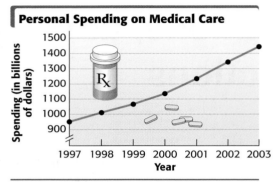

**FIGURE 1**

The line graph in Figure 1 presents information based on a method for locating a point in a plane developed by René Descartes, a 17th-century French mathematician. Legend has it that Descartes, who was lying in bed ill, was watching a fly crawl about on the ceiling near a corner of the room. It occurred to him that the location of the fly on the ceiling could be described by determining its distances from the two adjacent walls. See the figure in the margin. We use this insight to plot points and graph linear equations in two variables whose graphs are straight lines.

Locating a fly on a ceiling

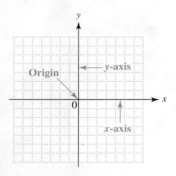

**FIGURE 2**

**OBJECTIVE 2** Plot ordered pairs. Each of the pairs of numbers $(3, 1)$, $(-5, 6)$, and $(4, -1)$ is an example of an **ordered pair**—that is, a pair of numbers written within parentheses in which the order of the numbers is important. We graph an ordered pair by using two perpendicular number lines that intersect at their 0 points, as shown in the plane in Figure 2. The common 0 point is called the **origin**. The position of any point in this plane is determined by referring to the horizontal number line, or **$x$-axis,** and the vertical number line, or **$y$-axis.** The first number in the ordered pair indicates the position relative to the $x$-axis, and the second number indicates the position relative to the $y$-axis. The $x$-axis and the $y$-axis make up a **rectangular** (or **Cartesian,** for Descartes) **coordinate system.**

To locate, or **plot,** the point on the graph that corresponds to the ordered pair $(3, 1)$, we move three units from 0 to the right along the $x$-axis and then one unit up parallel to the $y$-axis. The point corresponding to the ordered pair $(3, 1)$ is labeled $A$ in Figure 3. Additional points are labeled $B$–$E$. The phrase "the point corresponding to the ordered pair $(3, 1)$" is often abbreviated as "the point $(3, 1)$." The numbers in an ordered pair are called the **coordinates** of the corresponding point.

We can relate this method of locating ordered pairs to the line graph in Figure 1. We move along the horizontal axis to a year and then up parallel to the vertical axis to approximate medical spending for that year. Thus, we can write the ordered pair $(2002, 1340)$ to indicate that in 2002 personal spending on medical care was about $1340 billion.

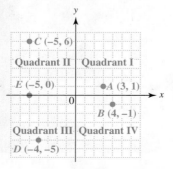

**FIGURE 3**

> **CAUTION** The parentheses used to represent an ordered pair are also used to represent an open interval (introduced in **Section 1.1**). The context of the discussion tells whether ordered pairs or open intervals are being represented.

The four regions of the graph, shown in Figure 3, are called **quadrants I, II, III,** and **IV,** reading counterclockwise from the upper right quadrant. The points on the *x*-axis and *y*-axis do not belong to any quadrant. For example, point *E* in Figure 3 belongs to no quadrant.

**NOW TRY** Exercises 3, 9, 13, 15, and 21.

**OBJECTIVE 3** Find ordered pairs that satisfy a given equation. Each solution of an equation with two variables, such as

$$2x + 3y = 6,$$

includes two numbers, one for each variable. To keep track of which number goes with which variable, we write the solutions as ordered pairs. (If *x* and *y* are used as the variables, the *x*-value is given first.) For example, we can show that $(6, -2)$ is a solution of $2x + 3y = 6$ by substitution.

$$2x + 3y = 6$$
$$2(6) + 3(-2) = 6 \quad ? \quad \text{Let } x = 6, y = -2.$$
$$12 - 6 = 6 \quad ?$$
$$6 = 6 \quad \text{True}$$

**Use parentheses to avoid errors.**

Because the ordered pair $(6, -2)$ makes the equation true, it is a solution. On the other hand, $(5, 1)$ is *not* a solution of the equation $2x + 3y = 6$, because

$$2x + 3y = 2(5) + 3(1)$$
$$= 10 + 3$$
$$= 13, \ \textbf{\textit{not}} \ \ 6.$$

To find ordered pairs that satisfy an equation, select any number for one of the variables, substitute it into the equation for that variable, and then solve for the other variable. Two other ordered pairs satisfying $2x + 3y = 6$ are $(0, 2)$ and $(3, 0)$.

*Since any real number could be selected for one variable and would lead to a real number for the other variable, linear equations in two variables have an infinite number of solutions.*

**EXAMPLE 1** Completing Ordered Pairs and Making a Table

In parts (a) and (b), complete each ordered pair for $2x + 3y = 6$. Then, in part (c), write the results as a table of ordered pairs.

**(a)** $(-3, \ )$
Replace *x* with $-3$ in the equation to find *y*.

$$2x + 3y = 6$$
$$2(-3) + 3y = 6 \quad \text{Let } x = -3.$$
$$-6 + 3y = 6$$
$$3y = 12$$
$$y = 4$$

The ordered pair is $(-3, 4)$.

**(b)** $( \ , -4)$
Replace *y* with $-4$ in the equation to find *x*.

$$2x + 3y = 6$$
$$2x + 3(-4) = 6 \quad \text{Let } y = -4.$$
$$2x - 12 = 6$$
$$2x = 18$$
$$x = 9$$

The ordered pair is $(9, -4)$.

**(c)** We write a table of these ordered pairs as shown.

| x | y | |
|---|---|---|
| −3 | 4 | ← Represents the ordered pair (−3, 4) |
| 9 | −4 | ← Represents the ordered pair (9, −4) |

**NOW TRY** Exercise 25(a).

**OBJECTIVE 4** Graph lines. The **graph of an equation** is the set of points corresponding to *all* ordered pairs that satisfy the equation. It gives a "picture" of the equation.

To graph an equation, we plot several ordered pairs that satisfy the equation until we have enough points to suggest the shape of the graph. For example, to graph $2x + 3y = 6$, we plot all ordered pairs found in Objective 3 and Example 1. These points, shown in a table of values and plotted in Figure 4(a), appear to lie on a straight line. If all the ordered pairs that satisfy the equation $2x + 3y = 6$ were graphed, they would form the straight line shown in Figure 4(b).

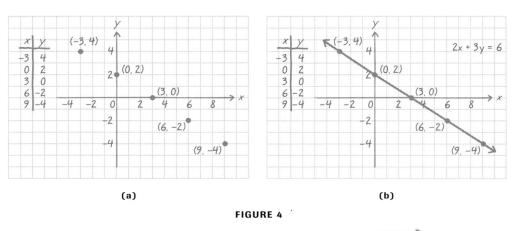

**FIGURE 4**

**NOW TRY** Exercise 25(b).

The equation $2x + 3y = 6$ is called a **first-degree equation,** because it has no term with a variable to a power greater than 1.

***The graph of any first-degree equation in two variables is a straight line.***

Since first-degree equations with two variables have straight-line graphs, they are called *linear equations in two variables*. (We discussed linear equations in one variable in **Chapter 2.**)

---
**Linear Equation in Two Variables**

A **linear equation in two variables** can be written in the form
$$Ax + By = C,$$
where $A$, $B$, and $C$ are real numbers ($A$ and $B$ not both 0). This form is called **standard form.**

---

**OBJECTIVE 5** Find *x*- and *y*-intercepts. A straight line is determined if any two different points on the line are known, so finding two different points is enough to graph the line. Two useful points for graphing are the *x*- and *y*-intercepts. The ***x*-intercept** is the point (if any) where the line intersects the *x*-axis; likewise, the ***y*-intercept**

is the point (if any) where the line intersects the *y*-axis.* In Figure 4(b), the *y*-value of the point where the line intersects the *x*-axis is 0. Similarly, the *x*-value of the point where the line intersects the *y*-axis is 0. This suggests a method for finding the *x*- and *y*-intercepts.

**Finding Intercepts**

When graphing the equation of a line,

let $y = 0$ to find the *x*-intercept;

let $x = 0$ to find the *y*-intercept.

**EXAMPLE 2** Finding Intercepts

Find the *x*- and *y*-intercepts of $4x - y = -3$ and graph the equation.

We find the *x*-intercept by letting $y = 0$.

$$4x - 0 = -3 \qquad \text{Let } y = 0.$$
$$4x = -3$$
$$x = -\frac{3}{4} \qquad \text{*x*-intercept is } \left(-\frac{3}{4}, 0\right).$$

For the *y*-intercept, we let $x = 0$.

$$4(0) - y = -3 \qquad \text{Let } x = 0.$$
$$-y = -3$$
$$y = 3 \qquad \text{*y*-intercept is } (0, 3).$$

The intercepts are the two points $\left(-\frac{3}{4}, 0\right)$ and $(0, 3)$. We show these ordered pairs in the table next to Figure 5 and use them to draw the graph.

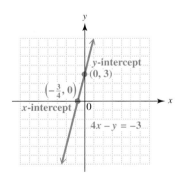

**FIGURE 5**

**NOW TRY** Exercise 37.

**NOTE** While two points, such as the two intercepts in Figure 5, are sufficient to graph a straight line, *it is a good idea to use a third point to guard against errors.* Verify by substitution that $(-2, -5)$ also lies on the graph of $4x - y = -3$.

---

*Some texts define an intercept as a number, not a point. For example, "*y*-intercept (0, 4)" would be given as "*y*-intercept 4."

**OBJECTIVE 6** Recognize equations of horizontal and vertical lines and lines passing through the origin. A graph can fail to have an *x*-intercept or a *y*-intercept. The next examples illustrate these special cases.

### EXAMPLE 3  Graphing a Horizontal Line

Graph $y = 2$.

Writing $y = 2$ as $0x + 1y = 2$ shows that any value of *x*, including $x = 0$, gives $y = 2$, making the *y*-intercept $(0, 2)$. Since *y* is always 2, there is no value of *x* corresponding to $y = 0$, so the graph has no *x*-intercept. The graph, shown with a table of ordered pairs in Figure 6, is a horizontal line.

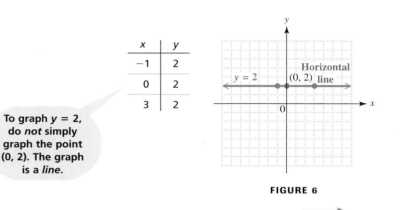

To graph $y = 2$, do *not* simply graph the point $(0, 2)$. The graph is a *line*.

**FIGURE 6**

NOW TRY Exercise 43.

### EXAMPLE 4  Graphing a Vertical Line

Graph $x + 1 = 0$.

The form $1x + 0y = -1$ shows that every value of *y* leads to $x = -1$, making the *x*-intercept $(-1, 0)$. No value of *y* makes $x = 0$, so the graph has no *y*-intercept. The only way a straight line can have no *y*-intercept is to be vertical, as shown in Figure 7.

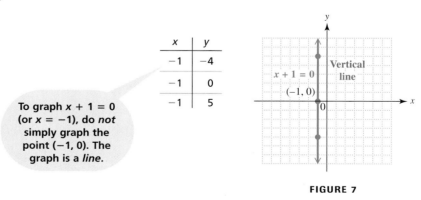

To graph $x + 1 = 0$ (or $x = -1$), do *not* simply graph the point $(-1, 0)$. The graph is a *line*.

**FIGURE 7**

NOW TRY Exercise 47.

Some lines have their *x*- and *y*-intercepts at the origin.

### EXAMPLE 5  Graphing a Line That Passes through the Origin

Graph $x + 2y = 0$.

Find the intercepts.

| *x*-intercept | *y*-intercept |
|---|---|
| $x + 2y = 0$ | $x + 2y = 0$ |
| $x + 2(0) = 0$    Let $y = 0$. | $0 + 2y = 0$    Let $x = 0$. |
| $x + 0 = 0$ | $y = 0$    *y*-intercept is (0, 0). |
| $x = 0$    *x*-intercept is (0, 0). | |

Both intercepts are the same point, (0, 0), which means that the graph passes through the origin. To find another point to graph the line, choose any nonzero number for $x$, say, $x = 4$, and solve for $y$.

$$x + 2y = 0$$
$$4 + 2y = 0 \quad \text{Let } x = 4.$$
$$2y = -4$$
$$y = -2$$

This gives the ordered pair $(4, -2)$. To find the additional point, we could have chosen any number (except 0) for $y$ instead of $x$.

The points $(0, 0)$ and $(4, -2)$ lead to the graph shown in Figure 8. As a check, verify that $(-2, 1)$ also lies on the line.

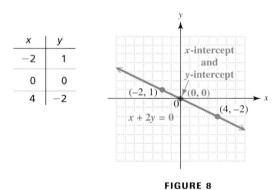

**FIGURE 8**

NOW TRY Exercise 49.

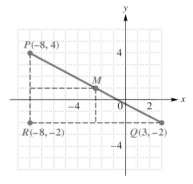

**FIGURE 9**

**OBJECTIVE 7** Use the midpoint formula. If the coordinates of the endpoints of a line segment are known, then the coordinates of the *midpoint* of the segment can be found. Figure 9 shows that segment $PQ$ has endpoints $P(-8, 4)$ and $Q(3, -2)$. $R$ is the point with the same *x*-coordinate as $P$ and the same *y*-coordinate as $Q$. So the coordinates of $R$ are $(-8, -2)$.

The *x*-coordinate of the midpoint $M$ of $PQ$ is the same as the *x*-coordinate of the midpoint of $RQ$. Since $RQ$ is horizontal, the *x*-coordinate of its midpoint is the *average* of the *x*-coordinates of its endpoints:

$$\frac{1}{2}(-8 + 3) = -2.5.$$

The $y$-coordinate of $M$ is the average of the $y$-coordinates of the midpoint of $PR$:

$$\frac{1}{2}(4 + (-2)) = 1.$$

The midpoint of $PQ$ is $M(-2.5, 1)$. This discussion leads to the *midpoint formula*.

> **Midpoint Formula**
>
> If the endpoints of a line segment $PQ$ are $(x_1, y_1)$ and $(x_2, y_2)$, its midpoint $M$ is
>
> $$\left( \frac{x_1 + x_2}{2}, \frac{y_1 + y_2}{2} \right).$$

The small numbers 1 and 2 in these ordered pairs are called *subscripts*. Read $(x_1, y_1)$ as "$x$-sub-one, $y$-sub-one."

**EXAMPLE 6** Finding the Coordinates of a Midpoint

Find the coordinates of the midpoint of line segment $PQ$ with endpoints $P(4, -3)$ and $Q(6, -1)$.

Use the midpoint formula with $x_1 = 4$, $x_2 = 6$, $y_1 = -3$, and $y_2 = -1$:

$$\left( \frac{4 + 6}{2}, \frac{-3 + (-1)}{2} \right) = \left( \frac{10}{2}, \frac{-4}{2} \right) = (5, -2). \leftarrow \text{Midpoint}$$

**NOW TRY** Exercise 55.

## 3.1 Exercises

**NOW TRY Exercise**

*In Exercises 1 and 2, answer each question by locating ordered pairs on the graphs. See Objective 1.*

1. The graph indicates U.S. federal government tax revenues in billions of dollars.

    (a) If the ordered pair $(x, y)$ represents a point on the graph, what does $x$ represent? What does $y$ represent?
    (b) Estimate revenue in 2002.
    (c) Write an ordered pair $(x, y)$ that gives approximate federal tax revenues in 2002.
    (d) What does the ordered pair (2000, 2030) mean in the context of this graph?

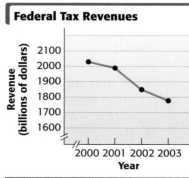

*Source:* U.S. Office of Management and Budget.

2. The graph shows the percent of women in mathematics or computer science professions.
   (a) If the ordered pair $(x, y)$ represents a point on the graph, what does $x$ represent? What does $y$ represent?
   (b) In what decade (10-yr period) did the percent of women in mathematics or computer science professions decrease?
   (c) Write an ordered pair $(x, y)$ that gives the approximate percent of women in mathematics or computer science professions in 1990.
   (d) What does the ordered pair (2000, 30) mean in the context of this graph?

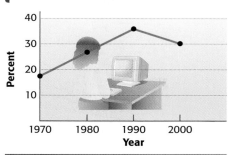

*Source:* U.S. Census Bureau and Bureau of Labor Statistics.

*Fill in the Blanks*

3. The point with coordinates (0, 0) is called the _____ of a rectangular coordinate system.
4. For any value of $x$, the point $(x, 0)$ lies on the _____-axis.
5. To find the $x$-intercept of a line, we let _____ equal 0 and solve for _____; to find the $y$-intercept, we let _____ equal 0 and solve for _____.
6. The equation _____ = 4 has a horizontal line as its graph.
         ($x$ or $y$)
7. To graph a straight line, we must find a minimum of _____ points.
8. The point (_____, 4) is on the graph of $2x - 3y = 0$.

*Name the quadrant, if any, in which each point is located.*

9. (a) $(1, 6)$    (b) $(-4, -2)$      10. (a) $(-2, -10)$    (b) $(4, 8)$
   (c) $(-3, 6)$    (d) $(7, -5)$            (c) $(-9, 12)$    (d) $(3, -9)$
   (e) $(-3, 0)$    (f) $(0, -0.5)$          (e) $(0, -8)$      (f) $(2.3, 0)$

11. *Concept Check* Use the given information to determine the quadrants in which the point $(x, y)$ may lie.
    (a) $xy > 0$    (b) $xy < 0$    (c) $\dfrac{x}{y} < 0$    (d) $\dfrac{x}{y} > 0$

12. *Concept Check* What must be true about the coordinates of any point that lies along an axis?

*Plot each point in a rectangular coordinate system. See Objective 2.*

13. $(2, 3)$      14. $(-1, 2)$      15. $(-3, -2)$      16. $(1, -4)$      17. $(0, 5)$
18. $(-2, -4)$    19. $(-2, 4)$      20. $(3, 0)$        21. $(-2, 0)$     22. $(3, -3)$

*In Exercises 23–32, (a) complete the given table for each equation and then (b) graph the equation. See Example 1 and Figure 4.*

**23.** $y = x - 4$

| x | y |
|---|---|
| 0 |   |
| 1 |   |
| 2 |   |
| 3 |   |
| 4 |   |

**24.** $y = x + 3$

| x | y |
|---|---|
| 0 |   |
| 1 |   |
| 2 |   |
| 3 |   |
| 4 |   |

**25.** $x - y = 3$

| x | y |
|---|---|
| 0 |   |
|   | 0 |
| 5 |   |
| 2 |   |

**26.** $x - y = 5$

| x | y |
|---|---|
| 0 |   |
|   | 0 |
| 1 |   |
| 3 |   |

**27.** $x + 2y = 5$

| x | y |
|---|---|
| 0 |   |
|   | 0 |
| 2 |   |
|   | 2 |

**28.** $x + 3y = -5$

| x | y |
|---|---|
| 0 |   |
|   | 0 |
| 1 |   |
|   | -1 |

**29.** $4x - 5y = 20$

| x | y |
|---|---|
| 0 |   |
|   | 0 |
| 2 |   |
|   | -3 |

**30.** $6x - 5y = 30$

| x | y |
|---|---|
| 0 |   |
|   | 0 |
| 3 |   |
|   | -2 |

**31.** $y = -2x + 3$

| x | y |
|---|---|
| 0 |   |
| 1 |   |
| 2 |   |
| 3 |   |

**32.** $y = -3x + 1$

| x | y |
|---|---|
| 0 |   |
| 1 |   |
| 2 |   |
| 3 |   |

**33.** *Fill in the Blanks*  Consider the patterns formed in the tables for Exercises 23 and 31. Fill in each blank with the appropriate number.

  (a) In Exercise 23, for every increase in x by 1 unit, y increases by _____ unit(s).
  (b) In Exercise 31, for every increase in x by 1 unit, y decreases by _____ unit(s).
  (c) On the basis of your observations in parts (a) and (b), make a conjecture about a similar pattern for $y = 2x + 4$. Then test your conjecture.

**34.** Explain why the graph of $x + y = k$ cannot pass through quadrant III if $k > 0$.

**35.** A student attempted to graph $4x + 5y = 0$ by finding intercepts. First she let $x = 0$ and found y; then she let $y = 0$ and found x. In both cases, the resulting point was (0, 0). She knew that she needed at least two points to graph the line, but was unsure what to do next because finding intercepts gave her only one point. Explain to her what to do next.

**36.** *Concept Check*  What is the equation of the x-axis? What is the equation of the y-axis?

*Find the x- and y-intercepts. Then graph each equation. See Examples 2–5.*

**37.** $2x + 3y = 12$  
**38.** $5x + 2y = 10$  
**39.** $x - 3y = 6$  
**40.** $x - 2y = -4$  
**41.** $\frac{2}{3}x - 3y = 7$  
**42.** $\frac{5}{7}x + \frac{6}{7}y = -2$  
**43.** $y = 5$  
**44.** $y = -3$  
**45.** $x = 2$  
**46.** $x = -3$  
**47.** $x + 4 = 0$  
**48.** $y + 2 = 0$  
**49.** $x + 5y = 0$  
**50.** $x - 3y = 0$  
**51.** $2x = 3y$  
**52.** $4y = 3x$  
**53.** $-\frac{2}{3}y = x$  
**54.** $3y = -\frac{4}{3}x$  

*Find the midpoint of each segment with the given endpoints. See Example 6.*

**55.** $(-8, 4)$ and $(-2, -6)$  
**56.** $(5, 2)$ and $(-1, 8)$  
**57.** $(3, -6)$ and $(6, 3)$  
**58.** $(-10, 4)$ and $(7, 1)$  
**59.** $(-9, 3)$ and $(9, 8)$  
**60.** $(4, -3)$ and $(-1, 3)$  
**61.** $(2.5, 3.1)$ and $(1.7, -1.3)$  
**62.** $(6.2, 5.8)$ and $(1.4, -0.6)$  
**63.** $\left(\frac{1}{2}, \frac{1}{3}\right)$ and $\left(\frac{3}{2}, \frac{5}{3}\right)$  
**64.** $\left(\frac{21}{4}, \frac{2}{5}\right)$ and $\left(\frac{7}{4}, \frac{3}{5}\right)$  
**65.** $\left(-\frac{1}{3}, \frac{2}{7}\right)$ and $\left(-\frac{1}{2}, \frac{1}{14}\right)$  
**66.** $\left(\frac{3}{5}, -\frac{1}{3}\right)$ and $\left(\frac{1}{2}, -\frac{7}{2}\right)$

*A linear equation can be used as a model to describe real data in some cases. Exercises 67 and 68 are based on this idea.*

**67.** The number of U.S. travelers to other countries for 2000–2003 is approximated by the linear equation

$$y = -1237x + 60{,}936,$$

where $y$ is the number of travelers, in thousands, in year $x$. In the equation, $x = 0$ corresponds to 2000, $x = 1$ corresponds to 2001, and so on. Use the equation to approximate the number of U.S. travelers to other countries in 2003. (*Source:* U.S. Department of Commerce.)

**68.** The total expenditures for dental services in the United States from 1990 through 2003 can be approximated by the linear equation

$$y = 3.32x + 28.7,$$

where $y$ is the expenditure in billions of dollars. In the equation, $x = 0$ corresponds to 1990, $x = 1$ corresponds to 1991, and so on. On the basis of this equation, find the total amount spent on dental services in the United States in 2002. (*Source:* U.S. Centers for Medicare and Medicaid Services.)

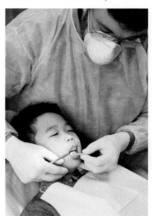

## 3.2 The Slope of a Line

**OBJECTIVES**

1. Find the slope of a line, given two points on the line.
2. Find the slope of a line, given an equation of the line.
3. Graph a line, given its slope and a point on the line.
4. Use slopes to determine whether two lines are parallel, perpendicular, or neither.
5. Solve problems involving average rate of change.

Slope (steepness) is used in many practical ways. The slope of a highway (sometimes called the *grade*) is often given as a percent. For example, a 10% $\left(\text{or } \frac{10}{100} = \frac{1}{10}\right)$ slope means that the highway rises 1 unit for every 10 horizontal units. Stairs and roofs have slopes too, as shown in Figure 10.

Slope is $\frac{1}{10}$.
(not to scale)

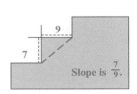

Slope is $\frac{7}{9}$.

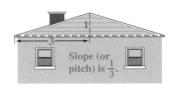

Slope (or pitch) is $\frac{1}{3}$.

**FIGURE 10**

In each example mentioned, slope is the ratio of vertical change, or **rise**, to horizontal change, or **run**. A simple way to remember this is to think, *"Slope is rise over run."*

**OBJECTIVE 1** Find the slope of a line, given two points on the line. To get a formal definition of the slope of a line, we designate two different points on the line. To differentiate between the points, we write them as $(x_1, y_1)$ and $(x_2, y_2)$. See Figure 11.

As we move along the line in Figure 11 from $(x_1, y_1)$ to $(x_2, y_2)$, the *y*-value changes (vertically) from $y_1$ to $y_2$, an amount equal to $y_2 - y_1$. As *y* changes from $y_1$ to $y_2$, the value of *x* changes (horizontally) from $x_1$ to $x_2$ by the amount $x_2 - x_1$. (The Greek letter **delta**, $\Delta$, is used in mathematics to denote "change in," so $\Delta y$ and $\Delta x$ represent the change in *y* and the change in *x*, respectively.) The ratio of the change in *y* to the change in *x* (the rise over the run) is called the *slope* of the line, with the letter *m* traditionally used for slope.

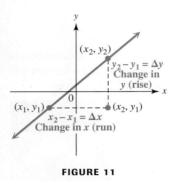

**FIGURE 11**

### Slope Formula

The **slope** of the line through the distinct points $(x_1, y_1)$ and $(x_2, y_2)$ is

$$m = \frac{\text{rise}}{\text{run}} = \frac{\text{change in } y}{\text{change in } x} = \frac{\Delta y}{\Delta x} = \frac{y_2 - y_1}{x_2 - x_1} \quad (x_1 \neq x_2).$$

**EXAMPLE 1** Finding the Slope of a Line

Find the slope of the line through the points $(2, -1)$ and $(-5, 3)$.

If $(2, -1) = (x_1, y_1)$ and $(-5, 3) = (x_2, y_2)$, then

$$m = \frac{y_2 - y_1}{x_2 - x_1} = \frac{3 - (-1)}{-5 - 2} = \frac{4}{-7} = -\frac{4}{7}.$$

Thus, the slope is $-\frac{4}{7}$. See Figure 12.

If the ordered pairs are interchanged so that $(2, -1) = (x_2, y_2)$ and $(-5, 3) = (x_1, y_1)$ in the slope formula, the slope is the same.

$$m = \frac{-1 - 3}{2 - (-5)} = \frac{-4}{7} = -\frac{4}{7}$$

*y*-values are in the *numerator*, *x*-values in the *denominator*.

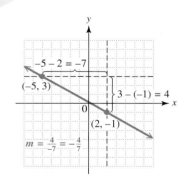

**FIGURE 12**

**NOW TRY** Exercise 23.

Example 1 suggests that the slope is the same no matter which point we consider first. Also, using similar triangles from geometry, we can show that the slope is the same no matter which two different points on the line we choose.

**CAUTION** *In calculating slope, be careful to subtract the y-values and the x-values in the same order.*

**Correct**

$$\frac{y_2 - y_1}{x_2 - x_1} \quad \text{or} \quad \frac{y_1 - y_2}{x_1 - x_2}$$

**Incorrect**

$$\frac{y_2 - y_1}{x_1 - x_2} \quad \text{or} \quad \frac{y_1 - y_2}{x_2 - x_1}$$

*The change in y is the numerator and the change in x is the denominator.*

**OBJECTIVE 2** Find the slope of a line, given an equation of the line. When an equation of a line is given, one way to find the slope is to use the definition of slope by first finding two different points on the line.

**EXAMPLE 2** Finding the Slope of a Line

Find the slope of the line $4x - y = -8$.
  The intercepts can be used as the two different points needed to find the slope. Let $y = 0$ to find that the x-intercept is $(-2, 0)$. Then let $x = 0$ to find that the y-intercept is $(0, 8)$. Use these two points in the slope formula. The slope is

$$m = \frac{\text{rise}}{\text{run}} = \frac{8 - 0}{0 - (-2)} = \frac{8}{2} = 4.$$

**NOW TRY** Exercise 41.

**EXAMPLE 3** Finding Slopes of Horizontal and Vertical Lines

Find the slope of each line.

**(a)** $y = 2$
  The graph of $y = 2$ is a horizontal line. To find the slope, select two different points on the line, such as $(3, 2)$ and $(-1, 2)$, and use the slope formula.

$$m = \frac{\text{rise}}{\text{run}} = \frac{2 - 2}{3 - (-1)} = \frac{0}{4} = 0$$

In this case, the *rise* is 0, so the slope is 0.

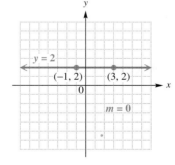

**(b)** $x + 1 = 0$
  The graph of $x + 1 = 0$, or $x = -1$, is a vertical line. Two points that satisfy the equation $x = -1$ are $(-1, 5)$ and $(-1, -4)$. If we use these two points to try to find the slope, we obtain

$$m = \frac{\text{rise}}{\text{run}} = \frac{-4 - 5}{-1 - (-1)} = \frac{-9}{0}.$$

Since division by 0 is undefined, the slope is undefined. This is why the definition of slope includes the restriction that $x_1 \neq x_2$.

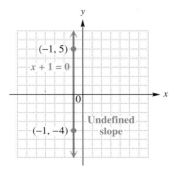

**NOW TRY** Exercises 47 and 49.

Example 3 illustrates the following important concepts.

> **Horizontal and Vertical Lines**
> - An equation of the form $y = b$ always intersects the $y$-axis; thus, the line with that equation is horizontal and has slope 0.
> - An equation of the form $x = a$ always intersects the $x$-axis; thus, the line with that equation is vertical and has undefined slope.

The slope of a line can also be found directly from its equation. Look again at the equation $4x - y = -8$ from Example 2. Solve this equation for $y$.

$$4x - y = -8 \quad \text{Equation from Example 2}$$
$$-y = -4x - 8 \quad \text{Subtract } 4x.$$
$$y = 4x + 8 \quad \text{Multiply by } -1.$$

Notice that the slope, 4, found with the slope formula in Example 2 is the same number as the coefficient of $x$ in the equation $y = 4x + 8$. We will see in the next section that this always happens, *as long as the equation is solved for y.*

### EXAMPLE 4  Finding the Slope from an Equation

Find the slope of the graph of $3x - 5y = 8$.
Solve the equation for $y$.

$$3x - 5y = 8$$
$$-5y = -3x + 8 \quad \text{Subtract } 3x.$$
$$y = \frac{3}{5}x - \frac{8}{5} \quad \text{Divide by } -5.$$

The slope is given by the coefficient of $x$, so the slope is $\frac{3}{5}$.

**NOW TRY** Exercise 43.

**OBJECTIVE 3** Graph a line, given its slope and a point on the line. Example 5 shows how to graph a straight line by using the geometric interpretation of slope and one point on the line.

### EXAMPLE 5  Using the Slope and a Point to Graph Lines

Graph each line.

**(a)** With slope $\frac{2}{3}$ and $y$-intercept $(0, -4)$

Begin by plotting the point $P(0, -4)$, as shown in Figure 13 on the next page. Then use the slope to find a second point. From the slope formula,

$$m = \frac{\text{change in } y}{\text{change in } x} = \frac{2}{3},$$

so move 2 units *up* and then 3 units to the *right* to locate another point on the graph, $R(3, -2)$. The line through $P(0, -4)$ and $R$ is the required graph.

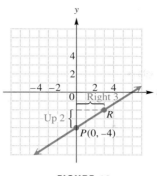

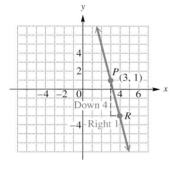

**FIGURE 13**  **FIGURE 14**

**(b)** Through (3, 1) with slope $-4$

Start by locating the point $P(3, 1)$ on a graph. Find a second point $R$ on the line by writing the slope $-4$ as $\frac{-4}{1}$ and using the slope formula.

$$m = \frac{\text{change in } y}{\text{change in } x} = \frac{-4}{1}$$

Move 4 units *down* from (3, 1), and then move 1 unit to the *right*. Draw a line through this second point $R$ and (3, 1), as shown in Figure 14.

The slope also could be written as

$$m = \frac{\text{change in } y}{\text{change in } x} = \frac{4}{-1}.$$

In this case, the second point $R$ is located 4 units *up* and 1 unit to the *left*. Verify that this approach also produces the line in Figure 14.

**NOW TRY** Exercises 55 and 57.

In Example 5(a), the slope of the line is the *positive* number $\frac{2}{3}$. The graph of the line in Figure 13 goes up (rises) from left to right. The line in Example 5(b) has *negative* slope $-4$. As Figure 14 shows, its graph goes down (falls) from left to right. These facts suggest the following generalization.

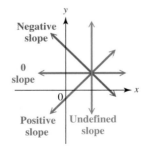

**FIGURE 15**

### Orientation of a Line in the Plane

A positive slope indicates that the line goes *up* (rises) from left to right.

A negative slope indicates that the line goes *down* (falls) from left to right.

Figure 15 shows lines of positive, 0, negative, and undefined slopes.

**OBJECTIVE 4** Use slopes to determine whether two lines are parallel, perpendicular, or neither. The slopes of a pair of parallel or perpendicular lines are related in a special way. Recall that the slope of a line measures the steepness of the line. Since parallel lines have equal steepness, their slopes must be equal; also, lines with the same slope are parallel.

### Slopes of Parallel Lines

Two nonvertical lines with the same slope are parallel.
Two nonvertical parallel lines have the same slope.

**EXAMPLE 6** Determining Whether Two Lines Are Parallel

Determine whether the lines $L_1$, through $(-2, 1)$ and $(4, 5)$, and $L_2$, through $(3, 0)$ and $(0, -2)$, are parallel.

The slope of $L_1$ is

$$m_1 = \frac{5 - 1}{4 - (-2)} = \frac{4}{6} = \frac{2}{3}.$$

The slope of $L_2$ is

$$m_2 = \frac{-2 - 0}{0 - 3} = \frac{-2}{-3} = \frac{2}{3}.$$

Because the slopes are equal, the two lines are parallel.

**NOW TRY** Exercise 65.

To see how the slopes of perpendicular lines are related, consider a nonvertical line with slope $\frac{a}{b}$. If this line is rotated 90°, the vertical change and the horizontal change are interchanged and the slope is $-\frac{b}{a}$, since the horizontal change is now negative. See Figure 16. Thus, the slopes of perpendicular lines have product $-1$ and are negative reciprocals of each other. For example, if the slopes of two lines are $\frac{3}{4}$ and $-\frac{4}{3}$, then the lines are perpendicular because

$$\frac{3}{4}\left(-\frac{4}{3}\right) = -1.$$

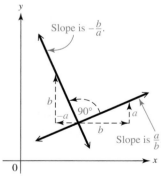

**FIGURE 16**

### Slopes of Perpendicular Lines

Two perpendicular lines, neither of which is parallel to an axis, have slopes that are negative reciprocals; that is, their product is $-1$. Also, lines with slopes that are negative reciprocals are perpendicular.

### EXAMPLE 7  Determining Whether Two Lines Are Perpendicular

Determine whether the lines with equations $2y = 3x - 6$ and $2x + 3y = -6$ are perpendicular.

Find the slope of each line by first solving each equation for $y$.

$$2y = 3x - 6 \qquad\qquad 2x + 3y = -6$$
$$y = \frac{3}{2}x - 3 \qquad\qquad 3y = -2x - 6$$
$$\uparrow \text{Slope} \qquad\qquad y = -\frac{2}{3}x - 2$$
$$\qquad\qquad\qquad\qquad\uparrow \text{Slope}$$

Since the product of the slopes of the two lines is $\frac{3}{2}\left(-\frac{2}{3}\right) = -1$, the lines are perpendicular.

**NOW TRY** Exercise 67.

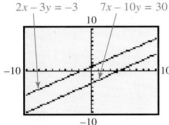

The graphs are *not* parallel, although they may appear to be.

**FIGURE 17**

We must be careful when interpreting calculator graphs of parallel and perpendicular lines. For example, the graphs of the equations in Figure 17 appear to be parallel. However, checking their slopes algebraically, we find that

$$2x - 3y = -3 \qquad\qquad 7x - 10y = 30$$
$$-3y = -2x - 3 \qquad\qquad -10y = -7x + 30$$
$$y = \frac{2}{3}x + 1 \qquad\qquad y = \frac{7}{10}x - 3$$

Since the slopes $\frac{2}{3}$ and $\frac{7}{10}$ are not equal, the lines are *not* parallel.

Figure 18(a) shows graphs of the perpendicular lines from Example 7. As graphed in the standard viewing window, the lines do not appear to be perpendicular. However, if we use a *square viewing window* as in Figure 18(b), we get a more realistic view. (Many graphing calculators can set a square window automatically. See your owner's manual.) *We cannot rely completely on what we see on a calculator screen—we must understand the mathematical concepts as well.*

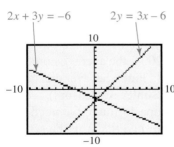

In the standard window, the lines *do not* appear to be perpendicular.

(a)

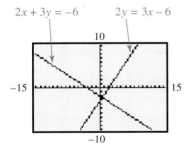

In the square window, the lines *do* appear to be perpendicular.

(b)

**FIGURE 18**

**OBJECTIVE 5** Solve problems involving average rate of change. We know that the slope of a line is the ratio of the vertical change in $y$ to the horizontal change in $x$. Thus, slope gives the *average rate of change* in $y$ per unit change in $x$, where the value of $y$ depends on the value of $x$. The next examples illustrate this idea. We assume a linear relationship between $x$ and $y$.

### EXAMPLE 8  Interpreting Slope as Average Rate of Change

The graph in Figure 19 approximates the average number of hours per year spent watching cable and satellite TV for each person in the United States during the years 2000 through 2004. Find the average rate of change in number of hours per year.

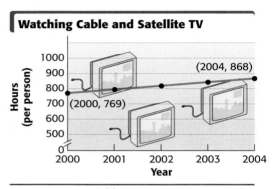

**Source:** Veronis Suhler Stevenson.

**FIGURE 19**

To determine the average rate of change, we need two pairs of data. From the graph, if $x = 2000$, then $y = 769$ and if $x = 2004$, then $y = 868$. Thus, we have the ordered pairs $(2000, 769)$ and $(2004, 868)$. By the slope formula,

$$\text{average rate of change} = \frac{868 - 769}{2004 - 2000} = \frac{99}{4} = 24.75.$$

A positive slope indicates an increase.

This means that the average time per person spent watching cable and satellite TV *increased* by about 25 hr per year from 2000 through 2004.

**NOW TRY** Exercise 85.

### EXAMPLE 9  Interpreting Slope as Average Rate of Change

During the year 2000, the average person in the United States spent 866 hr watching broadcast TV. In 2004, the average number of hours per person spent watching broadcast TV was 678. Find the average rate of change in number of hours per year. (*Source:* Veronis Suhler Stevenson.)

To use the slope formula, we need two ordered pairs. Here, we let one ordered pair be $(2000, 866)$ and the other be $(2004, 678)$.

$$\text{average rate of change} = \frac{678 - 866}{2004 - 2000} = \frac{-188}{4} = -47$$

A negative slope indicates a decrease.

The graph in Figure 20 confirms that the line through the ordered pairs falls from left to right and, therefore, has negative slope. Thus, the average time per person spent watching broadcast TV *decreased* by 47 hr per year from 2000 through 2004.

**FIGURE 20**

**NOW TRY** Exercise 87.

## 3.2 Exercises

**NOW TRY** **Exercise**

1. *Multiple Choice* A ski slope drops 30 ft for every horizontal 100 ft. Which of the following express its slope? (There are several correct choices.)

   **A.** 0.3  **B.** $\dfrac{3}{10}$  **C.** $3\dfrac{1}{3}$

   **D.** $\dfrac{30}{100}$  **E.** $\dfrac{10}{3}$  **F.** 30

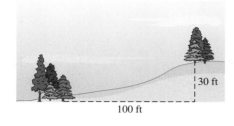

2. *Concept Check* A hill has slope 0.05. How many feet in the vertical direction correspond to a run of 50 ft?

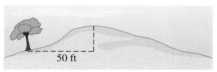

NOT TO SCALE

3. *Matching* Match each situation in (a)–(d) with the most appropriate graph in A–D.

   (a) Sales rose sharply during the first quarter, leveled off during the second quarter, and then rose slowly for the rest of the year.
   (b) Sales fell sharply during the first quarter and then rose slowly during the second and third quarters before leveling off for the rest of the year.
   (c) Sales rose sharply during the first quarter and then fell to the original level during the second quarter before rising steadily for the rest of the year.
   (d) Sales fell during the first two quarters of the year, leveled off during the third quarter, and rose during the fourth quarter.

**A.**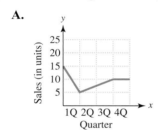

**B.**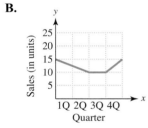

**C.**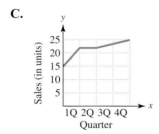

**D.**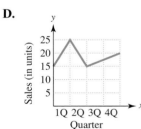

*Determine the slope of each line segment in the given figure.*

4. *AB*
5. *BC*
6. *CD*
7. *DE*
8. *EF*
9. *FG*

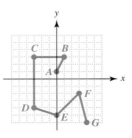

**10.** *Concept Check* If B and D were joined by a line segment in the figure for Exercises 4–9, what would be the slope of the segment?

*Calculate the value of each slope m, if possible, by using the slope formula. See Example 1.*

**11.** $m = \dfrac{6 - 2}{5 - 3}$

**12.** $m = \dfrac{5 - 7}{-4 - 2}$

**13.** $m = \dfrac{4 - (-1)}{-3 - (-5)}$

**14.** $m = \dfrac{-6 - 0}{0 - (-3)}$

**15.** $m = \dfrac{-5 - (-5)}{3 - 2}$

**16.** $m = \dfrac{-2 - (-2)}{4 - (-3)}$

**17.** $m = \dfrac{3 - 8}{-2 - (-2)}$

**18.** $m = \dfrac{5 - 6}{-8 - (-8)}$

**19.** $m = \dfrac{\frac{4}{3} + \frac{1}{2}}{\frac{1}{6} - \frac{1}{6}}$

**20.** *Multiple Choice* Which of the following forms of the slope formula are correct? Explain.

**A.** $m = \dfrac{y_1 - y_2}{x_2 - x_1}$ **B.** $m = \dfrac{y_1 - y_2}{x_1 - x_2}$ **C.** $m = \dfrac{x_2 - x_1}{y_2 - y_1}$ **D.** $m = \dfrac{y_2 - y_1}{x_2 - x_1}$

*Find the slope of the line through each pair of points. See Example 1.*

**21.** $(-2, -3)$ and $(-1, 5)$  **22.** $(-4, 3)$ and $(-3, 4)$  **23.** $(-4, 1)$ and $(2, 6)$

**24.** $(-3, -3)$ and $(5, 6)$  **25.** $(2, 4)$ and $(-4, 4)$  **26.** $(-6, 3)$ and $(2, 3)$

**27.** $(1.5, 2.6)$ and $(0.5, 3.6)$  **28.** $(3.4, 4.2)$ and $(1.4, 10.2)$

**29.** $\left(\dfrac{1}{6}, \dfrac{1}{2}\right)$ and $\left(\dfrac{5}{6}, \dfrac{9}{2}\right)$  **30.** $\left(\dfrac{3}{4}, \dfrac{1}{3}\right)$ and $\left(\dfrac{5}{4}, \dfrac{10}{3}\right)$

**31.** $\left(-\dfrac{2}{9}, \dfrac{5}{18}\right)$ and $\left(\dfrac{1}{18}, -\dfrac{5}{9}\right)$  **32.** $\left(-\dfrac{4}{5}, \dfrac{9}{10}\right)$ and $\left(-\dfrac{3}{10}, \dfrac{1}{5}\right)$

*Find the slope of each line.*

**33.**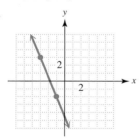

**34.**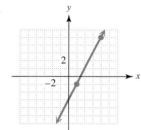

**35.**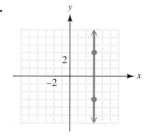

**36.** *Concept Check* Let $k$ be the number of letters in your last name. Sketch the graph of $y = k$. What is the slope of this line?

*Multiple Choice* On the basis of the figure shown here, determine which line satisfies the given description.

**37.** The line has positive slope.

**38.** The line has negative slope.

**39.** The line has slope 0.

**40.** The line has undefined slope.

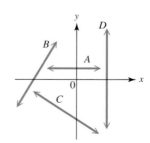

*Find the slope of the line and sketch the graph. See Examples 1–4.*

**41.** $x + 2y = 4$   **42.** $x + 3y = -6$   **43.** $5x - 2y = 10$
**44.** $4x - y = 4$   **45.** $y = 4x$   **46.** $y = -3x$
**47.** $x - 3 = 0$   **48.** $x + 2 = 0$   **49.** $y = -5$
**50.** $y = -4$   **51.** $2y = 3$   **52.** $3x = 4$

*Graph the line described. See Example 5.*

**53.** Through $(-4, 2)$; $m = \dfrac{1}{2}$   **54.** Through $(-2, -3)$; $m = \dfrac{5}{4}$

**55.** $y$-intercept $(0, -2)$; $m = -\dfrac{2}{3}$   **56.** $y$-intercept $(0, -4)$; $m = -\dfrac{3}{2}$

**57.** Through $(-1, -2)$; $m = 3$   **58.** Through $(-2, -4)$; $m = 4$
**59.** $m = 0$; through $(2, -5)$   **60.** $m = 0$; through $(5, 3)$
**61.** Undefined slope; through $(-3, 1)$   **62.** Undefined slope; through $(-4, 1)$

**63.** *Fill in the Blanks*  If a line has slope $-\dfrac{4}{9}$, then any line parallel to it has slope _____, and any line perpendicular to it has slope _____.

**64.** *Fill in the Blanks*  If a line has slope $0.2$, then any line parallel to it has slope _____, and any line perpendicular to it has slope _____.

*Decide whether each pair of lines is* parallel, perpendicular, *or* neither. *See Examples 6 and 7.*

**65.** The line through $(15, 9)$ and $(12, -7)$ and the line through $(8, -4)$ and $(5, -20)$

**66.** The line through $(4, 6)$ and $(-8, 7)$ and the line through $(-5, 5)$ and $(7, 4)$

**67.** $x + 4y = 7$ and $4x - y = 3$   **68.** $2x + 5y = -7$ and $5x - 2y = 1$
**69.** $4x - 3y = 6$ and $3x - 4y = 2$   **70.** $2x + y = 6$ and $x - y = 4$
**71.** $x = 6$ and $6 - x = 8$   **72.** $3x = y$ and $2y - 6x = 5$
**73.** $4x + y = 0$ and $5x - 8 = 2y$   **74.** $2x + 5y = -8$ and $6 + 2x = 5y$
**75.** $2x = y + 3$ and $2y + x = 3$   **76.** $4x - 3y = 8$ and $4y + 3x = 12$

*Find and interpret the average rate of change illustrated in each graph.*

**77.** 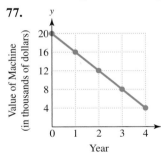   **78.** 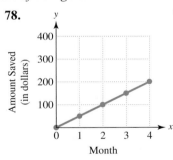   **79.**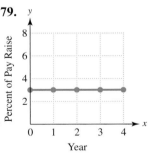

**80.** *Fill in the Blanks*  If the graph of a linear equation rises from left to right, then the average rate of change is _____. If the graph of a linear equation falls from
(positive/negative)
left to right, then the average rate of change is _____.
(positive/negative)

*Concept Check* Solve each problem.

81. When designing the arena now known as TD Banknorth Garden in Boston, architects designed the ramps leading up to the entrances so that circus elephants would be able to walk up the ramps. The maximum grade (or slope) that an elephant will walk on is 13%. Suppose that such a ramp was constructed with a horizontal run of 150 ft. What would be the maximum vertical rise the architects could use?

82. The upper deck at U.S. Cellular Field (formerly Comiskey Park) in Chicago has produced, among other complaints, displeasure with its steepness. It is 160 ft from home plate to the front of the upper deck and 250 ft from home plate to the back. The top of the upper deck is 63 ft above the bottom. What is its slope? (Consider the slope as a positive number here.)

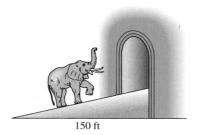
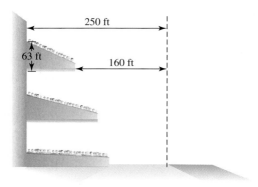

*Solve each problem. See Examples 8 and 9.*

83. The table gives the number of cellular telephone subscribers (in thousands) from 1999 through 2004.

    (a) Find the average rate of change in subscribers for 1999–2000, 2000–2001, and so on.
    (b) Is the average rate of change in subscribers in successive years approximately the same? If the ordered pairs in the table were plotted, could an approximately straight line be drawn through them?

**CELLULAR TELEPHONE SUBSCRIBERS**

| Year | Subscribers (in thousands) |
|---|---|
| 1999 | 86,047 |
| 2000 | 109,478 |
| 2001 | 128,375 |
| 2002 | 140,767 |
| 2003 | 158,722 |
| 2004 | 182,140 |

*Source:* CTIA: The Wireless Association.

84. The table gives book publishers' approximate net dollar sales (in millions) from 1995 through 2000.

    (a) Find the average rate of change for 1995–1996, 1995–1999, and 1998–2000.
    (b) What do you notice about your answers in part (a)? What does this tell you?

**BOOK PUBLISHERS' SALES**

| Year | Sales (in millions) |
|---|---|
| 1995 | 19,000 |
| 1996 | 20,000 |
| 1997 | 21,000 |
| 1998 | 22,000 |
| 1999 | 23,000 |
| 2000 | 24,000 |

*Source:* Book Industry Study Group.

**85.** Personal spending on recreation in the United States (in billions of dollars) in recent years is closely approximated by the graph.

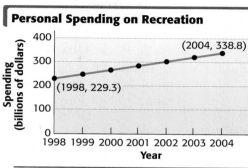

*Source:* U.S. Department of Commerce.

(a) Use the given ordered pairs to determine the average rate of change in these expenditures per year.

(b) Explain how a positive slope is interpreted in this situation.

**86.** The graph provides a good approximation of the number of drive-in theaters in the United States from 2000 through 2004.

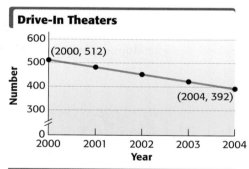

*Source:* Neilsen EDI.

(a) Use the given ordered pairs to find the average rate of change in the number of drive-in theaters per year during this period.

(b) Explain how a negative slope is interpreted in this situation.

**87.** When introduced in 1997, a DVD player sold for about $500. Five years later, the average price was $155. Find and interpret the average rate of change in price per year. (*Source: The Gazette,* Cedar Rapids, Iowa, June 22, 2002.)

**88.** In 1997, when DVD players entered the market, 0.349 million (that is, 349,000) were sold. Five years later, sales of DVD players reached 15.5 million (estimated). Find and interpret the average rate of change in sales, in millions per year. Round your answer to the nearest hundredth. (*Source: The Gazette,* Cedar Rapids, Iowa, June 22, 2002.)

## 3.3 Linear Equations in Two Variables

**OBJECTIVES**

1. Write an equation of a line, given its slope and y-intercept.
2. Graph a line, using its slope and y-intercept.
3. Write an equation of a line, given its slope and a point on the line.
4. Write an equation of a line, given two points on the line.
5. Write an equation of a line parallel or perpendicular to a given line.
6. Write an equation of a line that models real data.

**OBJECTIVE 1** Write an equation of a line, given its slope and y-intercept. In the previous section, we found the slope of a line from the equation of the line by solving the equation for $y$. For example, we found that the slope of the line with equation $y = 4x + 8$ is 4, the coefficient of $x$. What does the number 8 represent?

To find out, suppose a line has slope $m$ and y-intercept $(0, b)$. We can find an equation of this line by choosing another point $(x, y)$ on the line, as shown in Figure 21. Using the slope formula gives

$$m = \frac{y - b}{x - 0}$$

$$m = \frac{y - b}{x}$$

$mx = y - b$     Multiply by $x$.

$mx + b = y$     Add $b$.

$y = mx + b$.     Rewrite.

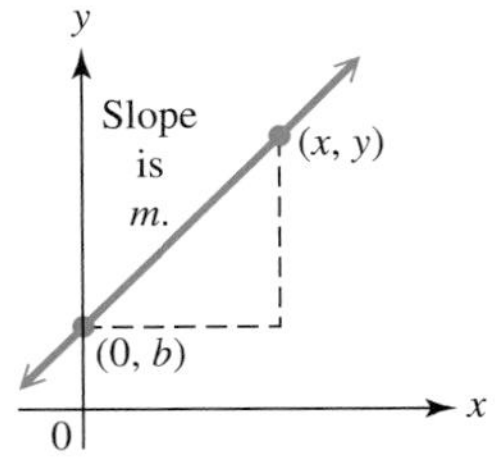

**FIGURE 21**

This last equation is called the *slope-intercept form* of the equation of a line, because we can identify the slope and y-intercept at a glance. Thus, in the line with equation $y = 4x + 8$, the number 8 indicates that the y-intercept is $(0, 8)$.

---

**Slope-Intercept Form**

The **slope-intercept form** of the equation of a line with slope $m$ and y-intercept $(0, b)$ is

$$y = mx + b.$$

↑    ↑
Slope   y-intercept is $(0, b)$.

---

**EXAMPLE 1** Using the Slope-Intercept Form to Find an Equation of a Line

Find an equation of the line with slope $-\frac{4}{5}$ and y-intercept $(0, -2)$.

Here, $m = -\frac{4}{5}$ and $b = -2$. Substitute these values into the slope-intercept form.

$y = mx + b$     Slope-intercept form

$y = -\frac{4}{5}x - 2$     $m = -\frac{4}{5}; b = -2$

**NOW TRY** Exercise 19.

**OBJECTIVE 2** Graph a line, using its slope and y-intercept. If the equation of a line is written in slope-intercept form, we can use the slope and y-intercept to obtain the graph of the equation. (We first saw this approach in Example 5(a) in **Section 3.2**.)

### EXAMPLE 2  Graphing Lines Using Slope and y-Intercept

Graph each line, using the slope and y-intercept.

**(a)** $y = 3x - 6$

Here, $m = 3$ and $b = -6$. Plot the y-intercept $(0, -6)$. The slope 3 can be interpreted as

$$m = \frac{\text{rise}}{\text{run}} = \frac{\text{change in } y}{\text{change in } x} = \frac{3}{1}.$$

From $(0, -6)$, move 3 units *up* and 1 unit to the *right,* and plot a second point at $(1, -3)$. Join the two points with a straight line to obtain the graph in Figure 22.

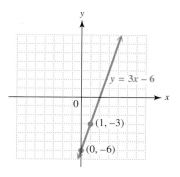

FIGURE 22

**(b)** $3y + 2x = 9$

Write the equation in slope-intercept form by solving for y.

$$3y + 2x = 9$$
$$3y = -2x + 9 \quad \text{Subtract } 2x.$$
$$y = -\frac{2}{3}x + 3 \quad \text{Slope-intercept form}$$

Slope ↑         ↑ y-intercept is (0, 3).

To graph this equation, plot the y-intercept $(0, 3)$. The slope can be interpreted as either $\frac{-2}{3}$ or $\frac{2}{-3}$. Using $\frac{-2}{3}$, begin at $(0, 3)$ and move 2 units *down* and 3 units to the *right* to locate the point $(3, 1)$. The line through these two points is the required graph. See Figure 23. (Verify that the point obtained with $\frac{2}{-3}$ as the slope is also on this line.)

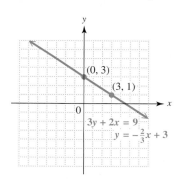

FIGURE 23

**NOW TRY** Exercise 25.

**OBJECTIVE 3** Write an equation of a line, given its slope and a point on the line.

Let $m$ represent the slope of a line and $(x_1, y_1)$ represent a given point on the line. Let $(x, y)$ represent any other point on the line. See Figure 24. Then, by the slope formula,

$$m = \frac{y - y_1}{x - x_1}$$
$$m(x - x_1) = y - y_1 \quad \text{Multiply each side by } x - x_1.$$
$$y - y_1 = m(x - x_1). \quad \text{Rewrite.}$$

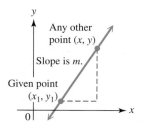

FIGURE 24

This last equation is the *point-slope form* of the equation of a line.

> **Point-Slope Form**
>
> The **point-slope form** of the equation of a line with slope $m$ passing through the point $(x_1, y_1)$ is
>
> $$y - y_1 = m(x - x_1).$$
>
> ↑ Given point ↑ (Slope ↓)

**EXAMPLE 3** Finding an Equation of a Line, Given the Slope and a Point

Find an equation of the line with slope $\frac{1}{3}$ and passing through the point $(-2, 5)$.

*Method 1:* Use the point-slope form of the equation of a line, with $(x_1, y_1) = (-2, 5)$ and $m = \frac{1}{3}$.

$$y - y_1 = m(x - x_1) \qquad \text{Point-slope form}$$
$$y - 5 = \frac{1}{3}[x - (-2)] \qquad y_1 = 5, m = \tfrac{1}{3}, x_1 = -2$$
$$y - 5 = \frac{1}{3}(x + 2)$$
$$3y - 15 = x + 2 \qquad \text{Multiply by 3.}$$
$$-x + 3y = 17 \qquad (*) \qquad \text{Subtract } x; \text{ add 15.}$$

By solving for $y$, we find the slope-intercept form to be $y = \frac{1}{3}x + \frac{17}{3}$.

*Method 2:* An alternative method for finding this equation uses the slope-intercept form, with $(x, y) = (-2, 5)$ and $m = \frac{1}{3}$.

$$y = mx + b \qquad \text{Slope-intercept form}$$
$$5 = \frac{1}{3}(-2) + b \qquad \text{Substitute for } y, m, \text{ and } x.$$
$$5 = -\frac{2}{3} + b$$
$$b = \frac{17}{3} \qquad \text{Solve for } b.$$

Knowing that $m = \frac{1}{3}$ and $b = \frac{17}{3}$ gives the equation $y = \frac{1}{3}x + \frac{17}{3}$.

In **Section 3.1**, we defined *standard form* for a linear equation as

$$Ax + By = C,$$

where $A$, $B$, and $C$ are real numbers. Most often, however, $A$, $B$, and $C$ are integers. In this case, let us agree that integers $A$, $B$, and $C$ have no common factor (except 1) and $A \geq 0$. For example, equation $(*)$ in Example 3, $-x + 3y = 17$, is written in standard form by multiplying each side by $-1$ to obtain $x - 3y = -17$.

**NOW TRY** Exercise 35.

**OBJECTIVE 4** Write an equation of a line, given two points on the line. To find an equation of a line when two points on the line are known, first use the slope formula to find the slope of the line. Then use the slope with either of the given points and the point-slope form of the equation of a line.

### EXAMPLE 4  Finding an Equation of a Line, Given Two Points

Find an equation of the line passing through the points $(-4, 3)$ and $(5, -7)$. Write the equation in standard form.

First find the slope by the slope formula.

$$m = \frac{-7 - 3}{5 - (-4)} = -\frac{10}{9}$$

Use either $(-4, 3)$ or $(5, -7)$ as $(x_1, y_1)$ in the point-slope form of the equation of a line. If you choose $(-4, 3)$, then $-4 = x_1$ and $3 = y_1$.

$y - y_1 = m(x - x_1)$    Point-slope form

$y - 3 = -\frac{10}{9}[x - (-4)]$    $y_1 = 3, m = -\frac{10}{9}, x_1 = -4$

$y - 3 = -\frac{10}{9}(x + 4)$

$9y - 27 = -10x - 40$    Multiply by 9; distributive property.

$10x + 9y = -13$    Standard form

Verify that if $(5, -7)$ were used, the same equation would result.

**NOW TRY** Exercise 51.

A horizontal line has slope 0. Using point-slope form, we find that the equation of a horizontal line through the point $(a, b)$ is

$y - y_1 = m(x - x_1)$    Point-slope form
$y - b = 0(x - a)$    $y_1 = b, m = 0, x_1 = a$
$y - b = 0$    Multiplication property of 0
$y = b.$    Add $b$.

Notice that point-slope form does not apply to a vertical line, since the slope of a vertical line is undefined. A vertical line through the point $(a, b)$ has equation $x = a$.

In summary, horizontal and vertical lines have the following special equations.

### Equations of Horizontal and Vertical Lines

The horizontal line through the point $(a, b)$ has equation **$y = b$**.

The vertical line through the point $(a, b)$ has equation **$x = a$**.

**NOW TRY** Exercises 43 and 45.

**OBJECTIVE 5** Write an equation of a line parallel or perpendicular to a given line. As mentioned in **Section 3.2**, parallel lines have the same slope and perpendicular lines have slopes that are negative reciprocals of each other.

### EXAMPLE 5  Finding Equations of Parallel or Perpendicular Lines

Find an equation of the line passing through the point $(-4, 5)$ and **(a)** parallel to the line $2x + 3y = 6$; **(b)** perpendicular to the line $2x + 3y = 6$. Write each equation in slope-intercept form.

**(a)** We find the slope of the line $2x + 3y = 6$ by solving for $y$.

$$2x + 3y = 6$$
$$3y = -2x + 6 \quad \text{Subtract } 2x.$$
$$y = -\frac{2}{3}x + 2 \quad \text{Divide by 3.}$$
        ↑
        Slope

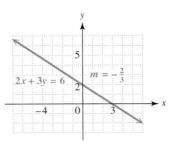

**FIGURE 25**

The slope of the line is given by the coefficient of $x$, so $m = -\frac{2}{3}$. See Figure 25.

The required equation of the line through $(-4, 5)$ and parallel to $2x + 3y = 6$ must also have slope $-\frac{2}{3}$. To find this equation, we use the point-slope form, with $(x_1, y_1) = (-4, 5)$ and $m = -\frac{2}{3}$.

$$y - 5 = -\frac{2}{3}[x - (-4)] \quad y_1 = 5,\ m = -\frac{2}{3},\ x_1 = -4$$

$$y - 5 = -\frac{2}{3}(x + 4)$$

$$y - 5 = -\frac{2}{3}x - \frac{8}{3} \quad \text{Distributive property}$$

$$y = -\frac{2}{3}x - \frac{8}{3} + \frac{15}{3} \quad \text{Add } 5 = \frac{15}{3}.$$

$$y = -\frac{2}{3}x + \frac{7}{3} \quad \text{Combine like terms.}$$

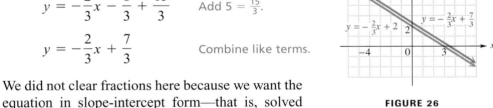

**FIGURE 26**

We did not clear fractions here because we want the equation in slope-intercept form—that is, solved for $y$. Both lines are shown in Figure 26.

**(b)** To be perpendicular to the line $2x + 3y = 6$, a line must have a slope that is the negative reciprocal of $-\frac{2}{3}$, which is $\frac{3}{2}$. We use $(-4, 5)$ and slope $\frac{3}{2}$ in the point-slope form to get the equation of the perpendicular line shown in Figure 27.

$$y - 5 = \frac{3}{2}[x - (-4)] \quad y_1 = 5,\ m = \frac{3}{2},\ x_1 = -4$$

$$y - 5 = \frac{3}{2}(x + 4)$$

$$y - 5 = \frac{3}{2}x + 6 \quad \text{Distributive property}$$

$$y = \frac{3}{2}x + 11 \quad \text{Add 5.}$$

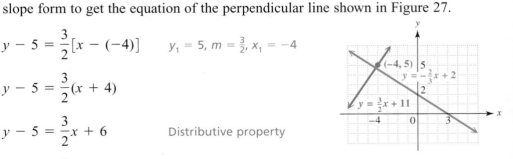

**FIGURE 27**

**NOW TRY** Exercises 63 and 67.

A summary of the various forms of linear equations follows.

### Forms of Linear Equations

| Equation | Description | When to Use |
| --- | --- | --- |
| $y = mx + b$ | **Slope-Intercept Form** <br> Slope is $m$. <br> $y$-intercept is $(0, b)$. | The slope and $y$-intercept can be easily identified and used to quickly graph the equation. |
| $y - y_1 = m(x - x_1)$ | **Point-Slope Form** <br> Slope is $m$. <br> Line passes through $(x_1, y_1)$. | This form is ideal for finding the equation of a line if the slope and a point on the line or two points on the line are known. |
| $Ax + By = C$ | **Standard Form** <br> ($A$, $B$, and $C$ integers, $A \geq 0$) <br> Slope is $-\frac{A}{B}$ ($B \neq 0$). <br> $x$-intercept is $\left(\frac{C}{A}, 0\right)$ ($A \neq 0$). <br> $y$-intercept is $\left(0, \frac{C}{B}\right)$ ($B \neq 0$). | The $x$- and $y$-intercepts can be found quickly and used to graph the equation. The slope must be calculated. |
| $y = b$ | **Horizontal Line** <br> Slope is 0. <br> $y$-intercept is $(0, b)$. | If the graph intersects only the $y$-axis, then $y$ is the only variable in the equation. |
| $x = a$ | **Vertical Line** <br> Slope is undefined. <br> $x$-intercept is $(a, 0)$. | If the graph intersects only the $x$-axis, then $x$ is the only variable in the equation. |

**OBJECTIVE 6** Write an equation of a line that models real data. We can use the information presented in this section to write equations of lines that mathematically describe, or *model*, real data if the given set of data changes at a fairly constant rate. In this case, the data fit a linear pattern, and the rate of change is the slope of the line.

**EXAMPLE 6** Determining a Linear Equation to Describe Real Data

Suppose it is time to fill your car with gasoline. At your local station, 89-octane gas is selling for $3.20 per gal.

**(a)** Write an equation that describes the cost $y$ to buy $x$ gallons of gas.

Experience has taught you that the total price you pay is determined by the number of gallons you buy multiplied by the price per gallon (in this case, $3.20). As you pump the gas, two sets of numbers spin by: the number of gallons pumped and the price for that number of gallons.

The table uses ordered pairs to illustrate this situation.

| Number of Gallons Pumped | Price of This Number of Gallons |
| --- | --- |
| 0 | 0($3.20) = $ 0.00 |
| 1 | 1($3.20) = $ 3.20 |
| 2 | 2($3.20) = $ 6.40 |
| 3 | 3($3.20) = $ 9.60 |
| 4 | 4($3.20) = $12.80 |

If we let $x$ denote the number of gallons pumped, then the total price $y$ in dollars can be found by the linear equation

Total price ⎯⎯⎯⎯ ⎯⎯⎯⎯ Number of gallons
$$y = 3.20x.$$

Theoretically, there are infinitely many ordered pairs $(x, y)$ that satisfy this equation, but here we are limited to nonnegative values for $x$, since we cannot have a negative number of gallons. In this situation, there is also a practical maximum value for $x$ that varies from one car to another. What determines this maximum value?

**(b)** You can also get a car wash at the gas station if you pay an additional $3.00. Write an equation that defines the price for gas and a car wash.

Since an additional $3.00 will be charged, you pay $3.20x + 3.00$ dollars for $x$ gallons of gas and a car wash, or

$$y = 3.2x + 3. \qquad \text{Delete unnecessary zeros.}$$

**(c)** Interpret the ordered pairs $(5, 19)$ and $(10, 35)$ in relation to the equation from part (b).

The ordered pair $(5, 19)$ indicates that the price of 5 gal of gas and a car wash is $19.00. Similarly, $(10, 35)$ indicates that the price of 10 gal of gas and a car wash is $35.00.

> **NOW TRY** Exercises 71 and 75.

In Example 6(a), the ordered pair $(0, 0)$ satisfied the equation, so the linear equation has the form $y = mx$, where $b = 0$. If a realistic situation involves an initial charge plus a charge per unit, as in Example 6(b), the equation has the form $y = mx + b$, where $b \neq 0$.

**EXAMPLE 7  Finding an Equation of a Line That Models Data**

Average annual tuition and fees for in-state students at public four-year colleges are shown in the table for selected years and graphed as ordered pairs of points in the *scatter diagram* in Figure 28, where $x = 0$ represents 1990, $x = 4$ represents 1994, and so on, and $y$ represents the cost in dollars.

| Year | Cost (in dollars) |
|------|-------------------|
| 1990 | 2035 |
| 1994 | 2820 |
| 1996 | 3151 |
| 1998 | 3486 |
| 2000 | 3774 |
| 2002 | 4273 |

*Source:* U.S. National Center for Education Statistics.

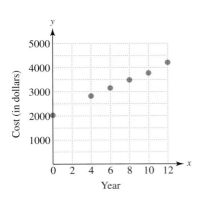

**FIGURE 28**

(a) Find an equation that models the data.

Since the points in Figure 28 lie approximately on a straight line, we can write a linear equation that models the relationship between year $x$ and cost $y$. We choose two data points, $(0, 2035)$ and $(12, 4273)$, to find the slope of the line.

$$m = \frac{4273 - 2035}{12 - 0} = \frac{2238}{12} = 186.5$$

*Start with the x- and y-values of the same point.*

The slope 186.5 indicates that the cost of tuition and fees increased by about $186.50 per year from 1990 to 2002. We use this slope, the $y$-intercept $(0, 2035)$, and the slope-intercept form to write an equation of the line. Thus,

$$y = 186.5x + 2035.$$

(b) Use the equation from part (a) to approximate the cost of tuition and fees at public four-year colleges in 2004.

The value $x = 14$ corresponds to the year 2004.

$$y = 186.5x + 2035$$
$$y = 186.5(14) + 2035 \quad \text{Substitute 14 for } x.$$
$$y = 4646$$

According to the model, average tuition and fees for in-state students at public four-year colleges in 2004 were about $4646.

**NOW TRY** Exercise 81.

---

**EXAMPLE 8** Finding an Equation of a Line That Models Data

Retail spending (in billions of dollars) on prescription drugs in the United States is shown in the graph in Figure 29.

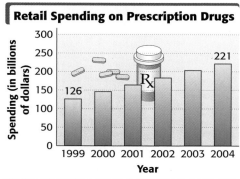

**FIGURE 29**

(a) Write an equation that models the data.

The data shown in the bar graph increase linearly; that is, we could draw a straight line through the tops of any two bars that would be close to the top of each bar.

We can use the data and the point-slope form of the equation of a line to get an equation that models the relationship between year $x$ and spending on prescription drugs, $y$. If we let $x = 9$ represent 1999, $x = 10$ represent 2000, and so on, the given data for 1999 and 2004 can be written as the ordered pairs $(9, 126)$ and $(14, 221)$. The slope of the line through these two points is

$$m = \frac{221 - 126}{14 - 9} = \frac{95}{5} = 19.$$

Thus, retail spending on prescription drugs increased by about $19 billion per year. Using this slope, one of the points, say, $(9, 126)$, and the point-slope form, we obtain the following:

| | |
|---|---|
| $y - y_1 = m(x - x_1)$ | Point-slope form |
| $y - 126 = 19(x - 9)$ | $(x_1, y_1) = (9, 126)$; $m = 19$ |
| $y - 126 = 19x - 171$ | Distributive property |
| $y = 19x - 45.$ | Slope-intercept form |

**Either point can be used here. (14, 221) provides the same answer.**

Thus, retail spending $y$ (in billions of dollars) on prescription drugs in the United States in year $x$ can be approximated by the equation $y = 19x - 45$.

**(b)** Use the equation from part (a) to predict retail spending on prescription drugs in the United States in 2007. (Assume a constant rate of change.)

Since $x = 9$ represents 1999 and 2007 is 8 yr after 1999, $x = 17$ represents 2007.

$y = 19x - 45$
$y = 19(17) - 45$  Substitute 17 for $x$.
$y = 278$

According to the model, $278 billion will be spent on prescription drugs in 2007.

 Exercise 83.

## 3.3 Exercises

 Exercise

*Concept Check* In Exercises 1–6, provide the appropriate response.

**1.** *Multiple Choice* The following equations all represent the same line. Which one is in standard form as defined in the text?

**A.** $3x - 2y = 5$   **B.** $2y = 3x - 5$   **C.** $\frac{3}{5}x - \frac{2}{5}y = 1$   **D.** $3x = 2y + 5$

**2.** *Multiple Choice* Which equation is in point-slope form?

**A.** $y = 6x + 2$   **B.** $4x + y = 9$   **C.** $y - 3 = 2(x - 1)$   **D.** $2y = 3x - 7$

**3.** *Multiple Choice* Which equation in Exercise 2 is in slope-intercept form?

**4.** Write the equation $y + 2 = -3(x - 4)$ in slope-intercept form.

**5.** Write the equation from Exercise 4 in standard form.

**6.** Write the equation $10x - 7y = 70$ in slope-intercept form.

*Matching* Match each equation with the graph that it most closely resembles. (Hint: Determine the signs of m and b to help you make your decision.)

7. $y = 2x + 3$
8. $y = -2x + 3$
9. $y = -2x - 3$
10. $y = 2x - 3$
11. $y = 2x$
12. $y = -2x$
13. $y = 3$
14. $y = -3$

A.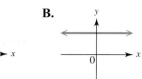
B.
C.
D.
E.
F.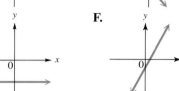
G.
H.
I.

*Find the equation in slope-intercept form of the line satisfying the given conditions. See Example 1.*

15. $m = 5; b = 15$
16. $m = -2; b = 12$
17. $m = -\frac{2}{3}; b = \frac{4}{5}$
18. $m = -\frac{5}{8}; b = -\frac{1}{3}$
19. Slope $\frac{2}{5}$; y-intercept $(0, 5)$
20. Slope $-\frac{3}{4}$; y-intercept $(0, 7)$

*Concept Check* Write an equation in slope-intercept form of the line shown in each graph. (Hint: Use the indicated points to find the slope.)

21.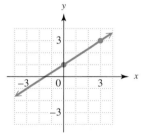
22.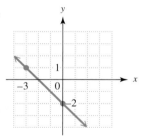

*For each equation, (a) write it in slope-intercept form, (b) give the slope of the line, (c) give the y-intercept, and (d) graph the line. See Example 2.*

23. $-x + y = 4$
24. $-x + y = 6$
25. $6x + 5y = 30$
26. $3x + 4y = 12$
27. $4x - 5y = 20$
28. $7x - 3y = 3$
29. $x + 2y = -4$
30. $x + 3y = -9$
31. $-4x + 3y = 12$

32. *Concept Check* Express the slope $m$ of the graph of $Ax + By = C (A \neq 0)$ in terms of $A$ and $B$.

*Find an equation of the line that satisfies the given conditions. (a) Write the equation in standard form. (b) Write the equation in slope-intercept form. See Example 3.*

33. Through $(5, 8)$; slope $-2$
34. Through $(12, 10)$; slope $1$
35. Through $(-2, 4)$; slope $-\frac{3}{4}$
36. Through $(-1, 6)$; slope $-\frac{5}{6}$

37. Through $(-5, 4)$; slope $\dfrac{1}{2}$
38. Through $(7, -2)$; slope $\dfrac{1}{4}$
39. $x$-intercept $(3, 0)$; slope $4$
40. $x$-intercept $(-2, 0)$; slope $-5$
41. Through $(2, 6.8)$; slope $1.4$
42. Through $(6, -1.2)$; slope $0.8$

*Find an equation of the line that satisfies the given conditions.*

43. Through $(9, 5)$; slope $0$
44. Through $(-4, -2)$; slope $0$
45. Through $(9, 10)$; undefined slope
46. Through $(-2, 8)$; undefined slope
47. Through $(0.5, 0.2)$; vertical
48. Through $\left(\dfrac{5}{8}, \dfrac{2}{9}\right)$; vertical
49. Through $(-7, 8)$; horizontal
50. Through $(2, 7)$; horizontal

*Find an equation of the line passing through the given points.* **(a)** *Write the equation in standard form.* **(b)** *Write the equation in slope-intercept form if possible. See Example 4.*

51. $(3, 4)$ and $(5, 8)$
52. $(5, -2)$ and $(-3, 14)$
53. $(6, 1)$ and $(-2, 5)$
54. $(-2, 5)$ and $(-8, 1)$
55. $\left(-\dfrac{2}{5}, \dfrac{2}{5}\right)$ and $\left(\dfrac{4}{3}, \dfrac{2}{3}\right)$
56. $\left(\dfrac{3}{4}, \dfrac{8}{3}\right)$ and $\left(\dfrac{2}{5}, \dfrac{2}{3}\right)$
57. $(2, 5)$ and $(1, 5)$
58. $(-2, 2)$ and $(4, 2)$
59. $(7, 6)$ and $(7, -8)$
60. $(13, 5)$ and $(13, -1)$
61. $\left(\dfrac{1}{2}, -3\right)$ and $\left(-\dfrac{2}{3}, -3\right)$
62. $\left(-\dfrac{4}{9}, -6\right)$ and $\left(\dfrac{12}{7}, -6\right)$

*Find an equation of the line that satisfies the given conditions.* **(a)** *Write the equation in slope-intercept form.* **(b)** *Write the eqution in standard form. See Example 5.*

63. Through $(7, 2)$; parallel to $3x - y = 8$
64. Through $(4, 1)$; parallel to $2x + 5y = 10$
65. Through $(-2, -2)$; parallel to $-x + 2y = 10$
66. Through $(-1, 3)$; parallel to $-x + 3y = 12$
67. Through $(8, 5)$; perpendicular to $2x - y = 7$
68. Through $(2, -7)$; perpendicular to $5x + 2y = 18$
69. Through $(-2, 7)$; perpendicular to $x = 9$
70. Through $(8, 4)$; perpendicular to $x = -3$

*Write an equation in the form $y = mx$ for each situation. Then give the three ordered pairs associated with the equation for x-values 0, 5, and 10. See Example 6(a).*

71. $x$ represents the number of hours traveling at 45 mph, and $y$ represents the distance traveled (in miles).

72. $x$ represents the number of compact discs sold at $16 each, and $y$ represents the total cost of the discs (in dollars).

73. $x$ represents the number of gallons of gas sold at $3.01 per gal, and $y$ represents the total cost of the gasoline (in dollars).

74. $x$ represents the number of days a DVD movie is rented at $3.50 per day, and $y$ represents the total charge for the rental (in dollars).

*For each situation, **(a)** write an equation in the form $y = mx + b$, **(b)** find and interpret the ordered pair associated with the equation for $x = 5$, and **(c)** answer the question. See Examples 6(b) and 6(c).*

75. A membership in the Midwest Athletic Club costs $99, plus $39 per month. (*Source:* Midwest Athletic Club.) Let $x$ represent the number of months selected. How much does the first year's membership cost?

76. For a family membership, the athletic club in Exercise 75 charges a membership fee of $159, plus $60 for each additional family member after the first. Let $x$ represent the number of additional family members. What is the membership fee for a four-person family?

77. A cell phone plan includes 450 anytime minutes for $35 per month, plus $19.95 for a Nokia 5165 cell phone and $25 for a one-time activation fee. (*Source:* U.S. Cellular.) Let $x$ represent the number of months of service. If you sign a 1-yr contract, how much will this cell phone package cost? (Assume that you never use more than the allotted number of minutes.)

78. Another cell phone plan includes 900 anytime minutes for $50 per month, plus a one-time activation fee of $25. A Nokia 5165 cell phone is included at no additional charge. (*Source:* U.S. Cellular.) Let $x$ represent the number of months of service. If you sign a 2-yr contract, how much will this cell phone plan cost? (Assume that you never use more than the allotted number of minutes.)

79. There is a $30 fee to rent a chain saw, plus $6 per day. Let $x$ represent the number of days the saw is rented and $y$ represent the charge to the user in dollars. If the total charge is $138, for how many days is the saw rented?

80. A rental car costs $50 plus $0.20 per mile. Let $x$ represent the number of miles driven and $y$ represent the total charge to the renter. How many miles was the car driven if the renter paid $84.60?

*Solve each problem. In part (a), give equations in slope-intercept form. See Examples 7 and 8. (Source for Exercises 81 and 82: Jupiter Media Metrix.)*

81. The percent of U.S. households that access the Internet by dial-up is shown in the graph, where the year 2000 corresponds to $x = 0$.

    (a) Use the ordered pairs from the graph to write an equation that models the data. What does the slope tell us in the context of this problem?
    (b) Use the equation from part (a) to predict the percent of U.S. households that were expected to access the Internet by dial-up in 2006. Round your answer to the nearest percent.

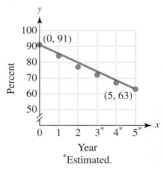

82. The percent of households that access the Internet by high-speed broadband is shown in the graph, where the year 2000 corresponds to $x = 0$.

    (a) Use the ordered pairs from the graph to write an equation that models the data. What does the slope tell us in the context of this problem?
    (b) Use the equation from part (a) to predict the percent of U.S. households that were expected to access the Internet by broadband in 2006. Round your answer to the nearest percent.

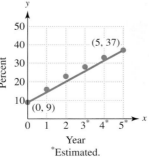

83. Median household income of African-Americans is shown in the bar graph.

    (a) Use the information given for the years 1995 and 2003, letting $x = 5$ represent 1995, $x = 13$ represent 2003, and $y$ represent the median income, to write an equation that models median household income.
    (b) Use the equation to approximate the median income for 1999. How does your result compare with the actual value, $27,910?

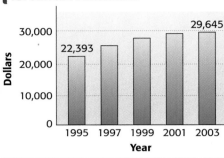

*Source:* U.S. Census Bureau.

84. The number of post offices in the United States is shown in the bar graph.

    (a) Use the information given for the years 1995 and 2000, letting $x = 5$ represent 1995, $x = 10$ represent 2000, and $y$ represent the number of post offices, to write an equation that models the data.
    (b) Use the equation to approximate the number of post offices in 1998. How does this result compare with the actual value, 27,952?

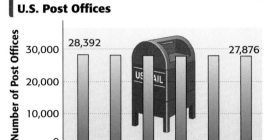

*Source:* U.S. Postal Service, *Annual Report of the Postmaster General.*

## 3.4 Linear Inequalities in Two Variables

**OBJECTIVES**

1. Graph linear inequalities in two variables.
2. Graph the intersection of two linear inequalities.
3. Graph the union of two linear inequalities.

**OBJECTIVE 1** Graph linear inequalities in two variables. In **Chapter 2**, we graphed linear inequalities in one variable on the number line. In this section, we graph linear inequalities in two variables on a rectangular coordinate system.

**Linear Inequality in Two Variables**

An inequality that can be written as

$$Ax + By < C \quad \text{or} \quad Ax + By > C,$$

where $A$, $B$, and $C$ are real numbers and $A$ and $B$ are not both 0, is a **linear inequality in two variables.**

The symbols $\leq$ and $\geq$ may replace $<$ and $>$ in the definition.

Consider the graph in Figure 30. The graph of the line $x + y = 5$ divides the points in the rectangular coordinate system into three sets:

1. Those points that lie on the line itself and satisfy the equation $x + y = 5$ [like $(0, 5)$, $(2, 3)$, and $(5, 0)$];
2. Those that lie in the half-plane above the line and satisfy the inequality $x + y > 5$ [like $(5, 3)$ and $(2, 4)$];
3. Those that lie in the half-plane below the line and satisfy the inequality $x + y < 5$ [like $(0, 0)$ and $(-3, -1)$].

The graph of the line $x + y = 5$ is called the **boundary line** for the inequalities $x + y > 5$ and $x + y < 5$. Graphs of linear inequalities in two variables are *regions* in the real number plane that may or may not include boundary lines.

To graph a linear inequality in two variables, follow these steps.

### Graphing a Linear Inequality

**Step 1** **Draw the graph of the straight line that is the boundary.** Make the line solid if the inequality involves $\leq$ or $\geq$. Make the line dashed if the inequality involves $<$ or $>$.

**Step 2** **Choose a test point.** Choose any point not on the line, and substitute the coordinates of that point in the inequality.

**Step 3** **Shade the appropriate region.** Shade the region that includes the test point if it satisfies the original inequality. Otherwise, shade the region on the other side of the boundary line.

**CAUTION** When drawing the boundary line in Step 1, be careful to draw a solid line if the inequality includes equality ($\leq$, $\geq$) or a dashed line if equality is not included ($<$, $>$). Students often make errors in this step.

### EXAMPLE 1  Graphing a Linear Inequality

Graph $3x + 2y \geq 6$.

**Step 1**   First graph the line $3x + 2y = 6$. The graph of this line, the boundary of the graph of the inequality, is shown in Figure 31.

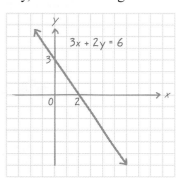

FIGURE 31

**Step 2**   The graph of the inequality $3x + 2y \geq 6$ includes the points of the line $3x + 2y = 6$ and either the points *above* the line $3x + 2y = 6$ or the points *below* that line. To decide which, select any point not on the boundary line $3x + 2y = 6$ as a test point. The origin, $(0, 0)$, is often a good choice because the substitution is easy. Substitute the values from the test point (0, 0) for $x$ and $y$ in the inequality $3x + 2y > 6$.

$$3(0) + 2(0) > 6 \quad ?$$
$$0 > 6 \quad \text{False}$$

**Step 3** Because the result is false, (0, 0) does *not* satisfy the inequality, so the solution set includes all points on the other side of the line. This region is shaded in Figure 32.

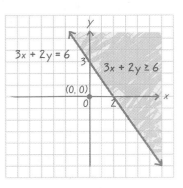

**FIGURE 32**

**NOW TRY** Exercise 7.

If the inequality is written in the form $y > mx + b$ or $y < mx + b$, then the inequality symbol indicates which half-plane to shade.

If $y > mx + b$, then shade **above** the boundary line;

if $y < mx + b$, then shade **below** the boundary line.

**This method works only if the inequality is solved for y.**

> **CAUTION** A common error in using the method just described is to use the original inequality symbol when deciding which half-plane to shade. Be sure to use the inequality symbol found in the inequality *after* it is solved for $y$.

**EXAMPLE 2** Graphing a Linear Inequality

Graph $x - 3y < 4$.

First graph the boundary line, shown in Figure 33. The points of the boundary line do not belong to the inequality $x - 3y < 4$ (because the inequality symbol is $<$, not $\leq$). For this reason, the line is dashed. Now solve the inequality for $y$.

$$x - 3y < 4$$
$$-3y < -x + 4 \qquad \text{Subtract } x.$$
$$y > \frac{1}{3}x - \frac{4}{3} \qquad \text{Multiply by } -\frac{1}{3}; \text{ change } < \text{ to } >.$$

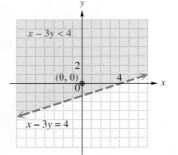

**FIGURE 33**

Because of the *is greater than* symbol that occurs **when the inequality is solved for y,** shade *above* the line. As a check, choose a test point not on the line, say, (0, 0), and substitute for *x* and *y* in the original inequality.

$$0 - 3(0) < 4 \quad ?$$
$$0 < 4 \quad \text{True}$$

This result agrees with the decision to shade above the line. The solution set, graphed in Figure 33, includes only those points in the shaded half-plane (not those on the line).

**NOW TRY** Exercise 9.

**OBJECTIVE 2** Graph the intersection of two linear inequalities. A pair of inequalities joined with the word *and* is interpreted as the intersection of the solution sets of the inequalities. ***The graph of the intersection of two or more inequalities is the region of the plane where all points satisfy all of the inequalities at the same time.***

**EXAMPLE 3** Graphing the Intersection of Two Inequalities

Graph $2x + 4y \geq 5$ and $x \geq 1$.

To begin, we graph each of the two inequalities $2x + 4y \geq 5$ and $x \geq 1$ separately, as shown in Figures 34(a) and (b). Then we use heavy shading to identify the intersection of the graphs, as shown in Figure 34(c).

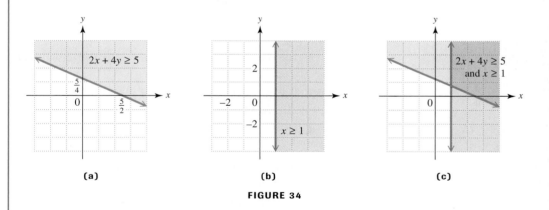

**FIGURE 34**

In practice, the graphs in Figures 34(a) and (b) are graphed on the same axes. To check the graph in Figure 34(c), we use a test point from each of the four regions formed by the intersection of the boundary lines. Verify that only ordered pairs in the heavily shaded region satisfy both inequalities.

**NOW TRY** Exercise 19.

**OBJECTIVE 3** Graph the union of two linear inequalities. When two inequalities are joined by the word *or,* we must find the union of the graphs of the inequalities. ***The graph of the union of two inequalities includes all of the points that satisfy either inequality.***

### EXAMPLE 4 Graphing the Union of Two Inequalities

Graph $2x + 4y \geq 5$ or $x \geq 1$.

The graphs of the two inequalities are shown in Figures 34(a) and (b) in Example 3. The graph of the union is shown in Figure 35.

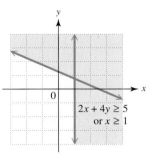

**FIGURE 35**

NOW TRY Exercise 29.

## 3.4 Exercises

*Fill in the Blanks* In Exercises 1–4, fill in the first blank with either *solid* or *dashed*. Fill in the second blank with either *above* or *below*.

1. The boundary of the graph of $y \leq -x + 2$ will be a _____ line, and the shading will be _____ the line.

2. The boundary of the graph of $y < -x + 2$ will be a _____ line, and the shading will be _____ the line.

3. The boundary of the graph of $y > -x + 2$ will be a _____ line, and the shading will be _____ the line.

4. The boundary of the graph of $y \geq -x + 2$ will be a _____ line, and the shading will be _____ the line.

5. How is the boundary line $Ax + By = C$ used in graphing either $Ax + By < C$ or $Ax + By > C$?

6. Describe the two methods discussed in the text for deciding which region is the solution set of a linear inequality in two variables.

*Graph each linear inequality in two variables. See Examples 1 and 2.*

7. $x + y \leq 2$
8. $x + y \leq -3$
9. $4x - y < 4$
10. $3x - y < 3$
11. $x + 3y \geq -2$
12. $x + 4y \geq -3$
13. $x + y > 0$
14. $x + 2y > 0$
15. $x - 3y \leq 0$
16. $x - 5y \leq 0$
17. $y < x$
18. $y \leq 4x$

*Graph each compound inequality. See Example 3.*

**19.** $x + y \leq 1$ and $x \geq 1$
**20.** $x - y \geq 2$ and $x \geq 3$
**21.** $2x - y \geq 2$ and $y < 4$
**22.** $3x - y \geq 3$ and $y < 3$
**23.** $x + y > -5$ and $y < -2$
**24.** $6x - 4y < 10$ and $y > 2$

*Use the method described in* **Section 2.6** *to write each inequality as a compound inequality, and graph its solution set in the rectangular coordinate plane.*

**25.** $|x| < 3$
**26.** $|y| < 5$
**27.** $|x + 1| < 2$
**28.** $|y - 3| < 2$

*Graph each compound inequality. See Example 4.*

**29.** $x - y \geq 1$ or $y \geq 2$
**30.** $x + y \leq 2$ or $y \geq 3$
**31.** $x - 2 > y$ or $x < 1$
**32.** $x + 3 < y$ or $x > 3$
**33.** $3x + 2y < 6$ or $x - 2y > 2$
**34.** $x - y \geq 1$ or $x + y \leq 4$

## 3.5 Introduction to Functions

**OBJECTIVES**

1. Distinguish between independent and dependent variables.
2. Define and identify relations and functions.
3. Find the domain and range.
4. Identify functions defined by graphs and equations.
5. Use function notation.
6. Graph linear and constant functions.

**OBJECTIVE 1** Distinguish between independent and dependent variables. We often describe one quantity in terms of another. Consider the following:

- The amount of your paycheck if you are paid hourly depends on the number of hours you worked.
- The cost at the gas station depends on the number of gallons of gas you pumped into your car.
- The distance traveled by a car moving at a constant speed depends on the time traveled.

We can use ordered pairs to represent these corresponding quantities. For example, we indicate the relationship between the amount of your paycheck and your hours worked by writing ordered pairs in which the first number represents hours worked and the second number represents the paycheck amount in dollars. Then the ordered pair (5, 40) indicates that when you work 5 hr, your paycheck is $40. Similarly, the ordered pairs (10, 80) and (20, 160) show that working 10 hr results in an $80 paycheck and working 20 hr results in a $160 paycheck. In this example, what would the ordered pair (40, 320) indicate?

Since the amount of your paycheck *depends* on the number of hours worked, your paycheck amount is called the *dependent variable,* and the number of hours worked is

called the *independent variable*. Generalizing, if the value of the variable $y$ depends on the value of the variable $x$, then $y$ is the **dependent variable** and $x$ is the **independent variable.**

$$\underset{(x,\ y)}{\underset{\uparrow\qquad\uparrow}{\text{Independent variable}\qquad\text{Dependent variable}}}$$

**OBJECTIVE 2** Define and identify relations and functions. Since we can write related quantities as ordered pairs, a set of ordered pairs such as

$$\{(5, 40), (10, 80), (20, 160), (40, 320)\}$$

is called a *relation*.

> **Relation**
>
> A **relation** is a set of ordered pairs.

A special kind of relation called a *function* is very important in mathematics and its applications.

> **Function**
>
> A **function** is a relation in which, for each value of the first component of the ordered pairs, there is *exactly one value* of the second component.

**EXAMPLE 1** Determining Whether Relations Are Functions

Tell whether each relation defines a function.

$$F = \{(1, 2), (-2, 4), (3, -1)\}$$
$$G = \{(-2, -1), (-1, 0), (0, 1), (1, 2), (2, 2)\}$$
$$H = \{(-4, 1), (-2, 1), (-2, 0)\}$$

Relations $F$ and $G$ are functions, because, for each different $x$-value, there is exactly one $y$-value. Notice that in $G$, the last two ordered pairs have the same $y$-value. (1 is paired with 2, and 2 is paired with 2.) This does not violate the definition of a function, since the first components ($x$-values) are different and each is paired with only one second component ($y$-value).

In relation $H$, however, the last two ordered pairs have the *same x*-value paired with *two different y*-values ($-2$ is paired with both 1 and 0), so $H$ is a relation, but not a function. **In a function, no two ordered pairs can have the same first component and different second components.**

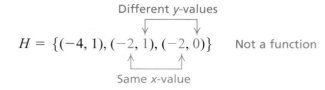

**NOW TRY** Exercises 5 and 7.

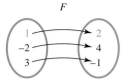

F is a function.

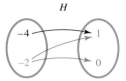

H is not a function.

**FIGURE 36**

Relations and functions can also be expressed as a correspondence or *mapping* from one set to another, as shown in Figure 36 for function F and relation H from Example 1. The arrow from 1 to 2 indicates that the ordered pair (1, 2) belongs to F. Each first component is paired with exactly one second component. In the mapping for set H, which is not a function, the first component −2 is paired with two different second components, 1 and 0.

Since relations and functions are sets of ordered pairs, we can represent them by tables and graphs. A table and graph for function F is shown in Figure 37.

Finally, we can describe a relation or function by using a rule that tells how to determine the dependent variable for a specific value of the independent variable. The rule may be given in words: the dependent variable is twice the independent variable. Usually the rule is given as an equation. For example, if the value of y is twice the value of x, the rule is

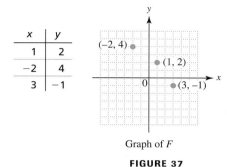

Graph of F

**FIGURE 37**

$$y = 2x.$$

↑ Dependent variable   ↑ Independent variable

Another way to think of a function relationship is to think of the independent variable as an input and the dependent variable as an output. This is illustrated by the input–output (function) machine for the function defined by $y = 2x$.

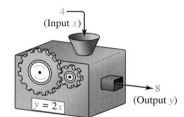

Function machine

*In a function, there is exactly one value of the dependent variable, the second component, for each value of the independent variable, the first component.* This is what makes functions so important in applications.

**OBJECTIVE 3** Find the domain and range.

**Domain and Range**

In a relation, the set of all values of the independent variable (x) is the **domain**. The set of all values of the dependent variable (y) is the **range**.

**EXAMPLE 2** Finding Domains and Ranges of Relations

Give the domain and range of each relation. Tell whether the relation defines a function.

**(a)** $\{(3, -1), (4, 2), (4, 5), (6, 8)\}$

The domain, the set of x-values, is $\{3, 4, 6\}$; the range, the set of y-values, is $\{-1, 2, 5, 8\}$. This relation is not a function because the same x-value 4 is paired with two different y-values, 2 and 5.

**(b)**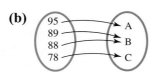

The domain of this relation is

$$\{95, 89, 88, 78\};$$

the range is

$$\{A, B, C\}.$$

This mapping defines a function: Each *x*-value corresponds to exactly one *y*-value.

**(c)**

| x  | y |
|----|---|
| −5 | 2 |
| 0  | 2 |
| 5  | 2 |

This is a table of ordered pairs, so the domain is the set of *x*-values $\{-5, 0, 5\}$ and the range is the set of *y*-values $\{2\}$. The table defines a function because each different *x*-value corresponds to exactly one *y*-value (even though it is the same *y*-value).

**NOW TRY** Exercises 11, 13, and 15.

A graph gives a picture of a relation and can be used to determine its domain and range.

**EXAMPLE 3** Finding Domains and Ranges from Graphs

Give the domain and range of each relation.

**(a)**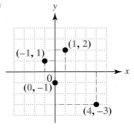

This relation includes the four ordered pairs that are graphed. The domain is the set of *x*-values,

$$\{-1, 0, 1, 4\}.$$

The range is the set of *y*-values,

$$\{-3, -1, 1, 2\}.$$

**(b)**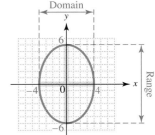

The *x*-values of the points on the graph include all numbers between $-4$ and 4, inclusive. The *y*-values include all numbers between $-6$ and 6, inclusive. In interval notation,

the domain is $[-4, 4]$;
the range is $[-6, 6]$.

**(c)**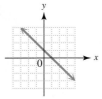

The arrowheads indicate that the line extends indefinitely left and right, as well as up and down. Therefore, both the domain and the range include all real numbers, written $(-\infty, \infty)$.

**(d)**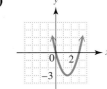

The arrowheads indicate that the graph extends indefinitely left and right, as well as upward. The domain is $(-\infty, \infty)$. Because there is a least *y*-value, $-3$, the range includes all numbers greater than or equal to $-3$, written $[-3, \infty)$.

**NOW TRY** Exercises 17 and 19.

Since relations are often defined by equations, such as $y = 2x + 3$ and $y^2 = x$, we must sometimes determine the domain of a relation from its equation. In this book, we assume the following agreement on the domain of a relation.

### Agreement on Domain

Unless specified otherwise, the domain of a relation is assumed to be all real numbers that produce real numbers when substituted for the independent variable.

To illustrate this agreement, since any real number can be used as a replacement for $x$ in $y = 2x + 3$, its domain is the set of all real numbers. As another example, the function defined by $y = \frac{1}{x}$ has all real numbers except 0 as domain, since $y$ is undefined if $x = 0$. Also, the domain of $y = \sqrt{x}$ is all nonnegative real numbers. *In general, the domain of a function defined by an algebraic expression is all real numbers, except those numbers that lead to division by 0 or an even root of a negative number.*

**OBJECTIVE 4** Identify functions defined by graphs and equations. Since each value of $x$ leads to only one value of $y$ in a function, any vertical line drawn through the graph of a function must intersect the graph in at most one point. This is the *vertical line test* for a function.

### Vertical Line Test

If every vertical line intersects the graph of a relation in no more than one point, then the relation is a function.

For example, the graph shown in Figure 38(a) is not the graph of a function, since a vertical line intersects the graph in more than one point. The graph in Figure 38(b) does represent a function.

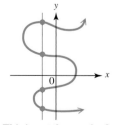

This is not the graph of a function. The same *x*-value corresponds to four different *y*-values.

(a)

This is the graph of a function. Each *x*-value corresponds to only one *y*-value.

(b)

**FIGURE 38**

## EXAMPLE 4  Using the Vertical Line Test

Use the vertical line test to determine whether each relation graphed in Example 3 is a function.

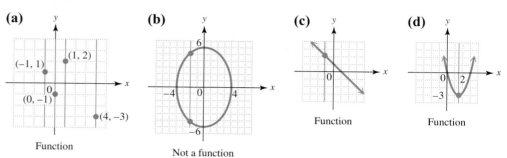

(a) Function  (b) Not a function  (c) Function  (d) Function

The graphs in (a), (c), and (d) represent functions. The graph of the relation in (b) fails the vertical line test, since the same *x*-value corresponds to two different *y*-values; therefore, it is not the graph of a function.

**NOW TRY** Exercise 21.

Graphs that do not represent functions are still relations. *Remember that all equations and graphs represent relations and that all relations have a domain and range.*

The vertical line test is a simple method for identifying a function defined by a graph. It is more difficult to decide whether a relation defined by an equation is a function. The next example gives some hints that may help.

## EXAMPLE 5  Identifying Functions from Their Equations

Decide whether each relation defines *y* as a function of *x*, and give the domain.

**(a)** $y = x + 4$

In the defining equation (or rule) $y = x + 4$, *y* is always found by adding 4 to *x*. Thus, each value of *x* corresponds to just one value of *y*, and the relation defines a function; *x* can be any real number, so the domain is $\{x \mid x \text{ is a real number}\}$, or $(-\infty, \infty)$.

**(b)** $y = \sqrt{2x - 1}$

For any choice of *x* in the domain, there is exactly one corresponding value for *y* (the radical is a nonnegative number), so this equation defines a function. Refer to the agreement on domain stated previously. Since the equation involves a square root, the quantity under the radical sign cannot be negative. Thus, numbers in the domain must satisfy

$$2x - 1 \geq 0$$
$$2x \geq 1$$
$$x \geq \frac{1}{2},$$

and the domain of the function is $\left[\frac{1}{2}, \infty\right)$.

**(c)** $y^2 = x$

The ordered pairs (16, 4) and (16, −4) both satisfy this equation. Since one value of *x*, 16, corresponds to two values of *y*, 4 and −4, this equation does not define a function. (When solved for *y*, *two* equations are obtained: $y = \sqrt{x}$ and $y = -\sqrt{x}$. See **Section 9.1**.) Because *x* is equal to the square of *y*, the values of *x* must always be nonnegative. The domain of the relation is $[0, \infty)$.

**(d)** $y \leq x - 1$

By definition, $y$ is a function of $x$ if every value of $x$ leads to exactly one value of $y$. In this example, a particular value of $x$, say, 1, corresponds to many values of $y$. The ordered pairs $(1, 0)$, $(1, -1)$, $(1, -2)$, $(1, -3)$, and so on all satisfy the inequality. Thus, this relation does not define a function. Any number can be used for $x$, so the domain is the set of real numbers, $(-\infty, \infty)$.

**(e)** $y = \dfrac{5}{x - 1}$

Given any value of $x$ in the domain, we find $y$ by subtracting 1 and then dividing the result into 5. This process produces exactly one value of $y$ for each value in the domain, so the given equation defines a function. The domain includes all real numbers except those which make the denominator 0. We find these numbers by setting the denominator equal to 0 and solving for $x$.

$$x - 1 = 0$$
$$x = 1$$

The domain includes all real numbers *except* 1, written as $(-\infty, 1) \cup (1, \infty)$.

**NOW TRY** Exercises 27, 29, and 35.

In summary, three variations of the definition of a function are given here.

---

**Variations of the Definition of a Function**

1. A **function** is a relation in which, for each value of the first component of the ordered pairs, there is exactly one value of the second component.
2. A **function** is a set of distinct ordered pairs in which no first component is repeated.
3. A **function** is a rule or correspondence that assigns exactly one range value to each domain value.

---

**OBJECTIVE 5** Use function notation. When a function $f$ is defined with a rule or an equation using $x$ and $y$ for the independent and dependent variables, we say, "$y$ is a function of $x$" to emphasize that $y$ depends on $x$. We use the notation

$$y = f(x),$$

called **function notation,** to express this and read $f(x)$ as **"$f$ of $x$."** (In this special notation, the parentheses do not indicate multiplication.) The letter $f$ is a name for this particular function. For example, if $y = 9x - 5$, we can name this function $f$ and write

$$f(x) = 9x - 5.$$

Note that **$f(x)$ is just another name for the dependent variable $y$.** For example, if $y = f(x) = 9x - 5$ and $x = 2$, then we find $y$, or $f(2)$, by replacing $x$ with 2.

$$y = f(2)$$
$$= 9 \cdot 2 - 5$$
$$= 18 - 5$$
$$= 13.$$

The statement "If $x = 2$, then $y = 13$" is represented by the ordered pair $(2, 13)$ and is abbreviated with function notation as

$$f(2) = 13.$$

Read $f(2)$ as "$f$ of 2" or "$f$ at 2." Also,

$$f(0) = 9 \cdot 0 - 5 = -5 \quad \text{and} \quad f(-3) = 9(-3) - 5 = -32.$$

These ideas and the symbols used to represent them can be illustrated as follows.

$$\underset{\underset{\text{Value of the function}}{\uparrow}}{y} = \underset{\text{Name of the function}}{\overset{\frown}{f(x)}} = \underset{\text{Name of the independent variable}}{\overset{\text{Defining expression}}{\overbrace{9x - 5}}}$$

**CAUTION** The symbol $f(x)$ *does not* indicate "$f$ times $x$," but represents the $y$-value associated with the indicated $x$-value. As just shown, $f(2)$ is the $y$-value that corresponds to the $x$-value 2.

### EXAMPLE 6  Using Function Notation

Let $f(x) = -x^2 + 5x - 3$. Find the following.

**(a)** $f(2)$ 

*Do not read this as "$f$ times 2." Read it as "$f$ of 2."*

$$\begin{aligned} f(x) &= -x^2 + 5x - 3 & &\text{The base in } -x^2 \text{ is } x, \text{ not } (-x). \\ f(2) &= -2^2 + 5 \cdot 2 - 3 & &\text{Replace } x \text{ with 2.} \\ &= -4 + 10 - 3 & &\text{Apply the exponent; multiply.} \\ &= 3 & &\text{Add and subtract.} \end{aligned}$$

Thus, $f(2) = 3$, and the ordered pair $(2, 3)$ belongs to $f$.

**(b)** $f(q)$

Replace $x$ with $q$.

$$\begin{aligned} f(x) &= -x^2 + 5x - 3 \\ f(q) &= -q^2 + 5q - 3 \end{aligned}$$

The replacement of one variable with another expression (such as $q$) is important in later courses.

> **NOW TRY** Exercises 41 and 49.

Sometimes letters other than $f$, such as $g$, $h$, or capital letters $F$, $G$, and $H$ are used to name functions.

### EXAMPLE 7  Using Function Notation

Let $g(x) = 2x + 3$. Find and simplify $g(a + 1)$.

$$\begin{aligned} g(x) &= 2x + 3 \\ g(a + 1) &= 2(a + 1) + 3 & &\text{Replace } x \text{ with } a + 1. \\ &= 2a + 2 + 3 \\ &= 2a + 5 \end{aligned}$$

> **NOW TRY** Exercise 53.

### EXAMPLE 8  Using Function Notation

For each function, find $f(3)$.

**(a)** $f(x) = 3x - 7$

$f(3) = 3(3) - 7$   Replace $x$ with 3.

$f(3) = 2$

**(b)** $f = \{(-3, 5), (0, 3), (3, 1), (6, -1)\}$

We want $f(3)$, the $y$-value of the ordered pair whose first component is $x = 3$. As indicated by the ordered pair $(3, 1)$, when $x = 3$, $y = 1$, so $f(3) = 1$.

**(c)**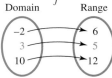

The domain element 3 is paired with 5 in the range, so $f(3) = 5$.

**(d)** The function graphed in Figure 39

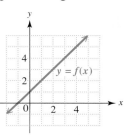

FIGURE 39      FIGURE 40

To evaluate $f(3)$, find 3 on the $x$-axis. See Figure 40. Then move up until the graph of $f$ is reached. Moving horizontally to the $y$-axis gives 4 for the corresponding $y$-value. Thus, $f(3) = 4$.

**NOW TRY** Exercises 61, 63, and 65.

If a function $f$ is defined by an equation with $x$ and $y$, and $y$ is not solved for $x$, use the following steps to find $f(x)$.

### Finding an Expression for $f(x)$

**Step 1** Solve the equation for $y$.

**Step 2** Replace $y$ with $f(x)$.

### EXAMPLE 9  Writing Equations Using Function Notation

Rewrite each equation using function notation $f(x)$. Then find $f(-2)$ and $f(a)$.

**(a)** $y = x^2 + 1$

This equation is already solved for $y$. Since $y = f(x)$,

$$f(x) = x^2 + 1.$$

To find $f(-2)$, let $x = -2$.

$$f(-2) = (-2)^2 + 1$$
$$= 4 + 1$$
$$= 5$$

Find $f(a)$ by letting $x = a$:  $f(a) = a^2 + 1$.

**(b)** $x - 4y = 5$

First solve $x - 4y = 5$ for $y$. Then replace $y$ with $f(x)$.

$$x - 4y = 5$$
$$x - 5 = 4y$$
$$y = \frac{x-5}{4} \quad \text{so} \quad f(x) = \frac{1}{4}x - \frac{5}{4}$$

Now find $f(-2)$ and $f(a)$.

$$f(-2) = \frac{1}{4}(-2) - \frac{5}{4} = -\frac{7}{4} \quad \text{Let } x = -2.$$

$$f(a) = \frac{1}{4}a - \frac{5}{4} \quad \text{Let } x = a.$$

**NOW TRY** Exercise 67.

**OBJECTIVE 6** Graph linear and constant functions. Linear equations (except for vertical lines with equations $x = a$) define *linear functions*.

---

**Linear Function**

A function that can be defined by

$$f(x) = ax + b$$

for real numbers $a$ and $b$ is a **linear function**. The value of $a$ is the slope $m$ of the graph of the function.

---

A linear function defined by $f(x) = b$ (whose graph is a horizontal line) is sometimes called a **constant function**. The domain of any linear function is $(-\infty, \infty)$. The range of a nonconstant linear function is $(-\infty, \infty)$, while the range of the constant function defined by $f(x) = b$ is $\{b\}$.

**EXAMPLE 10** Graphing Linear and Constant Functions

Graph each function. Give the domain and range.

**(a)** $f(x) = \frac{1}{4}x - \frac{5}{4}$

Recall from **Section 3.3** that the graph of $y = \frac{1}{4}x - \frac{5}{4}$ has slope $m = \frac{1}{4}$ and $y$-intercept $\left(0, -\frac{5}{4}\right)$. So here we have

$$f(x) = \frac{1}{4}x - \frac{5}{4}.$$

Slope ↑           ↑ $y$-intercept is $\left(0, -\frac{5}{4}\right)$.

To graph this function, plot the $y$-intercept $\left(0, -\frac{5}{4}\right)$ and use the definition of slope as $\frac{\text{rise}}{\text{run}}$ to find a second point on the line. Since the slope is $\frac{1}{4}$, move 1 unit up from

$\left(0, -\frac{5}{4}\right)$ and 4 units to the right. Draw the straight line through the points to obtain the graph shown in Figure 41. The domain and range are both $(-\infty, \infty)$.

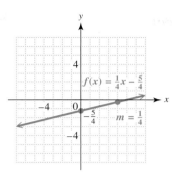

**FIGURE 41**   **FIGURE 42**

**(b)** $f(x) = 4$

This is a constant function. Its graph is the horizontal line containing all points with $y$-coordinate equal to 4. See Figure 42. The domain is $(-\infty, \infty)$ and the range is $\{4\}$.

**NOW TRY** Exercises 75 and 81.

## 3.5 Exercises

1. In your own words, define a function and give an example.

2. In your own words, define the domain of a function and give an example.

3. *Concept Check* In an ordered pair of a relation, is the first element the independent or the dependent variable?

4. *Concept Check* Give an example of a relation that is not a function and that has domain $\{-3, 2, 6\}$ and range $\{4, 6\}$. (There are many possible correct answers.)

*Tell whether each relation defines a function. See Example 1.*

5. $\{(5, 1), (3, 2), (4, 9), (7, 6)\}$
6. $\{(8, 0), (5, 4), (9, 3), (3, 8)\}$
7. $\{(2, 4), (0, 2), (2, 5)\}$
8. $\{(9, -2), (-3, 5), (9, 2)\}$
9. $\{(-3, 1), (4, 1), (-2, 7)\}$
10. $\{(-12, 5), (-10, 3), (8, 3)\}$

*Decide whether each relation defines a function, and give the domain and range. See Examples 1–4.*

11. $\{(1, 1), (1, -1), (0, 0), (2, 4), (2, -4)\}$
12. $\{(2, 5), (3, 7), (4, 9), (5, 11)\}$

13.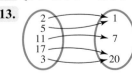
14.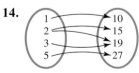

15. 
| x | y |
|---|---|
| 1 | 5 |
| 1 | 2 |
| 1 | -1 |
| 1 | -4 |

16. 
| x | y |
|---|---|
| 4 | -3 |
| 2 | -3 |
| 0 | -3 |
| -2 | -3 |

**17.** 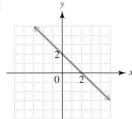  **18.**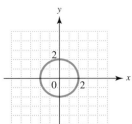

**19.** 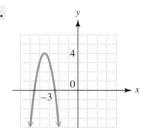  **20.**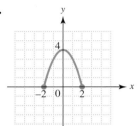

**21.** 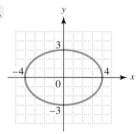  **22.**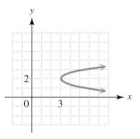

*Decide whether each relation defines y as a function of x. (Solve for y first if necessary.) Give the domain. See Example 5.*

**23.** $y = x^2$   **24.** $y = x^3$   **25.** $x = y^6$   **26.** $x = y^4$

**27.** $y = 2x - 6$   **28.** $y = -6x + 8$   **29.** $x + y < 4$   **30.** $x - y < 3$

**31.** $y = \sqrt{x}$   **32.** $y = -\sqrt{x}$   **33.** $xy = 1$   **34.** $xy = -3$

**35.** $y = \sqrt{4x + 2}$   **36.** $y = \sqrt{9 - 2x}$   **37.** $y = \dfrac{2}{x - 4}$   **38.** $y = \dfrac{-7}{x + 2}$

**39.** *Multiple Choice* Choose the correct response: The notation $f(3)$ means

   **A.** the variable $f$ times 3, or $3f$.
   **B.** the value of the dependent variable when the independent variable is 3.
   **C.** the value of the independent variable when the dependent variable is 3.
   **D.** $f$ equals 3.

**40.** *Fill in the Blanks* Give an example of a function from everyday life. (Hint: _____ depends on _____, so _____ is a function of _____.)

Let $f(x) = -3x + 4$ and $g(x) = -x^2 + 4x + 1$. Find the following. See Examples 6 and 7.

**41.** $f(0)$   **42.** $f(-3)$   **43.** $g(-2)$   **44.** $g(10)$

**45.** $f\left(\dfrac{1}{3}\right)$   **46.** $f\left(\dfrac{7}{3}\right)$   **47.** $g(0.5)$   **48.** $g(1.5)$

**49.** $f(p)$   **50.** $g(k)$   **51.** $f(-x)$   **52.** $g(-x)$

**53.** $f(x + 2)$   **54.** $f(x - 2)$   **55.** $g(\pi)$   **56.** $g(e)$

**57.** $f(x + h)$   **58.** $f(x + h) - f(x)$   **59.** $f(4) - g(4)$   **60.** $f(10) - g(10)$

*For each function, find (a) $f(2)$ and (b) $f(-1)$. See Example 8.*

**61.** $f = \{(-1, 3), (4, 7), (0, 6), (2, 2)\}$   **62.** $f = \{(2, 5), (3, 9), (-1, 11), (5, 3)\}$

**63.** 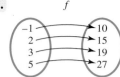      **64.**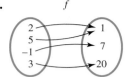

**65.** 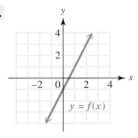      **66.**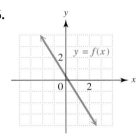

*An equation that defines y as a function f of x is given.* **(a)** *Solve for y in terms of x, and replace y with the function notation f(x).* **(b)** *Find f(3). See Example 9.*

**67.** $x + 3y = 12$      **68.** $x - 4y = 8$      **69.** $y + 2x^2 = 3$

**70.** $y - 3x^2 = 2$      **71.** $4x - 3y = 8$      **72.** $-2x + 5y = 9$

**73.** *Fill in the Blanks*

The equation $2x + y = 4$ has a straight _____ as its graph. One point that lies on the graph is (3, _____). If we solve the equation for y and use function notation, we obtain $f(x) =$ _____. For this function, $f(3) =$ _____, meaning that the point (_____, _____) lies on the graph of the function.

**74.** *Multiple Choice* Which of the following defines y as a linear function of x?

**A.** $y = \dfrac{1}{4}x - \dfrac{5}{4}$     **B.** $y = \dfrac{1}{x}$     **C.** $y = x^2$     **D.** $y = \sqrt{x}$

*Graph each linear function. Give the domain and range. See Example 10.*

**75.** $f(x) = -2x + 5$      **76.** $g(x) = 4x - 1$      **77.** $h(x) = \dfrac{1}{2}x + 2$

**78.** $F(x) = -\dfrac{1}{4}x + 1$      **79.** $G(x) = 2x$      **80.** $H(x) = -3x$

**81.** $g(x) = -4$      **82.** $f(x) = 5$      **83.** $f(x) = 0$

**84.** *Concept Check* What is the name that is usually given to the graph in Exercise 83?

*Solve each problem.*

**85.** A package weighing x pounds costs $f(x)$ dollars to mail to a given location, where

$$f(x) = 2.75x.$$

(a) Evaluate $f(3)$.

(b) In your own words, describe what 3 and the value $f(3)$ mean in part (a), using the terminology *independent variable* and *dependent variable*.

(c) How much would it cost to mail a 5-lb package? Interpret this question and its answer, using function notation.

**86.** Suppose that a taxicab driver charges $2.50 per mile.

(a) Fill in the table with the correct response for the price $f(x)$ he charges for a trip of x miles.

| x | f(x) |
|---|------|
| 0 |      |
| 1 |      |
| 2 |      |
| 3 |      |

(b) *Fill in the Blank* The linear function that gives a rule for the amount charged is $f(x) =$ _____.

(c) Graph this function for the domain $\{0, 1, 2, 3\}$.

87. Forensic scientists use the lengths of certain bones to calculate the height of a person. Two bones often used are the tibia ($t$), the bone from the ankle to the knee, and the femur ($r$), the bone from the knee to the hip socket. A person's height ($h$) is determined from the lengths of these bones by using functions defined by the following formulas. All measurements are in centimeters.

For men:   $h(r) = 69.09 + 2.24r$   or   $h(t) = 81.69 + 2.39t$
For women: $h(r) = 61.41 + 2.32r$   or   $h(t) = 72.57 + 2.53t$

(a) Find the height of a man with a femur measuring 56 cm.
(b) Find the height of a man with a tibia measuring 40 cm.
(c) Find the height of a woman with a femur measuring 50 cm.
(d) Find the height of a woman with a tibia measuring 36 cm.

88. Federal regulations set standards for the size of the quarters of marine mammals. A pool to house sea otters must have a volume of "the square of the sea otter's average adult length (in meters) multiplied by 3.14 and by 0.91 meter." If $x$ represents the sea otter's average adult length and $f(x)$ represents the volume (in cubic meters) of the corresponding pool size, this formula can be written as

$$f(x) = 0.91(3.14)x^2.$$

Find the volume of the pool for each adult sea otter length (in meters). Round answers to the nearest hundredth.

(a) 0.8   (b) 1.0   (c) 1.2   (d) 1.5

89. Refer to the graph to answer the questions.

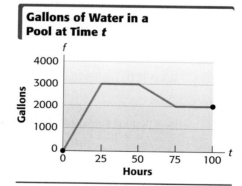

(a) What numbers are possible values of the independent variable? the dependent variable?
(b) For how long is the water level increasing? decreasing?
(c) How many gallons of water are in the pool after 90 hr?
(d) Call this function $f$. What is $f(0)$? What does it mean?
(e) What is $f(25)$? What does it mean?

90. The graph shows the daily number of megawatts of electricity used on a record-breaking summer day in Sacramento, California.

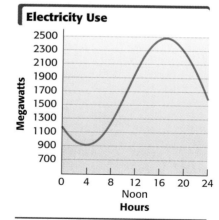

*Source:* Sacramento Municipal Utility District.

(a) Why is this the graph of a function?
(b) What is the domain?
(c) Estimate the number of megawatts used at 8 A.M.
(d) At what time was the most electricity used? the least electricity?
(e) Call this function $f$. What is $f(12)$? What does it mean?

# 3 SUMMARY

## KEY TERMS

**3.1** ordered pair
origin
$x$-axis
$y$-axis
rectangular (Cartesian) coordinate system
plot
coordinate
quadrant

graph of an equation
first-degree equation
linear equation in two variables
standard form
$x$-intercept
$y$-intercept
**3.2** rise
run

slope
**3.3** slope-intercept form
point-slope form
**3.4** linear inequality in two variables
boundary line
**3.5** dependent variable
independent variable
relation

function
domain
range
function notation
linear function
constant function

## NEW SYMBOLS

$(a, b)$ ordered pair

$x_1$ a specific value of the variable $x$ (read "$x$-sub-one"), usually used to distinguish that value from a different value of $x$

$\Delta$ Greek letter delta
$m$ slope

$f(x)$ function of $x$ (read "$f$ of $x$")

## QUICK REVIEW

| CONCEPTS | EXAMPLES |
|---|---|
| **3.1 The Rectangular Coordinate System**<br>**Finding Intercepts**<br>To find the $x$-intercept, let $y = 0$ and solve for $x$.<br>To find the $y$-intercept, let $x = 0$ and solve for $y$. | Find the intercepts of the graph of $2x + 3y = 12$.<br><br>$x$-intercept<br>$2x + 3(0) = 12$<br>$2x = 12$<br>$x = 6$<br>The $x$-intercept is $(6, 0)$.<br><br>$y$-intercept<br>$2(0) + 3y = 12$<br>$3y = 12$<br>$y = 4$<br>The $y$-intercept is $(0, 4)$. |
| **3.2 The Slope of a Line**<br>If $x_2 \neq x_1$, then<br>$$m = \frac{\text{rise}}{\text{run}} = \frac{\text{change in } y}{\text{change in } x} = \frac{\Delta y}{\Delta x} = \frac{y_2 - y_1}{x_2 - x_1}.$$<br><br>A vertical line has undefined slope.<br>A horizontal line has 0 slope. | Find the slope of the graph of $2x + 3y = 12$.<br>Use the intercepts $(6, 0)$ and $(0, 4)$ and the slope formula.<br>$$m = \frac{4 - 0}{0 - 6} = \frac{4}{-6} = -\frac{2}{3} \quad x_1 = 6, y_1 = 0, x_2 = 0, y_2 = 4$$<br><br>The graph of the line $x = 3$ has undefined slope.<br>The graph of the line $y = -5$ has slope $m = 0$. |

*(continued)*

| CONCEPTS | EXAMPLES |
|---|---|
| Parallel lines have equal slopes. | The lines $y = 2x + 3$ and $4x - 2y = 6$ are parallel; both have $m = 2$. $$y = 2x + 3 \qquad 4x - 2y = 6$$ $$m = 2 \qquad -2y = -4x + 6$$ $$y = 2x - 3$$ $$m = 2$$ |
| The slopes of perpendicular lines are negative reciprocals with a product of $-1$. | The lines $y = 3x - 1$ and $x + 3y = 4$ are perpendicular; their slopes are negative reciprocals. $$y = 3x - 1 \qquad x + 3y = 4$$ $$m = 3 \qquad 3y = -x + 4$$ $$y = -\frac{1}{3}x + \frac{4}{3}$$ $$m = -\frac{1}{3}$$ |

## 3.3 Linear Equations in Two Variables

| | |
|---|---|
| **Slope-Intercept Form** $y = mx + b$ | $y = 2x + 3 \qquad m = 2$, $y$-intercept is $(0, 3)$. |
| **Point-Slope Form** $y - y_1 = m(x - x_1)$ | $y - 3 = 4(x - 5) \qquad (5, 3)$ is on the line, $m = 4$. |
| **Standard Form** $Ax + By = C$ ($A, B, C$ integers, $A \geq 0$) | $2x - 5y = 8 \qquad$ Standard form |
| **Horizontal Line** $y = b$ | $y = 4 \qquad$ Horizontal line |
| **Vertical Line** $x = a$ | $x = -1 \qquad$ Vertical line |

## 3.4 Linear Inequalities In Two Variables

**Graphing a Linear Inequality**

**Step 1** Draw the graph of the line that is the boundary. Make the line solid if the inequality involves $\leq$ or $\geq$. Make the line dashed if the inequality involves $<$ or $>$.

**Step 2** Choose any point not on the line as a test point. Substitute the coordinates into the inequality.

**Step 3** Shade the region that includes the test point if the test point satisfies the original inequality. Otherwise, shade the region on the other side of the boundary line.

Graph $2x - 3y \leq 6$.

Draw the graph of $2x - 3y = 6$. Use a solid line because of the inclusion of equality in the symbol $\leq$.

Choose $(0, 0)$, for example.
$2(0) - 3(0) = 0$, and $0 \leq 6 \qquad$ True

Shade the side of the line that includes $(0, 0)$.

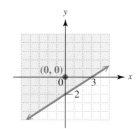

*(continued)*

| CONCEPTS | EXAMPLES |
|---|---|
| **3.5 Introduction to Functions** | |
| A **function** is a set of ordered pairs such that, for each first component, there is one and only one second component. The set of first components is called the **domain**, and the set of second components is called the **range.** | $y = f(x) = x^2$ defines a function $f$ with domain $(-\infty, \infty)$ and range $[0, \infty)$. |
| To evaluate a function $f$, where $f(x)$ defines the range value for a given value of $x$ in the domain, substitute the value wherever $x$ appears. | If $f(x) = x^2 - 7x + 12$, then $$f(1) = 1^2 - 7(1) + 12 = 6.$$ |
| To write an equation that defines a function $f$ in function notation, do the following: **Step 1** Solve the equation for $y$. **Step 2** Replace $y$ with $f(x)$. | Write $2x + 3y = 12$ using notation for a function $f$. $3y = -2x + 12$  Subtract $2x$. $y = -\frac{2}{3}x + 4$  Divide by 3. $f(x) = -\frac{2}{3}x + 4$  Function notation |

# 3 REVIEW EXERCISES

*Complete the table of ordered pairs for each equation. Then graph the equation.*

**1.** $3x + 2y = 10$

| x | y |
|---|---|
| 0 |   |
|   | 0 |
| 2 |   |
|   | -2 |

**2.** $x - y = 8$

| x | y |
|---|---|
| 2 |   |
|   | -3 |
| 3 |   |
|   | -2 |

*Find the x- and y-intercepts and then graph each equation.*

**3.** $4x - 3y = 12$

**4.** $5x + 7y = 28$

**5.** $2x + 5y = 20$

**6.** $x - 4y = 8$

**7.** Explain how the signs of the x- and y-coordinates of a point determine the quadrant in which the point lies.

*Find the slope of each line.*

**8.** Through $(-1, 2)$ and $(4, -5)$

**9.** Through $(0, 3)$ and $(-2, 4)$

**10.** $y = 2x + 3$

**11.** $3x - 4y = 5$

**12.** $x = 5$

**13.** Parallel to $3y = 2x + 5$

**14.** Perpendicular to $3x - y = 4$

**15.** Through $(-1, 5)$ and $(-1, -4)$

**16.** 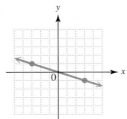     **17.**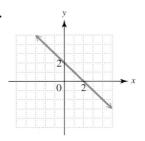

*Tell whether each line has* positive, negative, 0, *or* undefined *slope.*

**18.**    **19.**    **20.**    **21.**

**22.** *Multiple Choice* If a walkway rises 2 ft for every 10 ft on the horizontal, which of the following express its slope (or grade)? (There are several correct choices.)

**A.** 0.2    **B.** $\dfrac{2}{10}$    **C.** $\dfrac{1}{5}$

**D.** 20%    **E.** 5    **F.** $\dfrac{20}{100}$

**G.** 500%    **H.** $\dfrac{10}{2}$    **I.** $-5$

**23.** *Concept Check* If the pitch of a roof is $\tfrac{1}{4}$, how many feet in the horizontal direction correspond to a rise of 3 ft?

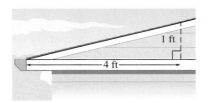

**24.** Family income in the United States has increased steadily for many years (primarily due to inflation). In 1980, the median family income was about $21,000 per year. In 2003, it was about $52,700 per year. Find the average rate of change of median family income to the nearest dollar over that period. (*Source:* U.S. Census Bureau.)

*Find an equation for each line.* **(a)** *Write the equation in slope-intercept form.* **(b)** *Write the equation in standard form.*

**25.** Slope $-\dfrac{1}{3}$; y-intercept $(0, -1)$     **26.** Slope 0; y-intercept $(0, -2)$

**27.** Slope $-\dfrac{4}{3}$; through $(2, 7)$     **28.** Slope 3; through $(-1, 4)$

**29.** Vertical; through $(2, 5)$     **30.** Through $(2, -5)$ and $(1, 4)$

**31.** Through $(-3, -1)$ and $(2, 6)$     **32.** The line pictured in Exercise 17

**33.** Parallel to $4x - y = 3$ and through $(7, -1)$     **34.** Perpendicular to $2x - 5y = 7$ and through $(4, 3)$

**35.** The Midwest Athletic Club (**Section 3.3,** Exercises 75 and 76) offers two special membership plans. (*Source:* Midwest Athletic Club.) For each plan, write a linear equation in slope-intercept form and give the cost *y* in dollars of a 1-yr membership. Let *x* represent the number of months.

(a) Executive VIP/Gold membership: $159 fee, plus $57 per month
(b) Executive Regular/Silver membership: $159 fee, plus $47 per month

**36.** The percent of tax returns filed electronically for the years 1996–2002 is shown in the graph.

(a) Use the information given for the years 1996 and 2002, letting $x = 6$ represent 1996, $x = 12$ represent 2002, and *y* represent the percent of returns filed electronically to find a linear equation that models the data. Write the equation in slope-intercept form. Interpret the slope of the graph.

(b) Use your equation from part (a) to predict the percent of tax returns that were filed electronically in 2005.

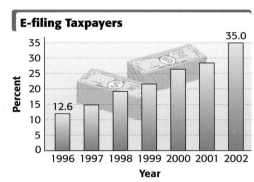

*Source:* Internal Revenue Service.

*Graph the solution set of each inequality or compound inequality.*

**37.** $3x - 2y \le 12$  **38.** $5x - y > 6$
**39.** $2x + y \le 1$ and $x \ge 2y$  **40.** $x \ge 2$ or $y \ge 2$

**41.** *Multiple Choice* Which one of the following has as its graph a dashed boundary line and shading below the line?

**A.** $y \ge 4x + 3$  **B.** $y > 4x + 3$  **C.** $y \le 4x + 3$  **D.** $y < 4x + 3$

*In Exercises 42–45, give the domain and range of each relation. Identify any functions.*

**42.** $\{(-4, 2), (-4, -2), (1, 5), (1, -5)\}$  **43.**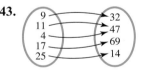

**44.** 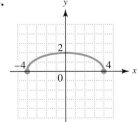  **45.**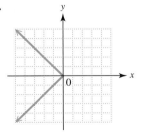

*Determine whether each equation or inequality defines y as a function of x. Give the domain in each case. Identify any linear functions.*

**46.** $y = 3x - 3$  **47.** $y < x + 2$  **48.** $y = |x|$
**49.** $y = \sqrt{4x + 7}$  **50.** $x = y^2$  **51.** $y = \dfrac{7}{x - 6}$

**52.** Explain the test that allows us to determine whether a graph is that of a function.

*Given $f(x) = -2x^2 + 3x - 6$, find each function value or expression.*

**53.** $f(0)$  **54.** $f(2.1)$  **55.** $f\left(-\dfrac{1}{2}\right)$  **56.** $f(k)$

**57.** The table shows life expectancy at birth in the United States for selected years.

(a) Does the table define a function?
(b) What are the domain and range?
(c) Call this function $f$. Give two ordered pairs that belong to $f$.
(d) Find $f(1973)$. What does this mean?
(e) If $f(x) = 75.5$, what does $x$ equal?

| Year | Life Expectancy at Birth (years) |
|---|---|
| 1943 | 63.3 |
| 1953 | 68.8 |
| 1963 | 69.9 |
| 1973 | 71.4 |
| 1983 | 74.6 |
| 1993 | 75.5 |
| 2003 | 77.6 |

*Source:* Centers for Disease Control and Prevention.

**58.** The equation $2x^2 - y = 0$ defines $y$ as a function $f$ of $x$. Write it using function notation, and find $f(3)$.

**59.** *Multiple Choice* Suppose that $2x - 5y = 7$ defines $y$ as a function $f$ of $x$. If $y = f(x)$, which one of the following defines the same function?

**A.** $f(x) = -\dfrac{2}{5}x + \dfrac{7}{5}$    **B.** $f(x) = -\dfrac{2}{5}x - \dfrac{7}{5}$

**C.** $f(x) = \dfrac{2}{5}x - \dfrac{7}{5}$    **D.** $f(x) = \dfrac{2}{5}x + \dfrac{7}{5}$

**60.** Can the graph of a linear function have an undefined slope? Explain.

# CHAPTER 3 TEST

**1.** Complete the table of ordered pairs for the equation $2x - 3y = 12$.

| x | y |
|---|---|
| 1 |   |
| 3 |   |
|   | -4 |

*Find the x- and y-intercepts, and graph each equation.*

**2.** $3x - 2y = 20$    **3.** $y = 5$    **4.** $x = 2$

**5.** Find the slope of the line through the points $(6, 4)$ and $(-4, -1)$.

**6.** *Concept Check* Describe how the graph of a line with undefined slope is situated in a rectangular coordinate system.

*Determine whether each pair of lines is* parallel, perpendicular, *or* neither.

**7.** $5x - y = 8$ and $5y = -x + 3$    **8.** $2y = 3x + 12$ and $3y = 2x - 5$

9. In 1980, there were 119,000 farms in Iowa. As of 2005, there were 89,000. Find and interpret the average rate of change in the number of farms per year. (*Source:* U.S. Department of Agriculture.)

*Find an equation of each line, and write it in* **(a)** *slope-intercept form if possible and* **(b)** *standard form.*

10. Through $(4, -1)$; $m = -5$
11. Through $(-3, 14)$; horizontal
12. Through $(-7, 2)$ and parallel to $3x + 5y = 6$
13. Through $(-7, 2)$ and perpendicular to $y = 2x$
14. Through $(-2, 3)$ and $(6, -1)$
15. Through $(5, -6)$; vertical
16. *Multiple Choice* Which one of the following has positive slope and negative *y*-coordinate for its *y*-intercept?

A. 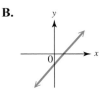  B. 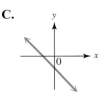  C. 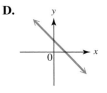  D.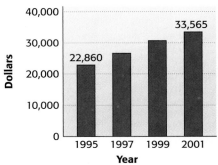

17. The bar graph shows median household income for Hispanics.

    (a) Use the information for the years 1995 and 2001 to find an equation that models the data. Let $x = 5$ represent 1995, $x = 11$ represent 2001, and $y$ represent the median income. Write the equation in slope-intercept form.
    (b) Use the equation from part (a) to approximate median household income for 1999 to the nearest dollar. How does your result compare against the actual value, $30,735?

**Median Household Income for Hispanics**

*Source:* U.S. Census Bureau.

*Graph each inequality or compound inequality.*

18. $3x - 2y > 6$
19. $y < 2x - 1$ and $x - y < 3$

20. *Multiple Choice* Which one of the following is the graph of a function?

A. 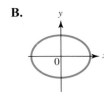  B. 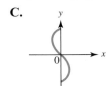  C.   D.

21. *Multiple Choice* Which of the following does not define $y$ as a function of $x$?

   **A.** $\{(0, 1), (-2, 3), (4, 8)\}$    **B.** $y = 2x - 6$    **C.** $y = \sqrt{x + 2}$    **D.**

   | x | y |
   |---|---|
   | 0 | 1 |
   | 3 | 2 |
   | 0 | 2 |
   | 6 | 3 |

*Give the domain and range of the relation shown in each of the following.*

22. Choice A of Exercise 20
23. Choice A of Exercise 21
24. For $f(x) = -x^2 + 2x - 1,$
    
    **(a)** find $f(1)$.    **(b)** find $f(a)$.

25. Graph the linear function defined by $f(x) = \frac{2}{3}x - 1$. What is its domain and range?

# 4

# Systems and Matrices

Have you ever wondered how an object moves in a video game? Computer graphics programs must perform many calculations before finally drawing an object on the screen.

Each corner, or *vertex,* of the object is assigned a set of coordinates. The object is moved across the screen by creating a *matrix* of the coordinates of the vertices, which is then manipulated using special *transformation matrices* to translate, rotate, or scale the original object. Multiplying the original matrix by the transformation matrix generates the new coordinates of the object. When the object is drawn on the screen using the new coordinates, it appears to have moved. You will learn how to multiply matrices in Section 4.4.

**4.1** Systems of Linear Equations in Two Variables

**4.2** Systems of Linear Equations in Three Variables

**4.3** Solving Systems of Linear Equations by Matrix Methods

**4.4** Properties of Matrices

**4.5** Matrix Inverses

**4.6** Determinants and Cramer's Rule

# 4.1 Systems of Linear Equations in Two Variables

**OBJECTIVES**

1. Decide whether an ordered pair is a solution of a linear system.
2. Solve linear systems by graphing.
3. Solve linear systems (with two equations and two variables) by substitution.
4. Solve linear systems (with two equations and two variables) by elimination.
5. Solve special systems.
6. Recognize how a graphing calculator is used to solve a linear system.

In recent years, the sale of digital cameras has increased, while that of conventional cameras has decreased. These trends can be seen in the graph in Figure 1. The two straight-line graphs intersect at the point in time when the two types of cameras had the *same* sales.

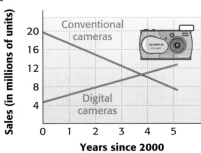

$2.5x + y = 19.4$
$-1.7x + y = 4.4$    Linear system of equations

(Here, $x = 0$ represents 2000, $x = 1$ represents 2001, and so on; $y$ represents sales in millions of units.)

**Source:** Consumer Electronics Association.

**FIGURE 1**

As shown beside Figure 1, we can use a linear equation to model the graph of digital camera sales (the blue equation) and another linear equation to model the graph of conventional camera sales (the red equation). Such a set of equations is called a **system of equations**—in this case, a **linear system of equations.** The point where the graphs in Figure 1 intersect is a solution of each of the individual equations. It is also the solution of the linear system of equations.

**OBJECTIVE 1** Decide whether an ordered pair is a solution of a linear system. The **solution set of a linear system** of equations contains all ordered pairs that satisfy all the equations of the system *at the same time.*

**EXAMPLE 1** Deciding Whether an Ordered Pair Is a Solution

Decide whether the given ordered pair is a solution of the given system.

(a) $x + y = 6$
$4x - y = 14$; $(4, 2)$

Replace $x$ with 4 and $y$ with 2 in each equation of the system.

$x + y = 6$           $4x - y = 14$
$4 + 2 = 6$   ?        $4(4) - 2 = 14$   ?
$6 = 6$    True         $14 = 14$    True

Since $(4, 2)$ makes *both* equations true, $(4, 2)$ is a solution of the system.

**(b)** $3x + 2y = 11$
$x + 5y = 36$ ; $(-1, 7)$

| $3x + 2y = 11$ | | $x + 5y = 36$ | |
|---|---|---|---|
| $3(-1) + 2(7) = 11$ | ? | $-1 + 5(7) = 36$ | ? |
| $-3 + 14 = 11$ | ? | $-1 + 35 = 36$ | ? |
| $11 = 11$ | True | $34 = 36$ | False |

*Use parentheses when substituting to avoid errors.*

The ordered pair $(-1, 7)$ is not a solution of the system, since it does not make *both* equations true.

**NOW TRY** Exercises 11 and 13.

**OBJECTIVE 2** Solve linear systems by graphing. One way to find the solution set of a linear system of equations is to graph each equation and find the point where the graphs intersect.

**EXAMPLE 2** Solving a System by Graphing

Solve the system of equations by graphing.

$$x + y = 5 \quad (1)$$
$$2x - y = 4 \quad (2)$$

When we graph these linear equations as shown in Figure 2, the graph suggests that the point of intersection is the ordered pair $(3, 2)$.

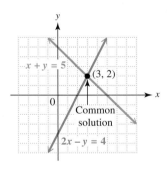

**FIGURE 2**

To be sure that $(3, 2)$ is a solution of *both* equations, we check by substituting 3 for $x$ and 2 for $y$ in each equation.

Check:
| $x + y = 5$ | (1) | $2x - y = 4$ | (2) |
|---|---|---|---|
| $3 + 2 = 5$ | ? | $2(3) - 2 = 4$ | ? |
| $5 = 5$ | True | $6 - 2 = 4$ | ? |
| | | $4 = 4$ | True |

Since $(3, 2)$ makes both equations true, $\{(3, 2)\}$ is the solution set of the system.

**NOW TRY** Exercise 15.

There are three possibilities for the solution set of a linear system in two variables.

> **Graphs of Linear Systems in Two Variables**
>
> 1. **The two graphs intersect in a single point.** The coordinates of this point give the only solution of the system. Since the system has a solution, it is **consistent.** The equations are *not* equivalent, so they are **independent.** See Figure 3(a).
> 2. **The graphs are parallel lines.** There is no solution common to both equations, so the solution set is $\emptyset$ and the system is **inconsistent.** Since the equations are *not* equivalent, they are **independent.** See Figure 3(b).
> 3. **The graphs are the same line.** Since any solution of one equation of the system is a solution of the other, the solution set is an infinite set of ordered pairs representing the points on the line. This type of system is **consistent** because there is a solution. The equations are equivalent, so they are **dependent.** See Figure 3(c).

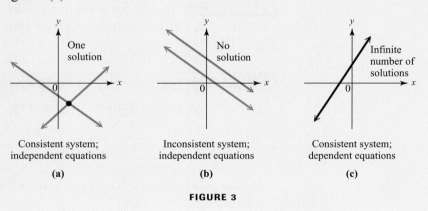

Consistent system; independent equations (a)

Inconsistent system; independent equations (b)

Consistent system; dependent equations (c)

**FIGURE 3**

**OBJECTIVE 3** Solve linear systems (with two equations and two variables) by substitution. Since it can be difficult to read exact coordinates, especially if they are not integers, from a graph, we usually use algebraic methods to solve systems. One such method, the **substitution method,** is most useful for solving linear systems in which one equation is solved or can be easily solved for one variable in terms of the other.

**EXAMPLE 3** Solving a System by Substitution

Solve the system

$$2x - y = 6 \quad (1)$$
$$x = y + 2. \quad (2)$$

Since equation (2) is solved for $x$, substitute $y + 2$ for $x$ in equation (1).

$$2x - y = 6 \quad (1)$$
$$2(y + 2) - y = 6 \quad \text{Let } x = y + 2.$$
$$2y + 4 - y = 6 \quad \text{Distributive property}$$
$$y + 4 = 6 \quad \text{Combine like terms.}$$
$$y = 2 \quad \text{Subtract 4.}$$

Be sure to use parentheses here.

We found $y$. Now find $x$ by substituting $2$ for $y$ in equation (2).

$$x = y + 2 = 2 + 2 = 4$$

Write the $x$-value first in the ordered pair.

Thus, $x = 4$ and $y = 2$, giving the ordered pair $(4, 2)$. Check this solution in both equations of the original system.

Check:
$$2x - y = 6 \quad (1) \qquad\qquad x = y + 2 \quad (2)$$
$$2(4) - 2 = 6 \quad ? \qquad\qquad 4 = 2 + 2 \quad ?$$
$$8 - 2 = 6 \quad ? \qquad\qquad 4 = 4 \quad \text{True}$$
$$6 = 6 \quad \text{True}$$

Since $(4, 2)$ makes both equations true, the solution set is $\{(4, 2)\}$.

**NOW TRY** Exercise 19.

The substitution method is summarized as follows.

---

### Solving a Linear System by Substitution

**Step 1** **Solve one of the equations for either variable.** If one of the equations has a variable term with coefficient 1 or −1, choose that equation since the substitution method is usually easier this way.

**Step 2** **Substitute** for that variable in the other equation. The result should be an equation with just one variable.

**Step 3** **Solve** the equation from Step 2.

**Step 4** **Find the other value.** Substitute the result from Step 3 into the equation from Step 1 to find the value of the other variable.

**Step 5** **Check** the solution in both of the original equations. Then write the solution set.

---

### EXAMPLE 4 Solving a System by Substitution

Solve the system
$$3x + 2y = 13 \quad (1)$$
$$4x - y = -1. \quad (2)$$

**Step 1** Solve one of the equations for either $x$ or $y$. Since the coefficient of $y$ in equation (2) is $-1$, it is easiest to solve for $y$ in equation (2).

$$4x - y = -1 \quad (2)$$
$$-y = -1 - 4x \qquad \text{Subtract } 4x.$$
$$y = 1 + 4x \qquad \text{Multiply by } .$$

**Step 2** Substitute $1 + 4x$ for $y$ in equation (1).

$$3x + 2y = 13 \quad (1)$$
$$3x + 2(1 + 4x) = 13 \qquad \text{Let } y = 1 + 4x.$$

**Step 3** Solve for $x$.

$$3x + 2(1 + 4x) = 13 \quad \text{Equation from Step 2}$$
$$3x + 2 + 8x = 13 \quad \text{Distributive property}$$
$$11x = 11 \quad \text{Combine like terms; subtract 2.}$$
$$x = 1 \quad \text{Divide by 11.}$$

**Step 4** Now find $y$. Since $y = 1 + 4x$ and $x = 1$,

$$y = 1 + 4(1) = 5. \quad \text{Let } x = 1.$$

**Step 5** Check the solution $(1, 5)$ in both equations (1) and (2).

| | |
|---|---|
| $3x + 2y = 13 \quad (1)$ | $4x - y = -1 \quad (2)$ |
| $3(1) + 2(5) = 13 \quad ?$ | $4(1) - 5 = -1 \quad ?$ |
| $3 + 10 = 13 \quad ?$ | $4 - 5 = -1 \quad ?$ |
| $13 = 13 \quad \text{True}$ | $-1 = -1 \quad \text{True}$ |

The solution set is $\{(1, 5)\}$.

**NOW TRY** Exercise 25.

---

**EXAMPLE 5** Solving a System with Fractional Coefficients

Solve the system

$$\frac{2}{3}x - \frac{1}{2}y = \frac{7}{6} \quad (1)$$
$$3x - y = 6. \quad (2)$$

This system will be easier to solve if we clear the fractions in equation (1).

$$6\left(\frac{2}{3}x - \frac{1}{2}y\right) = 6\left(\frac{7}{6}\right) \quad \text{Multiply (1) by the LCD, 6.}$$

*Remember to multiply each term by 6.*

$$6 \cdot \frac{2}{3}x - 6 \cdot \frac{1}{2}y = 6 \cdot \frac{7}{6} \quad \text{Distributive property}$$
$$4x - 3y = 7 \quad (3)$$

Now the system consists of equations (2) and (3). Solve equation (2) for $y$.

$$3x - y = 6 \quad (2)$$
$$-y = 6 - 3x \quad \text{Subtract } 3x.$$
$$y = 3x - 6 \quad \text{Multiply by } -1; \text{ rewrite.}$$

Substitute $3x - 6$ for $y$ in equation (3).

$$4x - 3y = 7 \quad (3)$$
$$4x - 3(3x - 6) = 7 \quad \text{Let } y = 3x - 6.$$
$$4x - 9x + 18 = 7 \quad \text{Distributive property}$$

*Be careful with signs.*

$$-5x + 18 = 7 \quad \text{Combine like terms.}$$
$$-5x = -11 \quad \text{Subtract 18.}$$
$$x = \frac{11}{5} \quad \text{Divide by } -5.$$

Since $y = 3x - 6$ and $x = \frac{11}{5}$,

$$y = 3\left(\frac{11}{5}\right) - 6 = \frac{33}{5} - \frac{30}{5} = \frac{3}{5}.$$

A check verifies that the solution set is $\left\{\left(\frac{11}{5}, \frac{3}{5}\right)\right\}$.

**NOW TRY** Exercise 29.

If an equation in a system contains decimal coefficients, it is best to first clear the decimals by multiplying by an appropriate power of 10, depending on the number of decimal places. Then solve the system. For example, we multiply *each side* of the equation

$$0.5x + 0.75y = 3.25$$

by $10^2$, or 100, to get the equivalent equation

$$50x + 75y = 325.$$

**OBJECTIVE 4** Solve linear systems (with two equations and two variables) by elimination. Another algebraic method, the **elimination method,** involves combining the two equations in a system so that one variable is eliminated. This is done using the following logic:

If $a = b$ and $c = d$, then $a + c = b + d$.

**EXAMPLE 6** Solving a System by Elimination

Solve the system

$$2x + 3y = -6 \quad (1)$$
$$4x - 3y = 6. \quad (2)$$

Notice that adding the equations together will eliminate the variable $y$.

$$\begin{aligned} 2x + 3y &= -6 \quad &(1) \\ \underline{4x - 3y} &= \underline{6} \quad &(2) \\ 6x \phantom{{}+3y} &= 0 \quad &\text{Add.} \\ x &= 0 \quad &\text{Solve for } x. \end{aligned}$$

To find $y$, substitute 0 for $x$ in either equation (1) or equation (2).

$$\begin{aligned} 2x + 3y &= -6 \quad &(1) \\ 2(0) + 3y &= -6 \quad &\text{Let } x = 0. \\ 0 + 3y &= -6 \quad &\text{Multiply.} \\ 3y &= -6 \quad &\text{Add.} \\ y &= -2 \quad &\text{Divide by 3.} \end{aligned}$$

The solution is $(0, -2)$. Check mentally by substituting 0 for $x$ and $-2$ for $y$ in both equations of the original system. The solution set is $\{(0, -2)\}$.

**NOW TRY** Exercise 35.

By adding the equations in Example 6, we eliminated the variable $y$ because the coefficients of the $y$-terms were opposites. In many cases the coefficients will *not* be opposites, and we must transform one or both equations so that the coefficients of one pair of variable terms are opposites.

The elimination method is summarized as follows.

> **Solving a Linear System by Elimination**
>
> *Step 1* **Write** both equations in standard form $Ax + By = C$.
>
> *Step 2* **Make the coefficients of one pair of variable terms opposites.** Multiply one or both equations by appropriate numbers so that the sum of the coefficients of either the $x$- or $y$-terms is 0.
>
> *Step 3* **Add** the new equations to eliminate a variable. The sum should be an equation with just one variable.
>
> *Step 4* **Solve** the equation from Step 3 for the remaining variable.
>
> *Step 5* **Find the other value.** Substitute the result of Step 4 into either of the original equations and solve for the other variable.
>
> *Step 6* **Check** the solution in both of the original equations. Then write the solution set.

**EXAMPLE 7** Solving a System by Elimination

Solve the system

$$5x - 2y = 4 \quad (1)$$
$$2x + 3y = 13. \quad (2)$$

*Step 1* Both equations are in standard form.

*Step 2* Suppose that you wish to eliminate the variable $x$. One way to do this is to multiply equation (1) by 2 and equation (2) by $-5$.

$\quad\quad 10x - 4y = 8 \quad$ 2 times each side of equation (1)
$\quad -10x - 15y = -65 \quad$ $-5$ times each side of equation (2)

*Step 3* Now add.

$$10x - 4y = 8$$
$$\underline{-10x - 15y = -65}$$
$$-19y = -57 \quad \text{Add.}$$

*Step 4* Solve for $y$. $\quad\quad y = 3 \quad$ Divide by $-19$.

*Step 5* To find $x$, substitute 3 for $y$ in either equation (1) or equation (2).

$\quad 2x + 3y = 13 \quad$ (2)
$\quad 2x + 3(3) = 13 \quad$ Let $y = 3$.
$\quad 2x + 9 = 13 \quad$ Multiply.
$\quad 2x = 4 \quad$ Subtract 9.
$\quad x = 2 \quad$ Divide by 2.

*Step 6* The solution is (2, 3). To check, substitute 2 for $x$ and 3 for $y$ in both equations (1) and (2).

$\quad 5x - 2y = 4 \quad$ (1) $\quad\quad 2x + 3y = 13 \quad$ (2)
$\quad 5(2) - 2(3) = 4 \quad$ ? $\quad\quad 2(2) + 3(3) = 13 \quad$ ?
$\quad 10 - 6 = 4 \quad$ ? $\quad\quad 4 + 9 = 13 \quad$ ?
$\quad 4 = 4 \quad$ True $\quad\quad 13 = 13 \quad$ True

The solution set is $\{(2, 3)\}$.

**NOW TRY** Exercise 39.

**OBJECTIVE 5** Solve special systems. As we saw in Figures 3(b) and (c), some systems of linear equations have no solution or an infinite number of solutions.

**EXAMPLE 8** Solving a System of Dependent Equations

Solve the system

$$2x - y = 3 \quad (1)$$
$$6x - 3y = 9. \quad (2)$$

We multiply equation (1) by $-3$ and then add the result to equation (2).

$$-6x + 3y = -9 \quad \text{$-3$ times each side of equation (1)}$$
$$\underline{6x - 3y = \phantom{-}9} \quad (2)$$
$$0 = 0 \quad \text{True}$$

Adding these equations gives the true statement $0 = 0$. In the original system, we could get equation (2) from equation (1) by multiplying equation (1) by 3. Because of this, equations (1) and (2) are equivalent and have the same graph, as shown in Figure 4. The equations are dependent. The solution set is the set of all points on the line with equation $2x - y = 3$, written in set-builder notation **(Section 1.1)** as

$$\{(x, y) \mid 2x - y = 3\}$$

and read "the set of all ordered pairs $(x, y)$, such that $2x - y = 3$."

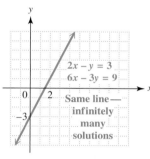

FIGURE 4

**NOW TRY** Exercise 41.

When a system has dependent equations and an infinite number of solutions, as in Example 8, either equation of the system could be used to write the solution set. We prefer to use the equation (in standard form) with coefficients that are integers having no common factor (except 1).

**EXAMPLE 9** Solving an Inconsistent System

Solve the system

$$x + 3y = 4 \quad (1)$$
$$-2x - 6y = 3. \quad (2)$$

Multiply equation (1) by 2, and then add the result to equation (2).

$$2x + 6y = 8 \quad \text{Equation (1) multiplied by 2}$$
$$\underline{-2x - 6y = 3} \quad (2)$$
$$0 = 11 \quad \text{False}$$

The result of the addition step is a false statement, which indicates that the system is inconsistent. As shown in Figure 5, the graphs of the equations of the system are parallel lines. There are no ordered pairs that satisfy both equations, so there is no solution for the system; the solution set is $\emptyset$.

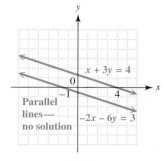

FIGURE 5

**NOW TRY** Exercise 45.

The results of Examples 8 and 9 are generalized as follows.

> **Special Cases of Linear Systems**
>
> If both variables are eliminated when a system of linear equations is solved,
> 1. there are infinitely many solutions if the resulting statement is *true;*
> 2. there is no solution if the resulting statement is *false.*

Slopes and y-intercepts can be used to decide whether the graphs of a system of equations are parallel lines or whether they coincide. In Example 8, writing each equation in slope-intercept form shows that both lines have slope 2 and y-intercept $(0, -3)$, so the graphs are the same line and the system has an infinite solution set.

In Example 9, both equations have slope $-\frac{1}{3}$, but y-intercepts $\left(0, \frac{4}{3}\right)$ and $\left(0, -\frac{1}{2}\right)$, showing that the graphs are two distinct parallel lines. Thus, the system has no solution.

**NOW TRY** Exercises 53 and 55.

**OBJECTIVE 6** Recognize how a graphing calculator is used to solve a linear system. In Example 2, we showed how to solve the system

$$x + y = 5$$
$$2x - y = 4$$

by graphing the two lines and finding their point of intersection. We can also do this with a graphing calculator.

**EXAMPLE 10** Finding the Solution Set of a System from a Graphing Calculator Screen

Solve the system

$$x + y = 5$$
$$2x - y = 4.$$

See Figure 6. The two lines were graphed by solving the first equation to get

$$y = 5 - x$$

and the second to get

$$y = 2x - 4.$$

The coordinates of their point of intersection are displayed at the bottom of the screen, indicating that the solution set is $\{(3, 2)\}$. (Compare this graph with the one in Figure 2.)

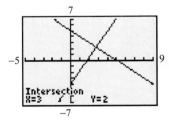

**FIGURE 6**

**NOW TRY** Exercise 67.

## 4.1 Exercises

**NOW TRY Exercise**

*Fill in the Blanks*

1. If $(3, -6)$ is a solution of a linear system in two variables, then substituting _____ for x and _____ for y leads to true statements in *both* equations.

2. A solution of a system of independent linear equations in two variables is a(n) _____.

3. If solving a system leads to a false statement such as $0 = 5$, the solution set is _____.

4. If solving a system leads to a true statement such as $0 = 0$, the system has _____ equations.

5. If the two lines forming a system have the same slope and different $y$-intercepts, the system has _____ solution(s).
(how many?)

6. If the two lines forming a system have different slopes, the system has _____ solution(s).
(how many?)

7. *Multiple Choice* Which ordered pair could be a solution of the graphed system of equations? Why?

   A. $(3, 3)$
   B. $(-3, 3)$
   C. $(-3, -3)$
   D. $(3, -3)$

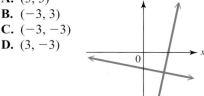

8. *Multiple Choice* Which ordered pair could be a solution of the graphed system of equations? Why?

   A. $(3, 0)$
   B. $(-3, 0)$
   C. $(0, 3)$
   D. $(0, -3)$

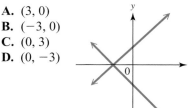

9. *Matching* Match each system with the correct graph.

   (a) $x + y = 6$
       $x - y = 0$

   (b) $x + y = -6$
       $x - y = 0$

   (c) $x + y = 0$
       $x - y = -6$

   (d) $x + y = 0$
       $x - y = 6$

   A. 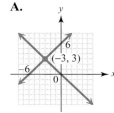    B. 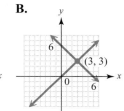    C. 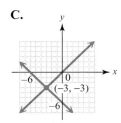    D.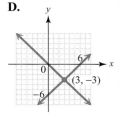

*Decide whether the given ordered pair is a solution of the given system. See Example 1.*

10. $x + y = 6$
    $x - y = 4$; $(5, 1)$

11. $x - y = 17$
    $x + y = -1$; $(8, -9)$

12. $2x - y = 8$
    $3x + 2y = 20$; $(5, 2)$

13. $3x - 5y = -12$
    $x - y = 1$; $(-1, 2)$

*Solve each system by graphing. See Example 2.*

14. $x + y = 4$
    $2x - y = 2$

15. $x + y = -5$
    $-2x + y = 1$

16. $x - 4y = -4$
    $3x + y = 1$

*Solve each system by substitution. If the system is inconsistent or has dependent equations, say so. See Examples 3–5, 8, and 9.*

17. $4x + y = 6$
    $y = 2x$

18. $2x - y = 6$
    $y = 5x$

19. $-x - 4y = -14$
    $y = 2x - 1$

20. $-3x - 5y = -17$
    $y = 4x + 8$

21. $3x - 4y = -22$
    $-3x + y = 0$

22. $-3x + y = -5$
    $x + 2y = 0$

23. $5x - 4y = 9$
    $3 - 2y = -x$

24. $6x - y = -9$
    $4 + 7x = -y$

25. $4x - 5y = -11$
    $x + 2y = 7$

26. $3x - y = 10$
    $2x + 5y = 1$

27. $x = 3y + 5$
    $x = \dfrac{3}{2}y$

28. $x = 6y - 2$
    $x = \dfrac{3}{4}y$

**29.** $\frac{1}{2}x + \frac{1}{3}y = 3$
$-3x + y = 0$

**30.** $\frac{1}{4}x - \frac{1}{5}y = 9$
$5x - y = 0$

**31.** $y = 2x$
$4x - 2y = 0$

**32.** $x = 3y$
$3x - 9y = 0$

**33.** $x = 5y$
$5x - 25y = 5$

**34.** $y = -4x$
$8x + 2y = 4$

*Solve each system by elimination. If the system is inconsistent or has dependent equations, say so. See Examples 5–9.*

**35.** $-2x + 3y = -16$
$2x - 5y = 24$

**36.** $6x + 5y = -7$
$-6x - 11y = 1$

**37.** $2x - 5y = 11$
$3x + y = 8$

**38.** $-2x + 3y = 1$
$-4x + y = -3$

**39.** $3x + 4y = -6$
$5x + 3y = 1$

**40.** $4x + 3y = 1$
$3x + 2y = 2$

**41.** $7x + 2y = 6$
$-14x - 4y = -12$

**42.** $x - 4y = 2$
$4x - 16y = 8$

**43.** $3x + 3y = 0$
$4x + 2y = 3$

**44.** $8x + 4y = 0$
$4x - 2y = 2$

**45.** $5x - 5y = 3$
$x - y = 12$

**46.** $2x - 3y = 7$
$-4x + 6y = 14$

**47.** $x + y = 0$
$2x - 2y = 0$

**48.** $3x + 3y = 0$
$-2x - y = 0$

**49.** $x - \frac{1}{2}y = 2$
$-x + \frac{2}{5}y = -\frac{8}{5}$

**50.** $\frac{3}{2}x + y = 3$
$\frac{2}{3}x + \frac{1}{3}y = 1$

**51.** $\frac{1}{2}x + \frac{1}{3}y = -\frac{1}{3}$
$\frac{1}{2}x + 2y = -7$

**52.** $\frac{1}{5}x + y = \frac{6}{5}$
$\frac{1}{10}x + \frac{1}{3}y = \frac{5}{6}$

*Concept Check* Write each equation in slope-intercept form and then tell how many solutions the system has. Do not actually solve.

**53.** $3x + 7y = 4$
$6x + 14y = 3$

**54.** $-x + 2y = 8$
$4x - 8y = 1$

**55.** $2x = -3y + 1$
$6x = -9y + 3$

**56.** $5x = -2y + 1$
$10x = -4y + 2$

**57.** Assuming that you want to minimize the amount of work required, tell whether you would use the substitution or elimination method to solve each system. Explain your answers. *Do not actually solve.*

**(a)** $6x - y = 5$
$y = 11x$

**(b)** $3x + y = -7$
$x - y = -5$

**(c)** $3x - 2y = 0$
$9x + 8y = 7$

*Solve each system by the method of your choice. See Examples 3–9. (For Exercises 58–60, see your answers to Exercise 57.)*

**58.** $6x - y = 5$
$y = 11x$

**59.** $3x + y = -7$
$x - y = -5$

**60.** $3x - 2y = 0$
$9x + 8y = 7$

**61.** $2x + 3y = 10$
$-3x + y = 18$

**62.** $3x - 5y = 7$
$2x + 3y = 30$

**63.** $\frac{1}{2}x - \frac{1}{8}y = -\frac{1}{4}$
$4x - y = -2$

**64.** $\frac{1}{6}x + \frac{1}{3}y = 8$
$\frac{1}{4}x + \frac{1}{2}y = 12$

**65.** $0.3x + 0.2y = 0.4$
$0.5x + 0.4y = 0.7$

**66.** $0.2x + 0.5y = 6$
$0.4x + y = 9$

*For each system, (a) solve by elimination or substitution and (b) use a graphing calculator to support your result. In part (b), be sure to solve each equation for y first. See Example 10.*

**67.** $x + y = 10$
$2x - y = 5$

**68.** $6x + y = 5$
$-x + y = -9$

**69.** $3x - 2y = 4$
$3x + y = -2$

**70.** $2x - 3y = 3$
$2x + 2y = 8$

*Answer the questions in Exercises 71 and 72 by observing the graphs provided.*

**71.** The figure shows graphs that represent supply and demand for a certain brand of low-fat frozen yogurt at various prices per half-gallon (in dollars).

(a) At what price does supply equal demand?
(b) For how many half-gallons does supply equal demand?
(c) What are the supply and demand at a price of $2 per half-gallon?

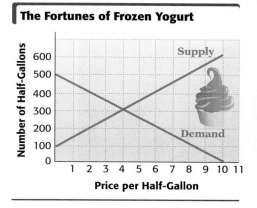

**72.** La Bronda Jones compared the monthly payments she would incur for two types of mortgages: fixed rate and variable rate. Her observations led to the following graphs.

(a) For which years would the monthly payment be more for the fixed-rate mortgage than for the variable-rate mortgage?
(b) In what year would the payments be the same, and what would those payments be?

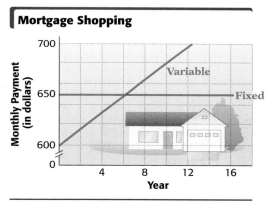

*Use the graph in Figure 1 at the beginning of this section (repeated here) to work Exercises 73–76.*

**73.** For which years during the period 2000–2004 were sales of digital cameras less than sales of conventional cameras?

**74.** Estimate the year in which sales for the two types of cameras were the same. About what was this sales figure?

**75.** If $x = 0$ represents 2000 and $x = 4$ represents 2004, sales ($y$) in millions of units can be modeled by the linear equations in the following system.

$2.5x + y = 19.4$   Conventional cameras
$-1.7x + y = 4.4$   Digital cameras

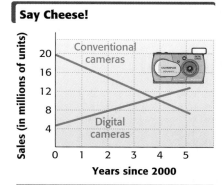

*Source:* Consumer Electronics Association.

Solve this system. Express values to the nearest tenth. Write the solution as an ordered pair of the form (year, sales).

**76.** Interpret your answer for Exercise 75. How does that answer compare with your estimate from Exercise 74?

## 4.2 Systems of Linear Equations in Three Variables

**OBJECTIVES**

1. Understand the geometry of systems of three equations in three variables.
2. Solve linear systems (with three equations and three variables) by elimination.
3. Solve linear systems (with three equations and three variables) in which some of the equations have missing terms.
4. Solve special systems.

A solution of an equation in three variables, such as

$$2x + 3y - z = 4, \quad \text{Linear equation in three variables}$$

is called an **ordered triple** and is written **(x, y, z)**. For example, the ordered triple $(0, 1, -1)$ is a solution of the preceding equation, because

$$2(0) + 3(1) - (-1) = 4$$

is a true statement. Verify that another solution of this equation is $(10, -3, 7)$.

We now extend the term *linear equation* to equations of the form

$$Ax + By + Cz + \cdots + Dw = K,$$

where not all the coefficients $A, B, C, \ldots, D$ equal 0. For example,

$$2x + 3y - 5z = 7 \quad \text{and} \quad x - 2y - z + 3u - 2w = 8$$

are linear equations, the first with three variables and the second with five.

**OBJECTIVE 1** Understand the geometry of systems of three equations in three variables.

Consider the solution of a system such as

$$4x + 8y + z = 2$$
$$x + 7y - 3z = -14 \quad \text{System of linear equations in three variables}$$
$$2x - 3y + 2z = 3.$$

Theoretically, a system of this type can be solved by graphing. However, the graph of a linear equation with three variables is a *plane*, not a line. Since visualizing a plane requires three-dimensional graphing, the method of graphing is not practical with these systems. However, it does illustrate the number of solutions possible for such systems, as shown in Figure 7.

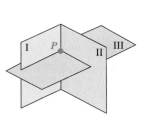

A single solution

(a)

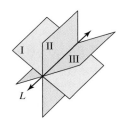

Points of a line in common

(b)

All points in common

(c)

No points in common

(d)

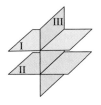

No points in common

(e)

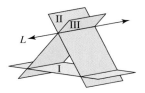

No points in common

(f)

No points in common

(g)

**FIGURE 7**

Figure 7 illustrates the following cases.

> **Graphs of Linear Systems in Three Variables**
>
> 1. **The three planes may meet at a single, common point** that is the solution of the system. See Figure 7(a).
> 2. **The three planes may have the points of a line in common,** so that the infinite set of points that satisfy the equation of the line is the solution of the system. See Figure 7(b).
> 3. **The three planes may coincide,** so that the solution of the system is the set of all points on a plane. See Figure 7(c).
> 4. **The planes may have no points common to all three,** so that there is no solution of the system. See Figures 7(d)–(g).

**OBJECTIVE 2** Solve linear systems (with three equations and three variables) by elimination. Since graphing to find the solution set of a system of three equations in three variables is impractical, these systems are solved with an extension of the elimination method from **Section 4.1,** summarized as follows.

> **Solving a Linear System in Three Variables**
>
> *Step 1* **Eliminate a variable.** Use the elimination method to eliminate any variable from any two of the original equations. The result is an equation in two variables.
>
> *Step 2* **Eliminate the same variable again.** Eliminate the *same* variable from any *other* two equations. The result is an equation in the same two variables as in Step 1.
>
> *Step 3* **Eliminate a different variable and solve.** Use the elimination method to eliminate a second variable from the two equations in two variables that result from Steps 1 and 2. The result is an equation in one variable which gives the value of that variable.
>
> *Step 4* **Find a second value.** Substitute the value of the variable found in Step 3 into either of the equations in two variables to find the value of the second variable.
>
> *Step 5* **Find a third value.** Use the values of the two variables from Steps 3 and 4 to find the value of the third variable by substituting into an appropriate equation.
>
> *Step 6* **Check** the solution in all of the original equations. Then write the solution set.

**EXAMPLE 1** Solving a System in Three Variables

Solve the system

$$4x + 8y + z = 2 \quad (1)$$
$$x + 7y - 3z = -14 \quad (2)$$
$$2x - 3y + 2z = 3. \quad (3)$$

$4x + 8y + z = 2$     (1)
$x + 7y - 3z = -14$     (2)
$2x - 3y + 2z = 3$     (3)

The original system of Example 1

**Step 1**    We must eliminate a variable from the sum of two equations. *The choice of which variable to eliminate is arbitrary.* Suppose we decide to eliminate $z$. We multiply equation (1) by 3 and then add the result to equation (2).

$$12x + 24y + 3z = 6 \quad \text{Multiply each side of (1) by 3.}$$
$$x + 7y - 3z = -14 \quad (2)$$
$$\overline{13x + 31y \phantom{ - 3z} = -8} \quad \text{Add. \quad (4)}$$

**Step 2**    Equation (4) has only two variables. To get another equation without $z$, we multiply equation (1) by $-2$ and add the result to equation (3). *It is essential at this point to eliminate the same variable, $z$.*

$$-8x - 16y - 2z = -4 \quad \text{Multiply each side of (1) by } -2.$$
$$2x - 3y + 2z = 3 \quad (3)$$
$$\overline{-6x - 19y \phantom{ + 2z} = -1} \quad \text{Add. \quad (5)}$$

> Make sure equation (5) has the same variables as equation (4).

**Step 3**    Now we solve the system of equations (4) and (5) for $x$ and $y$. This step is possible only if the *same* variable was eliminated in Steps 1 and 2.

$$78x + 186y = -48 \quad \text{Multiply each side of (4) by 6.}$$
$$-78x - 247y = -13 \quad \text{Multiply each side of (5) by 13.}$$
$$\overline{-61y = -61} \quad \text{Add.}$$
$$y = 1$$

**Step 4**    We substitute 1 for $y$ in either equation (4) or equation (5) to find $x$.

$$-6x - 19y = -1 \quad (5)$$
$$-6x - 19(1) = -1 \quad \text{Let } y = 1.$$
$$-6x - 19 = -1 \quad \text{Multiply.}$$
$$-6x = 18 \quad \text{Add 19.}$$
$$x = -3 \quad \text{Divide by } -6.$$

**Step 5**    We substitute $-3$ for $x$ and 1 for $y$ in any one of the three original equations to find $z$. Choosing (1) gives

$$4x + 8y + z = 2 \quad (1)$$
$$4(-3) + 8(1) + z = 2 \quad \text{Let } x = -3 \text{ and } y = 1.$$
$$-4 + z = 2$$
$$z = 6.$$

> Write the values of $x$, $y$, and $z$ in the correct order.

**Step 6**    It appears that the ordered triple $(-3, 1, 6)$ is the only solution of the system. We must check that this solution satisfies all three equations of the system.

*Check:*    $4x + 8y + z = 2 \quad (1)$
$$4(-3) + 8(1) + 6 = 2 \quad ?$$
$$-12 + 8 + 6 = 2 \quad ?$$
$$2 = 2 \quad \text{True}$$

Because $(-3, 1, 6)$ also satisfies equations (2) and (3), the solution set is $\{(-3, 1, 6)\}$.

**NOW TRY** Exercise 3.

**OBJECTIVE 3** Solve linear systems (with three equations and three variables) in which some of the equations have missing terms. When this situation occurs, one elimination step can be omitted.

### EXAMPLE 2 Solving a System of Equations with Missing Terms

Solve the system

$$6x - 12y = -5 \quad (1)$$
$$8y + z = 0 \quad (2)$$
$$9x - z = 12. \quad (3)$$

Since equation (3) is missing the variable $y$, use equations (1) and (2) to eliminate $y$.

$$12x - 24y = -10 \quad \text{Multiply each side of (1) by 2.}$$
$$\underline{\phantom{12x -\ } 24y + 3z = \phantom{-}0} \quad \text{Multiply each side of (2) by 3.}$$
$$12x \phantom{- 24y} + 3z = -10 \quad \text{Add.} \quad (4)$$

Use this result, together with equation (3), to eliminate $z$. Multiply equation (3) by 3.

$$27x - 3z = 36 \quad \text{Multiply each side of (3) by 3.}$$
$$\underline{12x + 3z = -10} \quad (4)$$
$$39x \phantom{+ 3z} = 26 \quad \text{Add.}$$
$$x = \frac{26}{39} \quad \text{Divide by 39.}$$
$$x = \frac{2}{3} \quad \text{Lowest terms}$$

Substituting this value for $x$ in equation (3) gives

$$9x - z = 12 \quad (3)$$
$$9\left(\frac{2}{3}\right) - z = 12 \quad \text{Let } x = \tfrac{2}{3}.$$
$$6 - z = 12 \quad \text{Multiply.}$$
$$z = -6. \quad \text{Subtract 6; multiply by } -1.$$

Substituting $-6$ for $z$ in equation (2) gives

$$8y + z = 0 \quad (2)$$
$$8y - 6 = 0 \quad \text{Let } z = -6.$$
$$8y = 6 \quad \text{Add 6.}$$
$$y = \frac{6}{8} \quad \text{Divide by 8.}$$
$$y = \frac{3}{4}. \quad \text{Lowest terms}$$

Check in each of the original equations of the system to verify that the solution set is $\left\{\left(\frac{2}{3}, \frac{3}{4}, -6\right)\right\}$.

**NOW TRY** Exercise 21.

**OBJECTIVE 4** Solve special systems. Linear systems with three variables may be inconsistent or may include dependent equations.

**EXAMPLE 3** Solving an Inconsistent System with Three Variables

Solve the system

$$2x - 4y + 6z = 5 \quad (1)$$
$$-x + 3y - 2z = -1 \quad (2)$$
$$x - 2y + 3z = 1. \quad (3)$$

Eliminate $x$ by adding equations (2) and (3) to get the equation

$$y + z = 0.$$

*Eliminate $x$ again,* using equations (1) and (3).

$$-2x + 4y - 6z = -2 \quad \text{Multiply each side of (3) by } -2.$$
$$\underline{2x - 4y + 6z = 5} \quad (1)$$
$$0 = 3 \quad \text{False}$$

The resulting false statement indicates that equations (1) and (3) have no common solution. Thus, the system is inconsistent and the solution set is ∅. The graph of this system would show the two planes parallel to one another.

**NOW TRY** Exercise 27.

**EXAMPLE 4** Solving a System of Dependent Equations with Three Variables

Solve the system

$$2x - 3y + 4z = 8 \quad (1)$$
$$-x + \frac{3}{2}y - 2z = -4 \quad (2)$$
$$6x - 9y + 12z = 24. \quad (3)$$

Multiplying each side of equation (1) by 3 gives equation (3). Multiplying each side of equation (2) by −6 also gives equation (3). Because of this, the equations are dependent. All three equations have the same graph, as illustrated in Figure 7(c). The solution set is written

$$\{(x, y, z) \mid 2x - 3y + 4z = 8\}. \quad \text{Set-builder notation}$$

Although any one of the three equations could be used to write the solution set, we use the equation in standard form with coefficients that are integers with no common factor (except 1), as we did in **Section 4.1**.

**NOW TRY** Exercise 33.

**EXAMPLE 5** Solving Another Special System

Solve the system

$$2x - y + 3z = 6 \quad (1)$$
$$x - \frac{1}{2}y + \frac{3}{2}z = 3 \quad (2)$$
$$4x - 2y + 6z = 1. \quad (3)$$

Multiplying each side of equation (2) by 2 gives equation (1), so these two equations are dependent. Equations (1) and (3) are not equivalent, however. Multiplying equation (3) by $\frac{1}{2}$ does *not* give equation (1). Instead, we obtain two equations with the same coefficients, but with different constant terms. The graphs of these two equations have no points in common (that is, the planes are parallel). Thus, the system is inconsistent and the solution set is $\emptyset$, as illustrated in Figure 7(g).

**NOW TRY** Exercise 37.

## 4.2 Exercises

**1.** *Multiple Choice* The two equations $\begin{array}{l} x + y + z = 6 \\ 2x - y + z = 3 \end{array}$ have a common solution of $(1, 2, 3)$. Which equation would complete a system of three linear equations in three variables having solution set $\{(1, 2, 3)\}$?

**A.** $3x + 2y - z = 1$  **B.** $3x + 2y - z = 4$
**C.** $3x + 2y - z = 5$  **D.** $3x + 2y - z = 6$

**2.** Explain what the following statement means: $\{(-1, 2, 3)\}$ is the solution set of the system

$$2x + y + z = 3$$
$$3x - y + z = -2$$
$$4x - y + 2z = 0.$$

*Solve each system of equations. See Example 1.*

**3.** $2x - 5y + 3z = -1$
$x + 4y - 2z = 9$
$x - 2y - 4z = -5$

**4.** $x + 3y - 6z = 7$
$2x - y + z = 1$
$x + 2y + 2z = -1$

**5.** $3x + 2y + z = 8$
$2x - 3y + 2z = -16$
$x + 4y - z = 20$

**6.** $-3x + y - z = -10$
$-4x + 2y + 3z = -1$
$2x + 3y - 2z = -5$

**7.** $2x + 5y + 2z = 0$
$4x - 7y - 3z = 1$
$3x - 8y - 2z = -6$

**8.** $5x - 2y + 3z = -9$
$4x + 3y + 5z = 4$
$2x + 4y - 2z = 14$

**9.** $x + 2y + z = 4$
$2x + y - z = -1$
$x - y - z = -2$

**10.** $x - 2y + 5z = -7$
$-2x - 3y + 4z = -14$
$-3x + 5y - z = -7$

**11.** $\frac{1}{3}x + \frac{1}{6}y - \frac{2}{3}z = -1$
$-\frac{3}{4}x - \frac{1}{3}y - \frac{1}{4}z = 3$
$\frac{1}{2}x + \frac{3}{2}y + \frac{3}{4}z = 21$

**12.** $\frac{2}{3}x - \frac{1}{4}y + \frac{5}{8}z = 0$
$\frac{1}{5}x + \frac{2}{3}y - \frac{1}{4}z = -7$
$-\frac{3}{5}x + \frac{4}{3}y - \frac{7}{8}z = -5$

**13.** $-x + 2y + 6z = 2$
$3x + 2y + 6z = 6$
$x + 4y - 3z = 1$

**14.** $2x + y + 2z = 1$
$x + 2y + z = 2$
$x - y - z = 0$

**15.** $x + y - z = -2$
$2x - y + z = -5$
$-x + 2y - 3z = -4$

**16.** $x + 2y + 3z = 1$
$-x - y + 3z = 2$
$-6x + y + z = -2$

*Solve each system of equations. See Example 2.*

**17.** $2x - 3y + 2z = -1$
$x + 2y + z = 17$
$2y - z = 7$

**18.** $2x - y + 3z = 6$
$x + 2y - z = 8$
$2y + z = 1$

**19.** $4x + 2y - 3z = 6$
$x - 4y + z = -4$
$-x + 2z = 2$

**20.** $2x + 3y - 4z = 4$
$x - 6y + z = -16$
$-x + 3z = 8$

**21.** $2x + y = 6$
$3y - 2z = -4$
$3x - 5z = -7$

**22.** $4x - 8y = -7$
$4y + z = 7$
$-8x + z = -4$

**23.** $-5x + 2y + z = 5$
$-3x - 2y - z = 3$
$-x + 6y = 1$

**24.** $x + y - z = 0$
$2y - z = 1$
$2x + 3y - 4z = -4$

**25.** $4x - z = -6$
$\frac{3}{5}y + \frac{1}{2}z = 0$
$\frac{1}{3}x + \frac{2}{3}z = -5$

**26.** $5x - 2z = 8$
$4y + 3z = -9$
$\frac{1}{2}x + \frac{2}{3}y = -1$

*Solve each system of equations. If the system is inconsistent or has dependent equations, say so. See Examples 1, 3, 4, and 5.*

**27.** $2x + 2y - 6z = 5$
$-3x + y - z = -2$
$-x - y + 3z = 4$

**28.** $-2x + 5y + z = -3$
$5x + 14y - z = -11$
$7x + 9y - 2z = -5$

**29.** $-5x + 5y - 20z = -40$
$x - y + 4z = 8$
$3x - 3y + 12z = 24$

**30.** $x + 4y - z = 3$
$-2x - 8y + 2z = -6$
$3x + 12y - 3z = 9$

**31.** $x + 5y - 2z = -1$
$-2x + 8y + z = -4$
$3x - y + 5z = 19$

**32.** $x + 3y + z = 2$
$4x + y + 2z = -4$
$5x + 2y + 3z = -2$

**33.** $2x + y - z = 6$
$4x + 2y - 2z = 12$
$-x - \frac{1}{2}y + \frac{1}{2}z = -3$

**34.** $2x - 8y + 2z = -10$
$-x + 4y - z = 5$
$\frac{1}{8}x - \frac{1}{2}y + \frac{1}{8}z = -\frac{5}{8}$

**35.** $x + y - 2z = 0$
$3x - y + z = 0$
$4x + 2y - z = 0$

**36.** $2x + 3y - z = 0$
$x - 4y + 2z = 0$
$3x - 5y - z = 0$

**37.** $x - 2y + \frac{1}{3}z = 4$
$3x - 6y + z = 12$
$-6x + 12y - 2z = -3$

**38.** $4x + y - 2z = 3$
$x + \frac{1}{4}y - \frac{1}{2}z = \frac{3}{4}$
$2x + \frac{1}{2}y - z = 1$

# 4.3 Solving Systems of Linear Equations by Matrix Methods

**OBJECTIVES**

1. Define a matrix.
2. Write the augmented matrix of a system.
3. Use row operations to solve a system with two equations.
4. Use row operations to solve a system with three equations.
5. Use row operations to solve special systems.

**OBJECTIVE 1** Define a matrix. A **matrix** is an ordered array of numbers, such as

$$\text{Rows} \begin{bmatrix} 2 & 3 & 5 \\ 7 & 1 & 2 \end{bmatrix}. \quad \text{Matrix}$$

with Columns indicated above.

The numbers are called **elements** of the matrix. Matrices (the plural of *matrix*) are named according to the number of **rows** and **columns** they contain. The rows are read horizontally, and the columns are read vertically. For example, the first row in the preceding matrix is 2  3  5 and the first column is $\begin{matrix} 2 \\ 7 \end{matrix}$. This matrix is a $2 \times 3$ (read "two by three") matrix, because it has 2 rows and 3 columns. The number of rows followed by the number of columns gives the **dimensions** of the matrix.

$$\begin{bmatrix} -1 & 0 \\ 1 & -2 \end{bmatrix} \quad \begin{matrix} 2 \times 2 \\ \text{matrix} \end{matrix} \qquad \begin{bmatrix} 8 & -1 & -3 \\ 2 & 1 & 6 \\ 0 & 5 & -3 \\ 5 & 9 & 7 \end{bmatrix} \quad \begin{matrix} 4 \times 3 \\ \text{matrix} \end{matrix}$$

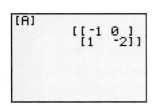

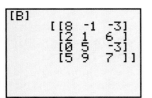

**FIGURE 8**

A **square matrix** is a matrix that has the same number of rows as columns. The $2 \times 2$ matrix above is a square matrix.

Figure 8 shows how a graphing calculator displays the preceding two matrices. Work with matrices is made much easier by the use of technology when it is available. Consult your owner's manual for details.

In this section, we discuss a matrix method of solving linear systems that is really just a very structured way of using the elimination method. The advantage of this new method is that it can be done by a graphing calculator or a computer.

**OBJECTIVE 2** Write the augmented matrix of a system. To solve a linear system using matrices, we begin by writing an *augmented matrix* of the system. An **augmented matrix** has a vertical bar that separates the columns of the matrix into two groups. For example, to solve the system

$$\begin{aligned} x - 3y &= 1 \\ 2x + y &= -5, \end{aligned} \quad \text{start with the augmented matrix} \quad \begin{bmatrix} 1 & -3 & | & 1 \\ 2 & 1 & | & -5 \end{bmatrix}.$$

Place the coefficients of the variables to the left of the bar, and the constants to the right. The bar separates the coefficients from the constants. *The matrix is just a shorthand way of writing the system of equations, so the rows of the augmented matrix can be treated the same as the equations of a system of equations.*

We know that exchanging the positions of two equations in a system does not change the system. Also, multiplying any equation in a system by a nonzero number does not change the system. Comparable changes to the augmented matrix of a system of equations produce new matrices that correspond to systems with the same solutions as the original system.

The following **row operations** produce new matrices that lead to systems having the same solutions as the original system.

> **Matrix Row Operations**
>
> 1. Any two rows of the matrix may be interchanged.
> 2. The elements of any row may be multiplied by any nonzero real number.
> 3. Any row may be changed by adding to the elements of the row the product of a real number and the corresponding elements of another row.

Examples of these row operations follow.

Row operation 1:

$$\begin{bmatrix} 2 & 3 & 9 \\ 4 & 8 & -3 \\ 1 & 0 & 7 \end{bmatrix} \text{ becomes } \begin{bmatrix} 1 & 0 & 7 \\ 4 & 8 & -3 \\ 2 & 3 & 9 \end{bmatrix}.$$

Interchange row 1 and row 3.

Row operation 2:

$$\begin{bmatrix} 2 & 3 & 9 \\ 4 & 8 & -3 \\ 1 & 0 & 7 \end{bmatrix} \text{ becomes } \begin{bmatrix} 6 & 9 & 27 \\ 4 & 8 & -3 \\ 1 & 0 & 7 \end{bmatrix}.$$

Multiply the numbers in row 1 by 3.

Row operation 3:

$$\begin{bmatrix} 2 & 3 & 9 \\ 4 & 8 & -3 \\ 1 & 0 & 7 \end{bmatrix} \text{ becomes } \begin{bmatrix} 0 & 3 & -5 \\ 4 & 8 & -3 \\ 1 & 0 & 7 \end{bmatrix}.$$

Multiply the numbers in row 3 by $-2$; add them to the corresponding numbers in row 1.

The third row operation corresponds to the way we eliminated a variable from a pair of equations in previous sections.

**OBJECTIVE 3** Use row operations to solve a system with two equations. Row operations can be used to rewrite a matrix until it is the matrix of a system whose solution is easy to find. The goal is a matrix in the form

$$\begin{bmatrix} 1 & a & | & b \\ 0 & 1 & | & c \end{bmatrix} \text{ or } \begin{bmatrix} 1 & a & b & | & c \\ 0 & 1 & d & | & e \\ 0 & 0 & 1 & | & f \end{bmatrix}$$

for systems with two and three equations, respectively. Notice that there are 1's down the diagonal from upper left to lower right and 0's below the 1's. A matrix written this way is said to be in **row echelon form.** When these matrices are rewritten as systems of equations, the value of one variable is known and the rest can be found by substitution. The following examples illustrate this method.

### EXAMPLE 1  Using Row Operations to Solve a System with Two Variables

Use row operations to solve the system

$$x - 3y = 1$$
$$2x + y = -5.$$

We start with the augmented matrix of the system.

$$\begin{bmatrix} 1 & -3 & | & 1 \\ 2 & 1 & | & -5 \end{bmatrix} \quad \text{Augmented matrix}$$

Now we use the various row operations to change this matrix into one that leads to a system that is easier to solve.

It is best to work by columns. We start with the first column and make sure that there is a 1 in the first row, first column, position. There already is a 1 in this position. Next, we get 0 in every position below the first. To get a 0 in row two, column one, we use the third row operation and add to the numbers in row two the result of multiplying each number in row one by $-2$. (We abbreviate this as $-2R_1 + R_2$.) Row one remains unchanged.

$$\begin{bmatrix} 1 & -3 & | & 1 \\ 2 + 1(-2) & 1 + (-3)(-2) & | & -5 + 1(-2) \end{bmatrix}$$

↑ Original number from row two    ↑ $-2$ times number from row one

$$\begin{bmatrix} 1 & -3 & | & 1 \\ 0 & 7 & | & -7 \end{bmatrix} \quad -2R_1 + R_2$$

The matrix now has a 1 in the first position of column one, with 0 in every position below the first.

Now we go to column two. The number 1 is needed in row two, column two. We get this 1 by using the second row operation, multiplying each number of row two by $\frac{1}{7}$.

Stop here—this matrix is in row echelon form.

$$\begin{bmatrix} 1 & -3 & | & 1 \\ 0 & 1 & | & -1 \end{bmatrix} \quad \tfrac{1}{7}R_2$$

This augmented matrix leads to the system of equations

$$\begin{array}{l} 1x - 3y = 1 \\ 0x + 1y = -1, \end{array} \quad \text{or} \quad \begin{array}{l} x - 3y = 1 \\ y = -1. \end{array}$$

From the second equation, $y = -1$, we substitute $-1$ for $y$ in the first equation to get

$$x - 3y = 1$$
$$x - 3(-1) = 1 \quad \text{Let } y = -1.$$
$$x + 3 = 1 \quad \text{Multiply.}$$
$$x = -2. \quad \text{Subtract 3.}$$

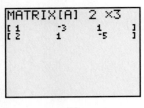

(a)

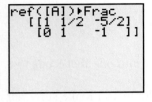

(b)

**FIGURE 9**

The solution set of the system is $\{(-2, -1)\}$. Check this solution by substitution in both equations of the system.

*Write the values of x and y in the correct order.*

**NOW TRY** Exercise 3.

If the augmented matrix of the system in Example 1 is entered as matrix [A] in a graphing calculator (Figure 9(a)) and the row echelon form of the matrix is found (Figure 9(b)), then the system becomes

$$x + \frac{1}{2}y = -\frac{5}{2}$$
$$y = -1.$$

While this system looks different from the one we obtained in Example 1, it is equivalent, since its solution set is also $\{(-2, -1)\}$.

**OBJECTIVE 4** Use row operations to solve a system with three equations. A linear system with three equations is solved in a similar way. We use row operations to get 1's down the diagonal from left to right and all 0's below each 1.

**EXAMPLE 2** Using Row Operations to Solve a System with Three Variables

Use row operations to solve the system

$$x - y + 5z = -6$$
$$3x + 3y - z = 10$$
$$x + 3y + 2z = 5.$$

Start by writing the augmented matrix of the system.

$$\begin{bmatrix} 1 & -1 & 5 & | & -6 \\ 3 & 3 & -1 & | & 10 \\ 1 & 3 & 2 & | & 5 \end{bmatrix} \quad \text{Augmented matrix}$$

This matrix already has 1 in row one, column one. Next get 0's in the rest of column one. First, add to row two the results of multiplying each number of row one by $-3$. This gives the matrix

$$\begin{bmatrix} 1 & -1 & 5 & | & -6 \\ 0 & 6 & -16 & | & 28 \\ 1 & 3 & 2 & | & 5 \end{bmatrix}. \quad -3R_1 + R_2$$

Now add to the numbers in row three the results of multiplying each number of row one by $-1$.

$$\begin{bmatrix} 1 & -1 & 5 & | & -6 \\ 0 & 6 & -16 & | & 28 \\ 0 & 4 & -3 & | & 11 \end{bmatrix} \quad -1R_1 + R_3$$

Introduce 1 in row two, column two, by multiplying each number in row two by $\frac{1}{6}$.

$$\begin{bmatrix} 1 & -1 & 5 & | & -6 \\ 0 & 1 & -\frac{8}{3} & | & \frac{14}{3} \\ 0 & 4 & -3 & | & 11 \end{bmatrix} \quad \frac{1}{6}R_2$$

To obtain 0 in row three, column two, add to row three the results of multiplying each number in row two by $-4$.

$$\begin{bmatrix} 1 & -1 & 5 & | & -6 \\ 0 & 1 & -\frac{8}{3} & | & \frac{14}{3} \\ 0 & 0 & \frac{23}{3} & | & -\frac{23}{3} \end{bmatrix} \quad -4R_2 + R_3$$

Finally, obtain 1 in row three, column three, by multiplying each number in row three by $\frac{3}{23}$.

$$\begin{bmatrix} 1 & -1 & 5 & | & -6 \\ 0 & 1 & -\frac{8}{3} & | & \frac{14}{3} \\ 0 & 0 & 1 & | & -1 \end{bmatrix} \quad \frac{3}{23}R_3$$

This final matrix gives the system of equations

$$x - y + 5z = -6$$
$$y - \frac{8}{3}z = \frac{14}{3}$$
$$z = -1.$$

Substitute $-1$ for $z$ in the second equation, $y - \frac{8}{3}z = \frac{14}{3}$, to find that $y = 2$. Finally, substitute 2 for $y$ and $-1$ for $z$ in the first equation, $x - y + 5z = -6$, to determine that $x = 1$. The solution set of the original system is $\{(1, 2, -1)\}$. Check by substitution.

**NOW TRY** Exercise 15.

**OBJECTIVE 5** Use row operations to solve special systems. In the final example, we show how to recognize inconsistent systems or systems with dependent equations when solving these systems with row operations.

**EXAMPLE 3** Recognizing Inconsistent Systems or Dependent Equations

Use row operations to solve each system.

**(a)** $\quad 2x - 3y = 8$
$\quad\;\, -6x + 9y = 4$

$$\begin{bmatrix} 2 & -3 & | & 8 \\ -6 & 9 & | & 4 \end{bmatrix} \quad \text{Write the augmented matrix.}$$

$$\begin{bmatrix} 1 & -\frac{3}{2} & | & 4 \\ -6 & 9 & | & 4 \end{bmatrix} \quad \frac{1}{2}R_1$$

$$\begin{bmatrix} 1 & -\frac{3}{2} & | & 4 \\ 0 & 0 & | & 28 \end{bmatrix} \quad 6R_1 + R_2$$

The corresponding system of equations is

$$x - \frac{3}{2}y = 4$$
$$0 = 28, \quad \text{False}$$

which has no solution and is inconsistent. The solution set is $\emptyset$.

**(b)** $-10x + 12y = 30$
$\phantom{-}5x - 6y = -15$

$$\begin{bmatrix} -10 & 12 & | & 30 \\ 5 & -6 & | & -15 \end{bmatrix} \quad \text{Write the augmented matrix.}$$

$$\begin{bmatrix} 1 & -\frac{6}{5} & | & -3 \\ 5 & -6 & | & -15 \end{bmatrix} \quad -\frac{1}{10}R_1$$

$$\begin{bmatrix} 1 & -\frac{6}{5} & | & -3 \\ 0 & 0 & | & 0 \end{bmatrix} \quad -5R_1 + R_2$$

The corresponding system is

$$x - \frac{6}{5}y = -3$$
$$0 = 0, \quad \text{True}$$

which has dependent equations. Using the second equation of the given system, which is in standard form, we write the solution set as

$$\{(x, y) \mid 5x - 6y = -15\}.$$

**NOW TRY** Exercises 11 and 13.

## 4.3 Exercises

1. *Concept Check* Consider the matrix $\begin{bmatrix} -2 & 3 & 1 \\ 0 & 5 & -3 \\ 1 & 4 & 8 \end{bmatrix}$ and answer the following.

   **(a)** What are the elements of the second row?
   **(b)** What are the elements of the third column?
   **(c)** Is this a square matrix? Explain why or why not.
   **(d)** Give the matrix obtained by interchanging the first and third rows.
   **(e)** Give the matrix obtained by multiplying the first row by $-\frac{1}{2}$.
   **(f)** Give the matrix obtained by multiplying the third row by 3 and adding to the first row.

2. *Concept Check* Give the dimensions of each matrix.

   **(a)** $\begin{bmatrix} 3 & -7 \\ 4 & 5 \\ -1 & 0 \end{bmatrix}$ **(b)** $\begin{bmatrix} 4 & 9 & 0 \\ -1 & 2 & -4 \end{bmatrix}$ **(c)** $\begin{bmatrix} 6 & 3 \\ -2 & 5 \\ 4 & 10 \\ 1 & -1 \end{bmatrix}$

**SECTION 4.3** Solving Systems of Linear Equations by Matrix Methods

*Complete the steps in the matrix solution of each system by filling in the boxes. Give the final system and the solution set. See Example 1.*

**3.** $\begin{aligned} 4x + 8y &= 44 \\ 2x - y &= -3 \end{aligned} \longrightarrow \begin{bmatrix} 4 & 8 & | & 44 \\ 2 & -1 & | & -3 \end{bmatrix}$

$\begin{bmatrix} 1 & \blacksquare & | & \blacksquare \\ 2 & -1 & | & -3 \end{bmatrix} \quad \frac{1}{4}R_1$

$\begin{bmatrix} 1 & 2 & | & 11 \\ 0 & \blacksquare & | & \blacksquare \end{bmatrix} \quad -2R_1 + R_2$

$\begin{bmatrix} 1 & 2 & | & 11 \\ 0 & 1 & | & \blacksquare \end{bmatrix} \quad -\frac{1}{5}R_2$

**4.** $\begin{aligned} 2x - 5y &= -1 \\ 3x + y &= 7 \end{aligned} \longrightarrow \begin{bmatrix} 2 & -5 & | & -1 \\ 3 & 1 & | & 7 \end{bmatrix}$

$\begin{bmatrix} 1 & -\frac{5}{2} & | & \blacksquare \\ 3 & 1 & | & 7 \end{bmatrix} \quad \frac{1}{2}R_1$

$\begin{bmatrix} 1 & -\frac{5}{2} & | & -\frac{1}{2} \\ 0 & \blacksquare & | & \blacksquare \end{bmatrix} \quad -3R_1 + R_2$

$\begin{bmatrix} 1 & -\frac{5}{2} & | & -\frac{1}{2} \\ 0 & 1 & | & \blacksquare \end{bmatrix} \quad \frac{2}{17}R_2$

*Use row operations to solve each system. See Examples 1 and 3.*

**5.** $\begin{aligned} x + y &= 5 \\ x - y &= 3 \end{aligned}$

**6.** $\begin{aligned} x + 2y &= 7 \\ x - y &= -2 \end{aligned}$

**7.** $\begin{aligned} 2x + 4y &= 6 \\ 3x - y &= 2 \end{aligned}$

**8.** $\begin{aligned} 4x + 5y &= -7 \\ x - y &= 5 \end{aligned}$

**9.** $\begin{aligned} 3x + 4y &= 13 \\ 2x - 3y &= -14 \end{aligned}$

**10.** $\begin{aligned} 5x + 2y &= 8 \\ 3x - y &= 7 \end{aligned}$

**11.** $\begin{aligned} -4x + 12y &= 36 \\ x - 3y &= 9 \end{aligned}$

**12.** $\begin{aligned} 2x - 4y &= 8 \\ -3x + 6y &= 5 \end{aligned}$

**13.** $\begin{aligned} 2x + y &= 4 \\ 4x + 2y &= 8 \end{aligned}$

**14.** $\begin{aligned} -3x - 4y &= 1 \\ 6x + 8y &= -2 \end{aligned}$

*Complete the steps in the matrix solution of each system by filling in the boxes. Give the final system and the solution set. See Example 2.*

**15.** $\begin{aligned} x + y - z &= -3 \\ 2x + y + z &= 4 \\ 5x - y + 2z &= 23 \end{aligned}$

$\begin{bmatrix} 1 & 1 & -1 & | & -3 \\ 2 & 1 & 1 & | & 4 \\ 5 & -1 & 2 & | & 23 \end{bmatrix}$

$\begin{bmatrix} 1 & 1 & -1 & | & -3 \\ 0 & \blacksquare & \blacksquare & | & \blacksquare \\ 0 & \blacksquare & \blacksquare & | & \blacksquare \end{bmatrix} \quad \begin{matrix} -2R_1 + R_2 \\ -5R_1 + R_3 \end{matrix}$

$\begin{bmatrix} 1 & 1 & -1 & | & -3 \\ 0 & 1 & \blacksquare & | & \blacksquare \\ 0 & -6 & 7 & | & 38 \end{bmatrix} \quad -1R_2$

$\begin{bmatrix} 1 & 1 & -1 & | & -3 \\ 0 & 1 & -3 & | & -10 \\ 0 & 0 & \blacksquare & | & \blacksquare \end{bmatrix} \quad 6R_2 + R_3$

$\begin{bmatrix} 1 & 1 & -1 & | & -3 \\ 0 & 1 & -3 & | & -10 \\ 0 & 0 & 1 & | & \blacksquare \end{bmatrix} \quad -\frac{1}{11}R_3$

**16.** $\begin{aligned} 2x + y + 2z &= 11 \\ 2x - y - z &= -3 \\ 3x + 2y + z &= 9 \end{aligned}$

$\begin{bmatrix} 2 & 1 & 2 & | & 11 \\ 2 & -1 & -1 & | & -3 \\ 3 & 2 & 1 & | & 9 \end{bmatrix}$

$\begin{bmatrix} 1 & \blacksquare & \blacksquare & | & \blacksquare \\ 2 & -1 & -1 & | & -3 \\ 3 & 2 & 1 & | & 9 \end{bmatrix} \quad \frac{1}{2}R_1$

$\begin{bmatrix} 1 & \frac{1}{2} & 1 & | & \frac{11}{2} \\ 0 & \blacksquare & \blacksquare & | & \blacksquare \\ 0 & \blacksquare & \blacksquare & | & \blacksquare \end{bmatrix} \quad \begin{matrix} -2R_1 + R_2 \\ -3R_1 + R_3 \end{matrix}$

$\begin{bmatrix} 1 & \frac{1}{2} & 1 & | & \frac{11}{2} \\ 0 & 1 & \blacksquare & | & \blacksquare \\ 0 & \frac{1}{2} & -2 & | & -\frac{15}{2} \end{bmatrix} \quad -\frac{1}{2}R_1$

$\begin{bmatrix} 1 & \frac{1}{2} & 1 & | & \frac{11}{2} \\ 0 & 1 & \frac{3}{2} & | & 7 \\ 0 & 0 & \blacksquare & | & \blacksquare \end{bmatrix} \quad -\frac{1}{2}R_2 + R_3$

$\begin{bmatrix} 1 & \frac{1}{2} & 1 & | & \frac{11}{2} \\ 0 & 1 & \frac{3}{2} & | & 7 \\ 0 & 0 & 1 & | & \blacksquare \end{bmatrix} \quad -\frac{4}{11}R_3$

*Use row operations to solve each system. See Examples 2 and 3.*

17. $x + y - 3z = 1$
    $2x - y + z = 9$
    $3x + y - 4z = 8$

18. $2x + 4y - 3z = -18$
    $3x + y - z = -5$
    $x - 2y + 4z = 14$

19. $x + y - z = 6$
    $2x - y + z = -9$
    $x - 2y + 3z = 1$

20. $x + 3y - 6z = 7$
    $2x - y + 2z = 0$
    $x + y + 2z = -1$

21. $x - y = 1$
    $y - z = 6$
    $x + z = -1$

22. $x + y = 1$
    $2x - z = 0$
    $y + 2z = -2$

23. $x - 2y + z = 4$
    $3x - 6y + 3z = 12$
    $-2x + 4y - 2z = -8$

24. $4x + 8y + 4z = 9$
    $x + 3y + 4z = 10$
    $5x + 10y + 5z = 12$

25. $x + 2y + 3z = -2$
    $2x + 4y + 6z = -5$
    $x - y + 2z = 6$

26. $x + 3y + z = 1$
    $2x + 6y + 2z = 2$
    $3x + 9y + 3z = 3$

*The augmented matrix of the system in Exercise 3 is shown in the graphing calculator screen on the left as matrix [A]. The screen in the middle shows the row echelon form for [A]. Compare it with the matrix shown in the answer section for Exercise 3. The screen on the right shows the "reduced" row echelon form, and from this it can be determined by inspection that the solution set of the system is $\{(1, 5)\}$.*

[A]
[[4  8  44]
 [2 -1  -3]]

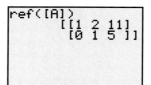

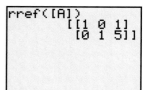

*Use a graphing calculator and either one of the two matrix methods illustrated to solve each system.*

27. $4x + y = 5$
    $2x + y = 3$

28. $5x + 3y = 7$
    $7x - 3y = -19$

29. $5x + y - 3z = -6$
    $2x + 3y + z = 5$
    $-3x - 2y + 4z = 3$

30. $x + y + z = 3$
    $3x - 3y - 4z = -1$
    $x + y + 3z = 11$

31. $x + z = -3$
    $y + z = 3$
    $x + y = 8$

32. $x - y = -1$
    $-y + z = -2$
    $x + z = -2$

# 4.4 Properties of Matrices

**OBJECTIVES**

1. Know the basic definitions for matrices.
2. Add and subtract matrices.
3. Multiply a matrix by a scalar.
4. Multiply matrices.
5. Use matrices in applications.

Matrices were used to solve systems of linear equations in **Section 4.3.** Here we discuss algebraic properties of matrices.

**OBJECTIVE 1** Know the basic definitions for matrices. We use capital letters to name matrices. Subscript notation is often used to name the elements of a matrix, as in the following matrix $A$.

$$A = \begin{bmatrix} a_{11} & a_{12} & a_{13} & \cdots & a_{1n} \\ a_{21} & a_{22} & a_{23} & \cdots & a_{2n} \\ a_{31} & a_{32} & a_{33} & \cdots & a_{3n} \\ \vdots & \vdots & \vdots & & \vdots \\ a_{m1} & a_{m2} & a_{m3} & \cdots & a_{mn} \end{bmatrix}$$

With this notation, the first-row, first-column element is $a_{11}$ (read "$a$-sub-one-one"); the second-row, third-column element is $a_{23}$; and in general, the $i$th-row, $j$th-column element is $a_{ij}$.

Certain matrices have special names. An $n \times n$ matrix is a **square matrix of order $n$.** Also, a matrix with just one row is a **row matrix**, and a matrix with just one column is a **column matrix.**

Two matrices are equal if they are the same size and if corresponding elements, position by position, are equal. Using this definition, the matrices

$$\begin{bmatrix} 2 & 1 \\ 3 & -5 \end{bmatrix} \quad \text{and} \quad \begin{bmatrix} 1 & 2 \\ -5 & 3 \end{bmatrix}$$

are *not* equal (even though they contain the same elements and are the same size), since the corresponding elements differ.

**EXAMPLE 1** Deciding Whether Two Matrices Are Equal

Find the values of the variables for which each statement is true.

(a) $\begin{bmatrix} 2 & 1 \\ p & q \end{bmatrix} = \begin{bmatrix} x & y \\ -1 & 0 \end{bmatrix}$

The only way this statement can be true is if $2 = x$, $1 = y$, $p = -1$, and $q = 0$.

(b) $\begin{bmatrix} x \\ y \end{bmatrix} = \begin{bmatrix} 1 \\ 4 \\ 0 \end{bmatrix}$

This statement can never be true, since the two matrices are different sizes. (One is $2 \times 1$ and the other is $3 \times 1$.)

**NOW TRY** Exercise 1.

**OBJECTIVE 2** Add and subtract matrices. Addition of matrices is defined as follows.

> **Addition of Matrices**
>
> To add two matrices of the same size, add corresponding elements. Only matrices of the same size can be added.

**EXAMPLE 2** Adding Matrices

Find each sum.

(a) $\begin{bmatrix} 5 & -6 \\ 8 & 9 \end{bmatrix} + \begin{bmatrix} -4 & 6 \\ 8 & -3 \end{bmatrix} = \begin{bmatrix} 5 + (-4) & -6 + 6 \\ 8 + 8 & 9 + (-3) \end{bmatrix} = \begin{bmatrix} 1 & 0 \\ 16 & 6 \end{bmatrix}$

(b) $\begin{bmatrix} 2 \\ 5 \\ 8 \end{bmatrix} + \begin{bmatrix} -6 \\ 3 \\ 12 \end{bmatrix} = \begin{bmatrix} -4 \\ 8 \\ 20 \end{bmatrix}$

(c) Because matrices $A = \begin{bmatrix} 5 & 8 \\ 6 & 2 \end{bmatrix}$ and $B = \begin{bmatrix} 3 & 9 & 1 \\ 4 & 2 & 5 \end{bmatrix}$ are different sizes, the sum $A + B$ cannot be found.

**NOW TRY** Exercises 15 and 21.

A matrix containing only zero elements is called a **zero matrix**. For example, $O = \begin{bmatrix} 0 & 0 & 0 \end{bmatrix}$ is the $1 \times 3$ zero matrix, while

$$O = \begin{bmatrix} 0 & 0 & 0 \\ 0 & 0 & 0 \end{bmatrix}$$

is the $2 \times 3$ zero matrix.

By the additive inverse property in **Chapter 1**, each real number has an additive inverse: if $a$ is a real number, there is a real number $-a$ such that

$$a + (-a) = 0 \quad \text{and} \quad -a + a = 0.$$

What about matrices? Given the matrix

$$A = \begin{bmatrix} -5 & 2 & -1 \\ 3 & 4 & -6 \end{bmatrix},$$

is there a matrix $-A$ such that

$$A + (-A) = O$$

where $O$ is the $2 \times 3$ zero matrix? The answer is yes: the matrix $-A$ has as elements the additive inverses of the elements of $A$. (Remember, each element of $A$ is a real number and therefore has an additive inverse.)

$$-A = \begin{bmatrix} 5 & -2 & 1 \\ -3 & -4 & 6 \end{bmatrix}$$

To check, test that $A + (-A)$ equals the zero matrix, $O$.

$$A + (-A) = \begin{bmatrix} -5 & 2 & -1 \\ 3 & 4 & -6 \end{bmatrix} + \begin{bmatrix} 5 & -2 & 1 \\ -3 & -4 & 6 \end{bmatrix} = \begin{bmatrix} 0 & 0 & 0 \\ 0 & 0 & 0 \end{bmatrix} = O$$

Matrix $-A$ is called the **additive inverse**, or **negative**, of matrix $A$. Every matrix has an additive inverse.

The real number $b$ is subtracted from the real number $a$, written $a - b$, by adding $a$ and the additive inverse of $b$. That is,

$$a - b = a + (-b).$$

The same definition works for subtraction of matrices.

### Subtraction of Matrices

If $A$ and $B$ are two matrices of the same size, then

$$A - B = A + (-B).$$

In practice, the difference of two matrices of the same size is found by subtracting corresponding elements.

### EXAMPLE 3  Subtracting Matrices

Find each difference, if possible.

(a) $\begin{bmatrix} -5 & 6 \\ 2 & 4 \end{bmatrix} - \begin{bmatrix} -3 & 2 \\ 5 & -8 \end{bmatrix} = \begin{bmatrix} -5 - (-3) & 6 - 2 \\ 2 - 5 & 4 - (-8) \end{bmatrix} = \begin{bmatrix} -2 & 4 \\ -3 & 12 \end{bmatrix}$

(b) $\begin{bmatrix} 8 & 6 & -4 \end{bmatrix} - \begin{bmatrix} 3 & 5 & -8 \end{bmatrix} = \begin{bmatrix} 5 & 1 & 4 \end{bmatrix}$

(c) The matrices $\begin{bmatrix} -2 & 5 \\ 0 & 1 \end{bmatrix}$ and $\begin{bmatrix} 3 \\ 5 \end{bmatrix}$ are different sizes and cannot be subtracted.

NOW TRY Exercise 17.

**OBJECTIVE 3** Multiply a matrix by a scalar. In work with matrices, a real number is called a **scalar** to distinguish it from a matrix. The product of a scalar $k$ and a matrix $X$ is the matrix $kX$, each of whose elements is $k$ times the corresponding element of $X$.

### EXAMPLE 4  Multiplying a Matrix by a Scalar

Find each product.

(a) $5 \begin{bmatrix} 2 & -3 \\ 0 & 4 \end{bmatrix} = \begin{bmatrix} 10 & -15 \\ 0 & 20 \end{bmatrix}$   (b) $\frac{3}{4} \begin{bmatrix} 20 & 36 \\ 12 & -16 \end{bmatrix} = \begin{bmatrix} 15 & 27 \\ 9 & -12 \end{bmatrix}$

NOW TRY Exercise 23.

### Properties of Scalar Multiplication

If $A$ and $B$ are matrices of the same size and $c$ and $d$ are real numbers, then

$$(c + d)A = cA + dA \qquad c(A)d = cd(A)$$
$$c(A + B) = cA + cB \qquad (cd)A = c(dA).$$

**OBJECTIVE 4** Multiply matrices. Multiplying two matrices is a little more complicated than scalar multiplication, but it is useful in applications. To find the product of

$$A = \begin{bmatrix} -3 & 4 & 2 \\ 5 & 0 & 4 \end{bmatrix} \quad \text{and} \quad B = \begin{bmatrix} -6 & 4 \\ 2 & 3 \\ 3 & -2 \end{bmatrix},$$

first locate *row* 1 of *A* and *column* 1 of *B*, shown shaded below.

$$A = \begin{bmatrix} -3 & 4 & 2 \\ 5 & 0 & 4 \end{bmatrix} \quad B = \begin{bmatrix} -6 & 4 \\ 2 & 3 \\ 3 & -2 \end{bmatrix}$$

Multiply corresponding elements, and find the sum of the products.

$$-3(-6) + 4(2) + 2(3) = 32$$

This result is the element for row 1, column 1 of the product matrix.

Now use *row* 1 of *A* and *column* 2 of *B* (shaded below) to determine the element in row 1, column 2 of the product matrix.

$$A = \begin{bmatrix} -3 & 4 & 2 \\ 5 & 0 & 4 \end{bmatrix} \quad B = \begin{bmatrix} -6 & 4 \\ 2 & 3 \\ 3 & -2 \end{bmatrix}$$

Multiply corresponding elements, and add the products.

$$-3(4) + 4(3) + 2(-2) = -4$$

Next, use *row* 2 of *A* and *column* 1 of *B*; this will give the row 2, column 1 entry of the product matrix.

$$\begin{bmatrix} -3 & 4 & 2 \\ 5 & 0 & 4 \end{bmatrix} \begin{bmatrix} -6 & 4 \\ 2 & 3 \\ 3 & -2 \end{bmatrix} \quad 5(-6) + 0(2) + 4(3) = -18$$

Finally, use *row* 2 of *A* and *column* 2 of *B* to find the entry for row 2, column 2 of the product matrix.

$$\begin{bmatrix} -3 & 4 & 2 \\ 5 & 0 & 4 \end{bmatrix} \begin{bmatrix} -6 & 4 \\ 2 & 3 \\ 3 & -2 \end{bmatrix} \quad 5(4) + 0(3) + 4(-2) = 12$$

The product matrix can now be written.

$$\begin{bmatrix} -3 & 4 & 2 \\ 5 & 0 & 4 \end{bmatrix} \begin{bmatrix} -6 & 4 \\ 2 & 3 \\ 3 & -2 \end{bmatrix} = \begin{bmatrix} 32 & -4 \\ -18 & 12 \end{bmatrix}$$

We see that the product of a $2 \times 3$ matrix and a $3 \times 2$ matrix is a $2 \times 2$ matrix.

By definition, the product *AB* of an $m \times n$ matrix *A* and an $n \times p$ matrix *B* is found as follows: Multiply each element of the first row of *A* by the corresponding element of the first column of *B*. The sum of these *n* products is the first-row, first-column element of *AB*. Also, the sum of the products found by multiplying the elements of the first row of *A* times the corresponding elements of the second column of *B* gives the first-row, second-column element of *AB*, and so on.

To find the *i*th-row, *j*th-column element of *AB*, multiply each element in the *i*th row of *A* by the corresponding element in the *j*th column of *B*. (Note the shaded areas in the matrices below.) The sum of these products will give the element of row $i$, column $j$ of $AB$.

$$A = \begin{bmatrix} a_{11} & a_{12} & a_{13} & \cdots & a_{1n} \\ a_{21} & a_{22} & a_{23} & \cdots & a_{2n} \\ \vdots & & & & \\ a_{i1} & a_{i2} & a_{i3} & \cdots & a_{in} \\ \vdots & & & & \\ a_{m1} & a_{m2} & a_{m3} & \cdots & a_{mn} \end{bmatrix} \qquad B = \begin{bmatrix} b_{11} & b_{12} & \cdots & b_{1j} & \cdots & b_{1p} \\ b_{21} & b_{22} & \cdots & b_{2j} & \cdots & b_{2p} \\ \vdots & & & \vdots & & \\ b_{n1} & b_{n2} & \cdots & b_{nj} & \cdots & b_{np} \end{bmatrix}$$

### Matrix Multiplication

If the number of columns of matrix $A$ is the same as the number of rows of matrix $B$, then entry $c_{ij}$ of the product matrix $C = AB$ is found as follows:

$$c_{ij} = a_{i1}b_{1j} + a_{i2}b_{2j} + \cdots + a_{in}b_{nj}.$$

The final product will have as many rows as $A$ and as many columns as $B$.

---

**EXAMPLE 5** Deciding Whether Two Matrices Can Be Multiplied

Suppose matrix $A$ is $3 \times 2$, while matrix $B$ is $2 \times 4$. Can the product $AB$ be calculated? What is the size of the product? Can the product $BA$ be calculated? What is the size of $BA$?

The following diagram helps answer the questions about the product $AB$.

```
    Matrix A              Matrix B
    3 × 2                 2 × 4
         ↑_____Must match_____↑
         ↑_____Size of AB_____↑
                 3 × 4
```

The product $AB$ exists, since the number of columns of $A$ equals the number of rows of $B$. (Both are 2.) The product is a $3 \times 4$ matrix. Make a similar diagram for $BA$.

```
    Matrix B              Matrix A
    2 × 4                 3 × 2
         ↑_____Different_____↑
```

The product $BA$ does not exist, since $B$ has 4 columns and $A$ has only 3 rows.

**NOW TRY** Exercise 29.

---

**EXAMPLE 6** Multiplying Two Matrices

Find $AB$ and $BA$, if possible, where

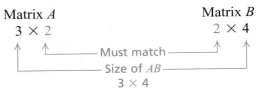

$$A = \begin{bmatrix} 1 & -3 \\ 7 & 2 \end{bmatrix} \quad \text{and} \quad B = \begin{bmatrix} 1 & 0 & -1 & 2 \\ 3 & 1 & 4 & -1 \end{bmatrix}.$$

First decide whether $AB$ can be found. Since $A$ is $2 \times 2$ and $B$ is $2 \times 4$, the product can be found and will be a $2 \times 4$ matrix. Now use the definition of matrix multiplication.

$$AB = \begin{bmatrix} 1 & -3 \\ 7 & 2 \end{bmatrix} \begin{bmatrix} 1 & 0 & -1 & 2 \\ 3 & 1 & 4 & -1 \end{bmatrix}$$

$$= \begin{bmatrix} 1(1)+(-3)3 & 1(0)+(-3)1 & 1(-1)+(-3)4 & 1(2)+(-3)(-1) \\ 7(1)+2(3) & 7(0)+2(1) & 7(-1)+2(4) & 7(2)+2(-1) \end{bmatrix}$$

$$= \begin{bmatrix} -8 & -3 & -13 & 5 \\ 13 & 2 & 1 & 12 \end{bmatrix}$$

Since $B$ is a $2 \times 4$ matrix and $A$ is a $2 \times 2$ matrix, the number of columns of $B$ (4) does not equal the number of rows of $A$ (2). Therefore, the product $BA$ cannot be found.

**NOW TRY** Exercise 33.

### EXAMPLE 7   Multiplying Square Matrices in Different Orders

If $A = \begin{bmatrix} 1 & 3 \\ -2 & 5 \end{bmatrix}$ and $B = \begin{bmatrix} -2 & 7 \\ 0 & 2 \end{bmatrix}$, then the definition of matrix multiplication can be used to show that

$$AB = \begin{bmatrix} -2 & 13 \\ 4 & -4 \end{bmatrix} \quad \text{and} \quad BA = \begin{bmatrix} -16 & 29 \\ -4 & 10 \end{bmatrix}.$$

**NOW TRY** Exercise 35.

**CAUTION** Examples 5 and 6 showed that the order in which two matrices are to be multiplied may determine whether their product can be found. Example 7 showed that even when both products $AB$ and $BA$ can be found, they may not be equal. In general, for matrices $A$ and $B$, $AB \neq BA$, so **matrix multiplication is not commutative.** Matrix multiplication does, however, satisfy the associative and distributive properties.

### Properties of Matrix Multiplication

If $A$, $B$, and $C$ are matrices such that all the following products and sums exist, then

$$(AB)C = A(BC)$$
$$A(B + C) = AB + AC$$
$$(B + C)A = BA + CA.$$

A graphing calculator with matrix capability will perform matrix multiplication, as well as other matrix operations. The three screens in Figure 10 show matrix multiplication using a calculator. Compare to the product $AB$ in Example 6.

```
[A]
      [[1  -3]
       [7   2]]
```

```
[B]
      [[1  0  -1   2]
       [3  1   4  -1]]
```

```
[A][B]
      [[-8  -3  -13   5]
       [13   2    1  12]]
```

**FIGURE 10**

## SECTION 4.4 Properties of Matrices

**OBJECTIVE 5** Use matrices in applications.

**EXAMPLE 8** Applying Matrix Multiplication

A contractor builds three kinds of houses, models A, B, and C, with a choice of two styles, colonial or ranch. Matrix $P$ below shows the number of each kind of house the contractor is planning to build for a new 100-home subdivision. The amounts for each of the main materials used depend on the style of the house. These amounts are shown in matrix $Q$ below, while matrix $R$ gives the cost in dollars for each kind of material. Concrete is measured here in cubic yards, lumber in 1000 board feet, brick in 1000s, and shingles in 100 square feet.

$$\begin{array}{c} \text{Model A} \\ \text{Model B} \\ \text{Model C} \end{array} \begin{array}{cc} \text{Colonial} & \text{Ranch} \\ \begin{bmatrix} 0 & 30 \\ 10 & 20 \\ 20 & 20 \end{bmatrix} \end{array} = P$$

$$\begin{array}{c} \text{Colonial} \\ \text{Ranch} \end{array} \begin{array}{cccc} \text{Concrete} & \text{Lumber} & \text{Brick} & \text{Shingles} \\ \begin{bmatrix} 10 & 2 & 0 & 2 \\ 50 & 1 & 20 & 2 \end{bmatrix} \end{array} = Q$$

$$\begin{array}{c} \text{Concrete} \\ \text{Lumber} \\ \text{Brick} \\ \text{Shingles} \end{array} \begin{array}{c} \text{Cost per unit} \\ \begin{bmatrix} 20 \\ 180 \\ 60 \\ 25 \end{bmatrix} \end{array} = R$$

**(a)** What is the total cost of materials for all houses of each model?

To find the materials cost of each model, first find matrix $PQ$, which will show the total amount of each material needed for all houses of each model.

$$PQ = \begin{bmatrix} 0 & 30 \\ 10 & 20 \\ 20 & 20 \end{bmatrix} \begin{bmatrix} 10 & 2 & 0 & 2 \\ 50 & 1 & 20 & 2 \end{bmatrix} = \begin{array}{c} \text{Concrete Lumber Brick Shingles} \\ \begin{bmatrix} 1500 & 30 & 600 & 60 \\ 1100 & 40 & 400 & 60 \\ 1200 & 60 & 400 & 80 \end{bmatrix} \end{array} \begin{array}{c} \text{Model A} \\ \text{Model B} \\ \text{Model C} \end{array}$$

Multiplying $PQ$ and the cost matrix $R$ gives the total cost of materials for each model.

$$(PQ)R = \begin{bmatrix} 1500 & 30 & 600 & 60 \\ 1100 & 40 & 400 & 60 \\ 1200 & 60 & 400 & 80 \end{bmatrix} \begin{bmatrix} 20 \\ 180 \\ 60 \\ 25 \end{bmatrix} = \begin{array}{c} \text{Cost} \\ \begin{bmatrix} 72{,}900 \\ 54{,}700 \\ 60{,}800 \end{bmatrix} \end{array} \begin{array}{c} \text{Model A} \\ \text{Model B} \\ \text{Model C} \end{array}$$

**(b)** How much of each of the four kinds of material must be ordered?

The totals of the columns of matrix $PQ$ will give a matrix whose elements represent the total amounts of each material needed for the subdivision. Call this matrix $T$ and write it as a row matrix.

$$T = \begin{bmatrix} 3800 & 130 & 1400 & 200 \end{bmatrix}$$

**(c)** What is the total cost of the materials?

The total cost of all the materials is given by the product of matrix $R$, the cost matrix, and matrix $T$, the total amounts matrix. To multiply these and get a $1 \times 1$ matrix, representing the total cost, requires multiplying a $1 \times 4$ matrix and a $4 \times 1$ matrix.

This is why in part (b) a row matrix was written rather than a column matrix. The total materials cost is given by *TR*, so

$$TR = \begin{bmatrix} 3800 & 130 & 1400 & 200 \end{bmatrix} \begin{bmatrix} 20 \\ 180 \\ 60 \\ 25 \end{bmatrix} = \begin{bmatrix} 188{,}400 \end{bmatrix}.$$

The total cost of the materials is $188,400.

**NOW TRY** Exercise 47.

To help keep track of the quantities a matrix represents, let matrix *P*, from Example 8, represent models/styles, matrix *Q* represent styles/materials, and matrix *R* represent materials/cost. In each case the meaning of the rows is written first and that of the columns second. When the product *PQ* was found in Example 8, the rows of the matrix represented models and the columns represented materials. Therefore, the matrix product *PQ* represents models/materials. The common quantity, styles, in both *P* and *Q* was eliminated in the product *PQ*. Do you see that the product *(PQ)R* represents models/cost?

In practical problems this notation helps to identify the order in which two matrices should be multiplied so that the results are meaningful. In Example 8(c), either product *RT* or product *TR* could have been found. However, since *T* represents subdivisions/materials and *R* represents materials/cost, only *TR* gave the required matrix representing subdivisions/cost.

## 4.4 Exercises

**NOW TRY** Exercise

*Find the values of each variable. See Example 1.*

**1.** $\begin{bmatrix} w & x \\ y & z \end{bmatrix} = \begin{bmatrix} 3 & 2 \\ -1 & 4 \end{bmatrix}$

**2.** $\begin{bmatrix} 0 & 5 & x \\ -1 & 3 & y+2 \\ 4 & 1 & z \end{bmatrix} = \begin{bmatrix} 0 & w+3 & 6 \\ -1 & 3 & 0 \\ 4 & 1 & 8 \end{bmatrix}$

**3.** $\begin{bmatrix} 2 & 5 & 6 \\ 1 & m & n \end{bmatrix} = \begin{bmatrix} z & y & w \\ 1 & 8 & -2 \end{bmatrix}$

**4.** $\begin{bmatrix} -7+z & 4r & 8s \\ 6p & 2 & 5 \end{bmatrix} + \begin{bmatrix} -9 & 8r & 3 \\ 2 & 5 & 4 \end{bmatrix} = \begin{bmatrix} 2 & 36 & 27 \\ 20 & 7 & 12a \end{bmatrix}$

**5.** $\begin{bmatrix} a+2 & 3z+1 & 5m \\ 8k & 0 & 3 \end{bmatrix} + \begin{bmatrix} 3a & 2z & 5m \\ 2k & 5 & 6 \end{bmatrix} = \begin{bmatrix} 10 & -14 & 80 \\ 10 & 5 & 9 \end{bmatrix}$

**6.** *Fill in the Blanks* A 3 × 8 matrix has _____ columns and _____ rows.

*Find the size of each matrix. Identify any square, column, or row matrices.*

**7.** $\begin{bmatrix} -4 & 8 \\ 2 & 3 \end{bmatrix}$

**8.** $\begin{bmatrix} -9 & 6 & 2 \\ 4 & 1 & 8 \end{bmatrix}$

**9.** $\begin{bmatrix} -6 & 8 & 0 & 0 \\ 4 & 1 & 9 & 2 \\ 3 & -5 & 7 & 1 \end{bmatrix}$

**10.** $\begin{bmatrix} 8 & -2 & 4 & 6 & 3 \end{bmatrix}$

**11.** $\begin{bmatrix} 2 \\ 4 \end{bmatrix}$

**12.** $\begin{bmatrix} -9 \end{bmatrix}$

13. Your friend missed the lecture on adding matrices. In your own words, explain to him how to add two matrices. Give an example.

14. Explain in your own words how to subtract two matrices. Give an example.

*Perform each operation in Exercises 15–22, whenever possible. See Examples 2 and 3.*

15. $\begin{bmatrix} 6 & -9 & 2 \\ 4 & 1 & 3 \end{bmatrix} + \begin{bmatrix} -8 & 2 & 5 \\ 6 & -3 & 4 \end{bmatrix}$

16. $\begin{bmatrix} 9 & 4 \\ -8 & 2 \end{bmatrix} + \begin{bmatrix} -3 & 2 \\ -4 & 7 \end{bmatrix}$

17. $\begin{bmatrix} -6 & 8 \\ 0 & 0 \end{bmatrix} - \begin{bmatrix} 0 & 0 \\ -4 & -2 \end{bmatrix}$

18. $\begin{bmatrix} 1 & -4 \\ 2 & -3 \\ -8 & 4 \end{bmatrix} - \begin{bmatrix} -6 & 9 \\ -2 & 5 \\ -7 & -12 \end{bmatrix}$

19. $\begin{bmatrix} 3x+y & x-2y & 2x \\ 5x & 3y & x+y \end{bmatrix} + \begin{bmatrix} 2x & 3y & 5x+y \\ 3x+2y & x & 2x \end{bmatrix}$

20. $\begin{bmatrix} 4k-8y \\ 6z-3x \\ 2k+5a \\ -4m+2n \end{bmatrix} - \begin{bmatrix} 5k+6y \\ 2z+5x \\ 4k+6a \\ 4m-2n \end{bmatrix}$

21. $\begin{bmatrix} 3 \\ 2 \end{bmatrix} + \begin{bmatrix} 2 & 3 \end{bmatrix}$

22. $\begin{bmatrix} 0 \\ 0 \end{bmatrix} - \begin{bmatrix} 0 & 0 & 0 \end{bmatrix}$

Let $A = \begin{bmatrix} -2 & 4 \\ 0 & 3 \end{bmatrix}$ and $B = \begin{bmatrix} -6 & 2 \\ 4 & 0 \end{bmatrix}$. *Find each of the following. See Example 4.*

23. $2A$

24. $-3B$

25. $2A - B$

26. $-2A + 4B$

27. $-A + \frac{1}{2}B$

28. $\frac{3}{4}A - B$

**Concept Check** *Decide whether each product can be found given*

$$A = \begin{bmatrix} 3 & 7 & 1 \\ -2 & 4 & 0 \end{bmatrix} \text{ and } B = \begin{bmatrix} 1 & 2 \\ 5 & 7 \end{bmatrix}.$$

*Give the size of each product if it exists. See Example 5.*

29. $AB$

30. $BA$

*Find each matrix product, whenever possible. See Examples 6 and 7.*

31. $\begin{bmatrix} 1 & 2 \\ 3 & 4 \end{bmatrix} \begin{bmatrix} -1 \\ 7 \end{bmatrix}$

32. $\begin{bmatrix} -1 & 5 \\ 7 & 0 \end{bmatrix} \begin{bmatrix} 6 \\ 2 \end{bmatrix}$

33. $\begin{bmatrix} 3 & -4 & 1 \\ 5 & 0 & 2 \end{bmatrix} \begin{bmatrix} -1 \\ 4 \\ 2 \end{bmatrix}$

34. $\begin{bmatrix} -6 & 3 & 5 \\ 2 & 9 & 1 \end{bmatrix} \begin{bmatrix} -2 \\ 0 \\ 3 \end{bmatrix}$

35. $\begin{bmatrix} 5 & 2 \\ -1 & 4 \end{bmatrix} \begin{bmatrix} 3 & -2 \\ 1 & 0 \end{bmatrix}$

36. $\begin{bmatrix} -4 & 0 \\ 1 & 3 \end{bmatrix} \begin{bmatrix} -2 & 4 \\ 0 & 1 \end{bmatrix}$

37. $\begin{bmatrix} 2 & 2 & -1 \\ 3 & 0 & 1 \end{bmatrix} \begin{bmatrix} 0 & 2 \\ -1 & 4 \\ 0 & 2 \end{bmatrix}$

38. $\begin{bmatrix} -9 & 2 & 1 \\ 3 & 0 & 0 \end{bmatrix} \begin{bmatrix} 2 \\ -1 \\ 4 \end{bmatrix}$

39. $\begin{bmatrix} -1 & 2 & 0 \\ 0 & 3 & 2 \\ 0 & 1 & 4 \end{bmatrix} \begin{bmatrix} 2 & -1 & 2 \\ 0 & 2 & 1 \\ 3 & 0 & -1 \end{bmatrix}$

40. $\begin{bmatrix} -2 & -3 & -4 \\ 2 & -1 & 0 \\ 4 & -2 & 3 \end{bmatrix} \begin{bmatrix} 0 & 1 & 4 \\ 1 & 2 & -1 \\ 3 & 2 & -2 \end{bmatrix}$

41. $\begin{bmatrix} -2 & 4 & 1 \end{bmatrix} \begin{bmatrix} 3 & -2 & 4 \\ 2 & 1 & 0 \\ 0 & -1 & 4 \end{bmatrix}$

42. $\begin{bmatrix} 0 & 3 & -4 \end{bmatrix} \begin{bmatrix} -2 & 6 & 3 \\ 0 & 4 & 2 \\ -1 & 1 & 4 \end{bmatrix}$

43. $\begin{bmatrix} -3 & 0 & 2 & 1 \\ 4 & 0 & 2 & 6 \end{bmatrix} \begin{bmatrix} -4 & 2 \\ 0 & 1 \end{bmatrix}$

44. $\begin{bmatrix} -1 & 2 & 4 & 1 \\ 0 & 2 & -3 & 5 \end{bmatrix} \begin{bmatrix} 1 & 2 & 4 \\ -2 & 5 & 1 \end{bmatrix}$

45. A hardware chain does an inventory of a particular size of screw and finds that its Adelphi store has 100 flat-head and 150 round-head screws, its Beltsville store has 125 flat and 50 round, and its College Park store has 175 flat and 200 round. Write this information first as a 3 × 2 matrix and then as a 2 × 3 matrix.

46. At the grocery store, Miguel bought 4 quarts of milk, 2 loaves of bread, 4 potatoes, and an apple. Mary bought 2 quarts of milk, a loaf of bread, 5 potatoes, and 4 apples. Write this information first as a 2 × 4 matrix and then as a 4 × 2 matrix.

47. Yummy Yogurt sells three types of yogurt, nonfat, regular, and super creamy, at three locations. Location I sells 50 gal of nonfat, 100 gal of regular, and 30 gal of super creamy each day. Location II sells 10 gal of nonfat and Location III sells 60 gal of nonfat each day. Daily sales of regular yogurt are 90 gal at Location II and 120 gal at Location III. At Location II, 50 gal of super creamy are sold each day, and 40 gal of super creamy are sold each day at Location III.

   (a) Write a 3 × 3 matrix that shows the sales figures for the three locations.
   (b) The income per gallon for nonfat, regular, and super creamy is $12, $10, and $15, respectively. Write a 1 × 3 or 3 × 1 matrix displaying the income.
   (c) Find a matrix product that gives the daily income at each of the three locations.
   (d) What is Yummy Yogurt's total daily income from the three locations?

48. The Bread Box, a neighborhood bakery, sells four main items: sweet rolls, bread, cakes, and pies. The amount of each ingredient (in cups, except for eggs) required for these items is given by matrix $A$.

$$\begin{array}{c} \\ \text{Rolls (dozen)} \\ \text{Bread (loaves)} \\ \text{Cakes} \\ \text{Pies (crust)} \end{array} \begin{array}{c} \text{Eggs} \quad \text{Flour} \quad \text{Sugar} \quad \text{Shortening} \quad \text{Milk} \\ \begin{bmatrix} 1 & 4 & \frac{1}{4} & \frac{1}{4} & 1 \\ 0 & 3 & 0 & \frac{1}{4} & 0 \\ 4 & 3 & 2 & 1 & 1 \\ 0 & 1 & 0 & \frac{1}{3} & 0 \end{bmatrix} = A \end{array}$$

The cost (in cents) for each ingredient when purchased in large lots or small lots is given in matrix $B$.

$$\begin{array}{c} \\ \text{Eggs} \\ \text{Flour} \\ \text{Sugar} \\ \text{Shortening} \\ \text{Milk} \end{array} \begin{array}{c} \text{Cost} \\ \text{Large lot} \quad \text{Small lot} \\ \begin{bmatrix} 5 & 5 \\ 8 & 10 \\ 10 & 12 \\ 12 & 15 \\ 5 & 6 \end{bmatrix} = B \end{array}$$

   (a) Use matrix multiplication to find a matrix giving the comparative cost per item for the two purchase options.
   (b) Suppose a day's orders consist of 20 dozen sweet rolls, 200 loaves of bread, 50 cakes, and 60 pies. Write the orders as a 1 × 4 matrix and, using matrix multiplication, write as a matrix the amount of each ingredient needed to fill the day's orders.
   (c) Use matrix multiplication to find a matrix giving the costs under the two purchase options to fill the day's orders.

## 4.5 Matrix Inverses

**OBJECTIVES**

1. Understand and write identity matrices.
2. Find multiplicative inverse matrices.
3. Use inverse matrices to solve systems of linear equations.

In **Section 4.4** we saw several parallels between the set of real numbers and the set of matrices. Another similarity is that both sets have identity and inverse elements for multiplication.

**OBJECTIVE 1** Understand and write identity matrices. By the identity property for real numbers, $a \cdot 1 = a$ and $1 \cdot a = a$ for any real number $a$. If there is to be a multiplicative **identity matrix** $I$, such that

$$AI = A \quad \text{and} \quad IA = A,$$

for any matrix $A$, then $A$ and $I$ must be square matrices of the same size. Otherwise it would not be possible to find both products.

### 2 × 2 Identity Matrix

If $I_2$ represents the 2 × 2 identity matrix, then

$$I_2 = \begin{bmatrix} 1 & 0 \\ 0 & 1 \end{bmatrix}.$$

To verify that $I_2$ is the 2 × 2 identity matrix, we must show that $AI = A$ and $IA = A$ for any 2 × 2 matrix. Let

$$A = \begin{bmatrix} x & y \\ z & w \end{bmatrix}.$$

Then

$$AI = \begin{bmatrix} x & y \\ z & w \end{bmatrix}\begin{bmatrix} 1 & 0 \\ 0 & 1 \end{bmatrix} = \begin{bmatrix} x \cdot 1 + y \cdot 0 & x \cdot 0 + y \cdot 1 \\ z \cdot 1 + w \cdot 0 & z \cdot 0 + w \cdot 1 \end{bmatrix} = \begin{bmatrix} x & y \\ z & w \end{bmatrix} = A,$$

and

$$IA = \begin{bmatrix} 1 & 0 \\ 0 & 1 \end{bmatrix}\begin{bmatrix} x & y \\ z & w \end{bmatrix} = \begin{bmatrix} 1 \cdot x + 0 \cdot z & 1 \cdot y + 0 \cdot w \\ 0 \cdot x + 1 \cdot z & 0 \cdot y + 1 \cdot w \end{bmatrix} = \begin{bmatrix} x & y \\ z & w \end{bmatrix} = A.$$

Generalizing from this example, there is an $n \times n$ identity matrix having 1s on the main diagonal and 0s elsewhere.

### $n \times n$ Identity Matrix

The $n \times n$ identity matrix is given by $I_n$, where

$$I_n = \begin{bmatrix} 1 & 0 & \cdots & 0 \\ 0 & 1 & \cdots & 0 \\ \vdots & \vdots & a_{ij} & \vdots \\ 0 & 0 & \cdots & 1 \end{bmatrix}.$$

The element $a_{ij} = 1$ when $i = j$ (the diagonal elements) and $a_{ij} = 0$ otherwise.

**EXAMPLE 1** Stating and Verifying the 3 × 3 Identity Matrix

Let $A = \begin{bmatrix} -2 & 4 & 0 \\ 3 & 5 & 9 \\ 0 & 8 & -6 \end{bmatrix}$. Give the 3 × 3 identity matrix $I$ and show that $AI = A$.

The 3 × 3 identity matrix is

$$I = \begin{bmatrix} 1 & 0 & 0 \\ 0 & 1 & 0 \\ 0 & 0 & 1 \end{bmatrix}.$$

By the definition of matrix multiplication,

$$AI = \begin{bmatrix} -2 & 4 & 0 \\ 3 & 5 & 9 \\ 0 & 8 & -6 \end{bmatrix} \begin{bmatrix} 1 & 0 & 0 \\ 0 & 1 & 0 \\ 0 & 0 & 1 \end{bmatrix} = \begin{bmatrix} -2 & 4 & 0 \\ 3 & 5 & 9 \\ 0 & 8 & -6 \end{bmatrix} = A.$$

**NOW TRY** Exercise 1.

The graphing calculator screen in Figure 11(a) shows identity matrices for $n = 2$ and $n = 3$. The screens in Figures 11(b) and 11(c) support the result in Example 1.

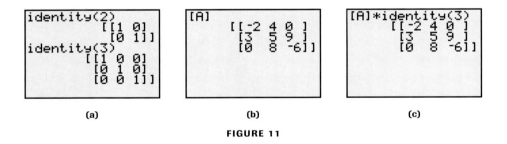

(a)    (b)    (c)

**FIGURE 11**

**OBJECTIVE 2** Find multiplicative inverse matrices. For every nonzero real number $a$, there is a multiplicative inverse $\frac{1}{a}$ such that

$$a \cdot \frac{1}{a} = 1 \quad \text{and} \quad \frac{1}{a} \cdot a = 1.$$

Recall that $\frac{1}{a}$ is also written $a^{-1}$. In a similar way, if $A$ is an $n \times n$ matrix, then its **multiplicative inverse**, written $A^{-1}$, must satisfy both

$$AA^{-1} = I_n \quad \text{and} \quad A^{-1}A = I_n.$$

This means that only a square matrix can have a multiplicative inverse.

**NOW TRY** Exercises 3 and 7.

**CAUTION** Although $a^{-1} = \frac{1}{a}$ for any nonzero real number $a$, if $A$ is a matrix, then

$$A^{-1} \neq \frac{1}{A}.$$

In fact, $\frac{1}{A}$ has no meaning, since 1 is a *number* and $A$ is a *matrix*.

We use the row operations introduced in **Section 4.3** to find the matrix $A^{-1}$. As an example, we find the inverse of

$$A = \begin{bmatrix} 2 & 4 \\ 1 & -1 \end{bmatrix}.$$

Let the unknown inverse matrix be

$$A^{-1} = \begin{bmatrix} x & y \\ z & w \end{bmatrix}.$$

By the definition of matrix inverse, $AA^{-1} = I_2$, or

$$AA^{-1} = \begin{bmatrix} 2 & 4 \\ 1 & -1 \end{bmatrix} \begin{bmatrix} x & y \\ z & w \end{bmatrix} = \begin{bmatrix} 1 & 0 \\ 0 & 1 \end{bmatrix}.$$

By matrix multiplication,

$$\begin{bmatrix} 2x + 4z & 2y + 4w \\ x - z & y - w \end{bmatrix} = \begin{bmatrix} 1 & 0 \\ 0 & 1 \end{bmatrix}.$$

Setting corresponding elements equal gives the system of equations

$$2x + 4z = 1 \quad (1)$$
$$2y + 4w = 0 \quad (2)$$
$$x - z = 0 \quad (3)$$
$$y - w = 1. \quad (4)$$

Since equations (1) and (3) involve only $x$ and $z$, while equations (2) and (4) involve only $y$ and $w$, these four equations lead to two systems of equations,

$$\begin{array}{ccc} 2x + 4z = 1 & & 2y + 4w = 0 \\ x - z = 0 & \text{and} & y - w = 1. \end{array}$$

Writing the two systems as augmented matrices gives

$$\begin{bmatrix} 2 & 4 & | & 1 \\ 1 & -1 & | & 0 \end{bmatrix} \quad \text{and} \quad \begin{bmatrix} 2 & 4 & | & 0 \\ 1 & -1 & | & 1 \end{bmatrix}.$$

Each of these systems can be solved using row operations. However, since the elements to the left of the vertical bar are identical, the two systems can be combined into one matrix,

$$\begin{bmatrix} 2 & 4 & | & 1 & 0 \\ 1 & -1 & | & 0 & 1 \end{bmatrix},$$

and solved simultaneously using matrix row operations. We need to change the numbers on the left of the vertical bar to the 2 × 2 identity matrix.

Interchange the two rows to get a 1 in the upper left corner.

$$\begin{bmatrix} 1 & -1 & | & 0 & 1 \\ 2 & 4 & | & 1 & 0 \end{bmatrix}$$

Multiply the first row by $-2$, and add the results to the second row to obtain

$$\begin{bmatrix} 1 & -1 & | & 0 & 1 \\ 0 & 6 & | & 1 & -2 \end{bmatrix}. \quad -2R_1 + R_2$$

Now, to get a 1 in the second-row, second-column position, multiply the second row by $\frac{1}{6}$.

$$\begin{bmatrix} 1 & -1 & | & 0 & 1 \\ 0 & 1 & | & \frac{1}{6} & -\frac{1}{3} \end{bmatrix} \quad \frac{1}{6}R_2$$

Finally, add the second row to the first row to get a 0 in the second column above the 1.

$$\begin{bmatrix} 1 & 0 & | & \frac{1}{6} & \frac{2}{3} \\ 0 & 1 & | & \frac{1}{6} & -\frac{1}{3} \end{bmatrix} \quad R_2 + R_1$$

The numbers in the first column to the right of the vertical bar give the values of $x$ and $z$. The second column gives the values of $y$ and $w$. That is,

$$\begin{bmatrix} 1 & 0 & | & x & y \\ 0 & 1 & | & z & w \end{bmatrix} = \begin{bmatrix} 1 & 0 & | & \frac{1}{6} & \frac{2}{3} \\ 0 & 1 & | & \frac{1}{6} & -\frac{1}{3} \end{bmatrix}$$

so that

$$A^{-1} = \begin{bmatrix} x & y \\ z & w \end{bmatrix} = \begin{bmatrix} \frac{1}{6} & \frac{2}{3} \\ \frac{1}{6} & -\frac{1}{3} \end{bmatrix}.$$

To check, multiply $A$ by $A^{-1}$. The result should be $I_2$.

$$AA^{-1} = \begin{bmatrix} 2 & 4 \\ 1 & -1 \end{bmatrix} \begin{bmatrix} \frac{1}{6} & \frac{2}{3} \\ \frac{1}{6} & -\frac{1}{3} \end{bmatrix} = \begin{bmatrix} \frac{1}{3} + \frac{2}{3} & \frac{4}{3} - \frac{4}{3} \\ \frac{1}{6} - \frac{1}{6} & \frac{2}{3} + \frac{1}{3} \end{bmatrix}$$

$$= \begin{bmatrix} 1 & 0 \\ 0 & 1 \end{bmatrix} = I_2$$

Finally,

$$A^{-1} = \begin{bmatrix} \frac{1}{6} & \frac{2}{3} \\ \frac{1}{6} & -\frac{1}{3} \end{bmatrix}.$$

The process for finding the multiplicative inverse $A^{-1}$ for any $n \times n$ matrix $A$ that has an inverse is summarized as follows.

## Finding an Inverse Matrix

To obtain $A^{-1}$ for any $n \times n$ matrix $A$ for which $A^{-1}$ exists, follow these steps.

**Step 1** Form the augmented matrix $[A \mid I_n]$, where $I_n$ is the $n \times n$ identity matrix.

**Step 2** Perform row operations on $[A \mid I_n]$ to obtain a matrix of the form $[I_n \mid B]$.

**Step 3** Matrix $B$ is $A^{-1}$.

To confirm that two $n \times n$ matrices $A$ and $B$ are inverses of each other, it is sufficient to show that $AB = I_n$. It is not necessary to show also that $BA = I_n$.

### EXAMPLE 2 Finding the Inverse of a 3 × 3 Matrix

Find $A^{-1}$ if $A = \begin{bmatrix} 1 & 0 & 1 \\ 2 & -2 & -1 \\ 3 & 0 & 0 \end{bmatrix}$.

Use row operations as follows.

**Step 1** Write the augmented matrix $[A \mid I_3]$.

$$\begin{bmatrix} 1 & 0 & 1 & | & 1 & 0 & 0 \\ 2 & -2 & -1 & | & 0 & 1 & 0 \\ 3 & 0 & 0 & | & 0 & 0 & 1 \end{bmatrix}$$

**Step 2** Since 1 is already in the upper left-hand corner as desired, begin by using a row operation that will result in a 0 for the first element in the second row. Multiply the elements of the first row by $-2$, and add the results to the second row.

$$\begin{bmatrix} 1 & 0 & 1 & | & 1 & 0 & 0 \\ 0 & -2 & -3 & | & -2 & 1 & 0 \\ 3 & 0 & 0 & | & 0 & 0 & 1 \end{bmatrix} \quad -2R_1 + R_2$$

To get 0 for the first element in the third row, multiply the elements of the first row by $-3$ and add to the third row.

$$\begin{bmatrix} 1 & 0 & 1 & | & 1 & 0 & 0 \\ 0 & -2 & -3 & | & -2 & 1 & 0 \\ 0 & 0 & -3 & | & -3 & 0 & 1 \end{bmatrix} \quad -3R_1 + R_3$$

To get 1 for the second element in the second row, multiply the elements of the second row by $-\frac{1}{2}$.

$$\begin{bmatrix} 1 & 0 & 1 & | & 1 & 0 & 0 \\ 0 & 1 & \frac{3}{2} & | & 1 & -\frac{1}{2} & 0 \\ 0 & 0 & -3 & | & -3 & 0 & 1 \end{bmatrix} \quad -\frac{1}{2}R_2$$

To get 1 for the third element in the third row, multiply the elements of the third row by $-\frac{1}{3}$.

$$\left[\begin{array}{ccc|ccc} 1 & 0 & 1 & 1 & 0 & 0 \\ 0 & 1 & \frac{3}{2} & 1 & -\frac{1}{2} & 0 \\ 0 & 0 & 1 & 1 & 0 & -\frac{1}{3} \end{array}\right] \quad -\frac{1}{3}R_3$$

To get 0 for the third element in the first row, multiply the elements of the third row by $-1$ and add to the first row.

$$\left[\begin{array}{ccc|ccc} 1 & 0 & 0 & 0 & 0 & \frac{1}{3} \\ 0 & 1 & \frac{3}{2} & 1 & -\frac{1}{2} & 0 \\ 0 & 0 & 1 & 1 & 0 & -\frac{1}{3} \end{array}\right] \quad -1R_3 + R_1$$

To get 0 for the third element in the second row, multiply the elements of the third row by $-\frac{3}{2}$ and add to the second row.

$$\left[\begin{array}{ccc|ccc} 1 & 0 & 0 & 0 & 0 & \frac{1}{3} \\ 0 & 1 & 0 & -\frac{1}{2} & -\frac{1}{2} & \frac{1}{2} \\ 0 & 0 & 1 & 1 & 0 & -\frac{1}{3} \end{array}\right] \quad -\frac{3}{2}R_3 + R_2$$

**Step 3** The last operation shows that the inverse is

$$A^{-1} = \left[\begin{array}{ccc} 0 & 0 & \frac{1}{3} \\ -\frac{1}{2} & -\frac{1}{2} & \frac{1}{2} \\ 1 & 0 & -\frac{1}{3} \end{array}\right].$$

Confirm this by forming the product $A^{-1}A$ or $AA^{-1}$, each of which should equal the matrix $I_3$.

**NOW TRY** Exercise 15.

As the examples indicate, the most efficient order in which to perform the row operations is to make changes column by column from left to right, so for each column the required 1 is the result of the first change. Next, perform the operations that obtain the 0s in that column. Then proceed to another column.

**EXAMPLE 3** Identifying a Matrix with No Inverse

Find $A^{-1}$, if possible, given $A = \begin{bmatrix} 2 & -4 \\ 1 & -2 \end{bmatrix}$.

Using row operations to change the first column of the augmented matrix

$$\left[\begin{array}{cc|cc} 2 & -4 & 1 & 0 \\ 1 & -2 & 0 & 1 \end{array}\right]$$

results in the following matrices:

$$\left[\begin{array}{cc|cc} 1 & -2 & \frac{1}{2} & 0 \\ 1 & -2 & 0 & 1 \end{array}\right] \quad \text{and} \quad \left[\begin{array}{cc|cc} 1 & -2 & \frac{1}{2} & 0 \\ 0 & 0 & -\frac{1}{2} & 1 \end{array}\right].$$

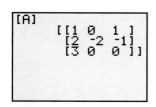

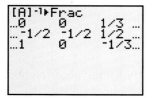

**FIGURE 12**

(We multiplied the elements in row one by $\frac{1}{2}$ in the first step.) At this point, the matrix should be changed so that the second-row, second-column element will be 1. Since that element is now 0, there is no way to complete the desired operations, so $A^{-1}$ does not exist for this matrix $A$. Just as there is no multiplicative inverse for the real number 0, not every matrix has a multiplicative inverse. Matrix $A$ is an example of such a matrix.

**NOW TRY** Exercise 13.

*If the inverse of a matrix exists, it is unique. That is, any given square matrix has no more than one inverse.*

A graphing calculator can find the inverse of a matrix. The screens in Figure 12 support the result of Example 2. The elements of the inverse are expressed as fractions.

**OBJECTIVE 3** Use inverse matrices to solve systems of linear equations. Matrix inverses can be used to solve square linear systems of equations. (A **square system** has the same number of equations as variables.) For example, given the linear system

$$a_{11}x + a_{12}y + a_{13}z = b_1$$
$$a_{21}x + a_{22}y + a_{23}z = b_2$$
$$a_{31}x + a_{32}y + a_{33}z = b_3,$$

the definition of matrix multiplication can be used to rewrite the system as

$$\begin{bmatrix} a_{11} & a_{12} & a_{13} \\ a_{21} & a_{22} & a_{23} \\ a_{31} & a_{32} & a_{33} \end{bmatrix} \cdot \begin{bmatrix} x \\ y \\ z \end{bmatrix} = \begin{bmatrix} b_1 \\ b_2 \\ b_3 \end{bmatrix}. \quad (1)$$

(To see this, multiply the matrices on the left.)

$$\text{If } A = \begin{bmatrix} a_{11} & a_{12} & a_{13} \\ a_{21} & a_{22} & a_{23} \\ a_{31} & a_{32} & a_{33} \end{bmatrix}, \quad X = \begin{bmatrix} x \\ y \\ z \end{bmatrix}, \quad \text{and} \quad B = \begin{bmatrix} b_1 \\ b_2 \\ b_3 \end{bmatrix},$$

the system given in (1) becomes

$$AX = B.$$

If $A^{-1}$ exists, then we can multiply both sides of $AX = B$ by $A^{-1}$ on the left to obtain

$$A^{-1}(AX) = A^{-1}B$$
$$(A^{-1}A)X = A^{-1}B \quad \text{Associative property}$$
$$I_3 X = A^{-1}B \quad \text{Inverse property}$$
$$X = A^{-1}B. \quad \text{Identity property}$$

Matrix $A^{-1}B$ gives the solution of the system.

**Solution of the Matrix Equation $AX = B$**

If $A$ is an $n \times n$ matrix with inverse $A^{-1}$, $X$ is an $n \times 1$ matrix of variables, and $B$ is an $n \times 1$ matrix, then the matrix equation

$$AX = B$$

has the solution

$$X = A^{-1}B.$$

This method of using matrix inverses to solve systems of equations is useful when the inverse is already known or when many systems of the form $AX = B$ must be solved and only $B$ changes.

### EXAMPLE 4 Solving Systems of Equations Using Matrix Inverses

Use the inverse of the coefficient matrix to solve each system.

**(a)** $2x - 3y = 4$
$\quad\; x + 5y = 2$

To represent the system as a matrix equation, use one matrix for the coefficients, one for the variables, and one for the constants, as follows.

$$A = \begin{bmatrix} 2 & -3 \\ 1 & 5 \end{bmatrix}, \quad X = \begin{bmatrix} x \\ y \end{bmatrix}, \quad \text{and} \quad B = \begin{bmatrix} 4 \\ 2 \end{bmatrix}$$

Write the system in matrix form as the equation $AX = B$.

$$AX = \begin{bmatrix} 2 & -3 \\ 1 & 5 \end{bmatrix} \begin{bmatrix} x \\ y \end{bmatrix} = \begin{bmatrix} 2x - 3y \\ x + 5y \end{bmatrix} = \begin{bmatrix} 4 \\ 2 \end{bmatrix} = B$$

To solve the system, first find $A^{-1}$.

$$A^{-1} = \begin{bmatrix} \frac{5}{13} & \frac{3}{13} \\ -\frac{1}{13} & \frac{2}{13} \end{bmatrix}$$

Next, find the product $A^{-1}B$.

$$A^{-1}B = \begin{bmatrix} \frac{5}{13} & \frac{3}{13} \\ -\frac{1}{13} & \frac{2}{13} \end{bmatrix} \begin{bmatrix} 4 \\ 2 \end{bmatrix} = \begin{bmatrix} 2 \\ 0 \end{bmatrix}$$

Since $X = A^{-1}B$,

$$X = \begin{bmatrix} x \\ y \end{bmatrix} = \begin{bmatrix} 2 \\ 0 \end{bmatrix}.$$

From the final matrix, the solution set of the system is $\{(2, 0)\}$.

**(b)** $2x - 3y = 1$
$\quad\; x + 5y = 20$

This system has the same matrix of coefficients as the system in part (a). Only matrix $B$ is different. Use $A^{-1}$ from part (a) and multiply by $B$ to obtain

$$X = A^{-1}B = \begin{bmatrix} \frac{5}{13} & \frac{3}{13} \\ -\frac{1}{13} & \frac{2}{13} \end{bmatrix} \begin{bmatrix} 1 \\ 20 \end{bmatrix} = \begin{bmatrix} 5 \\ 3 \end{bmatrix},$$

giving the solution set $\{(5, 3)\}$.

**NOW TRY** Exercise 25.

## 4.5 Exercises

**NOW TRY Exercise**

**1.** Let $A = \begin{bmatrix} 4 & -2 \\ 3 & 1 \end{bmatrix}$. Give the 2 × 2 identity matrix $I$ and show that $AI = A$. See Example 1.

**2.** In Exercise 1, verify that $IA = A$.

*Decide whether the given matrices are inverses of each other. (Check to see if their product is the identity matrix $I_n$.)*

**3.** $\begin{bmatrix} 5 & 7 \\ 2 & 3 \end{bmatrix}$ and $\begin{bmatrix} 3 & -7 \\ -2 & 5 \end{bmatrix}$

**4.** $\begin{bmatrix} 2 & 3 \\ 1 & 1 \end{bmatrix}$ and $\begin{bmatrix} -1 & 3 \\ 1 & -2 \end{bmatrix}$

**5.** $\begin{bmatrix} -1 & 2 \\ 3 & -5 \end{bmatrix}$ and $\begin{bmatrix} -5 & -2 \\ -3 & -1 \end{bmatrix}$

**6.** $\begin{bmatrix} 2 & 1 \\ 3 & 2 \end{bmatrix}$ and $\begin{bmatrix} 2 & 1 \\ -3 & 2 \end{bmatrix}$

**7.** $\begin{bmatrix} 0 & 1 & 0 \\ 0 & 0 & -2 \\ 1 & -1 & 0 \end{bmatrix}$ and $\begin{bmatrix} 1 & 0 & 1 \\ 1 & 0 & 0 \\ 0 & -1 & 0 \end{bmatrix}$

**8.** $\begin{bmatrix} 1 & 2 & 0 \\ 0 & 1 & 0 \\ 0 & 1 & 0 \end{bmatrix}$ and $\begin{bmatrix} 1 & -2 & 0 \\ 0 & 1 & 0 \\ 0 & -1 & 1 \end{bmatrix}$

**9.** $\begin{bmatrix} -1 & -1 & -1 \\ 4 & 5 & 0 \\ 0 & 1 & -3 \end{bmatrix}$ and $\begin{bmatrix} 15 & 4 & -5 \\ -12 & -3 & 4 \\ -4 & -1 & 1 \end{bmatrix}$

**10.** $\begin{bmatrix} 1 & 3 & 3 \\ 1 & 4 & 3 \\ 1 & 3 & 4 \end{bmatrix}$ and $\begin{bmatrix} 7 & -3 & -3 \\ -1 & 1 & 0 \\ -1 & 0 & 1 \end{bmatrix}$

*Find the inverse, if it exists, for each matrix. See Examples 2 and 3.*

**11.** $\begin{bmatrix} -1 & -2 \\ 3 & 4 \end{bmatrix}$

**12.** $\begin{bmatrix} 3 & -1 \\ -5 & 2 \end{bmatrix}$

**13.** $\begin{bmatrix} 5 & 10 \\ -3 & -6 \end{bmatrix}$

**14.** $\begin{bmatrix} -6 & 4 \\ -3 & 2 \end{bmatrix}$

**15.** $\begin{bmatrix} 1 & 0 & 1 \\ 0 & -1 & 0 \\ 2 & 1 & 1 \end{bmatrix}$

**16.** $\begin{bmatrix} 1 & 0 & 0 \\ 0 & -1 & 0 \\ 1 & 0 & 1 \end{bmatrix}$

**17.** $\begin{bmatrix} 1 & 3 & 3 \\ 1 & 4 & 3 \\ 1 & 3 & 4 \end{bmatrix}$

**18.** $\begin{bmatrix} -2 & 2 & 4 \\ -3 & 4 & 5 \\ 1 & 0 & 2 \end{bmatrix}$

**19.** $\begin{bmatrix} 2 & 2 & -4 \\ 2 & 6 & 0 \\ -3 & -3 & 5 \end{bmatrix}$

**20.** $\begin{bmatrix} 2 & 4 & 6 \\ -1 & -4 & -3 \\ 0 & 1 & -1 \end{bmatrix}$

**21.** $\begin{bmatrix} 1 & 1 & 0 & 2 \\ 2 & -1 & 1 & -1 \\ 3 & 3 & 2 & -2 \\ 1 & 2 & 1 & 0 \end{bmatrix}$

**22.** $\begin{bmatrix} 1 & -2 & 3 & 0 \\ 0 & 1 & -1 & 1 \\ -2 & 2 & -2 & 4 \\ 0 & 2 & -3 & 1 \end{bmatrix}$

*Solve each system by using the inverse of the coefficient matrix. See Example 4.*

**23.** $-x + y = 1$
    $2x - y = 1$

**24.** $x + y = 5$
    $x - y = -1$

**25.** $2x - y = -8$
    $3x + y = -2$

**26.** $x + 3y = -12$
    $2x - y = 11$

**27.** $2x + 3y = -10$
    $3x + 4y = -12$

**28.** $2x - 3y = 10$
    $2x + 2y = 5$

*Solve each system of equations by using the inverse of the coefficient matrix. The inverses were found in Exercises 17–20. See Example 4.*

29. $x + 3y + 3z = 1$
    $x + 4y + 3z = 0$
    $x + 3y + 4z = -1$

30. $-2x + 2y + 4z = 3$
    $-3x + 4y + 5z = 1$
    $x + 2z = 2$

31. $2x + 2y - 4z = 12$
    $2x + 6y = 16$
    $-3x - 3y + 5z = -20$

32. $2x + 4y + 6z = 4$
    $-x - 4y - 3z = 8$
    $y - z = -4$

*Solve each system of equations by using the inverse of the coefficient matrix. The inverses were found in Exercises 21 and 22.*

33. $x + y + 2w = 3$
    $2x - y + z - w = 3$
    $3x + 3y + 2z - 2w = 5$
    $x + 2y + z = 3$

34. $x - 2y + 3z = 1$
    $y - z + w = -1$
    $-2x + 2y - 2z + 4w = 2$
    $2y - 3z + w = -3$

*Solve each problem.*

35. The amount of plate-glass sales $S$ (in millions of dollars) can be affected by the number of new building contracts $B$ issued (in millions) and automobiles $A$ produced (in millions). A plate-glass company in California wants to forecast future sales by using the past three years of sales. The totals for three years are given in the table.

| S | A | B |
|---|---|---|
| 602.7 | 5.543 | 37.14 |
| 656.7 | 6.933 | 41.30 |
| 778.5 | 7.638 | 45.62 |

To describe the relationship between these variables, the equation

$$S = a + bA + cB$$

was used, where the coefficients $a$, $b$, and $c$ are constants that must be determined. (*Source:* Makridakis, S., and S. Wheelwright, *Forecasting Methods for Management,* John Wiley & Sons, 1989.)

(a) Substitute the values for $S$, $A$, and $B$ for each year from the table into the equation $S = a + bA + cB$, and obtain three linear equations involving $a$, $b$, and $c$.

(b) Use a graphing calculator to solve this linear system for $a$, $b$, and $c$. Use matrix inverse methods.

(c) Write the equation for $S$ using these values for the coefficients.

(d) For the next year it is estimated that $A = 7.752$ and $B = 47.38$. Predict $S$. (The actual value for $S$ was 877.6.)

(e) It is predicted that in six years $A = 8.9$ and $B = 66.25$. Find the value of $S$ in this situation and discuss its validity.

36. The number of automobile tire sales is dependent on several variables. In one study the relationship between annual tire sales $S$ (in thousands of dollars), automobile registrations $R$ (in millions), and personal disposable income $I$ (in millions of dollars) was investigated. The results for three years are given in the table.

| S | R | I |
|---|---|---|
| 10,170 | 112.9 | 307.5 |
| 15,305 | 132.9 | 621.63 |
| 21,289 | 155.2 | 1937.13 |

To describe the relationship between these variables, mathematicians often use the equation
$$S = a + bR + cI,$$
where the coefficients $a$, $b$, and $c$ are constants that must be determined before the equation can be used. (*Source:* Jarrett, J., *Business Forecasting Methods,* Basil Blackwell, 1991.)

(a) Substitute the values for $S$, $R$, and $I$ for each year from the table into the equation $S = a + bR + cI$, and obtain three linear equations involving $a$, $b$, and $c$.
(b) Use a graphing calculator to solve this linear system for $a$, $b$, and $c$. Use matrix inverse methods.
(c) Write the equation for $S$ using these values for the coefficients.
(d) If $R = 117.6$ and $I = 310.73$, predict $S$. (The actual value for $S$ was 11,314.)
(e) If $R = 143.8$ and $I = 829.06$, predict $S$. (The actual value for $S$ was 18,481.)

## 4.6 Determinants and Cramer's Rule

**OBJECTIVES**

1. Evaluate 2 × 2 determinants.
2. Use expansion by minors to evaluate 3 × 3 determinants.
3. Understand the derivation of Cramer's rule.
4. Apply Cramer's rule to solve linear systems.

Recall from **Section 4.3** that an ordered array of numbers within square brackets is called a *matrix* (plural *matrices*). Matrices are named according to the number of rows and columns they contain. A *square matrix* has the same number of rows and columns.

$$\text{Rows} \rightarrow \begin{bmatrix} 2 & 3 & 5 \\ 7 & 1 & 2 \end{bmatrix} \quad \begin{array}{c} 2 \times 3 \\ \text{matrix} \end{array} \qquad \begin{bmatrix} -1 & 0 \\ 1 & -2 \end{bmatrix} \quad \begin{array}{c} 2 \times 2 \\ \text{square matrix} \end{array}$$

Associated with every *square matrix* is a real number called the **determinant** of the matrix. A determinant is symbolized by the entries of the matrix placed between two vertical lines, such as

$$\begin{vmatrix} 2 & 3 \\ 7 & 1 \end{vmatrix} \quad \begin{array}{c} 2 \times 2 \\ \text{determinant} \end{array} \qquad \begin{vmatrix} 7 & 4 & 3 \\ 0 & 1 & 5 \\ 6 & 0 & 1 \end{vmatrix}. \quad \begin{array}{c} 3 \times 3 \\ \text{determinant} \end{array}$$

Like matrices, determinants are named according to the number of rows and columns they contain.

**OBJECTIVE 1** Evaluate 2 × 2 determinants. As mentioned above, the value of a determinant is a *real number*. We use the following rule to evaluate a 2 × 2 determinant.

---

**Value of a 2 × 2 Determinant**

$$\begin{vmatrix} a & b \\ c & d \end{vmatrix} = ad - bc$$

**EXAMPLE 1** Evaluating a 2 × 2 Determinant

Evaluate the determinant.

$$\begin{vmatrix} -1 & -3 \\ 4 & -2 \end{vmatrix}$$

Here $a = -1$, $b = -3$, $c = 4$, and $d = -2$, so

$$\begin{vmatrix} -1 & -3 \\ 4 & -2 \end{vmatrix} = -1(-2) - (-3)4 = 2 + 12 = 14.$$

**NOW TRY** Exercise 3.

A 3 × 3 determinant is evaluated in a similar way.

**Value of a 3 × 3 Determinant**

$$\begin{vmatrix} a_1 & b_1 & c_1 \\ a_2 & b_2 & c_2 \\ a_3 & b_3 & c_3 \end{vmatrix} = (a_1 b_2 c_3 + b_1 c_2 a_3 + c_1 a_2 b_3) \\ - (a_3 b_2 c_1 + b_3 c_2 a_1 + c_3 a_2 b_1)$$

To calculate a 3 × 3 determinant, we rearrange terms using the distributive property.

$$\begin{vmatrix} a_1 & b_1 & c_1 \\ a_2 & b_2 & c_2 \\ a_3 & b_3 & c_3 \end{vmatrix} = a_1(b_2 c_3 - b_3 c_2) - a_2(b_1 c_3 - b_3 c_1) + a_3(b_1 c_2 - b_2 c_1) \quad (1)$$

Each quantity in parentheses represents a 2 × 2 determinant that is the part of the 3 × 3 determinant remaining when the row and column of the multiplier are eliminated, as shown below.

$a_1(b_2 c_3 - b_3 c_2)$ $\quad \begin{vmatrix} a_1 & b_1 & c_1 \\ a_2 & b_2 & c_2 \\ a_3 & b_3 & c_3 \end{vmatrix}$ Eliminate the 1st row and 1st column.

$a_2(b_1 c_3 - b_3 c_1)$ $\quad \begin{vmatrix} a_1 & b_1 & c_1 \\ a_2 & b_2 & c_2 \\ a_3 & b_3 & c_3 \end{vmatrix}$ Eliminate the 2nd row and 1st column.

$a_3(b_1 c_2 - b_2 c_1)$ $\quad \begin{vmatrix} a_1 & b_1 & c_1 \\ a_2 & b_2 & c_2 \\ a_3 & b_3 & c_3 \end{vmatrix}$ Eliminate the 3rd row and 1st column.

These 2 × 2 determinants are called **minors** of the elements in the 3 × 3 determinant. In the determinant above, the minors of $a_1$, $a_2$, and $a_3$ are, respectively,

$$\begin{vmatrix} b_2 & c_2 \\ b_3 & c_3 \end{vmatrix}, \quad \begin{vmatrix} b_1 & c_1 \\ b_3 & c_3 \end{vmatrix}, \quad \text{and} \quad \begin{vmatrix} b_1 & c_1 \\ b_2 & c_2 \end{vmatrix}. \quad \text{Minors}$$

**OBJECTIVE 2** Use expansion by minors to evaluate 3 × 3 determinants. We evaluate a 3 × 3 determinant by multiplying each element in the first column by its minor and combining the products as indicated in equation (1). This procedure is called **expansion of the determinant by minors** about the first column.

## EXAMPLE 2 Evaluating a 3 × 3 Determinant

Evaluate the determinant using expansion by minors about the first column.

$$\begin{vmatrix} 1 & 3 & -2 \\ -1 & -2 & -3 \\ 1 & 1 & 2 \end{vmatrix}$$

In this determinant, $a_1 = 1$, $a_2 = -1$, and $a_3 = 1$. Multiply each of these numbers by its minor, and combine the three terms using the definition. Notice that the second term in the definition is *subtracted*.

$$\begin{vmatrix} 1 & 3 & -2 \\ -1 & -2 & -3 \\ 1 & 1 & 2 \end{vmatrix} = 1\begin{vmatrix} -2 & -3 \\ 1 & 2 \end{vmatrix} - (-1)\begin{vmatrix} 3 & -2 \\ 1 & 2 \end{vmatrix} + 1\begin{vmatrix} 3 & -2 \\ -2 & -3 \end{vmatrix}$$

**Use parentheses and brackets to avoid errors.**

$$= 1[-2(2) - (-3)1] + 1[3(2) - (-2)1]$$
$$+ 1[3(-3) - (-2)(-2)]$$
$$= 1(-1) + 1(8) + 1(-13)$$
$$= -1 + 8 - 13$$
$$= -6$$

**NOW TRY** Exercise 9.

To obtain equation (1), we could have rearranged terms in the definition of the determinant and used the distributive property to factor out the three elements of the second or third column or of any of the three rows. Therefore, expanding by minors about any row or any column results in the same value for a 3 × 3 determinant. To determine the correct signs for the terms of other expansions, the following **array of signs** is helpful.

### Array of Signs for a 3 × 3 Determinant

$$\begin{matrix} + & - & + \\ - & + & - \\ + & - & + \end{matrix}$$

The signs alternate for each row and column beginning with a + in the first row, first column position. For example, if the expansion is to be about the second column, the first term would have a minus sign associated with it, the second term a plus sign, and the third term a minus sign.

## EXAMPLE 3 Evaluating a 3 × 3 Determinant

Evaluate the determinant of Example 2 using expansion by minors about the second column.

$$\begin{vmatrix} 1 & 3 & -2 \\ -1 & -2 & -3 \\ 1 & 1 & 2 \end{vmatrix} = -3\begin{vmatrix} -1 & -3 \\ 1 & 2 \end{vmatrix} + (-2)\begin{vmatrix} 1 & -2 \\ 1 & 2 \end{vmatrix} - 1\begin{vmatrix} 1 & -2 \\ -1 & -3 \end{vmatrix}$$
$$= -3(1) - 2(4) - 1(-5)$$
$$= -3 - 8 + 5$$
$$= -6 \qquad \text{The result is the same as in Example 2.}$$

**NOW TRY** Exercise 15.

**OBJECTIVE 3** **Understand the derivation of Cramer's rule.** We can use determinants to solve a system of equations of the form

$$a_1x + b_1y = c_1 \quad (1)$$
$$a_2x + b_2y = c_2. \quad (2)$$

The result will be a formula that can be used to solve any system of two equations with two variables. To get this general solution, we eliminate $y$ and solve for $x$ by first multiplying each side of equation (1) by $b_2$ and each side of equation (2) by $-b_1$. Then we add these results and solve for $x$.

$$\begin{array}{ll} a_1b_2x + b_1b_2y = c_1b_2 & \text{Multiply equation (1) by } b_2. \\ -a_2b_1x - b_1b_2y = -c_2b_1 & \text{Multiply equation (2) by } -b_1. \\ \hline (a_1b_2 - a_2b_1)x = c_1b_2 - c_2b_1 & \text{Add.} \end{array}$$

$$x = \frac{c_1b_2 - c_2b_1}{a_1b_2 - a_2b_1} \quad (\text{if } a_1b_2 - a_2b_1 \neq 0)$$

To solve for $y$, we multiply each side of equation (1) by $-a_2$ and each side of equation (2) by $a_1$ and add.

$$\begin{array}{ll} -a_1a_2x - a_2b_1y = -a_2c_1 & \text{Multiply equation (1) by } -a_2. \\ a_1a_2x + a_1b_2y = a_1c_2 & \text{Multiply equation (2) by } a_1. \\ \hline (a_1b_2 - a_2b_1)y = a_1c_2 - a_2c_1 & \text{Add.} \end{array}$$

$$y = \frac{a_1c_2 - a_2c_1}{a_1b_2 - a_2b_1} \quad (\text{if } a_1b_2 - a_2b_1 \neq 0)$$

We can write both numerators and the common denominator of these values for $x$ and $y$ as determinants because

$$a_1c_2 - a_2c_1 = \begin{vmatrix} a_1 & c_1 \\ a_2 & c_2 \end{vmatrix}, \quad c_1b_2 - c_2b_1 = \begin{vmatrix} c_1 & b_1 \\ c_2 & b_2 \end{vmatrix}, \text{ and } a_1b_2 - a_2b_1 = \begin{vmatrix} a_1 & b_1 \\ a_2 & b_2 \end{vmatrix}.$$

Using these results, the solutions for $x$ and $y$ become

$$x = \frac{\begin{vmatrix} c_1 & b_1 \\ c_2 & b_2 \end{vmatrix}}{\begin{vmatrix} a_1 & b_1 \\ a_2 & b_2 \end{vmatrix}} \quad \text{and} \quad y = \frac{\begin{vmatrix} a_1 & c_1 \\ a_2 & c_2 \end{vmatrix}}{\begin{vmatrix} a_1 & b_1 \\ a_2 & b_2 \end{vmatrix}}, \quad \begin{vmatrix} a_1 & b_1 \\ a_2 & b_2 \end{vmatrix} \neq 0.$$

For convenience, we denote the three determinants in the solution as

$$\begin{vmatrix} a_1 & b_1 \\ a_2 & b_2 \end{vmatrix} = D, \quad \begin{vmatrix} c_1 & b_1 \\ c_2 & b_2 \end{vmatrix} = D_x, \quad \text{and} \quad \begin{vmatrix} a_1 & c_1 \\ a_2 & c_2 \end{vmatrix} = D_y.$$

Notice that the elements of $D$ are the four coefficients of the variables in the given system; the elements of $D_x$ are obtained by replacing the coefficients of $x$ by the respective constants; the elements of $D_y$ are obtained by replacing the coefficients of $y$ by the respective constants.

These results are summarized as **Cramer's rule.**

**Cramer's Rule for 2 × 2 Systems**

Given the system $\begin{aligned} a_1x + b_1y &= c_1 \\ a_2x + b_2y &= c_2 \end{aligned}$ with $a_1b_2 - a_2b_1 = D \neq 0$, then

$$x = \frac{\begin{vmatrix} c_1 & b_1 \\ c_2 & b_2 \end{vmatrix}}{\begin{vmatrix} a_1 & b_1 \\ a_2 & b_2 \end{vmatrix}} = \frac{D_x}{D} \quad \text{and} \quad y = \frac{\begin{vmatrix} a_1 & c_1 \\ a_2 & c_2 \end{vmatrix}}{\begin{vmatrix} a_1 & b_1 \\ a_2 & b_2 \end{vmatrix}} = \frac{D_y}{D}.$$

**OBJECTIVE 4** Apply Cramer's rule to solve linear systems. To use Cramer's rule to solve a system of equations, we find the three determinants, $D$, $D_x$, and $D_y$, and then write the necessary quotients for $x$ and $y$.

**CAUTION** As indicated in the box, *Cramer's rule does not apply if* $D = a_1b_2 - a_2b_1 = 0$. When $D = 0$, the system is inconsistent or has dependent equations. For this reason, it is a good idea to evaluate $D$ first.

**EXAMPLE 4** Using Cramer's Rule to Solve a 2 × 2 System

Use Cramer's rule to solve the system

$$5x + 7y = -1$$
$$6x + 8y = 1.$$

By Cramer's rule, $x = \frac{D_x}{D}$ and $y = \frac{D_y}{D}$. If $D \neq 0$, then we find $D_x$ and $D_y$.

$$D = \begin{vmatrix} 5 & 7 \\ 6 & 8 \end{vmatrix} = 5(8) - 7(6) = -2$$

$$D_x = \begin{vmatrix} -1 & 7 \\ 1 & 8 \end{vmatrix} = -1(8) - 7(1) = -15$$

$$D_y = \begin{vmatrix} 5 & -1 \\ 6 & 1 \end{vmatrix} = 5(1) - (-1)6 = 11$$

From Cramer's rule,

$$x = \frac{D_x}{D} = \frac{-15}{-2} = \frac{15}{2} \quad \text{and} \quad y = \frac{D_y}{D} = \frac{11}{-2} = -\frac{11}{2}.$$

The solution set is $\left\{\left(\frac{15}{2}, -\frac{11}{2}\right)\right\}$, as can be verified by checking in the given system.

**NOW TRY** Exercise 23.

We can extend Cramer's rule to systems of three equations with three variables.

**Cramer's Rule for 3 × 3 Systems**

Given the system

$$a_1 x + b_1 y + c_1 z = d_1$$
$$a_2 x + b_2 y + c_2 z = d_2$$
$$a_3 x + b_3 y + c_3 z = d_3$$

with

$$D_x = \begin{vmatrix} d_1 & b_1 & c_1 \\ d_2 & b_2 & c_2 \\ d_3 & b_3 & c_3 \end{vmatrix}, \quad D_y = \begin{vmatrix} a_1 & d_1 & c_1 \\ a_2 & d_2 & c_2 \\ a_3 & d_3 & c_3 \end{vmatrix},$$

$$D_z = \begin{vmatrix} a_1 & b_1 & d_1 \\ a_2 & b_2 & d_2 \\ a_3 & b_3 & d_3 \end{vmatrix}, \quad D = \begin{vmatrix} a_1 & b_1 & c_1 \\ a_2 & b_2 & c_2 \\ a_3 & b_3 & c_3 \end{vmatrix} \neq 0,$$

then

$$x = \frac{D_x}{D}, \quad y = \frac{D_y}{D}, \quad \text{and} \quad z = \frac{D_z}{D}.$$

**EXAMPLE 5  Using Cramer's Rule to Solve a 3 × 3 System**

Use Cramer's rule to solve the system

$$x + y - z + 2 = 0$$
$$2x - y + z + 5 = 0$$
$$x - 2y + 3z - 4 = 0.$$

To use Cramer's rule, we first rewrite the system in the form

$$x + y - z = -2$$
$$2x - y + z = -5$$
$$x - 2y + 3z = 4.$$

We expand by minors about row 1 to find $D$.

$$D = \begin{vmatrix} 1 & 1 & -1 \\ 2 & -1 & 1 \\ 1 & -2 & 3 \end{vmatrix}$$

$$= 1 \begin{vmatrix} -1 & 1 \\ -2 & 3 \end{vmatrix} - 1 \begin{vmatrix} 2 & 1 \\ 1 & 3 \end{vmatrix} + (-1) \begin{vmatrix} 2 & -1 \\ 1 & -2 \end{vmatrix}$$

$$= 1(-1) - 1(5) - 1(-3)$$

$$= -3$$

Expanding $D_x$ by minors about row 1 gives

$$D_x = \begin{vmatrix} -2 & 1 & -1 \\ -5 & -1 & 1 \\ 4 & -2 & 3 \end{vmatrix}$$

$$= -2 \begin{vmatrix} -1 & 1 \\ -2 & 3 \end{vmatrix} - 1 \begin{vmatrix} -5 & 1 \\ 4 & 3 \end{vmatrix} + (-1) \begin{vmatrix} -5 & -1 \\ 4 & -2 \end{vmatrix}$$

$$= -2(-1) - 1(-19) - 1(14)$$

$$= 7.$$

Verify that $D_y = -22$ and $D_z = -21$. Thus,

$$x = \frac{D_x}{D} = \frac{7}{-3} = -\frac{7}{3}, \qquad y = \frac{D_y}{D} = \frac{-22}{-3} = \frac{22}{3}, \qquad z = \frac{D_z}{D} = \frac{-21}{-3} = 7.$$

Check that the solution set is $\left\{\left(-\frac{7}{3}, \frac{22}{3}, 7\right)\right\}$.

**NOW TRY** Exercise 29.

---

**EXAMPLE 6** Determining When Cramer's Rule Does Not Apply

Use Cramer's rule, if possible, to solve the system

$$\begin{aligned} 2x - 3y + 4z &= 8 \\ 6x - 9y + 12z &= 24 \\ x + 2y - 3z &= 5. \end{aligned}$$

First, find $D$.

$$D = \begin{vmatrix} 2 & -3 & 4 \\ 6 & -9 & 12 \\ 1 & 2 & -3 \end{vmatrix}$$

$$= 2\begin{vmatrix} -9 & 12 \\ 2 & -3 \end{vmatrix} - 6\begin{vmatrix} -3 & 4 \\ 2 & -3 \end{vmatrix} + 1\begin{vmatrix} -3 & 4 \\ -9 & 12 \end{vmatrix}$$

$$= 2(3) - 6(1) + 1(0)$$

$$= 0$$

Since $D = 0$ here, Cramer's rule does not apply and we must use another method to solve the system. Multiplying each side of the first equation by 3 shows that the first two equations have the same solution set, so this system has dependent equations and an infinite solution set.

**NOW TRY** Exercise 31.

---

## 4.6 Exercises

1. *True or False* Decide whether each statement is *true* or *false*. If a statement is false, explain why.

   (a) A matrix is an array of numbers, while a determinant is a single number.
   
   (b) A square matrix has the same number of rows as columns.
   
   (c) The determinant $\begin{vmatrix} a & b \\ c & d \end{vmatrix}$ is equal to $ad + bc$.
   
   (d) The value of $\begin{vmatrix} 0 & 0 \\ x & y \end{vmatrix}$ is 0 for any replacements for $x$ and $y$.

2. *Multiple Choice* Which one of the following is the expression for the determinant

   $$\begin{vmatrix} -2 & -3 \\ 4 & -6 \end{vmatrix}?$$

   **A.** $-2(-6) + (-3)4$  **B.** $-2(-6) - 3(4)$
   **C.** $-3(4) - (-2)(-6)$  **D.** $-2(-6) - (-3)4$

*Evaluate each determinant. See Example 1.*

**3.** $\begin{vmatrix} -2 & 5 \\ -1 & 4 \end{vmatrix}$    **4.** $\begin{vmatrix} 3 & -6 \\ 2 & -2 \end{vmatrix}$    **5.** $\begin{vmatrix} 1 & -2 \\ 7 & 0 \end{vmatrix}$

**6.** $\begin{vmatrix} -5 & -1 \\ 1 & 0 \end{vmatrix}$    **7.** $\begin{vmatrix} 0 & 4 \\ 0 & 4 \end{vmatrix}$    **8.** $\begin{vmatrix} 8 & -3 \\ 0 & 0 \end{vmatrix}$

*Evaluate each determinant by expansion by minors about the first column. See Example 2.*

**9.** $\begin{vmatrix} -1 & 2 & 4 \\ -3 & -2 & -3 \\ 2 & -1 & 5 \end{vmatrix}$    **10.** $\begin{vmatrix} 2 & -3 & -5 \\ 1 & 2 & 2 \\ 5 & 3 & -1 \end{vmatrix}$

**11.** $\begin{vmatrix} 1 & 0 & -2 \\ 0 & 2 & 3 \\ 1 & 0 & 5 \end{vmatrix}$    **12.** $\begin{vmatrix} 2 & -1 & 0 \\ 0 & -1 & 1 \\ 1 & 2 & 0 \end{vmatrix}$

**13.** Explain in your own words how to evaluate a 2 × 2 determinant. Illustrate with an example.

*Evaluate each determinant by expansion by minors about any row or column. (Hint: The work is easier if you choose a row or a column with 0s.) See Example 3.*

**14.** $\begin{vmatrix} 4 & 4 & 2 \\ 1 & -1 & -2 \\ 1 & 0 & 2 \end{vmatrix}$    **15.** $\begin{vmatrix} 3 & -1 & 2 \\ 1 & 5 & -2 \\ 0 & 2 & 0 \end{vmatrix}$

**16.** $\begin{vmatrix} 3 & 5 & -2 \\ 1 & -4 & 1 \\ 3 & 1 & -2 \end{vmatrix}$    **17.** $\begin{vmatrix} 0 & 0 & 3 \\ 4 & 0 & -2 \\ 2 & -1 & 3 \end{vmatrix}$

**18.** $\begin{vmatrix} 3 & 0 & -2 \\ 1 & -4 & 1 \\ 3 & 1 & -2 \end{vmatrix}$    **19.** $\begin{vmatrix} 1 & 1 & 2 \\ 5 & 5 & 7 \\ 3 & 3 & 1 \end{vmatrix}$

**20.** *Matching*  Consider the system

$$4x + 3y - 2z = 1$$
$$7x - 4y + 3z = 2$$
$$-2x + y - 8z = 0.$$

Match each determinant in parts (a)–(d) with its correct representation from choices A–D.

**(a)** $D$    **(b)** $D_x$    **(c)** $D_y$    **(d)** $D_z$

**A.** $\begin{vmatrix} 1 & 3 & -2 \\ 2 & -4 & 3 \\ 0 & 1 & -8 \end{vmatrix}$    **B.** $\begin{vmatrix} 4 & 3 & 1 \\ 7 & -4 & 2 \\ -2 & 1 & 0 \end{vmatrix}$    **C.** $\begin{vmatrix} 4 & 1 & -2 \\ 7 & 2 & 3 \\ -2 & 0 & -8 \end{vmatrix}$    **D.** $\begin{vmatrix} 4 & 3 & -2 \\ 7 & -4 & 3 \\ -2 & 1 & -8 \end{vmatrix}$

**21.** For the system

$$x + 3y - 6z = 7$$
$$2x - y + z = 1$$
$$x + 2y + 2z = -1,$$

$D = -43$, $D_x = -43$, $D_y = 0$, and $D_z = 43$. What is the solution set of the system?

*Use Cramer's rule to solve each linear system in two variables. See Example 4.*

22. $3x + 5y = -5$
    $-2x + 3y = 16$

23. $5x + 2y = -3$
    $4x - 3y = -30$

24. $8x + 3y = 1$
    $6x - 5y = 2$

25. $3x - y = 9$
    $2x + 5y = 8$

26. $2x + 3y = 4$
    $5x + 6y = 7$

27. $4x + 5y = 6$
    $7x + 8y = 9$

*Use Cramer's rule where applicable to solve each linear system in three variables. See Examples 5 and 6.*

28. $2x + 3y + 2z = 15$
    $x - y + 2z = 5$
    $x + 2y - 6z = -26$

29. $x - y + 6z = 19$
    $3x + 3y - z = 1$
    $x + 9y + 2z = -19$

30. $2x - 3y + 4z = 8$
    $6x - 9y + 12z = 24$
    $-4x + 6y - 8z = -16$

31. $7x + y - z = 4$
    $2x - 3y + z = 2$
    $-6x + 9y - 3z = -6$

32. $3x + 5z = 0$
    $2x + 3y = 1$
    $-y + 2z = -11$

33. $-x + 2y = 4$
    $3x + y = -5$
    $2x + z = -1$

34. $x - 3y = 13$
    $2y + z = 5$
    $-x + z = -7$

35. $-5x - y = -10$
    $3x + 2y + z = -3$
    $-y - 2z = -13$

36. Under what conditions can a system *not* be solved using Cramer's rule?

# 4 SUMMARY

## KEY TERMS

**4.1** system of equations
system of linear equations
solution set of a linear system
consistent system
independent equations
inconsistent system
dependent equations

elimination method
substitution method
**4.2** ordered triple
**4.3** matrix
element of a matrix
row
column
dimensions of a matrix
square matrix
augmented matrix

row operations
row echelon form
**4.4** square matrix of order $n$
row matrix
column matrix
zero matrix
additive inverse (or negative) of a matrix
scalar

**4.5** identity matrix
multiplicative inverse of a matrix
square system
**4.6** determinant
minor
expansion by minors
array of signs
Cramer's rule

## NEW SYMBOLS

$(x, y, z)$  ordered triple

$\begin{bmatrix} a & b & c \\ d & e & f \end{bmatrix}$  matrix with two rows, three columns

$a_{ij}$  the element in row $i$, column $j$ of a matrix

$I_n$  $n \times n$ identity matrix

$A^{-1}$  multiplicative inverse of matrix $A$

$|A|$  determinant of matrix $A$

$D, D_x, D_y, D_z$  determinants used in Cramer's rule

## QUICK REVIEW

| CONCEPTS | EXAMPLES |
|---|---|

### 4.1 Systems of Linear Equations in Two Variables

**Solving a Linear System by Substitution**

**Step 1** Solve one of the equations for either variable.

**Step 2** Substitute for that variable in the other equation. The result should be an equation with just one variable.

**Step 3** Solve the equation from Step 2.

**Step 4** Find the value of the other variable by substituting the result from Step 3 into the equation from Step 1.

**Step 5** Check the solution in both of the original equations. Then write the solution set.

Solve by substitution: $\begin{aligned} 4x - y &= 7 \quad (1) \\ 3x + 2y &= 30. \quad (2) \end{aligned}$

Solve for $y$ in equation (1): $y = 4x - 7$.

Substitute $4x - 7$ for $y$ in equation (2), and solve for $x$.

$3x + 2y = 30$  (2)
$3x + 2(4x - 7) = 30$  Let $y = 4x - 7$.
$3x + 8x - 14 = 30$  Distributive property
$11x = 44$  Combine terms; add 14.
$x = 4$  Divide by 11.

Substitute 4 for $x$ in the equation $y = 4x - 7$ to find that $y = 4(4) - 7 = 9$.

Check to see that $\{(4, 9)\}$ is the solution set.

**Solving a Linear System by Elimination**

**Step 1** Write both equations in standard form.

**Step 2** Make the coefficients of one pair of variable terms opposites.

**Step 3** Add the new equations. The sum should be an equation with just one variable.

**Step 4** Solve the equation from Step 3.

Solve by elimination: $\begin{aligned} 5x + y &= 2 \quad (1) \\ 2x - 3y &= 11. \quad (2) \end{aligned}$

To eliminate $y$, multiply equation (1) by 3 and add the result to equation (2).

$\begin{aligned} 15x + 3y &= 6 \quad &\text{3 times equation (1)} \\ 2x - 3y &= 11 \quad &(2) \\ \hline 17x &= 17 \quad &\text{Add.} \\ x &= 1 \quad &\text{Divide by 17.} \end{aligned}$

*(continued)*

| CONCEPTS | EXAMPLES |
|---|---|
| **Step 5** Find the value of the other variable by substituting the result of Step 4 into either of the original equations. | Let $x = 1$ in equation (1), and solve for $y$. $$5(1) + y = 2$$ $$y = -3$$ |
| **Step 6** Check the solution in both of the original equations. Then write the solution set. | Check to verify that $\{(1, -3)\}$ is the solution set. |

## 4.2 Systems of Linear Equations in Three Variables

| | |
|---|---|
| **Solving a Linear System in Three Variables** | Solve the system |
| **Step 1** Use the elimination method to eliminate any variable from any two of the original equations. | $$x + 2y - z = 6 \quad (1)$$ $$x + y + z = 6 \quad (2)$$ $$2x + y - z = 7. \quad (3)$$ Add equations (1) and (2); $z$ is eliminated and the result is $2x + 3y = 12$. |
| **Step 2** Eliminate the *same* variable from any *other* two equations. | Eliminate $z$ again by adding equations (2) and (3) to get $3x + 2y = 13$. Now solve the system $$2x + 3y = 12 \quad (4)$$ $$3x + 2y = 13. \quad (5)$$ |
| **Step 3** Eliminate a second variable from the two equations in two variables that result from Steps 1 and 2. The result is an equation in one variable that gives the value of that variable. | To eliminate $x$, multiply equation (4) by $-3$ and equation (5) by 2. $$-6x - 9y = -36$$ $$\underline{6x + 4y = 26}$$ $$-5y = -10$$ $$y = 2$$ |
| **Step 4** Substitute the value of the variable found in Step 3 into either of the equations in two variables to find the value of the second variable. | Let $y = 2$ in equation (4). $$2x + 3(2) = 12$$ $$2x + 6 = 12$$ $$2x = 6$$ $$x = 3$$ |
| **Step 5** Use the values of the two variables from Steps 3 and 4 to find the value of the third variable by substituting into an appropriate equation. | Let $y = 2$ and $x = 3$ in any of the original equations to find that $z = 1$. |
| **Step 6** Check the solution in all of the original equations. Then write the solution set. | *Check.* The solution set is $\{(3, 2, 1)\}$. |

## 4.3 Solving Systems of Linear Equations by Matrix Methods

**Matrix Row Operations**

1. Any two rows of the matrix may be interchanged.

$$\begin{bmatrix} 1 & 5 & 7 \\ 3 & 9 & -2 \\ 0 & 6 & 4 \end{bmatrix} \text{ becomes } \begin{bmatrix} 3 & 9 & -2 \\ 1 & 5 & 7 \\ 0 & 6 & 4 \end{bmatrix} \quad \text{Interchange } R_1 \text{ and } R_2.$$

2. The elements of any row may be multiplied by any nonzero real number.

$$\begin{bmatrix} 1 & 5 & 7 \\ 3 & 9 & -2 \\ 0 & 6 & 4 \end{bmatrix} \text{ becomes } \begin{bmatrix} 1 & 5 & 7 \\ 1 & 3 & -\frac{2}{3} \\ 0 & 6 & 4 \end{bmatrix} \quad \tfrac{1}{3} R_2$$

3. Any row may be changed by adding to the elements of the row the product of a real number and the elements of another row.

$$\begin{bmatrix} 1 & 5 & 7 \\ 3 & 9 & -2 \\ 0 & 6 & 4 \end{bmatrix} \text{ becomes } \begin{bmatrix} 1 & 5 & 7 \\ 0 & -6 & -23 \\ 0 & 6 & 4 \end{bmatrix} \quad -3R_1 + R_2$$

*(continued)*

| CONCEPTS | EXAMPLES |
|---|---|
| A system can be solved by matrix methods. Write the augmented matrix and use row operations to obtain a matrix in row echelon form. | Solve using row operations: $\begin{array}{l} x + 3y = 7 \\ 2x + y = 4 \end{array}$ <br><br> $\begin{bmatrix} 1 & 3 & \vert & 7 \\ 2 & 1 & \vert & 4 \end{bmatrix}$ Augmented matrix <br><br> $\begin{bmatrix} 1 & 3 & \vert & 7 \\ 0 & -5 & \vert & -10 \end{bmatrix}$ $-2R_1 + R_2$ <br><br> $\begin{bmatrix} 1 & 3 & \vert & 7 \\ 0 & 1 & \vert & 2 \end{bmatrix}$ $-\frac{1}{5}R_2$ $\xrightarrow{\text{implies}}$ $\begin{array}{l} x + 3y = 7 \\ y = 2 \end{array}$ <br><br> When $y = 2$, $x + 3(2) = 7$, so $x = 1$. The solution set is $\{(1, 2)\}$. |

## 4.4 Properties of Matrices

### Addition and Subtraction of Matrices

To add (subtract) matrices of the same size, add (subtract) the corresponding elements.

Find the sum or difference.

$$\begin{bmatrix} 2 & 3 & -1 \\ 0 & 4 & 9 \end{bmatrix} + \begin{bmatrix} -8 & 12 & 1 \\ 5 & 3 & -3 \end{bmatrix} = \begin{bmatrix} -6 & 15 & 0 \\ 5 & 7 & 6 \end{bmatrix}$$

$$\begin{bmatrix} 5 & -1 \\ -8 & 8 \end{bmatrix} - \begin{bmatrix} -2 & 4 \\ 3 & -6 \end{bmatrix} = \begin{bmatrix} 7 & -5 \\ -11 & 14 \end{bmatrix}$$

### Scalar Multiplication

To multiply a matrix by a scalar, multiply each element of the matrix by the scalar.

Find the scalar product.

$$3 \begin{bmatrix} 6 & 2 \\ 1 & -2 \\ 0 & 8 \end{bmatrix} = \begin{bmatrix} 18 & 6 \\ 3 & -6 \\ 0 & 24 \end{bmatrix}$$

### Multiplication of Matrices

The product $AB$ of an $m \times n$ matrix $A$ and an $n \times p$ matrix $B$ is found as follows. To get the $i$th row, $j$th column element of the $m \times p$ matrix $AB$, multiply each element in the $i$th row of $A$ by the corresponding element in the $j$th column of $B$. The sum of these products will give the element of row $i$, column $j$ of $AB$.

Find the matrix product.

$$\begin{bmatrix} 1 & -2 & 3 \\ 5 & 0 & 4 \\ -8 & 7 & -7 \end{bmatrix} \begin{bmatrix} 1 \\ -2 \\ 3 \end{bmatrix} = \begin{bmatrix} 14 \\ 17 \\ -43 \end{bmatrix}$$

## 4.5 Matrix Inverses

### Finding an Inverse Matrix

To obtain $A^{-1}$ for any $n \times n$ matrix $A$ for which $A^{-1}$ exists, follow these steps.

**Step 1** Form the augmented matrix $[A \vert I_n]$, where $I_n$ is the $n \times n$ identity matrix.

**Step 2** Perform row transformations on $[A \vert I_n]$ to get a matrix of the form $[I_n \vert B]$.

**Step 3** Matrix $B$ is $A^{-1}$.

Find $A^{-1}$ if $A = \begin{bmatrix} 5 & 2 \\ 2 & 1 \end{bmatrix}$.

$$\begin{bmatrix} 5 & 2 & \vert & 1 & 0 \\ 2 & 1 & \vert & 0 & 1 \end{bmatrix}$$

$$\begin{bmatrix} 1 & 0 & \vert & 1 & -2 \\ 2 & 1 & \vert & 0 & 1 \end{bmatrix} \quad -2R_2 + R_1$$

$$\begin{bmatrix} 1 & 0 & \vert & 1 & -2 \\ 0 & 1 & \vert & -2 & 5 \end{bmatrix} \quad -2R_1 + R_2$$

$$\underbrace{\phantom{\begin{matrix}1 & 0\\0 & 1\end{matrix}}}_{I_n} \quad \underbrace{\phantom{\begin{matrix}1 & -2\\-2 & 5\end{matrix}}}_{A^{-1}}$$

Therefore, $A^{-1} = \begin{bmatrix} 1 & -2 \\ -2 & 5 \end{bmatrix}$.

*(continued)*

## CONCEPTS

### 4.6 Determinants and Cramer's Rule

**Determinant of a $2 \times 2$ Matrix**

If $A = \begin{bmatrix} a_{11} & a_{12} \\ a_{21} & a_{22} \end{bmatrix}$, then

$$|A| = \begin{vmatrix} a_{11} & a_{12} \\ a_{21} & a_{22} \end{vmatrix} = a_{11}a_{22} - a_{21}a_{12}.$$

**Determinant of a $3 \times 3$ Matrix**

If $A = \begin{bmatrix} a_{11} & a_{12} & a_{13} \\ a_{21} & a_{22} & a_{23} \\ a_{31} & a_{32} & a_{33} \end{bmatrix}$, then

$$|A| = \begin{vmatrix} a_{11} & a_{12} & a_{13} \\ a_{21} & a_{22} & a_{23} \\ a_{31} & a_{32} & a_{33} \end{vmatrix} = (a_{11}a_{22}a_{33} + a_{12}a_{23}a_{31} + a_{13}a_{21}a_{32}) \\ - (a_{31}a_{22}a_{13} + a_{32}a_{23}a_{11} + a_{33}a_{21}a_{12}).$$

In practice, we usually evaluate determinants by expansion by minors.

**Cramer's Rule for $2 \times 2$ Systems**

Given the system

$$a_1 x + b_1 y = c_1 \\ a_2 x + b_2 y = c_2,$$

if $D \neq 0$, then the system has the unique solution

$$x = \frac{D_x}{D} \quad \text{and} \quad y = \frac{D_y}{D},$$

where $D = \begin{vmatrix} a_1 & b_1 \\ a_2 & b_2 \end{vmatrix}$, $D_x = \begin{vmatrix} c_1 & b_1 \\ c_2 & b_2 \end{vmatrix}$, and

$D_y = \begin{vmatrix} a_1 & c_1 \\ a_2 & c_2 \end{vmatrix}.$

**Cramer's Rule for $3 \times 3$ Systems**

Given the system

$$a_1 x + b_1 y + c_1 z = d_1 \\ a_2 x + b_2 y + c_2 z = d_2 \\ a_3 x + b_3 y + c_3 z = d_3$$

with

$$D_x = \begin{vmatrix} d_1 & b_1 & c_1 \\ d_2 & b_2 & c_2 \\ d_3 & b_3 & c_3 \end{vmatrix}, \quad D_y = \begin{vmatrix} a_1 & d_1 & c_1 \\ a_2 & d_2 & c_2 \\ a_3 & d_3 & c_3 \end{vmatrix},$$

$$D_z = \begin{vmatrix} a_1 & b_1 & d_1 \\ a_2 & b_2 & d_2 \\ a_3 & b_3 & d_3 \end{vmatrix}, \text{ and } D = \begin{vmatrix} a_1 & b_1 & c_1 \\ a_2 & b_2 & c_2 \\ a_3 & b_3 & c_3 \end{vmatrix},$$

if $D \neq 0$, then the system has the unique solution

$$x = \frac{D_x}{D}, \quad y = \frac{D_y}{D}, \quad \text{and} \quad z = \frac{D_z}{D}.$$

## EXAMPLES

Evaluate.

$$\begin{bmatrix} 3 & 5 \\ -2 & 6 \end{bmatrix} = 3(6) - (-2)5 = 28$$

Evaluate by expanding about the second column.

$$\begin{vmatrix} 2 & -3 & -2 \\ -1 & -4 & -3 \\ -1 & 0 & 2 \end{vmatrix} = -(-3)\begin{vmatrix} -1 & -3 \\ -1 & 2 \end{vmatrix} + (-4)\begin{vmatrix} 2 & -2 \\ -1 & 2 \end{vmatrix}$$

$$- 0\begin{vmatrix} 2 & -2 \\ -1 & -3 \end{vmatrix}$$

$$= 3(-5) - 4(2) - 0(-8)$$
$$= -15 - 8 + 0$$
$$= -23$$

Solve using Cramer's rule.

$$x - 2y = -1 \\ 2x + 5y = 16$$

$$x = \frac{\begin{vmatrix} -1 & -2 \\ 16 & 5 \end{vmatrix}}{\begin{vmatrix} 1 & -2 \\ 2 & 5 \end{vmatrix}} = \frac{-5 + 32}{5 + 4} = \frac{27}{9} = 3$$

$$y = \frac{\begin{vmatrix} 1 & -1 \\ 2 & 16 \end{vmatrix}}{\begin{vmatrix} 1 & -2 \\ 2 & 5 \end{vmatrix}} = \frac{16 + 2}{5 + 4} = \frac{18}{9} = 2$$

The solution set is $\{(3, 2)\}$.

Solve using Cramer's rule.

$$3x + 2y + z = -5 \\ x - y + 3z = -5 \\ 2x + 3y + z = 0$$

Using the method of expansion by minors, it can be shown that $D_x = 45$, $D_y = -30$, $D_z = 0$, and $D = -15$. Thus,

$$x = \frac{D_x}{D} = \frac{45}{-15} = -3, \quad y = \frac{D_y}{D} = \frac{-30}{-15} = 2,$$

$$z = \frac{D_z}{D} = \frac{0}{-15} = 0.$$

The solution set is $\{(-3, 2, 0)\}$.

# CHAPTER 4 REVIEW EXERCISES

1. The graph shows the trends during the years 1975 through 2002 relating to bachelor's degrees awarded in the United States.

   (a) Between what years shown on the horizontal axis did the number of degrees for men equal that for women?

   (b) When the number of degrees for men was equal to that for women, what was that number (approximately)?

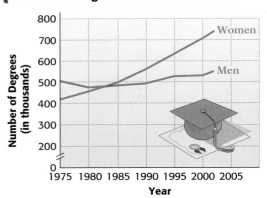

**Bachelor's Degrees in the United States**

*Source:* U.S. National Center for Education Statistics, *Digest of Education Statistics*, annual.

2. Solve the system by graphing.

   $$x + 3y = 8$$
   $$2x - y = 2$$

3. *Multiple Choice* Which one of the following ordered pairs is not a solution of the equation $3x + 2y = 6$?

   **A.** $(2, 0)$   **B.** $(0, 3)$   **C.** $(4, -3)$   **D.** $(3, -2)$

4. *Concept Check* Suppose that two linear equations are graphed on the same set of coordinate axes. Sketch what the graph might look like if the system has the given description.

   (a) The system has a single solution.   (b) The system has no solution.

   (c) The system has infinitely many solutions.

*Solve each system by the substitution method. If a system is inconsistent or has dependent equations, say so.*

5. $3x + y = -4$
   $x = \frac{2}{3}y$

6. $9x - y = -4$
   $y = x + 4$

7. $-5x + 2y = -2$
   $x + 6y = 26$

*Solve each system of equations by the elimination method. If a system is inconsistent or has dependent equations, say so.*

8. $5x + y = 12$
   $2x - 2y = 0$

9. $x - 4y = -4$
   $3x + y = 1$

10. $6x + 5y = 4$
    $-4x + 2y = 8$

11. $\frac{1}{6}x + \frac{1}{6}y = -\frac{1}{2}$
    $x - y = -9$

12. $-3x + y = 6$
    $y = 6 + 3x$

13. $5x - 4y = 2$
    $-10x + 8y = 7$

14. Without doing any algebraic work, but answering on the basis of your knowledge of the graphs of the two lines, explain why the system

$$y = 3x + 2$$
$$y = 3x - 4$$

has ∅ as its solution set.

*Solve each system. If a system is inconsistent or has dependent equations, say so.*

15. $\begin{aligned} 2x + 3y - z &= -16 \\ x + 2y + 2z &= -3 \\ -3x + y + z &= -5 \end{aligned}$

16. $\begin{aligned} 4x - y &= 2 \\ 3y + z &= 9 \\ x + 2z &= 7 \end{aligned}$

17. $\begin{aligned} 3x - y - z &= -8 \\ 4x + 2y + 3z &= 15 \\ -6x + 2y + 2z &= 10 \end{aligned}$

*Solve each system of equations by using row operations.*

18. $\begin{aligned} 2x + 5y &= -4 \\ 4x - y &= 14 \end{aligned}$

19. $\begin{aligned} 6x + 3y &= 9 \\ -7x + 2y &= 17 \end{aligned}$

20. $\begin{aligned} x + 2y - z &= 1 \\ 3x + 4y + 2z &= -2 \\ -2x - y + z &= -1 \end{aligned}$

21. $\begin{aligned} x + 3y &= 7 \\ 3x + z &= 2 \\ y - 2z &= 4 \end{aligned}$

*Find the value of each variable.*

22. $\begin{bmatrix} 5 & x+2 \\ -6y & z \end{bmatrix} = \begin{bmatrix} a & 3x-1 \\ 5y & 9 \end{bmatrix}$

23. $\begin{bmatrix} -6+k & 2 & a+3 \\ -2+m & 3p & 2r \end{bmatrix} + \begin{bmatrix} 3-2k & 5 & 7 \\ 5 & 8p & 5r \end{bmatrix} = \begin{bmatrix} 5 & y & 6a \\ 2m & 11 & -35 \end{bmatrix}$

*Perform each operation, whenever possible.*

24. $\begin{bmatrix} 3 \\ 2 \\ 5 \end{bmatrix} - \begin{bmatrix} 8 \\ -4 \\ 6 \end{bmatrix} + \begin{bmatrix} 1 \\ 0 \\ 2 \end{bmatrix}$

25. $4\begin{bmatrix} 3 & -4 & 2 \\ 5 & -1 & 6 \end{bmatrix} + \begin{bmatrix} -3 & 2 & 5 \\ 1 & 0 & 4 \end{bmatrix}$

26. $\begin{bmatrix} 2 & 5 & 8 \\ 1 & 9 & 2 \end{bmatrix} - \begin{bmatrix} 3 & 4 \\ 7 & 1 \end{bmatrix}$

27. $\begin{bmatrix} -3 & 4 \\ 2 & 8 \end{bmatrix} \begin{bmatrix} -1 & 0 \\ 2 & 5 \end{bmatrix}$

28. $\begin{bmatrix} -1 & 0 \\ 2 & 5 \end{bmatrix} \begin{bmatrix} -3 & 4 \\ 2 & 8 \end{bmatrix}$

29. $\begin{bmatrix} 1 & 2 \\ 3 & 0 \\ -6 & 5 \end{bmatrix} \begin{bmatrix} 4 & 8 \\ -1 & 2 \end{bmatrix}$

30. $\begin{bmatrix} 3 & 2 & -1 \\ 4 & 0 & 6 \end{bmatrix} \begin{bmatrix} -2 & 0 \\ 0 & 2 \\ 3 & 1 \end{bmatrix}$

31. $\begin{bmatrix} 1 & -2 & 4 & 2 \\ 0 & 1 & -1 & 8 \end{bmatrix} \begin{bmatrix} -1 \\ 2 \\ 0 \\ 1 \end{bmatrix}$

32. $\begin{bmatrix} -2 & 5 & 5 \\ 0 & 1 & 4 \\ 3 & -4 & -1 \end{bmatrix} \begin{bmatrix} 1 & 0 & -1 \\ -1 & 0 & 0 \\ 1 & 1 & -1 \end{bmatrix}$

*Find the inverse of each matrix that has an inverse.*

33. $\begin{bmatrix} -4 & 2 \\ 0 & 3 \end{bmatrix}$

34. $\begin{bmatrix} 2 & 1 \\ 5 & 3 \end{bmatrix}$

35. $\begin{bmatrix} 2 & 3 & 5 \\ -2 & -3 & -5 \\ 1 & 4 & 2 \end{bmatrix}$

36. $\begin{bmatrix} 2 & -1 & 0 \\ 1 & 0 & 1 \\ 1 & -2 & 0 \end{bmatrix}$

*Use the method of matrix inverses to solve each system.*

**37.** $2x + y = 5$
$3x - 2y = 4$

**38.** $3x + 2y + z = -5$
$x - y + 3z = -5$
$2x + 3y + z = 0$

**39.** $x + y + z = 1$
$2x - y = -2$
$3y + z = 2$

*Evaluate each determinant.*

**40.** $\begin{vmatrix} -2 & 4 \\ 0 & 3 \end{vmatrix}$

**41.** $\begin{vmatrix} -1 & 8 \\ 2 & 9 \end{vmatrix}$

**42.** $\begin{vmatrix} -1 & 2 & 3 \\ 4 & 0 & 3 \\ 5 & -1 & 2 \end{vmatrix}$

**43.** $\begin{vmatrix} -2 & 4 & 1 \\ 3 & 0 & 2 \\ -1 & 0 & 3 \end{vmatrix}$

*Solve each system by Cramer's rule, if possible. Identify any inconsistent systems or systems with infinitely many solutions. (Use another method if Cramer's rule cannot be used.)*

**44.** $3x + 7y = 2$
$5x - y = -22$

**45.** $3x + y = -1$
$5x + 4y = 10$

**46.** $6x + y = -3$
$12x + 2y = 1$

**47.** $3x + 2y + z = 2$
$4x - y + 3z = -16$
$x + 3y - z = 12$

**48.** $x + y = -1$
$2y + z = 5$
$3x - 2z = -28$

# 4 TEST

*If the rates of growth between 1990 and 2000 continue, the populations of Houston, Phoenix, Dallas, and Philadelphia will follow the trends indicated in the graph. Use the graph to work Exercises 1 and 2.*

1. **(a)** Which of these cities will experience population growth?
   **(b)** Which city will experience population decline?
   **(c)** Rank the city populations from least to greatest for the year 2000.

2. **(a)** In which year will the population of Dallas equal that of Philadelphia? About what will this population be?
   **(b)** Write as an ordered pair (year, population in millions) the point at which Houston and Phoenix will have the same population.

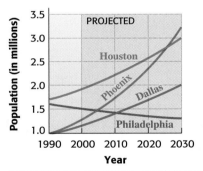

**The Growth Game**
Size of cities if the rate of population growth from 1990 to 2000 continues:

*Source:* U.S. Census Bureau, *Chronicle* research.

**3.** Use a graph to solve the system $\begin{aligned} x + y &= 7 \\ x - y &= 5. \end{aligned}$

*Solve each system by substitution or elimination. If a system is inconsistent or has dependent equations, say so.*

**4.** $2x - 3y = 24$
$y = -\frac{2}{3}x$

**5.** $3x - y = -8$
$2x + 6y = 3$

**6.** $12x - 5y = 8$
$3x = \frac{5}{4}y + 2$

7. $3x + y = 12$
   $2x - y = 3$

8. $-5x + 2y = -4$
   $6x + 3y = -6$

9. $3x + 4y = 8$
   $8y = 7 - 6x$

10. $3x + 5y + 3z = 2$
    $6x + 5y + z = 0$
    $3x + 10y - 2z = 6$

11. $4x + y + z = 11$
    $x - y - z = 4$
    $y + 2z = 0$

*Solve each system using row operations.*

12. $3x + 2y = 4$
    $5x + 5y = 9$

13. $x + 3y + 2z = 11$
    $3x + 7y + 4z = 23$
    $5x + 3y - 5z = -14$

14. Find the value of each variable in the equation $\begin{bmatrix} 5 & x+6 \\ 0 & 4 \end{bmatrix} = \begin{bmatrix} y-2 & 4-x \\ 0 & w+7 \end{bmatrix}$.

*Perform each operation when possible.*

15. $3\begin{bmatrix} 2 & 3 \\ 1 & -4 \\ 5 & 9 \end{bmatrix} - \begin{bmatrix} -2 & 6 \\ 3 & -1 \\ 0 & 8 \end{bmatrix}$

16. $\begin{bmatrix} 1 \\ 2 \end{bmatrix} + \begin{bmatrix} 4 \\ -6 \end{bmatrix} + \begin{bmatrix} 2 & 8 \\ -7 & 5 \end{bmatrix}$

17. $\begin{bmatrix} 2 & 1 & -3 \\ 4 & 0 & 5 \end{bmatrix} \begin{bmatrix} 1 & 3 \\ 2 & 4 \\ 3 & -2 \end{bmatrix}$

18. $\begin{bmatrix} 2 & -4 \\ 3 & 5 \end{bmatrix} \begin{bmatrix} 4 \\ 2 \\ 7 \end{bmatrix}$

19. *Multiple Choice* Which of the following properties does not apply to multiplication of matrices?

   **A.** commutative  **B.** associative  **C.** distributive  **D.** identity

*Find the inverse, if it exists, of each matrix.*

20. $\begin{bmatrix} -8 & 5 \\ 3 & -2 \end{bmatrix}$

21. $\begin{bmatrix} 4 & 12 \\ 2 & 6 \end{bmatrix}$

22. $\begin{bmatrix} 1 & 3 & 4 \\ 2 & 7 & 8 \\ -2 & -5 & -7 \end{bmatrix}$

*Use matrix inverses to solve each system.*

23. $2x + y = -6$
    $3x - y = -29$

24. $x + y = 5$
    $y - 2z = 23$
    $x + 3z = -27$

*Evaluate each determinant.*

25. $\begin{vmatrix} 6 & 8 \\ 2 & -7 \end{vmatrix}$

26. $\begin{vmatrix} 2 & 0 & 8 \\ -1 & 7 & 9 \\ 12 & 5 & -3 \end{vmatrix}$

*Solve each system by Cramer's rule.*

27. $2x - 3y = -33$
    $4x + 5y = 11$

28. $x + y - z = -4$
    $2x - 3y - z = 5$
    $x + 2y + 2z = 3$

# Exponents, Polynomials, and Polynomial Functions

In 1980, MasterCard International Incorporated first began offering debit cards in an effort to challenge Visa, the leader in credit card transactions at that time. Now immensely popular, debit cards draw money from consumers' bank accounts rather than from established lines of credit. It is estimated that 354 million debit cards were in use in the United States in 2006. (*Source: Statistical Abstract of the United States.*)

We introduced the concept of a function in Section 3.5 and extend our work to include *polynomial functions* in this chapter. In Exercise 11 of Section 5.3, we use a polynomial function to model the number of bank debit cards issued.

**5.1** Integer Exponents and Scientific Notation

**5.2** Adding and Subtracting Polynomials

**5.3** Polynomial Functions, Graphs, and Composition

**5.4** Multiplying Polynomials

**5.5** Dividing Polynomials

# 5.1 Integer Exponents and Scientific Notation

**OBJECTIVES**

1. Use the product rule for exponents.
2. Define 0 and negative exponents.
3. Use the quotient rule for exponents.
4. Use the power rules for exponents.
5. Simplify exponential expressions.
6. Use the rules for exponents with scientific notation.

Recall that we use exponents to write products of repeated factors. For example,

$$2^5 \text{ is defined as } 2 \cdot 2 \cdot 2 \cdot 2 \cdot 2 = 32.$$

The number 5, the *exponent*, shows that the *base* 2 appears as a factor five times. The quantity $2^5$ is called an *exponential* or a *power*. We read $2^5$ as "2 to the fifth power" or "2 to the fifth."

**OBJECTIVE 1** Use the product rule for exponents. There are rules that simplify work with exponents. For example, the product $2^5 \cdot 2^3$ can be simplified as follows.

$$2^5 \cdot 2^3 = (2 \cdot 2 \cdot 2 \cdot 2 \cdot 2)(2 \cdot 2 \cdot 2) = 2^8$$

with $5 + 3 = 8$.

This result, that products of exponential expressions with the *same base* are found by adding exponents, is generalized as the **product rule for exponents.**

---

**Product Rule for Exponents**

If $m$ and $n$ are natural numbers and $a$ is any real number, then

$$a^m \cdot a^n = a^{m+n}.$$

That is, when multiplying powers of like bases, keep the same base and add the exponents.

---

To see that the product rule is true, use the definition of an exponent as follows.

$$a^m = \underbrace{a \cdot a \cdot a \cdot \ldots \cdot a}_{a \text{ appears as a factor } m \text{ times.}} \qquad a^n = \underbrace{a \cdot a \cdot a \cdot \ldots \cdot a}_{a \text{ appears as a factor } n \text{ times.}}$$

From this, $a^m \cdot a^n = \underbrace{a \cdot a \cdot a \cdot \ldots \cdot a}_{m \text{ factors}} \cdot \underbrace{a \cdot a \cdot a \cdot \ldots \cdot a}_{n \text{ factors}}$

$$= \underbrace{a \cdot a \cdot a \cdot \ldots \cdot a}_{(m+n) \text{ factors}}$$

$$a^m \cdot a^n = a^{m+n}.$$

---

**EXAMPLE 1** Using the Product Rule for Exponents

Apply the product rule for exponents, if possible, in each case.

(a) $3^4 \cdot 3^7 = 3^{4+7} = 3^{11}$    *Do not multiply the bases! Keep the same base.*

(b) $5^3 \cdot 5 = 5^3 \cdot 5^1 = 5^{3+1} = 5^4$

(c) $y^3 \cdot y^8 \cdot y^2 = y^{3+8+2} = y^{13}$

(d) $(5y^2)(-3y^4) = 5(-3)y^2y^4$    Associative and commutative properties

$\phantom{(5y^2)(-3y^4)} = -15y^{2+4}$    Product rule

$\phantom{(5y^2)(-3y^4)} = -15y^6$

**(e)** $(7p^3q)(2p^5q^2) = 7(2)p^3p^5q^1q^2 = 14p^8q^3$

**(f)** $x^2 \cdot y^4$    The product rule does not apply because the bases are not the same.

**NOW TRY** Exercises 7, 11, 13, and 15.

**CAUTION** Be careful in problems like Example 1(a) not to multiply the bases. Notice that $3^4 \cdot 3^7 = 3^{11}$, *not* $9^{11}$. *Remember to keep the same base and add the exponents.*

**OBJECTIVE 2** Define 0 and negative exponents.   Suppose we multiply $4^2$ by $4^0$. Extending the product rule, we should have

$$4^2 \cdot 4^0 = 4^{2+0} = 4^2.$$

For the product rule to hold, $4^0$ must equal 1, so we define $a^0$ that way for any nonzero real number $a$.

### Zero Exponent

If $a$ is any nonzero real number, then

$$a^0 = 1.$$

*The expression $0^0$ is undefined.**

**EXAMPLE 2** Using 0 as an Exponent

Evaluate.

**(a)** $6^0 = 1$   *The base is 6, not $-6$*

**(b)** $(-6)^0 = 1$   *Here the base is $-6$*

**(c)** $-6^0 = -(6^0) = -(1) = -1$

**(d)** $-(-6)^0 = -1$

**(e)** $5^0 + 12^0 = 1 + 1 = 2$

**(f)** $(8k)^0 = 1, \quad k \neq 0$

**NOW TRY** Exercises 17 and 25.

To define a negative exponent, we extend the product rule. For example,

$$8^2 \cdot 8^{-2} = 8^{2+(-2)} = 8^0 = 1.$$

Here $8^{-2}$ is the reciprocal of $8^2$. But $\frac{1}{8^2}$ is the reciprocal of $8^2$, and a number can have only one reciprocal. Therefore, $8^{-2} = \frac{1}{8^2}$. We generalize this result as follows.

### Negative Exponent

For any natural number $n$ and any nonzero real number $a$,

$$a^{-n} = \frac{1}{a^n}.$$

---

*In advanced mathematics, $0^0$ is called an *indeterminate form.*

With this definition, the expression $a^n$ is meaningful for any integer exponent $n$ and any nonzero real number $a$. The product rule is valid for integers $m$ and $n$.

**CAUTION** *A negative exponent does not indicate a negative number;* negative exponents lead to reciprocals. For example,

$$3^{-2} = \frac{1}{3^2} = \frac{1}{9} \quad \text{Not negative} \qquad \bigg| \qquad -3^{-2} = -\frac{1}{3^2} = -\frac{1}{9}. \quad \text{Negative}$$

### EXAMPLE 3 Using Negative Exponents

In parts (a)–(f), write each exponential with only positive exponents.

(a) $2^{-3} = \dfrac{1}{2^3}$

(b) $6^{-1} = \dfrac{1}{6^1} = \dfrac{1}{6}$

(c) $(5z)^{-3} = \dfrac{1}{(5z)^3}, \quad z \neq 0$
  ↑
  Base is $5z$.

(d) $5z^{-3} = 5\left(\dfrac{1}{z^3}\right) = \dfrac{5}{z^3}, \quad z \neq 0$
  ↑
  Base is $z$.

(e) $-m^{-2} = -\dfrac{1}{m^2}, \quad m \neq 0$
  (What is the base here?)

(f) $(-m)^{-2} = \dfrac{1}{(-m)^2}, \quad m \neq 0$
  (What is the base here?)

In parts (g) and (h), evaluate.

(g) $3^{-1} + 4^{-1} = \dfrac{1}{3} + \dfrac{1}{4} = \dfrac{4}{12} + \dfrac{3}{12} = \dfrac{7}{12}$ $\quad \frac{1}{3} \cdot \frac{4}{4} = \frac{4}{12}; \frac{1}{4} \cdot \frac{3}{3} = \frac{3}{12}$

*The product rule does not apply here.*

(h) $5^{-1} - 2^{-1} = \dfrac{1}{5} - \dfrac{1}{2} = \dfrac{2}{10} - \dfrac{5}{10} = -\dfrac{3}{10}$

**NOW TRY** Exercises 33, 37, 39, 41, and 45.

**CAUTION** In Example 3(g), note that $3^{-1} + 4^{-1} \neq (3 + 4)^{-1}$. The expression on the left is equal to $\frac{7}{12}$, as shown in the example, while the expression on the right is $7^{-1} = \frac{1}{7}$. Similar reasoning can be applied to part (h).

### EXAMPLE 4 Using Negative Exponents

Evaluate.

(a) $\dfrac{1}{2^{-3}} = \dfrac{1}{\frac{1}{2^3}} = 1 \div \dfrac{1}{2^3} = 1 \cdot \dfrac{2^3}{1} = 2^3 = 8$

Multiply by the reciprocal of the divisor.

**(b)** $\dfrac{2^{-3}}{3^{-2}} = \dfrac{\frac{1}{2^3}}{\frac{1}{3^2}} = \dfrac{1}{2^3} \div \dfrac{1}{3^2} = \dfrac{1}{2^3} \cdot \dfrac{3^2}{1} = \dfrac{3^2}{2^3} = \dfrac{9}{8}$

**NOW TRY** Exercises 51 and 53.

Example 4 suggests the following generalizations.

### Special Rules for Negative Exponents

If $a \neq 0$ and $b \neq 0$, then
$$\dfrac{1}{a^{-n}} = a^n \quad \text{and} \quad \dfrac{a^{-n}}{b^{-m}} = \dfrac{b^m}{a^n}.$$

**OBJECTIVE 3** Use the quotient rule for exponents. A quotient, such as $\dfrac{a^8}{a^3}$, can be simplified in much the same way as a product. (In all quotients of this type, assume that the denominator is not 0.) Using the definition of an exponent gives

$$\dfrac{a^8}{a^3} = \dfrac{a \cdot a \cdot a \cdot a \cdot a \cdot a \cdot a \cdot a}{a \cdot a \cdot a} = a \cdot a \cdot a \cdot a \cdot a = a^5.$$

Notice that $8 - 3 = 5$. In the same way,

$$\dfrac{a^3}{a^8} = \dfrac{a \cdot a \cdot a}{a \cdot a \cdot a \cdot a \cdot a \cdot a \cdot a \cdot a} = \dfrac{1}{a^5} = a^{-5}.$$

Here $3 - 8 = -5$. These examples suggest the **quotient rule for exponents.**

### Quotient Rule for Exponents

If $a$ is any nonzero real number and $m$ and $n$ are integers, then
$$\dfrac{a^m}{a^n} = a^{m-n}.$$

That is, when dividing powers of like bases, keep the same base and subtract the exponent of the denominator from the exponent of the numerator.

**EXAMPLE 5** Using the Quotient Rule for Exponents

Apply the quotient rule for exponents, if possible, and write each result with only positive exponents.

**(a)** $\dfrac{3^7}{3^2} = 3^{7-2} = 3^5$ (Numerator exponent, Denominator exponent, Minus sign)

**(b)** $\dfrac{p^6}{p^2} = p^{6-2} = p^4, \quad p \neq 0$

**(c)** $\dfrac{k^7}{k^{12}} = k^{7-12} = k^{-5} = \dfrac{1}{k^5}, \quad k \neq 0$

**(d)** $\dfrac{2^7}{2^{-3}} = 2^{7-(-3)} = 2^{7+3} = 2^{10}$

Use parentheses to avoid errors.

**(e)** $\dfrac{8^{-2}}{8^5} = 8^{-2-5} = 8^{-7} = \dfrac{1}{8^7}$

**(f)** $\dfrac{6}{6^{-1}} = \dfrac{6^1}{6^{-1}} = 6^{1-(-1)} = 6^2$

**(g)** $\dfrac{z^{-5}}{z^{-8}} = z^{-5-(-8)} = z^3, \quad z \neq 0$

*Be careful with signs.*

**(h)** $\dfrac{a^3}{b^4}, \quad b \neq 0$

This expression cannot be simplified further.

The quotient rule does not apply because the bases are different.

**NOW TRY** Exercises 63, 67, 75, and 77.

**CAUTION** Be careful when working with quotients that involve negative exponents in the denominator. Be sure to write the numerator exponent, then a minus sign, and then the denominator exponent. Use parentheses.

**OBJECTIVE 4** Use the power rules for exponents. We can simplify $(3^4)^2$ as

$$(3^4)^2 = 3^4 \cdot 3^4 = 3^{4+4} = 3^8,$$

where $4 \cdot 2 = 8$. This example suggests the first **power rule for exponents.** The other two power rules can be demonstrated with similar examples.

### Power Rules for Exponents

If $a$ and $b$ are real numbers and $m$ and $n$ are integers, then

**(a)** $(a^m)^n = a^{mn}$, **(b)** $(ab)^m = a^m b^m$, and **(c)** $\left(\dfrac{a}{b}\right)^m = \dfrac{a^m}{b^m} \quad (b \neq 0)$.

That is,

**(a)** To raise a power to a power, multiply exponents.

**(b)** To raise a product to a power, raise each factor to that power.

**(c)** To raise a quotient to a power, raise the numerator and the denominator to that power.

**EXAMPLE 6** Using the Power Rules for Exponents

Simplify, using the power rules.

**(a)** $(p^8)^3 = p^{8 \cdot 3} = p^{24}$

**(b)** $\left(\dfrac{2}{3}\right)^4 = \dfrac{2^4}{3^4} = \dfrac{16}{81}$

**(c)** $(3y)^4 = 3^4 y^4 = 81 y^4$

**(d)** $(6p^7)^2 = 6^2 p^{7 \cdot 2} = 6^2 p^{14} = 36 p^{14}$

**(e)** $\left(\dfrac{-2m^5}{z}\right)^3 = \dfrac{(-2)^3 m^{5 \cdot 3}}{z^3} = \dfrac{(-2)^3 m^{15}}{z^3} = \dfrac{-8 m^{15}}{z^3}, \quad z \neq 0$

**NOW TRY** Exercises 79, 81, 83, and 87.

The reciprocal of $a^n$ is $\dfrac{1}{a^n} = \left(\dfrac{1}{a}\right)^n$. Also, by definition, $a^n$ and $a^{-n}$ are reciprocals, since

$$a^n \cdot a^{-n} = a^n \cdot \dfrac{1}{a^n} = 1.$$

Thus, since both are reciprocals of $a^n$,

$$a^{-n} = \left(\dfrac{1}{a}\right)^n.$$

Some examples of this result are

$$6^{-3} = \left(\dfrac{1}{6}\right)^3 \quad \text{and} \quad \left(\dfrac{1}{3}\right)^{-2} = 3^2.$$

This discussion can be generalized as follows.

### More Special Rules for Negative Exponents

If $a \neq 0$ and $b \neq 0$ and $n$ is an integer, then

$$a^{-n} = \left(\dfrac{1}{a}\right)^n \quad \text{and} \quad \left(\dfrac{a}{b}\right)^{-n} = \left(\dfrac{b}{a}\right)^n.$$

That is, any nonzero number raised to the negative $n$th power is equal to the reciprocal of that number raised to the $n$th power.

### EXAMPLE 7 Using Negative Exponents with Fractions

Write with only positive exponents and then evaluate.

**(a)** $\left(\dfrac{3}{7}\right)^{-2} = \left(\dfrac{7}{3}\right)^2 = \dfrac{49}{9}$

**(b)** $\left(\dfrac{4}{5}\right)^{-3} = \left(\dfrac{5}{4}\right)^3 = \dfrac{125}{64}$

Change the fraction to its reciprocal and change the sign of the exponent.

**NOW TRY** Exercise 55.

The definitions and rules of this section are summarized here.

### Definitions and Rules for Exponents

For all integers $m$ and $n$ and all real numbers $a$ and $b$, the following rules apply.

**Product Rule** $\quad a^m \cdot a^n = a^{m+n}$

**Quotient Rule** $\quad \dfrac{a^m}{a^n} = a^{m-n} \quad (a \neq 0)$

**Zero Exponent** $\quad a^0 = 1 \quad (a \neq 0)$

**Negative Exponent** $\quad a^{-n} = \dfrac{1}{a^n} \quad (a \neq 0)$

*(continued)*

### Definitions and Rules for Exponents (continued)

**Power Rules**  $(a^m)^n = a^{mn}$  $\qquad$ $(ab)^m = a^m b^m$

$$\left(\frac{a}{b}\right)^m = \frac{a^m}{b^m} \quad (b \neq 0)$$

**Special Rules**  $\dfrac{1}{a^{-n}} = a^n \quad (a \neq 0)$  $\qquad$ $\dfrac{a^{-n}}{b^{-m}} = \dfrac{b^m}{a^n} \quad (a, b \neq 0)$

$$a^{-n} = \left(\frac{1}{a}\right)^n \quad (a \neq 0) \qquad \left(\frac{a}{b}\right)^{-n} = \left(\frac{b}{a}\right)^n \quad (a, b \neq 0)$$

**OBJECTIVE 5** Simplify exponential expressions.

**EXAMPLE 8** Using the Definitions and Rules for Exponents

Simplify so that no negative exponents appear in the final result. Assume that all variables represent nonzero real numbers.

**(a)** $3^2 \cdot 3^{-5} = 3^{2+(-5)} = 3^{-3} = \dfrac{1}{3^3}, \quad \text{or} \quad \dfrac{1}{27}$

**(b)** $x^{-3} x^{-4} x^2 = x^{-3+(-4)+2} = x^{-5} = \dfrac{1}{x^5}$

**(c)** $(4^{-2})^{-5} = 4^{(-2)(-5)} = 4^{10}$  $\qquad$ **(d)** $(x^{-4})^6 = x^{(-4)6} = x^{-24} = \dfrac{1}{x^{24}}$

**(e)** $\dfrac{x^{-4} y^2}{x^2 y^{-5}} = \dfrac{x^{-4}}{x^2} \cdot \dfrac{y^2}{y^{-5}}$  $\qquad$ **(f)** $(2^3 x^{-2})^{-2} = (2^3)^{-2} \cdot (x^{-2})^{-2}$

$\qquad\qquad = x^{-4-2} \cdot y^{2-(-5)}$  $\qquad\qquad\qquad\qquad = 2^{-6} x^4$

$\qquad\qquad = x^{-6} y^7$  $\qquad\qquad\qquad\qquad\qquad\qquad = \dfrac{x^4}{2^6}, \quad \text{or} \quad \dfrac{x^4}{64}$

$\qquad\qquad = \dfrac{y^7}{x^6}$

**(g)** $\left(\dfrac{3x^2}{y}\right)^2 \left(\dfrac{4x^3}{y^{-2}}\right)^{-1} = \dfrac{3^2 (x^2)^2}{y^2} \cdot \dfrac{y^{-2}}{4x^3}$ $\qquad$ Combination of rules

$\qquad\qquad\qquad\qquad\qquad = \dfrac{9x^4}{y^2} \cdot \dfrac{y^{-2}}{4x^3}$ $\qquad\qquad$ Power rule

$\qquad\qquad\qquad\qquad\qquad = \dfrac{9}{4} x^{4-3} y^{-2-2}$ $\qquad\qquad$ Quotient rule

$\qquad\qquad\qquad\qquad\qquad = \dfrac{9x}{4y^4}$

**NOW TRY** Exercises 91, 93, 107, and 117.

**OBJECTIVE 6** Use the rules for exponents with scientific notation. Scientists often use numbers that are very large or very small. For example, the number of one-celled organisms that will sustain a whale for a few hours is 400,000,000,000,000, and the shortest wavelength of visible light is approximately 0.0000004 m. It is simpler to write these numbers in *scientific notation*.

In scientific notation, a number is written with the decimal point after the first nonzero digit and multiplied by a power of 10, as indicated in the following definition.

### Scientific Notation

A number is written in **scientific notation** when it is expressed in the form

$$a \times 10^n$$

where $1 \leq |a| < 10$ and $n$ is an integer.

For example, in scientific notation,

$$8000 = 8 \times 1000 = 8 \times 10^3.$$

In scientific notation it is customary to use a times sign ($\times$) instead of a multiplication dot. The following numbers are *not* in scientific notation.

$$0.230 \times 10^4 \qquad 46.5 \times 10^{-3}$$

0.230 is less than 1.    46.5 is greater than 10.

To write a number in scientific notation, use the following steps. (If the number is negative, go through these steps, and then attach a negative sign to the result.)

### Converting to Scientific Notation

*Step 1* **Position the decimal point.** Place a caret, ^, to the right of the first nonzero digit, where the decimal point will be placed.

*Step 2* **Determine the numeral for the exponent.** Count the number of digits from the decimal point to the caret. This number gives the absolute value of the exponent on 10.

*Step 3* **Determine the sign for the exponent.** Decide whether multiplying by $10^n$ should make the result of Step 1 greater or less. The exponent should be positive to make the result greater; it should be negative to make the result less.

It is helpful to remember that, for $n \geq 1$, $10^{-n} < 1$ and $10^n \geq 10$.

### EXAMPLE 9  Writing Numbers in Scientific Notation

Write each number in scientific notation.

**(a)** 820,000

*Step 1* Place a caret to the right of the 8 (the first nonzero digit) to mark the new location of the decimal point.

$$8\!\wedge\!20{,}000$$

*Step 2* Count from the decimal point, which is understood to be after the last 0, to the caret.

$$8\!\wedge\!20{,}000. \quad \leftarrow \text{Decimal point}$$
Count 5 places.

*Step 3* Since 8.2 is to be made greater, the exponent on 10 is positive.

$$820{,}000 = 8.2 \times 10^5$$

**(b)** 0.0000072

Count from left to right.

$$0.000007.2$$
6 places

Since the number 7.2 is to be made less, the exponent on 10 is negative.

$$0.0000072 = 7.2 \times 10^{-6}$$

**(c)** $-0.0000462 = -4.62 \times 10^{-5}$

Count 5 places.

**NOW TRY** Exercises 121, 125, and 127.

To convert scientific notation to standard notation, just work in reverse.

### Converting from Scientific Notation

Multiplying a number by a positive power of 10 makes the number greater, so move the decimal point to the right if $n$ is positive in $10^n$.

Multiplying by a negative power of 10 makes a number less, so move the decimal point to the left if $n$ is negative.

If $n$ is 0, leave the decimal point where it is.

**EXAMPLE 10  Converting from Scientific Notation to Standard Notation**

Write each number in standard notation.

**(a)** $6.93 \times 10^7$

$$6.93{00000}$$  Attach 0's as necessary.
7 places

We moved the decimal point 7 places to the right. (We had to attach five 0's.)

$$6.93 \times 10^7 = 69,300,000$$

**(b)** $4.7 \times 10^{-6}$

$$000004.7$$  Attach 0's as necessary.
6 places

Add a leading 0 since the decimal is between 0 and 1.

We moved the decimal point 6 places to the left.

$$4.7 \times 10^{-6} = .0000047, \quad \text{or} \quad 0.0000047$$

**(c)** $-1.083 \times 10^0 = -1.083 \times 1 = -1.083$

**NOW TRY** Exercises 129, 131, and 133.

When problems require operations with numbers that are very large or very small, it is often advantageous to write the numbers in scientific notation first and then perform the calculations, using the rules for exponents.

## EXAMPLE 11  Using Scientific Notation in Computation

Evaluate $\dfrac{1{,}920{,}000 \times 0.0015}{0.000032 \times 45{,}000}$.

First, express all numbers in scientific notation.

$$\dfrac{1{,}920{,}000 \times 0.0015}{0.000032 \times 45{,}000} = \dfrac{1.92 \times 10^{6} \times 1.5 \times 10^{-3}}{3.2 \times 10^{-5} \times 4.5 \times 10^{4}}$$

$$= \dfrac{1.92 \times 1.5 \times 10^{6} \times 10^{-3}}{3.2 \times 4.5 \times 10^{-5} \times 10^{4}} \quad \text{Commutative and associative properties}$$

$$= \dfrac{1.92 \times 1.5 \times 10^{3}}{3.2 \times 4.5 \times 10^{-1}} \quad \text{Product rule}$$

$$= \dfrac{1.92 \times 1.5}{3.2 \times 4.5} \times 10^{4} \quad \text{Quotient rule}$$

$$= 0.2 \times 10^{4}$$

**Don't stop here!**

$$= (2 \times 10^{-1}) \times 10^{4} \quad \text{Write 0.2 in scientific notation.}$$

$$= 2 \times 10^{3} \quad \text{Product rule}$$

$$= 2000 \quad \text{Standard notation}$$

**NOW TRY** Exercise 143.

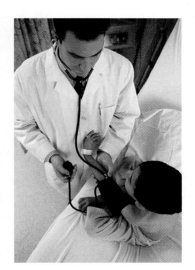

## EXAMPLE 12  Using Scientific Notation to Solve Problems

In 1990, the national health care expenditure was $696 billion. By 2003, this figure had risen by a factor of 2.4. The expenditure had risen by almost $2\tfrac{1}{2}$ times in only 13 yr. (*Source:* U.S. Centers for Medicare & Medicaid Services.)

**(a)** Write the 1990 health care expenditure in scientific notation.

$$696 \text{ billion} = 696 \times 10^{9} \quad \text{1 billion} = 1{,}000{,}000{,}000 = 10^{9}$$

$$= (6.96 \times 10^{2}) \times 10^{9} \quad \text{Write 696 in scientific notation.}$$

$$= 6.96 \times 10^{11} \quad \text{Product rule}$$

In 1990, the expenditure was $6.96 \times 10^{11}$.

**(b)** What was the expenditure in 2003?
Multiply the result in part (a) by 2.4.

$$(6.96 \times 10^{11}) \times 2.4 = (2.4 \times 6.96) \times 10^{11} \quad \text{Commutative and associative properties}$$

$$= 16.7 \times 10^{11} \quad \text{Round to the nearest tenth.}$$

$$= (1.67 \times 10^{1}) \times 10^{11} \quad \text{Write 16.7 in scientific notation.}$$

$$= 1.67 \times 10^{12} \quad \text{Product rule}$$

The 2003 expenditure was about $1,670,000,000,000, almost $1.7 trillion.

**NOW TRY** Exercise 145.

## 5.1 Exercises

*Concept Check* Decide whether each expression has been simplified correctly. If not, correct it.

1. $(ab)^2 = ab^2$
2. $y^2 \cdot y^6 = y^{12}$
3. $\left(\dfrac{4}{a}\right)^3 = \dfrac{4^3}{a}$ $(a \neq 0)$
4. $xy^0 = 0$ $(y \neq 0)$

5. State the product rule for exponents in your own words. Give an example.
6. *Concept Check* Your friend evaluated $4^5 \cdot 4^2$ as $16^7$. **WHAT WENT WRONG?** Give the correct answer.

*Apply the product rule for exponents, if possible, in each case. See Example 1.*

7. $13^4 \cdot 13^8$
8. $9^6 \cdot 9^4$
9. $x^3 \cdot x^5 \cdot x^9$
10. $y^4 \cdot y^5 \cdot y^6$
11. $(-3w^5)(9w^3)$
12. $(-5x^2)(3x^4)$
13. $(2x^2y^5)(9xy^3)$
14. $(8s^4t)(3s^3t^5)$
15. $r^2 \cdot s^4$
16. $p^3 \cdot q^2$

*Matching* In Exercises 17 and 18, match the expression in Column I with its equivalent expression in Column II. Choices may be used once, more than once, or not at all.* See Example 2.

17.      I      II
(a) $9^0$    A. $0$
(b) $-9^0$    B. $1$
(c) $(-9)^0$    C. $-1$
(d) $-(-9)^0$    D. $9$
     E. $-9$

18.      I      II
(a) $2x^0$    A. $0$
(b) $-2x^0$    B. $1$
(c) $(2x)^0$    C. $-1$
(d) $(-2x)^0$    D. $2$
     E. $-2$

*Evaluate. Assume that all variables represent nonzero real numbers.* See Example 2.

19. $25^0$
20. $14^0$
21. $-7^0$
22. $-10^0$
23. $(-15)^0$
24. $(-20)^0$
25. $3^0 + (-3)^0$
26. $5^0 + (-5)^0$
27. $-3^0 + 3^0$
28. $-5^0 + 5^0$
29. $-4^0 - m^0$
30. $-8^0 - k^0$

*Matching* In Exercises 31 and 32, match the expression in Column I with its equivalent expression in Column II. Choices may be used once, more than once, or not at all.* See Example 3.

31.      I      II
(a) $4^{-2}$    A. $16$
(b) $-4^{-2}$    B. $\dfrac{1}{16}$
(c) $(-4)^{-2}$    C. $-16$
(d) $-(-4)^{-2}$    E. $-\dfrac{1}{16}$

32.      I      II
(a) $5^{-3}$    A. $125$
(b) $-5^{-3}$    B. $-125$
(c) $(-5)^{-3}$    C. $\dfrac{1}{125}$
(d) $-(-5)^{-3}$    D. $-\dfrac{1}{125}$

*Write each expression with only positive exponents. Assume that all variables represent nonzero real numbers. In Exercises 45–48, simplify each expression. See Example 3.*

33. $5^{-4}$
34. $7^{-2}$
35. $8^{-1}$
36. $12^{-1}$
37. $(4x)^{-2}$
38. $(5t)^{-3}$
39. $4x^{-2}$
40. $5t^{-3}$

---

*The authors thank Mitchel Levy of Broward Community College for his suggestions for these exercises.

**41.** $-a^{-3}$  **42.** $-b^{-4}$  **43.** $(-a)^{-4}$  **44.** $(-b)^{-6}$

**45.** $5^{-1} + 6^{-1}$  **46.** $2^{-1} + 8^{-1}$  **47.** $8^{-1} - 3^{-1}$  **48.** $6^{-1} - 4^{-1}$

**49.** *Concept Check* Consider the expressions $-a^n$ and $(-a)^n$. In some cases they are equal, and in some cases they are not. Using $n = 2, 3, 4, 5,$ and $6$ and $a = 2$, draw a conclusion as to when they are equal and when they are opposites.

**50.** *Concept Check* Your friend thinks that $(-3)^{-2}$ is a negative number. Why is she incorrect?

*Evaluate each expression. See Examples 4 and 7.*

**51.** $\dfrac{1}{4^{-2}}$  **52.** $\dfrac{1}{3^{-3}}$  **53.** $\dfrac{2^{-2}}{3^{-3}}$  **54.** $\dfrac{3^{-3}}{2^{-2}}$

**55.** $\left(\dfrac{2}{3}\right)^{-3}$  **56.** $\left(\dfrac{3}{2}\right)^{-3}$  **57.** $\left(\dfrac{4}{5}\right)^{-2}$  **58.** $\left(\dfrac{5}{4}\right)^{-2}$

*Matching* In Exercises 59 and 60, match the expression in Column I with its equivalent expression in Column II. Choices may be used once, more than once, or not at all.*

|  | I |  | II |  | I |  | II |
|---|---|---|---|---|---|---|---|
| **59.** (a) | $\left(\dfrac{1}{3}\right)^{-1}$ | A. | $\dfrac{1}{3}$ | **60.** (a) | $\left(\dfrac{2}{5}\right)^{-2}$ | A. | $\dfrac{25}{4}$ |
| (b) | $\left(-\dfrac{1}{3}\right)^{-1}$ | B. | $3$ | (b) | $\left(-\dfrac{2}{5}\right)^{-2}$ | B. | $-\dfrac{25}{4}$ |
| (c) | $-\left(\dfrac{1}{3}\right)^{-1}$ | C. | $-\dfrac{1}{3}$ | (c) | $-\left(\dfrac{2}{5}\right)^{-2}$ | C. | $\dfrac{4}{25}$ |
| (d) | $-\left(-\dfrac{1}{3}\right)^{-1}$ | D. | $-3$ | (d) | $-\left(-\dfrac{2}{5}\right)^{-2}$ | D. | $-\dfrac{4}{25}$ |

**61.** State the quotient rule for exponents in your own words. Give an example.

**62.** State the three power rules for exponents in your own words. Give examples.

*Write each result with only positive exponents. Assume that all variables represent nonzero real numbers. See Example 5.*

**63.** $\dfrac{4^8}{4^6}$  **64.** $\dfrac{5^9}{5^7}$  **65.** $\dfrac{x^{12}}{x^8}$  **66.** $\dfrac{y^{14}}{y^{10}}$

**67.** $\dfrac{r^7}{r^{10}}$  **68.** $\dfrac{y^8}{y^{12}}$  **69.** $\dfrac{6^4}{6^{-2}}$  **70.** $\dfrac{7^5}{7^{-3}}$

**71.** $\dfrac{6^{-3}}{6^7}$  **72.** $\dfrac{5^{-4}}{5^2}$  **73.** $\dfrac{7}{7^{-1}}$  **74.** $\dfrac{8}{8^{-1}}$

**75.** $\dfrac{r^{-3}}{r^{-6}}$  **76.** $\dfrac{s^{-4}}{s^{-8}}$  **77.** $\dfrac{x^3}{y^2}$  **78.** $\dfrac{y^5}{t^3}$

*Simplify each expression. Assume that all variables represent nonzero real numbers. See Example 6.*

**79.** $(x^3)^6$  **80.** $(y^5)^4$  **81.** $\left(\dfrac{3}{5}\right)^3$  **82.** $\left(\dfrac{4}{3}\right)^2$

**83.** $(4t)^3$  **84.** $(5t)^4$  **85.** $(-6x^2)^3$  **86.** $(-2x^5)^5$

**87.** $\left(\dfrac{-4m^2}{t}\right)^3$  **88.** $\left(\dfrac{-5n^4}{r^2}\right)^3$

---

*The authors thank Mitchel Levy of Broward Community College for his suggestions for these exercises.

*Simplify each expression so that no negative exponents appear in the final result. Assume that all variables represent nonzero real numbers. See Examples 1–8.*

**89.** $3^5 \cdot 3^{-6}$
**90.** $4^4 \cdot 4^{-6}$
**91.** $a^{-3}a^2a^{-4}$
**92.** $k^{-5}k^{-3}k^4$
**93.** $(k^2)^{-3}k^4$
**94.** $(x^3)^{-4}x^5$
**95.** $-4r^{-2}(r^4)^2$
**96.** $-2m^{-1}(m^3)^2$
**97.** $(5a^{-1})^4(a^2)^{-3}$
**98.** $(3p^{-4})^2(p^3)^{-1}$
**99.** $(z^{-4}x^3)^{-1}$
**100.** $(y^{-2}z^4)^{-3}$
**101.** $7k^2(-2k)(4k^{-5})^0$
**102.** $3a^2(-5a^{-6})(-2a)^0$
**103.** $\dfrac{(p^{-2})^0}{5p^{-4}}$
**104.** $\dfrac{(m^4)^0}{9m^{-3}}$
**105.** $\dfrac{(3pq)q^2}{6p^2q^4}$
**106.** $\dfrac{(-8xy)y^3}{4x^5y^4}$
**107.** $\dfrac{4a^5(a^{-1})^3}{(a^{-2})^{-2}}$
**108.** $\dfrac{12k^{-2}(k^{-3})^{-4}}{6k^5}$
**109.** $\dfrac{(-y^{-4})^2}{6(y^{-5})^{-1}}$
**110.** $\dfrac{2(-m^{-1})^{-4}}{9(m^{-3})^2}$
**111.** $\dfrac{(2k)^2m^{-5}}{(km)^{-3}}$
**112.** $\dfrac{(3rs)^{-2}}{3^2r^2s^{-4}}$
**113.** $\dfrac{(2k)^2k^3}{k^{-1}k^{-5}}(5k^{-2})^{-3}$
**114.** $\dfrac{(3r^2)^2r^{-5}}{r^{-2}r^3}(2r^{-6})^2$
**115.** $\left(\dfrac{3k^{-2}}{k^4}\right)^{-1} \cdot \dfrac{2}{k}$
**116.** $\left(\dfrac{7m^{-2}}{m^{-3}}\right)^{-2} \cdot \dfrac{m^3}{4}$
**117.** $\left(\dfrac{2p}{q^2}\right)^3\left(\dfrac{3p^4}{q^{-4}}\right)^{-1}$
**118.** $\left(\dfrac{5z^3}{2a^2}\right)^{-3}\left(\dfrac{8a^{-1}}{15z^{-2}}\right)^{-3}$
**119.** $\dfrac{2^2y^4(y^{-3})^{-1}}{2^5y^{-2}}$
**120.** $\dfrac{3^{-1}m^4(m^2)^{-1}}{3^2m^{-2}}$

*Write each number in scientific notation. See Example 9.*

**121.** 530
**122.** 1600
**123.** 0.830
**124.** 0.0072
**125.** 0.00000692
**126.** 0.875
**127.** $-38,500$
**128.** $-976,000,000$

*Write each number in standard notation. See Example 10.*

**129.** $7.2 \times 10^4$
**130.** $8.91 \times 10^2$
**131.** $2.54 \times 10^{-3}$
**132.** $5.42 \times 10^{-4}$
**133.** $-6 \times 10^4$
**134.** $-9 \times 10^3$
**135.** $1.2 \times 10^{-5}$
**136.** $2.7 \times 10^{-6}$

*Evaluate. See Example 11.*

**137.** $\dfrac{12 \times 10^4}{2 \times 10^6}$
**138.** $\dfrac{16 \times 10^5}{4 \times 10^8}$
**139.** $\dfrac{3 \times 10^{-2}}{12 \times 10^3}$
**140.** $\dfrac{5 \times 10^{-3}}{25 \times 10^2}$
**141.** $\dfrac{0.05 \times 1600}{0.0004}$
**142.** $\dfrac{0.003 \times 40,000}{0.00012}$
**143.** $\dfrac{20,000 \times 0.018}{300 \times 0.0004}$
**144.** $\dfrac{840,000 \times 0.03}{0.00021 \times 600}$

*Solve each problem. See Example 12.*

**145.** On March 1, 2009, the population of the United States was estimated to be 305.9 million. (*Source:* U.S. Census Bureau.)

(a) Write this population in scientific notation.

(b) Write $1 trillion—that is, $1,000,000,000,000—in scientific notation.

(c) Using your answers from parts (a) and (b), calculate how much each person in the United States in the year 2009 would have had to contribute in order to make one person in the nation a trillionaire. Write this amount in standard notation to the nearest dollar.

**146.** In the early years of the Powerball Lottery, a player would choose five numbers from 1 through 49 and one number from 1 through 42. It can be shown that there are about $8.009 \times 10^7$ different ways to do this. Suppose that a group of 2000 persons decided to purchase tickets for all these numbers and each ticket cost $1.00. How much should each person have expected to pay? (*Source:* www.powerball.com)

**147.** The speed of light is approximately $3 \times 10^{10}$ cm per sec. How long will it take light to travel $9 \times 10^{12}$ cm?

**148.** The average distance from Earth to the Sun is $9.3 \times 10^7$ mi. How long would it take a rocket, traveling at $2.9 \times 10^3$ mph, to reach the Sun?

**149.** A *light-year* is the distance that light travels in one year. Find the number of miles in a light-year if light travels $1.86 \times 10^5$ mi per sec.

**150.** Use the information given in the previous two exercises to find the number of minutes necessary for light from the Sun to reach Earth.

**151. (a)** The planet Mercury has an average distance from the Sun of $3.6 \times 10^7$ mi, while the average distance of Venus to the Sun is $6.7 \times 10^7$ mi. How long would it take a spacecraft traveling at $1.55 \times 10^3$ mph to travel from Venus to Mercury? (Give your answer in hours, in standard notation.)

**(b)** Use the information from part (a) to find the number of days it would take the spacecraft to travel from Venus to Mercury. Round your answer to the nearest whole number of days.

**152.** When the distance between the centers of the moon and Earth is $4.60 \times 10^8$ m, an object on the line joining the centers exerts the same gravitational force on each of those bodies when it is $4.14 \times 10^8$ m from the center of Earth. How far is the object from the center of the moon at that point?

# 5.2 Adding and Subtracting Polynomials

**OBJECTIVES**

1. Know the basic definitions for polynomials.
2. Find the degree of a polynomial.
3. Add and subtract polynomials.

**OBJECTIVE 1** Know the basic definitions for polynomials. Just as whole numbers are the basis of arithmetic, *polynomials* are fundamental in algebra. To understand polynomials, we review several words from **Section 1.4**. A **term** is a number, a variable, or the product or quotient of a number and one or more variables raised to powers. Examples of terms include

$$4x, \quad \frac{1}{2}m^5 \left(\text{or } \frac{m^5}{2}\right), \quad -7z^9, \quad 6x^2z, \quad \frac{5}{3x^2}, \quad \text{and} \quad 9. \quad \text{Terms}$$

The number in the product is called the **numerical coefficient,** or just the **coefficient.** In the term $8x^3$, the coefficient is $8$. In the term $-4p^5$, it is $-4$. The coefficient of the term $k$ is understood to be 1 (since $k$ can be written as $1k$). The coefficient of $-r$ is $-1$. In the term $\frac{x}{3}$, the coefficient is $\frac{1}{3}$, since $\frac{x}{3} = \frac{1x}{3} = \frac{1}{3}x$. More generally, any factor in a term is the coefficient of the product of the remaining factors. For example, $3x^2$ is the coefficient of $y$ in the term $3x^2y$, and $3y$ is the coefficient of $x^2$ in $3x^2y$.

Any combination of variables or constants (numerical values) joined by the basic operations of addition, subtraction, multiplication, and division (except by 0), or raising to powers or taking roots, formed according to the rules of algebra, is called an **algebraic expression.** The simplest kind of algebraic expression is a *polynomial*.

> **Polynomial**
>
> A **polynomial** is a term or a finite sum of terms in which all variables have whole number exponents and no variables appear in denominators.

Examples of polynomials include

$$3x - 5, \quad 4m^3 - 5m^2p + 8, \quad \text{and} \quad -5t^2s^3. \quad \text{Polynomials}$$

Even though the expression $3x - 5$ involves subtraction, it is a sum of terms, since it could be written as $3x + (-5)$.

Some examples of expressions that are not polynomials are

$$x^{-1} + 3x^{-2}, \quad \sqrt{9 - x}, \quad \text{and} \quad \frac{1}{x}. \quad \text{Not polynomials}$$

The first of these is not a polynomial because it has a negative integer exponent, the second because it involves a variable under a radical, and the third because it contains a variable in the denominator.

Most of the polynomials used in this book contain only one variable. A polynomial containing only the variable $x$ is called a **polynomial in $x$.** A polynomial in one variable is written in **descending powers** of the variable if the exponents on the variable decrease from left to right. For example,

$$x^5 - 6x^2 + 12x - 5$$

is a polynomial in descending powers of $x$. The term $-5$ in this polynomial can be thought of as $-5x^0$, since $-5x^0 = -5(1) = -5$.

**EXAMPLE 1** Writing Polynomials in Descending Powers

Write each polynomial in descending powers of the variable.

**(a)** $y - 6y^3 + 8y^5 - 9y^4 + 12$ is written as $8y^5 - 9y^4 - 6y^3 + y + 12$.

**(b)** $-2 + m + 6m^2 - 4m^3$ is written as $-4m^3 + 6m^2 + m - 2$.

**NOW TRY** Exercise 1.

Some polynomials with a specific number of terms are so common that they are given special names. A polynomial with exactly three terms is a **trinomial,** and a polynomial with exactly two terms is a **binomial**. A single-term polynomial is a **monomial**. The table gives examples.

| Type of Polynomial | Examples |
|---|---|
| Monomial | $5x$,  $7m^9$,  $-8$,  $x^2y^2$ |
| Binomial | $3x^2 - 6$,  $11y + 8$,  $5a^2b + 3a$ |
| Trinomial | $y^2 + 11y + 6$,  $8p^3 - 7p + 2m$,  $-3 + 2k^5 + 9z^4$ |
| None of these | $p^3 - 5p^2 + 2p - 5$,  $-9z^3 + 5c^3 + 2m^5 + 11r^2 - 7r$ |

**OBJECTIVE 2** Find the degree of a polynomial. The **degree of a term** with one variable is the exponent on the variable. For example, the degree of $2x^3$ is 3, the degree of $-x^4$ is 4, and the degree of $17x$ (that is, $17x^1$) is 1. The degree of a term with more than one variable is defined to be the sum of the exponents on the variables. For example, the degree of $5x^3y^7$ is 10, because $3 + 7 = 10$.

The greatest degree of any term in a polynomial is called the **degree of the polynomial**. In most cases, we will be interested in finding the degree of a polynomial in one variable. For example, $4x^3 - 2x^2 - 3x + 7$ has degree 3, because the greatest degree of any term is 3 (the degree of $4x^3$).

The table shows several polynomials and their degrees.

| Polynomial | Degree |
|---|---|
| $9x^2 - 5x + 8$ | 2 |
| $17m^9 + 18m^{14} - 9m^3$ | 14 |
| $5x$ | 1, because $5x = 5x^1$ |
| $-2$ | 0, because $-2 = -2x^0$ (Any nonzero constant has degree 0.) |
| $5a^2b^5$ | 7, because $2 + 5 = 7$ |
| $x^3y^9 + 12xy^4 + 7xy$ | 12, because the degrees of the terms are 12, 5, and 2, and 12 is the greatest. |

**NOTE** The number 0 has no degree, since 0 times a variable to any power is 0.

**NOW TRY** Exercises 21, 25, and 27.

**OBJECTIVE 3** Add and subtract polynomials. We use the distributive property to simplify polynomials by combining terms. For example,

$$x^3 + 4x^2 + 5x^2 - 1 = x^3 + (4 + 5)x^2 - 1 \quad \text{Distributive property}$$
$$= x^3 + 9x^2 - 1.$$

Notice that the terms in the polynomial $4x + 5x^2$ cannot be combined. As these examples suggest, only terms containing exactly the same variables to the same powers may be combined. Recall that such terms are called **like terms.**

**CAUTION** *Remember that only like terms can be combined.*

### EXAMPLE 2  Combining Like Terms

Combine like terms.

(a) $-5y^3 + 8y^3 - y^3 = (-5 + 8 - 1)y^3$     Distributive property
$= 2y^3$

(b) $6x + 5y - 9x + 2y = 6x - 9x + 5y + 2y$     Associative and commutative properties
$= -3x + 7y$     Combine like terms.

Since $-3x$ and $7y$ are unlike terms, no further simplification is possible.

(c) $5x^2y - 6xy^2 + 9x^2y + 13xy^2 = 5x^2y + 9x^2y - 6xy^2 + 13xy^2$
$= 14x^2y + 7xy^2$

**NOW TRY** Exercises 31, 37, and 43.

We use the following rule to add two polynomials.

### Adding Polynomials

To add two polynomials, combine like terms.

### EXAMPLE 3  Adding Polynomials

Add $(3a^5 - 9a^3 + 4a^2) + (-8a^5 + 8a^3 + 2)$.

Use the commutative and associative properties to rearrange the polynomials so that like terms are together. Then use the distributive property to combine like terms.

$(3a^5 - 9a^3 + 4a^2) + (-8a^5 + 8a^3 + 2)$
$= 3a^5 - 8a^5 - 9a^3 + 8a^3 + 4a^2 + 2$
$= -5a^5 - a^3 + 4a^2 + 2$     Combine like terms.

We can add these same two polynomials vertically by placing like terms in columns.

$$\begin{array}{r} 3a^5 - 9a^3 + 4a^2 \phantom{+2} \\ -8a^5 + 8a^3 \phantom{+4a^2} + 2 \\ \hline -5a^5 - a^3 + 4a^2 + 2 \end{array}$$

**NOW TRY** Exercises 51 and 65.

In **Section 1.2,** we defined subtraction of real numbers as

$$a - b = a + (-b).$$

That is, we add the first number and the negative (or opposite) of the second. We can give a similar definition for subtraction of polynomials by defining the **negative of a polynomial** as that polynomial with the sign of every coefficient changed.

### Subtracting Polynomials

To subtract two polynomials, add the first polynomial and the negative of the *second* polynomial.

### EXAMPLE 4 Subtracting Polynomials

Subtract $(-6m^2 - 8m + 5) - (-5m^2 + 7m - 8)$.
Change every sign in the second polynomial and add.

$(-6m^2 - 8m + 5) - (-5m^2 + 7m - 8)$
$= -6m^2 - 8m + 5 + 5m^2 - 7m + 8$
$= -6m^2 + 5m^2 - 8m - 7m + 5 + 8$     Rearrange terms.
$= -m^2 - 15m + 13$     Combine like terms.

Check by adding the sum, $-m^2 - 15m + 13$, to the second polynomial. The result should be the first polynomial.

To subtract these two polynomials vertically, write the first polynomial above the second, lining up like terms in columns.

$$-6m^2 - 8m + 5$$
$$-5m^2 + 7m - 8$$

Change all the signs in the second polynomial and add.

$-6m^2 - 8m + 5$
$+5m^2 - 7m + 8$     Change all signs.
$-m^2 - 15m + 13$     Add in columns.

**NOW TRY** Exercises 61 and 69.

## 5.2 Exercises

*Write each polynomial in descending powers of the variable. See Example 1.*

**1.** $2x^3 + x - 3x^2 + 4$    **2.** $3y^2 + y^4 - 2y^3 + y$    **3.** $4p^3 - 8p^5 + p^7$

**4.** $q^2 + 3q^4 - 2q + 1$    **5.** $-m^3 + 5m^2 + 3m^4 + 10$    **6.** $4 - x + 3x^2$

*Give the numerical coefficient and the degree of each term.*

**7.** $7z$    **8.** $3r$    **9.** $-15p^2$    **10.** $-27k^3$    **11.** $x^4$

**12.** $y^6$    **13.** $\dfrac{t}{6}$    **14.** $\dfrac{m}{4}$    **15.** $-mn^5$    **16.** $-a^5 b$

*Identify each polynomial as a* monomial, binomial, trinomial, *or none of these. Also, give the degree.*

**17.** $25$    **18.** $5$    **19.** $7m - 22$

**20.** $6x + 15$    **21.** $-7y^6 + 11y^8$    **22.** $12k^2 - 9k^5$

**23.** $-5m^3 + 6m - 9m^2$    **24.** $4z^2 - 11z + 2$

**25.** $-6p^4 q - 3p^3 q^2 + 2pq^3 - q^4$    **26.** $8s^3 t - 4s^2 t^2 + 2st^3 + 9$

**27.** *Multiple Choice* Which one of the following is a trinomial in descending powers, having degree 6?

   **A.** $5x^6 - 4x^5 + 12$    **B.** $6x^5 - x^6 + 4$
   **C.** $2x + 4x^2 - x^6$    **D.** $4x^6 - 6x^4 + 9x^2 - 8$

**28.** *Concept Check* Give an example of a polynomial of four terms in the variable $x$, having degree 5, written in descending powers, and lacking a fourth-degree term.

*Combine like terms. See Example 2.*

**29.** $5z^4 + 3z^4$    **30.** $8r^5 - 2r^5$    **31.** $-m^3 + 2m^3 + 6m^3$

**32.** $3p^4 + 5p^4 - 2p^4$    **33.** $x + x + x + x + x$    **34.** $z - z - z + z$

35. $m^4 - 3m^2 + m$  36. $5a^5 + 2a^4 - 9a^3$  37. $5t + 4s - 6t + 9s$
38. $8p - 9q - 3p + q$  39. $2k + 3k^2 + 5k^2 - 7$  40. $4x^2 + 2x - 6x^2 - 6$
41. $n^4 - 2n^3 + n^2 - 3n^4 + n^3$  42. $2q^3 + 3q^2 - 4q - q^3 + 5q^2$
43. $3ab^2 + 7a^2b - 5ab^2 + 13a^2b$  44. $6m^2n - 8mn^2 + 3mn^2 - 7m^2n$
45. $4 - (2 + 3m) + 6m + 9$  46. $8a - (3a + 4) - (5a - 3)$
47. $6 + 3p - (2p + 1) - (2p + 9)$  48. $4x - 8 - (-1 + x) - (11x + 5)$

49. Define *polynomial* in your own words. Give examples. Include the words *term, monomial, binomial,* and *trinomial* in your explanation.

50. Write a paragraph explaining how to add and subtract polynomials. Give examples.

*Add or subtract as indicated. See Examples 3 and 4.*

51. $(5x^2 + 7x - 4) + (3x^2 - 6x + 2)$  52. $(4k^3 + k^2 + k) + (2k^3 - 4k^2 - 3k)$
53. $(6t^2 - 4t^4 - t) + (3t^4 - 4t^2 + 5)$  54. $(3p^2 + 2p - 5) + (7p^2 - 4p^3 + 3p)$
55. $(y^3 + 3y + 2) + (4y^3 - 3y^2 + 2y - 1)$  56. $(2x^5 - 2x^4 + x^3 - 1) + (x^4 - 3x^3 + 2)$
57. $(3r + 8) - (2r - 5)$  58. $(2d + 7) - (3d - 1)$
59. $(2a^2 + 3a - 1) - (4a^2 + 5a + 6)$  60. $(q^4 - 2q^2 + 10) - (3q^4 + 5q^2 - 5)$
61. $(z^5 + 3z^2 + 2z) - (4z^5 + 2z^2 - 5z)$  62. $(5t^3 - 3t^2 + 2t) - (4t^3 + 2t^2 + 3t)$

63. Add.
$21p - 8$
$\underline{-9p + 4}$

64. Add.
$15m - 9$
$\underline{4m + 12}$

65. Add.
$-12p^2 + 4p - 1$
$\underline{3p^2 + 7p - 8}$

66. Add.
$-6y^3 + 8y + 5$
$\underline{9y^3 + 4y - 6}$

67. Subtract.
$12a + 15$
$\underline{7a - 3}$

68. Subtract.
$-3b + 6$
$\underline{2b - 8}$

69. Subtract.
$6m^2 - 11m + 5$
$\underline{-8m^2 + 2m - 1}$

70. Subtract.
$-4z^2 + 2z - 1$
$\underline{3z^2 - 5z + 2}$

71. Add.
$12z^2 - 11z + 8$
$5z^2 + 16z - 2$
$\underline{-4z^2 + 5z - 9}$

72. Add.
$-6m^3 + 2m^2 + 5m$
$8m^3 + 4m^2 - 6m$
$\underline{-3m^3 + 2m^2 - 7m}$

73. Add.
$6y^3 - 9y^2 \phantom{+ 5y} + 8$
$\underline{4y^3 + 2y^2 + 5y}$

74. Add.
$-7r^8 + 2r^6 - r^5$
$\underline{\phantom{-7r^8 +} 3r^6 \phantom{- r^5} + 5}$

75. Subtract.
$-5a^4 \phantom{+ 6a^3} + 8a^2 - 9$
$\underline{\phantom{-5a^4 +} 6a^3 - a^2 + 2}$

76. Subtract.
$\phantom{m^4 -} -2m^3 + 8m^2 \phantom{+ 2m}$
$\underline{m^4 - \phantom{2}m^3 \phantom{+ 8m^2} + 2m}$

*Perform the indicated operations. See Examples 2–4.*

77. Subtract $4y^2 - 2y + 3$ from $7y^2 - 6y + 5$.
78. Subtract $-(-4x + 2z^2 + 3m)$ from $[(2z^2 - 3x + m) + (z^2 - 2m)]$.
79. $(-4m^2 + 3n^2 - 5n) - [(3m^2 - 5n^2 + 2n) + (-3m^2) + 4n^2]$
80. $[-(4m^2 - 8m + 4m^3) - (3m^2 + 2m + 5m^3)] + m^2$
81. $[-(y^4 - y^2 + 1) - (y^4 + 2y^2 + 1)] + (3y^4 - 3y^2 - 2)$
82. $[2p - (3p - 6)] - [(5p - (8 - 9p)) + 4p]$
83. $-[3z^2 + 5z - (2z^2 - 6z)] + [(8z^2 - [5z - z^2]) + 2z^2]$
84. $5k - (5k - [2k - (4k - 8k)]) + 11k - (9k - 12k)$

# 5.3 Polynomial Functions, Graphs, and Composition

**OBJECTIVES**

1. Recognize and evaluate polynomial functions.
2. Use a polynomial function to model data.
3. Add and subtract polynomial functions.
4. Find the composition of functions.
5. Graph basic polynomial functions.

**OBJECTIVE 1** Recognize and evaluate polynomial functions. In **Chapter 3**, we studied linear (first-degree polynomial) functions, defined as $f(x) = ax + b$. Now we consider more general polynomial functions.

**Polynomial Function**

A **polynomial function of degree $n$** is defined by

$$f(x) = a_n x^n + a_{n-1} x^{n-1} + \cdots + a_1 x + a_0,$$

for real numbers $a_n, a_{n-1}, \ldots, a_1,$ and $a_0$, where $a_n \neq 0$ and $n$ is a whole number.

Another way of describing a polynomial function is to say that it is a function defined by a polynomial in one variable, consisting of one or more terms. It is usually written in descending powers of the variable, and its degree is the degree of the polynomial that defines it. Suppose we consider the polynomial $3x^2 - 5x + 7$, so that

$$f(x) = 3x^2 - 5x + 7.$$

If $x = -2$, then $f(x) = 3x^2 - 5x + 7$ takes on the value

$$f(-2) = 3(-2)^2 - 5(-2) + 7 \quad \text{Let } x = -2.$$
$$= 3(4) + 10 + 7$$
$$= 29.$$

Thus, $f(-2) = 29$, and the ordered pair $(-2, 29)$ belongs to $f$.

**EXAMPLE 1** Evaluating Polynomial Functions

Let $f(x) = 4x^3 - x^2 + 5$. Find each value.

**(a)** $f(3)$

*Read this as "f of 3," not "f times 3."*

$$f(x) = 4x^3 - x^2 + 5$$
$$f(3) = 4(3)^3 - 3^2 + 5 \quad \text{Substitute 3 for } x.$$
$$= 4(27) - 9 + 5 \quad \text{Order of operations}$$
$$= 108 - 9 + 5 \quad \text{Multiply.}$$
$$= 104 \quad \text{Subtract; add.}$$

**(b)** $f(-4) = 4(-4)^3 - (-4)^2 + 5 \quad \text{Let } x = -4.$
$$= 4(-64) - 16 + 5$$
$$= -267$$

*Be careful with signs.*

**NOW TRY** Exercise 7.

While $f$ is the most common letter used to represent functions, recall that other letters, such as $g$ and $h$, are also used. The capital letter $P$ is often used for polynomial functions. Note that the function defined as $P(x) = 4x^3 - x^2 + 5$ yields the same ordered pairs as the function $f$ in Example 1.

**OBJECTIVE 2** Use a polynomial function to model data. Polynomial functions can be used to approximate data. Such functions are usually valid for small intervals, and they allow us to predict (with caution) what might happen for values just outside the intervals. These intervals are often periods of years, as shown in Example 2.

**EXAMPLE 2** Using a Polynomial Model to Approximate Data

The number of U.S. households estimated to see and pay at least one bill on-line each month during the years 2000 through 2006 can be modeled by the polynomial function defined by

$$P(x) = 0.808x^2 + 2.625x + 0.502,$$

where $x = 0$ corresponds to the year 2000, $x = 1$ corresponds to 2001, and so on, and $P(x)$ is in millions. Use this function to approximate the number of households that paid at least one bill on-line each month in 2005.

Since $x = 5$ corresponds to 2005, we must find $P(5)$.

$$P(x) = 0.808x^2 + 2.625x + 0.502$$
$$P(5) = 0.808(5)^2 + 2.625(5) + 0.502 \quad \text{Let } x = 5.$$
$$= 33.827 \quad \text{Evaluate.}$$

According to this model, in 2005 about 33.83 million households were expected to pay at least one bill on-line each month.

**NOW TRY** Exercise 9.

**OBJECTIVE 3** Add and subtract polynomial functions. The operations of addition, subtraction, multiplication, and division are also defined for functions. For example, businesses use the equation "profit equals revenue minus cost," written in function notation as

$$\underset{\substack{\uparrow \\ \text{Profit} \\ \text{function}}}{P(x)} = \underset{\substack{\uparrow \\ \text{Revenue} \\ \text{function}}}{R(x)} - \underset{\substack{\uparrow \\ \text{Cost} \\ \text{function}}}{C(x)},$$

where $x$ is the number of items produced and sold. Thus, the profit function is found by subtracting the cost function from the revenue function.

We define the following **operations on functions.**

---

**Adding and Subtracting Functions**

If $f(x)$ and $g(x)$ define functions, then

$$(f + g)(x) = f(x) + g(x) \quad \text{Sum function}$$

and

$$(f - g)(x) = f(x) - g(x). \quad \text{Difference function}$$

In each case, the domain of the new function is the intersection of the domains of $f(x)$ and $g(x)$.

### EXAMPLE 3  Adding and Subtracting Functions

For the polynomial functions defined by

$$f(x) = x^2 - 3x + 7 \quad \text{and} \quad g(x) = -3x^2 - 7x + 7,$$

find **(a)** the sum and **(b)** the difference.

**(a)**
$$\begin{aligned}
(f + g)(x) &= f(x) + g(x) &&\text{Use the definition.} \\
&= (x^2 - 3x + 7) + (-3x^2 - 7x + 7) &&\text{Substitute.} \\
&= -2x^2 - 10x + 14 &&\text{Add the polynomials.}
\end{aligned}$$

*This notation does not indicate the distributive property.*

**(b)**
$$\begin{aligned}
(f - g)(x) &= f(x) - g(x) &&\text{Use the definition.} \\
&= (x^2 - 3x + 7) - (-3x^2 - 7x + 7) &&\text{Substitute.} \\
&= (x^2 - 3x + 7) + (3x^2 + 7x - 7) &&\text{Change subtraction to addition.} \\
&= 4x^2 + 4x &&\text{Add.}
\end{aligned}$$

**NOW TRY** Exercise 15.

### EXAMPLE 4  Adding and Subtracting Functions

Find each of the following for the functions defined by

$$f(x) = 10x^2 - 2x \quad \text{and} \quad g(x) = 2x.$$

**(a)** $(f + g)(2)$

$$\begin{aligned}
(f + g)(2) &= f(2) + g(2) &&\text{Use the definition.} \\
&= \overbrace{[10(2)^2 - 2(2)]}^{f(x) = 10x^2 - 2x} + \overbrace{2(2)}^{g(x) = 2x} &&\text{Substitute.} \\
&= [40 - 4] + 4 &&\text{Order of operations} \\
&= 40
\end{aligned}$$

*This is a key step.*

Alternatively, we could first find $(f + g)(x)$.

$$\begin{aligned}
(f + g)(x) &= f(x) + g(x) &&\text{Use the definition.} \\
&= (10x^2 - 2x) + 2x &&\text{Substitute.} \\
&= 10x^2 &&\text{Combine like terms.}
\end{aligned}$$

Then,

$$(f + g)(2) = 10(2)^2 = 40. \qquad \text{The result is the same.}$$

**(b)** $(f - g)(x)$ and $(f - g)(1)$

$$\begin{aligned}
(f - g)(x) &= f(x) - g(x) &&\text{Use the definition.} \\
&= (10x^2 - 2x) - 2x &&\text{Substitute.} \\
&= 10x^2 - 4x &&\text{Combine like terms.}
\end{aligned}$$

Then,

$$(f - g)(1) = 10(1)^2 - 4(1) = 6. \qquad \text{Substitute.}$$

Confirm that $f(1) - g(1)$ gives the same result.

**NOW TRY** Exercises 17 and 19.

**OBJECTIVE 4** Find the composition of functions. The diagram in Figure 1 shows a function $f$ that assigns, to each element $x$ of set $X$, some element $y$ of set $Y$. Suppose that a function $g$ takes each element of set $Y$ and assigns a value $z$ of set $Z$. Then $f$ and $g$ together assign an element $x$ in $X$ to an element $z$ in $Z$. The result of this process is a new function $h$ that takes an element $x$ in $X$ and assigns it an element $z$ in $Z$.

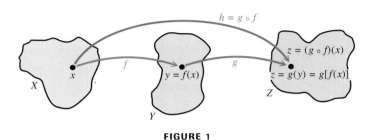

**FIGURE 1**

This function $h$ is called the *composition* of functions $g$ and $f$, written $g \circ f$.

---

**Composition of Functions**

If $f$ and $g$ are functions, then the **composite function**, or **composition**, of $g$ and $f$ is defined by

$$(g \circ f)(x) = g(f(x))$$

for all $x$ in the domain of $f$ such that $f(x)$ is in the domain of $g$.

---

Read $g \circ f$ as "$g$ of $f$".

As a real-life example of how composite functions occur, consider the following retail situation.

*A $40 pair of blue jeans is on sale for 25% off. If you purchase the jeans before noon, the retailer offers an additional 10% off. What is the final sale price of the blue jeans?*

You might be tempted to say that the blue jeans are $25\% + 10\% = 35\%$ off and calculate $\$40(0.35) = \$14$, giving a final sale price of $\$40 - \$14 = \$26$ for the jeans. ***This is not correct.*** To find the final sale price, we must first find the price after taking 25% off, and then take an additional 10% off that price.

$\$40(0.25) = \$10$, giving a sale price of $\$40 - \$10 = \$30$.  Take 25% off original price.

$\$30(0.10) = \$3$, giving a ***final sale price*** of $\$30 - \$3 = \$27$.  Take additional 10% off.

This is the idea behind composition of functions.

As another example of composition, suppose an oil well off the Louisiana coast is leaking, with the leak spreading oil in a circular layer over the surface. See Figure 2. At any time $t$, in minutes, after the beginning of the leak, the radius of the circular oil slick is given by $r(t) = 5t$ feet. Since $A(r) = \pi r^2$ gives the area of a circle of radius $r$, the area can be expressed as a function of time by substituting $5t$ for $r$ in $A(r) = \pi r^2$ to get

$$A(r(t)) = \pi(5t)^2 = 25\pi t^2.$$

**FIGURE 2**

The function $A(r(t))$ is a composite function of the functions $A$ and $r$.

---

**EXAMPLE 5  Finding a Composite Function**

Let $f(x) = x^2$ and $g(x) = x + 3$. Find $(f \circ g)(4)$.

$(f \circ g)(4) = f(g(4))$     Definition

*Evaluate the "inside" function value first.*

$\quad\quad\quad\quad = f(4 + 3)$     Use the rule for g(x); g(4) = 4 + 3.

$\quad\quad\quad\quad = f(7)$     Add.

*Now evaluate the "outside" function.*

$\quad\quad\quad\quad = 7^2$     Use the rule for f(x); f(7) = 7².

$\quad\quad\quad\quad = 49$

**NOW TRY** Exercise 35.

---

Notice in Example 5 that if we interchange the order of the functions, the composition of $g$ and $f$ is defined by $g(f(x))$. Once again, letting $x = 4$, we have

$(g \circ f)(4) = g(f(4))$     Definition

$\quad\quad\quad\quad = g(4^2)$     Use the rule for f(x); f(4) = 4².

$\quad\quad\quad\quad = g(16)$     Square 4.

$\quad\quad\quad\quad = 16 + 3$     Use the rule for g(x); g(16) = 16 + 3.

$\quad\quad\quad\quad = 19.$

Here we see that $(f \circ g)(4) \neq (g \circ f)(4)$ because $49 \neq 19$. In general,

$$(f \circ g)(x) \neq (g \circ f)(x).$$

---

**EXAMPLE 6  Finding Composite Functions**

Let $f(x) = 4x - 1$ and $g(x) = x^2 + 5$. Find the following.

**(a)** $(f \circ g)(2)$

$(f \circ g)(2) = f(g(2))$

$\quad\quad\quad\quad = f(2^2 + 5)$     g(x) = x² + 5

$\quad\quad\quad\quad = f(9)$     Work inside the parentheses.

$\quad\quad\quad\quad = 4(9) - 1$     f(x) = 4x − 1

$\quad\quad\quad\quad = 35$     Multiply; subtract.

**(b)** $(f \circ g)(x)$

Here, use $g(x)$ as the input for the function $f$.

$(f \circ g)(x) = f(g(x))$

$\quad\quad\quad\quad = 4(g(x)) - 1$     Use the rule for f(x); f(x) = 4x − 1.

$\quad\quad\quad\quad = 4(x^2 + 5) - 1$     g(x) = x² + 5

$\quad\quad\quad\quad = 4x^2 + 20 - 1$     Distributive property

$\quad\quad\quad\quad = 4x^2 + 19$     Combine terms.

**(c)** Find $(f \circ g)(2)$ again, this time using the rule obtained in part (b).

$$(f \circ g)(x) = 4x^2 + 19 \quad \text{From part (b)}$$
$$(f \circ g)(2) = 4(2)^2 + 19 \quad \text{Let } x = 2.$$
$$= 4(4) + 19$$
$$= 16 + 19$$
$$= 35$$

The result, 35, is the same as the result in part (a).

**NOW TRY** Exercises 37 and 41.

**OBJECTIVE 5** Graph basic polynomial functions. Functions were introduced in **Section 3.5.** Recall that each input (or *x*-value) of a function results in one output (or *y*-value). The simplest polynomial function is the **identity function,** defined by $f(x) = x$. The domain (set of *x*-values) of this function is all real numbers, $(-\infty, \infty)$, and the function pairs each real number with itself. Therefore, the range (set of *y*-values) is also $(-\infty, \infty)$. The graph of the function is a straight line, as first seen in **Chapter 3.** (Notice that a *linear function* is a specific kind of polynomial function.) Figure 3 shows the graph of $f(x) = x$ and a table of selected ordered pairs.

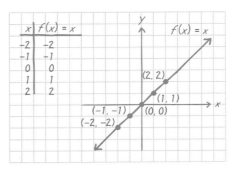

**FIGURE 3**

Another polynomial function, defined by $f(x) = x^2$, is the **squaring function.** For this function, every real number is paired with its square. The input can be any real number, so the domain is $(-\infty, \infty)$. Since the square of any real number is nonnegative, the range is $[0, \infty)$. The graph of the squaring function is a *parabola*. Figure 4 shows the graph and a table of selected ordered pairs.

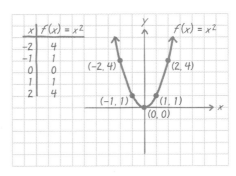

**FIGURE 4**

The **cubing function** is defined by $f(x) = x^3$. This function pairs every real number with its cube. The domain and the range are both $(-\infty, \infty)$. The graph of the cubing function is neither a line nor a parabola. See Figure 5 and the table of ordered pairs.

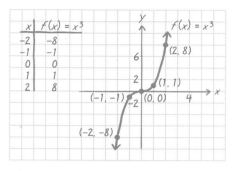

**FIGURE 5**

---

**EXAMPLE 7** Graphing Variations of the Identity, Squaring, and Cubing Functions

Graph each function by creating a table of ordered pairs. Give the domain and range of each function by observing its graph.

**(a)** $f(x) = 2x$

To find each range value, multiply the domain value by 2. Plot the points and join them with a straight line. See Figure 6. Both the domain and the range are $(-\infty, \infty)$.

| x | f(x) = 2x |
|---|---|
| −2 | −4 |
| −1 | −2 |
| 0 | 0 |
| 1 | 2 |
| 2 | 4 |

| x | f(x) = −x² |
|---|---|
| −2 | −4 |
| −1 | −1 |
| 0 | 0 |
| 1 | −1 |
| 2 | −4 |

**FIGURE 6**  **FIGURE 7**

**(b)** $f(x) = -x^2$

For each input $x$, square it and then take its opposite. Plotting and joining the points gives a parabola that opens down. It is a *reflection* of the graph of the squaring function across the $x$-axis. See the table and Figure 7. The domain is $(-\infty, \infty)$ and the range is $(-\infty, 0]$.

**(c)** $f(x) = x^3 - 2$

For this function, cube the input and then subtract 2 from the result. The graph is that of the cubing function *shifted* 2 units down. See the table and Figure 8 on the next page. The domain and range are both $(-\infty, \infty)$.

| x | $f(x) = x^3 - 2$ |
|---|---|
| -2 | -10 |
| -1 | -3 |
| 0 | -2 |
| 1 | -1 |
| 2 | 6 |

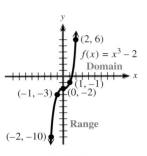

**FIGURE 8**

NOW TRY Exercises 55, 57, and 59.

## 5.3 Exercises

*For each polynomial function, find* **(a)** $f(-1)$ *and* **(b)** $f(2)$. *See Example 1.*

1. $f(x) = 6x - 4$
2. $f(x) = -2x + 5$
3. $f(x) = x^2 - 3x + 4$
4. $f(x) = 3x^2 + x - 5$
5. $f(x) = 5x^4 - 3x^2 + 6$
6. $f(x) = -4x^4 + 2x^2 - 1$
7. $f(x) = -x^2 + 2x^3 - 8$
8. $f(x) = -x^2 - x^3 + 11x$

*Solve each problem. See Example 2.*

9. The number of airports in the United States during the period from 1980 through 2003 can be approximated by the polynomial function defined by

$$f(x) = -2.19x^2 + 245.7x + 15{,}163,$$

where $x = 0$ represents 1980, $x = 1$ represents 1981, and so on. Use this function to approximate the number of airports in each given year. (*Source:* U.S. Federal Aviation Administration.)

**(a)** 1980  **(b)** 1990  **(c)** 2003

10. The percent of births to unmarried women during the period from 1990 through 2002 can be approximated by the polynomial function defined by

$$f(x) = 0.010x^3 - 0.245x^2 + 2.09x + 26.6,$$

where $x = 0$ represents 1990, $x = 1$ represents 1991, and so on. Use this function to approximate the percent (to the nearest tenth) of births to unmarried women in each given year. (*Source:* National Center for Health Statistics.)

**(a)** 1990  **(b)** 1995  **(c)** 2002

11. The number of bank debit cards issued during the period from 1990 through 2000 can be modeled by the polynomial function defined by

$$P(x) = -0.31x^3 + 5.8x^2 - 15x + 9,$$

where $x = 0$ corresponds to the year 1990, $x = 1$ corresponds to 1991, and so on, and $P(x)$ is in millions. Use this function to approximate the number of bank debit cards issued in each given year. Round answers to the nearest million. (*Source: Statistical Abstract of the United States.*)

(a) 1990  (b) 1996  (c) 1999

12. The total number of cases on the docket of the U.S. Supreme Court during the period from 1980 to 2000 can be modeled by the polynomial function defined by

$$P(x) = -0.0325x^3 + 4.524x^2 + 102.5x + 5128,$$

where $x = 0$ represents 1980, $x = 1$ represents 1981, and so on. Use this function to approximate the number of cases on the docket of the U.S. Supreme Court in each given year. (*Source:* Office of the Clerk, Supreme Court of the United States.)

(a) 1980  (b) 1990  (c) 2000

*For each pair of functions, find* (a) $(f + g)(x)$ *and* (b) $(f - g)(x)$. *See Example 3.*

13. $f(x) = 5x - 10, g(x) = 3x + 7$
14. $f(x) = -4x + 1, g(x) = 6x + 2$
15. $f(x) = 4x^2 + 8x - 3, g(x) = -5x^2 + 4x - 9$
16. $f(x) = 3x^2 - 9x + 10, g(x) = -4x^2 + 2x + 12$

Let $f(x) = x^2 - 9$, $g(x) = 2x$, and $h(x) = x - 3$. Find each of the following. See Example 4.

17. $(f + g)(x)$
18. $(f - g)(x)$
19. $(f + g)(3)$
20. $(f - g)(-3)$
21. $(f - h)(x)$
22. $(f + h)(x)$
23. $(f - h)(-3)$
24. $(f + h)(-2)$
25. $(g + h)(-10)$
26. $(g - h)(10)$
27. $(g - h)(-3)$
28. $(g + h)\left(\frac{1}{3}\right)$
29. $(g + h)\left(\frac{1}{4}\right)$
30. $(g + h)\left(-\frac{1}{4}\right)$
31. $(g + h)\left(-\frac{1}{2}\right)$

32. Is the equation defining the sum of two functions, $(f + g)(x) = f(x) + g(x)$, an application of the distributive property? Explain.

33. Construct two polynomial functions defined by $f(x)$, a polynomial of degree 3, and $g(x)$, a polynomial of degree 4. Find $(f - g)(x)$ and $(g - f)(x)$. Use your answers to decide whether subtraction of polynomial functions is a commutative operation. Explain.

34. *Concept Check* Find two polynomial functions defined by $f(x)$ and $g(x)$ such that

$$(f + g)(x) = 3x^3 - x + 3.$$

Let $f(x) = x^2 + 4$, $g(x) = 2x + 3$, and $h(x) = x + 5$. Find each value or expression. See Examples 5 and 6.

35. $(h \circ g)(4)$
36. $(f \circ g)(4)$
37. $(g \circ f)(6)$
38. $(h \circ f)(6)$
39. $(f \circ h)(-2)$
40. $(h \circ g)(-2)$
41. $(f \circ g)(x)$
42. $(g \circ h)(x)$
43. $(f \circ h)(x)$
44. $(g \circ f)(x)$
45. $(h \circ g)(x)$
46. $(h \circ f)(x)$
47. $(f \circ h)\left(\frac{1}{2}\right)$
48. $(h \circ f)\left(\frac{1}{2}\right)$
49. $(f \circ g)\left(-\frac{1}{2}\right)$
50. $(g \circ f)\left(-\frac{1}{2}\right)$

*Solve each problem.*

**51.** The function defined by $f(x) = 12x$ computes the number of inches in $x$ feet, and the function defined by $g(x) = 5280x$ computes the number of feet in $x$ miles. What is $(f \circ g)(x)$ and what does it compute?

**52.** The perimeter $x$ of a square with sides of length $s$ is given by the formula $x = 4s$.

(a) Solve for $s$ in terms of $x$.
(b) If $y$ represents the area of this square, write $y$ as a function of the perimeter $x$.
(c) Use the composite function of part (b) to find the area of a square with perimeter 6.

**53.** When a thermal inversion layer is over a city (as happens often in Los Angeles), pollutants cannot rise vertically, but are trapped below the layer and must disperse horizontally. Assume that a factory smokestack begins emitting a pollutant at 8 A.M. Assume that the pollutant disperses horizontally over a circular area. Suppose that $t$ represents the time, in hours, since the factory began emitting pollutants ($t = 0$ represents 8 A.M.), and assume that the radius of the circle of pollution is $r(t) = 2t$ miles. Let $A(r) = \pi r^2$ represent the area of a circle of radius $r$. Find and interpret $(A \circ r)(t)$.

**54.** An oil well off the Gulf Coast is leaking, with the leak spreading oil over the surface as a circle. At any time $t$, in minutes, after the beginning of the leak, the radius of the circular oil slick on the surface is $r(t) = 4t$ feet. Let $A(r) = \pi r^2$ represent the area of a circle of radius $r$. Find and interpret $(A \circ r)(t)$.

*Graph each function. Give the domain and range. See Example 7.*

**55.** $f(x) = -2x + 1$     **56.** $f(x) = 3x + 2$     **57.** $f(x) = -3x^2$

**58.** $f(x) = \dfrac{1}{2}x^2$     **59.** $f(x) = x^3 + 1$     **60.** $f(x) = -x^3 + 2$

## 5.4 Multiplying Polynomials

**OBJECTIVES**

1. Multiply terms.
2. Multiply any two polynomials.
3. Multiply binomials.
4. Find the product of the sum and difference of two terms.
5. Find the square of a binomial.
6. Multiply polynomial functions.

**OBJECTIVE 1** Multiply terms. Recall that the product of $3x^4$ and $5x^3$ is found by using the commutative and associative properties, along with the rules for exponents.

$$(3x^4)(5x^3) = 3 \cdot 5 \cdot x^4 \cdot x^3$$
$$= 15x^{4+3}$$
$$= 15x^7$$

**EXAMPLE 1** Multiplying Monomials

Find each product.

(a) $-4a^3(3a^5) = -4(3)a^3 \cdot a^5 = -12a^8$

(b) $2m^2z^4(8m^3z^2) = 2(8)m^2 \cdot m^3 \cdot z^4 \cdot z^2 = 16m^5z^6$

**NOW TRY** Exercises 5 and 7.

**OBJECTIVE 2** **Multiply any two polynomials.** We use the distributive property to extend this process to find the product of any two polynomials.

### EXAMPLE 2 Multiplying Polynomials

Find each product.

(a) $-2(8x^3 - 9x^2) = -2(8x^3) - 2(-9x^2)$   Distributive property
$= -16x^3 + 18x^2$

(b) $5x^2(-4x^2 + 3x - 2) = 5x^2(-4x^2) + 5x^2(3x) + 5x^2(-2)$
$= -20x^4 + 15x^3 - 10x^2$

(c) $(3x - 4)(2x^2 + x)$

Use the distributive property to multiply each term of $2x^2 + x$ by $3x - 4$.

$$(3x - 4)(2x^2 + x) = (3x - 4)(2x^2) + (3x - 4)(x)$$

Here, $3x - 4$ was treated as a single expression so that the distributive property could be used. Now use the distributive property two more times.

$= 3x(2x^2) + (-4)(2x^2) + (3x)(x) + (-4)(x)$
$= 6x^3 - 8x^2 + 3x^2 - 4x$
$= 6x^3 - 5x^2 - 4x$   Combine like terms.

(d) $2x^2(x + 1)(x - 3) = 2x^2[(x + 1)(x) + (x + 1)(-3)]$   Distributive property
$= 2x^2[x^2 + x - 3x - 3]$   Distributive property
$= 2x^2(x^2 - 2x - 3)$   Combine like terms.
$= 2x^4 - 4x^3 - 6x^2$   Distributive property

**NOW TRY** Exercises 11, 13, and 19.

It is often easier to multiply polynomials by writing them vertically.

### EXAMPLE 3 Multiplying Polynomials Vertically

Find each product.

(a) $(5a - 2b)(3a + b)$

$$\begin{array}{r} 5a - 2b \\ 3a + b \\ \hline 5ab - 2b^2 \\ 15a^2 - 6ab \phantom{- 2b^2} \\ \hline 15a^2 - ab - 2b^2 \end{array}$$

← Multiply $b(5a - 2b)$.
← Multiply $3a(5a - 2b)$.
Combine like terms.

(b) $(3m^3 - 2m^2 + 4)(3m - 5)$

$$\begin{array}{r} 3m^3 - 2m^2 + 4 \\ 3m - 5 \\ \hline -15m^3 + 10m^2 \phantom{+ 12m} - 20 \\ 9m^4 - 6m^3 \phantom{+ 10m^2} + 12m \phantom{- 20} \\ \hline 9m^4 - 21m^3 + 10m^2 + 12m - 20 \end{array}$$

$-5(3m^3 - 2m^2 + 4)$
$3m(3m^3 - 2m^2 + 4)$
Combine like terms.

Be sure to write like terms in columns.

**NOW TRY** Exercises 23 and 25.

**OBJECTIVE 3** Multiply binomials. When working with polynomials, the product of two binomials occurs repeatedly. There is a shortcut method for finding these products. Recall that a binomial has just two terms, such as $3x - 4$ or $2x + 3$. We can find the product of these binomials by using the distributive property as follows:

$$(3x - 4)(2x + 3) = 3x(2x + 3) - 4(2x + 3)$$
$$= 3x(2x) + 3x(3) - 4(2x) - 4(3)$$
$$= 6x^2 + 9x - 8x - 12.$$

Before combining like terms to find the simplest form of the answer, we check the origin of each of the four terms in the sum $6x^2 + 9x - 8x - 12$. First, $6x^2$ is the product of the two *first* terms.

$$(3x - 4)(2x + 3) \qquad 3x(2x) = 6x^2 \qquad \text{First terms}$$

To get $9x$, the *outer* terms are multiplied.

$$(3x - 4)(2x + 3) \qquad 3x(3) = 9x \qquad \text{Outer terms}$$

The term $-8x$ comes from the *inner* terms.

$$(3x - 4)(2x + 3) \qquad -4(2x) = -8x \qquad \text{Inner terms}$$

Finally, $-12$ comes from the *last* terms.

$$(3x - 4)(2x + 3) \qquad -4(3) = -12 \qquad \text{Last terms}$$

The product is found by combining these four results.

$$(3x - 4)(2x + 3) = 6x^2 + 9x + (-8x) + (-12)$$
$$= 6x^2 + x - 12$$

To keep track of the order of multiplying terms, we use the initials FOIL (First, Outer, Inner, Last). All the steps of the FOIL method can be done as follows. Try to do as many of these steps as possible mentally.

$$(3x - 4)(2x + 3)$$

First, Last, Inner, Outer

$$(3x - 4)(2x + 3)$$

$6x^2$, $-12$, $-8x$, $9x$, $x$  Add.

**CAUTION** The FOIL method is an extension of the distributive property, and the acronym *"FOIL" applies only to multiplying two binomials.*

---

**EXAMPLE 4** Using the FOIL Method

Use the FOIL method to find each product.

**(a)** $(4m - 5)(3m + 1)$

*First terms* $\quad (4m - 5)(3m + 1) \qquad 4m(3m) = 12m^2$

*Outer terms* $\quad (4m - 5)(3m + 1) \qquad 4m(1) = 4m$

*Inner terms*  $(4m - 5)(3m + 1)$   $-5(3m) = -15m$

*Last terms*   $(4m - 5)(3m + 1)$   $-5(1) = -5$

Thus,  $\quad\quad\quad\quad\overset{\text{F}\quad\quad\text{O}\quad\quad\text{I}\quad\quad\text{L}}{(4m - 5)(3m + 1) = 12m^2 + 4m - 15m - 5}$
$\quad\quad\quad\quad\quad\quad\quad\quad\quad\quad\quad = 12m^2 - 11m - 5.$   Combine like terms.

The procedure can be written in compact form as follows.

$$\begin{array}{c} 12m^2 \quad\quad -5 \\ (4m - 5)(3m + 1) \\ -15m \\ 4m \\ \hline -11m \quad \text{Add.} \end{array}$$

Combine these four results to get $12m^2 - 11m - 5$.

$\quad\quad\quad\quad\quad\quad\overset{\text{First}\quad\text{Outer}\quad\text{Inner}\quad\text{Last}}{\downarrow\quad\quad\downarrow\quad\quad\downarrow\quad\quad\downarrow}$

**(b)** $(6a - 5b)(3a + 4b) = 18a^2 + 24ab - 15ab - 20b^2$
$\quad\quad\quad\quad\quad\quad\quad\quad = 18a^2 + 9ab - 20b^2$   Combine like terms.

**(c)** $(2k + 3z)(5k - 3z) = 10k^2 - 6kz + 15kz - 9z^2$   FOIL
$\quad\quad\quad\quad\quad\quad\quad\quad = 10k^2 + 9kz - 9z^2$

**NOW TRY** Exercises 35 and 39.

**OBJECTIVE 4** Find the product of the sum and difference of two terms. Some types of binomial products occur frequently. The product of the sum and difference of the same two terms, $x$ and $y$, is

$\quad\quad (x + y)(x - y) = x^2 - xy + xy - y^2$   FOIL
$\quad\quad\quad\quad\quad\quad\quad = x^2 - y^2.$   Combine like terms.

**Product of the Sum and Difference of Two Terms**

The **product of the sum and difference of the two terms** $x$ **and** $y$ is the difference of the squares of the terms.

$$(x + y)(x - y) = x^2 - y^2$$

**EXAMPLE 5**  Multiplying the Sum and Difference of Two Terms

Find each product.

**(a)** $(p + 7)(p - 7) = p^2 - 7^2$  $\quad\quad$ **(b)** $(2r + 5)(2r - 5) = (2r)^2 - 5^2$
$\quad\quad\quad\quad\quad\quad = p^2 - 49$ $\quad\quad\quad\quad\quad\quad\quad\quad\quad\quad = 2^2 r^2 - 25$
$\quad\quad\quad\quad\quad\quad\quad\quad\quad\quad\quad\quad\quad\quad\quad\quad\quad\quad\quad\quad = 4r^2 - 25$

**(c)** $(6m + 5n)(6m - 5n) = (6m)^2 - (5n)^2$ $\quad$ **(d)** $2x^3(x + 3)(x - 3) = 2x^3(x^2 - 9)$
$\quad\quad\quad\quad\quad\quad\quad\quad = 36m^2 - 25n^2$ $\quad\quad\quad\quad\quad\quad\quad\quad\quad\quad = 2x^5 - 18x^3$

**NOW TRY** Exercises 47, 51, and 57.

**OBJECTIVE 5** Find the square of a binomial. Another special binomial product is the *square of a binomial*. To find the square of $x + y$, or $(x + y)^2$, multiply $x + y$ by itself.

$$(x + y)(x + y) = x^2 + xy + xy + y^2$$
$$= x^2 + 2xy + y^2$$

A similar result is true for the square of a difference.

> **Square of a Binomial**
>
> The **square of a binomial** is the sum of the square of the first term, twice the product of the two terms, and the square of the last term.
>
> $$(x + y)^2 = x^2 + 2xy + y^2$$
> $$(x - y)^2 = x^2 - 2xy + y^2$$

**EXAMPLE 6** Squaring Binomials

Find each product.

(a) $(m + 7)^2 = m^2 + 2 \cdot m \cdot 7 + 7^2$   $(x + y)^2 = x^2 + 2xy + y^2$
$= m^2 + 14m + 49$

(b) $(p - 5)^2 = p^2 - 2 \cdot p \cdot 5 + 5^2$   $(x - y)^2 = x^2 - 2xy + y^2$
$= p^2 - 10p + 25$

(c) $(2p + 3v)^2 = (2p)^2 + 2(2p)(3v) + (3v)^2$
$= 4p^2 + 12pv + 9v^2$

(d) $(3r - 5s)^2 = (3r)^2 - 2(3r)(5s) + (5s)^2$
$= 9r^2 - 30rs + 25s^2$

**NOW TRY** Exercises 59 and 63.

**CAUTION** As the products in the formula for the square of a binomial show,

$$(x + y)^2 \neq x^2 + y^2.$$

More generally,

$$(x + y)^n \neq x^n + y^n \quad (n \neq 1).$$

**EXAMPLE 7** Multiplying More Complicated Binomials

Find each product.

(a) $[(3p - 2) + 5q][(3p - 2) - 5q]$
$= (3p - 2)^2 - (5q)^2$     Product of sum and difference of terms
$= 9p^2 - 12p + 4 - 25q^2$   Square both quantities.

**(b)** $[(2z + r) + 1]^2 = (2z + r)^2 + 2(2z + r)(1) + 1^2$    Square of a binomial

$\qquad\qquad\qquad\quad = 4z^2 + 4zr + r^2 + 4z + 2r + 1$    Square again; use the distributive property.

**(c)** $(x + y)^3 = (x + y)^2(x + y)$

$\qquad\qquad\;\; = (x^2 + 2xy + y^2)(x + y)$    Square $x + y$.

$\qquad\qquad\;\; = x^3 + 2x^2y + xy^2 + x^2y + 2xy^2 + y^3$    Distributive property

$\qquad\qquad\;\; = x^3 + 3x^2y + 3xy^2 + y^3$    Combine like terms.

This does not equal $x^3 + y^3$.

**(d)** $(2a + b)^4 = (2a + b)^2(2a + b)^2$

$\qquad\qquad\quad = (4a^2 + 4ab + b^2)(4a^2 + 4ab + b^2)$    Square $2a + b$.

$\qquad\qquad\quad = 16a^4 + 16a^3b + 4a^2b^2 + 16a^3b + 16a^2b^2$

$\qquad\qquad\qquad\;\; + 4ab^3 + 4a^2b^2 + 4ab^3 + b^4$

$\qquad\qquad\quad = 16a^4 + 32a^3b + 24a^2b^2 + 8ab^3 + b^4$

**NOW TRY** Exercises 71, 75, 79, and 83.

**OBJECTIVE 6** Multiply polynomial functions. In **Section 5.3,** we introduced operations on functions and saw how functions can be added and subtracted. Functions can also be multiplied.

---

**Multiplying Functions**

If $f(x)$ and $g(x)$ define functions, then

$$(fg)(x) = f(x) \cdot g(x). \quad \text{Product function}$$

The domain of the product function is the intersection of the domains of $f(x)$ and $g(x)$.

---

**EXAMPLE 8** Multiplying Polynomial Functions

For $f(x) = 3x + 4$ and $g(x) = 2x^2 + x$, find $(fg)(x)$ and $(fg)(-1)$.

$\qquad(fg)(x) = f(x) \cdot g(x)$    Use the definition.

$\qquad\qquad\quad = (3x + 4)(2x^2 + x)$

$\qquad\qquad\quad = 6x^3 + 3x^2 + 8x^2 + 4x$    FOIL

$\qquad\qquad\quad = 6x^3 + 11x^2 + 4x$    Combine like terms.

Then

$\qquad(fg)(-1) = 6(-1)^3 + 11(-1)^2 + 4(-1)$    Let $x = -1$.

$\qquad\qquad\qquad = -6 + 11 - 4$

$\qquad\qquad\qquad = 1.$

Be careful with signs

Confirm that $f(-1) \cdot g(-1)$ is equal to $(fg)(-1)$.

**NOW TRY** Exercises 99 and 101.

## 5.4 Exercises

**Matching** Match each product in Column I with the correct polynomial in Column II.

| I | II |
|---|---|
| 1. $(2x - 5)(3x + 4)$ | A. $6x^2 + 23x + 20$ |
| 2. $(2x + 5)(3x + 4)$ | B. $6x^2 + 7x - 20$ |
| 3. $(2x - 5)(3x - 4)$ | C. $6x^2 - 7x - 20$ |
| 4. $(2x + 5)(3x - 4)$ | D. $6x^2 - 23x + 20$ |

*Find each product. See Examples 1–3.*

5. $-8m^3(3m^2)$
6. $4p^2(-5p^4)$
7. $14x^2y^3(-2x^5y)$
8. $-5m^3n^4(4m^2n^5)$
9. $3x(-2x + 5)$
10. $5y(-6y - 1)$
11. $-q^3(2 + 3q)$
12. $-3a^4(4 - a)$
13. $6k^2(3k^2 + 2k + 1)$
14. $5r^3(2r^2 - 3r - 4)$
15. $(2m + 3)(3m^2 - 4m - 1)$
16. $(4z - 2)(z^2 + 3z + 5)$
17. $m(m + 5)(m - 8)$
18. $p(p - 6)(p + 4)$
19. $4z(2z + 1)(3z - 4)$
20. $2y(8y - 3)(2y + 1)$
21. $4x^3(x - 3)(x + 2)$
22. $2y^5(y - 8)(y + 2)$
23. $(2y + 3)(3y - 4)$
24. $(5m - 3)(2m + 6)$

25. $\begin{array}{r} -b^2 + 3b + 3 \\ 2b + 4 \end{array}$

26. $\begin{array}{r} -r^2 - 4r + 8 \\ 3r - 2 \end{array}$

27. $\begin{array}{r} 5m - 3n \\ 5m + 3n \end{array}$

28. $\begin{array}{r} 2k + 6q \\ 2k - 6q \end{array}$

29. $\begin{array}{r} 2z^3 - 5z^2 + 8z - 1 \\ 4z + 3 \end{array}$

30. $\begin{array}{r} 3z^4 - 2z^3 + z - 5 \\ 2z - 5 \end{array}$

31. $\begin{array}{r} 2p^2 + 3p + 6 \\ 3p^2 - 4p - 1 \end{array}$

32. $\begin{array}{r} 5y^2 - 2y + 4 \\ 2y^2 + y + 3 \end{array}$

*Use the FOIL method to find each product. See Example 4.*

33. $(m + 5)(m - 8)$
34. $(p - 6)(p + 4)$
35. $(4k + 3)(3k - 2)$
36. $(5w + 2)(2w + 5)$
37. $(z - w)(3z + 4w)$
38. $(s + t)(2s - 5t)$
39. $(6c - d)(2c + 3d)$
40. $(2m - n)(3m + 5n)$
41. $(0.2x + 1.3)(0.5x - 0.1)$
42. $(0.5y - 0.4)(0.1y + 2.1)$
43. $\left(3w + \dfrac{1}{4}z\right)(w - 2z)$
44. $\left(5r - \dfrac{2}{3}y\right)(r + 5y)$

45. Describe the FOIL method in your own words.

46. Explain why the product of the sum and difference of two terms is not a trinomial.

*Find each product. See Example 5.*

47. $(2p - 3)(2p + 3)$
48. $(3x - 8)(3x + 8)$
49. $(5m - 1)(5m + 1)$
50. $(6y + 3)(6y - 3)$
51. $(3a + 2c)(3a - 2c)$
52. $(5r - 4s)(5r + 4s)$
53. $\left(4x - \dfrac{2}{3}\right)\left(4x + \dfrac{2}{3}\right)$
54. $\left(3t + \dfrac{5}{4}\right)\left(3t - \dfrac{5}{4}\right)$
55. $(4m + 7n^2)(4m - 7n^2)$
56. $(2k^2 + 6h)(2k^2 - 6h)$
57. $3y(5y^3 + 2)(5y^3 - 2)$
58. $4x(3x^3 + 4)(3x^3 - 4)$

*Find each product. See Example 6.*

59. $(y - 5)^2$
60. $(a - 3)^2$
61. $(2p + 7)^2$
62. $(3z + 8)^2$
63. $(4n + 3m)^2$
64. $(5r + 7s)^2$
65. $\left(k - \dfrac{5}{7}p\right)^2$
66. $\left(q - \dfrac{3}{4}r\right)^2$
67. $(0.2x - 1.4y)^2$

68. Explain why $(x + y)^2$ and $x^2 + y^2$ are not equivalent.
69. *Concept Check* Find the product $101 \cdot 99$, using the special product rule $(x + y)(x - y) = x^2 - y^2$.
70. *Concept Check* Repeat Exercise 69 for the product $202 \cdot 198$.

*Find each product. See Example 7.*

71. $[(5x + 1) + 6y]^2$
72. $[(3m - 2) + p]^2$
73. $[(2a + b) - 3]^2$
74. $[(4k + h) - 4]^2$
75. $[(2a + b) - 3][(2a + b) + 3]$
76. $[(m + p) + 5][(m + p) - 5]$
77. $[(2h - k) + j][(2h - k) - j]$
78. $[(3m - y) + z][(3m - y) - z]$
79. $(y + 2)^3$
80. $(z - 3)^3$
81. $(5r - s)^3$
82. $(x + 3y)^3$
83. $(q - 2)^4$
84. $(r + 3)^4$

*In Exercises 85–88, two expressions are given. Replace x with 3 and y with 4 to show that, in general, the two expressions do not equal each other.*

85. $(x + y)^2$;  $x^2 + y^2$
86. $(x + y)^3$;  $x^3 + y^3$
87. $(x + y)^4$;  $x^4 + y^4$
88. $(x + y)^5$;  $x^5 + y^5$

*Find the area of each figure. Express it as a polynomial in descending powers of the variable x. Refer to the formulas on the inside covers of this book if necessary.*

89.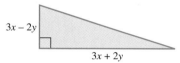
90.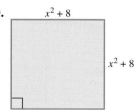
91.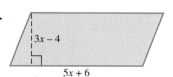
92.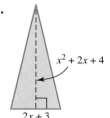

*For each pair of functions, find the product $(fg)(x)$. See Example 8.*

93. $f(x) = 2x, g(x) = 5x - 1$
94. $f(x) = 3x, g(x) = 6x - 8$
95. $f(x) = x + 1, g(x) = 2x - 3$
96. $f(x) = x - 7, g(x) = 4x + 5$
97. $f(x) = 2x - 3, g(x) = 4x^2 + 6x + 9$
98. $f(x) = 3x + 4, g(x) = 9x^2 - 12x + 16$

*Let $f(x) = x^2 - 9$, $g(x) = 2x$, and $h(x) = x - 3$. Find each of the following. See Example 8.*

99. $(fg)(x)$
100. $(fh)(x)$
101. $(fg)(2)$
102. $(fh)(1)$
103. $(gh)(x)$
104. $(fh)(-1)$
105. $(gh)(-3)$
106. $(fg)(-2)$
107. $(fg)\left(-\dfrac{1}{2}\right)$
108. $(fg)\left(-\dfrac{1}{3}\right)$
109. $(fh)\left(-\dfrac{1}{4}\right)$
110. $(fh)\left(-\dfrac{1}{5}\right)$

## 5.5 Dividing Polynomials

**OBJECTIVES**
1. Divide a polynomial by a monomial.
2. Divide a polynomial by a polynomial of two or more terms.
3. Divide polynomial functions.

**OBJECTIVE 1** Divide a polynomial by a monomial. We now discuss polynomial division, beginning with division by a monomial. (Recall that a monomial is a single term, such as $8x$, $-9m^4$, or $11y^2$.)

### Dividing a Polynomial by a Monomial

To divide a polynomial by a monomial, divide each term in the polynomial by the monomial, and then write each quotient in lowest terms.

**EXAMPLE 1** Dividing a Polynomial by a Monomial

Divide.

(a) $\dfrac{15x^2 - 12x + 6}{3} = \dfrac{15x^2}{3} - \dfrac{12x}{3} + \dfrac{6}{3}$  Divide each term by 3.

$= 5x^2 - 4x + 2$  Write in lowest terms.

Check this answer by multiplying it by the divisor, 3.

$$3(5x^2 - 4x + 2) = 15x^2 - 12x + 6$$

Divisor  Quotient    Original polynomial

(b) $\dfrac{5m^3 - 9m^2 + 10m}{5m^2} = \dfrac{5m^3}{5m^2} - \dfrac{9m^2}{5m^2} + \dfrac{10m}{5m^2}$  Divide each term by $5m^2$.

$= m - \dfrac{9}{5} + \dfrac{2}{m}$  Write in lowest terms.

Check: $5m^2\left(m - \dfrac{9}{5} + \dfrac{2}{m}\right) = 5m^3 - 9m^2 + 10m$, the original polynomial.

The result $m - \dfrac{9}{5} + \dfrac{2}{m}$ is not a polynomial. (Why?) The quotient of two polynomials need not be a polynomial.

(c) $\dfrac{8xy^2 - 9x^2y + 6x^2y^2}{x^2y^2} = \dfrac{8xy^2}{x^2y^2} - \dfrac{9x^2y}{x^2y^2} + \dfrac{6x^2y^2}{x^2y^2}$

$= \dfrac{8}{x} - \dfrac{9}{y} + 6$

**NOW TRY** Exercises 5, 9, and 11.

**OBJECTIVE 2** Divide a polynomial by a polynomial of two or more terms. The process for dividing one polynomial by another polynomial that is not a monomial is similar to that for dividing whole numbers.

### EXAMPLE 2  Dividing a Polynomial by a Polynomial

Divide $\dfrac{2m^2 + m - 10}{m - 2}$.

Write the problem as if dividing whole numbers, making sure that both polynomials are written in descending powers of the variables.

$$m - 2 \overline{\smash{\big)}\, 2m^2 + m - 10}$$

Divide the first term of $2m^2 + m - 10$ by the first term of $m - 2$. Since $\dfrac{2m^2}{m} = 2m$, place this result above the division line.

$$\begin{array}{r} 2m \\ m - 2 \overline{\smash{\big)}\, 2m^2 + m - 10} \end{array}$$ ← Result of $\frac{2m^2}{m}$

Multiply $m - 2$ and $2m$, and write the result below $2m^2 + m - 10$.

$$\begin{array}{r} 2m \phantom{{}+ m - 10} \\ m - 2 \overline{\smash{\big)}\, 2m^2 + m - 10} \\ \underline{2m^2 - 4m} \phantom{{}- 10} \end{array}$$ ← $2m(m - 2) = 2m^2 - 4m$

Now subtract $2m^2 - 4m$ from $2m^2 + m$. Do this by mentally changing the signs on $2m^2 - 4m$ and *adding*.

$$\begin{array}{r} 2m \phantom{{}+ m - 10} \\ m - 2 \overline{\smash{\big)}\, 2m^2 + m - 10} \\ \underline{2m^2 - 4m} \phantom{{}- 10} \\ 5m \phantom{{}- 10} \end{array}$$

To subtract, add the opposite. ← Subtract. The difference is $5m$.

Bring down $-10$ and continue by dividing $5m$ by $m$.

$$\begin{array}{r} 2m + 5 \phantom{0} \\ m - 2 \overline{\smash{\big)}\, 2m^2 + m - 10} \\ \underline{2m^2 - 4m} \phantom{{}- 10} \\ 5m - 10 \\ \underline{5m - 10} \\ 0 \end{array}$$ ← $\frac{5m}{m} = 5$

← Bring down $-10$.
← $5(m - 2) = 5m - 10$
← Subtract. The difference is 0.

Finally, $(2m^2 + m - 10) \div (m - 2) = 2m + 5$. Check by multiplying $m - 2$ and $2m + 5$. The result should be $2m^2 + m - 10$.

**NOW TRY** Exercise 17.

### EXAMPLE 3  Dividing a Polynomial with a Missing Term

Divide $3x^3 - 2x + 5$ by $x - 3$.

Make sure that $3x^3 - 2x + 5$ is in descending powers of the variable. Add a term with 0 coefficient as a placeholder for the missing $x^2$-term.

Missing term

$$x - 3 \overline{\smash{\big)}\, 3x^3 + 0x^2 - 2x + 5}$$

Start with $\dfrac{3x^3}{x} = 3x^2$.

$$
\begin{array}{r}
3x^2\phantom{xxxxxxxxx} \\
x-3\overline{\smash{)}3x^3 + 0x^2 - 2x + 5} \\
\underline{3x^3 - 9x^2\phantom{xxxxxxx}}
\end{array}
$$

← $\dfrac{3x^3}{x} = 3x^2$
← $3x^2(x-3)$

Subtract by mentally changing the signs on $3x^3 - 9x^2$ and adding.

$$
\begin{array}{r}
3x^2\phantom{xxxxxxxxx} \\
x-3\overline{\smash{)}3x^3 + 0x^2 - 2x + 5} \\
\underline{3x^3 - 9x^2\phantom{xxxxxxx}} \\
9x^2\phantom{xxxxxxx}
\end{array}
$$

← Subtract.

Bring down the next term.

$$
\begin{array}{r}
3x^2\phantom{xxxxxxxxx} \\
x-3\overline{\smash{)}3x^3 + 0x^2 - 2x + 5} \\
\underline{3x^3 - 9x^2\phantom{xxxxxxx}} \\
9x^2 - 2x\phantom{xxx}
\end{array}
$$

← Bring down $-2x$.

In the next step, $\dfrac{9x^2}{x} = 9x$.

$$
\begin{array}{r}
3x^2 + \phantom{0}9x\phantom{xxxxx} \\
x-3\overline{\smash{)}3x^3 + 0x^2 - \phantom{0}2x + 5} \\
\underline{3x^3 - 9x^2\phantom{xxxxxxxx}} \\
9x^2 - \phantom{0}2x\phantom{xxx} \\
\underline{9x^2 - 27x\phantom{xxx}} \\
25x + 5
\end{array}
$$

← $\dfrac{9x^2}{x} = 9x$
← $9x(x-3)$
← Subtract; bring down 5.

Finally, $\dfrac{25x}{x} = 25$.

$$
\begin{array}{r}
3x^2 + \phantom{0}9x + 25\phantom{xx} \\
x-3\overline{\smash{)}3x^3 + 0x^2 - \phantom{0}2x + \phantom{0}5} \\
\underline{3x^3 - 9x^2\phantom{xxxxxxxxxx}} \\
9x^2 - \phantom{0}2x\phantom{xxxxx} \\
\underline{9x^2 - 27x\phantom{xxxxx}} \\
25x + \phantom{0}5 \\
\underline{25x - 75} \\
80
\end{array}
$$

← $\dfrac{25x}{x} = 25$

← $25(x-3)$
← Remainder

Write the remainder, 80, as the numerator of the fraction $\dfrac{80}{x-3}$. In summary,

$$\dfrac{3x^3 - 2x + 5}{x-3} = 3x^2 + 9x + 25 + \dfrac{80}{x-3}.$$

Be sure to add $\dfrac{\text{remainder}}{\text{divisor}}$ here. Don't forget the + sign.

Check by multiplying $x - 3$ and $3x^2 + 9x + 25$ and adding 80 to the result. You should get $3x^3 - 2x + 5$.

**NOW TRY** Exercise 37.

**CAUTION** Remember to include $\frac{\text{remainder}}{\text{divisor}}$ as part of the answer. *Don't forget to insert a plus sign between the polynomial quotient and this fraction.*

### EXAMPLE 4  Dividing by a Polynomial with a Missing Term

Divide $6r^4 + 9r^3 + 2r^2 - 8r + 7$ by $3r^2 - 2$.

The polynomial $3r^2 - 2$ has a missing term. Write it as $3r^2 + 0r - 2$.

$$
\begin{array}{r}
2r^2 + 3r + 2 \phantom{0}\\
3r^2 + 0r - 2 \overline{\smash{\big)}\, 6r^4 + 9r^3 + 2r^2 - 8r + 7} \\
\underline{6r^4 + 0r^3 - 4r^2 \phantom{- 8r + 7}} \\
9r^3 + 6r^2 - 8r \phantom{+ 7} \\
\underline{9r^3 + 0r^2 - 6r \phantom{+ 7}} \\
6r^2 - 2r + 7 \\
\underline{6r^2 + 0r - 4} \\
-2r + 11 \phantom{0}
\end{array}
$$

Missing term ⟵ (pointing to $0r$)

Stop when the degree of the remainder is less than the degree of the divisor.

⟵ Remainder

The degree of the remainder, $-2r + 11$, is less than the degree of the divisor, $3r^2 - 2$, so the division process is now finished. The result is written

$$2r^2 + 3r + 2 + \frac{-2r + 11}{3r^2 - 2}.$$

**NOW TRY** Exercise 43.

**CAUTION** Remember the following steps when dividing a polynomial by a polynomial of two or more terms.

1. Be sure the terms in both polynomials are in descending powers.
2. Write any missing terms with 0 placeholders.

### EXAMPLE 5  Performing a Division with a Fractional Coefficient in the Quotient

Divide $2p^3 + 5p^2 + p - 2$ by $2p + 2$.

$\frac{3p^2}{2p} = \frac{3}{2}p$

$$
\begin{array}{r}
p^2 + \dfrac{3}{2}p - 1 \phantom{0}\\
2p + 2 \overline{\smash{\big)}\, 2p^3 + 5p^2 + \phantom{3}p - 2} \\
\underline{2p^3 + 2p^2 \phantom{+ 3p - 2}} \\
3p^2 + \phantom{3}p \phantom{- 2} \\
\underline{3p^2 + 3p \phantom{- 2}} \\
-2p - 2 \\
\underline{-2p - 2} \\
0
\end{array}
$$

Since the remainder is 0, the quotient is $p^2 + \frac{3}{2}p - 1$.

**NOW TRY** Exercise 45.

**OBJECTIVE 3** Divide polynomial functions. In the preceding sections, we used operations on functions to add, subtract, and multiply polynomial functions. We now define the quotient of two functions.

> **Dividing Functions**
>
> If $f(x)$ and $g(x)$ define functions, then
>
> $$\left(\frac{f}{g}\right)(x) = \frac{f(x)}{g(x)}. \quad \text{Quotient function}$$
>
> The domain of the quotient function is the intersection of the domains of $f(x)$ and $g(x)$, excluding any values of $x$ for which $g(x) = 0$.

**EXAMPLE 6** Dividing Polynomial Functions

For $f(x) = 2x^2 + x - 10$ and $g(x) = x - 2$, find $\left(\frac{f}{g}\right)(x)$ and $\left(\frac{f}{g}\right)(-3)$. What value of $x$ is not in the domain of the quotient function?

$$\left(\frac{f}{g}\right)(x) = \frac{f(x)}{g(x)} = \frac{2x^2 + x - 10}{x - 2}$$

This quotient was found in Example 2, with $m$ replacing $x$. The result here is $2x + 5$, so

$$\left(\frac{f}{g}\right)(x) = 2x + 5, \quad x \neq 2.$$

The number 2 is not in the domain because it causes the denominator $g(x) = x - 2$ to equal 0. Then

$$\left(\frac{f}{g}\right)(-3) = 2(-3) + 5 = -1. \quad \text{Let } x = -3.$$

Verify that the same value is found by evaluating $\frac{f(-3)}{g(-3)}$.

**NOW TRY** Exercises 61 and 63.

## 5.5 Exercises

**NOW TRY Exercise**

*Fill in the Blanks* Complete each statement with the correct word(s).

1. We find the quotient of two monomials by using the _____ rule for _____.

2. To divide a polynomial by a monomial, divide each _____ of the polynomial by the _____.

3. When dividing polynomials that are not monomials, first write them in _____ powers.

4. If a polynomial in a division problem has a missing term, insert a term with coefficient equal to _____ as a placeholder.

*Divide. See Example 1.*

5. $\dfrac{15x^3 - 10x^2 + 5}{5}$

6. $\dfrac{27m^4 - 18m^3 + 9m}{9}$

7. $\dfrac{9y^2 + 12y - 15}{3y}$

8. $\dfrac{80r^2 - 40r + 10}{10r}$

9. $\dfrac{15m^3 + 25m^2 + 30m}{5m^2}$

10. $\dfrac{64x^3 - 72x^2 + 12x}{8x^3}$

11. $\dfrac{14m^2n^2 - 21mn^3 + 28m^2n}{14m^2n}$

12. $\dfrac{24h^2k + 56hk^2 - 28hk}{16h^2k^2}$

13. $\dfrac{8wxy^2 + 3wx^2y + 12w^2xy}{4wx^2y}$

14. $\dfrac{12ab^2c + 10a^2bc + 18abc^2}{6a^2bc}$

*Complete the division.*

15. 
$$\begin{array}{r} r^2\phantom{aaaaaaaaaa} \\ 3r-1\overline{\smash{\big)}\,3r^3 - 22r^2 + 25r - 6} \\ \underline{3r^3 - \phantom{2}r^2}\phantom{aaaaaaaaaa} \\ -21r^2\phantom{aaaaaaaa} \end{array}$$

16. 
$$\begin{array}{r} 3b^2\phantom{aaaaaaaaaa} \\ 2b-5\overline{\smash{\big)}\,6b^3 - \phantom{2}7b^2 - 4b - 40} \\ \underline{6b^3 - 15b^2}\phantom{aaaaaaaaaa} \\ 8b^2\phantom{aaaaaaaa} \end{array}$$

*Divide. See Examples 2–5.*

17. $\dfrac{y^2 + y - 20}{y + 5}$

18. $\dfrac{y^2 + 3y - 18}{y + 6}$

19. $\dfrac{q^2 + 4q - 32}{q - 4}$

20. $\dfrac{q^2 + 2q - 35}{q - 5}$

21. $\dfrac{3t^2 + 17t + 10}{3t + 2}$

22. $\dfrac{2k^2 - 3k - 20}{2k + 5}$

23. $\dfrac{p^2 + 2p + 20}{p + 6}$

24. $\dfrac{x^2 + 11x + 16}{x + 8}$

25. $\dfrac{3m^3 + 5m^2 - 5m + 1}{3m - 1}$

26. $\dfrac{8z^3 - 6z^2 - 5z + 3}{4z + 3}$

27. $\dfrac{m^3 - 2m^2 - 9}{m - 3}$

28. $\dfrac{p^3 + 3p^2 - 4}{p + 2}$

29. $(2z^3 - 5z^2 + 6z - 15) \div (2z - 5)$

30. $(3p^3 + p^2 + 18p + 6) \div (3p + 1)$

31. $(4x^3 + 9x^2 - 10x + 3) \div (4x + 1)$

32. $(10z^3 - 26z^2 + 17z - 13) \div (5z - 3)$

33. $\dfrac{6x^3 - 19x^2 + 14x - 15}{3x^2 - 2x + 4}$

34. $\dfrac{8m^3 - 18m^2 + 37m - 13}{2m^2 - 3m + 6}$

35. $(x^3 + 2x - 3) \div (x - 1)$

36. $(2x^3 - 11x^2 + 25) \div (x - 5)$

37. $(3x^3 - x + 4) \div (x - 2)$

38. $(3k^3 + 9k - 14) \div (k - 2)$

39. $\dfrac{4k^4 + 6k^3 + 3k - 1}{2k^2 + 1}$

40. $\dfrac{9k^4 + 12k^3 - 4k - 1}{3k^2 - 1}$

41. $\dfrac{6y^4 + 4y^3 + 4y - 6}{3y^2 + 2y - 3}$

42. $\dfrac{8t^4 + 6t^3 + 12t - 32}{4t^2 + 3t - 8}$

43. $(x^4 - 4x^3 + 5x^2 - 3x + 2) \div (x^2 + 3)$

44. $(3t^4 + 5t^3 - 8t^2 - 13t + 2) \div (t^2 - 5)$

45. $(2p^3 + 7p^2 + 9p + 3) \div (2p + 2)$

46. $(3x^3 + 4x^2 + 7x + 4) \div (3x + 3)$

47. $(3a^2 - 11a + 17) \div (2a + 6)$

48. $(5t^2 + 19t + 7) \div (4t + 12)$

49. $\dfrac{p^3 - 1}{p - 1}$

50. $\dfrac{8a^3 + 1}{2a + 1}$

*Solve each problem.*

**51.** The volume of a box is $(2p^3 + 15p^2 + 28p)$. The height is $p$ and the length is $(p + 4)$. Give an expression in $p$ that represents the width.

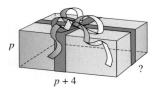

**52.** Suppose that a car travels a distance of $(2m^3 + 15m^2 + 35m + 36)$ miles in $(2m + 9)$ hours. Give an expression in $m$ that represents the rate of the car in mph.

**53.** For $P(x) = x^3 - 4x^2 + 3x - 5$, find $P(-1)$. Then divide $P(x)$ by $D(x) = x + 1$. Compare the remainder with $P(-1)$. What do these results suggest?

**54.** *Concept Check* Let $P(x) = 4x^3 - 8x^2 + 13x - 2$ and $D(x) = 2x - 1$. Use division to find polynomials $Q(x)$ and $R(x)$ such that $P(x) = Q(x) \cdot D(x) + R(x)$.

*For each pair of functions, find the quotient $\left(\frac{f}{g}\right)(x)$ and give any x-values that are not in the domain of the quotient function. See Example 6.*

**55.** $f(x) = 10x^2 - 2x, g(x) = 2x$

**56.** $f(x) = 18x^2 - 24x, g(x) = 3x$

**57.** $f(x) = 2x^2 - x - 3, g(x) = x + 1$

**58.** $f(x) = 4x^2 - 23x - 35, g(x) = x - 7$

**59.** $f(x) = 8x^3 - 27, g(x) = 2x - 3$

**60.** $f(x) = 27x^3 + 64, g(x) = 3x + 4$

*Let $f(x) = x^2 - 9$, $g(x) = 2x$, and $h(x) = x - 3$. Find each of the following. See Example 6.*

**61.** $\left(\dfrac{f}{g}\right)(x)$

**62.** $\left(\dfrac{f}{h}\right)(x)$

**63.** $\left(\dfrac{f}{g}\right)(2)$

**64.** $\left(\dfrac{f}{h}\right)(1)$

**65.** $\left(\dfrac{h}{g}\right)(x)$

**66.** $\left(\dfrac{g}{h}\right)(x)$

**67.** $\left(\dfrac{h}{g}\right)(3)$

**68.** $\left(\dfrac{g}{h}\right)(-1)$

**69.** $\left(\dfrac{f}{g}\right)\left(\dfrac{1}{2}\right)$

**70.** $\left(\dfrac{f}{g}\right)\left(\dfrac{3}{2}\right)$

**71.** $\left(\dfrac{h}{g}\right)\left(-\dfrac{1}{2}\right)$

**72.** $\left(\dfrac{h}{g}\right)\left(-\dfrac{3}{2}\right)$

# 5 SUMMARY

## KEY TERMS

**5.2** term
numerical coefficient (coefficient)
algebraic expression
polynomial
polynomial in $x$
descending powers
trinomial
binomial
monomial
degree of a term
degree of a polynomial
negative of a polynomial
**5.3** polynomial function
composition of functions
identity function
squaring function
cubing function

## NEW SYMBOLS

$(f \circ g)(x) = f(g(x))$   composite function

## QUICK REVIEW

### CONCEPTS — EXAMPLES

### 5.1 Integer Exponents and Scientific Notation

**Definitions and Rules for Exponents**

For all integers $m$ and $n$ and all real numbers $a$ and $b$,

**Product Rule:** $a^m \cdot a^n = a^{m+n}$

**Quotient Rule:** $\dfrac{a^m}{a^n} = a^{m-n}$ $(a \neq 0)$

**Zero Exponent:** $a^0 = 1$ $(a \neq 0)$

**Negative Exponent:** $a^{-n} = \dfrac{1}{a^n}$ $(a \neq 0)$

**Power Rules:** $(a^m)^n = a^{mn}$

$(ab)^m = a^m b^m$

$\left(\dfrac{a}{b}\right)^n = \dfrac{a^n}{b^n}$ $(b \neq 0)$.

**Special Rules for Negative Exponents:**

$\dfrac{1}{a^{-n}} = a^n$   $(a \neq 0)$

$\dfrac{a^{-n}}{b^{-m}} = \dfrac{b^m}{a^n}$   $(a, b \neq 0)$

$a^{-n} = \left(\dfrac{1}{a}\right)^n$   $(a \neq 0)$

$\left(\dfrac{a}{b}\right)^{-n} = \left(\dfrac{b}{a}\right)^n$   $(a, b \neq 0)$

Apply the rules for exponents.

$3^4 \cdot 3^2 = 3^6$

$\dfrac{2^5}{2^3} = 2^2$

$27^0 = 1, \quad (-5)^0 = 1$

$5^{-2} = \dfrac{1}{5^2}$

$(6^3)^4 = 6^{12}$

$(5p)^4 = 5^4 p^4$

$\left(\dfrac{2}{3}\right)^5 = \dfrac{2^5}{3^5}$

$\dfrac{1}{3^{-2}} = 3^2$

$\dfrac{5^{-3}}{4^{-6}} = \dfrac{4^6}{5^3}$

$4^{-3} = \left(\dfrac{1}{4}\right)^3$

$\left(\dfrac{4}{7}\right)^{-2} = \left(\dfrac{7}{4}\right)^2$

**Scientific Notation**

A number is in scientific notation when it is written as a product of a number between 1 and 10 (inclusive of 1) and an integer power of 10.

Write 23,500,000,000 in scientific notation.

$$23,500,000,000 = 2.35 \times 10^{10}$$

Write $4.3 \times 10^{-6}$ in standard notation.

$$4.3 \times 10^{-6} = 0.0000043$$

*(continued)*

## CONCEPTS

### 5.2 Adding and Subtracting Polynomials

Add or subtract polynomials by combining like terms.

### 5.3 Polynomial Functions, Graphs, and Composition

**Composition of $f$ and $g$**

$$(f \circ g)(x) = f(g(x))$$

The graph of $f(x) = x$ is a line, and the graph of $f(x) = x^2$ is a parabola. These graphs define the identity and squaring functions, respectively.

The graph of $f(x) = x^3$ defines the cubing function.

### 5.4 Multiplying Polynomials

To multiply two polynomials, multiply each term of one by each term of the other.

To multiply two binomials, use the **FOIL method.** Multiply the **First** terms, the **Outer** terms, the **Inner** terms, and the **Last** terms. Then add these products.

**Special Products**

$$(x + y)(x - y) = x^2 - y^2$$
$$(x + y)^2 = x^2 + 2xy + y^2$$
$$(x - y)^2 = x^2 - 2xy + y^2$$

### 5.5 Dividing Polynomials

**Dividing by a Monomial**

To divide a polynomial by a monomial, divide each term in the polynomial by the monomial, and then write each fraction in lowest terms.

## EXAMPLES

Add. $(x^2 - 2x + 3) + (2x^2 - 8) = 3x^2 - 2x - 5$

Subtract. $(5x^4 + 3x^2) - (7x^4 + x^2 - x) = -2x^4 + 2x^2 + x$

If $f(x) = x^2$ and $g(x) = 2x + 1$, then

$$(f \circ g)(x) = f(g(x)) = f(2x + 1)$$
$$= (2x + 1)^2$$

and

$$(g \circ f)(x) = g(f(x)) = g(x^2)$$
$$= 2x^2 + 1.$$

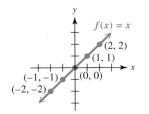

Identity Function

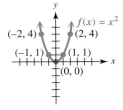

Squaring Function

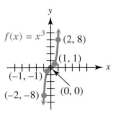

Cubing Function

Multiply. $(x^3 + 3x)(4x^2 - 5x + 2)$
$$= 4x^5 + 12x^3 - 5x^4 - 15x^2 + 2x^3 + 6x$$
$$= 4x^5 - 5x^4 + 14x^3 - 15x^2 + 6x$$

Multiply. $(2x + 3)(x - 7) = 2x(x) + 2x(-7) + 3x + 3(-7)$
$$= 2x^2 - 14x + 3x - 21$$
$$= 2x^2 - 11x - 21$$

Multiply.
$$(3m + 8)(3m - 8) = 9m^2 - 64$$
$$(5a + 3b)^2 = 25a^2 + 30ab + 9b^2$$
$$(2k - 1)^2 = 4k^2 - 4k + 1$$

Divide. $\dfrac{2x^3 - 4x^2 + 6x - 8}{2x} = \dfrac{2x^3}{2x} - \dfrac{4x^2}{2x} + \dfrac{6x}{2x} - \dfrac{8}{2x}$

$$= x^2 - 2x + 3 - \dfrac{4}{x}$$

*(continued)*

## CONCEPTS

**Dividing by a Polynomial**
Use the long division process. The process ends when the remainder is 0 or when the degree of the remainder is less than the degree of the divisor.

## EXAMPLES

Divide. $\dfrac{m^3 - m^2 + 2m + 5}{m + 1}$

$$
\begin{array}{r}
m^2 - 2m + 4\phantom{)} \\
m + 1 \overline{\smash{)}\,m^3 - m^2 + 2m + 5} \\
\underline{m^3 + m^2\phantom{ + 2m + 5}} \\
-2m^2 + 2m\phantom{ + 5} \\
\underline{-2m^2 - 2m\phantom{ + 5}} \\
4m + 5 \\
\underline{4m + 4} \\
1 \leftarrow \text{Remainder}
\end{array}
$$

The answer is $m^2 - 2m + 4 + \dfrac{1}{m + 1}$.

# 5 REVIEW EXERCISES

*Simplify. Write answers with only positive exponents. Assume that all variables represent nonzero real numbers.*

1. $4^3$
2. $\left(\dfrac{1}{3}\right)^4$
3. $(-5)^3$
4. $\dfrac{2}{(-3)^{-2}}$
5. $\left(\dfrac{2}{3}\right)^{-4}$
6. $\left(\dfrac{5}{4}\right)^{-2}$
7. $5^{-1} + 6^{-1}$
8. $(5 + 6)^{-1}$
9. $-3^0 + 3^0$
10. $(3^{-4})^2$
11. $(x^{-4})^{-2}$
12. $(xy^{-3})^{-2}$
13. $(z^{-3})^3 z^{-6}$
14. $(5m^{-3})^2 (m^4)^{-3}$
15. $\dfrac{(3r)^2 r^4}{r^{-2} r^{-3}} (9r^{-3})^{-2}$
16. $\left(\dfrac{5z^{-3}}{z^{-1}}\right) \dfrac{5}{z^2}$
17. $\left(\dfrac{6m^{-4}}{m^{-9}}\right)^{-1} \left(\dfrac{m^{-2}}{16}\right)$
18. $\left(\dfrac{3r^5}{5r^{-3}}\right)^{-2} \left(\dfrac{9r^{-1}}{2r^{-5}}\right)^3$
19. $(-3x^4 y^3)(4x^{-2} y^5)$
20. $\dfrac{6m^{-4} n^3}{-3mn^2}$
21. $\dfrac{(5p^{-2} q)(4p^5 q^{-3})}{2p^{-5} q^5}$
22. $\left(\dfrac{a^{-2} b^{-1}}{3a^2}\right)^{-2} \left(\dfrac{b^{-2} \cdot 3a^4}{2b^{-3}}\right)^{-2} \left(\dfrac{a^{-4} b^5}{a^3}\right)^{-2}$

23. Explain the difference between the expressions $(-6)^0$ and $-6^0$.

24. By choosing a specific value for $a$, give an example to show that, in general, $(2a)^{-3}$ is not equal to $\dfrac{2}{a^3}$.

25. *Concept Check* Is $\left(\dfrac{a}{b}\right)^{-1} = \dfrac{a^{-1}}{b^{-1}}$ true for all $a$ and $b \neq 0$? If not, explain.

26. *Concept Check* Is $(ab)^{-1} = ab^{-1}$ true for all $a$ and $b \neq 0$? If not, explain.

27. By choosing specific values for $x$ and $y$, give an example to show that, in general, $(x^2 + y^2)^2 \neq x^4 + y^4$.

*Write in scientific notation.*

28. 13,450

29. 0.0000000765

30. 0.138

*Write each number in standard notation.*

31. $5.8 \times 10^{-3}$

32. $1.21 \times 10^6$

*Find each value. Give answers in both scientific notation and standard notation.*

33. $\dfrac{6 \times 10^{-2}}{4 \times 10^{-5}}$

34. $\dfrac{16 \times 10^4}{8 \times 10^8}$

35. $\dfrac{0.0009 \times 12{,}000{,}000}{400{,}000}$

36. $\dfrac{0.0000000164}{0.0004}$

37. The population of Fresno, California, is approximately $3.45 \times 10^5$. The population density is 5449 per mi². 

  (a) Write the population density in scientific notation.
  (b) To the nearest square mile, what is the area of Fresno?

*Give the numerical coefficient of each term.*

38. $14p^5$

39. $-z$

40. $\dfrac{x}{10}$

41. $504p^3 r^5$

*For each polynomial, (a) write it in descending powers, (b) identify it as a monomial, binomial, trinomial, or none of these, and (c) give its degree.*

42. $9k + 11k^3 - 3k^2$

43. $14m^6 + 9m^7$

44. $-5y^4 + 3y^3 + 7y^2 - 2y$

45. $-7q^5 r^3$

46. *Concept Check* Give an example of a polynomial in the variable $x$ such that the polynomial has degree 5, is lacking a third-degree term, and is in descending powers of the variable.

*Add or subtract as indicated.*

47. Add.
$$\begin{array}{r} 3x^2 - 5x + 6 \\ -4x^2 + 2x - 5 \end{array}$$

48. Subtract.
$$\begin{array}{r} -5y^3 \phantom{+4y^2} + 8y - 3 \\ 4y^2 + 2y + 9 \end{array}$$

49. $(4a^3 - 9a + 15) - (-2a^3 + 4a^2 + 7a)$

50. $(3y^2 + 2y - 1) + (5y^2 - 11y + 6)$

51. Find the perimeter of the triangle.

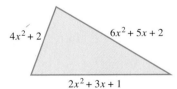

52. Find each of the following for the polynomial function defined by
$$f(x) = -2x^2 + 5x + 7.$$
  (a) $f(-2)$   (b) $f(3)$

53. Find each of the following for the polynomial functions defined by
$$f(x) = 2x + 3 \text{ and } g(x) = 5x^2 - 3x + 2.$$
  (a) $(f+g)(x)$   (b) $(f-g)(x)$   (c) $(f+g)(-1)$   (d) $(f-g)(-1)$

**54.** Find each of the following for the polynomial functions defined by
$$f(x) = 3x^2 + 2x - 1 \quad \text{and} \quad g(x) = 5x + 7.$$
**(a)** $(g \circ f)(3)$     **(b)** $(f \circ g)(3)$     **(c)** $(f \circ g)(-2)$
**(d)** $(g \circ f)(-2)$     **(e)** $(f \circ g)(x)$     **(f)** $(g \circ f)(x)$

**55.** The number of people, in millions, enrolled in health maintenance organizations (HMOs) during the period from 1995 through 2003 can be modeled by the polynomial function defined by
$$f(x) = -0.574x^2 + 6.01x + 15.9,$$
where $x = 0$ corresponds to 1995, $x = 1$ corresponds to 1996, and so on. Use this model to approximate the number of people enrolled in each given year. (*Source:* Interstudy; U.S. National Center for Health Statistics.)

**(a)** 1995     **(b)** 2000     **(c)** 2003

*Graph each polynomial function defined as follows.*

**56.** $f(x) = -2x + 5$     **57.** $f(x) = x^2 - 6$     **58.** $f(x) = -x^3 + 1$

*Find each product.*

**59.** $-6k(2k^2 + 7)$     **60.** $(3m - 2)(5m + 1)$

**61.** $(3w - 2t)(2w - 3t)$     **62.** $(2p^2 + 6p)(5p^2 - 4)$

**63.** $(3q^2 + 2q - 4)(q - 5)$     **64.** $(3z^3 - 2z^2 + 4z - 1)(3z - 2)$

**65.** $(6r^2 - 1)(6r^2 + 1)$     **66.** $\left(z + \dfrac{3}{5}\right)\left(z - \dfrac{3}{5}\right)$

**67.** $(4m + 3)^2$     **68.** $t(3t + 2)^2$

*Divide.*

**69.** $\dfrac{4y^3 - 12y^2 + 5y}{4y}$     **70.** $\dfrac{x^3 - 9x^2 + 26x - 30}{x - 5}$

**71.** $\dfrac{2p^3 + 9p^2 + 27}{2p - 3}$     **72.** $\dfrac{5p^4 + 15p^3 - 33p^2 - 9p + 18}{5p^2 - 3}$

# 5 TEST

**1.** *Matching* Match each expression (a)–(i) in Column I with its equivalent expression A–I in Column II. Choices may be used once, more than once, or not at all.

| I | | II | |
|---|---|---|---|
| **(a)** $7^{-2}$ | **(b)** $7^0$ | **A.** 1 | **B.** $\dfrac{1}{9}$ |
| **(c)** $-7^0$ | **(d)** $(-7)^0$ | **C.** $\dfrac{1}{49}$ | **D.** $-1$ |
| **(e)** $-7^2$ | **(f)** $7^{-1} + 2^{-1}$ | **E.** $-49$ | **F.** $\dfrac{9}{14}$ |
| **(g)** $(7 + 2)^{-1}$ | **(h)** $\dfrac{7^{-1}}{2^{-1}}$ | **G.** $\dfrac{2}{7}$ | **H.** 0 |
| **(i)** $(-7)^{-2}$ | | **I.** none of these | |

*Simplify. Write answers with only positive exponents. Assume that all variables represent nonzero real numbers.*

**2.** $(3x^{-2}y^3)^{-2}(4x^3y^{-4})$

**3.** $\dfrac{36r^{-4}(r^2)^{-3}}{6r^4}$

**4.** $\left(\dfrac{4p^2}{q^4}\right)^3 \left(\dfrac{6p^8}{q^{-8}}\right)^{-2}$

**5.** $(-2x^4y^{-3})^0(-4x^{-3}y^{-8})^2$

**6.** Write $9.1 \times 10^{-7}$ in standard form.

**7.** Use scientific notation to simplify $\dfrac{2{,}500{,}000 \times 0.00003}{0.05 \times 5{,}000{,}000}$. Write the answer in both scientific notation and standard notation.

**8.** Find each of the following for the functions defined by
$$f(x) = -2x^2 + 5x - 6 \quad \text{and} \quad g(x) = 7x - 3.$$
**(a)** $f(4)$    **(b)** $(f+g)(x)$    **(c)** $(f-g)(x)$    **(d)** $(f-g)(-2)$

**9.** Find each of the following for the functions defined by
$$f(x) = 3x + 5 \quad \text{and} \quad g(x) = x^2 + 2.$$
**(a)** $(f \circ g)(-2)$    **(b)** $(f \circ g)(x)$    **(c)** $(g \circ f)(x)$

*Graph each polynomial function.*

**10.** $f(x) = -2x^2 + 3$      **11.** $f(x) = -x^3 + 3$

**12.** The number of medical doctors, in thousands, in the United States during the period from 1990 through 2002 can be modeled by the polynomial function defined by
$$f(x) = -0.141x^2 + 21.5x + 616,$$
where $x = 0$ corresponds to 1990, $x = 1$ corresponds to 1991, and so on. Use this model to approximate the number of doctors to the nearest thousand in each given year. (*Source: American Medical Association.*)

**(a)** 1990    **(b)** 1996    **(c)** 2002

*Perform the indicated operations.*

**13.** $(4x^3 - 3x^2 + 2x - 5) - (3x^3 + 11x + 8) + (x^2 - x)$

**14.** $(5x - 3)(2x + 1)$      **15.** $(2m - 5)(3m^2 + 4m - 5)$

**16.** $(6x + y)(6x - y)$      **17.** $(3k + q)^2$

**18.** $[2y + (3z - x)][2y - (3z - x)]$      **19.** $\dfrac{16p^3 - 32p^2 + 24p}{4p^2}$

**20.** $(x^3 + 3x^2 - 4) \div (x - 1)$

**21.** If $f(x) = x^2 + 3x + 2$ and $g(x) = x + 1$, find each of the following.

**(a)** $(fg)(x)$    **(b)** $(fg)(-2)$

**22.** Use $f(x)$ and $g(x)$ from Exercise 21 to find each of the following.

**(a)** $\left(\dfrac{f}{g}\right)(x)$    **(b)** $\left(\dfrac{f}{g}\right)(-2)$

# 6

# Factoring

*Factoring* is used to solve *quadratic equations,* which have many useful applications. An important one is to express the distance a falling or projected object travels in a specific time. Such equations are used in astronomy and the space program to describe the motion of objects in space.

In Section 6.5, we use the concepts of this chapter to explore how to find the heights of objects after they are projected or dropped.

**6.1** Greatest Common Factors; Factoring by Grouping

**6.2** Factoring Trinomials

**6.3** Special Factoring

**6.4** A General Approach to Factoring

**6.5** Solving Equations by Factoring

## 6.1 Greatest Common Factors; Factoring by Grouping

**OBJECTIVES**
1. Factor out the greatest common factor.
2. Factor by grouping.

Writing a polynomial as the product of two or more simpler polynomials is called **factoring** the polynomial. For example, the product of $3x$ and $5x - 2$ is $15x^2 - 6x$, and $15x^2 - 6x$ can be factored as the product $3x(5x - 2)$.

$$3x(5x - 2) = 15x^2 - 6x \quad \text{Multiplying}$$
$$15x^2 - 6x = 3x(5x - 2) \quad \text{Factoring}$$

Notice that both multiplying and factoring use the distributive property, but in opposite directions. *Factoring "undoes," or reverses, multiplying.*

**OBJECTIVE 1** Factor out the greatest common factor. The first step in factoring a polynomial is to find the *greatest common factor* for the terms of the polynomial. The **greatest common factor (GCF)** is the largest term that is a factor of all terms in the polynomial. For example, the greatest common factor for $8x + 12$ is 4, since 4 is the largest term that is a factor of (divides into) both $8x$ and 12. Using the distributive property gives

$$8x + 12 = 4(2x) + 4(3)$$
$$= 4(2x + 3).$$

As a check, multiply 4 and $2x + 3$. The result should be $8x + 12$. Using the distributive property this way is called **factoring out the greatest common factor**.

**EXAMPLE 1** Factoring Out the Greatest Common Factor

Factor out the greatest common factor.

**(a)** $9z - 18$
Since 9 is the GCF, factor 9 from each term.
$$9z - 18 = 9 \cdot z - 9 \cdot 2$$
$$= 9(z - 2)$$
*Check:* $9(z - 2) = 9z - 18$ Original polynomial

*Always check by multiplying.*

**(b)** $56m + 35p = 7(8m + 5p)$

**(c)** $2y + 5$   There is no common factor other than 1.

**(d)** $12 + 24z = 12 \cdot 1 + 12 \cdot 2z$   Identity property
$$= 12(1 + 2z) \qquad \text{12 is the GCF.}$$

*Remember to write the 1.*

*Check:* $12(1 + 2z) = 12(1) + 12(2z)$   Distributive property
$$= 12 + 24z \qquad \text{Original polynomial}$$

**NOW TRY** Exercises 1, 3, and 5.

SECTION 6.1 Greatest Common Factors; Factoring by Grouping **287**

**EXAMPLE 2** Factoring Out the Greatest Common Factor

Factor out the greatest common factor.

**(a)** $9x^2 + 12x^3$

The numerical part of the GCF is 3. For the variable parts, $x^2$ and $x^3$, use the least exponent that appears on $x$; here, the least exponent is 2. The GCF is $3x^2$.

$$9x^2 + 12x^3 = 3x^2(3) + 3x^2(4x) \qquad \text{GCF} = 3x^2$$
$$= 3x^2(3 + 4x)$$

**(b)** $32p^4 - 24p^3 + 40p^5 = 8p^3(4p) + 8p^3(-3) + 8p^3(5p^2) \qquad \text{GCF} = 8p^3$
$$= 8p^3(4p - 3 + 5p^2)$$

**(c)** $3k^4 - 15k^7 + 24k^9 = 3k^4(1 - 5k^3 + 8k^5)$

*Remember the 1.*

**(d)** $24m^3n^2 - 18m^2n + 6m^4n^3$

The numerical part of the GCF is 6. Here, 2 is the least exponent that appears on $m$, while 1 is the least exponent on $n$. The GCF is $6m^2n$.

$$24m^3n^2 - 18m^2n + 6m^4n^3 = 6m^2n(4mn) + 6m^2n(-3) + 6m^2n(m^2n^2)$$
$$= 6m^2n(4mn - 3 + m^2n^2)$$

**(e)** $25x^2y^3 + 30y^5 - 15x^4y^7 = 5y^3(5x^2 + 6y^2 - 3x^4y^4)$

*In each case, remember to check the factored form by multiplying.*

**NOW TRY** Exercises 7, 11, and 15.

**EXAMPLE 3** Factoring Out a Binomial Factor

Factor out the greatest common factor.

**(a)** $(x - 5)(x + 6) + (x - 5)(2x + 5)$

The greatest common factor here is the binomial $x - 5$.

$$(x - 5)(x + 6) + (x - 5)(2x + 5) = (x - 5)[(x + 6) + (2x + 5)]$$
$$= (x - 5)(x + 6 + 2x + 5)$$
$$= (x - 5)(3x + 11)$$

**(b)** $z^2(m + n)^2 + x^2(m + n)^2 = (m + n)^2(z^2 + x^2)$

**(c)** $p(r + 2s)^2 - q(r + 2s)^3 = (r + 2s)^2[p - q(r + 2s)]$
$$= (r + 2s)^2(p - qr - 2qs)$$

*Be careful with signs.*

**(d)** $(p - 5)(p + 2) - (p - 5)(3p + 4)$
$$= (p - 5)[(p + 2) - (3p + 4)] \qquad \text{Factor out the common factor.}$$
$$= (p - 5)[p + 2 - 3p - 4] \qquad \text{Distributive property}$$
$$= (p - 5)[-2p - 2] \qquad \text{Combine like terms.}$$
$$= (p - 5)[-2(p + 1)] \qquad \text{Look for a common factor.}$$
$$= -2(p - 5)(p + 1) \qquad \text{Commutative property}$$

**NOW TRY** Exercises 21 and 25.

**EXAMPLE 4** Factoring Out a Negative Common Factor

Factor $-a^3 + 3a^2 - 5a$ in two ways.
First, $a$ could be used as the common factor, giving

$-a^3 + 3a^2 - 5a = a(-a^2) + a(3a) + a(-5)$   Factor out $a$.
$\phantom{-a^3 + 3a^2 - 5a} = a(-a^2 + 3a - 5)$.

Because of the leading negative sign, $-a$ could also be used as the common factor.

$-a^3 + 3a^2 - 5a = -a(a^2) + (-a)(-3a) + (-a)(5)$   Factor out $-a$.
$\phantom{-a^3 + 3a^2 - 5a} = -a(a^2 - 3a + 5)$

Sometimes there may be a reason to prefer one of these forms, but either is correct.

**NOW TRY** Exercise 33.

**OBJECTIVE 2** Factor by grouping. Sometimes the *individual terms* of a polynomial have a greatest common factor of 1, but it still may be possible to factor the polynomial by using a process called *factoring by grouping*. *We usually factor by grouping when a polynomial has more than three terms.*

**EXAMPLE 5** Factoring by Grouping

Factor $ax - ay + bx - by$.
*Group* the terms as follows:

Terms with common factor $a$   Terms with common factor $b$

$(ax - ay) + (bx - by)$.

Then factor $ax - ay$ as $a(x - y)$ and factor $bx - by$ as $b(x - y)$.

$ax - ay + bx - by = (ax - ay) + (bx - by)$   Group the terms.
$\phantom{ax - ay + bx - by} = a(x - y) + b(x - y)$   Factor each group.
$\phantom{ax - ay + bx - by} = (x - y)(a + b)$   The common factor is $x - y$.

*Check* by multiplying.

**NOW TRY** Exercise 39.

**EXAMPLE 6** Factoring by Grouping

Factor $3x - 3y - ax + ay$.
Grouping terms gives

$(3x - 3y) + (-ax + ay) = 3(x - y) + a(-x + y)$.

The factors $(x - y)$ and $(-x + y)$ are opposites, so if we factor out $-a$ instead of $a$ in the second group of terms, we get

$(3x - 3y) + (-ax + ay) = 3(x - y) - a(x - y)$   Be careful with signs.
$\phantom{(3x - 3y) + (-ax + ay)} = (x - y)(3 - a)$.

Check:  $(x - y)(3 - a) = 3x - ax - 3y + ay$   FOIL
$\phantom{(x - y)(3 - a)} = 3x - 3y - ax + ay$   Original polynomial

**NOW TRY** Exercise 43.

Use the following steps to factor by grouping.

**Factoring by Grouping**

*Step 1* **Group terms.** Collect the terms into groups so that each group has a common factor.

*Step 2* **Factor within the groups.** Factor out the common factor in each group.

*Step 3* **Factor the entire polynomial.** If each group now has a common factor, factor it out. If not, try a different grouping.

*Always check the factored form by multiplying.*

**EXAMPLE 7** Factoring by Grouping

Factor $6ax + 12bx + a + 2b$.

$$6ax + 12bx + a + 2b = (6ax + 12bx) + (a + 2b) \quad \text{Group terms.}$$

Now factor $6x$ from the first group, and use the identity property of multiplication to introduce the factor 1 in the second group.

Remember to write the 1.

$$(6ax + 12bx) + (a + 2b) = 6x(a + 2b) + 1(a + 2b)$$
$$= (a + 2b)(6x + 1) \quad \text{Factor out } a + 2b.$$

Check: $(a + 2b)(6x + 1) = 6ax + a + 12bx + 2b \quad$ FOIL
$\phantom{(a + 2b)(6x + 1)} = 6ax + 12bx + a + 2b \quad$ Original polynomial

**NOW TRY** Exercise 47.

**EXAMPLE 8** Rearranging Terms before Factoring by Grouping

Factor $p^2q^2 - 10 - 2q^2 + 5p^2$.

Neither the first two terms nor the last two terms have a common factor except the identity element 1. Rearrange and group the terms as follows:

$$p^2q^2 - 10 - 2q^2 + 5p^2$$
$$= (p^2q^2 - 2q^2) + (5p^2 - 10) \quad \text{Rearrange and group the terms.}$$
$$= q^2(p^2 - 2) + 5(p^2 - 2) \quad \text{Factor out the common factors.}$$

Don't stop here. $\quad = (p^2 - 2)(q^2 + 5). \quad \text{Factor out } p^2 - 2.$

Check: $(p^2 - 2)(q^2 + 5) = p^2q^2 + 5p^2 - 2q^2 - 10 \quad$ FOIL
$\phantom{(p^2 - 2)(q^2 + 5)} = p^2q^2 - 10 - 2q^2 + 5p^2 \quad$ Original polynomial

**NOW TRY** Exercise 53.

**CAUTION** In Example 8, do not stop at the step

$$q^2(p^2 - 2) + 5(p^2 - 2).$$

This expression is *not in factored form,* because it is a *sum* of two terms, $q^2(p^2 - 2)$ and $5(p^2 - 2)$, not a *product*.

## 6.1 Exercises

*Factor out the greatest common factor. Simplify the factors, if possible. See Examples 1–4.*

1. $12m - 60$
2. $15r - 45$
3. $4 + 20z$
4. $9 + 27x$
5. $8y - 15$
6. $7x - 40$
7. $8k^3 + 24k$
8. $9z^4 + 81z$
9. $-4p^3q^4 - 2p^2q^5$
10. $-3z^5w^2 - 18z^3w^4$
11. $21x^5 + 35x^4 - 14x^3$
12. $6k^3 - 36k^4 - 48k^5$
13. $10t^5 - 8t^4 - 4t^3$
14. $6p^3 - 3p^2 - 9p^4$
15. $15a^2c^3 - 25ac^2 + 5ac$
16. $15y^3z^3 + 27y^2z^4 - 36yz^5$
17. $16z^2n^6 + 64zn^7 - 32z^3n^3$
18. $5r^3s^5 + 10r^2s^2 - 15r^4s^2$
19. $14a^3b^2 + 7a^2b - 21a^5b^3 + 42ab^4$
20. $12km^3 - 24k^3m^2 + 36k^2m^4 - 60k^4m^3$
21. $(m - 4)(m + 2) + (m - 4)(m + 3)$
22. $(z - 5)(z + 7) + (z - 5)(z + 9)$
23. $(2z - 1)(z + 6) - (2z - 1)(z - 5)$
24. $(3x + 2)(x - 4) - (3x + 2)(x + 8)$
25. $5(2 - x)^2 - 2(2 - x)^3$
26. $2(5 - x)^3 - 3(5 - x)^2$
27. $4(3 - x)^2 - (3 - x)^3 + 3(3 - x)$
28. $2(t - s) + 4(t - s)^2 - (t - s)^3$
29. $15(2z + 1)^3 + 10(2z + 1)^2 - 25(2z + 1)$
30. $6(a + 2b)^2 - 4(a + 2b)^3 + 12(a + 2b)^4$
31. $5(m + p)^3 - 10(m + p)^2 - 15(m + p)^4$
32. $-9a^2(p + q) - 3a^3(p + q)^2 + 6a(p + q)^3$

*Factor each polynomial twice. First use a common factor with a positive coefficient, and then use a common factor with a negative coefficient. See Example 4.*

33. $-r^3 + 3r^2 + 5r$
34. $-t^4 + 8t^3 - 12t$
35. $-12s^5 + 48s^4$
36. $-16y^4 + 64y^3$
37. $-2x^5 + 6x^3 + 4x^2$
38. $-5a^3 + 10a^4 - 15a^5$

*Factor by grouping. See Examples 5–8.*

39. $mx + qx + my + qy$
40. $2k + 2h + jk + jh$
41. $10m + 2n + 5mk + nk$
42. $3ma + 3mb + 2ab + 2b^2$
43. $4 - 2q - 6p + 3pq$
44. $20 + 5m + 12n + 3mn$
45. $p^2 - 4zq + pq - 4pz$
46. $r^2 - 9tw + 3rw - 3rt$
47. $2xy + 3y + 2x + 3$
48. $7ab + 35bc + a + 5c$
49. $m^3 + 4m^2 - 6m - 24$
50. $2a^3 + a^2 - 14a - 7$
51. $-3a^3 - 3ab^2 + 2a^2b + 2b^3$
52. $-16m^3 + 4m^2p^2 - 4mp + p^3$
53. $4 + xy - 2y - 2x$
54. $10ab - 21 - 6b + 35a$
55. $8 + 9y^4 - 6y^3 - 12y$
56. $x^3y^2 - 3 - 3y^2 + x^3$
57. $1 - a + ab - b$
58. $2ab^2 - 8b^2 + a - 4$

59. *Concept Check* When directed to completely factor the polynomial
$$4x^2y^5 - 8xy^3,$$
a student wrote
$$2xy^3(2xy^2 - 4).$$
When the teacher did not give him full credit, he complained because when his answer is multiplied out, the result is the original polynomial. **WHAT WENT WRONG?** Give the correct answer.

60. *Multiple Choice* Which choice is an example of a polynomial in factored form?

    **A.** $3x^2y^3 + 6x^2(2x + y)$
    **B.** $5(x + y)^2 - 10(x + y)^3$
    **C.** $(-2 + 3x)(5y^2 + 4y + 3)$
    **D.** $(3x + 4)(5x - y) - (3x + 4)(2x - 1)$

## 6.2 Factoring Trinomials

**OBJECTIVES**

1. Factor trinomials when the coefficient of the squared term is 1.
2. Factor trinomials when the coefficient of the squared term is not 1.
3. Use an alternative method for factoring trinomials.
4. Factor by substitution.

**OBJECTIVE 1** Factor trinomials when the coefficient of the squared term is 1. We begin by finding the product of $x + 3$ and $x - 5$.

$$(x + 3)(x - 5) = x^2 - 5x + 3x - 15$$
$$= x^2 - 2x - 15$$

We see by this result that the factored form of $x^2 - 2x - 15$ is $(x + 3)(x - 5)$.

$$\text{Factored form} \xrightarrow{\text{Multiplication}} (x + 3)(x - 5) = x^2 - 2x - 15 \xleftarrow{\text{Factoring}} \text{Product}$$

Since multiplying and factoring are operations that "undo" each other, factoring trinomials involves using FOIL backwards. As shown here, the $x^2$-term came from multiplying $x$ and $x$, and $-15$ came from multiplying 3 and $-5$.

Product of $x$ and $x$ is $x^2$.
$$(x + 3)(x - 5) = x^2 - 2x - 15$$
Product of 3 and $-5$ is $-15$.

We find the $-2x$ in $x^2 - 2x - 15$ by multiplying the outer terms, then the inner terms, and adding.

Outer terms: $x(-5) = -5x$
$$(x + 3)(x - 5)$$
Inner terms: $3 \cdot x = 3x$
Add to get $-2x$.

Based on this example, use the following steps to factor a trinomial $x^2 + bx + c$, where 1 is the coefficient of the squared term. (A procedure for factoring a trinomial when the coefficient of the squared term is *not* 1 follows later in this section.)

### Factoring $x^2 + bx + c$

*Step 1* **Find pairs whose product is $c$.** Find all pairs of integers whose product is $c$, the third term of the trinomial.

*Step 2* **Find the pair whose sum is $b$.** Choose the pair whose sum is $b$, the coefficient of the middle term.

If there are no such integers, the polynomial cannot be factored.

A polynomial that cannot be factored with integer coefficients is a **prime polynomial**.

> **EXAMPLE 1** Factoring Trinomials in $x^2 + bx + c$ Form
>
> Factor each polynomial.
>
> **(a)** $y^2 + 2y - 35$
>
> | **Step 1** Find pairs of numbers whose product is $-35$. | **Step 2** Write sums of those numbers. |
> |---|---|
> | $-35(1)$ | $-35 + 1 = -34$ |
> | $35(-1)$ | $35 + (-1) = 34$ |
> | $7(-5)$ | $7 + (-5) = 2$ ← Coefficient of the middle term |
> | $5(-7)$ | $5 + (-7) = -2$ |
>
> The required numbers are 7 and $-5$, so
> $$y^2 + 2y - 35 = (y + 7)(y - 5).$$
> *Check* by finding the product of $y + 7$ and $y - 5$.
>
> **(b)** $r^2 + 8r + 12$
>
> Look for two numbers with a product of 12 and a sum of 8. Of all pairs of numbers having a product of 12, only the pair 6 and 2 has a sum of 8. Therefore,
> $$r^2 + 8r + 12 = (r + 6)(r + 2).$$
> Because of the commutative property, it would be equally correct to write $(r + 2)(r + 6)$. ***Check by using FOIL to multiply the factored form.***
>
> **NOW TRY** Exercises 5 and 7.

> **EXAMPLE 2** Recognizing a Prime Polynomial
>
> Factor $m^2 + 6m + 7$.
>
> Look for two numbers whose product is 7 and whose sum is 6. Only two pairs of integers, 7 and 1 and $-7$ and $-1$, give a product of 7. Neither of these pairs has a sum of 6, so $m^2 + 6m + 7$ cannot be factored with integer coefficients and is prime.
>
> **NOW TRY** Exercise 9.

We use a similar process to factor a trinomial that has more than one variable.

> **EXAMPLE 3** Factoring a Trinomial in Two Variables
>
> Factor $p^2 + 6ap - 16a^2$.
>
> Look for two expressions whose product is $-16a^2$ and whose sum is $6a$. The quantities $8a$ and $-2a$ have the necessary product and sum, so
> $$p^2 + 6ap - 16a^2 = (p + 8a)(p - 2a).$$
> Check:  $(p + 8a)(p - 2a) = p^2 - 2ap + 8ap - 16a^2$   FOIL
> $\phantom{Check:  (p + 8a)(p - 2a)} = p^2 + 6ap - 16a^2$   Original polynomial
>
> **NOW TRY** Exercise 11.

A trinomial may have a common factor that should be factored out first.

**SECTION 6.2** Factoring Trinomials **293**

> **EXAMPLE 4** Factoring a Trinomial with a Common Factor
>
> Factor $16y^3 - 32y^2 - 48y$.
> Start by factoring out the greatest common factor, $16y$.
> $$16y^3 - 32y^2 - 48y = 16y(y^2 - 2y - 3)$$
> To factor $y^2 - 2y - 3$, look for two integers whose product is $-3$ and whose sum is $-2$. The necessary integers are $-3$ and $1$, so
> $$16y^3 - 32y^2 - 48y = 16y(y - 3)(y + 1).$$
> *Remember to include the GCF.*
>
> **NOW TRY** Exercise 39.

**CAUTION** When factoring, always look for a common factor first. Remember to write the common factor as part of the answer.

**OBJECTIVE 2** Factor trinomials when the coefficient of the squared term is not 1. We can use a generalization of the method shown in Objective 1 to factor a trinomial of the form $ax^2 + bx + c$, where $a \neq 1$. To factor $3x^2 + 7x + 2$, for example, we first identify the values $a$, $b$, and $c$:

$$ax^2 + bx + c$$
$$3x^2 + 7x + 2, \quad \text{so} \quad a = 3, \ b = 7, \ c = 2.$$

The product $ac$ is $3 \cdot 2 = 6$, so we must find integers having a product of 6 and a sum of 7 (since the middle term has coefficient 7). The necessary integers are 1 and 6, so we write $7x$ as $1x + 6x$, or $x + 6x$, giving

$$3x^2 + 7x + 2 = 3x^2 + \underline{x + 6x} + 2.$$
$$x + 6x = 7x$$

*Check by multiplying.*
$$= (3x^2 + x) + (6x + 2) \quad \text{Group terms.}$$
$$= x(3x + 1) + 2(3x + 1) \quad \text{Factor by grouping.}$$
$$= (3x + 1)(x + 2) \quad \text{Factor out the common factor.}$$

> **EXAMPLE 5** Factoring a Trinomial in $ax^2 + bx + c$ Form
>
> Factor $12r^2 - 5r - 2$.
> Since $a = 12$, $b = -5$, and $c = -2$, the product $ac$ is $-24$. The two integers whose product is $-24$ and whose sum is $-5$ are $-8$ and $3$.
> $$12r^2 - 5r - 2 = 12r^2 + 3r - 8r - 2 \quad \text{Write } -5r \text{ as } 3r - 8r.$$
> $$= 3r(4r + 1) - 2(4r + 1) \quad \text{Factor by grouping.}$$
> $$= (4r + 1)(3r - 2) \quad \text{Factor out the common factor.}$$
> *Check by multiplying.*
>
> **NOW TRY** Exercise 19.

**OBJECTIVE 3** Use an alternative method for factoring trinomials. When the product $ac$ is large, trying repeated combinations and using FOIL is helpful.

> **EXAMPLE 6** Factoring Trinomials in $ax^2 + bx + c$ Form
>
> Factor each trinomial.
>
> **(a)** $3x^2 + 7x + 2$
>
> To factor this polynomial, we must find the correct numbers to put in the blanks.
>
> $$3x^2 + 7x + 2 = (\underline{\phantom{xx}}x + \underline{\phantom{xx}})(\underline{\phantom{xx}}x + \underline{\phantom{xx}})$$
>
> Addition signs are used, since all the signs in the polynomial indicate addition. The first two expressions have a product of $3x^2$, so they must be $3x$ and $1x$, or $x$.
>
> $$3x^2 + 7x + 2 = (3x + \underline{\phantom{xx}})(x + \underline{\phantom{xx}})$$
>
> The product of the two last terms must be 2, so the numbers must be 2 and 1. There is a choice. The 2 could be used with the $3x$ or with the $x$. Only one of these choices can give the correct middle term, $7x$. We use FOIL to try each one.
>
> $$\overset{3x}{(3x + 2)(x + 1)} \qquad \overset{6x}{(3x + 1)(x + 2)}$$
> $$\underset{2x}{} \qquad \underset{x}{}$$
>
> $3x + 2x = 5x$ $\qquad$ $6x + x = 7x$
> Wrong middle term $\qquad$ Correct middle term
>
> Therefore, $3x^2 + 7x + 2 = (3x + 1)(x + 2)$. (Compare with the solution obtained by factoring by grouping on the preceding page.)
>
> **(b)** $12r^2 - 5r - 2$
>
> To reduce the number of trials, we note that the trinomial has no common factor (except 1). This means that neither of its factors can have a common factor. We should keep this in mind as we choose factors. We try 4 and 3 for the two first terms.
>
> $$12r^2 - 5r - 2 = (4r \underline{\phantom{xx}})(3r \underline{\phantom{xx}})$$
>
> The factors of $-2$ are $-2$ and 1 or $-1$ and 2. We try both possibilities to see if we obtain the correct middle term, $-5r$.
>
> $(4r - 2)(3r + 1)$ $\qquad$ $\overset{8r}{(4r - 1)(3r + 2)}$
> Wrong: $4r - 2$ has a $\qquad\qquad\quad$ $\underset{-3r}{}$
> common factor of 2, $\qquad\qquad\quad$ $8r - 3r = 5r$
> which cannot be correct, $\qquad\quad$ Wrong middle term
> since 2 is not a factor
> of $12r^2 - 5r - 2$.
>
> The middle term on the right is $5r$, instead of the $-5r$ that is needed. We get $-5r$ by interchanging the signs of the second terms in the factors.
>
> $$\overset{-8r}{(4r + 1)(3r - 2)}$$
> $$\underset{3r}{}$$
> $-8r + 3r = -5r$
> Correct middle term
>
> Thus, $12r^2 - 5r - 2 = (4r + 1)(3r - 2)$. (Compare with Example 5.)

**NOW TRY** Exercise 21.

As shown in Example 6(b), if the terms of a polynomial have no common factor (except 1), then none of the terms of its factors can have a common factor. Remembering this will eliminate some potential factors.

We summarize this alternative method of factoring a trinomial in the form
$$ax^2 + bx + c, \text{ where } a \neq 1.$$

> **Factoring $ax^2 + bx + c$**
>
> **Step 1** **Find pairs whose product is $a$.** Write all pairs of integer factors of $a$, the coefficient of the squared term.
>
> **Step 2** **Find pairs whose product is $c$.** Write all pairs of integer factors of $c$, the last term.
>
> **Step 3** **Choose inner and outer terms.** Use FOIL and various combinations of the factors from Steps 1 and 2 until the necessary middle term is found.
>
> If no such combinations exist, the trinomial is prime.

### EXAMPLE 7 Factoring a Trinomial in Two Variables

Factor $18m^2 - 19mx - 12x^2$.

There is no common factor (except 1). Follow the steps for factoring a trinomial. There are many possible factors of both 18 and $-12$. Try 6 and 3 for 18 and $-3$ and 4 for $-12$.

$(6m - 3x)(3m + 4x)$   |   $(6m + 4x)(3m - 3x)$
Wrong: common factor   |   Wrong: common factors

Since 6 and 3 do not work as factors of 18, try 9 and 2 instead, with 3 and $-4$ as factors of $-12$.

$(9m + 3x)(2m - 4x)$   |   $(9m - 4x)(2m + 3x)$
Wrong: common factors

$27mx + (-8mx) = 19mx$
Wrong middle term

The result on the right differs from the correct middle term only in sign, so interchange the signs of the second terms in the factors.

$$18m^2 - 19mx - 12x^2 = (9m + 4x)(2m - 3x)$$

Check by using FOIL to multiply the factors.

**NOW TRY** Exercise 23.

### EXAMPLE 8 Factoring $ax^2 + bx + c, a < 0$

Factor $-3x^2 + 16x + 12$.

While we could factor directly, it is helpful to first factor out $-1$ so that the coefficient of the $x^2$-term is positive.

$$-3x^2 + 16x + 12 = -1(3x^2 - 16x - 12) \quad \text{Factor out } -1.$$
$$= -1(3x + 2)(x - 6) \quad \text{Factor the trinomial.}$$
$$= -(3x + 2)(x - 6)$$

This factored form can be written in other ways. Two of them are

$$(-3x - 2)(x - 6) \quad \text{and} \quad (3x + 2)(-x + 6).$$

Verify that these both give the original trinomial when multiplied.

**NOW TRY** Exercise 33.

**EXAMPLE 9** Factoring a Trinomial with a Common Factor

Factor $16y^3 + 24y^2 - 16y$.

$$16y^3 + 24y^2 - 16y = 8y(2y^2 + 3y - 2) \quad \text{GCF} = 8y$$
$$= 8y(2y - 1)(y + 2) \quad \text{Factor the trinomial.}$$

*Remember the common factor.*

**NOW TRY** Exercise 31.

**OBJECTIVE 4** Factor by substitution. Sometimes we can factor a more complicated polynomial by substituting a variable for an expression.

**EXAMPLE 10** Factoring a Polynomial by Substitution

Factor $2(x + 3)^2 + 5(x + 3) - 12$.

Since the binomial $x + 3$ appears to powers 2 and 1, we let the substitution variable represent $x + 3$. We may choose any letter we wish except $x$. We choose $t$ to represent $x + 3$.

$$2(x + 3)^2 + 5(x + 3) - 12 = 2t^2 + 5t - 12 \quad \text{Let } t = x + 3.$$
$$= (2t - 3)(t + 4) \quad \text{Factor.}$$
$$= [2(x + 3) - 3][(x + 3) + 4] \quad \text{Replace } t \text{ with } x + 3.$$
$$= (2x + 6 - 3)(x + 7) \quad \text{Simplify.}$$
$$= (2x + 3)(x + 7)$$

**NOW TRY** Exercise 49.

**CAUTION** *Remember to make the final substitution* of $x + 3$ for $t$ in Example 10.

**EXAMPLE 11** Factoring a Trinomial in $ax^4 + bx^2 + c$ Form

Factor $6y^4 + 7y^2 - 20$.

The variable $y$ appears to powers in which the larger exponent is twice the smaller exponent. We can let a substitution variable equal the smaller power. Here, we let $t = y^2$.

$$6y^4 + 7y^2 - 20 = 6(y^2)^2 + 7y^2 - 20$$
$$= 6t^2 + 7t - 20 \quad \text{Substitute.}$$
$$= (3t - 4)(2t + 5) \quad \text{Factor.}$$
$$= (3y^2 - 4)(2y^2 + 5) \quad t = y^2$$

*Don't stop here. Replace $t$ with $y^2$.*

**NOW TRY** Exercise 59.

## 6.2 Exercises

*Multiple Choice* Answer each question.

1. Which is *not* a valid way of starting the process of factoring $12x^2 + 29x + 10$?
   - **A.** $(12x\quad)(x\quad)$
   - **B.** $(4x\quad)(3x\quad)$
   - **C.** $(6x\quad)(2x\quad)$
   - **D.** $(8x\quad)(4x\quad)$

2. Which is the completely factored form of $2x^6 - 5x^5 - 3x^4$?
   - **A.** $x^4(2x + 1)(x - 3)$
   - **B.** $x^4(2x - 1)(x + 3)$
   - **C.** $(2x^5 + x^4)(x - 3)$
   - **D.** $x^3(2x^2 + x)(x - 3)$

3. Which is *not* a factored form of $-x^2 + 16x - 60$?
   - **A.** $(x - 10)(-x + 6)$
   - **B.** $(-x - 10)(x + 6)$
   - **C.** $(-x + 10)(x - 6)$
   - **D.** $-(x - 10)(x - 6)$

4. Which is the completely factored form of $4x^2 - 4x - 24$?
   - **A.** $4(x - 2)(x + 3)$
   - **B.** $4(x + 2)(x + 3)$
   - **C.** $4(x + 2)(x - 3)$
   - **D.** $4(x - 2)(x - 3)$

*Factor each trinomial. See Examples 1–9.*

5. $y^2 + 7y - 30$
6. $z^2 + 2z - 24$
7. $p^2 + 15p + 56$
8. $k^2 - 11k + 30$
9. $m^2 - 11m + 60$
10. $p^2 - 12p - 27$
11. $a^2 - 2ab - 35b^2$
12. $z^2 + 8zw + 15w^2$
13. $y^2 - 3yq - 15q^2$
14. $k^2 - 11hk + 28h^2$
15. $x^2y^2 + 11xy + 18$
16. $p^2q^2 - 5pq - 18$
17. $-6m^2 - 13m + 15$
18. $-15y^2 + 17y + 18$
19. $10x^2 + 3x - 18$
20. $8k^2 + 34k + 35$
21. $20k^2 + 47k + 24$
22. $27z^2 + 42z - 5$
23. $15a^2 - 22ab + 8b^2$
24. $15p^2 + 24pq + 8q^2$
25. $36m^2 - 60m + 25$
26. $25r^2 - 90r + 81$
27. $40x^2 + xy + 6y^2$
28. $14c^2 - 17cd - 6d^2$
29. $6x^2z^2 + 5xz - 4$
30. $8m^2n^2 - 10mn + 3$
31. $24x^2 + 42x + 15$
32. $36x^2 + 18x - 4$
33. $-15a^2 - 70a + 120$
34. $-12a^2 - 10a + 42$
35. $-11x^3 + 110x^2 - 264x$
36. $-9k^3 - 36k^2 + 189k$
37. $2x^3y^3 - 48x^2y^4 + 288xy^5$
38. $6m^3n^2 - 24m^2n^3 - 30mn^4$
39. $6a^3 + 12a^2 - 90a$
40. $3m^4 + 6m^3 - 72m^2$
41. $13y^3 + 39y^2 - 52y$
42. $4p^3 + 24p^2 - 64p$
43. $12p^3 - 12p^2 + 3p$
44. $45t^3 + 60t^2 + 20t$

45. *Concept Check* When a student was given the polynomial $4x^2 + 2x - 20$ to factor completely on a test, the student lost some credit when her answer was $(4x + 10)(x - 2)$. She complained to her teacher that when we multiply $(4x + 10)(x - 2)$, we get the original polynomial. **WHAT WENT WRONG?** Give the correct answer.

46. When factoring the polynomial $-4x^2 - 29x + 24$, Terry obtained $(-4x + 3)(x + 8)$, while John got $(4x - 3)(-x - 8)$. Who is correct? Explain your answer.

*Factor each trinomial. See Example 10.*

47. $12p^6 - 32p^3r + 5r^2$
48. $2y^6 + 7xy^3 + 6x^2$
49. $10(k + 1)^2 - 7(k + 1) + 1$
50. $4(m - 5)^2 - 4(m - 5) - 15$
51. $3(m + p)^2 - 7(m + p) - 20$
52. $4(x - y)^2 - 23(x - y) - 6$

*Factor each trinomial.* (Hint: *Factor out the GCF first.*)

53. $a^2(a + b) - ab(a + b)^2 - 6b^2(a + b)^2$

54. $m^2(m - p) + mp(m - p) - 2p^2(m - p)$

55. $p^2(p + q) + 4pq(p + q) + 3q^2(p + q)$

56. $2k^2(5 - y) - 7k(5 - y) + 5(5 - y)$

57. $z^2(z - x) - zx(x - z) - 2x^2(z - x)$

58. $r^2(r - s) - 5rs(s - r) - 6s^2(r - s)$

*Factor each trinomial. See Example 11.*

**59.** $p^4 - 10p^2 + 16$      **60.** $k^4 + 10k^2 + 9$      **61.** $2x^4 - 9x^2 - 18$

**62.** $6z^4 + z^2 - 1$      **63.** $16x^4 + 16x^2 + 3$      **64.** $9r^4 + 9r^2 + 2$

## 6.3 Special Factoring

**OBJECTIVES**

1. Factor a difference of squares.
2. Factor a perfect square trinomial.
3. Factor a difference of cubes.
4. Factor a sum of cubes.

**OBJECTIVE 1** Factor a difference of squares. The special products introduced in Section 5.4 are used in reverse when factoring. Recall that the product of the sum and difference of two terms leads to a **difference of squares.**

**Difference of Squares**

$$x^2 - y^2 = (x + y)(x - y)$$

**EXAMPLE 1** Factoring Differences of Squares

Factor each polynomial.

**(a)** $t^2 - 36 = t^2 - 6^2$     $36 = 6^2$
$= (t + 6)(t - 6)$     Factor the difference of squares.

**(b)** $4a^2 - 64$
There is a common factor of 4.

$4a^2 - 64 = 4(a^2 - 16)$     Factor out the common factor.
$= 4(a + 4)(a - 4)$     Factor the difference of squares.

$\quad\quad\quad A^2 \;-\; B^2 \;=\; (A \;+\; B)\;(A \;-\; B)$
$\quad\quad\quad\;\downarrow\quad\quad\;\downarrow\quad\quad\quad\downarrow\quad\;\downarrow\quad\;\downarrow$
**(c)** $16m^2 - 49p^2 = (4m)^2 - (7p)^2 = (4m + 7p)(4m - 7p)$

$\quad\quad\quad A^2 \;-\; B^2 \;=\; (A \;+\; B)\;(A \;-\; B)$
$\quad\quad\quad\;\downarrow\quad\quad\;\downarrow\quad\quad\quad\downarrow\quad\;\downarrow\quad\;\downarrow$
**(d)** $81k^2 - (a + 2)^2 = (9k)^2 - (a + 2)^2 = (9k + \overline{a + 2})(9k - \overline{(a + 2)})$
$\quad\quad\quad\quad\quad\quad\quad\quad\quad\quad\quad\quad\quad\;= (9k + a + 2)(9k - a - 2)$

We could have used the method of substitution here.

**(e)** $x^4 - 81 = (x^2 + 9)(x^2 - 9)$     Factor the difference of squares.
$\quad\quad\quad\quad= (x^2 + 9)(x + 3)(x - 3)$     Factor $x^2 - 9$.

**NOW TRY** Exercises 7, 9, 13, 15, and 19.

**CAUTION** *Assuming no greatest common factor except 1, it is not possible to factor (with real numbers) a sum of squares* such as $x^2 + 9$ in Example 1(e). In particular, $x^2 + y^2 \neq (x + y)^2$, as shown next.

**OBJECTIVE 2** Factor a perfect square trinomial. Two other special products from **Section 5.4** lead to the following rules for factoring.

### Perfect Square Trinomial

$$x^2 + 2xy + y^2 = (x + y)^2$$
$$x^2 - 2xy + y^2 = (x - y)^2$$

Because the trinomial $x^2 + 2xy + y^2$ is the square of $x + y$, it is called a **perfect square trinomial**. In this pattern, both the first and the last terms of the trinomial must be perfect squares. In the factored form $(x + y)^2$, twice the product of the first and the last terms must give the middle term of the trinomial. You should understand these patterns in words, since they occur with different symbols (other than $x$ and $y$).

$4m^2 + 20m + 25$  $\qquad$  $p^2 - 8p + 64$

Perfect square trinomial; $4m^2 = (2m)^2$, $25 = 5^2$, and $2(2m)(5) = 20m$. $\qquad$ Not a perfect square trinomial; middle term would have to be $16p$ or $-16p$.

**EXAMPLE 2** Factoring Perfect Square Trinomials

Factor each polynomial.

**(a)** $144p^2 - 120p + 25$

Here, $144p^2 = (12p)^2$ and $25 = 5^2$. The sign on the middle term is $-$, so if $144p^2 - 120p + 25$ is a perfect square trinomial, the factored form will have to be
$$(12p - 5)^2.$$
Take twice the product of the two terms to see if this is correct.
$$2(12p)(-5) = -120p$$
This is the middle term of the given trinomial, so
$$144p^2 - 120p + 25 = (12p - 5)^2.$$

**(b)** $4m^2 + 20mn + 49n^2$

If this is a perfect square trinomial, it will equal $(2m + 7n)^2$. By the pattern in the box, if multiplied out, this squared binomial has a middle term of
$$2(2m)(7n) = 28mn,$$
which *does not equal* $20mn$. Verify that this trinomial cannot be factored by the methods of the previous section either. It is prime.

**(c)** $(r + 5)^2 + 6(r + 5) + 9 = [(r + 5) + 3]^2$
$\qquad\qquad\qquad\qquad\qquad\quad = (r + 8)^2 \qquad$ $2(r + 5)(3) = 6(r + 5)$, the middle term

**(d)** $m^2 - 8m + 16 - p^2$

Since there are four terms, use factoring by grouping. The first three terms here form a perfect square trinomial. Group them together, and factor as follows.

$$m^2 - 8m + 16 - p^2 = (m^2 - 8m + 16) - p^2$$
$$= (m - 4)^2 - p^2 \quad \text{Factor the perfect square trinomial.}$$
$$= (m - 4 + p)(m - 4 - p) \quad \text{Factor the difference of squares.}$$

**NOW TRY** Exercises 23, 25, and 33.

Perfect square trinomials, of course, can be factored by the general methods shown earlier for other trinomials. The patterns given here provide "shortcuts."

**OBJECTIVE 3** Factor a difference of cubes. A **difference of cubes**, $x^3 - y^3$, can be factored as follows.

**Difference of Cubes**

$$x^3 - y^3 = (x - y)(x^2 + xy + y^2)$$

We could check this pattern by finding the product of $x - y$ and $x^2 + xy + y^2$.

**EXAMPLE 3** Factoring Differences of Cubes

Factor each polynomial.

$$A^3 - B^3 = (A - B)(A^2 + A \cdot B + B^2)$$

**(a)** $m^3 - 8 = m^3 - 2^3 = (m - 2)(m^2 + m \cdot 2 + 2^2)$
$$= (m - 2)(m^2 + 2m + 4)$$

Check: $(m - 2)(m^2 + 2m + 4)$

Opposite of the product of the cube roots gives the middle term.

**(b)** $27x^3 - 8y^3 = (3x)^3 - (2y)^3$
$$= (3x - 2y)[(3x)^2 + (3x)(2y) + (2y)^2]$$
$$= (3x - 2y)(9x^2 + 6xy + 4y^2)$$

**(c)** $1000k^3 - 27n^3 = (10k)^3 - (3n)^3$
$$= (10k - 3n)[(10k)^2 + (10k)(3n) + (3n)^2]$$
$$= (10k - 3n)(100k^2 + 30kn + 9n^2)$$

**NOW TRY** Exercises 37 and 51.

**OBJECTIVE 4** Factor a sum of cubes. While an expression of the form $x^2 + y^2$ cannot be factored with real numbers, a **sum of cubes** is factored as follows.

### Sum of Cubes

$$x^3 + y^3 = (x + y)(x^2 - xy + y^2)$$

To verify this result, find the product of $x + y$ and $x^2 - xy + y^2$. Compare this pattern with the pattern for factoring a difference of cubes.

Notice that the sign of the second term in the binomial factor of a sum or difference of cubes is *always the same* as the sign in the original polynomial. In the trinomial factor, the first and last terms are *always positive;* the sign of the middle term is *the opposite of* the sign of the second term in the binomial factor.

#### EXAMPLE 4 Factoring Sums of Cubes

Factor each polynomial.

(a) $r^3 + 27 = r^3 + 3^3$
$= (r + 3)(r^2 - 3r + 3^2)$
$= (r + 3)(r^2 - 3r + 9)$

(b) $27z^3 + 125 = (3z)^3 + 5^3$
$= (3z + 5)[(3z)^2 - (3z)(5) + 5^2]$
$= (3z + 5)(9z^2 - 15z + 25)$

(c) $125t^3 + 216s^6 = (5t)^3 + (6s^2)^3$
$= (5t + 6s^2)[(5t)^2 - (5t)(6s^2) + (6s^2)^2]$
$= (5t + 6s^2)(25t^2 - 30ts^2 + 36s^4)$

(d) $3x^3 + 192 = 3(x^3 + 64)$     Factor out the common factor.
$= 3(x + 4)(x^2 - 4x + 16)$     Factor the sum of cubes.

**Remember the common factor.**

(e) $(x + 2)^3 + t^3 = [(x + 2) + t][(x + 2)^2 - (x + 2)t + t^2]$
$= (x + 2 + t)(x^2 + 4x + 4 - xt - 2t + t^2)$

**NOW TRY** Exercises 41, 53, 55, and 57.

**CAUTION** A common error when factoring $x^3 + y^3$ or $x^3 - y^3$ is to think that the $xy$-term has a coefficient of 2. Since there is no coefficient of 2, expressions of the form $x^2 + xy + y^2$ and $x^2 - xy + y^2$ usually cannot be factored further.

The special types of factoring are summarized here. ***These should be memorized.***

### Special Types of Factoring

**Difference of Squares**     $x^2 - y^2 = (x + y)(x - y)$

**Perfect Square Trinomial**  $x^2 + 2xy + y^2 = (x + y)^2$
$x^2 - 2xy + y^2 = (x - y)^2$

**Difference of Cubes**       $x^3 - y^3 = (x - y)(x^2 + xy + y^2)$

**Sum of Cubes**              $x^3 + y^3 = (x + y)(x^2 - xy + y^2)$

## 6.3 Exercises

*Multiple choice* Answer each question.

1. Which of the following binomials are differences of squares?
   A. $64 - m^2$   B. $2x^2 - 25$   C. $k^2 + 9$   D. $4z^4 - 49$

2. Which of the following binomials are sums or differences of cubes?
   A. $64 + y^3$   B. $125 - p^6$   C. $9x^3 + 125$   D. $(x + y)^3 - 1$

3. Which of the following trinomials are perfect squares?
   A. $x^2 - 8x - 16$   B. $4m^2 + 20m + 25$
   C. $9z^4 + 30z^2 + 25$   D. $25a^2 - 45a + 81$

4. Of the 12 polynomials listed in Exercises 1–3, which ones can be factored by the methods of this section?

5. The binomial $9x^2 + 81$ is an example of a sum of two squares that can be factored. Under what conditions can the sum of two squares be factored?

6. *Fill in the Blanks* Insert the correct signs in the blanks.
   (a) $8 + t^3 = (2 \_\_ t)(4 \_\_ 2t \_\_ t^2)$   (b) $z^3 - 1 = (z \_\_ 1)(z^2 \_\_ z \_\_ 1)$

*Factor each polynomial. See Examples 1–4.*

7. $p^2 - 16$
8. $k^2 - 9$
9. $25x^2 - 4$
10. $36m^2 - 25$
11. $18a^2 - 98b^2$
12. $32c^2 - 98d^2$
13. $64m^4 - 4y^4$
14. $243x^4 - 3t^4$
15. $(y + z)^2 - 81$
16. $(h + k)^2 - 9$
17. $16 - (x + 3y)^2$
18. $64 - (r + 2t)^2$
19. $p^4 - 256$
20. $a^4 - 625$
21. $k^2 - 6k + 9$
22. $x^2 + 10x + 25$
23. $4z^2 + 4zw + w^2$
24. $9y^2 + 6yz + z^2$
25. $16m^2 - 8m + 1 - n^2$
26. $25c^2 - 20c + 4 - d^2$
27. $4r^2 - 12r + 9 - s^2$
28. $9a^2 - 24a + 16 - b^2$
29. $x^2 - y^2 + 2y - 1$
30. $-k^2 - h^2 + 2kh + 4$
31. $98m^2 + 84mn + 18n^2$
32. $80z^2 - 40zw + 5w^2$
33. $(p + q)^2 + 2(p + q) + 1$
34. $(x + y)^2 + 6(x + y) + 9$
35. $(a - b)^2 + 8(a - b) + 16$
36. $(m - n)^2 + 4(m - n) + 4$
37. $x^3 - 27$
38. $y^3 - 64$
39. $t^3 - 216$
40. $m^3 - 512$
41. $x^3 + 64$
42. $r^3 + 343$
43. $1000 + y^3$
44. $729 + x^3$
45. $8x^3 + 1$
46. $27y^3 + 1$
47. $125x^3 - 216$
48. $8w^3 - 125$
49. $x^3 - 8y^3$
50. $z^3 - 125p^3$
51. $64g^3 - 27h^3$
52. $27a^3 - 8b^3$
53. $343p^3 + 125q^3$
54. $512t^3 + 27s^3$
55. $24n^3 + 81p^3$
56. $250x^3 + 16y^3$
57. $(y + z)^3 + 64$
58. $(p - q)^3 + 125$
59. $m^6 - 125$
60. $27r^6 + 1$
61. $1000x^9 - 27$
62. $729p^9 - 64$
63. $125y^6 + z^3$

## 6.4 A General Approach to Factoring

**OBJECTIVES**

1. Factor out any common factor.
2. Factor binomials.
3. Factor trinomials.
4. Factor polynomials of more than three terms.

In this section, we summarize and apply the factoring methods presented in preceding sections. A polynomial is completely factored when it is in the following form.

1. The polynomial is written as a product of prime polynomials with integer coefficients.
2. None of the polynomial factors can be factored further, except that a monomial factor need not be factored completely.

### Factoring a Polynomial

*Step 1* **Factor out any common factor.**

*Step 2* **If the polynomial is a binomial,** check to see if it is the difference of squares, the difference of cubes, or the sum of cubes.

**If the polynomial is a trinomial,** check to see if it is a perfect square trinomial. If it is not, factor as in **Section 6.2**.

**If the polynomial has more than three terms,** try to factor by grouping.

*Step 3* **Check the factored form by multiplying.**

**OBJECTIVE 1** Factor out any common factor. This step is always the same, regardless of the number of terms in the polynomial.

**EXAMPLE 1** Factoring Out a Common Factor

Factor each polynomial.

(a) $9p + 45 = 9(p + 5)$      (b) $8m^2p^2 + 4mp = 4mp(2mp + 1)$

(c) $5x(a + b) - y(a + b) = (a + b)(5x - y)$

**NOW TRY** Exercises 13 and 23.

**OBJECTIVE 2** Factor binomials. Use one of the following rules.

### Factoring a Binomial

For a **binomial** (two terms), check for the following:

**Difference of squares**     $x^2 - y^2 = (x + y)(x - y)$

**Difference of cubes**       $x^3 - y^3 = (x - y)(x^2 + xy + y^2)$

**Sum of cubes**             $x^3 + y^3 = (x + y)(x^2 - xy + y^2).$

**EXAMPLE 2** Factoring Binomials

Factor each binomial if possible.

(a) $64m^2 - 9n^2 = (8m)^2 - (3n)^2$      Difference of squares

                $= (8m + 3n)(8m - 3n)$

(b) $8p^3 - 27 = (2p)^3 - 3^3$      Difference of cubes

             $= (2p - 3)[(2p)^2 + (2p)(3) + 3^2]$

             $= (2p - 3)(4p^2 + 6p + 9)$

(c) $1000m^3 + 1 = (10m)^3 + 1^3$     Sum of cubes
$= (10m + 1)[(10m)^2 - (10m)(1) + 1^2]$
$= (10m + 1)(100m^2 - 10m + 1)$

(d) $25m^2 + 121$ is prime. It is the sum of squares.

**NOW TRY** Exercises 7, 11, 29, and 31.

**OBJECTIVE 3** Factor trinomials.   Consider the following when factoring trinomials.

### Factoring a Trinomial

For a **trinomial** (three terms), decide whether it is a perfect square trinomial of the form

$$x^2 + 2xy + y^2 = (x + y)^2 \quad \text{or} \quad x^2 - 2xy + y^2 = (x - y)^2.$$

If not, use the methods of **Section 6.2**.

### EXAMPLE 3  Factoring Trinomials

Factor each trinomial.

(a) $p^2 + 10p + 25 = (p + 5)^2$     Perfect square trinomial
(b) $49z^2 - 42z + 9 = (7z - 3)^2$     Perfect square trinomial
(c) $y^2 - 5y - 6 = (y - 6)(y + 1)$

The numbers $-6$ and $1$ have a product of $-6$ and a sum of $-5$.

(d) $r^2 + 18r + 72 = (r + 6)(r + 12)$
(e) $2k^2 - k - 6 = (2k + 3)(k - 2)$     Use either method from **Section 6.2**.
(f) $28z^2 + 6z - 10 = 2(14z^2 + 3z - 5)$     Factor out the common factor.
$= 2(7z + 5)(2z - 1)$

*Remember the common factor.*

**NOW TRY** Exercises 9, 19, and 41.

**OBJECTIVE 4** Factor polynomials of more than three terms.   Try factoring by grouping.

### EXAMPLE 4  Factoring Polynomials with More than Three Terms

Factor each polynomial.

(a) $xy^2 - y^3 + x^3 - x^2y = (xy^2 - y^3) + (x^3 - x^2y)$
$= y^2(x - y) + x^2(x - y)$
$= (x - y)(y^2 + x^2)$

(b) $20k^3 + 4k^2 - 45k - 9 = (20k^3 + 4k^2) - (45k + 9)$     *Be careful with signs.*
$= 4k^2(5k + 1) - 9(5k + 1)$
$= (5k + 1)(4k^2 - 9)$     $5k + 1$ is a common factor.
$= (5k + 1)(2k + 3)(2k - 3)$     Difference of squares

**(c)** $4a^2 + 4a + 1 - b^2 = (4a^2 + 4a + 1) - b^2$    Associative property
$= (2a + 1)^2 - b^2$    Perfect square trinomial
$= (2a + 1 + b)(2a + 1 - b)$    Difference of squares

**(d)** $8m^3 + 4m^2 - n^3 - n^2$

Notice that the terms must be rearranged before grouping, since

$$(8m^3 + 4m^2) - (n^3 + n^2) = 4m^2(2m + 1) - n^2(n + 1),$$

which cannot be factored further. Factor the polynomial as follows:

$8m^3 + 4m^2 - n^3 - n^2 = (8m^3 - n^3) + (4m^2 - n^2)$    Group the cubes and squares.
$= (2m - n)(4m^2 + 2mn + n^2) + (2m - n)(2m + n)$    Factor each group.
$= (2m - n)(4m^2 + 2mn + n^2 + 2m + n).$    Factor out the common factor $2m - n$.

**NOW TRY** Exercises 21 and 45.

## 6.4 Exercises

*Factor each polynomial. See Examples 1–4.*

1. $100a^2 - 9b^2$
2. $10r^2 + 13r - 3$
3. $3p^4 - 3p^3 - 90p^2$
4. $k^4 - 16$
5. $3a^2pq + 3abpq - 90b^2pq$
6. $49z^2 - 16$
7. $225p^2 + 256$
8. $18m^3n + 3m^2n^2 - 6mn^3$
9. $6b^2 - 17b - 3$
10. $k^2 - 6k - 16$
11. $x^3 - 1000$
12. $6t^2 + 19tu - 77u^2$
13. $4(p + 2) + m(p + 2)$
14. $40p - 32r$
15. $9m^2 - 45m + 18m^3$
16. $4k^2 + 28kr + 49r^2$
17. $54m^3 - 2000$
18. $mn - 2n + 5m - 10$
19. $9m^2 - 30mn + 25n^2$
20. $2a^2 - 7a - 4$
21. $kq - 9q + kr - 9r$
22. $56k^3 - 875$
23. $16z^3x^2 - 32z^2x$
24. $9r^2 + 100$
25. $x^2 + 2x - 35$
26. $9 - a^2 + 2ab - b^2$
27. $x^4 - 625$
28. $2m^2 - mn - 15n^2$
29. $p^3 + 1$
30. $48y^2z^3 - 28y^3z^4$
31. $64m^2 - 625$
32. $14z^2 - 3zk - 2k^2$
33. $12z^3 - 6z^2 + 18z$
34. $225k^2 - 36r^2$
35. $256b^2 - 400c^2$
36. $z^2 - zp - 20p^2$
37. $1000z^3 + 512$
38. $64m^2 - 25n^2$
39. $10r^2 + 23rs - 5s^2$
40. $12k^2 - 17kq - 5q^2$
41. $24p^3q + 52p^2q^2 + 20pq^3$
42. $32x^2 + 16x^3 - 24x^5$
43. $48k^4 - 243$
44. $14x^2 - 25xq - 25q^2$
45. $m^3 + m^2 - n^3 - n^2$
46. $64x^3 + y^3 - 16x^2 + y^2$
47. $x^2 - 4m^2 - 4mn - n^2$
48. $4r^2 - s^2 - 2st - t^2$
49. $18p^5 - 24p^3 + 12p^6$
50. $k^2 - 6k + 16$
51. $2x^2 - 2x - 40$
52. $27x^3 - 3y^3$
53. $(2m + n)^2 - (2m - n)^2$
54. $(3k + 5)^2 - 4(3k + 5) + 4$
55. $50p^2 - 162$
56. $y^2 + 3y - 10$
57. $12m^2rx + 4mnrx + 40n^2rx$
58. $18p^2 + 53pr - 35r^2$

59. $21a^2 - 5ab - 4b^2$
60. $x^2 - 2xy + y^2 - 4$
61. $x^2 - y^2 - 4$
62. $(5r + 2s)^2 - 6(5r + 2s) + 9$
63. $(p + 8q)^2 - 10(p + 8q) + 25$
64. $z^4 - 9z^2 + 20$
65. $21m^4 - 32m^2 - 5$
66. $(x - y)^3 - (27 - y)^3$
67. $(r + 2t)^3 + (r - 3t)^3$
68. $16x^3 + 32x^2 - 9x - 18$
69. $x^5 + 3x^4 - x - 3$
70. $x^{16} - 1$
71. $m^2 - 4m + 4 - n^2 + 6n - 9$
72. $x^2 + 4 + x^2y + 4y$

## 6.5 Solving Equations by Factoring

**OBJECTIVES**

1. Learn and use the zero-factor property.
2. Solve applied problems that require the zero-factor property.

In **Chapter 2**, we developed methods for solving linear, or first-degree, equations. Solving higher degree polynomial equations requires other methods, one of which involves factoring.

**OBJECTIVE 1** Learn and use the zero-factor property. Solving equations by factoring depends on a special property of the number 0, called the **zero-factor property**.

> **Zero-Factor Property**
>
> If two numbers have a product of 0, then at least one of the numbers must be 0. That is,
>
> if $ab = 0$, then either $a = 0$ or $b = 0$.

To prove the zero-factor property, we first assume that $a \neq 0$. (If $a$ does equal 0, then the property is proved already.) If $a \neq 0$, then $\frac{1}{a}$ exists, and both sides of $ab = 0$ can be multiplied by $\frac{1}{a}$ to get

$$\frac{1}{a} \cdot ab = \frac{1}{a} \cdot 0$$
$$b = 0.$$

Thus, if $a \neq 0$, then $b = 0$, and the property is proved.

**CAUTION** If $ab = 0$, then $a = 0$ or $b = 0$. However, if $ab = 6$, for example, it is not necessarily true that $a = 6$ or $b = 6$; in fact, it is very likely that neither $a = 6$ nor $b = 6$. *The zero-factor property works only for a product equal to 0.*

### EXAMPLE 1  Using the Zero-Factor Property to Solve an Equation

Solve $(x + 6)(2x - 3) = 0$.

Here, the product of $x + 6$ and $2x - 3$ is 0. By the zero-factor property, this can be true only if

$$x + 6 = 0 \quad \text{or} \quad 2x - 3 = 0. \quad \text{Zero-factor property}$$

Solve each of these equations.

$$x + 6 = 0 \quad \text{or} \quad 2x - 3 = 0$$
$$x = -6 \quad \text{or} \quad 2x = 3$$
$$x = \frac{3}{2}$$

*Check* the two solutions $-6$ and $\frac{3}{2}$ by substitution into the *original* equation.

If $x = -6$, then
$$(x + 6)(2x - 3) = 0$$
$$(-6 + 6)[2(-6) - 3] = 0 \quad ?$$
$$0(-15) = 0 \quad ?$$
$$0 = 0. \quad \text{True}$$

If $x = \frac{3}{2}$, then
$$(x + 6)(2x - 3) = 0$$
$$\left(\frac{3}{2} + 6\right)\left(2 \cdot \frac{3}{2} - 3\right) = 0 \quad ?$$
$$\frac{15}{2}(0) = 0 \quad ?$$
$$0 = 0. \quad \text{True}$$

Both solutions check; the solution set is $\left\{-6, \frac{3}{2}\right\}$.

**NOW TRY** Exercise 5.

Since the product $(x + 6)(2x - 3)$ equals $2x^2 + 9x - 18$, the equation of Example 1 has a term with a squared variable and is an example of a *quadratic equation*. ***A quadratic equation has degree 2.***

### Quadratic Equation

An equation that can be written in the form

$$ax^2 + bx + c = 0,$$

where $a$, $b$, and $c$ are real numbers, with $a \neq 0$, is a **quadratic equation**. This form is called **standard form.**

Quadratic equations are discussed in more detail in **Chapter 9.**

The steps for solving a quadratic equation by factoring are summarized here.

### Solving a Quadratic Equation by Factoring

**Step 1** **Write in standard form.** Rewrite the equation if necessary so that one side is 0.

**Step 2** **Factor** the polynomial.

**Step 3** **Use the zero-factor property.** Set each variable factor equal to 0.

**Step 4** **Find the solution(s).** Solve each equation formed in Step 3.

**Step 5** **Check** each solution in the *original* equation.

**EXAMPLE 2** Solving Quadratic Equations by Factoring

Solve each equation.

**(a)** $2x^2 + 3x = 2$

**Step 1**
$$2x^2 + 3x = 2$$
$$2x^2 + 3x - 2 = 0 \quad \text{Standard form}$$

**Step 2** $\quad (2x - 1)(x + 2) = 0 \quad$ Factor.

**Step 3** $\quad 2x - 1 = 0 \quad \text{or} \quad x + 2 = 0 \quad$ Zero-factor property

**Step 4** $\quad 2x = 1 \quad \text{or} \quad x = -2 \quad$ Solve each equation.

$$x = \frac{1}{2}$$

**Step 5** *Check* each solution in the original equation.

If $x = \frac{1}{2}$, then
$$2x^2 + 3x = 2$$
$$2\left(\frac{1}{2}\right)^2 + 3\left(\frac{1}{2}\right) = 2 \quad ?$$
$$2\left(\frac{1}{4}\right) + \frac{3}{2} = 2 \quad ?$$
$$\frac{1}{2} + \frac{3}{2} = 2 \quad ?$$
$$2 = 2. \quad \text{True}$$

If $x = -2$, then
$$2x^2 + 3x = 2$$
$$2(-2)^2 + 3(-2) = 2 \quad ?$$
$$2(4) - 6 = 2 \quad ?$$
$$8 - 6 = 2 \quad ?$$
$$2 = 2. \quad \text{True}$$

We write solutions in the order they appear on a number line.

Because both solutions check, the solution set is $\left\{-2, \frac{1}{2}\right\}$.

**(b)**
$$4x^2 - 4x + 1 = 0 \quad \text{Standard form}$$
$$(2x - 1)^2 = 0 \quad \text{Factor.}$$
$$2x - 1 = 0 \quad \text{Zero-factor property}$$
$$2x = 1 \quad \text{Add 1.}$$
$$x = \frac{1}{2} \quad \text{Divide by 2.}$$

There is only one solution, because the trinomial is a perfect square. The solution set is $\left\{\frac{1}{2}\right\}$.

**NOW TRY** Exercises 11 and 29.

**EXAMPLE 3** Solving a Quadratic Equation with a Missing Constant Term

Solve $4z^2 - 20z = 0$.

This quadratic equation has a missing term. Comparing it with the standard form $ax^2 + bx + c = 0$ shows that $c = 0$. The zero-factor property can still be used.

$$4z^2 - 20z = 0$$
$$4z(z - 5) = 0 \quad \text{Factor.}$$

Set each *variable* factor equal to 0.

$$4z = 0 \quad \text{or} \quad z - 5 = 0 \quad \text{Zero-factor property}$$
$$z = 0 \quad \text{or} \quad z = 5 \quad \text{Solve each equation.}$$

**Check:** If $z = 0$, then

$4z^2 - 20z = 0$

$4(0)^2 - 20(0) = 0$ ?

$0 - 0 = 0.$  True

If $z = 5$, then

$4z^2 - 20z = 0$

$4(5)^2 - 20(5) = 0$ ?

$100 - 100 = 0.$  True

The solution set is $\{0, 5\}$.

**NOW TRY** Exercise 19.

**CAUTION** Remember to include 0 as a solution in Example 3.

### EXAMPLE 4  Solving a Quadratic Equation with a Missing Linear Term

Solve $3m^2 - 108 = 0$.

*The factor 3 does not lead to a solution.*

$3m^2 - 108 = 0$

$3(m^2 - 36) = 0$    Factor out 3.

$3(m + 6)(m - 6) = 0$    Factor $m^2 - 36$.

$m + 6 = 0$   or   $m - 6 = 0$    Zero-factor property

$m = -6$   or   $m = 6$

*Check* that the solution set is $\{-6, 6\}$.

**NOW TRY** Exercise 23.

**CAUTION** The factor 3 in Example 4 is not a *variable* factor, so it does *not* lead to a solution of the equation. In Example 3, however, the factor $4z$ is a variable factor and leads to the solution 0.

### EXAMPLE 5  Solving an Equation That Requires Rewriting

Solve $(2q + 1)(q + 1) = 2(1 - q) + 6$.

$(2q + 1)(q + 1) = 2(1 - q) + 6$

$2q^2 + 3q + 1 = 2 - 2q + 6$    Multiply on each side.

$2q^2 + 5q - 7 = 0$    Standard form

$(2q + 7)(q - 1) = 0$    Factor.

$2q + 7 = 0$   or   $q - 1 = 0$    Zero-factor property

$q = -\dfrac{7}{2}$   or   $q = 1$    Solve each equation.

**Check:** $(2q + 1)(q + 1) = 2(1 - q) + 6$

$\left[2\left(-\dfrac{7}{2}\right) + 1\right]\left(-\dfrac{7}{2} + 1\right) = 2\left[1 - \left(-\dfrac{7}{2}\right)\right] + 6$   ?    Let $q = -\dfrac{7}{2}$.

$(-7 + 1)\left(-\dfrac{5}{2}\right) = 2\left(\dfrac{9}{2}\right) + 6$   ?    Simplify; $1 = \dfrac{2}{2}$.

$(-6)\left(-\dfrac{5}{2}\right) = 9 + 6$   ?

$15 = 15$    True

*Check* that 1 is a solution. The solution set is $\left\{-\dfrac{7}{2}, 1\right\}$.

**NOW TRY** Exercise 35.

### EXAMPLE 6  Solving an Equation of Degree 3

Solve $-x^3 + x^2 = -6x$.

Start by adding $6x$ to each side to get 0 on the right side.

$$-x^3 + x^2 + 6x = 0$$
$$x^3 - x^2 - 6x = 0 \quad \text{Multiply each side by } -1.$$
$$x(x^2 - x - 6) = 0 \quad \text{Factor out } x.$$
$$x(x - 3)(x + 2) = 0 \quad \text{Factor the trinomial.}$$

Use the zero-factor property, extended to include the three *variable* factors.

**Remember to set x equal to 0.**

$$x = 0 \quad \text{or} \quad x - 3 = 0 \quad \text{or} \quad x + 2 = 0$$
$$x = 3 \quad \text{or} \quad x = -2$$

*Check* that the solution set is $\{-2, 0, 3\}$.

**NOW TRY** Exercise 39.

**OBJECTIVE 2** Solve applied problems that require the zero-factor property. The next example shows an application that leads to a quadratic equation. We continue to use the six-step problem-solving method introduced in **Section 2.3**.

### EXAMPLE 7  Using a Quadratic Equation in an Application

A piece of sheet metal is in the shape of a parallelogram. The longer sides of the parallelogram are each 8 m longer than the distance between them. The area of the piece is 48 m². Find the length of the longer sides and the distance between them.

*Step 1*  **Read** the problem again. There will be two answers.

*Step 2*  **Assign a variable.**

  Let  $x =$ the distance between the longer sides;
  $x + 8 =$ the length of each longer side. (See Figure 1.)

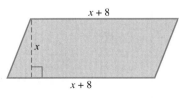

**FIGURE 1**

*Step 3*  **Write an equation.** The area of a parallelogram is given by $A = bh$, where $b$ is the length of the longer side and $h$ is the distance between the longer sides. Here, $b = x + 8$ and $h = x$.

$$A = bh$$
$$48 = (x + 8)x \quad \text{Let } A = 48, b = x + 8, h = x.$$

**Step 4** Solve.
$$48 = (x + 8)x$$
$$48 = x^2 + 8x \quad \text{Distributive property}$$
$$x^2 + 8x - 48 = 0 \quad \text{Standard form}$$
$$(x + 12)(x - 4) = 0 \quad \text{Factor.}$$
$$x + 12 = 0 \quad \text{or} \quad x - 4 = 0 \quad \text{Zero-factor property}$$
$$x = -12 \quad \text{or} \quad x = 4$$

**Step 5** State the answer. *A distance cannot be negative, so reject $-12$ as a solution.* The only possible solution is 4, so the distance between the longer sides is 4 m. The length of the longer sides is $4 + 8 = 12$ m.

**Step 6** Check. The length of the longer sides is 8 m more than the distance between them, and the area is $4 \cdot 12 = 48$ m², as required, so the answer checks.

**NOW TRY** Exercise 51.

**CAUTION** When applications lead to quadratic equations, a solution of the equation may not satisfy the physical requirements of the problem, as in Example 7. Reject such solutions.

A function defined by a quadratic polynomial is called a *quadratic function*. (See **Chapter 10.**) The next example uses such a function.

**EXAMPLE 8** Using a Quadratic Function in an Application

Quadratic functions are used to describe the height a falling object or a projected object reaches in a specific time. For example, if a small rocket is launched vertically upward from ground level with an initial velocity of 128 ft per sec, then its height in feet after $t$ seconds is a function defined by

$$h(t) = -16t^2 + 128t$$

if air resistance is neglected. After how many seconds will the rocket be 220 ft above the ground?

We must let $h(t) = 220$ and solve for $t$.

$$220 = -16t^2 + 128t \quad \text{Let } h(t) = 220.$$
$$16t^2 - 128t + 220 = 0 \quad \text{Standard form}$$
$$4t^2 - 32t + 55 = 0 \quad \text{Divide by 4.}$$
$$(2t - 5)(2t - 11) = 0 \quad \text{Factor.}$$
$$2t - 5 = 0 \quad \text{or} \quad 2t - 11 = 0 \quad \text{Zero-factor property}$$
$$t = 2.5 \quad \text{or} \quad t = 5.5$$

The rocket will reach a height of 220 ft twice: on its way up at 2.5 sec and again on its way down at 5.5 sec.

**NOW TRY** Exercise 57.

## 6.5 Exercises

1. Explain in your own words how the zero-factor property is used in solving a quadratic equation.

2. *Multiple Choice* One of the following equations is *not* in proper form for using the zero-factor property. Which one is it? Tell why it is not in proper form.
   A. $(x + 2)(x - 6) = 0$
   B. $x(3x - 7) = 0$
   C. $3t(t + 8)(t - 9) = 0$
   D. $y(y - 3) + 6(y - 3) = 0$

*Solve each equation. See Examples 1–5.*

3. $(x + 10)(x - 5) = 0$
4. $(x + 7)(x + 3) = 0$
5. $(2k - 5)(3k + 8) = 0$
6. $(3q - 4)(2q + 5) = 0$
7. $m^2 - 3m - 10 = 0$
8. $x^2 + x - 12 = 0$
9. $z^2 + 9z + 18 = 0$
10. $x^2 - 18x + 80 = 0$
11. $2x^2 = 7x + 4$
12. $2x^2 = 3 - x$
13. $15k^2 - 7k = 4$
14. $3c^2 + 3 = -10c$
15. $2x^2 - 12 - 4x = x^2 - 3x$
16. $3p^2 + 9p + 30 = 2p^2 - 2p$
17. $(5z + 1)(z + 3) = -2(5z + 1)$
18. $(3x + 1)(x - 3) = 2 + 3(x + 5)$
19. $4p^2 + 16p = 0$
20. $2a^2 - 8a = 0$
21. $6m^2 - 36m = 0$
22. $-3m^2 + 27m = 0$
23. $4p^2 - 16 = 0$
24. $9x^2 - 81 = 0$
25. $-3m^2 + 27 = 0$
26. $-2a^2 + 8 = 0$
27. $-x^2 = 9 - 6x$
28. $-m^2 - 8m = 16$
29. $9k^2 + 24k + 16 = 0$
30. $4m^2 - 20m + 25 = 0$
31. $(x - 3)(x + 5) = -7$
32. $(x + 8)(x - 2) = -21$
33. $(2x + 1)(x - 3) = 6x + 3$
34. $(3x + 2)(x - 3) = 7x - 1$
35. $(x + 3)(x - 6) = (2x + 2)(x - 6)$
36. $(2x + 1)(x + 5) = (x + 11)(x + 3)$

*Solve each equation. See Example 6.*

37. $2x^3 - 9x^2 - 5x = 0$
38. $6x^3 - 13x^2 - 5x = 0$
39. $x^3 - 2x^2 = 3x$
40. $y^3 - 6y^2 = -8y$
41. $9t^3 = 16t$
42. $25x^3 = 64x$
43. $2r^3 + 5r^2 - 2r - 5 = 0$
44. $2p^3 + p^2 - 98p - 49 = 0$
45. $x^3 - 6x^2 - 9x + 54 = 0$
46. $x^3 - 3x^2 - 4x + 12 = 0$

47. *Concept Check* A student tried to solve the equation in Exercise 41 by first dividing each side by $t$, obtaining $9t^2 = 16$. She then solved the resulting equation by the zero-factor property to get the solution set $\left\{-\frac{4}{3}, \frac{4}{3}\right\}$. *WHAT WENT WRONG?* Give the correct solution set.

48. *Multiple Choice* Without actually solving each equation, determine which one of the following has 0 in its solution set.
   A. $4x^2 - 25 = 0$
   B. $x^2 + 2x - 3 = 0$
   C. $6x^2 + 9x + 1 = 0$
   D. $x^3 + 4x^2 = 3x$

*Solve each problem. See Examples 7 and 8.*

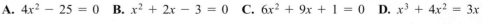

49. A garden has an area of 320 ft². Its length is 4 ft more than its width. What are the dimensions of the garden? (*Hint:* $320 = 16 \cdot 20$)

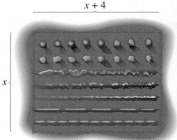

50. A square mirror has sides measuring 2 ft less than the sides of a square painting. If the difference between their areas is 32 ft², find the lengths of the sides of the mirror and the painting.

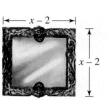

51. The base of a parallelogram is 7 ft more than the height. If the area of the parallelogram is 60 ft², what are the measures of the base and the height?

52. A sign has the shape of a triangle. The length of the base is 3 m less than the height. What are the measures of the base and the height if the area is 44 m²?

53. A farmer has 300 ft of fencing and wants to enclose a rectangular area of 5000 ft². What dimensions should she use? (*Hint:* $5000 = 50 \cdot 100$)

54. A rectangular landfill has an area of 30,000 ft². Its length is 200 ft more than its width. What are the dimensions of the landfill? (*Hint:* $30{,}000 = 300 \cdot 100$)

55. A box with no top is to be constructed from a piece of cardboard whose length measures 6 in. more than its width. The box is to be formed by cutting squares that measure 2 in. on each side from the four corners and then folding up the sides. If the volume of the box will be 110 in.³, what are the dimensions of the piece of cardboard?

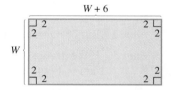

56. The surface area of the box with open top shown in the figure is 161 in.². Find the dimensions of the base. (*Hint:* The surface area of the box is a function defined by $S(x) = x^2 + 16x$.)

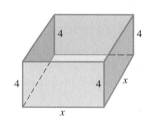

57. If an object is projected upward with an initial velocity of 64 ft per sec from a height of 80 ft, then its height in feet $t$ seconds after it is projected is a function defined by

$$f(t) = -16t^2 + 64t + 80.$$

How long after it is projected will it hit the ground? (*Hint:* When it hits the ground, its height is 0 ft.)

58. Refer to Example 8. After how many seconds will the rocket be 240 ft above the ground? 112 ft above the ground?

59. If a baseball is dropped from a helicopter 625 ft above the ground, then its distance in feet from the ground $t$ seconds later is a function defined by

$$f(t) = -16t^2 + 625.$$

How long after it is dropped will it hit the ground?

60. If a rock is dropped from a building 576 ft high, then its distance in feet from the ground $t$ seconds later is a function defined by

$$f(t) = -16t^2 + 576.$$

How long after it is dropped will it hit the ground?

# 6 SUMMARY

## KEY TERMS

**6.1** factoring
greatest common factor (GCF)

**6.2** prime polynomial
**6.3** difference of squares
perfect square trinomial

difference of cubes
sum of cubes
**6.5** quadratic equation

standard form of a quadratic equation

## QUICK REVIEW

### CONCEPTS — EXAMPLES

### 6.1 Greatest Common Factors; Factoring by Grouping

**The Greatest Common Factor**

The product of the largest common numerical factor and each common variable raised to the least exponent that appears on that variable in any term is the greatest common factor of the terms of the polynomial.

Factor $4x^2y - 50xy^2$.
The greatest common factor is $2xy$.
$$4x^2y - 50xy^2 = 2xy(2x - 25y)$$

**Factoring by Grouping**

*Step 1* Group the terms so that each group has a common factor.

*Step 2* Factor out the common factor in each group.

*Step 3* If the groups now have a common factor, factor it out. If not, try a different grouping.

*Always check the factored form by multiplying.*

Factor by grouping.
$$5a - 5b - ax + bx = (5a - 5b) + (-ax + bx)$$
$$= 5(a - b) - x(a - b)$$
$$= (a - b)(5 - x)$$

*Be careful with signs.*

### 6.2 Factoring Trinomials

To factor a trinomial, choose factors of the first term and factors of the last term. Then place them within a pair of parentheses of this form:

$$(\quad)(\quad).$$

Try various combinations of the factors until the correct middle term of the trinomial is found.

Factor $15x^2 + 14x - 8$.
The factors of 15 are 5 and 3, and 15 and 1.
The factors of $-8$ are $-4$ and 2, 4 and $-2$, $-1$ and 8, and 1 and $-8$.

Various combinations lead to the correct factorization
$$15x^2 + 14x - 8 = (5x - 2)(3x + 4).$$

*Check by multiplying.*

### 6.3 Special Factoring

**Difference of Squares**
$$x^2 - y^2 = (x + y)(x - y)$$

**Perfect Square Trinomials**
$$x^2 + 2xy + y^2 = (x + y)^2$$
$$x^2 - 2xy + y^2 = (x - y)^2$$

**Difference of Cubes**
$$x^3 - y^3 = (x - y)(x^2 + xy + y^2)$$

**Sum of Cubes**
$$x^3 + y^3 = (x + y)(x^2 - xy + y^2)$$

Factor.
$$4m^2 - 25n^2 = (2m)^2 - (5n)^2$$
$$= (2m + 5n)(2m - 5n)$$

$$9y^2 + 6y + 1 = (3y + 1)^2$$
$$16p^2 - 56p + 49 = (4p - 7)^2$$

$$8 - 27a^3 = (2 - 3a)(4 + 6a + 9a^2)$$

$$64z^3 + 1 = (4z + 1)(16z^2 - 4z + 1)$$

*(continued)*

─────── CONCEPTS ───────                    ─────── EXAMPLES ───────

**6.4 A General Approach to Factoring**

*Step 1* Factor out any common factors.

*Step 2* For a binomial, check for the difference of squares, the difference of cubes, or the sum of cubes.

For a trinomial, see if it is a perfect square. If not, factor as in **Section 6.2**.

For more than three terms, try factoring by grouping.

*Step 3* Check the factored form by multiplying.

Factor.
$$ak^3 + 2ak^2 - 9ak - 18a$$
$$= a(k^3 + 2k^2 - 9k - 18) \quad \text{Factor out the common factor.}$$
$$= a[(k^3 + 2k^2) - (9k + 18)] \quad \text{Factor by grouping.}$$
$$= a[k^2(k + 2) - 9(k + 2)]$$
$$= a[(k + 2)(k^2 - 9)]$$
$$= a(k + 2)(k - 3)(k + 3) \quad \text{Factor the difference of squares.}$$

**6.5 Solving Equations by Factoring**

*Step 1* Rewrite the equation if necessary so that one side is 0.

*Step 2* Factor the polynomial.

*Step 3* Set each factor equal to 0.

*Step 4* Solve each equation from Step 3.

*Step 5* Check each solution.

Solve.
$$2x^2 + 5x = 3$$
$$2x^2 + 5x - 3 = 0 \quad \text{Standard form}$$
$$(2x - 1)(x + 3) = 0 \quad \text{Factor.}$$
$$2x - 1 = 0 \quad \text{or} \quad x + 3 = 0 \quad \text{Zero-factor property}$$
$$2x = 1 \qquad\qquad x = -3$$
$$x = \frac{1}{2}$$

A check verifies that the solution set is $\left\{-3, \frac{1}{2}\right\}$.

# 6 REVIEW EXERCISES

*Factor out the greatest common factor.*

**1.** $12p^2 - 6p$
**2.** $21x^2 + 35x$
**3.** $12q^2b + 8qb^2 - 20q^3b^2$
**4.** $6r^3t - 30r^2t^2 + 18rt^3$
**5.** $(x + 3)(4x - 1) - (x + 3)(3x + 2)$
**6.** $(z + 1)(z - 4) + (z + 1)(2z + 3)$

*Factor by grouping.*

**7.** $4m + nq + mn + 4q$
**8.** $x^2 + 5y + 5x + xy$
**9.** $2m + 6 - am - 3a$
**10.** $x^2 + 3x - 3y - xy$

*Factor completely.*

**11.** $3p^2 - p - 4$
**12.** $6k^2 + 11k - 10$
**13.** $12r^2 - 5r - 3$
**14.** $10m^2 + 37m + 30$
**15.** $10k^2 - 11kh + 3h^2$
**16.** $9x^2 + 4xy - 2y^2$
**17.** $24x - 2x^2 - 2x^3$
**18.** $6b^3 - 9b^2 - 15b$
**19.** $y^4 + 2y^2 - 8$
**20.** $2k^4 - 5k^2 - 3$

**21.** $p^2(p + 2)^2 + p(p + 2)^2 - 6(p + 2)^2$   **22.** $3(r + 5)^2 - 11(r + 5) - 4$

**23.** *Concept Check* When asked to factor $x^2y^2 - 6x^2 + 5y^2 - 30$, a student gave the following incorrect answer: $x^2(y^2 - 6) + 5(y^2 - 6)$. *WHAT WENT WRONG?* What is the correct answer?

**24.** If the area of this rectangle is represented by $4p^2 + 3p - 1$, what is the width in terms of $p$?

*Factor completely.*

**25.** $16x^2 - 25$  **26.** $9t^2 - 49$  **27.** $36m^2 - 25n^2$

**28.** $x^2 + 14x + 49$  **29.** $9k^2 - 12k + 4$  **30.** $r^3 + 27$

**31.** $125x^3 - 1$  **32.** $m^6 - 1$  **33.** $x^8 - 1$

**34.** $x^2 + 6x + 9 - 25y^2$  **35.** $(a + b)^3 - (a - b)^3$  **36.** $x^5 - x^3 - 8x^2 + 8$

*Solve each equation.*

**37.** $x^2 - 8x + 16 = 0$  **38.** $(5x + 2)(x + 1) = 0$

**39.** $p^2 - 5p + 6 = 0$  **40.** $q^2 + 2q = 8$

**41.** $6z^2 = 5z + 50$  **42.** $6r^2 + 7r = 3$

**43.** $8k^2 + 14k + 3 = 0$  **44.** $-4m^2 + 36 = 0$

**45.** $6x^2 + 9x = 0$  **46.** $(2x + 1)(x - 2) = -3$

**47.** $(r + 2)(r - 2) = (r - 2)(r + 3) - 2$  **48.** $2x^3 - x^2 - 28x = 0$

**49.** $-t^3 - 3t^2 + 4t + 12 = 0$  **50.** $(r + 2)(5r^2 - 9r - 18) = 0$

*Solve each problem.*

**51.** A triangular wall brace has the shape of a right triangle. One of the perpendicular sides is 1 ft longer than twice the other. The area enclosed by the triangle is 10.5 ft². Find the shorter of the perpendicular sides.

**52.** A rectangular parking lot has a length 20 ft more than its width. Its area is 2400 ft². What are the dimensions of the lot?

The area is 10.5 ft².

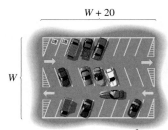

The area is 2400 ft².

*A rock is projected directly upward from ground level. After $t$ seconds, its height is given by $f(t) = -16t^2 + 256t$ (if air resistance is neglected).*

**53.** When will the rock return to the ground?

**54.** After how many seconds will it be 240 ft above the ground?

**55.** Why does the question in Exercise 54 have two answers?

# 6 TEST

*Factor.*

1. $11z^2 - 44z$
2. $10x^2y^5 - 5x^2y^3 - 25x^5y^3$
3. $3x + by + bx + 3y$
4. $-2x^2 - x + 36$
5. $6x^2 + 11x - 35$
6. $4p^2 + 3pq - q^2$
7. $16a^2 + 40ab + 25b^2$
8. $x^2 + 2x + 1 - 4z^2$
9. $a^3 + 2a^2 - ab^2 - 2b^2$
10. $9k^2 - 121j^2$
11. $y^3 - 216$
12. $6k^4 - k^2 - 35$
13. $27x^6 + 1$

14. Explain why $(x^2 + 2y)p + 3(x^2 + 2y)$ is not in factored form. Then factor the polynomial.

15. *Multiple Choice* Which one of the following is *not* a factored form of $-x^2 - x + 12$?

   A. $(3 - x)(x + 4)$
   B. $-(x - 3)(x + 4)$
   C. $(-x + 3)(x + 4)$
   D. $(x - 3)(-x + 4)$

*Solve each equation.*

16. $3x^2 + 8x = -4$
17. $3x^2 - 5x = 0$
18. $5m(m - 1) = 2(1 - m)$

*Solve each problem.*

19. The area of the rectangle shown is 40 in.$^2$. Find the length and the width of the rectangle.

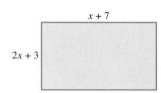

The area is 40 in.$^2$.

20. A ball is projected upward from ground level. After $t$ seconds, its height in feet is a function defined by $f(t) = -16t^2 + 96t$. After how many seconds will it reach a height of 128 ft?

# Rational Expressions and Functions

Americans have been car crazy ever since the first automobiles hit the road early in the 20th century. Today, there are about 213.5 million vehicles in the United States driving on 3.4 million miles of paved roadways. There is even a museum devoted exclusively to our four-wheeled passion and its influence on our lives and culture. The Museum of Automobile History in Syracuse, New York, features some 200 years of automobile memorabilia, including rare advertising pieces, designer drawings, and Hollywood movie posters. (*Source: Home and Away,* May/June 2002.)

In Exercise 85, Section 7.2, we use a *rational expression* to determine the cost of restoring a vintage automobile.

**7.1** Rational Expressions and Functions; Multiplying and Dividing

**7.2** Adding and Subtracting Rational Expressions

**7.3** Complex Fractions

**7.4** Equations with Rational Expressions and Graphs

**7.5** Applications of Rational Expressions

**7.6** Variation

# 7.1 Rational Expressions and Functions; Multiplying and Dividing

**OBJECTIVES**

1. Define rational expressions.
2. Define rational functions and describe their domains.
3. Write rational expressions in lowest terms.
4. Multiply rational expressions.
5. Find reciprocals of rational expressions.
6. Divide rational expressions.

**OBJECTIVE 1** Define rational expressions. In arithmetic, a rational number is the quotient of two integers, with the denominator not 0. In algebra, a **rational expression,** or *algebraic fraction,* is the quotient of two polynomials, again with the denominator not 0. For example,

$$\frac{x}{y}, \quad \frac{-a}{4}, \quad \frac{m+4}{m-2}, \quad \frac{8x^2 - 2x + 5}{4x^2 + 5x}, \quad \text{and} \quad x^5 \left(\text{or } \frac{x^5}{1}\right) \qquad \text{Rational expressions}$$

are all rational expressions. Rational expressions are elements of the set

$$\left\{ \frac{P}{Q} \,\middle|\, P \text{ and } Q \text{ are polynomials, with } Q \neq 0 \right\}.$$

**OBJECTIVE 2** Define rational functions and describe their domains. A function that is defined by a quotient of polynomials is called a **rational function** and has the form

$$f(x) = \frac{P(x)}{Q(x)}, \quad \text{where } Q(x) \neq 0.$$

The domain of a rational function consists of all real numbers except those that make $Q(x)$—that is, the denominator—equal to 0. For example, the domain of

$$f(x) = \frac{2}{\underbrace{x-5}_{\text{Cannot equal 0}}}$$

includes all real numbers except 5, because 5 would make the denominator equal to 0.

Figure 1 shows a graph of the function defined by $f(x) = \frac{2}{x-5}$. Notice that the graph does not exist when $x = 5$. It does not intersect the dashed vertical line whose equation is $x = 5$. This line is an *asymptote*. We discuss graphs of rational functions in more detail in **Section 7.4.**

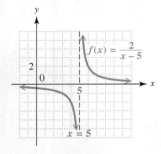

**FIGURE 1**

**EXAMPLE 1** Finding Numbers That Are Not in the Domains of Rational Functions

For each rational function, find all numbers that are not in the domain. Then give the domain in set notation.

**(a)** $f(x) = \dfrac{3}{7x - 14}$

The only values that cannot be used are those that make the denominator 0. To find these values, set the denominator equal to 0 and solve the resulting equation.

$$7x - 14 = 0$$
$$7x = 14 \qquad \text{Add 14.}$$
$$x = 2 \qquad \text{Divide by 7.}$$

The number 2 cannot be used as a replacement for $x$; the domain of $f$ includes all real numbers except 2. This is written $\{x \mid x \neq 2\}$.

**(b)** $g(x) = \dfrac{3 + x}{x^2 - 4x + 3}$

> Values that make the denominator 0 must be *excluded*.

$$x^2 - 4x + 3 = 0 \quad \text{Set the denominator equal to 0.}$$
$$(x - 1)(x - 3) = 0 \quad \text{Factor.}$$
$$x - 1 = 0 \quad \text{or} \quad x - 3 = 0 \quad \text{Zero-factor property}$$
$$x = 1 \quad \text{or} \quad x = 3 \quad \text{Solve each equation.}$$

The domain of $g$ includes all real numbers except 1 and 3, written $\{x \mid x \neq 1, 3\}$.

**(c)** $h(x) = \dfrac{8x + 2}{3}$

The denominator, 3, can never be 0, so the domain of $h$ includes all real numbers, written $(-\infty, \infty)$.

**(d)** $f(x) = \dfrac{2}{x^2 + 4}$

Setting $x^2 + 4$ equal to 0 leads to $x^2 = -4$. There is no real number whose square is $-4$. Therefore, any real number can be used, and as in part (c), the domain of $f$ includes all real numbers $(-\infty, \infty)$.

**NOW TRY** Exercises 11, 15, 17, and 19.

**OBJECTIVE 3** Write rational expressions in lowest terms. In arithmetic, we write the fraction $\tfrac{15}{20}$ in lowest terms by dividing the numerator and denominator by 5 to get $\tfrac{3}{4}$. We write rational expressions in lowest terms in a similar way, using the **fundamental property of rational numbers**.

> **Fundamental Property of Rational Numbers**
>
> If $\tfrac{a}{b}$ is a rational number and if $c$ is any nonzero real number, then
>
> $$\dfrac{a}{b} = \dfrac{ac}{bc}.$$
>
> That is, the numerator and denominator of a rational number may either be multiplied or divided by the same nonzero number without changing the value of the rational number.

Because $\tfrac{c}{c}$ is equivalent to 1, the fundamental property is based on the identity property of multiplication.

A rational expression is a quotient of two polynomials. Since the value of a polynomial is a real number for every value of the variable for which it is defined, any statement that applies to rational numbers will also apply to rational expressions.

We use the following steps to write rational expressions in lowest terms.

> **Writing a Rational Expression in Lowest Terms**
>
> *Step 1* **Factor** both numerator and denominator to find their greatest common factor (GCF).
>
> *Step 2* **Apply the fundamental property.**

### EXAMPLE 2 Writing Rational Expressions in Lowest Terms

Write each rational expression in lowest terms.

(a) $\dfrac{(x+5)(x+2)}{(x+2)(x-3)}$

$= \dfrac{(x+5)(x+2)}{(x-3)(x+2)}$     Commutative property

$= \dfrac{x+5}{x-3} \cdot 1$     Fundamental property

$= \dfrac{x+5}{x-3}$     Lowest terms

(b) $\dfrac{a^2 - a - 6}{a^2 + 5a + 6}$

$= \dfrac{(a-3)(a+2)}{(a+3)(a+2)}$     Factor the numerator and the denominator.

$= \dfrac{a-3}{a+3} \cdot 1$     Fundamental property

$= \dfrac{a-3}{a+3}$     Lowest terms

(c) $\dfrac{y^2 - 4}{2y + 4}$

$= \dfrac{(y+2)(y-2)}{2(y+2)}$     Factor; the numerator is a difference of squares.

$= \dfrac{y-2}{2}$     Lowest terms

(d) $\dfrac{x^3 - 27}{x - 3}$

$= \dfrac{(x-3)(x^2 + 3x + 9)}{x - 3}$     Factor the difference of cubes.

$= x^2 + 3x + 9$     Lowest terms

(e) $\dfrac{pr + qr + ps + qs}{pr + qr - ps - qs}$

$= \dfrac{(pr + qr) + (ps + qs)}{(pr + qr) - (ps + qs)}$     Group terms.

$= \dfrac{r(p+q) + s(p+q)}{r(p+q) - s(p+q)}$     Factor within groups.

$= \dfrac{(p+q)(r+s)}{(p+q)(r-s)}$     Factor by grouping.

$= \dfrac{r+s}{r-s}$     Lowest terms

**(f)** $\dfrac{8+k}{16}$   *Be careful. The numerator cannot be factored.*

This expression cannot be simplified further and is in lowest terms.

**NOW TRY** Exercises 27, 31, 35, 39, 43, and 47.

---

**CAUTION** Be careful! When you use the fundamental property of rational numbers, *only common factors may be divided.* For example,

$$\frac{y-2}{2} \neq y \quad \text{and} \quad \frac{y-2}{2} \neq y-1$$

because the 2 in $y-2$ is not a *factor* of the numerator. **Remember to factor before writing a fraction in lowest terms.** To see this, replace $y$ with a number and evaluate the fraction. For example, if $y = 5$, then

$$\frac{y-2}{2} = \frac{5-2}{2} = \frac{3}{2}, \quad \text{which does not equal } y \text{ or } y-1.$$

---

In the rational expression from Example 2(b), namely,

$$\frac{a^2 - a - 6}{a^2 + 5a + 6}, \quad \text{or} \quad \frac{(a-3)(a+2)}{(a+3)(a+2)},$$

$a$ can take any value except $-3$ or $-2$, since these values make the denominator 0. In the simplified rational expression

$$\frac{a-3}{a+3},$$

$a$ cannot equal $-3$. Because of this,

$$\frac{a^2 - a - 6}{a^2 + 5a + 6} = \frac{a-3}{a+3}$$

for all values of $a$ except $-3$ or $-2$. From now on, such statements of equality will be made with the understanding that they apply only to those real numbers that make neither denominator equal 0. We will no longer state such restrictions.

---

**EXAMPLE 3** Writing Rational Expressions in Lowest Terms

Write each rational expression in lowest terms.

**(a)** $\dfrac{m-3}{3-m}$

Here, the numerator and denominator are opposites. The given expression can be written in lowest terms by writing the denominator as $-1(m-3)$, giving

$$\frac{m-3}{3-m} = \frac{m-3}{-1(m-3)} = \frac{1}{-1} = -1.$$

The numerator could have been rewritten instead to get the same result.

**(b)** $\dfrac{r^2 - 16}{4 - r}$

$= \dfrac{(r + 4)(r - 4)}{4 - r}$  Factor the difference of squares.

$= \dfrac{(r + 4)(r - 4)}{-1(r - 4)}$  Write $4 - r$ as $-1(r - 4)$.

$= \dfrac{r + 4}{-1}$  Fundamental property

$= -(r + 4),$ or $-r - 4$  Lowest terms

**NOW TRY** Exercises 49 and 51.

As shown in Examples 3(a) and (b), the quotient $\dfrac{a}{-a}$ ($a \neq 0$) can be simplified as

$$\dfrac{a}{-a} = \dfrac{a}{-1(a)} = \dfrac{1}{-1} = -1.$$

The following statement summarizes this result.

### Quotient of Opposites

In general, if the numerator and the denominator of a rational expression are opposites, then the expression equals $-1$.

Based on this result, the following are true:

$$\dfrac{q - 7}{7 - q} = -1 \quad \text{and} \quad \dfrac{-5a + 2b}{5a - 2b} = -1.$$

Numerator and denominator in each expression are opposites.

However, the expression

$$\dfrac{r - 2}{r + 2} \quad \text{Numerator and denominator are } not \text{ opposites.}$$

cannot be simplified further.

**OBJECTIVE 4** **Multiply rational expressions.** To multiply rational expressions, follow these steps. (In practice, we usually simplify before multiplying.)

### Multiplying Rational Expressions

*Step 1* **Factor** all numerators and denominators as completely as possible.

*Step 2* **Apply the fundamental property.**

*Step 3* **Multiply** remaining factors in the numerator and remaining factors in the denominator. Leave the denominator in factored form.

*Step 4* **Check** to be sure that the product is in lowest terms.

### EXAMPLE 4  Multiplying Rational Expressions

Multiply.

(a) $\dfrac{5p - 5}{p} \cdot \dfrac{3p^2}{10p - 10}$

$= \dfrac{5(p - 1)}{p} \cdot \dfrac{3p \cdot p}{2 \cdot 5(p - 1)}$  Factor.

$= \dfrac{1}{1} \cdot \dfrac{3p}{2}$  Lowest terms

$= \dfrac{3p}{2}$  Multiply.

(b) $\dfrac{k^2 + 2k - 15}{k^2 - 4k + 3} \cdot \dfrac{k^2 - k}{k^2 + k - 20}$

$= \dfrac{(k + 5)(k - 3)}{(k - 3)(k - 1)} \cdot \dfrac{k(k - 1)}{(k + 5)(k - 4)}$  Factor.

$= \dfrac{k}{k - 4}$  Fundamental property

(c) $(p - 4) \cdot \dfrac{3}{5p - 20}$

$= \dfrac{p - 4}{1} \cdot \dfrac{3}{5p - 20}$  Write $p - 4$ as $\dfrac{p - 4}{1}$.

$= \dfrac{p - 4}{1} \cdot \dfrac{3}{5(p - 4)}$  Factor.

$= \dfrac{3}{5}$  Fundamental property

(d) $\dfrac{x^2 + 2x}{x + 1} \cdot \dfrac{x^2 - 1}{x^3 + x^2}$

$= \dfrac{x(x + 2)}{x + 1} \cdot \dfrac{(x + 1)(x - 1)}{x^2(x + 1)}$  Factor.

$= \dfrac{(x + 2)(x - 1)}{x(x + 1)}$  Multiply; lowest terms

(e) $\dfrac{x - 6}{x^2 - 12x + 36} \cdot \dfrac{x^2 - 3x - 18}{x^2 + 7x + 12}$

$= \dfrac{x - 6}{(x - 6)^2} \cdot \dfrac{(x + 3)(x - 6)}{(x + 3)(x + 4)}$  Factor.

$= \dfrac{1}{x + 4}$  Lowest terms

*Remember to include 1 in the numerator when all other factors are eliminated.*

**NOW TRY** Exercises 71, 73, 77, and 83.

| Rational Expression | Reciprocal |
|---|---|
| $3$, or $\frac{3}{1}$ | $\frac{1}{3}$ |
| $\frac{5}{k}$ | $\frac{k}{5}$ |
| $\frac{m^2 - 9m}{2}$ | $\frac{2}{m^2 - 9m}$ |
| $\frac{0}{4}$ | undefined |

*Reciprocals have a product of 1.*

**OBJECTIVE 5** Find reciprocals of rational expressions. The rational numbers $\frac{a}{b}$ and $\frac{c}{d}$ are reciprocals of each other if they have a product of 1. The **reciprocal** of a rational expression is defined in the same way: Two rational expressions are reciprocals of each other if they have a product of 1. ***Recall that 0 has no reciprocal.*** The table shows several rational expressions and their reciprocals.

The examples in the table suggest the following procedure.

### Finding the Reciprocal

To find the reciprocal of a nonzero rational expression, invert the rational expression.

**OBJECTIVE 6** Divide rational expressions. Dividing rational expressions is like dividing rational numbers.

### Dividing Rational Expressions

To divide two rational expressions, *multiply* the first (the *dividend*) by the reciprocal of the second (the *divisor*).

**EXAMPLE 5** Dividing Rational Expressions

Divide.

(a) $\dfrac{2z}{9} \div \dfrac{5z^2}{18}$

$= \dfrac{2z}{9} \cdot \dfrac{18}{5z^2}$  Multiply by the reciprocal of the divisor.

$= \dfrac{2z}{9} \cdot \dfrac{2 \cdot 9}{5z \cdot z}$  Factor.

$= \dfrac{4}{5z}$  Multiply; lowest terms

(b) $\dfrac{m^2 pq^3}{mp^4} \div \dfrac{m^5 p^2 q}{mpq^2}$

$= \dfrac{m^2 pq^3}{mp^4} \cdot \dfrac{mpq^2}{m^5 p^2 q}$  Multiply by the reciprocal.

$= \dfrac{m^3 p^2 q^5}{m^6 p^6 q}$  Properties of exponents

$= \dfrac{q^4}{m^3 p^4}$  Properties of exponents

(c) $\dfrac{8k-16}{3k} \div \dfrac{3k-6}{4k^2}$

$= \dfrac{8k-16}{3k} \cdot \dfrac{4k^2}{3k-6}$     Multiply by the reciprocal.

$= \dfrac{8(k-2)}{3k} \cdot \dfrac{4k \cdot k}{3(k-2)}$     Factor.

$= \dfrac{32k}{9}$     Multiply; lowest terms

(d) $\dfrac{5m^2+17m-12}{3m^2+7m-20} \div \dfrac{5m^2+2m-3}{15m^2-34m+15}$

$= \dfrac{5m^2+17m-12}{3m^2+7m-20} \cdot \dfrac{15m^2-34m+15}{5m^2+2m-3}$     Definition of division

$= \dfrac{(5m-3)(m+4)}{(m+4)(3m-5)} \cdot \dfrac{(3m-5)(5m-3)}{(5m-3)(m+1)}$     Factor.

$= \dfrac{5m-3}{m+1}$     Lowest terms

**NOW TRY** Exercises 63, 79, and 91.

## 7.1 Exercises

**NOW TRY Exercise**

*Matching* Rational expressions often can be written in lowest terms in seemingly different ways. For example,

$$\dfrac{y-3}{-5} \quad \text{and} \quad \dfrac{-y+3}{5}$$

look different, but we get the second quotient by multiplying the first by $-1$ in both the numerator and denominator. To practice recognizing equivalent rational expressions, match the expressions in Exercises 1–6 with their equivalents in choices A–F.

1. $\dfrac{x-3}{x+4}$    2. $\dfrac{x+3}{x-4}$    3. $\dfrac{x-3}{x-4}$    4. $\dfrac{x+3}{x+4}$    5. $\dfrac{3-x}{x+4}$    6. $\dfrac{x+3}{4-x}$

A. $\dfrac{-x-3}{4-x}$    B. $\dfrac{-x-3}{-x-4}$    C. $\dfrac{3-x}{-x-4}$    D. $\dfrac{-x+3}{-x+4}$    E. $\dfrac{x-3}{-x-4}$    F. $\dfrac{-x-3}{x-4}$

7. In Example 1(a), we showed that the domain of the rational function defined by $f(x) = \dfrac{3}{7x-14}$ does not include 2. Explain in your own words why this is so. In general, how do we find the value or values excluded from the domain of a rational function?

8. The domain of the rational function defined by $g(x) = \dfrac{x+1}{x^2+3}$ includes all real numbers. Explain.

*For each function, find all numbers that are not in the domain. Then give the domain in set notation. See Example 1.*

9. $f(x) = \dfrac{x}{x - 7}$

10. $f(x) = \dfrac{x}{x + 3}$

11. $f(x) = \dfrac{6x - 5}{7x + 1}$

12. $f(x) = \dfrac{8x - 3}{2x + 7}$

13. $f(x) = \dfrac{12x + 3}{x}$

14. $f(x) = \dfrac{9x + 8}{x}$

15. $f(x) = \dfrac{3x + 1}{2x^2 + x - 6}$

16. $f(x) = \dfrac{2x + 4}{3x^2 + 11x - 42}$

17. $f(x) = \dfrac{x + 2}{14}$

18. $f(x) = \dfrac{x - 9}{26}$

19. $f(x) = \dfrac{2x^2 - 3x + 4}{3x^2 + 8}$

20. $f(x) = \dfrac{9x^2 - 8x + 3}{4x^2 + 1}$

21. (a) *Concept Check* Identify the two *terms* in the numerator and the two *terms* in the denominator of the rational expression $\dfrac{x^2 + 4x}{x + 4}$.

(b) Describe the steps you would use to write this rational expression in lowest terms. (*Hint:* It simplifies to $x$.)

*Multiple Choice  Answer each question.*

22. Which one of the following rational expressions can be simplified?

   A. $\dfrac{x^2 + 2}{x^2}$   B. $\dfrac{x^2 + 2}{2}$   C. $\dfrac{x^2 + y^2}{y^2}$   D. $\dfrac{x^2 - 5x}{x}$

23. Which one of the following rational expressions is *not* equivalent to $\dfrac{x - 3}{4 - x}$?

   A. $\dfrac{3 - x}{x - 4}$   B. $\dfrac{x + 3}{4 + x}$   C. $-\dfrac{3 - x}{4 - x}$   D. $-\dfrac{x - 3}{x - 4}$

24. Which two of the following rational expressions equal $-1$?

   A. $\dfrac{2x + 3}{2x - 3}$   B. $\dfrac{2x - 3}{3 - 2x}$   C. $\dfrac{2x + 3}{3 + 2x}$   D. $\dfrac{2x + 3}{-2x - 3}$

*Write each rational expression in lowest terms. See Example 2.*

25. $\dfrac{x^2(x + 1)}{x(x + 1)}$

26. $\dfrac{y^3(y - 4)}{y^2(y - 4)}$

27. $\dfrac{(x + 4)(x - 3)}{(x + 5)(x + 4)}$

28. $\dfrac{(2x + 7)(x - 1)}{(2x + 3)(2x + 7)}$

29. $\dfrac{4x(x + 3)}{8x^2(x - 3)}$

30. $\dfrac{5y^2(y + 8)}{15y(y - 8)}$

31. $\dfrac{3x + 7}{3}$

32. $\dfrac{4x - 9}{4}$

33. $\dfrac{6m + 18}{7m + 21}$

34. $\dfrac{5r - 20}{3r - 12}$

35. $\dfrac{t^2 - 9}{3t + 9}$

36. $\dfrac{m^2 - 25}{4m - 20}$

37. $\dfrac{2t + 6}{t^2 - 9}$

38. $\dfrac{5s - 25}{s^2 - 25}$

39. $\dfrac{x^2 + 2x - 15}{x^2 + 6x + 5}$

40. $\dfrac{y^2 - 5y - 14}{y^2 + y - 2}$

41. $\dfrac{8x^2 - 10x - 3}{8x^2 - 6x - 9}$

42. $\dfrac{12x^2 - 4x - 5}{8x^2 - 6x - 5}$

43. $\dfrac{a^3 + b^3}{a + b}$

44. $\dfrac{r^3 - s^3}{r - s}$

45. $\dfrac{2c^2 + 2cd - 60d^2}{2c^2 - 12cd + 10d^2}$

46. $\dfrac{3s^2 - 9st - 54t^2}{3s^2 - 6st - 72t^2}$

47. $\dfrac{ac - ad + bc - bd}{ac - ad - bc + bd}$

48. $\dfrac{2xy + 2xw + y + w}{2xy + y - 2xw - w}$

*Write each rational expression in lowest terms. See Example 3.*

49. $\dfrac{7-b}{b-7}$

50. $\dfrac{r-13}{13-r}$

51. $\dfrac{x^2-y^2}{y-x}$

52. $\dfrac{m^2-n^2}{n-m}$

53. $\dfrac{(a-3)(x+y)}{(3-a)(x-y)}$

54. $\dfrac{(8-p)(x+2)}{(p-8)(x-2)}$

55. $\dfrac{5k-10}{20-10k}$

56. $\dfrac{7x-21}{63-21x}$

57. $\dfrac{a^2-b^2}{a^2+b^2}$

58. $\dfrac{p^2+q^2}{p^2-q^2}$

✎ 59. Explain in a few words how to multiply rational expressions. Give an example.

✎ 60. Explain in a few words how to divide rational expressions. Give an example.

*Multiply or divide as indicated. See Examples 4 and 5.*

61. $\dfrac{x^3}{3y} \cdot \dfrac{9y^2}{x^5}$

62. $\dfrac{a^4}{5b^2} \cdot \dfrac{25b^4}{a^3}$

63. $\dfrac{5a^4b^2}{16a^2b} \div \dfrac{25a^2b}{60a^3b^2}$

64. $\dfrac{s^3t^2}{10s^2t^4} \div \dfrac{8s^4t^2}{5t^6}$

65. $\dfrac{(-3mn)^2 \cdot 64(m^2n)^3}{16m^2n^4(mn^2)^3} \div \dfrac{24(m^2n^2)^4}{(3m^2n^3)^2}$

66. $\dfrac{(-4a^2b^3)^2 \cdot 9(a^2b^4)^2}{(2a^2b^3)^4 \cdot (3a^3b)^2} \div \dfrac{(ab)^4}{(a^2b^3)^2}$

67. $\dfrac{(x+2)(x+1)}{(x+3)(x-2)} \cdot \dfrac{(x+3)(x+4)}{(x+2)(x+1)}$

68. $\dfrac{(x+3)(x-4)}{(x-4)(x+2)} \cdot \dfrac{(x+5)(x-6)}{(x+3)(x-6)}$

69. $\dfrac{(2x+3)(x-4)}{(x+8)(x-4)} \div \dfrac{(x-4)(x+2)}{(x-4)(x+8)}$

70. $\dfrac{(6x+5)(x-3)}{(x+9)(x-1)} \div \dfrac{(x-3)(2x+7)}{(x-1)(x+9)}$

71. $\dfrac{4x}{8x+4} \cdot \dfrac{14x+7}{6}$

72. $\dfrac{12x-20}{5x} \cdot \dfrac{6}{9x-15}$

73. $\dfrac{p^2-25}{4p} \cdot \dfrac{2}{5-p}$

74. $\dfrac{a^2-1}{4a} \cdot \dfrac{2}{1-a}$

75. $(7k+7) \div \dfrac{4k+4}{5}$

76. $(8y-16) \div \dfrac{3y-6}{10}$

77. $(z^2-1) \cdot \dfrac{1}{1-z}$

78. $(y^2-4) \div \dfrac{2-y}{8y}$

79. $\dfrac{4x-20}{5x} \div \dfrac{2x-10}{7x^3}$

80. $\dfrac{k^2-4}{3k^2} \div \dfrac{2-k}{11k}$

81. $\dfrac{12x-10y}{3x+2y} \cdot \dfrac{6x+4y}{10y-12x}$

82. $\dfrac{9s-12t}{2s+2t} \cdot \dfrac{3s+3t}{4t-3s}$

83. $\dfrac{x^2-25}{x^2+x-20} \cdot \dfrac{x^2+7x+12}{x^2-2x-15}$

84. $\dfrac{t^2-49}{t^2+4t-21} \cdot \dfrac{t^2+8t+15}{t^2-2t-35}$

85. $\dfrac{a^3-b^3}{a^2-b^2} \div \dfrac{2a-2b}{2a+2b}$

86. $\dfrac{x^3+y^3}{2x+2y} \div \dfrac{x^2-y^2}{2x-2y}$

87. $\dfrac{8x^3-27}{2x^2-18} \cdot \dfrac{2x+6}{8x^2+12x+18}$

88. $\dfrac{64x^3+1}{4x^2-100} \cdot \dfrac{4x+20}{64x^2-16x+4}$

89. $\dfrac{a^3-8b^3}{a^2-ab-6b^2} \cdot \dfrac{a^2+ab-12b^2}{a^2+2ab-8b^2}$

90. $\dfrac{p^3-27q^3}{p^2+pq-12q^2} \cdot \dfrac{p^2-2pq-24q^2}{p^2-5pq-6q^2}$

91. $\dfrac{6x^2+5x-6}{12x^2-11x+2} \div \dfrac{4x^2-12x+9}{8x^2-14x+3}$

92. $\dfrac{8a^2-6a-9}{6a^2-5a-6} \div \dfrac{4a^2+11a+6}{9a^2+12a+4}$

# 7.2 Adding and Subtracting Rational Expressions

**OBJECTIVES**

1. Add and subtract rational expressions with the same denominator.
2. Find a least common denominator.
3. Add and subtract rational expressions with different denominators.

**OBJECTIVE 1** Add and subtract rational expressions with the same denominator. The following steps, used to add or subtract rational numbers, are also used to add or subtract rational expressions.

### Adding or Subtracting Rational Expressions

**Step 1** **If the denominators are the same,** add or subtract the numerators. Place the result over the common denominator.

**If the denominators are different,** first find the least common denominator. Write all rational expressions with this least common denominator, and then add or subtract the numerators. Place the result over the common denominator.

**Step 2** **Simplify.** Write all answers in lowest terms.

**EXAMPLE 1** Adding and Subtracting Rational Expressions with the Same Denominators

Add or subtract as indicated.

**(a)** $\dfrac{3y}{5} + \dfrac{x}{5}$

$= \dfrac{3y + x}{5}$ ← Add the numerators.
  ← Keep the common denominator.

**(b)** $\dfrac{7}{2r^2} - \dfrac{11}{2r^2}$

$= \dfrac{7 - 11}{2r^2}$   Subtract the numerators; keep the common denominator.

$= \dfrac{-4}{2r^2}$

$= -\dfrac{2}{r^2}$   Lowest terms

**(c)** $\dfrac{m}{m^2 - p^2} + \dfrac{p}{m^2 - p^2}$

$= \dfrac{m + p}{m^2 - p^2}$   Add the numerators; keep the common denominator.

$= \dfrac{m + p}{(m + p)(m - p)}$   Factor.

$= \dfrac{1}{m - p}$   Lowest terms

Remember to write 1 in the numerator.

(d) $\dfrac{4}{x^2 + 2x - 8} + \dfrac{x}{x^2 + 2x - 8}$

$= \dfrac{4 + x}{x^2 + 2x - 8}$   Add.

$= \dfrac{4 + x}{(x - 2)(x + 4)}$   Factor.

$= \dfrac{1}{x - 2}$   Lowest terms

**NOW TRY** Exercises 3, 7, 11, and 13.

**OBJECTIVE 2** Find a least common denominator. We add or subtract rational expressions with different denominators by first writing them with a common denominator, usually the **least common denominator (LCD).**

---

**Finding the Least Common Denominator**

*Step 1* **Factor** each denominator.

*Step 2* **Find the least common denominator.** The LCD is the product of all of the different factors from each denominator, with each factor raised to the *greatest* power that occurs in any denominator.

---

**EXAMPLE 2** Finding Least Common Denominators

Assume that the given expressions are denominators of fractions. Find the LCD for each group.

(a) $5xy^2$, $2x^3y$

Each denominator is already factored.

$$5xy^2 = 5 \cdot x \cdot y^2$$
$$2x^3y = 2 \cdot x^3 \cdot y$$

LCD $= 5 \cdot 2 \cdot x^3 \cdot y^2$   ← Greatest exponent on $x$ is 3.
                                       ← Greatest exponent on $y$ is 2.

$= 10x^3y^2$

(b) $k - 3$, $k$

Each denominator is already factored. The LCD, an expression divisible by *both* $k - 3$ and $k$, is

Don't forget $k$.   $k(k - 3)$.

It is usually best to leave a least common denominator in factored form.

(c) $y^2 - 2y - 8$, $y^2 + 3y + 2$

$y^2 - 2y - 8 = (y - 4)(y + 2)$ ⎫
$y^2 + 3y + 2 = (y + 2)(y + 1)$ ⎬ Factor.

The LCD, divisible by both polynomials, is $(y - 4)(y + 2)(y + 1)$.

**(d)** $8z - 24$, $5z^2 - 15z$

$$\left. \begin{array}{l} 8z - 24 = 8(z - 3) \\ 5z^2 - 15z = 5z(z - 3) \end{array} \right\} \text{Factor.}$$

The LCD is $8 \cdot 5z \cdot (z - 3) = 40z(z - 3)$.

**(e)** $m^2 + 5m + 6$, $m^2 + 4m + 4$, $2m^2 + 4m - 6$

$$\left. \begin{array}{l} m^2 + 5m + 6 = (m + 3)(m + 2) \\ m^2 + 4m + 4 = (m + 2)^2 \\ 2m^2 + 4m - 6 = 2(m + 3)(m - 1) \end{array} \right\} \text{Factor.}$$

The LCD is $2(m + 3)(m + 2)^2(m - 1)$.

**NOW TRY** Exercises 19, 21, 31, and 35.

**OBJECTIVE 3** Add and subtract rational expressions with different denominators. Before adding or subtracting two rational expressions, we write each expression with the least common denominator by multiplying its numerator and denominator by the factors needed to get the LCD. This procedure is valid because we are multiplying each rational expression by a form of 1, the identity element for multiplication.

Consider the sum $\frac{7}{15} + \frac{5}{12}$. The LCD for 15 and 12 is 60. Multiply $\frac{7}{15}$ by $\frac{4}{4}$ (a form of 1) and multiply $\frac{5}{12}$ by $\frac{5}{5}$, so that each fraction has denominator 60, and then add the numerators.

$$\frac{7}{15} + \frac{5}{12} = \frac{7 \cdot 4}{15 \cdot 4} + \frac{5 \cdot 5}{12 \cdot 5} \quad \text{Fundamental property}$$

$$= \frac{28}{60} + \frac{25}{60}$$

$$= \frac{28 + 25}{60} \quad \text{Add the numerators; keep the common denominator.}$$

$$= \frac{53}{60}$$

**EXAMPLE 3** Adding and Subtracting Rational Expressions with Different Denominators

Add or subtract as indicated.

**(a)** $\dfrac{5}{2p} + \dfrac{3}{8p}$    The LCD for $2p$ and $8p$ is $8p$.

$$= \frac{5 \cdot 4}{2p \cdot 4} + \frac{3}{8p} \quad \text{Fundamental property}$$

$$= \frac{20}{8p} + \frac{3}{8p}$$

$$= \frac{20 + 3}{8p} \quad \text{Add the numerators; keep the common denominator.}$$

$$= \frac{23}{8p}$$

**(b)** $\dfrac{6}{r} - \dfrac{5}{r-3}$  The LCD is $r(r-3)$.

$= \dfrac{6(r-3)}{r(r-3)} - \dfrac{r \cdot 5}{r(r-3)}$  Fundamental property

$= \dfrac{6r - 18}{r(r-3)} - \dfrac{5r}{r(r-3)}$  Distributive and commutative properties

$= \dfrac{6r - 18 - 5r}{r(r-3)}$  Subtract numerators.

$= \dfrac{r - 18}{r(r-3)}$  Combine terms in the numerator.

**NOW TRY** Exercises 39 and 49.

---

**CAUTION** One of the most common sign errors in algebra occurs when one is subtracting a rational expression with two or more terms in the numerator. Remember that in this situation *the subtraction sign must be distributed to every term in the numerator of the fraction that follows it.* Study Example 4 carefully to see how this is done.

---

**EXAMPLE 4** Using the Distributive Property When Subtracting Rational Expressions

Subtract.

**(a)** $\dfrac{7x}{3x+1} - \dfrac{x-2}{3x+1}$

The denominators are the same for both rational expressions. The subtraction sign must be applied to *both* terms in the numerator of the second rational expression.

$$\dfrac{7x}{3x+1} - \dfrac{x-2}{3x+1}$$

*Use parentheses to avoid errors.*

$= \dfrac{7x - (x - 2)}{3x + 1}$  Subtract the numerators; keep the common denominator.

*Be careful with signs.*

$= \dfrac{7x - x + 2}{3x + 1}$  Distributive property

$= \dfrac{6x + 2}{3x + 1}$  Combine terms in the numerator.

$= \dfrac{2(3x + 1)}{3x + 1}$  Factor the numerator.

$= 2$  Lowest terms

**(b)** $\dfrac{1}{q-1} - \dfrac{1}{q+1}$

$= \dfrac{1(q+1)}{(q-1)(q+1)} - \dfrac{1(q-1)}{(q+1)(q-1)}$  The LCD is $(q-1)(q+1)$; fundamental property

$= \dfrac{(q+1) - (q-1)}{(q-1)(q+1)}$  Subtract.

*Be careful with signs.*

$= \dfrac{q+1-q+1}{(q-1)(q+1)}$  Distributive property

$= \dfrac{2}{(q-1)(q+1)}$  Combine terms in the numerator.

**NOW TRY** Exercises 53 and 59.

In some problems, rational expressions to be added or subtracted have denominators that are opposites of each other, such as

$$\dfrac{y}{y-2} + \dfrac{8}{2-y}.$$  Denominators are opposites.

The next example illustrates how to proceed in such a problem.

**EXAMPLE 5** Adding Rational Expressions with Denominators That Are Opposites

Add.

$\dfrac{y}{y-2} + \dfrac{8}{2-y}$

$= \dfrac{y}{y-2} + \dfrac{8(-1)}{(2-y)(-1)}$  Multiply the second expression by $\dfrac{-1}{-1}$.

$= \dfrac{y}{y-2} + \dfrac{-8}{y-2}$  The LCD is $y-2$.

$= \dfrac{y-8}{y-2}$  Add the numerators.

We could use $2-y$ as the common denominator and rewrite the first expression.

$\dfrac{y}{y-2} + \dfrac{8}{2-y}$

$= \dfrac{y(-1)}{(y-2)(-1)} + \dfrac{8}{2-y}$

$= \dfrac{-y+8}{2-y},$ or $\dfrac{8-y}{2-y}$  This is an equivalent form of the answer.

**NOW TRY** Exercise 55.

**EXAMPLE 6** Adding and Subtracting Three Rational Expressions

Add and subtract as indicated.

$$\frac{3}{x-2} + \frac{5}{x} - \frac{6}{x^2-2x}$$

$$= \frac{3}{x-2} + \frac{5}{x} - \frac{6}{x(x-2)} \qquad \text{Factor the third denominator.}$$

$$= \frac{3x}{x(x-2)} + \frac{5(x-2)}{x(x-2)} - \frac{6}{x(x-2)} \qquad \text{The LCD is } x(x-2);\text{ fundamental property}$$

$$= \frac{3x + 5(x-2) - 6}{x(x-2)} \qquad \text{Add and subtract the numerators.}$$

$$= \frac{3x + 5x - 10 - 6}{x(x-2)} \qquad \text{Distributive property}$$

$$= \frac{8x - 16}{x(x-2)} \qquad \text{Combine terms in the numerator.}$$

$$= \frac{8(x-2)}{x(x-2)} \qquad \text{Factor the numerator.}$$

$$= \frac{8}{x} \qquad \text{Lowest terms}$$

**NOW TRY** Exercise 61.

**EXAMPLE 7** Subtracting Rational Expressions

Subtract.

$$\frac{m+4}{m^2 - 2m - 3} - \frac{2m-3}{m^2 - 5m + 6}$$

$$= \frac{m+4}{(m-3)(m+1)} - \frac{2m-3}{(m-3)(m-2)} \qquad \text{Factor each denominator.}$$

$$= \frac{(m+4)(m-2)}{(m-3)(m+1)(m-2)} - \frac{(2m-3)(m+1)}{(m-3)(m-2)(m+1)} \qquad \text{The LCD is } (m-3) \cdot (m+1)(m-2).$$

$$= \frac{(m+4)(m-2) - (2m-3)(m+1)}{(m-3)(m+1)(m-2)} \qquad \text{Subtract.}$$

*Note the careful use of parentheses.*

$$= \frac{m^2 + 2m - 8 - (2m^2 - m - 3)}{(m-3)(m+1)(m-2)} \qquad \text{Multiply in the numerator.}$$

*Be careful with signs.*

$$= \frac{m^2 + 2m - 8 - 2m^2 + m + 3}{(m-3)(m+1)(m-2)} \qquad \text{Distributive property}$$

$$= \frac{-m^2 + 3m - 5}{(m-3)(m+1)(m-2)} \qquad \text{Combine terms in the numerator.}$$

If we try to factor the numerator, we find that this rational expression is in lowest terms.

**NOW TRY** Exercise 75.

### EXAMPLE 8 Adding Rational Expressions

Add.

$$\frac{5}{x^2 + 10x + 25} + \frac{2}{x^2 + 7x + 10}$$

$$= \frac{5}{(x + 5)^2} + \frac{2}{(x + 5)(x + 2)} \qquad \text{Factor each denominator.}$$

$$= \frac{5(x + 2)}{(x + 5)^2(x + 2)} + \frac{2(x + 5)}{(x + 5)^2(x + 2)} \qquad \text{The LCD is } (x + 5)^2(x + 2); \text{ fundamental property}$$

$$= \frac{5(x + 2) + 2(x + 5)}{(x + 5)^2(x + 2)} \qquad \text{Add.}$$

$$= \frac{5x + 10 + 2x + 10}{(x + 5)^2(x + 2)} \qquad \text{Distributive property}$$

$$= \frac{7x + 20}{(x + 5)^2(x + 2)} \qquad \text{Combine terms in the numerator.}$$

**NOW TRY** Exercise 83.

## 7.2 Exercises

*Add or subtract as indicated. Write all answers in lowest terms. See Example 1.*

1. $\dfrac{7}{t} + \dfrac{2}{t}$

2. $\dfrac{5}{r} + \dfrac{9}{r}$

3. $\dfrac{6x}{7} + \dfrac{y}{7}$

4. $\dfrac{12t}{5} + \dfrac{s}{5}$

5. $\dfrac{11}{5x} - \dfrac{1}{5x}$

6. $\dfrac{7}{4y} - \dfrac{3}{4y}$

7. $\dfrac{9}{4x^3} - \dfrac{17}{4x^3}$

8. $\dfrac{6}{5y^4} - \dfrac{21}{5y^4}$

9. $\dfrac{5x + 4}{6x + 5} + \dfrac{x + 1}{6x + 5}$

10. $\dfrac{6y + 12}{4y + 3} + \dfrac{2y - 6}{4y + 3}$

11. $\dfrac{x^2}{x + 5} - \dfrac{25}{x + 5}$

12. $\dfrac{y^2}{y + 6} - \dfrac{36}{y + 6}$

13. $\dfrac{-3p + 7}{p^2 + 7p + 12} + \dfrac{8p + 13}{p^2 + 7p + 12}$

14. $\dfrac{5x + 6}{x^2 + x - 20} + \dfrac{4 - 3x}{x^2 + x - 20}$

15. $\dfrac{a^3}{a^2 + ab + b^2} - \dfrac{b^3}{a^2 + ab + b^2}$

16. $\dfrac{p^3}{p^2 - pq + q^2} + \dfrac{q^3}{p^2 - pq + q^2}$

✍ 17. Write a step-by-step method for adding or subtracting rational expressions that have a common denominator. Illustrate with an example.

✍ 18. Write a step-by-step method for adding or subtracting rational expressions that have different denominators. Give an example.

*Assume that the expressions given are denominators of fractions. Find the least common denominator (LCD) for each group. See Example 2.*

19. $18x^2y^3$,  $24x^4y^5$

20. $24a^3b^4$,  $18a^5b^2$

21. $z - 2$,  $z$

22. $k + 3$,  $k$

23. $2y + 8$,  $y + 4$

24. $3r - 21$,  $r - 7$

25. $x^2 - 81$,  $x^2 + 18x + 81$

26. $y^2 - 16$,  $y^2 - 8y + 16$

27. $m + n$,  $m - n$,  $m^2 - n^2$

28. $r + s$,  $r - s$,  $r^2 - s^2$

**29.** $x^2 - 3x - 4, \quad x + x^2$

**30.** $y^2 - 8y + 12, \quad y^2 - 6y$

**31.** $2t^2 + 7t - 15, \quad t^2 + 3t - 10$

**32.** $s^2 - 3s - 4, \quad 3s^2 + s - 2$

**33.** $2y + 6, \quad y^2 - 9, \quad y$

**34.** $9x + 18, \quad x^2 - 4, \quad x$

**35.** $2x - 6, \quad x^2 - x - 6, \quad x^2 + 4x + 4$

**36.** $3a - 3b, \quad a^2 + ab - 2b^2, \quad a^2 - 2ab + b^2$

**37.** *Concept Check* Consider the following *incorrect* work.

$$\frac{x}{x+2} - \frac{4x-1}{x+2}$$
$$= \frac{x - 4x - 1}{x+2}$$
$$= \frac{-3x - 1}{x+2}$$

**WHAT WENT WRONG?** Work the problem correctly.

**38.** One student added two rational expressions and obtained the answer $\frac{3}{5-y}$. Another student obtained the answer $\frac{-3}{y-5}$ for the same problem. Is it possible that both answers are correct? Explain.

*Add or subtract as indicated. Write all answers in lowest terms. See Examples 3–6.*

**39.** $\dfrac{8}{t} + \dfrac{7}{3t}$

**40.** $\dfrac{5}{x} + \dfrac{9}{4x}$

**41.** $\dfrac{5}{12x^2y} - \dfrac{11}{6xy}$

**42.** $\dfrac{7}{18a^3b^2} - \dfrac{2}{9ab}$

**43.** $\dfrac{4}{15a^4b^5} + \dfrac{3}{20a^2b^6}$

**44.** $\dfrac{5}{12x^5y^2} + \dfrac{5}{18x^4y^5}$

**45.** $\dfrac{2r}{7p^3q^4} + \dfrac{3s}{14p^4q}$

**46.** $\dfrac{4t}{9a^8b^7} + \dfrac{5s}{27a^4b^3}$

**47.** $\dfrac{1}{a^3b^2} - \dfrac{2}{a^4b} + \dfrac{3}{a^5b^7}$

**48.** $\dfrac{5}{t^4u^7} - \dfrac{3}{t^5u^9} + \dfrac{6}{t^{10}u}$

**49.** $\dfrac{1}{x-1} - \dfrac{1}{x}$

**50.** $\dfrac{3}{x-3} - \dfrac{1}{x}$

**51.** $\dfrac{3a}{a+1} + \dfrac{2a}{a-3}$

**52.** $\dfrac{2x}{x+4} + \dfrac{3x}{x-7}$

**53.** $\dfrac{17y+3}{9y+7} - \dfrac{-10y-18}{9y+7}$

**54.** $\dfrac{7x+8}{3x+2} - \dfrac{x+4}{3x+2}$

**55.** $\dfrac{2}{4-x} + \dfrac{5}{x-4}$

**56.** $\dfrac{3}{2-t} + \dfrac{1}{t-2}$

**57.** $\dfrac{w}{w-z} - \dfrac{z}{z-w}$

**58.** $\dfrac{a}{a-b} - \dfrac{b}{b-a}$

**59.** $\dfrac{1}{x+1} - \dfrac{1}{x-1}$

**60.** $\dfrac{-2}{x-1} + \dfrac{2}{x+1}$

**61.** $\dfrac{4x}{x-1} - \dfrac{2}{x+1} - \dfrac{4}{x^2-1}$

**62.** $\dfrac{4}{x+3} - \dfrac{x}{x-3} - \dfrac{18}{x^2-9}$

**63.** $\dfrac{15}{y^2+3y} + \dfrac{2}{y} + \dfrac{5}{y+3}$

**64.** $\dfrac{7}{t-2} - \dfrac{6}{t^2-2t} - \dfrac{3}{t}$

**65.** $\dfrac{5}{x-2} + \dfrac{1}{x} + \dfrac{2}{x^2-2x}$

**66.** $\dfrac{5x}{x-3} + \dfrac{2}{x} + \dfrac{6}{x^2-3x}$

**67.** $\dfrac{3x}{x+1} + \dfrac{4}{x-1} - \dfrac{6}{x^2-1}$

**68.** $\dfrac{5x}{x+3} + \dfrac{x+2}{x} - \dfrac{6}{x^2+3x}$

**69.** $\dfrac{4}{x+1} + \dfrac{1}{x^2-x+1} - \dfrac{12}{x^3+1}$

**70.** $\dfrac{5}{x+2} + \dfrac{2}{x^2-2x+4} - \dfrac{60}{x^3+8}$

71. $\dfrac{2x+4}{x+3} + \dfrac{3}{x} - \dfrac{6}{x^2+3x}$

72. $\dfrac{4x+1}{x+5} - \dfrac{2}{x} + \dfrac{10}{x^2+5x}$

73. $\dfrac{3}{(p-2)^2} - \dfrac{5}{p-2} + 4$

74. $\dfrac{8}{(3r-1)^2} + \dfrac{2}{3r-1} - 6$

*Add or subtract as indicated. Write all answers in lowest terms. See Examples 7 and 8.* \*

75. $\dfrac{3}{x^2-5x+6} - \dfrac{2}{x^2-4x+4}$

76. $\dfrac{2}{m^2-4m+4} + \dfrac{3}{m^2+m-6}$

77. $\dfrac{5x}{x^2+xy-2y^2} - \dfrac{3x}{x^2+5xy-6y^2}$

78. $\dfrac{6x}{6x^2+5xy-4y^2} - \dfrac{2y}{9x^2-16y^2}$

79. $\dfrac{5x-y}{x^2+xy-2y^2} - \dfrac{3x+2y}{x^2+5xy-6y^2}$

80. $\dfrac{6x+5y}{6x^2+5xy-4y^2} - \dfrac{x+2y}{9x^2-16y^2}$

81. $\dfrac{r+s}{3r^2+2rs-s^2} - \dfrac{s-r}{6r^2-5rs+s^2}$

82. $\dfrac{3y}{y^2+yz-2z^2} + \dfrac{4y-1}{y^2-z^2}$

83. $\dfrac{3}{x^2+4x+4} + \dfrac{7}{x^2+5x+6}$

84. $\dfrac{5}{x^2+6x+9} - \dfrac{2}{x^2+4x+3}$

*Work each problem.*

85. A **Concours D'elegance** is a competition in which a maximum of 100 points is awarded to a car on the basis of its general attractiveness. The function defined by the rational expression

$$c(x) = \dfrac{1010}{49(101-x)} - \dfrac{10}{49}$$

approximates the cost, in thousands of dollars, of restoring a car so that it will win $x$ points.

(a) Simplify the expression for $c(x)$ by performing the indicated subtraction.

(b) Use the simplified expression to determine how much it would cost to win 95 points.

86. A **cost–benefit model** expresses the cost of an undertaking in terms of the benefits received. One cost–benefit model gives the cost in thousands of dollars to remove $x$ percent of a certain pollutant as

$$c(x) = \dfrac{6.7x}{100-x}.$$

Another model produces the relationship

$$c(x) = \dfrac{6.5x}{102-x}.$$

(a) What is the cost found by averaging the two models? (*Hint:* The average of two quantities is half their sum.)

(b) Using the two given models and your answer to part (a), find the cost to the nearest dollar to remove 95% ($x = 95$) of the pollutant.

(c) Average the two costs in part (b) from the given models. What do you notice about this result compared with the cost obtained by using the average of the two models?

---

\*The authors wish to thank Joyce Nemeth of Broward Community College for her suggestions regarding some of these exercises.

## 7.3 Complex Fractions

**OBJECTIVES**

1. Simplify complex fractions by simplifying the numerator and denominator (Method 1).
2. Simplify complex fractions by multiplying by a common denominator (Method 2).
3. Compare the two methods of simplifying complex fractions.
4. Simplify rational expressions with negative exponents.

A **complex fraction** is an expression having a fraction in the numerator, denominator, or both. Examples of complex fractions include

$$\frac{1 + \frac{1}{x}}{2}, \quad \frac{\frac{4}{y}}{6 - \frac{3}{y}}, \quad \text{and} \quad \frac{\frac{m^2 - 9}{m + 1}}{\frac{m + 3}{m^2 - 1}}.$$  Complex fractions

**OBJECTIVE 1** Simplify complex fractions by simplifying the numerator and denominator (Method 1). This is the first of two different methods for simplifying complex fractions.

### Simplifying a Complex Fraction: Method 1

*Step 1* Simplify the numerator and denominator separately.

*Step 2* Divide by multiplying the numerator by the reciprocal of the denominator.

*Step 3* Simplify the resulting fraction if possible.

In Step 2, we are treating the complex fraction as a quotient of two rational expressions and dividing. Before performing this step, be sure that both numerator and denominator are single fractions.

### EXAMPLE 1  Simplifying Complex Fractions by Method 1

Use Method 1 to simplify each complex fraction.

(a) $\dfrac{\frac{x + 1}{x}}{\frac{x - 1}{2x}}$   Both the numerator and the denominator are already simplified. (Step 1)

$= \dfrac{x + 1}{x} \div \dfrac{x - 1}{2x}$   Write as a division problem.

$= \dfrac{x + 1}{x} \cdot \dfrac{2x}{x - 1}$   Multiply by the reciprocal of $\frac{x - 1}{2x}$. (Step 2)

$= \dfrac{2x(x + 1)}{x(x - 1)}$   Multiply.

$= \dfrac{2(x + 1)}{x - 1}$   Simplify. (Step 3)

**(b)** $\dfrac{2 + \dfrac{1}{y}}{3 - \dfrac{2}{y}}$  Simplify the numerator and denominator. (Step 1)

$= \dfrac{\dfrac{2y}{y} + \dfrac{1}{y}}{\dfrac{3y}{y} - \dfrac{2}{y}}$

$= \dfrac{\dfrac{2y + 1}{y}}{\dfrac{3y - 2}{y}}$

$= \dfrac{2y + 1}{y} \div \dfrac{3y - 2}{y}$  Write as a division problem.

$= \dfrac{2y + 1}{y} \cdot \dfrac{y}{3y - 2}$  Multiply by the reciprocal of $\dfrac{3y-2}{y}$. (Step 2)

$= \dfrac{2y + 1}{3y - 2}$  Multiply and simplify. (Step 3)

**NOW TRY** Exercises 5 and 9.

**OBJECTIVE 2** Simplify complex fractions by multiplying by a common denominator (Method 2). The second method for simplifying complex fractions uses the identity property for multiplication.

### Simplifying a Complex Fraction: Method 2

**Step 1** Multiply the numerator and denominator of the complex fraction by the least common denominator of the fractions in the numerator and the fractions in the denominator of the complex fraction.

**Step 2** Simplify the resulting fraction if possible.

### EXAMPLE 2  Simplifying Complex Fractions by Method 2

Use Method 2 to simplify each complex fraction.

**(a)** $\dfrac{2 + \dfrac{1}{y}}{3 - \dfrac{2}{y}}$

$= \dfrac{2 + \dfrac{1}{y}}{3 - \dfrac{2}{y}} \cdot 1$  Identity property of multiplication

$$= \frac{\left(2 + \dfrac{1}{y}\right) \cdot y}{\left(3 - \dfrac{2}{y}\right) \cdot y}$$ 

The LCD of all the fractions is $y$; Multiply the numerator and denominator by $y$, since $\dfrac{y}{y} = 1$.

$$= \frac{2 \cdot y + \dfrac{1}{y} \cdot y}{3 \cdot y - \dfrac{2}{y} \cdot y}$$ 

Distributive property

$$= \frac{2y + 1}{3y - 2}$$ 

Multiply.

Compare this method of solution with that used in Example 1(b).

**(b)** $\dfrac{2p + \dfrac{5}{p-1}}{3p - \dfrac{2}{p}}$

$$= \frac{\left(2p + \dfrac{5}{p-1}\right) \cdot p(p-1)}{\left(3p - \dfrac{2}{p}\right) \cdot p(p-1)}$$ 

Multiply the numerator and denominator by the LCD, $p(p-1)$.

$$= \frac{2p[p(p-1)] + \dfrac{5}{p-1} \cdot p(p-1)}{3p[p(p-1)] - \dfrac{2}{p} \cdot p(p-1)}$$ 

Distributive property

$$= \frac{2p[p(p-1)] + 5p}{3p[p(p-1)] - 2(p-1)}$$ 

Multiply.

$$= \frac{2p^3 - 2p^2 + 5p}{3p^3 - 3p^2 - 2p + 2}$$ 

Multiply again.

This rational expression is in lowest terms.

**NOW TRY** Exercises 9 (using Method 2) and 23.

**OBJECTIVE 3** Compare the two methods of simplifying complex fractions. Choosing whether to use Method 1 or Method 2 to simplify a complex fraction is usually a matter of preference. Some students prefer one method over the other, while other students feel comfortable with both methods and rely on practice with many examples to determine which method they will use on a particular problem. In the next example, we illustrate how to simplify a complex fraction by both methods so that you can observe the processes and decide for yourself the pros and cons of each method.

**342** CHAPTER 7 Rational Expressions and Functions

**EXAMPLE 3** Simplifying Complex Fractions by Both Methods

Use both Method 1 and Method 2 to simplify each complex fraction.

**Method 1**

(a) $\dfrac{\dfrac{2}{x-3}}{\dfrac{5}{x^2-9}}$

$= \dfrac{\dfrac{2}{x-3}}{\dfrac{5}{(x-3)(x+3)}}$

$= \dfrac{2}{x-3} \div \dfrac{5}{(x-3)(x+3)}$

$= \dfrac{2}{x-3} \cdot \dfrac{(x-3)(x+3)}{5}$

$= \dfrac{2(x+3)}{5}$

(b) $\dfrac{\dfrac{1}{x}+\dfrac{1}{y}}{\dfrac{1}{x^2}-\dfrac{1}{y^2}}$

$= \dfrac{\dfrac{y}{xy}+\dfrac{x}{xy}}{\dfrac{y^2}{x^2y^2}-\dfrac{x^2}{x^2y^2}}$

$= \dfrac{\dfrac{y+x}{xy}}{\dfrac{y^2-x^2}{x^2y^2}}$

$= \dfrac{y+x}{xy} \div \dfrac{y^2-x^2}{x^2y^2}$

$= \dfrac{y+x}{xy} \cdot \dfrac{x^2y^2}{(y-x)(y+x)}$

$= \dfrac{xy}{y-x}$

**Method 2**

(a) $\dfrac{\dfrac{2}{x-3}}{\dfrac{5}{x^2-9}}$

$= \dfrac{\dfrac{2}{x-3} \cdot (x-3)(x+3)}{\dfrac{5}{(x-3)(x+3)} \cdot (x-3)(x+3)}$

$= \dfrac{2(x+3)}{5}$

(b) $\dfrac{\dfrac{1}{x}+\dfrac{1}{y}}{\dfrac{1}{x^2}-\dfrac{1}{y^2}}$

$= \dfrac{\left(\dfrac{1}{x}+\dfrac{1}{y}\right) \cdot x^2y^2}{\left(\dfrac{1}{x^2}-\dfrac{1}{y^2}\right) \cdot x^2y^2}$

$= \dfrac{xy^2 + x^2y}{y^2 - x^2}$

$= \dfrac{xy(y+x)}{(y+x)(y-x)}$

$= \dfrac{xy}{y-x}$

**NOW TRY** Exercises 17 and 19.

**SECTION 7.3** Complex Fractions **343**

**OBJECTIVE 4** Simplify rational expressions with negative exponents. Rational expressions and complex fractions sometimes involve negative exponents. To simplify such expressions, we begin by rewriting the expressions with only positive exponents.

**EXAMPLE 4** Simplifying Rational Expressions with Negative Exponents

Simplify each expression, using only positive exponents in the answer.

(a) $\dfrac{m^{-1} + p^{-2}}{2m^{-2} - p^{-1}}$

$= \dfrac{\dfrac{1}{m} + \dfrac{1}{p^2}}{\dfrac{2}{m^2} - \dfrac{1}{p}}$  Write with positive exponents.

*Be careful! The base is $m$, not $2m$; $2m^{-2} = \dfrac{2}{m^2}$.*

$= \dfrac{m^2 p^2 \left(\dfrac{1}{m} + \dfrac{1}{p^2}\right)}{m^2 p^2 \left(\dfrac{2}{m^2} - \dfrac{1}{p}\right)}$  Simplify by Method 2, multiplying the numerator and denominator by the LCD, $m^2 p^2$.

$= \dfrac{m^2 p^2 \cdot \dfrac{1}{m} + m^2 p^2 \cdot \dfrac{1}{p^2}}{m^2 p^2 \cdot \dfrac{2}{m^2} - m^2 p^2 \cdot \dfrac{1}{p}}$  Distributive property

$= \dfrac{mp^2 + m^2}{2p^2 - m^2 p}$  Lowest terms

(b) $\dfrac{x^{-2} - 2y^{-1}}{y - 2x^2}$

*The 2 does not go in the denominator of this fraction.*

$= \dfrac{\dfrac{1}{x^2} - \dfrac{2}{y}}{y - 2x^2}$  Write with positive exponents.

$= \dfrac{\left(\dfrac{1}{x^2} - \dfrac{2}{y}\right) x^2 y}{(y - 2x^2) x^2 y}$  Use Method 2; multiply by the LCD, $x^2 y$.

$= \dfrac{y - 2x^2}{(y - 2x^2) x^2 y}$  Use the distributive property in the numerator.

$= \dfrac{1}{x^2 y}$  Lowest terms

*Remember to write 1 in the numerator.*

**NOW TRY** Exercises 29 and 31.

## 7.3 Exercises

1. Explain in your own words the two methods of simplifying complex fractions.
2. Method 2 of simplifying complex fractions says that we can multiply both the numerator and the denominator of the complex fraction by the same nonzero expression. What property of real numbers from **Section 1.4** justifies this method?

*Use either method to simplify each complex fraction. See Examples 1–3.*

3. $\dfrac{\dfrac{12}{x-1}}{\dfrac{6}{x}}$

4. $\dfrac{\dfrac{24}{t+4}}{\dfrac{6}{t}}$

5. $\dfrac{\dfrac{k+1}{2k}}{\dfrac{3k-1}{4k}}$

6. $\dfrac{\dfrac{1-r}{4r}}{\dfrac{1+r}{8r}}$

7. $\dfrac{\dfrac{4z^2x^4}{9}}{\dfrac{12x^2z^5}{15}}$

8. $\dfrac{\dfrac{3y^2x^3}{8}}{\dfrac{9y^3x^4}{16}}$

9. $\dfrac{6+\dfrac{1}{x}}{7-\dfrac{3}{x}}$

10. $\dfrac{4-\dfrac{1}{p}}{9+\dfrac{5}{p}}$

11. $\dfrac{\dfrac{3}{x}+\dfrac{3}{y}}{\dfrac{3}{x}-\dfrac{3}{y}}$

12. $\dfrac{\dfrac{4}{t}-\dfrac{4}{s}}{\dfrac{4}{t}+\dfrac{4}{s}}$

13. $\dfrac{\dfrac{8x-24y}{10}}{\dfrac{x-3y}{5x}}$

14. $\dfrac{\dfrac{20x-10y}{12y}}{\dfrac{2x-y}{6y^2}}$

15. $\dfrac{\dfrac{x^2-16y^2}{xy}}{\dfrac{1}{y}-\dfrac{4}{x}}$

16. $\dfrac{\dfrac{4t^2-9s^2}{st}}{\dfrac{2}{s}-\dfrac{3}{t}}$

17. $\dfrac{\dfrac{6}{y-4}}{\dfrac{12}{y^2-16}}$

18. $\dfrac{\dfrac{8}{t+7}}{\dfrac{24}{t^2-49}}$

19. $\dfrac{\dfrac{1}{b^2}-\dfrac{1}{a^2}}{\dfrac{1}{b}-\dfrac{1}{a}}$

20. $\dfrac{\dfrac{1}{x^2}-\dfrac{1}{y^2}}{\dfrac{1}{x}+\dfrac{1}{y}}$

21. $\dfrac{x+y}{\dfrac{1}{y}+\dfrac{1}{x}}$

22. $\dfrac{s-r}{\dfrac{1}{r}-\dfrac{1}{s}}$

23. $\dfrac{y-\dfrac{y-3}{3}}{\dfrac{4}{9}+\dfrac{2}{3y}}$

24. $\dfrac{p-\dfrac{p+2}{4}}{\dfrac{3}{4}-\dfrac{5}{2p}}$

25. $\dfrac{\dfrac{x+2}{x}+\dfrac{1}{x+2}}{\dfrac{5}{x}+\dfrac{x}{x+2}}$

26. $\dfrac{\dfrac{y+3}{y}-\dfrac{4}{y-1}}{\dfrac{y}{y-1}+\dfrac{1}{y}}$

*Simplify each expression, using only positive exponents in your answer. See Example 4.*

27. $\dfrac{1}{x^{-2}+y^{-2}}$

28. $\dfrac{1}{p^{-2}-q^{-2}}$

29. $\dfrac{x^{-2}+y^{-2}}{x^{-1}+y^{-1}}$

30. $\dfrac{x^{-1}-y^{-1}}{x^{-2}-y^{-2}}$

31. $\dfrac{x^{-1}+2y^{-1}}{2y+4x}$

32. $\dfrac{a^{-2}-4b^{-2}}{3b-6a}$

# 7.4 Equations with Rational Expressions and Graphs

**OBJECTIVES**
1. Determine the domain of the variable in a rational equation.
2. Solve rational equations.
3. Recognize the graph of a rational function.

At the beginning of this chapter, we defined the domain of a rational expression as the set of all possible values of the variable. Any value that makes the denominator 0 is excluded. A **rational equation** is an equation that contains at least one rational expression with a variable in the denominator.

**OBJECTIVE 1** Determine the domain of the variable in a rational equation. The **domain of the variable in a rational equation** is the intersection (overlap) of the domains of the rational expressions in the equation.

**EXAMPLE 1** Determining the Domains of the Variables in Rational Equations

Find the domain of the variables in each equation.

(a) $\dfrac{2}{x} - \dfrac{3}{2} = \dfrac{7}{2x}$

The domains of the three rational expressions of the equation $\dfrac{2}{x} - \dfrac{3}{2} = \dfrac{7}{2x}$ are, in order, $\{x \mid x \neq 0\}$, $(-\infty, \infty)$, and $\{x \mid x \neq 0\}$. The intersection of these three domains is all real numbers except 0, which may be written $\{x \mid x \neq 0\}$.

(b) $\dfrac{2}{x-3} - \dfrac{3}{x+3} = \dfrac{12}{x^2-9}$

The domains of the three expressions are, respectively, $\{x \mid x \neq 3\}$, $\{x \mid x \neq -3\}$, and $\{x \mid x \neq \pm 3\}$. The domain of the variables in the equation is the intersection of the three domains: all real numbers except 3 and $-3$, written $\{x \mid x \neq \pm 3\}$.

**NOW TRY** Exercises 1 and 5.

**OBJECTIVE 2** Solve rational equations. The easiest way to solve most rational equations is to multiply all terms in the equation by the least common denominator. This step will clear the equation of all denominators, as the next examples show. **We can do this only with equations, not expressions.**

Because the first step in solving a rational equation is to multiply both sides of the equation by a common denominator, it is *necessary* to either check the proposed solutions or verify that they are in the domain.

**CAUTION** When both sides of an equation are multiplied by a *variable* expression, the resulting proposed solutions may not satisfy the original equation. ***You must either determine and observe the domain or check all proposed solutions in the original equation. It is wise to do both.***

**EXAMPLE 2** Solving a Rational Equation

Solve $\dfrac{2}{x} - \dfrac{3}{2} = \dfrac{7}{2x}$.

The domain, which excludes 0, was found in Example 1(a).

$$2x\left(\dfrac{2}{x} - \dfrac{3}{2}\right) = 2x\left(\dfrac{7}{2x}\right) \quad \text{Multiply by the LCD, } 2x.$$

$$2x\left(\dfrac{2}{x}\right) - 2x\left(\dfrac{3}{2}\right) = 2x\left(\dfrac{7}{2x}\right) \quad \text{Distributive property}$$

$$4 - 3x = 7 \quad \text{Multiply.}$$

$$-3x = 3 \quad \text{Subtract 4.}$$

Proposed solution → $x = -1$ $\quad$ Divide by

Check:
$$\dfrac{2}{x} - \dfrac{3}{2} = \dfrac{7}{2x} \quad \text{Original equation}$$

$$\dfrac{2}{-1} - \dfrac{3}{2} = \dfrac{7}{2(-1)} \quad ? \quad \text{Let } x = -1.$$

$$-\dfrac{7}{2} = -\dfrac{7}{2} \quad \text{True}$$

The solution set is $\{-1\}$.

**NOW TRY** Exercise 15.

**EXAMPLE 3** Solving a Rational Equation with No Solution

Solve $\dfrac{2}{x-3} - \dfrac{3}{x+3} = \dfrac{12}{x^2-9}$.

Using the result from Example 1(b), we know that the domain excludes 3 and $-3$. We multiply each side by the LCD, $(x+3)(x-3)$.

$$(x+3)(x-3)\left(\dfrac{2}{x-3} - \dfrac{3}{x+3}\right) = (x+3)(x-3)\left(\dfrac{12}{x^2-9}\right)$$

$$2(x+3) - 3(x-3) = 12 \quad \text{Distributive property}$$

$$2x + 6 - 3x + 9 = 12 \quad \text{Distributive property}$$

$$-x + 15 = 12 \quad \text{Combine like terms.}$$

$$-x = -3 \quad \text{Subtract 15.}$$

Proposed solution → $x = 3$ $\quad$ Divide by $-1$.

Since the proposed solution, 3, is not in the domain, it cannot be an actual solution of the equation. Substituting 3 into the original equation shows why.

$$\dfrac{2}{x-3} - \dfrac{3}{x+3} = \dfrac{12}{x^2-9} \quad \text{Original equation}$$

$$\dfrac{2}{3-3} - \dfrac{3}{3+3} = \dfrac{12}{3^2-9} \quad ? \quad \text{Let } x = 3.$$

$$\dfrac{2}{0} - \dfrac{3}{6} = \dfrac{12}{0} \quad ?$$

Division by 0 is undefined. The equation has no solution and the solution set is $\emptyset$.

**NOW TRY** Exercise 29.

### EXAMPLE 4 Solving a Rational Equation

Solve $\dfrac{3}{p^2 + p - 2} - \dfrac{1}{p^2 - 1} = \dfrac{7}{2(p^2 + 3p + 2)}$.

Factor each denominator to find the LCD, $2(p - 1)(p + 2)(p + 1)$. The domain excludes 1, $-2$, and $-1$. Multiply each side by the LCD.

$$2(p - 1)(p + 2)(p + 1)\left[\dfrac{3}{(p + 2)(p - 1)} - \dfrac{1}{(p + 1)(p - 1)}\right]$$
$$= 2(p - 1)(p + 2)(p + 1)\left[\dfrac{7}{2(p + 2)(p + 1)}\right]$$

| | |
|---|---|
| $2 \cdot 3(p + 1) - 2(p + 2) = 7(p - 1)$ | Distributive property |
| $6p + 6 - 2p - 4 = 7p - 7$ | Distributive property |
| $4p + 2 = 7p - 7$ | Combine like terms. |
| $9 = 3p$ | Subtract $4p$; add 7. |
| Proposed solution $\rightarrow 3 = p$ | Divide by 3. |

Note that 3 is in the domain. Substitute 3 for $p$ in the original equation to check that the solution set is $\{3\}$.

**NOW TRY** Exercise 39.

### EXAMPLE 5 Solving a Rational Equation That Leads to a Quadratic Equation

Solve $\dfrac{2}{3x + 1} = \dfrac{1}{x} - \dfrac{6x}{3x + 1}$.

Since the denominator $3x + 1$ cannot equal 0, $-\dfrac{1}{3}$ is excluded from the domain, as is 0.

| | |
|---|---|
| $x(3x + 1)\left(\dfrac{2}{3x + 1}\right) = x(3x + 1)\left(\dfrac{1}{x} - \dfrac{6x}{3x + 1}\right)$ | Multiply by the LCD, |
| $2x = 3x + 1 - 6x^2$ | Distributive property |
| $6x^2 - x - 1 = 0$ | Standard form |
| $(3x + 1)(2x - 1) = 0$ | Factor. |
| $3x + 1 = 0$ or $2x - 1 = 0$ | Zero-factor property |
| $x = -\dfrac{1}{3}$ or $x = \dfrac{1}{2}$ | Proposed solutions |

Because $-\dfrac{1}{3}$ is not in the domain of the equation, it is not a solution. Check that the solution set is $\left\{\dfrac{1}{2}\right\}$.

**NOW TRY** Exercise 35.

**OBJECTIVE 3** Recognize the graph of a rational function. As mentioned in **Section 7.1,** a function defined by a quotient of polynomials is a *rational function.* Because one or more values of $x$ are excluded from the domain of many rational functions, their graphs are often **discontinuous.** That is, there will be one or more breaks in the graph. For example, consider the graph of the simple rational function defined by

$$f(x) = \dfrac{1}{x}.$$

The domain of this function includes all real numbers except 0. Thus, there will be no point on the graph with $x = 0$. The vertical line with equation $x = 0$ is called a *vertical asymptote* of the graph. If the $y$-values of a rational function approach $\infty$ or $-\infty$ as the $x$-values approach a real number $a$, the vertical line $x = a$ is called a **vertical asymptote** of the graph of the function.

We show some typical ordered pairs for $f(x) = \frac{1}{x}$ in the table for both negative and positive $x$-values.

| $x$ | $-3$ | $-2$ | $-1$ | $-0.5$ | $-0.25$ | $-0.1$ | $0.1$ | $0.25$ | $0.5$ | $1$ | $2$ | $3$ |
|---|---|---|---|---|---|---|---|---|---|---|---|---|
| $y$ | $-\frac{1}{3}$ | $-\frac{1}{2}$ | $-1$ | $-2$ | $-4$ | $-10$ | $10$ | $4$ | $2$ | $1$ | $\frac{1}{2}$ | $\frac{1}{3}$ |

Notice that the closer positive values of $x$ are to 0, the larger $y$ is. Similarly, as negative values of $x$ get closer to 0, $y$-values are negative and get larger in absolute value. Using this observation and the fact that the domain excludes 0, and plotting some of the points just found, produces the graph in Figure 2.

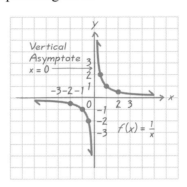

**FIGURE 2**

The graph of $g(x) = \frac{-2}{x - 3}$ is shown in Figure 3. Some ordered pairs that belong to the function are listed in the table.

| $x$ | $-2$ | $-1$ | $0$ | $1$ | $2$ | $2.5$ | $2.75$ | $3.25$ | $3.5$ | $4$ | $5$ | $6$ | $7$ |
|---|---|---|---|---|---|---|---|---|---|---|---|---|---|
| $y$ | $\frac{2}{5}$ | $\frac{1}{2}$ | $\frac{2}{3}$ | $1$ | $2$ | $4$ | $8$ | $-8$ | $-4$ | $-2$ | $-1$ | $-\frac{2}{3}$ | $-\frac{1}{2}$ |

There is no point on the graph for $x = 3$, because 3 is excluded from the domain. The dashed line $x = 3$ represents the asymptote and is not part of the graph. As suggested by the points from the table, the graph gets closer to the vertical asymptote $x = 3$ as the $x$-values get closer to 3.

**NOW TRY** Exercise 45.

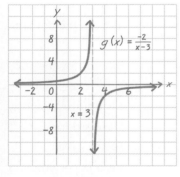

**FIGURE 3**

## 7.4 Exercises

**NOW TRY Exercise**

*As explained in this section, any values that would cause a denominator to equal 0 must be excluded from the domain and, consequently, as solutions of an equation that has variable expressions in the denominators. **(a)** Without actually solving the equation, list all possible numbers that would have to be rejected if they appeared as proposed solutions. **(b)** Then give the domain, using set notation. See Example 1.*

1. $\dfrac{1}{3x} + \dfrac{1}{2x} = \dfrac{x}{3}$

2. $\dfrac{5}{6x} - \dfrac{8}{2x} = \dfrac{x}{4}$

3. $\dfrac{1}{x + 1} - \dfrac{1}{x - 2} = 0$

4. $\dfrac{3}{x + 4} - \dfrac{2}{x - 9} = 0$

5. $\dfrac{1}{x^2 - 16} - \dfrac{2}{x - 4} = \dfrac{1}{x + 4}$

6. $\dfrac{2}{x^2 - 25} - \dfrac{1}{x + 5} = \dfrac{1}{x - 5}$

7. $\dfrac{2}{x^2 - x} + \dfrac{1}{x + 3} = \dfrac{4}{x - 2}$

8. $\dfrac{3}{x^2 + x} - \dfrac{1}{x + 5} = \dfrac{2}{x - 7}$

9. $\dfrac{6}{4x + 7} - \dfrac{3}{x} = \dfrac{5}{6x - 13}$

10. $\dfrac{4}{3x - 5} + \dfrac{2}{x} = \dfrac{9}{4x + 13}$

11. $\dfrac{3x + 1}{x - 4} = \dfrac{6x + 5}{2x - 7}$

12. $\dfrac{4x - 1}{2x + 3} = \dfrac{12x - 25}{6x - 2}$

13. Suppose that in solving $\dfrac{x + 7}{4} - \dfrac{x + 3}{3} = \dfrac{x}{12}$, all of your algebraic steps are correct. Is there a possibility that your proposed solution will have to be rejected? Explain.

14. Consider the equation in Exercise 13.
   (a) Solve it.  (b) Check your solution, showing all steps.

*Solve each equation. See Examples 2–5.*

15. $\dfrac{3}{4x} = \dfrac{5}{2x} - \dfrac{7}{4}$

16. $\dfrac{6}{5x} + \dfrac{8}{45} = \dfrac{2}{3x}$

17. $x - \dfrac{24}{x} = -2$

18. $p + \dfrac{15}{p} = -8$

19. $\dfrac{x - 4}{x + 6} = \dfrac{2x + 3}{2x - 1}$

20. $\dfrac{5x - 8}{x + 2} = \dfrac{5x - 1}{x + 3}$

21. $\dfrac{3x + 1}{x - 4} = \dfrac{6x + 5}{2x - 7}$

22. $\dfrac{4x - 1}{2x + 3} = \dfrac{12x - 25}{6x - 2}$

23. $\dfrac{1}{y - 1} + \dfrac{5}{12} = \dfrac{-2}{3y - 3}$

24. $\dfrac{4}{m + 2} - \dfrac{11}{9} = \dfrac{1}{3m + 6}$

25. $\dfrac{7}{6x + 3} - \dfrac{1}{3} = \dfrac{2}{2x + 1}$

26. $\dfrac{3}{4m + 2} = \dfrac{17}{2} - \dfrac{7}{2m + 1}$

27. $\dfrac{3}{k + 2} - \dfrac{2}{k^2 - 4} = \dfrac{1}{k - 2}$

28. $\dfrac{3}{x - 2} + \dfrac{21}{x^2 - 4} = \dfrac{14}{x + 2}$

29. $\dfrac{1}{y + 2} + \dfrac{3}{y + 7} = \dfrac{5}{y^2 + 9y + 14}$

30. $\dfrac{1}{t + 3} + \dfrac{4}{t + 5} = \dfrac{2}{t^2 + 8t + 15}$

31. $\dfrac{9}{x} + \dfrac{4}{6x - 3} = \dfrac{2}{6x - 3}$

32. $\dfrac{5}{n} + \dfrac{4}{6 - 3n} = \dfrac{2n}{6 - 3n}$

33. $\dfrac{6}{w + 3} + \dfrac{-7}{w - 5} = \dfrac{-48}{w^2 - 2w - 15}$

34. $\dfrac{2}{r - 5} + \dfrac{3}{2r + 1} = \dfrac{22}{2r^2 - 9r - 5}$

35. $\dfrac{x}{x - 3} + \dfrac{4}{x + 3} = \dfrac{18}{x^2 - 9}$

36. $\dfrac{2x}{x - 3} + \dfrac{4}{x + 3} = \dfrac{-24}{x^2 - 9}$

37. $\dfrac{1}{x + 4} + \dfrac{x}{x - 4} = \dfrac{-8}{x^2 - 16}$

38. $\dfrac{5}{x - 4} - \dfrac{3}{x - 1} = \dfrac{x^2 - 1}{x^2 - 5x + 4}$

39. $\dfrac{2}{k^2 + k - 6} + \dfrac{1}{k^2 - k - 2} = \dfrac{4}{k^2 + 4k + 3}$

40. $\dfrac{5}{p^2 + 3p + 2} - \dfrac{3}{p^2 - 4} = \dfrac{1}{p^2 - p - 2}$

41. $\dfrac{5x + 14}{x^2 - 9} = \dfrac{-2x^2 - 5x + 2}{x^2 - 9} + \dfrac{2x + 4}{x - 3}$

42. $\dfrac{4x - 7}{4x^2 - 9} = \dfrac{-2x^2 + 5x - 4}{4x^2 - 9} + \dfrac{x + 1}{2x + 3}$

43. *Concept Check* Professor Dan Abbey asked the following question on a test: What is the solution set of $\dfrac{x + 3}{x + 3} = 1$? Only one student answered it correctly.
   (a) What is the solution set?
   (b) Many students answered $(-\infty, \infty)$. Why is this not correct?

44. Without actually solving the equation, explain why the solution set must be ∅.

$$\frac{1}{x^2+2} + \frac{2}{x^2+1} = -4$$

*Graph each rational function. Give the equation of the vertical asymptote. See Figures 2 and 3.*

**45.** $f(x) = \dfrac{2}{x}$   **46.** $f(x) = \dfrac{3}{x}$   **47.** $f(x) = \dfrac{1}{x-2}$   **48.** $f(x) = \dfrac{1}{x+2}$

*Solve each problem.*

**49.** The average number of vehicles waiting in line to enter a sports arena parking area is modeled by the function defined by

$$w(x) = \frac{x^2}{2(1-x)},$$

where $x$ is a quantity between 0 and 1 known as the **traffic intensity.** (*Source:* Mannering, F., and W. Kilareski, *Principles of Highway Engineering and Traffic Control,* John Wiley and Sons, 1990.) For each traffic intensity, find the average number of vehicles waiting (to the nearest tenth).

(a) 0.1   (b) 0.8   (c) 0.9
(d) What happens to waiting time as traffic intensity increases?

**50.** The percent of deaths caused by smoking is modeled by the rational function defined by

$$p(x) = \frac{x-1}{x},$$

where $x$ is the number of times a smoker is more likely to die of lung cancer than a nonsmoker is. This is called the **incidence rate.** (*Source:* Walker, A., *Observation and Inference: An Introduction to the Methods of Epidemiology,* Epidemiology Resources Inc., 1991.) For example, $x = 10$ means that a smoker is 10 times more likely than a nonsmoker to die of lung cancer.

(a) Find $p(x)$ if $x$ is 10.
(b) For what values of $x$ is $p(x) = 80\%$? (*Hint:* Change 80% to a decimal.)
(c) Can the incidence rate equal 0? Explain.

**51.** The force required to keep a 2000-lb car going 30 mph from skidding on a curve, where $r$ is the radius of the curve in feet, is given by

$$F(r) = \frac{225{,}000}{r}.$$

(a) What radius must a curve have if a force of 450 lb is needed to keep the car from skidding?
(b) As the radius of the curve is lengthened, how is the force affected?

**52.** The amount of heating oil produced (in gallons per day) by an oil refinery is modeled by the rational function defined by

$$f(x) = \frac{125{,}000 - 25x}{125 + 2x},$$

where $x$ is the amount of gasoline produced (in hundreds of gallons per day). Suppose the refinery must produce 300 gal of heating oil per day to meet the needs of its customers.

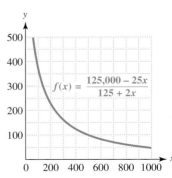

(a) How much gasoline will be produced per day?
(b) The graph of $f$ is shown in the figure. Use it to decide what happens to the amount of gasoline ($x$) produced as the amount of heating oil ($y$) produced increases.

# 7.5 Applications of Rational Expressions

**OBJECTIVES**

1. Find the value of an unknown variable in a formula.
2. Solve a formula for a specified variable.
3. Solve applications by using proportions.
4. Solve applications about distance, rate, and time.
5. Solve applications about work rates.

**OBJECTIVE 1** Find the value of an unknown variable in a formula. Formulas may contain rational expressions, as does $t = \frac{d}{r}$ and $\frac{1}{f} = \frac{1}{p} + \frac{1}{q}$.

### EXAMPLE 1 Finding the Value of a Variable in a Formula

In physics, the focal length $f$ of a lens is given by the formula

$$\frac{1}{f} = \frac{1}{p} + \frac{1}{q}.$$

In the formula, $p$ is the distance from the object to the lens and $q$ is the distance from the lens to the image. See Figure 4. Find $q$ if $p = 20$ cm and $f = 10$ cm.

Focal Length of Camera Lens

**FIGURE 4**

$$\frac{1}{f} = \frac{1}{p} + \frac{1}{q}$$

$$\frac{1}{10} = \frac{1}{20} + \frac{1}{q} \qquad \text{Let } f = 10, p = 20.$$

$$20q \cdot \frac{1}{10} = 20q\left(\frac{1}{20} + \frac{1}{q}\right) \qquad \text{Multiply by the LCD, } 20q.$$

$$2q = q + 20$$

$$q = 20 \qquad \text{Subtract } q.$$

The distance from the lens to the image is 20 cm.

**NOW TRY** Exercise 5.

**OBJECTIVE 2** Solve a formula for a specified variable. Recall that the goal in solving for a specified variable is to isolate it on one side of the equals sign.

### EXAMPLE 2 Solving a Formula for a Specified Variable

Solve $\frac{1}{f} = \frac{1}{p} + \frac{1}{q}$ for $p$.

$$fpq \cdot \frac{1}{f} = fpq\left(\frac{1}{p} + \frac{1}{q}\right) \qquad \text{Multiply by the LCD, } fpq.$$

$$pq = fq + fp \qquad \text{Distributive property}$$

Transform so that the terms with $p$ (the specified variable) are on the same side.

$$pq - fp = fq \qquad \text{Subtract } fp.$$

$$p(q - f) = fq \qquad \text{Factor out } p.$$

*This is a key step.*
$$p = \frac{fq}{q - f} \qquad \text{Divide by } q - f.$$

**NOW TRY** Exercise 11.

351

### EXAMPLE 3  Solving a Formula for a Specified Variable

Solve $I = \dfrac{nE}{R + nr}$ for $n$.

$$(R + nr)I = (R + nr)\dfrac{nE}{R + nr} \quad \text{Multiply by } R + nr.$$

$$RI + nrI = nE$$

$$RI = nE - nrI \quad \text{Subtract } nrI.$$

$$RI = n(E - rI) \quad \text{Factor out } n.$$

$$\dfrac{RI}{E - rI} = n \quad \text{Divide by } E - rI.$$

**NOW TRY** Exercise 19.

**CAUTION** Refer to the steps in Examples 2 and 3 that factor out the desired variable. The variable for which you are solving *must* be a factor on only one side of the equation, so that each side can be divided by the remaining factor in the last step.

We now solve problems that translate into equations with rational expressions, using the six-step problem-solving method from **Chapter 2.**

**OBJECTIVE 3** Solve applications by using proportions. A **ratio** is a comparison of two quantities. The ratio of $a$ to $b$ may be written in any of the following ways:

$$a \text{ to } b, \quad a:b, \quad \text{or} \quad \dfrac{a}{b}. \quad \text{Ratio of } a \text{ to } b$$

Ratios are usually written as quotients in algebra. A **proportion** is a statement that two ratios are equal, such as

$$\dfrac{a}{b} = \dfrac{c}{d}. \quad \text{Proportion}$$

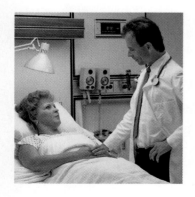

### EXAMPLE 4  Solving a Proportion

In 2003, about 15 of every 100 Americans had no health insurance. The population at that time was about 288 million. How many million Americans had no health insurance? (*Source:* U.S. Census Bureau.)

*Step 1* **Read** the problem.

*Step 2* **Assign a variable.** Let $x =$ the number of Americans (in millions) who had no health insurance.

*Step 3* **Write an equation.** To get an equation, set up a proportion. The ratio 15 to 100 should equal the ratio $x$ to 288.

$$\dfrac{15}{100} = \dfrac{x}{288} \quad \text{Write a proportion.}$$

**Step 4** Solve. $\dfrac{15}{100} = \dfrac{x}{288}$

$$28{,}800\left(\dfrac{15}{100}\right) = 28{,}800\left(\dfrac{x}{288}\right) \quad \text{Multiply by a common denominator.}$$

$$4320 = 100x \quad \text{Simplify.}$$

$$x = 43.2 \quad \text{Divide by 100.}$$

**Step 5** **State the answer.** There were 43.2 million Americans with no health insurance in 2003.

**Step 6** **Check** that the ratio of 43.2 million to 288 million equals $\dfrac{15}{100}$.

**NOW TRY** Exercise 31.

---

**EXAMPLE 5** Solving a Proportion Involving Rates

Marissa's car uses 10 gal of gasoline to travel 210 mi. She has 5 gal of gasoline in the car, and she wants to know how much more gasoline she will need to drive 640 mi. If we assume the car continues to use gasoline at the same rate, how many more gallons will she need?

**Step 1** **Read** the problem.

**Step 2** **Assign a variable.** Let $x =$ the additional number of gallons of gas needed.

**Step 3** **Write an equation.** To get an equation, set up a proportion.

$$\text{gallons} \rightarrow \dfrac{10}{210} = \dfrac{5 + x}{640} \leftarrow \text{gallons}$$
$$\text{miles} \rightarrow \phantom{\dfrac{10}{210}} \phantom{=} \phantom{\dfrac{5+x}{640}} \leftarrow \text{miles}$$

**Step 4** **Solve.** We could multiply both sides by the LCD $10 \cdot 21 \cdot 64$. Instead we use an alternative method that involves *cross products*: For $\dfrac{a}{b} = \dfrac{c}{d}$ to be true, then the cross products $ad$ and $bc$ must be equal. Thus,

$$10 \cdot 640 = 210(5 + x) \quad \text{If } \dfrac{a}{b} = \dfrac{c}{d}, \text{ then } ad = bc.$$

$$6400 = 1050 + 210x \quad \text{Multiply; distributive property}$$

$$5350 = 210x \quad \text{Subtract 1050.}$$

$$25.5 \approx x. \quad \text{Divide by 210; round to the nearest tenth.}$$

**Step 5** **State the answer.** Marissa will need about 25.5 more gallons of gas.

**Step 6** **Check** the answer in the words of the problem. The 25.5 gal plus the 5 gal equals 30.5 gal.

$$\dfrac{30.5}{640} \approx 0.048 \quad \text{and} \quad \dfrac{10}{210} \approx 0.048$$

Since the rates are equal, the solution is correct.

**NOW TRY** Exercise 37.

---

**OBJECTIVE 4** Solve applications about distance, rate, and time. The next examples use the distance formula $d = rt$ introduced in **Chapter 2**. A familiar example of a rate is speed, which is the ratio of distance to time, or $r = \dfrac{d}{t}$.

> **EXAMPLE 6** Solving a Problem about Distance, Rate, and Time
>
> A paddlewheeler goes 10 mi against the current in a river in the same time that it goes 15 mi with the current. If the speed of the current is 3 mph, find the speed of the boat in still water.
>
> *Step 1* **Read** the problem. We must find the speed of the boat in still water.
>
> *Step 2* **Assign a variable.** Let $x$ = the speed of the boat in still water.
>
> When the boat is traveling *against* the current, the current slows the boat down, and the speed of the boat is the *difference* between its speed in still water and the speed of the current—that is, $(x - 3)$ mph.
>
> When the boat is traveling *with* the current, the current speeds the boat up, and the speed of the boat is the *sum* of its speed in still water and the speed of the current—that is, $(x + 3)$ mph.
>
> Thus, $x - 3$ = the speed of the boat *against* the current,
> and $x + 3$ = the speed of the boat *with* the current.
>
> Because the time is the same going against the current as with the current, find time in terms of distance and rate (speed) for each situation. Start with the distance formula, $d = rt$, and divide each side by $r$ to get $t = \frac{d}{r}$. Against the current, the distance is 10 mi and the rate is $(x - 3)$ mph, giving
>
> $$t = \frac{d}{r} = \frac{10}{x - 3}. \quad \text{Time } \textit{against} \text{ the current}$$
>
> With the current, the distance is 15 mi and the rate is $(x + 3)$ mph, so
>
> $$t = \frac{d}{r} = \frac{15}{x + 3}. \quad \text{Time } \textit{with} \text{ the current}$$
>
> This information is summarized in the following table.
>
> |  | Distance | Rate | Time |
> |---|---|---|---|
> | Against Current | 10 | $x - 3$ | $\frac{10}{x-3}$ |
> | With Current | 15 | $x + 3$ | $\frac{15}{x+3}$ |
>
> Times are equal.
>
> *Step 3* **Write an equation.** Because the times are equal,
>
> $$\frac{10}{x - 3} = \frac{15}{x + 3}.$$
>
> *Step 4* **Solve** this equation. The LCD is $(x + 3)(x - 3)$.
>
> $$(x + 3)(x - 3)\left(\frac{10}{x - 3}\right) = (x + 3)(x - 3)\left(\frac{15}{x + 3}\right) \quad \text{Multiply by the LCD.}$$
>
> $$10(x + 3) = 15(x - 3)$$
> $$10x + 30 = 15x - 45 \quad \text{Distributive property}$$
> $$75 = 5x \quad \text{Subtract } 10x; \text{ add } 45.$$
> $$15 = x \quad \text{Divide by 5.}$$
>
> *Step 5* **State the answer.** The speed of the boat in still water is 15 mph.
>
> *Step 6* **Check** the answer: $\frac{10}{15 - 3} = \frac{15}{15 + 3}$ is true.

**NOW TRY** Exercise 41.

**SECTION 7.5** Applications of Rational Expressions **355**

> **EXAMPLE 7** Solving a Problem about Distance, Rate, and Time
>
> At O'Hare International Airport in Chicago, Cheryl and Bill are walking to the gate (at the same speed) to catch their flight to Akron, Ohio. Since Bill wants a window seat, he steps onto the moving sidewalk and continues to walk while Cheryl uses the stationary sidewalk. If the sidewalk moves at 1 m per sec and Bill saves 50 sec covering the 300-m distance, what is their walking speed?
>
> *Step 1* **Read** the problem. We must find their walking speed.
>
> *Step 2* **Assign a variable.** Let $x$ represent their walking speed in meters per second. Thus, Cheryl travels at $x$ meters per second and Bill travels at $(x + 1)$ meters per second. Since Bill's time is 50 sec less than Cheryl's time, express their times in terms of the known distances and the variable rates. As in Example 6, start with $d = rt$ and divide each side by $r$ to get $t = \frac{d}{r}$. For Cheryl, the distance is 300 m and the rate is $x$ m per sec. Cheryl's time is
>
> $$t = \frac{d}{r} = \frac{300}{x}.$$
>
> Bill travels 300 m at a rate of $(x + 1)$ m per sec, so his time is
>
> $$t = \frac{d}{r} = \frac{300}{x + 1}.$$
>
> This information is summarized in the following table.
>
> |        | Distance | Rate  | Time              |
> | ------ | -------- | ----- | ----------------- |
> | Cheryl | 300      | $x$   | $\frac{300}{x}$   |
> | Bill   | 300      | $x+1$ | $\frac{300}{x+1}$ |
>
> *Step 3* **Write an equation,** using the times from the table.
>
> Bill's time is Cheryl's time less 50 seconds.
>
> $$\frac{300}{x+1} = \frac{300}{x} - 50$$
>
> *Step 4* **Solve.**
>
> $$x(x+1)\left(\frac{300}{x+1}\right) = x(x+1)\left(\frac{300}{x} - 50\right) \quad \text{Multiply by the LCD, } x(x+1).$$
>
> $$300x = 300(x+1) - 50x(x+1) \quad \text{Multiply; distributive property}$$
> $$300x = 300x + 300 - 50x^2 - 50x \quad \text{Distributive property}$$
> $$0 = 50x^2 + 50x - 300 \quad \text{Standard form}$$
> $$0 = x^2 + x - 6 \quad \text{Divide by 50.}$$
> $$0 = (x+3)(x-2) \quad \text{Factor.}$$
> $$x + 3 = 0 \quad \text{or} \quad x - 2 = 0 \quad \text{Zero-factor property}$$
> $$x = -3 \quad \text{or} \quad x = 2 \quad \text{Solve each equation.}$$
>
> Discard the negative answer, since speed cannot be negative.
>
> *Step 5* **State the answer.** Their walking speed is 2 m per sec.
>
> *Step 6* **Check** the solution in the words of the original problem.

**NOW TRY** Exercise 47.

**CAUTION** The method of solving a rational equation by using cross products, introduced in Example 5, cannot be used to solve the equation in Example 7 (as it appears in Step 3). The method can be used only when there is a single rational expression on each side.

**OBJECTIVE 5** **Solve applications about work rates.** Problems about work are closely related to distance problems.

To solve a work problem, we begin by using this fact to express all rates of work.

### Rate of Work

If a job can be accomplished in $t$ units of time, then the rate of work is

$$\frac{1}{t} \text{ job per unit of time.}$$

See if you can identify the problem-solving steps in the work problem that follows.

**EXAMPLE 8** Solving a Problem about Work

Letitia and Kareem are working on a neighborhood cleanup. Kareem can clean up all the trash in the area in 7 hr, while Letitia can do the same job in 5 hr. How long will it take them if they work together?

Let $x =$ the number of hours it will take the two people working together. Just as we made a table for the distance formula, $d = rt$, we make a table here for $A = rt$, with $A = 1$. Since $A = 1$, the rate for each person will be $\frac{1}{t}$, where $t$ is the time it takes the person to complete the job alone. For example, since Kareem can clean up all the trash in 7 hr, his rate is $\frac{1}{7}$ of the job per hour. Similarly, Letitia's rate is $\frac{1}{5}$ of the job per hour. Fill in the table as shown.

|  | Rate | Time Working Together | Fractional Part of the Job Done |
|---|---|---|---|
| Kareem | $\frac{1}{7}$ | $x$ | $\frac{1}{7}x$ |
| Letitia | $\frac{1}{5}$ | $x$ | $\frac{1}{5}x$ |

Since together they complete 1 job, the sum of the fractional parts accomplished by them should equal 1.

$$\underbrace{\frac{1}{7}x}_{\text{Part done by Kareem}} + \underbrace{\frac{1}{5}x}_{\text{part done by Letitia}} = \underbrace{1}_{\text{1 whole job.}}$$

$$35\left(\frac{1}{7}x + \frac{1}{5}x\right) = 35 \cdot 1 \quad \text{Multiply by the LCD, 35.}$$

$$5x + 7x = 35 \quad \text{Distributive property}$$

$$12x = 35 \quad \text{Combine like terms.}$$

$$x = \frac{35}{12} \quad \text{Divide by 12.}$$

Working together, Kareem and Letitia can do the job in $\frac{35}{12}$ hr, or 2 hr, 55 min. Check this result in the original problem.

**NOW TRY** Exercise 49.

There is another way to approach problems about rates of work. For instance, in Example 8, $x$ represents the number of hours it will take the two people working together to complete the entire job. In 1 hr, $\frac{1}{x}$ of the entire job will be completed. In 1 hr, Kareem completes $\frac{1}{7}$ of the job and Letitia completes $\frac{1}{5}$ of the job. The sum of their rates should equal $\frac{1}{x}$. This reasoning gives the equation

$$\frac{1}{7} + \frac{1}{5} = \frac{1}{x}.$$

When each side of this equation is multiplied by $35x$, the result is $5x + 7x = 35$, the same equation we got in Example 8 in the third line from the bottom. Thus, the solution of the equation is the same with either approach.

## 7.5 Exercises

*Multiple Choice* Exercises 1–4 present a familiar formula. Give the letter of the choice that is an equivalent form of the formula.

1. $p = br$ (percent)

    **A.** $b = \dfrac{p}{r}$   **B.** $r = \dfrac{b}{p}$   **C.** $b = \dfrac{r}{p}$   **D.** $p = \dfrac{r}{b}$

2. $V = LWH$ (geometry)

    **A.** $H = \dfrac{LW}{V}$   **B.** $L = \dfrac{V}{WH}$   **C.** $L = \dfrac{WH}{V}$   **D.** $W = \dfrac{H}{VL}$

3. $m = \dfrac{F}{a}$ (physics)

    **A.** $a = mF$   **B.** $F = \dfrac{m}{a}$   **C.** $F = \dfrac{a}{m}$   **D.** $F = ma$

4. $I = \dfrac{E}{R}$ (electricity)

    **A.** $R = \dfrac{I}{E}$   **B.** $R = IE$   **C.** $E = \dfrac{I}{R}$   **D.** $E = RI$

*Solve each problem. See Example 1.*

5. In work with electric circuits, the formula
$$\frac{1}{a} = \frac{1}{b} + \frac{1}{c}$$
occurs. Find $b$ if $a = 8$ and $c = 12$.

6. A gas law in chemistry says that
$$\frac{PV}{T} = \frac{pv}{t}.$$
Suppose that $T = 300$, $t = 350$, $V = 9$, $P = 50$, and $v = 8$. Find $p$.

7. A formula from anthropology says that
$$c = \frac{100b}{L}.$$
Find $L$ if $c = 80$ and $b = 5$.

8. The gravitational force between two masses is given by
$$F = \frac{GMm}{d^2}.$$
Find $M$ if $F = 10$, $G = 6.67 \times 10^{-11}$, $m = 1$, and $d = 3 \times 10^{-6}$.

*Solve each formula for the specified variable. See Examples 2 and 3.*

9. $F = \dfrac{GMm}{d^2}$ for $G$ (physics)

10. $F = \dfrac{GMm}{d^2}$ for $M$ (physics)

11. $\dfrac{1}{a} = \dfrac{1}{b} + \dfrac{1}{c}$ for $a$ (electricity)

12. $\dfrac{1}{a} = \dfrac{1}{b} + \dfrac{1}{c}$ for $b$ (electricity)

13. $\dfrac{PV}{T} = \dfrac{pv}{t}$ for $v$ (chemistry)

14. $\dfrac{PV}{T} = \dfrac{pv}{t}$ for $T$ (chemistry)

15. $I = \dfrac{nE}{R + nr}$ for $r$ (engineering)

16. $a = \dfrac{V - v}{t}$ for $V$ (physics)

17. $A = \dfrac{1}{2}h(b + B)$ for $b$ (mathematics)

18. $S = \dfrac{n}{2}(a + \ell)d$ for $n$ (mathematics)

19. $\dfrac{E}{e} = \dfrac{R + r}{r}$ for $r$ (engineering)

20. $y = \dfrac{x + z}{a - x}$ for $x$

21. *Concept Check* In solving the equation $m = \dfrac{ab}{a - b}$ for $a$, what is the first step?

22. *Concept Check* Suppose you are asked to solve the equation
$$rp - rq = p + q$$
for $r$. What is the first step?

*Solve each problem mentally. Use proportions in Exercises 23 and 24.*

23. In a mathematics class, 3 of every 4 students are girls. If there are 20 students in the class, how many are girls? How many are boys?

24. In a certain southern state, sales tax on a purchase of $1.50 is $0.12. What is the sales tax on a purchase of $6.00?

25. If Marin can mow her yard in 2 hr, what is her rate (in terms of the proportion of the job per hour)?

26. A van traveling from Atlanta to Detroit averages 50 mph and takes 14 hr to make the trip. How far is it from Atlanta to Detroit?

*Use a proportion to solve each problem. See Examples 4 and 5.*

27. On a map of the United States, the distance between Seattle and Durango is 4.125 in. The two cities are actually 1238 miles apart. On this same map, what would be the distance between Chicago and El Paso, two cities that are actually 1606 mi apart? (*Source:* Universal Map Atlas.)

28. On a map of the United States, the distance between Reno and Phoenix is 2.5 in. The two cities are actually 768 miles apart. On this same map, what would be the distance between St. Louis and Jacksonville, two cities that are actually 919 mi apart? (*Source:* Universal Map Atlas.)

29. On a world globe, the distance between New York and Cairo, two cities that are actually 5619 mi apart, is 8.5 in. On this same globe, how far apart are Madrid and Rio de Janeiro, two cities that are actually 5045 mi apart? (*Sources:* Author's globe, *The World Almanac and Book of Facts.*)

30. On a world globe, the distance between San Francisco and Melbourne, two cities that are actually 7856 mi apart, is 11.875 in. On this same globe, how far apart are Mexico City and Singapore, two cities that are actually 10,327 mi apart? (*Sources:* Author's globe, *The World Almanac and Book of Facts.*)

31. During the 2001–2002 school year, the average ratio of teachers to students in private elementary schools was approximately 1 to 14. If one such private school had 554 students, how many teachers would be at the school if this ratio was valid for that school? Round your answer to the nearest whole number. (*Source:* U.S. National Center for Education Statistics.)

**32.** On June 28, 2006, the Boston Red Sox were in first place in the Eastern Division of the American League, having won 47 of their first 75 regular season games. If the team continued to win the same fraction of its games, how many games would the Red Sox win for the complete 162-game season? Round your answer to the nearest whole number. (*Source:* www.mlb.com)

**33.** To estimate the deer population of a forest preserve, wildlife biologists caught, tagged, and then released 42 deer. A month later, they returned and caught a sample of 75 deer and found that 15 of them were tagged. Based on this experiment, approximately how many deer lived in the forest preserve?

**34.** Suppose that in the experiment in Exercise 33, when the biologists returned, they found only 5 tagged deer in their sample of 75. What would be the estimate of the deer population?

**35.** Biologists tagged 500 fish in a lake on January 1. On February 1, they returned and collected a random sample of 400 fish, 8 of which had been previously tagged. On the basis of this experiment, approximately how many fish does the lake have?

**36.** Suppose that in the experiment of Exercise 35, 10 of the previously tagged fish were collected on February 1. What would be the estimate of the fish population?

**37.** Bruce Johnston's Shelby Cobra uses 5 gal of gasoline to drive 156 miles. He has 3 gal of gasoline in the car, and he wants to know how much more gasoline he will need to drive 300 mi. If we assume that the car continues to use gasoline at the same rate, how many more gallons will he need?

**38.** Mike Love's T-bird uses 6 gal of gasoline to drive 141 miles. He has 4 gal of gasoline in the car, and he wants to know how much more gasoline he will need to drive 250 mi. If we assume that the car continues to use gasoline at the same rate, how many more gallons will he need?

*In geometry, two triangles with corresponding angle measures that are equal, called* **similar triangles,** *have corresponding sides proportional. For example, in the figure, angle A = angle D, angle B = angle E, and angle C = angle F, so the triangles are similar. Then the following ratios of corresponding sides are equal.*

$$\frac{4}{6} = \frac{6}{9} = \frac{2x+1}{2x+5}$$

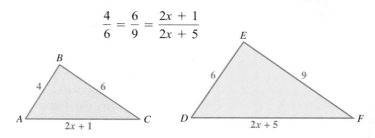

**39.** Solve for *x*, using the given proportion to find the lengths of the third sides of the triangles.

**40.** Suppose the following triangles are similar. Find *y* and the lengths of the two longest sides of each triangle.

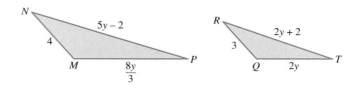

*Solve each problem. See Examples 6 and 7.*

**41.** Kellen's boat goes 12 mph. Find the rate of the current of the river if she can go 6 mi upstream in the same amount of time she can go 10 mi downstream.

|  | Distance | Rate | Time |
|---|---|---|---|
| Downstream | 10 | $12 + x$ |  |
| Upstream | 6 | $12 - x$ |  |

**42.** Kasey can travel 8 mi upstream in the same time it takes her to go 12 mi downstream. Her boat goes 15 mph in still water. What is the rate of the current?

|  | Distance | Rate | Time |
|---|---|---|---|
| Downstream |  |  |  |
| Upstream |  |  |  |

**43.** On his drive from Montpelier, Vermont, to Columbia, South Carolina, Dylan Davis averaged 51 mph. If he had been able to average 60 mph, he would have reached his destination 3 hr earlier. What is the driving distance between Montpelier and Columbia?

**44.** Rebecca Song is a college student who lives in an off-campus apartment. Some days she rides her bike to campus, while other days she walks. When she rides her bike, she gets to her first classroom building 36 min faster than when she walks. If her average walking speed is 3 mph and her average biking speed is 12 mph, how far is it from her apartment to the classroom building?

**45.** A plane averaged 500 mph on a trip going east, but only 350 mph on the return trip. The total flying time in both directions was 8.5 hr. What was the one-way distance?

**46.** Leah drove from her apartment to her parents' house for the weekend. Driving to their house on Saturday morning, she was able to average 60 mph because traffic was light. However, returning on Sunday night, she was able to average only 45 mph on the same route, because traffic was heavy. The drive on Sunday took her 1.5 hr longer than the drive on Saturday. What is the distance between Leah's apartment and her parents' house?

**47.** On the first part of a trip to Carmel traveling on the freeway, Marge averaged 60 mph. On the rest of the trip, which was 10 mi longer than the first part, she averaged 50 mph. Find the total distance to Carmel if the second part of the trip took 30 min more than the first part.

**48.** While on vacation, Jim and Annie decided to drive all day. During the first part of their trip on the highway, they averaged 60 mph. When they got to Houston, traffic caused them to average only 30 mph. The distance they drove in Houston was 100 mi less than their distance on the highway. What was their total driving distance if they spent 50 min more on the highway than they did in Houston?

*Solve each problem. See Example 8.*

**49.** Butch and Peggy want to pick up the mess that their grandson, Grant, has made in his playroom. Butch can do it in 15 min working alone. Peggy, working alone, can clean it in 12 min. How long will it take them if they work together?

|  | Rate | Time Working Together | Fractional Part of the Job Done |
|---|---|---|---|
| Butch | $\frac{1}{15}$ | $x$ |  |
| Peggy | $\frac{1}{12}$ | $x$ |  |

**50.** Lou can groom Jay Beckenstein's dogs in 8 hr, but it takes his business partner, Janet, only 5 hr to groom the same dogs. How long will it take them to groom Jay's dogs if they work together?

|  | Rate | Time Working Together | Fractional Part of the Job Done |
|---|---|---|---|
| Lou | $\frac{1}{8}$ | $x$ |  |
| Janet | $\frac{1}{5}$ | $x$ |  |

**51.** Jerry and Kuba are laying a hardwood floor. Working alone, Jerry can do the job in 20 hr. If the two of them work together, they can complete the job in 12 hr. How long would it take Kuba to lay the floor working alone?

52. Mrs. Disher is a high school mathematics teacher. She can grade a set of chapter tests in 5 hr working alone. If her student teacher Mr. Howes helps her, it will take 3 hr to grade the tests. How long would it take Mr. Howes to grade the tests if he worked alone?

53. Dixie can paint a room in 3 hr working alone. Trixie can paint the same room in 6 hr working alone. How long after Dixie starts to paint the room will it be finished if Trixie joins her 1 hr later?

54. Chipper can wash a car in 30 min working alone. Dalie can do the same job in 45 min working alone. How long after Chipper starts to wash the car will it be finished if Dalie joins him 5 min later?

55. If a vat of acid can be filled by an inlet pipe in 10 hr and emptied by an outlet pipe in 20 hr, how long will it take to fill the vat if both pipes are open?

56. A winery has a vat to hold Chardonnay. An inlet pipe can fill the vat in 9 hr, while an outlet pipe can empty it in 12 hr. How long will it take to fill the vat if both the outlet and the inlet pipes are open?

57. Suppose that Hortense and Mort can clean their entire house in 7 hr, while their toddler, Mimi, just by being around, can completely mess it up in only 2 hr. If Hortense and Mort clean the house while Mimi is at her grandma's and then start cleaning up after Mimi the minute she gets home, how long does it take from the time Mimi gets home until the whole place is a shambles?

58. An inlet pipe can fill an artificial lily pond in 60 min, while an outlet pipe can empty it in 80 min. Through an error, both pipes are left open. How long will it take for the pond to fill?

## 7.6 Variation

**OBJECTIVES**

1. Write an equation expressing direct variation.
2. Find the constant of variation, and solve direct variation problems.
3. Solve inverse variation problems.
4. Solve joint variation problems.
5. Solve combined variation problems.

Certain types of functions are common, especially in business and the physical sciences. These are functions in which $y$ depends on a multiple of $x$ or $y$ depends on a number divided by $x$. In such situations, $y$ is said to *vary directly as $x$* (in the first case) or *vary inversely as $x$* (in the second case). For example, by the distance formula, the distance traveled varies directly as the rate (or speed) and the time. Formulas for area and volume are other familiar examples of *direct variation*.

By contrast, the force required to keep a car from skidding on a curve varies inversely as the radius of the curve. Another example of *inverse variation* is how travel time is inversely proportional to rate or speed.

**OBJECTIVE 1** Write an equation expressing direct variation. The circumference of a circle is given by the formula $C = 2\pi r$, where $r$ is the radius of the circle. See Figure 5. The circumference is always a constant multiple of the radius. ($C$ is always found by multiplying $r$ by the constant $2\pi$.) Thus,

as the *radius increases,* the *circumference increases.*

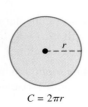

**FIGURE 5**

The following is also true:

As the *radius decreases*, the *circumference decreases*.

Because of these relationships, the circumference is said to *vary directly* as the radius.

> **Direct Variation**
>
> **y varies directly as x** if there exists a real number $k$ such that
> $$y = kx.$$

Also, $y$ is said to be **proportional to** $x$. The number $k$ is called the **constant of variation**. In direct variation, for $k > 0$, as the value of $x$ increases, the value of $y$ also increases. Similarly, as $x$ decreases, $y$ decreases.

**OBJECTIVE 2** Find the constant of variation, and solve direct variation problems.
The direct variation equation $y = kx$ defines a linear function, where the constant of variation $k$ is the slope of the line. For example, we wrote the equation

$$y = 3.20x$$

to describe the cost $y$ to buy $x$ gallons of gasoline in Example 6 of **Section 3.3.** The cost varies directly as, or is proportional to, the number of gallons of gasoline purchased. That is, as the number of gallons of gasoline increases, the cost increases; also, as the number of gallons of gasoline decreases, the cost decreases. The constant of variation $k$ is 3.20, the cost of 1 gal of gasoline.

**EXAMPLE 1** Finding the Constant of Variation and the Variation Equation

Ariel Mejia is paid an hourly wage. One week she worked 43 hr and was paid $795.50. How much does she earn per hour?

Let $h$ represent the number of hours she works and $P$ represent her corresponding pay. Then $P$ varies directly as $h$, so

$$P = kh.$$

Here, $k$ represents Abby's hourly wage. Since $P = 795.50$ when $h = 43$,

$$795.50 = 43k \quad \text{Substitute for } P \text{ and } h.$$
$$k = 18.50. \quad \text{Use a calculator.}$$

Her hourly wage is $18.50, and $P$ and $h$ are related by

$$P = 18.50h.$$

**NOW TRY** Exercise 29.

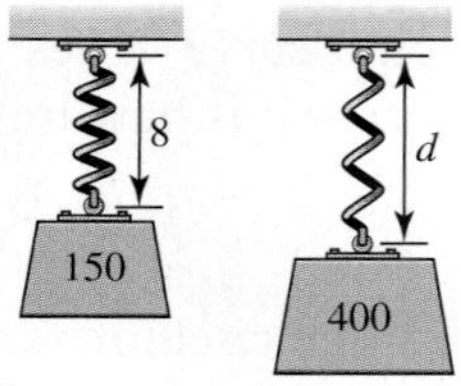

**FIGURE 6**

**EXAMPLE 2** Solving a Direct Variation Problem

Hooke's law for an elastic spring states that the distance a spring stretches is directly proportional to the force applied. If a force of 150 newtons* stretches a certain spring 8 cm, how much will a force of 400 newtons stretch the spring? See Figure 6.

---

*A newton is a unit of measure of force used in physics.

If $d$ is the distance the spring stretches and $f$ is the force applied, then $d = kf$ for some constant $k$. Since a force of 150 newtons stretches the spring 8 cm, use these values to find $k$.

$$d = kf \qquad \text{Variation equation}$$
$$8 = k \cdot 150 \qquad \text{Let } d = 8 \text{ and } f = 150.$$
$$k = \frac{8}{150} \qquad \text{Solve for } k.$$
$$k = \frac{4}{75} \qquad \text{Lowest terms}$$

Substitute $\frac{4}{75}$ for $k$ in the variation equation $d = kf$ to get

$$d = \frac{4}{75}f.$$

For a force of 400 newtons,

$$d = \frac{4}{75}(400) = \frac{64}{3}. \qquad \text{Let } f = 400.$$

The spring will stretch $\frac{64}{3}$ cm, or $21\frac{1}{3}$ cm, if a force of 400 newtons is applied.

**NOW TRY** Exercise 31.

In summary, use the following steps to solve a variation problem.

### Solving a Variation Problem

*Step 1* Write the variation equation.
*Step 2* Substitute the initial values and solve for $k$.
*Step 3* Rewrite the variation equation with the value of $k$ from Step 2.
*Step 4* Substitute the remaining values, solve for the unknown, and find the required answer.

The direct variation equation $y = kx$ is a linear equation. However, other kinds of variation involve other types of equations. For example, one variable can be proportional to a power of another variable.

### Direct Variation as a Power

**$y$ varies directly as the $n$th power of $x$** if there exists a real number $k$ such that
$$y = kx^n.$$

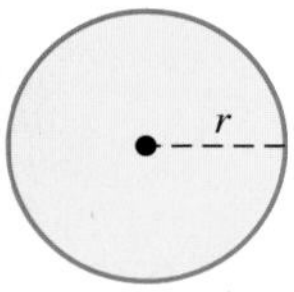

$A = \pi r^2$

**FIGURE 7**

An example of direct variation as a power is the formula for the area of a circle, $A = \pi r^2$. See Figure 7. Here, $\pi$ is the constant of variation, and the area varies directly as the *square* of the radius.

**EXAMPLE 3** Solving a Direct Variation Problem

The distance a body falls from rest varies directly as the square of the time it falls (disregarding air resistance). If a skydiver falls 64 ft in 2 sec, how far will she fall in 8 sec?

**Step 1** If $d$ represents the distance the skydiver falls and $t$ the time it takes to fall, then $d$ is a function of $t$, and, for some constant $k$,
$$d = kt^2.$$

**Step 2** To find the value of $k$, use the fact that the skydiver falls 64 ft in 2 sec.

$d = kt^2$     Variation equation
$64 = k(2)^2$     Let $d = 64$ and $t = 2$.
$k = 16$     Find $k$.

**Step 3** Using 16 for $k$, we find that the variation equation is
$$d = 16t^2.$$

**Step 4** Now let $t = 8$ to find the number of feet the skydiver will fall in 8 sec.
$$d = 16(8)^2 = 1024 \quad \text{Let } t = 8.$$

The skydiver will fall 1024 ft in 8 sec.

**NOW TRY** Exercise 35.

As pressure increases, volume decreases.

**FIGURE 8**

**OBJECTIVE 3** Solve inverse variation problems. In direct variation, where $k > 0$, as $x$ increases, $y$ increases. Similarly, as $x$ decreases, $y$ decreases. Another type of variation is *inverse variation*. With inverse variation, where $k > 0$, as one variable increases, the other variable decreases. For example, in a closed space, volume decreases as pressure increases, as illustrated by a trash compactor. See Figure 8. As the compactor presses down, the pressure on the trash increases; in turn, the trash occupies a smaller space.

**Inverse Variation**

**$y$ varies inversely as $x$** if there exists a real number $k$ such that
$$y = \frac{k}{x}.$$

Also, **$y$ varies inversely as the $n$th power of $x$** if there exists a real number $k$ such that
$$y = \frac{k}{x^n}.$$

The inverse variation equation also defines a function. Since $x$ is in the denominator, these functions are rational functions. Another example of inverse variation comes from the distance formula. In its usual form, the formula is
$$d = rt.$$
Dividing each side by $r$ gives
$$t = \frac{d}{r}.$$
Here, $t$ (time) varies inversely as $r$ (rate or speed), with $d$ (distance) serving as the constant of variation. For example, if the distance between Chicago and Des Moines is 300 mi, then
$$t = \frac{300}{r},$$

and the values of $r$ and $t$ might be any of the following.

$$r = 50, t = 6$$
$$r = 60, t = 5$$  As $r$ increases, $t$ decreases.
$$r = 75, t = 4$$

$$r = 30, t = 10$$
$$r = 25, t = 12$$  As $r$ decreases, $t$ increases.
$$r = 20, t = 15$$

If we *increase* the rate (speed) at which we drive, time *decreases*. If we *decrease* the rate (speed) at which we drive, time *increases*.

### EXAMPLE 4 Solving an Inverse Variation Problem

In the manufacture of a certain medical syringe, the cost of producing the syringe varies inversely as the number produced. If 10,000 syringes are produced, the cost is $2 per syringe. Find the cost per syringe of producing 25,000 syringes.

Let  $x =$ the number of syringes produced,
and  $c =$ the cost per syringe.

Here, as production increases, cost decreases, and as production decreases, cost increases. Since $c$ varies inversely as $x$, there is a constant $k$ such that

$$c = \frac{k}{x}.$$

Find $k$ by replacing $c$ with 2 and $x$ with 10,000.

$$2 = \frac{k}{10,000}$$

$$20,000 = k \qquad \text{Multiply by 10,000.}$$

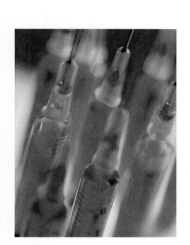

Since $c = \frac{k}{x}$,

$$c = \frac{20,000}{25,000} = 0.80. \qquad \text{Let } k = 20,000 \text{ and } x = 25,000.$$

The cost per syringe to make 25,000 syringes is $0.80.

 Exercise 37.

### EXAMPLE 5 Solving an Inverse Variation Problem

The weight of an object above Earth varies inversely as the square of its distance from the center of Earth. A space shuttle in an elliptical orbit has a maximum distance from the center of Earth (*apogee*) of 6700 mi. Its minimum distance from the center of Earth (*perigee*) is 4090 mi. See Figure 9. If an astronaut in the shuttle weighs 57 lb at its apogee, what does the astronaut weigh at its perigee?

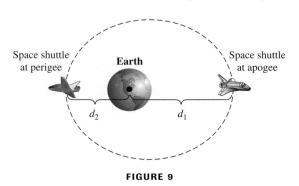

**FIGURE 9**

If $w$ is the weight and $d$ is the distance from the center of Earth, then

$$w = \frac{k}{d^2}$$

for some constant $k$. At the apogee, the astronaut weighs 57 lb, and the distance from the center of Earth is 6700 mi. Use these values to find $k$.

$$57 = \frac{k}{(6700)^2} \qquad \text{Let } w = 57 \text{ and } d = 6700.$$
$$k = 57(6700)^2$$

Then the weight at the perigee with $d = 4090$ mi is

$$w = \frac{57(6700)^2}{(4090)^2} \approx 153 \text{ lb.} \qquad \text{Use a calculator.}$$

**NOW TRY** Exercise 41.

**OBJECTIVE 4** Solve joint variation problems. It is common for one variable to depend on several others. If one variable varies directly as the *product* of several other variables (perhaps raised to powers), the first variable is said to *vary jointly* as the others.

### Joint Variation

*y* **varies jointly as** *x* **and** *z* if there exists a real number $k$ such that

$$y = kxz.$$

**CAUTION** Note that *and* in the expression "*y* varies directly as *x and z*" translates as the product $y = kxz$. The word *and* does not indicate addition here.

**EXAMPLE 6** Solving a Joint Variation Problem

The interest on a loan or an investment is given by the formula $I = prt$. Here, for a given principal $p$, the interest earned, $I$, varies jointly as the interest rate $r$ and the time $t$ the principal is left earning interest. If an investment earns \$100 interest at 5% for 2 yr, how much interest will the same principal earn at 4.5% for 3 yr?

We use the formula $I = prt$, where $p$ is the constant of variation because it is the same for both investments. For the first investment,

$$I = prt$$
$$100 = p(0.05)(2) \qquad \text{Let } I = 100, r = 0.05, \text{ and } t = 2.$$
$$100 = 0.1p$$
$$p = 1000. \qquad \text{Divide by 0.1; rewrite.}$$

Now we find $I$ when $p = 1000$, $r = 0.045$, and $t = 3$.

$$I = 1000(0.045)(3) = 135 \qquad \text{Let } p = 1000, r = 0.045, \text{ and } t = 3.$$

The interest will be \$135.

**NOW TRY** Exercise 43.

**OBJECTIVE 5** Solve combined variation problems. There are many combinations of direct and inverse variation. Example 7 shows a typical **combined variation** problem.

### EXAMPLE 7  Solving a Combined Variation Problem

Body mass index, or BMI, is used by physicians to assess a person's level of fatness. A BMI from 19 through 25 is considered desirable. BMI varies directly as an individual's weight in pounds and inversely as the square of the individual's height in inches. A person who weighs 118 lb and is 64 in. tall has a BMI of 20. (The BMI is rounded to the nearest whole number.) Find the BMI of a person who weighs 165 lb and is 70 in. tall. (*Source: The Washington Post.*)

Let $B$ represent the BMI, $w$ the weight, and $h$ the height. Then

$$B = \frac{kw}{h^2}.$$

 — BMI varies directly as the weight.
 — BMI varies inversely as the square of the height.

To find $k$, let $B = 20$, $w = 118$, and $h = 64$.

$$20 = \frac{k(118)}{64^2} \qquad B = \frac{kw}{h^2}$$

$$k = \frac{20(64^2)}{118} \qquad \text{Multiply by } 64^2; \text{ divide by } 118.$$

$$k \approx 694 \qquad \text{Use a calculator.}$$

Now find $B$ when $k = 694$, $w = 165$, and $h = 70$.

$$B = \frac{694(165)}{70^2} \approx 23 \qquad \text{Nearest whole number}$$

The person's BMI is 23.

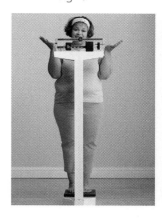

NOW TRY Exercise 51.

## 7.6 Exercises

NOW TRY Exercise

*Concept Check* Use personal experience or intuition to determine whether the situation suggests *direct* or *inverse variation*.

1. The number of raffle tickets you buy and your probability of winning that raffle
2. The rate and the distance traveled by a pickup truck in 2 hr
3. The amount of pressure put on the accelerator of a car and the speed of the car
4. The number of days from now until December 25 and the magnitude of the frenzy of Christmas shopping
5. Your age and the probability that you believe in Santa Claus
6. The surface area of a balloon and its diameter
7. The number of days until the end of the baseball season and the number of home runs that Albert Pujols has
8. The amount of gasoline you pump and the amount you pay

*Concept Check* Determine whether each equation represents *direct, inverse, joint,* or *combined variation*.

9. $y = \dfrac{3}{x}$
10. $y = \dfrac{8}{x}$
11. $y = 10x^2$
12. $y = 2x^3$
13. $y = 3xz^4$
14. $y = 6x^3z^2$
15. $y = \dfrac{4x}{wz}$
16. $y = \dfrac{6x}{st}$

**17.** *Fill in the Blanks* For $k > 0$, if $y$ varies directly as $x$, then when $x$ increases, $y$ _____, and when $x$ decreases, $y$ _____.

**18.** *Fill in the Blanks* For $k > 0$, if $y$ varies inversely as $x$, then when $x$ increases, $y$ _____, and when $x$ decreases, $y$ _____.

*Concept Check* Solve each problem.

**19.** If $x$ varies directly as $y$, and $x = 9$ when $y = 3$, find $x$ when $y = 12$.

**20.** If $x$ varies directly as $y$, and $x = 10$ when $y = 7$, find $y$ when $x = 50$.

**21.** If $a$ varies directly as the square of $b$, and $a = 4$ when $b = 3$, find $a$ when $b = 2$.

**22.** If $h$ varies directly as the square of $m$, and $h = 15$ when $m = 5$, find $h$ when $m = 7$.

**23.** If $z$ varies inversely as $w$, and $z = 10$ when $w = 0.5$, find $z$ when $w = 8$.

**24.** If $t$ varies inversely as $s$, and $t = 3$ when $s = 5$, find $s$ when $t = 5$.

**25.** If $m$ varies inversely as $p^2$, and $m = 20$ when $p = 2$, find $m$ when $p = 5$.

**26.** If $a$ varies inversely as $b^2$, and $a = 48$ when $b = 4$, find $a$ when $b = 7$.

**27.** $p$ varies jointly as $q$ and $r^2$, and $p = 200$ when $q = 2$ and $r = 3$. Find $p$ when $q = 5$ and $r = 2$.

**28.** $f$ varies jointly as $g^2$ and $h$, and $f = 50$ when $g = 4$ and $h = 2$. Find $f$ when $g = 3$ and $h = 6$.

*Solve each problem involving variation. See Examples 1–7.*

**29.** Ben bought 15 gal of gasoline and paid $43.79. To the nearest tenth of a cent, what is the price of gasoline per gallon?

**30.** Sara gives horseback rides at Shadow Mountain Ranch. A 2.5-hr ride costs $75.00. What is the price per hour?

**31.** The weight of an object on Earth is directly proportional to the weight of that same object on the moon. A 200-lb astronaut would weigh 32 lb on the moon. How much would a 50-lb dog weigh on the moon?

**32.** The pressure exerted by a certain liquid at a given point is directly proportional to the depth of the point beneath the surface of the liquid. The pressure at 30 m is 80 newtons. What pressure is exerted at 50 m?

**33.** The volume of a can of pork and beans is directly proportional to the height of the can. If the volume of the can is 300 in.³ when its height is 10.62 in., find the volume of a can with height 15.92 in.

**34.** The force required to compress a spring is directly proportional to the change in length of the spring. If a force of 20 newtons is required to compress a certain spring 2 cm, how much force is required to compress the spring from 20 cm to 8 cm?

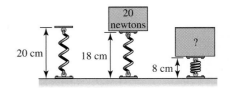

**35.** For a body falling freely from rest (disregarding air resistance), the distance the body falls varies directly as the square of the time. If an object is dropped from the top of a tower 576 ft high and hits the ground in 6 sec, how far did it fall in the first 4 sec?

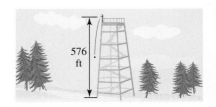

36. The amount of water emptied by a pipe varies directly as the square of the diameter of the pipe. For a certain constant water flow, a pipe emptying into a canal will allow 200 gal of water to escape in an hour. The diameter of the pipe is 6 in. How much water would a 12-in. pipe empty into the canal in an hour, assuming the same water flow?

37. Over a specified distance, speed varies inversely with time. If a Dodge Viper on a test track goes a certain distance in one-half minute at 160 mph, what speed is needed to go the same distance in three-fourths minute?

38. For a constant area, the length of a rectangle varies inversely as the width. The length of a rectangle is 27 ft when the width is 10 ft. Find the width of a rectangle with the same area if the length is 18 ft.

39. The frequency of a vibrating string varies inversely as its length. That is, a longer string vibrates fewer times in a second than a shorter string. Suppose a piano string 2 ft long vibrates 250 cycles per sec. What frequency would a string 5 ft long have?

40. The current in a simple electrical circuit varies inversely as the resistance. If the current is 20 amps when the resistance is 5 ohms, find the current when the resistance is 7.5 ohms.

41. The amount of light (measured in foot-candles) produced by a light source varies inversely as the square of the distance from the source. If the illumination produced 1 m from a light source is 768 foot-candles, find the illumination produced 6 m from the same source.

42. The force with which Earth attracts an object above Earth's surface varies inversely as the square of the distance of the object from the center of Earth. If an object 4000 mi from the center of Earth is attracted with a force of 160 lb, find the force of attraction if the object were 6000 mi from the center of Earth.

43. For a given interest rate, simple interest varies jointly as principal and time. If $2000 left in an account for 4 yr earned interest of $280, how much interest would be earned in 6 yr?

44. The collision impact of an automobile varies jointly as its mass and the square of its speed. Suppose a 2000-lb car traveling at 55 mph has a collision impact of 6.1. What is the collision impact of the same car at 65 mph?

45. The weight of a bass varies jointly as its girth and the square of its length. (*Girth* is the distance around the body of the fish.) A prize-winning bass weighed in at 22.7 lb and measured 36 in. long with a 21-in. girth. How much would a bass 28 in. long with an 18-in. girth weigh?

46. The weight of a trout varies jointly as its length and the square of its girth. One angler caught a trout that weighed 10.5 lb and measured 26 in. long with an 18-in. girth. Find the weight of a trout that is 22 in. long with a 15-in. girth.

47. The force needed to keep a car from skidding on a curve varies inversely as the radius of the curve and jointly as the weight of the car and the square of the speed. If 242 lb of force keeps a 2000-lb car from skidding on a curve of radius 500 ft at 30 mph, what force would keep the same car from skidding on a curve of radius 750 ft at 50 mph?

48. The maximum load that a cylindrical column with a circular cross section can hold varies directly as the fourth power of the diameter of the cross section and inversely as the square of the height. A 9-m column 1 m in diameter will support 8 metric tons. How many metric tons can be supported by a column 12 m high and $\frac{2}{3}$ m in diameter?

Load = 8 metric tons

49. The number of long-distance phone calls between two cities during a certain period varies jointly as the populations of the cities, $p_1$ and $p_2$, and inversely as the distance between them. If 80,000 calls are made between two cities 400 mi apart, with populations of 70,000 and 100,000, how many calls are made between cities with populations of 50,000 and 75,000 that are 250 mi apart?

50. Natural gas provides 35.8% of U.S. energy. (*Source:* U.S. Energy Department.) The volume of gas varies inversely as the pressure and directly as the temperature. [Temperature must be measured in *kelvins* (K), a unit of measurement used in physics.] If a certain gas occupies a volume of 1.3 L at 300 K and a pressure of 18 newtons, find the volume at 340 K and a pressure of 24 newtons.

51. A body mass index from 27 through 29 carries a slight risk of weight-related health problems, while one of 30 or more indicates a great increase in risk. Use your own height and weight and the information in Example 7 to determine your BMI and whether you are at risk.

52. The maximum load of a horizontal beam that is supported at both ends varies directly as the width and the square of the height and inversely as the length between the supports. A beam 6 m long, 0.1 m wide, and 0.06 m high supports a load of 360 kg. What is the maximum load supported by a beam 16 m long, 0.2 m wide, and 0.08 m high?

53. Explain the difference between inverse variation and direct variation.

54. What is meant by the constant of variation in a direct variation problem? If you were to graph the linear equation $y = kx$ for some nonnegative constant $k$, what role would $k$ play in the graph?

# 7 SUMMARY

## KEY TERMS

**7.1** rational expression
rational function

**7.2** least common denominator (LCD)

**7.3** complex fraction

**7.4** rational equation
domain of the variable in a rational equation
discontinuous
vertical asymptote

**7.5** ratio
proportion

**7.6** vary directly
proportional
constant of variation

vary inversely
vary jointly
combined variation

## QUICK REVIEW

### CONCEPTS

### EXAMPLES

### 7.1 Rational Expressions and Functions; Multiplying and Dividing

**Rational Functions**

A function of the form

$$f(x) = \frac{P(x)}{Q(x)}, \quad \text{where } Q(x) \neq 0,$$

is a rational function. Its domain consists of all real numbers except those that make $Q(x) = 0$.

Find the domain.

$$f(x) = \frac{2x + 1}{3x + 6}$$

Solve $3x + 6 = 0$ to find $x = -2$. This is the only real number excluded from the domain. The domain is $\{x \mid x \neq -2\}$.

**Fundamental Property of Rational Numbers**

If $\frac{a}{b}$ is a rational number and if $c$ is any nonzero real number, then

$$\frac{a}{b} = \frac{ac}{bc}.$$

$$\frac{3}{4} = \frac{3 \cdot 5}{4 \cdot 5} = \frac{15}{20}$$

**Writing a Rational Expression in Lowest Terms**

*Step 1* Factor the numerator and the denominator completely.

*Step 2* Apply the fundamental property.

Write in lowest terms.

$$\frac{2x + 8}{x^2 - 16}$$

$$= \frac{2(x + 4)}{(x - 4)(x + 4)} \quad \text{Factor.}$$

$$= \frac{2}{x - 4} \quad \text{Lowest terms}$$

**Multiplying Rational Expressions**

*Step 1* Factor numerators and denominators.

*Step 2* Apply the fundamental property.

*Step 3* Multiply the remaining factors in the numerator and in the denominator.

*Step 4* Check that the product is in lowest terms.

Multiply.

$$\frac{x^2 + 2x + 1}{x^2 - 1} \cdot \frac{5}{3x + 3}$$

$$= \frac{(x + 1)^2}{(x - 1)(x + 1)} \cdot \frac{5}{3(x + 1)} \quad \text{Factor.}$$

$$= \frac{5}{3(x - 1)} \quad \text{Multiply; lowest terms}$$

*(continued)*

371

## CONCEPTS

**Dividing Rational Expressions**

Multiply the first rational expression by the reciprocal of the second.

## EXAMPLES

Divide.

$$\frac{2x+5}{x-3} \div \frac{2x^2+3x-5}{x^2-9}$$

$$= \frac{2x+5}{x-3} \cdot \frac{x^2-9}{2x^2+3x-5} \quad \text{Multiply by the reciprocal.}$$

$$= \frac{2x+5}{x-3} \cdot \frac{(x+3)(x-3)}{(2x+5)(x-1)} \quad \text{Factor.}$$

$$= \frac{x+3}{x-1} \quad \text{Multiply; lowest terms}$$

## 7.2 Adding and Subtracting Rational Expressions

**Adding or Subtracting Rational Expressions**

**Step 1** If the denominators are the same, add or subtract the numerators. Place the result over the common denominator.

If the denominators are different, write all rational expressions with the LCD. Then add or subtract the numerators, and place the result over the common denominator.

**Step 2** Make sure that the answer is in lowest terms.

Subtract.

$$\frac{1}{x+6} - \frac{3}{x+2}$$

$$= \frac{x+2}{(x+6)(x+2)} - \frac{3(x+6)}{(x+6)(x+2)}$$

$$= \frac{x+2-3(x+6)}{(x+6)(x+2)}$$

$$= \frac{x+2-3x-18}{(x+6)(x+2)} \quad \text{Be careful with signs.}$$

$$= \frac{-2x-16}{(x+6)(x+2)}$$

## 7.3 Complex Fractions

**Simplifying a Complex Fraction**

**Method 1** Simplify the numerator and denominator separately as much as possible. Then multiply the numerator by the reciprocal of the denominator. Write the answer in lowest terms.

Simplify the complex fraction.

**Method 1**

$$\frac{\dfrac{1}{x^2} - \dfrac{1}{y^2}}{\dfrac{1}{x} + \dfrac{1}{y}}$$

$$= \frac{\dfrac{y^2}{x^2y^2} - \dfrac{x^2}{x^2y^2}}{\dfrac{y}{xy} + \dfrac{x}{xy}}$$

$$= \frac{\dfrac{y^2 - x^2}{x^2y^2}}{\dfrac{y+x}{xy}}$$

$$= \frac{y^2 - x^2}{x^2y^2} \div \frac{y+x}{xy}$$

$$= \frac{(y+x)(y-x)}{x^2y^2} \cdot \frac{xy}{y+x}$$

$$= \frac{y-x}{xy}$$

*(continued)*

| CONCEPTS | EXAMPLES |
|---|---|
| **Method 2** Multiply the numerator and denominator of the complex fraction by the least common denominator of all fractions appearing in the complex fraction. Then simplify the result. | **Method 2** $\dfrac{\dfrac{1}{x^2} - \dfrac{1}{y^2}}{\dfrac{1}{x} + \dfrac{1}{y}}$ $= \dfrac{x^2 y^2 \left(\dfrac{1}{x^2} - \dfrac{1}{y^2}\right)}{x^2 y^2 \left(\dfrac{1}{x} + \dfrac{1}{y}\right)}$ $= \dfrac{y^2 - x^2}{xy^2 + x^2 y}$ $= \dfrac{(y - x)(y + x)}{xy(y + x)}$ $= \dfrac{y - x}{xy}$ |

## 7.4 Equations with Rational Expressions and Graphs

**Solving an Equation with Rational Expressions**
To solve a rational equation, first determine the domain of the variable. Then multiply all the terms in the equation by the least common denominator. Solve the resulting equation. Each proposed solution *must* be checked to see that it is in the domain.

Solve.
$$\frac{3x + 2}{x - 2} + \frac{2}{x(x - 2)} = \frac{-1}{x}$$

Note that 0 and 2 are excluded from the domain.

$x(3x + 2) + 2 = -(x - 2)$    Multiply by $x(x - 2)$.
$3x^2 + 2x + 2 = -x + 2$    Distributive property
$3x^2 + 3x = 0$    Add $x$; subtract 2.
$3x(x + 1) = 0$    Factor.
$3x = 0$   or   $x + 1 = 0$    Zero-factor property
$x = 0$   or   $x = -1$

Of the two proposed solutions, 0 must be discarded because it is not in the domain. The solution set is $\{-1\}$.

The graph of a rational function (written in lowest terms) may have one or more breaks. At such points, the graph will approach an asymptote.

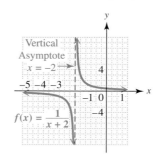

## 7.5 Applications of Rational Expressions

To solve a formula for a particular variable, isolate that variable on one side.

Solve for $L$.
$$c = \frac{100b}{L}$$
$cL = 100b$    Multiply by $L$.
$L = \dfrac{100b}{c}$    Divide by $c$.

*(continued)*

| CONCEPTS | EXAMPLES |
|---|---|
| To solve a motion problem, use the formula $$d = rt$$ or one of its equivalents, $$t = \frac{d}{r} \quad \text{or} \quad r = \frac{d}{t}.$$ | Solve. <br> A canal has a current of 2 mph. Find the speed of Amy's boat in still water if it goes 11 mi downstream in the same time that it goes 8 mi upstream. <br> Let $x$ represent the speed of the boat in still water. |

|  | Distance | Rate | Time |
|---|---|---|---|
| Downstream | 11 | $x + 2$ | $\frac{11}{x+2}$ |
| Upstream | 8 | $x - 2$ | $\frac{8}{x-2}$ |

Because the times are the same, the equation is

$$\frac{11}{x+2} = \frac{8}{x-2}. \qquad \text{Use } t = \frac{d}{r}.$$

The LCD is $(x + 2)(x - 2)$.

$$\begin{aligned} 11(x - 2) &= 8(x + 2) &&\text{Multiply by the LCD.} \\ 11x - 22 &= 8x + 16 &&\text{Distributive property} \\ 3x &= 38 &&\text{Subtract } 8x; \text{ add } 22. \\ x &= 12\tfrac{2}{3} &&\text{Divide by 3.} \end{aligned}$$

The speed in still water is $12\tfrac{2}{3}$ mph.

To solve a work problem, use the fact that if a complete job is done in $t$ units of time, the rate of work is $\frac{1}{t}$ job per unit of time. The amount of work accomplished is $A = rt$, so if one job is accomplished in time $t$, use the formula

$$1 = rt.$$

## 7.6 Variation

If there is some constant $k$ such that

$y = kx^n,$ then $y$ varies directly as $x^n$.

$y = \dfrac{k}{x^n},$ then $y$ varies inversely as $x^n$.

$y = kxz,$ then $y$ varies jointly as $x$ and $z$.

The area of a circle varies directly as the square of the radius.

$$A = kr^2 \qquad \text{Here, } k = \pi.$$

Pressure varies inversely as volume.

$$p = \frac{k}{V}$$

For a given principal, interest varies jointly as interest rate and time.

$$I = krt \qquad k \text{ is the given principal.}$$

# 7 REVIEW EXERCISES

*(a) Find all real numbers that are excluded from the domain. (b) Give the domain in set notation.*

1. $f(x) = \dfrac{-7}{3x + 18}$

2. $f(x) = \dfrac{5x + 17}{x^2 - 7x + 10}$

3. $f(x) = \dfrac{9}{x^2 - 18x + 81}$

*Write in lowest terms.*

4. $\dfrac{12x^2 + 6x}{24x + 12}$

5. $\dfrac{25m^2 - n^2}{25m^2 - 10mn + n^2}$

6. $\dfrac{r - 2}{4 - r^2}$

7. What is meant by the reciprocal of a rational expression?

*Multiply or divide. Write the answer in lowest terms.*

8. $\dfrac{(2y + 3)^2}{5y} \cdot \dfrac{15y^3}{4y^2 - 9}$

9. $\dfrac{w^2 - 16}{w} \cdot \dfrac{3}{4 - w}$

10. $\dfrac{z^2 - z - 6}{z - 6} \cdot \dfrac{z^2 - 6z}{z^2 + 2z - 15}$

11. $\dfrac{m^3 - n^3}{m^2 - n^2} \div \dfrac{m^2 + mn + n^2}{m + n}$

*Assume that each expression is the denominator of a rational expression. Find the least common denominator for each group.*

12. $32b^3, \quad 24b^5$

13. $9r^2, \quad 3r + 1, \quad 9$

14. $6x^2 + 13x - 5, \quad 9x^2 + 9x - 4$

15. $3x - 12, \quad x^2 - 2x - 8, \quad x^2 - 8x + 16$

*Add or subtract as indicated.*

16. $\dfrac{5}{3x^6 y^5} - \dfrac{8}{9x^4 y^7}$

17. $\dfrac{5y + 13}{y + 1} - \dfrac{1 - 7y}{y + 1}$

18. $\dfrac{6}{5a + 10} + \dfrac{7}{6a + 12}$

19. $\dfrac{3r}{10r^2 - 3rs - s^2} + \dfrac{2r}{2r^2 + rs - s^2}$

20. *Concept Check* Two students worked the following problem:

$$\text{Add: } \dfrac{2}{y - x} + \dfrac{1}{x - y}.$$

One student got a sum of $\dfrac{1}{y - x}$ and the other got a sum of $\dfrac{-1}{x - y}$. Which student(s) got the correct answer?

*Simplify each complex fraction.*

21. $\dfrac{\dfrac{3}{t} + 2}{\dfrac{4}{t} - 7}$

22. $\dfrac{\dfrac{2}{m - 3n}}{\dfrac{1}{3n - m}}$

23. $\dfrac{\dfrac{3}{p} - \dfrac{2}{q}}{\dfrac{9q^2 - 4p^2}{qp}}$

24. $\dfrac{x^{-2} - y^{-2}}{x^{-1} - y^{-1}}$

*Solve each equation.*

25. $\dfrac{1}{t + 4} + \dfrac{1}{2} = \dfrac{3}{2t + 8}$

26. $\dfrac{-5m}{m + 1} + \dfrac{m}{3m + 3} = \dfrac{56}{6m + 6}$

27. $\dfrac{2}{k - 1} - \dfrac{4k + 1}{k^2 - 1} = \dfrac{-1}{k + 1}$

28. $\dfrac{5}{x + 2} + \dfrac{3}{x + 3} = \dfrac{x}{x^2 + 5x + 6}$

**29.** *Concept Check*  After solving the equation

$$\frac{3}{x-3} - \frac{2}{x-2} = \frac{3}{x^2 - 5x + 6},$$

a student got $x = 3$ as her final step. She could not understand why the answer in the back of the book was "∅," because she checked her algebra several times and was sure that all her algebraic work was correct. Was she wrong, or was the answer in the back of the book wrong?

**30.** Explain the difference between the following:

Simplify the expression $\frac{4}{x} + \frac{1}{2} - \frac{1}{3}$;  Solve the equation $\frac{4}{x} + \frac{1}{2} = \frac{1}{3}$.

**31.** *Multiple Choice*  Which graph has a vertical asymptote, and what is its equation?

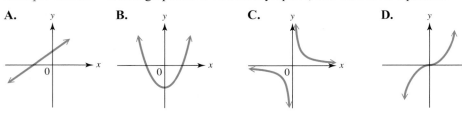

**A.**   **B.**   **C.**   **D.**

**32.** The equation from a law from physics is in the form $\frac{1}{A} = \frac{1}{B} + \frac{1}{C}$. Find $A$ if $B = 30$ and $C = 10$.

*Solve each formula for the specified variable.*

**33.** $F = \dfrac{GMm}{d^2}$  for $m$  (physics)

**34.** $\mu = \dfrac{Mv}{M + m}$  for $M$  (electronics)

*Solve each problem.*

**35.** An article in *Scientific American* predicts that, in the year 2050, about 23,200 of the 58,000 passenger-km per day in North America will be provided by high-speed trains. If the traffic volume in a typical region of North America is 15,000, how many passenger-kilometers per day will high-speed trains provide there? (*Source:* Schafer, Andreas, and David Victor, "The Past and Future of Global Mobility," *Scientific American,* October 1997.)

**36.** A river has a current of 4 km per hr. Find the speed of Lynn McTernan's boat in still water if it goes 40 km downstream in the same time that it takes to go 24 km upstream.

|  | d | r | t |
|---|---|---|---|
| Upstream | 24 | $x - 4$ |  |
| Downstream | 40 |  |  |

**37.** A sink can be filled by a cold-water tap in 8 min and by a hot-water tap in 12 min. How long would it take to fill the sink with both taps open?

**38.** Melena Fenn and Jeff Houck need to sort a pile of bottles at the recycling center. Working alone, Melena could do the entire job in 9 hr, while Jeff could do the entire job in 6 hr. How long will it take them if they work together?

**39.** *Multiple Choice*  In which one of the following does $y$ vary inversely as $x$?

**A.** $y = 2x$   **B.** $y = \dfrac{x}{3}$   **C.** $y = \dfrac{3}{x}$   **D.** $y = x^2$

*Solve each problem.*

**40.** For the subject in a photograph to appear in the same perspective in the photograph as in real life, the viewing distance must be properly related to the amount of enlargement. For a particular camera, the viewing distance varies directly as the amount of enlargement. A picture that is taken with this camera and enlarged 5 times should be viewed from a distance of 250 mm. Suppose a print 8.6 times the size of the negative is made. From what distance should it be viewed?

**41.** The frequency (number of vibrations per second) of a vibrating guitar string varies inversely as its length. That is, a longer string vibrates fewer times in a second than a shorter string. Suppose a guitar string 0.65 m long vibrates 4.3 times per sec. What frequency would a string 0.5 m long have?

**42.** The volume of a rectangular box of a given height is proportional to its width and length. A box with width 2 ft and length 4 ft has volume 12 ft³. Find the volume of a box with the same height, but that is 3 ft wide and 5 ft long.

## 7 TEST

**1.** Find all real numbers excluded from the domain of $f(x) = \dfrac{x+3}{3x^2 + 2x - 8}$. Then give the domain in set notation.

**2.** Write $\dfrac{6x^2 - 13x - 5}{9x^3 - x}$ in lowest terms.

*Multiply or divide.*

**3.** $\dfrac{(x+3)^2}{4} \cdot \dfrac{6}{2x+6}$

**4.** $\dfrac{y^2 - 16}{y^2 - 25} \cdot \dfrac{y^2 + 2y - 15}{y^2 - 7y + 12}$

**5.** $\dfrac{3-t}{5} \div \dfrac{t-3}{10}$

**6.** $\dfrac{x^2 - 9}{x^3 + 3x^2} \div \dfrac{x^2 + x - 12}{x^3 + 9x^2 + 20x}$

**7.** Find the least common denominator for the following group of denominators: $t^2 + t - 6$, $t^2 + 3t$, $t^2$.

*Add or subtract as indicated.*

**8.** $\dfrac{7}{6t^2} - \dfrac{1}{3t}$

**9.** $\dfrac{3}{7a^4b^3} + \dfrac{5}{21a^5b^2}$

**10.** $\dfrac{9}{x^2 - 6x + 9} + \dfrac{2}{x^2 - 9}$

**11.** $\dfrac{6}{x+4} + \dfrac{1}{x+2} - \dfrac{3x}{x^2 + 6x + 8}$

*Simplify each complex fraction.*

**12.** $\dfrac{\dfrac{12}{r+4}}{\dfrac{11}{6r+24}}$

**13.** $\dfrac{\dfrac{1}{a} - \dfrac{1}{b}}{\dfrac{a}{b} - \dfrac{b}{a}}$

**14.** $\dfrac{2x^{-2} + y^{-2}}{x^{-1} - y^{-1}}$

15. One of the following is an expression to be simplified by algebraic operations, and the other is an equation to be solved. Identify each, and then simplify the one that is an expression and solve the one that is an equation.

    (a) $\dfrac{2x}{3} + \dfrac{x}{4} - \dfrac{11}{2}$
    
    (b) $\dfrac{2x}{3} + \dfrac{x}{4} = \dfrac{11}{2}$

*Solve each equation.*

16. $\dfrac{1}{x} - \dfrac{4}{3x} = \dfrac{1}{x-2}$

17. $\dfrac{y}{y+2} - \dfrac{1}{y-2} = \dfrac{8}{y^2-4}$

18. Checking the solution(s) of an equation in earlier chapters verified that the algebraic steps were performed correctly. When an equation includes a term with a variable denominator, what additional reason *requires* that the solutions be checked?

19. Solve for the variable $\ell$ in this formula from mathematics: $S = \dfrac{n}{2}(a + \ell)$.

20. Graph the function defined by $f(x) = \dfrac{-2}{x+1}$. Give the equation of its vertical asymptote.

*Solve each problem.*

21. Wayne can do a job in 9 hr, while Susan can do the same job in 5 hr. How long would it take them to do the job if they worked together?

22. The rate of the current in a stream is 3 mph. Nana's boat can go 36 mi downstream in the same time that it takes to go 24 mi upstream. Find the rate of her boat in still water.

23. Biologists collected a sample of 600 fish from Lake Linda on May 1 and tagged each of them. When they returned on June 1, a new sample of 800 fish was collected and 10 of these had been previously tagged. Use this experiment to determine the approximate fish population of Lake Linda.

24. In biology, the function defined by

    $$g(x) = \dfrac{5x}{2+x}$$

    gives the growth rate of a population for $x$ units of available food. (*Source:* Smith, J. Maynard, *Models in Ecology,* Cambridge University Press, 1974.)

    (a) What amount of food (in appropriate units) would produce a growth rate of 3 units of growth per unit of food?
    
    (b) What is the growth rate if no food is available?

25. The current in a simple electrical circuit is inversely proportional to the resistance. If the current is 80 amps when the resistance is 30 ohms, find the current when the resistance is 12 ohms.

26. The force of the wind blowing on a vertical surface varies jointly as the area of the surface and the square of the velocity. If a wind blowing at 40 mph exerts a force of 50 lb on a surface of 500 ft², how much force will a wind of 80 mph place on a surface of 2 ft²?

# Roots, Radicals, and Root Functions

Tom Skilling is the chief meteorologist for the *Chicago Tribune*. He writes a column titled "Ask Tom Why," in which readers question him on a variety of topics. In the Saturday, August 17, 2002 issue, reader Ted Fleischaker wrote,

*I cannot remember the formula to calculate the distance to the horizon. I have a stunning view from my 14th-floor condo, 150 ft above the ground. How far can I see?*

Skilling's answer in Section 8.3, Exercise 125, provides a formula for finding the distance to the horizon. The formula includes a *square root,* one of the topics of this chapter.

**8.1** Radical Expressions and Graphs

**8.2** Rational Exponents

**8.3** Simplifying Radical Expressions

**8.4** Adding and Subtracting Radical Expressions

**8.5** Multiplying and Dividing Radical Expressions

**8.6** Solving Equations with Radicals

**8.7** Complex Numbers

# 8.1 Radical Expressions and Graphs

**OBJECTIVES**
1. Find roots of numbers.
2. Find principal roots.
3. Graph functions defined by radical expressions.
4. Find $n$th roots of $n$th powers.
5. Use a calculator to find roots.

**OBJECTIVE 1** Find roots of numbers. Recall from **Section 1.3** that $6^2 = 36$; that is, 6 *squared* is 36. The opposite (or inverse) of *squaring* a number is taking its *square root*. Thus,

*It is customary to write $\sqrt{\phantom{x}}$, rather than $\sqrt[2]{\phantom{x}}$.*

$\sqrt{36} = 6$, because $6^2 = 36$.

We now extend our discussion of roots to *cube roots* $\sqrt[3]{\phantom{x}}$, *fourth roots* $\sqrt[4]{\phantom{x}}$, and higher roots. In general, $\sqrt[n]{a}$ is a number whose $n$th power equals $a$. That is,

$$\sqrt[n]{a} = b \quad \text{means} \quad b^n = a.$$

The number $a$ is the **radicand**, $n$ is the **index** or **order**, and the expression $\sqrt[n]{a}$ is a **radical**.

### EXAMPLE 1 Simplifying Higher Roots

Simplify.

(a) $\sqrt[3]{64} = 4$, because $4^3 = 64$.  (b) $\sqrt[3]{125} = 5$, because $5^3 = 125$.
(c) $\sqrt[4]{16} = 2$, because $2^4 = 16$.  (d) $\sqrt[5]{32} = 2$, because $2^5 = 32$.
(e) $\sqrt[3]{\dfrac{8}{27}} = \dfrac{2}{3}$, because $\left(\dfrac{2}{3}\right)^3 = \dfrac{8}{27}$.
(f) $\sqrt[4]{0.0016} = 0.2$, because $(0.2)^4 = 0.0016$.

**NOW TRY** Exercises 5, 15, 33, and 39.

**OBJECTIVE 2** Find principal roots. If $n$ is even, positive numbers have two $n$th roots. For example, both 4 and $-4$ are square roots of 16, and 2 and $-2$ are fourth roots of 16. In such cases, the notation $\sqrt[n]{a}$ represents the positive root, called the **principal root**, and $-\sqrt[n]{a}$ represents the negative root.

### $n$th Root

1. If $n$ is **even** and $a$ is **positive or 0,** then

    $\sqrt[n]{a}$ represents the **principal $n$th root** of $a$,

    and $-\sqrt[n]{a}$ represents the **negative $n$th root** of $a$.

2. If $n$ is **even** and $a$ is **negative,** then

    $\sqrt[n]{a}$ is not a real number.

3. If $n$ is **odd,** then

    there is exactly one real $n$th root of $a$, written $\sqrt[n]{a}$.

If $n$ is even, the two $n$th roots of $a$ are often written together as $\pm \sqrt[n]{a}$, with $\pm$ read "positive or negative," or "plus or minus."

### EXAMPLE 2  Finding Roots

Find each root.

**(a)** $\sqrt{100} = 10$

Because the radicand, 100, is *positive,* there are two square roots: 10 and $-10$. We want the principal root, which is 10.

**(b)** $-\sqrt{100} = -10$

Here, we want the negative square root, $-10$.

**(c)** $\sqrt[4]{81} = 3$   Principal 4th root   **(d)** $-\sqrt[4]{81} = -3$   Negative 4th root

Parts (a)–(d) illustrate Case 1 in the preceding box.

**(e)** $\sqrt[4]{-81}$

The index is *even* and the radicand is *negative,* so $\sqrt[4]{-81}$ is not a real number. This is Case 2 in the preceding box.

**(f)** $\sqrt[3]{8} = 2$, because $2^3 = 8$.   **(g)** $\sqrt[3]{-8} = -2$, because $(-2)^3 = -8$.

Parts (f) and (g) illustrate Case 3 in the box. The index is *odd,* so each radical represents exactly one $n$th root (regardless of whether the radicand is positive, negative, or 0).

**NOW TRY** Exercises 13, 17, 21, and 25.

**OBJECTIVE 3** Graph functions defined by radical expressions. A **radical expression** is an algebraic expression that contains radicals.

$$3 - \sqrt{x}, \quad \sqrt[3]{x}, \quad \text{and} \quad \sqrt{2x-1} \quad \text{Radical expressions}$$

In earlier chapters, we graphed functions defined by polynomial and rational expressions. Now we examine the graphs of functions defined by the basic radical expressions $f(x) = \sqrt{x}$ and $f(x) = \sqrt[3]{x}$.

Figure 1 shows the graph of the **square root function,** together with a table of selected points. Only nonnegative values can be used for $x$, so the domain is $[0, \infty)$. Because $\sqrt{x}$ is the principal square root of $x$, it always has a nonnegative value, so the range is also $[0, \infty)$.

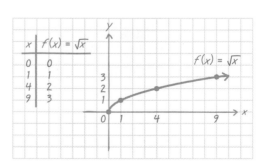

**FIGURE 1**

Figure 2 on the next page shows the graph of the **cube root function** and a table of selected points. Since any real number (positive, negative, or 0) can be used for $x$ in the cube root function, $\sqrt[3]{x}$ can be positive, negative, or 0. Thus, both the domain and the range of the cube root function are $(-\infty, \infty)$.

**382** CHAPTER 8 Roots, Radicals, and Root Functions

| x | $f(x) = \sqrt[3]{x}$ |
|---|---|
| −8 | −2 |
| −1 | −1 |
| 0 | 0 |
| 1 | 1 |
| 8 | 2 |

**FIGURE 2**

### EXAMPLE 3 Graphing Functions Defined with Radicals

Graph each function by creating a table of values. Give the domain and range.

**(a)** $f(x) = \sqrt{x - 3}$

A table of values is given with Figure 3. The *x*-values were chosen in such a way that the function values are all integers. For the radicand to be nonnegative, we must have $x - 3 \geq 0$, or $x \geq 3$. Therefore, the domain is $[3, \infty)$. Function values are positive or 0, so the range is $[0, \infty)$. The graph is shown in Figure 3.

| x | $f(x) = \sqrt{x - 3}$ |
|---|---|
| 3 | $\sqrt{3 - 3} = 0$ |
| 4 | $\sqrt{4 - 3} = 1$ |
| 7 | $\sqrt{7 - 3} = 2$ |

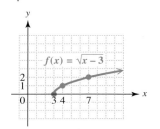

| x | $f(x) = \sqrt[3]{x} + 2$ |
|---|---|
| −8 | $\sqrt[3]{-8} + 2 = 0$ |
| −1 | $\sqrt[3]{-1} + 2 = 1$ |
| 0 | $\sqrt[3]{0} + 2 = 2$ |
| 1 | $\sqrt[3]{1} + 2 = 3$ |
| 8 | $\sqrt[3]{8} + 2 = 4$ |

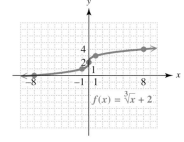

**FIGURE 3**    **FIGURE 4**

**(b)** $f(x) = \sqrt[3]{x} + 2$

See Figure 4. Both the domain and range are $(-\infty, \infty)$.

**NOW TRY** Exercises 41 and 45.

### OBJECTIVE 4 Find *n*th roots of *n*th powers.

What does $\sqrt{a^2}$ equal? Your first answer might be *a*, but this is not necessarily true. For example, consider the following:

If $a = 6$, then $\sqrt{a^2} = \sqrt{6^2} = \sqrt{36} = 6$.

If $a = -6$, then $\sqrt{a^2} = \sqrt{(-6)^2} = \sqrt{36} = 6$. ← Instead of −6, we get 6, the *absolute value* of −6.

Since the symbol $\sqrt{a^2}$ represents the *nonnegative* square root, we express $\sqrt{a^2}$ with absolute value bars as $|a|$ in case *a* is a negative number.

---

$\sqrt{a^2}$

For any real number *a*,    $\sqrt{a^2} = |a|$.

That is, the principal square root of $a^2$ is the absolute value of *a*.

### EXAMPLE 4  Simplifying Square Roots by Using Absolute Value

Find each square root.

(a) $\sqrt{7^2} = |7| = 7$

(b) $\sqrt{(-7)^2} = |-7| = 7$

(c) $\sqrt{k^2} = |k|$

(d) $\sqrt{(-k)^2} = |-k| = |k|$

**NOW TRY** Exercises 49, 51, 59, and 61.

We can generalize this idea to any $n$th root.

#### $\sqrt[n]{a^n}$

If $n$ is an **even** positive integer, then $\sqrt[n]{a^n} = |a|$.

If $n$ is an **odd** positive integer, then $\sqrt[n]{a^n} = a$.

That is, use absolute value when $n$ is even; absolute value is not necessary when $n$ is odd.

### EXAMPLE 5  Simplifying Higher Roots by Using Absolute Value

Simplify each root.

(a) $\sqrt[6]{(-3)^6} = |-3| = 3$     $n$ is even; use absolute value.

(b) $\sqrt[5]{(-4)^5} = -4$     $n$ is odd.

(c) $-\sqrt[4]{(-9)^4} = -|-9| = -9$     $n$ is even; use absolute value.

(d) $-\sqrt{m^4} = -|m^2| = -m^2$

No absolute value bars are needed here, because $m^2$ is nonnegative for any real number value of $m$.

(e) $\sqrt[3]{a^{12}} = a^4$, because $a^{12} = (a^4)^3$.

(f) $\sqrt[4]{x^{12}} = |x^3|$

We use absolute value bars to guarantee that the result is not negative (because $x^3$ can be either positive or negative, depending on $x$). Also, $|x^3|$ can be written as $x^2 \cdot |x|$.

**NOW TRY** Exercises 53, 55, 57, and 65.

**OBJECTIVE 5** Use a calculator to find roots. While numbers such as $\sqrt{9}$ and $\sqrt[3]{-8}$ are rational, radicals are often irrational numbers. To find approximations of roots such as $\sqrt{15}$, $\sqrt[3]{10}$, and $\sqrt[4]{2}$, we usually use scientific or graphing calculators. Using a calculator, we find that

$$\sqrt{15} \approx 3.872983346, \quad \sqrt[3]{10} \approx 2.15443469, \quad \text{and} \quad \sqrt[4]{2} \approx 1.189207115,$$

where the symbol $\approx$ means "is approximately equal to." In this book, we usually show approximations rounded to three decimal places. Thus, we would write

$$\sqrt{15} \approx 3.873, \quad \sqrt[3]{10} \approx 2.154, \quad \text{and} \quad \sqrt[4]{2} \approx 1.189.$$

Figure 5 shows how the preceding approximations are displayed on a TI-83/84 Plus graphing calculator. In Figure 5(a), eight or nine decimal places are shown, while in Figure 5(b), the number of decimal places is fixed at three.

There is a simple way to check that a calculator approximation is "in the ballpark." Because 16 is a little larger than 15, $\sqrt{16} = 4$ should be a little larger than $\sqrt{15}$. Thus, 3.873 is reasonable as an approximation for $\sqrt{15}$.

```
√(15)
        3.872983346
³√(10)
        2.15443469
4*√2
        1.189207115
```
(a)

```
√(15)
            3.873
³√(10)
            2.154
4*√2
            1.189
```
(b)

**FIGURE 5**

> **EXAMPLE 6** Finding Approximations for Roots
>
> Use a calculator to verify that each approximation is correct.
>
> (a) $\sqrt{39} \approx 6.245$    (b) $-\sqrt{72} \approx -8.485$
> (c) $\sqrt[3]{93} \approx 4.531$    (d) $\sqrt[4]{39} \approx 2.499$
>
> **NOW TRY** Exercises 69, 73, 75, and 77.

## 8.1 Exercises

**NOW TRY Exercise**

*Matching* Match each expression from Column I with the equivalent choice from Column II. Answers may be used more than once. See Examples 1 and 2.

| I | | II | |
|---|---|---|---|
| 1. $-\sqrt{16}$ | 2. $\sqrt{-16}$ | A. 3 | B. $-2$ |
| 3. $\sqrt[3]{-27}$ | 4. $\sqrt[5]{-32}$ | C. 2 | D. $-3$ |
| 5. $\sqrt[4]{81}$ | 6. $\sqrt[3]{8}$ | E. $-4$ | F. Not a real number |

*Multiple Choice* Choose the closest approximation of each square root.

7. $\sqrt{123.5}$
   A. 9    B. 10    C. 11    D. 12

8. $\sqrt{67.8}$
   A. 7    B. 8    C. 9    D. 10

*Multiple Choice* Refer to the rectangle to answer the questions in Exercises 9 and 10.

9. Which one of the following is the best estimate of its area?
   A. 2500    B. 250    C. 50    D. 100

10. Which one of the following is the best estimate of its perimeter?
    A. 15    B. 250    C. 100    D. 30

11. *Concept Check* Consider the expression $-\sqrt{-a}$. Decide whether it is *positive*, *negative*, 0, or *not a real number* if
    (a) $a > 0$,    (b) $a < 0$,    (c) $a = 0$.

12. *Concept Check* If $n$ is odd, under what conditions is $\sqrt[n]{a}$
    (a) positive,    (b) negative,    (c) 0?

*Find each root that is a real number. See Examples 1 and 2.*

13. $-\sqrt{81}$
14. $-\sqrt{121}$
15. $\sqrt[3]{216}$
16. $\sqrt[3]{343}$
17. $\sqrt[3]{-64}$
18. $\sqrt[3]{-125}$
19. $-\sqrt[3]{512}$
20. $-\sqrt[3]{1000}$
21. $\sqrt[4]{1296}$
22. $\sqrt[4]{625}$
23. $-\sqrt[4]{16}$
24. $-\sqrt[4]{256}$
25. $\sqrt[4]{-625}$
26. $\sqrt[4]{-256}$
27. $\sqrt[6]{64}$
28. $\sqrt[6]{729}$
29. $\sqrt[6]{-32}$
30. $\sqrt[8]{-1}$
31. $\sqrt{\dfrac{64}{81}}$
32. $\sqrt{\dfrac{100}{9}}$
33. $\sqrt[3]{\dfrac{64}{27}}$
34. $\sqrt[4]{\dfrac{81}{16}}$
35. $-\sqrt[6]{\dfrac{1}{64}}$
36. $-\sqrt[5]{\dfrac{1}{32}}$
37. $\sqrt{0.49}$
38. $\sqrt{0.81}$
39. $\sqrt[3]{0.001}$
40. $\sqrt[3]{0.125}$

*Graph each function and give its domain and range. See Example 3.*

41. $f(x) = \sqrt{x+3}$
42. $f(x) = \sqrt{x-5}$
43. $f(x) = \sqrt{x} - 2$
44. $f(x) = \sqrt{x} + 4$
45. $f(x) = \sqrt[3]{x} - 3$
46. $f(x) = \sqrt[3]{x} + 1$
47. $f(x) = \sqrt[3]{x-3}$
48. $f(x) = \sqrt[3]{x+1}$

*Simplify each root. See Examples 4 and 5.*

49. $\sqrt{12^2}$
50. $\sqrt{19^2}$
51. $\sqrt{(-10)^2}$
52. $\sqrt{(-13)^2}$
53. $\sqrt[6]{(-2)^6}$
54. $\sqrt[6]{(-4)^6}$
55. $\sqrt[5]{(-9)^5}$
56. $\sqrt[5]{(-8)^5}$
57. $-\sqrt[6]{(-5)^6}$
58. $-\sqrt[6]{(-7)^6}$
59. $\sqrt{x^2}$
60. $-\sqrt{x^2}$
61. $\sqrt{(-z)^2}$
62. $\sqrt{(-q)^2}$
63. $\sqrt[3]{x^3}$
64. $-\sqrt[3]{x^3}$
65. $\sqrt[3]{x^{15}}$
66. $\sqrt[3]{m^9}$
67. $\sqrt[6]{x^{30}}$
68. $\sqrt[4]{k^{20}}$

*Find a decimal approximation for each radical. Round the answer to three decimal places. See Example 6.*

69. $\sqrt{9483}$
70. $\sqrt{6825}$
71. $\sqrt{284.361}$
72. $\sqrt{846.104}$
73. $-\sqrt{82}$
74. $-\sqrt{91}$
75. $\sqrt[3]{423}$
76. $\sqrt[3]{555}$
77. $\sqrt[4]{100}$
78. $\sqrt[4]{250}$
79. $\sqrt[5]{23.8}$
80. $\sqrt[5]{98.4}$

*Solve each problem.*

81. According to an article in *The World Scanner Report*, the distance $D$, in miles, to the horizon from an observer's point of view over water or "flat" earth is given by

$$D = \sqrt{2H},$$

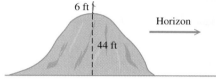

where $H$ is the height of the point of view, in feet. If a person whose eyes are 6 ft above ground level is standing at the top of a hill 44 ft above "flat" earth, approximately how far to the horizon will she be able to see?

82. The time for one complete swing of a simple pendulum is

$$t = 2\pi\sqrt{\frac{L}{g}},$$

where $t$ is time in seconds, $L$ is the length of the pendulum in feet, and $g$, the acceleration due to gravity, is about 32 ft per sec². Find the time of a complete swing of a 2-ft pendulum to the nearest tenth of a second.

83. **Heron's formula** gives a method of finding the area of a triangle if the lengths of its sides are known. Suppose that $a$, $b$, and $c$ are the lengths of the sides. Let $s$ denote one-half of the perimeter of the triangle (called the *semiperimeter*); that is,

$$s = \frac{1}{2}(a + b + c).$$

Then the area of the triangle is

$$A = \sqrt{s(s - a)(s - b)(s - c)}.$$

Find the area of the Bermuda Triangle if the "sides" of this triangle measure approximately 850 mi, 925 mi, and 1300 mi. Give your answer to the nearest thousand square miles.

84. The Vietnam Veterans Memorial in Washington, DC, is in the shape of an unenclosed isosceles triangle with equal sides of length 246.75 ft. If the triangle were enclosed, the third side would have length 438.14 ft. Use Heron's formula from the previous exercise to find the area of this enclosure to the nearest hundred square feet. (*Source:* Information pamphlet obtained at the Vietnam Veterans Memorial.)

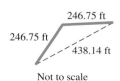

## 8.2 Rational Exponents

**OBJECTIVES**
1. Use exponential notation for nth roots.
2. Define and use expressions of the form $a^{m/n}$.
3. Convert between radicals and rational exponents.
4. Use the rules for exponents with rational exponents.

**OBJECTIVE 1** Use exponential notation for *n*th roots. We now look at exponents that are rational numbers of the form $\frac{1}{n}$, or $1/n$, where $n$ is a natural number.

Consider the product $(3^{1/2})^2 = 3^{1/2} \cdot 3^{1/2}$. Using the rules of exponents from **Section 5.1,** we can simplify this product as follows:

$$(3^{1/2})^2 = 3^{1/2} \cdot 3^{1/2}$$
$$= 3^{1/2+1/2} \quad \text{Product rule:}$$
$$= 3^1 \quad \text{Add exponents.}$$
$$= 3.$$

Also, by definition, $(\sqrt{3})^2 = \sqrt{3} \cdot \sqrt{3} = 3$. Since both $(3^{1/2})^2$ and $(\sqrt{3})^2$ are equal to 3, it seems reasonable to define

$$3^{1/2} = \sqrt{3}.$$

This suggests the following generalization.

---

**$a^{1/n}$**

If $\sqrt[n]{a}$ is a real number, then $\quad a^{1/n} = \sqrt[n]{a}.$

---

For example,

$$4^{1/2} = \sqrt{4}, \quad 8^{1/3} = \sqrt[3]{8}, \quad \text{and} \quad 16^{1/4} = \sqrt[4]{16}.$$

*Notice that the denominator of the rational exponent is the index of the radical.*

---

**EXAMPLE 1** Evaluating Exponentials of the Form $a^{1/n}$

Evaluate each exponential.

> The denominator is the index.

> The denominator is the index. $\sqrt{\phantom{x}}$ means $\sqrt[2]{\phantom{x}}$.

(a) $64^{1/3} = \sqrt[3]{64} = 4$ 　　　(b) $100^{1/2} = \sqrt{100} = 10$

(c) $-256^{1/4} = -\sqrt[4]{256} = -4$

(d) $(-256)^{1/4} = \sqrt[4]{-256}$ is not a real number, because the radicand, $-256$, is negative and the index is even.

(e) $(-32)^{1/5} = \sqrt[5]{-32} = -2$ 　　　(f) $\left(\frac{1}{8}\right)^{1/3} = \sqrt[3]{\frac{1}{8}} = \frac{1}{2}$

**NOW TRY** Exercises 11, 13, 19, and 21.

---

**CAUTION** Notice the difference between parts (c) and (d) in Example 1. The radical in part (c) is the **negative fourth root of a positive number,** while the radical in part (d) is the **principal fourth root of a negative number, which is not a real number.**

**OBJECTIVE 2** Define and use expressions of the form $a^{m/n}$. We know that $8^{1/3} = \sqrt[3]{8}$. How should we define a number like $8^{2/3}$? For past rules of exponents to be valid,

$$8^{2/3} = 8^{(1/3)2} = (8^{1/3})^2.$$

Since $8^{1/3} = \sqrt[3]{8}$,

$$8^{2/3} = (\sqrt[3]{8})^2 = 2^2 = 4.$$

Generalizing from this example, we define $a^{m/n}$ as follows.

---

**$a^{m/n}$**

If $m$ and $n$ are positive integers with $m/n$ in lowest terms, then

$$a^{m/n} = (a^{1/n})^m,$$

provided that $a^{1/n}$ is a real number. If $a^{1/n}$ is not a real number, then $a^{m/n}$ is not a real number.

---

**EXAMPLE 2** Evaluating Exponentials of the Form $a^{m/n}$

Evaluate each exponential.

Think: $36^{1/2} = \sqrt{36} = 6$

Think: $125^{1/3} = \sqrt[3]{125} = 5$

**(a)** $36^{3/2} = (36^{1/2})^3 = 6^3 = 216$  **(b)** $125^{2/3} = (125^{1/3})^2 = 5^2 = 25$

Be careful. The base is 4.

**(c)** $-4^{5/2} = -(4^{5/2}) = -(4^{1/2})^5 = -(2)^5 = -32$

Because the base here is 4, the negative sign is *not* affected by the exponent.

**(d)** $(-27)^{2/3} = [(-27)^{1/3}]^2 = (-3)^2 = 9$

Notice how the $-$ sign is used in parts (c) and (d). In part (c), we first evaluate the exponential and then find its negative. In part (d), the $-$ sign is part of the base, $-27$.

**(e)** $(-100)^{3/2} = [(-100)^{1/2}]^3$, which is not a real number, since $(-100)^{1/2}$, or $\sqrt{-100}$, is not a real number.

**NOW TRY** Exercises 23, 27, and 29.

When a rational exponent is negative, the earlier interpretation of negative exponents is applied.

---

**$a^{-m/n}$**

If $a^{m/n}$ is a real number, then

$$a^{-m/n} = \frac{1}{a^{m/n}} \quad (a \neq 0).$$

**EXAMPLE 3** Evaluating Exponentials with Negative Rational Exponents

Evaluate each exponential.

(a) $16^{-3/4} = \dfrac{1}{16^{3/4}} = \dfrac{1}{(16^{1/4})^3} = \dfrac{1}{(\sqrt[4]{16})^3} = \dfrac{1}{2^3} = \dfrac{1}{8}$

*The denominator of 3/4 is the index and the numerator is the exponent.*

(b) $25^{-3/2} = \dfrac{1}{25^{3/2}} = \dfrac{1}{(25^{1/2})^3} = \dfrac{1}{(\sqrt{25})^3} = \dfrac{1}{5^3} = \dfrac{1}{125}$

(c) $\left(\dfrac{8}{27}\right)^{-2/3} = \dfrac{1}{\left(\dfrac{8}{27}\right)^{2/3}} = \dfrac{1}{\left(\sqrt[3]{\dfrac{8}{27}}\right)^2} = \dfrac{1}{\left(\dfrac{2}{3}\right)^2} = \dfrac{1}{\dfrac{4}{9}} = \dfrac{9}{4}$

$\dfrac{1}{\frac{4}{9}} = 1 \div \dfrac{4}{9} = 1 \cdot \dfrac{9}{4}$

We could also use the rule $\left(\dfrac{b}{a}\right)^{-m} = \left(\dfrac{a}{b}\right)^m$ here, as follows:

$\left(\dfrac{8}{27}\right)^{-2/3} = \left(\dfrac{27}{8}\right)^{2/3} = \left(\sqrt[3]{\dfrac{27}{8}}\right)^2 = \left(\dfrac{3}{2}\right)^2 = \dfrac{9}{4}.$

*Take the reciprocal only of the base, not the exponent.*

**NOW TRY** Exercises 31 and 35.

**CAUTION** Be careful to distinguish between exponential expressions like the following:

$16^{-1/4} = \dfrac{1}{2}, \quad -16^{1/4} = -2, \quad \text{and} \quad -16^{-1/4} = -\dfrac{1}{2}.$

*A negative exponent does not necessarily lead to a negative result. Negative exponents lead to reciprocals, which may be positive.*

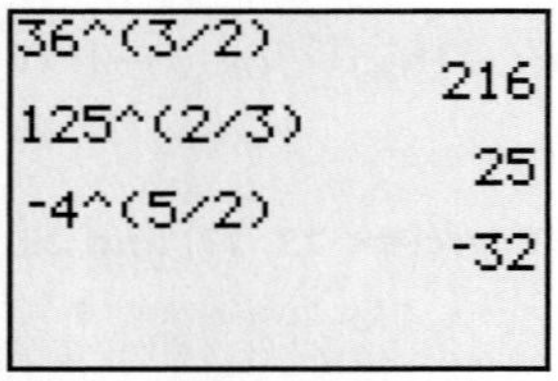

```
36^(3/2)
            216
125^(2/3)
             25
-4^(5/2)
            -32
```
**FIGURE 6**

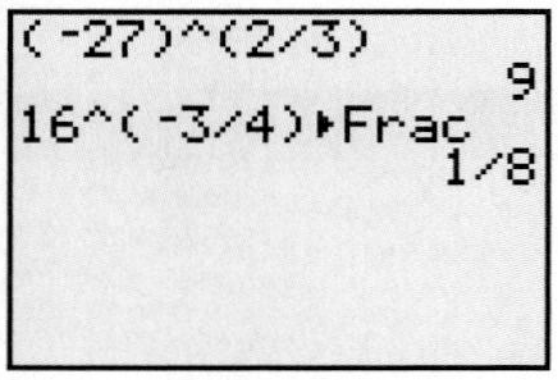

```
(-27)^(2/3)
              9
16^(-3/4)▶Frac
            1/8
```
**FIGURE 7**

The screens in Figures 6 and 7 illustrate how a graphing calculator performs some of the evaluations seen in Examples 2 and 3.

We obtain an alternative definition of $a^{m/n}$ by using the power rule for exponents a little differently than in the earlier definition. If all indicated roots are real numbers, then
$$a^{m/n} = a^{m(1/n)} = (a^m)^{1/n},$$
so
$$a^{m/n} = (a^m)^{1/n}.$$

$a^{m/n}$

If all indicated roots are real numbers, then
$$a^{m/n} = (a^{1/n})^m = (a^m)^{1/n}.$$

We can now evaluate an expression such as $27^{2/3}$ in two ways:

$$27^{2/3} = (27^{1/3})^2 = 3^2 = 9$$

or $\quad 27^{2/3} = (27^2)^{1/3} = 729^{1/3} = 9.\quad$ The result is the same.

***In most cases, it is easier to use $(a^{1/n})^m$.***
This rule can also be expressed with radicals as follows.

> **Radical Form of $a^{m/n}$**
>
> If all indicated roots are real numbers, then
> $$a^{m/n} = \sqrt[n]{a^m} = (\sqrt[n]{a})^m.$$
>
> That is, raise $a$ to the $m$th power and then take the $n$th root, or take the $n$th root of $a$ and then raise to the $m$th power.

For example,
$$8^{2/3} = \sqrt[3]{8^2} = \sqrt[3]{64} = 4, \quad \text{and} \quad 8^{2/3} = (\sqrt[3]{8})^2 = 2^2 = 4,$$
so
$$8^{2/3} = \sqrt[3]{8^2} = (\sqrt[3]{8})^2.$$

**OBJECTIVE 3** **Convert between radicals and rational exponents.** Using the definition of rational exponents, we can simplify many problems involving radicals by converting the radicals to numbers with rational exponents. After simplifying, we convert the answer back to radical form.

> **EXAMPLE 4** **Converting between Rational Exponents and Radicals**
>
> Write each exponential as a radical. Assume that all variables represent positive real numbers. Use the definition that takes the root first.
>
> **(a)** $13^{1/2} = \sqrt{13}$ 　　　　　　　　　**(b)** $6^{3/4} = (\sqrt[4]{6})^3$
> **(c)** $9m^{5/8} = 9(\sqrt[8]{m})^5$ 　　　　　　**(d)** $6x^{2/3} - (4x)^{3/5} = 6(\sqrt[3]{x})^2 - (\sqrt[5]{4x})^3$
> **(e)** $r^{-2/3} = \dfrac{1}{r^{2/3}} = \dfrac{1}{(\sqrt[3]{r})^2}$
> **(f)** $(a^2 + b^2)^{1/2} = \sqrt{a^2+b^2}$ 　　$\sqrt{a^2+b^2} \ne a+b$
>
> In (g)–(i), write each radical as an exponential. Simplify. Assume that all variables represent positive real numbers.
>
> **(g)** $\sqrt{10} = 10^{1/2}$ 　　　　　　　　　**(h)** $\sqrt[4]{3^8} = 3^{8/4} = 3^2 = 9$
> **(i)** $\sqrt[6]{z^6} = z$, since $z$ is positive.
>
> **NOW TRY** Exercises 37, 39, 41, 53, and 55.

**OBJECTIVE 4** **Use the rules for exponents with rational exponents.** The definition of rational exponents allows us to apply the rules for exponents from **Section 5.1.**

### Rules for Rational Exponents

Let $r$ and $s$ be rational numbers. For all real numbers $a$ and $b$ for which the indicated expressions exist,

$$a^r \cdot a^s = a^{r+s} \qquad a^{-r} = \frac{1}{a^r} \qquad \frac{a^r}{a^s} = a^{r-s} \qquad \left(\frac{a}{b}\right)^{-r} = \frac{b^r}{a^r}$$

$$(a^r)^s = a^{rs} \qquad (ab)^r = a^r b^r \qquad \left(\frac{a}{b}\right)^r = \frac{a^r}{b^r} \qquad a^{-r} = \left(\frac{1}{a}\right)^r.$$

### EXAMPLE 5  Applying Rules for Rational Exponents

Write with only positive exponents. Assume that all variables represent positive real numbers.

**(a)** $2^{1/2} \cdot 2^{1/4} = 2^{1/2 + 1/4} = 2^{3/4}$  Product rule

**(b)** $\dfrac{5^{2/3}}{5^{7/3}} = 5^{2/3 - 7/3} = 5^{-5/3} = \dfrac{1}{5^{5/3}}$  Quotient rule

**(c)** $\dfrac{(x^{1/2} y^{2/3})^4}{y} = \dfrac{(x^{1/2})^4 (y^{2/3})^4}{y}$  Power rule

$\phantom{(c)}\quad = \dfrac{x^2 y^{8/3}}{y^1}$  Power rule

$\phantom{(c)}\quad = x^2 y^{8/3 - 1}$  Quotient rule

$\phantom{(c)}\quad = x^2 y^{5/3}$

**(d)** $\left(\dfrac{x^4 y^{-6}}{x^{-2} y^{1/3}}\right)^{-2/3} = \dfrac{(x^4)^{-2/3} (y^{-6})^{-2/3}}{(x^{-2})^{-2/3} (y^{1/3})^{-2/3}}$

$\phantom{(d)}\quad = \dfrac{x^{-8/3} y^4}{x^{4/3} y^{-2/9}}$  Power rule

$\phantom{(d)}\quad = x^{-8/3 - 4/3} y^{4 - (-2/9)}$  Quotient rule

$\phantom{(d)}\quad = x^{-4} y^{38/9}$  *Use parentheses to avoid errors.*

$\phantom{(d)}\quad = \dfrac{y^{38/9}}{x^4}$  Definition of negative exponent

The same result is obtained if we simplify within the parentheses first, leading to $(x^6 y^{-19/3})^{-2/3}$. Then apply the power rule. (Show that the result is the same.)

**(e)** $m^{3/4}(m^{5/4} - m^{1/4}) = m^{3/4} \cdot m^{5/4} - m^{3/4} \cdot m^{1/4}$  Distributive property

$\phantom{(e)}\quad = m^{3/4 + 5/4} - m^{3/4 + 1/4}$  Product rule

$\phantom{(e)}\quad = m^{8/4} - m^{4/4}$

$\phantom{(e)}\quad = m^2 - m$

*Do not make the common mistake of multiplying exponents in the first step.*

**NOW TRY** Exercises 61, 63, 71, 81, and 83.

**CAUTION** Use the rules of exponents in problems like those in Example 5. Do not convert the expressions to radical form.

### EXAMPLE 6   Applying Rules for Rational Exponents

Write all radicals as exponentials, and then apply the rules for rational exponents. Leave answers in exponential form. Assume that all variables represent positive real numbers.

(a) $\sqrt[3]{x^2} \cdot \sqrt[4]{x} = x^{2/3} \cdot x^{1/4}$   Convert to rational exponents.
$= x^{2/3+1/4}$   Product rule
$= x^{8/12+3/12}$   Write exponents with a common denominator.
$= x^{11/12}$

(b) $\dfrac{\sqrt{x^3}}{\sqrt[3]{x^2}} = \dfrac{x^{3/2}}{x^{2/3}} = x^{3/2 - 2/3} = x^{5/6}$   $\dfrac{3}{2} - \dfrac{2}{3} = \dfrac{9}{6} - \dfrac{4}{6} = \dfrac{5}{6}$

(c) $\sqrt{\sqrt[4]{z}} = \sqrt{z^{1/4}} = (z^{1/4})^{1/2} = z^{1/8}$

**NOW TRY** Exercises 89, 91, and 95.

## 8.2 Exercises

*Matching*   Match each expression from Column I with the equivalent choice from Column II.

| I | | II | |
|---|---|---|---|
| **1.** $2^{1/2}$ | **2.** $(-27)^{1/3}$ | **A.** $-4$ | **B.** 8 |
| **3.** $-16^{1/2}$ | **4.** $(-16)^{1/2}$ | **C.** $\sqrt{2}$ | **D.** $-\sqrt{6}$ |
| **5.** $(-32)^{1/5}$ | **6.** $(-32)^{2/5}$ | **E.** $-3$ | **F.** $\sqrt{6}$ |
| **7.** $4^{3/2}$ | **8.** $6^{2/4}$ | **G.** 4 | **H.** $-2$ |
| **9.** $-6^{2/4}$ | **10.** $36^{0.5}$ | **I.** 6 | **J.** Not a real number |

*Evaluate each exponential. See Examples 1–3.*

**11.** $169^{1/2}$   **12.** $121^{1/2}$   **13.** $729^{1/3}$   **14.** $512^{1/3}$

**15.** $16^{1/4}$   **16.** $625^{1/4}$   **17.** $\left(\dfrac{64}{81}\right)^{1/2}$   **18.** $\left(\dfrac{8}{27}\right)^{1/3}$

**19.** $(-27)^{1/3}$   **20.** $(-32)^{1/5}$   **21.** $(-144)^{1/2}$   **22.** $(-36)^{1/2}$

**23.** $100^{3/2}$   **24.** $64^{3/2}$   **25.** $81^{3/4}$   **26.** $216^{2/3}$

**27.** $-16^{5/2}$   **28.** $-32^{3/5}$   **29.** $(-8)^{4/3}$   **30.** $(-243)^{2/5}$

**31.** $32^{-3/5}$   **32.** $27^{-4/3}$   **33.** $64^{-3/2}$   **34.** $81^{-3/2}$

**35.** $\left(\dfrac{125}{27}\right)^{-2/3}$   **36.** $\left(\dfrac{64}{125}\right)^{-2/3}$

*Write with radicals. Assume that all variables represent positive real numbers. See Example 4.*

**37.** $10^{1/2}$   **38.** $3^{1/2}$   **39.** $8^{3/4}$

**40.** $7^{2/3}$   **41.** $(9q)^{5/8} - (2x)^{2/3}$   **42.** $(3p)^{3/4} + (4x)^{1/3}$

**43.** $(2m)^{-3/2}$   **44.** $(5y)^{-3/5}$   **45.** $(2y + x)^{2/3}$

**46.** $(r + 2z)^{3/2}$   **47.** $(3m^4 + 2k^2)^{-2/3}$   **48.** $(5x^2 + 3z^3)^{-5/6}$

**49.** Show that, in general, $\sqrt{a^2 + b^2} \neq a + b$ by replacing $a$ with 3 and $b$ with 4.

**50.** Suppose someone claims that $\sqrt[n]{a^n + b^n}$ must equal $a + b$, since, when $a = 1$ and $b = 0$, a true statement results: $\sqrt[n]{a^n + b^n} = \sqrt[n]{1^n + 0^n} = \sqrt[n]{1^n} = 1 = 1 + 0 = a + b$. Explain why this is faulty reasoning.

*Simplify by first converting to rational exponents. Assume that all variables represent positive real numbers. See Example 4.*

**51.** $\sqrt{2^{12}}$  **52.** $\sqrt{5^{10}}$  **53.** $\sqrt[3]{4^9}$  **54.** $\sqrt[4]{6^8}$  **55.** $\sqrt{x^{20}}$

**56.** $\sqrt{r^{50}}$  **57.** $\sqrt[3]{x} \cdot \sqrt{x}$  **58.** $\sqrt[4]{y} \cdot \sqrt[5]{y^2}$  **59.** $\dfrac{\sqrt[3]{t^4}}{\sqrt[5]{t^4}}$  **60.** $\dfrac{\sqrt[4]{w^3}}{\sqrt[6]{w}}$

*Simplify each expression. Write all answers with positive exponents. Assume that all variables represent positive real numbers. See Example 5.*

**61.** $3^{1/2} \cdot 3^{3/2}$   **62.** $6^{4/3} \cdot 6^{2/3}$   **63.** $\dfrac{64^{5/3}}{64^{4/3}}$

**64.** $\dfrac{125^{7/3}}{125^{5/3}}$   **65.** $y^{7/3} \cdot y^{-4/3}$   **66.** $r^{-8/9} \cdot r^{17/9}$

**67.** $x^{2/3} \cdot x^{-1/4}$   **68.** $x^{2/5} \cdot x^{-1/3}$   **69.** $\dfrac{k^{1/3}}{k^{2/3} \cdot k^{-1}}$

**70.** $\dfrac{z^{3/4}}{z^{5/4} \cdot z^{-2}}$   **71.** $\dfrac{(x^{1/4} y^{2/5})^{20}}{x^2}$   **72.** $\dfrac{(r^{1/5} s^{2/3})^{15}}{r^2}$

**73.** $\dfrac{(x^{2/3})^2}{(x^2)^{7/3}}$   **74.** $\dfrac{(p^3)^{1/4}}{(p^{5/4})^2}$   **75.** $\dfrac{m^{3/4} n^{-1/4}}{(m^2 n)^{1/2}}$

**76.** $\dfrac{(a^2 b^5)^{-1/4}}{(a^{-3} b^2)^{1/6}}$   **77.** $\dfrac{p^{1/5} p^{7/10} p^{1/2}}{(p^3)^{-1/5}}$   **78.** $\dfrac{z^{1/3} z^{-2/3} z^{1/6}}{(z^{-1/6})^3}$

**79.** $\left(\dfrac{b^{-3/2}}{c^{-5/3}}\right)^2 (b^{-1/4} c^{-1/3})^{-1}$   **80.** $\left(\dfrac{m^{-2/3}}{a^{-3/4}}\right)^4 (m^{-3/8} a^{1/4})^{-2}$   **81.** $\left(\dfrac{p^{-1/4} q^{-3/2}}{3^{-1} p^{-2} q^{-2/3}}\right)^{-2}$

**82.** $\left(\dfrac{2^{-2} w^{-3/4} x^{-5/8}}{w^{3/4} x^{-1/2}}\right)^{-3}$   **83.** $p^{2/3}(p^{1/3} + 2p^{4/3})$   **84.** $z^{5/8}(3z^{5/8} + 5z^{11/8})$

**85.** $k^{1/4}(k^{3/2} - k^{1/2})$   **86.** $r^{3/5}(r^{1/2} + r^{3/4})$   **87.** $6a^{7/4}(a^{-7/4} + 3a^{-3/4})$

**88.** $4m^{5/3}(m^{-2/3} - 4m^{-5/3})$

*Write with rational exponents, and then apply the properties of exponents. Assume that all radicands represent positive real numbers. Give answers in exponential form. See Example 6.*

**89.** $\sqrt[5]{x^3} \cdot \sqrt[4]{x}$   **90.** $\sqrt[6]{y^5} \cdot \sqrt[3]{y^2}$   **91.** $\dfrac{\sqrt{x^5}}{\sqrt{x^8}}$   **92.** $\dfrac{\sqrt[3]{k^5}}{\sqrt[3]{k^7}}$

**93.** $\sqrt{y} \cdot \sqrt[3]{yz}$   **94.** $\sqrt[3]{xz} \cdot \sqrt{z}$   **95.** $\sqrt[4]{\sqrt[3]{m}}$   **96.** $\sqrt[3]{\sqrt{k}}$

**97.** $\sqrt{\sqrt[3]{\sqrt[4]{x}}}$   **98.** $\sqrt[3]{\sqrt[5]{\sqrt{y}}}$

The **windchill factor** is a measure of the cooling effect that the wind has on a person's skin. It calculates the equivalent cooling temperature if there were no wind. The National Weather Service uses the formula

$$\text{Windchill temperature} = 35.74 + 0.6215T - 35.75V^{4/25} + 0.4275TV^{4/25},$$

where $T$ is the temperature in °F and $V$ is the wind speed in miles per hour, to calculate windchill. The chart gives the windchill factor for various wind speeds and temperatures at which frostbite is a risk, and how quickly it may occur.

| | Temperature (°F) | | | | | | | | |
|---|---|---|---|---|---|---|---|---|---|
| Calm | 40 | 30 | 20 | 10 | 0 | −10 | −20 | −30 | −40 |
| 5 | 36 | 25 | 13 | 1 | −11 | −22 | −34 | −46 | −57 |
| 10 | 34 | 21 | 9 | −4 | −16 | −28 | −41 | −53 | −66 |
| 15 | 32 | 19 | 6 | −7 | −19 | −32 | −45 | −58 | −71 |
| 20 | 30 | 17 | 4 | −9 | −22 | −35 | −48 | −61 | −74 |
| 25 | 29 | 16 | 3 | −11 | −24 | −37 | −51 | −64 | −78 |
| 30 | 28 | 15 | 1 | −12 | −26 | −39 | −53 | −67 | −80 |
| 35 | 28 | 14 | 0 | −14 | −27 | −41 | −55 | −69 | −82 |
| 40 | 27 | 13 | −1 | −15 | −29 | −43 | −57 | −71 | −84 |

Wind speed (mph)

Frostbites times: ☐ 30 minutes ☐ 10 minutes ☐ 5 minutes

*Source*: National Oceanic and Atmospheric Administration, National Weather Service.

*Use the formula and a calculator to determine the windchill to the nearest tenth of a degree, given the following conditions. Compare your answers with the appropriate entries in the table.*

**99.** 30°F, 15-mph wind

**100.** 10°F, 30-mph wind

## 8.3 Simplifying Radical Expressions

**OBJECTIVES**

1. Use the product rule for radicals.
2. Use the quotient rule for radicals.
3. Simplify radicals.
4. Simplify products and quotients of radicals with different indexes.
5. Use the Pythagorean theorem.
6. Use the distance formula.

**OBJECTIVE 1** Use the product rule for radicals. We now develop rules for multiplying and dividing radicals that have the same index. For example, is the product of two $n$th-root radicals equal to the $n$th root of the product of the radicands? Are the expressions $\sqrt{36 \cdot 4}$ and $\sqrt{36} \cdot \sqrt{4}$ equal?

$$\sqrt{36 \cdot 4} = \sqrt{144} = 12$$
$$\sqrt{36} \cdot \sqrt{4} = 6 \cdot 2 = 12$$

The result is the same.

This is an example of the **product rule for radicals**.

**Product Rule for Radicals**

If $\sqrt[n]{a}$ and $\sqrt[n]{b}$ are real numbers and $n$ is a natural number, then

$$\sqrt[n]{a} \cdot \sqrt[n]{b} = \sqrt[n]{ab}.$$

That is, the product of two $n$th roots is the $n$th root of the product.

We justify the product rule by using the rules for rational exponents. Since $\sqrt[n]{a} = a^{1/n}$ and $\sqrt[n]{b} = b^{1/n}$,

$$\sqrt[n]{a} \cdot \sqrt[n]{b} = a^{1/n} \cdot b^{1/n} = (ab)^{1/n} = \sqrt[n]{ab}.$$

**CAUTION** *Use the product rule only when the radicals have the same index.*

**EXAMPLE 1** Using the Product Rule

Multiply. Assume that all variables represent positive real numbers.

(a) $\sqrt{5} \cdot \sqrt{7} = \sqrt{5 \cdot 7} = \sqrt{35}$   (b) $\sqrt{2} \cdot \sqrt{19} = \sqrt{2 \cdot 19} = \sqrt{38}$

(c) $\sqrt{11} \cdot \sqrt{p} = \sqrt{11p}$   (d) $\sqrt{7} \cdot \sqrt{11xyz} = \sqrt{77xyz}$

**NOW TRY** Exercises 7, 9, and 11.

**EXAMPLE 2** Using the Product Rule

Multiply. Assume that all variables represent positive real numbers.

Remember to write the index.

(a) $\sqrt[3]{3} \cdot \sqrt[3]{12} = \sqrt[3]{3 \cdot 12} = \sqrt[3]{36}$   (b) $\sqrt[4]{8y} \cdot \sqrt[4]{3r^2} = \sqrt[4]{24yr^2}$

(c) $\sqrt[6]{10m^4} \cdot \sqrt[6]{5m} = \sqrt[6]{50m^5}$

(d) $\sqrt[4]{2} \cdot \sqrt[5]{2}$ cannot be simplified by using the product rule for radicals, because the indexes (4 and 5) are different.

**NOW TRY** Exercises 13, 15, 17, and 19.

**OBJECTIVE 2** Use the quotient rule for radicals. The **quotient rule for radicals** is similar to the product rule.

### Quotient Rule for Radicals

If $\sqrt[n]{a}$ and $\sqrt[n]{b}$ are real numbers, $b \neq 0$, and $n$ is a natural number, then

$$\sqrt[n]{\frac{a}{b}} = \frac{\sqrt[n]{a}}{\sqrt[n]{b}}.$$

That is, the $n$th root of a quotient is the quotient of the $n$th roots.

**EXAMPLE 3** Using the Quotient Rule

Simplify. Assume that all variables represent positive real numbers.

(a) $\sqrt{\frac{16}{25}} = \frac{\sqrt{16}}{\sqrt{25}} = \frac{4}{5}$   (b) $\sqrt{\frac{7}{36}} = \frac{\sqrt{7}}{\sqrt{36}} = \frac{\sqrt{7}}{6}$

(c) $\sqrt[3]{-\frac{8}{125}} = \sqrt[3]{\frac{-8}{125}} = \frac{\sqrt[3]{-8}}{\sqrt[3]{125}} = \frac{-2}{5} = -\frac{2}{5}$   $-\frac{a}{b} = \frac{-a}{b}$

(d) $\sqrt[3]{\frac{7}{216}} = \frac{\sqrt[3]{7}}{\sqrt[3]{216}} = \frac{\sqrt[3]{7}}{6}$

Think: $\sqrt[3]{m^6} = m^{6/3} = m^2$

(e) $\sqrt[5]{\frac{x}{32}} = \frac{\sqrt[5]{x}}{\sqrt[5]{32}} = \frac{\sqrt[5]{x}}{2}$   (f) $-\sqrt[3]{\frac{m^6}{125}} = -\frac{\sqrt[3]{m^6}}{\sqrt[3]{125}} = -\frac{m^2}{5}$

**NOW TRY** Exercises 23, 25, 31, 33, and 35.

**SECTION 8.3** Simplifying Radical Expressions  **395**

**OBJECTIVE 3** Simplify radicals. We use the product and quotient rules to simplify radicals. A radical is **simplified** if the following four conditions are met.

### Conditions for a Simplified Radical

1. The radicand has no factor raised to a power greater than or equal to the index.
2. The radicand has no fractions.
3. No denominator contains a radical.
4. Exponents in the radicand and the index of the radical have no common factor (except 1).

### EXAMPLE 4 Simplifying Roots of Numbers

Simplify.

**(a)** $\sqrt{24}$

Check to see whether 24 is divisible by a perfect square (the square of a natural number), such as 4, 9, . . . . The largest perfect square that divides into 24 is 4. Write 24 as the product of 4 and 6, and then use the product rule.

$$\sqrt{24} = \sqrt{4 \cdot 6} = \sqrt{4} \cdot \sqrt{6} = 2\sqrt{6}$$

**(b)** $\sqrt{108}$

As shown on the left, the number 108 is divisible by the perfect square 36. If this is not obvious, try factoring 108 into its prime factors, as shown on the right.

$$\sqrt{108} = \sqrt{36 \cdot 3} \qquad\qquad \sqrt{108} = \sqrt{2^2 \cdot 3^3}$$
$$= \sqrt{36} \cdot \sqrt{3} \qquad\qquad\quad = \sqrt{2^2 \cdot 3^2 \cdot 3}$$
$$= 6\sqrt{3} \qquad\qquad\qquad = 2 \cdot 3 \cdot \sqrt{3} \qquad \text{Product rule;}$$
$$\qquad\qquad\qquad\qquad\qquad\qquad\qquad\qquad \sqrt{2^2} = 2, \sqrt{3^2} = 3$$
$$\qquad\qquad\qquad\qquad\qquad = 6\sqrt{3}$$

**(c)** $\sqrt{10}$

No perfect square (other than 1) divides into 10, so $\sqrt{10}$ cannot be simplified further.

**(d)** $\sqrt[3]{16}$

The largest perfect *cube* that divides into 16 is 8, so write 16 as $8 \cdot 2$ (or factor 16 into prime factors).

$$\sqrt[3]{16} = \sqrt[3]{8 \cdot 2} = \sqrt[3]{8} \cdot \sqrt[3]{2} = 2\sqrt[3]{2}$$

**(e)** $-\sqrt[4]{162} = -\sqrt[4]{81 \cdot 2}$   81 is a perfect 4th power.

*Remember the negative sign in each line.* $= -\sqrt[4]{81} \cdot \sqrt[4]{2}$   Product rule

$\qquad\qquad\qquad\qquad = -3\sqrt[4]{2}$

**NOW TRY** Exercises 39, 41, 47, 49, and 55.

**CAUTION** *Be careful with which factors belong outside the radical sign and which belong inside.* Note in Example 4(b) how $2 \cdot 3$ is written outside because $\sqrt{2^2} = 2$ and $\sqrt{3^2} = 3$, while the remaining 3 is left inside the radical.

**EXAMPLE 5** Simplifying Radicals Involving Variables

Simplify. Assume that all variables represent positive real numbers.

(a) $\sqrt{16m^3} = \sqrt{16m^2 \cdot m}$  Factor.
$= \sqrt{16m^2} \cdot \sqrt{m}$  Product rule
$= 4m\sqrt{m}$

Absolute value bars are not needed around the $m$ in color because all the variables represent *positive* real numbers.

(b) $\sqrt{200k^7q^8} = \sqrt{10^2 \cdot 2 \cdot (k^3)^2 \cdot k \cdot (q^4)^2}$  Factor.
$= 10k^3q^4\sqrt{2k}$  Remove perfect square factors.

(c) $\sqrt[3]{-8x^4y^5} = \sqrt[3]{(-8x^3y^3)(xy^2)}$  Choose $-8x^3y^3$ as the perfect cube that divides into $-8x^4y^5$.
$= \sqrt[3]{-8x^3y^3} \cdot \sqrt[3]{xy^2}$  Product rule
$= -2xy\sqrt[3]{xy^2}$

(d) $-\sqrt[4]{32y^9} = -\sqrt[4]{(16y^8)(2y)}$  $16y^8$ is the largest 4th power that divides into $32y^9$.
$= -\sqrt[4]{16y^8} \cdot \sqrt[4]{2y}$  Product rule
$= -2y^2\sqrt[4]{2y}$

> **NOW TRY** Exercises 75, 79, 83, and 87.

The conditions for a simplified radical given earlier state that an exponent in the radicand and the index of the radical should have no common factor (except 1). The next example shows how to simplify radicals with such common factors.

**EXAMPLE 6** Simplifying Radicals by Using Smaller Indexes

Simplify. Assume that all variables represent positive real numbers.

(a) $\sqrt[9]{5^6}$

We write this radical by using rational exponents and then write the exponent in lowest terms. We then express the answer as a radical.

$$\sqrt[9]{5^6} = (5^6)^{1/9} = 5^{6/9} = 5^{2/3} = \sqrt[3]{5^2}, \text{ or } \sqrt[3]{25}$$

(b) $\sqrt[4]{p^2} = (p^2)^{1/4} = p^{2/4} = p^{1/2} = \sqrt{p}$  (Recall the assumption that $p > 0$.)

> **NOW TRY** Exercises 93 and 97.

These examples suggest the following rule.

> If $m$ is an integer, $n$ and $k$ are natural numbers, and all indicated roots exist, then
> $$\sqrt[kn]{a^{km}} = \sqrt[n]{a^m}.$$

**OBJECTIVE 4** Simplify products and quotients of radicals with different indexes.
Since the product and quotient rules for radicals apply only when they have the same index, we multiply and divide radicals with different indexes by using rational exponents.

### EXAMPLE 7  Multiplying Radicals with Different Indexes

Simplify $\sqrt{7} \cdot \sqrt[3]{2}$.

Because the different indexes, 2 and 3, have a least common index of 6, use rational exponents to write each radical as a sixth root.

$$\sqrt{7} = 7^{1/2} = 7^{3/6} = \sqrt[6]{7^3} = \sqrt[6]{343}$$
$$\sqrt[3]{2} = 2^{1/3} = 2^{2/6} = \sqrt[6]{2^2} = \sqrt[6]{4}$$

Therefore,

$$\sqrt{7} \cdot \sqrt[3]{2} = \sqrt[6]{343} \cdot \sqrt[6]{4} = \sqrt[6]{1372}. \quad \text{Product rule}$$

**NOW TRY** Exercise 99.

Results such as the one in Example 7 can be supported with a calculator, as shown in Figure 8. Notice that the calculator gives the same approximation for the initial product and the final radical that we obtained.

**FIGURE 8**

**CAUTION** The computation in Figure 8 is not *proof* that the two expressions are equal. The algebra in Example 7, however, is valid proof of their equality.

**OBJECTIVE 5** Use the Pythagorean theorem.   The **Pythagorean theorem** relates the lengths of the three sides of a right triangle.

### Pythagorean Theorem

If $c$ is the length of the longest side of a right triangle and $a$ and $b$ are the lengths of the shorter sides, then

$$c^2 = a^2 + b^2.$$

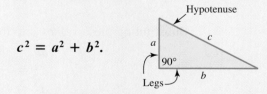

The longest side is the **hypotenuse** and the two shorter sides are the **legs** of the triangle. The hypotenuse is the side opposite the right angle.

In **Section 9.1** we will see that an equation such as $x^2 = 7$ has two solutions: $\sqrt{7}$ (the principal, or positive, square root of 7) and $-\sqrt{7}$. Similarly, $c^2 = 52$ has two solutions, $\pm\sqrt{52} = \pm 2\sqrt{13}$. In applications we often choose only the positive square root, as seen in the example that follows.

### EXAMPLE 8  Using the Pythagorean Theorem

Use the Pythagorean theorem to find the length of the hypotenuse of the triangle in Figure 9.

$c^2 = a^2 + b^2$   Pythagorean theorem
$c^2 = 4^2 + 6^2$   Let $a = 4$ and $b = 6$.

Substitute carefully.
$c^2 = 52$
$c = \sqrt{52}$   Choose the principal root.
$c = \sqrt{4 \cdot 13}$   Factor.
$c = \sqrt{4} \cdot \sqrt{13}$   Product rule
$c = 2\sqrt{13}$

The length of the hypotenuse is $2\sqrt{13}$.

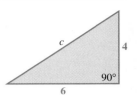

FIGURE 9

**NOW TRY** Exercise 105.

**CAUTION** When using the Pythagorean theorem, be sure that the length of the hypotenuse is substituted for $c$ and that the lengths of the legs are substituted for $a$ and $b$. Errors often occur because values are substituted incorrectly.

### OBJECTIVE 6  Use the distance formula.

An important result in algebra is derived from the Pythagorean theorem. The *distance formula* allows us to find the distance between two points in the coordinate plane, or the length of the line segment joining those two points. Figure 10 shows the points $(3, -4)$ and $(-5, 3)$. The vertical line through $(-5, 3)$ and the horizontal line through $(3, -4)$ intersect at the point $(-5, -4)$. Thus, the point $(-5, -4)$ becomes the vertex of the right angle in a right triangle. By the Pythagorean theorem, the square of the length of the hypotenuse $d$ of the right triangle in Figure 10 is equal to the sum of the squares of the lengths of the two legs $a$ and $b$:

$$d^2 = a^2 + b^2.$$

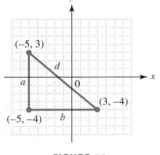

FIGURE 10

The length $a$ is the difference between the $y$-coordinates of the endpoints. Since the $x$-coordinate of both points in Figure 10 is $-5$, the side is vertical, and we can find $a$ by finding the difference between the $y$-coordinates. We subtract $-4$ from 3 to get a positive value for $a$.

$$a = 3 - (-4) = 7$$

Similarly, we find $b$ by subtracting $-5$ from 3.

$$b = 3 - (-5) = 8$$

Substituting these values into the formula, we obtain

$d^2 = a^2 + b^2$
$d^2 = 7^2 + 8^2$   Let $a = 7$ and $b = 8$.
$d^2 = 49 + 64$
$d^2 = 113$
$d = \sqrt{113}$.

We choose the principal root, since distance cannot be negative. Therefore, the distance between $(-5, 3)$ and $(3, -4)$ is $\sqrt{113}$. (It is customary to leave the distance in radical form. Do not use a calculator to get an approximation, unless you are specifically directed to do so.)

This result can be generalized. Figure 11 shows the two points $(x_1, y_1)$ and $(x_2, y_2)$. Notice that the distance between $(x_1, y_1)$ and $(x_2, y_1)$ is given by

$$a = |x_2 - x_1|,$$

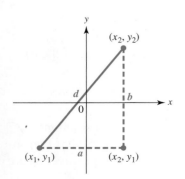

FIGURE 11

and the distance between $(x_2, y_2)$ and $(x_2, y_1)$ is given by
$$b = |y_2 - y_1|.$$
From the Pythagorean theorem,
$$d^2 = a^2 + b^2$$
$$d^2 = (x_2 - x_1)^2 + (y_2 - y_1)^2.$$
Choosing the principal square root gives the **distance formula.**

### Distance Formula

The distance between the points $(x_1, y_1)$ and $(x_2, y_2)$ is
$$d = \sqrt{(x_2 - x_1)^2 + (y_2 - y_1)^2}.$$

**EXAMPLE 9** Using the Distance Formula

Find the distance between the points $(-3, 5)$ and $(6, 4)$.

Designating the points as $(x_1, y_1)$ and $(x_2, y_2)$ is arbitrary. We choose $(x_1, y_1) = (-3, 5)$ and $(x_2, y_2) = (6, 4)$.

$$d = \sqrt{(x_2 - x_1)^2 + (y_2 - y_1)^2}$$
$$= \sqrt{[6 - (-3)]^2 + (4 - 5)^2} \quad x_2 = 6, y_2 = 4, x_1 = -3, y_1 = 5$$

*Start with the x-value and y-value of the same point.*

$$= \sqrt{9^2 + (-1)^2}$$
$$= \sqrt{82}$$

**NOW TRY** Exercise 111.

## 8.3 Exercises

**NOW TRY Exercise**

*Concept Check* Decide whether each statement is true or false by using the product rule explained in this section. Then support your answer by finding a calculator approximation for each expression.

1. $2\sqrt{12} = \sqrt{48}$  2. $\sqrt{72} = 2\sqrt{18}$  3. $3\sqrt{8} = 2\sqrt{18}$  4. $5\sqrt{72} = 6\sqrt{50}$

5. *Multiple Choice* Which one of the following is *not* equal to $\sqrt{\frac{1}{2}}$? (Do not use calculator approximations.)

   A. $\sqrt{0.5}$  B. $\sqrt{\frac{2}{4}}$  C. $\sqrt{\frac{3}{6}}$  D. $\frac{\sqrt{4}}{\sqrt{16}}$

6. Use the $\pi$ key on your calculator to get a value for $\pi$. Now find an approximation for $\sqrt[4]{\frac{2143}{22}}$. Does the result mean that $\pi$ is actually equal to $\sqrt[4]{\frac{2143}{22}}$? Why or why not?

*Multiply, using the product rule. Assume that all variables represent positive real numbers. See Examples 1 and 2.*

7. $\sqrt{5} \cdot \sqrt{6}$  8. $\sqrt{10} \cdot \sqrt{3}$  9. $\sqrt{14} \cdot \sqrt{x}$  10. $\sqrt{23} \cdot \sqrt{t}$

11. $\sqrt{14} \cdot \sqrt{3pqr}$  12. $\sqrt{7} \cdot \sqrt{5xt}$  13. $\sqrt[3]{7x} \cdot \sqrt[3]{2y}$  14. $\sqrt[3]{9x} \cdot \sqrt[3]{4y}$

15. $\sqrt[4]{11} \cdot \sqrt[4]{3}$  16. $\sqrt[4]{6} \cdot \sqrt[4]{9}$  17. $\sqrt[4]{2x} \cdot \sqrt[4]{3x^2}$  18. $\sqrt[4]{3y^2} \cdot \sqrt[4]{6y}$

19. $\sqrt[3]{7} \cdot \sqrt[4]{3}$  20. $\sqrt[5]{8} \cdot \sqrt[6]{12}$

✏ 21. Explain the product rule for radicals in your own words. Give examples.

✏ 22. Explain the quotient rule for radicals in your own words. Give examples.

*Simplify each radical. Assume that all variables represent positive real numbers. See Example 3.*

23. $\sqrt{\dfrac{64}{121}}$
24. $\sqrt{\dfrac{16}{49}}$
25. $\sqrt{\dfrac{3}{25}}$
26. $\sqrt{\dfrac{13}{49}}$

27. $\sqrt{\dfrac{x}{25}}$
28. $\sqrt{\dfrac{k}{100}}$
29. $\sqrt{\dfrac{p^6}{81}}$
30. $\sqrt{\dfrac{w^{10}}{36}}$

31. $\sqrt[3]{-\dfrac{27}{64}}$
32. $\sqrt[3]{-\dfrac{216}{125}}$
33. $\sqrt[3]{\dfrac{r^2}{8}}$
34. $\sqrt[3]{\dfrac{t}{125}}$

35. $-\sqrt[4]{\dfrac{81}{x^4}}$
36. $-\sqrt[4]{\dfrac{625}{y^4}}$
37. $\sqrt[5]{\dfrac{1}{x^{15}}}$
38. $\sqrt[5]{\dfrac{32}{y^{20}}}$

*Express each radical in simplified form. See Example 4.*

39. $\sqrt{12}$
40. $\sqrt{18}$
41. $\sqrt{288}$
42. $\sqrt{72}$
43. $-\sqrt{32}$

44. $-\sqrt{48}$
45. $-\sqrt{28}$
46. $-\sqrt{24}$
47. $\sqrt{30}$
48. $\sqrt{46}$

49. $\sqrt[3]{128}$
50. $\sqrt[3]{24}$
51. $\sqrt[3]{-16}$
52. $\sqrt[3]{-250}$
53. $\sqrt[3]{40}$

54. $\sqrt[3]{375}$
55. $-\sqrt[4]{512}$
56. $-\sqrt[4]{1250}$
57. $\sqrt[5]{64}$
58. $\sqrt[5]{128}$

✏ 59. A student claimed that $\sqrt[3]{14}$ is not in simplified form, since $14 = 8 + 6$, and 8 is a perfect cube. Was his reasoning correct? Why or why not?

✏ 60. Explain in your own words why $\sqrt[3]{k^4}$ is not a simplified radical.

*Express each radical in simplified form. Assume that all variables represent positive real numbers. See Example 5.*

61. $\sqrt{72k^2}$
62. $\sqrt{18m^2}$
63. $\sqrt{144x^3y^9}$

64. $\sqrt{169s^5t^{10}}$
65. $\sqrt{121x^6}$
66. $\sqrt{256z^{12}}$

67. $-\sqrt[3]{27t^{12}}$
68. $-\sqrt[3]{64y^{18}}$
69. $-\sqrt{100m^8z^4}$

70. $-\sqrt{25t^6s^{20}}$
71. $-\sqrt[3]{-125a^6b^9c^{12}}$
72. $-\sqrt[3]{-216y^{15}x^6z^3}$

73. $\sqrt[4]{\dfrac{1}{16}r^8t^{20}}$
74. $\sqrt[4]{\dfrac{81}{256}t^{12}u^8}$
75. $\sqrt{50x^3}$
76. $\sqrt{300z^3}$

77. $-\sqrt{500r^{11}}$
78. $-\sqrt{200p^{13}}$
79. $\sqrt{13x^7y^8}$
80. $\sqrt{23k^9p^{14}}$

81. $\sqrt[3]{8z^6w^9}$
82. $\sqrt[3]{64a^{15}b^{12}}$
83. $\sqrt[3]{-16z^5t^7}$
84. $\sqrt[3]{-81m^4n^{10}}$

85. $\sqrt[4]{81x^{12}y^{16}}$
86. $\sqrt[4]{81t^8u^{28}}$
87. $-\sqrt[4]{162r^{15}s^{10}}$
88. $-\sqrt[4]{32k^5m^{10}}$

89. $\sqrt{\dfrac{y^{11}}{36}}$
90. $\sqrt{\dfrac{v^{13}}{49}}$
91. $\sqrt[3]{\dfrac{x^{16}}{27}}$
92. $\sqrt[3]{\dfrac{y^{17}}{125}}$

*Simplify each radical. Assume that $x \geq 0$. See Example 6.*

93. $\sqrt[4]{48^2}$
94. $\sqrt[4]{50^2}$
95. $\sqrt[4]{25}$

96. $\sqrt[6]{8}$
97. $\sqrt[10]{x^{25}}$
98. $\sqrt[12]{x^{44}}$

*Simplify by first writing the radicals as radicals with the same index. Then multiply. Assume that all variables represent positive real numbers. See Example 7.*

99. $\sqrt[3]{4} \cdot \sqrt{3}$
100. $\sqrt[3]{5} \cdot \sqrt{6}$
101. $\sqrt[3]{3} \cdot \sqrt[3]{4}$

102. $\sqrt[5]{7} \cdot \sqrt[7]{5}$
103. $\sqrt{x} \cdot \sqrt[3]{x}$
104. $\sqrt[3]{y} \cdot \sqrt[4]{y}$

*Find the unknown length in each right triangle. Simplify the answer if possible. See Example 8.*

**105.**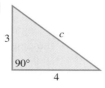

**106.**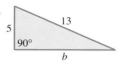

**107.**

**108.**

*Find the distance between each pair of points. See Example 9.*

**109.** $(6, 13)$ and $(1, 1)$     **110.** $(8, 13)$ and $(2, 5)$     **111.** $(-6, 5)$ and $(3, -4)$

**112.** $(-1, 5)$ and $(-7, 7)$     **113.** $(-8, 2)$ and $(-4, 1)$     **114.** $(-1, 2)$ and $(5, 3)$

**115.** $(4.7, 2.3)$ and $(1.7, -1.7)$     **116.** $(-2.9, 18.2)$ and $(2.1, 6.2)$

**117.** $(\sqrt{2}, \sqrt{6})$ and $(-2\sqrt{2}, 4\sqrt{6})$     **118.** $(\sqrt{7}, 9\sqrt{3})$ and $(-\sqrt{7}, 4\sqrt{3})$

**119.** $(x + y, y)$ and $(x - y, x)$     **120.** $(c, c - d)$ and $(d, c + d)$

**121.** *Concept Check* As given in the text, the distance formula is expressed with a radical. Write the distance formula with rational exponents.

**122.** An alternative form of the distance formula is
$$d = \sqrt{(x_1 - x_2)^2 + (y_1 - y_2)^2}.$$
Compare this with the form given in this section, and explain why the two forms are equivalent.

*Find the perimeter of each triangle. (Hint: For Exercise 123, $\sqrt{k} + \sqrt{k} = 2\sqrt{k}$.)*

**123.**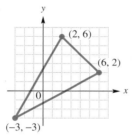

**124.**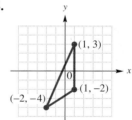

*Solve each problem.*

**125.** The following letter appeared in the column "Ask Tom Why," written by Tom Skilling of the *Chicago Tribune*:

*Dear Tom,*

*I cannot remember the formula to calculate the distance to the horizon. I have a stunning view from my 14th-floor condo, 150 ft above the ground. How far can I see?*

*Ted Fleischaker; Indianapolis, Ind.*

Skilling's answer was as follows:

To find the distance to the horizon in miles, take the square root of the height of your view in feet and multiply that result by 1.224. Your answer will be the number of miles to the horizon. (*Source: Chicago Tribune,* August 17, 2002.)

Assuming that Ted's eyes are 6 ft above the ground, the total height from the ground is $150 + 6 = 156$ ft. To the nearest tenth of a mile, how far can he see to the horizon?

126. The length of the diagonal of a box is given by
$$D = \sqrt{L^2 + W^2 + H^2},$$
where $L$, $W$, and $H$ are, respectively, the length, width, and height of the box. Find the length of the diagonal $D$ of a box that is 4 ft long, 2 ft wide, and 3 ft high. Give the exact value, and then round to the nearest tenth of a foot.

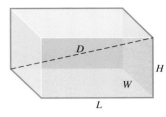

127. A Sanyo color television, model AVM-2755, has a rectangular screen with a 21.7-in. width. Its height is 16 in. What is the measure of the diagonal of the screen, to the nearest tenth of an inch? (*Source:* Actual measurements of the author's television.)

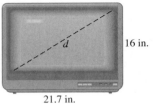

128. The illumination $I$, in foot-candles, produced by a light source is related to the distance $d$, in feet, from the light source by the equation
$$d = \sqrt{\frac{k}{I}},$$
where $k$ is a constant. If $k = 640$, how far from the light source will the illumination be 2 foot-candles? Give the exact value, and then round to the nearest tenth of a foot.

## 8.4 Adding and Subtracting Radical Expressions

**OBJECTIVE**

1. Simplify radical expressions involving addition and subtraction.

**OBJECTIVE 1** Simplify radical expressions involving addition and subtraction. An expression such as $4\sqrt{2} + 3\sqrt{2}$ can be simplified by the distributive property.

$$4\sqrt{2} + 3\sqrt{2}$$
$$= (4 + 3)\sqrt{2} = 7\sqrt{2}$$
This is similar to simplifying $4x + 3x$ to $7x$.

$$2\sqrt{3} - 5\sqrt{3}$$
$$= (2 - 5)\sqrt{3} = -3\sqrt{3}$$
This is similar to simplifying $2x - 5x$ to $-3x$.

**CAUTION** *Only radical expressions with the same index and the same radicand may be combined.* Expressions such as $5\sqrt{3} + 2\sqrt{2}$ or $3\sqrt{3} + 2\sqrt[3]{3}$ cannot be simplified by combining terms.

**EXAMPLE 1** Adding and Subtracting Radicals

Add or subtract to simplify each radical expression.

(a) $3\sqrt{24} + \sqrt{54}$

Simplify each radical; then use the distributive property to combine terms.

$$3\sqrt{24} + \sqrt{54}$$
$$= 3\sqrt{4} \cdot \sqrt{6} + \sqrt{9} \cdot \sqrt{6} \quad \text{Product rule}$$
$$= 3 \cdot 2\sqrt{6} + 3\sqrt{6} \quad \sqrt{4} = 2; \sqrt{9} = 3$$
$$= 6\sqrt{6} + 3\sqrt{6}$$
$$= 9\sqrt{6} \quad \text{Combine like terms.}$$

**(b)** $2\sqrt{20x} - \sqrt{45x}, \quad x \geq 0$

$= 2\sqrt{4} \cdot \sqrt{5x} - \sqrt{9} \cdot \sqrt{5x}$     Product rule

$= 2 \cdot 2\sqrt{5x} - 3\sqrt{5x}$

$= 4\sqrt{5x} - 3\sqrt{5x}$

$= \sqrt{5x}$     Combine like terms.

**(c)** $2\sqrt{3} - 4\sqrt{5}$

Here the radicands differ and are already simplified, so $2\sqrt{3} - 4\sqrt{5}$ cannot be simplified further.

**NOW TRY** Exercises 7, 15, and 19.

**CAUTION** Do not confuse the product rule with combining like terms. *The root of a sum does not equal the sum of the roots.* For example,

$$\sqrt{9 + 16} \neq \sqrt{9} + \sqrt{16}$$

since $\sqrt{9 + 16} = \sqrt{25} = 5$, but $\sqrt{9} + \sqrt{16} = 3 + 4 = 7$.

### EXAMPLE 2   Adding and Subtracting Radicals with Higher Indexes

Add or subtract to simplify each radical expression. Assume that all variables represent positive real numbers.

**(a)** $2\sqrt[3]{16} - 5\sqrt[3]{54}$

$= 2\sqrt[3]{8 \cdot 2} - 5\sqrt[3]{27 \cdot 2}$     Factor.

$= 2\sqrt[3]{8} \cdot \sqrt[3]{2} - 5\sqrt[3]{27} \cdot \sqrt[3]{2}$     Product rule

$= 2 \cdot 2 \cdot \sqrt[3]{2} - 5 \cdot 3 \cdot \sqrt[3]{2}$

$= 4\sqrt[3]{2} - 15\sqrt[3]{2}$

$= (4 - 15)\sqrt[3]{2}$     Distributive property

$= -11\sqrt[3]{2}$     Combine like terms.

**(b)** $2\sqrt[3]{x^2y} + \sqrt[3]{8x^5y^4}$

$= 2\sqrt[3]{x^2y} + \sqrt[3]{(8x^3y^3)x^2y}$     Factor.

$= 2\sqrt[3]{x^2y} + \sqrt[3]{8x^3y^3} \cdot \sqrt[3]{x^2y}$     Product rule

$= 2\sqrt[3]{x^2y} + 2xy\sqrt[3]{x^2y}$

$= (2 + 2xy)\sqrt[3]{x^2y}$     Distributive property

*This result cannot be simplified further.*

*Be careful! The indexes are different.*

**(c)** $5\sqrt{4x^3} + 3\sqrt[3]{64x^4}$

$= 5\sqrt{4x^2 \cdot x} + 3\sqrt[3]{64x^3 \cdot x}$     Factor.

$= 5\sqrt{4x^2} \cdot \sqrt{x} + 3\sqrt[3]{64x^3} \cdot \sqrt[3]{x}$     Product rule

$= 5 \cdot 2x\sqrt{x} + 3 \cdot 4x\sqrt[3]{x}$

$= 10x\sqrt{x} + 12x\sqrt[3]{x}$

*Remember to write the correct index.*

The radicands are both $x$, but since the indexes are different, this expression cannot be simplified further.

**NOW TRY** Exercises 25, 37, and 41.

**CAUTION** *Remember to write the index when working with cube roots, fourth roots, and so on.*

### EXAMPLE 3  Adding and Subtracting Radicals with Fractions

Perform the indicated operations. Assume that all variables represent positive real numbers.

(a) $2\sqrt{\dfrac{75}{16}} + 4\dfrac{\sqrt{8}}{\sqrt{32}}$

$= 2\dfrac{\sqrt{25 \cdot 3}}{\sqrt{16}} + 4\dfrac{\sqrt{4 \cdot 2}}{\sqrt{16 \cdot 2}}$   Quotient rule

$= 2\left(\dfrac{5\sqrt{3}}{4}\right) + 4\left(\dfrac{2\sqrt{2}}{4\sqrt{2}}\right)$   Product rule

$= \dfrac{5\sqrt{3}}{2} + 2$   Multiply; $\dfrac{\sqrt{2}}{\sqrt{2}} = 1$.

$= \dfrac{5\sqrt{3}}{2} + \dfrac{4}{2}$   Write with a common denominator.

$= \dfrac{5\sqrt{3} + 4}{2}$   Add fractions.

(b) $10\sqrt[3]{\dfrac{5}{x^6}} - 3\sqrt[3]{\dfrac{4}{x^9}}$

$= 10\dfrac{\sqrt[3]{5}}{\sqrt[3]{x^6}} - 3\dfrac{\sqrt[3]{4}}{\sqrt[3]{x^9}}$   Quotient rule

$= \dfrac{10\sqrt[3]{5}}{x^2} - \dfrac{3\sqrt[3]{4}}{x^3}$   Simplify denominators.

$= \dfrac{10\sqrt[3]{5} \cdot x}{x^2 \cdot x} - \dfrac{3\sqrt[3]{4}}{x^3}$   Write with a common denominator.

$= \dfrac{10x\sqrt[3]{5} - 3\sqrt[3]{4}}{x^3}$   Subtract fractions.

**NOW TRY** Exercises 51 and 57.

## 8.4 Exercises

1. *Multiple Choice* Which one of the following sums could be simplified without first simplifying the individual radical expressions?

   **A.** $\sqrt{50} + \sqrt{32}$   **B.** $3\sqrt{6} + 9\sqrt{6}$   **C.** $\sqrt[3]{32} - \sqrt[3]{108}$   **D.** $\sqrt[5]{6} - \sqrt[5]{192}$

2. *Concept Check* Let $a = 1$ and let $b = 64$.
   
   (a) Evaluate $\sqrt{a} + \sqrt{b}$. Then find $\sqrt{a+b}$. Are they equal?
   
   (b) Evaluate $\sqrt[3]{a} + \sqrt[3]{b}$. Then find $\sqrt[3]{a+b}$. Are they equal?
   
   (c) *Fill in the Blank* In general, $\sqrt[n]{a} + \sqrt[n]{b} \neq$ _____, based on the observations in parts (a) and (b) of this exercise.

**SECTION 8.4** Adding and Subtracting Radical Expressions **405**

✏️ **3.** Even though the root indexes of the terms are not equal, the sum $\sqrt{64} + \sqrt[3]{125} + \sqrt[4]{16}$ can be simplified quite easily. What is this sum? Why can we add these terms so easily?

✏️ **4.** Explain why $28 - 4\sqrt{2}$ is not equal to $24\sqrt{2}$. (This is a common error among algebra students.)

*Simplify. Assume that all variables represent positive real numbers. See Examples 1 and 2.*

**5.** $\sqrt{36} - \sqrt{100}$      **6.** $\sqrt{25} - \sqrt{81}$      **7.** $-2\sqrt{48} + 3\sqrt{75}$

**8.** $4\sqrt{32} - 2\sqrt{8}$      **9.** $\sqrt[3]{16} + 4\sqrt[3]{54}$      **10.** $3\sqrt[3]{24} - 2\sqrt[3]{192}$

**11.** $\sqrt[4]{32} + 3\sqrt[4]{2}$                      **12.** $\sqrt[4]{405} - 2\sqrt[4]{5}$

**13.** $6\sqrt{18} - \sqrt{32} + 2\sqrt{50}$           **14.** $5\sqrt{8} + 3\sqrt{72} - 3\sqrt{50}$

**15.** $5\sqrt{6} + 2\sqrt{10}$                   **16.** $3\sqrt{11} - 5\sqrt{13}$

**17.** $2\sqrt{5} + 3\sqrt{20} + 4\sqrt{45}$           **18.** $5\sqrt{54} - 2\sqrt{24} - 2\sqrt{96}$

**19.** $\sqrt{72x} - \sqrt{8x}$                  **20.** $4\sqrt{18k} - \sqrt{72k}$

**21.** $3\sqrt{72m^2} - 5\sqrt{32m^2} - 3\sqrt{18m^2}$     **22.** $9\sqrt{27p^2} - 14\sqrt{108p^2} + 2\sqrt{48p^2}$

**23.** $2\sqrt[3]{16} - \sqrt[3]{54}$                 **24.** $15\sqrt[3]{81} - 4\sqrt[3]{24}$

**25.** $2\sqrt[3]{27x} - 2\sqrt[3]{8x}$              **26.** $6\sqrt[3]{128m} + 3\sqrt[3]{16m}$

**27.** $\sqrt[3]{x^2y} - \sqrt[3]{8x^2y}$             **28.** $3\sqrt[3]{x^2y^2} - 2\sqrt[3]{64x^2y^2}$

**29.** $3x\sqrt[3]{xy^2} - 2\sqrt[3]{8x^4y^2}$         **30.** $6q^2\sqrt[3]{5q} - 2q\sqrt[3]{40q^4}$

**31.** $5\sqrt[4]{32} + 3\sqrt[4]{162}$             **32.** $2\sqrt[4]{512} + 4\sqrt[4]{32}$

**33.** $3\sqrt[4]{x^5y} - 2x\sqrt[4]{xy}$           **34.** $2\sqrt[4]{m^9p^6} - 3m^2p\sqrt[4]{mp^2}$

**35.** $2\sqrt[4]{32a^3} + 5\sqrt[4]{2a^3}$          **36.** $-\sqrt[4]{16r} + 5\sqrt[4]{r}$

**37.** $\sqrt[3]{64xy^2} + \sqrt[3]{27x^4y^5}$        **38.** $\sqrt[4]{625s^3t} - \sqrt[4]{81s^7t^5}$

**39.** $4\sqrt[3]{x} - 6\sqrt{x}$                   **40.** $3\sqrt{7z} - 9\sqrt[3]{7z}$

**41.** $2\sqrt[3]{8x^4} + 3\sqrt[4]{16x^5}$          **42.** $3\sqrt[3]{64m^4} + 5\sqrt[4]{81m^5}$

*Simplify. Assume that all variables represent positive real numbers. See Example 3.*

**43.** $\sqrt{8} - \dfrac{\sqrt{64}}{\sqrt{16}}$      **44.** $\sqrt{48} - \dfrac{\sqrt{81}}{\sqrt{9}}$      **45.** $\dfrac{2\sqrt{5}}{3} + \dfrac{\sqrt{5}}{6}$

**46.** $\dfrac{4\sqrt{3}}{3} + \dfrac{2\sqrt{3}}{9}$      **47.** $\sqrt{\dfrac{8}{9}} + \sqrt{\dfrac{18}{36}}$      **48.** $\sqrt{\dfrac{12}{16}} + \sqrt{\dfrac{48}{64}}$

**49.** $\dfrac{\sqrt{32}}{3} + \dfrac{2\sqrt{2}}{3} - \dfrac{\sqrt{2}}{\sqrt{9}}$    **50.** $\dfrac{\sqrt{27}}{2} - \dfrac{3\sqrt{3}}{2} + \dfrac{\sqrt{3}}{\sqrt{4}}$    **51.** $3\sqrt{\dfrac{50}{9}} + 8\dfrac{\sqrt{2}}{\sqrt{8}}$

**52.** $9\sqrt{\dfrac{48}{25}} - 2\dfrac{\sqrt{2}}{\sqrt{98}}$    **53.** $\sqrt{\dfrac{25}{x^8}} - \sqrt{\dfrac{9}{x^6}}$    **54.** $\sqrt{\dfrac{100}{y^4}} + \sqrt{\dfrac{81}{y^{10}}}$

**55.** $3\sqrt[3]{\dfrac{m^5}{27}} - 2m\sqrt[3]{\dfrac{m^2}{64}}$    **56.** $2a\sqrt[4]{\dfrac{a}{16}} - 5a\sqrt[4]{\dfrac{a}{81}}$    **57.** $3\sqrt[3]{\dfrac{2}{x^6}} - 4\sqrt[3]{\dfrac{5}{x^9}}$

**58.** $-4\sqrt[3]{\dfrac{4}{t^9}} + 3\sqrt[3]{\dfrac{9}{t^{12}}}$

*Multiple Choice*   *Solve each problem.*

**59.** A rectangular yard has a length of $\sqrt{192}$ m and a width of $\sqrt{48}$ m. Choose the best estimate of its dimensions. Then estimate the perimeter.

    **A.** 14 m by 7 m      **B.** 5 m by 7 m      **C.** 14 m by 8 m      **D.** 15 m by 8 m

**406** CHAPTER 8 Roots, Radicals, and Root Functions

**60.** If the sides of a triangle are $\sqrt{65}$ in., $\sqrt{35}$ in., and $\sqrt{26}$ in., which one of the following is the best estimate of its perimeter?

**A.** 20 in.   **B.** 26 in.   **C.** 19 in.   **D.** 24 in.

*Solve each problem. Give answers as simplified radical expressions.*

**61.** Find the perimeter of the triangle.

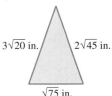

**62.** Find the perimeter of the rectangle.

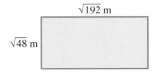

**63.** What is the perimeter of the computer graphic?

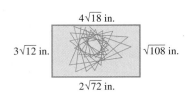

**64.** Find the area of the trapezoid.

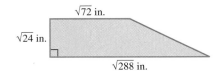

## 8.5 Multiplying and Dividing Radical Expressions

**OBJECTIVES**

1. Multiply radicals.
2. Rationalize denominators with one radical term.
3. Rationalize denominators with binomials involving radicals.
4. Write radical quotients in lowest terms.

**OBJECTIVE 1** Multiply radicals. We multiply binomial expressions involving radicals by using the FOIL method from **Section 5.4.** For example, we find the product of the binomials $\sqrt{5} + 3$ and $\sqrt{6} + 1$ as follows:

$$(\sqrt{5} + 3)(\sqrt{6} + 1) = \overbrace{\sqrt{5} \cdot \sqrt{6}}^{\text{First}} + \overbrace{\sqrt{5} \cdot 1}^{\text{Outer}} + \overbrace{3 \cdot \sqrt{6}}^{\text{Inner}} + \overbrace{3 \cdot 1}^{\text{Last}}$$
$$= \sqrt{30} + \sqrt{5} + 3\sqrt{6} + 3.$$

This result cannot be simplified further.

**EXAMPLE 1** Multiplying Binomials Involving Radicals

Multiply, using the FOIL method.

(a) $(7 - \sqrt{3})(\sqrt{5} + \sqrt{2}) = \overset{F}{7\sqrt{5}} + \overset{O}{7\sqrt{2}} - \overset{I}{\sqrt{3} \cdot \sqrt{5}} - \overset{L}{\sqrt{3} \cdot \sqrt{2}}$
$= 7\sqrt{5} + 7\sqrt{2} - \sqrt{15} - \sqrt{6}$

(b) $(\sqrt{10} + \sqrt{3})(\sqrt{10} - \sqrt{3})$
$= \sqrt{10} \cdot \sqrt{10} - \sqrt{10} \cdot \sqrt{3} + \sqrt{10} \cdot \sqrt{3} - \sqrt{3} \cdot \sqrt{3}$   FOIL
$= 10 - 3$
$= 7$

The product $(\sqrt{10} + \sqrt{3})(\sqrt{10} - \sqrt{3}) = (\sqrt{10})^2 - (\sqrt{3})^2$ is the difference of squares:

$$(x + y)(x - y) = x^2 - y^2.$$

Here, $x = \sqrt{10}$ and $y = \sqrt{3}$.

**(c)** $(\sqrt{7} - 3)^2 = (\sqrt{7} - 3)(\sqrt{7} - 3)$
$= \sqrt{7} \cdot \sqrt{7} - 3\sqrt{7} - 3\sqrt{7} + 3 \cdot 3$
$= 7 - 6\sqrt{7} + 9$   *Be careful! These terms cannot be combined.*
$= 16 - 6\sqrt{7}$

**(d)** $(5 - \sqrt[3]{3})(5 + \sqrt[3]{3}) = 5 \cdot 5 + 5\sqrt[3]{3} - 5\sqrt[3]{3} - \sqrt[3]{3} \cdot \sqrt[3]{3}$
$= 25 - \sqrt[3]{3^2}$   *Remember to write the index 3 in each radical.*
$= 25 - \sqrt[3]{9}$

**(e)** $(\sqrt{k} + \sqrt{y})(\sqrt{k} - \sqrt{y}) = (\sqrt{k})^2 - (\sqrt{y})^2$   Difference of squares
$= k - y$,   $k \geq 0$ and $y \geq 0$

**NOW TRY** Exercises 13, 17, 23, 27, and 39.

In Example 1(c), we could have used the formula for the square of a binomial,

$$(x - y)^2 = x^2 - 2xy + y^2,$$

to obtain the same result:

$(\sqrt{7} - 3)^2 = (\sqrt{7})^2 - 2(\sqrt{7})(3) + 3^2$
$= 7 - 6\sqrt{7} + 9$
$= 16 - 6\sqrt{7}.$

**OBJECTIVE 2** Rationalize denominators with one radical term. As defined earlier, a simplified radical expression has no radical in the denominator. The origin of this agreement no doubt occurred before the days of high-speed calculation, when computation was a tedious process performed by hand. To see this, consider the radical expression $\frac{1}{\sqrt{2}}$. To find a decimal approximation by hand, it is necessary to divide 1 by a decimal approximation for $\sqrt{2}$, such as 1.414. It is much easier if the divisor is a whole number. This can be accomplished by multiplying $\frac{1}{\sqrt{2}}$ by 1 in the form $\frac{\sqrt{2}}{\sqrt{2}}$. *Multiplying by 1 in any form does not change the value of the original expression.*

$$\frac{1}{\sqrt{2}} \cdot \frac{\sqrt{2}}{\sqrt{2}} = \frac{\sqrt{2}}{2} \qquad \frac{\sqrt{2}}{\sqrt{2}} = 1$$

Now the computation requires dividing 1.414 by 2 to obtain 0.707, a much easier task.

With current technology, either form of this fraction can be approximated with the same number of keystrokes. See Figure 12, which shows how a calculator gives the same approximation for both forms of the expression.

A common way of "standardizing" the form of a radical expression is to have the denominator contain no radicals. The process of removing radicals from a denominator so that the denominator contains only rational numbers is called **rationalizing the denominator.**

```
1/√(2)
        .7071067812
√(2)/2
        .7071067812
```

**FIGURE 12**

**EXAMPLE 2** Rationalizing Denominators with Square Roots

Rationalize each denominator.

(a) $\dfrac{3}{\sqrt{7}}$

Multiply the numerator and denominator by $\sqrt{7}$. This is, in effect, multiplying by 1.

$$\dfrac{3}{\sqrt{7}} = \dfrac{3 \cdot \sqrt{7}}{\sqrt{7} \cdot \sqrt{7}} = \dfrac{3\sqrt{7}}{7}$$

In the denominator, $\sqrt{7} \cdot \sqrt{7} = \sqrt{7 \cdot 7} = \sqrt{49} = 7$. The final denominator is now a rational number.

(b) $\dfrac{5\sqrt{2}}{\sqrt{5}} = \dfrac{5\sqrt{2} \cdot \sqrt{5}}{\sqrt{5} \cdot \sqrt{5}} = \dfrac{5\sqrt{10}}{5} = \sqrt{10}$

(c) $\dfrac{-6}{\sqrt{12}}$

Less work is involved if we simplify the radical in the denominator first.

$$\dfrac{-6}{\sqrt{12}} = \dfrac{-6}{\sqrt{4 \cdot 3}} = \dfrac{-6}{2\sqrt{3}} = \dfrac{-3}{\sqrt{3}}$$

Now we rationalize the denominator.

$$\dfrac{-3}{\sqrt{3}} = \dfrac{-3 \cdot \sqrt{3}}{\sqrt{3} \cdot \sqrt{3}} = \dfrac{-3\sqrt{3}}{3} = -\sqrt{3}$$

**NOW TRY** Exercises 43, 49, and 51.

**EXAMPLE 3** Rationalizing Denominators in Roots of Fractions

Simplify each radical. In part (b), $p > 0$.

(a) $-\sqrt{\dfrac{18}{125}} = -\dfrac{\sqrt{18}}{\sqrt{125}}$    Quotient rule

$= -\dfrac{\sqrt{9 \cdot 2}}{\sqrt{25 \cdot 5}}$    Factor.

$= -\dfrac{3\sqrt{2}}{5\sqrt{5}}$    Product rule

$= -\dfrac{3\sqrt{2} \cdot \sqrt{5}}{5\sqrt{5} \cdot \sqrt{5}}$    Multiply by $\dfrac{\sqrt{5}}{\sqrt{5}}$.

$= -\dfrac{3\sqrt{10}}{5 \cdot 5}$    Product rule

$= -\dfrac{3\sqrt{10}}{25}$

**(b)** $\sqrt{\dfrac{50m^4}{p^5}} = \dfrac{\sqrt{50m^4}}{\sqrt{p^5}}$   Quotient rule

$= \dfrac{5m^2\sqrt{2}}{p^2\sqrt{p}}$   Product rule

$= \dfrac{5m^2\sqrt{2} \cdot \sqrt{p}}{p^2\sqrt{p} \cdot \sqrt{p}}$   Multiply by $\dfrac{\sqrt{p}}{\sqrt{p}}$.

$= \dfrac{5m^2\sqrt{2p}}{p^2 \cdot p}$   Product rule

$= \dfrac{5m^2\sqrt{2p}}{p^3}$

**NOW TRY** Exercises 55 and 63.

---

**EXAMPLE 4** Rationalizing Denominators with Cube and Fourth Roots

Simplify.

**(a)** $\sqrt[3]{\dfrac{27}{16}}$

Use the quotient rule, and simplify the numerator and denominator.

$$\sqrt[3]{\dfrac{27}{16}} = \dfrac{\sqrt[3]{27}}{\sqrt[3]{16}} = \dfrac{3}{\sqrt[3]{8} \cdot \sqrt[3]{2}} = \dfrac{3}{2\sqrt[3]{2}}$$

To get a rational denominator, multiply the numerator and denominator by a number that will result in a perfect *cube* in the radicand in the denominator. Since $2 \cdot 4 = 8$, a perfect cube, multiply the numerator and denominator by $\sqrt[3]{4}$.

$$\sqrt[3]{\dfrac{27}{16}} = \dfrac{3}{2\sqrt[3]{2}} = \dfrac{3 \cdot \sqrt[3]{4}}{2\sqrt[3]{2} \cdot \sqrt[3]{4}} = \dfrac{3\sqrt[3]{4}}{2\sqrt[3]{8}} = \dfrac{3\sqrt[3]{4}}{2 \cdot 2} = \dfrac{3\sqrt[3]{4}}{4}$$

**(b)** $\sqrt[4]{\dfrac{5x}{z}} = \dfrac{\sqrt[4]{5x}}{\sqrt[4]{z}} \cdot \dfrac{\sqrt[4]{z^3}}{\sqrt[4]{z^3}} = \dfrac{\sqrt[4]{5xz^3}}{\sqrt[4]{z^4}} = \dfrac{\sqrt[4]{5xz^3}}{z}$,  $x \geq 0, z > 0$

**NOW TRY** Exercises 71 and 81.

---

**CAUTION** It is easy to make mistakes in problems like Example 4(a). A typical error is to multiply the numerator and denominator by $\sqrt[3]{2}$, forgetting that

$$\sqrt[3]{2} \cdot \sqrt[3]{2} \neq 2.$$

You need *three* factors of 2 to obtain $2^3$ under the radical.

$$\sqrt[3]{2} \cdot \sqrt[3]{2} \cdot \sqrt[3]{2} = 2$$

---

**OBJECTIVE 3** Rationalize denominators with binomials involving radicals. Recall the special product

$$(x + y)(x - y) = x^2 - y^2.$$

To rationalize a denominator that contains a binomial expression (one that contains exactly two terms) involving radicals, such as

$$\dfrac{3}{1 + \sqrt{2}},$$

we must use *conjugates*. The conjugate of $1 + \sqrt{2}$ is $1 - \sqrt{2}$. In general, $x + y$ and $x - y$ are **conjugates**.

### Rationalizing a Binomial Denominator

Whenever a radical expression has a sum or difference with square root radicals in the denominator, rationalize the denominator by multiplying both the numerator and denominator by the *conjugate* of the denominator.

For the expression $\frac{3}{1+\sqrt{2}}$, we rationalize the denominator by multiplying both the numerator and denominator by $1 - \sqrt{2}$, the conjugate of the denominator.

$$\frac{3}{1+\sqrt{2}} = \frac{3(1-\sqrt{2})}{(1+\sqrt{2})(1-\sqrt{2})}$$

$$= \frac{3(1-\sqrt{2})}{-1} \quad (1+\sqrt{2})(1-\sqrt{2}) = 1^2 - (\sqrt{2})^2 = 1 - 2 = -1.$$

$$= \frac{3}{-1}(1-\sqrt{2}) \quad \text{The denominator is now a rational number.}$$

$$= -3(1-\sqrt{2})$$

$$= -3 + 3\sqrt{2}$$

**EXAMPLE 5** Rationalizing Binomial Denominators

Rationalize each denominator.

**(a)** $\dfrac{5}{4-\sqrt{3}}$

To rationalize the denominator, multiply both the numerator and denominator by the conjugate of the denominator, $4 + \sqrt{3}$.

$$\frac{5}{4-\sqrt{3}} = \frac{5(4+\sqrt{3})}{(4-\sqrt{3})(4+\sqrt{3})}$$

$$= \frac{5(4+\sqrt{3})}{16-3}$$

$$= \frac{5(4+\sqrt{3})}{13}$$

Notice that the numerator is left in factored form. This makes it easier to determine whether the expression is written in lowest terms.

**(b)** $\dfrac{\sqrt{2}-\sqrt{3}}{\sqrt{5}+\sqrt{3}} = \dfrac{(\sqrt{2}-\sqrt{3})(\sqrt{5}-\sqrt{3})}{(\sqrt{5}+\sqrt{3})(\sqrt{5}-\sqrt{3})}$  Multiply numerator and denominator by $\sqrt{5} - \sqrt{3}$.

$$= \frac{\sqrt{10} - \sqrt{6} - \sqrt{15} + 3}{5 - 3}$$

$$= \frac{\sqrt{10} - \sqrt{6} - \sqrt{15} + 3}{2}$$

**(c)** $\dfrac{3}{\sqrt{5m}-\sqrt{p}} = \dfrac{3(\sqrt{5m}+\sqrt{p})}{(\sqrt{5m}-\sqrt{p})(\sqrt{5m}+\sqrt{p})}$

$$= \frac{3(\sqrt{5m}+\sqrt{p})}{5m - p}, \quad 5m \neq p, m > 0, p > 0$$

**NOW TRY** Exercises 83, 89, and 93.

## SECTION 8.5 Multiplying and Dividing Radical Expressions

**OBJECTIVE 4** Write radical quotients in lowest terms.

**EXAMPLE 6** Writing Radical Quotients in Lowest Terms

Write each quotient in lowest terms.

(a) $\dfrac{6 + 2\sqrt{5}}{4}$

Factor the numerator and denominator, and then write in lowest terms.

$$\dfrac{6 + 2\sqrt{5}}{4} = \dfrac{2(3 + \sqrt{5})}{2 \cdot 2} = \dfrac{3 + \sqrt{5}}{2} \qquad \text{Factor first; then divide out the common factor.}$$

Here is an alternative method for writing this quotient in lowest terms:

$$\dfrac{6 + 2\sqrt{5}}{4} = \dfrac{6}{4} + \dfrac{2\sqrt{5}}{4} = \dfrac{3}{2} + \dfrac{\sqrt{5}}{2} = \dfrac{3 + \sqrt{5}}{2}.$$

(b) $\dfrac{5y - \sqrt{8y^2}}{6y} = \dfrac{5y - 2y\sqrt{2}}{6y}, \quad y > 0 \qquad$ Product rule

$\qquad\qquad\quad = \dfrac{y(5 - 2\sqrt{2})}{6y} \qquad$ Factor the numerator.

$\qquad\qquad\quad = \dfrac{5 - 2\sqrt{2}}{6} \qquad$ Divide out the common factor.

**NOW TRY** Exercises 103 and 105.

**CAUTION** *Be careful to factor before writing a quotient in lowest terms.*

## 8.5 Exercises

*Matching* Match each part of a rule for a special product in Column I with the other part in Column II. Assume that all variables represent positive real numbers.

| I | II |
|---|---|
| **1.** $(x + \sqrt{y})(x - \sqrt{y})$ | **A.** $x - y$ |
| **2.** $(\sqrt{x} + y)(\sqrt{x} - y)$ | **B.** $x + 2y\sqrt{x} + y^2$ |
| **3.** $(\sqrt{x} + \sqrt{y})(\sqrt{x} - \sqrt{y})$ | **C.** $x - y^2$ |
| **4.** $(\sqrt{x} + \sqrt{y})^2$ | **D.** $x - 2\sqrt{xy} + y$ |
| **5.** $(\sqrt{x} - \sqrt{y})^2$ | **E.** $x^2 - y$ |
| **6.** $(\sqrt{x} + y)^2$ | **F.** $x + 2\sqrt{xy} + y$ |

*Multiply, and then simplify each product. Assume that all variables represent positive real numbers. See Example 1.*

**7.** $\sqrt{6}(3 + \sqrt{2})$  **8.** $\sqrt{2}(\sqrt{32} - \sqrt{9})$  **9.** $5(\sqrt{72} - \sqrt{8})$

**10.** $\sqrt{3}(\sqrt{12} + 2)$  **11.** $(\sqrt{7} + 3)(\sqrt{7} - 3)$  **12.** $(\sqrt{3} - 5)(\sqrt{3} + 5)$

13. $(\sqrt{2} - \sqrt{3})(\sqrt{2} + \sqrt{3})$
14. $(\sqrt{7} + \sqrt{3})(\sqrt{7} - \sqrt{3})$
15. $(\sqrt{8} - \sqrt{2})(\sqrt{8} + \sqrt{2})$
16. $(\sqrt{20} - \sqrt{5})(\sqrt{20} + \sqrt{5})$
17. $(\sqrt{2} + 1)(\sqrt{3} - 1)$
18. $(\sqrt{3} + 3)(\sqrt{5} - 2)$
19. $(\sqrt{11} - \sqrt{7})(\sqrt{2} + \sqrt{5})$
20. $(\sqrt{6} + \sqrt{2})(\sqrt{3} + \sqrt{2})$
21. $(2\sqrt{3} + \sqrt{5})(3\sqrt{3} - 2\sqrt{5})$
22. $(\sqrt{7} - \sqrt{11})(2\sqrt{7} + 3\sqrt{11})$
23. $(\sqrt{5} + 2)^2$
24. $(\sqrt{11} - 1)^2$
25. $(\sqrt{21} - \sqrt{5})^2$
26. $(\sqrt{6} - \sqrt{2})^2$
27. $(2 + \sqrt[3]{6})(2 - \sqrt[3]{6})$
28. $(\sqrt[3]{3} + 6)(\sqrt[3]{3} - 6)$
29. $(2 + \sqrt[3]{2})(4 - 2\sqrt[3]{2} + \sqrt[3]{4})$
30. $(\sqrt[3]{3} - 1)(\sqrt[3]{9} + \sqrt[3]{3} + 1)$
31. $(3\sqrt{x} - \sqrt{5})(2\sqrt{x} + 1)$
32. $(4\sqrt{p} + \sqrt{7})(\sqrt{p} - 9)$
33. $(3\sqrt{r} - \sqrt{s})(3\sqrt{r} + \sqrt{s})$
34. $(\sqrt{k} + 4\sqrt{m})(\sqrt{k} - 4\sqrt{m})$
35. $(\sqrt[3]{2y} - 5)(4\sqrt[3]{2y} + 1)$
36. $(\sqrt[3]{9z} - 2)(5\sqrt[3]{9z} + 7)$
37. $(\sqrt{3x} + 2)(\sqrt{3x} - 2)$
38. $(\sqrt{6y} - 4)(\sqrt{6y} + 4)$
39. $(2\sqrt{x} + \sqrt{y})(2\sqrt{x} - \sqrt{y})$
40. $(\sqrt{p} + 5\sqrt{s})(\sqrt{p} - 5\sqrt{s})$
41. $[(\sqrt{2} + \sqrt{3}) - \sqrt{6}][(\sqrt{2} + \sqrt{3}) + \sqrt{6}]$
42. $[(\sqrt{5} - \sqrt{2}) - \sqrt{3}][(\sqrt{5} - \sqrt{2}) + \sqrt{3}]$

*Rationalize the denominator in each expression. Assume that all variables represent positive real numbers. See Examples 2 and 3.*

43. $\dfrac{7}{\sqrt{7}}$
44. $\dfrac{11}{\sqrt{11}}$
45. $\dfrac{15}{\sqrt{3}}$
46. $\dfrac{12}{\sqrt{6}}$
47. $\dfrac{\sqrt{3}}{\sqrt{2}}$
48. $\dfrac{\sqrt{7}}{\sqrt{6}}$
49. $\dfrac{9\sqrt{3}}{\sqrt{5}}$
50. $\dfrac{3\sqrt{2}}{\sqrt{11}}$
51. $\dfrac{-6}{\sqrt{18}}$
52. $\dfrac{-5}{\sqrt{24}}$
53. $\sqrt{\dfrac{7}{2}}$
54. $\sqrt{\dfrac{10}{3}}$
55. $-\sqrt{\dfrac{7}{50}}$
56. $-\sqrt{\dfrac{13}{75}}$
57. $\sqrt{\dfrac{24}{x}}$
58. $\sqrt{\dfrac{52}{y}}$
59. $\dfrac{-8\sqrt{3}}{\sqrt{k}}$
60. $\dfrac{-4\sqrt{13}}{\sqrt{m}}$
61. $-\sqrt{\dfrac{150m^5}{n^3}}$
62. $-\sqrt{\dfrac{98r^3}{s^5}}$
63. $\sqrt{\dfrac{288x^7}{y^9}}$
64. $\sqrt{\dfrac{242t^9}{u^{11}}}$
65. $\dfrac{5\sqrt{2m}}{\sqrt{y^3}}$
66. $\dfrac{2\sqrt{5r}}{\sqrt{m^3}}$
67. $-\sqrt{\dfrac{48k^2}{z}}$
68. $-\sqrt{\dfrac{75m^3}{p}}$

*Simplify. Assume that all variables represent positive real numbers. See Example 4.*

69. $\sqrt[3]{\dfrac{2}{3}}$
70. $\sqrt[3]{\dfrac{4}{5}}$
71. $\sqrt[3]{\dfrac{4}{9}}$
72. $\sqrt[3]{\dfrac{5}{16}}$
73. $\sqrt[3]{\dfrac{9}{32}}$
74. $\sqrt[3]{\dfrac{10}{9}}$
75. $-\sqrt[3]{\dfrac{2p}{r^2}}$
76. $-\sqrt[3]{\dfrac{6x}{y^2}}$
77. $\sqrt[3]{\dfrac{x^6}{y}}$
78. $\sqrt[3]{\dfrac{m^9}{q}}$
79. $\sqrt[4]{\dfrac{16}{x}}$
80. $\sqrt[4]{\dfrac{81}{y}}$
81. $\sqrt[4]{\dfrac{2y}{z}}$
82. $\sqrt[4]{\dfrac{7t}{s^2}}$

*Rationalize the denominator in each expression. Assume that all variables represent positive real numbers and no denominators are 0. See Example 5.*

83. $\dfrac{3}{4 + \sqrt{5}}$  84. $\dfrac{4}{3 - \sqrt{7}}$  85. $\dfrac{\sqrt{8}}{3 - \sqrt{2}}$  86. $\dfrac{\sqrt{27}}{2 + \sqrt{3}}$

87. $\dfrac{2}{3\sqrt{5} + 2\sqrt{3}}$  88. $\dfrac{-1}{3\sqrt{2} - 2\sqrt{7}}$  89. $\dfrac{\sqrt{2} - \sqrt{3}}{\sqrt{6} - \sqrt{5}}$  90. $\dfrac{\sqrt{5} + \sqrt{6}}{\sqrt{3} - \sqrt{2}}$

91. $\dfrac{m - 4}{\sqrt{m} + 2}$  92. $\dfrac{r - 9}{\sqrt{r} - 3}$  93. $\dfrac{4}{\sqrt{x} - 2\sqrt{y}}$  94. $\dfrac{5}{3\sqrt{r} + \sqrt{s}}$

95. $\dfrac{\sqrt{x} - \sqrt{y}}{\sqrt{x} + \sqrt{y}}$  96. $\dfrac{\sqrt{a} + \sqrt{b}}{\sqrt{a} - \sqrt{b}}$  97. $\dfrac{5\sqrt{k}}{2\sqrt{k} + \sqrt{q}}$  98. $\dfrac{3\sqrt{x}}{\sqrt{x} - 2\sqrt{y}}$

*Write each expression in lowest terms. Assume that all variables represent positive real numbers. See Example 6.*

99. $\dfrac{30 - 20\sqrt{6}}{10}$  100. $\dfrac{24 + 12\sqrt{5}}{12}$  101. $\dfrac{3 - 3\sqrt{5}}{3}$

102. $\dfrac{-5 + 5\sqrt{2}}{5}$  103. $\dfrac{16 - 4\sqrt{8}}{12}$  104. $\dfrac{12 - 9\sqrt{72}}{18}$

105. $\dfrac{6p + \sqrt{24p^3}}{3p}$  106. $\dfrac{11y - \sqrt{242y^5}}{22y}$

*Rationalize each denominator. Assume that all radicals represent real numbers and no denominators are 0.*

107. $\dfrac{1}{\sqrt{x + y}}$  108. $\dfrac{5}{\sqrt{m - n}}$  109. $\dfrac{p}{\sqrt{p + 2}}$  110. $\dfrac{3q}{\sqrt{5 + q}}$

## 8.6 Solving Equations with Radicals

**OBJECTIVES**

1. Solve radical equations by using the power rule.
2. Solve radical equations that require additional steps.
3. Solve radical equations with indexes greater than 2.
4. Solve radical equations by using a graphing calculator.
5. Use the power rule to solve a formula for a specified variable.

An equation that includes one or more radical expressions with a variable is called a **radical equation.**

$\sqrt{x - 4} = 8$,  $\sqrt{5x + 12} = 3\sqrt{2x - 1}$,  and  $\sqrt[3]{6 + x} = 27$   Radical equations

**OBJECTIVE 1** Solve radical equations by using the power rule. The equation $x = 1$ has only one solution. Its solution set is $\{1\}$. If we square both sides of this equation, we get $x^2 = 1$. This new equation has *two* solutions: $-1$ and $1$. Notice that the solution of the original equation is also a solution of the squared equation. However, the squared equation has another solution, $-1$, that is *not* a solution of the original equation. When solving equations with radicals, we use this idea of raising both sides to a power. It is an application of the **power rule.**

**Power Rule for Solving an Equation with Radicals**

If both sides of an equation are raised to the same power, all solutions of the original equation are also solutions of the new equation.

***Read the power rule carefully; it does not say that all solutions of the new equation are solutions of the original equation. They may or may not be.*** Solutions that do not satisfy the original equation are called **extraneous solutions;** they must be discarded.

> **CAUTION** When the power rule is used to solve an equation, *every solution of the new equation must be checked in the original equation.*

**EXAMPLE 1** Using the Power Rule

Solve $\sqrt{3x + 4} = 8$.

Use the power rule and square both sides to obtain

$$(\sqrt{3x + 4})^2 = 8^2$$
$$3x + 4 = 64$$
$$3x = 60 \quad \text{Subtract 4.}$$
$$x = 20. \quad \text{Divide by 3.}$$

Check: $\sqrt{3x + 4} = 8$     Original equation
$\sqrt{3 \cdot 20 + 4} = 8$  ?   Let $x = 20$.
$\sqrt{64} = 8$  ?
$8 = 8$     True

Since 20 satisfies the *original* equation, the solution set is {20}.

**NOW TRY** Exercise 9.

Use the following steps to solve equations with radicals.

### Solving an Equation with Radicals

*Step 1* **Isolate the radical.** Make sure that one radical term is alone on one side of the equation.

*Step 2* **Apply the power rule.** Raise both sides of the equation to a power that is the same as the index of the radical.

*Step 3* **Solve** the resulting equation; if it still contains a radical, repeat Steps 1 and 2.

*Step 4* **Check** all proposed solutions in the original equation.

**EXAMPLE 2** Using the Power Rule

Solve $\sqrt{5q - 1} + 3 = 0$.

*Step 1* To isolate the radical on one side, subtract 3 from each side.

$$\sqrt{5q - 1} = -3$$

*Step 2* Now square both sides.

$$(\sqrt{5q - 1})^2 = (-3)^2$$

**Step 3**
$$5q - 1 = 9$$
$$5q = 10 \quad \text{Add 1.}$$
$$q = 2 \quad \text{Divide by 5.}$$

**Step 4** Check:
$$\sqrt{5q - 1} + 3 = 0 \quad \text{Original equation}$$
$$\sqrt{5 \cdot 2 - 1} + 3 = 0 \quad ? \quad \text{Let } q = 2.$$
$$3 + 3 = 0 \quad \text{False}$$

*Be sure to check the proposed solution.*

This false result shows that 2 is *not* a solution of the original equation; it is extraneous. The solution set is ∅.

**NOW TRY** Exercise 11.

We could have determined after Step 1 that the equation in Example 2 has no solution because the expression on the left cannot be negative.

**OBJECTIVE 2** Solve radical equations that require additional steps. The next examples involve finding the square of a binomial. Recall from **Section 5.4** that

$$(x + y)^2 = x^2 + 2xy + y^2.$$

**EXAMPLE 3** Using the Power Rule; Squaring a Binomial

Solve $\sqrt{4 - x} = x + 2$.

**Step 1** The radical is alone on the left side of the equation.

**Step 2** Square both sides; the square of $x + 2$ is $(x + 2)^2 = x^2 + 4x + 4$.

$$(\sqrt{4 - x})^2 = (x + 2)^2$$
$$4 - x = x^2 + 4x + 4$$

*Remember the middle term.*

— Twice the product of 2 and $x$

**Step 3** The new equation is quadratic, so get 0 on one side.

$$0 = x^2 + 5x \quad \text{Subtract 4; add } x.$$
$$0 = x(x + 5) \quad \text{Factor.}$$
$$x = 0 \quad \text{or} \quad x + 5 = 0 \quad \text{Zero-factor property}$$
$$x = -5$$

*Set each factor equal to 0.*

**Step 4** Check each proposed solution in the original equation.

Check: If $x = 0$, then

$$\sqrt{4 - x} = x + 2$$
$$\sqrt{4 - 0} = 0 + 2 \quad ?$$
$$\sqrt{4} = 2 \quad ?$$
$$2 = 2. \quad \text{True}$$

If $x = -5$, then

$$\sqrt{4 - x} = x + 2$$
$$\sqrt{4 - (-5)} = -5 + 2 \quad ?$$
$$\sqrt{9} = -3 \quad ?$$
$$3 = -3. \quad \text{False}$$

The solution set is {0}. The other proposed solution, −5, is extraneous.

**NOW TRY** Exercise 27.

**EXAMPLE 4** Using the Power Rule; Squaring a Binomial

Solve $\sqrt{x^2 - 4x + 9} = x - 1$.

Squaring both sides gives $(x - 1)^2 = x^2 - 2(x)(1) + 1^2$ on the right.

$$(\sqrt{x^2 - 4x + 9})^2 = (x - 1)^2 \qquad \text{Remember the middle term.}$$

$$x^2 - 4x + 9 = x^2 - 2x + 1$$

↑ Twice the product of $x$ and $-1$

$$-2x = -8 \qquad \text{Subtract } x^2 \text{ and } 9; \text{ add } 2x.$$

$$x = 4 \qquad \text{Divide by } -2.$$

Check: 
$$\sqrt{x^2 - 4x + 9} = x - 1 \qquad \text{Original equation}$$
$$\sqrt{4^2 - 4 \cdot 4 + 9} = 4 - 1 \quad ? \qquad \text{Let } x = 4.$$
$$3 = 3 \qquad \text{True}$$

The solution set of the original equation is $\{4\}$.

**NOW TRY** Exercise 29.

---

**EXAMPLE 5** Using the Power Rule; Squaring Twice

Solve $\sqrt{5x + 6} + \sqrt{3x + 4} = 2$.

Start by isolating one radical on one side of the equation. Do this by subtracting $\sqrt{3x + 4}$ from each side.

$$\sqrt{5x + 6} = 2 - \sqrt{3x + 4} \qquad \text{Subtract } \sqrt{3x + 4}.$$
$$(\sqrt{5x + 6})^2 = (2 - \sqrt{3x + 4})^2 \qquad \text{Square both sides.}$$
$$5x + 6 = 4 - 4\sqrt{3x + 4} + (3x + 4)$$

↑ Twice the product of 2 and

**Remember the middle term.**

This equation still contains a radical, so isolate the radical term on the right and square both sides again.

$$5x + 6 = 8 + 3x - 4\sqrt{3x + 4}$$
$$2x - 2 = -4\sqrt{3x + 4} \qquad \text{Subtract 8 and } 3x.$$

*Divide each term by 2.*
$$x - 1 = -2\sqrt{3x + 4} \qquad \text{Divide by 2.}$$
$$(x - 1)^2 = (-2\sqrt{3x + 4})^2 \qquad \text{Square both sides again.}$$
$$x^2 - 2x + 1 = (-2)^2(\sqrt{3x + 4})^2 \qquad \text{On the right, } (ab)^2 = a^2b^2.$$
$$x^2 - 2x + 1 = 4(3x + 4) \qquad \text{Apply the exponents.}$$
$$x^2 - 2x + 1 = 12x + 16 \qquad \text{Distributive property}$$
$$x^2 - 14x - 15 = 0 \qquad \text{Standard form}$$
$$(x - 15)(x + 1) = 0 \qquad \text{Factor.}$$
$$x - 15 = 0 \quad \text{or} \quad x + 1 = 0 \qquad \text{Zero-factor property}$$
$$x = 15 \quad \text{or} \quad x = -1$$

*Check:* If $x = 15$, then

$$\sqrt{5x + 6} + \sqrt{3x + 4} = 2 \quad \text{Original equation}$$
$$\sqrt{5(15) + 6} + \sqrt{3(15) + 4} = 2 \quad ? \quad \text{Let}$$
$$\sqrt{81} + \sqrt{49} = 2 \quad ?$$
$$9 + 7 = 2 \quad ?$$
$$16 = 2. \quad \text{False}$$

Thus, 15 is an extraneous solution and must be discarded. Confirm that the proposed solution $-1$ checks, so the solution set is $\{-1\}$.

**NOW TRY** Exercise 51.

**OBJECTIVE 3** Solve radical equations with indexes greater than 2. The power rule also works for root indexes and powers greater than 2.

**EXAMPLE 6** Using the Power Rule for a Root Index Greater than 2

Solve $\sqrt[3]{z + 5} = \sqrt[3]{2z - 6}$.

Raise both sides to the third power.

$$(\sqrt[3]{z + 5})^3 = (\sqrt[3]{2z - 6})^3$$
$$z + 5 = 2z - 6$$
$$11 = z \quad \text{Subtract } z; \text{ add 6.}$$

*Check:*
$$\sqrt[3]{z + 5} = \sqrt[3]{2z - 6} \quad \text{Original equation}$$
$$\sqrt[3]{11 + 5} = \sqrt[3]{2 \cdot 11 - 6} \quad ? \quad \text{Let}$$
$$\sqrt[3]{16} = \sqrt[3]{16} \quad \text{True}$$

The solution set is $\{11\}$.

**NOW TRY** Exercise 37.

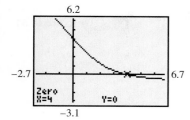

FIGURE 13

**OBJECTIVE 4** Solve radical equations by using a graphing calculator. In Example 4, we solved the equation $\sqrt{x^2 - 4x + 9} = x - 1$ by using algebraic methods. If we write this equation with one side equal to 0, we get

$$\sqrt{x^2 - 4x + 9} - x + 1 = 0.$$

Using a graphing calculator to graph the function defined by

$$f(x) = \sqrt{x^2 - 4x + 9} - x + 1,$$

we obtain the graph shown in Figure 13. Notice that its *zero* (x-value of the x-intercept) is 4, which is the solution we found in Example 4.

In Example 3, we found that the single solution of $\sqrt{4 - x} = x + 2$ is 0. An extraneous solution is $-5$. If we graph $f(x) = \sqrt{4 - x}$ and $g(x) = x + 2$ in the same window, we find that the x-coordinate of the point of intersection of the two graphs is 0, which is the solution of the equation. See Figure 14.

We solved the equation in Example 3 by squaring both sides, obtaining $4 - x = x^2 + 4x + 4$. In Figure 15 on the next page, we show that the two functions defined by $f(x) = 4 - x$ and $g(x) = x^2 + 4x + 4$ have two points of intersection.

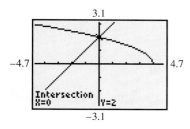

FIGURE 14

The extraneous solution −5 that we found in Example 3 shows up as an *x*-value of one of these points of intersection. However, our check showed that −5 was not a solution of the *original* equation (before the squaring step). Here we see a graphical interpretation of the extraneous solution.

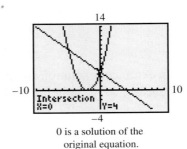
0 is a solution of the original equation.

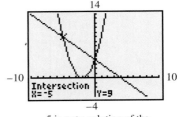
−5 is *not* a solution of the original equation.

**FIGURE 15**

**NOW TRY** Exercise 59.

**OBJECTIVE 5** Use the power rule to solve a formula for a specified variable.

**EXAMPLE 7** Solving a Formula from Electronics for a Variable

An important property of a radio-frequency transmission line is its **characteristic impedance,** represented by Z and measured in ohms. If L and C are the inductance and capacitance, respectively, per unit of length of the line, then these quantities are related by the formula $Z = \sqrt{\frac{L}{C}}$. Solve this formula for C.

$$Z = \sqrt{\frac{L}{C}} \quad \text{Given formula}$$

Our goal is to isolate C on one side of the equals sign.

$$Z^2 = \frac{L}{C} \quad \text{Square both sides.}$$

$$CZ^2 = L \quad \text{Multiply by C.}$$

$$C = \frac{L}{Z^2} \quad \text{Divide by}$$

**NOW TRY** Exercise 61.

## 8.6 Exercises

**NOW TRY** Exercise

*Concept Check* Check each equation to see if the given value for x is a solution.

1. $\sqrt{3x + 18} = x$
   (a) 6  (b) −3

2. $\sqrt{3x - 3} = x - 1$
   (a) 1  (b) 4

3. $\sqrt{x + 2} = \sqrt{9x - 2} - 2\sqrt{x - 1}$
   (a) 2  (b) 7

4. $\sqrt{8x - 3} = 2x$
   (a) $\frac{3}{2}$  (b) $\frac{1}{2}$

5. Is 9 a solution of the equation $\sqrt{x} = -3$? If not, what is the solution of this equation? Explain.

6. Before even attempting to solve $\sqrt{3x + 18} = x$, how can you be sure that the equation cannot have a negative solution?

*Solve each equation. See Examples 1–4.*

7. $\sqrt{r-2} = 3$
8. $\sqrt{q+1} = 7$
9. $\sqrt{6k-1} = 1$
10. $\sqrt{7m-3} = 5$
11. $\sqrt{4r+3} + 1 = 0$
12. $\sqrt{5k-3} + 2 = 0$
13. $\sqrt{3k+1} - 4 = 0$
14. $\sqrt{5z+1} - 11 = 0$
15. $4 - \sqrt{x-2} = 0$
16. $9 - \sqrt{4k+1} = 0$
17. $\sqrt{9a-4} = \sqrt{8a+1}$
18. $\sqrt{4p-2} = \sqrt{3p+5}$
19. $2\sqrt{x} = \sqrt{3x+4}$
20. $2\sqrt{m} = \sqrt{5m-16}$
21. $3\sqrt{z-1} = 2\sqrt{2z+2}$
22. $5\sqrt{4a+1} = 3\sqrt{10a+25}$
23. $k = \sqrt{k^2 + 4k - 20}$
24. $p = \sqrt{p^2 - 3p + 18}$
25. $a = \sqrt{a^2 + 3a + 9}$
26. $z = \sqrt{z^2 - 4z - 8}$
27. $\sqrt{9-x} = x + 3$
28. $\sqrt{5-x} = x + 1$
29. $\sqrt{k^2 + 2k + 9} = k + 3$
30. $\sqrt{a^2 - 3a + 3} = a - 1$
31. $\sqrt{r^2 + 9r + 3} = -r$
32. $\sqrt{p^2 - 15p + 15} = p - 5$
33. $\sqrt{z^2 + 12z - 4} + 4 - z = 0$
34. $\sqrt{m^2 + 3m + 12} - m - 2 = 0$

35. *Concept Check* In solving the equation $\sqrt{3x+4} = 8 - x$, a student wrote the following for her first step:

$$3x + 4 = 64 + x^2.$$

**WHAT WENT WRONG?** Solve the given equation correctly.

36. *Concept Check* In solving the equation $\sqrt{5x+6} - \sqrt{x+3} = 3$, a student wrote the following for his first step:

$$(5x + 6) + (x + 3) = 9.$$

**WHAT WENT WRONG?** Solve the given equation correctly.

*Solve each equation. See Examples 5 and 6.*

37. $\sqrt[3]{2x+5} = \sqrt[3]{6x+1}$
38. $\sqrt[3]{p-1} = 2$
39. $\sqrt[3]{a^2 + 5a + 1} = \sqrt[3]{a^2 + 4a}$
40. $\sqrt[3]{r^2 + 2r + 8} = \sqrt[3]{r^2}$
41. $\sqrt[3]{2m-1} = \sqrt[3]{m+13}$
42. $\sqrt[3]{2k-11} - \sqrt[3]{5k+1} = 0$
43. $\sqrt[4]{a+8} = \sqrt[4]{2a}$
44. $\sqrt[4]{z+11} = \sqrt[4]{2z+6}$
45. $\sqrt[3]{x-8} + 2 = 0$
46. $\sqrt[3]{r+1} + 1 = 0$
47. $\sqrt[4]{2k-5} + 4 = 0$
48. $\sqrt[4]{8z-3} + 2 = 0$
49. $\sqrt{k+2} - \sqrt{k-3} = 1$
50. $\sqrt{r+6} - \sqrt{r-2} = 2$
51. $\sqrt{2r+11} - \sqrt{5r+1} = -1$
52. $\sqrt{3x-2} - \sqrt{x+3} = 1$
53. $\sqrt{3p+4} - \sqrt{2p-4} = 2$
54. $\sqrt{4x+5} - \sqrt{2x+2} = 1$
55. $\sqrt{3-3p} - 3 = \sqrt{3p+2}$
56. $\sqrt{4x+7} - 4 = \sqrt{4x-1}$
57. $\sqrt{2\sqrt{x}+11} = \sqrt{4x+2}$
58. $\sqrt{1 + \sqrt{24-10x}} = \sqrt{3x+5}$

59. Use a graphing calculator to solve $\sqrt{3-3x} = 3 + \sqrt{3x+2}$. What is the domain of $y = \sqrt{3-3x} - 3 - \sqrt{3x+2}$?

60. Use a graphing calculator with a viewing window of $[-1, 4]$ by $[-1, 3]$ to solve $\sqrt{2\sqrt{7x+2}} = \sqrt{3x+2}$. What is the domain of $f(x) = \sqrt{2\sqrt{7x+2}} - \sqrt{3x+2}$?

*Solve each formula from electricity and radio for the indicated variable. See Example 7.*
*(Source: Cooke, Nelson M., and Joseph B. Orleans, Mathematics Essential to Electricity and Radio, McGraw-Hill, 1943.)*

61. $V = \sqrt{\dfrac{2K}{m}}$ for $K$
62. $V = \sqrt{\dfrac{2K}{m}}$ for $m$
63. $f = \dfrac{1}{2\pi\sqrt{LC}}$ for $L$
64. $r = \sqrt{\dfrac{Mm}{F}}$ for $F$

## 8.7 Complex Numbers

**OBJECTIVES**

1. Simplify numbers of the form $\sqrt{-b}$, where $b > 0$.
2. Recognize complex numbers.
3. Add and subtract complex numbers.
4. Multiply complex numbers.
5. Divide complex numbers.
6. Find powers of $i$.

As we saw in **Section 1.1,** the set of real numbers includes many other number sets (the rational numbers, integers, and natural numbers, for example). In this section, a new set of numbers is introduced that includes the set of real numbers, as well as numbers that are even roots of negative numbers, such as $\sqrt{-2}$.

**OBJECTIVE 1** Simplify numbers of the form $\sqrt{-b}$, where $b > 0$. The equation $x^2 + 1 = 0$ has no real number solution, since any solution must be a number whose square is $-1$. In the set of real numbers, all squares are nonnegative numbers because the product of two positive numbers or two negative numbers is positive and $0^2 = 0$. To provide a solution of the equation $x^2 + 1 = 0$, we introduce a new number $i$.

**Imaginary Unit $i$**

The **imaginary unit $i$** is defined as

$$i = \sqrt{-1}, \quad \text{where} \quad i^2 = -1.$$

That is, $i$ is the principal square root of $-1$.

This definition of $i$ makes it possible to define any square root of a negative real number as follows.

**$\sqrt{-b}$**

For any positive real number $b$, $\quad \sqrt{-b} = i\sqrt{b}.$

**EXAMPLE 1** Simplifying Square Roots of Negative Numbers

Write each number as a product of a real number and $i$.

**(a)** $\sqrt{-100} = i\sqrt{100} = 10i$  
**(b)** $-\sqrt{-36} = -i\sqrt{36} = -6i$  
**(c)** $\sqrt{-2} = i\sqrt{2}$  
**(d)** $\sqrt{-54} = i\sqrt{54} = i\sqrt{9 \cdot 6} = 3i\sqrt{6}$

**NOW TRY** Exercises 7, 9, 11, and 13.

**CAUTION** It is easy to mistake $\sqrt{2}i$ for $\sqrt{2i}$, with the $i$ under the radical. For this reason, we usually write $\sqrt{2}i$ as $i\sqrt{2}$, as in the definition of $\sqrt{-b}$.

When finding a product such as $\sqrt{-4} \cdot \sqrt{-9}$, we cannot use the product rule for radicals because it applies only to *nonnegative* radicands. **For this reason, we change $\sqrt{-b}$ to the form $i\sqrt{b}$ before performing any multiplications or divisions.** For example,

$$\sqrt{-4} \cdot \sqrt{-9} = i\sqrt{4} \cdot i\sqrt{9} \quad \sqrt{-b} = i\sqrt{b}$$

First write all square roots in terms of $i$.

$$= i \cdot 2 \cdot i \cdot 3$$
$$= 6i^2$$
$$= 6(-1) \quad \text{Substitute: } i^2 = -1.$$
$$= -6.$$

**CAUTION** Using the product rule for radicals *before* using the definition of $\sqrt{-b}$ gives a *wrong* answer. The preceding example shows that

$$\sqrt{-4} \cdot \sqrt{-9} = -6, \quad \text{Correct}$$

but

$$\sqrt{-4(-9)} = \sqrt{36} = 6, \quad \text{Incorrect}$$

so

$$\sqrt{-4} \cdot \sqrt{-9} \neq \sqrt{-4(-9)}.$$

---

**EXAMPLE 2** Multiplying Square Roots of Negative Numbers

Multiply.

(a) $\sqrt{-3} \cdot \sqrt{-7} = i\sqrt{3} \cdot i\sqrt{7}$     $\sqrt{-b} = i\sqrt{b}$

First write all square roots in terms of *i*.
$= i^2\sqrt{3 \cdot 7}$     Product rule
$= (-1)\sqrt{21}$     Substitute: $i^2 = -1$.
$= -\sqrt{21}$

(b) $\sqrt{-2} \cdot \sqrt{-8} = i\sqrt{2} \cdot i\sqrt{8}$     $\sqrt{-b} = i\sqrt{b}$
$= i^2\sqrt{2 \cdot 8}$     Product rule
$= (-1)\sqrt{16}$     $i^2 = -1$
$= -4$

(c) $\sqrt{-5} \cdot \sqrt{6} = i\sqrt{5} \cdot \sqrt{6} = i\sqrt{30}$

**NOW TRY** Exercises 15, 17, and 19.

---

The methods used to find products also apply to quotients. First write all square roots in terms of *i*.

**EXAMPLE 3** Dividing Square Roots of Negative Numbers

Divide.

(a) $\dfrac{\sqrt{-75}}{\sqrt{-3}} = \dfrac{i\sqrt{75}}{i\sqrt{3}} = \sqrt{\dfrac{75}{3}} = \sqrt{25} = 5$

(b) $\dfrac{\sqrt{-32}}{\sqrt{8}} = \dfrac{i\sqrt{32}}{\sqrt{8}} = i\sqrt{\dfrac{32}{8}} = i\sqrt{4} = 2i$

**NOW TRY** Exercises 21 and 23.

---

**OBJECTIVE 2** Recognize complex numbers. With the imaginary unit *i* and the real numbers, a new set of numbers, the *complex numbers,* are defined as follows.

**Complex Number**

If *a* and *b* are real numbers, then any number of the form *a* + *bi* is called a **complex number**. In the complex number *a* + *bi*, the number *a* is called the **real part** and *b* is called the **imaginary part**.*

---

*Some texts define *bi* as the imaginary part of the complex number *a* + *bi*.

For a complex number $a + bi$, if $b = 0$, then $a + bi = a$, which is a real number. *Thus, the set of real numbers is a subset of the set of complex numbers.* If $a = 0$ and $b \neq 0$, the complex number is said to be a **pure imaginary number.** For example, $3i$ is a pure imaginary number. A number such as $7 + 2i$ is a **nonreal complex number.**

A complex number written in the form $a + bi$ is in **standard form.** In this section, most answers will be given in standard form, but if $a$ or $b$ is 0, we consider answers such as $a$ and $bi$ to be in standard form.

The relationships among the various sets of numbers are shown in Figure 16.

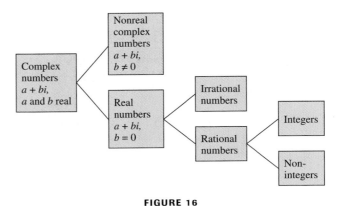

**FIGURE 16**

**OBJECTIVE 3** Add and subtract complex numbers. The commutative, associative, and distributive properties for real numbers are also valid for complex numbers. Thus, to add complex numbers, we add their real parts and add their imaginary parts.

### EXAMPLE 4 Adding Complex Numbers

Add.

**(a)** $(2 + 3i) + (6 + 4i) = (2 + 6) + (3 + 4)i$  Properties of real numbers
$= 8 + 7i$  Add real parts; add imaginary parts.

**(b)** $(4 + 2i) + (3 - i) + (-6 + 3i) = [4 + 3 + (-6)] + [2 + (-1) + 3]i$
$= 1 + 4i$

**NOW TRY** Exercises 25 and 33.

We subtract complex numbers by subtracting their real parts and subtracting their imaginary parts.

### EXAMPLE 5 Subtracting Complex Numbers

Subtract.

**(a)** $(6 + 5i) - (3 + 2i) = (6 - 3) + (5 - 2)i$  Properties of real numbers
$= 3 + 3i$  Subtract real parts; subtract imaginary parts.

**(b)** $(7 - 3i) - (8 - 6i) = (7 - 8) + [-3 - (-6)]i$
$= -1 + 3i$

**(c)** $(-9 + 4i) - (-9 + 8i) = (-9 + 9) + (4 - 8)i$
$= 0 - 4i$
$= -4i$

> **NOW TRY** Exercises 29 and 31.

**OBJECTIVE 4** **Multiply complex numbers.** We multiply complex numbers as we multiply polynomials. Complex numbers of the form $a + bi$ have the same form as binomials, so when applicable, we multiply two complex numbers in standard form by using the FOIL method for multiplying binomials. (Recall from **Section 5.4** that FOIL stands for *First, Outer, Inner, Last.*)

**EXAMPLE 6** **Multiplying Complex Numbers**

Multiply.

**(a)** $4i(2 + 3i) = 4i(2) + 4i(3i)$  Distributive property
$= 8i + 12i^2$  Multiply.
$= 8i + 12(-1)$  Substitute: $i^2 = -1$.
$= -12 + 8i$  Standard form

**(b)** $(3 + 5i)(4 - 2i) = \underbrace{3(4)}_{\text{First}} + \underbrace{3(-2i)}_{\text{Outer}} + \underbrace{5i(4)}_{\text{Inner}} + \underbrace{5i(-2i)}_{\text{Last}}$  (FOIL)
$= 12 - 6i + 20i - 10i^2$
$= 12 + 14i - 10(-1)$  Combine imaginary terms; $i^2 = -1$.
$= 12 + 14i + 10$
$= 22 + 14i$  Combine real terms.

**(c)** $(2 + 3i)(1 - 5i) = 2(1) + 2(-5i) + 3i(1) + 3i(-5i)$  FOIL
$= 2 - 10i + 3i - 15i^2$
$= 2 - 7i - 15(-1)$   *Use parentheses around $-1$ to avoid errors.*
$= 2 - 7i + 15$
$= 17 - 7i$

> **NOW TRY** Exercises 43 and 45.

The two complex numbers $a + bi$ and $a - bi$ are called **complex conjugates,** or simply *conjugates,* of each other. *The product of a complex number and its conjugate is always a real number,* as shown here:

$(a + bi)(a - bi) = a^2 - abi + abi - b^2i^2$
$= a^2 - b^2(-1)$
$(a + bi)(a - bi) = a^2 + b^2.$   *The product eliminates $i$.*

For example, $(3 + 7i)(3 - 7i) = 3^2 + 7^2 = 9 + 49 = 58$.

**OBJECTIVE 5** **Divide complex numbers.** The quotient of two complex numbers should be a complex number. To write the quotient as a complex number, we need to eliminate $i$ in the denominator. We use conjugates and a process like that for rationalizing a denominator to do this.

**EXAMPLE 7** Dividing Complex Numbers

Find each quotient.

(a) $\dfrac{8 + 9i}{5 + 2i}$

Multiply both the numerator and denominator by the conjugate of the denominator. The conjugate of $5 + 2i$ is $5 - 2i$.

$$\dfrac{8 + 9i}{5 + 2i} = \dfrac{(8 + 9i)(5 - 2i)}{(5 + 2i)(5 - 2i)} \qquad \dfrac{5 - 2i}{5 - 2i} = 1$$

$$= \dfrac{40 - 16i + 45i - 18i^2}{5^2 + 2^2}$$

$$= \dfrac{58 + 29i}{29} \qquad -18i^2 = -18(-1) = 18;\ \text{combine terms.}$$

Factor first; then divide out the common factor.
$$= \dfrac{29(2 + i)}{29} \qquad \text{Factor the numerator.}$$

$$= 2 + i \qquad \text{Lowest terms}$$

(b) $\dfrac{1 + i}{i} = \dfrac{(1 + i)(-i)}{i(-i)}$    Multiply numerator and denominator by $-i$, the conjugate of $i$.

$$= \dfrac{-i - i^2}{-i^2}$$

$$= \dfrac{-i - (-1)}{-(-1)} \qquad \text{Substitute: } i^2 = -1.$$

Use parentheses to avoid errors.
$$= \dfrac{-i + 1}{1}$$

$$= 1 - i$$

**NOW TRY** Exercises 59 and 65.

**OBJECTIVE 6** Find powers of $i$. Because $i^2$ is defined to be $-1$, we can find larger powers of $i$, as shown in the following examples:

$$i^3 = i \cdot i^2 = i(-1) = -i \qquad\qquad i^6 = i^2 \cdot i^4 = (-1) \cdot 1 = -1$$
$$i^4 = i^2 \cdot i^2 = (-1)(-1) = 1 \qquad\quad i^7 = i^3 \cdot i^4 = (-i) \cdot 1 = -i$$
$$i^5 = i \cdot i^4 = i \cdot 1 = i \qquad\qquad\qquad i^8 = i^4 \cdot i^4 = 1 \cdot 1 = 1.$$

Notice that the powers of $i$ rotate through the four numbers $i$, $-1$, $-i$, and $1$. Larger powers of $i$ can be simplified by using the fact that $i^4 = 1$. For example,

$$i^{75} = (i^4)^{18} \cdot i^3 = 1^{18} \cdot i^3 = 1 \cdot i^3 = i^3 = -i.$$

**EXAMPLE 8** Simplifying Powers of $i$

Find each power of $i$.

(a) $i^{12} = (i^4)^3 = 1^3 = 1$    (b) $i^{39} = i^{36} \cdot i^3 = (i^4)^9 \cdot i^3 = 1^9 \cdot (-i) = -i$

(c) $i^{-2} = \dfrac{1}{i^2} = \dfrac{1}{-1} = -1$    (d) $i^{-1} = \dfrac{1}{i} = \dfrac{1(-i)}{i(-i)} = \dfrac{-i}{-i^2} = \dfrac{-i}{-(-1)} = \dfrac{-i}{1} = -i$

**NOW TRY** Exercises 67 and 75.

## 8.7 Exercises

**Concept Check** Decide whether each expression is equal to $1, -1, i,$ or $-i$.

1. $\sqrt{-1}$
2. $-\sqrt{-1}$
3. $i^2$
4. $-i^2$
5. $\dfrac{1}{i}$
6. $(-i)^2$

*Write each number as a product of a real number and i. Simplify all radical expressions. See Example 1.*

7. $\sqrt{-169}$
8. $\sqrt{-225}$
9. $-\sqrt{-144}$
10. $-\sqrt{-196}$
11. $\sqrt{-5}$
12. $\sqrt{-21}$
13. $\sqrt{-48}$
14. $\sqrt{-96}$

*Multiply or divide as indicated. See Examples 2 and 3.*

15. $\sqrt{-7} \cdot \sqrt{-15}$
16. $\sqrt{-3} \cdot \sqrt{-19}$
17. $\sqrt{-4} \cdot \sqrt{-25}$
18. $\sqrt{-9} \cdot \sqrt{-81}$
19. $\sqrt{-3} \cdot \sqrt{11}$
20. $\sqrt{-10} \cdot \sqrt{2}$
21. $\dfrac{\sqrt{-300}}{\sqrt{-100}}$
22. $\dfrac{\sqrt{-40}}{\sqrt{-10}}$
23. $\dfrac{\sqrt{-75}}{\sqrt{3}}$
24. $\dfrac{\sqrt{-160}}{\sqrt{10}}$

*Add or subtract as indicated. Write your answers in the form $a + bi$. See Examples 4 and 5.*

25. $(3 + 2i) + (-4 + 5i)$
26. $(7 + 15i) + (-11 + 14i)$
27. $(5 - i) + (-5 + i)$
28. $(-2 + 6i) + (2 - 6i)$
29. $(4 + i) - (-3 - 2i)$
30. $(9 + i) - (3 + 2i)$
31. $(-3 - 4i) - (-1 - 4i)$
32. $(-2 - 3i) - (-5 - 3i)$
33. $(-4 + 11i) + (-2 - 4i) + (7 + 6i)$
34. $(-1 + i) + (2 + 5i) + (3 + 2i)$
35. $[(7 + 3i) - (4 - 2i)] + (3 + i)$
36. $[(7 + 2i) + (-4 - i)] - (2 + 5i)$

37. *Fill in the Blank* Because $(4 + 2i) - (3 + i) = 1 + i$, using the definition of subtraction, we can check this to find that $(1 + i) + (3 + i) = $ _____.

38. *Fill in the Blank* Because $\dfrac{-5}{2 - i} = -2 - i$, using the definition of division, we can check this to find that $(-2 - i)(2 - i) = $ _____.

*Multiply. See Example 6.*

39. $(3i)(27i)$
40. $(5i)(125i)$
41. $(-8i)(-2i)$
42. $(-32i)(-2i)$
43. $5i(-6 + 2i)$
44. $3i(4 + 9i)$
45. $(4 + 3i)(1 - 2i)$
46. $(7 - 2i)(3 + i)$
47. $(4 + 5i)^2$
48. $(3 + 2i)^2$
49. $2i(-4 - i)^2$
50. $3i(-3 - i)^2$
51. $(12 + 3i)(12 - 3i)$
52. $(6 + 7i)(6 - 7i)$
53. $(4 + 9i)(4 - 9i)$
54. $(7 + 2i)(7 - 2i)$

55. *Concept Check* What is the conjugate of $a + bi$?

56. *Fill in the Blank* If we multiply $a + bi$ by its conjugate, we get _____, which is always a real number.

*Find each quotient. See Example 7.*

**57.** $\dfrac{2}{1-i}$

**58.** $\dfrac{29}{5+2i}$

**59.** $\dfrac{-7+4i}{3+2i}$

**60.** $\dfrac{-38-8i}{7+3i}$

**61.** $\dfrac{8i}{2+2i}$

**62.** $\dfrac{-8i}{1+i}$

**63.** $\dfrac{2-3i}{2+3i}$

**64.** $\dfrac{-1+5i}{3+2i}$

**65.** $\dfrac{3+i}{i}$

**66.** $\dfrac{5-i}{-i}$

*Find each power of i. See Example 8.*

**67.** $i^{18}$

**68.** $i^{26}$

**69.** $i^{89}$

**70.** $i^{48}$

**71.** $i^{38}$

**72.** $i^{102}$

**73.** $i^{43}$

**74.** $i^{83}$

**75.** $i^{-5}$

**76.** $i^{-17}$

**77.** A student simplified $i^{-18}$ as follows:

$$i^{-18} = i^{-18} \cdot i^{20} = i^{-18+20} = i^2 = -1.$$

Explain the mathematical justification for this correct work.

**78.** Explain why

$$(46 + 25i)(3 - 6i) \quad \text{and} \quad (46 + 25i)(3 - 6i)i^{12}$$

must be equal. (Do not actually perform the computation.)

# 8 SUMMARY

## KEY TERMS

**8.1** radicand
index (order)
radical
principal root
radical expression

square root function
cube root function
**8.3** simplified radical
**8.5** rationalizing the denominator

conjugates
**8.6** radical equation
extraneous solution
**8.7** complex number
real part

imaginary part
pure imaginary number
standard form
complex conjugates

## NEW SYMBOLS

$\sqrt{\phantom{a}}$   radical sign

$\sqrt[n]{a}$   radical; principal $n$th root of $a$

$\pm$   "positive or negative" or "plus or minus"

$\approx$   is approximately equal to

$a^{1/n}$   $a$ to the power $\dfrac{1}{n}$

$a^{m/n}$   $a$ to the power $\dfrac{m}{n}$

$i$   imaginary unit

## QUICK REVIEW

| CONCEPTS | EXAMPLES |
|---|---|
| **8.1 Radical Expressions and Graphs**<br>$\sqrt[n]{a} = b$ means $b^n = a$.<br>$\sqrt[n]{a}$ is the principal $n$th root of $a$.<br>$\sqrt[n]{a^n} = \lvert a \rvert$ if $n$ is even; $\sqrt[n]{a^n} = a$ if $n$ is odd.<br>**Functions Defined by Radical Expressions**<br>The square root function defined by $f(x) = \sqrt{x}$ and the cube root function defined by $f(x) = \sqrt[3]{x}$ are two important functions defined by radical expressions. | The two square roots of 64 are $\sqrt{64} = 8$ (the principal square root) and $-\sqrt{64} = -8$.<br>$\sqrt[4]{(-2)^4} = \lvert -2 \rvert = 2 \qquad \sqrt[3]{-27} = -3$<br>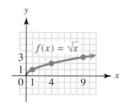 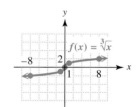 |
| **8.2 Rational Exponents**<br>$a^{1/n} = \sqrt[n]{a}$ whenever $\sqrt[n]{a}$ exists.<br>If $m$ and $n$ are positive integers with $\dfrac{m}{n}$ in lowest terms, then $a^{m/n} = (a^{1/n})^m$, provided that $a^{1/n}$ is a real number.<br>All of the usual definitions and rules for exponents are valid for rational exponents. | $81^{1/2} = \sqrt{81} = 9 \qquad -64^{1/3} = -\sqrt[3]{64} = -4$<br>$8^{5/3} = (8^{1/3})^5 = 2^5 = 32$<br>$5^{-1/2} \cdot 5^{1/4} = 5^{-1/2+1/4} = 5^{-1/4} = \dfrac{1}{5^{1/4}} \qquad (y^{2/5})^{10} = y^4$<br>$\dfrac{x^{-1/3}}{x^{-1/2}} = x^{-1/3-(-1/2)} = x^{-1/3+1/2} = x^{1/6}, \quad x > 0$ |
| **8.3 Simplifying Radical Expressions**<br>**Product and Quotient Rules for Radicals**<br>If $\sqrt[n]{a}$ and $\sqrt[n]{b}$ are real numbers and $n$ is a natural number, then<br>$\sqrt[n]{a} \cdot \sqrt[n]{b} = \sqrt[n]{ab}$<br>and $\sqrt[n]{\dfrac{a}{b}} = \dfrac{\sqrt[n]{a}}{\sqrt[n]{b}}, \quad b \neq 0.$ | $\sqrt{3} \cdot \sqrt{7} = \sqrt{21}$<br>$\sqrt[5]{x^3 y} \cdot \sqrt[5]{xy^2} = \sqrt[5]{x^4 y^3}$<br>$\dfrac{\sqrt{x^5}}{\sqrt{x^4}} = \sqrt{\dfrac{x^5}{x^4}} = \sqrt{x}, \quad x > 0$ |

*(continued)*

## CONCEPTS

**Conditions for a Simplified Radical**
1. The radicand has no factor raised to a power greater than or equal to the index.
2. The radicand has no fractions.
3. No denominator contains a radical.
4. Exponents in the radicand and the index of the radical have no common factor (except 1).

**Pythagorean Theorem**
If $c$ is the length of the longest side of a right triangle and $a$ and $b$ are the lengths of the shorter sides, then $c^2 = a^2 + b^2$. The longest side is the hypotenuse and the two shorter sides are the legs of the triangle. The hypotenuse is opposite the right angle.

**Distance Formula**
The distance between $(x_1, y_1)$ and $(x_2, y_2)$ is

$$d = \sqrt{(x_2 - x_1)^2 + (y_2 - y_1)^2}.$$

## EXAMPLES

$\sqrt{18} = \sqrt{9 \cdot 2} = 3\sqrt{2}$

$\sqrt[3]{54x^5y^3} = \sqrt[3]{27x^3y^3 \cdot 2x^2} = 3xy\sqrt[3]{2x^2}$

$\sqrt{\dfrac{7}{4}} = \dfrac{\sqrt{7}}{\sqrt{4}} = \dfrac{\sqrt{7}}{2}$

$\sqrt[9]{x^3} = x^{3/9} = x^{1/3},$ or $\sqrt[3]{x}$

Find $b$ for the triangle in the figure.

$10^2 + b^2 = (2\sqrt{61})^2$
$b^2 = 4(61) - 100$
$b^2 = 144$
$b = 12$

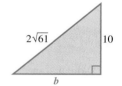

The distance between $(3, -2)$ and $(-1, 1)$ is

$\sqrt{(-1 - 3)^2 + [1 - (-2)]^2} = \sqrt{(-4)^2 + 3^2}$
$= \sqrt{16 + 9}$
$= \sqrt{25}$
$= 5.$

Start with the *x*-value and *y*-value of the same point.

### 8.4 Adding and Subtracting Radical Expressions

*Only radical expressions with the same index and the same radicand may be combined.*

$3\sqrt{17} + 2\sqrt{17} - 8\sqrt{17} = (3 + 2 - 8)\sqrt{17}$
$= -3\sqrt{17}$

$\sqrt[3]{2} - \sqrt[3]{250} = \sqrt[3]{2} - 5\sqrt[3]{2}$  $\sqrt[3]{250} = \sqrt[3]{125 \cdot 2} = 5\sqrt[3]{2}$
$= -4\sqrt[3]{2}$

$\left.\begin{array}{l}\sqrt{15} + \sqrt{30} \\ \sqrt{3} + \sqrt[3]{9}\end{array}\right\}$ cannot be simplified further

### 8.5 Multiplying and Dividing Radical Expressions

Multiply binomial radical expressions by using the FOIL method. Special products from **Section 5.4** may apply.

$(\sqrt{2} + \sqrt{7})(\sqrt{3} - \sqrt{6})$
$= \sqrt{6} - 2\sqrt{3} + \sqrt{21} - \sqrt{42}$   $\sqrt{12} = 2\sqrt{3}$
$(\sqrt{5} - \sqrt{10})(\sqrt{5} + \sqrt{10}) = 5 - 10 = -5$
$(\sqrt{3} - \sqrt{2})^2 = 3 - 2\sqrt{3} \cdot \sqrt{2} + 2 = 5 - 2\sqrt{6}$

Rationalize the denominator by multiplying both the numerator and the denominator by the same expression.

$\dfrac{\sqrt{7}}{\sqrt{5}} = \dfrac{\sqrt{7} \cdot \sqrt{5}}{\sqrt{5} \cdot \sqrt{5}} = \dfrac{\sqrt{35}}{5}$

$\dfrac{\sqrt[3]{2}}{\sqrt[3]{4}} = \dfrac{\sqrt[3]{2} \cdot \sqrt[3]{2}}{\sqrt[3]{4} \cdot \sqrt[3]{2}} = \dfrac{\sqrt[3]{4}}{\sqrt[3]{8}} = \dfrac{\sqrt[3]{4}}{2}$

$\dfrac{4}{\sqrt{5} - \sqrt{2}} = \dfrac{4(\sqrt{5} + \sqrt{2})}{(\sqrt{5} - \sqrt{2})(\sqrt{5} + \sqrt{2})}$
$= \dfrac{4(\sqrt{5} + \sqrt{2})}{5 - 2} = \dfrac{4(\sqrt{5} + \sqrt{2})}{3}$

$\dfrac{5 + 15\sqrt{6}}{10} = \dfrac{5(1 + 3\sqrt{6})}{5 \cdot 2} = \dfrac{1 + 3\sqrt{6}}{2}$

Factor first; then divide out the common factor.

*(continued)*

## CONCEPTS

### 8.6 Solving Equations with Radicals

**Solving an Equation with Radicals**

**Step 1** Isolate one radical on one side of the equation.

**Step 2** Raise both sides of the equation to a power that is the same as the index of the radical.

**Step 3** Solve the resulting equation; if it still contains a radical, repeat Steps 1 and 2.

**Step 4** Check all proposed solutions in the *original* equation.

### 8.7 Complex Numbers

$i = \sqrt{-1}$, where $i^2 = -1$.
For any positive number $b$, $\sqrt{-b} = i\sqrt{b}$.
*To multiply radicals with negative radicands, first change each factor to the form $i\sqrt{b}$ and then multiply. The same procedure applies to quotients.*

**Adding and Subtracting Complex Numbers**
Add (or subtract) the real parts and add (or subtract) the imaginary parts.

**Multiplying Complex Numbers**
Multiply complex numbers by using the FOIL method.

**Dividing Complex Numbers**
Divide complex numbers by multiplying the numerator and the denominator by the conjugate of the denominator.

## EXAMPLES

Solve $\sqrt{2x + 3} - x = 0$.

| | |
|---|---|
| $\sqrt{2x + 3} = x$ | Add $x$. |
| $(\sqrt{2x + 3})^2 = x^2$ | Square both sides. |
| $2x + 3 = x^2$ | |
| $x^2 - 2x - 3 = 0$ | Standard form |
| $(x - 3)(x + 1) = 0$ | Zero-factor property |
| $x - 3 = 0$ or $x + 1 = 0$ | Solve each equation. |
| $x = 3$ or $x = -1$ | |

A check shows that 3 is a solution, but $-1$ is extraneous. The solution set is $\{3\}$.

$\sqrt{-25} = i\sqrt{25} = 5i$

First write all square roots in terms of $i$.
$$\sqrt{-3} \cdot \sqrt{-27} = i\sqrt{3} \cdot i\sqrt{27}$$
$$= i^2\sqrt{81}$$
$$= -1 \cdot 9$$
$$= -9$$

$$\frac{\sqrt{-18}}{\sqrt{-2}} = \frac{i\sqrt{18}}{i\sqrt{2}} = \sqrt{\frac{18}{2}} = \sqrt{9} = 3$$

$(5 + 3i) + (8 - 7i) = 13 - 4i$
$(5 + 3i) - (8 - 7i) = -3 + 10i$

| | | |
|---|---|---|
| $(2 + i)(5 - 3i)$ | $= 10 - 6i + 5i - 3i^2$ | FOIL |
| | $= 10 - i - 3(-1)$ | $i^2 = -1$ |
| | $= 10 - i + 3$ | |
| | $= 13 - i$ | |

$$\frac{20}{3 + i} = \frac{20(3 - i)}{(3 + i)(3 - i)} = \frac{20(3 - i)}{9 - i^2}$$
$$= \frac{20(3 - i)}{10} = 2(3 - i) = 6 - 2i$$

# 8 REVIEW EXERCISES

*Find each root.*

1. $\sqrt{1764}$
2. $-\sqrt{289}$
3. $\sqrt[3]{216}$
4. $\sqrt[3]{-125}$
5. $-\sqrt[3]{27}$
6. $\sqrt[5]{-32}$

7. *Concept Check* Under what conditions is $\sqrt[n]{a}$ not a real number?

8. Simplify each radical so that no radicals appear. Assume that $x$ represents any real number.
   (a) $\sqrt{x^2}$   (b) $-\sqrt{x^2}$   (c) $\sqrt[3]{x^3}$

*Use a calculator to find a decimal approximation for each number. Give the answer to the nearest thousandth.*

9. $-\sqrt{47}$
10. $\sqrt[3]{-129}$
11. $\sqrt[4]{605}$
12. $500^{-3/4}$
13. $-500^{4/3}$
14. $-28^{-1/2}$

*Graph each function. Give the domain and range.*

15. $f(x) = \sqrt{x-1}$
16. $f(x) = \sqrt[3]{x} + 4$

17. *Multiple Choice* What is the best estimate of the area of the triangle shown here?

    A. 3600   B. 30   C. 60   D. 360

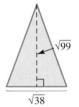

18. *Fill in the Blanks* One way to evaluate $8^{2/3}$ is to first find the _____ root of _____, which is _____. Then raise that result to the _____ power, to get an answer of _____. Therefore, $8^{2/3} = $ _____.

19. *Multiple Choice* Which one of the following is a positive number?
    A. $(-27)^{2/3}$   B. $(-64)^{5/3}$   C. $(-100)^{1/2}$   D. $(-32)^{1/5}$

20. *Concept Check* If $a$ is a negative number and $n$ is odd, then what must be true about $m$ for $a^{m/n}$ to be
    (a) positive   (b) negative?

21. *Concept Check* If $a$ is negative and $n$ is even, then what can be said about $a^{1/n}$?

*Simplify. If the expression does not represent a real number, say so.*

22. $49^{1/2}$
23. $-121^{1/2}$
24. $16^{5/4}$
25. $-8^{2/3}$
26. $-\left(\dfrac{36}{25}\right)^{3/2}$
27. $\left(-\dfrac{1}{8}\right)^{-5/3}$
28. $\left(\dfrac{81}{10{,}000}\right)^{-3/4}$
29. $(-16)^{3/4}$

30. Solve the formula of the Pythagorean theorem, $a^2 + b^2 = c^2$, for $b$, where $b > 0$.

31. Explain the relationship between the expressions $a^{m/n}$ and $\sqrt[n]{a^m}$. Give an example.

*Write each expression as a radical.*

32. $(m + 3n)^{1/2}$
33. $(3a + b)^{-5/3}$

*Write each expression with a rational exponent.*

34. $\sqrt{7^9}$
35. $\sqrt[5]{p^4}$

*Use the rules for exponents to simplify each expression. Write the answer with only positive exponents. Assume that all variables represent positive real numbers.*

**36.** $5^{1/4} \cdot 5^{7/4}$ 

**37.** $\dfrac{96^{2/3}}{96^{-1/3}}$ 

**38.** $\dfrac{(a^{1/3})^4}{a^{2/3}}$

**39.** $\dfrac{y^{-1/3} \cdot y^{5/6}}{y}$ 

**40.** $\left(\dfrac{z^{-1}x^{-3/5}}{2^{-2}z^{-1/2}x}\right)^{-1}$ 

**41.** $r^{-1/2}(r + r^{3/2})$

*Simplify by first writing each radical in exponential form. Leave the answer in exponential form. Assume that all variables represent positive real numbers.*

**42.** $\sqrt[8]{s^4}$ 

**43.** $\sqrt[6]{r^9}$ 

**44.** $\dfrac{\sqrt{p^5}}{p^2}$

**45.** $\sqrt[4]{k^3} \cdot \sqrt{k^3}$ 

**46.** $\sqrt[3]{m^5} \cdot \sqrt[3]{m^8}$ 

**47.** $\sqrt[4]{\sqrt[3]{z}}$

**48.** $\sqrt{\sqrt{\sqrt{x}}}$ 

**49.** $\sqrt[3]{\sqrt[5]{x}}$ 

**50.** $\sqrt{\sqrt[6]{\sqrt[3]{x}}}$

**51.** By the product rule for exponents, we know that $2^{1/4} \cdot 2^{1/5} = 2^{9/20}$. However, there is no exponent rule for simplifying $3^{1/4} \cdot 2^{1/5}$. Why?

*Simplify each radical. Assume that all variables represent positive real numbers.*

**52.** $\sqrt{6} \cdot \sqrt{11}$ 

**53.** $\sqrt{5} \cdot \sqrt{r}$ 

**54.** $\sqrt[3]{6} \cdot \sqrt[3]{5}$ 

**55.** $\sqrt[4]{7} \cdot \sqrt[4]{3}$

**56.** $\sqrt{20}$ 

**57.** $\sqrt{75}$ 

**58.** $-\sqrt{125}$ 

**59.** $\sqrt[3]{-108}$

**60.** $\sqrt{100y^7}$ 

**61.** $\sqrt[3]{64p^4q^6}$ 

**62.** $\sqrt[3]{108a^8b^5}$ 

**63.** $\sqrt[3]{632r^8t^4}$

**64.** $\sqrt{\dfrac{y^3}{144}}$ 

**65.** $\sqrt[3]{\dfrac{m^{15}}{27}}$ 

**66.** $\sqrt[3]{\dfrac{r^2}{8}}$ 

**67.** $\sqrt[4]{\dfrac{a^9}{81}}$

*Simplify each radical expression.*

**68.** $\sqrt[6]{15^3}$ 

**69.** $\sqrt[4]{p^6}$ 

**70.** $\sqrt[3]{2} \cdot \sqrt[5]{5}$ 

**71.** $\sqrt{x} \cdot \sqrt[5]{x}$

**72.** Find the missing length in the right triangle. Simplify the answer if applicable.

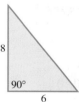

**73.** Find the distance between the points $(-4, 7)$ and $(10, 6)$.

*Perform the indicated operations. Assume that all variables represent positive real numbers.*

**74.** $2\sqrt{8} - 3\sqrt{50}$ 

**75.** $8\sqrt{80} - 3\sqrt{45}$ 

**76.** $-\sqrt{27y} + 2\sqrt{75y}$

**77.** $2\sqrt{54m^3} + 5\sqrt{96m^3}$ 

**78.** $3\sqrt[3]{54} + 5\sqrt[3]{16}$ 

**79.** $-6\sqrt[4]{32} + \sqrt[4]{512}$

**80.** $\dfrac{3}{\sqrt{16}} - \dfrac{\sqrt{5}}{2}$ 

**81.** $\dfrac{4}{\sqrt{25}} + \dfrac{\sqrt{5}}{4}$

*In Exercises 82 and 83, leave answers as simplified radicals.*

**82.** Find the perimeter of a rectangular electronic billboard having sides of lengths shown in the figure.

**83.** Find the perimeter of a triangular electronic highway road sign having the dimensions shown in the figure.

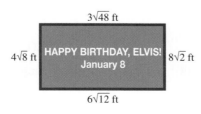

*Multiply.*

**84.** $(\sqrt{3}+1)(\sqrt{3}-2)$

**85.** $(\sqrt{7}+\sqrt{5})(\sqrt{7}-\sqrt{5})$

**86.** $(3\sqrt{2}+1)(2\sqrt{2}-3)$

**87.** $(\sqrt{13}-\sqrt{2})^2$

**88.** $(\sqrt[3]{2}+3)(\sqrt[3]{4}-3\sqrt[3]{2}+9)$

**89.** $(\sqrt[3]{4y}-1)(\sqrt[3]{4y}+3)$

**90.** Use a calculator to show that the answer to Exercise 87, $15 - 2\sqrt{26}$, is not equal to $13\sqrt{26}$.

**91.** *Concept Check* A friend wants to rationalize the denominator of the fraction $\dfrac{5}{\sqrt[3]{6}}$, and she decides to multiply the numerator and denominator by $\sqrt[3]{6}$. Why will her plan *not* work?

*Rationalize each denominator. Assume that all variables represent positive real numbers.*

**92.** $\dfrac{\sqrt{6}}{\sqrt{5}}$

**93.** $\dfrac{-6\sqrt{3}}{\sqrt{2}}$

**94.** $\dfrac{3\sqrt{7p}}{\sqrt{y}}$

**95.** $\sqrt{\dfrac{11}{8}}$

**96.** $-\sqrt[3]{\dfrac{9}{25}}$

**97.** $\sqrt[3]{\dfrac{108m^3}{n^5}}$

**98.** $\dfrac{1}{\sqrt{2}+\sqrt{7}}$

**99.** $\dfrac{-5}{\sqrt{6}-3}$

*Write in lowest terms.*

**100.** $\dfrac{2-2\sqrt{5}}{8}$

**101.** $\dfrac{4-8\sqrt{8}}{12}$

**102.** $\dfrac{-18+\sqrt{27}}{6}$

*Solve each equation.*

**103.** $\sqrt{8x+9}=5$

**104.** $\sqrt{2z-3}-3=0$

**105.** $\sqrt{3m+1}-2=-3$

**106.** $\sqrt{7z+1}=z+1$

**107.** $3\sqrt{m}=\sqrt{10m-9}$

**108.** $\sqrt{p^2+3p+7}=p+2$

**109.** $\sqrt{a+2}-\sqrt{a-3}=1$

**110.** $\sqrt[3]{5m-1}=\sqrt[3]{3m-2}$

**111.** $\sqrt[3]{2x^2+3x-7}=\sqrt[3]{2x^2+4x+6}$

**112.** $\sqrt[3]{3y^2-4y+6}=\sqrt[3]{3y^2-2y+8}$

**113.** $\sqrt[3]{1-2k}-\sqrt[3]{-k-13}=0$

**114.** $\sqrt[3]{11-2t}-\sqrt[3]{-1-5t}=0$

**115.** $\sqrt[4]{x-1}+2=0$

**116.** $\sqrt[4]{2k+3}+1=0$

**117.** $\sqrt[4]{x+7}=\sqrt[4]{2x}$

**118.** $\sqrt[4]{x+8}=\sqrt[3]{3x}$

**119.** Carpenters stabilize wall frames with a diagonal brace, as shown in the figure. The length of the brace is given by $L=\sqrt{H^2+W^2}$.

  **(a)** Solve this formula for $H$.

  **(b)** If the bottom of the brace is attached 9 ft from the corner and the brace is 12 ft long, how far up the corner post should it be nailed? Give your answer to the nearest tenth of a foot.

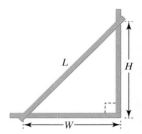

*Write each expression as a product of a real number and i.*

**120.** $\sqrt{-25}$

**121.** $\sqrt{-200}$

**122.** *Concept Check* If $a$ is a positive real number, is $-\sqrt{-a}$ a real number?

*Perform the indicated operations. Give answers in standard form.*

**123.** $(-2+5i)+(-8-7i)$

**124.** $(5+4i)-(-9-3i)$

**125.** $\sqrt{-5}\cdot\sqrt{-7}$

126. $\sqrt{-25} \cdot \sqrt{-81}$

127. $\dfrac{\sqrt{-72}}{\sqrt{-8}}$

128. $(2 + 3i)(1 - i)$

129. $(6 - 2i)^2$

130. $\dfrac{3 - i}{2 + i}$

131. $\dfrac{5 + 14i}{2 + 3i}$

*Find each power of i.*

132. $i^{11}$

133. $i^{36}$

134. $i^{-10}$

135. $i^{-8}$

# 8 TEST

*Evaluate.*

1. $-\sqrt{841}$

2. $\sqrt[3]{-512}$

3. $125^{1/3}$

4. *Multiple Choice* For $\sqrt{146.25}$, which choice gives the best estimate?
   A. 10   B. 11   C. 12   D. 13

*Use a calculator to approximate each root to the nearest thousandth.*

5. $\sqrt{478}$

6. $\sqrt[3]{-832}$

7. Graph the function defined by $f(x) = \sqrt{x + 6}$, and give the domain and range.

*Simplify each expression. Assume that all variables represent positive real numbers.*

8. $\left(\dfrac{16}{25}\right)^{-3/2}$

9. $(-64)^{-4/3}$

10. $\dfrac{3^{2/5} x^{-1/4} y^{2/5}}{3^{-8/5} x^{7/4} y^{1/10}}$

11. $\left(\dfrac{x^{-4} y^{-6}}{x^{-2} y^3}\right)^{-2/3}$

12. $7^{3/4} \cdot 7^{-1/4}$

13. $\sqrt[3]{a^4} \cdot \sqrt[3]{a^7}$

14. Use the Pythagorean theorem to find the exact length of side $b$ in the figure.

15. Find the distance between the points $(-4, 2)$ and $(2, 10)$.

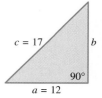

$c = 17$, $b$, $90°$, $a = 12$

*Simplify each expression. Assume that all variables represent positive real numbers.*

16. $\sqrt{54 x^5 y^6}$

17. $\sqrt[4]{32 a^7 b^{13}}$

18. $\sqrt{2} \cdot \sqrt[3]{5}$   (Express as a radical.)

19. $3\sqrt{20} - 5\sqrt{80} + 4\sqrt{500}$

20. $\sqrt[3]{16 t^3 s^5} - \sqrt[3]{54 t^6 s^2}$

21. $(7\sqrt{5} + 4)(2\sqrt{5} - 1)$

22. $(\sqrt{3} - 2\sqrt{5})^2$

23. $\dfrac{-5}{\sqrt{40}}$

24. $\dfrac{2}{\sqrt[3]{5}}$

25. $\dfrac{-4}{\sqrt{7} + \sqrt{5}}$

26. Write $\dfrac{6 + \sqrt{24}}{2}$ in lowest terms.

**27.** The following formula is used in physics, relating the velocity $V$ of sound to the temperature $T$:

$$V = \frac{V_0}{\sqrt{1 - kT}}.$$

    **(a)** Find an approximation of $V$ to the nearest tenth if $V_0 = 50$, $k = 0.01$, and $T = 30$. Use a calculator.

    **(b)** Solve the formula for $T$.

*Solve each equation.*

**28.** $\sqrt[3]{5x} = \sqrt[3]{2x - 3}$            **29.** $x + \sqrt{x + 6} = 9 - x$

**30.** $\sqrt{x + 4} - \sqrt{1 - x} = -1$

*Perform the indicated operations. Give the answers in standard form.*

**31.** $(-2 + 5i) - (3 + 6i) - 7i$      **32.** $(1 + 5i)(3 + i)$

**33.** $\dfrac{7 + i}{1 - i}$

**34.** Simplify $i^{37}$.

**35.** *True or False*

    **(a)** $i^2 = -1$     **(b)** $i = \sqrt{-1}$     **(c)** $i = -1$     **(d)** $\sqrt{-3} = i\sqrt{3}$

# Quadratic Equations and Inequalities

According to the U.S. Securities and Exchange Commission, federal bankruptcy laws determine how companies suffering from extensive debt can go out of business or recover from that debt. *Chapter 11* of the bankruptcy code allows a company to reorganize its affairs and become profitable again. Under Chapter 11, the company continues to operate with the stipulation that major decisions be approved by a bankruptcy court. Under *Chapter 7*, a company ceases operations completely. A trustee liquidates, or sells, the assets of the company and the money is then used to pay off debts to creditors and investors. The order in which they are paid is determined by bankruptcy laws. In general, investors who have taken the least risk are paid off first. (*Source:* www.sec.gov/investor/pubs/bankrupt.htm)

In Example 6 of Section 9.4, we use a *quadratic* (*second-degree*) *function* to model bankruptcies during the last decade of the twentieth century.

**9.1** The Square Root Property and Completing the Square

**9.2** The Quadratic Formula

**9.3** Equations Quadratic in Form

**9.4** Formulas and Further Applications

**9.5** Quadratic and Rational Inequalities

## 9.1 The Square Root Property and Completing the Square

**OBJECTIVES**

1. Review the zero-factor property.
2. Learn the square root property.
3. Solve quadratic equations of the form $(ax + b)^2 = c$ by using the square root property.
4. Solve quadratic equations by completing the square.
5. Solve quadratic equations with solutions that are not real numbers.

We introduced quadratic equations in **Section 6.5**. Recall that a *quadratic equation* is defined as follows.

> **Quadratic Equation**
>
> An equation that can be written in the form
> $$ax^2 + bx + c = 0,$$
> where $a$, $b$, and $c$ are real numbers, with $a \neq 0$, is a **quadratic equation**. The given form is called **standard form**.

A quadratic equation is a *second-degree equation,* that is, an equation with a squared variable term and no terms of greater degree. For example,

$$4x^2 + 4x - 5 = 0 \quad \text{and} \quad 3x^2 = 4x - 8$$

are quadratic equations, with the first equation in standard form.

**OBJECTIVE 1** Review the zero-factor property. In **Section 6.5** we used factoring and the zero-factor property to solve quadratic equations.

> **Zero-Factor Property**
>
> If two numbers have a product of 0, then at least one of the numbers must be 0. That is, if $ab = 0$, then $a = 0$ or $b = 0$.

We solved a quadratic equation such as $3x^2 - 5x - 28 = 0$ using the zero-factor property as follows.

$$3x^2 - 5x - 28 = 0$$
$$(3x + 7)(x - 4) = 0 \quad \text{Factor.}$$
$$3x + 7 = 0 \quad \text{or} \quad x - 4 = 0 \quad \text{Zero-factor property}$$
$$3x = -7 \quad \text{or} \quad x = 4 \quad \text{Solve each equation.}$$
$$x = -\frac{7}{3}$$

The solution set is $\left\{-\frac{7}{3}, 4\right\}$.

**OBJECTIVE 2** Learn the square root property. Although factoring is the simplest way to solve quadratic equations, not every quadratic equation can be solved easily by factoring. In this section and the next, we develop three other methods of solving quadratic equations based on the following property.

> **Square Root Property**
>
> If $x$ and $k$ are complex numbers and $x^2 = k$, then
> $$x = \sqrt{k} \quad \text{or} \quad x = -\sqrt{k}.$$

## SECTION 9.1 The Square Root Property and Completing the Square

The following steps justify the square root property.

$$x^2 = k$$
$$x^2 - k = 0 \quad \text{Subtract } k.$$
$$(x - \sqrt{k})(x + \sqrt{k}) = 0 \quad \text{Factor.}$$
$$x - \sqrt{k} = 0 \quad \text{or} \quad x + \sqrt{k} = 0 \quad \text{Zero-factor property}$$
$$\boldsymbol{x = \sqrt{k}} \quad \text{or} \quad \boldsymbol{x = -\sqrt{k}} \quad \text{Solve each equation.}$$

**CAUTION** Remember that if $k \neq 0$, using the square root property always produces *two* square roots, one positive and one negative.

### EXAMPLE 1  Using the Square Root Property

Solve each equation.

**(a)** $x^2 = 5$

By the square root property,

$$x = \sqrt{5} \quad \text{or} \quad x = -\sqrt{5},$$

*Don't forget the negative solution.*

and the solution set is $\{\sqrt{5}, -\sqrt{5}\}$.

**(b)**
$$4x^2 - 48 = 0$$
$$4x^2 = 48 \quad \text{Add 48.}$$
$$x^2 = 12 \quad \text{Divide by 4.}$$
$$x = \sqrt{12} \quad \text{or} \quad x = -\sqrt{12} \quad \text{Square root property}$$
$$x = 2\sqrt{3} \quad \text{or} \quad x = -2\sqrt{3} \quad \sqrt{12} = \sqrt{4} \cdot \sqrt{3} = 2\sqrt{3}$$

The solutions are $2\sqrt{3}$ and $-2\sqrt{3}$. Check each in the original equation.

*Check:*
$$4x^2 - 48 = 0 \quad \text{Original equation}$$

| | |
|---|---|
| $4(2\sqrt{3})^2 - 48 = 0 \quad ?$ | $4(-2\sqrt{3})^2 - 48 = 0 \quad ?$ |
| $4(12) - 48 = 0 \quad ?$ | $4(12) - 48 = 0 \quad ?$ |
| $48 - 48 = 0 \quad ?$ | $48 - 48 = 0 \quad ?$ |
| $0 = 0 \quad$ True | $0 = 0 \quad$ True |

The solution set is $\{2\sqrt{3}, -2\sqrt{3}\}$.

**NOW TRY** Exercises 7 and 13.

Recall that solutions such as those in Example 1 are sometimes abbreviated with the symbol $\pm$ (read "positive or negative," or "plus or minus"). With this symbol the solutions in Example 1 would be written $\pm\sqrt{5}$ and $\pm 2\sqrt{3}$.

### EXAMPLE 2  Using the Square Root Property in an Application

Galileo Galilei (1564–1642) developed a formula for freely falling objects described by

$$d = 16t^2,$$

where $d$ is the distance in feet that an object falls (disregarding air resistance) in $t$ seconds, regardless of weight. Galileo dropped objects from the Leaning Tower of Pisa to develop this formula. If the Leaning Tower is about 180 ft tall, use Galileo's formula to determine how long it would take an object dropped from the tower to fall to the ground. (*Source: Microsoft Encarta Encyclopedia.*)

We substitute 180 for $d$ in Galileo's formula.

$$d = 16t^2$$
$$180 = 16t^2 \quad \text{Let } d = 180.$$
$$11.25 = t^2 \quad \text{Divide by 16.}$$
$$t = \sqrt{11.25} \quad \text{or} \quad t = -\sqrt{11.25} \quad \text{Square root property}$$

Since time cannot be negative, we discard the negative solution. In applied problems, we usually prefer approximations to exact values. Using a calculator, $\sqrt{11.25} \approx 3.4$ so $t \approx 3.4$. The object would fall to the ground in about 3.4 sec.

**NOW TRY** Exercise 27.

**OBJECTIVE 3** Solve quadratic equations of the form $(ax + b)^2 = c$ by using the square root property. To solve more complicated equations by using the square root property, such as

$$(x - 5)^2 = 36,$$

substitute $(x - 5)^2$ for $x^2$ and 36 for $k$ in the square root property to obtain

$$x - 5 = \sqrt{36} \quad \text{or} \quad x - 5 = -\sqrt{36}$$
$$x - 5 = 6 \quad \text{or} \quad x - 5 = -6$$
$$x = 11 \quad \text{or} \quad x = -1.$$

Check: $\quad (x - 5)^2 = 36 \quad$ Original equation

| $(11 - 5)^2 = 36$ ? | $(-1 - 5)^2 = 36$ ? |
| $6^2 = 36$ ? | $(-6)^2 = 36$ ? |
| $36 = 36$ True | $36 = 36$ True |

Since both solutions satisfy the original equation, the solution set is $\{-1, 11\}$.

**EXAMPLE 3** Using the Square Root Property

Solve $(2x - 3)^2 = 18$.

$$2x - 3 = \sqrt{18} \quad \text{or} \quad 2x - 3 = -\sqrt{18} \quad \text{Square root property}$$
$$2x = 3 + \sqrt{18} \quad \text{or} \quad 2x = 3 - \sqrt{18} \quad \text{Add 3.}$$
$$x = \frac{3 + \sqrt{18}}{2} \quad \text{or} \quad x = \frac{3 - \sqrt{18}}{2} \quad \text{Divide by 2.}$$
$$x = \frac{3 + 3\sqrt{2}}{2} \quad \text{or} \quad x = \frac{3 - 3\sqrt{2}}{2} \quad \sqrt{18} = \sqrt{9 \cdot 2} = 3\sqrt{2}$$

We show the check for the first solution. The check for the other solution is similar.

Check: $\quad (2x - 3)^2 = 18 \quad$ Original equation

$$\left[ 2\left(\frac{3 + 3\sqrt{2}}{2}\right) - 3 \right]^2 = 18 \quad ? \quad \text{Let } x = \frac{3 + 3\sqrt{2}}{2}.$$
$$(3 + 3\sqrt{2} - 3)^2 = 18 \quad ? \quad \text{Multiply.}$$
$$(3\sqrt{2})^2 = 18 \quad ? \quad \text{Simplify.}$$
$$18 = 18 \quad \text{True}$$

The solution set is $\left\{ \dfrac{3 + 3\sqrt{2}}{2}, \dfrac{3 - 3\sqrt{2}}{2} \right\}$.

**NOW TRY** Exercise 23.

**OBJECTIVE 4** Solve quadratic equations by completing the square. We can use the square root property to solve *any* quadratic equation by writing it in the form $(x + k)^2 = n$. That is, we must write the left side of the equation as a perfect square trinomial that can be factored as $(x + k)^2$, the square of a binomial, and the right side must be a constant. Rewriting a quadratic equation in this form is called **completing the square.**

Recall that the perfect square trinomial

$$x^2 + 10x + 25$$

can be factored as $(x + 5)^2$. In the trinomial, the coefficient of $x$ (the first-degree term) is 10 and the constant term is 25. Notice that if we take half of 10 and square it, we get the constant term, 25.

$$\left[\underbrace{\frac{1}{2}(10)}_{\text{Coefficient of } x}\right]^2 = 5^2 = \underbrace{25}_{\text{Constant}}$$

Similarly, in

$$x^2 + 12x + 36, \qquad \left[\frac{1}{2}(12)\right]^2 = 6^2 = 36,$$

and in

$$m^2 - 6m + 9, \qquad \left[\frac{1}{2}(-6)\right]^2 = (-3)^2 = 9.$$

This relationship is true in general and is the idea behind completing the square.

**EXAMPLE 4** Solving a Quadratic Equation by Completing the Square

Solve $x^2 + 8x + 10 = 0$.

This quadratic equation cannot be solved by factoring, and it is not in the correct form to solve using the square root property. To solve it by completing the square, we need a perfect square trinomial on the left side of the equation. To get this form, we first subtract 10 from each side.

$$x^2 + 8x + 10 = 0 \qquad \text{Original equation}$$
$$x^2 + 8x = -10 \qquad \text{Subtract 10.}$$

We must add a constant to get a perfect square trinomial on the left.

$$\underbrace{x^2 + 8x + \underline{\ ?\ }}_{\text{Needs to be a perfect square trinomial}}$$

To find this constant, we apply the ideas preceding this example. Take half the coefficient of the first-degree term and square the result.

$$\left[\frac{1}{2}(8)\right]^2 = 4^2 = 16 \quad \leftarrow \text{Desired constant}$$

We add this constant, 16, to *each* side of the equation.

*This is a key step.* $\qquad x^2 + 8x + 16 = -10 + 16$

Next we factor the perfect square trinomial on the left and add on the right.

$$(x + 4)^2 = 6$$

We can now solve this equation using the square root property.

$$x + 4 = \sqrt{6} \quad \text{or} \quad x + 4 = -\sqrt{6}$$
$$x = -4 + \sqrt{6} \quad \text{or} \quad x = -4 - \sqrt{6}$$

*Check:*

$$x^2 + 8x + 10 = 0 \quad \text{Original equation}$$
$$(-4 + \sqrt{6})^2 + 8(-4 + \sqrt{6}) + 10 = 0 \quad ? \quad \text{Let } x = -4 + \sqrt{6}.$$
$$16 - 8\sqrt{6} + 6 - 32 + 8\sqrt{6} + 10 = 0 \quad ?$$
$$0 = 0 \quad \text{True}$$

*Remember the middle term when squaring* $-4 + \sqrt{6}$.

The check of the other solution is similar. The solution set is

$$\{-4 + \sqrt{6}, -4 - \sqrt{6}\}.$$

**NOW TRY** Exercise 49.

The procedure from Example 4 can be generalized.

### Completing the Square

To solve $ax^2 + bx + c = 0$ ($a \neq 0$) by completing the square, use these steps.

**Step 1** **Be sure the second-degree (squared) term has coefficient 1.** If the coefficient of the squared term is 1, proceed to Step 2. If the coefficient of the squared term is not 1 but some other nonzero number $a$, divide each side of the equation by $a$.

**Step 2** **Write the equation in correct form** so that terms with variables are on one side of the equals sign and the constant is on the other side.

(Steps 1 and 2 can be done in either order. With some equations, it is more convenient to do Step 2 first.)

**Step 3** **Square half the coefficient of the first-degree (linear) term.**

**Step 4** **Add the square to each side.**

**Step 5** **Factor the perfect square trinomial.** One side should now be a perfect square trinomial. Factor it as the square of a binomial. Simplify the other side.

**Step 6** **Solve the equation.** Apply the square root property to complete the solution.

**EXAMPLE 5** Solving a Quadratic Equation with $a = 1$ by Completing the Square

Solve $x^2 + 5x - 1 = 0$.

Follow the steps in the box. Since the coefficient of the squared term is 1, begin with Step 2.

**Step 2** $\qquad\qquad x^2 + 5x = 1 \qquad$ Add 1 to each side.

**Step 3** Take half the coefficient of the first-degree term and square the result.

$$\left[\frac{1}{2}(5)\right]^2 = \left(\frac{5}{2}\right)^2 = \frac{25}{4}$$

**Step 4** $\qquad\qquad x^2 + 5x + \dfrac{25}{4} = 1 + \dfrac{25}{4} \qquad$ Add the square to each side of the equation.

**Step 5** $\qquad\qquad \left(x + \dfrac{5}{2}\right)^2 = \dfrac{29}{4} \qquad$ Factor on the left; add on the right.

**Step 6** $x + \dfrac{5}{2} = \sqrt{\dfrac{29}{4}}$ or $x + \dfrac{5}{2} = -\sqrt{\dfrac{29}{4}}$   Square root property

$x + \dfrac{5}{2} = \dfrac{\sqrt{29}}{2}$ or $x + \dfrac{5}{2} = -\dfrac{\sqrt{29}}{2}$   $\sqrt{\dfrac{a}{b}} = \dfrac{\sqrt{a}}{\sqrt{b}}$

$x = -\dfrac{5}{2} + \dfrac{\sqrt{29}}{2}$ or $x = -\dfrac{5}{2} - \dfrac{\sqrt{29}}{2}$   Add $-\dfrac{5}{2}$.

$x = \dfrac{-5 + \sqrt{29}}{2}$ or $x = \dfrac{-5 - \sqrt{29}}{2}$   $\dfrac{a}{c} + \dfrac{b}{c} = \dfrac{a+b}{c}$

Check that the solution set is $\left\{\dfrac{-5 + \sqrt{29}}{2}, \dfrac{-5 - \sqrt{29}}{2}\right\}$.

**NOW TRY** Exercise 51.

**EXAMPLE 6** Solving a Quadratic Equation with $a \neq 1$ by Completing the Square

Solve $2x^2 - 4x - 5 = 0$.
   Divide each side by 2 to get 1 as the coefficient of the second-degree term.

$x^2 - 2x - \dfrac{5}{2} = 0$   Step 1

$x^2 - 2x = \dfrac{5}{2}$   Step 2

$\left[\dfrac{1}{2}(-2)\right]^2 = (-1)^2 = 1$   Step 3

$x^2 - 2x + 1 = \dfrac{5}{2} + 1$   Step 4

$(x - 1)^2 = \dfrac{7}{2}$   Step 5

$x - 1 = \sqrt{\dfrac{7}{2}}$ or $x - 1 = -\sqrt{\dfrac{7}{2}}$   Step 6

$x = 1 + \sqrt{\dfrac{7}{2}}$ or $x = 1 - \sqrt{\dfrac{7}{2}}$   Add 1.

$x = 1 + \dfrac{\sqrt{14}}{2}$ or $x = 1 - \dfrac{\sqrt{14}}{2}$   Rationalize denominators.

Add the two terms in each solution as follows.

$1 + \dfrac{\sqrt{14}}{2} = \dfrac{2}{2} + \dfrac{\sqrt{14}}{2} = \dfrac{2 + \sqrt{14}}{2}$   $1 = \dfrac{2}{2}$

$1 - \dfrac{\sqrt{14}}{2} = \dfrac{2}{2} - \dfrac{\sqrt{14}}{2} = \dfrac{2 - \sqrt{14}}{2}$.

Check that the solution set is $\left\{\dfrac{2 + \sqrt{14}}{2}, \dfrac{2 - \sqrt{14}}{2}\right\}$.

**NOW TRY** Exercise 57.

**OBJECTIVE 5** Solve quadratic equations with solutions that are not real numbers. So far, all the equations we have solved using the square root property have had two real solutions. In the equation $x^2 = k$, if $k < 0$, there will be two nonreal complex solutions.

**EXAMPLE 7** Solving Quadratic Equations with Nonreal Complex Solutions

Solve each equation.

(a) $x^2 = -15$

$\quad x = \sqrt{-15} \quad$ or $\quad x = -\sqrt{-15} \quad$ Square root property
$\quad x = i\sqrt{15} \quad$ or $\quad x = -i\sqrt{15} \quad \sqrt{-1} = i$

The solution set is $\{i\sqrt{15}, -i\sqrt{15}\}$.

(b) $(x + 2)^2 = -16$

$\quad x + 2 = \sqrt{-16} \quad$ or $\quad x + 2 = -\sqrt{-16} \quad$ Square root property
$\quad x + 2 = 4i \quad$ or $\quad x + 2 = -4i \quad \sqrt{-16} = 4i$
$\quad x = -2 + 4i \quad$ or $\quad x = -2 - 4i \quad$ Add $-2$.

The solution set is $\{-2 + 4i, -2 - 4i\}$.

(c) $x^2 + 2x + 7 = 0$

Solve by completing the square.

$\quad x^2 + 2x = -7 \quad$ Subtract 7.
$\quad x^2 + 2x + 1 = -7 + 1 \quad \left[\frac{1}{2}(2)\right]^2 = 1$; add 1 to each side.
$\quad (x + 1)^2 = -6 \quad$ Factor on the left; add on the right.
$\quad x + 1 = \pm i\sqrt{6} \quad$ Square root property
$\quad x = -1 \pm i\sqrt{6} \quad$ Subtract 1.

The solution set is $\{-1 + i\sqrt{6}, -1 - i\sqrt{6}\}$.

**NOW TRY** Exercises 65, 67, and 71.

## 9.1 Exercises

1. *Concept Check* A student was asked to solve the quadratic equation $x^2 = 16$ and did not get full credit for the solution set $\{4\}$. **WHAT WENT WRONG?** Give the correct solution set.

2. *Concept Check* Why can't the zero-factor property be used to solve every quadratic equation?

3. Give a one-sentence description or explanation of each phrase.

   (a) Quadratic equation in standard form  (b) Zero-factor property
   (c) Square root property

4. *Concept Check* A student tried to solve $x^2 - x - 2 = 5$ as follows.

$\quad x^2 - x - 2 = 5$
$\quad (x - 2)(x + 1) = 5 \quad$ Factor.
$\quad x - 2 = 5 \quad$ or $\quad x + 1 = 5 \quad$ Zero-factor property
$\quad x = 7 \quad$ or $\quad x = 4 \quad$ Solve each equation.

This method is incorrect. **WHAT WENT WRONG?**

*Use the square root property to solve each equation. See Examples 1 and 3.*

5. $x^2 = 81$
6. $x^2 = 225$
7. $x^2 = 17$
8. $x^2 = 19$
9. $x^2 = 32$
10. $x^2 = 54$
11. $x^2 - 20 = 0$
12. $p^2 - 50 = 0$
13. $3n^2 - 72 = 0$
14. $5z^2 - 200 = 0$
15. $(x + 2)^2 = 25$
16. $(t + 8)^2 = 9$
17. $(x - 4)^2 = 3$
18. $(x + 3)^2 = 11$
19. $(t + 5)^2 = 48$
20. $(m - 6)^2 = 27$
21. $(3x - 1)^2 = 7$
22. $(2x + 4)^2 = 10$

**23.** $(4p + 1)^2 = 24$  **24.** $(5t - 2)^2 = 12$  **25.** $(2 - 5t)^2 = 12$

**26.** Explain why Exercises 24 and 25 have the same solution set.

*Solve Exercises 27 and 28 using Galileo's formula, $d = 16t^2$. Round answers to the nearest tenth. See Example 2.*

**27.** Mount Rushmore National Memorial in South Dakota features a sculpture of four of America's favorite presidents carved into the rim of the mountain, 500 ft above the valley floor. How long would it take a rock dropped from the top of the sculpture to fall to the ground? (*Source: Microsoft Encarta Encyclopedia.*)

**28.** The Gateway Arch in St. Louis, Missouri, is 630 ft tall. How long would it take an object dropped from the top of it to fall to the ground? (*Source: Home & Away,* November/December 2000.)

**29.** *Concept Check* Of the two equations
$$(2x + 1)^2 = 5 \quad \text{and} \quad x^2 + 4x = 12,$$
which one is more suitable for solving by the square root property? Which one is more suitable for solving by completing the square?

**30.** Why would most students find the equation $x^2 + 4x = 20$ easier to solve by completing the square than the equation $5x^2 + 2x = 3$?

*Concept Check Decide what number must be added to make each expression a perfect square trinomial. Then factor the trinomial.*

**31.** $x^2 + 6x +$   **32.** $x^2 + 14x +$   **33.** $p^2 - 12p +$
**34.** $x^2 + 20x +$   **35.** $q^2 + 9q +$   **36.** $t^2 - 13t +$
**37.** $x^2 + \dfrac{1}{4}x +$   **38.** $x^2 + \dfrac{1}{2}x +$   **39.** $x^2 - 0.8x +$

**40.** *Concept Check* What would be the first step in solving $2x^2 + 8x = 9$ by completing the square?

*Determine the number that will complete the square to solve each equation after the constant term has been written on the right side. Do not actually solve. See Examples 4–6.*

**41.** $x^2 + 4x - 2 = 0$   **42.** $t^2 + 2t - 1 = 0$   **43.** $x^2 + 10x + 18 = 0$
**44.** $x^2 + 8x + 11 = 0$   **45.** $3w^2 - w - 24 = 0$   **46.** $4z^2 - z - 39 = 0$

*Solve each equation by completing the square. See Examples 4–6.*

**47.** $x^2 - 2x - 24 = 0$   **48.** $m^2 - 4m - 32 = 0$   **49.** $x^2 + 4x - 2 = 0$
**50.** $t^2 + 2t - 1 = 0$   **51.** $x^2 + 7x - 1 = 0$   **52.** $x^2 + 13x - 3 = 0$
**53.** $3w^2 - w = 24$   **54.** $4z^2 - z = 39$   **55.** $2k^2 + 5k - 2 = 0$
**56.** $3r^2 + 2r - 2 = 0$   **57.** $5x^2 - 10x + 2 = 0$   **58.** $2x^2 - 16x + 25 = 0$
**59.** $9x^2 - 24x = -13$   **60.** $25n^2 - 20n = 1$   **61.** $z^2 - \dfrac{4}{3}z = -\dfrac{1}{9}$
**62.** $p^2 - \dfrac{8}{3}p = -1$   **63.** $0.1x^2 - 0.2x - 0.1 = 0$   **64.** $0.1p^2 - 0.4p + 0.1 = 0$

*Find the nonreal complex solutions of each equation. See Example 7.*

**65.** $x^2 = -12$  **66.** $x^2 = -18$  **67.** $(r - 5)^2 = -4$
**68.** $(t + 6)^2 = -9$  **69.** $(6k - 1)^2 = -8$  **70.** $(4m - 7)^2 = -27$
**71.** $m^2 + 4m + 13 = 0$  **72.** $t^2 + 6t + 10 = 0$  **73.** $3r^2 + 4r + 4 = 0$
**74.** $4x^2 + 5x + 5 = 0$  **75.** $-m^2 - 6m - 12 = 0$  **76.** $-k^2 - 5k - 10 = 0$

## 9.2 The Quadratic Formula

**OBJECTIVES**

1. Derive the quadratic formula.
2. Solve quadratic equations by using the quadratic formula.
3. Use the discriminant to determine the number and type of solutions.

In this section, we complete the square to solve the general quadratic equation
$$ax^2 + bx + c = 0,$$
where $a$, $b$, and $c$ are complex numbers and $a \neq 0$. The solution of this general equation gives a formula for finding the solution of any specific quadratic equation.

**OBJECTIVE 1** Derive the quadratic formula. To solve $ax^2 + bx + c = 0$ by completing the square (assuming $a > 0$), we follow the steps given in **Section 9.1**.

$$ax^2 + bx + c = 0$$

$$x^2 + \frac{b}{a}x + \frac{c}{a} = 0 \qquad \text{Divide by } a. \text{ (Step 1)}$$

$$x^2 + \frac{b}{a}x = -\frac{c}{a} \qquad \text{Subtract } \frac{c}{a}. \text{ (Step 2)}$$

$$\left[\frac{1}{2}\left(\frac{b}{a}\right)\right]^2 = \left(\frac{b}{2a}\right)^2 = \frac{b^2}{4a^2} \qquad \text{(Step 3)}$$

$$x^2 + \frac{b}{a}x + \frac{b^2}{4a^2} = -\frac{c}{a} + \frac{b^2}{4a^2} \qquad \text{Add } \frac{b^2}{4a^2} \text{ to each side. (Step 4)}$$

Write the left side as a perfect square and rearrange the right side.

$$\left(x + \frac{b}{2a}\right)^2 = \frac{b^2}{4a^2} + \frac{-c}{a} \qquad \text{(Step 5)}$$

$$\left(x + \frac{b}{2a}\right)^2 = \frac{b^2}{4a^2} + \frac{-4ac}{4a^2} \qquad \text{Write with a common denominator.}$$

$$\left(x + \frac{b}{2a}\right)^2 = \frac{b^2 - 4ac}{4a^2} \qquad \text{Add fractions.}$$

$$x + \frac{b}{2a} = \sqrt{\frac{b^2 - 4ac}{4a^2}} \quad \text{or} \quad x + \frac{b}{2a} = -\sqrt{\frac{b^2 - 4ac}{4a^2}} \qquad \text{Square root property (Step 6)}$$

Since $\sqrt{\dfrac{b^2 - 4ac}{4a^2}} = \dfrac{\sqrt{b^2 - 4ac}}{\sqrt{4a^2}} = \dfrac{\sqrt{b^2 - 4ac}}{2a},$

the right side of each equation can be expressed as

$$x + \frac{b}{2a} = \frac{\sqrt{b^2 - 4ac}}{2a} \quad \text{or} \quad x + \frac{b}{2a} = \frac{-\sqrt{b^2 - 4ac}}{2a}$$

$$x = \frac{-b}{2a} + \frac{\sqrt{b^2 - 4ac}}{2a} \quad \text{or} \quad x = \frac{-b}{2a} - \frac{\sqrt{b^2 - 4ac}}{2a}$$

$$x = \frac{-b + \sqrt{b^2 - 4ac}}{2a} \quad \text{or} \quad x = \frac{-b - \sqrt{b^2 - 4ac}}{2a}.$$

If $a < 0$, the same two solutions are obtained. The result is the **quadratic formula,** which is abbreviated as follows.

### Quadratic Formula

The solutions of the equation $ax^2 + bx + c = 0 \; (a \neq 0)$ are given by

$$x = \frac{-b \pm \sqrt{b^2 - 4ac}}{2a}.$$

**CAUTION** In the quadratic formula, $x = \dfrac{-b \pm \sqrt{b^2 - 4ac}}{2a}$, *the square root is added to or subtracted from the value of $-b$ before dividing by $2a$.*

**OBJECTIVE 2** Solve quadratic equations by using the quadratic formula. To use the quadratic formula, first write the given equation in standard form

$$ax^2 + bx + c = 0.$$

Then identify the values of $a$, $b$, and $c$ and substitute them into the formula.

### EXAMPLE 1  Using the Quadratic Formula (Rational Solutions)

Solve $6x^2 - 5x - 4 = 0$.

Here $a$, the coefficient of the second-degree term, is 6, while $b$, the coefficient of the first-degree term, is $-5$, and the constant $c$ is $-4$. Substitute these values into the quadratic formula.

$$x = \frac{-b \pm \sqrt{b^2 - 4ac}}{2a} \quad \text{Quadratic formula}$$

$$x = \frac{-(-5) \pm \sqrt{(-5)^2 - 4(6)(-4)}}{2(6)} \quad a = 6, b = -5, c = -4$$

Use parentheses and substitute carefully to avoid errors.

$$x = \frac{5 \pm \sqrt{25 + 96}}{12}$$

$$x = \frac{5 \pm \sqrt{121}}{12}$$

$$x = \frac{5 \pm 11}{12}$$

There are two solutions, one from the $+$ sign and one from the $-$ sign.

$$x = \frac{5 + 11}{12} = \frac{16}{12} = \frac{4}{3} \quad \text{or} \quad x = \frac{5 - 11}{12} = \frac{-6}{12} = -\frac{1}{2}$$

Check each solution in the original equation. The solution set is $\left\{-\frac{1}{2}, \frac{4}{3}\right\}$.

**NOW TRY** Exercise 5.

We could have used factoring to solve the equation in Example 1.

$$6x^2 - 5x - 4 = 0$$

$(3x - 4)(2x + 1) = 0$      Factor.

$3x - 4 = 0$ or $2x + 1 = 0$      Zero-factor property

$3x = 4$ or $2x = -1$      Solve each equation.

$x = \frac{4}{3}$ or $x = -\frac{1}{2}$      Same solutions as in Example 1

When solving quadratic equations, it is a good idea to try factoring first. If the equation cannot be factored or if factoring is difficult, then use the quadratic formula. Later in this section, we will show a way to determine whether factoring can be used to solve a quadratic equation.

**EXAMPLE 2**   Using the Quadratic Formula (Irrational Solutions)

Solve $4x^2 = 8x - 1$.

Write the equation in standard form as $4x^2 - 8x + 1 = 0$.

$$x = \frac{-b \pm \sqrt{b^2 - 4ac}}{2a} \quad \text{Quadratic formula}$$

$$x = \frac{-(-8) \pm \sqrt{(-8)^2 - 4(4)(1)}}{2(4)} \quad a = 4, b = -8, c = 1$$

$$= \frac{8 \pm \sqrt{64 - 16}}{8}$$

$$= \frac{8 \pm \sqrt{48}}{8}$$

$$= \frac{8 \pm 4\sqrt{3}}{8} \quad \sqrt{48} = \sqrt{16} \cdot \sqrt{3} = 4\sqrt{3}$$

$$= \frac{4(2 \pm \sqrt{3})}{4(2)} \quad \text{Factor.}$$

Factor first; then divide out the common factor.

$$= \frac{2 \pm \sqrt{3}}{2} \quad \text{Lowest terms}$$

The solution set is $\left\{\frac{2 + \sqrt{3}}{2}, \frac{2 - \sqrt{3}}{2}\right\}$.

**NOW TRY** Exercise 9.

**CAUTION** Remember the following:

1. *Every quadratic equation must be expressed in standard form $ax^2 + bx + c = 0$ before we begin to solve it,* whether we use factoring or the quadratic formula.
2. *When writing solutions in lowest terms, be sure to factor first; then divide out the common factor,* as shown in the last two steps in Example 2.

**EXAMPLE 3** Using the Quadratic Formula (Nonreal Complex Solutions)

Solve $(9x + 3)(x - 1) = -8$.

Write the equation in standard form.

$$(9x + 3)(x - 1) = -8$$
$$9x^2 - 6x - 3 = -8 \quad \text{Multiply.}$$
$$9x^2 - 6x + 5 = 0 \quad \text{Standard form}$$

From the equation $9x^2 - 6x + 5 = 0$, we identify $a = 9$, $b = -6$, and $c = 5$.

$$x = \frac{-(-6) \pm \sqrt{(-6)^2 - 4(9)(5)}}{2(9)} \quad \text{Substitute into the quadratic formula.}$$

$$= \frac{6 \pm \sqrt{-144}}{18}$$

$$= \frac{6 \pm 12i}{18} \quad \sqrt{-144} = 12i$$

$$= \frac{6(1 \pm 2i)}{6(3)} \quad \text{Factor.}$$

$$= \frac{1 \pm 2i}{3} \quad \text{Lowest terms}$$

The solution set, written in standard form $a + bi$, is $\left\{\frac{1}{3} + \frac{2}{3}i, \frac{1}{3} - \frac{2}{3}i\right\}$.

**NOW TRY** Exercise 37.

**OBJECTIVE 3** Use the discriminant to determine the number and type of solutions. The solutions of the quadratic equation $ax^2 + bx + c = 0$ are given by

$$x = \frac{-b \pm \sqrt{b^2 - 4ac}}{2a}. \quad \leftarrow \text{Discriminant}$$

If $a$, $b$, and $c$ are integers, the type of solutions of a quadratic equation—that is, rational, irrational, or nonreal complex—is determined by the expression under the radical sign, $b^2 - 4ac$. Because it distinguishes among the three types of solutions, $b^2 - 4ac$ is called the *discriminant*. By calculating the discriminant before solving a quadratic equation, we can predict whether the solutions will be rational numbers, irrational numbers, or nonreal complex numbers.

### Discriminant

The **discriminant** of $ax^2 + bx + c = 0$ is $b^2 - 4ac$. If $a$, $b$, and $c$ are integers, then the number and type of solutions are determined as follows.

| Discriminant | Number and Type of Solutions |
| --- | --- |
| Positive, and the square of an integer | Two rational solutions |
| Positive, but not the square of an integer | Two irrational solutions |
| Zero | One rational solution |
| Negative | Two nonreal complex solutions |

Calculating the discriminant can also help you decide whether to solve a quadratic equation by factoring or by using the quadratic formula. *If the discriminant is a perfect square (including 0), then the equation can be solved by factoring. Otherwise, the quadratic formula should be used.*

### EXAMPLE 4  Using the Discriminant

Find the discriminant. Use it to predict the number and type of solutions for each equation. Tell whether the equation can be solved by factoring or whether the quadratic formula should be used.

**(a)** $6x^2 - x - 15 = 0$

We find the discriminant by evaluating $b^2 - 4ac$.

$$b^2 - 4ac = (-1)^2 - 4(6)(-15) \quad a = 6, b = -1, c = -15$$
$$= 1 + 360$$
$$= 361$$

Use parentheses and substitute carefully.

A calculator shows that $361 = 19^2$, a perfect square. Since $a$, $b$, and $c$ are integers and the discriminant is a perfect square, there will be two rational solutions and the equation can be solved by factoring.

**(b)** $3x^2 - 4x = 5$

Write the equation in standard form as $3x^2 - 4x - 5 = 0$.

$$b^2 - 4ac = (-4)^2 - 4(3)(-5) \quad a = 3, b = -4, c = -5$$
$$= 16 + 60$$
$$= 76$$

Because 76 is positive but not the square of an integer and $a$, $b$, and $c$ are integers, the equation will have two irrational solutions and is best solved using the quadratic formula.

**(c)** $4x^2 + x + 1 = 0$

Since $a = 4$, $b = 1$, and $c = 1$, the discriminant is

$$b^2 - 4ac = 1^2 - 4(4)(1)$$
$$= 1 - 16$$
$$= -15.$$

Because the discriminant is negative and $a$, $b$, and $c$ are integers, this quadratic equation will have two nonreal complex solutions. The quadratic formula should be used to solve it.

**(d)** $4x^2 + 9 = 12x$

Write the equation as $4x^2 - 12x + 9 = 0$. The discriminant is

$$b^2 - 4ac = (-12)^2 - 4(4)(9) \quad a = 4, b = -12, c = 9$$
$$= 144 - 144$$
$$= 0.$$

Because the discriminant is 0, the quantity under the radical in the quadratic formula is 0, and there is only one rational solution. Again, the equation can be solved by factoring.

**NOW TRY** Exercises 39, 41, and 43.

### EXAMPLE 5  Using the Discriminant

Find $k$ so that $9x^2 + kx + 4 = 0$ will have only one rational solution.

The equation will have only one rational solution if the discriminant is 0. Since $a = 9$, $b = k$, and $c = 4$, the discriminant is

$$b^2 - 4ac = k^2 - 4(9)(4) = k^2 - 144.$$

Set the discriminant equal to 0 and solve for $k$.

$$k^2 - 144 = 0$$
$$k^2 = 144 \quad \text{Subtract 144.}$$
$$k = 12 \quad \text{or} \quad k = -12 \quad \text{Square root property}$$

The equation will have only one rational solution if $k = 12$ or $k = -12$.

**NOW TRY** Exercise 55.

## 9.2 Exercises

*Concept Check*  Answer each question in Exercises 1–4.

**1.** The Cadillac Bar in Houston, Texas, encourages patrons to write (tasteful) messages on the walls. One person attempted to write the quadratic formula, as shown here.

$$x = \frac{-b\sqrt{b^2 - 4ac}}{2a}$$

Was this correct? If not, correct it.

**2.** An early version of Microsoft *Word* for Windows included the 1.0 edition of *Equation Editor*. The documentation used the following for the quadratic formula.

$$x = -b \pm \frac{\sqrt{b^2 - 4ac}}{2a}$$

Was this correct? If not, correct it.

**3.** A student claimed that the equation $2x^2 - 5 = 0$ cannot be solved using the quadratic formula because there is no first-degree $x$-term. Was the student correct? If not, give the values of $a$, $b$, and $c$.

**4.** A student attempted to solve $5x^2 - 5x + 1 = 0$ as follows.

$$x = \frac{-(-5) \pm \sqrt{(-5)^2 - 4(5)(1)}}{2(5)} = \frac{5 \pm \sqrt{5}}{10} = \frac{1}{2} \pm \sqrt{5}$$

Solution set: $\left\{\dfrac{1}{2} \pm \sqrt{5}\right\}$

This is incorrect.  **WHAT WENT WRONG?**  Give the correct solution set.

*Use the quadratic formula to solve each equation. (All solutions for these equations are real numbers.) See Examples 1 and 2.*

5. $x^2 - 8x + 15 = 0$
6. $x^2 + 3x - 28 = 0$
7. $2x^2 + 4x + 1 = 0$
8. $2x^2 + 3x - 1 = 0$
9. $2x^2 - 2x = 1$
10. $9x^2 + 6x = 1$
11. $x^2 + 18 = 10x$
12. $x^2 - 4 = 2x$
13. $4k^2 + 4k - 1 = 0$
14. $4r^2 - 4r - 19 = 0$
15. $2 - 2x = 3x^2$
16. $26r - 2 = 3r^2$
17. $\dfrac{x^2}{4} - \dfrac{x}{2} = 1$
18. $p^2 + \dfrac{p}{3} = \dfrac{1}{6}$
19. $-2t(t + 2) = -3$
20. $-3x(x + 2) = -4$
21. $(r - 3)(r + 5) = 2$
22. $(k + 1)(k - 7) = 1$
23. $(x + 2)(x - 3) = 1$
24. $(x - 5)(x + 2) = 6$
25. $p = \dfrac{5(5 - p)}{3(p + 1)}$
26. $x = \dfrac{2(x + 3)}{x + 5}$
27. $(2x + 1)^2 = x + 4$
28. $(2x - 1)^2 = x + 2$

*Use the quadratic formula to solve each equation. (All solutions for these equations are nonreal complex numbers.) See Example 3.*

29. $x^2 - 3x + 6 = 0$
30. $x^2 - 5x + 20 = 0$
31. $r^2 - 6r + 14 = 0$
32. $t^2 + 4t + 11 = 0$
33. $4x^2 - 4x = -7$
34. $9x^2 - 6x = -7$
35. $x(3x + 4) = -2$
36. $z(2z + 3) = -2$
37. $(2x - 1)(8x - 4) = -1$
38. $(x - 1)(9x - 3) = -2$

*Multiple Choice* Use the discriminant to determine whether the solutions for each equation are

A. two rational numbers;  B. one rational number;
C. two irrational numbers;  D. two nonreal complex numbers.

*Do not actually solve. See Example 4.*

39. $25x^2 + 70x + 49 = 0$
40. $4k^2 - 28k + 49 = 0$
41. $x^2 + 4x + 2 = 0$
42. $9x^2 - 12x - 1 = 0$
43. $3x^2 = 5x + 2$
44. $4x^2 = 4x + 3$
45. $3m^2 - 10m + 15 = 0$
46. $18x^2 + 60x + 82 = 0$
47. $0.5x^2 + 10x + 50 = 0$

48. Using the discriminant, which equations in Exercises 39–47 can be solved by factoring?

*Based on your answer in Exercise 48, solve the equation given in each exercise.*

49. Exercise 39   50. Exercise 40   51. Exercise 43   52. Exercise 44

53. Find the discriminant for each quadratic equation. Use it to tell whether the equation can be solved by factoring or whether the quadratic formula should be used. Then solve each equation.

   (a) $3k^2 + 13k = -12$   (b) $2x^2 + 19 = 14x$

54. *Concept Check* Is it possible for the solution of a quadratic equation with integer coefficients to include just one irrational number? Why or why not?

*Find the value of a, b, or c so that each equation will have exactly one rational solution. See Example 5.*

55. $p^2 + bp + 25 = 0$
56. $r^2 - br + 49 = 0$
57. $am^2 + 8m + 1 = 0$
58. $at^2 + 24t + 16 = 0$
59. $9x^2 - 30x + c = 0$
60. $4m^2 + 12m + c = 0$

61. One solution of $4x^2 + bx - 3 = 0$ is $-\dfrac{5}{2}$. Find $b$ and the other solution.
62. One solution of $3x^2 - 7x + c = 0$ is $\dfrac{1}{3}$. Find $c$ and the other solution.

## 9.3 Equations Quadratic in Form

**OBJECTIVES**

1. Solve an equation with fractions by writing it in quadratic form.
2. Use quadratic equations to solve applied problems.
3. Solve an equation with radicals by writing it in quadratic form.
4. Solve an equation that is quadratic in form by substitution.

We have introduced four methods for solving quadratic equations written in standard form $ax^2 + bx + c = 0$. The following table lists some advantages and disadvantages of each method.

**METHODS FOR SOLVING QUADRATIC EQUATIONS**

| Method | Advantages | Disadvantages |
| --- | --- | --- |
| Factoring | This is usually the fastest method. | Not all polynomials are factorable; some factorable polynomials are difficult to factor. |
| Square root property | This is the simplest method for solving equations of the form $(ax + b)^2 = c$. | Few equations are given in this form. |
| Completing the square | This method can always be used, although most people prefer the quadratic formula. | It requires more steps than other methods. |
| Quadratic formula | This method can always be used. | It is more difficult than factoring because of the square root, although calculators can simplify its use. |

**OBJECTIVE 1** Solve an equation with fractions by writing it in quadratic form. A variety of nonquadratic equations can be written in the form of a quadratic equation and solved by using one of the methods in the table. As you solve the equations in this section, try to decide which method is best for each equation.

**EXAMPLE 1** Solving an Equation with Fractions that Leads to a Quadratic Equation

Solve $\dfrac{1}{x} + \dfrac{1}{x-1} = \dfrac{7}{12}$.

Clear fractions by multiplying each term by the least common denominator, $12x(x - 1)$. (Note that the domain must be restricted to $x \neq 0, x \neq 1$.)

$$12x(x-1)\dfrac{1}{x} + 12x(x-1)\dfrac{1}{x-1} = 12x(x-1)\dfrac{7}{12}$$

$$12(x - 1) + 12x = 7x(x - 1)$$

$12x - 12 + 12x = 7x^2 - 7x$     Distributive property

$24x - 12 = 7x^2 - 7x$     Combine terms.

$7x^2 - 31x + 12 = 0$     Standard form

$(7x - 3)(x - 4) = 0$     Factor.

$7x - 3 = 0$   or   $x - 4 = 0$     Zero-factor property

$7x = 3$   or   $x = 4$     Solve for $x$.

$x = \dfrac{3}{7}$

The solution set is $\left\{\dfrac{3}{7}, 4\right\}$.

**NOW TRY** Exercise 19.

**OBJECTIVE 2** Use quadratic equations to solve applied problems. Earlier we solved distance-rate-time (or motion) problems that led to linear equations or rational equations. Now we solve motion problems that lead to quadratic equations. We continue to use the six-step problem-solving method from **Section 2.3.**

**EXAMPLE 2** Solving a Motion Problem

A riverboat for tourists averages 12 mph in still water. It takes the boat 1 hr, 4 min to go 6 mi upstream and return. Find the speed of the current. See Figure 1.

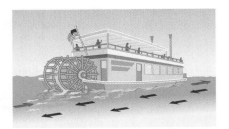

Riverboat, traveling *upstream*—
the current slows it down

**FIGURE 1**

*Step 1* **Read** the problem carefully.

*Step 2* **Assign a variable.** Let $x =$ the speed of the current. The current slows down the boat when it is going upstream, so the rate (or speed) of the boat going upstream is its speed in still water less the speed of the current, or $(12 - x)$ mph. Similarly, the current speeds up the boat as it travels downstream, so its speed downstream is $(12 + x)$ mph. Thus,

$$12 - x = \text{the rate upstream};$$
$$12 + x = \text{the rate downstream}.$$

This information can be used to complete a table. We use the distance formula, $d = rt$, solved for time $t$, $t = \frac{d}{r}$, to write expressions for $t$.

|  | $d$ | $r$ | $t$ |
|---|---|---|---|
| Upstream | 6 | $12 - x$ | $\dfrac{6}{12 - x}$ |
| Downstream | 6 | $12 + x$ | $\dfrac{6}{12 + x}$ |

Times in hours

*Step 3* **Write an equation.** The total time of 1 hr, 4 min can be written as

$$1 + \frac{4}{60} = 1 + \frac{1}{15} = \frac{16}{15} \text{ hr.}$$

Because the time upstream plus the time downstream equals $\frac{16}{15}$ hr,

Time upstream + Time downstream = Total time

$$\frac{6}{12 - x} + \frac{6}{12 + x} = \frac{16}{15}.$$

***Step 4*** **Solve** the equation. Multiply each side by $15(12 - x)(12 + x)$, the LCD, and solve the resulting quadratic equation.

$$15(12 + x)6 + 15(12 - x)6 = 16(12 - x)(12 + x)$$
$$90(12 + x) + 90(12 - x) = 16(144 - x^2)$$
$$1080 + 90x + 1080 - 90x = 2304 - 16x^2 \quad \text{Distributive property}$$
$$2160 = 2304 - 16x^2 \quad \text{Combine terms.}$$
$$16x^2 = 144$$
$$x^2 = 9 \quad \text{Divide by 16.}$$
$$x = 3 \quad \text{or} \quad x = -3 \quad \text{Square root property}$$

***Step 5*** **State the answer.** The current speed cannot be $-3$, so the answer is 3 mph.

***Step 6*** **Check** that this value satisfies the original problem.

**NOW TRY** Exercise 33.

In **Section 7.5** we solved problems about work rates. Recall that a person's work rate is $\frac{1}{t}$ part of the job per hour, where $t$ is the time in hours required to do the complete job. Thus, the part of the job the person will do in $x$ hours is $\frac{1}{t}x$.

### EXAMPLE 3  Solving a Work Problem

It takes two carpet layers 4 hr to carpet a room. If each worked alone, one of them could do the job in 1 hr less time than the other. How long would it take each carpet layer to complete the job alone?

***Step 1*** **Read** the problem again. There will be two answers.

***Step 2*** **Assign a variable.** Let $x =$ the number of hours for the slower carpet layer to complete the job alone. Then the faster carpet layer could do the entire job in $(x - 1)$ hours. The slower person's rate is $\frac{1}{x}$, and the faster person's rate is $\frac{1}{x-1}$. Together, they do the job in 4 hr. Complete a table.

|  | Rate | Time Working Together | Fractional Part of the Job Done |
|---|---|---|---|
| Slower Worker | $\frac{1}{x}$ | 4 | $\frac{1}{x}(4)$ |
| Faster Worker | $\frac{1}{x-1}$ | 4 | $\frac{1}{x-1}(4)$ |

Sum is 1 whole job.

***Step 3*** **Write an equation.** The sum of the fractional parts done by the workers should equal 1 (the whole job).

Part done by slower worker + Part done by faster worker = 1 whole job

$$\frac{4}{x} + \frac{4}{x-1} = 1$$

***Step 4*** **Solve** the equation from Step 3.

$$x(x-1)\left(\frac{4}{x} + \frac{4}{x-1}\right) = x(x-1)(1) \quad \text{Multiply by the LCD.}$$

$$4(x-1) + 4x = x(x-1) \quad \text{Distributive property}$$

$$4x - 4 + 4x = x^2 - x \quad \text{Distributive property}$$

$$x^2 - 9x + 4 = 0 \quad \text{Standard form}$$

This equation cannot be solved by factoring, so use the quadratic formula.

$$x = \frac{-(-9) \pm \sqrt{(-9)^2 - 4(1)(4)}}{2(1)} = \frac{9 \pm \sqrt{65}}{2} \quad a = 1, b = -9, c = 4$$

$$x = \frac{9 + \sqrt{65}}{2} \approx 8.5 \quad \text{or} \quad x = \frac{9 - \sqrt{65}}{2} \approx 0.5 \quad \text{Use a calculator.}$$

***Step 5*** **State the answer.** Only the solution 8.5 makes sense in the original problem, because $x - 1 = 0.5 - 1 = -0.5$ cannot represent the time for the faster worker. Thus, the slower worker could do the job in about 8.5 hr and the faster in about $8.5 - 1 = 7.5$ hr.

***Step 6*** **Check** that these results satisfy the original problem.

**NOW TRY** Exercise 39.

**OBJECTIVE 3** Solve an equation with radicals by writing it in quadratic form.

**EXAMPLE 4** Solving Radical Equations That Lead to Quadratic Equations

Solve each equation.

**(a)** $x = \sqrt{6x - 8}$

This equation is not quadratic. However, squaring both sides of the equation gives a quadratic equation that can be solved by factoring.

$$x^2 = 6x - 8 \quad \text{Square each side.}$$

$$x^2 - 6x + 8 = 0 \quad \text{Standard form}$$

$$(x - 4)(x - 2) = 0 \quad \text{Factor.}$$

$$x - 4 = 0 \quad \text{or} \quad x - 2 = 0 \quad \text{Zero-factor property}$$

$$x = 4 \quad \text{or} \quad x = 2 \quad \text{Proposed solutions}$$

Recall from **Section 8.6** that squaring both sides of an equation can introduce extraneous solutions that do not satisfy the original equation. *All proposed solutions must be checked in the original (not the squared) equation.*

*Check:* If $x = 4$, then
$$x = \sqrt{6x - 8}$$
$$4 = \sqrt{6(4) - 8} \quad ?$$
$$4 = \sqrt{16} \quad ?$$
$$4 = 4. \quad \text{True}$$

If $x = 2$, then
$$x = \sqrt{6x - 8}$$
$$2 = \sqrt{6(2) - 8} \quad ?$$
$$2 = \sqrt{4} \quad ?$$
$$2 = 2. \quad \text{True}$$

Both solutions check, so the solution set is $\{2, 4\}$.

**(b)**
$$x + \sqrt{x} = 6$$
$$\sqrt{x} = 6 - x \quad \text{Isolate the radical on one side.}$$
$$x = 36 - 12x + x^2 \quad \text{Square each side.}$$
$$x^2 - 13x + 36 = 0 \quad \text{Standard form}$$
$$(x - 4)(x - 9) = 0 \quad \text{Factor.}$$
$$x - 4 = 0 \quad \text{or} \quad x - 9 = 0 \quad \text{Zero-factor property}$$
$$x = 4 \quad \text{or} \quad x = 9 \quad \text{Proposed solutions}$$

Check both proposed solutions in the *original* equation.

If $x = 4$, then
$$x + \sqrt{x} = 6$$
$$4 + \sqrt{4} = 6 \quad ?$$
$$6 = 6. \quad \text{True}$$

If $x = 9$, then
$$x + \sqrt{x} = 6$$
$$9 + \sqrt{9} = 6 \quad ?$$
$$12 = 6. \quad \text{False}$$

Only the solution 4 checks, so the solution set is $\{4\}$.

**NOW TRY** Exercises 43 and 49.

**OBJECTIVE 4** Solve an equation that is quadratic in form by substitution. A non-quadratic equation that can be written in the form $au^2 + bu + c = 0$, for $a \neq 0$ and an algebraic expression $u$, is called **quadratic in form.**

**EXAMPLE 5** Solving Equations That Are Quadratic in Form

Solve each equation.

**(a)** $x^4 - 13x^2 + 36 = 0$

Because $x^4 = (x^2)^2$, we can write this equation in quadratic form with $u = x^2$ and $u^2 = x^4$. (Any letter except $x$ could be used instead of $u$.)

$$x^4 - 13x^2 + 36 = 0$$
$$(x^2)^2 - 13x^2 + 36 = 0 \quad x^4 = (x^2)^2$$
$$u^2 - 13u + 36 = 0 \quad \text{Let } u = x^2.$$
$$(u - 4)(u - 9) = 0 \quad \text{Factor.}$$
$$u - 4 = 0 \quad \text{or} \quad u - 9 = 0 \quad \text{Zero-factor property}$$

Don't stop here.
$$u = 4 \quad \text{or} \quad u = 9 \quad \text{Solve.}$$
$$x^2 = 4 \quad \text{or} \quad x^2 = 9 \quad \text{Substitute } x^2 \text{ for } u.$$
$$x = \pm 2 \quad \text{or} \quad x = \pm 3 \quad \text{Square root property}$$

The equation $x^4 - 13x^2 + 36 = 0$, a fourth-degree equation, has four solutions.* The solution set is $\{-3, -2, 2, 3\}$. Each solution can be verified by substituting it into the original equation for $x$.

---

*In general, an equation in which an $n$th-degree polynomial equals 0 has $n$ complex solutions, although some of them may be repeated.

**(b)**
$$4x^6 + 1 = 5x^3$$
$$4(x^3)^2 + 1 = 5x^3 \qquad x^6 = (x^3)^2$$
$$4u^2 + 1 = 5u \qquad \text{Let } u = x^3.$$
$$4u^2 - 5u + 1 = 0 \qquad \text{Standard form}$$
$$(4u - 1)(u - 1) = 0 \qquad \text{Factor.}$$
$$4u - 1 = 0 \quad \text{or} \quad u - 1 = 0 \qquad \text{Zero-factor property}$$
$$u = \frac{1}{4} \quad \text{or} \quad u = 1 \qquad \text{Solve.}$$

*This is a key step.* $\quad x^3 = \frac{1}{4} \quad \text{or} \quad x^3 = 1 \qquad u = x^3$

From these two equations,

$$x = \sqrt[3]{\frac{1}{4}} = \frac{\sqrt[3]{1}}{\sqrt[3]{4}} = \frac{1}{\sqrt[3]{4}} \cdot \frac{\sqrt[3]{2}}{\sqrt[3]{2}} = \frac{\sqrt[3]{2}}{2} \quad \text{or} \quad x = \sqrt[3]{1} = 1.$$

There are nonreal complex solutions for this equation, but finding them involves trigonometry. The real number solution set of $4x^6 + 1 = 5x^3$ is $\left\{\dfrac{\sqrt[3]{2}}{2}, 1\right\}$.

**(c)**
$$x^4 = 6x^2 - 3$$
$$x^4 - 6x^2 + 3 = 0 \qquad \text{Standard form}$$
$$(x^2)^2 - 6x^2 + 3 = 0 \qquad x^4 = (x^2)^2$$
$$u^2 - 6u + 3 = 0 \qquad \text{Let } u = x^2.$$

Since this equation cannot be solved by factoring, use the quadratic formula.

$$u = \frac{-(-6) \pm \sqrt{(-6)^2 - 4(1)(3)}}{2(1)} \qquad a = 1, b = -6, c = 3$$

$$u = \frac{6 \pm \sqrt{24}}{2}$$

$$u = \frac{6 \pm 2\sqrt{6}}{2} \qquad \sqrt{24} = \sqrt{4} \cdot \sqrt{6} = 2\sqrt{6}$$

$$u = \frac{2(3 \pm \sqrt{6})}{2} \qquad \text{Factor.}$$

$$u = 3 \pm \sqrt{6} \qquad \text{Lowest terms}$$

*Find both square roots in each case.* $\quad x^2 = 3 + \sqrt{6} \quad \text{or} \quad x^2 = 3 - \sqrt{6} \qquad u = x^2$

$$x = \pm\sqrt{3 + \sqrt{6}} \quad \text{or} \quad x = \pm\sqrt{3 - \sqrt{6}}$$

The solution set contains four numbers:

$$\left\{\sqrt{3 + \sqrt{6}}, -\sqrt{3 + \sqrt{6}}, \sqrt{3 - \sqrt{6}}, -\sqrt{3 - \sqrt{6}}\right\}.$$

**NOW TRY** Exercises 55, 83, and 87.

**EXAMPLE 6** Solving Equations That Are Quadratic in Form

Solve each equation.

**(a)** $2(4x - 3)^2 + 7(4x - 3) + 5 = 0$

Because of the repeated quantity $4x - 3$, this equation is quadratic in form with $u = 4x - 3$.

$$2(4x - 3)^2 + 7(4x - 3) + 5 = 0$$
$$2u^2 + 7u + 5 = 0 \qquad \text{Let } u = 4x - 3.$$

$$(2u + 5)(u + 1) = 0 \qquad \text{Factor.}$$
$$2u + 5 = 0 \quad \text{or} \quad u + 1 = 0 \qquad \text{Zero-factor property}$$

**Don't stop here.** $\quad u = -\dfrac{5}{2} \quad \text{or} \quad u = -1 \qquad \text{Solve for } u.$

$$4x - 3 = -\dfrac{5}{2} \quad \text{or} \quad 4x - 3 = -1 \qquad \text{Substitute } 4x - 3 \text{ for } u.$$

$$4x = \dfrac{1}{2} \quad \text{or} \quad 4x = 2 \qquad \text{Solve for } x.$$

$$x = \dfrac{1}{8} \quad \text{or} \quad x = \dfrac{1}{2}$$

Check that the solution set of the original equation is $\left\{\dfrac{1}{8}, \dfrac{1}{2}\right\}$.

**(b)** $2x^{2/3} - 11x^{1/3} + 12 = 0$

Let $x^{1/3} = u$; then $x^{2/3} = (x^{1/3})^2 = u^2$. Substitute into the given equation.

$$2u^2 - 11u + 12 = 0 \qquad \text{Let } x^{1/3} = u;\ x^{2/3} = u^2.$$
$$(2u - 3)(u - 4) = 0 \qquad \text{Factor.}$$
$$2u - 3 = 0 \quad \text{or} \quad u - 4 = 0 \qquad \text{Zero-factor property}$$
$$u = \dfrac{3}{2} \quad \text{or} \quad u = 4$$
$$x^{1/3} = \dfrac{3}{2} \quad \text{or} \quad x^{1/3} = 4 \qquad u = x^{1/3}$$
$$(x^{1/3})^3 = \left(\dfrac{3}{2}\right)^3 \quad \text{or} \quad (x^{1/3})^3 = 4^3 \qquad \text{Cube each side.}$$
$$x = \dfrac{27}{8} \quad \text{or} \quad x = 64$$

Check that the solution set is $\left\{\dfrac{27}{8}, 64\right\}$.

**NOW TRY** Exercises 63 and 69.

---

**CAUTION** A common error when solving problems like those in Examples 5 and 6 is to stop too soon. *Once you have solved for u, remember to substitute and solve for the values of the original variable.*

---

## 9.3 Exercises

**NOW TRY Exercise**

*Concept Check* Refer to the box at the beginning of this section. Decide whether *factoring, the square root property,* or the *quadratic formula* is most appropriate for solving each quadratic equation. Do not actually solve the equations.

1. $(2x + 3)^2 = 4$
2. $4x^2 - 3x = 1$
3. $x^2 + 5x - 8 = 0$
4. $2x^2 + 3x = 1$
5. $3x^2 = 2 - 5x$
6. $x^2 = 5$

*Concept Check* Write a sentence describing the first step you would take to solve each equation. Do not actually solve.

7. $\dfrac{14}{x} = x - 5$
8. $\sqrt{1 + x} + x = 5$
9. $(x^2 + x)^2 - 8(x^2 + x) + 12 = 0$
10. $3x = \sqrt{16 - 10x}$

**11.** *Concept Check* A student solved the equation $x = \sqrt{3x+4}$ as follows.

$$x = \sqrt{3x+4}$$
$$x^2 = 3x+4 \quad \text{Square both sides.}$$
$$x^2 - 3x - 4 = 0$$
$$(x-4)(x+1) = 0$$
$$x-4=0 \quad \text{or} \quad x+1=0$$
$$x = 4 \quad \text{or} \quad x = -1$$

Solution set: $\{4, -1\}$

This solution set is incorrect. **WHAT WENT WRONG?** Give the correct solution set.

**12.** *Concept Check* A student solved the following equation as shown.

$$2(x-1)^2 - 3(x-1) + 1 = 0$$
$$2u^2 - 3u + 1 = 0 \quad \text{Let } u = x-1.$$
$$(2u-1)(u-1) = 0$$
$$2u-1 = 0 \quad \text{or} \quad u-1 = 0$$
$$u = \frac{1}{2} \quad \text{or} \quad u = 1$$

Solution set: $\{\frac{1}{2}, 1\}$

This solution set is incorrect. **WHAT WENT WRONG?** Give the correct solution set.

*Solve each equation. Check your solutions. See Example 1.*

**13.** $\dfrac{14}{x} = x - 5$

**14.** $\dfrac{-12}{x} = x + 8$

**15.** $1 - \dfrac{3}{x} - \dfrac{28}{x^2} = 0$

**16.** $4 - \dfrac{7}{r} - \dfrac{2}{r^2} = 0$

**17.** $3 - \dfrac{1}{t} = \dfrac{2}{t^2}$

**18.** $1 + \dfrac{2}{k} = \dfrac{3}{k^2}$

**19.** $\dfrac{1}{x} + \dfrac{2}{x+2} = \dfrac{17}{35}$

**20.** $\dfrac{2}{m} + \dfrac{3}{m+9} = \dfrac{11}{4}$

**21.** $\dfrac{2}{x+1} + \dfrac{3}{x+2} = \dfrac{7}{2}$

**22.** $\dfrac{4}{3-p} + \dfrac{2}{5-p} = \dfrac{26}{15}$

**23.** $\dfrac{3}{2x} - \dfrac{1}{2(x+2)} = 1$

**24.** $\dfrac{4}{3x} - \dfrac{1}{2(x+1)} = 1$

**25.** $3 = \dfrac{1}{t+2} + \dfrac{2}{(t+2)^2}$

**26.** $1 + \dfrac{2}{3z+2} = \dfrac{15}{(3z+2)^2}$

**27.** $\dfrac{6}{p} = 2 + \dfrac{p}{p+1}$

**28.** $\dfrac{k}{2-k} + \dfrac{2}{k} = 5$

**29.** $1 - \dfrac{1}{2x+1} - \dfrac{1}{(2x+1)^2} = 0$

**30.** $1 - \dfrac{1}{3x-2} - \dfrac{1}{(3x-2)^2} = 0$

*Concept Check* Answer each question.

**31.** A boat goes 20 mph in still water, and the rate of the current is $t$ mph.

  **(a)** What is the rate of the boat when it travels upstream?
  **(b)** What is the rate of the boat when it travels downstream?

**32.** If it takes $m$ hours to grade a set of papers, what is the grader's rate (in job per hour)?

*Solve each problem. See Examples 2 and 3.*

**33.** On a windy day Eduardo found that he could go 16 mi downstream and then 4 mi back upstream at top speed in a total of 48 min. What was the top speed of Eduardo's boat if the current was 15 mph?

**34.** Latarsha flew her plane for 6 hr at a constant speed. She traveled 810 mi with the wind, then turned around and traveled 720 mi against the wind. The wind speed was a constant 15 mph. Find the speed of the plane.

|  | d | r | t |
|---|---|---|---|
| Upstream | 4 | x − 15 | |
| Downstream | 16 | | |

|  | d | r | t |
|---|---|---|---|
| With Wind | 810 | | |
| Against Wind | 720 | | |

35. In California, the distance from Jackson to Lodi is about 40 mi, as is the distance from Lodi to Manteca. Rico drove from Jackson to Lodi during the rush hour, stopped in Lodi for a high energy drink, and then drove on to Manteca at 10 mph faster. Driving time for the entire trip was 88 min. Find his speed from Jackson to Lodi. (*Source: State Farm Road Atlas.*)

36. In Canada, Medicine Hat and Cranbrook are 300 km apart. Harry rides his Harley 20 km per hr faster than Yoshi rides his Yamaha. Find Harry's average speed if he travels from Cranbrook to Medicine Hat in $1\frac{1}{4}$ hr less time than Yoshi. (*Source: State Farm Road Atlas.*)

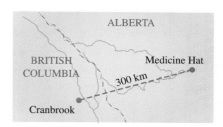

37. Working together, two people can cut a large lawn in 2 hr. One person can do the job alone in 1 hr less time than the other. How long (to the nearest tenth) would it take the faster person to do the job? (*Hint: x* is the time of the faster person.)

| | Rate | Time Working Together | Fractional Part of the Job Done |
|---|---|---|---|
| Faster Worker | $\frac{1}{x}$ | 2 | |
| Slower Worker | | 2 | |

38. A janitorial service provides two people to clean an office building. Working together, the two can clean the building in 5 hr. One person is new to the job and would take 2 hr longer than the other person to clean the building alone. How long (to the nearest tenth) would it take the new worker to clean the building alone?

| | Rate | Time Working Together | Fractional Part of the Job Done |
|---|---|---|---|
| Faster Worker | | | |
| Slower Worker | | | |

39. Rusty and Nancy Brauner are planting flats of spring flowers. Working alone, Rusty would take 2 hr longer than Nancy to plant the flowers. Working together, they do the job in 12 hr. How long would it have taken each person working alone?

40. Jay Beckenstein can work through a stack of invoices in 1 hr less time than Colleen Manley Jones can. Working together they take $1\frac{1}{2}$ hr. How long would it take each person working alone?

41. Two pipes together can fill a large tank in 2 hr. One of the pipes, used alone, takes 3 hr longer than the other to fill the tank. How long would each pipe take to fill the tank alone?

42. A washing machine can be filled in 6 min if both the hot and cold water taps are fully opened. Filling the washer with hot water alone takes 9 min longer than filling it with cold water alone. How long does it take to fill the washer with cold water?

*Solve each equation. Check your solutions. See Example 4.*

**43.** $x = \sqrt{7x - 10}$

**44.** $z = \sqrt{5z - 4}$

**45.** $2x = \sqrt{11x + 3}$

**46.** $4x = \sqrt{6x + 1}$

**47.** $3x = \sqrt{16 - 10x}$

**48.** $4t = \sqrt{8t + 3}$

**49.** $k + \sqrt{k} = 12$

**50.** $p - 2\sqrt{p} = 8$

**51.** $m = \sqrt{\dfrac{6 - 13m}{5}}$

**52.** $r = \sqrt{\dfrac{20 - 19r}{6}}$

**53.** $-x = \sqrt{\dfrac{8 - 2x}{3}}$

**54.** $-x = \sqrt{\dfrac{3x + 7}{4}}$

*Solve each equation. Check your solutions. See Examples 5 and 6.*

**55.** $x^4 - 29x^2 + 100 = 0$

**56.** $x^4 - 37x^2 + 36 = 0$

**57.** $4k^4 - 13k^2 + 9 = 0$

**58.** $9x^4 - 25x^2 + 16 = 0$

**59.** $x^4 + 48 = 16x^2$

**60.** $z^4 = 17z^2 - 72$

**61.** $(x + 3)^2 + 5(x + 3) + 6 = 0$

**62.** $(k - 4)^2 + (k - 4) - 20 = 0$

**63.** $3(m + 4)^2 - 8 = 2(m + 4)$

**64.** $(t + 5)^2 + 6 = 7(t + 5)$

**65.** $2 + \dfrac{5}{3k - 1} = \dfrac{-2}{(3k - 1)^2}$

**66.** $3 - \dfrac{7}{2p + 2} = \dfrac{6}{(2p + 2)^2}$

**67.** $2 - 6(m - 1)^{-2} = (m - 1)^{-1}$

**68.** $3 - 2(x - 1)^{-1} = (x - 1)^{-2}$

**69.** $x^{2/3} + x^{1/3} - 2 = 0$

**70.** $x^{2/3} - 2x^{1/3} - 3 = 0$

**71.** $r^{2/3} + r^{1/3} - 12 = 0$

**72.** $3x^{2/3} - x^{1/3} - 24 = 0$

**73.** $4k^{4/3} - 13k^{2/3} + 9 = 0$

**74.** $9t^{4/3} - 25t^{2/3} + 16 = 0$

**75.** $2(1 + \sqrt{r})^2 = 13(1 + \sqrt{r}) - 6$

**76.** $(k^2 + k)^2 + 12 = 8(k^2 + k)$

**77.** $2x^4 + x^2 - 3 = 0$

**78.** $4k^4 + 5k^2 + 1 = 0$

*The equations in Exercises 79–88 are not grouped by type. Decide which method of solution applies, and then solve each equation. Give only real solutions. See Examples 1 and 4–6.*

**79.** $12x^4 - 11x^2 + 2 = 0$

**80.** $\left(x - \dfrac{1}{2}\right)^2 + 5\left(x - \dfrac{1}{2}\right) - 4 = 0$

**81.** $\sqrt{2x + 3} = 2 + \sqrt{x - 2}$

**82.** $\sqrt{m + 1} = -1 + \sqrt{2m}$

**83.** $2m^6 + 11m^3 + 5 = 0$

**84.** $8x^6 + 513x^3 + 64 = 0$

**85.** $6 = 7(2w - 3)^{-1} + 3(2w - 3)^{-2}$

**86.** $m^6 - 10m^3 = -9$

**87.** $2x^4 - 9x^2 = -2$

**88.** $8x^4 + 1 = 11x^2$

## 9.4 Formulas and Further Applications

**OBJECTIVES**

1. Solve formulas for variables involving squares and square roots.
2. Solve applied problems using the Pythagorean theorem.
3. Solve applied problems using area formulas.
4. Solve applied problems using quadratic functions as models.

**OBJECTIVE 1** Solve formulas for variables involving squares and square roots. The methods presented earlier can be used to solve such formulas.

**EXAMPLE 1** Solving for Variables Involving Squares or Square Roots

Solve each formula for the given variable. Keep $\pm$ in the answer in part (a).

**(a)** $w = \dfrac{kFr}{v^2}$ for $v$

$$w = \dfrac{kFr}{v^2}$$ *The goal is to isolate $v$ on one side.*

$$v^2 w = kFr \quad \text{Multiply by } v^2.$$

$$v^2 = \dfrac{kFr}{w} \quad \text{Divide by } w.$$

$$v = \pm\sqrt{\dfrac{kFr}{w}} \quad \text{Square root property}$$

$$v = \dfrac{\pm\sqrt{kFr}}{\sqrt{w}} \cdot \dfrac{\sqrt{w}}{\sqrt{w}} \quad \text{Rationalize the denominator.}$$

$$v = \dfrac{\pm\sqrt{kFrw}}{w}$$

**(b)** $d = \sqrt{\dfrac{4A}{\pi}}$ for $A$

$$d = \sqrt{\dfrac{4A}{\pi}}$$ *The goal is to isolate $A$ on one side.*

$$d^2 = \dfrac{4A}{\pi} \quad \text{Square both sides.}$$

$$\pi d^2 = 4A \quad \text{Multiply by } \pi.$$

$$\dfrac{\pi d^2}{4} = A, \quad \text{or} \quad A = \dfrac{\pi d^2}{4} \quad \text{Divide by 4.}$$

**NOW TRY** Exercises 9 and 19.

**EXAMPLE 2** Solving for a Variable that Appears in First- and Second-Degree Terms

Solve $s = 2t^2 + kt$ for $t$.

Since the given equation has terms with $t^2$ and $t$, write it in standard form $ax^2 + bx + c = 0$, with $t$ as the variable instead of $x$.

$$2t^2 + kt - s = 0$$

Now use the quadratic formula with $a = 2$, $b = k$, and $c = -s$.

$$t = \frac{-k \pm \sqrt{k^2 - 4(2)(-s)}}{2(2)} \quad \text{Solve for } t.$$

$$t = \frac{-k \pm \sqrt{k^2 + 8s}}{4}$$

The solutions are $t = \dfrac{-k + \sqrt{k^2 + 8s}}{4}$ and $t = \dfrac{-k - \sqrt{k^2 + 8s}}{4}$.

**NOW TRY** Exercise 15.

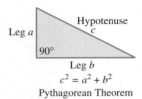
Pythagorean Theorem

**OBJECTIVE 2** Solve applied problems using the Pythagorean theorem. The Pythagorean theorem, illustrated by the figure in the margin, was introduced in **Section 8.3** and is used to solve applications involving right triangles. Such problems often require solving quadratic equations.

**EXAMPLE 3** Using the Pythagorean Theorem

Two cars left an intersection at the same time, one heading due north, the other due west. Some time later, they were exactly 100 mi apart. The car headed north had gone 20 mi farther than the car headed west. How far had each car traveled?

*Step 1* **Read** the problem carefully.

*Step 2* **Assign a variable.**

Let $x =$ the distance traveled by the car headed west.

Then $x + 20 =$ the distance traveled by the car headed north.

See Figure 2. The cars are 100 mi apart, so the hypotenuse of the right triangle equals 100.

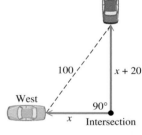
**FIGURE 2**

*Step 3* **Write an equation.** Use the Pythagorean theorem.

$$c^2 = a^2 + b^2$$
$$100^2 = x^2 + (x + 20)^2 \quad (x + y)^2 = x^2 + 2xy + y^2$$

*Step 4* **Solve.**
$\quad 10{,}000 = x^2 + x^2 + 40x + 400$    Square the binomial.
$\quad\quad\quad\;\; 0 = 2x^2 + 40x - 9600$    Standard form
$\quad\quad\quad\;\; 0 = x^2 + 20x - 4800$    Divide by 2.
$\quad\quad\quad\;\; 0 = (x + 80)(x - 60)$    Factor.
$\quad x + 80 = 0 \quad \text{or} \quad x - 60 = 0$    Zero-factor property
$\quad\quad\;\; x = -80 \quad \text{or} \quad\quad x = 60$

*Step 5* **State the answer.** Since distance cannot be negative, discard the negative solution. The required distances are 60 mi and $60 + 20 = 80$ mi.

*Step 6* **Check.** Since $60^2 + 80^2 = 100^2$, the answer is correct.

**NOW TRY** Exercise 31.

**OBJECTIVE 3** Solve applied problems using area formulas.

## EXAMPLE 4  Solving an Area Problem

A rectangular reflecting pool in a park is 20 ft wide and 30 ft long. The park gardener wants to plant a strip of grass of uniform width around the edge of the pool. She has enough seed to cover 336 ft². How wide will the strip be?

*Step 1*  **Read** the problem carefully.

*Step 2*  **Assign a variable.** The pool is shown in Figure 3.

Let $x$ = the unknown width of the grass strip.

Then $20 + 2x$ = the width of the large rectangle (the width of the pool plus two grass strips),

and $30 + 2x$ = the length of the large rectangle.

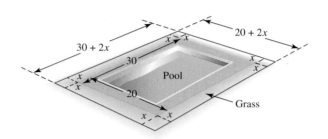

**FIGURE 3**

*Step 3*  **Write an equation.** The area of the large rectangle is given by the product of its length and width, $(30 + 2x)(20 + 2x)$. The area of the pool is $30 \cdot 20 = 600$ ft². The area of the large rectangle minus the area of the pool should equal the area of the grass strip. Since the area of the grass strip is to be 336 ft², the equation is $(30 + 2x)(20 + 2x) - 600 = 336$.

$$\underbrace{(30 + 2x)(20 + 2x)}_{\text{Area of rectangle}} - \underbrace{600}_{\text{Area of pool}} = \underbrace{336}_{\text{Area of grass}}$$

*Step 4*  **Solve.**

$600 + 100x + 4x^2 - 600 = 336$     Multiply.

$4x^2 + 100x - 336 = 0$     Standard form

$x^2 + 25x - 84 = 0$     Divide by 4.

$(x + 28)(x - 3) = 0$     Factor.

$x = -28$   or   $x = 3$     Zero-factor property

*Step 5*  **State the answer.** The width cannot be $-28$ ft, so the grass strip should be 3 ft wide.

*Step 6*  **Check.** If $x = 3$, then the area of the large rectangle (which includes the grass strip) is

$(30 + 2 \cdot 3)(20 + 2 \cdot 3) = 36 \cdot 26 = 936$ ft².     Area of pool and strip

The area of the pool is $30 \cdot 20 = 600$ ft². So, the area of the grass strip is $936 - 600 = 336$ ft², as required. The answer is correct.

**NOW TRY** Exercise 37.

**OBJECTIVE 4** Solve applied problems using quadratic functions as models. Some applied problems can be modeled by *quadratic functions*, which for real numbers $a$, $b$, and $c$, can be written in the form

$$f(x) = ax^2 + bx + c, \quad \text{with } a \neq 0.$$

**EXAMPLE 5** Solving an Applied Problem Using a Quadratic Function

If an object is projected upward from the top of a 144-ft building at 112 ft per sec, its position (in feet above the ground) is given by

$$s(t) = -16t^2 + 112t + 144,$$

where $t$ is time in seconds after it was projected. When does it hit the ground?

When the object hits the ground, its distance above the ground is 0. We must find the value of $t$ that makes $s(t) = 0$.

$$0 = -16t^2 + 112t + 144 \qquad \text{Let } s(t) = 0.$$
$$0 = t^2 - 7t - 9 \qquad \text{Divide by } -16.$$
$$t = \frac{-(-7) \pm \sqrt{(-7)^2 - 4(1)(-9)}}{2(1)} \qquad \text{Substitute into the quadratic formula.}$$
$$t = \frac{7 \pm \sqrt{85}}{2} \approx \frac{7 \pm 9.2}{2} \qquad \text{Use a calculator.}$$

The solutions are $t \approx 8.1$ or $t \approx -1.1$. Time cannot be negative, so we discard the negative solution. The object hits the ground about 8.1 sec after it is projected.

**NOW TRY** Exercise 43.

**EXAMPLE 6** Using a Quadratic Function to Model Company Bankruptcy Filings

The number of publicly traded companies filing for bankruptcy was high in the early 1990s due to an economic recession. The number then declined during the middle 1990s, and in recent years has increased again. The quadratic function defined by

$$f(x) = 3.37x^2 - 28.6x + 133$$

approximates the number of company bankruptcy filings during the years 1990–2001, where $x$ is the number of years since 1990. (*Source:* www.BankruptcyData.com)

**(a)** Use the model to approximate the number of company bankruptcy filings in 1995. For 1995, $x = 5$, so find $f(5)$.

$$f(5) = 3.37(5)^2 - 28.6(5) + 133 \qquad \text{Let } x = 5.$$
$$= 74.25$$

There were about 74 company bankruptcy filings in 1995.

**(b)** In what year did company bankruptcy filings reach 150? Find the value of $x$ that makes $f(x) = 150$.

$$f(x) = 3.37x^2 - 28.6x + 133$$
$$150 = 3.37x^2 - 28.6x + 133 \qquad \text{Let } f(x) = 150.$$
$$0 = 3.37x^2 - 28.6x - 17 \qquad \text{Standard form}$$

Now use $a = 3.37$, $b = -28.6$, and $c = -17$ in the quadratic formula.

$$x = \frac{-(-28.6) \pm \sqrt{(-28.6)^2 - 4(3.37)(-17)}}{2(3.37)}$$

$x \approx 9.0$ or $x \approx -0.56$   Use a calculator.

The positive solution is $x \approx 9$, so company bankruptcy filings reached 150 in the year $1990 + 9 = 1999$. (Reject the negative solution since the model is not valid for negative values of $x$.) Note that company bankruptcy filings doubled from about 74 in 1995 to 150 in 1999.

**NOW TRY** Exercises 53 and 55.

## 9.4 Exercises

*Concept Check* Answer each question in Exercises 1–4.

1. In solving a formula that has the specified variable in the denominator, what is the first step?
2. What is the first step in solving a formula like $gw^2 = 2r$ for $w$?
3. What is the first step in solving a formula like $gw^2 = kw + 24$ for $w$?
4. Why is it particularly important to check all proposed solutions to an applied problem against the information in the original problem?

*In Exercises 5 and 6, solve for $m$ in terms of the other variables ($m > 0$).*

5.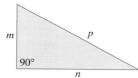

6.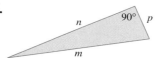

*Solve each equation for the indicated variable. (Leave $\pm$ in your answers.) See Examples 1 and 2.*

7. $d = kt^2$ for $t$
8. $s = kwd^2$ for $d$
9. $I = \dfrac{ks}{d^2}$ for $d$
10. $R = \dfrac{k}{d^2}$ for $d$
11. $F = \dfrac{kA}{v^2}$ for $v$
12. $L = \dfrac{kd^4}{h^2}$ for $h$
13. $V = \dfrac{1}{3}\pi r^2 h$ for $r$
14. $V = \pi(r^2 + R^2)h$ for $r$
15. $At^2 + Bt = -C$ for $t$
16. $S = 2\pi rh + \pi r^2$ for $r$
17. $D = \sqrt{kh}$ for $h$
18. $F = \dfrac{k}{\sqrt{d}}$ for $d$
19. $p = \sqrt{\dfrac{k\ell}{g}}$ for $\ell$
20. $p = \sqrt{\dfrac{k\ell}{g}}$ for $g$
21. $S = 4\pi r^2$ for $r$

22. Refer to Example 2 of this section. Suppose that $k$ and $s$ both represent positive numbers.
    (a) Which one of the two solutions given is positive?
    (b) Which one is negative?   (c) How can you tell?

*Solve each equation for the indicated variable.*

23. $p = \dfrac{E^2 R}{(r + R)^2}$ for $R$   $(E > 0)$
24. $S(6S - t) = t^2$ for $S$

**25.** $10p^2c^2 + 7pcr = 12r^2$ for $r$

**26.** $S = vt + \frac{1}{2}gt^2$ for $t$

**27.** $LI^2 + RI + \frac{1}{c} = 0$ for $I$

**28.** $P = EI - RI^2$ for $I$

*Solve each problem. When appropriate, round answers to the nearest tenth. See Example 3.*

**29.** Find the lengths of the sides of the triangle.

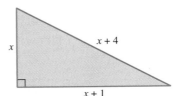

**30.** Find the lengths of the sides of the triangle.

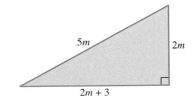

**31.** Two ships leave port at the same time, one heading due south and the other heading due east. Several hours later, they are 170 mi apart. If the ship traveling south traveled 70 mi farther than the other ship, how many miles did they each travel?

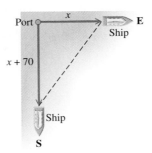

**32.** Emily McCants is flying a kite that is 30 ft farther above her hand than its horizontal distance from her. The string from her hand to the kite is 150 ft long. How high is the kite?

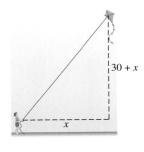

**33.** A game board is in the shape of a right triangle. The hypotenuse is 2 inches longer than the longer leg, and the longer leg is 1 inch less than twice as long as the shorter leg. How long is each side of the game board?

**34.** Sedona Levy is planting a vegetable garden in the shape of a right triangle. The longer leg is 3 ft longer than the shorter leg, and the hypotenuse is 3 ft longer than the longer leg. Find the lengths of the three sides of the garden.

**35.** The diagonal of a rectangular rug measures 26 ft, and the length is 4 ft more than twice the width. Find the length and width of the rug.

**36.** A 13-ft ladder is leaning against a house. The distance from the bottom of the ladder to the house is 7 ft less than the distance from the top of the ladder to the ground. How far is the bottom of the ladder from the house?

*Solve each problem. See Example 4.*

**37.** A club swimming pool is 30 ft wide and 40 ft long. The club members want an exposed aggregate border in a strip of uniform width around the pool. They have enough material for 296 ft². How wide can the strip be?

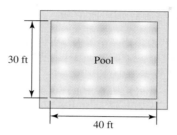

**38.** A couple wants to buy a rug for a room that is 20 ft long and 15 ft wide. They want to leave an even strip of flooring uncovered around the edges of the room. How wide a strip will they have if they buy a rug with an area of 234 ft²?

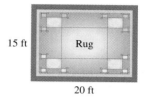

**39.** A rectangle has a length 2 m less than twice its width. When 5 m are added to the width, the resulting figure is a square with an area of 144 m². Find the dimensions of the original rectangle.

**40.** Ahmad's backyard measures 20 m by 30 m. He wants to put a flower garden in the middle of the backyard, leaving a strip of grass of uniform width around the flower garden. Ahmad must have 184 m² of grass. Under these conditions, what will the length and width of the garden be?

**41.** A rectangular piece of sheet metal has a length that is 4 in. less than twice the width. A square piece 2 in. on a side is cut from each corner. The sides are then turned up to form an uncovered box of volume 256 in.³. Find the length and width of the original piece of metal.

**42.** Another rectangular piece of sheet metal is 2 in. longer than it is wide. A square piece 3 in. on a side is cut from each corner. The sides are then turned up to form an uncovered box of volume 765 in.³. Find the dimensions of the original piece of metal.

*Solve each problem. When appropriate, round answers to the nearest tenth. See Example 5.*

**43.** An object is projected directly upward from the ground. After $t$ seconds its distance in feet above the ground is

$$s(t) = 144t - 16t^2.$$

After how many seconds will the object be 128 ft above the ground? (*Hint:* Look for a common factor before solving the equation.)

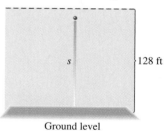

**44.** When does the object in Exercise 43 strike the ground?

**45.** A ball is projected upward from the ground. Its distance in feet from the ground in $t$ seconds is given by

$$s(t) = -16t^2 + 128t.$$

At what times will the ball be 213 ft from the ground?

**46.** A toy rocket is launched from ground level. Its distance in feet from the ground in $t$ seconds is given by

$$s(t) = -16t^2 + 208t.$$

At what times will the rocket be 550 ft from the ground?

**47.** The function defined by

$$D(t) = 13t^2 - 100t$$

gives the distance in feet a car going approximately 68 mph will skid in $t$ seconds. Find the time it would take for the car to skid 180 ft.

**48.** The function given in Exercise 47 becomes $D(t) = 13t^2 - 73t$ for a car going 50 mph. Find the time it takes for this car to skid 218 ft.

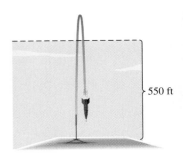

A rock is projected upward from ground level, and its distance in feet from the ground in $t$ seconds is given by $s(t) = -16t^2 + 160t$. Use algebra and a short explanation to answer Exercises 49 and 50.

**49.** After how many seconds does it reach a height of 400 ft? How would you describe in words its position at this height?

**50.** After how many seconds does it reach a height of 425 ft? How would you interpret the mathematical result here?

*Recall from geometry that corresponding sides of similar triangles are proportional. Use this fact to find the lengths of the indicated sides of each pair of similar triangles. Check all possible solutions in both triangles. Sides of a triangle cannot be negative (and are not drawn to scale here).*

**51.** Side $AC$

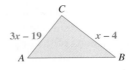

**52.** Side $RQ$

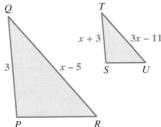

The total number of miles traveled by all motor vehicles in the United States for the years 1994–2003 are shown in the bar graph and can be modeled by the quadratic function defined by

$$f(x) = -1.705x^2 + 75.93x + 2351.$$

Here, $x = 0$ represents 1994, $x = 1$ represents 1995, and so on. Use the graph and the model to work Exercises 53–56. See Example 6.

**53.** (a) Use the graph to estimate miles traveled in 2000 to the nearest ten billion.
(b) Use the model to approximate miles traveled in 2000 to the nearest ten billion. How does this result compare to your estimate from part (a)?

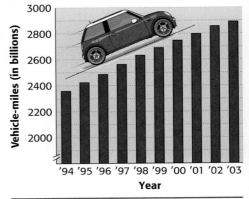

*Source:* U.S. Federal Highway Administration.

**54.** Based on the model, in what year did miles traveled reach 2600 billion? (Round down to the nearest year.) How does this result compare to the vehicle-miles shown in the graph?

**55.** Based on the model, in what year did miles traveled reach 2800 billion? (Round down to the nearest year.) How does this result compare to the vehicle-miles shown in the graph?

**56.** If these data were modeled by a *linear* function defined by $f(x) = ax + b$, would the value of $a$ be positive or negative? Explain.

# 9.5 Quadratic and Rational Inequalities

**OBJECTIVES**

1. Solve quadratic inequalities.
2. Solve polynomial inequalities of degree 3 or greater.
3. Solve rational inequalities.

Now we combine the methods of solving linear inequalities with the methods of solving quadratic equations to solve *quadratic inequalities*.

### Quadratic Inequality

A **quadratic inequality** can be written in the form
$$ax^2 + bx + c < 0 \quad \text{or} \quad ax^2 + bx + c > 0,$$
where $a$, $b$, and $c$ are real numbers, with $a \neq 0$.

As before, the symbols $<$ and $>$ may be replaced with $\leq$ and $\geq$.

**OBJECTIVE 1** Solve quadratic inequalities.

**EXAMPLE 1** Solving a Quadratic Inequality Using Test Numbers

Solve $x^2 - x - 12 > 0$.

Solve the quadratic equation $x^2 - x - 12 = 0$ by factoring.
$$(x - 4)(x + 3) = 0$$
$$x - 4 = 0 \quad \text{or} \quad x + 3 = 0$$
$$x = 4 \quad \text{or} \quad x = -3$$

The numbers 4 and $-3$ divide a number line into the three intervals shown in Figure 4. ***Be careful to put the lesser number on the left.***

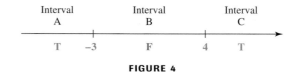

**FIGURE 4**

The numbers 4 and $-3$ are the only numbers that make the expression $x^2 - x - 12$ equal to 0. All other numbers make the expression either positive or negative. The sign of the expression can change from positive to negative or from negative to positive only at a number that makes it 0. Therefore, if one number in an interval satisfies the inequality, then all the numbers in that interval will satisfy the inequality.

To see if the numbers in Interval A satisfy the inequality, choose any number from Interval A in Figure 4 (that is, any number less than $-3$). Substitute this test number for $x$ in the original inequality $x^2 - x - 12 > 0$. If the result is *true*, then all numbers in Interval A satisfy the inequality.

Try $-5$ from Interval A. Substitute $-5$ for $x$.

$$x^2 - x - 12 > 0 \quad \text{Original inequality}$$
$$(-5)^2 - (-5) - 12 > 0 \quad ?$$
$$25 + 5 - 12 > 0 \quad ?$$
$$18 > 0 \quad \text{True}$$

**Use parentheses to avoid sign errors.**

Because $-5$ from Interval A satisfies the inequality, all numbers from Interval A are solutions.

Now try 0 from Interval B. If $x = 0$, then

$$0^2 - 0 - 12 > 0 \quad ?$$
$$-12 > 0. \quad \text{False}$$

The numbers in Interval B are *not* solutions. Verify that the test number 5 satisfies the inequality, so the numbers in Interval C are also solutions.

Based on these results (shown by the colored letters in Figure 4), the solution set includes the numbers in Intervals A and C, as shown on the graph in Figure 5. The solution set is written in interval notation as

$$(-\infty, -3) \cup (4, \infty).$$

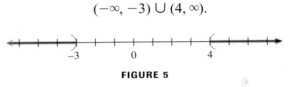

**FIGURE 5**

**NOW TRY** Exercise 11.

If the inequality in Example 1 had used $\geq$, the solution set would have included the solutions of the *equation* and been written in interval notation as $(-\infty, -3] \cup [4, \infty)$.

In summary, follow these steps to solve a quadratic inequality.

---

**Solving a Quadratic Inequality**

*Step 1* **Write the inequality as an equation and solve it.**

*Step 2* **Use the solutions from Step 1 to determine intervals.** Graph the numbers found in Step 1 on a number line. These numbers divide the number line into intervals.

*Step 3* **Find the intervals that satisfy the inequality.** Substitute a test number from each interval into the original inequality to determine the intervals that satisfy the inequality. All numbers in those intervals are in the solution set. A graph of the solution set will usually look like one of these. (Square brackets might be used instead of parentheses.)

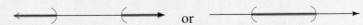

*Step 4* **Consider the endpoints separately.** The numbers from Step 1 are included in the solution set if the inequality symbol is $\leq$ or $\geq$; they are not included if it is $<$ or $>$.

---

Special cases of quadratic inequalities may occur, as in the next example.

### EXAMPLE 2   Solving Special Cases

Solve each inequality.

**(a)** $(2x - 3)^2 > -1$

Because $(2x - 3)^2$ is never negative, it is always greater than $-1$. Thus, the solution for $(2x - 3)^2 > -1$ is the set of all real numbers, $(-\infty, \infty)$.

**(b)** $(2x - 3)^2 < -1$

Using the same reasoning as in part (a), there is no solution for this inequality. The solution set is $\emptyset$.

**NOW TRY** Exercises 25 and 27.

**OBJECTIVE 2** Solve polynomial inequalities of degree 3 or greater. Higher-degree polynomial inequalities that can be factored are solved in the same way as quadratic inequalities.

### EXAMPLE 3   Solving a Third-Degree Polynomial Inequality

Solve $(x - 1)(x + 2)(x - 4) \leq 0$.

This is a *cubic* (third-degree) inequality rather than a quadratic inequality, but it can be solved using the method shown in the box by extending the zero-factor property to more than two factors. Begin by setting the factored polynomial *equal* to 0 and solving the equation. (Step 1)

$$(x - 1)(x + 2)(x - 4) = 0$$

$x - 1 = 0$  or  $x + 2 = 0$  or  $x - 4 = 0$

$x = 1$  or  $x = -2$  or  $x = 4$

Locate the numbers $-2$, 1, and 4 on a number line, as in Figure 6, to determine the Intervals A, B, C, and D. (Step 2)

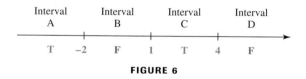

**FIGURE 6**

Substitute a test number from each interval in the *original* inequality to determine which intervals satisfy the inequality. (Step 3) Use a table to organize this information.

| Interval | Test Number | Test of Inequality | True or False? |
|---|---|---|---|
| A | $-3$ | $-28 \leq 0$ | T |
| B | 0 | $8 \leq 0$ | F |
| C | 2 | $-8 \leq 0$ | T |
| D | 5 | $28 \leq 0$ | F |

Verify the information given in the table and graphed in Figure 7 on the next page. The numbers in Intervals A and C are in the solution set, which is written in interval notation as

$$(-\infty, -2] \cup [1, 4].$$

The three endpoints are included since the inequality symbol is ≤. (Step 4)

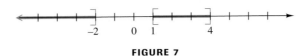

**FIGURE 7**

NOW TRY Exercise 29.

**OBJECTIVE 3** Solve rational inequalities. Inequalities that involve rational expressions, called **rational inequalities,** are solved similarly using the following steps.

### Solving a Rational Inequality

*Step 1* **Write the inequality** so that 0 is on one side and there is a single fraction on the other side.

*Step 2* **Determine the numbers** that make the numerator or denominator equal to 0.

*Step 3* **Divide a number line into intervals.** Use the numbers from Step 2.

*Step 4* **Find the intervals that satisfy the inequality.** Test a number from each interval by substituting it into the *original* inequality.

*Step 5* **Consider the endpoints separately.** Exclude any values that make the denominator 0.

**CAUTION** As indicated in Step 5, any number that makes the denominator 0 *must* be excluded from the solution set.

### EXAMPLE 4 Solving a Rational Inequality

Solve $\dfrac{-1}{x-3} > 1$.

Write the inequality so that 0 is on one side. (Step 1)

$$\dfrac{-1}{x-3} - 1 > 0 \qquad \text{Subtract 1.}$$

$$\dfrac{-1}{x-3} - \dfrac{x-3}{x-3} > 0 \qquad \text{Use } x - 3 \text{ as the common denominator.}$$

**Be careful with signs.**

$$\dfrac{-1-x+3}{x-3} > 0 \qquad \text{Write the left side as a single fraction.}$$

$$\dfrac{-x+2}{x-3} > 0 \qquad \text{Combine terms.}$$

The sign of the rational expression $\dfrac{-x+2}{x-3}$ will change from positive to negative or negative to positive only at those numbers that make the numerator or denominator 0. The number 2 makes the numerator 0, and 3 makes the denominator 0. (Step 2) These two numbers, 2 and 3, divide a number line into three intervals. See Figure 8. (Step 3)

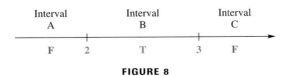

**FIGURE 8**

Testing a number from each interval in the *original* inequality, $\frac{-1}{x-3} > 1$, gives the results shown in the table. (Step 4)

| Interval | Test Number | Test of Inequality | True or False? |
| --- | --- | --- | --- |
| A | 0 | $\frac{1}{3} > 1$ | F |
| B | 2.5 | $2 > 1$ | T |
| C | 4 | $-1 > 1$ | F |

The solution set is the interval $(2, 3)$. This interval does not include 3 since it would make the denominator of the original equality 0; 2 is not included either since the inequality symbol is $>$. (Step 5) A graph of the solution set is given in Figure 9.

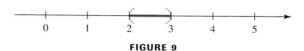

**FIGURE 9**

NOW TRY Exercise 37.

### EXAMPLE 5  Solving a Rational Inequality

Solve $\frac{x-2}{x+2} \leq 2$.

Write the inequality so that 0 is on one side. (Step 1)

$$\frac{x-2}{x+2} - 2 \leq 0 \quad \text{Subtract 2.}$$

$$\frac{x-2}{x+2} - \frac{2(x+2)}{x+2} \leq 0 \quad \text{Use } x + 2 \text{ as the common denominator.}$$

**Be careful with signs.**

$$\frac{x - 2 - 2x - 4}{x+2} \leq 0 \quad \text{Write as a single fraction.}$$

$$\frac{-x - 6}{x+2} \leq 0 \quad \text{Combine terms.}$$

The number $-6$ makes the numerator 0, and $-2$ makes the denominator 0. (Step 2) These two numbers determine three intervals. (Step 3) Test one number from each interval (Step 4) to see that the solution set is

$$(-\infty, -6] \cup (-2, \infty).$$

The number $-6$ satisfies the original inequality, but $-2$ cannot be used as a solution since it makes the denominator 0. (Step 5) A graph of the solution set is shown in Figure 10.

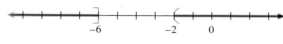

**FIGURE 10**

NOW TRY Exercise 43.

## 9.5 Exercises

*True or False* Choose a test number from the given interval and decide whether the inequality is true or false.

1. $x^2 - 2x - 3 > 0$; interval: $(-1, 3)$
2. $x^2 + 2x - 3 < 0$; interval: $(-\infty, -3)$
3. $2x^2 - 7x - 15 > 0$; interval: $(5, \infty)$
4. $2x^2 + x - 10 < 0$; interval: $\left(-\frac{5}{2}, 2\right)$

5. Explain how to determine whether to include or exclude endpoints when solving a quadratic or greater-degree inequality.

6. *Concept Check* The solution set of the inequality $x^2 + x - 12 < 0$ is the interval $(-4, 3)$. Without actually performing any work, give the solution set of the inequality $x^2 + x - 12 \geq 0$.

*Solve each inequality and graph the solution set. See Example 1. (Hint: In Exercises 23 and 24, use the quadratic formula.)*

7. $(x + 1)(x - 5) > 0$
8. $(x + 6)(x - 2) > 0$
9. $(x + 4)(x - 6) < 0$
10. $(x + 4)(x - 8) < 0$
11. $x^2 - 4x + 3 \geq 0$
12. $x^2 - 3x - 10 \geq 0$
13. $10x^2 + 9x \geq 9$
14. $3x^2 + 10x \geq 8$
15. $4x^2 - 9 \leq 0$
16. $9x^2 - 25 \leq 0$
17. $6x^2 + x \geq 1$
18. $4x^2 + 7x \geq -3$
19. $z^2 - 4z \geq 0$
20. $x^2 + 2x < 0$
21. $3k^2 - 5k \leq 0$
22. $2z^2 + 3z > 0$
23. $x^2 - 6x + 6 \geq 0$
24. $3k^2 - 6k + 2 \leq 0$

*Solve each inequality. See Example 2.*

25. $(4 - 3x)^2 \geq -2$
26. $(6x + 7)^2 \geq -1$
27. $(3x + 5)^2 \leq -4$
28. $(8x + 5)^2 \leq -5$

*Solve each inequality and graph the solution set. See Example 3.*

29. $(x - 1)(x - 2)(x - 4) < 0$
30. $(2x + 1)(3x - 2)(4x + 7) < 0$
31. $(x - 4)(2x + 3)(3x - 1) \geq 0$
32. $(x + 2)(4x - 3)(2x + 7) \geq 0$

*Solve each inequality and graph the solution set. See Examples 4 and 5.*

33. $\dfrac{x - 1}{x - 4} > 0$
34. $\dfrac{x + 1}{x - 5} > 0$
35. $\dfrac{2x + 3}{x - 5} \leq 0$
36. $\dfrac{3x + 7}{x - 3} \leq 0$
37. $\dfrac{8}{x - 2} \geq 2$
38. $\dfrac{20}{x - 1} \geq 1$
39. $\dfrac{3}{2x - 1} < 2$
40. $\dfrac{6}{x - 1} < 1$
41. $\dfrac{x - 3}{x + 2} \geq 2$
42. $\dfrac{m + 4}{m + 5} \geq 2$
43. $\dfrac{x - 8}{x - 4} < 3$
44. $\dfrac{2t - 3}{t + 1} > 4$
45. $\dfrac{4k}{2k - 1} < k$
46. $\dfrac{r}{r + 2} < 2r$
47. $\dfrac{2x - 3}{x^2 + 1} \geq 0$
48. $\dfrac{9x - 8}{4x^2 + 25} < 0$
49. $\dfrac{(3x - 5)^2}{x + 2} > 0$
50. $\dfrac{(5x - 3)^2}{2x + 1} \leq 0$

# 9 SUMMARY

## KEY TERMS

**9.1** quadratic equation

**9.2** quadratic formula  
discriminant

**9.3** quadratic in form

**9.5** quadratic inequality  
rational inequality

## QUICK REVIEW

--- CONCEPTS --- | --- EXAMPLES ---

### 9.1 The Square Root Property and Completing the Square

**Square Root Property**

If $x$ and $k$ are complex numbers and $x^2 = k$, then

$$x = \sqrt{k} \quad \text{or} \quad x = -\sqrt{k}.$$

**Completing the Square**

To solve $ax^2 + bx + c = 0 \ (a \neq 0)$:

**Step 1** If $a \neq 1$, divide each side by $a$.

**Step 2** Write the equation with the variable terms on one side and the constant on the other.

**Step 3** Take half the coefficient of $x$ and square it.

**Step 4** Add the square to each side.

**Step 5** Factor the perfect square trinomial, and write it as the square of a binomial. Simplify the other side.

**Step 6** Use the square root property to complete the solution.

Solve $(x - 1)^2 = 8$.

$x - 1 = \sqrt{8} \quad \text{or} \quad x - 1 = -\sqrt{8}$

$x = 1 + 2\sqrt{2} \quad \text{or} \quad x = 1 - 2\sqrt{2}$

The solution set is $\{1 + 2\sqrt{2}, 1 - 2\sqrt{2}\}$.

Solve $2x^2 - 4x - 18 = 0$.

$x^2 - 2x - 9 = 0 \qquad$ Divide by 2.

$x^2 - 2x = 9 \qquad$ Add 9.

$\left[\frac{1}{2}(-2)\right]^2 = (-1)^2 = 1$

$x^2 - 2x + 1 = 9 + 1 \qquad$ Add 1.

$(x - 1)^2 = 10 \qquad$ Factor.

$x - 1 = \sqrt{10} \quad \text{or} \quad x - 1 = -\sqrt{10} \qquad$ Square root property

$x = 1 + \sqrt{10} \quad \text{or} \quad x = 1 - \sqrt{10}$

The solution set is $\{1 + \sqrt{10}, 1 - \sqrt{10}\}$.

### 9.2 The Quadratic Formula

**Quadratic Formula**

The solutions of $ax^2 + bx + c = 0 \ (a \neq 0)$ are given by

$$x = \frac{-b \pm \sqrt{b^2 - 4ac}}{2a}.$$

Solve $3x^2 + 5x + 2 = 0$.

$$x = \frac{-5 \pm \sqrt{5^2 - 4(3)(2)}}{2(3)} = \frac{-5 \pm 1}{6}$$

$x = \frac{-5 + 1}{6} = -\frac{2}{3} \quad \text{or} \quad x = \frac{-5 - 1}{6} = -1$

The solution set is $\left\{-1, -\frac{2}{3}\right\}$.

*(continued)*

## CONCEPTS

### The Discriminant

If $a$, $b$, and $c$ are integers, then the discriminant, $b^2 - 4ac$, of $ax^2 + bx + c = 0$ determines the number and type of solutions as follows.

| Discriminant | Number and Type of Solutions |
|---|---|
| Positive, the square of an integer | Two rational solutions |
| Positive, not the square of an integer | Two irrational solutions |
| Zero | One rational solution |
| Negative | Two nonreal complex solutions |

### 9.3 Equations Quadratic in Form

A nonquadratic equation that can be written in the form

$$au^2 + bu + c = 0,$$

for $a \neq 0$ and an algebraic expression $u$, is called quadratic in form. Substitute $u$ for the expression, solve for $u$, and then solve for the variable in the expression.

### 9.4 Formulas and Further Applications

To solve a formula for a squared variable, proceed as follows.

**(a) If the variable appears only to the second power:** Isolate the squared variable on one side of the equation, and then use the square root property.

**(b) If the variable appears to the first and second powers:** Write the equation in standard form, and then use the quadratic formula.

## EXAMPLES

For $x^2 + 3x - 10 = 0$, the discriminant is

$3^2 - 4(1)(-10) = 49.$   Two rational solutions

For $4x^2 + x + 1 = 0$, the discriminant is

$1^2 - 4(4)(1) = -15.$   Two nonreal complex solutions

Solve $3(x + 5)^2 + 7(x + 5) + 2 = 0$.

$3u^2 + 7u + 2 = 0$   Let $u = x + 5$.

$(3u + 1)(u + 2) = 0$   Factor.

$u = -\dfrac{1}{3}$   or   $u = -2$

$x + 5 = -\dfrac{1}{3}$   or   $x + 5 = -2$   $x + 5 = u$

$x = -\dfrac{16}{3}$   or   $x = -7$

The solution set is $\left\{-7, -\dfrac{16}{3}\right\}$.

Solve $A = \dfrac{2mp}{r^2}$ for $r$.

$r^2 A = 2mp$   Multiply by $r^2$.

$r^2 = \dfrac{2mp}{A}$   Divide by $A$.

$r = \pm\sqrt{\dfrac{2mp}{A}}$   Square root property

$r = \dfrac{\pm\sqrt{2mpA}}{A}$   Rationalize the denominator.

Solve $x^2 + rx = t$ for $x$.

$x^2 + rx - t = 0$   Standard form

$x = \dfrac{-r \pm \sqrt{r^2 - 4(1)(-t)}}{2(1)}$   $a = 1, b = r, c = -t$

$x = \dfrac{-r \pm \sqrt{r^2 + 4t}}{2}$

*(continued)*

## CONCEPTS

### 9.5 Quadratic and Rational Inequalities

**Solving a Quadratic (or Higher-Degree Polynomial) Inequality**

**Step 1** Write the inequality as an equation and solve.

**Step 2** Use the numbers found in Step 1 to divide a number line into intervals.

**Step 3** Substitute a test number from each interval into the original inequality to determine the intervals that belong to the solution set.

**Step 4** Consider the endpoints separately.

**Solving a Rational Inequality**

**Step 1** Write the inequality so that 0 is on one side and there is a single fraction on the other side.

**Step 2** Determine the numbers that make the numerator or denominator 0.

**Step 3** Use the numbers from Step 2 to divide a number line into intervals.

**Step 4** Substitute a test number from each interval into the original inequality to determine the intervals that belong to the solution set.

**Step 5** Consider the endpoints separately.

## EXAMPLES

Solve $2x^2 + 5x + 2 < 0$.

$$2x^2 + 5x + 2 = 0$$
$$(2x + 1)(x + 2) = 0$$
$$x = -\tfrac{1}{2} \quad \text{or} \quad x = -2$$

Intervals: $(-\infty, -2)$, $\left(-2, -\tfrac{1}{2}\right)$, $\left(-\tfrac{1}{2}, \infty\right)$

Test values: $-3, -1, 0$

$x = -3$ makes the original inequality false; $x = -1$ makes it true; $x = 0$ makes it false. Choose the interval(s) which yield(s) a true statement. The solution set is the interval $\left(-2, -\tfrac{1}{2}\right)$.

Solve $\dfrac{x}{x + 2} \geq 4$.

$$\dfrac{x}{x + 2} - 4 \geq 0 \quad \text{Subtract 4.}$$

$$\dfrac{x}{x + 2} - \dfrac{4(x + 2)}{x + 2} \geq 0 \quad \text{Write with a common denominator.}$$

$$\dfrac{-3x - 8}{x + 2} \geq 0 \quad \text{Subtract fractions.}$$

$-\tfrac{8}{3}$ makes the numerator 0; $-2$ makes the denominator 0.

$-4$ from A makes the original inequality false; $-\tfrac{7}{3}$ from B makes it true; 0 from C makes it false.

The solution set is the interval $\left[-\tfrac{8}{3}, -2\right)$. The endpoint $-2$ is not included since it makes the denominator 0.

---

# 9 REVIEW EXERCISES

*Exercises marked * require knowledge of the complex number system.*

*Solve each equation by using the square root property or completing the square.*

**1.** $t^2 = 121$     **2.** $p^2 = 3$     **3.** $(2x + 5)^2 = 100$

***4.** $(3k - 2)^2 = -25$     **5.** $x^2 + 4x = 15$     **6.** $2m^2 - 3m = -1$

7. *Concept Check* A student gave the following "solution" to the equation $x^2 = 12$.

$$x^2 = 12$$
$$x = \sqrt{12} \quad \text{Square root property}$$
$$x = 2\sqrt{3}$$

Solution set: $\{2\sqrt{3}\}$

The answer is not correct. **WHAT WENT WRONG?** Give the correct solution set.

8. Navy Pier Center in Chicago, Illinois, features a 150-ft tall Ferris wheel. Use Galileo's formula $d = 16t^2$ to find how long it would take an MP-3 player dropped from the top of the Ferris wheel to fall to the ground. Round your answer to the nearest tenth of a second. (*Source: Microsoft Encarta Encyclopedia.*)

*Multiple Choice* Use the discriminant to predict whether the solutions to each equation are

**A.** *two rational numbers;*      **B.** *one rational number;*
**C.** *two irrational numbers;*      **D.** *two nonreal complex numbers.*

9. $x^2 + 5x + 2 = 0$      10. $4t^2 = 3 - 4t$
11. $4x^2 = 6x - 8$      12. $9z^2 + 30z + 25 = 0$

*Solve each equation by using the quadratic formula.*

13. $2x^2 + x - 21 = 0$     14. $k^2 + 5k = 7$     15. $(t + 3)(t - 4) = -2$

*16. $2x^2 + 3x + 4 = 0$     *17. $3p^2 = 2(2p - 1)$     18. $m(2m - 7) = 3m^2 + 3$

*Solve each equation. Check your solutions.*

19. $\dfrac{15}{x} = 2x - 1$      20. $\dfrac{1}{n} + \dfrac{2}{n+1} = 2$

21. $-2r = \sqrt{\dfrac{48 - 20r}{2}}$      22. $8(3x + 5)^2 + 2(3x + 5) - 1 = 0$

23. $2x^{2/3} - x^{1/3} - 28 = 0$      24. $p^4 - 10p^2 + 9 = 0$

*Solve each problem. Round answers to the nearest tenth, as necessary.*

25. Phong paddled his canoe 20 mi upstream, then paddled back. If the speed of the current was 3 mph and the total trip took 7 hr, what was Phong's speed?

26. Maureen O'Connor drove 8 mi to pick up her friend Laurie, and then drove 11 mi to a mall at a speed 15 mph faster. If Maureen's total travel time was 24 min, what was her speed on the trip to pick up Laurie?

27. An old machine processes a batch of checks in 1 hr more time than a new one. How long would it take the old machine to process a batch of checks that the two machines together process in 2 hr?

28. Greg Tobin can process a stack of invoices 1 hr faster than Carter Fenton can. Working together, they take 1.5 hr. How long would it take each person working alone?

*Solve each formula for the indicated variable. (Give answers with $\pm$.)*

29. $k = \dfrac{rF}{wv^2}$ for $v$      30. $p = \sqrt{\dfrac{yz}{6}}$ for $y$      31. $mt^2 = 3mt + 6$ for $t$

*Solve each problem. Round answers to the nearest tenth, as necessary.*

**32.** A large machine requires a part in the shape of a right triangle with a hypotenuse 9 ft less than twice the length of the longer leg. The shorter leg must be $\frac{3}{4}$ the length of the longer leg. Find the lengths of the three sides of the part.

**33.** A square has an area of 256 cm². If the same amount is removed from one dimension and added to the other, the resulting rectangle has an area 16 cm² less. Find the dimensions of the rectangle.

**34.** Nancy wants to buy a mat for a photograph that measures 14 in. by 20 in. She wants to have an even border around the picture when it is mounted on the mat. If the area of the mat she chooses is 352 in.², how wide will the border be?

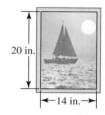

**35.** A searchlight moves horizontally back and forth along a wall with the distance of the light from a starting point at $t$ minutes given by the quadratic function defined by

$$f(t) = 100t^2 - 300t.$$

How long will it take before the light returns to the starting point?

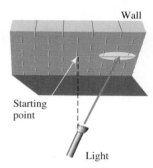

**36.** Aria (formerly the Lewis Tower), built in Philadelphia, Pennsylvania, in 1929, is 400 ft high. Suppose that a ball is projected upward from the top of the building, and its position in feet above the ground is given by the quadratic function defined by

$$f(t) = -16t^2 + 45t + 400,$$

where $t$ is the number of seconds elapsed. How long will it take for the ball to reach a height of 200 ft above the ground? (*Source: World Almanac and Book of Facts.*)

**37.** The Alberta Stock Exchange building in Calgary, Alberta, is 407 ft high. Suppose that a ball is projected upward from the top of the building, and its position in feet above the ground is given by the quadratic function defined by

$$s(t) = -16t^2 + 75t + 407,$$

where $t$ is the number of seconds elapsed. How long will it take for the ball to reach a height of 450 ft above the ground? (*Source: World Almanac and Book of Facts.*)

**38.** The manager of a restaurant has determined that the demand for frozen yogurt is $\frac{25}{p}$ units per day, where $p$ is the price (in dollars) per unit. The supply is $70p + 15$ units per day. Find the price at which supply and demand are equal.

**39.** Use the formula $A = P(1 + r)^2$ to find the interest rate $r$ at which a principal $P$ of $10,000 will increase to $10,920.25 in 2 yr.

**40.** *True or False*  Determine whether a test value from the interval $(-7, 4)$ makes the inequality $x^2 + 3x - 28 > 0$ *true* or *false*.

*Solve each inequality and graph the solution set.*

**41.** $x^2 + x \leq 12$

**42.** $(x - 4)(2x + 3) > 0$

**43.** $(4m + 3)^2 \leq -4$

**44.** $(x + 2)(x - 3)(x + 5) \leq 0$

**45.** $\dfrac{3t + 4}{t - 2} \leq 1$

**46.** $\dfrac{6}{2z - 1} < 2$

# CHAPTER 9 TEST

*Problems marked * require knowledge of the complex number system.*

*Solve each equation by using the square root property or completing the square.*

**1.** $t^2 = 54$

**2.** $(7x + 3)^2 = 25$

**3.** $2x^2 + 4x = 8$

*Solve by using the quadratic formula.*

**4.** $2x^2 - 3x - 1 = 0$

***5.** $3t^2 - 4t = -5$

**6.** $(2x + 1)(3x + 1) = 4$

***7.** *Multiple Choice*  If $k$ is a negative number, then which one of the following equations will have two nonreal complex solutions?

**A.** $x^2 = 4k$
**B.** $x^2 = -4k$
**C.** $(x + 2)^2 = -k$
**D.** $x^2 + k = 0$

**8.** What is the discriminant for $2x^2 - 8x - 3 = 0$? How many and what type of solutions does this equation have? (Do not actually solve.)

*Solve by any method.*

**9.** $3 - \dfrac{16}{x} - \dfrac{12}{x^2} = 0$

**10.** $4x^2 + 7x - 3 = 0$

**11.** $3x = \sqrt{\dfrac{9x + 2}{2}}$

**12.** $9x^4 + 4 = 37x^2$

**13.** $12 = (2n + 1)^2 + (2n + 1)$

**14.** Solve $S = 4\pi r^2$ for $r$. (Leave $\pm$ in your answer.)

*Solve each problem.*

**15.** Andrew and Kent do word processing. For a certain prospectus, Kent can prepare it 2 hr faster than Andrew can. If they work together, they can do the entire prospectus in 5 hr. How long will it take each of them working alone to prepare the prospectus? Round your answers to the nearest tenth of an hour.

**16.** Abby Tanenbaum paddled her canoe 10 mi upstream and then paddled back to her starting point. If the rate of the current was 3 mph and the entire trip took $3\tfrac{1}{2}$ hr, what was Abby's rate?

**17.** Tyler McGinnis has a pool 24 ft long and 10 ft wide. He wants to construct a concrete walk around the pool. If he plans for the walk to be of uniform width and cover 152 ft², what will the width of the walk be?

**18.** At a point 30 m from the base of a tower, the distance to the top of the tower is 2 m more than twice the height of the tower. Find the height of the tower.

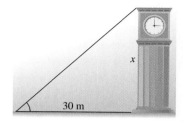

*Solve, and graph each solution set.*

**19.** $2x^2 + 7x > 15$

**20.** $\dfrac{5}{t-4} \leq 1$

# Additional Graphs of Functions and Relations

Since 1980, the number of multiple births in the United States has increased 59%, primarily due to greater use of fertility drugs and greater numbers of births to women over age 40. The number of higher-order multiple births—that is, births involving triplets or more—has increased over 400%. One of the most publicized higher-order multiple births occurred November 19, 1997, with the birth of the McCaughey septuplets in Des Moines, Iowa. All seven premature babies survived, a first in medical history. (*Source:* American College of Obstetricians and Gynecologists; *The Gazette,* November 19, 2003.)

In Example 6 of Section 10.2, we determine a *quadratic function* that models the number of higher-order multiple births in the United States.

**10.1** Review of Operations and Composition

**10.2** Graphs of Quadratic Functions

**10.3** More about Parabolas and Their Applications

**10.4** Symmetry; Increasing and Decreasing Functions

**10.5** Piecewise Linear Functions

## 10.1 Review of Operations and Composition

**OBJECTIVES**

1. Review how functions are formed using operations on functions.
2. Find a difference quotient.
3. Form composite functions and find their domains.

**OBJECTIVE 1** Review how functions are formed using operations on functions. In **Section 5.3** we defined the sum, difference, product, and quotient of functions. We review these definitions here.

### Operations on Functions

If $f$ and $g$ are functions, then for all values of $x$ for which both $f(x)$ and $g(x)$ exist,

$$(f + g)(x) = f(x) + g(x), \quad \text{Sum function}$$
$$(f - g)(x) = f(x) - g(x), \quad \text{Difference function}$$
$$(fg)(x) = f(x) \cdot g(x), \quad \text{Product function}$$

and

$$\left(\frac{f}{g}\right)(x) = \frac{f(x)}{g(x)}, \quad g(x) \neq 0. \quad \text{Quotient function}$$

**EXAMPLE 1** Using the Operations on Functions

Let $f(x) = x^2 + 1$ and $g(x) = 3x + 5$. Find each of the following.

**(a)** $(f + g)(1)$
Since $f(1) = 1^2 + 1 = 2$ and $g(1) = 3(1) + 5 = 8$,
$$(f + g)(1) = f(1) + g(1)$$
$$= 2 + 8$$
$$= 10.$$

**(b)** $(f - g)(-3) = f(-3) - g(-3)$

This is a key step.
$$= \overbrace{[(-3)^2 + 1]}^{f(x) = x^2 + 1} - \overbrace{[3(-3) + 5]}^{g(x) = 3x + 5}$$
$$= 10 - (-4)$$
$$= 14$$

**(c)** $(fg)(5) = f(5) \cdot g(5)$
$$= [5^2 + 1] \cdot [3(5) + 5]$$
$$= 26 \cdot 20$$
$$= 520$$

**(d)** $\left(\dfrac{f}{g}\right)(0) = \dfrac{f(0)}{g(0)} = \dfrac{0^2 + 1}{3 \cdot 0 + 5} = \dfrac{1}{5}$

**NOW TRY** Exercises 1, 3, and 5.

**EXAMPLE 2** Using the Operations on Functions

Let $f(x) = 8x - 9$ and $g(x) = \sqrt{2x - 1}$. Find each of the following, and give their domains.

**(a)** $(f + g)(x) = f(x) + g(x) = 8x - 9 + \sqrt{2x - 1}$

**(b)** $(f - g)(x) = f(x) - g(x) = 8x - 9 - \sqrt{2x - 1}$

**(c)** $(fg)(x) = f(x) \cdot g(x) = (8x - 9)\sqrt{2x - 1}$

**(d)** $\left(\dfrac{f}{g}\right)(x) = \dfrac{f(x)}{g(x)} = \dfrac{8x - 9}{\sqrt{2x - 1}}$

In parts (a)–(d), the domain of $f$ is the set of all real numbers, $(-\infty, \infty)$ while the domain of $g$, where $g(x) = \sqrt{2x - 1}$, includes just those real numbers that make $2x - 1 \geq 0$; the domain of $g$ is the interval $\left[\frac{1}{2}, \infty\right)$. The domain of $f + g, f - g$, and $fg$ is thus $\left[\frac{1}{2}, \infty\right)$. With $\frac{f}{g}$, the denominator cannot be 0, so the value $\frac{1}{2}$ is excluded from the domain. The domain of $\frac{f}{g}$ is $\left(\frac{1}{2}, \infty\right)$.

**NOW TRY** Exercise 13.

The domains of $f + g, f - g, fg$, and $\frac{f}{g}$ are summarized below. (Recall that the intersection of two sets is the set of all elements belonging to *both* sets.)

**Domains of $f + g$, $f - g$, $fg$, $\frac{f}{g}$**

For functions $f$ and $g$, the domains of $f + g, f - g$, and $fg$ include all real numbers in the intersection of the domains of $f$ and $g$, while the domain of $\frac{f}{g}$ includes those real numbers in the intersection of the domains of $f$ and $g$ for which $g(x) \neq 0$.

**OBJECTIVE 2** Find a difference quotient. If the coordinates of point $P$ are $(x, f(x))$ and the coordinates of point $Q$ are $(x + h, f(x + h))$, then the slope of the line $PQ$ is

$$m = \dfrac{f(x + h) - f(x)}{(x + h) - x} = \boldsymbol{\dfrac{f(x + h) - f(x)}{h}}, \quad h \neq 0. \quad \text{Difference quotient}$$

The boldface expression is called the **difference quotient** and is important in the study of calculus.

**EXAMPLE 3** Finding Difference Quotients

For $f(x) = x^2 - 2x$, find the following and simplify each expression.

**(a)** $f(x + h) - f(x)$

First find $f(x + h)$. Replace $x$ in $f(x) = x^2 - 2x$ with $x + h$.

$$f(x + h) = (x + h)^2 - 2(x + h)$$
$$= x^2 + 2xh + h^2 - 2x - 2h$$

Now subtract $f(x)$ from this expression for $f(x + h)$.

$$f(x + h) - f(x) = (x^2 + 2xh + h^2 - 2x - 2h) - (x^2 - 2x)$$
$$= 2xh + h^2 - 2h$$

**(b)** $\dfrac{f(x + h) - f(x)}{h}$

Use the result from part (a) to find the difference quotient.

$$\dfrac{f(x + h) - f(x)}{h} = \dfrac{2xh + h^2 - 2h}{h} = 2x + h - 2 \quad \text{Difference quotient}$$

**NOW TRY** Exercise 21.

**OBJECTIVE 3** Form composite functions and find their domains. The diagram in Figure 1 shows a function $f$ that assigns to each element $x$ of set $X$ some element $y$ of set $Y$. Suppose that a function $g$ takes each element of set $Y$ and assigns a value $z$ of set $Z$. Using both $f$ and $g$, then, an element $x$ in $X$ is assigned to an element $z$ in $Z$. The result of this process is a new function $h$, which takes an element $x$ in $X$ and assigns it an element $z$ in $Z$.

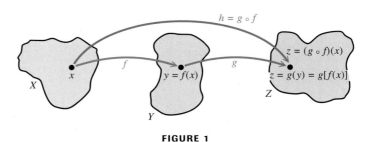

**FIGURE 1**

This function $h$ is called the *composition* of functions $g$ and $f$, written $g \circ f$, and is defined as follows.

### Composition of Functions

If $f$ and $g$ are functions, then the **composite function**, or **composition**, of $g$ and $f$ is defined by

$$(g \circ f)(x) = g(f(x))$$

for all $x$ in the domain of $f$ such that $f(x)$ is in the domain of $g$.

Read $g \circ f$ as "*g* of *f*."

As a real-life example of how composite functions occur, suppose an oil well off the Louisiana coast is leaking, with the leak spreading oil in a circular layer over the surface. See Figure 2. At any time $t$, in minutes, after the beginning of the leak, the radius of the circular oil slick is given by $r(t) = 5t$ feet. Since $A(r) = \pi r^2$ gives the area of a circle of radius $r$, the area can be expressed as a function of time by substituting $5t$ for $r$ in $A(r) = \pi r^2$ to get

$$A(r) = \pi r^2$$
$$A(r(t)) = \pi(5t)^2 = 25\pi t^2.$$

See the table in the margin. The function $A(r(t))$ is a composite function of the functions $A$ and $r$.

**FIGURE 2**

| $t$ | $r(t)$ | $A(r(t))$ |
|---|---|---|
| 1 | 5 | $\pi(5)^2 = 25\pi$ |
| 2 | 10 | $\pi(10)^2 = 100\pi$ |
| 3 | 15 | $\pi(15)^2 = 225\pi$ |
| $t$ | $5t$ | $\pi(5t)^2 = 25\pi t^2$ |

### EXAMPLE 4 Evaluating Composite Functions

Given $f(x) = 2x - 1$ and $g(x) = \dfrac{4}{x-1}$, find each of the following.

**(a)** $(f \circ g)(2)$

By definition, $(f \circ g)(2) = f(g(2))$, so we first find $g(2)$. Since $g(x) = \dfrac{4}{x-1}$,

$$g(2) = \frac{4}{2-1} = \frac{4}{1} = 4.$$

Now find $(f \circ g)(2) = f(g(2)) = f(4)$. Since $f(x) = 2x - 1$,
$$f(g(2)) = f(4) = 2(4) - 1 = 7.$$

**(b)** $(g \circ f)(-3)$

$$f(-3) = 2(-3) - 1 = -7$$

$$(g \circ f)(-3) = g(f(-3)) = g(-7) = \frac{4}{-7-1} = \frac{4}{-8} = -\frac{1}{2}$$

**NOW TRY** Exercises 7 and 9.

### EXAMPLE 5 Finding Composite Functions

Let $f(x) = 4x + 1$ and $g(x) = 2x^2 + 5x$. Find each of the following, and give their domains.

**(a)** $(g \circ f)(x)$

By definition, $(g \circ f)(x) = g(f(x))$. Using the given functions,

$$(g \circ f)(x) = g(f(x))$$
$$= g(4x + 1) \qquad f(x) = 4x + 1$$
$$= 2(4x + 1)^2 + 5(4x + 1) \qquad g(x) = 2x^2 + 5x$$
$$= 2(16x^2 + 8x + 1) + 20x + 5 \qquad \text{Multiply.}$$
$$= 32x^2 + 16x + 2 + 20x + 5 \qquad \text{Distributive property}$$
$$= 32x^2 + 36x + 7. \qquad \text{Combine terms.}$$

*Substitute $4x + 1$ for $x$ in the rule for $g$.*

**(b)** $(f \circ g)(x)$

If we use the preceding definition with $f$ and $g$ interchanged, $(f \circ g)(x)$ becomes $f(g(x))$.

$$(f \circ g)(x) = f(g(x))$$
$$= f(2x^2 + 5x) \qquad g(x) = 2x^2 + 5x$$
$$= 4(2x^2 + 5x) + 1 \qquad f(x) = 4x + 1$$
$$= 8x^2 + 20x + 1 \qquad \text{Distributive property}$$

*Substitute $2x^2 + 5x$ for $x$ in the rule for $f$.*

The domain of both composite functions is the set of all real numbers, $(-\infty, \infty)$.

**NOW TRY** Exercise 25.

As Example 5 shows, it is *not always true* that $f \circ g = g \circ f$.

---

**CAUTION** In general, the composite function $f \circ g$ is not the same as the product $fg$. For example, with $f$ and $g$ defined as in Example 5,

$$(f \circ g)(x) = 8x^2 + 20x + 1$$

but

$$(fg)(x) = (4x + 1)(2x^2 + 5x) = 8x^3 + 22x^2 + 5x.$$

---

### EXAMPLE 6 Finding Composite Functions and Their Domains

Let $f(x) = \frac{1}{x}$ and $g(x) = \sqrt{3 - x}$. Find $f \circ g$ and $g \circ f$. Give the domain of each.

First find $f \circ g$.

$$(f \circ g)(x) = f(g(x))$$
$$= f(\sqrt{3 - x})$$
$$= \frac{1}{\sqrt{3 - x}}$$

The radical expression $\sqrt{3 - x}$ is a positive real number only when $3 - x > 0$, or $x < 3$, so the domain of $f \circ g$ is the interval $(-\infty, 3)$.

Use the same functions to find $g \circ f$, as follows.

$$(g \circ f)(x) = g(f(x))$$
$$= g\left(\frac{1}{x}\right) \quad f(x) = \frac{1}{x}$$
$$= \sqrt{3 - \frac{1}{x}} \quad g(x) = \sqrt{3-x}$$
$$= \sqrt{\frac{3x-1}{x}} \quad \text{Write as a single fraction.}$$

The domain of $g \circ f$ is the set of all real numbers $x$ such that $x \neq 0$ and $3 - f(x) \geq 0$. By the methods of **Section 9.5,** the domain of $g \circ f$ is the set $(-\infty, 0) \cup \left[\frac{1}{3}, \infty\right)$.

**NOW TRY** Exercise 27.

### EXAMPLE 7  Finding the Functions That Form a Given Composite

Find functions $f$ and $g$ such that
$$(f \circ g)(x) = (x^2 - 5)^3 - 4(x^2 - 5) + 3.$$

Note the repeated quantity $x^2 - 5$. If $g(x) = x^2 - 5$ and $f(x) = x^3 - 4x + 3$, then
$$(f \circ g)(x) = f(g(x))$$
$$= f(x^2 - 5)$$
$$= (x^2 - 5)^3 - 4(x^2 - 5) + 3.$$

There are other pairs of functions $f$ and $g$ that also work. For instance,
$$f(x) = (x-5)^3 - 4(x-5) + 3 \quad \text{and} \quad g(x) = x^2.$$

**NOW TRY** Exercise 51.

## 10.1 Exercises

**NOW TRY** Exercise

Let $f(x) = 4x^2 - 2x$ and let $g(x) = 8x + 1$. Find each given value. See Examples 1 and 4.

1. $(f + g)(3)$
2. $(f - g)(-5)$
3. $(fg)(4)$
4. $(gf)(-3)$
5. $\left(\frac{g}{f}\right)(-1)$
6. $\left(\frac{f}{g}\right)(4)$
7. $(f \circ g)(2)$
8. $(f \circ g)(-5)$
9. $(g \circ f)(2)$
10. $(g \circ f)(-5)$
11. $(f \circ g)(-2)$
12. $(g \circ f)(-2)$

For each pair of functions $f$ and $g$, find **(a)** $f + g$, **(b)** $f - g$, **(c)** $fg$, and **(d)** $\frac{f}{g}$. Give the domain for each of these in Exercises 13–16. See Example 2.

13. $f(x) = 4x - 1, \quad g(x) = 6x + 3$
14. $f(x) = 9 - 2x, \quad g(x) = -5x + 2$
15. $f(x) = 3x^2 - 2x, \quad g(x) = x^2 - 2x + 1$
16. $f(x) = 6x^2 - 11x, \quad g(x) = x^2 - 4x - 5$
17. $f(x) = \sqrt{2x + 5}, \quad g(x) = \sqrt{4x + 9}$
18. $f(x) = \sqrt{11x - 3}, \quad g(x) = \sqrt{2x - 15}$

19. If $f$ and $g$ are functions, explain in your own words how to find the function values for $f + g$. Give an example.

20. If $f$ and $g$ are functions, explain in your own words how to find the function values for $fg$. Give an example.

*For each function, find and simplify* **(a)** $f(x + h) - f(x)$ *and* **(b)** $\dfrac{f(x + h) - f(x)}{h}$. *See Example 3.*

21. $f(x) = 2x^2 - 1$
22. $f(x) = 3x^2 + 2$
23. $f(x) = x^2 + 4x$
24. $f(x) = x^2 - 5x$

*Find $f \circ g$ and $g \circ f$ and their domains for each pair of functions. See Examples 5 and 6.*

25. $f(x) = 5x + 3$, $g(x) = -x^2 + 4x + 3$
26. $f(x) = 4x^2 + 2x + 8$, $g(x) = x + 5$
27. $f(x) = \dfrac{1}{x}$, $g(x) = x^2$
28. $f(x) = \dfrac{2}{x^4}$, $g(x) = 2 - x$
29. $f(x) = \sqrt{x + 2}$, $g(x) = 8x - 6$
30. $f(x) = 9x - 11$, $g(x) = 2\sqrt{x + 2}$
31. $f(x) = \dfrac{1}{x - 5}$, $g(x) = \dfrac{2}{x}$
32. $f(x) = \dfrac{8}{x - 6}$, $g(x) = \dfrac{4}{3x}$

33. Describe the steps required to find the composite function $f \circ g$. Give an example.

34. Composition is an operation that is unique to functions. Is composition of functions commutative? That is, does $f \circ g = g \circ f$ for all functions $f$ and $g$? Explain, using an example.

*The graphs of functions $f$ and $g$ are shown. Use these graphs to find each indicated value.*

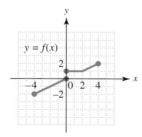
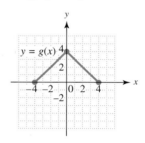

35. $f(1) + g(1)$
36. $f(4) - g(3)$
37. $f(-2) \cdot g(4)$
38. $\dfrac{f(4)}{g(2)}$
39. $(f \circ g)(2)$
40. $(g \circ f)(2)$
41. $(g \circ f)(-4)$
42. $(f \circ g)(-2)$

*The tables give some selected ordered pairs for functions $f$ and $g$.*

| $x$ | 3 | 4 | 6 |
|---|---|---|---|
| $f(x)$ | 1 | 3 | 9 |

| $x$ | 2 | 7 | 1 | 9 |
|---|---|---|---|---|
| $g(x)$ | 3 | 6 | 9 | 12 |

*Find each of the following.*

43. $(f \circ g)(2)$
44. $(f \circ g)(7)$
45. $(g \circ f)(3)$
46. $(g \circ f)(6)$
47. $(f \circ f)(4)$
48. $(g \circ g)(1)$

49. Why can you not determine $(f \circ g)(1)$ given the information in the tables for Exercises 43–48?

50. Extend the concept of composition of functions to evaluate $(g \circ (f \circ g))(7)$ using the tables for Exercises 43–48.

*A function $h$ is given. Find functions $f$ and $g$ such that $h(x) = (f \circ g)(x)$. Many such pairs of functions exist. See Example 7.*

51. $h(x) = (6x - 2)^2$
52. $h(x) = \sqrt{x^2 - 1}$
53. $h(x) = \dfrac{1}{x^2 + 2}$
54. $h(x) = (x + 2)^3 - 3(x + 2)^2$

*Solve each problem.*

**55.** The function defined by $f(x) = 3x$ computes the number of feet in $x$ yards and the function defined by $g(x) = 1760x$ computes the number of yards in $x$ miles. What is $(f \circ g)(x)$ and what does it compute?

**56.** The perimeter $x$ of a square with sides of length $t$ is given by the formula $x = 4t$.
   (a) Solve for $t$ in terms of $x$.
   (b) If $y$ represents the area of this square, write $y$ as a function of the perimeter $x$.
   (c) Use the composite function of part (b) to find the area of a square with perimeter 12.

**57.** Suppose the demand for a certain brand of vacuum cleaner is given by

$$D(p) = \frac{-p^2}{100} + 500,$$

where $p$ is the price in dollars. If the price in terms of the cost, $c$, is expressed as

$$p(c) = 2c - 10,$$

find $D(c)$, the demand in terms of the cost.

**58.** Suppose the population $P$ of a certain species of fish depends on the number $x$ (in hundreds) of a smaller kind of fish that serves as its food supply, where

$$P(x) = 2x^2 + 1.$$

Suppose, also, that the number $x$ (in hundreds) of the smaller species of fish depends on the amount $a$ (in appropriate units) of its food supply, a kind of plankton, where

$$x = f(a) = 3a + 2.$$

Find $(P \circ f)(a)$, the relationship between the population $P$ of the large fish and the amount $a$ of plankton available.

## 10.2 Graphs of Quadratic Functions

**OBJECTIVES**

1. Graph a quadratic function.
2. Graph parabolas with horizontal and vertical shifts.
3. Use the coefficient of $x^2$ to predict the shape and direction in which a parabola opens.
4. Find a quadratic function to model data.
5. Use the geometric definition of a parabola.

**OBJECTIVE 1** Graph a quadratic function. Figure 3 gives a graph of the simplest *quadratic function*, defined by $y = x^2$.

| x | y |
|---|---|
| -2 | 4 |
| -1 | 1 |
| 0 | 0 |
| 1 | 1 |
| 2 | 4 |

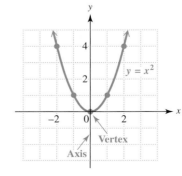

**FIGURE 3**

As mentioned in **Section 5.3,** this graph is called a **parabola.** The point (0, 0), the lowest point on the curve, is the **vertex** of this parabola. The vertical line through the vertex is the **axis** of the parabola, here $x = 0$. A parabola is **symmetric about its axis;** that is, if the graph were folded along the axis, the two portions of the curve would coincide. As Figure 3 suggests, $x$ can be any real number, so the domain of the function defined by $y = x^2$ is $(-\infty, \infty)$. Since $y$ is always nonnegative, the range is $[0, \infty)$.

In **Section 9.4,** we solved applications modeled by quadratic functions. We consider graphs of more general quadratic functions as defined here.

> **Quadratic Function**
>
> A function that can be written in the form
> $$f(x) = ax^2 + bx + c$$
> for real numbers $a$, $b$, and $c$, with $a \neq 0$, is a **quadratic function.**

*The graph of any quadratic function is a parabola with a vertical axis.*

We use the variable $y$ and function notation $f(x)$ interchangeably when discussing parabolas. Although we use the letter $f$ most often to name quadratic functions, other letters can be used. In this section we use the capital letter $F$ to distinguish between different parabolas graphed on the same coordinate axes.

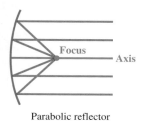

Parabolic reflector

**FIGURE 4**

Parabolas, which are a type of *conic section* (**Chapter 13**), have many applications. Cross sections of satellite dishes and automobile headlights form parabolas. See Figure 4. The cables that are used to support suspension bridges are shaped like parabolas.

We can use operations on and composition of functions to get variations of $f(x) = x^2$.

1. $f(x) = ax^2$     Multiply $g(x) = a$ and $h(x) = x^2$ to get $f(x)$.
2. $f(x) = x^2 + k$     Add $h(x) = x^2$ and $g(x) = k$ to get $f(x)$.
3. $f(x) = (x - h)^2$     Form the composite function $H(g(x))$ with $H(x) = x^2$ and $g(x) = x - h$ to get $f(x)$.
4. $f(x) = a(x - h)^2 + k$     Form $H(g(x))$ with $H(x) = ax^2 + k$ and $g(x) = x - h$ to get $f(x)$.

Each of these functions has a graph that is a parabola, but in each case the graph is modified from that of $f(x) = x^2$.

**OBJECTIVE 2** Graph parabolas with horizontal and vertical shifts. Parabolas need not have their vertices at the origin, as does the graph of $f(x) = x^2$. For example, to graph a parabola of the form

$$F(x) = x^2 + k,$$

start by selecting sample values of $x$ like those that were used to graph $f(x) = x^2$. The corresponding values of $F(x)$ in $F(x) = x^2 + k$ differ by $k$ from those of $f(x) = x^2$. Thus, the graph of $F(x) = x^2 + k$ is *shifted,* or *translated,* $k$ units vertically compared with that of $f(x) = x^2$.

**EXAMPLE 1** Graphing a Parabola with a Vertical Shift

Graph $F(x) = x^2 - 2$.

This graph has the same shape as that of $f(x) = x^2$, but since $k$ here is $-2$, the graph is shifted 2 units down, with vertex $(0, -2)$. Every function value is 2 less than the corresponding function value of $f(x) = x^2$. Plotting points on both sides of the vertex gives the graph in Figure 5 on the next page.

Notice that since the parabola is symmetric about its axis $x = 0$, the plotted points are "mirror images" of each other. Since $x$ can be any real number, the domain is still $(-\infty, \infty)$; the value of $y$ (or $F(x)$) is always greater than or equal to $-2$, so the range is $[-2, \infty)$.

The graph of $f(x) = x^2$ is shown in Figure 5 for comparison.

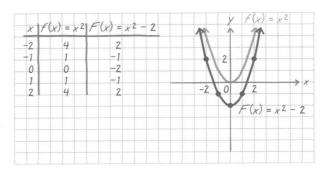

**FIGURE 5**

NOW TRY Exercise 23.

### Vertical Shift

The graph of $F(x) = x^2 + k$ is a parabola with the same shape as the graph of $f(x) = x^2$. The parabola is shifted $k$ units up if $k > 0$ and $|k|$ units down if $k < 0$. The vertex is $(0, k)$.

The graph of the function defined by

$$F(x) = (x - h)^2$$

is also a parabola with the same shape as that of $f(x) = x^2$. Because $(x - h)^2 \geq 0$ for all $x$, the vertex of $F(x) = (x - h)^2$ is the lowest point on the parabola. The lowest point occurs here when $F(x)$ is 0. To get $F(x)$ equal to 0, let $x = h$ so the vertex of $F(x) = (x - h)^2$ is $(h, 0)$. Thus, the graph of $F(x) = (x - h)^2$ is shifted $h$ units horizontally compared with that of $f(x) = x^2$.

**EXAMPLE 2** Graphing a Parabola with a Horizontal Shift

Graph $F(x) = (x - 2)^2$.

If $x = 2$, then $F(x) = 0$, giving the vertex $(2, 0)$. The graph of $F(x) = (x - 2)^2$ has the same shape as that of $f(x) = x^2$ but is shifted 2 units to the right. Plot points on one side of the vertex and use symmetry about the axis $x = 2$ to find corresponding points on the other side. See Figure 6. The domain is $(-\infty, \infty)$; the range is $[0, \infty)$.

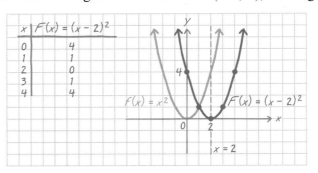

**FIGURE 6**

NOW TRY Exercise 27.

### Horizontal Shift

The graph of $F(x) = (x - h)^2$ is a parabola with the same shape as the graph of $f(x) = x^2$. The parabola is shifted $h$ units horizontally: $h$ units to the right if $h > 0$, and $|h|$ units to the left if $h < 0$. The vertex is $(h, 0)$.

**CAUTION** Errors frequently occur when horizontal shifts are involved. To determine the direction and magnitude of a horizontal shift, find the value that would cause the expression $x - h$ to equal 0.

### EXAMPLE 3 Graphing a Parabola with Horizontal and Vertical Shifts

Graph $F(x) = (x + 3)^2 - 2$.

This graph has the same shape as that of $f(x) = x^2$ but is shifted 3 units to the left (since $x + 3 = 0$ if $x = -3$) and 2 units down (because of the $-2$). See Figure 7. The vertex is $(-3, -2)$, with axis $x = -3$, domain $(-\infty, \infty)$, and range $[-2, \infty)$.

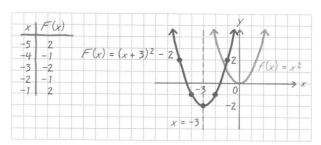

**FIGURE 7**

**NOW TRY** Exercise 29.

The characteristics of the graph of a parabola of the form $F(x) = (x - h)^2 + k$ are summarized as follows.

### Vertex and Axis of a Parabola

The graph of $F(x) = (x - h)^2 + k$ is a parabola with the same shape as the graph of $f(x) = x^2$ but with vertex $(h, k)$. The axis is the vertical line $x = h$.

**OBJECTIVE 3** Use the coefficient of $x^2$ to predict the shape and direction in which a parabola opens. Not all parabolas open up, and not all parabolas have the same shape as the graph of $f(x) = x^2$.

### EXAMPLE 4 Graphing a Parabola That Opens Down

Graph $f(x) = -\frac{1}{2}x^2$.

This parabola is shown in Figure 8 on the next page. The coefficient $-\frac{1}{2}$ affects the shape of the graph; the $\frac{1}{2}$ makes the parabola wider $\left(\text{since the values of } \frac{1}{2}x^2 \text{ increase more slowly than those of } x^2\right)$, and the negative sign makes the parabola open down.

The graph is not shifted in any direction; the vertex is still (0, 0) and the axis is $x = 0$. Unlike the parabolas graphed in Examples 1–3, the vertex here has the *largest* function value of any point on the graph. The domain is $(-\infty, \infty)$; the range is $(-\infty, 0]$.

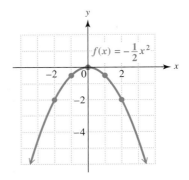

**FIGURE 8**

NOW TRY Exercise 21.

Some general principles concerning the graph of $F(x) = a(x - h)^2 + k$ follow.

### General Principles

1. The graph of the quadratic function defined by
$$F(x) = a(x - h)^2 + k, \quad a \neq 0,$$
is a parabola with vertex $(h, k)$ and the vertical line $x = h$ as axis.
2. The graph opens up if $a$ is positive and down if $a$ is negative.
3. The graph is wider than that of $f(x) = x^2$ if $0 < |a| < 1$. The graph is narrower than that of $f(x) = x^2$ if $|a| > 1$.

### EXAMPLE 5 Using the General Principles to Graph a Parabola

Graph $F(x) = -2(x + 3)^2 + 4$.

The parabola opens down (because $a < 0$) and is narrower than the graph of $f(x) = x^2$, since $|-2| = 2 > 1$, causing values of $F(x)$ to decrease more quickly than those of $f(x) = -x^2$. This parabola has vertex $(-3, 4)$, as shown in Figure 9. To complete the graph, we plotted the ordered pairs $(-4, 2)$ and, by symmetry, $(-2, 2)$. Symmetry can be used to find additional ordered pairs that satisfy the equation. The domain is $(-\infty, \infty)$; the range is $(-\infty, 4]$.

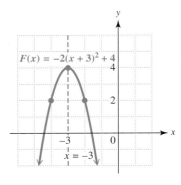

**FIGURE 9**

NOW TRY Exercise 33.

### OBJECTIVE 4 Find a quadratic function to model data.

**EXAMPLE 6** Finding a Quadratic Function to Model the Rise in Multiple Births

The number of higher-order multiple births in the United States is rising. Let $x$ represent the number of years since 1970 and $y$ represent the rate of higher-order multiples born per 100,000 births since 1971. The data are shown in the following table. Find a quadratic function that models the data.

| Year | x | y |
|---|---|---|
| 1971 | 1 | 29.1 |
| 1976 | 6 | 35.0 |
| 1981 | 11 | 40.0 |
| 1986 | 16 | 47.0 |
| 1991 | 21 | 100.0 |
| 1996 | 26 | 152.6 |
| 2001 | 31 | 185.6 |

*Source:* National Center for Health Statistics.

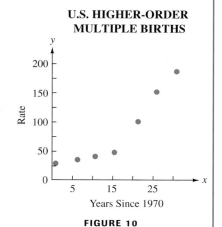

**U.S. HIGHER-ORDER MULTIPLE BIRTHS**

**FIGURE 10**

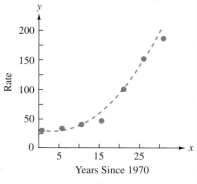

**U.S. HIGHER-ORDER MULTIPLE BIRTHS**

**FIGURE 11**

A scatter diagram of the ordered pairs $(x, y)$ is shown in Figure 10. The general shape suggested by the scatter diagram indicates that a parabola should approximate these points, as shown by the dashed curve in Figure 11. The equation for such a parabola would have a positive coefficient for $x^2$ since the graph opens up. To find a quadratic function of the form

$$y = ax^2 + bx + c$$

that models, or *fits,* these data, we choose three representative ordered pairs and use them to write a system of three equations. Using (1, 29.1), (11, 40), and (21, 100), we substitute the $x$- and $y$-values from the ordered pairs into the quadratic form $y = ax^2 + bx + c$ to get the following three equations.

$$a(1)^2 + b(1) + c = 29.1 \quad \text{or} \quad a + b + c = 29.1 \quad (1)$$
$$a(11)^2 + b(11) + c = 40 \quad \text{or} \quad 121a + 11b + c = 40 \quad (2)$$
$$a(21)^2 + b(21) + c = 100 \quad \text{or} \quad 441a + 21b + c = 100 \quad (3)$$

We can find the values of $a$, $b$, and $c$ by solving this system of three equations in three variables using the methods of **Section 4.2.** Multiplying equation (1) by $-1$ and adding the result to equation (2) gives

$$120a + 10b = 10.9. \quad (4)$$

Multiplying equation (2) by $-1$ and adding the result to equation (3) gives

$$320a + 10b = 60. \quad (5)$$

We eliminate $b$ from this system of two equations in two variables by multiplying equation (4) by $-1$ and adding the result to equation (5).

$$200a = 49.1$$
$$a = 0.2455 \quad \text{Use a calculator.}$$

We substitute 0.2455 for $a$ in equation (4) or (5) to find that $b = -1.856$. Substituting the values of $a$ and $b$ into equation (1) gives $c = 30.7105$. Using these values of $a$, $b$, and $c$, our model is defined by

$$y = 0.2455x^2 - 1.856x + 30.7105.$$

**NOW TRY** Exercise 41.

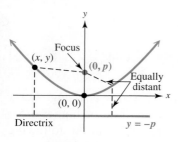

FIGURE 12

**OBJECTIVE 5** Use the geometric definition of a parabola. Geometrically, a parabola is defined as the set of all points in a plane that are equally distant from a fixed point and a fixed line not containing the point. The point is the **focus** and the line is the **directrix.** The line through the focus and perpendicular to the directrix is the axis of the parabola. The point on the axis that is equally distant from the focus and the directrix is the vertex of the parabola. See Figure 12.

The parabola in Figure 12 has the point $(0, p)$ as focus and the line $y = -p$ as directrix. The vertex is $(0, 0)$. Let $(x, y)$ be any point on the parabola. The distance from $(x, y)$ to the directrix is $|y - (-p)|$, while the distance from $(x, y)$ to $(0, p)$ is $\sqrt{(x - 0)^2 + (y - p)^2}$. Since $(x, y)$ is equally distant from the directrix and the focus,

$$|y - (-p)| = \sqrt{(x - 0)^2 + (y - p)^2}$$
$$(y + p)^2 = x^2 + (y - p)^2 \quad \text{Square both sides.}$$
$$y^2 + 2py + p^2 = x^2 + y^2 - 2py + p^2 \quad \text{Square of a binomial}$$
$$4py = x^2, \quad \text{Combine terms.}$$

*Remember the middle term.*

which is the equation of the parabola with focus $(0, p)$ and directrix $y = -p$. Solving $4py = x^2$ for $y$ gives

$$y = \frac{1}{4p}x^2,$$

so $\frac{1}{4p} = a$ when the equation is written as $y = ax^2 + bx + c$.

This result could be extended to a parabola with vertex at $(h, k)$, focus $|p|$ units above $(h, k)$, and directrix $|p|$ units below $(h, k)$, or to a parabola with vertex at $(h, k)$, focus $|p|$ units below $(h, k)$, and directrix $|p|$ units above $(h, k)$.

**EXAMPLE 7** Using the Geometric Definition of a Parabola

Use the geometric definition to find the equation of the parabola with focus at $(0, -3)$ and directrix $y = 3$.

Since the vertex is halfway between the focus and the directrix, the vertex is $(0, 0)$, as shown in Figure 13. Here $p = -3$, so the distance between the focus and the vertex is $|p| = 3$. The equation is

$$y = \frac{1}{4p}x^2 = \frac{1}{4(-3)}x^2 = -\frac{1}{12}x^2, \quad \text{or} \quad -12y = x^2.$$

**NOW TRY** Exercise 45.

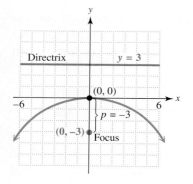

FIGURE 13

## 10.2 Exercises

1. *Matching* Match each quadratic function with its graph from choices A–D.

   (a) $f(x) = (x + 2)^2 - 1$

   (b) $f(x) = (x + 2)^2 + 1$

   (c) $f(x) = (x - 2)^2 - 1$

   (d) $f(x) = (x - 2)^2 + 1$

   A.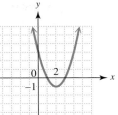

   B.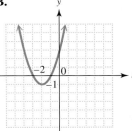

   C.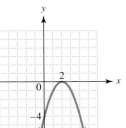

   D.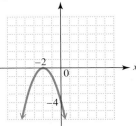

2. *Matching* Match each quadratic function with its graph from choices A–D.

   (a) $f(x) = -x^2 + 2$

   (b) $f(x) = -x^2 - 2$

   (c) $f(x) = -(x + 2)^2$

   (d) $f(x) = -(x - 2)^2$

   A.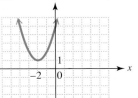

   B.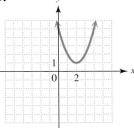

   C.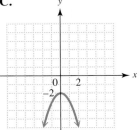

   D.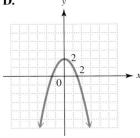

*Identify the vertex of each parabola. See Examples 1–4.*

3. $f(x) = -3x^2$

4. $f(x) = \dfrac{1}{2}x^2$

5. $f(x) = x^2 + 4$

6. $f(x) = x^2 - 4$

7. $f(x) = (x - 1)^2$

8. $f(x) = (x + 3)^2$

9. $f(x) = (x + 3)^2 - 4$

10. $f(x) = (x - 5)^2 - 8$

11. $f(x) = -(x - 5)^2 + 6$

12. (a) How are the graphs of the parabolas in Exercises 9 and 10 shifted compared to the graph of $f(x) = x^2$?

    (b) What does the value of $a$ in $F(x) = a(x - h)^2 + k$ tell you about the graph of the function compared to the graph of $f(x) = x^2$?

*For each quadratic function, tell whether the graph opens up or down and whether the graph is wider, narrower, or the same shape as the graph of $f(x) = x^2$. See Examples 4 and 5.*

**13.** $f(x) = -\dfrac{2}{5}x^2$  **14.** $f(x) = -2x^2$

**15.** $f(x) = 3x^2 + 1$  **16.** $f(x) = \dfrac{2}{3}x^2 - 4$

**17.** *Concept Check* For $f(x) = a(x - h)^2 + k$, in what quadrant is the vertex if

(a) $h > 0, k > 0$;  (b) $h > 0, k < 0$;  (c) $h < 0, k > 0$;  (d) $h < 0, k < 0$?

**18.** *Concept Check* Answer each question.

(a) What is the value of $h$ if the graph of $f(x) = a(x - h)^2 + k$ has vertex on the $y$-axis?
(b) What is the value of $k$ if the graph of $f(x) = a(x - h)^2 + k$ has vertex on the $x$-axis?

**19.** *Matching* Match each quadratic function with the description of the parabola that is its graph.

(a) $f(x) = (x - 4)^2 - 2$  **A.** Vertex $(2, -4)$, opens down
(b) $f(x) = (x - 2)^2 - 4$  **B.** Vertex $(2, -4)$, opens up
(c) $f(x) = -(x - 4)^2 - 2$  **C.** Vertex $(4, -2)$, opens down
(d) $f(x) = -(x - 2)^2 - 4$  **D.** Vertex $(4, -2)$, opens up

**20.** Explain in your own words the meaning of each term.

(a) Vertex of a parabola  (b) Axis of a parabola

*Graph each parabola. Plot at least two points as well as the vertex. Give the vertex, axis, domain, and range in Exercises 27–36. See Examples 1–5.*

**21.** $f(x) = -2x^2$  **22.** $f(x) = \dfrac{1}{3}x^2$  **23.** $f(x) = x^2 - 1$

**24.** $f(x) = x^2 + 3$  **25.** $f(x) = -x^2 + 2$  **26.** $f(x) = 2x^2 - 2$

**27.** $f(x) = (x - 4)^2$  **28.** $f(x) = -2(x + 1)^2$

**29.** $f(x) = (x + 2)^2 - 1$  **30.** $f(x) = (x - 1)^2 + 2$

**31.** $f(x) = 2(x - 2)^2 - 4$  **32.** $f(x) = -3(x - 2)^2 + 1$

**33.** $f(x) = -\dfrac{1}{2}(x + 1)^2 + 2$  **34.** $f(x) = -\dfrac{2}{3}(x + 2)^2 + 1$

**35.** $f(x) = 2(x - 2)^2 - 3$  **36.** $f(x) = \dfrac{4}{3}(x - 3)^2 - 2$

*Concept Check* In Exercises 37–40, tell whether a linear or quadratic function would be a more appropriate model for each set of graphed data. If linear, tell whether the slope should be positive or negative. If quadratic, tell whether the coefficient $a$ of $x^2$ should be positive or negative. See Example 6.

**37.**

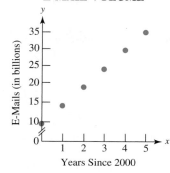

AVERAGE DAILY E-MAIL VOLUME

*Source:* General Accounting Office.

**38.**

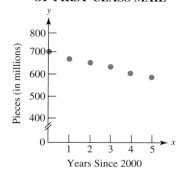

AVERAGE DAILY VOLUME OF FIRST-CLASS MAIL

*Source:* General Accounting Office.

**39. INCREASES IN WHOLESALE DRUG PRICES**

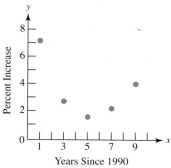

Years Since 1990

*Source:* IMS Health, Retail and Provider Perspective.

**40. SOCIAL SECURITY ASSETS***

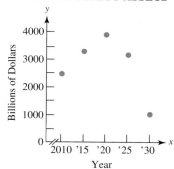

*Projected

*Source:* Social Security Administration.

*Solve each problem. See Example 6.*

**41.** The number of publicly traded companies filing for bankruptcy for selected years between 1990 and 2000 are shown in the table. In the year column, 0 represents 1990, 2 represents 1992, and so on.

| Year | Number of Bankruptcies |
|---|---|
| 0 | 115 |
| 2 | 91 |
| 4 | 70 |
| 6 | 84 |
| 8 | 120 |
| 10 | 176 |

*Source:* www.BankruptcyData.com

(a) Use the ordered pairs (year, number of bankruptcies) to make a scatter diagram of the data.

(b) Use the scatter diagram to decide whether a linear or quadratic function would better model the data. If quadratic, should the coefficient $a$ of $x^2$ be positive or negative?

(c) Use the ordered pairs (0, 115), (4, 70), and (8, 120) to find a quadratic function that models the data. Round the values of $a$, $b$, and $c$ in your model to three decimal places, as necessary.

(d) Use your model from part (c) to approximate the number of company bankruptcy filings in 2002. Round your answer to the nearest whole number.

(e) The number of company bankruptcy filings through August 16, 2002 was 129. Based on this, is your estimate from part (d) reasonable? Explain.

**42.** The percent of U.S. high school students in grades 9–12 who smoke is shown in the table for selected years. In the year column, 1 represents 1991, 3 represents 1993, and so on.

| Year | Percent of Students |
|---|---|
| 1 | 28 |
| 3 | 31 |
| 5 | 35 |
| 7 | 36 |
| 9 | 35 |
| 11 | 29 |
| 13 | 22 |

*Source:* National Center for Health Statistics.

(a) Use the ordered pairs (year, percent of students) to make a scatter diagram of the data.

(b) Would a linear or quadratic function better model the data?

(c) Should the coefficient $a$ of $x^2$ in a quadratic model be positive or negative?

(d) Use the ordered pairs (1, 28), (7, 36), and (11, 29) to find a quadratic function that models the data. Round the values of $a$, $b$, and $c$ in your model to the nearest tenth, as necessary.

(e) Use your model from part (d) to approximate the percent of high school students who smoked during 1995 and 2003 to the nearest percent. How well does the model approximate the actual data from the table?

**43.** In Example 6, we determined that the quadratic function defined by

$$y = 0.2455x^2 - 1.856x + 30.7105$$

modeled the rate (per 100,000) of higher-order multiple births, where $x$ represents the number of years since 1970.

(a) Use this model to approximate the rate of higher-order births in 2002 to the nearest tenth.

(b) The actual rate of higher-order births in 2002 was 184.0. (*Source:* National Center for Health Statistics.) How does the approximation using the model compare to the actual rate for 2002?

**44.** Should the model from Exercise 43 be used to approximate the rate of higher-order multiple births in years after 2002? Explain.

*Use the geometric definition of a parabola to find the equation of each parabola described. See Example 7.*

**45.** Focus $(0, 2)$, directrix $y = -2$

**46.** Focus $\left(0, -\frac{1}{2}\right)$, directrix $y = \frac{1}{2}$

**47.** Focus $(0, -1)$, directrix $y = 1$

**48.** Focus $\left(0, \frac{1}{4}\right)$, directrix $y = -\frac{1}{4}$

**49.** Focus $(0, 4)$, directrix $y = 2$

**50.** Focus $(0, -1)$, directrix $y = -2$

## 10.3 More about Parabolas and Their Applications

**OBJECTIVES**

1. Find the vertex of a vertical parabola.
2. Graph a quadratic function.
3. Use the discriminant to find the number of $x$-intercepts of a parabola with a vertical axis.
4. Use quadratic functions to solve problems involving maximum or minimum value.
5. Graph parabolas with horizontal axes.

**OBJECTIVE 1** Find the vertex of a vertical parabola. When the equation of a parabola is given in the form $f(x) = ax^2 + bx + c$, we need to locate the vertex to sketch an accurate graph. There are two ways to do this:

1. Complete the square as shown in Examples 1 and 2, or
2. Use a formula derived by completing the square, as shown in Example 3.

**EXAMPLE 1** Completing the Square to Find the Vertex

Find the vertex of the graph of $f(x) = x^2 - 4x + 5$.

We need to express $x^2 - 4x + 5$ in the form $(x - h)^2 + k$ by completing the square on $x^2 - 4x$, as in **Section 9.1**. The process is slightly different here because we want to keep $f(x)$ alone on one side of the equation. Instead of adding the appropriate number to each side, we *add and subtract* it on the right. This is equivalent to adding 0.

$$f(x) = x^2 - 4x + 5$$
$$= (x^2 - 4x \quad) + 5 \qquad \text{Group the variable terms.}$$
$$\left[\frac{1}{2}(-4)\right]^2 = (-2)^2 = 4$$
$$= (x^2 - 4x + 4 - 4) + 5 \qquad \text{Add and subtract 4.}$$
$$= (x^2 - 4x + 4) - 4 + 5 \qquad \text{Bring } -4 \text{ outside the parentheses.}$$
$$f(x) = (x - 2)^2 + 1 \qquad \text{Factor; combine terms.}$$

The vertex of this parabola is $(2, 1)$.

**NOW TRY** Exercise 5.

> **EXAMPLE 2** Completing the Square to Find the Vertex When $a \neq 1$
>
> Find the vertex of the graph of $f(x) = -3x^2 + 6x - 1$.
>
> We must complete the square on $-3x^2 + 6x$. Because the $x^2$-term has a coefficient other than 1, we factor that coefficient out of the first two terms.
>
> $$f(x) = -3x^2 + 6x - 1$$
> $$= -3(x^2 - 2x) - 1 \qquad \text{Factor out } -3.$$
>
> $\left[\frac{1}{2}(-2)\right]^2 = (-1)^2 = 1$
>
> $$= -3(x^2 - 2x + 1 - 1) - 1 \qquad \text{Add and subtract 1 within the parentheses.}$$
>
> Bring $-1$ outside the parentheses; **be sure to multiply it by $-3$.**
>
> $$= -3(x^2 - 2x + 1) + (-3)(-1) - 1 \qquad \text{Distributive property}$$
> $$= -3(x^2 - 2x + 1) + 3 - 1 \qquad \text{This is a key step.}$$
> $$f(x) = -3(x - 1)^2 + 2 \qquad \text{Factor; combine terms.}$$
>
> The vertex is $(1, 2)$.
>
> **NOW TRY** Exercise 7.

To derive a formula for the vertex of the graph of the quadratic function defined by $f(x) = ax^2 + bx + c$ ($a \neq 0$), complete the square.

$$f(x) = ax^2 + bx + c \qquad \text{Standard form}$$

$$= a\left(x^2 + \frac{b}{a}x\right) + c \qquad \text{Factor } a \text{ from the first two terms.}$$

$\left[\frac{1}{2}\left(\frac{b}{a}\right)\right]^2 = \left(\frac{b}{2a}\right)^2 = \frac{b^2}{4a^2}$

$$= a\left(x^2 + \frac{b}{a}x + \frac{b^2}{4a^2} - \frac{b^2}{4a^2}\right) + c \qquad \text{Add and subtract } \tfrac{b^2}{4a^2}.$$

$$= a\left(x^2 + \frac{b}{a}x + \frac{b^2}{4a^2}\right) + a\left(-\frac{b^2}{4a^2}\right) + c \qquad \text{Distributive property}$$

$$= a\left(x^2 + \frac{b}{a}x + \frac{b^2}{4a^2}\right) - \frac{b^2}{4a} + c$$

$$= a\left(x + \frac{b}{2a}\right)^2 + \frac{4ac - b^2}{4a} \qquad \text{Factor; combine terms.}$$

$$f(x) = a\underbrace{\left[x - \left(\frac{-b}{2a}\right)\right]^2}_{h} + \underbrace{\frac{4ac - b^2}{4a}}_{k} \qquad f(x) = (x - h)^2 + k$$

Thus, the vertex $(h, k)$ can be expressed in terms of $a$, $b$, and $c$. However, it is not necessary to remember this expression for $k$, since it can be found by replacing $x$ with $\frac{-b}{2a}$. Using function notation, if $y = f(x)$, then the $y$-value of the vertex is $f\left(\frac{-b}{2a}\right)$.

### Vertex Formula

The graph of the quadratic function defined by $f(x) = ax^2 + bx + c \; (a \neq 0)$ has vertex

$$\left( \frac{-b}{2a}, f\left(\frac{-b}{2a}\right) \right),$$

and the axis of the parabola is the line

$$x = \frac{-b}{2a}.$$

**EXAMPLE 3** Using the Formula to Find the Vertex

Use the vertex formula to find the vertex of the graph of $f(x) = x^2 - x - 6$.
 For this function, $a = 1$, $b = -1$, and $c = -6$. The $x$-coordinate of the vertex of the parabola is given by

$$\frac{-b}{2a} = \frac{-(-1)}{2(1)} = \frac{1}{2}.$$

The $y$-coordinate is $f\left(\frac{-b}{2a}\right) = f\left(\frac{1}{2}\right)$.

$$f\left(\frac{1}{2}\right) = \left(\frac{1}{2}\right)^2 - \frac{1}{2} - 6 = \frac{1}{4} - \frac{1}{2} - 6 = -\frac{25}{4}$$

The vertex is $\left(\frac{1}{2}, -\frac{25}{4}\right)$.

**NOW TRY** Exercise 9.

**OBJECTIVE 2** Graph a quadratic function. We give a general approach for graphing any quadratic function here.

### Graphing a Quadratic Function $y = f(x)$

**Step 1** **Determine whether the graph opens up or down.** If $a > 0$, the parabola opens up; if $a < 0$, it opens down.

**Step 2** **Find the vertex.** Use either the vertex formula or completing the square.

**Step 3** **Find any intercepts.** To find the $x$-intercepts (if any), solve $f(x) = 0$. To find the $y$-intercept, evaluate $f(0)$.

**Step 4** **Complete the graph.** Plot the points found so far. Find and plot additional points as needed, using symmetry about the axis.

**EXAMPLE 4** Using the Steps to Graph a Quadratic Function

Graph the quadratic function defined by

$$f(x) = x^2 - x - 6.$$

**Step 1** From the equation, $a = 1$, so the graph of the function opens up.

**Step 2** The vertex, $\left(\frac{1}{2}, -\frac{25}{4}\right)$, was found in Example 3 by using the vertex formula.

**Step 3** Find any intercepts. Since the vertex, $\left(\frac{1}{2}, -\frac{25}{4}\right)$, is in quadrant IV and the graph opens up, there will be two x-intercepts. To find them, let $f(x) = 0$ and solve.

$$f(x) = x^2 - x - 6$$
$$0 = x^2 - x - 6 \quad \text{Let } f(x) = 0.$$
$$0 = (x - 3)(x + 2) \quad \text{Factor.}$$
$$x - 3 = 0 \quad \text{or} \quad x + 2 = 0 \quad \text{Zero-factor property}$$
$$x = 3 \quad \text{or} \quad x = -2$$

The x-intercepts are $(3, 0)$ and $(-2, 0)$.
Now find the y-intercept by evaluating $f(0)$.

$$f(x) = x^2 - x - 6$$
$$f(0) = 0^2 - 0 - 6 \quad \text{Let } x = 0.$$
$$f(0) = -6$$

The y-intercept is $(0, -6)$.

**Step 4** Plot the points found so far and additional points as needed using symmetry about the axis, $x = \frac{1}{2}$. The graph is shown in Figure 14. The domain is $(-\infty, \infty)$, and the range is $\left[-\frac{25}{4}, \infty\right)$.

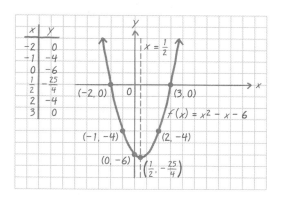

**FIGURE 14**

NOW TRY Exercise 23.

**OBJECTIVE 3** Use the discriminant to find the number of x-intercepts of a parabola with a vertical axis. Recall from **Section 9.2** that $b^2 - 4ac$ is called the *discriminant* of the quadratic equation $ax^2 + bx + c = 0$ and that we can use it to determine the number of real solutions of a quadratic equation.

In a similar way, we can use the discriminant of a quadratic *function* to determine the number of x-intercepts of its graph. See Figure 15. If the discriminant is positive, the parabola will have two x-intercepts. If the discriminant is 0, there will be only one x-intercept, and it will be the vertex of the parabola. If the discriminant is negative, the graph will have no x-intercepts.

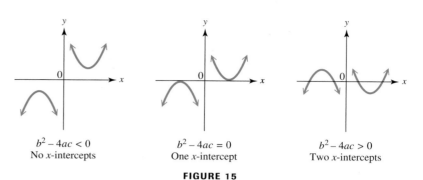

**FIGURE 15**

### EXAMPLE 5  Using the Discriminant to Determine the Number of x-Intercepts

Find the discriminant and use it to determine the number of x-intercepts of the graph of each quadratic function.

**(a)** $f(x) = 2x^2 + 3x - 5$

The discriminant is $b^2 - 4ac$. Here $a = 2$, $b = 3$, and $c = -5$, so

$$b^2 - 4ac = 9 - 4(2)(-5) = 49.$$

Since the discriminant is positive, the parabola has two x-intercepts.

**(b)** $f(x) = -3x^2 - 1$

Here, $a = -3$, $b = 0$, and $c = -1$. The discriminant is

$$b^2 - 4ac = 0 - 4(-3)(-1) = -12.$$

The discriminant is negative, so the graph has no x-intercepts.

**(c)** $f(x) = 9x^2 + 6x + 1$

Here, $a = 9$, $b = 6$, and $c = 1$. The discriminant is

$$b^2 - 4ac = 36 - 4(9)(1) = 0.$$

The parabola has only one x-intercept (its vertex).

**NOW TRY** Exercises 11 and 13.

**OBJECTIVE 4** Use quadratic functions to solve problems involving maximum or minimum value. The vertex of the graph of a quadratic function is either the highest or the lowest point on the parabola. The y-value of the vertex gives the maximum or minimum value of y, while the x-value tells where that maximum or minimum occurs. In many applied problems we must find the greatest or least value of some quantity. When we can express that quantity in terms of a quadratic function, the value of k in the vertex (h, k) gives that optimum value.

### EXAMPLE 6  Finding the Maximum Area of a Rectangular Region

A farmer has 120 ft of fencing. He wants to put a fence around a rectangular field next to a building. Find the maximum area he can enclose and the dimensions of the field when the area is maximized.

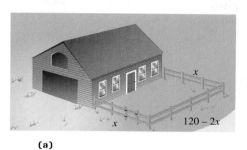
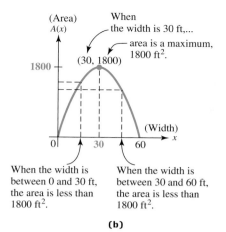

**FIGURE 16**

Let $x =$ the width of the field. See Figure 16(a). Then

$x + x + \text{length} = 120$     Sum of the sides is 120 ft.

$2x + \text{length} = 120$     Combine terms.

$\text{length} = 120 - 2x.$     Subtract 2x.

The area is given by the product of the width and length, so

$$A(x) = x(120 - 2x)$$
$$= 120x - 2x^2.$$

To determine the maximum area, find the vertex of the parabola given by $A(x) = 120x - 2x^2$ using the vertex formula. Writing the equation in standard form as $A(x) = -2x^2 + 120x$ gives $a = -2$, $b = 120$, and $c = 0$, so

$$h = \frac{-b}{2a} = \frac{-120}{2(-2)} = \frac{-120}{-4} = 30;$$

$$A(30) = -2(30)^2 + 120(30) = -2(900) + 3600 = 1800.$$

The graphical interpretation in Figure 16(b) shows that the graph is a parabola that opens down and that its vertex is $(30, 1800)$. Thus, the maximum area will be 1800 ft². This area will occur if $x$, the width of the field, is 30 ft and the length is $120 - 2(30) = 60$ ft.

**NOW TRY** Exercise 33.

> **CAUTION** *Be careful when interpreting the meanings of the coordinates of the vertex.* The first coordinate, $x$, gives the value for which the *function value* is a maximum or a minimum. Be sure to read the problem carefully to determine whether you are asked to find the value of the independent variable, the function value, or both.

**EXAMPLE 7** Finding the Maximum Height Attained by a Projectile

If air resistance is neglected, a projectile on Earth shot straight upward with an initial velocity of 40 m per sec will be at a height $s$ in meters given by

$$s(t) = -4.9t^2 + 40t,$$

where $t$ is the number of seconds elapsed after projection. After how many seconds will it reach its maximum height, and what is this maximum height?

For this function, $a = -4.9$, $b = 40$, and $c = 0$. Use the vertex formula.

$$h = \frac{-b}{2a} = \frac{-40}{2(-4.9)} \approx 4.1 \quad \text{Use a calculator.}$$

This indicates that the maximum height is attained at 4.1 sec. To find this maximum height, calculate $s(4.1)$.

$$s(4.1) = -4.9(4.1)^2 + 40(4.1)$$
$$\approx 81.6 \quad \text{Use a calculator.}$$

The projectile will attain a maximum height of approximately 81.6 m.

**NOW TRY** Exercise 35.

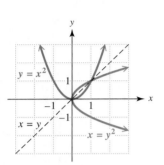

FIGURE 17

**OBJECTIVE 5** Graph parabolas with horizontal axes. The directrix of a parabola could be the *vertical* line $x = -p$, where $p > 0$, with focus on the $x$-axis at $(p, 0)$, producing a parabola opening to the right. This parabola is the graph of the relation

$$y^2 = 4px \quad \text{or} \quad x = \frac{1}{4p}y^2.$$

For example, we obtain the equation $x = y^2$ from $y = x^2$ by interchanging $x$ and $y$. Choosing values of $y$ and finding the corresponding values of $x$ gives the parabola in Figure 17. The graph of $x = y^2$, shown in red, is symmetric with respect to the line $y = 0$ and has vertex $(0, 0)$. For comparison, the graph of $y = x^2$ is shown in blue. These graphs are mirror images of each other with respect to the line $y = x$. From the graph, the domain of $x = y^2$ is $[0, \infty)$, and the range is $(-\infty, \infty)$.

If $x$ and $y$ are interchanged in the equation $y = ax^2 + bx + c$, the equation becomes $x = ay^2 + by + c$.

**Graph of a Parabola with Horizontal Axis**

The graph of

$$x = ay^2 + by + c \quad \text{or} \quad x = a(y - k)^2 + h$$

is a parabola with vertex $(h, k)$ and the horizontal line $y = k$ as axis. The graph opens to the right if $a > 0$ and to the left if $a < 0$.

**EXAMPLE 8** Graphing a Parabola with Horizontal Axis

Graph $x = (y - 2)^2 - 3$. Give the vertex, axis, domain, and range.

This graph has its vertex at $(-3, 2)$, since the roles of $x$ and $y$ are interchanged. It opens to the right because $a = 1$ and $1 > 0$, and it has the same shape as $y = x^2$. Plotting a few additional points gives the graph in Figure 18. Note that the graph is symmetric about its axis, $y = 2$. The domain is $[-3, \infty)$, and the range is $(-\infty, \infty)$.

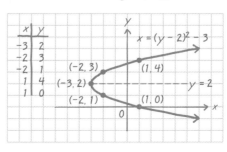

FIGURE 18

**NOW TRY** Exercise 27.

When a quadratic equation is given in the form $x = ay^2 + by + c$, completing the square on $y$ allows us to find the vertex.

**EXAMPLE 9** Completing the Square to Graph a Parabola with Horizontal Axis

Graph $x = -2y^2 + 4y - 3$. Give the vertex, axis, domain, and range of the relation.

$$x = -2y^2 + 4y - 3$$
$$= -2(y^2 - 2y) - 3 \qquad \text{Factor out } -2.$$
$$= -2(y^2 - 2y + 1 - 1) - 3 \qquad \text{Complete the square within the parentheses; add and subtract 1.}$$
$$= -2(y^2 - 2y + 1) + (-2)(-1) - 3 \qquad \text{Distributive property}$$

Be careful here.

$$x = -2(y - 1)^2 - 1 \qquad \text{Factor; simplify.}$$

Because of the negative coefficient $(-2)$ in $x = -2(y - 1)^2 - 1$, the graph opens to the left (the negative $x$-direction). The graph is narrower than the graph of $y = x^2$ because $|-2| > 1$. As shown in Figure 19, the vertex is $(-1, 1)$ and the axis is $y = 1$. The domain is $(-\infty, -1]$, and the range is $(-\infty, \infty)$.

**NOW TRY** Exercise 31.

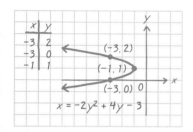

**FIGURE 19**

**CAUTION** Only quadratic equations solved for $y$ (whose graphs are vertical parabolas) are examples of functions. *The horizontal parabolas in Examples 8 and 9 are not graphs of functions,* because they do not satisfy the vertical line test.

In summary, the graphs of parabolas studied in **Sections 10.2 and 10.3** fall into the following categories.

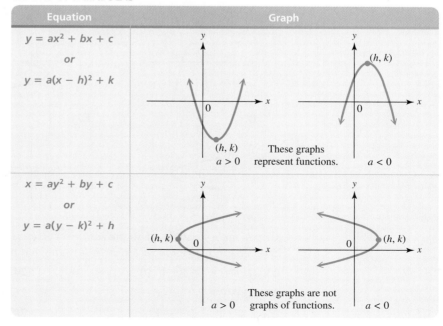

## 10.3 Exercises

*Concept Check*  In Exercises 1–4, answer each question.

1. How can you determine just by looking at the equation of a parabola whether it has a vertical or a horizontal axis?
2. Why can't the graph of a quadratic function be a parabola with a horizontal axis?
3. How can you determine the number of $x$-intercepts of the graph of a quadratic function without graphing the function?
4. If the vertex of the graph of a quadratic function is $(1, -3)$, and the graph opens down, how many $x$-intercepts does the graph have?

*Find the vertex of each parabola. See Examples 1–3.*

5. $f(x) = x^2 + 8x + 10$
6. $f(x) = x^2 + 10x + 23$
7. $f(x) = -2x^2 + 4x - 5$
8. $f(x) = -3x^2 + 12x - 8$
9. $f(x) = x^2 + x - 7$
10. $f(x) = x^2 - x + 5$

*Find the vertex of each parabola. For each equation, decide whether the graph opens up, down, to the left, or to the right, and whether it is wider, narrower, or the same shape as the graph of $y = x^2$. If it is a parabola with vertical axis, find the discriminant and use it to determine the number of $x$-intercepts. See Examples 1–3, 5, 8, and 9.*

11. $f(x) = 2x^2 + 4x + 5$
12. $f(x) = 3x^2 - 6x + 4$
13. $f(x) = -x^2 + 5x + 3$
14. $x = -y^2 + 7y + 2$
15. $x = \frac{1}{3}y^2 + 6y + 24$
16. $x = \frac{1}{2}y^2 + 10y - 5$

*Matching*  Use the concepts of this section to match each equation in Exercises 17–22 with its graph in A–F.

17. $y = 2x^2 + 4x - 3$
18. $y = -x^2 + 3x + 5$
19. $y = -\frac{1}{2}x^2 - x + 1$
20. $x = y^2 + 6y + 3$
21. $x = -y^2 - 2y + 4$
22. $x = 3y^2 + 6y + 5$

A.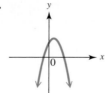
B.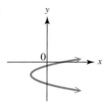
C.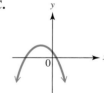
D.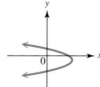
E.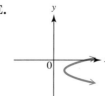
F.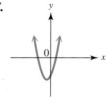

*Graph each parabola. (Use the results of Exercises 5–8 to help graph the parabolas in Exercises 23–26.) Give the vertex, axis, domain, and range. See Examples 4, 8, and 9.*

23. $f(x) = x^2 + 8x + 10$
24. $f(x) = x^2 + 10x + 23$
25. $f(x) = -2x^2 + 4x - 5$
26. $f(x) = -3x^2 + 12x - 8$
27. $x = (y + 2)^2 + 1$
28. $x = (y + 3)^2 - 2$

**29.** $x = -\dfrac{1}{5}y^2 + 2y - 4$  **30.** $x = -\dfrac{1}{2}y^2 - 4y - 6$

**31.** $x = 3y^2 + 12y + 5$  **32.** $x = 4y^2 + 16y + 11$

*Solve each problem. See Examples 6 and 7.*

**33.** Morgan's Department Store wants to construct a rectangular parking lot on land bordered on one side by a highway. It has 280 ft of fencing that is to be used to fence off the other three sides. What should be the dimensions of the lot if the enclosed area is to be a maximum? What is the maximum area?

**34.** Emmylou has 100 ft of fencing material to enclose a rectangular exercise run for her dog. One side of the run will border her house, so she will only need to fence three sides. What dimensions will give the enclosure the maximum area? What is the maximum area?

**35.** If an object on Earth is projected upward with an initial velocity of 32 ft per sec, then its height after $t$ seconds is given by

$$s(t) = -16t^2 + 32t.$$

Find the maximum height attained by the object and the number of seconds it takes to hit the ground.

**36.** A projectile on Earth is fired straight upward so that its distance (in feet) above the ground $t$ seconds after firing is given by

$$s(t) = -16t^2 + 400t.$$

Find the maximum height it reaches and the number of seconds it takes to reach that height.

**37.** The percent of births in the United States to teenage mothers in the years 1990–2002 can be modeled by the quadratic function defined by

$$f(x) = -0.0334x^2 + 0.2351x + 12.79,$$

where $x = 0$ represents 1990, $x = 1$ represents 1991, and so on. (*Source:* U.S. National Center for Health Statistics.)

(a) Since the coefficient of $x^2$ in the model is negative, the graph of this quadratic function is a parabola that opens down. Will the $y$-value of the vertex of this graph be a maximum or a minimum?

(b) In what year during this period was the percent of births in the United States to teenage mothers a maximum? (Round down to the nearest year.) Use the actual $y$-value of the vertex, to the nearest tenth, to find this percent.

**38.** The number of tickets sold (in millions) to the top 50 rock concerts in the years 1998–2002 is modeled by the quadratic function defined by

$$f(x) = -0.386x^2 + 1.28x + 11.3,$$

where $x = 0$ represents 1998, $x = 1$ represents 1999, and so on. (*Source:* Pollstar Online.)

(a) Since the coefficient of $x^2$ in the model is negative, the graph of this quadratic function is a parabola that opens down. Will the $y$-value of the vertex of this graph be a maximum or a minimum?

(b) In what year was the maximum number of tickets sold? (Round down to the nearest year.) Use the actual $x$-value of the vertex, to the nearest tenth, to find this number.

**39.** A charter flight charges a fare of $200 per person, plus $4 per person for each unsold seat on the plane. If the plane holds 100 passengers and if $x$ represents the number of unsold seats, find the following.

(a) A function defined by $R(x)$ that describes the total revenue received for the flight (*Hint:* Multiply the number of people flying, $100 - x$, by the price per ticket, $200 + 4x$.)

(b) The graph of the function from part (a)

(c) The number of unsold seats that will produce the maximum revenue

(d) The maximum revenue

**40.** For a trip to a resort, a charter bus company charges a fare of $48 per person, plus $2 per person for each unsold seat on the bus. If the bus has 42 seats and $x$ represents the number of unsold seats, find the following.

(a) A function defined by $R(x)$ that describes the total revenue from the trip (*Hint:* Multiply the total number riding, $42 - x$, by the price per ticket, $48 + 2x$.)
(b) The graph of the function from part (a)
(c) The number of unsold seats that produces the maximum revenue
(d) The maximum revenue

## 10.4 Symmetry; Increasing and Decreasing Functions

**OBJECTIVES**

1. Understand how multiplying a function by a real number $a$ affects its graph.
2. Test for symmetry with respect to an axis.
3. Test for symmetry with respect to the origin.
4. Decide if a function is increasing or decreasing on an interval.

**OBJECTIVE 1** Understand how multiplying a function by a real number $a$ affects its graph. In **Section 10.2,** we saw that the value of $a$ affects the graph of $g(x) = ax^2$ in several ways. If $a$ is positive, then the graph opens up; if $a$ is negative, then the graph opens down—that is, the graph of $g(x) = -x^2$ is the same as the graph of $f(x) = x^2$ reflected about the $x$-axis. Also, if $0 < |a| < 1$, then the graph is wider than the graph of $f(x) = x^2$; if $|a| > 1$, then the graph is narrower than that of $f(x) = x^2$. Figure 8 in **Section 10.2** demonstrates these effects with the graph of $f(x) = -\frac{1}{2}x^2$.

The same effects are true with the graphs of any function, as follows.

> **Reflection, Stretching, and Shrinking**
>
> The graph of $g(x) = a \cdot f(x)$ has the same general shape as the graph of $f(x)$.
> It is reflected about the $x$-axis if $a$ is negative;
> It is stretched vertically compared to the graph of $f(x)$ if $|a| > 1$;
> It is shrunken vertically compared to the graph of $f(x)$ if $0 < |a| < 1$.

**EXAMPLE 1** Comparing the Graph of $g(x) = a \cdot f(x)$ with the Graph of $f(x)$

Graph each function.

(a) $g(x) = 2\sqrt{x}$

We graphed $f(x) = \sqrt{x}$ in **Section 8.1.** Because the coefficient 2 is positive and greater than 1, the graph of $g(x) = 2\sqrt{x}$ will have the same general shape as the graph of $f(x)$ but is stretched vertically. See Figure 20.

(b) $h(x) = -\sqrt{x}$

Because of the negative sign, the graph of $h(x) = -\sqrt{x}$ will have the same general shape as the graph of $f(x) = \sqrt{x}$ but is reflected about the $x$-axis. See Figure 21.

| x | $\sqrt{x}$ | $2\sqrt{x}$ |
|---|---|---|
| 0 | 0 | 0 |
| 1 | 1 | 2 |
| 4 | 2 | 4 |
| 9 | 3 | 6 |

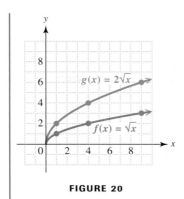

FIGURE 20

| x | $\sqrt{x}$ | $-\sqrt{x}$ |
|---|---|---|
| 0 | 0 | 0 |
| 1 | 1 | -1 |
| 4 | 2 | -2 |
| 9 | 3 | -3 |

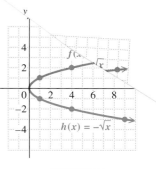

FIGURE 21

**NOW TRY** Exercise 1.

The parabolas graphed in the previous two sections were symmetric with respect to the axis of the parabola, a vertical or horizontal line through the vertex. The graphs of many other relations also are symmetric with respect to a line or a point. As we saw when graphing parabolas, symmetry is helpful in drawing graphs.

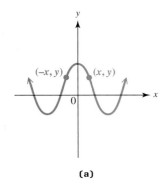

(a)

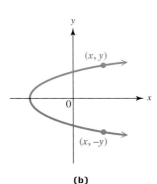

(b)

FIGURE 22

**OBJECTIVE 2** Test for symmetry with respect to an axis. The graph in Figure 22(a) is cut in half by the y-axis with each half the mirror image of the other half. Such a graph is *symmetric with respect to the y-axis.* As suggested by Figure 22(a), for a graph to be symmetric with respect to the y-axis, the point $(-x, y)$ must be on the graph whenever the point $(x, y)$ is on the graph.

Similarly, if the graph in Figure 22(b) were folded in half along the x-axis, the portion from the top would exactly match the portion from the bottom. Such a graph is *symmetric with respect to the x-axis.* As the graph suggests, symmetry with respect to the x-axis means that the point $(x, -y)$ must be on the graph whenever the point $(x, y)$ is on the graph.

The following tests tell when a graph is symmetric with respect to the x-axis or y-axis.

**Symmetry with Respect to an Axis**

The graph of a relation is **symmetric with respect to the y-axis** if the replacement of x with $-x$ results in an equivalent equation.

The graph of a relation is **symmetric with respect to the x-axis** if the replacement of y with $-y$ results in an equivalent equation.

**EXAMPLE 2** Testing for Symmetry with Respect to an Axis

Test for symmetry with respect to the x-axis and the y-axis.

**(a)** $y = x^2 + 4$

Replace x with $-x$.

$$y = x^2 + 4 \quad \text{becomes} \quad y = (-x)^2 + 4 = x^2 + 4.$$

The result is equivalent to the original equation, so the graph (shown in Figure 23) is symmetric with respect to the y-axis. The graph is *not* symmetric with respect to the x-axis, since replacing y with $-y$ gives

$$-y = x^2 + 4 \quad \text{or} \quad y = -x^2 - 4,$$

which is not equivalent to the original equation.

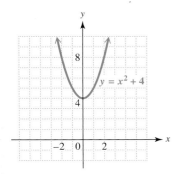

FIGURE 23

**(b)** $2x + y = 4$

Replace $x$ with $-x$ and then replace $y$ with $-y$; in neither case does an equivalent equation result. This graph is neither symmetric with respect to the $x$-axis nor to the $y$-axis.

**(c)** $x = |y|$

Replacing $x$ with $-x$ gives $-x = |y|$, which is not equivalent to the original equation. The graph is not symmetric with respect to the $y$-axis. Replacing $y$ with $-y$ gives $x = |-y| = |y|$. Thus, the graph is symmetric with respect to the $x$-axis.

> **NOW TRY** Exercises 3(a) and 3(b).

**OBJECTIVE 3** Test for symmetry with respect to the origin. Another kind of symmetry occurs when a graph can be rotated 180° about the origin and have the result coincide exactly with the original graph. Symmetry of this type is called *symmetry with respect to the origin.* It can be shown that rotating a graph 180° is equivalent to saying that the point $(-x, -y)$ is on the graph whenever the point $(x, y)$ is on the graph. Figure 24 shows two graphs that are symmetric with respect to the origin.

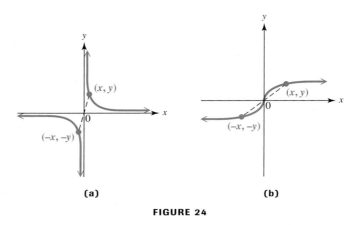

**FIGURE 24**

A test for this type of symmetry follows.

### Symmetry with Respect to the Origin

The graph of a relation is **symmetric with respect to the origin** if the replacement of both $x$ with $-x$ and $y$ with $-y$ results in an equivalent equation.

---

**EXAMPLE 3** Testing for Symmetry with Respect to the Origin

For each equation, decide if the graph is symmetric with respect to the origin.

**(a)** $3x = 5y$

Replace $x$ with $-x$ and $y$ with $-y$ in the equation.

$$3(-x) = 5(-y) \quad \text{Substitute } -x \text{ for } x \text{ and } -y \text{ for } y.$$
$$-3x = -5y \quad \text{Multiply.}$$
$$3x = 5y \quad \text{Multiply each side by } -1.$$

*Use parentheses here.*

Since this equation is equivalent to the original equation, the graph has symmetry with respect to the origin. See Figure 25(a).

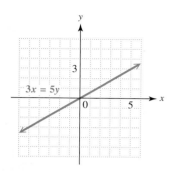

(a)

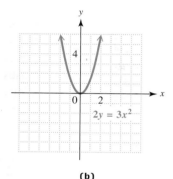

(b)

**FIGURE 25**

**(b)** $2y = 3x^2$

Substituting $-x$ for $x$ and $-y$ for $y$ gives

$$2(-y) = 3(-x)^2 \quad \text{Replace } y \text{ with } -y \text{ and } x \text{ with } -x.$$
$$-2y = 3x^2, \quad \text{Multiply.}$$

Use parentheses here.

which is not equivalent to the original equation. As Figure 25(b) shows, the graph is not symmetric with respect to the origin.

**NOW TRY** Exercise 3(c).

A summary of the tests for symmetry follows.

**TESTS FOR SYMMETRY**

| | Symmetric with Respect to | | |
|---|---|---|---|
| | x-axis | y-axis | Origin |
| Test | Replace $y$ with $-y$. | Replace $x$ with $-x$. | Replace $x$ with $-x$ and replace $y$ with $-y$. |
| Example | | | |

**NOW TRY** Exercises 9, 11, and 13.

**OBJECTIVE 4** Decide if a function is increasing or decreasing on an interval. Intuitively, a function is said to be *increasing* if its graph goes up from left to right. The function graphed in Figure 26(a) is an increasing function. On the other hand, a function is *decreasing* if its graph goes down from left to right, like the function in Figure 26(b). The function graphed in Figure 26(c) is neither an increasing function nor a decreasing function. However, it is increasing on the interval $(-\infty, -1]$ and decreasing on the interval $[-1, \infty)$.

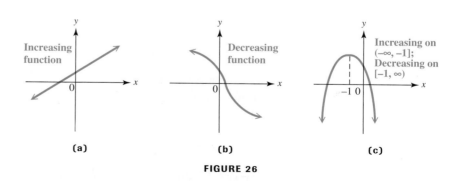

**FIGURE 26**

In the following definition of a function increasing or decreasing on an interval, $I$ represents any interval of real numbers.

### Increasing and Decreasing Functions

Let $f$ be a function, with $x_1$ and $x_2$ in an interval $I$ in the domain of $f$. Then

$f$ is **increasing** on $I$ if $f(x_1) < f(x_2)$ whenever $x_1 < x_2$;

$f$ is **decreasing** on $I$ if $f(x_1) > f(x_2)$ whenever $x_1 < x_2$.

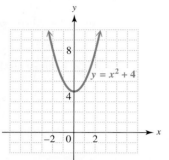

There can be confusion regarding whether endpoints of an interval should be included when determining intervals over which a function is increasing or decreasing. For example, consider the graph of $y = f(x) = x^2 + 4$, shown in the margin. Is the function increasing on $[0, \infty)$, or just $(0, \infty)$? The definition of increasing and decreasing allows us to include 0 as a part of the interval $I$ over which this function is increasing, because if we let $x_1 = 0$, then $f(0) < f(x_2)$ whenever $0 < x_2$. Thus, $f(x) = x^2 + 4$ is increasing on $[0, \infty)$. A similar discussion can be used to show that this function is decreasing on $(-\infty, 0]$. Do not confuse these concepts by saying that $f$ both increases and decreases at the point $(0, 0)$. *The concepts of increasing and decreasing functions apply to intervals of the domain, not to individual points.*

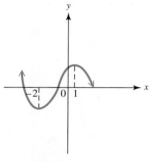

FIGURE 27

### EXAMPLE 4  Determining Increasing or Decreasing Intervals

Give the intervals where each function is increasing or decreasing.

(a) The function graphed in Figure 27 is decreasing on $(-\infty, -2]$ and $[1, \infty)$. The function is increasing on $[-2, 1]$.

(b) The function graphed in Figure 28 is increasing on $(-\infty, 2]$. On the interval $[2, \infty)$, the function is neither increasing nor decreasing.

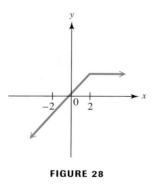

FIGURE 28

**NOW TRY** Exercises 21 and 23.

A function that is neither increasing nor decreasing on an interval is said to be **constant** on that interval. The function graphed in Figure 28 is constant on the interval $[2, \infty)$.

**CAUTION** *When identifying intervals over which a function is increasing, decreasing, or constant, remember that we are interested in identifying domain intervals. Range values do not appear in these stated intervals.*

## 10.4 Exercises

**NOW TRY Exercise**

*Concept Check* For Exercises 1 and 2, see Example 1.

1. Use the graph of $y = f(x)$ in the figure to obtain the graph of each equation. Describe how each graph is related to the graph of $y = f(x)$.

   **(a)** $y = -f(x)$     **(b)** $y = 2f(x)$

2. Use the graph of $y = g(x)$ in the figure to obtain the graph of each equation. Describe how each graph is related to the graph of $y = g(x)$.

   **(a)** $y = \dfrac{1}{2}g(x)$     **(b)** $y = -g(x)$

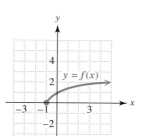
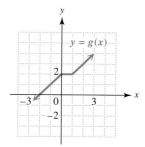

3. State how you would go about testing an equation in $x$ and $y$ to see if its graph is symmetric with respect to **(a)** the $x$-axis, **(b)** the $y$-axis, and **(c)** the origin.

4. Explain why the graph of a function cannot be symmetric with respect to the $x$-axis.

*Plot each point, and then use the same axes to plot the points that are symmetric to the given point with respect to the following: **(a)** x-axis, **(b)** y-axis, **(c)** origin.*

5. $(-4, -2)$     6. $(-8, 3)$     7. $(-8, 0)$     8. $(0, -3)$

*Use the tests for symmetry to decide whether the graph of each relation is symmetric with respect to the x-axis, the y-axis, or the origin. Remember that more than one of these symmetries may apply and that perhaps none apply. See Examples 2 and 3. (Do not graph.)*

9. $x^2 + y^2 = 5$     10. $y^2 = 4 - x^2$     11. $y = x^2 - 8x$

12. $y = 4x - x^2$     13. $y = |x|$     14. $y = |x| + 1$

15. $y = x^3$     16. $y = -x^3$     17. $f(x) = \dfrac{1}{1 + x^2}$

18. $f(x) = \dfrac{-1}{x^2 + 9}$     19. $xy = 2$     20. $xy = -6$

*For each function, give the interval where f is decreasing and the interval where f is increasing. In Exercises 25 and 26, first graph the function. See Example 4.*

21.
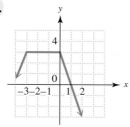

22.
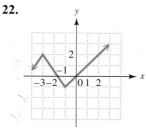

**23.** 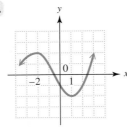 **24.**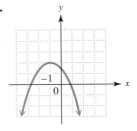

**25.** $f(x) = 2x^2 + 1$   **26.** $f(x) = 3 - x^2$

*Concept Check* Decide whether each figure is symmetric with respect to **(a)** the given line, and **(b)** the given point.

**27.**  **28.** 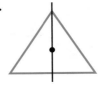 **29.** **30.**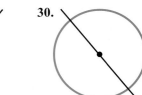

*Concept Check* Assume that for $y = f(x)$, $f(2) = 3$. For each given statement, find another value for the function.

**31.** The graph of $y = f(x)$ is symmetric with respect to the origin.

**32.** The graph of $y = f(x)$ is symmetric with respect to the $y$-axis.

**33.** The graph of $y = f(x)$ is symmetric with respect to the line $x = 3$.

**34.** A graph that is symmetric with respect to both the $x$-axis and the $y$-axis is also symmetric with respect to the origin. Explain why.

*Concept Check* Complete the left half of the graph of $y = f(x)$ based on the given assumption.

**35.** For all $x$, $f(-x) = f(x)$.

**36.** For all $x$, $f(-x) = -f(x)$.

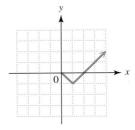

## 10.5 Piecewise Linear Functions

**OBJECTIVES**
1. Graph absolute value functions.
2. Graph other piecewise linear functions.
3. Graph step functions.

A function defined by different linear equations over different intervals of its domain is called a **piecewise linear function**.

**OBJECTIVE 1** Graph absolute value functions. An example of a function with a graph that includes portions of two lines is the **absolute value function**, defined by $f(x) = |x|$.

## Absolute Value Function

$$f(x) = |x| = \begin{cases} x & \text{if } x \geq 0 \\ -x & \text{if } x < 0 \end{cases}$$

Its graph, along with a table of selected ordered pairs, is shown in Figure 29. Its domain is $(-\infty, \infty)$, and its range is $[0, \infty)$.

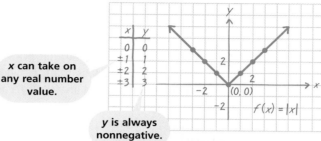

*x* can take on any real number value.

*y* is always nonnegative.

**FIGURE 29**

### EXAMPLE 1 Graphing Absolute Value Functions

Graph each function.

**(a)** $f(x) = -|x|$

As shown in **Section 10.4**, the negative sign indicates that the graph of $f(x)$ is the reflection of the graph of $y = |x|$ about the *x*-axis. The domain is $(-\infty, \infty)$, and the range is $(-\infty, 0]$. As shown in Figure 30, on the interval $(-\infty, 0]$, the graph is the same as the graph of $y = x$; on the interval $(0, \infty)$, it is the graph of $y = -x$.

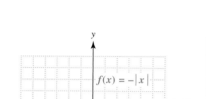

**FIGURE 30**

**(b)** $f(x) = |3x + 4| + 1$

In **Section 10.2**, we saw that the graph of $y = a(x - h)^2 + k$ is a parabola that, depending on the absolute value of *a*, is stretched or shrunken compared with the graph of $y = x^2$, and that is shifted *h* units horizontally and *k* units vertically. The same idea applies here. If we write

$$y = |3x + 4| + 1 \quad \text{as} \quad y = a|x - h| + k,$$

its graph will compare similarly with the graph of $y = |x|$.

$$y = |3x + 4| + 1$$

$$y = \left|3\left(x + \frac{4}{3}\right)\right| + 1 \quad \text{Factor 3 from the absolute value expression.}$$

$$y = |3|\left|x + \frac{4}{3}\right| + 1 \quad |ab| = |a| \cdot |b|; \text{ write each factor in absolute value bars.}$$

$$y = 3\left|x + \frac{4}{3}\right| + 1 \quad |3| = 3$$

In this form, we see that the graph is narrower than the graph of $y = |x|$ (that is, stretched vertically), with the "vertex" at the point $\left(-\frac{4}{3}, 1\right)$. See Figure 31. The axis of symmetry is $x = -\frac{4}{3}$. The coefficient of *x*, 3, determines the slopes of the two partial lines that form the graph. One has slope 3, and the other has slope $-3$. Because the absolute value of the slopes is greater than 1, the lines are steeper than the lines that form the graph of $y = |x|$.

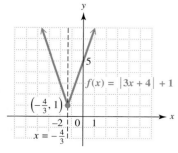

**FIGURE 31**

**NOW TRY** Exercises 5 and 11.

**518** CHAPTER 10 Additional Graphs of Functions and Relations

> **OBJECTIVE 2** Graph other piecewise linear functions. The different parts of piecewise linear functions may have completely different equations.
>
> **EXAMPLE 2** Graphing Piecewise Linear Functions
>
> Graph each function.
>
> **(a)** $f(x) = \begin{cases} x + 1 & \text{if } x \leq 2 \\ -2x + 7 & \text{if } x > 2 \end{cases}$
>
> Graph the function over each interval of the domain separately. If $x \leq 2$, this portion of the graph has an endpoint at $x = 2$. Find the $y$-value by substituting 2 for $x$ in $y = x + 1$ to get $y = 3$. Another point is needed to complete this portion of the graph. Choose an $x$-value less than 2. Choosing $x = -1$ gives $y = -1 + 1 = 0$. Draw the graph through $(2, 3)$ and $(-1, 0)$ as a partial line with an endpoint at $(2, 3)$.
>
> Graph the function over the interval $x > 2$ similarly. This line will have an open endpoint when $x = 2$ and $y = -2(2) + 7 = 3$. Choosing $x = 4$ gives $y = -2(4) + 7 = -1$. The partial line through $(2, 3)$ and $(4, -1)$ completes the graphs. The two parts meet at $(2, 3)$. See Figure 32.
>
> **(b)** $f(x) = \begin{cases} 2x + 3 & \text{if } x \leq 1 \\ -x + 4 & \text{if } x > 1 \end{cases}$
>
> Graph the function over each interval of the domain separately. If $x \leq 1$, the graph has an endpoint at $x = 1$. Substitute 1 for $x$ in $y = 2x + 3$ to get the ordered pair $(1, 5)$. For another point on this portion of the graph, choose a number less than 1, say $x = -2$. This gives the ordered pair $(-2, -1)$. Draw the partial line through these points with an endpoint at $(1, 5)$. Graph the function over the interval $x > 1$ similarly. This line has an open endpoint at $(1, 3)$ and passes through $(4, 0)$. The completed graph is shown in Figure 33.

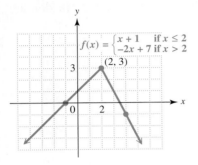

**FIGURE 32**

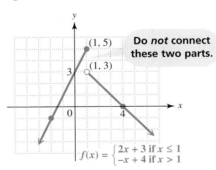

**FIGURE 33**

**NOW TRY** Exercises 15 and 17.

> **CAUTION** In Example 2, we did not graph the entire lines but only those portions with domain intervals as given. Graphs of these functions should not be two intersecting lines.
>
> **EXAMPLE 3** Applying a Piecewise Linear Function
>
> The table and the graph in Figure 34 show the number of cable TV stations from 1984 through 1999. Write equations for each part of the graph and use them to define a function that models the number of cable TV stations from 1984 through 1999. Let $x = 0$ represent 1984, $x = 1$ represent 1985, and so on.

| Year | Number |
|------|--------|
| 1984 | 48     |
| 1992 | 87     |
| 1999 | 181    |

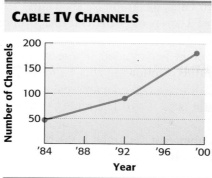

*Source:* National Cable Television Association.

**FIGURE 34**

The data in the table can be used to write the ordered pairs (0, 48) and (8, 87). The slope of the line through these points is

$$m = \frac{87 - 48}{8 - 0} = \frac{39}{8} = 4.875.$$

Using the ordered pair (0, 48) and $m = 4.875$ in the point-slope form of the equation of a line gives an equation of the first line.

$$\begin{aligned} y - y_1 &= m(x - x_1) &&\text{Point-slope form} \\ y - 48 &= 4.875(x - 0) &&\text{Substitute for } y_1, m, \text{ and } x_1. \\ y &= 4.875x + 48 &&\text{Solve for } y. \end{aligned}$$

Similarly, the equation of the other line is $y = 13.429x - 20.432$ (with the slope rounded to 3 decimal places). Verify this. Thus, the number of cable TV channels can be modeled by the function defined by

$$f(x) = \begin{cases} 4.875x + 48 & \text{if } 0 \leq x \leq 8 \\ 13.429x - 20.432 & \text{if } 8 < x \leq 15. \end{cases}$$

**NOW TRY** Exercise 31.

**OBJECTIVE 3** Graph step functions. The **greatest integer function**, written $f(x) = [\![x]\!]$, is defined as follows.

$[\![x]\!]$ **denotes the greatest integer that is less than or equal to $x$.**

For example,

$$[\![8]\!] = 8, \quad [\![7.45]\!] = 7, \quad [\![\pi]\!] = 3, \quad [\![-1]\!] = -1, \quad \text{and} \quad [\![-2.6]\!] = -3.$$

**EXAMPLE 4** Graphing the Greatest Integer Function

Graph $f(x) = [\![x]\!]$. Give the domain and range.

For $[\![x]\!]$,

$$\begin{aligned} &\text{if } -1 \leq x < 0, &&\text{then} &&[\![x]\!] = -1; \\ &\text{if } \phantom{-}0 \leq x < 1, &&\text{then} &&[\![x]\!] = 0; \\ &\text{if } \phantom{-}1 \leq x < 2, &&\text{then} &&[\![x]\!] = 1, \end{aligned}$$

and so on. Thus, the graph, as shown in Figure 35, consists of a series of horizontal line segments. In each one, the left endpoint is included and the right endpoint is excluded. These segments continue infinitely following this pattern to the left and right. Since $x$ can take any real number value, the domain is $(-\infty, \infty)$. The range is the set of integers $\{\ldots, -4, -3, -2, -1, 0, 1, 2, 3, 4, \ldots\}$. The appearance of the graph is the reason that this function is called a **step function.**

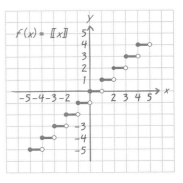

**FIGURE 35**

The graph of a step function also may be shifted. For example, the graph of $h(x) = [\![x - 2]\!]$ is the same as the graph of $f(x) = [\![x]\!]$ shifted two units to the right. Similarly, the graph of $g(x) = [\![x]\!] + 2$ is the graph of $f(x)$ shifted two units up.

**NOW TRY** Exercise 25.

**EXAMPLE 5** Graphing the Greatest Integer Function $f(x) = [\![ax + b]\!]$

Graph $f(x) = [\![\frac{1}{2}x + 1]\!]$.

If $x$ is in the interval $[0, 2)$, then $y = 1$. For $x$ in $[2, 4)$, $y = 2$, and so on. The graph is shown in Figure 36. Again, the domain of the function is $(-\infty, \infty)$. The range is the set of integers, $\{\ldots, -2, -1, 0, 1, 2, \ldots\}$. (As usual, we show only a portion of the graph.)

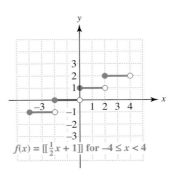

$f(x) = [\![\frac{1}{2}x + 1]\!]$ for $-4 \leq x < 4$

**FIGURE 36**

**NOW TRY** Exercise 27.

**EXAMPLE 6** Applying a Greatest Integer Function

An overnight delivery service charges $25 for a package weighing up to 2 lb. For each additional pound or fraction of a pound there is an additional charge of $3. Let $D(x)$ represent the cost to send a package weighing $x$ pounds. Graph $D(x)$ for $x$ in the interval $(0, 6]$.

For $x$ in the interval $(0, 2]$, $\quad y = 25$.

For $x$ in the interval $(2, 3]$, $\quad y = 25 + 3 = 28$.

For $x$ in the interval $(3, 4]$, $\quad y = 28 + 3 = 31$, and so on.

The graph, which is that of a step function, is shown in Figure 37.

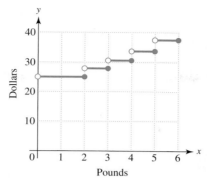

FIGURE 37

NOW TRY Exercise 33.

## 10.5 Exercises

NOW TRY Exercise

*Matching* Without actually plotting points, match each function with its graph from choices A–D.

**1.** $f(x) = |x - 2| + 2$

**2.** $f(x) = |x + 2| + 2$

**3.** $f(x) = |x - 2| - 2$

**4.** $f(x) = |x + 2| - 2$

**A.**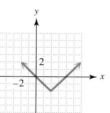

**B.**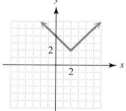

**C.**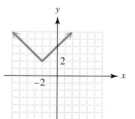

**D.**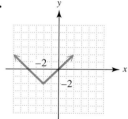

*Sketch the graph of each function defined by the absolute value expression. See Example 1.*

**5.** $f(x) = |x + 1|$

**6.** $f(x) = |x - 1|$

**7.** $f(x) = |2 - x|$

**8.** $f(x) = |-3 - x|$

**9.** $y = |x| + 4$

**10.** $y = 2|x| - 1$

**11.** $y = 3|x - 2| - 1$

**12.** $y = \frac{1}{2}|x + 3| + 1$

*For each piecewise linear function, find (a) $f(-5)$, (b) $f(-1)$, (c) $f(0)$, (d) $f(3)$, and (e) $f(5)$. See Example 2.*

**13.** $f(x) = \begin{cases} 2x & \text{if } x \leq -1 \\ x - 1 & \text{if } x > -1 \end{cases}$

**14.** $f(x) = \begin{cases} 3x + 5 & \text{if } x \leq 0 \\ 4 - 2x & \text{if } 0 < x < 2 \\ x & \text{if } x \geq 2 \end{cases}$

*Graph each piecewise linear function. See Example 2.*

**15.** $f(x) = \begin{cases} x - 1 & \text{if } x \leq 3 \\ 2 & \text{if } x > 3 \end{cases}$

**16.** $f(x) = \begin{cases} 6 - x & \text{if } x \leq 3 \\ 3x - 6 & \text{if } x > 3 \end{cases}$

**17.** $f(x) = \begin{cases} 4 - x & \text{if } x < 2 \\ 1 + 2x & \text{if } x \geq 2 \end{cases}$

**18.** $f(x) = \begin{cases} -2 & \text{if } x \geq 1 \\ 2 & \text{if } x < 1 \end{cases}$

**19.** $f(x) = \begin{cases} 2x + 1 & \text{if } x \geq 0 \\ x & \text{if } x < 0 \end{cases}$

**20.** $f(x) = \begin{cases} 5x - 4 & \text{if } x \geq 1 \\ x & \text{if } x < 1 \end{cases}$

*Graph each piecewise function. (Hint: At least one part is not linear.)*

**21.** $f(x) = \begin{cases} 2 + x & \text{if } x < -4 \\ -x^2 & \text{if } x \geq -4 \end{cases}$

**22.** $f(x) = \begin{cases} -2x & \text{if } x \leq 2 \\ -x^2 & \text{if } x > 2 \end{cases}$

**23.** $f(x) = \begin{cases} |x| & \text{if } x > -2 \\ x^2 - 2 & \text{if } x \leq -2 \end{cases}$

**24.** $f(x) = \begin{cases} |x| - 1 & \text{if } x > -1 \\ x^2 - 1 & \text{if } x \leq -1 \end{cases}$

*Graph each step function defined by the greatest integer expressions. See Examples 4 and 5.*

**25.** $f(x) = [\![-x]\!]$

**26.** $f(x) = [\![2x]\!]$

**27.** $f(x) = [\![2x - 1]\!]$

**28.** $f(x) = [\![3x + 1]\!]$

**29.** $f(x) = [\![3x]\!]$

**30.** $f(x) = [\![3x]\!] + 1$

*Work each problem. See Examples 3 and 6.*

**31.** The light vehicle market share (in percent) in the United States for domestic cars is shown in the graph.

*Source:* J.D. Power & Associates.

Let $x = 3$ represent 1993, $x = 6$ represent 1996, and so on. Use the points on the graph to write equations for the line segments in the intervals $[3, 6]$ and $(6, 9]$. Then define $f(x)$ for the piecewise linear function.

**32.** To rent a midsized car from Avis costs $30 per day or fraction of a day. If you pick up the car in Lansing and drop it in West Lafayette, there is a fixed $50 dropoff charge. Let $C(x)$ represent the cost of renting the car for $x$ days, taking it from Lansing to West Lafayette. Find each of the following.

(a) $C\left(\dfrac{3}{4}\right)$  (b) $C\left(\dfrac{9}{10}\right)$  (c) $C(1)$  (d) $C\left(1\dfrac{5}{8}\right)$  (e) $C(2.4)$

(f) Graph $y = C(x)$.

**33.** Suppose a chain-saw rental firm charges a fixed $4 sharpening fee plus $7 per day or fraction of a day. Let $S(x)$ represent the cost of renting a saw for $x$ days. Find each of the following.

(a) $S(1)$  (b) $S(1.25)$  (c) $S(3.5)$

(d) Graph $y = S(x)$.  (e) Give the domain and range of $S$.

34. When a diabetic takes long-acting insulin, the insulin reaches its peak effect on the blood sugar level in about 3 hr. This effect remains fairly constant for 5 hr, then declines, and is very low until the next injection. In a typical patient, the level of insulin might be given by the following function.

$$i(t) = \begin{cases} 40t + 100 & \text{if } 0 \le t \le 3 \\ 220 & \text{if } 3 < t \le 8 \\ -80t + 860 & \text{if } 8 < t \le 10 \\ 60 & \text{if } 10 < t \le 24 \end{cases}$$

Here $i(t)$ is the blood sugar level, in appropriate units, at time $t$ measured in hours from the time of the injection. Chuck takes his insulin at 6 A.M. Find the blood sugar level at each of the following times.

(a) 7 A.M.   (b) 9 A.M.   (c) 10 A.M.   (d) noon
(e) 3 P.M.   (f) 5 P.M.   (g) midnight   (h) Graph $y = i(t)$.

# 10 SUMMARY

## KEY TERMS

**10.1** difference quotient
composite function (composition)
**10.2** parabola
vertex
axis
quadratic function

focus
directrix
**10.4** symmetric with respect to the *y*-axis
symmetric with respect to the *x*-axis

symmetric with respect to the origin
increasing function
decreasing function
**10.5** piecewise linear function

absolute value function
greatest integer function
step function

## NEW SYMBOLS

$f \circ g$  composite function

$[\![x]\!]$  greatest integer function

## QUICK REVIEW

── CONCEPTS ── | ── EXAMPLES ──

### 10.1 Review of Operations and Composition

**Operations on Functions**

$(f + g)(x) = f(x) + g(x)$  Sum
$(f - g)(x) = f(x) - g(x)$  Difference
$(fg)(x) = f(x) \cdot g(x)$  Product
$\left(\dfrac{f}{g}\right)(x) = \dfrac{f(x)}{g(x)}, \; g(x) \neq 0$  Quotient

**Composition of Functions**

If $f$ and $g$ are functions, then the composite function of $g$ and $f$ is

$(g \circ f)(x) = g(f(x))$.

If $f(x) = 3x^2 + 2$ and $g(x) = \sqrt{x}$, then

$(f + g)(x) = 3x^2 + 2 + \sqrt{x}$
$(f - g)(x) = 3x^2 + 2 - \sqrt{x}$
$(fg)(x) = (3x^2 + 2)(\sqrt{x})$
$\left(\dfrac{f}{g}\right)(x) = \dfrac{3x^2 + 2}{\sqrt{x}}, \; x > 0.$

If $g(x) = \sqrt{x}$ and $f(x) = x^2 - 1$, then the composite function of $g$ and $f$ is

$(g \circ f)(x) = g(f(x)) = \sqrt{x^2 - 1}.$

### 10.2 Graphs of Quadratic Functions

1. The graph of the quadratic function defined by $F(x) = a(x - h)^2 + k, \; a \neq 0$, is a parabola with vertex at $(h, k)$ and the vertical line $x = h$ as axis.

2. The graph opens up if $a$ is positive and down if $a$ is negative.

3. The graph is wider than the graph of $f(x) = x^2$ if $0 < |a| < 1$ and narrower if $|a| > 1$.

The parabola with focus at $(0, p)$ and directrix $y = -p$ has equation

$$y = \dfrac{1}{4p}x^2.$$

Graph $f(x) = -(x + 3)^2 + 1$.

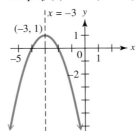

The graph opens down since $a < 0$. It is shifted 3 units left and 1 unit up, so the vertex is $(-3, 1)$, with axis $x = -3$. The domain is $(-\infty, \infty)$; the range is $(-\infty, 1]$.

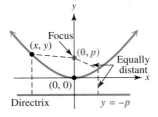

*(continued)*

― CONCEPTS ― | ― EXAMPLES ―

## 10.3 More about Parabolas and Their Applications

The vertex of the graph of $f(x) = ax^2 + bx + c$, $a \neq 0$, may be found by completing the square.

The vertex has coordinates $\left(\dfrac{-b}{2a}, f\left(\dfrac{-b}{2a}\right)\right)$.

**Graphing a Quadratic Function**

*Step 1* Determine whether the graph opens up or down.
*Step 2* Find the vertex.
*Step 3* Find the x-intercepts (if any). Find the y-intercept.
*Step 4* Find and plot additional points as needed.

Graph $f(x) = x^2 + 4x + 3$.

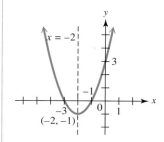

The graph opens up since $a > 0$. The vertex is $(-2, -1)$. The solutions of $x^2 + 4x + 3 = 0$ are $-1$ and $-3$, so the x-intercepts are $(-1, 0)$ and $(-3, 0)$. Since $f(0) = 3$, the y-intercept is $(0, 3)$. The domain is $(-\infty, \infty)$; the range is $[-1, \infty)$.

**Horizontal Parabolas**

The graph of $x = ay^2 + by + c$ is a horizontal parabola, opening to the right if $a > 0$ or to the left if $a < 0$. Horizontal parabolas do not represent functions.

Graph $x = 2y^2 + 6y + 5$.

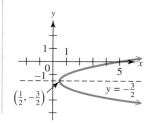

The graph opens to the right since $a > 0$. The vertex is $\left(\tfrac{1}{2}, -\tfrac{3}{2}\right)$. The axis is $y = -\tfrac{3}{2}$.

The domain is $\left[\tfrac{1}{2}, \infty\right)$; the range is $(-\infty, \infty)$.

## 10.4 Symmetry; Increasing and Decreasing Functions

The graph of $g(x) = a \cdot f(x)$ has the same general shape as the graph of $f(x)$.

It is reflected about the x-axis if $a$ is negative.

It is stretched vertically compared to the graph of $f(x)$ if $|a| > 1$.

It is shrunken vertically compared to the graph of $f(x)$ if $0 < |a| < 1$.

Let $g(x) = -\tfrac{1}{2}\sqrt{x}$; the graph of $g(x)$ is shown with the graph of $f(x) = \sqrt{x}$.

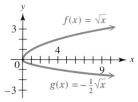

**Symmetry**

To decide whether the graph of a relation is symmetric with respect to the following, perform the indicated test.

**(a)** The x-axis
Replace $y$ with $-y$.

**(b)** The y-axis
Replace $x$ with $-x$.

**(c)** The origin
Replace $x$ with $-x$ and $y$ with $-y$.

The symmetry holds if the resulting equation is equivalent to the original equation.

Test each relation for symmetry.

**(a)** $x = y^2 - 5$
$x = (-y)^2 - 5 = y^2 - 5$
The graph is symmetric with respect to the x-axis.

**(b)** $y = -2x^2 + 1$
$y = -2(-x)^2 + 1 = -2x^2 + 1$
The graph is symmetric with respect to the y-axis.

**(c)** $\quad x^2 + y^2 = 4$
$(-x)^2 + (-y)^2 = 4$
$x^2 + y^2 = 4$

The graph is symmetric with respect to the origin (and to the x-axis and y-axis).

*(continued)*

## CONCEPTS

**Increasing and Decreasing Functions**

A function $f$ is increasing on an interval if $f(x_1) < f(x_2)$ whenever $x_1 < x_2$.

A function $f$ is decreasing on an interval if $f(x_1) > f(x_2)$ whenever $x_1 < x_2$.

If $f$ is neither increasing nor decreasing on an interval, it is constant there.

### 10.5 Piecewise Linear Functions

**Absolute Value Function**

$$f(x) = |x| = \begin{cases} x & \text{if } x \geq 0 \\ -x & \text{if } x < 0 \end{cases}$$

The graph of $f(x) = |ax + b| + c$ has "vertex" at $\left(-\frac{b}{a}, c\right)$ and it is symmetric with respect to a vertical axis through the "vertex."

**Other Piecewise Linear Functions**

Graph each portion with an open or solid endpoint as appropriate.

**Greatest Integer Function**

$$f(x) = [\![x]\!]$$

$[\![x]\!]$ is the greatest integer less than or equal to $x$.

## EXAMPLES

$f$ is increasing on $(-\infty, a]$.

$f$ is decreasing on $[a, b]$.

$f$ is constant on $[b, \infty)$.

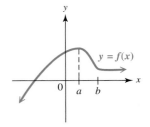

Graph $f(x) = 2|x - 1| + 3$.

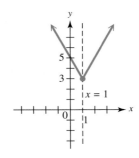

Graph $f(x) = \begin{cases} x - 2 & \text{if } x \geq 1 \\ 3x & \text{if } x < 1 \end{cases}$.

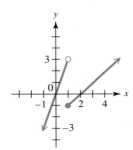

Graph $f(x) = [\![2x - 1]\!]$.

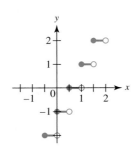

# 10 REVIEW EXERCISES

*Given $f(x) = x^2 - 2x$ and $g(x) = 5x + 3$, find the following. Give the domain of each.*

1. $(f + g)(x)$
2. $(f - g)(x)$
3. $(fg)(x)$
4. $\left(\dfrac{f}{g}\right)(x)$
5. $(g \circ f)(x)$
6. $(f \circ g)(x)$

*For $f(x) = 2x - 3$ and $g(x) = \sqrt{x}$, find the following.*

7. $(f - g)(4)$
8. $\left(\dfrac{f}{g}\right)(9)$
9. $(fg)(5)$
10. $(f + g)(5)$
11. $(fg)(2b)$
12. $(g \circ f)(2)$

13. *Concept Check* After working Exercise 12, find $(f \circ g)(2)$. Are your answers equal? Is composition of functions a commutative operation?

14. Explain in your own words why 5 is not in the domain of $f(x) = \sqrt{9 - 2x}$.

*Identify the vertex of each parabola.*

15. $y = 6 - 2x^2$
16. $f(x) = -(x - 1)^2$
17. $f(x) = (x - 3)^2 + 7$
18. $y = -3x^2 + 4x - 2$
19. $x = (y - 3)^2 - 4$

*Graph each parabola. Give the vertex, axis, domain, and range.*

20. $f(x) = -5x^2$
21. $f(x) = 3x^2 - 2$
22. $y = (x + 2)^2$
23. $y = 2(x - 2)^2 - 3$
24. $f(x) = -2x^2 + 8x - 5$
25. $y = x^2 + 3x + 2$
26. $x = (y - 1)^2 + 2$
27. $x = 2(y + 3)^2 - 4$
28. $x = -\dfrac{1}{2}y^2 + 6y - 14$

*Solve each problem.*

29. Find the length and width of a rectangle having a perimeter of 200 m if the area is to be a maximum. What is the maximum area?

30. The height (in feet) of a projectile $t$ seconds after being fired from Earth into the air is given by
$$f(t) = -16t^2 + 160t.$$
Find the number of seconds required for the projectile to reach maximum height. What is the maximum height?

*Use the tests for symmetry to determine any symmetries of the graph of each relation. Do not graph.*

31. $2x^2 - y^2 = 4$
32. $3x^2 + 4y^2 = 12$
33. $2x - y^2 = 8$
34. $y = 2x^2 + 3$
35. $y = 2\sqrt{x} - 4$
36. $y = \dfrac{1}{x^2}$

37. What is wrong with this statement? "A function whose graph is a circle centered at the origin has symmetry with respect to both axes and the origin."

38. *Concept Check* Suppose that a circle has its center at the origin. Is it symmetric with respect to **(a)** the $x$-axis, **(b)** the $y$-axis, **(c)** the origin?

39. *Concept Check* Suppose that a linear function in the form $f(x) = mx + b$ has $m < 0$. Is it increasing or decreasing over all real numbers?

*Give the intervals where each function is increasing, decreasing, or constant.*

**40.** 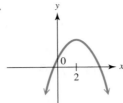   **41.**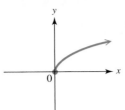

**42.** 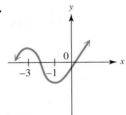   **43.**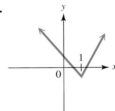

*Graph each function.*

**44.** $f(x) = 2|x| + 3$

**45.** $f(x) = |x - 2|$

**46.** $f(x) = -|x - 1|$

**47.** $f(x) = \begin{cases} 2x + 1 & \text{if } x \leq -1 \\ x + 3 & \text{if } x > -1 \end{cases}$

**48.** $f(x) = \begin{cases} -x & \text{if } x \leq 0 \\ x^2 & \text{if } x > 0 \end{cases}$

**49.** $f(x) = -[\![x]\!]$

**50.** $f(x) = [\![x + 1]\!]$

**51.** Describe how the graph of $y = 2|x + 4| - 3$ relates to the graph of $y = |x|$.

**52.** Taxi rates in a small town are 90¢ for the first $\frac{1}{9}$ mi and 10¢ for each additional $\frac{1}{9}$ mi or fraction of $\frac{1}{9}$ mi. Let $C(x)$ be the cost for a taxi ride of $\frac{x}{9}$ mi. Find

(a) $C(1)$   (b) $C(2.3)$   (c) $C(8)$.
(d) Graph $y = C(x)$.   (e) Give the domain and range of $C$.

# 10 TEST

*Let f and g be functions defined by $f(x) = 4x + 2$ and $g(x) = -x^2 + 3$. Find each function value.*

**1.** $g(1)$

**2.** $(f + g)(-2)$

**3.** $\left(\dfrac{f}{g}\right)(3)$

**4.** $(f \circ g)(2)$

**5.** For the functions defined in the directions for Exercises 1–4, find and simplify $(f - g)(x)$ and give its domain.

**6.** *Multiple Choice* Which one of the following most closely resembles the graph of $f(x) = a(x - h)^2 + k$ if $a < 0, h > 0$, and $k < 0$?

**A.** 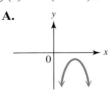   **B.** 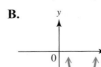   **C.** 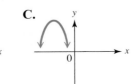   **D.**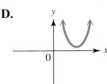

*Graph each parabola. Identify the vertex, axis, domain, and range.*

**7.** $f(x) = \dfrac{1}{2}x^2 - 2$  **8.** $f(x) = -x^2 + 4x - 1$  **9.** $x = -(y-2)^2 + 2$

**10.** The percent increase for in-state tuition at Iowa public universities during the years 1992 through 2002 can be modeled by the quadratic function defined by

$$f(x) = 0.156x^2 - 2.05x + 10.2,$$

where $x = 2$ represents 1992, $x = 3$ represents 1993, and so on. (*Source:* Iowa Board of Regents.)

**(a)** Based on this model, by what percent (to the nearest tenth) did tuition increase in 2001?

**(b)** In what year was the minimum tuition increase? (Round down to the nearest year.) To the nearest tenth, by what percent did tuition increase that year?

**11.** Palo Alto College is planning to construct a rectangular parking lot on land bordered on one side by a highway. The plan is to use 640 ft of fencing to fence off the other three sides. What should the dimensions of the lot be if the enclosed area is to be a maximum? What is the maximum area?

*Use the tests for symmetry to determine the symmetries, if any, for each relation.*

**12.** $f(x) = -x^2 + 1$  **13.** $x = y^2 + 7$  **14.** $x^2 + y^2 = 4$

**15.** *Matching*  Match each function (a)–(d) with its graph from choices A–D.

  **(a)** $f(x) = |x - 2|$  **(b)** $f(x) = |x + 2|$  **(c)** $f(x) = |x| + 2$  **(d)** $f(x) = |x| - 2$

**A.**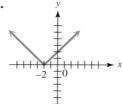

**B.**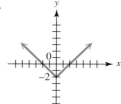

**C.**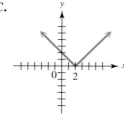

**D.**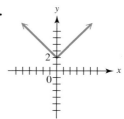

**16.** Give the intervals over which the function is increasing, decreasing, and constant.

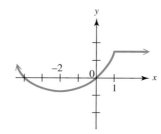

*Graph each relation.*

**17.** $f(x) = |x - 3| + 4$

**18.** $f(x) = [\![2x]\!]$

**19.** $f(x) = \begin{cases} -x & \text{if } x \leq 2 \\ x - 4 & \text{if } x > 2 \end{cases}$

**20.** Assume that postage rates are 44¢ for the first ounce, plus 17¢ for each additional ounce, and that each letter carries one 44¢ stamp and as many 17¢ stamps as necessary. Graph the function defined by $p(x)$ = the number of stamps on a letter weighing $x$ ounces. Use the interval $(0, 5]$.

# Inverse, Exponential, and Logarithmic Functions

In 2001, Apple Computer Inc., introduced the iPod. Since then, the company has sold over 40 million of the popular music players, in spite of warnings by experts that listening to the devices at high volumes may put people at increased risk of hearing loss. In 2006, a federal class-action lawsuit was filed against the company, accusing it of not taking adequate steps to protect the hearing of iPod users. As a result, Apple issued a software update that allows listeners to set maximum volume limits on some of the newer iPod models. (*Source: Sacramento Bee, USA Today.*)

In Example 4 of Section 11.5, we use a *logarithmic function* to calculate the volume level, in *decibels,* of an iPod.

**11.1** Inverse Functions

**11.2** Exponential Functions

**11.3** Logarithmic Functions

**11.4** Properties of Logarithms

**11.5** Common and Natural Logarithms

**11.6** Exponential and Logarithmic Equations; Further Applications

# 11.1 Inverse Functions

**OBJECTIVES**

1. Decide whether a function is one-to-one and, if it is, find its inverse.
2. Use the horizontal line test to determine whether a function is one-to-one.
3. Find the equation of the inverse of a function.
4. Graph $f^{-1}$ from the graph of $f$.

In this chapter we will study two important types of functions, *exponential* and *logarithmic*. These functions are related in a special way: They are *inverses* of one another. We begin by discussing inverse functions in general.

**OBJECTIVE 1** Decide whether a function is one-to-one and, if it is, find its inverse. Suppose we define the function

$$G = \{(-2, 2), (-1, 1), (0, 0), (1, 3), (2, 5)\}.$$

We can form another set of ordered pairs from $G$ by interchanging the $x$- and $y$-values of each pair in $G$. We can call this set $F$, so

$$F = \{(2, -2), (1, -1), (0, 0), (3, 1), (5, 2)\}.$$

To show that these two sets are related, $F$ is called the *inverse* of $G$. For a function $f$ to have an inverse, $f$ must be a *one-to-one function*.

---

**One-to-One Function**

In a **one-to-one function,** each $x$-value corresponds to only one $y$-value, and each $y$-value corresponds to only one $x$-value.

---

The function shown in Figure 1(a) is not one-to-one because the $y$-value 7 corresponds to *two* $x$-values, 2 and 3. That is, the ordered pairs (2, 7) and (3, 7) both belong to the function. The function in Figure 1(b) is one-to-one.

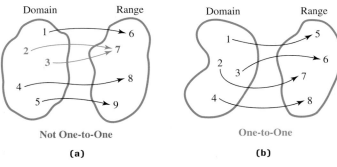

**FIGURE 1**

The *inverse* of any one-to-one function $f$ is found by interchanging the components of the ordered pairs of $f$. The inverse of $f$ is written $f^{-1}$. Read $f^{-1}$ as **"the inverse of $f$"** or **"$f$-inverse."**

---

**CAUTION** The symbol $f^{-1}(x)$ does ***not*** represent $\dfrac{1}{f(x)}$.

---

532

The definition of the inverse of a function follows.

### Inverse of a Function

The **inverse** of a one-to-one function $f$, written $f^{-1}$, is the set of all ordered pairs of the form $(y, x)$, where $(x, y)$ belongs to $f$. Since the inverse is formed by interchanging $x$ and $y$, the domain of $f$ becomes the range of $f^{-1}$ and the range of $f$ becomes the domain of $f^{-1}$.

For inverses $f$ and $f^{-1}$, it follows that

$$(f \circ f^{-1})(x) = x \quad \text{and} \quad (f^{-1} \circ f)(x) = x.$$

**EXAMPLE 1** Finding the Inverses of One-to-One Functions

Decide whether each function is one-to-one. If it is, find the inverse.

**(a)** $F = \{(-2, 1), (-1, 0), (0, 1), (1, 2), (2, 2)\}$

Each $x$-value in $F$ corresponds to just one $y$-value. However, the $y$-value 1 corresponds to two $x$-values, $-2$ and 0. Also, the $y$-value 2 corresponds to both 1 and 2. Because some $y$-values correspond to more than one $x$-value, $F$ is not one-to-one and does not have an inverse.

**(b)** $G = \{(3, 1), (0, 2), (2, 3), (4, 0)\}$

Every $x$-value in $G$ corresponds to only one $y$-value, and every $y$-value corresponds to only one $x$-value, so $G$ is a one-to-one function. The inverse function is found by interchanging the $x$- and $y$-values in each ordered pair.

$$G^{-1} = \{(1, 3), (2, 0), (3, 2), (0, 4)\}$$

Notice how the domain and range of $G$ become the range and domain, respectively, of $G^{-1}$.

**(c)** The Pollutant Standard Index (PSI) is an indicator of air quality. If the PSI exceeds 100 on a particular day, then that day is classified as unhealthy. The table shows the number of unhealthy days in Chicago for the years 1991–2002, based on new standards set in 1998.

| Year | Number of Unhealthy Days | Year | Number of Unhealthy Days |
|---|---|---|---|
| 1991 | 24 | 1997 | 10 |
| 1992 | 5 | 1998 | 12 |
| 1993 | 4 | 1999 | 19 |
| 1994 | 13 | 2000 | 2 |
| 1995 | 24 | 2001 | 22 |
| 1996 | 7 | 2002 | 21 |

*Source:* U.S. Environmental Protection Agency, Office of Air Quality Planning and Standards.

Let $f$ be the function defined in the table, with the years forming the domain and the numbers of unhealthy days forming the range. Then $f$ is not one-to-one, because in two different years (1991 and 1995), the number of unhealthy days was the same, 24.

**NOW TRY** Exercises 1, 9, and 11.

**OBJECTIVE 2** Use the horizontal line test to determine whether a function is one-to-one. It may be difficult to decide whether a function is one-to-one just by looking at the equation that defines the function. However, by graphing the function and observing the graph, we can use the *horizontal line test* to tell whether the function is one-to-one.

> **Horizontal Line Test**
>
> A function is one-to-one if every horizontal line intersects the graph of the function at most once.

The horizontal line test follows from the definition of a one-to-one function. Any two points that lie on the same horizontal line have the same $y$-coordinate. No two ordered pairs that belong to a one-to-one function may have the same $y$-coordinate, and, therefore, no horizontal line will intersect the graph of a one-to-one function more than once.

**EXAMPLE 2** Using the Horizontal Line Test

Use the horizontal line test to determine whether each graph is the graph of a one-to-one function.

(a)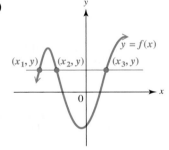

(b)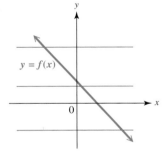

Because a horizontal line intersects the graph in more than one point (actually three points), the function is not one-to-one.

Every horizontal line will intersect the graph in exactly one point. This function is one-to-one.

**NOW TRY** Exercise 7.

**OBJECTIVE 3** Find the equation of the inverse of a function. The inverse of a one-to-one function is found by interchanging the $x$- and $y$-values of each of its ordered pairs. The equation of the inverse of a function defined by $y = f(x)$ is found in the same way.

### Finding the Equation of the Inverse of $y = f(x)$

For a one-to-one function $f$ defined by an equation $y = f(x)$, find the defining equation of the inverse as follows.

**Step 1** Interchange $x$ and $y$.
**Step 2** Solve for $y$.
**Step 3** Replace $y$ with $f^{-1}(x)$.

### EXAMPLE 3  Finding Equations of Inverses

Decide whether each equation defines a one-to-one function. If so, find the equation that defines the inverse.

**(a)** $f(x) = 2x + 5$

The graph of $y = 2x + 5$ is a nonvertical line, so by the horizontal line test, $f$ is a one-to-one function. To find the inverse, let $y = f(x)$ so that

$$y = 2x + 5$$
$$x = 2y + 5 \quad \text{Interchange } x \text{ and } y. \text{ (Step 1)}$$
$$2y = x - 5 \quad \text{Solve for } y. \text{ (Step 2)}$$
$$y = \frac{x - 5}{2}$$
$$f^{-1}(x) = \frac{x - 5}{2}, \quad \text{Replace } y \text{ with } f^{-1}(x). \text{ (Step 3)}$$

which can be written

$$f^{-1}(x) = \frac{x}{2} - \frac{5}{2}, \quad \text{or} \quad f^{-1}(x) = \frac{1}{2}x - \frac{5}{2}. \qquad \frac{a - b}{c} = \frac{a}{c} - \frac{b}{c}$$

Thus, $f^{-1}$ is a linear function. In the function defined by $y = 2x + 5$, the value of $y$ is found by starting with a value of $x$, multiplying by 2, and adding 5. The equation $f^{-1}(x) = \frac{x - 5}{2}$ for the inverse has us *subtract* 5 and then *divide* by 2. This shows how an inverse is used to "undo" what a function does to the variable $x$.

**(b)** $y = x^2 + 2$

This equation has a vertical parabola as its graph, so some horizontal lines will intersect the graph at two points. For example, both $x = 3$ and $x = -3$ correspond to $y = 11$. Because of the $x^2$-term, there are many pairs of $x$-values that correspond to the same $y$-value. This means that the function defined by $y = x^2 + 2$ is not one-to-one and does not have an inverse.

If this is not noticed, then following the steps for finding the equation of an inverse leads to

$$y = x^2 + 2$$
$$x = y^2 + 2 \quad \text{Interchange } x \text{ and } y.$$
$$y^2 = x - 2 \quad \text{Solve for } y.$$
$$y = \pm\sqrt{x - 2}. \quad \text{Square root property}$$

The last step shows that there are two $y$-values for each choice of $x > 2$, so the given function is not one-to-one and does not have an inverse.

**(c)** $f(x) = (x - 2)^3$

Refer to **Section 5.3** to see that the graphs of cubing functions are one-to-one.

$$y = (x - 2)^3 \quad \text{Replace } f(x) \text{ with } y.$$
$$x = (y - 2)^3 \quad \text{Interchange } x \text{ and } y.$$
$$\sqrt[3]{x} = \sqrt[3]{(y - 2)^3} \quad \text{Take the cube root on each side.}$$
$$\sqrt[3]{x} = y - 2$$
$$y = \sqrt[3]{x} + 2 \quad \text{Solve for } y.$$
$$f^{-1}(x) = \sqrt[3]{x} + 2 \quad \text{Replace } y \text{ with } f^{-1}(x).$$

**NOW TRY** Exercises 13, 17, and 19.

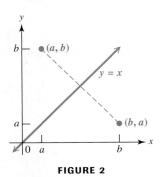

FIGURE 2

**OBJECTIVE 4** Graph $f^{-1}$ from the graph of $f$. One way to graph the inverse of a function $f$ whose equation is known is to find some ordered pairs that belong to $f$, interchange $x$ and $y$ to get ordered pairs that belong to $f^{-1}$, plot those points, and sketch the graph of $f^{-1}$ through the points. A simpler way is to select points on the graph of $f$ and use symmetry to find corresponding points on the graph of $f^{-1}$.

For example, suppose the point $(a, b)$ shown in Figure 2 belongs to a one-to-one function $f$. Then the point $(b, a)$ belongs to $f^{-1}$. The line segment connecting $(a, b)$ and $(b, a)$ is perpendicular to, and cut in half by, the line $y = x$. The points $(a, b)$ and $(b, a)$ are "mirror images" of each other with respect to $y = x$. For this reason **we can find the graph of $f^{-1}$ from the graph of $f$ by locating the mirror image of each point in $f$ with respect to the line $y = x$.**

**EXAMPLE 4** Graphing the Inverse

Graph the inverses of the functions $f$ (shown in blue) in Figure 3.

In Figure 3 the graphs of two functions $f$ are shown in blue. Their inverses are shown in red. In each case, the graph of $f^{-1}$ is a reflection of the graph of $f$ with respect to the line $y = x$.

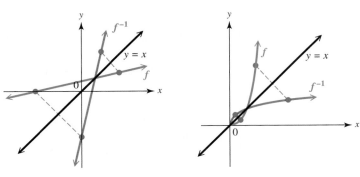

FIGURE 3

**NOW TRY** Exercises 25 and 29.

## 11.1 Exercises

*In Exercises 1–4, write a few sentences of explanation. See Example 1.*

1. A new study found that the trans fat content in fast-food products varied widely around the world, based on the type of frying oil used, as shown in the table.

    If the set of countries is the domain and the set of trans fat percentages is the range of the function consisting of the eight pairs listed, is it a one-to-one function? Why or why not?

| Country | Percentage of Trans Fat in McDonald's Chicken |
|---|---|
| Scotland | 14 |
| France | 11 |
| United States | 11 |
| Peru | 9 |
| Hungary | 8 |
| Poland | 8 |
| Russia | 5 |
| Denmark | 1 |

*Source: New England Journal of Medicine.*

2. The table shows the number of uncontrolled hazardous waste sites in 2004 that require further investigation to determine whether remedies are needed under the Superfund program. The eight states listed are ranked in the top ten in the United States.

    If this correspondence is considered to be a function that pairs each state with its number of uncontrolled waste sites, is it one-to-one? If not, explain why.

| State | Number of Sites |
|---|---|
| New Jersey | 114 |
| Pennsylvania | 96 |
| California | 95 |
| New York | 91 |
| Michigan | 69 |
| Florida | 52 |
| Illinois | 47 |
| Washington | 47 |

*Source:* U.S. Environmental Protection Agency.

3. The road mileage between Denver, Colorado, and several selected U.S. cities is shown in the table. If we consider this as a function that pairs each city with a distance, is it a one-to-one function? How could we change the answer to this question by adding 1 mile to one of the distances shown?

| City | Distance to Denver (in miles) |
|---|---|
| Atlanta | 1398 |
| Dallas | 781 |
| Indianapolis | 1058 |
| Kansas City, MO | 600 |
| Los Angeles | 1059 |
| San Francisco | 1235 |

4. Suppose you consider the set of ordered pairs $(x, y)$ such that $x$ represents a person in your mathematics class and $y$ represents that person's mother. Explain how this function might not be a one-to-one function.

*In Exercises 5–8, choose the correct response from the given list.*

**5.** *Multiple Choice* If a function is made up of ordered pairs in such a way that the same $y$-value appears in a correspondence with two different $x$-values, then

   **A.** the function is one-to-one
   **B.** the function is not one-to-one
   **C.** its graph does not pass the vertical line test
   **D.** it has an inverse function associated with it.

**6.** Which equation defines a one-to-one function? Explain why the others do not, using specific examples.

   **A.** $f(x) = x$   **B.** $f(x) = x^2$   **C.** $f(x) = |x|$   **D.** $f(x) = -x^2 + 2x - 1$

**7.** *Multiple Choice* Only one of the graphs illustrates a one-to-one function. Which one is it? (See Example 2.)

**A.** 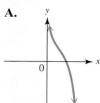  **B.** 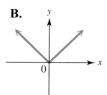  **C.** 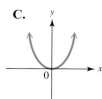  **D.**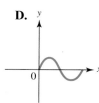

**8.** *Multiple Choice* If a function $f$ is one-to-one and the point $(p, q)$ lies on the graph of $f$, then which point *must* lie on the graph of $f^{-1}$?

   **A.** $(-p, q)$   **B.** $(-q, -p)$   **C.** $(p, -q)$   **D.** $(q, p)$

*If the function is one-to-one, find its inverse. See Examples 1–3.*

**9.** $\{(3, 6), (2, 10), (5, 12)\}$

**10.** $\left\{(-1, 3), (0, 5), (5, 0), \left(7, -\dfrac{1}{2}\right)\right\}$

**11.** $\{(-1, 3), (2, 7), (4, 3), (5, 8)\}$

**12.** $\{(-8, 6), (-4, 3), (0, 6), (5, 10)\}$

**13.** $f(x) = 2x + 4$

**14.** $f(x) = 3x + 1$

**15.** $g(x) = \sqrt{x - 3}, \quad x \geq 3$

**16.** $g(x) = \sqrt{x + 2}, \quad x \geq -2$

**17.** $f(x) = 3x^2 + 2$

**18.** $f(x) = -4x^2 - 1$

**19.** $f(x) = x^3 - 4$

**20.** $f(x) = x^3 - 3$

Let $f(x) = 2^x$. We will see in the next section that this function is one-to-one. Find each value, always working part (a) before part (b).

**21.** (a) $f(3)$   (b) $f^{-1}(8)$     **22.** (a) $f(4)$   (b) $f^{-1}(16)$

**23.** (a) $f(0)$   (b) $f^{-1}(1)$     **24.** (a) $f(-2)$   (b) $f^{-1}\left(\dfrac{1}{4}\right)$

*The graphs of some functions are given in Exercises 25–30.* **(a)** *Use the horizontal line test to determine whether the function is one-to-one.* **(b)** *If the function is one-to-one, then graph the inverse of the function. (Remember that if $f$ is one-to-one and $(a, b)$ is on the graph of $f$, then $(b, a)$ is on the graph of $f^{-1}$.) See Example 4.*

**25.** 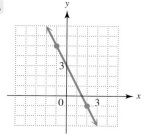   **26.** 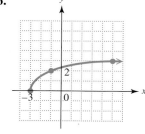   **27.**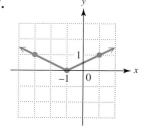

28. 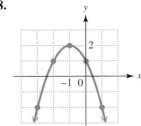   29. 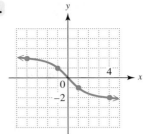   30.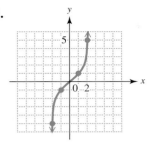

*Each function defined in Exercises 31–38 is a one-to-one function. Graph the function as a solid line (or curve) and then graph its inverse on the same set of axes as a dashed line (or curve). In Exercises 35–38 you are given a table to complete so that graphing the function will be easier. See Example 4.*

**31.** $f(x) = 2x - 1$  **32.** $f(x) = 2x + 3$  **33.** $g(x) = -4x$  **34.** $g(x) = -2x$

**35.** $f(x) = \sqrt{x}$,  **36.** $f(x) = -\sqrt{x}$,  **37.** $f(x) = x^3 - 2$  **38.** $f(x) = x^3 + 3$
  $x \geq 0$       $x \geq 0$

| x | f(x) |
|---|------|
| 0 |      |
| 1 |      |
| 4 |      |

| x | f(x) |
|---|------|
| 0 |      |
| 1 |      |
| 4 |      |

| x | f(x) |
|---|------|
| -1 |     |
| 0 |      |
| 1 |      |
| 2 |      |

| x | f(x) |
|---|------|
| -2 |     |
| -1 |     |
| 0 |      |
| 1 |      |

## 11.2 Exponential Functions

**OBJECTIVES**

1. Define an exponential function.
2. Graph an exponential function.
3. Solve exponential equations of the form $a^x = a^k$ for x.
4. Use exponential functions in applications involving growth or decay.

**OBJECTIVE 1** Define an exponential function. In **Section 8.2** we showed how to evaluate $2^x$ for rational values of $x$. For example,

$$2^3 = 8, \quad 2^{-1} = \frac{1}{2}, \quad 2^{1/2} = \sqrt{2}, \quad \text{and} \quad 2^{3/4} = \sqrt[4]{2^3} = \sqrt[4]{8}.$$

In more advanced courses it is shown that $2^x$ exists for all real number values of $x$, both rational and irrational. (Later in this chapter, we will see how to approximate the value of $2^x$ for irrational $x$.) The following definition of an exponential function assumes that $a^x$ exists for all real numbers $x$.

> **Exponential Function**
>
> For $a > 0$, $a \neq 1$, and all real numbers $x$,
>
> $$f(x) = a^x$$
>
> defines the **exponential function with base** $a$.

*The two restrictions on a in the definition of an exponential function $f(x) = a^x$ are important.*

1. The restriction $a > 0$ is necessary so that the function can be defined for all real numbers $x$. For example, letting $a$ be negative ($a = -2$, for instance) and letting $x = \frac{1}{2}$ would give the expression $(-2)^{1/2}$, which is not real.
2. The restriction $a \neq 1$ is necessary because 1 raised to any power is equal to 1, and the function would then be the linear function defined by $f(x) = 1$.

**540** CHAPTER 11 Inverse, Exponential, and Logarithmic Functions

> **OBJECTIVE 2** Graph an exponential function. We graph an exponential function by finding several ordered pairs that belong to the function, plotting these points, and connecting them with a smooth curve.

> **EXAMPLE 1** Graphing an Exponential Function with $a > 1$
>
> Graph $f(x) = 2^x$.
>
> Choose some values of $x$, and find the corresponding values of $f(x)$. Plotting these points and drawing a smooth curve through them gives the darker graph shown in Figure 4. This graph is typical of the graphs of exponential functions of the form $F(x) = a^x$, where $a > 1$. **The larger the value of $a$, the faster the graph rises.** Compare the lighter graph of $F(x) = 5^x$ with the graph of $f(x) = 2^x$ in Figure 4.
>
>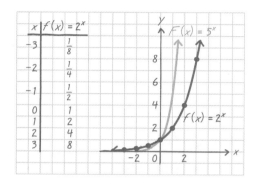
>
> **FIGURE 4**
>
> By the vertical line test, the graphs in Figure 4 represent functions. As these graphs suggest, the domain of an exponential function includes all real numbers. Because $y$ is always positive, the range is $(0, \infty)$. Figure 4 also shows an important characteristic of exponential functions where $a > 1$: **As $x$ gets larger, $y$ increases at a faster and faster rate.**
>
> **NOW TRY** Exercise 5.

> **CAUTION** The graph of an exponential function *approaches* the $x$-axis, but does *not* touch it.

> **EXAMPLE 2** Graphing an Exponential Function with $0 < a < 1$
>
> Graph $g(x) = \left(\dfrac{1}{2}\right)^x$.
>
> Again, find some points on the graph. The graph, shown in Figure 5 on the next page, is very similar to that of $f(x) = 2^x$ (Figure 4) with the same domain and range, except that here **as $x$ gets larger, $y$ decreases.** This graph is typical of the graph of a function of the form $F(x) = a^x$, where $0 < a < 1$.

| x | $g(x) = \left(\frac{1}{2}\right)^x$ |
|---|---|
| -3 | 8 |
| -2 | 4 |
| -1 | 2 |
| 0 | 1 |
| 1 | $\frac{1}{2}$ |
| 2 | $\frac{1}{4}$ |
| 3 | $\frac{1}{8}$ |

**FIGURE 5**

**NOW TRY** Exercise 7.

Based on Examples 1 and 2, we make the following generalizations about the graphs of exponential functions of the form $F(x) = a^x$.

### Graph of $F(x) = a^x$

1. The graph contains the point (0, 1).
2. When $a > 1$, the graph will *rise* from left to right. (See Figure 4.) When $0 < a < 1$, the graph will *fall* from left to right. (See Figure 5.) In both cases, the graph goes from the second quadrant to the first.
3. The graph will approach the x-axis but never touch it. (Recall from **Section 7.4** that such a line is called an *asymptote*.)
4. The domain is $(-\infty, \infty)$, and the range is $(0, \infty)$.

**EXAMPLE 3** Graphing a More Complicated Exponential Function

Graph $f(x) = 3^{2x-4}$.

Find some ordered pairs.

If $x = 0$, then $y = 3^{2(0)-4} = 3^{-4} = \frac{1}{81}$.

If $x = 2$, then $y = 3^{2(2)-4} = 3^0 = 1$.

These ordered pairs, $\left(0, \frac{1}{81}\right)$ and (2, 1), along with the other ordered pairs shown in the table, lead to the graph in Figure 6. The graph is similar to the graph of $f(x) = 3^x$ except that it is shifted to the right and rises more rapidly.

| x | y |
|---|---|
| 0 | $\frac{1}{81}$ |
| 1 | $\frac{1}{9}$ |
| 2 | 1 |
| 3 | 9 |

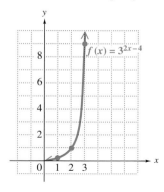

**FIGURE 6**

**NOW TRY** Exercise 11.

**OBJECTIVE 3** Solve exponential equations of the form $a^x = a^k$ for $x$. Until this chapter, we have solved only equations that had the variable as a base, like $x^2 = 8$; all exponents have been constants. An **exponential equation** is an equation that has a variable in an exponent, such as

$$9^x = 27.$$

By the horizontal line test, the exponential function defined by $F(x) = a^x$ is a one-to-one function, so we can use the following property to solve many exponential equations.

> **Property for Solving an Exponential Equation**
>
> For $a > 0$ and $a \neq 1$, if $a^x = a^y$ then $x = y$.

This property would not necessarily be true if $a = 1$.

To solve an exponential equation using this property, follow these steps.

> **Solving an Exponential Equation**
>
> *Step 1* **Each side must have the same base.** If the two sides of the equation do not have the same base, express each as a power of the same base if possible.
>
> *Step 2* **Simplify exponents** if necessary, using the rules of exponents.
>
> *Step 3* **Set exponents equal** using the property given in this section.
>
> *Step 4* **Solve** the equation obtained in Step 3.

These steps cannot be applied to an exponential equation like $3^x = 12$ because Step 1 cannot easily be done. A method for solving such equations is given in **Section 11.6**.

**EXAMPLE 4** Solving an Exponential Equation

Solve the equation $9^x = 27$.

$$9^x = 27$$
$$(3^2)^x = 3^3 \quad \text{Write with the same base; } 9 = 3^2 \text{ and } 27 = 3^3. \text{ (Step 1)}$$
$$3^{2x} = 3^3 \quad \text{Power rule for exponents (Step 2)}$$
$$2x = 3 \quad \text{If } a^x = a^y, \text{ then } x = y. \text{ (Step 3)}$$
$$x = \frac{3}{2} \quad \text{Solve for } x. \text{ (Step 4)}$$

*Check* that the solution set is $\left\{\frac{3}{2}\right\}$ by substituting $\frac{3}{2}$ for $x$:

$$9^x = 9^{3/2} = (9^{1/2})^3 = 3^3 = 27, \quad \text{as required.}$$

**NOW TRY** Exercise 17.

### EXAMPLE 5  Solving Exponential Equations

Solve each equation.

(a) $4^{3x-1} = 16^{x+2}$  *Be careful multiplying the exponents.*

$4^{3x-1} = (4^2)^{x+2}$   Write with the same base; $16 = 4^2$.

$4^{3x-1} = 4^{2x+4}$   Power rule for exponents

$3x - 1 = 2x + 4$   Set exponents equal.

$x = 5$   Subtract $2x$; add 1.

Verify that the solution set is $\{5\}$.

(b) $6^x = \dfrac{1}{216}$

$6^x = \dfrac{1}{6^3}$   $216 = 6^3$

$6^x = 6^{-3}$   Write with the same base; $\dfrac{1}{6^3} = 6^{-3}$.

$x = -3$   Set exponents equal.

Check that the solution set is $\{-3\}$ by substituting $-3$ for $x$:

$$6^x = 6^{-3} = \dfrac{1}{6^3} = \dfrac{1}{216}, \text{ as required.}$$

(c) $\left(\dfrac{2}{3}\right)^x = \dfrac{9}{4}$

$\left(\dfrac{2}{3}\right)^x = \left(\dfrac{4}{9}\right)^{-1}$   $\dfrac{9}{4} = \left(\dfrac{4}{9}\right)^{-1}$

$\left(\dfrac{2}{3}\right)^x = \left[\left(\dfrac{2}{3}\right)^2\right]^{-1}$   Write with the same base.

$\left(\dfrac{2}{3}\right)^x = \left(\dfrac{2}{3}\right)^{-2}$   Power rule for exponents

$x = -2$   Set exponents equal.

Check that the solution set is $\{-2\}$.

**NOW TRY** Exercises 19, 21, and 25.

---

**OBJECTIVE 4** Use exponential functions in applications involving growth or decay.

**EXAMPLE 6** Solving an Application Involving Exponential Growth

The graph in Figure 7 shows the concentration of carbon dioxide (in parts per million) in the air. This concentration is increasing exponentially.

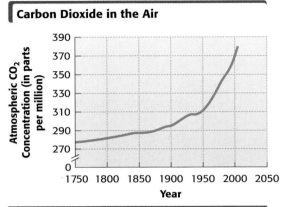

**Source:** *Sacramento Bee*; National Oceanic and Atmospheric Administration.

**FIGURE 7**

The data are approximated by the function defined by $f(x) = 266(1.001)^x$, where $x$ is the number of years since 1750. Use this function and a calculator to approximate the concentration of carbon dioxide in parts per million for each year.

**(a)** 1900

Because $x$ represents the number of years since 1750, $x = 1900 - 1750 = 150$. Thus, evaluate $f(150)$.

$$f(x) = 266(1.001)^x$$
$$f(150) = 266(1.001)^{150} \quad \text{Let } x = 150.$$
$$\approx 309 \text{ parts per million} \quad \text{Use a calculator.}$$

**(b)** 1950

$$f(200) = 266(1.001)^{200} \quad x = 1950 - 1750 = 200$$
$$\approx 325 \text{ parts per million}$$

**NOW TRY** Exercise 37.

**EXAMPLE 7** Applying an Exponential Decay Function

The atmospheric pressure (in millibars) at a given altitude $x$, in meters, can be approximated by the function defined by

$$f(x) = 1038(1.000134)^{-x},$$

for values of $x$ between 0 and 10,000. Because the base is greater than 1 and the coefficient of $x$ in the exponent is negative, the function values decrease as $x$ increases. This means that as the altitude increases, the atmospheric pressure decreases. (*Source:* Miller, A. and J. Thompson, *Elements of Meteorology,* Fourth Edition, Charles E. Merrill Publishing Company, 1993.)

**(a)** According to this function, what is the pressure at ground level?

At ground level, $x = 0$, so

$$f(0) = 1038(1.000134)^{-0} = 1038(1) = 1038.$$

The pressure is 1038 millibars.

**(b)** What is the pressure at 5000 m?

$$f(5000) = 1038(1.000134)^{-5000}$$
$$\approx 531 \qquad \text{Use a calculator.}$$

The pressure is approximately 531 millibars.

**NOW TRY** Exercise 39.

## 11.2 Exercises

**NOW TRY Exercise**

**1.** *Multiple Choice* Which point lies on the graph of $f(x) = 2^x$?

   **A.** $(1, 0)$    **B.** $(2, 1)$    **C.** $(0, 1)$    **D.** $\left(\sqrt{2}, \frac{1}{2}\right)$

**2.** *Multiple Choice* Which statement is true?

   **A.** The $y$-intercept of the graph of $f(x) = 10^x$ is $(0, 10)$.
   **B.** For any $a > 1$, the graph of $f(x) = a^x$ falls from left to right.
   **C.** The point $\left(\frac{1}{2}, \sqrt{5}\right)$ lies on the graph of $f(x) = 5^x$.
   **D.** The graph of $y = 4^x$ rises at a faster rate than the graph of $y = 10^x$.

**3.** *Multiple Choice* The asymptote of the graph of $F(x) = a^x$

   **A.** is the $x$-axis.       **B.** is the $y$-axis.
   **C.** has equation $x = 1$.   **D.** has equation $y = 1$.

**4.** *Multiple Choice* Which equation is graphed here?

   **A.** $y = 1000\left(\frac{1}{2}\right)^{0.3x}$    **B.** $y = 1000\left(\frac{1}{2}\right)^{x}$
   **C.** $y = 1000(2)^{0.3x}$      **D.** $y = 1000^x$

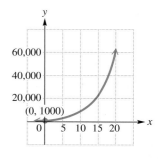

*Graph each exponential function. See Examples 1–3.*

**5.** $f(x) = 3^x$         **6.** $f(x) = 5^x$         **7.** $g(x) = \left(\frac{1}{3}\right)^x$

**8.** $g(x) = \left(\frac{1}{5}\right)^x$    **9.** $y = 4^{-x}$        **10.** $y = 6^{-x}$

**11.** $y = 2^{2x-2}$       **12.** $y = 2^{2x+1}$

**13.** *Concept Check*

   **(a)** *Fill in the Blanks* For an exponential function defined by $f(x) = a^x$, if $a > 1$, the graph _____ from left to right. If $0 < a < 1$, the graph _____
                                 (rises/falls)                                (rises/falls)
from left to right.

   **(b)** Based on your answers in part (a), make a conjecture (an educated guess) concerning whether an exponential function defined by $f(x) = a^x$ is one-to-one. Then decide whether it has an inverse based on the concepts of **Section 11.1**.

**14.** In your own words, describe the characteristics of the graph of an exponential function. Use the exponential function defined by $f(x) = 3^x$ (Exercise 5) and the words *asymptote, domain,* and *range* in your explanation.

*Solve each equation. See Examples 4 and 5.*

15. $6^x = 36$
16. $8^x = 64$
17. $100^x = 1000$
18. $8^x = 4$
19. $16^{2x+1} = 64^{x+3}$
20. $9^{2x-8} = 27^{x-4}$
21. $5^x = \dfrac{1}{125}$
22. $3^x = \dfrac{1}{81}$
23. $5^x = 0.2$
24. $10^x = 0.1$
25. $\left(\dfrac{3}{2}\right)^x = \dfrac{8}{27}$
26. $\left(\dfrac{4}{3}\right)^x = \dfrac{27}{64}$

*Use the exponential key of a calculator to find an approximation to the nearest thousandth.*

27. $12^{2.6}$
28. $13^{1.8}$
29. $0.5^{3.921}$
30. $0.6^{4.917}$
31. $2.718^{2.5}$
32. $2.718^{-3.1}$

*The graph shown here accompanied the article "Is Our World Warming?" which appeared in* National Geographic. *It shows projected temperature increases using two graphs: one an exponential-type curve, and the other linear. From the graph, approximate the increase* **(a)** *for the exponential curve and* **(b)** *for the linear graph for each year.*

33. 2000
34. 2010
35. 2020
36. 2040

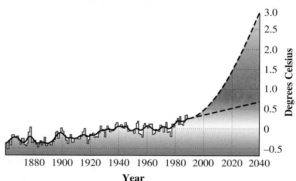

**IS OUR WORLD WARMING?**

Graph, "Zero Equals Average Global Temperature for the Period 1950–1979." Dale D. Glasgow, © National Geographic Society. Reprinted by permission.

*Solve each problem. See Examples 6 and 7.*

37. Based on figures from 1970 through 2002, the worldwide carbon monoxide emissions in thousands of tons are approximated by the exponential function defined by

$$f(x) = 220{,}717(1.0217)^{-x},$$

where $x = 0$ corresponds to 1970, $x = 5$ corresponds to 1975, and so on. (*Source:* U.S. Environmental Protection Agency.)

(a) Use this model to approximate the emissions in 1970.
(b) Use this model to approximate the emissions in 1995.
(c) In 2002, the actual amount of emissions was 112,049 million tons. How does this compare to the number that the model provides?

38. Based on figures from 1980 through 2003, the municipal solid waste generated in millions of tons can be approximated by the exponential function defined by

$$f(x) = 158.64(1.0189)^x,$$

where $x = 0$ corresponds to 1980, $x = 5$ corresponds to 1985, and so on. (*Source:* U.S. Environmental Protection Agency.)

(a) Use the model to approximate the number of tons of this waste in 1980.
(b) Use the model to approximate the number of tons of this waste in 1995.
(c) In 2003, the actual number of millions of tons of this waste was 236.2. How does this compare to the number that the model provides?

**39.** A small business estimates that the value $V(t)$ of a copy machine is decreasing according to the function defined by

$$V(t) = 5000(2)^{-0.15t},$$

where $t$ is the number of years that have elapsed since the machine was purchased, and $V(t)$ is in dollars.

(a) What was the original value of the machine?
(b) What is the value of the machine 5 yr after purchase? Give your answer to the nearest dollar.
(c) What is the value of the machine 10 yr after purchase? Give your answer to the nearest dollar.
(d) Graph the function.

**40.** The amount of radioactive material in an ore sample is given by the function defined by

$$A(t) = 100(3.2)^{-0.5t},$$

where $A(t)$ is the amount present, in grams, of the sample $t$ months after the initial measurement.

(a) How much was present at the initial measurement? (*Hint:* $t = 0$.)
(b) How much was present 2 months later?
(c) How much was present 10 months later?
(d) Graph the function.

**41.** Refer to the function in Exercise 39. When will the value of the machine be $2500? (*Hint:* Let $V(t) = 2500$, divide both sides by 5000, and use the method of Example 4.)

**42.** Refer to the function in Exercise 39. When will the value of the machine be $1250?

## 11.3 Logarithmic Functions

**OBJECTIVES**

1. Define a logarithm.
2. Convert between exponential and logarithmic forms.
3. Solve logarithmic equations of the form $\log_a b = k$ for $a$, $b$, or $k$.
4. Define and graph logarithmic functions.
5. Use logarithmic functions in applications involving growth or decay.

The graph of $y = 2^x$ is the curve shown in blue in Figure 8. Because $y = 2^x$ defines a one-to-one function, it has an inverse. Interchanging $x$ and $y$ gives

$$x = 2^y, \quad \text{the inverse of} \quad y = 2^x.$$

As we saw in **Section 11.1,** the graph of the inverse is found by reflecting the graph of $y = 2^x$ about the line $y = x$. The graph of $x = 2^y$ is shown as a red curve in Figure 8.

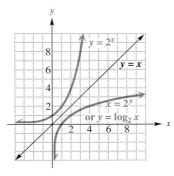

**FIGURE 8**

**OBJECTIVE 1** **Define a logarithm.** We cannot solve the equation $x = 2^y$ for the dependent variable $y$ with the methods presented up to now. The following definition is used to solve $x = 2^y$ for $y$.

### Logarithm

For all positive numbers $a$, with $a \neq 1$, and all positive numbers $x$,

$$y = \log_a x \quad \text{means the same as} \quad x = a^y.$$

*This key statement should be memorized.* The abbreviation **log** is used for the word **logarithm.** Read $\log_a x$ as **"the logarithm of $x$ to the base $a$"** or **"the base $a$ logarithm of $x$."** To remember the location of the base and the exponent in each form, refer to the following diagrams.

$$\text{Logarithmic form: } y = \log_a x \qquad \text{Exponential form: } x = a^y$$

with Exponent pointing to $y$ (log form) and to the exponent in $a^y$; Base pointing to $a$ in both.

In working with logarithmic form and exponential form, remember the following.

### Meaning of $\log_a x$

A logarithm is an exponent. The expression $\log_a x$ represents the exponent to which the base $a$ must be raised to obtain $x$.

**OBJECTIVE 2** **Convert between exponential and logarithmic forms.** We can use the definition of logarithm to write exponential statements in logarithmic form and logarithmic statements in exponential form. The following table shows several pairs of equivalent statements.

| Exponential Form | Logarithmic Form |
|---|---|
| $3^2 = 9$ | $\log_3 9 = 2$ |
| $\left(\frac{1}{5}\right)^{-2} = 25$ | $\log_{1/5} 25 = -2$ |
| $10^5 = 100{,}000$ | $\log_{10} 100{,}000 = 5$ |
| $4^{-3} = \frac{1}{64}$ | $\log_4 \frac{1}{64} = -3$ |

**NOW TRY** Exercises 3 and 11.

**OBJECTIVE 3** **Solve logarithmic equations of the form $\log_a b = k$ for $a$, $b$, or $k$.** A **logarithmic equation** is an equation with a logarithm in at least one term. We solve logarithmic equations of the form $\log_a b = k$ for any of the three variables by first writing the equation in exponential form.

### EXAMPLE 1  Solving Logarithmic Equations

Solve each equation.

**(a)** $\log_4 x = -2$

By the definition of logarithm, $\log_4 x = -2$ is equivalent to $x = 4^{-2}$. Solve this exponential equation.

$$x = 4^{-2} = \frac{1}{16}$$

The solution set is $\left\{\frac{1}{16}\right\}$.

**(b)** $\log_{1/2}(3x + 1) = 2$

$3x + 1 = \left(\frac{1}{2}\right)^2$    Write in exponential form.

$3x + 1 = \frac{1}{4}$    Apply the exponent.

$12x + 4 = 1$    Multiply each term by 4.

$12x = -3$    Subtract 4.

$x = -\frac{1}{4}$    Divide by 12.

The solution set is $\left\{-\frac{1}{4}\right\}$.

**(c)** $\log_x 3 = 2$

$x^2 = 3$    Write in exponential form.

$x = \pm\sqrt{3}$    Take square roots.

Only the *principal* square root satisfies the equation since the base must be a positive number. The solution set is $\{\sqrt{3}\}$.

**(d)** $\log_{49} \sqrt[3]{7} = x$

$49^x = \sqrt[3]{7}$    Write in exponential form.

$(7^2)^x = 7^{1/3}$    Write with the same base.

$7^{2x} = 7^{1/3}$    Power rule for exponents

$2x = \frac{1}{3}$    Set exponents equal.

$x = \frac{1}{6}$    Divide by 2.

The solution set is $\left\{\frac{1}{6}\right\}$.

**NOW TRY** Exercises 21, 25, 37, and 39.

For any real number $b$, we know that $b^1 = b$ and for $b \neq 0$, $b^0 = 1$. Writing these two statements in logarithmic form gives the following two properties of logarithms.

---

**Properties of Logarithms**

For any positive real number $b$, with $b \neq 1$,

$$\log_b b = 1 \quad \text{and} \quad \log_b 1 = 0.$$

**EXAMPLE 2** Using Properties of Logarithms

Use the preceding two properties of logarithms to evaluate each logarithm.

(a) $\log_7 7 = 1$  (b) $\log_{\sqrt{2}} \sqrt{2} = 1$
(c) $\log_9 1 = 0$  (d) $\log_{0.2} 1 = 0$

**NOW TRY** Exercise 19.

**OBJECTIVE 4** Define and graph logarithmic functions. Now we define the logarithmic function with base $a$.

**Logarithmic Function**

If $a$ and $x$ are positive numbers, with $a \neq 1$, then

$$G(x) = \log_a x$$

defines the **logarithmic function with base $a$**.

**EXAMPLE 3** Graphing a Logarithmic Function with $a > 1$

Graph $f(x) = \log_2 x$.

By writing $y = f(x) = \log_2 x$ in exponential form as $x = 2^y$, we can identify ordered pairs that satisfy the equation. It is easier to choose values for $y$ and find the corresponding values of $x$. Plotting the points in the table of ordered pairs and connecting them with a smooth curve gives the graph in Figure 9. This graph is typical of logarithmic functions with base $a > 1$.

Be careful to write the x- and y-values in the correct order.

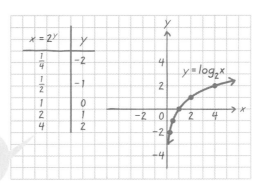

**FIGURE 9**

As the table and graph suggest, $x$ is always positive, so the domain of a logarithmic function is $(0, \infty)$. The range includes all real numbers, $(-\infty, \infty)$.

**NOW TRY** Exercise 41.

### EXAMPLE 4  Graphing a Logarithmic Function with $0 < a < 1$

Graph $g(x) = \log_{1/2} x$.

We write $y = g(x) = \log_{1/2} x$ in exponential form as $x = \left(\frac{1}{2}\right)^y$, then choose values for $y$ and find the corresponding values of $x$. See the table of ordered pairs.

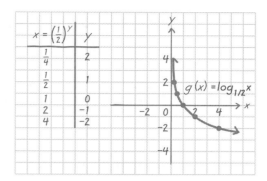

**FIGURE 10**

Plotting these points and connecting them with a smooth curve gives the graph in Figure 10. This graph, which is similar to that of $f(x) = \log_2 x$ (Figure 9) with the same domain and range, is typical of logarithmic functions with $0 < a < 1$.

**NOW TRY Exercise 43.**

Based on the graphs of the functions defined by $y = \log_2 x$ in Figure 9 and $y = \log_{1/2} x$ in Figure 10, we make the following generalizations about the graphs of logarithmic functions of the form $G(x) = \log_a x$.

---

**Graph of $G(x) = \log_a x$**

1. The graph contains the point $(1, 0)$.
2. When $a > 1$, the graph will *rise* from left to right, from the fourth quadrant to the first. (See Figure 9.) When $0 < a < 1$, the graph will *fall* from left to right, from the first quadrant to the fourth. (See Figure 10.)
3. The graph will approach the $y$-axis but never touch it. (The $y$-axis is an asymptote.)
4. The domain is $(0, \infty)$, and the range is $(-\infty, \infty)$.

---

Compare these generalizations to the similar ones for exponential functions found in **Section 11.2.**

**OBJECTIVE 5** Use logarithmic functions in applications involving growth or decay. Logarithmic functions, like exponential functions, can be applied to growth or decay of real-world phenomena.

### EXAMPLE 5  Solving an Application of a Logarithmic Function

The function defined by

$$f(x) = 27 + 1.105 \log_{10}(x + 1)$$

approximates the barometric pressure in inches of mercury at a distance of $x$ miles from the eye of a typical hurricane. (*Source:* Miller, A. and R. Anthes, *Meteorology*, Fifth Edition, Charles E. Merrill Publishing Company, 1985.)

**(a)** Approximate the pressure 9 mi from the eye of the hurricane.
    Let $x = 9$, and find $f(9)$.

$$\begin{aligned} f(9) &= 27 + 1.105 \log_{10}(9 + 1) &&\text{Let } x = 9. \\ &= 27 + 1.105 \log_{10} 10 &&\text{Add inside parentheses.} \\ &= 27 + 1.105(1) &&\log_{10} 10 = 1 \\ &= 28.105 \end{aligned}$$

The pressure 9 mi from the eye of the hurricane is 28.105 in.

**(b)** Approximate the pressure 99 mi from the eye of the hurricane.

$$\begin{aligned} f(99) &= 27 + 1.105 \log_{10}(99 + 1) &&\text{Let } x = 99. \\ &= 27 + 1.105 \log_{10} 100 &&\text{Add inside parentheses.} \\ &= 27 + 1.105(2) &&\log_{10} 100 = 2 \\ &= 29.21 \end{aligned}$$

The pressure 99 mi from the eye of the hurricane is 29.21 in.

**NOW TRY** Exercise 53.

## 11.3 Exercises

**1.** *Matching* Match the logarithmic equation in Column I with the corresponding exponential equation from Column II.

| I | II |
|---|---|
| (a) $\log_{1/3} 3 = -1$ | A. $8^{1/3} = \sqrt[3]{8}$ |
| (b) $\log_5 1 = 0$ | B. $\left(\dfrac{1}{3}\right)^{-1} = 3$ |
| (c) $\log_2 \sqrt{2} = \dfrac{1}{2}$ | C. $4^1 = 4$ |
| (d) $\log_{10} 1000 = 3$ | D. $2^{1/2} = \sqrt{2}$ |
| (e) $\log_8 \sqrt[3]{8} = \dfrac{1}{3}$ | E. $5^0 = 1$ |
| (f) $\log_4 4 = 1$ | F. $10^3 = 1000$ |

**2.** *Matching* Use the definition of logarithm to match the logarithm in Column I with its value in Column II. (*Example:* $\log_3 9$ is equal to 2 because 2 is the exponent to which 3 must be raised in order to obtain 9.)

| I | II |
|---|---|
| (a) $\log_4 16$ | A. $-2$ |
| (b) $\log_3 81$ | B. $-1$ |
| (c) $\log_3 \left(\dfrac{1}{3}\right)$ | C. $2$ |
| (d) $\log_{10} 0.01$ | D. $0$ |
| (e) $\log_5 \sqrt{5}$ | E. $\dfrac{1}{2}$ |
| (f) $\log_{13} 1$ | F. $4$ |

*Write in logarithmic form. See the table in Objective 2.*

**3.** $4^5 = 1024$

**4.** $3^6 = 729$

**5.** $\left(\dfrac{1}{2}\right)^{-3} = 8$

**6.** $\left(\dfrac{1}{6}\right)^{-3} = 216$

**7.** $10^{-3} = 0.001$

**8.** $36^{1/2} = 6$

**9.** $\sqrt[4]{625} = 5$

**10.** $\sqrt[3]{343} = 7$

*Write in exponential form. See the table in Objective 2.*

**11.** $\log_4 64 = 3$  **12.** $\log_2 512 = 9$  **13.** $\log_{10} \dfrac{1}{10{,}000} = -4$

**14.** $\log_{100} 100 = 1$  **15.** $\log_6 1 = 0$  **16.** $\log_\pi 1 = 0$

**17.** $\log_9 3 = \dfrac{1}{2}$  **18.** $\log_{64} 2 = \dfrac{1}{6}$

**19.** *Matching* Match each logarithm in Column I with its value in Column II. See Example 2.

| I | II |
| --- | --- |
| (a) $\log_8 8$ | A. $-1$ |
| (b) $\log_{16} 1$ | B. $0$ |
| (c) $\log_{0.3} 1$ | C. $1$ |
| (d) $\log_{\sqrt{7}} \sqrt{7}$ | D. $0.1$ |

**20.** When a student asked his teacher to explain how to evaluate $\log_9 3$ without showing any work, his teacher told him, "Think radically." Explain what the teacher meant by this hint.

*Solve each equation. See Examples 1 and 2.*

**21.** $x = \log_{27} 3$  **22.** $x = \log_{125} 5$  **23.** $\log_x 9 = \dfrac{1}{2}$

**24.** $\log_x 5 = \dfrac{1}{2}$  **25.** $\log_x 125 = -3$  **26.** $\log_x 64 = -6$

**27.** $\log_{12} x = 0$  **28.** $\log_4 x = 0$  **29.** $\log_x x = 1$

**30.** $\log_x 1 = 0$  **31.** $\log_x \dfrac{1}{25} = -2$  **32.** $\log_x \dfrac{1}{10} = -1$

**33.** $\log_8 32 = x$  **34.** $\log_{81} 27 = x$  **35.** $\log_\pi \pi^4 = x$

**36.** $\log_{\sqrt{2}} \sqrt{2^9} = x$  **37.** $\log_6 \sqrt{216} = x$  **38.** $\log_4 \sqrt{64} = x$

**39.** $\log_4(2x + 4) = 3$  **40.** $\log_3(2x + 7) = 4$

*If the point $(p, q)$ is on the graph of $f(x) = a^x$ (for $a > 0$ and $a \neq 1$), then the point $(q, p)$ is on the graph of $f^{-1}(x) = \log_a x$. Use this fact, and refer to the graphs required in Exercises 5–8 in Section 11.2 to graph each logarithmic function. See Examples 3 and 4.*

**41.** $y = \log_3 x$  **42.** $y = \log_5 x$  **43.** $y = \log_{1/3} x$  **44.** $y = \log_{1/5} x$

**45.** Explain why 1 is not allowed as a base for a logarithmic function.

**46.** Compare the summary of facts about the graph of $F(x) = a^x$ in **Section 11.2** with the similar summary of facts about the graph of $G(x) = \log_a x$ in this section. Make a list of the facts that reinforce the concept that $F$ and $G$ are inverse functions.

**47.** *Fill in the Blanks* The domain of $F(x) = a^x$ is $(-\infty, \infty)$, while the range is $(0, \infty)$. Therefore, since $G(x) = \log_a x$ defines the inverse of $F$, the domain of $G$ is _____, while the range of $G$ is _____.

**48.** *Concept Check* The graphs of both $F(x) = 3^x$ and $G(x) = \log_3 x$ rise from left to right. Which one rises at a faster rate?

*Concept Check* Use the graph at the right to predict the value of $f(t)$ for the given value of $t$.

**49.** $t = 0$

**50.** $t = 10$

**51.** $t = 60$

**52.** Show that the points determined in Exercises 49–51 lie on the graph of $f(t) = 8 \log_5(2t + 5)$.

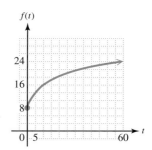

*Solve each problem. See Example 5.*

**53.** According to selected figures from 1981 through 2003, the number of billion cubic feet of natural gas gross withdrawals from crude oil wells in the United States can be approximated by the function defined by

$$f(x) = 3800 + 585 \log_2 x,$$

where $x = 1$ corresponds to 1981, $x = 2$ to 1982, and so on. (*Source:* Energy Information Administration.) Use this function to approximate the number of cubic feet withdrawn in each of the following years.

**(a)** 1982  **(b)** 1988  **(c)** 1996

**54.** According to selected figures from the last two decades of the 20th century, the number of trillion cubic feet of dry natural gas consumed worldwide can be approximated by the function defined by

$$f(x) = 51.47 + 6.044 \log_2 x,$$

where $x = 1$ corresponds to 1980, $x = 2$ to 1981, and so on. (*Source:* Energy Information Administration.) Use the function to approximate consumption in each year.

**(a)** 1980  **(b)** 1987  **(c)** 1995

**55.** Sales (in thousands of units) of a new product are approximated by the function defined by

$$S(t) = 100 + 30 \log_3(2t + 1),$$

where $t$ is the number of years after the product is introduced.

**(a)** What were the sales after 1 yr?
**(b)** What were the sales after 13 yr?
**(c)** Graph $y = S(t)$.

**56.** A study showed that the number of mice in an old abandoned house was approximated by the function defined by

$$M(t) = 6 \log_4(2t + 4),$$

where $t$ is measured in months and $t = 0$ corresponds to January 1998. Find the number of mice in the house in

**(a)** January 1998  **(b)** July 1998  **(c)** July 2000.
**(d)** Graph the function.

*In the United States, the intensity of an earthquake is rated using the **Richter scale**. The Richter scale rating of an earthquake of intensity x is given by*

$$R = \log_{10} \frac{x}{x_0},$$

*where $x_0$ is the intensity of an earthquake of a certain (small) size. The figure here shows Richter scale ratings for major Southern California earthquakes since 1930. As the figure indicates, earthquakes "come in bunches."*

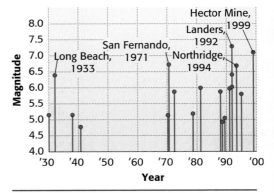

*Source:* Caltech; U.S. Geological Survey.

**57.** The 1994 Northridge earthquake had a Richter scale rating of 6.7; the 1992 Landers earthquake had a rating of 7.3. How much more powerful was the Landers earthquake than the Northridge earthquake?

**58.** Compare the smallest rated earthquake in the figure (at 4.8) with the Landers quake. How much more powerful was the Landers quake?

## 11.4 Properties of Logarithms

**OBJECTIVES**

1. Use the product rule for logarithms.
2. Use the quotient rule for logarithms.
3. Use the power rule for logarithms.
4. Use properties to write alternative forms of logarithmic expressions.

Logarithms have been used as an aid to numerical calculation for several hundred years. Today the widespread use of calculators has made the use of logarithms for calculation obsolete. However, logarithms are still very important in applications and in further work in mathematics.

**OBJECTIVE 1** Use the product rule for logarithms. One way in which logarithms simplify problems is by changing a problem of multiplication into one of addition. We know that $\log_2 4 = 2$, $\log_2 8 = 3$, and $\log_2 32 = 5$. Since $2 + 3 = 5$,

$$\log_2 32 = \log_2 4 + \log_2 8$$
$$\log_2(4 \cdot 8) = \log_2 4 + \log_2 8.$$

This is true in general.

---

**Product Rule for Logarithms**

If $x$, $y$, and $b$ are positive real numbers, where $b \neq 1$, then

$$\log_b xy = \log_b x + \log_b y.$$

That is, the logarithm of a product is the sum of the logarithms of the factors.

---

To prove this rule, let $m = \log_b x$ and $n = \log_b y$, and recall that

$$\log_b x = m \quad \text{means} \quad b^m = x.$$
$$\log_b y = n \quad \text{means} \quad b^n = y.$$

Now consider the product $xy$.

| | |
|---|---|
| $xy = b^m \cdot b^n$ | Substitute. |
| $xy = b^{m+n}$ | Product rule for exponents |
| $\log_b xy = m + n$ | Convert to logarithmic form. |
| $\log_b xy = \log_b x + \log_b y$ | Substitute. |

The last statement is the result we wished to prove.

---

**EXAMPLE 1** Using the Product Rule

Use the product rule to rewrite each logarithm. Assume $x > 0$.

**(a)** $\log_5 (6 \cdot 9) = \log_5 6 + \log_5 9$

**(b)** $\log_7 8 + \log_7 12 = \log_7(8 \cdot 12)$
$\qquad\qquad\qquad\qquad = \log_7 96$

**(c)** $\log_3(3x) = \log_3 3 + \log_3 x$     Product rule
$\qquad\qquad = 1 + \log_3 x$     $\log_3 3 = 1$

**(d)** $\log_4 x^3 = \log_4(x \cdot x \cdot x)$     $x^3 = x \cdot x \cdot x$
$\qquad\qquad = \log_4 x + \log_4 x + \log_4 x$     Product rule
$\qquad\qquad = 3 \log_4 x$

**NOW TRY** Exercises 7 and 21.

555

**OBJECTIVE 2** Use the quotient rule for logarithms. The rule for division is similar to the rule for multiplication.

> **Quotient Rule for Logarithms**
>
> If $x$, $y$, and $b$ are positive real numbers, where $b \neq 1$, then
>
> $$\log_b \frac{x}{y} = \log_b x - \log_b y.$$
>
> That is, the logarithm of a quotient is the difference between the logarithm of the numerator and the logarithm of the denominator.

The proof of this rule is very similar to the proof of the product rule.

**EXAMPLE 2** Using the Quotient Rule

Use the quotient rule to rewrite each logarithm. Assume $x > 0$.

(a) $\log_4 \dfrac{7}{9} = \log_4 7 - \log_4 9$

(b) $\log_5 6 - \log_5 x = \log_5 \dfrac{6}{x}$

(c) $\log_3 \dfrac{27}{5} = \log_3 27 - \log_3 5$
$= 3 - \log_3 5$

**NOW TRY** Exercises 9 and 23.

> **CAUTION** There is no property of logarithms to rewrite the logarithm of a **sum or difference.** For example, we **cannot** write $\log_b(x + y)$ in terms of $\log_b x$ and $\log_b y$. Also,
>
> $$\log_b \frac{x}{y} \neq \frac{\log_b x}{\log_b y}.$$

**OBJECTIVE 3** Use the power rule for logarithms. An exponential expression such as $2^3$ means $2 \cdot 2 \cdot 2$; the base is used as a factor 3 times. Similarly, the product rule can be extended to rewrite the logarithm of a power as the product of the exponent and the logarithm of the base. For example, by the product rule for logarithms,

$\log_5 2^3 = \log_5(2 \cdot 2 \cdot 2)$        $\log_2 7^4 = \log_2(7 \cdot 7 \cdot 7 \cdot 7)$
$\phantom{\log_5 2^3} = \log_5 2 + \log_5 2 + \log_5 2$        $\phantom{\log_2 7^4} = \log_2 7 + \log_2 7 + \log_2 7 + \log_2 7$
$\phantom{\log_5 2^3} = 3 \log_5 2.$        $\phantom{\log_2 7^4} = 4 \log_2 7.$

Furthermore, we saw in Example 1(d) that $\log_4 x^3 = 3 \log_4 x$. These examples suggest the following rule.

## Power Rule for Logarithms

If $x$ and $b$ are positive real numbers, where $b \neq 1$, and if $r$ is any real number, then

$$\log_b x^r = r \log_b x.$$

That is, the logarithm of a number to a power equals the exponent times the logarithm of the number.

As further examples of this rule,

$$\log_b m^5 = 5 \log_b m \quad \text{and} \quad \log_3 5^4 = 4 \log_3 5.$$

To prove the power rule, let $\log_b x = m$.

$\log_b x = m$
$b^m = x$      Convert to exponential form.
$(b^m)^r = x^r$      Raise to the power $r$.
$b^{mr} = x^r$      Power rule for exponents
$\log_b x^r = mr$      Convert to logarithmic form.
$\log_b x^r = rm$      Commutative property
$\log_b x^r = r \log_b x$      $m = \log_b x$

This is the statement to be proved.

As a special case of the power rule, let $r = \frac{1}{p}$, so

$$\log_b \sqrt[p]{x} = \log_b x^{1/p} = \frac{1}{p} \log_b x.$$

For example, using this result, with $x > 0$,

$$\log_b \sqrt[5]{x} = \log_b x^{1/5} = \frac{1}{5} \log_b x \quad \text{and} \quad \log_b \sqrt[3]{x^4} = \log_b x^{4/3} = \frac{4}{3} \log_b x.$$

Another special case is

$$\log_b \frac{1}{x} = \log_b x^{-1} = -\log_b x.$$

### EXAMPLE 3   Using the Power Rule

Use the power rule to rewrite each logarithm. Assume $b > 0$, $x > 0$, and $b \neq 1$.

(a) $\log_5 4^2 = 2 \log_5 4$      (b) $\log_b x^5 = 5 \log_b x$

(c) $\log_b \sqrt{7} = \log_b 7^{1/2}$    $\sqrt{x} = x^{1/2}$      (d) $\log_2 \sqrt[5]{x^2} = \log_2 x^{2/5}$    $\sqrt[5]{x^2} = x^{2/5}$

$\phantom{(c)\ \log_b \sqrt{7} } = \frac{1}{2} \log_b 7$    Power rule      $\phantom{(d)\ \log_2 \sqrt[5]{x^2}} = \frac{2}{5} \log_2 x$    Power rule

**NOW TRY** Exercise 11.

Two special properties involving both exponential and logarithmic expressions come directly from the fact that logarithmic and exponential functions are inverses of each other.

## Special Properties

If $b > 0$ and $b \neq 1$, then

$$b^{\log_b x} = x, \quad x > 0 \quad \text{and} \quad \log_b b^x = x.$$

To prove the first statement, let $y = \log_b x$.

$$y = \log_b x$$
$$b^y = x \quad \text{Convert to exponential form.}$$
$$b^{\log_b x} = x \quad \text{Replace } y \text{ with } \log_b x.$$

The proof of the second statement is similar.

**EXAMPLE 4** Using the Special Properties

Find each value.

(a) $\log_5 5^4 = 4$, since $\log_b b^x = x$.  (b) $\log_3 9 = \log_3 3^2 = 2$
(c) $4^{\log_4 10} = 10$

**NOW TRY** Exercises 3 and 5.

Here is a summary of the properties of logarithms.

---

**Properties of Logarithms**

If $x$, $y$, and $b$ are positive real numbers, where $b \neq 1$, and $r$ is any real number, then

**Product Rule**  $\quad \log_b xy = \log_b x + \log_b y$

**Quotient Rule**  $\quad \log_b \dfrac{x}{y} = \log_b x - \log_b y$

**Power Rule**  $\quad \log_b x^r = r \log_b x$

**Special Properties**  $\quad b^{\log_b x} = x \quad \text{and} \quad \log_b b^x = x.$

---

**OBJECTIVE 4** Use properties to write alternative forms of logarithmic expressions.

**EXAMPLE 5** Writing Logarithms in Alternative Forms

Use the properties of logarithms to rewrite each expression if possible. Assume that all variables represent positive real numbers.

(a) $\log_4 4x^3$
$= \log_4 4 + \log_4 x^3 \quad$ Product rule
$= 1 + 3 \log_4 x \quad$ $\log_4 4 = 1$; power rule

(b) $\log_7 \sqrt{\dfrac{m}{n}}$

$= \log_7 \left(\dfrac{m}{n}\right)^{1/2} \quad$ Write the radical expression with a rational exponent.

$= \dfrac{1}{2} \log_7 \dfrac{m}{n} \quad$ Power rule

$= \dfrac{1}{2}(\log_7 m - \log_7 n) \quad$ Quotient rule

(c) $\log_5 \dfrac{a^2}{bc}$

$= \log_5 a^2 - \log_5 bc$     Quotient rule

$= 2 \log_5 a - \log_5 bc$     Power rule

$= 2 \log_5 a - (\log_5 b + \log_5 c)$     Product rule

$= 2 \log_5 a - \log_5 b - \log_5 c$

*Use parentheses to avoid errors.*

(d) $4 \log_b m - \log_b n$

$= \log_b m^4 - \log_b n$     Power rule

$= \log_b \dfrac{m^4}{n}$     Quotient rule

(e) $\log_b(x + 1) + \log_b(2x + 1) - \dfrac{2}{3}\log_b x$

$= \log_b(x + 1) + \log_b(2x + 1) - \log_b x^{2/3}$     Power rule

$= \log_b \dfrac{(x + 1)(2x + 1)}{x^{2/3}}$     Product and quotient rules

$= \log_b \dfrac{2x^2 + 3x + 1}{x^{2/3}}$     Multiply in the numerator.

(f) $\log_8(2p + 3r)$ cannot be rewritten using the properties of logarithms. There is no property of logarithms to rewrite the logarithm of a sum.

**NOW TRY** Exercises 13, 15, 27, and 31.

In the next example, we use numerical values for $\log_2 5$ and $\log_2 3$. While we use the equals sign to give these values, they are actually just approximations since most logarithms of this type are irrational numbers. We use $=$ with the understanding that the values are correct to four decimal places.

**EXAMPLE 6** Using the Properties of Logarithms with Numerical Values

Given that $\log_2 5 = 2.3219$ and $\log_2 3 = 1.5850$, evaluate the following.

(a) $\log_2 15 = \log_2(3 \cdot 5)$

$= \log_2 3 + \log_2 5$     Product rule

$= 1.5850 + 2.3219$     Substitute the given values.

$= 3.9069$     Add.

(b) $\log_2 0.6 = \log_2 \dfrac{3}{5}$     $0.6 = \tfrac{6}{10} = \tfrac{3}{5}$

$= \log_2 3 - \log_2 5$     Quotient rule

$= 1.5850 - 2.3219$     Substitute the given values.

$= -0.7369$     Subtract.

(c) $\log_2 27 = \log_2 3^3$

$= 3 \log_2 3$     Power rule

$= 3(1.5850)$     Substitute the given value.

$= 4.7550$     Multiply.

**NOW TRY** Exercises 33, 35, and 43.

**EXAMPLE 7** Deciding Whether Statements about Logarithms Are True

Decide whether each statement is *true* or *false*.

(a) $\log_2 8 - \log_2 4 = \log_2 4$
Evaluate both sides.

Left side: $\log_2 8 - \log_2 4 = \log_2 2^3 - \log_2 2^2 = 3 - 2 = 1$
Right side: $\log_2 4 = \log_2 2^2 = 2$

The statement is false because $1 \neq 2$.

(b) $\log_3(\log_2 8) = \dfrac{\log_7 49}{\log_8 64}$
Evaluate both sides.

Left side: $\log_3(\log_2 8) = \log_3 3 = 1$

Right side: $\dfrac{\log_7 49}{\log_8 64} = \dfrac{\log_7 7^2}{\log_8 8^2} = \dfrac{2}{2} = 1$

The statement is true because $1 = 1$.

**NOW TRY** Exercises 45 and 51.

## 11.4 Exercises

**NOW TRY Exercise**

*Fill in the Blanks* Use the indicated rule of logarithms to complete each equation. See Examples 1–4.

1. $\log_{10}(3 \cdot 4) = $ _____ (product rule)

2. $\log_{10} \dfrac{3}{4} = $ _____ (quotient rule)

3. $3^{\log_3 4} = $ _____ (special property)

4. $\log_{10} 3^4 = $ _____ (power rule)

5. $\log_3 3^4 = $ _____ (special property)

6. Evaluate $\log_2(8 + 8)$. Then evaluate $\log_2 8 + \log_2 8$. Are the results the same? How could you change the operation in the first expression to make the two expressions equal?

*Use the properties of logarithms to express each logarithm as a sum or difference of logarithms, or as a single number if possible. Assume that all variables represent positive real numbers. See Examples 1–5.*

7. $\log_7(4 \cdot 5)$

8. $\log_8(9 \cdot 11)$

9. $\log_5 \dfrac{8}{3}$

10. $\log_3 \dfrac{7}{5}$

11. $\log_4 6^2$

12. $\log_5 7^4$

13. $\log_3 \dfrac{\sqrt[3]{4}}{x^2 y}$

14. $\log_7 \dfrac{\sqrt[3]{13}}{pq^2}$

15. $\log_3 \sqrt{\dfrac{xy}{5}}$

16. $\log_6 \sqrt{\dfrac{pq}{7}}$

17. $\log_2 \dfrac{\sqrt[3]{x} \cdot \sqrt[5]{y}}{r^2}$

18. $\log_4 \dfrac{\sqrt[4]{z} \cdot \sqrt[5]{w}}{s^2}$

✏️ **19.** A student erroneously wrote $\log_a(x + y) = \log_a x + \log_a y$. When his teacher explained that this was indeed wrong, the student claimed that he had used the distributive property. Write a few sentences explaining why the distributive property does not apply in this case.

✏️ **20.** Write a few sentences explaining how the rules for multiplying and dividing powers of the same base are similar to the rules for finding logarithms of products and quotients.

*Use the properties of logarithms to write each expression as a single logarithm. Assume that all variables are defined in such a way that the variable expressions are positive and that bases are positive numbers not equal to 1. See Examples 1–5.*

**21.** $\log_b x + \log_b y$

**22.** $\log_b 2 + \log_b z$

**23.** $\log_a m - \log_a n$

**24.** $\log_b x - \log_b y$

**25.** $(\log_a r - \log_a s) + 3 \log_a t$

**26.** $(\log_a p - \log_a q) + 2 \log_a r$

**27.** $3 \log_a 5 - 4 \log_a 3$

**28.** $3 \log_a 5 + \dfrac{1}{2} \log_a 9$

**29.** $\log_{10}(x + 3) + \log_{10}(x - 3)$

**30.** $\log_{10}(y + 4) + \log_{10}(y - 4)$

**31.** $3 \log_p x + \dfrac{1}{2} \log_p y - \dfrac{3}{2} \log_p z - 3 \log_p a$

**32.** $\dfrac{1}{3} \log_b x + \dfrac{2}{3} \log_b y - \dfrac{3}{4} \log_b s - \dfrac{2}{3} \log_b t$

*To four decimal places, the values of $\log_{10} 2$ and $\log_{10} 9$ are*

$$\log_{10} 2 = 0.3010 \qquad \log_{10} 9 = 0.9542.$$

*Evaluate each logarithm by applying the appropriate rule or rules from this section. DO NOT USE A CALCULATOR. See Example 6.*

**33.** $\log_{10} 18$

**34.** $\log_{10} \dfrac{9}{2}$

**35.** $\log_{10} \dfrac{2}{9}$

**36.** $\log_{10} 4$

**37.** $\log_{10} 36$

**38.** $\log_{10} 162$

**39.** $\log_{10} 3$

**40.** $\log_{10} \sqrt[5]{2}$

**41.** $\log_{10} \sqrt[4]{9}$

**42.** $\log_{10} \dfrac{1}{9}$

**43.** $\log_{10} 9^5$

**44.** $\log_{10} 2^{19}$

*True or False* Decide whether each statement is true or false. See Example 7.

**45.** $\log_2 (8 + 32) = \log_2 8 + \log_2 32$

**46.** $\log_2(64 - 16) = \log_2 64 - \log_2 16$

**47.** $\log_3 7 + \log_3 7^{-1} = 0$

**48.** $\log_9 14 - \log_{14} 9 = 0$

**49.** $\log_6 60 - \log_6 10 = 1$

**50.** $\log_3 8 + \log_3 \dfrac{1}{8} = 0$

**51.** $\dfrac{\log_{10} 7}{\log_{10} 14} = \dfrac{1}{2}$

**52.** $\dfrac{\log_{10} 10}{\log_{10} 100} = \dfrac{1}{10}$

# 11.5 Common and Natural Logarithms

**OBJECTIVES**

1. Evaluate common logarithms using a calculator.
2. Use common logarithms in applications.
3. Evaluate natural logarithms using a calculator.
4. Use natural logarithms in applications.
5. Use the change-of-base rule.

As mentioned earlier, logarithms are important in many applications of mathematics to everyday problems, particularly in biology, engineering, economics, and social science. In this section we find numerical approximations for logarithms. Traditionally, base 10 logarithms were used most often because our number system is base 10. Logarithms to base 10 are called **common logarithms**, and $\log_{10} x$ is abbreviated as simply **log x**, where the base is understood to be 10.

**OBJECTIVE 1** Evaluate common logarithms using a calculator. We use calculators to evaluate common logarithms. In the next example we give the results of evaluating some common logarithms using a calculator with a (LOG) key. (This may be a second function key on some calculators.) For simple scientific calculators, just enter the number, then press the (LOG) key. For graphing calculators, these steps are reversed. We give most approximations for logarithms to four decimal places.

**EXAMPLE 1** Evaluating Common Logarithms

Evaluate each logarithm using a calculator.

(a) $\log 327.1 \approx 2.5147$  (b) $\log 437,000 \approx 5.6405$
(c) $\log 0.0615 \approx -1.2111$

Figure 11 shows how a graphing calculator displays these common logarithms to four decimal places.

```
log(327.1)
          2.5147
log(437000)
          5.6405
log(.0615)
         -1.2111
```

**FIGURE 11**

In part (c), notice that $\log 0.0615 \approx -1.2111$, a negative result. *The common logarithm of a number between 0 and 1 is always negative* because the logarithm is the exponent on 10 that produces the number. In this case, we have

$$10^{-1.2111} \approx 0.0615.$$

If the exponent (the logarithm) were positive, the result would be greater than 1 because $10^0 = 1$. The graph in Figure 12 illustrates these concepts.

**NOW TRY** Exercises 7, 9, and 11.

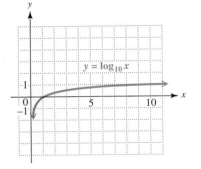

**FIGURE 12**

**OBJECTIVE 2** Use common logarithms in applications. In chemistry, pH is a measure of the acidity or alkalinity of a solution; pure water, for example, has pH 7. In general, acids have pH numbers less than 7 and alkaline solutions have pH values greater than 7, as shown in Figure 13 on the next page. The **pH** of a solution is defined as

$$\mathbf{pH} = -\log[\mathbf{H_3O^+}],$$

where $[H_3O^+]$ is the hydronium ion concentration in moles per liter. It is customary to round pH values to the nearest tenth.

**FIGURE 13** pH Scale

### EXAMPLE 2  Using pH in an Application

Wetlands are classified as *bogs, fens, marshes,* and *swamps,* on the basis of pH values. A pH value between 6.0 and 7.5, such as that of Summerby Swamp in Michigan's Hiawatha National Forest, indicates that the wetland is a "rich fen." When the pH is between 4.0 and 6.0, the wetland is a "poor fen," and if the pH falls to 3.0 or less, it is a "bog." (*Source:* Mohlenbrock, R., "Summerby Swamp, Michigan," *Natural History,* March 1994.)

Suppose that the hydronium ion concentration of a sample of water from a wetland is $6.3 \times 10^{-3}$. How would this wetland be classified?

$$\begin{aligned} pH &= -\log(6.3 \times 10^{-3}) && \text{Definition of pH} \\ &= -(\log 6.3 + \log 10^{-3}) && \text{Product rule} \\ &= -[0.7993 - 3(1)] && \text{Use a calculator to find log 6.3.} \\ &= -0.7993 + 3 && \text{Distributive property} \\ &\approx 2.2 \end{aligned}$$

Since the pH is less than 3.0, the wetland is a bog.

**NOW TRY** Exercise 29.

### EXAMPLE 3  Finding Hydronium Ion Concentration

Find the hydronium ion concentration of drinking water with pH 6.5.

$$\begin{aligned} pH &= -\log[H_3O^+] \\ 6.5 &= -\log[H_3O^+] && \text{Let pH = 6.5.} \\ \log[H_3O^+] &= -6.5 && \text{Multiply by } -1. \end{aligned}$$

Solve for $[H_3O^+]$ by writing the equation in exponential form using base 10.

$$\begin{aligned} [H_3O^+] &= 10^{-6.5} \\ [H_3O^+] &\approx 3.2 \times 10^{-7} && \text{Use a calculator.} \end{aligned}$$

**NOW TRY** Exercise 35.

The loudness of sound is measured in a unit called a **decibel**, abbreviated **dB**. To measure with this unit, we first assign an intensity of $I_0$ to a very faint sound, called the **threshold sound.** If a particular sound has intensity $I$, then the decibel level of this louder sound is

$$D = 10 \log\left(\frac{I}{I_0}\right).$$

| Decibel Level | Example |
|---|---|
| 60 | Normal conversation |
| 90 | Rush hour traffic, lawn mower |
| 100 | Garbage truck, chain saw, pneumatic drill |
| 120 | Rock concert, thunderclap |
| 140 | Gunshot blast, jet engine |
| 180 | Rocket launching pad |

*Source:* Deafness Research Foundation.

The table in the margin gives average decibel levels for some common sounds. Any sound over 85 dB exceeds what hearing experts consider safe. Permanent hearing damage can be suffered at levels above 150 dB.

### EXAMPLE 4   Measuring the Loudness of Sound

If music delivered through the earphones of an iPod has intensity $I$ of $3.162 \times 10^{11} I_0$, find the average decibel level. *(Source: Sacramento Bee.)*

$$D = 10 \log\left(\frac{I}{I_0}\right)$$

$$D = 10 \log\left(\frac{3.162 \times 10^{11} I_0}{I_0}\right) \quad \text{Substitute the given value for } I.$$

$$D = 10 \log(3.162 \times 10^{11})$$

$$D \approx 115 \quad \text{Use a calculator.}$$

**NOW TRY** Exercise 39.

**OBJECTIVE 3** Evaluate natural logarithms using a calculator. The most important logarithms used in applications are **natural logarithms**, which have as base the number $e$. The number $e$ is a fundamental number in our universe. For this reason $e$, like $\pi$, is called a **universal constant.** The letter $e$ is used to honor Leonhard Euler, who published extensive results on the number in 1748. Since it is an irrational number, its decimal expansion never terminates and never repeats.

The first few digits of the decimal value of $e$ are **2.718281828.** On a calculator, use $\boxed{e^x}$ or the two keys $\boxed{\text{INV}}$ and $\boxed{\ln x}$ to approximate powers of $e$, such as

$$e^2 \approx 7.389056099, \quad e^3 \approx 20.08553692, \quad \text{and} \quad e^{0.6} \approx 1.8221188.$$

Logarithms to base $e$ are called natural logarithms because they occur in natural situations that involve growth or decay. The base $e$ logarithm of $x$ is written **ln $x$** (read "el en $x$"). The graph of $y = \ln x$ is given in Figure 14.

A calculator key labeled $\boxed{\ln x}$ is used to evaluate natural logarithms. If your scientific calculator has an $\boxed{e^x}$ key, but not a key labeled $\boxed{\ln x}$, find a natural logarithm by entering the number and pressing the $\boxed{\text{INV}}$ key and then the $\boxed{e^x}$ key. This works because $y = e^x$ defines the inverse function of $y = \ln x$ (or $y = \log_e x$).

**Leonhard Euler (1707–1783)**

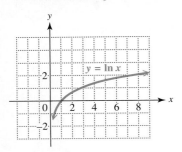

**FIGURE 14**

### EXAMPLE 5   Evaluating Natural Logarithms

Evaluate each logarithm using a calculator.

**(a)** $\ln 0.5841 \approx -0.5377$

As with common logarithms, *a number between 0 and 1 has a negative natural logarithm.*

**(b)** $\ln 192.7 \approx 5.2611$  **(c)** $\ln 10.84 \approx 2.3832$

See Figure 15.

```
ln(.5841)
           -.5377
ln(192.7)
           5.2611
ln(10.84)
           2.3832
```

**FIGURE 15**

**NOW TRY** Exercises 15, 17, and 19.

**OBJECTIVE 4** Use natural logarithms in applications.

### EXAMPLE 6  Applying a Natural Logarithmic Function

The altitude in meters that corresponds to an atmospheric pressure of $x$ millibars is given by the logarithmic function defined by

$$f(x) = 51{,}600 - 7457 \ln x.$$

(*Source:* Miller, A. and J. Thompson, *Elements of Meteorology,* Fourth Edition, Charles E. Merrill Publishing Company, 1993.) Use this function to find the altitude when atmospheric pressure is 400 millibars.

Let $x = 400$ and substitute in the expression for $f(x)$.

$$f(400) = 51{,}600 - 7457 \ln 400$$
$$\approx 6900$$

Atmospheric pressure is 400 millibars at approximately 6900 m.

**NOW TRY** Exercise 41.

In Example 6, the final answer was obtained using a calculator *without* rounding the intermediate values. In general, it is best to wait until the final step to round the answer; otherwise, a buildup of round-off error may cause the final answer to have an incorrect final decimal place digit.

**OBJECTIVE 5** Use the change-of-base rule. We have used a calculator to approximate the values of common logarithms (base 10) and natural logarithms (base $e$). However, some applications involve logarithms to other bases. For example, the percentage of women who had a baby in the last year and returned to work is given by

$$f(x) = 38.83 + 4.208 \log_2 x,$$

for year $x$ since 1980. (*Source:* U.S. Census Bureau.) To use this function, we need to find a base 2 logarithm. The following rule is used to convert logarithms from one base to another.

**Change-of-Base Rule**

If $a > 0$, $a \neq 1$, $b > 0$, $b \neq 1$, and $x > 0$, then

$$\log_a x = \frac{\log_b x}{\log_b a}.$$

Any positive number other than 1 can be used for base $b$ in the change-of-base rule, but usually the only practical bases are $e$ and 10 because calculators give logarithms only for these two bases.

To derive the change-of-base rule, let $\log_a x = m$.

$$\log_a x = m$$
$$a^m = x \quad \text{Change to exponential form.}$$

Since logarithmic functions are one-to-one, if all variables are positive and if $x = y$, then $\log_b x = \log_b y$.

$$\log_b(a^m) = \log_b x$$
$$m \log_b a = \log_b x \quad \text{Power rule}$$
$$(\log_a x)(\log_b a) = \log_b x \quad \text{Substitute for } m.$$
$$\log_a x = \frac{\log_b x}{\log_b a} \quad \text{Divide by } \log_b a.$$

The last step gives the change-of-base rule.

**EXAMPLE 7** Using the Change-of-Base Rule

Find $\log_5 12$.

Use common logarithms and the change-of-base rule.

$$\log_5 12 = \frac{\log 12}{\log 5} \approx 1.5440 \quad \text{Use a calculator.}$$

> **NOW TRY** Exercise 47.

Either common or natural logarithms can be used when applying the change-of-base rule. Verify that the same value is found in Example 7 if natural logarithms are used.

**EXAMPLE 8** Using the Change-of-Base Rule in an Application

Use natural logarithms in the change-of-base rule and the function

$$f(x) = 38.83 + 4.208 \log_2 x$$

(given earlier) to find the percent of women who returned to work after having a baby in 1995. In the equation, $x = 0$ represents 1980.

$$f(15) = 38.83 + 4.208 \log_2 15 \quad 1995 - 1980 = 15, \text{ so let } x = 15.$$

$$= 38.83 + 4.208 \left(\frac{\ln 15}{\ln 2}\right) \quad \text{Change-of-base rule}$$

$$\approx 55.3\% \quad \text{Use a calculator.}$$

This is very close to the actual value of 55%.

> **NOW TRY** Exercise 57.

## 11.5 Exercises

**NOW TRY Exercise**

*Multiple Choice* Choose the correct response in Exercises 1–4.

1. What is the base in the expression $\log x$?

   **A.** $e$   **B.** 1   **C.** 10   **D.** $x$

2. What is the base in the expression $\ln x$?

   **A.** $e$   **B.** 1   **C.** 10   **D.** $x$

3. Since $10^0 = 1$ and $10^1 = 10$, between what two consecutive integers is the value of $\log 5.6$?

   **A.** 5 and 6   **B.** 10 and 11   **C.** 0 and 1   **D.** $-1$ and 0

4. Since $e^1 \approx 2.718$ and $e^2 \approx 7.389$, between what two consecutive integers is the value of $\ln 5.6$?

   **A.** 5 and 6   **B.** 2 and 3   **C.** 1 and 2   **D.** 0 and 1

5. *Concept Check* Without using a calculator, give the value of $\log 10^{19.2}$.

6. *Concept Check* Without using a calculator, give the value of $\ln e^{\sqrt{2}}$.

*You will need a calculator for the remaining exercises in this set.*

*Find each logarithm. Give approximations to four decimal places. See Examples 1 and 5.*

**7.** $\log 43$   **8.** $\log 98$   **9.** $\log 328.4$

10. log 457.2  
11. log 0.0326  
12. log 0.1741  
13. log(4.76 × 10⁹)  
14. log(2.13 × 10⁴)  
15. ln 7.84  
16. ln 8.32  
17. ln 0.0556  
18. ln 0.0217  
19. ln 388.1  
20. ln 942.6  
21. ln(8.59 × $e^2$)  
22. ln(7.46 × $e^3$)  
23. ln 10  
24. log $e$  

25. Use your calculator to find approximations of the following logarithms:
   (a) log 356.8   (b) log 35.68   (c) log 3.568.
   (d) Observe your answers and make a conjecture concerning the decimal values of the common logarithms of numbers greater than 1 that have the same digits.

26. Let $k$ represent the number of letters in your last name.
   (a) Use your calculator to find log $k$.
   (b) Raise 10 to the power indicated by the number in part (a). What is your result?
   (c) Use the concepts of **Section 11.1** to explain why you obtained the answer you found in part (b). Would it matter what number you used for $k$ to observe the same result?

27. Try to find log(−1) using a calculator. (If you have a graphing calculator, it should be in real number mode.) What happens? Explain.

*Suppose that water from a wetland area is sampled and found to have the given hydronium ion concentration. Is the wetland a rich fen, a poor fen, or a bog? See Example 2.*

28. $2.5 \times 10^{-5}$   29. $2.5 \times 10^{-2}$   30. $2.5 \times 10^{-7}$

*Find the pH of the substance with the given hydronium ion concentration. See Example 2.*

31. Ammonia, $2.5 \times 10^{-12}$   32. Sodium bicarbonate, $4.0 \times 10^{-9}$
33. Grapes, $5.0 \times 10^{-5}$   34. Tuna, $1.3 \times 10^{-6}$

*Find the hydronium ion concentration of the substance with the given pH. See Example 3.*

35. Human blood plasma, 7.4   36. Human gastric contents, 2.0
37. Spinach, 5.4   38. Bananas, 4.6

*Solve each problem. See Examples 4 and 6.*

39. Consumers can now enjoy movies at home in elaborate home-theater systems. Find the average decibel level

$$D = 10 \log \left(\frac{I}{I_0}\right)$$

for each popular movie with the given intensity $I$.
   (a) *Spider-Man 2;* $5.012 \times 10^{10} I_0$
   (b) *Finding Nemo;* $10^{10} I_0$
   (c) *Saving Private Ryan;* $6{,}310{,}000{,}000 I_0$

40. The time $t$ in years for an amount increasing at a rate of $r$ (in decimal form) to double is given by

$$t(r) = \frac{\ln 2}{\ln(1 + r)}.$$

This is called **doubling time.** Find the doubling time to the nearest tenth for an investment at each interest rate.
   (a) 2% (or 0.02)   (b) 5% (or 0.05)   (c) 8% (or 0.08)

**41.** The number of years, $N(r)$, since two independently evolving languages split off from a common ancestral language is approximated by
$$N(r) = -5000 \ln r,$$
where $r$ is the percent of words (in decimal form) from the ancestral language common to both languages now. Find the number of years (to the nearest hundred years) since the split for each percent of common words.

**(a)** 85% (or 0.85)   **(b)** 35% (or 0.35)   **(c)** 10% (or 0.10)

**42.** The concentration of a drug injected into the bloodstream decreases with time. The intervals of time $T$ when the drug should be administered are given by
$$T = \frac{1}{k} \ln \frac{C_2}{C_1},$$
where $k$ is a constant determined by the drug in use, $C_2$ is the concentration at which the drug is harmful, and $C_1$ is the concentration below which the drug is ineffective. (*Source:* Horelick, Brindell and Sinan Koont, "Applications of Calculus to Medicine: Prescribing Safe and Effective Dosage," *UMAP Module 202,* 1977.) Thus, if $T = 4$, the drug should be administered every 4 hr. For a certain drug, $k = \frac{1}{3}$, $C_2 = 5$, and $C_1 = 2$. How often should the drug be administered? (*Hint:* Round down.)

**43.** The growth of outpatient surgeries as a percent of total surgeries at hospitals is approximated by
$$f(x) = -1317 + 304 \ln x,$$
where $x$ is the number of years since 1900. (*Source:* American Hospital Association.)

**(a)** What does this function predict for the percent of outpatient surgeries in 1998?
**(b)** When did outpatient surgeries reach 50%? (*Hint:* Substitute for $y$, then write the equation in exponential form to solve it.)

**44.** In the central Sierra Nevada of California, the percent of moisture that falls as snow rather than rain is approximated reasonably well by
$$f(x) = 86.3 \ln x - 680,$$
where $x$ is the altitude in feet.

**(a)** What percent of the moisture at 5000 ft falls as snow?
**(b)** What percent at 7500 ft falls as snow?

**45.** The **cost-benefit equation**
$$T = -0.642 - 189 \ln(1 - p)$$
describes the approximate tax $T$, in dollars per ton, that would result in a $p$% (in decimal form) reduction in carbon dioxide emissions.

**(a)** What tax will reduce emissions 25%?
**(b)** Explain why the equation is not valid for $p = 0$ or $p = 1$.

**46.** The age in years of a female blue whale of length $L$ in feet is approximated by
$$t = -2.57 \ln\left(\frac{87 - L}{63}\right).$$

**(a)** How old is a female blue whale that measures 80 ft?
**(b)** The equation that defines $t$ has domain $24 < L < 87$. Explain why.

*Use the change-of-base rule (with either common or natural logarithms) to find each logarithm to four decimal places. See Example 7.*

**47.** $\log_3 12$  **48.** $\log_4 18$  **49.** $\log_5 3$
**50.** $\log_7 4$  **51.** $\log_3 \sqrt{2}$  **52.** $\log_6 \sqrt[3]{5}$
**53.** $\log_\pi e$  **54.** $\log_\pi 10$  **55.** $\log_e 12$

**56.** Explain why the answer to Exercise 55 is the same one that you get when you use a calculator to approximate $\ln 12$.

*Solve each application of a logarithmic function (from Exercises 53 and 54 of **Section 11.3**). See Example 8.*

**57.** According to selected figures from 1981 through 2003, the number of billion cubic feet of natural gas gross withdrawals from crude oil wells in the United States can be approximated by the function defined by

$$f(x) = 3800 + 585 \log_2 x,$$

where $x = 1$ represents 1981, $x = 2$ represents 1982, and so on. (*Source:* Energy Information Administration.) Use this function to approximate the number of cubic feet withdrawn in 2003.

**58.** According to selected figures from the last two decades of the 20th century, the number of trillion cubic feet of dry natural gas consumed worldwide can be approximated by the function defined by

$$f(x) = 51.47 + 6.044 \log_2 x,$$

where $x = 1$ represents 1980, $x = 2$ represents 1981, and so on. (*Source:* Energy Information Administration.) Use this function to approximate consumption in 2003.

## 11.6 Exponential and Logarithmic Equations; Further Applications

**OBJECTIVES**

1. Solve equations involving variables in the exponents.
2. Solve equations involving logarithms.
3. Solve applications of compound interest.
4. Solve applications involving base $e$ exponential growth and decay.
5. Use a graphing calculator to solve exponential and logarithmic equations.

We solved exponential and logarithmic equations in **Sections 11.2 and 11.3**. General methods for solving these equations depend on the following properties.

> **Properties for Solving Exponential and Logarithmic Equations**
>
> For all real numbers $b > 0$, $b \neq 1$, and any real numbers $x$ and $y$:
>
> 1. If $x = y$, then $b^x = b^y$.
> 2. If $b^x = b^y$, then $x = y$.
> 3. If $x = y$, and $x > 0, y > 0$, then $\log_b x = \log_b y$.
> 4. If $x > 0, y > 0$, and $\log_b x = \log_b y$, then $x = y$.

We used Property 2 to solve exponential equations in **Section 11.2**.

**OBJECTIVE 1** Solve equations involving variables in the exponents. The first two examples illustrate the method for solving exponential equations using Property 3.

**EXAMPLE 1** Solving an Exponential Equation

Solve $3^x = 12$.

$$3^x = 12$$
$$\log 3^x = \log 12 \quad \text{Property 3 (common logs)}$$
$$x \log 3 = \log 12 \quad \text{Power rule}$$

Exact solution $\longrightarrow \quad x = \dfrac{\log 12}{\log 3} \quad$ Divide by log 3.

Decimal approximation $\longrightarrow \quad x \approx 2.262 \quad$ Use a calculator.

The solution set is $\{2.262\}$. Check with a calculator that $3^{2.262} \approx 12$.

**NOW TRY** Exercise 1.

**CAUTION** Be careful: $\frac{\log 12}{\log 3}$ is *not* equal to log 4. Check to see that log 4 ≈ 0.6021, but $\frac{\log 12}{\log 3} \approx 2.262$.

When an exponential equation has $e$ as the base, as in the next example, it is easiest to use base $e$ logarithms.

### EXAMPLE 2  Solving an Exponential Equation with Base e

Solve $e^{0.003x} = 40$.

$$\ln e^{0.003x} = \ln 40 \quad \text{Property 3 (natural logs)}$$
$$0.003x \ln e = \ln 40 \quad \text{Power rule}$$
$$0.003x = \ln 40 \quad \ln e = \ln e^1 = 1$$
$$x = \frac{\ln 40}{0.003} \quad \text{Divide by 0.003.}$$
$$x \approx 1230 \quad \text{Use a calculator.}$$

The solution set is {1230}. Check that $e^{0.003(1230)} \approx 40$.

**NOW TRY** Exercise 11.

### General Method for Solving an Exponential Equation

Take logarithms to the same base on both sides, and then use the power rule of logarithms or the special property $\log_b b^x = x$. (See Examples 1 and 2.)

As a special case, if both sides can be written as exponentials with the same base, do so, and set the exponents equal. (See **Section 11.2**.)

**OBJECTIVE 2**  Solve equations involving logarithms. The properties of logarithms from **Section 11.4** are useful here, as is using the definition of a logarithm to change the equation to exponential form.

### EXAMPLE 3  Solving a Logarithmic Equation

Solve $\log_2(x + 5)^3 = 4$. Give the exact solution.

$$\log_2(x + 5)^3 = 4$$
$$(x + 5)^3 = 2^4 \quad \text{Convert to exponential form.}$$
$$(x + 5)^3 = 16$$
$$x + 5 = \sqrt[3]{16} \quad \text{Take the cube root on each side.}$$
$$x = -5 + \sqrt[3]{16} \quad \text{Subtract 5.}$$
$$x = -5 + 2\sqrt[3]{2} \quad \sqrt[3]{16} = \sqrt[3]{8 \cdot 2} = \sqrt[3]{8} \cdot \sqrt[3]{2} = 2\sqrt[3]{2}$$

*Check:* $\log_2(x+5)^3 = 4$     Original equation

$\log_2(-5 + 2\sqrt[3]{2} + 5)^3 = 4$ ?    Let $x = -5 + 2\sqrt[3]{2}$.

$\log_2(2\sqrt[3]{2})^3 = 4$ ?    Work inside the parentheses.

$\log_2 16 = 4$ ?    $(2\sqrt[3]{2})^3 = 2^3(\sqrt[3]{2})^3 = 8 \cdot 2 = 16$

$2^4 = 16$ ?    Write in exponential form.

$16 = 16$    True

A true statement results, so the solution set is $\{-5 + 2\sqrt[3]{2}\}$.

**NOW TRY** Exercise 25.

---

**CAUTION** Recall that the domain of $y = \log_b x$ is $(0, \infty)$. For this reason, *always check that the solution of an equation with logarithms yields only logarithms of positive numbers in the original equation.*

---

**EXAMPLE 4** Solving a Logarithmic Equation

Solve $\log_2(x + 1) - \log_2 x = \log_2 7$.

$\log_2(x + 1) - \log_2 x = \log_2 7$

*Transform the left side to an expression with only one logarithm.*

$\log_2 \dfrac{x+1}{x} = \log_2 7$    Quotient rule

$\dfrac{x+1}{x} = 7$    Property 4

$x + 1 = 7x$    Multiply by $x$.

$1 = 6x$    Subtract $x$.

$\dfrac{1}{6} = x$    Divide by 6.

Since we cannot take the logarithm of a *nonpositive* number, both $x + 1$ and $x$ must be positive here. If $x = \tfrac{1}{6}$, then this condition is satisfied.

*Check:* $\log_2(x+1) - \log_2 x = \log_2 7$    Original equation

$\log_2\left(\dfrac{1}{6} + 1\right) - \log_2 \dfrac{1}{6} = \log_2 7$ ?    Let $x = \tfrac{1}{6}$.

$\log_2 \dfrac{7}{6} - \log_2 \dfrac{1}{6} = \log_2 7$ ?

$\log_2 \dfrac{\frac{7}{6}}{\frac{1}{6}} = \log_2 7$ ?    Quotient rule

$\dfrac{\frac{7}{6}}{\frac{1}{6}} = \dfrac{7}{6} \div \dfrac{1}{6} = \dfrac{7}{6} \cdot \dfrac{6}{1} = 7$

$\log_2 7 = \log_2 7$    True

A true statement results, so the solution set is $\left\{\tfrac{1}{6}\right\}$.

**NOW TRY** Exercise 31.

### EXAMPLE 5  Solving a Logarithmic Equation

Solve $\log x + \log(x - 21) = 2$.

$$\log x + \log(x - 21) = 2$$
$$\log x(x - 21) = 2 \qquad \text{Product rule}$$

The base is 10.
$$x(x - 21) = 10^2 \qquad \text{Write in exponential form.}$$
$$x^2 - 21x = 100 \qquad \text{Distributive property; multiply.}$$
$$x^2 - 21x - 100 = 0 \qquad \text{Standard form}$$
$$(x - 25)(x + 4) = 0 \qquad \text{Factor.}$$
$$x - 25 = 0 \quad \text{or} \quad x + 4 = 0 \qquad \text{Zero-factor property}$$
$$x = 25 \quad \text{or} \quad x = -4 \qquad \text{Solve each equation.}$$

The value $-4$ must be rejected as a solution since it leads to the logarithm of a negative number in the original equation:

$$\log(-4) + \log(-4 - 21) = 2. \qquad \text{The left side is undefined.}$$

Check that the only solution is 25, so the solution set is $\{25\}$.

**NOW TRY** Exercise 35.

**CAUTION** *Do not reject a potential solution just because it is nonpositive. Reject any value that leads to the logarithm of a nonpositive number.*

In summary, we use the following steps to solve a logarithmic equation.

### Solving a Logarithmic Equation

**Step 1** **Transform the equation so that a single logarithm appears on one side.** Use the product rule or quotient rule of logarithms to do this.

**Step 2** **(a) Use Property 4.** If $\log_b x = \log_b y$, then $x = y$. (See Example 4.)

**(b) Write the equation in exponential form.** If $\log_b x = k$, then $x = b^k$. (See Examples 3 and 5.)

**OBJECTIVE 3** Solve applications of compound interest. So far in this book, we have solved simple interest problems using the formula $I = prt$. In most cases, interest paid or charged is *compound interest* (interest paid on both principal and interest). The formula for compound interest is an application of exponential functions.

### Compound Interest Formula (for a Finite Number of Periods)

If a principal of $P$ dollars is deposited at an annual rate of interest $r$ compounded (paid) $n$ times per year, then the account will contain

$$A = P\left(1 + \frac{r}{n}\right)^{nt}$$

dollars after $t$ years. (In this formula, $r$ is expressed as a decimal.)

## SECTION 11.6 Exponential and Logarithmic Equations; Further Applications

**EXAMPLE 6** Solving a Compound Interest Problem for *A*

How much money will there be in an account at the end of 5 yr if $1000 is deposited at 6% compounded quarterly? (Assume no withdrawals are made.)

Because interest is compounded quarterly, $n = 4$. The other given values are $P = 1000$, $r = 0.06$ (because 6% = 0.06), and $t = 5$.

$$A = P\left(1 + \frac{r}{n}\right)^{nt}$$

$$A = 1000\left(1 + \frac{0.06}{4}\right)^{4 \cdot 5} \quad \text{Substitute the given values.}$$

$$A = 1000(1.015)^{20}$$

$$A = 1346.86 \quad \text{Use a calculator; round to the nearest cent.}$$

The account will contain $1346.86. (The actual amount of interest earned is $1346.86 − $1000 = $346.86. Why?)

**NOW TRY** Exercise 41(a).

**EXAMPLE 7** Solving a Compound Interest Problem for *t*

Suppose inflation is averaging 3% per year. How many years will it take for prices to double?

We want the number of years $t$ for $1 to grow to $2 at a rate of 3% per year. In the compound interest formula, we let $A = 2$, $P = 1$, $r = 0.03$, and $n = 1$.

$$2 = 1\left(1 + \frac{0.03}{1}\right)^{1t} \quad \text{Substitute in the compound interest formula.}$$

$$2 = (1.03)^t \quad \text{Simplify.}$$

$$\log 2 = \log(1.03)^t \quad \text{Property 3}$$

$$\log 2 = t \log(1.03) \quad \text{Power rule}$$

$$t = \frac{\log 2}{\log 1.03} \quad \text{Divide by log 1.03.}$$

$$t \approx 23.45 \quad \text{Use a calculator.}$$

Prices will double in about 23 yr. (This is called the **doubling time** of the money.) To check, verify that $1.03^{23.45} \approx 2$.

**NOW TRY** Exercise 41(b).

Interest can be compounded annually, semiannually, quarterly, daily, and so on. The number of compounding periods can get larger and larger. If the value of $n$ is allowed to approach infinity, we have an example of *continuous compounding*. However, the compound interest formula above cannot be used for continuous compounding since there is no finite value for $n$. The formula for continuous compounding is an example of exponential growth involving the number $e$.

---

**Continuous Compound Interest Formula**

If a principal of $P$ dollars is deposited at an annual rate of interest $r$ compounded continuously for $t$ years, the final amount on deposit is

$$A = Pe^{rt}.$$

### EXAMPLE 8  Solving a Continuous Compound Interest Problem

In Example 6 we found that $1000 invested for 5 yr at 6% interest compounded quarterly would grow to $1346.86.

**(a)** How much would this same investment grow to if interest were compounded continuously?

$$A = Pe^{rt} \quad \text{Formula for continuous compounding}$$
$$= 1000e^{0.06(5)} \quad \text{Let } P = 1000, r = 0.06, \text{ and } t = 5.$$
$$= 1000e^{0.30}$$
$$= 1349.86 \quad \text{Use a calculator; round to the nearest cent.}$$

Continuous compounding would cause the investment to grow to $1349.86. This is $3.00 more than the amount the investment grew to in Example 6, when interest was compounded quarterly.

**(b)** How long would it take for the initial investment to double its original amount?

We must find the value of $t$ that will cause $A$ to be $2(\$1000) = \$2000$. Once again, we use the continuous compounding formula with $P = 1000$ and $r = 0.06$.

$$A = Pe^{rt}$$
$$2000 = 1000e^{0.06t} \quad \text{Let } A = 2P = 2000.$$
$$2 = e^{0.06t} \quad \text{Divide by 1000.}$$
$$\ln 2 = \ln e^{0.06t} \quad \text{Take natural logarithms.}$$
$$\ln 2 = 0.06t \quad \ln e^k = k$$
$$t = \frac{\ln 2}{0.06} \quad \text{Divide by 0.06.}$$
$$t \approx 11.55 \quad \text{Use a calculator.}$$

It would take about 11.55 yr for the original investment to double.

**NOW TRY** Exercise 43.

---

**OBJECTIVE 4** Solve applications involving base $e$ exponential growth and decay.
One of the most common applications of exponential functions depends on the fact that in many situations involving growth or decay of a population, the amount or number of some quantity present at time $t$ can be closely approximated by

$$y = y_0 e^{kt},$$

where $y_0$ is the amount or number present at time $t = 0$, $k$ is a constant, and $e$ is the base of natural logarithms.

### EXAMPLE 9  Applying a Base $e$ Exponential Function

Hybrid cars, powered by a combination of a battery and gasoline, emit less pollution and get considerably better gas mileage than conventional cars of comparable size. The first hybrid car came on the U.S. market in 1999. Since then, sales of hybrid cars have been increasing rapidly. The number of hybrid cars sold in the United States over the period 2000–2005 can be modeled by the function defined by

$$f(x) = 10{,}014 e^{0.5761x},$$

where $x = 0$ represents 2000, $x = 1$ represents 2001, and so on. (*Source:* www.hybridcars.com) Use this function to approximate the number of hybrid cars sold in 2002.

Here, $x = 2$ represents 2002. Evaluate $f(2)$ using a calculator.

$$f(2) = 10{,}014e^{0.5761(2)} \approx 31{,}696$$

In 2002, about 31,696 hybrid cars were sold in the United States.

**NOW TRY Exercise 51.**

### EXAMPLE 10  Solving an Application Involving Exponential Decay

Carbon 14 is a radioactive form of carbon that is found in all living plants and animals. After a plant or animal dies, the radioactive carbon 14 disintegrates according to the function defined by

$$y = y_0 e^{-0.000121t},$$

where $t$ is time in years, $y$ is the amount of the sample at time $t$, and $y_0$ is the initial amount present at $t = 0$.

**(a)** If an initial sample contains $y_0 = 10$ g of carbon 14, how many grams will be present after 3000 yr?

Let $y_0 = 10$ and $t = 3000$ in the formula, and use a calculator.

$$y = 10e^{-0.000121(3000)} \approx 6.96 \text{ g}$$

**(b)** How long would it take for the initial sample to decay to half of its original amount? (This is called the **half-life**.)

Let $y = \frac{1}{2}(10) = 5$, and solve for $t$.

$$5 = 10e^{-0.000121t} \quad \text{Substitute.}$$

$$\frac{1}{2} = e^{-0.000121t} \quad \text{Divide by 10.}$$

$$\ln \frac{1}{2} = -0.000121t \quad \text{Take natural logarithms; } \ln e^k = k.$$

$$t = \frac{\ln \frac{1}{2}}{-0.000121} \quad \text{Divide by } -0.000121.$$

$$t \approx 5728 \quad \text{Use a calculator.}$$

The half-life is just over 5700 yr.

**NOW TRY Exercise 55.**

**OBJECTIVE 5** Use a graphing calculator to solve exponential and logarithmic equations. Recall that the $x$-intercepts of the graph of a function $f$ correspond to the real solutions of the equation $f(x) = 0$. In Example 1, we solved the equation $3^x = 12$ algebraically using rules for logarithms and found the solution set to be $\{2.262\}$. This can be supported graphically by showing that the $x$-intercept of the graph of the function defined by $y = 3^x - 12$ corresponds to this solution. See Figure 16 on the next page.

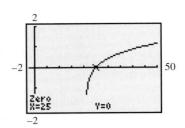

**FIGURE 16**  **FIGURE 17**

In Example 5, we solved $\log x + \log(x - 21) = 2$ to find the solution set $\{25\}$. (We rejected the apparent solution $-4$ since it led to the logarithm of a negative number.) Figure 17 shows that the $x$-intercept of the graph of the function defined by $y = \log x + \log(x - 21) - 2$ supports this result.

## 11.6 Exercises

**NOW TRY Exercise**

*You will need a calculator for many of the exercises in this set.*

*Solve each equation. Give solutions to three decimal places. See Example 1.*

1. $7^x = 5$
2. $4^x = 3$
3. $9^{-x+2} = 13$
4. $6^{-t+1} = 22$
5. $3^{2x} = 14$
6. $5^{0.3x} = 11$
7. $2^{x+3} = 5^x$
8. $6^{m+3} = 4^m$
9. $2^{x+3} = 3^{x-4}$

*Solve each equation. Use natural logarithms. When appropriate, give solutions to three decimal places. See Example 2.*

10. $e^{0.006x} = 30$
11. $e^{0.012x} = 23$
12. $e^{-0.103x} = 7$
13. $e^{-0.205x} = 9$
14. $\ln e^x = 4$
15. $\ln e^{3x} = 9$
16. $\ln e^{0.04x} = \sqrt{3}$
17. $\ln e^{0.45x} = \sqrt{7}$
18. $\ln e^{2x} = \pi$

19. Solve one of the equations in Exercises 10–13 using common logarithms rather than natural logarithms. (You should get the same solution.) Explain why using natural logarithms is a better choice.

20. *Concept Check* If you were asked to solve
$$10^{0.0025x} = 75,$$
would natural or common logarithms be a better choice? Why?

*Solve each equation. Give the exact solution. See Example 3.*

21. $\log_3(6x + 5) = 2$
22. $\log_5(12x - 8) = 3$
23. $\log_2(2x - 1) = 5$
24. $\log_6(4x + 2) = 2$
25. $\log_7(x + 1)^3 = 2$
26. $\log_4(x - 3)^3 = 4$

27. *Concept Check* Suppose that in solving a logarithmic equation having the term $\log(x - 3)$ you obtain an apparent solution of 2. All algebraic work is correct. Why must you reject 2 as a solution of the equation?

28. *Concept Check* Suppose that in solving a logarithmic equation having the term $\log(3 - x)$ you obtain an apparent solution of $-4$. All algebraic work is correct. Should you reject $-4$ as a solution of the equation? Why or why not?

*Solve each equation. Give exact solutions. See Examples 4 and 5.*

**29.** $\log(6x + 1) = \log 3$

**30.** $\log(7 - x) = \log 12$

**31.** $\log_5(3t + 2) - \log_5 t = \log_5 4$

**32.** $\log_2(x + 5) - \log_2(x - 1) = \log_2 3$

**33.** $\log 4x - \log(x - 3) = \log 2$

**34.** $\log(-x) + \log 3 = \log(2x - 15)$

**35.** $\log_2 x + \log_2(x - 7) = 3$

**36.** $\log(2x - 1) + \log 10x = \log 10$

**37.** $\log 5x - \log(2x - 1) = \log 4$

**38.** $\log_3 x + \log_3(2x + 5) = 1$

**39.** $\log_2 x + \log_2(x - 6) = 4$

**40.** $\log_2 x + \log_2(x + 4) = 5$

*Solve each problem. See Examples 6–8.*

**41.** **(a)** How much money will there be in an account at the end of 6 yr if $2000 is deposited at 4% compounded quarterly? (Assume no withdrawals are made.)
   **(b)** To one decimal place, how long will it take for the account to grow to $3000?

**42.** **(a)** How much money will there be in an account at the end of 7 yr if $3000 is deposited at 3.5% compounded quarterly? (Assume no withdrawals are made.)
   **(b)** To one decimal place, when will the account grow to $5000?

**43.** **(a)** What will be the amount $A$ in an account with initial principal $4000 if interest is compounded continuously at an annual rate of 3.5% for 6 yr?
   **(b)** To one decimal place, how long will it take for the initial amount to double?

**44.** Refer to Exercise 42. Does the money grow to a larger value under those conditions or when invested for 7 yr at 3% compounded continuously?

**45.** Find the amount of money in an account after 12 yr if $5000 is deposited at 7% annual interest compounded as follows.

   **(a)** Annually  **(b)** Semiannually  **(c)** Quarterly
   **(d)** Daily (Use $n = 365$.)  **(e)** Continuously

**46.** How much money will be in an account at the end of 8 yr if $4500 is deposited at 6% annual interest compounded as follows?

   **(a)** Annually  **(b)** Semiannually  **(c)** Quarterly
   **(d)** Daily (Use $n = 365$.)  **(e)** Continuously

**47.** How much money must be deposited today to amount to $1850 in 40 yr at 6.5% compounded continuously?

**48.** How much money must be deposited today to amount to $1000 in 10 yr at 5% compounded continuously?

*Solve each problem. See Examples 9 and 10.*

**49.** The total volume in millions of tons of materials recovered from municipal solid waste collections in the United States during the period 1980–2003 can be approximated by the function defined by

$$f(x) = 15.80e^{0.0708x},$$

where $x = 0$ corresponds to 1980, $x = 1$ to 1981, and so on. Approximate the volume recovered each year. (*Source:* Environmental Protection Agency.)

   **(a)** 1980  **(b)** 1985  **(c)** 1995  **(d)** 2003

**50.** Worldwide emissions in millions of metric tons of the greenhouse gas carbon dioxide from fossil fuel consumption during the period 1980–2003 can be modeled by the function defined by

$$f(x) = 18,315e^{0.01338x},$$

where $x = 0$ corresponds to 1980, $x = 1$ to 1981, and so on. Approximate the emissions for each year. (*Source:* U.S. Department of Energy.)

   **(a)** 1980  **(b)** 1995  **(c)** 2000  **(d)** 2003

**51.** Consumer expenditures on all types of books in the United States for the years 1995–2004 can be modeled by the function defined by

$$B(x) = 27{,}190e^{0.0448x},$$

where $x = 0$ represents 1995, $x = 1$ represents 1996, and so on, and $B(x)$ is in millions of dollars. Approximate consumer expenditures for 2004. (*Source:* Book Industry Study Group.)

**52.** Based on selected figures obtained during the years 1980–2003, the total number of bachelor's degrees earned in the United States can be modeled by the function defined by

$$D(x) = 918{,}030e^{0.0154x},$$

where $x = 0$ corresponds to 1980, $x = 10$ corresponds to 1990, and so on. Approximate the number of bachelor's degrees earned in 2003. (*Source:* U.S. National Center for Education Statistics.)

**53.** Suppose that the amount, in grams, of plutonium 241 present in a given sample is determined by the function defined by

$$A(t) = 2.00e^{-0.053t},$$

where $t$ is measured in years. Find the amount present in the sample after the given number of years.

**(a)** 4  **(b)** 10  **(c)** 20  **(d)** What was the initial amount present?

**54.** Suppose that the amount, in grams, of radium 226 present in a given sample is determined by the function defined by

$$A(t) = 3.25e^{-0.00043t},$$

where $t$ is measured in years. Find the amount present in the sample after the given number of years.

**(a)** 20  **(b)** 100  **(c)** 500  **(d)** What was the initial amount present?

**55.** A sample of 400 g of lead 210 decays to polonium 210 according to the function defined by

$$A(t) = 400e^{-0.032t},$$

where $t$ is time in years.

**(a)** How much lead will be left in the sample after 25 yr?
**(b)** How long will it take the initial sample to decay to half of its original amount?

**56.** The concentration of a drug in a person's system decreases according to the function defined by

$$C(t) = 2e^{-0.125t},$$

where $C(t)$ is in appropriate units, and $t$ is in hours.

**(a)** How much of the drug will be in the system after 1 hr?
**(b)** Find the time that it will take for the concentration to be half of its original amount.

**57.** Refer to Exercise 49. Assuming that the function continued to apply past 2003, in what year could we have expected the volume of materials recovered to have reached 100 million tons? (*Source:* Environmental Protection Agency.)

**58.** Refer to Exercise 50. Assuming that the function continued to apply past 2003, in what year can we expect worldwide carbon dioxide emissions from fossil fuel consumption to reach 28,000 million metric tons? Round up to the nearest year. (*Source:* U.S. Department of Energy.)

# 11 SUMMARY

## KEY TERMS

**11.1** one-to-one function
inverse of a function

**11.2** exponential function
asymptote
exponential equation

**11.3** logarithm
logarithmic equation
logarithmic function with base $a$

**11.5** common logarithm
natural logarithm
universal constant

## NEW SYMBOLS

$f^{-1}(x)$ the inverse of $f(x)$
$\log_a x$ the logarithm of $x$ to the base $a$

$\log x$ common (base 10) logarithm of $x$

$\ln x$ natural (base $e$) logarithm of $x$

$e$ a constant, approximately 2.7182818

## QUICK REVIEW

| CONCEPTS | EXAMPLES |
|---|---|
| **11.1 Inverse Functions** | |

### Horizontal Line Test

A function is one-to-one if every horizontal line intersects the graph of the function at most once.

Find $f^{-1}$ if $f(x) = 2x - 3$.

The graph of $f$ is a straight line, so $f$ is one-to-one by the horizontal line test.

### Inverse Functions

For a one-to-one function $f$ defined by an equation $y = f(x)$, the equation that defines the inverse function $f^{-1}$ is found by interchanging $x$ and $y$, solving for $y$, and replacing $y$ with $f^{-1}(x)$.

To find $f^{-1}(x)$, interchange $x$ and $y$ in the equation $y = 2x - 3$.
$$x = 2y - 3$$
Solve for $y$ to get
$$y = \frac{x + 3}{2}.$$
Therefore,
$$f^{-1}(x) = \frac{x + 3}{2}, \text{ or } f^{-1}(x) = \frac{1}{2}x + \frac{3}{2}.$$

In general, the graph of $f^{-1}$ is the mirror image of the graph of $f$ with respect to the line $y = x$.

The graphs of a function $f$ and its inverse $f^{-1}$ are shown here.

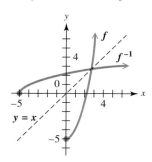

*(continued)*

| CONCEPTS | EXAMPLES |
|---|---|

## 11.2 Exponential Functions

For $a > 0$, $a \neq 1$, $F(x) = a^x$ defines the exponential function with base $a$.

**Graph of $F(x) = a^x$**
1. The graph contains the point $(0, 1)$.
2. When $a > 1$, the graph rises from left to right. When $0 < a < 1$, the graph falls from left to right.
3. The $x$-axis is an asymptote.
4. The domain is $(-\infty, \infty)$; the range is $(0, \infty)$.

$F(x) = 3^x$ defines the exponential function with base 3.

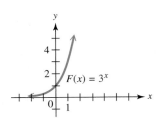

## 11.3 Logarithmic Functions

$y = \log_a x$ means $x = a^y$.

For $b > 0$, $b \neq 1$, $\log_b b = 1$ and $\log_b 1 = 0$.

For $a > 0$, $a \neq 1$, $x > 0$, $G(x) = \log_a x$ defines the logarithmic function with base $a$.

**Graph of $G(x) = \log_a x$**
1. The graph contains the point $(1, 0)$.
2. When $a > 1$, the graph rises from left to right. When $0 < a < 1$, the graph falls from left to right.
3. The $y$-axis is an asymptote.
4. The domain is $(0, \infty)$; the range is $(-\infty, \infty)$.

$y = \log_2 x$ means $x = 2^y$.

$\log_3 3 = 1 \qquad \log_5 1 = 0$

$G(x) = \log_3 x$ defines the logarithmic function with base 3.

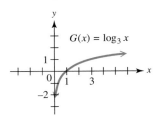

## 11.4 Properties of Logarithms

**Product Rule** $\quad \log_a xy = \log_a x + \log_a y$

**Quotient Rule** $\quad \log_a \dfrac{x}{y} = \log_a x - \log_a y$

**Power Rule** $\quad \log_a x^r = r \log_a x$

**Special Properties**
$$b^{\log_b x} = x \quad \text{and} \quad \log_b b^x = x$$

$\log_2 3m = \log_2 3 + \log_2 m$

$\log_5 \dfrac{9}{4} = \log_5 9 - \log_5 4$

$\log_{10} 2^3 = 3 \log_{10} 2$

$6^{\log_6 10} = 10 \qquad \log_3 3^4 = 4$

## 11.5 Common and Natural Logarithms

**Common logarithms (base 10)** are used in applications such as pH, sound level, and intensity of an earthquake. Use the $\boxed{\text{LOG}}$ key of a calculator to evaluate common logarithms.

Use the formula $\text{pH} = -\log[H_3 O^+]$ to find the pH (to one decimal place) of grapes with hydronium ion concentration $5.0 \times 10^{-5}$.

$$\begin{aligned} \text{pH} &= -\log(5.0 \times 10^{-5}) && \text{Substitute.} \\ &= -(\log 5.0 + \log 10^{-5}) && \text{Property of logarithms} \\ &\approx 4.3 && \text{Evaluate.} \end{aligned}$$

*(continued)*

| CONCEPTS | EXAMPLES |
|---|---|
| **Natural logarithms (base $e$)** are often used in applications of growth and decay, such as time for money invested to double, decay of chemical compounds, and biological growth. Use the $\boxed{\ln x}$ key or both the $\boxed{\text{INV}}$ and $\boxed{e^x}$ keys to evaluate natural logarithms. | Use the formula for doubling time (in years) $t(r) = \frac{\ln 2}{\ln(1+r)}$ to find the doubling time to the nearest tenth at an interest rate of 4%.<br><br>$t(0.04) = \frac{\ln 2}{\ln(1+0.04)}$  Substitute.<br><br>$\approx 17.7$  Evaluate.<br><br>The doubling time is about 17.7 yr. |
| **Change-of-Base Rule**<br>If $a > 0, a \neq 1, b > 0, b \neq 1, x > 0$, then<br>$$\log_a x = \frac{\log_b x}{\log_b a}.$$ | $\log_3 17 = \frac{\ln 17}{\ln 3} = \frac{\log 17}{\log 3} \approx 2.5789$ |

## 11.6 Exponential and Logarithmic Equations; Further Applications

| | |
|---|---|
| To solve exponential equations, use these properties $(b > 0, b \neq 1)$.<br>**1.** If $b^x = b^y$, then $x = y$. | Solve.  $2^{3x} = 2^5$<br>$3x = 5$  Set exponents equal.<br>$x = \frac{5}{3}$  Divide by 3.<br>The solution set is $\{\frac{5}{3}\}$. |
| **2.** If $x = y, x > 0, y > 0$, then $\log_b x = \log_b y$. | Solve.  $5^m = 8$<br>$\log 5^m = \log 8$  Take common logarithms.<br>$m \log 5 = \log 8$  Power rule<br>$m = \frac{\log 8}{\log 5} \approx 1.2920$  Divide by log 5.<br>The solution set is $\{1.2920\}$. |
| To solve logarithmic equations, use these properties, where $b > 0, b \neq 1, x > 0, y > 0$. First use the properties of **Section 11.4,** if necessary, to write the equation in the proper form.<br>**1.** If $\log_b x = \log_b y$, then $x = y$. | Solve.  $\log_3 2x = \log_3(x+1)$<br>$2x = x + 1$<br>$x = 1$  Subtract $x$.<br>This value checks, so the solution set is $\{1\}$. |
| **2.** If $\log_b x = y$, then $b^y = x$. | Solve.  $\log_2(3a - 1) = 4$<br>$3a - 1 = 2^4$  Exponential form<br>$3a - 1 = 16$  Apply the exponent.<br>$3a = 17$  Add 1.<br>$a = \frac{17}{3}$  Divide by 3.<br>The solution set is $\{\frac{17}{3}\}$. |

# 11 REVIEW EXERCISES

*Determine whether each graph is the graph of a one-to-one function.*

1.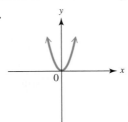

2.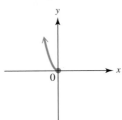

3. The table lists caffeine amounts in several popular 12-oz sodas. If the set of sodas is the domain and the set of caffeine amounts is the range of the function consisting of the six pairs listed, is it a one-to-one function? Why or why not?

| Soda | Caffeine (mg) |
| --- | --- |
| Mountain Dew | 55 |
| Diet Coke | 45 |
| Dr. Pepper | 41 |
| Sunkist Orange Soda | 41 |
| Diet Pepsi-Cola | 36 |
| Coca-Cola Classic | 34 |

*Source:* National Soft Drink Association.

*Determine whether each function is one-to-one. If it is, find its inverse.*

4. $f(x) = -3x + 7$

5. $f(x) = \sqrt[3]{6x - 4}$

6. $f(x) = -x^2 + 3$

*Each function graphed is one-to-one. Graph its inverse.*

7.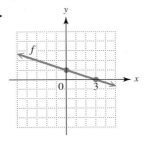

8.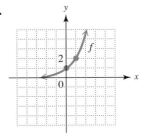

*Graph each function.*

9. $f(x) = 3^x$

10. $f(x) = \left(\dfrac{1}{3}\right)^x$

11. $y = 3^{x+1}$

12. $y = 2^{2x+3}$

*Solve each equation.*

13. $4^{3x} = 8^{x+4}$

14. $\left(\dfrac{1}{27}\right)^{x-1} = 9^{2x}$

15. The gross wastes generated in plastics, in millions of tons, from 1980 through 2003 can be approximated by the exponential function defined by

$$W(x) = 7.77(1.059)^x,$$

where $x = 0$ corresponds to 1980, $x = 5$ to 1985, and so on. Use this function to approximate the plastic waste amounts for each year. (*Source:* U.S. Environmental Protection Agency.)

(a) 1985   (b) 1995   (c) 2000

*Graph each function.*

16. $g(x) = \log_3 x$   (*Hint:* See Exercise 9.)   17. $g(x) = \log_{1/3} x$   (*Hint:* See Exercise 10.)

*Solve each equation.*

18. $\log_8 64 = x$   19. $\log_2 \sqrt{8} = x$   20. $\log_x\left(\dfrac{1}{49}\right) = -2$

21. $\log_4 x = \dfrac{3}{2}$   22. $\log_k 4 = 1$   23. $\log_b b^2 = 2$

24. In your own words, explain the meaning of $\log_b a$.

25. *Concept Check*   Based on the meaning of $\log_b a$, what is the simplest form of $b^{\log_b a}$?

26. A company has found that total sales, in thousands of dollars, are given by the function defined by

$$S(x) = 100 \log_2(x + 2),$$

where $x$ is the number of weeks after a major advertising campaign was introduced.

(a) What were the total sales 6 weeks after the campaign was introduced?
(b) Graph the function.

*Apply the properties of logarithms to express each logarithm as a sum or difference of logarithms. Assume that all variables represent positive real numbers.*

27. $\log_2 3xy^2$   28. $\log_4 \dfrac{\sqrt{x} \cdot w^2}{z}$

*Apply the properties of logarithms to write each expression as a single logarithm. Assume that all variables represent positive real numbers, $b \neq 1$.*

29. $\log_b 3 + \log_b x - 2 \log_b y$   30. $\log_3(x + 7) - \log_3(4x + 6)$

*Evaluate each logarithm. Give approximations to four decimal places.*

31. $\log 28.9$   32. $\log 0.257$   33. $\ln 28.9$   34. $\ln 0.257$

*Use the change-of-base rule (with either common or natural logarithms) to find each logarithm. Give approximations to four decimal places.*

35. $\log_{16} 13$   36. $\log_4 12$

*Use the formula* $\text{pH} = -\log[H_3O^+]$ *to find the pH of each substance with the given hydronium ion concentration.*

37. Milk, $4.0 \times 10^{-7}$   38. Crackers, $3.8 \times 10^{-9}$

39. If orange juice has pH 4.6, what is its hydronium ion concentration?

**40.** Suppose the quantity, measured in grams, of a radioactive substance present at time $t$ is given by

$$Q(t) = 500e^{-0.05t},$$

where $t$ is measured in days. Find the quantity present at the following times.

(a) $t = 0$    (b) $t = 4$

**41. Section 11.5,** Exercise 40 introduced the doubling function defined by

$$t(r) = \frac{\ln 2}{\ln(1 + r)},$$

that gives the number of years required to double your money when it is invested at interest rate $r$ (in decimal form) compounded annually. How long does it take to double your money at each rate? Round answers to the nearest year.

(a) 4%    (b) 6%    (c) 10%    (d) 12%

(e) Compare each answer in parts (a)–(d) with these numbers:

$$\frac{72}{4}, \frac{72}{6}, \frac{72}{10}, \frac{72}{12}.$$

What do you find?

*Solve each equation. Give solutions to three decimal places.*

**42.** $3^x = 9.42$      **43.** $2^{x-1} = 15$      **44.** $e^{0.06x} = 3$

*Solve each equation. Give exact solutions.*

**45.** $\log_3(9x + 8) = 2$      **46.** $\log_5(y + 6)^3 = 2$

**47.** $\log_3(p + 2) - \log_3 p = \log_3 2$      **48.** $\log(2x + 3) = 1 + \log x$

**49.** $\log_4 x + \log_4(8 - x) = 2$      **50.** $\log_2 x + \log_2(x + 15) = \log_2 16$

**51.** *Concept Check* Consider the following "solution" of the equation $\log x^2 = 2$. **WHAT WENT WRONG?** Give the correct solution set.

| | |
|---|---|
| $\log x^2 = 2$ | Original equation |
| $2 \log x = 2$ | Power rule for logarithms |
| $\log x = 1$ | Divide both sides by 2. |
| $x = 10^1$ | Write in exponential form. |
| $x = 10$ | $10^1 = 10$ |

Solution set: $\{10\}$

*Solve each problem. Use a calculator as necessary.*

**52.** If $20,000 is deposited at 7% annual interest compounded quarterly, how much will be in the account after 5 yr, assuming no withdrawals are made?

**53.** How much will $10,000 compounded continuously at 6% annual interest amount to in 3 yr?

**54.** Which is a better plan?

    *Plan A:* Invest $1000 at 4% compounded quarterly for 3 yr

    *Plan B:* Invest $1000 at 3.9% compounded monthly for 3 yr

**55.** What is the half-life of the radioactive substance described in Exercise 40?

**56.** A machine purchased for business use **depreciates**, or loses value, over a period of years. The value of the machine at the end of its useful life is called its **scrap value**. By one method of depreciation (where it is assumed a constant percentage of the value depreciates annually), the scrap value, $S$, is given by

$$S = C(1 - r)^n,$$

where $C$ is the original cost, $n$ is the useful life in years, and $r$ is the constant percent of depreciation.

(a) Find the scrap value of a machine costing $30,000, having a useful life of 12 yr and a constant annual rate of depreciation of 15%.

(b) A machine has a "half-life" of 6 yr. Find the constant annual rate of depreciation.

57. Recall from Exercise 41 in **Section 11.5** that the number of years, $N(r)$, since two independently evolving languages split off from a common ancestral language is approximated by

$$N(r) = -5000 \ln r,$$

where $r$ is the percent of words from the ancestral language common to both languages now. Find $r$ if the split occurred 2000 yr ago.

58. *Multiple Choice* Which one of the following is *not* equal to the solution of $7^x = 23$?

**A.** $\dfrac{\log 23}{\log 7}$  **B.** $\dfrac{\ln 23}{\ln 7}$  **C.** $\log_7 23$  **D.** $\log_{23} 7$

# 11 TEST

1. Decide whether each function is one-to-one.

    (a) $f(x) = x^2 + 9$    (b)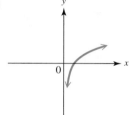

2. Find $f^{-1}(x)$ for the one-to-one function defined by $f(x) = \sqrt[3]{x + 7}$.

3. Graph the inverse of $f$, given the graph of $f$ at the right.

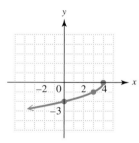

*Graph each function.*

4. $f(x) = 6^x$                5. $g(x) = \log_6 x$

6. Explain how the graph of the function in Exercise 5 can be obtained from the graph of the function in Exercise 4.

*Solve each equation. Give the exact solution.*

7. $5^x = \dfrac{1}{625}$            8. $2^{3x-7} = 8^{2x+2}$

9. A recent report predicts that the U.S. Hispanic population will increase from 26.7 million in 1995 to 96.5 million in 2050. (*Source:* U.S. Census Bureau.) Assuming an exponential growth pattern, the population is approximated by

$$f(x) = 26.7e^{0.023x},$$

where $x$ represents the number of years since 1995. Use this function to estimate the Hispanic population in each year.

(a) 2010   (b) 2015

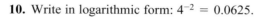

10. Write in logarithmic form: $4^{-2} = 0.0625$.

11. Write in exponential form: $\log_7 49 = 2$.

*Solve each equation.*

12. $\log_{1/2} x = -5$
13. $x = \log_9 3$
14. $\log_x 16 = 4$

15. *Fill in the Blanks*  The value of $\log_2 32$ is _____. This means that if we raise _____ to the _____ power, the result is _____.

*Use properties of logarithms to write each expression as a sum or difference of logarithms. Assume that variables represent positive real numbers.*

16. $\log_3 x^2 y$
17. $\log_5\left(\dfrac{\sqrt{x}}{yz}\right)$

*Use properties of logarithms to write each expression as a single logarithm. Assume that variables represent positive real numbers, $b \neq 1$.*

18. $3 \log_b s - \log_b t$
19. $\dfrac{1}{4} \log_b r + 2 \log_b s - \dfrac{2}{3} \log_b t$

20. Use a calculator to approximate each logarithm to four decimal places.

(a) $\log 23.1$   (b) $\ln 0.82$

21. Use the change-of-base rule to express $\log_3 19$

(a) in terms of common logarithms   (b) in terms of natural logarithms
(c) correct to four decimal places.

22. Solve $3^x = 78$, giving the correct solution to four decimal places.

23. Solve $\log_8(x + 5) + \log_8(x - 2) = 1$.

24. Suppose that $10,000 is invested at 4.5% annual interest, compounded quarterly. How much will be in the account in 5 yr if no money is withdrawn?

25. Suppose that $15,000 is invested at 5% annual interest, compounded continuously.

(a) How much will be in the account in 5 yr if no money is withdrawn?
(b) How long will it take for the initial principal to double?

# 12

# Polynomial and Rational Functions

Although the population of the United States has increased by about 26% since 1982, the time spent waiting in traffic has increased by 236%. Today, the average driver spends over 51 hours a year stuck in traffic. Traffic congestion costs Americans $70 billion a year due to wasted fuel and lost time from work. According to the Texas Transportation Institute, the five U.S. cities with the longest commutes in terms of extra travel time to and from work because of traffic are Los Angeles, San Francisco, Denver, Miami, and Chicago.

Traffic is subject to a *nonlinear effect*. According to Joe Sussman, an engineer from the Massachusetts Institute of Technology, you can put more cars on the road up to a point. Then, if traffic intensity increases even slightly, congestion and waiting time increase dramatically. This nonlinear effect can also occur when you are waiting to enter a parking ramp. (*Source:* Longman, Phillip J., "American Gridlock," *U.S. News & World Report,* May 28, 2001.)

In Section 12.4, we use a *rational function* to model this phenomenon.

**12.1** Synthetic Division

**12.2** Zeros of Polynomial Functions

**12.3** Graphs and Applications of Polynomial Functions

**12.4** Graphs and Applications of Rational Functions

## 12.1 Synthetic Division

**OBJECTIVES**

1. Know the meaning of the division algorithm.
2. Use synthetic division to divide by a polynomial of the form $x - k$.
3. Use the remainder theorem to evaluate a polynomial.
4. Decide whether a given number is a zero of a polynomial function.

To prepare for later work in this chapter, we now examine a shortcut method for division of polynomials, called *synthetic division*. Polynomial division was first covered in **Section 5.5,** and you may wish to review it at this time.

**OBJECTIVE 1** Know the meaning of the division algorithm. The following algorithm will be used in this chapter.

> **Division Algorithm**
>
> Let $f(x)$ and $g(x)$ be polynomials with $g(x)$ of lower degree than $f(x)$ and $g(x)$ of degree 1 or more. There exist unique polynomials $q(x)$ and $r(x)$ such that
> $$f(x) = g(x) \cdot q(x) + r(x),$$
> where either $r(x) = 0$ or the degree of $r(x)$ is less than the degree of $g(x)$.

In Example 4 of **Section 5.5,** we saw how to perform long division of two polynomials. Using $x$ as the variable (rather than $r$), we found that the following is true:

$$\underbrace{6x^4 + 9x^3 + 2x^2 - 8x + 7}_{\substack{f(x) \\ \text{Dividend} \\ \text{(Original polynomial)}}} = \underbrace{(3x^2 - 2)}_{\substack{g(x) \\ \text{Divisor}}} \underbrace{(2x^2 + 3x + 2)}_{\substack{q(x) \\ \text{Quotient}}} + \underbrace{(-2x + 11)}_{\substack{r(x) \\ \text{Remainder}}}.$$

**OBJECTIVE 2** Use synthetic division to divide by a polynomial of the form $x - k$. Often, when one polynomial is divided by a second, the second polynomial has the form $x - k$, where the coefficient of the $x$-term is 1. To see a shortcut for these divisions, look first below left, where the division of $3x^3 - 2x + 5$ by $x - 3$ is shown. Notice that we inserted 0 for the missing $x^2$-term.

$$
\begin{array}{r}
3x^2 + 9x + 25 \\
x - 3 \overline{\smash{)}3x^3 + 0x^2 - 2x + 5} \\
\underline{3x^3 - 9x^2\phantom{ + 00x + 00}} \\
9x^2 - 2x\phantom{ + 00} \\
\underline{9x^2 - 27x\phantom{ + 00}} \\
25x + 5 \\
\underline{25x - 75} \\
80
\end{array}
\qquad
\begin{array}{r}
3 \phantom{x} 9 \phantom{x} 25 \\
1 - 3 \overline{\smash{)}3 \phantom{xx} 0 \phantom{x} -2 \phantom{x} 5} \\
3 \phantom{x} -9 \phantom{xxxxxx} \\
9 \phantom{x} -2 \phantom{xxx} \\
9 \phantom{x} -27 \phantom{x} \\
25 \phantom{x} 5 \\
25 \phantom{x} -75 \\
80
\end{array}
$$

On the right, exactly the same division is shown written without the variables. This is why it is *essential* to use 0 as a placeholder in synthetic division. All the numbers in color on the right are repetitions of the numbers directly above them, so we omit them, as shown below.

$$
\begin{array}{r}
3 \phantom{x} 9 \phantom{x} 25 \\
1 - 3 \overline{\smash{)}3 \phantom{xx} 0 \phantom{x} -2 \phantom{x} 5} \\
-9 \phantom{xxxxxx} \\
9 \phantom{x} -2 \phantom{xxx} \\
-27 \phantom{xx} \\
25 \phantom{x} 5 \\
-75 \\
80
\end{array}
\qquad
\begin{array}{r}
3 \phantom{x} 9 \phantom{x} 25 \\
1 - 3 \overline{\smash{)}3 \phantom{xx} 0 \phantom{x} -2 \phantom{x} 5} \\
-9 \phantom{xxxxxx} \\
9 \phantom{xxxxxx} \\
-27 \phantom{xx} \\
25 \phantom{xx} \\
-75 \\
80
\end{array}
$$

The numbers in color on the left are again repetitions of the numbers directly above them; they too are omitted, as shown on the right.

Now we can condense the problem. If we bring the 3 in the dividend down to the beginning of the bottom row, the top row can be omitted, since it duplicates the bottom row.

$$1 - 3 \overline{)\begin{array}{cccc} 3 & 0 & -2 & 5 \\ & -9 & -27 & -75 \\ \hline 3 & 9 & 25 & 80 \end{array}}$$

Finally, we omit the 1 at the upper left. Also, to simplify the arithmetic, we replace subtraction in the second row by addition. To compensate for this, we change the $-3$ at the upper left to its additive inverse, 3.

Additive inverse → $3 \overline{)\begin{array}{cccc} 3 & 0 & -2 & 5 \\ & 9 & 27 & 75 \\ \hline 3 & 9 & 25 & 80 \end{array}}$ ← Signs changed
← Remainder

The quotient is read from the bottom row.

$$3x^2 + 9x + 25 + \frac{80}{x - 3}$$

Remember to add $\frac{\text{remainder}}{\text{divisor}}$.

The first three numbers in the bottom row are the coefficients of the quotient polynomial with degree 1 less than the degree of the dividend. The last number gives the remainder.

### Synthetic Division

This shortcut procedure is called **synthetic division.** It is used only when dividing a polynomial by a binomial of the form $x - k$.

### EXAMPLE 1 Using Synthetic Division

Use synthetic division to divide $5x^2 + 16x + 15$ by $x + 2$.

We change $x + 2$ into the form $x - k$ by writing it as

$$x + 2 = x - (-2),$$

where $k = -2$. Now write the coefficients of $5x^2 + 16x + 15$, placing $-2$ to the left.

$x + 2$ leads to $-2$. → $-2 \overline{)\begin{array}{ccc} 5 & 16 & 15 \end{array}}$ ← Coefficients

Bring down the 5, and multiply: $-2 \cdot 5 = -10$.

$$-2 \overline{)\begin{array}{ccc} 5 & 16 & 15 \\ & -10 & \\ \hline 5 & & \end{array}}$$

Add 16 and $-10$, getting 6, and multiply 6 and $-2$ to get $-12$.

$$-2 \overline{)\begin{array}{ccc} 5 & 16 & 15 \\ & -10 & -12 \\ \hline 5 & 6 & \end{array}}$$

Add 15 and $-12$, getting 3.

$$-2 \overline{)\begin{array}{ccc} 5 & 16 & 15 \\ & -10 & -12 \\ \hline 5 & 6 & 3 \end{array}}$$ ← Remainder

The result is read from the bottom row.

$$\frac{5x^2 + 16x + 15}{x + 2} = 5x + 6 + \frac{3}{x + 2}$$

Add $\frac{\text{remainder}}{\text{divisor}}$.

NOW TRY Exercise 7.

The result of the division in Example 1 can be written as

$$5x^2 + 16x + 15 = (x + 2)(5x + 6) + 3$$

by multiplying both sides by the denominator $x + 2$. The following theorem is a generalization of the division process illustrated above.

> **Division Algorithm for Divisor $x - k$**
>
> For any polynomial $f(x)$ and any complex number $k$, there exists a unique polynomial $q(x)$ and number $r$ such that
>
> $$f(x) = (x - k)q(x) + r.$$

For example, in the synthetic division above,

$$\underbrace{5x^2 + 16x + 15}_{f(x)} = \underbrace{(x + 2)}_{(x-k)} \cdot \underbrace{(5x + 6)}_{q(x)} + \underbrace{3}_{+\,r}.$$

This theorem is a special case of the division algorithm given earlier. Here $g(x)$ is the first-degree polynomial $x - k$.

**EXAMPLE 2  Using Synthetic Division with a Missing Term**

Use synthetic division to find $(-4x^5 + x^4 + 6x^3 + 2x^2 + 50) \div (x - 2)$. Use the steps given above, first inserting a 0 for the missing $x$-term.

```
2) -4    1    6    2    0   50
        -8  -14  -16  -28  -56
    ─────────────────────────────
        -4   -7   -8  -14  -28   -6
```

Use 0 as coefficient for the missing $x$-term.

Read the result from the bottom row.

$$\frac{-4x^5 + x^4 + 6x^3 + 2x^2 + 50}{x - 2} = -4x^4 - 7x^3 - 8x^2 - 14x - 28 + \frac{-6}{x - 2}$$

**NOW TRY** Exercise 13.

**OBJECTIVE 3**  Use the remainder theorem to evaluate a polynomial.  We can use synthetic division to evaluate polynomials. For example, in the synthetic division of Example 2, where the polynomial was divided by $x - 2$, the remainder was $-6$.

Replacing $x$ in the polynomial with 2 gives

$$-4x^5 + x^4 + 6x^3 + 2x^2 + 50 = -4 \cdot 2^5 + 2^4 + 6 \cdot 2^3 + 2 \cdot 2^2 + 50$$
$$= -4 \cdot 32 + 16 + 6 \cdot 8 + 2 \cdot 4 + 50$$
$$= -128 + 16 + 48 + 8 + 50$$
$$= -6,$$

the same number as the remainder; that is, dividing by $x - 2$ produced a remainder equal to the result when $x$ is replaced with 2.

By the division algorithm, $f(x) = (x - k)q(x) + r$. This equality is true for all complex values of $x$, so it is true for $x = k$. Replacing $x$ with $k$ gives

$$f(k) = (k - k)q(k) + r$$
$$f(k) = r.$$

This proves the following **remainder theorem,** which gives an alternative method of evaluating functions defined by polynomials.

> **Remainder Theorem**
>
> If the polynomial $f(x)$ is divided by $x - k$, then the remainder is equal to $f(k)$.

**EXAMPLE 3** Using the Remainder Theorem

Let $f(x) = 2x^3 - 5x^2 - 3x + 11$. Find $f(-2)$.

Use the remainder theorem; divide $f(x)$ by $x - (-2)$.

$$\text{Value of } k \rightarrow -2\overline{)\begin{array}{cccc} 2 & -5 & -3 & 11 \\ & -4 & 18 & -30 \\ \hline 2 & -9 & 15 & -19 \end{array}} \leftarrow \text{Remainder}$$

Thus, $f(-2) = -19$.

**NOW TRY** Exercise 31.

**OBJECTIVE 4** Decide whether a given number is a zero of a polynomial function. The function defined by $f(x) = 2x^3 - 5x^2 - 3x + 11$ in Example 3 is an example of a *polynomial function*. We extend the definition given in **Section 5.3** to include complex numbers as coefficients.

> **Polynomial Function**
>
> A **polynomial function of degree $n$** is a function defined by
>
> $$f(x) = a_n x^n + a_{n-1} x^{n-1} + \cdots + a_1 x + a_0,$$
>
> for complex numbers $a_n, a_{n-1}, \ldots, a_1,$ and $a_0$, where $a_n \neq 0$.

We are often required to find values of $x$ that make $f(x)$ equal 0. A **zero** of a polynomial function $f$ is a value of $k$ such that $f(k) = 0$.

The remainder theorem gives a quick way to decide if a number $k$ is a zero of a polynomial function defined by $f(x)$. Use synthetic division to find $f(k)$; if the remainder is 0, then $f(k) = 0$ and $k$ is a zero of $f(x)$. A zero of $f(x)$ is called a **root** or **solution** of the equation $f(x) = 0$.

**EXAMPLE 4** Deciding Whether a Number is a Zero

Decide whether the given number is a zero of the given polynomial function.

**(a)** $-5;\ f(x) = 2x^4 + 12x^3 + 6x^2 - 5x + 75$

Use synthetic division and the remainder theorem.

$$\text{Proposed zero} \rightarrow -5\overline{)\begin{array}{ccccc} 2 & 12 & 6 & -5 & 75 \\ & -10 & -10 & 20 & -75 \\ \hline 2 & 2 & -4 & 15 & 0 \end{array}} \leftarrow \text{Remainder}$$

Since the remainder is 0, $f(-5) = 0$, and $-5$ is a zero of $f(x) = 2x^4 + 12x^3 + 6x^2 - 5x + 75$.

**(b)** $-4$;  $f(x) = x^4 + x^2 - 3x + 1$

```
         -4 )1    0    1    -3    1
                -4   16   -68   284
                ─────────────────────
                 1   -4   17   -71   285
```
Use 0 as coefficient for the missing $x^3$-term.

The remainder is not 0, so $-4$ is not a zero of $f(x) = x^4 + x^2 - 3x + 1$. In fact, $f(-4) = 285$.

**(c)** $1 + 2i$;  $f(x) = x^4 - 2x^3 + 4x^2 + 2x - 5$

Use synthetic division and operations with complex numbers.

```
   1 + 2i )1    -2         4         2        -5
                1 + 2i    -5    -1 - 2i        5
                ──────────────────────────────────
           1   -1 + 2i    -1     1 - 2i        0
```

$(1 + 2i)(-1 + 2i)$
$= -1 + 2i - 2i + (2i)^2$
$= -1 + 4(-1) = -5$

Since the remainder is 0, $1 + 2i$ is a zero of the given polynomial function.

**NOW TRY** Exercises 41, 43, and 47.

The synthetic division in Example 4(a) shows that $x - (-5)$ divides the polynomial with 0 remainder. Thus $x - (-5) = x + 5$ is a *factor* of the polynomial and

$$2x^4 + 12x^3 + 6x^2 - 5x + 75 = (x + 5)(2x^3 + 2x^2 - 4x + 15).$$

The second factor is the quotient polynomial found in the last row of the synthetic division.

## 12.1 Exercises

**NOW TRY Exercise**

1. Explain why synthetic division is used.
2. *Concept Check* What type of polynomial divisors may be used with synthetic division?

*Use synthetic division to find each quotient. See Examples 1 and 2.*

3. $\dfrac{x^2 - 6x + 5}{x - 1}$

4. $\dfrac{x^2 - 4x - 21}{x + 3}$

5. $\dfrac{4m^2 + 19m - 5}{m + 5}$

6. $\dfrac{3k^2 - 5k - 12}{k - 3}$

7. $\dfrac{2a^2 + 8a + 13}{a + 2}$

8. $\dfrac{4y^2 - 5y - 20}{y - 4}$

9. $(p^2 - 3p + 5) \div (p + 1)$

10. $(z^2 + 4z - 6) \div (z - 5)$

11. $\dfrac{4a^3 - 3a^2 + 2a - 3}{a - 1}$

12. $\dfrac{5p^3 - 6p^2 + 3p + 14}{p + 1}$

13. $(x^5 - 2x^3 + 3x^2 - 4x - 2) \div (x - 2)$

14. $(2y^5 - 5y^4 - 3y^2 - 6y - 23) \div (y - 3)$

15. $(-4r^6 - 3r^5 - 3r^4 + 5r^3 - 6r^2 + 3r + 3) \div (r - 1)$

16. $(2t^6 - 3t^5 + 2t^4 - 5t^3 + 6t^2 - 3t - 2) \div (t - 2)$

17. $(-3y^5 + 2y^4 - 5y^3 - 6y^2 - 1) \div (y + 2)$

18. $(m^6 + 2m^4 - 5m + 11) \div (m - 2)$

19. $\dfrac{y^3 + 1}{y - 1}$

20. $\dfrac{z^4 + 81}{z - 3}$

*Express each polynomial function in the form $f(x) = (x - k)q(x) + r$ for the given value of $k$.*

**21.** $f(x) = 2x^3 + x^2 + x - 8$;  $k = -1$
**22.** $f(x) = 2x^3 + 3x^2 - 16x + 10$;  $k = -4$
**23.** $f(x) = -x^3 + 2x^2 + 4$;  $k = -2$
**24.** $f(x) = -4x^3 + 2x^2 - 3x - 10$;  $k = 2$
**25.** $f(x) = 4x^4 - 3x^3 - 20x^2 - x$;  $k = 3$
**26.** $f(x) = 2x^4 + x^3 - 15x^2 + 3x$;  $k = -3$

*For each polynomial function, use the remainder theorem and synthetic division to find $f(k)$. See Example 3.*

**27.** $k = 3$;  $f(x) = x^2 - 4x + 5$
**28.** $k = -2$;  $f(x) = x^2 + 5x + 6$
**29.** $k = 2$;  $f(x) = 2x^2 - 3x - 3$
**30.** $k = 4$;  $f(x) = -x^3 + 8x^2 + 63$
**31.** $k = -1$;  $f(x) = x^3 - 4x^2 + 2x + 1$
**32.** $k = 2$;  $f(x) = 2x^3 - 3x^2 - 5x + 4$
**33.** $k = 3$;  $f(x) = 2x^5 - 10x^3 - 19x^2 - 45$
**34.** $k = 4$;  $f(x) = x^4 + 6x^3 + 9x^2 + 3x - 3$
**35.** $k = -8$;  $f(x) = x^6 + 7x^5 - 5x^4 + 22x^3 - 16x^2 + x + 19$
**36.** $k = -\dfrac{1}{2}$;  $f(x) = 6x^3 - 31x^2 - 15x$
**37.** $k = 2 + i$;  $f(x) = x^2 - 5x + 1$
**38.** $k = 3 - 2i$;  $f(x) = x^2 - x + 3$

**39.** Explain why a 0 remainder in synthetic division of $f(x)$ by $x - k$ indicates that $k$ is a solution of the equation $f(x) = 0$.

**40.** Explain why it is important to insert 0s as placeholders for missing terms before performing synthetic division.

*Use synthetic division to decide whether the given number is a zero of the given polynomial function. See Example 4.*

**41.** 3;  $f(x) = 2x^3 - 6x^2 - 9x + 4$
**42.** $-6$;  $f(x) = 2x^3 + 9x^2 - 16x + 12$
**43.** $-5$;  $f(x) = x^3 + 7x^2 + 10x$
**44.** $-2$;  $f(x) = 2x^3 - 3x^2 - 5x$
**45.** $\dfrac{2}{5}$;  $f(x) = 5x^4 + 2x^3 - x + 15$
**46.** $\dfrac{1}{2}$;  $f(x) = 2x^4 - 3x^2 + 4$
**47.** $2 - i$;  $f(x) = x^2 + 3x + 4$
**48.** $1 - 2i$;  $f(x) = x^2 - 3x + 5$

## 12.2 Zeros of Polynomial Functions

**OBJECTIVES**

1. Use the factor theorem.
2. Use the rational zeros theorem.
3. Find polynomial functions that satisfy given conditions.
4. Apply the conjugate zeros theorem.

In this section we build upon some of the ideas presented in the previous section to learn about finding zeros of polynomial functions.

**OBJECTIVE 1** Use the factor theorem. By the remainder theorem, if $f(k) = 0$, then the remainder when $f(x)$ is divided by $x - k$ is 0. This means that $x - k$ is a factor of $f(x)$. Conversely, if $x - k$ is a factor of $f(x)$, then $f(k)$ must equal 0. This is summarized in the following **factor theorem**.

**Factor Theorem**

The polynomial $x - k$ is a factor of the polynomial $f(x)$ if and only if $f(k) = 0$.

> **EXAMPLE 1** Deciding Whether $x - k$ Is a Factor of $f(x)$
>
> Is $x - 1$ a factor of $f(x) = 2x^4 + 3x^2 - 5x + 7$?
>
> By the factor theorem, $x - 1$ will be a factor of $f(x)$ only if $f(1) = 0$. Use synthetic division and the remainder theorem to decide.
>
> Use 0 as coefficient for the missing $x^3$-term.
>
> $$\begin{array}{r|rrrrr} 1) & 2 & 0 & 3 & -5 & 7 \\ & & 2 & 2 & 5 & 0 \\ \hline & 2 & 2 & 5 & 0 & 7 \end{array} \leftarrow \text{Remainder}$$
>
> Since the remainder is 7, $f(1) = 7$ and not 0, so $x - 1$ is not a factor of $f(x)$.
>
> **NOW TRY** Exercises 7 and 9.

The factor theorem can be used to factor a polynomial of higher degree into linear factors. Linear factors are factors of the form $ax - b$.

> **EXAMPLE 2** Factoring a Polynomial Given a Zero
>
> Factor $f(x) = 6x^3 + 19x^2 + 2x - 3$ into linear factors given that $-3$ is a zero of $f$.
>
> Since $-3$ is a zero of $f$, $x - (-3) = x + 3$ is a factor.
>
> $$\begin{array}{r|rrrr} -3) & 6 & 19 & 2 & -3 \\ & & -18 & -3 & 3 \\ \hline & 6 & 1 & -1 & 0 \end{array}$$
>
> Use synthetic division to divide $f(x)$ by $x + 3$.
>
> The quotient is $6x^2 + x - 1$, so
>
> $$f(x) = (x + 3)(6x^2 + x - 1)$$
> $$= (x + 3)(2x + 1)(3x - 1), \quad \text{Factor } 6x^2 + x - 1.$$
>
> where all factors are linear.
>
> **NOW TRY** Exercise 11.

**OBJECTIVE 2** Use the rational zeros theorem. The **rational zeros theorem** gives a method to determine all possible candidates for rational zeros of a polynomial function with integer coefficients.

> **Rational Zeros Theorem**
>
> Let $f(x) = a_n x^n + a_{n-1} x^{n-1} + \cdots + a_1 x + a_0$, where $a_n \neq 0$, define a polynomial function with integer coefficients. If $\frac{p}{q}$ is a rational number written in lowest terms, and if $\frac{p}{q}$ is a zero of $f$, then $p$ is a factor of the constant term $a_0$, and $q$ is a factor of the leading coefficient $a_n$.

### EXAMPLE 3 Using the Rational Zeros Theorem

Do each of the following for the polynomial function defined by
$$f(x) = 6x^4 + 7x^3 - 12x^2 - 3x + 2.$$

**(a)** List all possible rational zeros.

For a rational number $\frac{p}{q}$ to be a zero, $p$ must be a factor of $a_0 = 2$ and $q$ must be a factor of $a_4 = 6$. Thus, $p$ can be $\pm 1$ or $\pm 2$, and $q$ can be $\pm 1$, $\pm 2$, $\pm 3$, or $\pm 6$. The possible rational zeros, $\frac{p}{q}$, are

$$\pm 1, \quad \pm 2, \quad \pm \frac{1}{2}, \quad \pm \frac{1}{3}, \quad \pm \frac{1}{6}, \quad \pm \frac{2}{3}.$$

**(b)** Find all rational zeros and factor $f(x)$.

Use the remainder theorem to show that 1 and $-2$ are zeros.

$$\begin{array}{r|rrrrr} 1) & 6 & 7 & -12 & -3 & 2 \\ & & 6 & 13 & 1 & -2 \\ \hline & 6 & 13 & 1 & -2 & 0 \end{array}$$

0 indicates that 1 is a zero.

The 0 remainder shows that 1 is a zero. Now, use the quotient polynomial $6x^3 + 13x^2 + x - 2$ and synthetic division to find that $-2$ is also a zero.

$$\begin{array}{r|rrrr} -2) & 6 & 13 & 1 & -2 \\ & & -12 & -2 & 2 \\ \hline & 6 & 1 & -1 & 0 \end{array}$$

The new quotient polynomial is $6x^2 + x - 1$. Factor to solve $6x^2 + x - 1 = 0$, obtaining $(3x - 1)(2x + 1) = 0$. The remaining two zeros are $\frac{1}{3}$ and $-\frac{1}{2}$.

Since the four zeros of $f(x) = 6x^4 + 7x^3 - 12x^2 - 3x + 2$ are $1, -2, \frac{1}{3}$, and $-\frac{1}{2}$, the factors of $f(x)$ are $x - 1, x + 2, x - \frac{1}{3}$, and $x + \frac{1}{2}$, and

$$f(x) = a(x - 1)(x + 2)\left(x - \frac{1}{3}\right)\left(x + \frac{1}{2}\right)$$

$$f(x) = 6(x - 1)(x + 2)\left(x - \frac{1}{3}\right)\left(x + \frac{1}{2}\right) \quad \text{The leading coefficient of } f(x) \text{ is } 6.$$

Factor 6 as $3 \cdot 2$ in the next line.

$$f(x) = (x - 1)(x + 2)(3)\left(x - \frac{1}{3}\right)(2)\left(x + \frac{1}{2}\right)$$

$$f(x) = (x - 1)(x + 2)(3x - 1)(2x + 1).$$

**NOW TRY** Exercise 25.

**CAUTION** The rational zeros theorem gives only *possible* rational zeros; it does not tell us whether these rational numbers are *actual* zeros. We must rely on other methods to determine whether they are indeed zeros. Furthermore, the function must have integer coefficients. To apply the rational zeros theorem to a polynomial with fractional coefficients, multiply through by the least common denominator of all the fractions. For example, any rational zeros of

$$p(x) = x^4 - \frac{1}{6}x^3 + \frac{2}{3}x^2 - \frac{1}{6}x - \frac{1}{3}$$

will also be rational zeros of

$$q(x) = 6x^4 - x^3 + 4x^2 - x - 2.$$

The function $q$ was obtained by multiplying the terms of $p$ by 6.

**Carl Friederich Gauss**
**(1777–1855)**

**OBJECTIVE 3** Find polynomial functions that satisfy given conditions. The next theorem says that every function defined by a polynomial of degree 1 or more has a zero, which means that every such polynomial can be factored. This was first proved by Carl F. Gauss as part of his doctoral dissertation completed in 1799. Gauss's proof used advanced mathematical concepts outside the field of algebra.

> **Fundamental Theorem of Algebra**
>
> Every function defined by a polynomial of degree 1 or more has at least one complex zero.

From the fundamental theorem, if $f(x)$ is of degree 1 or more, then there is some number $k_1$ such that $f(k_1) = 0$. By the factor theorem,

$$f(x) = (x - k_1) \cdot q_1(x)$$

for some polynomial $q_1(x)$. If $q_1(x)$ is of degree 1 or more, the fundamental theorem and the factor theorem can be used to factor $q_1(x)$ in the same way. There is some number $k_2$ such that $q_1(k_2) = 0$, so that

$$q_1(x) = (x - k_2)q_2(x)$$

and
$$f(x) = (x - k_1)(x - k_2)q_2(x).$$

Assuming that $f(x)$ has degree $n$ and repeating this process $n$ times gives

$$f(x) = a(x - k_1)(x - k_2) \cdots (x - k_n),$$

where $a$ is the leading coefficient of $f(x)$. Each of these factors leads to a zero of $f(x)$, so $f(x)$ has the $n$ zeros $k_1, k_2, \ldots, k_n$. This result suggests the next theorem.

> **Number of Zeros Theorem**
>
> A function defined by a polynomial of degree $n$ has at most $n$ distinct zeros.

The theorem says that there exist *at most* $n$ distinct zeros. For example, the polynomial function defined by

$$f(x) = x^3 + 3x^2 + 3x + 1 = (x + 1)^3$$

is of degree 3 but has only one distinct zero, $-1$. Actually, the zero $-1$ occurs three times, since there are three factors of $x + 1$; this zero is called a **zero of multiplicity** 3.

**EXAMPLE 4** Finding a Polynomial Function That Satisfies Given Conditions (Real Zeros)

Find a function $f$ defined by a polynomial of degree 3 that satisfies the given conditions.

**(a)** Zeros of $-1$, 2, and 4; $f(1) = 3$

These three zeros give $x - (-1) = x + 1$, $x - 2$, and $x - 4$ as factors of $f(x)$. Since $f(x)$ is to be of degree 3, these are the only possible factors by the theorem just stated. Therefore, $f(x)$ has the form

$$f(x) = a(x + 1)(x - 2)(x - 4)$$

for some real number $a$. To find $a$, use the fact that $f(1) = 3$.

$$f(1) = a(1 + 1)(1 - 2)(1 - 4) = 3$$
$$a(2)(-1)(-3) = 3$$
$$6a = 3$$
$$a = \frac{1}{2}$$

Thus $\quad f(x) = \frac{1}{2}(x + 1)(x - 2)(x - 4),$

or $\quad f(x) = \frac{1}{2}x^3 - \frac{5}{2}x^2 + x + 4.\quad$ Multiply.

**(b)** $-2$ is a zero of multiplicity 3;  $f(-1) = 4$
The polynomial function defined by $f(x)$ has the form

$$f(x) = a(x + 2)(x + 2)(x + 2)$$
$$= a(x + 2)^3.$$

Since $f(-1) = 4$,

$$f(-1) = a(-1 + 2)^3 = 4$$
$$a(1)^3 = 4$$
$$a = 4,$$

and $\quad f(x) = 4(x + 2)^3 = 4x^3 + 24x^2 + 48x + 32.$

**NOW TRY** Exercise 45.

**OBJECTIVE 4** Apply the conjugate zeros theorem. The remainder theorem can be used to show that both $2 + i$ and $2 - i$ are zeros of $f(x) = x^3 - x^2 - 7x + 15$. In general, if $a + bi$ is a zero of a polynomial function with *real* coefficients, then so is $a - bi$. This fact is called the **conjugate zeros theorem.**

### Conjugate Zeros Theorem

If $f(x)$ is a polynomial function *having only real coefficients* and if $a + bi$ is a zero of $f(x)$, where $a$ and $b$ are real numbers, then $a - bi$ is also a zero of $f(x)$.

**CAUTION** It is *essential* that the polynomial function have only real coefficients. For example, $f(x) = x - (1 + i)$ has $1 + i$ as a zero, but the conjugate $1 - i$ is not a zero.

**EXAMPLE 5** Finding a Polynomial Function That Satisfies Given Conditions (Complex Zeros)

Find a polynomial function of least possible degree having only real coefficients and zeros 3 and $2 + i$.

The complex number $2 - i$ also must be a zero, so the polynomial function has at least three zeros, 3, $2 + i$, and $2 - i$. For the polynomial to be of least possible degree, these must be the only zeros. By the factor theorem there must be three factors, $x - 3$, $x - (2 + i)$, and $x - (2 - i)$. A polynomial function of least possible degree is

$$f(x) = (x - 3)[x - (2 + i)][x - (2 - i)]$$
$$= (x - 3)(x - 2 - i)(x - 2 + i)$$
$$= x^3 - 7x^2 + 17x - 15.$$

Constant multiples such as $2(x^3 - 7x^2 + 17x - 15)$ or $\sqrt{5}(x^3 - 7x^2 + 17x - 15)$ also satisfy the given conditions on zeros. The information on zeros given in the problem is not enough to give a specific value for the leading coefficient.

**NOW TRY** Exercise 41.

The theorem on conjugate zeros can help predict the number of real zeros of polynomial functions with real coefficients. A polynomial function with real coefficients of odd degree $n$, where $n \geq 1$, must have at least one real zero (since zeros of the form $a + bi$, where $b \neq 0$, occur in conjugate pairs). A polynomial function with real coefficients of even degree $n$ may have no real zeros.

**EXAMPLE 6** Finding All Zeros of a Polynomial Function Given One Zero

Find all zeros of $f(x) = x^4 - 7x^3 + 18x^2 - 22x + 12$, given that $1 - i$ is a zero.

Since the polynomial function has only real coefficients and since $1 - i$ is a zero, by the conjugate zeros theorem $1 + i$ is also a zero. To find the remaining zeros, we first divide the original polynomial by $x - (1 - i)$.

$$\begin{array}{r|rrrrr}
1-i) & 1 & -7 & 18 & -22 & 12 \\
 &  & 1-i & -7+5i & 16-6i & -12 \\
\hline
 & 1 & -6-i & 11+5i & -6-6i & 0
\end{array}$$

By the factor theorem, since $x = 1 - i$ is a zero of $f(x)$, $x - (1 - i)$ is a factor, and $f(x)$ can be written as

$$f(x) = [x - (1 - i)][x^3 + (-6 - i)x^2 + (11 + 5i)x + (-6 - 6i)].$$

We know that $x = 1 + i$ is also a zero of $f(x)$, so

$$f(x) = [x - (1 - i)][x - (1 + i)]q(x)$$

for some polynomial $q(x)$. Thus,

$$x^3 + (-6 - i)x^2 + (11 + 5i)x + (-6 - 6i) = [x - (1 + i)]q(x).$$

We use synthetic division to find $q(x)$.

$$\begin{array}{r|rrrr}
1+i) & 1 & -6-i & 11+5i & -6-6i \\
 &  & 1+i & -5-5i & 6+6i \\
\hline
 & 1 & -5 & 6 & 0
\end{array}$$

Since $q(x) = x^2 - 5x + 6$, $f(x)$ can be written as

$$f(x) = [x - (1 - i)][x - (1 + i)](x^2 - 5x + 6).$$

Factoring the quadratic polynomial $x^2 - 5x + 6$ shows that the factors are $x - 2$ and $x - 3$, which leads to the zeros 2 and 3. Thus, the four zeros of $f(x)$ are $1 - i$, $1 + i$, 2, and 3.

**NOW TRY** Exercise 17.

## 12.2 Exercises

*True or False* Decide whether each statement is true or false.

1. Given that $x - 1$ is a factor of $f(x) = x^6 - x^4 + 2x^2 - 2$, we are assured that $f(1) = 0$.
2. Given that $f(1) = 0$ for $f(x) = x^6 - x^4 + 2x^2 - 2$, we are assured that $x - 1$ is a factor of $f(x)$.
3. For the function defined by $f(x) = (x + 2)^4(x - 3)$, 2 is a zero of multiplicity 4.
4. Given that $2 + 3i$ is a zero of $f(x) = x^2 - 4x + 13$, we are assured that $2 - 3i$ is also a zero.

*Use the factor theorem to decide whether the second polynomial is a factor of the first. See Example 1.*

5. $4x^2 + 2x + 54;\quad x - 4$
6. $5x^2 - 14x + 10;\quad x + 2$
7. $x^3 + 2x^2 - 3;\quad x - 1$
8. $2x^3 + x + 2;\quad x + 1$
9. $2x^4 + 5x^3 - 2x^2 + 5x + 6;\quad x + 3$
10. $5x^4 + 16x^3 - 15x^2 + 8x + 16;\quad x + 4$

*Factor $f(x)$ into linear factors given that $k$ is a zero of $f(x)$. See Example 2.*

11. $f(x) = 2x^3 - 3x^2 - 17x + 30;\quad k = 2$
12. $f(x) = 2x^3 - 3x^2 - 5x + 6;\quad k = 1$
13. $f(x) = 6x^3 + 13x^2 - 14x + 3;\quad k = -3$
14. $f(x) = 6x^3 + 17x^2 - 63x + 10;\quad k = -5$

*For each polynomial function, one zero is given. Find all others. See Examples 2 and 6.*

15. $f(x) = x^3 - x^2 - 4x - 6;\quad 3$
16. $f(x) = x^3 + 4x^2 - 5;\quad 1$
17. $f(x) = x^3 - 7x^2 + 17x - 15;\quad 2 - i$
18. $f(x) = 4x^3 + 6x^2 - 2x - 1;\quad \dfrac{1}{2}$
19. $f(x) = x^4 + 5x^2 + 4;\quad -i$
20. $f(x) = x^4 + 10x^3 + 27x^2 + 10x + 26;\quad i$

*For each polynomial function (a) list all possible rational zeros, (b) find all rational zeros, and (c) factor $f(x)$. See Example 3.*

21. $f(x) = x^3 - 2x^2 - 13x - 10$
22. $f(x) = x^3 + 5x^2 + 2x - 8$
23. $f(x) = x^3 + 6x^2 - x - 30$
24. $f(x) = x^3 - x^2 - 10x - 8$
25. $f(x) = 6x^3 + 17x^2 - 31x - 12$
26. $f(x) = 15x^3 + 61x^2 + 2x - 8$
27. $f(x) = 12x^3 + 20x^2 - x - 6$
28. $f(x) = 12x^3 + 40x^2 + 41x + 12$

*For each polynomial function, find all zeros and their multiplicities.*

29. $f(x) = (x + 4)^2(x^2 - 7)(x + 1)$
30. $f(x) = (x + 1)^2(x - 1)^3(x^2 - 10)$
31. $f(x) = 3(x - 2)(x + 3)(x^2 - 1)$
32. $f(x) = 5x^2(x + 1 - \sqrt{2})(2x + 5)$
33. $f(x) = (x^2 + x - 2)^5(x - 1 + \sqrt{3})^2$
34. $f(x) = (7x - 2)^3(x^2 + 9)^2$

*For each of the following, find a polynomial function of least possible degree with only real coefficients and having the given zeros. See Examples 4 and 5.*

35. $3 + i$ and $3 - i$
36. $7 - 2i$ and $7 + 2i$
37. $1 + \sqrt{2}, 1 - \sqrt{2}$, and 3
38. $1 - \sqrt{3}, 1 + \sqrt{3}$, and 1
39. $-2 + i, -2 - i, 3$, and $-3$
40. $3 + 2i, -1$, and 2
41. 2 and $3i$
42. $-1$ and $6 - 3i$
43. $1 + 2i$, 2 (multiplicity 2)
44. $2 + i$, $-3$ (multiplicity 2)

*Find a polynomial function of degree 3 with only real coefficients that satisfies the given conditions. See Examples 4 and 5.*

**45.** Zeros of $-3$, 1, and 4;   $f(2) = 30$
**46.** Zeros of 1, $-1$, and 0;   $f(2) = 3$
**47.** Zeros of $-2$, 1, and 0;   $f(-1) = -1$
**48.** Zeros of 2, $-3$, and 5;   $f(3) = 6$
**49.** Zeros of 5, $i$, and $-i$;   $f(2) = 5$
**50.** Zeros of $-2$, $i$, and $-i$;   $f(-3) = 30$

*Concept Check*   Work Exercises 51 and 52.

**51.** Show that $-2$ is a zero of multiplicity 2 of $f(x) = x^4 + 2x^3 - 7x^2 - 20x - 12$ and find all other complex zeros. Then write $f(x)$ in factored form.

**52.** Show that $-1$ is a zero of multiplicity 3 of $f(x) = x^5 - 4x^3 - 2x^2 + 3x + 2$ and find all other complex zeros. Then write $f(x)$ in factored form.

Descartes's rule of signs can help determine the number of positive and the number of negative real zeros of a polynomial function.

> **Descartes's Rule of Signs**
>
> Let $f(x)$ define a polynomial function with real coefficients and a nonzero constant term, with terms in descending powers of $x$.
>
> **(a)** The number of positive real zeros of $f$ either equals the number of variations in sign occurring in the coefficients of $f(x)$ or is less than the number of variations by a positive even integer.
>
> **(b)** The number of negative real zeros of $f$ either equals the number of variations in sign occurring in the coefficients of $f(-x)$ or is less than the number of variations by a positive even integer.

In the theorem, a *variation in sign* is a change from positive to negative or negative to positive in successive terms of the polynomial. Missing terms (those with 0 coefficients) are counted as no change in sign and can be ignored. For example, in the polynomial function defined by $f(x) = x^4 - 6x^3 + 8x^2 + 2x - 1$, $f(x)$ has three variations in sign:

$$+x^4 \underbrace{- 6x^3}_{1} \underbrace{+ 8x^2}_{2} + 2x \underbrace{- 1}_{3}.$$

Thus, by Descartes's rule of signs, $f$ has either 3 or $3 - 2 = 1$ positive real zeros. Since

$$f(-x) = (-x)^4 - 6(-x)^3 + 8(-x)^2 + 2(-x) - 1$$
$$= x^4 + 6x^3 + 8x^2 - 2x - 1$$

has only one variation in sign, $f$ has only one negative real zero.

*Use Descartes's rule of signs to determine the possible number of positive real zeros and the possible number of negative real zeros for each function.*

**53.** $f(x) = 2x^3 - 4x^2 + 2x + 7$
**54.** $f(x) = x^3 + 2x^2 + x - 10$
**55.** $f(x) = 5x^4 + 3x^2 + 2x - 9$
**56.** $f(x) = 3x^4 + 2x^3 - 8x^2 - 10x - 1$
**57.** $f(x) = x^5 + 3x^4 - x^3 + 2x + 3$
**58.** $f(x) = 2x^5 - x^4 + x^3 - x^2 + x + 5$

## 12.3 Graphs and Applications of Polynomial Functions

**OBJECTIVES**

1. Graph functions of the form $f(x) = a(x - h)^n + k$.
2. Graph general polynomial functions.
3. Use the intermediate value and boundedness theorems.
4. Approximate real zeros of polynomial functions using a graphing calculator.
5. Solve applications using polynomial functions as models.

We have already discussed the graphs of polynomial functions of degree 0 to 2. In this section we show how to graph polynomial functions of degree 3 or more. The domains will be restricted to real numbers, since we will be graphing on the real number plane.

The concepts presented here allow you to understand the ideas of finding real zeros of polynomial functions. Once these ideas are mastered, you may wish to investigate the use of computers and graphing calculators to find real zeros of polynomial functions. Learning the methods of this section first will underscore the power of technology.

**OBJECTIVE 1** Graph functions of the form $f(x) = a(x - h)^n + k$.

**EXAMPLE 1** Graphing Functions of the Form $f(x) = ax^n$

Graph each function $f$ defined as follows.

**(a)** $f(x) = x^3$

Choose several values for $x$, and find the corresponding values of $f(x)$, or $y$, as shown in the left table beside Figure 1. Plot the resulting ordered pairs and connect the points with a smooth curve. The graph of $f(x) = x^3$ is shown in blue in Figure 1.

$f(x) = x^3$

| x | f(x) |
|---|---|
| −2 | −8 |
| −1 | −1 |
| 0 | 0 |
| 1 | 1 |
| 2 | 8 |

$g(x) = x^5$

| x | g(x) |
|---|---|
| −1.5 | −7.6 |
| −1 | −1 |
| 0 | 0 |
| 1 | 1 |
| 1.5 | 7.6 |

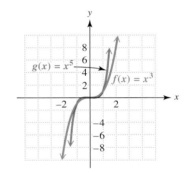

**FIGURE 1**

**(b)** $g(x) = x^5$

Work as in part (a) of this example to obtain the graph shown in red in Figure 1. Notice that the graphs of $f(x) = x^3$ and $g(x) = x^5$ are both symmetric with respect to the origin.

**(c)** $f(x) = x^4$, $g(x) = x^6$

Some typical ordered pairs for the graphs of $f(x) = x^4$ and $g(x) = x^6$ are given in the tables in Figure 2. These graphs are symmetric with respect to the $y$-axis, as is the graph of $f(x) = ax^2$ for a nonzero real number $a$.

$f(x) = x^4$

| x | f(x) |
|---|---|
| −2 | 16 |
| −1 | 1 |
| 0 | 0 |
| 1 | 1 |
| 2 | 16 |

$g(x) = x^6$

| x | g(x) |
|---|---|
| −1.5 | 11.4 |
| −1 | 1 |
| 0 | 0 |
| 1 | 1 |
| 1.5 | 11.4 |

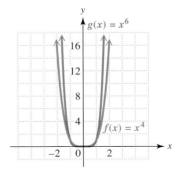

**FIGURE 2**

601

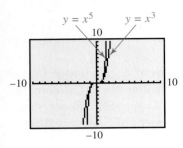
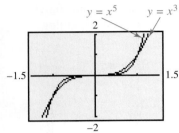
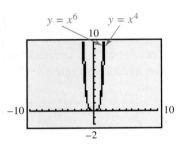
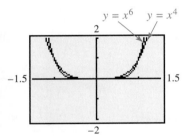

**FIGURE 3**

The ZOOM feature of a graphing calculator is useful with graphs like those in Example 1 to show the difference between the graphs of $y = x^3$ and $y = x^5$ and between $y = x^4$ and $y = x^6$ for values of $x$ in the interval $[-1.5, 1.5]$. See Figure 3. In each case the first window shows $x$ in $[-10, 10]$ and the second window shows $x$ in $[-1.5, 1.5]$.

The value of $a$ in $f(x) = ax^n$ determines the width of the graph. When $|a| > 1$, the graph is stretched vertically, making it narrower, while when $0 < |a| < 1$, the graph is shrunk or compressed vertically, so the graph is broader. The graph of $f(x) = -ax^n$ is reflected across the x-axis compared to the graph of $f(x) = ax^n$.

**NOW TRY** Exercises 1 and 3.

Compared with the graph $f(x) = ax^n$, the graph of **$f(x) = ax^n + k$** is translated (shifted) $k$ units up if $k > 0$ and $|k|$ units down if $k < 0$. Also, the graph of **$f(x) = a(x - h)^n$** is translated $h$ units to the right if $h > 0$ and $|h|$ units to the left if $h < 0$, when compared with the graph of $f(x) = ax^n$.

The graph of **$f(x) = a(x - h)^n + k$** shows a combination of these translations. The effects are the same as those we saw with quadratic functions in **Chapter 10**.

**EXAMPLE 2** Examining Vertical and Horizontal Translations

Graph each function.

**(a)** $f(x) = x^5 - 2$

The graph will be the same as that of $f(x) = x^5$ but translated 2 units down. See Figure 4.

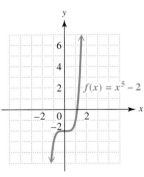

**FIGURE 4**

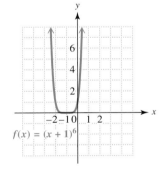

**FIGURE 5**

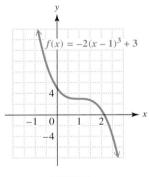

**FIGURE 6**

**(b)** $f(x) = (x + 1)^6$

This function $f$ has a graph like that of $f(x) = x^6$, but since $x + 1 = x - (-1)$, it is translated 1 unit to the left as shown in Figure 5.

**(c)** $f(x) = -2(x - 1)^3 + 3$

The negative sign in $-2$ causes the graph to be reflected about the x-axis when compared with the graph of $f(x) = x^3$. Because $|-2| > 1$, the graph is stretched vertically as compared to the graph of $f(x) = x^3$. As shown in Figure 6, the graph is also translated 1 unit to the right and 3 units up.

**NOW TRY** Exercises 5, 7, and 9.

**OBJECTIVE 2** Graph general polynomial functions. The domain of every polynomial function is the set of all real numbers. The range of a polynomial function of odd degree is also the set of all real numbers. Some typical graphs of polynomial functions of odd degree are shown in Figure 7(a). These graphs suggest that for every polynomial function $f$ of odd degree there is at least one real value of $x$ that makes $f(x) = 0$. The zeros are the $x$-intercepts of the graph.

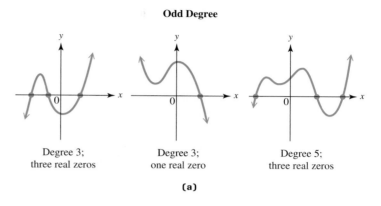

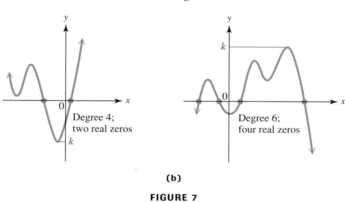

**FIGURE 7**

A polynomial function of even degree will have a range of the form $(-\infty, k]$ or $[k, \infty)$ for some real number $k$. Figure 7(b) shows two typical graphs of polynomial functions of even degree.

As mentioned in **Section 12.2,** a zero $k$ of a polynomial function has *multiplicity* that corresponds to the exponent of the factor $x - k$. For example, in

$$f(x) = (x - 1)^2(x + 4)^3,$$

the zero 1 has multiplicity 2 and the zero $-4$ has multiplicity 3. Determining whether a zero has even or odd multiplicity aids in sketching the graph near that zero. If the zero has odd multiplicity, the graph will cross the $x$-axis at the corresponding $x$-intercept. If the zero has even multiplicity, the graph will be tangent to the $x$-axis at the corresponding $x$-intercept (that is, it will touch but not cross the $x$-axis). See Figure 8.

The graph crosses the $x$-axis at $(c, 0)$ if $c$ is a zero of odd multiplicity.

The graph is tangent to the $x$-axis at $(c, 0)$ if $c$ is a zero of even multiplicity.

**FIGURE 8**

The graphs in Figures 7(a) and 7(b) show that polynomial functions often have **turning points** where the function changes from increasing to decreasing or from decreasing to increasing.

> **Turning Points**
>
> A polynomial function of degree $n$ has at most $n - 1$ turning points with at least one turning point between each pair of successive zeros.

The end behavior of a polynomial graph is determined by the term of greatest degree, called the **dominating term.** That is, a polynomial of the form $f(x) = a_n x^n + a_{n-1} x^{n-1} + \cdots + a_0$ has the same end behavior as $f(x) = a_n x^n$. For instance, $f(x) = 2x^3 - 8x^2 + 9$ has the same end behavior as $f(x) = 2x^3$. It is large and positive for large positive values of $x$ and large and negative for negative values of $x$ with large absolute value. The arrowheads at the ends of the graph look like those of the first graph in Figure 7(a); the right one points up and the left one points down. The end behavior of polynomial graphs is summarized in the following table.

END BEHAVIOR OF THE GRAPH OF $f(x) = a_n x^n + a_{n-1} x^{n-1} + \cdots + a_0$

| Degree | Sign of $a_n$ | Left Arrow | Right Arrow | Example |
|---|---|---|---|---|
| Odd | Positive | Down | Up | First graph of Figure 7(a) |
| Odd | Negative | Up | Down | Second graph of Figure 7(a) |
| Even | Positive | Up | Up | First graph of Figure 7(b) |
| Even | Negative | Down | Down | Second graph of Figure 7(b) |

We have discussed several characteristics of the graphs of polynomial functions that are useful in graphing the function. We now define what we mean by a **comprehensive graph** of a polynomial function.

> **Comprehensive Graph of a Polynomial Function**
>
> A **comprehensive graph** of a polynomial function will show the following characteristics.
>
> 1. all $x$-intercepts (zeros)
> 2. the $y$-intercept
> 3. all turning points
> 4. enough of the domain to show the end behavior

If the zeros of a polynomial function are known, its graph can be approximated without plotting very many points, as shown in the next example.

**EXAMPLE 3** Graphing a Polynomial Function

Graph the polynomial function defined by
$$f(x) = 2x^3 + 5x^2 - x - 6,$$
given that 1 is a zero.

Since 1 is a zero, we know that $x - 1$ is a factor of $f(x)$. To find the remaining quadratic factor, use synthetic division.

$$\begin{array}{r|rrr} 1) & 2 & 5 & -1 & -6 \\ & & 2 & 7 & 6 \\ \hline & 2 & 7 & 6 & 0 \end{array}$$

The final line indicates that $2x^2 + 7x + 6$ is a factor. This factors further as $(2x + 3)(x + 2)$, so the factored form of $f(x)$ is

$$f(x) = (x - 1)(2x + 3)(x + 2).$$

The three zeros are $1$, $-\frac{3}{2}$, and $-2$. To sketch the graph, we note that these zeros divide the $x$-axis into four intervals:

$$(-\infty, -2), \quad \left(-2, -\frac{3}{2}\right), \quad \left(-\frac{3}{2}, 1\right), \quad \text{and} \quad (1, \infty).$$

These intervals are shown in Figure 9.

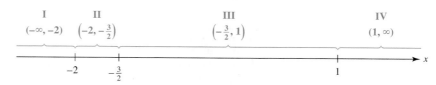

**FIGURE 9**

In any of these regions, the values of $f(x)$ are either always positive or always negative. To find the sign of $f(x)$ in each region, select an $x$-value in each region and substitute it into the expression for $f(x)$ to determine if the values of the function are positive or negative in that region. A typical selection of test points and the results of the tests are shown.

| Region | Test Point | Value of $f(x)$ | Sign of $f(x)$ |
| --- | --- | --- | --- |
| I $(-\infty, -2)$ | $-3$ | $-12$ | Negative |
| II $\left(-2, -\frac{3}{2}\right)$ | $-\frac{7}{4}$ | $\frac{11}{32}$ | Positive |
| III $\left(-\frac{3}{2}, 1\right)$ | $0$ | $-6$ | Negative |
| IV $(1, \infty)$ | $2$ | $28$ | Positive |

Plot the three zeros and the test points and join them with a smooth curve to get the graph. As expected, because each zero has odd multiplicity (that is, 1), the graph crosses the $x$-axis each time. The graph in Figure 10(a) shows that this function has two turning points, the maximum number for a third-degree polynomial function. When the values of $f(x)$ are negative, the graph is below the $x$-axis, and when $f(x)$ takes on positive values, the graph is above the $x$-axis, as shown in Figure 10(a). The sketch could be improved by plotting additional points in each region. Notice that the left arrowhead points down and the right one points up. This end behavior is correct because the first term is $2x^3$. Figure 10(b) shows how a graphing calculator graphs this function.

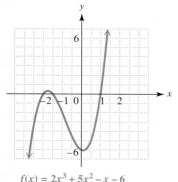

$f(x) = 2x^3 + 5x^2 - x - 6$
$= (x - 1)(2x + 3)(x + 2)$

(a)

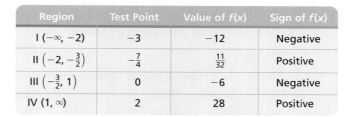

(b)

**FIGURE 10**

**NOW TRY** Exercise 25.

In summary, *there are important relationships among the following ideas:*

1. the *x*-intercepts of the graph of $y = f(x)$;
2. the zeros of the function $f$;
3. the solutions of the equation $f(x) = 0$.

For example, the graph of the function from Example 3, defined by

$$f(x) = 2x^3 + 5x^2 - x - 6$$
$$= (x - 1)(2x + 3)(x + 2), \quad \text{Factored form}$$

has *x*-intercepts $(1, 0)$, $\left(-\tfrac{3}{2}, 0\right)$, and $(-2, 0)$, as shown in Figure 10. Since $1, -\tfrac{3}{2}$, and $-2$ are the *x*-values for which the function is 0, they are the zeros of $f$. Futhermore, $1, -\tfrac{3}{2}$, and $-2$ are solutions of the equation $2x^3 + 5x^2 - x - 6 = 0$. This discussion is summarized as follows.

### *x*-Intercepts, Zeros, and Solutions

If the point $(a, 0)$ is an *x*-intercept of the graph of $y = f(x)$, then $a$ is a zero of $f$ and $a$ is a solution of the equation $f(x) = 0$.

**OBJECTIVE 3** Use the intermediate value and boundedness theorems. As Example 3 shows, one key to graphing a polynomial function is locating its zeros. In the special case where the zeros are rational numbers, the zeros can be found by the rational zeros theorem of **Section 12.2**. Occasionally, irrational zeros can be found by inspection. For instance, $f(x) = x^3 - 2$ has the irrational zero $\sqrt[3]{2}$. Two theorems presented in this section apply to the zeros of every polynomial function with real coefficients.

The first theorem uses the fact that graphs of polynomial functions are unbroken curves, with no gaps or sudden jumps. The proof requires advanced methods, so it is not given here. Figure 11 illustrates the theorem.

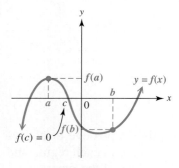

**FIGURE 11**

### Intermediate Value Theorem (as applied to locating zeros)

If $f(x)$ defines a polynomial function with only real coefficients, and if for real numbers $a$ and $b$ the values $f(a)$ and $f(b)$ are opposite in sign, then there exists at least one real zero between $a$ and $b$.

This theorem helps to identify intervals where zeros of polynomial functions are located. If $f(a)$ and $f(b)$ are opposite in sign, then 0 is between $f(a)$ and $f(b)$, and there must be a number $c$ between $a$ and $b$ where $f(c) = 0$.

**CAUTION** *Be careful how you interpret the intermediate value theorem.* If $f(a)$ and $f(b)$ are *not* opposite in sign, it does not necessarily mean that there is no zero between $a$ and $b$. For example, in Figure 12, $f(a)$ and $f(b)$ are both negative, but $-3$ and $-1$, which are between $a$ and $b$, are zeros of $f(x)$.

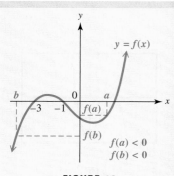

**FIGURE 12**

### EXAMPLE 4 Using the Intermediate Value Theorem to Locate a Zero

Show that $f(x) = x^3 - 2x^2 - x + 1$ has a real zero between 2 and 3.

Use synthetic division to find $f(2)$ and $f(3)$.

$$
\begin{array}{r|rrrr}
2) & 1 & -2 & -1 & 1 \\
   &   & 2  & 0  & -2 \\ \hline
   & 1 & 0  & -1 & -1 = f(2)
\end{array}
\qquad
\begin{array}{r|rrrr}
3) & 1 & -2 & -1 & 1 \\
   &   & 3  & 3  & 6 \\ \hline
   & 1 & 1  & 2  & 7 = f(3)
\end{array}
$$

The results show that $f(2) = -1$ and $f(3) = 7$. Since $f(2)$ is negative but $f(3)$ is positive, by the intermediate value theorem there must be a real zero between 2 and 3.

**NOW TRY** Exercise 47(a).

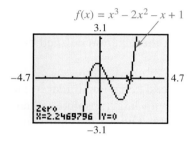

**FIGURE 13**

A graphing calculator can locate the zero established in Example 4. See Figure 13, which indicates that this zero is approximately 2.2469796. (Notice that there are two other zeros as well.)

The intermediate value theorem for polynomials is helpful in limiting the search for real zeros to smaller and smaller intervals. In Example 4 the theorem was used to verify that there is a real zero between 2 and 3. The theorem could then be used repeatedly to locate the zero more accurately. The next theorem, the **boundedness theorem,** shows how the bottom row of a synthetic division can be used to place upper and lower bounds on the possible real zeros of a polynomial function.

### Boundedness Theorem

Let $f(x)$ be a polynomial function of degree $n \geq 1$ with real coefficients and with a positive leading coefficient. If $f(x)$ is divided synthetically by $x - c$ and

(a) if $c > 0$ and all numbers in the bottom row of the synthetic division are nonnegative, then $f(x)$ has no zero greater than $c$;

(b) if $c < 0$ and the numbers in the bottom row of the synthetic division alternate in sign (with 0 considered positive or negative, as needed), then $f(x)$ has no zero less than $c$.

### EXAMPLE 5 Using the Boundedness Theorem

Show that the real zeros of $f(x) = 2x^4 - 5x^3 + 3x + 1$ satisfy the following conditions.

**(a)** No real zero is greater than 3.

Since $f(x)$ has real coefficients and the leading coefficient, 2, is positive, the boundedness theorem can be used. Divide $f(x)$ synthetically by $x - 3$.

$$
\begin{array}{r|rrrrr}
3) & 2 & -5 & 0 & 3 & 1 \\
   &   & 6  & 3 & 9 & 36 \\ \hline
   & 2 & 1  & 3 & 12 & 37
\end{array}
$$

Since $3 > 0$ and all numbers in the last row of the synthetic division are nonnegative, $f(x)$ has no real zero greater than 3.

**(b)** No real zero is less than $-1$.

$$-1 \overline{)\begin{array}{rrrrr} 2 & -5 & 0 & 3 & 1 \\ & -2 & 7 & -7 & 4 \\ \hline 2 & -7 & 7 & -4 & 5 \end{array}} \quad \text{Divide } f(x) \text{ by } x + 1.$$

Here $-1 < 0$ and the numbers in the last row alternate in sign, so $f(x)$ has no zero less than $-1$.

**NOW TRY** Exercises 51 and 53.

**OBJECTIVE 4** Approximate real zeros of polynomial functions using a graphing calculator.

**EXAMPLE 6** Approximating Real Zeros of a Polynomial Function

Use a graphing calculator to approximate the real zeros of the function defined by

$$f(x) = x^4 - 6x^3 + 8x^2 + 2x - 1.$$

The greatest degree term is $x^4$, so the graph will have end behavior similar to the graph of $f(x) = x^4$, which is positive for all values of $x$ with large absolute values. That is, the end behavior is upward at the left and the right. There are at most four real zeros, since the polynomial is fourth-degree.

Since $f(0) = -1$, the $y$-intercept is $(0, -1)$. Because the end behavior is positive on the left and the right and the $y$-value of the $y$-intercept is negative, by the intermediate value theorem $f$ has at least one zero on either side of $x = 0$. The calculator graph in Figure 14 supports these facts. We can see that there are four zeros, and the table in Figure 14 indicates that they are between $-1$ and $0$, $0$ and $1$, $2$ and $3$, and $3$ and $4$, because there is a change of sign in $f(x)$ in each case.

Using the capability of the calculator, we can find the zeros to a great degree of accuracy. Figure 15 shows that the negative zero is approximately $-0.4142136$. To the nearest hundredth, the four zeros are $-0.41$, $0.27$, $2.41$, and $3.73$.

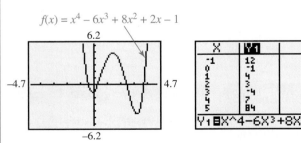

**FIGURE 14**

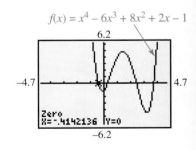

**FIGURE 15**

**NOW TRY** Exercise 59.

**OBJECTIVE 5** Solve applications using polynomial functions as models. In **Section 10.3**, we saw that a quadratic function of the form $f(x) = ax^2 + bx + c$ can be written in the form $f(x) = a(x - h)^2 + k$ by completing the square. This latter form allows us to identify the vertex of the parabola, $(h, k)$. We use this fact in the next example.

## EXAMPLE 7 Finding a Polynomial Model

The table lists the total (cumulative) number of AIDS cases diagnosed in the United States up to 1993. For example, a total of 22,620 AIDS cases were diagnosed between 1981 and 1985.

| Year | AIDS Cases | Year | AIDS Cases |
|---|---|---|---|
| 1982 | 1563 | 1988 | 105,489 |
| 1983 | 4647 | 1989 | 147,170 |
| 1984 | 10,845 | 1990 | 193,245 |
| 1985 | 22,620 | 1991 | 248,023 |
| 1986 | 41,662 | 1992 | 315,329 |
| 1987 | 70,222 | 1993 | 361,508 |

*Source:* U.S. Dept. of Health and Human Services, Centers for Disease Control and Prevention, *HIV/AIDS Surveillance,* March 1994.

Figure 16(a) shows these data plotted using a graphing calculator, with $x = 2$ corresponding to 1982, $x = 3$ corresponding to 1983, and so on.

**(a)** Find a quadratic function of the form $f(x) = a(x - h)^2 + k$ that models these data, using (2, 1563) as the vertex and the point (13, 361,508) to determine $a$.

$$f(x) = a(x - h)^2 + k \quad \text{Given form}$$
$$f(x) = a(x - 2)^2 + 1563 \quad (h, k) = (2, 1563)$$

Let $x = 13$ and $f(13) = 361{,}508$ to find $a$.

$$361{,}508 = a(13 - 2)^2 + 1563$$
$$361{,}508 = 121a + 1563$$
$$359{,}945 = 121a$$
$$a = 2975 \quad \text{Nearest whole number}$$

The desired function is $f(x) = 2975(x - 2)^2 + 1563$. It is graphed with the data points in Figure 16(b).

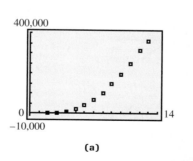
(a)

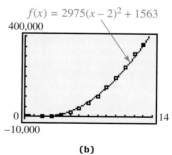
(b)

**FIGURE 16**

**(b)** Use the model from part (a) to approximate the number of cases for 1989, and compare the result to the actual data from the table.

$$f(x) = 2975(x - 2)^2 + 1563$$
$$f(9) = 2975(9 - 2)^2 + 1563 \quad \text{For 1989, } x = 9.$$
$$= 147{,}338$$

The number 147,338 compares very closely with the actual figure 147,170 given in the table.

**NOW TRY** Exercise 71.

## 12.3 Exercises

*Sketch the graph of each function. See Examples 1 and 2.*

**1.** $f(x) = \dfrac{1}{4}x^6$     **2.** $f(x) = 2x^4$     **3.** $f(x) = -\dfrac{5}{4}x^5$     **4.** $f(x) = -\dfrac{2}{3}x^5$

**5.** $f(x) = \dfrac{1}{2}x^3 + 1$     **6.** $f(x) = -x^4 + 2$     **7.** $f(x) = -(x+1)^3$

**8.** $f(x) = \dfrac{1}{3}(x+3)^4$     **9.** $f(x) = (x-1)^4 + 2$     **10.** $f(x) = (x+2)^3 - 1$

*Determine the maximum possible number of turning points of the graph of each function.*

**11.** $f(x) = x^3 - 3x^2 - 6x + 8$     **12.** $f(x) = x^3 + 4x^2 - 11x - 30$

**13.** $f(x) = 2x^4 - 9x^3 + 5x^2 + 57x - 45$     **14.** $f(x) = 4x^4 + 27x^3 - 42x^2 - 445x - 300$

**15.** $f(x) = -x^4 - 4x^3 + 3x^2 + 18x$     **16.** $f(x) = -x^4 + 2x^3 + 8x^2$

*Concept Check*   The graphs of four polynomial functions are shown in A–D. They represent the graphs of functions defined by these four equations, but not necessarily in the order listed.

$$y = x^3 - 3x^2 - 6x + 8 \qquad y = x^4 + 7x^3 - 5x^2 - 75x$$
$$y = -x^3 + 9x^2 - 27x + 17 \qquad y = -x^5 + 36x^3 - 22x^2 - 147x - 90.$$

**A.**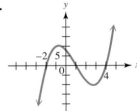

**B.**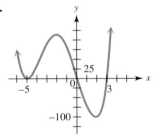

**C.**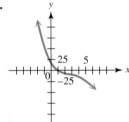

**D.**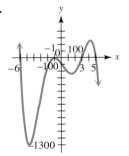

*Apply the concepts of this section to answer each question in Exercises 17–24.*

**17.** *Multiple Choice*   Which one of the graphs is that of $y = x^3 - 3x^2 - 6x + 8$?

**18.** *Multiple Choice*   Which one of the graphs is that of $y = x^4 + 7x^3 - 5x^2 - 75x$?

**19.** How many real zeros does the graph in C have?

**20.** *Multiple Choice*   Which one of C and D is the graph of $y = -x^3 + 9x^2 - 27x + 17$? (*Hint:* Look at the *y*-intercept.)

**21.** *Multiple Choice*   Which of the graphs cannot be that of a cubic polynomial function?

**22.** How many positive real zeros does the function graphed in D have?

**23.** How many negative real zeros does the function graphed in A have?

**24.** *Multiple Choice*   Which one of the graphs is that of a function whose range is *not* $(-\infty, \infty)$?

Graph each polynomial function. Factor first if the expression is not in factored form. See Example 3.

25. $f(x) = x^3 + 5x^2 + 2x - 8$
26. $f(x) = x^3 + 3x^2 - 13x - 15$
27. $f(x) = 2x(x - 3)(x + 2)$
28. $f(x) = x^2(x + 1)(x - 1)$
29. $f(x) = x^2(x - 2)(x + 3)^2$
30. $f(x) = x^2(x - 5)(x + 3)(x - 1)$
31. $f(x) = (3x - 1)(x + 2)^2$
32. $f(x) = (4x + 3)(x + 2)^2$
33. $f(x) = x^3 + 5x^2 - x - 5$
34. $f(x) = x^3 + x^2 - 36x - 36$
35. $f(x) = x^3 - x^2 - 2x$
36. $f(x) = 3x^4 + 5x^3 - 2x^2$
37. $f(x) = 2x^3(x^2 - 4)(x - 1)$
38. $f(x) = x^2(x - 3)^3(x + 1)$

For each polynomial function, one zero is given. Find all rational zeros and factor the polynomial. Then graph the function. See Example 3.

39. $f(x) = 2x^3 - 5x^2 - x + 6$; zero: $-1$
40. $f(x) = 3x^3 + x^2 - 10x - 8$; zero: 2
41. $f(x) = x^3 + x^2 - 8x - 12$; zero: 3
42. $f(x) = x^3 + 6x^2 - 32$; zero: $-4$
43. $f(x) = -x^3 - x^2 + 8x + 12$; zero: $-2$
44. $f(x) = -x^3 + 10x^2 - 33x + 36$; zero: 3
45. $f(x) = x^4 - 18x^2 + 81$; zero: $-3$ (multiplicity 2)
46. $f(x) = x^4 - 8x^2 + 16$; zero: 2 (multiplicity 2)

For each of the following, (a) show that the polynomial function has a zero between the two given integers; (b) use a graphing calculator to evaluate all zeros to the nearest thousandth. See Examples 4 and 6.

47. $f(x) = x^4 + x^3 - 6x^2 - 20x - 16$; between $-2$ and $-1$
48. $f(x) = x^4 - 2x^3 - 2x^2 - 18x + 5$; between 0 and 1
49. $f(x) = x^4 - 4x^3 - 20x^2 + 32x + 12$; between $-4$ and $-3$
50. $f(x) = x^4 - 4x^3 - 44x^2 + 160x - 80$; between 2 and 3

Use the boundedness theorem to show that the real zeros of each polynomial function satisfy the given conditions. See Example 5.

51. $f(x) = x^4 - x^3 + 3x^2 - 8x + 8$; no real zero greater than 2
52. $f(x) = 2x^5 - x^4 + 2x^3 - 2x^2 + 4x - 4$; no real zero greater than 1
53. $f(x) = x^4 + x^3 - x^2 + 3$; no real zero less than $-2$
54. $f(x) = x^5 + 2x^3 - 2x^2 + 5x + 5$; no real zero less than $-1$
55. $f(x) = 3x^4 + 2x^3 - 4x^2 + x - 1$; no real zero greater than 1
56. $f(x) = 3x^4 + 2x^3 - 4x^2 + x - 1$; no real zero less than $-2$

*Concept Check* In Exercises 57 and 58, find a cubic polynomial having the graph shown.

57.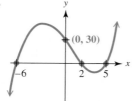
58.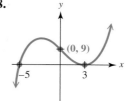

*Use a graphing calculator to approximate all real zeros of each function to the nearest hundredth. See Example 6.*

**59.** $f(x) = 0.86x^3 - 5.24x^2 + 3.55x + 7.84$  **60.** $f(x) = -2.47x^3 - 6.58x^2 - 3.33x + 0.14$
**61.** $f(x) = \sqrt{7}x^3 + \sqrt{5}x^2 + \sqrt{17}$  **62.** $f(x) = \sqrt{10}x^3 - \sqrt{11}x - \sqrt{8}$
**63.** $f(x) = 2.45x^4 - 3.22x^3 + 0.47x^2 - 6.54x - 3$
**64.** $f(x) = \sqrt{17}x^4 - \sqrt{22}x^2 + 1$

 *A graphing calculator can be used to find the coordinates of the turning points of the graph of a polynomial function. Use this capability of your calculator to find the coordinates of the turning point in the given interval. Give your answers to the nearest hundredth.*

**65.** $f(x) = x^3 + 4x^2 - 8x - 8$, $[-3.8, -3]$ **66.** $f(x) = x^3 + 4x^2 - 8x - 8$, $[0.3, 1]$
**67.** $f(x) = 2x^3 - 5x^2 - x + 1$, $[-1, 0]$ **68.** $f(x) = 2x^3 - 5x^2 - x + 1$, $[1.4, 2]$
**69.** $f(x) = x^4 - 7x^3 + 13x^2 + 6x - 28$, $[-1, 0]$
**70.** $f(x) = x^3 - x + 3$, $[-1, 0]$

*Solve each problem. See Example 7.*

**71.** The table lists the total (cumulative) number of known deaths caused by AIDS in the United States up to 1993.

(a) Plot the data. Let $x = 2$ correspond to the year 1982.
(b) Find a quadratic function of the form $g(x) = a(x - h)^2 + k$ that models the data. Use $(2, 620)$ as the vertex and the ordered pair $(13, 220{,}592)$ to approximate the value of $a$. Graph this function over the data points.
(c) Use the function from part (b) to approximate the number of deaths for 1987. Compare the result to the actual figure given in the table.

| Year | Deaths | Year | Deaths |
|---|---|---|---|
| 1982 | 620 | 1988 | 61,723 |
| 1983 | 2122 | 1989 | 89,172 |
| 1984 | 5600 | 1990 | 119,821 |
| 1985 | 12,529 | 1991 | 154,567 |
| 1986 | 24,550 | 1992 | 191,508 |
| 1987 | 40,820 | 1993 | 220,592 |

*Source:* U.S. Dept. of Health and Human Services, Centers for Disease Control and Prevention, *HIV/AIDS Surveillance,* March 1994.

**72.** Repeat Example 7, but use the ordered pair $(10, 193{,}245)$ to approximate $a$.

(a) Find the function of the form $f(x) = a(x - h)^2 + k$ that models the data.
(b) Use this new function to approximate the number of cases for 1989. Compare this result to the one found in part (b) of Example 7. Which model is more accurate for 1989?

**73.** A piece of rectangular sheet metal is 20 in. wide. It is to be made into a rain gutter by turning up the edges to form parallel sides. Let $x$ represent the length of each of the parallel sides.

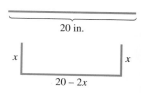

(a) Give the restrictions on $x$.
(b) Describe a function $A$ that gives the area of a cross section of the gutter.
(c) For what value of $x$ will $A$ be a maximum (and thus maximize the amount of water that the gutter will hold)? What is this maximum area?
(d) For what values of $x$ will the area of a cross section be less than 40 in.²?

**74.** A certain right triangle has area 84 in.² One leg of the triangle measures 1 in. less than the hypotenuse. Let $x$ represent the length of the hypotenuse.

(a) Express the length of the leg mentioned above in terms of $x$.
(b) Express the length of the other leg in terms of $x$.
(c) Write an equation based on the information determined thus far. Square both sides and then write the equation with one side as a polynomial with integer coefficients, in descending powers, and the other side equal to 0.
(d) Solve the equation in part (c) graphically. Find the lengths of the three sides of the triangle.

**75.** Find the value of $x$ in the figure that will maximize the area of rectangle $ABCD$.

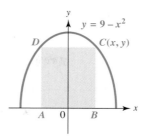

**76.** A storage tank for butane gas is to be built in the shape of a right circular cylinder of altitude 12 ft, with a half sphere attached to each end. If $x$ represents the radius of each half sphere, what radius should be used to cause the volume of the tank to be $144\pi$ ft$^3$?

## 12.4 Graphs and Applications of Rational Functions

Rational functions were introduced in **Section 7.1.** We review the definition.

**OBJECTIVES**

1. Graph rational functions using reflection and translation.
2. Find asymptotes of the graph of a rational function.
3. Graph rational functions where the degree of the numerator is less than the degree of the denominator.
4. Graph rational functions where the degrees of the numerator and the denominator are equal.
5. Graph rational functions where the degree of the numerator is greater than the degree of the denominator.
6. Graph a rational function that is not in lowest terms.
7. Graph rational functions and solve applications using a graphing calculator.

**Rational Function**

A function of the form

$$f(x) = \frac{P(x)}{Q(x)}$$

where $P(x)$ and $Q(x)$ are polynomial functions, $Q(x) \neq 0$, is called a **rational function.**

$$f(x) = \frac{1}{x}, \quad f(x) = \frac{x+1}{2x^2+5x-3}, \quad f(x) = \frac{3(x+1)(x-2)}{(x+4)^2} \quad \text{Rational functions}$$

Since any values of $x$ for which $Q(x) = 0$ are excluded from the domain of a rational function, this type of function often has a **discontinuous graph,** that is, a graph with one or more breaks in it.

**OBJECTIVE 1** Graph rational functions using reflection and translation. The simplest rational function with a variable denominator is

$$f(x) = \frac{1}{x}.$$

As we saw in **Section 7.4,** the domain of this function is the set of all real numbers except 0. The number 0 cannot be used as a value of $x$, but for graphing it is helpful to find the values of $f(x)$ for some values of $x$ close to 0. The table shows what happens to $f(x)$ as $x$ gets closer and closer to 0 from either side.

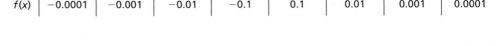

| $x$ | −1 | −0.1 | −0.01 | −0.001 | 0.001 | 0.01 | 0.1 | 1 |
|---|---|---|---|---|---|---|---|---|
| $f(x)$ | −1 | −10 | −100 | −1000 | 1000 | 100 | 10 | 1 |

$|f(x)|$ gets larger and larger.

The table suggests that $|f(x)|$ gets larger and larger as $x$ gets closer and closer to 0, which is written in symbols as

$$|f(x)| \to \infty \text{ as } x \to 0.$$

(The symbol $x \to 0$ means that $x$ approaches as close as desired to 0, without necessarily ever being equal to 0.) Since $x$ cannot equal 0, the graph of $f(x) = \frac{1}{x}$ will never intersect the vertical line $x = 0$. This line is called a **vertical asymptote**.

On the other hand, as $|x|$ gets larger and larger, the values of $f(x) = \frac{1}{x}$ get closer and closer to 0, as shown in the table.

| $x$ | −10,000 | −1000 | −100 | −10 | 10 | 100 | 1000 | 10,000 |
|---|---|---|---|---|---|---|---|---|
| $f(x)$ | −0.0001 | −0.001 | −0.01 | −0.1 | 0.1 | 0.01 | 0.001 | 0.0001 |

Letting $|x|$ get larger and larger without bound (written $|x| \to \infty$) causes the graph of $y = f(x) = \frac{1}{x}$ to move closer and closer to the horizontal line $y = 0$. This line is called a **horizontal asymptote**.

If the point $(a, b)$ lies on the graph of $f(x) = \frac{1}{x}$, then so does the point $(-a, -b)$. Therefore, the graph of $f$ is symmetric with respect to the origin. Choosing some positive values of $x$ and finding the corresponding values of $f(x)$ gives the first-quadrant part of the graph shown in Figure 17. The other part of the graph (in the third quadrant) can be found by symmetry.

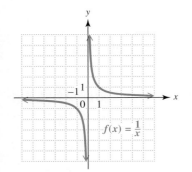

FIGURE 17

---

**EXAMPLE 1** Graphing a Rational Function Using Reflection

Graph $f(x) = -\dfrac{2}{x}$.

The expression on the right side of the equation can be rewritten so that

$$f(x) = -2 \cdot \frac{1}{x}.$$

Compared to $f(x) = \frac{1}{x}$, the graph will be reflected about the $x$-axis (because of the negative sign), and each point will be twice as far from the $x$-axis. The $x$-axis and $y$-axis remain the horizontal and vertical asymptotes. See Figure 18.

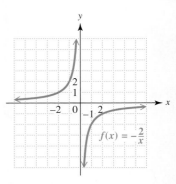

FIGURE 18

NOW TRY Exercise 7.

## EXAMPLE 2  Graphing a Rational Function Using Translation

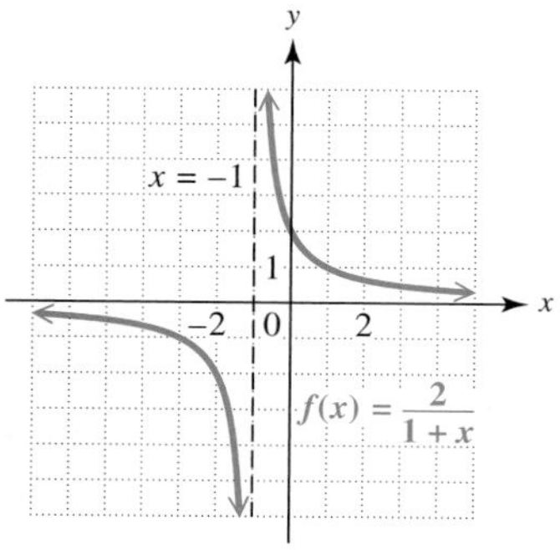

Graph $f(x) = \dfrac{2}{1+x}$.

The domain of this function is the set of all real numbers except $-1$. As shown in Figure 19, the graph is that of $f(x) = \dfrac{1}{x}$, translated 1 unit to the left, with each $y$-value doubled. This can be seen by writing the expression as

$$f(x) = 2 \cdot \dfrac{1}{x-(-1)}.$$

*Write in the form $x - k$.*

The line $x = -1$ is the vertical asymptote; the line $y = 0$ (the $x$-axis) remains the horizontal asymptote.

**FIGURE 19**

**NOW TRY** Exercise 9.

**OBJECTIVE 2** Find asymptotes of the graph of a rational function. The preceding examples suggest the following definitions of vertical and horizontal asymptotes.

### Vertical and Horizontal Asymptotes

Let $P(x)$ and $Q(x)$ define polynomials. For the rational function defined by $f(x) = \dfrac{P(x)}{Q(x)}$, written in lowest terms, and for real numbers $a$ and $b$:

1. If $|f(x)| \to \infty$ as $x \to a$, then the line $x = a$ is a **vertical asymptote**.
2. If $f(x) \to b$ as $|x| \to \infty$, then the line $y = b$ is a **horizontal asymptote**.

Locating asymptotes is important when graphing rational functions. Vertical asymptotes are found by setting the denominator of a rational function equal to 0 and then solving. Horizontal asymptotes (and, in some cases, *oblique asymptotes*) are found by considering what happens to $f(x)$ as $|x| \to \infty$.

## EXAMPLE 3  Finding the Equations of Asymptotes

For each rational function, find all asymptotes.

**(a)** $f(x) = \dfrac{x+1}{2x^2+5x-3}$

To find the vertical asymptotes, set the denominator equal to 0 and solve.

$(2x-1)(x+3) = 0$  Factor $2x^2 + 5x - 3$.

$2x - 1 = 0$ or $x + 3 = 0$  Zero-factor property

$x = \dfrac{1}{2}$ or $x = -3$  Solve each equation.

The equations of the vertical asymptotes are $x = \dfrac{1}{2}$ and $x = -3$.

To find the equation of the horizontal asymptote, divide each term by the greatest power of $x$ in the expression. In this case it is $x^2$, since 2 is the greatest exponent on $x$. This gives

$$f(x) = \dfrac{\dfrac{x}{x^2} + \dfrac{1}{x^2}}{\dfrac{2x^2}{x^2} + \dfrac{5x}{x^2} - \dfrac{3}{x^2}} = \dfrac{\dfrac{1}{x} + \dfrac{1}{x^2}}{2 + \dfrac{5}{x} - \dfrac{3}{x^2}}.$$

*Divide every term by $x^2$.*   *Simplify each term as needed.*

As $|x|$ gets larger and larger, the quotients $\frac{1}{x}, \frac{1}{x^2}, \frac{5}{x}$, and $\frac{3}{x^2}$ all approach 0, and the value of $f(x)$ approaches

$$\frac{0+0}{2+0-0} = \frac{0}{2} = 0.$$

The line $y = 0$ (that is, the $x$-axis) is therefore the horizontal asymptote.

**(b)** $f(x) = \dfrac{2x+1}{x-3}$

Set the denominator $x - 3$ equal to 0 to find that the vertical asymptote has the equation $x = 3$. To find the horizontal asymptote, divide each term in the rational expression by $x$, since the greatest power of $x$ in the expression is 1.

$$f(x) = \frac{2x+1}{x-3} = \frac{\frac{2x}{x}+\frac{1}{x}}{\frac{x}{x}-\frac{3}{x}} = \frac{2+\frac{1}{x}}{1-\frac{3}{x}}$$

As $|x|$ gets larger and larger, both $\frac{1}{x}$ and $\frac{3}{x}$ approach 0, and $f(x)$ approaches

$$\frac{2+0}{1-0} = \frac{2}{1} = 2,$$

so the line $y = 2$ is the horizontal asymptote.

**(c)** $f(x) = \dfrac{x^2+1}{x-2}$

Setting the denominator $x - 2$ equal to 0 shows that the vertical asymptote has the equation $x = 2$. If we divide by the greatest power of $x$ as before ($x^2$ in this case), we see that there is no horizontal asymptote because

$$f(x) = \frac{\frac{x^2}{x^2}+\frac{1}{x^2}}{\frac{x}{x^2}-\frac{2}{x^2}} = \frac{1+\frac{1}{x^2}}{\frac{1}{x}-\frac{2}{x^2}}$$

does not approach any real number as $|x| \to \infty$, since $\frac{1}{0}$ is undefined. This happens whenever the degree of the numerator is greater than the degree of the denominator. In such cases we divide the denominator into the numerator to write the expression in another form. Using synthetic division gives

$$\begin{array}{r} 2\overline{)1 \quad 0 \quad 1} \\ \underline{\quad\quad 2 \quad 4} \\ 1 \quad 2 \quad 5. \end{array}$$

The function can now be written as

$$f(x) = \frac{x^2+1}{x-2} = x + 2 + \frac{5}{x-2}.$$

For very large values of $|x|$, $\frac{5}{x-2}$ is close to 0, and the graph approaches the line $y = x + 2$. This line is an **oblique asymptote** (neither vertical nor horizontal) for the function.

**NOW TRY** Exercises 17, 21, and 25.

The results of Example 3 can be summarized as follows.

### Determining Asymptotes

To find asymptotes of a rational function defined by a rational expression written *in lowest terms,* use the following procedures.

1. **Vertical Asymptotes**
   Find any vertical asymptotes by setting the denominator equal to 0 and solving for $x$. If $a$ is a zero of the denominator, then the line $x = a$ is a vertical asymptote.

2. **Other Asymptotes**
   Determine any other asymptotes. There are three possibilities:
   
   (a) If the numerator has lesser degree than the denominator, there is a horizontal asymptote, $y = 0$ (the $x$-axis).
   
   (b) If the numerator and denominator have the same degree, and the function is of the form
   $$f(x) = \frac{a_n x^n + \cdots + a_0}{b_n x^n + \cdots + b_0}, \quad \text{where } b_n \neq 0,$$
   dividing by $x^n$ in the numerator and denominator produces the horizontal asymptote $y = \frac{a_n}{b_n}$.
   
   (c) If the numerator is of degree exactly one more than the denominator, there may be an oblique asymptote. To find it, divide the numerator by the denominator and disregard any remainder. Set the rest of the quotient equal to $y$ to obtain the equation of the asymptote.

The following procedure can be used to graph rational functions written in lowest terms.

### Guidelines for Graphing Rational Functions

Let $f(x) = \frac{P(x)}{Q(x)}$ be a rational function where $P(x)$ and $Q(x)$ are polynomials and the rational expression is written in lowest terms. Sketch its graph as follows.

*Step 1* **Find any vertical asymptotes.**

*Step 2* **Find any horizontal or oblique asymptote.**

*Step 3* **Find the $y$-intercept** by evaluating $f(0)$.

*Step 4* **Find the $x$-intercepts,** if any, by solving $f(x) = 0$. (These will be the zeros of the numerator, $P(x)$.)

*Step 5* **Determine whether the graph will intersect its nonvertical asymptote** by solving $f(x) = b$, where $b$ is the $y$-value of the nonvertical asymptote.

*Step 6* **Plot a few selected points, as necessary.** Choose an $x$-value in each interval of the domain as determined by the vertical asymptotes and $x$-intercepts.

*Step 7* **Complete the sketch.**

**OBJECTIVE 3** Graph rational functions where the degree of the numerator is less than the degree of the denominator.

**EXAMPLE 4** Graphing a Rational Function; Degree of Numerator < Degree of Denominator

Graph $f(x) = \dfrac{x+1}{2x^2 + 5x - 3}$.

**Step 1** As shown in Example 3(a), the vertical asymptotes have equations $x = \frac{1}{2}$ and $x = -3$.

**Step 2** Again, as shown in Example 3(a), the horizontal asymptote is the $x$-axis.

**Step 3** The $y$-intercept is $\left(0, -\frac{1}{3}\right)$, since

$$f(0) = \dfrac{0+1}{2(0)^2 + 5(0) - 3} = -\dfrac{1}{3}.$$

**Step 4** The $x$-intercept is found by solving $f(x) = 0$.

$$\dfrac{x+1}{2x^2 + 5x - 3} = 0$$

$$x + 1 = 0 \quad \text{Multiply by } 2x^2 + 5x - 3.$$

$$x = -1$$

The $x$-intercept is $(-1, 0)$.

**Step 5** To determine whether the graph intersects its horizontal asymptote, solve

$$f(x) = 0. \quad \leftarrow y\text{-value of horizontal asymptote}$$

Since the horizontal asymptote is the $x$-axis, the solution of this equation was found in Step 4. The graph intersects its horizontal asymptote at $(-1, 0)$.

**Step 6** Plot a point in each of the intervals determined by the $x$-intercepts and vertical asymptotes, $(-\infty, -3)$, $(-3, -1)$, $\left(-1, \frac{1}{2}\right)$, and $\left(\frac{1}{2}, \infty\right)$, to get an idea of how the graph behaves in each region.

**Step 7** Complete the sketch. The graph is shown in Figure 20.

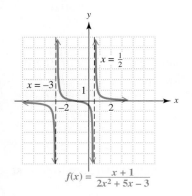

**FIGURE 20**

**NOW TRY** Exercise 35.

**OBJECTIVE 4** Graph rational functions where the degrees of the numerator and the denominator are equal. In the remaining examples, we will not specifically number the steps.

**EXAMPLE 5** Graphing a Rational Function; Degree of Numerator = Degree of Denominator

Graph $f(x) = \dfrac{2x+1}{x-3}$.

As shown in Example 3(b), the equation of the vertical asymptote is $x = 3$ and the equation of the horizontal asymptote is $y = 2$. Since $f(0) = -\frac{1}{3}$, the $y$-intercept is $\left(0, -\frac{1}{3}\right)$. The solution of $f(x) = 0$ is $-\frac{1}{2}$, so the only $x$-intercept is $\left(-\frac{1}{2}, 0\right)$.

The graph does not intersect its horizontal asymptote, since $f(x) = 2$ has no solution. (Verify this.) The points $(-4, 1)$ and $\left(6, \frac{13}{3}\right)$ are on the graph and can be used to complete the sketch, as shown in Figure 21.

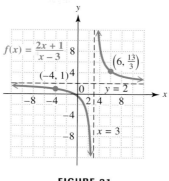

**FIGURE 21**

**NOW TRY** Exercise 39.

### EXAMPLE 6  Graphing a Rational Function; Graph Intersects the Asymptote

Graph $f(x) = \dfrac{3(x + 1)(x - 2)}{(x + 4)^2}$.

The only vertical asymptote is the line $x = -4$. To find any horizontal asymptotes, we multiply the factors in the numerator and denominator.

$$f(x) = \frac{3x^2 - 3x - 6}{x^2 + 8x + 16}$$

As explained in the guidelines for determining asymptotes, the equation of the horizontal asymptote can be shown to be

$$y = \frac{3}{1} \quad \begin{array}{l}\leftarrow \text{Leading coefficient of numerator}\\ \leftarrow \text{Leading coefficient of denominator}\end{array}$$

or $y = 3$. The $y$-intercept is $\left(0, -\frac{3}{8}\right)$, and the $x$-intercepts are $(-1, 0)$ and $(2, 0)$. By setting $f(x) = 3$ and solving, we find the point where the graph intersects the horizontal asymptote.

$$f(x) = \frac{3x^2 - 3x - 6}{x^2 + 8x + 16}$$

*This is the $y$-value of the horizontal asymptote.*

$$3 = \frac{3x^2 - 3x - 6}{x^2 + 8x + 16} \quad \text{Let } f(x) = 3.$$

$$3x^2 + 24x + 48 = 3x^2 - 3x - 6 \quad \text{Multiply by } x^2 + 8x + 16.$$

$$24x + 48 = -3x - 6 \quad \text{Subtract } 3x^2.$$

$$27x = -54 \quad \text{Add } 3x; \text{ subtract } 48.$$

$$x = -2 \quad \text{Divide by 27.}$$

The graph intersects its horizontal asymptote at $(-2, 3)$.

Some other points that lie on the graph are $(-10, 9)$, $(-3, 30)$, and $\left(5, \frac{2}{3}\right)$. These can be used to complete the graph, shown in Figure 22.

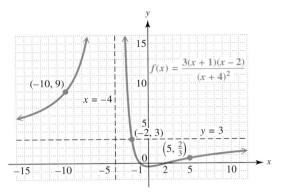

**FIGURE 22**

NOW TRY Exercise 45.

**OBJECTIVE 5** Graph rational functions where the degree of the numerator is greater than the degree of the denominator.

**EXAMPLE 7** Graphing a Rational Function; Degree of Numerator > Degree of Denominator

Graph $f(x) = \dfrac{x^2 + 1}{x - 2}$.

As shown in Example 3(c), the vertical asymptote has the equation $x = 2$, and the graph has an oblique asymptote with equation $y = x + 2$. The $y$-intercept is $\left(0, -\frac{1}{2}\right)$, and the graph has no $x$-intercepts, since the numerator, $x^2 + 1$, has no real zeros. Using the intercepts, asymptotes, the points $\left(4, \frac{17}{2}\right)$ and $\left(-1, -\frac{2}{3}\right)$, and the general behavior of the graph near its asymptotes, we obtain the graph shown in Figure 23.

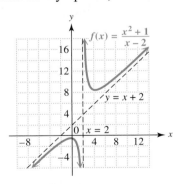

**FIGURE 23**

NOW TRY Exercise 47.

**OBJECTIVE 6** Graph a rational function that is not in lowest terms. As mentioned earlier, a rational function must be in lowest terms before we can use the methods discussed in this section to determine its graph. A rational function that has a common variable factor in the numerator and denominator will have a "hole," or **point of discontinuity,** in its graph.

### EXAMPLE 8  Graphing a Rational Function That is Not in Lowest Terms

Graph $f(x) = \dfrac{x^2 - 4}{x - 2}$.

Start by noticing that the domain of this function does not include 2. The rational expression $\dfrac{x^2 - 4}{x - 2}$ can be written in lowest terms by factoring the numerator, and using the fundamental property.

$$\frac{x^2 - 4}{x - 2} = \frac{(x + 2)(x - 2)}{x - 2} = x + 2 \quad (x \neq 2)$$

Therefore, the graph of this function will be the same as the graph of $y = x + 2$ (a straight line), with the exception of the point with $x$-value 2. Instead of an asymptote, there is a hole in the graph at (2, 4). See Figure 24.

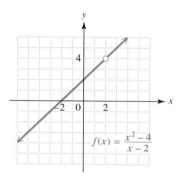

**FIGURE 24**

NOW TRY Exercise 51.

**OBJECTIVE 7** Graph rational functions and solve applications using a graphing calculator. Special care must be taken in interpreting rational function graphs generated by a graphing calculator. If the calculator is in connected mode, it may show a vertical line for $x$-values that produce vertical asymptotes. While this may be interpreted as a graph of the asymptote, a more realistic graph can be obtained by using dot mode.

In Figure 25, there are three "views": the first shows the graph of $f(x) = \dfrac{1}{x + 3}$ in connected mode, the second shows the same window but in dot mode, and the third shows a carefully chosen window, where the calculator is in connected mode but does not show the vertical line.

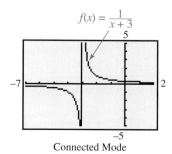
Connected Mode

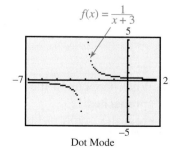
Dot Mode

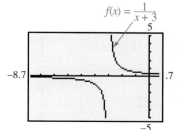
Carefully Chosen Window, Connected Mode

**FIGURE 25**

In Example 8, if the window of a graphing calculator is set so that 2 is an *x*-value for the location of the cursor, then the display will show an unlit pixel at 2. To see this, look carefully in Figure 26 at the point on the screen where $x = 2$. However, such points will often *not* be evident from calculator graphs—once again showing us a reason for studying the concepts along with the technology.

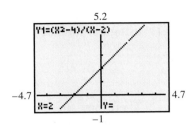

There is a tiny gap in the graph at $x = 2$.

**FIGURE 26**

Rational functions have a variety of applications, including traffic intensity as mentioned in the chapter opener.

**EXAMPLE 9** Applying Rational Functions to Traffic Intensity

Suppose that vehicles arrive randomly at a parking ramp with an average rate of 2.6 vehicles per minute. The parking attendant can admit 3.2 vehicles per minute. However, since arrivals are random, lines form at various times. (*Source:* Mannering, F., and W. Kilareski, *Principles of Highway Engineering and Traffic Control,* 2nd ed., John Wiley & Sons, 1998.)

**(a)** The traffic intensity *x* is defined as the ratio of the average arrival rate to the average admittance rate. Determine *x* for this parking ramp.

The average arrival rate is 2.6 vehicles and the average admittance rate is 3.2 vehicles, so

$$x = \frac{2.6}{3.2} = 0.8125.$$

**(b)** The average number of vehicles waiting in line to enter the ramp is given by the rational function defined by

$$f(x) = \frac{x^2}{2(1 - x)},$$

where $0 \leq x < 1$ is the traffic intensity. Graph this function using a graphing calculator, and compute $f(0.8125)$ for this parking ramp.

Figure 27 shows the graph. Find $f(0.8125)$ by substitution.

$$f(0.8125) = \frac{0.8125^2}{2(1 - 0.8125)}$$

$$\approx 1.76 \text{ vehicles}$$

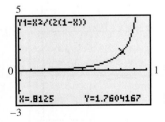

**FIGURE 27**

**(c)** What happens to the number of vehicles waiting in line as the traffic intensity approaches 1?

From the graph in Figure 27, we see that as *x* approaches 1, $y = f(x)$ gets very large. This is not surprising, because it is what we would expect.

**NOW TRY** Exercise 57.

## 12.4 Exercises

**Multiple Choice** *Use the graphs of the rational functions in A–D to answer the questions in Exercises 1–6. Give all possible answers; there may be more than one correct choice.*

A.
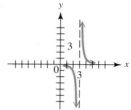

B.
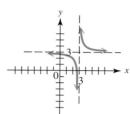

C.
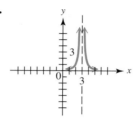

D.
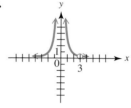

1. Which choices have domain $(-\infty, 3) \cup (3, \infty)$?
2. Which choices have range $(-\infty, 3) \cup (3, \infty)$?
3. Which choices have range $(-\infty, 0) \cup (0, \infty)$?
4. If $f$ represents the function, only one choice has a single solution to the equation $f(x) = 3$. Which one is it?
5. Which choices have the $x$-axis as a horizontal asymptote?
6. Which choices are symmetric with respect to a vertical line?

*Use reflections and/or translations to graph each rational function. See Examples 1 and 2.*

7. $f(x) = -\dfrac{3}{x}$

8. $f(x) = \dfrac{2}{x}$

9. $f(x) = \dfrac{1}{x + 2}$

10. $f(x) = \dfrac{1}{x - 3}$

11. $f(x) = \dfrac{1}{x} + 1$

12. $f(x) = \dfrac{1}{x} - 2$

13. *Concept Check* Sketch the following graphs and compare them with the graph of $f(x) = \dfrac{1}{x^2}$.

   (a) $f(x) = \dfrac{1}{(x - 3)^2}$
   (b) $f(x) = -\dfrac{2}{x^2}$
   (c) $f(x) = \dfrac{-2}{(x - 3)^2}$

14. Describe in your own words what is meant by an asymptote of the graph of a rational function.

*Give the equations of any vertical, horizontal, or oblique asymptotes for the graph of each rational function. See Example 3.*

15. $f(x) = \dfrac{-8}{3x - 7}$

16. $f(x) = \dfrac{-5}{4x - 9}$

17. $f(x) = \dfrac{x + 3}{3x^2 + x - 10}$

18. $f(x) = \dfrac{x - 8}{2x^2 - 9x - 18}$

19. $f(x) = \dfrac{2 - x}{x + 2}$

20. $f(x) = \dfrac{x - 4}{5 - x}$

21. $f(x) = \dfrac{3x - 5}{2x + 9}$

22. $f(x) = \dfrac{4x + 3}{3x - 7}$

23. $f(x) = \dfrac{2}{x^2 - 4x + 3}$

24. $f(x) = \dfrac{-5}{x^2 - 3x - 10}$

25. $f(x) = \dfrac{x^2 - 1}{x + 3}$

26. $f(x) = \dfrac{x^2 - 4}{x - 1}$

27. $f(x) = \dfrac{(x - 3)(x + 1)}{(x + 2)(2x - 5)}$

28. $f(x) = \dfrac{(x + 2)(x - 4)}{(6x - 1)(x - 5)}$

29. *Multiple Choice* Which choice has a graph that does not have a vertical asymptote?

   **A.** $f(x) = \dfrac{1}{x^2 + 2}$   **B.** $f(x) = \dfrac{1}{x^2 - 2}$   **C.** $f(x) = \dfrac{3}{x^2}$   **D.** $f(x) = \dfrac{2x + 1}{x - 8}$

30. *Multiple Choice* Which choice has a graph that does not have a horizontal asymptote?

   **A.** $f(x) = \dfrac{2x - 7}{x + 3}$   **B.** $f(x) = \dfrac{3x}{x^2 - 9}$

   **C.** $f(x) = \dfrac{x^2 - 9}{x + 3}$   **D.** $f(x) = \dfrac{x + 5}{(x + 2)(x - 3)}$

*Graph each rational function. See Examples 4–8.*

31. $f(x) = \dfrac{4}{5 + 3x}$

32. $f(x) = \dfrac{1}{(x - 2)(x + 4)}$

33. $f(x) = \dfrac{3}{(x + 4)^2}$

34. $f(x) = \dfrac{3x}{(x + 1)(x - 2)}$

35. $f(x) = \dfrac{2x + 1}{(x + 2)(x + 4)}$

36. $f(x) = \dfrac{5x}{x^2 - 1}$

37. $f(x) = \dfrac{-x}{x^2 - 4}$

38. $f(x) = \dfrac{3x}{x - 1}$

39. $f(x) = \dfrac{x + 1}{x - 4}$

40. $f(x) = \dfrac{4x}{1 - 3x}$

41. $f(x) = \dfrac{x - 5}{x + 3}$

42. $f(x) = \dfrac{x}{x^2 - 9}$

43. $f(x) = \dfrac{3x}{x^2 - 16}$

44. $f(x) = \dfrac{2x^2 + 3}{x - 4}$

45. $f(x) = \dfrac{(x - 3)(x + 1)}{(x - 1)^2}$

46. $f(x) = \dfrac{x^2 - x}{x + 2}$

47. $f(x) = \dfrac{x^2 + 1}{x + 3}$

48. $f(x) = \dfrac{x(x + 2)}{2x - 1}$

49. $f(x) = \dfrac{x(x - 2)}{(x + 3)^2}$

50. $f(x) = \dfrac{(x - 5)(x - 2)}{x^2 + 9}$

51. $f(x) = \dfrac{x^2 - 9}{x + 3}$

52. $f(x) = \dfrac{x^2 - 16}{x - 4}$

53. $f(x) = \dfrac{25 - x^2}{x - 5}$

54. $f(x) = \dfrac{36 - x^2}{x - 6}$

55. Suppose that a friend tells you that the graph of

$$f(x) = \frac{x^2 - 25}{x + 5}$$

has a vertical asymptote with equation $x = -5$. Is this correct? If not, describe the behavior of the graph at $x = -5$.

56. *Matching* The graphs in A–D show the four ways that a rational function can approach the vertical line $x = 2$ as an asymptote. Match each graph in A–D with the appropriate rational function in (a)–(d).

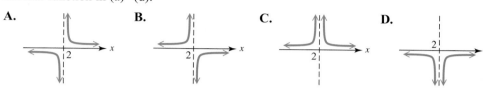

(a) $f(x) = \dfrac{1}{(x-2)^2}$   (b) $f(x) = \dfrac{1}{x-2}$   (c) $f(x) = \dfrac{-1}{x-2}$   (d) $f(x) = \dfrac{-1}{(x-2)^2}$

57. Refer to Example 9. Let the average number of vehicles arriving at the gate of an amusement park per minute equal $k$, and let the average number of vehicles admitted by the park attendants equal $r$. Then the average waiting time $T$ (in minutes) for each vehicle arriving at the park is given by the rational function defined by

$$T(r) = \frac{2r - k}{2r^2 - 2kr}, \quad \text{where } r > k.$$

(*Source:* Mannering, F., and W. Kilareski, *Principles of Highway Engineering and Traffic Control*, 2nd ed., John Wiley & Sons, 1998.)

(a) It is known from experience that on Saturday afternoon $k = 25$. Use graphing to estimate the admittance rate $r$ that is necessary to keep the average waiting time $T$ for each vehicle to 30 sec.

(b) If one park attendant can serve 5.3 vehicles per minute, how many park attendants will be needed to keep the average wait to 30 sec?

58. The rational function defined by

$$d(x) = \frac{8710x^2 - 69{,}400x + 470{,}000}{1.08x^2 - 324x + 82{,}200}$$

can be used to accurately model the braking distance for automobiles traveling at $x$ mph, where $20 \le x \le 70$. (*Source:* Mannering, F., and W. Kilareski, *Principles of Highway Engineering and Traffic Control*, 2nd ed., John Wiley & Sons, 1998.)

(a) Use graphing to estimate $x$ when $d(x) = 300$.
(b) Complete the table for each value of $x$.

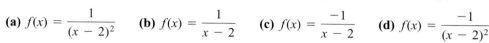

| x | 20 | 25 | 30 | 35 | 40 | 45 | 50 | 55 | 60 | 65 | 70 |
|---|----|----|----|----|----|----|----|----|----|----|----|
| d(x) | | | | | | | | | | | |

(c) If a car doubles its speed, does the stopping distance double or more than double? Explain.
(d) Suppose the stopping distance doubled whenever the speed doubled. What type of relationship would exist between the stopping distance and the speed?

# 12 SUMMARY

## KEY TERMS

**12.1** division algorithm
synthetic division
polynomial function of degree $n$
zero (root)

**12.2** multiplicity (of a zero)
**12.3** turning points
dominating term
comprehensive graph

**12.4** rational function
discontinuous graph
vertical asymptote
horizontal asymptote
oblique asymptote
point of discontinuity

## NEW SYMBOLS

$x \rightarrow a$    $x$ approaches $a$

## QUICK REVIEW

### CONCEPTS

### 12.1 Synthetic Division

Synthetic division provides a shortcut method for dividing a polynomial by a binomial of the form $x - k$.

### EXAMPLES

Use synthetic division to divide
$$f(x) = 2x^3 - 3x + 2$$
by $x - 1$. Write the result in the form $f(x) = g(x) \cdot q(x) + r(x)$.

```
1) 2    0   -3    2
         2    2   -1
   ─────────────────
      2   2   -1 | 1
```
Use 0 as coefficient for the missing $x^2$-term.

Coefficients of the quotient    Remainder

$2x^3 - 3x + 2 = (x - 1) \cdot (2x^2 + 2x - 1) + 1$

### 12.2 Zeros of Polynomial Functions

**Factor Theorem**

The polynomial $x - k$ is a factor of the polynomial $f(x)$ if and only if $f(k) = 0$.

For the polynomial function defined by
$$f(x) = x^3 + x + 2,$$
$f(-1) = 0$. Therefore, $x - (-1)$, or $x + 1$, is a factor of $f(x)$.
Also, since $x - 1$ is a factor of $g(x) = x^3 - 1$, $g(1) = 0$.

**Rational Zeros Theorem**

Let $f(x) = a_n x^n + a_{n-1} x^{n-1} + \cdots + a_1 x + a_0$, where $a_n \neq 0$, define a polynomial function with integer coefficients. If $\frac{p}{q}$ is a rational number written in lowest terms and if $\frac{p}{q}$ is a zero of $f$, then $p$ is a factor of the constant term $a_0$ and $q$ is a factor of the leading coefficient $a_n$.

The only rational numbers that can possibly be zeros of
$$f(x) = 2x^3 - 9x^2 - 4x - 5$$
are $\pm 1$, $\pm \frac{1}{2}$, $\pm 5$, and $\pm \frac{5}{2}$.
By synthetic division, it can be shown that the only rational zero of $f(x)$ is 5.

```
5) 2   -9   -4   -5
        10    5    5
   ─────────────────
   2    1    1    0  ← f(5)
```

*(continued)*

| CONCEPTS | EXAMPLES |
|---|---|

**Fundamental Theorem of Algebra**

Every function defined by a polynomial of degree 1 or more has at least one complex zero.

$f(x) = x^3 + x + 2$ has at least 1 and at most 3 zeros.

**Number of Zeros Theorem**

A function defined by a polynomial of degree $n$ has at most $n$ distinct zeros.

**Conjugate Zeros Theorem**

If $f(x)$ is a polynomial function having only real coefficients and if $a + bi$ is a zero of $f(x)$, then $a - bi$ is also a zero of $f(x)$.

Since $1 + 2i$ is a zero of
$$f(x) = x^3 - 5x^2 + 11x - 15,$$
its conjugate $1 - 2i$ is a zero as well.

## 12.3 Graphs and Applications of Polynomial Functions

**Graphing Using Translations**

The graph of the function defined by
$$f(x) = a(x - h)^n + k$$
can be found by considering the effects of the constants $a$, $h$, and $k$ on the graph of $y = x^n$.

Sketch the graph of $f(x) = -(x + 2)^4 + 1$.

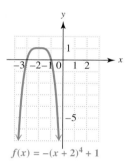

**Turning Points**

A polynomial function of degree $n$ has at most $n - 1$ turning points.

The graph of
$$f(x) = 4x^5 - 2x^3 + 3x^2 + x - 10$$
has at most $5 - 1 = 4$ turning points.

**Graphing Polynomial Functions**

To graph a polynomial function $f$, where $f(x)$ is factorable, find $x$-intercepts and $y$-intercepts. Choose a value in each region determined by the $x$-intercepts to decide whether the graph is above or below the $x$-axis.

Sketch the graph of $f(x) = (x + 2)(x - 1)(x + 3)$.

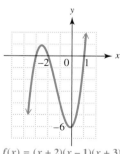

**Intermediate Value Theorem for Polynomials**

If $f(x)$ defines a polynomial function with *real coefficients*, and if for real numbers $a$ and $b$ the values of $f(a)$ and $f(b)$ are opposite in sign, then there exists at least one real zero between $a$ and $b$.

For the polynomial function defined by
$$f(x) = -x^4 + 2x^3 + 3x^2 + 6,$$
$$f(3.1) = 2.0599 \quad \text{and} \quad f(3.2) = -2.6016.$$

Since $f(3.1) > 0$ and $f(3.2) < 0$, there exists at least one real zero between 3.1 and 3.2.

*(continued)*

## CONCEPTS

**Boundedness Theorem**

Let $f(x)$ be a polynomial function with *real coefficients* and with a *positive* leading coefficient. If $f(x)$ is divided synthetically by $x - c$, and

(a) if $c > 0$ and all numbers in the bottom row of the synthetic division are nonnegative, then $f(x)$ has no zero greater than $c$;

(b) if $c < 0$ and the numbers in the bottom row of the synthetic division alternate in sign (with 0 considered positive or negative, as needed), then $f(x)$ has no zero less than $c$.

## EXAMPLES

Show that $f(x) = x^3 - x^2 - 8x + 12$ has no zero greater than 4 and no zero less than $-4$.

$$\begin{array}{r|rrrr} 4) & 1 & -1 & -8 & 12 \\ & & 4 & 12 & 16 \\ \hline & 1 & 3 & 4 & 28 \end{array} \leftarrow \text{All positive}$$

$$\begin{array}{r|rrrr} -4) & 1 & -1 & -8 & 12 \\ & & -4 & 20 & -48 \\ \hline & 1 & -5 & 12 & -36 \end{array} \leftarrow \text{Alternating signs}$$

### 12.4 Graphs and Applications of Rational Functions

**Graphing Rational Functions**

To graph a rational function in lowest terms, find asymptotes and intercepts. Determine whether the graph intersects the nonvertical asymptote. Plot a few points, as necessary, to complete the sketch.

Graph the rational function defined by $f(x) = \dfrac{x^2 - 1}{(x + 3)(x - 2)}$.

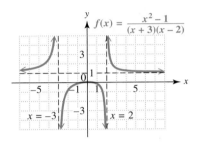

**Point of Discontinuity**

If a rational function is not written in lowest terms, there may be a "hole" in the graph instead of an asymptote.

Graph the rational function defined by $f(x) = \dfrac{x^2 - 1}{x + 1}$.

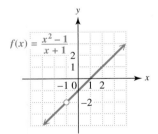

# 12 REVIEW EXERCISES

*Use synthetic division to perform each division.*

1. $\dfrac{3x^2 - x - 2}{x - 1}$
2. $\dfrac{10x^2 - 3x - 15}{x + 2}$
3. $(2x^3 - 5x^2 + 12) \div (x - 3)$
4. $(-x^4 + 19x^2 + 18x + 15) \div (x + 4)$

*Use synthetic division to decide whether $-5$ is a solution of each equation.*

5. $2x^3 + 8x^2 - 14x - 20 = 0$
6. $-3x^4 + 2x^3 + 5x^2 - 9x + 1 = 0$

*Use synthetic division to evaluate $f(k)$ for the given value of $k$.*

7. $f(x) = 3x^3 - 5x^2 + 4x - 1$; $k = -1$
8. $f(x) = x^4 - 2x^3 - 9x - 5$; $k = 3$
9. Use synthetic division to write $\dfrac{2x^3 + x - 6}{x + 2}$ in the form $f(x) = g(x) \cdot q(x) + r(x)$.

10. Explain how the rational zeros theorem can also be used to determine the answer in Exercise 6.

*Find all rational zeros of each polynomial function in Exercises 11–14.*

11. $f(x) = 2x^3 - 9x^2 - 6x + 5$
12. $f(x) = 3x^3 - 10x^2 - 27x + 10$
13. $f(x) = x^3 - \dfrac{17}{6}x^2 - \dfrac{13}{3}x - \dfrac{4}{3}$
14. $f(x) = 8x^4 - 14x^3 - 29x^2 - 4x + 3$

15. Is $-1$ a zero of $f$, if $f(x) = 2x^4 + x^3 - 4x^2 + 3x + 1$?
16. Is $-2$ a zero of $f$, if $f(x) = 2x^4 + x^3 - 4x^2 + 3x + 1$?
17. Is $x + 1$ a factor of $f(x) = x^3 + 2x^2 + 3x - 1$?
18. Is $x + 1$ a factor of $f(x) = 2x^3 - x^2 + x + 4$?

19. Find a function $f$ defined by a polynomial of degree 3 with real coefficients, having $-2$, 1, and 4 as zeros, and such that $f(2) = 16$.

20. Find a function $f$ defined by a polynomial of degree 4 with real coefficients, having 1, $-1$, and $3i$ as zeros, and such that $f(2) = 39$.

21. Find a least possible degree polynomial $f(x)$ with real coefficients defining a function having zeros 2, $-2$, and $-i$.

22. Find a least possible degree polynomial $f(x)$ with real coefficients defining a function having zeros 2, $-3$, and $5i$.

23. Find a polynomial $f(x)$ of least possible degree with real coefficients defining a function having $-3$ and $1 - i$ as zeros.

24. Find all zeros of $f$, if $f(x) = x^4 - 3x^3 - 8x^2 + 22x - 24$, given $1 - i$ is a zero, and factor $f(x)$.

25. Is it possible for a polynomial function of degree 4 with real coefficients to have three zeros that are not real and one zero that is real? Explain.

26. Suppose that a polynomial function has six real zeros. Is it possible for the function to be of degree 5? Explain.

27. *Fill in the Blanks* For the polynomial function defined by $f(x) = x^3 - 3x^2 - 7x + 12$, $f(4) = 0$. Therefore, we can say that 4 is a(n) _____ of the function, 4 is a(n) _____ of the equation $x^3 - 3x^2 - 7x + 12 = 0$, and that $(4, 0)$ is a(n) _____ of the graph of the function.

**28.** *Multiple Choice* Which one of the following is not a polynomial function?

  **A.** $f(x) = x^2$   **B.** $f(x) = 2x + 5$   **C.** $f(x) = \dfrac{1}{x}$   **D.** $f(x) = x^{100}$

*Give the maximum possible number of turning points of the graph of each polynomial function.*

**29.** $f(x) = x^3 - 9x$

**30.** $f(x) = 4x^4 - 6x^2 + 2$

*Graph each polynomial function.*

**31.** $f(x) = x^3 + 5$

**32.** $f(x) = 1 - x^4$

**33.** $f(x) = x^2(2x + 1)(x - 2)$

**34.** $f(x) = 2x^3 + 13x^2 + 15x$

**35.** $f(x) = 12x^3 - 13x^2 - 5x + 6$, given that 1 is a zero

**36.** $f(x) = x^4 - 2x^3 - 5x^2 + 6x$, given that 0 and 1 are zeros

*Show that each polynomial function has real zeros satisfying the given conditions.*

**37.** $f(x) = 3x^3 - 8x^2 + x + 2$, zeros in $[-1, 0]$ and $[2, 3]$

**38.** $f(x) = 4x^3 - 37x^2 + 50x + 60$, zeros in $[2, 3]$ and $[7, 8]$

**39.** $f(x) = x^3 + 2x^2 - 22x - 8$, zeros in $[-1, 0]$ and $[-6, -5]$

**40.** $f(x) = 2x^4 - x^3 - 21x^2 + 51x - 36$, no zero greater than 4

**41.** $f(x) = 6x^4 + 13x^3 - 11x^2 - 3x + 5$, no zero greater than 1 or less than $-3$

**42.** Use a graphing calculator to approximate the real zeros for each polynomial function. Round to the nearest thousandth for irrational zeros.

  **(a)** $f(x) = 2x^3 - 11x^2 - 2x + 2$
  **(b)** $f(x) = x^4 - 4x^3 - 5x^2 + 14x - 15$
  **(c)** $f(x) = x^3 + 3x^2 - 4x - 2$

*Graph each polynomial function.*

**43.** $f(x) = 2x^3 - 11x^2 - 2x + 2$ (See Exercise 42(a).)

**44.** $f(x) = x^4 - 4x^3 - 5x^2 + 14x - 15$ (See Exercise 42(b).)

**45.** $f(x) = x^3 + 3x^2 - 4x - 2$ (See Exercise 42(c).)

**46.** $f(x) = 2x^4 - 3x^3 + 4x^2 + 5x - 1$

*Graph each rational function.*

**47.** $f(x) = \dfrac{8}{x}$

**48.** $f(x) = \dfrac{2}{3x - 1}$

**49.** $f(x) = \dfrac{4x - 2}{3x + 1}$

**50.** $f(x) = \dfrac{6x}{(x - 1)(x + 2)}$

**51.** $f(x) = \dfrac{2x}{x^2 - 1}$

**52.** $f(x) = \dfrac{x^2 + 4}{x + 2}$

**53.** $f(x) = \dfrac{x^2 - 1}{x}$

**54.** $f(x) = \dfrac{x^2 + 6x + 5}{x - 3}$

**55.** $f(x) = \dfrac{4x^2 - 9}{2x + 3}$

# 12 TEST

*Use synthetic division in Exercises 1–5.*

1. Find the quotient when $2x^3 - 3x - 10$ is divided by $x - 2$.

2. Write $\dfrac{x^4 + 2x^3 - x^2 + 3x - 5}{x + 1}$ in the form $f(x) = g(x) \cdot q(x) + r(x)$.

3. Determine whether $x + 3$ is a factor of $x^4 - 2x^3 - 15x^2 - 4x - 12$.

4. Evaluate $f(-4)$ if $f(x) = 3x^3 - 4x^2 - 5x + 9$.

5. Is 3 a zero of $f(x) = 6x^4 - 11x^3 - 35x^2 + 34x + 24$? Why or why not?

6. Find a function $f$ defined by a polynomial having real coefficients, with degree 4, zeros 2, $-1$, and $i$, and such that $f(3) = 80$.

7. For the polynomial function defined by
$$f(x) = 6x^3 - 25x^2 + 12x + 7,$$
   **(a)** list all possible rational zeros;
   **(b)** find all rational zeros.

8. Show that the polynomial function defined by
$$f(x) = 2x^4 - 3x^3 + 4x^2 - 5x - 1$$
has no real zeros greater than 2 or less than $-1$.

9. For the polynomial function defined by
$$f(x) = 2x^3 - x + 3,$$
   **(a)** use the intermediate value theorem to show that there is a zero between $-2$ and $-1$;
   **(b)** use a graphing calculator to find this zero to the nearest thousandth.

10. For the graph of the polynomial function defined by
$$f(x) = 2x^3 - 9x^2 + 4x + 8,$$
without actually graphing,
   **(a)** determine the maximum possible number of $x$-intercepts;
   **(b)** determine the maximum possible number of turning points.

*Graph each polynomial function.*

11. $f(x) = (x - 1)^4$

12. $f(x) = x(x + 1)(x - 2)$

13. $f(x) = 2x^3 - 7x^2 + 2x + 3$, given that 3 is a zero

14. $f(x) = x^4 - 5x^2 + 6$

15. The polynomial function defined by
$$f(x) = -0.184x^3 + 1.45x^2 + 10.7x - 27.9$$
models the average temperature at Trout Lake, Canada, in degrees Fahrenheit, where $x = 1$ corresponds to January and $x = 12$ to December. What is the average temperature in June?

*Graph each rational function.*

16. $f(x) = \dfrac{-2}{x + 3}$

17. $f(x) = \dfrac{3x - 1}{x - 2}$

18. $f(x) = \dfrac{x^2 - 1}{x^2 - 9}$

19. What is the equation of the oblique asymptote of the graph of $f(x) = \dfrac{2x^2 + x - 6}{x - 1}$?

20. *Multiple Choice* Which one of the functions defined below has a graph with no $x$-intercepts?

   A. $f(x) = (x - 2)(x + 3)^2$

   B. $f(x) = \dfrac{x + 7}{x - 2}$

   C. $f(x) = x^3 - x$

   D. $f(x) = \dfrac{1}{x^2 + 4}$

# Conic Sections and Nonlinear Systems

In this chapter, we study a group of curves known as *conic sections*. One conic section, the *ellipse,* has a special reflecting property responsible for "whispering galleries." In a whispering gallery, a person whispering at a certain point in the room can be heard clearly at another point across the room.

The Old House Chamber of the U.S. Capitol, now called Statuary Hall, is a whispering gallery. History has it that John Quincy Adams, whose desk was positioned at exactly the right point beneath the ellipsoidal ceiling, often pretended to sleep there as he listened to political opponents whispering strategies across the room. (*Source: We, the People, The Story of the United States Capitol,* 1991.)

In Section 13.1, we investigate ellipses.

**13.1** The Circle and the Ellipse

**13.2** The Hyperbola and Functions Defined by Radicals

**13.3** Nonlinear Systems of Equations

# 13.1 The Circle and the Ellipse

**OBJECTIVES**

1. Find an equation of a circle given the center and radius.
2. Determine the center and radius of a circle given its equation.
3. Recognize an equation of an ellipse.
4. Graph ellipses.

When an infinite cone is intersected by a plane, the resulting figure is called a **conic section.** The parabola is one example of a conic section; circles, ellipses, and hyperbolas may also result. See Figure 1.

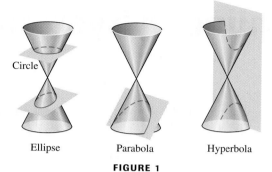

**FIGURE 1**

**OBJECTIVE 1** Find an equation of a circle given the center and radius. A **circle** is the set of all points in a plane that lie a fixed distance from a fixed point. The fixed point is called the **center,** and the fixed distance is called the **radius.** We use the distance formula from **Section 8.3** to find an equation of a circle.

**EXAMPLE 1** Finding an Equation of a Circle and Graphing It

Find an equation of the circle with radius 3 and center at (0, 0), and graph it.
  If the point $(x, y)$ is on the circle, then the distance from $(x, y)$ to the center (0, 0) is 3.

$$\sqrt{(x_2 - x_1)^2 + (y_2 - y_1)^2} = d \quad \text{Distance formula}$$
$$\sqrt{(x - 0)^2 + (y - 0)^2} = 3 \quad \text{Let } x_1 = 0, y_1 = 0, \text{ and } d = 3.$$
$$x^2 + y^2 = 9 \quad \text{Square each side.}$$

An equation of this circle is $x^2 + y^2 = 9$. The graph is shown in Figure 2.

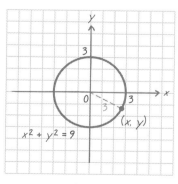

**FIGURE 2**

**NOW TRY** Exercise 1.

634

### EXAMPLE 2 Finding an Equation of a Circle and Graphing It

Find an equation of the circle with center at $(4, -3)$ and radius 5, and graph it.

$$\sqrt{(x - 4)^2 + [y - (-3)]^2} = 5 \quad \text{Distance formula}$$

$$(x - 4)^2 + (y + 3)^2 = 25 \quad \text{Square each side.}$$

To graph the circle, plot the center $(4, -3)$, then move 5 units right, left, up, and down from the center, plotting points as you move. Draw a smooth curve through these four points, sketching one quarter of the circle at a time. The graph of this circle is shown in Figure 3.

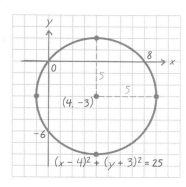

**FIGURE 3**

NOW TRY Exercise 7.

Examples 1 and 2 suggest the form of an equation of a circle with radius $r$ and center at $(h, k)$. If $(x, y)$ is a point on the circle, then the distance from the center $(h, k)$ to the point $(x, y)$ is $r$. By the distance formula,

$$\sqrt{(x - h)^2 + (y - k)^2} = r.$$

Squaring both sides gives the **center-radius form** of the equation of a circle.

---

**Equation of a Circle (Center-Radius Form)**

$$(x - h)^2 + (y - k)^2 = r^2$$

is an equation of the circle with radius $r$ and center at $(h, k)$.

---

### EXAMPLE 3 Using the Center-Radius Form of the Equation of a Circle

Find an equation of the circle with center at $(-1, 2)$ and radius $\sqrt{7}$.
Use the center-radius form, with $h = -1$, $k = 2$, and $r = \sqrt{7}$.

$$(x - h)^2 + (y - k)^2 = r^2$$

$$[x - (-1)]^2 + (y - 2)^2 = (\sqrt{7})^2$$

*Pay attention to signs here.*   $(x + 1)^2 + (y - 2)^2 = 7$

NOW TRY Exercise 9.

**OBJECTIVE 2** Determine the center and radius of a circle given its equation. In the equation found in Example 2, multiplying out $(x - 4)^2$ and $(y + 3)^2$ gives

$$(x - 4)^2 + (y + 3)^2 = 25$$
$$x^2 - 8x + 16 + y^2 + 6y + 9 = 25$$
$$x^2 + y^2 - 8x + 6y = 0.$$

This general form suggests that an equation with both $x^2$- and $y^2$-terms with equal coefficients may represent a circle. The next example shows how to tell, by completing the square. This procedure was introduced in **Section 9.1**.

**EXAMPLE 4** Completing the Square to Find the Center and Radius

Find the center and radius of the circle $x^2 + y^2 + 2x + 6y - 15 = 0$, and graph it.

Since the equation has $x^2$- and $y^2$-terms with equal coefficients, its graph might be that of a circle. To find the center and radius, complete the squares on $x$ and $y$.

$$x^2 + y^2 + 2x + 6y = 15 \qquad \text{Transform so that the constant is on the right.}$$

$$(x^2 + 2x \quad) + (y^2 + 6y \quad) = 15 \qquad \text{Write in anticipation of completing the square.}$$

Add 1 and 9 on **both** sides of the equation.
$$\left[\frac{1}{2}(2)\right]^2 = 1 \qquad \left[\frac{1}{2}(6)\right]^2 = 9 \qquad \text{Square half the coefficient of each middle term.}$$

$$(x^2 + 2x + 1) + (y^2 + 6y + 9) = 15 + 1 + 9 \qquad \text{Complete the squares on both } x \text{ and } y.$$

$$(x + 1)^2 + (y + 3)^2 = 25 \qquad \text{Factor on the left; add on the right.}$$

$$[x - (-1)]^2 + [y - (-3)]^2 = 5^2 \qquad \text{Center-radius form}$$

The final equation

$$[x - (-1)]^2 + [y - (-3)]^2 = 5^2$$

or

$$(x + 1)^2 + (y + 3)^2 = 5^2$$

shows that the graph is a circle with center at $(-1, -3)$ and radius 5. The graph is shown in Figure 4.

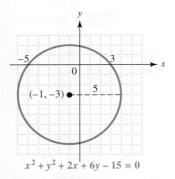

$x^2 + y^2 + 2x + 6y - 15 = 0$

**FIGURE 4**

**NOW TRY** Exercise 11.

If the procedure of Example 4 leads to an equation of the form $(x - h)^2 + (y - k)^2 = 0$, then the graph is the single point $(h, k)$. If the constant on the right side is negative, then the equation has no graph.

**OBJECTIVE 3** Recognize an equation of an ellipse. An **ellipse** is the set of all points in a plane the *sum* of whose distances from two fixed points is constant. These fixed points are called **foci** (singular: *focus*). Figure 5 shows an ellipse whose foci are $(c, 0)$ and $(-c, 0)$, with $x$-intercepts $(a, 0)$ and $(-a, 0)$ and $y$-intercepts $(0, b)$ and $(0, -b)$. It is shown in more advanced courses that $c^2 = a^2 - b^2$ for an ellipse of this type. The origin is the **center** of the ellipse.

An ellipse has the following equation.

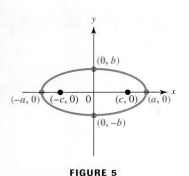

**FIGURE 5**

## Equation of an Ellipse

The ellipse whose $x$-intercepts are $(a, 0)$ and $(-a, 0)$ and whose $y$-intercepts are $(0, b)$ and $(0, -b)$ has an equation of the form

$$\frac{x^2}{a^2} + \frac{y^2}{b^2} = 1.$$

A circle is a special case of an ellipse, where $a^2 = b^2$.

When a ray of light or sound emanating from one focus of an ellipse bounces off the ellipse, it passes through the other focus. See Figure 6. As mentioned in the chapter introduction, this reflecting property is responsible for whispering galleries. John Quincy Adams was able to listen in on his opponents' conversations because his desk was positioned at one of the foci beneath the ellipsoidal ceiling and his opponents were located across the room at the other focus.

The paths of Earth and other planets around the Sun are approximately ellipses; the Sun is at one focus and a point in space is at the other.

Elliptical bicycle gears are designed to respond to the legs' natural strengths and weaknesses. At the top and bottom of the powerstroke, where the legs have the least leverage, the gear offers little resistance, but as the gear rotates, the resistance increases. This allows the legs to apply more power where it is most naturally available. See Figure 7.

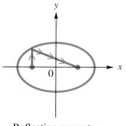

Reflecting property of an ellipse

**FIGURE 6**

**FIGURE 7**

**OBJECTIVE 4** Graph ellipses.

### EXAMPLE 5  Graphing Ellipses

Graph each ellipse.

**(a)** $\dfrac{x^2}{49} + \dfrac{y^2}{36} = 1$

Here, $a^2 = 49$, so $a = 7$, and the $x$-intercepts for this ellipse are $(7, 0)$ and $(-7, 0)$. Similarly, $b^2 = 36$, so $b = 6$, and the $y$-intercepts are $(0, 6)$ and $(0, -6)$. Plotting the intercepts and sketching the ellipse through them gives the graph in Figure 8.

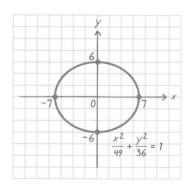

**FIGURE 8**

**(b)** $\dfrac{x^2}{36} + \dfrac{y^2}{121} = 1$

The $x$-intercepts for this ellipse are $(6, 0)$ and $(-6, 0)$, and the $y$-intercepts are $(0, 11)$ and $(0, -11)$. Join these with the smooth curve of an ellipse. The graph has been sketched in Figure 9.

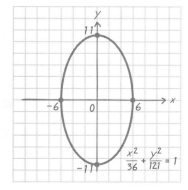

**FIGURE 9**

**NOW TRY** Exercises 29 and 35.

**EXAMPLE 6** Graphing an Ellipse Shifted Horizontally and Vertically

Graph $\dfrac{(x-2)^2}{25} + \dfrac{(y+3)^2}{49} = 1$.

Just as $(x-2)^2$ and $(y+3)^2$ would indicate that the center of a circle would be $(2, -3)$, so it is with this ellipse. Figure 10 shows that the graph goes through the four points $(2, 4), (7, -3), (2, -10),$ and $(-3, -3)$. The $x$-values of these points are found by adding $\pm a = \pm 5$ to 2, and the $y$-values come from adding $\pm b = \pm 7$ to $-3$.

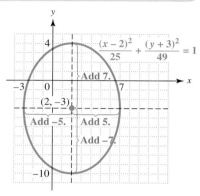

**FIGURE 10**

NOW TRY Exercise 37.

## 13.1 Exercises

**1.** See Example 1. Consider the circle whose equation is $x^2 + y^2 = 25$.
   **(a)** What are the coordinates of its center?   **(b)** What is its radius?
   **(c)** Sketch its graph.

**2.** Why does a set of points defined by a circle *not* satisfy the definition of a function?

*Matching*  Match each equation with the correct graph.

**3.** $(x-3)^2 + (y-2)^2 = 25$

**4.** $(x-3)^2 + (y+2)^2 = 25$

**5.** $(x+3)^2 + (y-2)^2 = 25$

**6.** $(x+3)^2 + (y+2)^2 = 25$

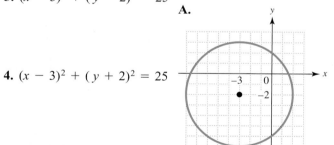

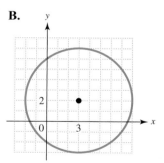

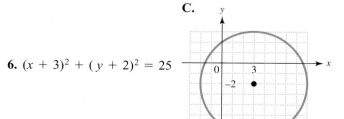

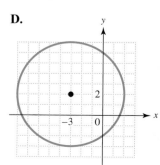

*Find the equation of a circle satisfying the given conditions. See Examples 2 and 3.*

**7.** Center: $(-4, 3)$; radius: 2     **8.** Center: $(5, -2)$; radius: 4

**9.** Center: $(-8, -5)$; radius: $\sqrt{5}$     **10.** Center: $(-12, 13)$; radius: $\sqrt{7}$

*Find the center and radius of each circle. (Hint: In Exercises 15 and 16, divide each side by a common factor.) See Example 4.*

**11.** $x^2 + y^2 + 4x + 6y + 9 = 0$     **12.** $x^2 + y^2 - 8x - 12y + 3 = 0$

**13.** $x^2 + y^2 + 10x - 14y - 7 = 0$     **14.** $x^2 + y^2 - 2x + 4y - 4 = 0$

**15.** $3x^2 + 3y^2 - 12x - 24y + 12 = 0$     **16.** $2x^2 + 2y^2 + 20x + 16y + 10 = 0$

*Graph each circle. Identify the center if it is not at the origin. See Examples 1, 2, and 4.*

**17.** $x^2 + y^2 = 9$     **18.** $x^2 + y^2 = 4$     **19.** $2y^2 = 10 - 2x^2$     **20.** $3x^2 = 48 - 3y^2$

**21.** $(x + 3)^2 + (y - 2)^2 = 9$     **22.** $(x - 1)^2 + (y + 3)^2 = 16$

**23.** $x^2 + y^2 - 4x - 6y + 9 = 0$     **24.** $x^2 + y^2 + 8x + 2y - 8 = 0$

**25.** $x^2 + y^2 + 6x - 6y + 9 = 0$     **26.** $x^2 + y^2 - 4x + 10y + 20 = 0$

**27.** A circle can be drawn on a piece of posterboard by fastening one end of a string with a thumbtack, pulling the string taut with a pencil, and tracing a curve, as shown in the figure. Explain why this method works.

**28.** An ellipse can be drawn on a piece of posterboard by fastening two ends of a length of string with thumbtacks, pulling the string taut with a pencil, and tracing a curve, as shown in the figure. Explain why this method works.

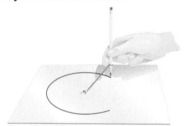

*Graph each ellipse. See Examples 5 and 6.*

**29.** $\dfrac{x^2}{9} + \dfrac{y^2}{25} = 1$     **30.** $\dfrac{x^2}{9} + \dfrac{y^2}{16} = 1$     **31.** $\dfrac{x^2}{36} = 1 - \dfrac{y^2}{16}$     **32.** $\dfrac{x^2}{9} = 1 - \dfrac{y^2}{4}$

**33.** $\dfrac{y^2}{25} = 1 - \dfrac{x^2}{49}$     **34.** $\dfrac{y^2}{9} = 1 - \dfrac{x^2}{16}$     **35.** $\dfrac{x^2}{16} + \dfrac{y^2}{4} = 1$     **36.** $\dfrac{x^2}{49} + \dfrac{y^2}{81} = 1$

**37.** $\dfrac{(x + 1)^2}{64} + \dfrac{(y - 2)^2}{49} = 1$     **38.** $\dfrac{(x - 4)^2}{9} + \dfrac{(y + 2)^2}{4} = 1$

**39.** $\dfrac{(x - 2)^2}{16} + \dfrac{(y - 1)^2}{9} = 1$     **40.** $\dfrac{(x + 3)^2}{25} + \dfrac{(y + 2)^2}{36} = 1$

**41.** Explain why a set of ordered pairs whose graph forms an ellipse does not satisfy the definition of a function.

**42.** (a) How many points are there on the graph of $(x - 4)^2 + (y - 1)^2 = 0$? Explain.
    (b) How many points are there on the graph of $(x - 4)^2 + (y - 1)^2 = -1$? Explain.

*Solve each problem.*

**43.** An arch has the shape of half an ellipse. The equation of the ellipse is
$$100x^2 + 324y^2 = 32,400,$$
where $x$ and $y$ are in meters.
(a) How high is the center of the arch?
(b) How wide is the arch across the bottom?

NOT TO SCALE

44. A one-way street passes under an overpass, which is in the form of the top half of an ellipse, as shown in the figure. Suppose that a truck 12 ft wide passes directly under the overpass. What is the maximum possible height of this truck?

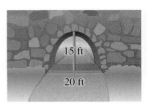

NOT TO SCALE

*In Exercises 45 and 46, see Figure 5 and use the fact that $c^2 = a^2 - b^2$ where $a^2 > b^2$.*

45. The orbit of Mars is an ellipse with the Sun at one focus. For $x$ and $y$ in millions of miles, the equation of the orbit is

$$\frac{x^2}{141.7^2} + \frac{y^2}{141.1^2} = 1.$$

(*Source:* Kaler, James B., *Astronomy!*, Addison-Wesley, 1997.)

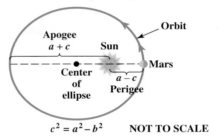

(a) Find the greatest distance (the **apogee**) from Mars to the Sun.
(b) Find the smallest distance (the **perigee**) from Mars to the Sun.

46. The orbit of Venus around the Sun (one of the foci) is an ellipse with equation

$$\frac{x^2}{5013} + \frac{y^2}{4970} = 1,$$

where $x$ and $y$ are measured in millions of miles. (*Source:* Kaler, James B., *Astronomy!*, Addison-Wesley, 1997.)

(a) Find the greatest distance between Venus and the Sun.
(b) Find the smallest distance between Venus and the Sun.

## 13.2 The Hyperbola and Functions Defined by Radicals

**OBJECTIVES**

1. Recognize the equation of a hyperbola.
2. Graph hyperbolas by using asymptotes.
3. Identify conic sections by their equations.
4. Graph certain square root functions.

**OBJECTIVE 1** Recognize the equation of a hyperbola. A **hyperbola** is the set of all points in a plane such that the absolute value of the *difference* of the distances from two fixed points (called *foci*) is constant. Figure 11 shows a hyperbola; using the distance formula and the definition above, we can show that this hyperbola has equation

$$\frac{x^2}{16} - \frac{y^2}{12} = 1.$$

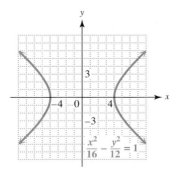

**FIGURE 11**

To graph hyperbolas centered at the origin, we need to find their intercepts. For the hyperbola in Figure 11, we proceed as follows.

### x-Intercepts

Let $y = 0$.

$\dfrac{x^2}{16} - \dfrac{0^2}{12} = 1$   Let $y = 0$.

$\dfrac{x^2}{16} = 1$

$x^2 = 16$   Multiply by 16.

$x = \pm 4$

The x-intercepts are $(4, 0)$ and $(-4, 0)$.

### y-Intercepts

Let $x = 0$.

$\dfrac{0^2}{16} - \dfrac{y^2}{12} = 1$   Let $x = 0$.

$-\dfrac{y^2}{12} = 1$

$y^2 = -12$   Multiply by $-12$.

Because there are no *real* solutions to the equation $y^2 = -12$, the graph has no y-intercepts.

The graph of $\dfrac{x^2}{16} - \dfrac{y^2}{12} = 1$ in Figure 11 on the preceding page has no y-intercepts. On the other hand, the hyperbola in Figure 12 has no x-intercepts. Its equation is $\dfrac{y^2}{25} - \dfrac{x^2}{9} = 1$, with y-intercepts $(0, 5)$ and $(0, -5)$.

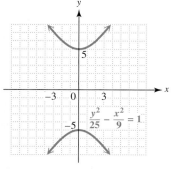

**FIGURE 12**

### Equations of Hyperbolas

A hyperbola with x-intercepts $(a, 0)$ and $(-a, 0)$ has an equation of the form

$$\dfrac{x^2}{a^2} - \dfrac{y^2}{b^2} = 1,$$

and a hyperbola with y-intercepts $(0, b)$ and $(0, -b)$ has an equation of the form

$$\dfrac{y^2}{b^2} - \dfrac{x^2}{a^2} = 1.$$

**OBJECTIVE 2** Graph hyperbolas by using asymptotes. The two branches of the graph of a hyperbola approach a pair of intersecting straight lines, which are its asymptotes. See Figure 13 on the next page. The asymptotes are useful for sketching the graph.

### Asymptotes of Hyperbolas

The extended diagonals of the rectangle with vertices (corners) at the points $(a, b)$, $(-a, b)$, $(-a, -b)$, and $(a, -b)$ are the **asymptotes** of the hyperbolas

$$\dfrac{x^2}{a^2} - \dfrac{y^2}{b^2} = 1 \quad \text{and} \quad \dfrac{y^2}{b^2} - \dfrac{x^2}{a^2} = 1.$$

This rectangle is called the **fundamental rectangle**. Using the methods of **Chapter 3**, we could show that the equations of these asymptotes are

$$y = \dfrac{b}{a}x \quad \text{and} \quad y = -\dfrac{b}{a}x.$$

To graph hyperbolas, follow these steps.

### Graphing a Hyperbola

*Step 1* **Find the intercepts.** Locate the intercepts at $(a, 0)$ and $(-a, 0)$ if the $x^2$-term has a positive coefficient or at $(0, b)$ and $(0, -b)$ if the $y^2$-term has a positive coefficient.

*Step 2* **Find the fundamental rectangle.** Locate the vertices of the fundamental rectangle at $(a, b)$, $(-a, b)$, $(-a, -b)$, and $(a, -b)$.

*Step 3* **Sketch the asymptotes.** The extended diagonals of the rectangle are the asymptotes of the hyperbola, and they have equations $y = \pm \frac{b}{a} x$.

*Step 4* **Draw the graph.** Sketch each branch of the hyperbola through an intercept and approaching (but not touching) the asymptotes.

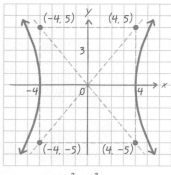

$\dfrac{x^2}{16} - \dfrac{y^2}{25} = 1$

**FIGURE 13**

### EXAMPLE 1  Graphing a Horizontal Hyperbola

Graph $\dfrac{x^2}{16} - \dfrac{y^2}{25} = 1$.

*Step 1*  Here $a = 4$ and $b = 5$. The $x$-intercepts are $(4, 0)$ and $(-4, 0)$.

*Step 2*  The four points $(4, 5)$, $(-4, 5)$, $(-4, -5)$, and $(4, -5)$ are the vertices of the fundamental rectangle, as shown in Figure 13.

*Steps 3 and 4*  The equations of the asymptotes are $y = \pm \frac{5}{4} x$, and the hyperbola approaches these lines as $x$ and $y$ get larger and larger in absolute value.

**NOW TRY** Exercise 7.

### EXAMPLE 2  Graphing a Vertical Hyperbola

Graph $\dfrac{y^2}{49} - \dfrac{x^2}{16} = 1$.

This hyperbola has $y$-intercepts $(0, 7)$ and $(0, -7)$. The asymptotes are the extended diagonals of the rectangle with vertices at $(4, 7)$, $(-4, 7)$, $(-4, -7)$, and $(4, -7)$. Their equations are $y = \pm \frac{7}{4} x$. See Figure 14.

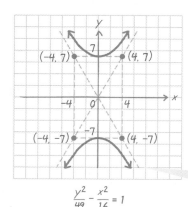

$\dfrac{y^2}{49} - \dfrac{x^2}{16} = 1$

**FIGURE 14**

When sketching the graph, be sure that the branches do *not* touch the asymptotes.

**NOW TRY** Exercise 9.

## SUMMARY OF CONIC SECTIONS

| Equation | Graph | Description | Identification |
|---|---|---|---|
| $y = ax^2 + bx + c$ or $y = a(x - h)^2 + k$ | Parabola | It opens up if $a > 0$, down if $a < 0$. The vertex is $(h, k)$. | It has an $x^2$-term. $y$ is not squared. |
| $x = ay^2 + by + c$ or $x = a(y - k)^2 + h$ | Parabola | It opens to the right if $a > 0$, to the left if $a < 0$. The vertex is $(h, k)$. | It has a $y^2$-term. $x$ is not squared. |
| $(x - h)^2 + (y - k)^2 = r^2$ | Circle | The center is $(h, k)$ and the radius is $r$. | $x^2$ and $y^2$-terms have the same positive coefficient. |
| $\dfrac{x^2}{a^2} + \dfrac{y^2}{b^2} = 1$ | Ellipse | The $x$-intercepts are $(a, 0)$ and $(-a, 0)$. The $y$-intercepts are $(0, b)$ and $(0, -b)$. | $x^2$- and $y^2$-terms have different positive coefficients. |
| $\dfrac{x^2}{a^2} - \dfrac{y^2}{b^2} = 1$ | Hyperbola | The $x$-intercepts are $(a, 0)$ and $(-a, 0)$. The asymptotes are found from $(a, b), (a, -b), (-a, -b),$ and $(-a, b)$. | $x^2$ has a positive coefficient. $y^2$ has a negative coefficient. |
| $\dfrac{y^2}{b^2} - \dfrac{x^2}{a^2} = 1$ | Hyperbola | The $y$-intercepts are $(0, b)$ and $(0, -b)$. The asymptotes are found from $(a, b), (a, -b), (-a, -b),$ and $(-a, b)$. | $y^2$ has a positive coefficient. $x^2$ has a negative coefficient. |

As with circles and ellipses, hyperbolas are graphed with a graphing calculator by first writing the equations of two functions whose union is equivalent to the equation of the hyperbola. A square window gives a truer shape for hyperbolas, too.

**OBJECTIVE 3** Identify conic sections by their equations. Rewriting a second-degree equation in one of the forms given for ellipses, hyperbolas, circles, or parabolas makes it possible to determine when the graph is one of these.

**EXAMPLE 3** Identifying the Graphs of Equations

Identify the graph of each equation.

**(a)** $9x^2 = 108 + 12y^2$

Both variables are squared, so the graph is either an ellipse or a hyperbola. (This situation also occurs for a circle, which is a special case of an ellipse.) To see whether the graph is an ellipse or a hyperbola, rewrite the equation so that the $x^2$- and $y^2$-terms are on one side of the equation and 1 is on the other.

$$9x^2 = 108 + 12y^2$$
$$9x^2 - 12y^2 = 108 \qquad \text{Subtract } 12y^2.$$
$$\frac{x^2}{12} - \frac{y^2}{9} = 1 \qquad \text{Divide by 108.}$$

Because of the minus sign, the graph of this equation is a hyperbola. The $x$-intercepts are $(\pm\sqrt{12}, 0)$, or $(\pm 2\sqrt{3}, 0)$.

**(b)** $x^2 = y - 3$

Only one of the two variables, $x$, is squared, so this is the vertical parabola

$$y = x^2 + 3.$$

The graph of this parabola opens up and has vertex $(0, 3)$.

**(c)** $x^2 = 9 - y^2$

Write the variable terms on the same side of the equation.

$$x^2 = 9 - y^2$$
$$x^2 + y^2 = 9 \qquad \text{Add } y^2.$$

The graph of this equation is a circle with center at the origin and radius 3.

**NOW TRY** Exercises 17, 19, and 21.

**OBJECTIVE 4** Graph certain square root functions. Recall that no vertical line will intersect the graph of a function in more than one point. Thus, horizontal parabolas, all circles and ellipses, and most hyperbolas discussed in this chapter are examples of graphs that do not satisfy the conditions of a function. However, by considering only a part of the graph of each of these we have the graph of a function, as seen in Figure 15.

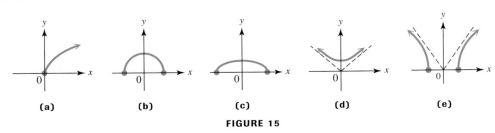

**FIGURE 15**

In parts (a), (b), (c), and (d) of Figure 15, the top portion of a conic section is shown (parabola, circle, ellipse, and hyperbola, respectively). In part (e), the top two portions of a hyperbola are shown. In each case, the graph is that of a function since the graph satisfies the conditions of the vertical line test.

In **Sections 8.1 and 10.4** we observed the square root function defined by $f(x) = \sqrt{x}$. To find equations for the types of graphs shown in Figure 15, we extend its definition.

**Square Root Function**

For an algebraic expression $u$, with $u \geq 0$, a function of the form

$$f(x) = \sqrt{u}$$

is called a **square root function**.

**EXAMPLE 4** Graphing a Semicircle

Graph $f(x) = \sqrt{25 - x^2}$. Give the domain and range.

Replace $f(x)$ with $y$ and square both sides to get the equation

$$y^2 = 25 - x^2 \quad \text{or} \quad x^2 + y^2 = 25.$$

This is the graph of a circle with center at $(0, 0)$ and radius 5. Since $f(x)$, or $y$, represents a principal square root in the original equation, $f(x)$ must be nonnegative. This restricts the graph to the upper half of the circle, as shown in Figure 16. Use the graph and the vertical line test to verify that it is indeed a function. The domain is $[-5, 5]$, and the range is $[0, 5]$.

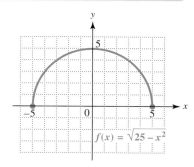

FIGURE 16

NOW TRY Exercise 25.

**EXAMPLE 5** Graphing a Portion of an Ellipse

Graph $\dfrac{y}{6} = -\sqrt{1 - \dfrac{x^2}{16}}$. Give the domain and range.

Square both sides to get an equation whose form is known.

$$\frac{y^2}{36} = 1 - \frac{x^2}{16}$$

$$\frac{x^2}{16} + \frac{y^2}{36} = 1 \qquad \text{Add } \tfrac{x^2}{16}.$$

This is the equation of an ellipse with $x$-intercepts $(4, 0)$ and $(-4, 0)$ and $y$-intercepts $(0, 6)$ and $(0, -6)$. Since $\dfrac{y}{6}$ equals a negative square root in the original equation, $y$ must be nonpositive, restricting the graph to the lower half of the ellipse, as shown in Figure 17. Verify that this is the graph of a function, using the vertical line test. The domain is $[-4, 4]$, and the range is $[-6, 0]$.

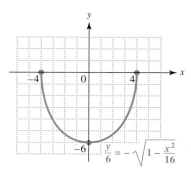

FIGURE 17

NOW TRY Exercise 27.

Root functions, since they are functions, can be entered and graphed directly with a graphing calculator.

## 13.2 Exercises

**Matching** Based on the discussions of ellipses in the previous section and of hyperbolas in this section, match each equation with its graph.

1. $\dfrac{x^2}{25} + \dfrac{y^2}{9} = 1$

2. $\dfrac{x^2}{9} + \dfrac{y^2}{25} = 1$

3. $\dfrac{x^2}{9} - \dfrac{y^2}{25} = 1$

4. $\dfrac{x^2}{25} - \dfrac{y^2}{9} = 1$

A.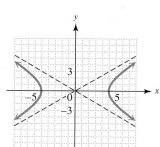

B.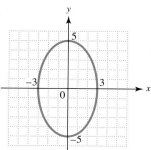

C.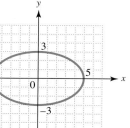

D.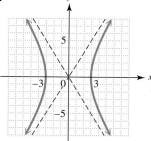

5. Write an explanation of how you can tell from the equation whether the branches of a hyperbola open up and down or left and right.

6. Describe how the fundamental rectangle is used to sketch a hyperbola.

*Graph each hyperbola. See Examples 1 and 2.*

7. $\dfrac{x^2}{16} - \dfrac{y^2}{9} = 1$

8. $\dfrac{y^2}{9} - \dfrac{x^2}{9} = 1$

9. $\dfrac{y^2}{4} - \dfrac{x^2}{25} = 1$

10. $\dfrac{x^2}{49} - \dfrac{y^2}{16} = 1$

11. $\dfrac{x^2}{25} - \dfrac{y^2}{36} = 1$

12. $\dfrac{y^2}{9} - \dfrac{x^2}{4} = 1$

13. $\dfrac{y^2}{16} - \dfrac{x^2}{16} = 1$

14. $\dfrac{x^2}{25} - \dfrac{y^2}{9} = 1$

*Identify the graph of each equation as a* parabola, circle, ellipse, *or* hyperbola, *and then sketch the graph. See Example 3.*

15. $x^2 - y^2 = 16$

16. $x^2 + y^2 = 16$

17. $4x^2 + y^2 = 16$

18. $y^2 = 36 - x^2$

19. $x^2 - 2y = 0$

20. $9x^2 + 25y^2 = 225$

21. $9x^2 = 144 + 16y^2$

22. $x^2 + 9y^2 = 9$

23. $y^2 = 4 + x^2$

24. State in your own words the major difference between the definitions of *ellipse* and *hyperbola*.

*Graph each function defined by a radical expression. Give the domain and range. See Examples 4 and 5.*

25. $f(x) = \sqrt{16 - x^2}$

26. $f(x) = \sqrt{9 - x^2}$

27. $f(x) = -\sqrt{36 - x^2}$

28. $f(x) = -\sqrt{25 - x^2}$

29. $\dfrac{y}{3} = \sqrt{1 + \dfrac{x^2}{9}}$

30. $y = \sqrt{\dfrac{x+4}{2}}$

In **Section 13.1,** Example 6, we saw that the center of an ellipse may be shifted away from the origin. The same process applies to hyperbolas. For example, the hyperbola shown at the right,

$$\frac{(x+5)^2}{4} - \frac{(y-2)^2}{9} = 1,$$

has the same graph as $\frac{x^2}{4} - \frac{y^2}{9} = 1$, but it is centered at $(-5, 2)$. Graph each hyperbola with center shifted away from the origin.

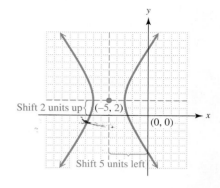

**31.** $\dfrac{(x-2)^2}{4} - \dfrac{(y+1)^2}{9} = 1$

**32.** $\dfrac{(x+3)^2}{16} - \dfrac{(y-2)^2}{25} = 1$

**33.** $\dfrac{y^2}{36} - \dfrac{(x-2)^2}{49} = 1$

**34.** $\dfrac{(y-5)^2}{9} - \dfrac{x^2}{25} = 1$

*Solve each problem.*

**35.** Two buildings in a sports complex are shaped and positioned like a portion of the branches of the hyperbola with equation

$$400x^2 - 625y^2 = 250{,}000,$$

where $x$ and $y$ are in meters.

(a) How far apart are the buildings at their closest point?

(b) Find the distance $d$ in the figure.

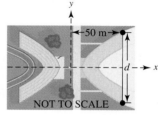

**36.** In rugby, after a *try* (similar to a touchdown in American football) the scoring team attempts a kick for extra points. The ball must be kicked from directly behind the point where the try was scored. The kicker can choose the distance but cannot move the ball sideways. It can be shown that the kicker's best choice is on the hyperbola with equation

$$\frac{x^2}{g^2} - \frac{y^2}{g^2} = 1,$$

where $2g$ is the distance between the goal posts. Since the hyperbola approaches its asymptotes, it is easier for the kicker to estimate points on the asymptotes instead of on the hyperbola. What are the asymptotes of this hyperbola? Why is it relatively easy to estimate them? (*Source:* Isaksen, Daniel C., "How to Kick a Field Goal," *The College Mathematics Journal,* September 1996.)

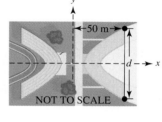

# 13.3 Nonlinear Systems of Equations

**OBJECTIVES**

1. Solve a nonlinear system by substitution.
2. Use the elimination method to solve a system with two second-degree equations.
3. Solve a system that requires a combination of methods.

An equation in which some terms have more than one variable or a variable of degree 2 or greater is called a **nonlinear equation**. A **nonlinear system of equations** includes at least one nonlinear equation.

When solving a nonlinear system, it helps to visualize the types of graphs of the equations of the system to determine the possible number of points of intersection. For example, if a system includes two equations where the graph of one is a parabola and the graph of the other is a line, then there may be zero, one, or two points of intersection, as illustrated in Figure 18.

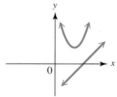  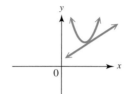  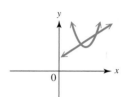

No points of intersection   One point of intersection   Two points of intersection

**FIGURE 18**

A similar situation exists for a system consisting of a circle and a line. See Figure 19.

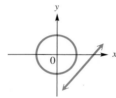  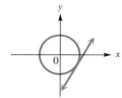  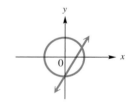

No points of intersection   One point of intersection   Two points of intersection

**FIGURE 19**

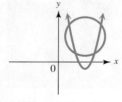

This system has four solutions, since there are four points of intersection.

**FIGURE 20**

If a system consists of two second-degree equations, then there may be zero, one, two, three, or four solutions. Figure 20 shows a case where a system consisting of a circle and a parabola has four solutions, all made up of ordered pairs of real numbers. (Exercises 7–14 are designed for you to visualize similar situations.)

**OBJECTIVE 1** Solve a nonlinear system by substitution. We solve nonlinear systems by the elimination method, the substitution method, or a combination of the two. The substitution method **(Section 4.1)** is usually appropriate when one equation is linear.

**EXAMPLE 1** Solving a Nonlinear System by Substitution

Solve the system

$$x^2 + y^2 = 9 \quad (1)$$
$$2x - y = 3. \quad (2)$$

The graph of (1) is a circle and the graph of (2) is a line. Recall from Figure 19 that the graphs could intersect in zero, one, or two points.

Solve the linear equation for one of the two variables, and then substitute the resulting expression into the nonlinear equation to obtain an equation in one variable.

$$2x - y = 3 \quad (2)$$
$$y = 2x - 3 \quad (3)$$

Substitute $2x - 3$ for $y$ in equation (1).

$$x^2 + y^2 = 9 \quad (1)$$
$$x^2 + (2x - 3)^2 = 9 \quad \text{Let } y = 2x - 3.$$
$$x^2 + 4x^2 - 12x + 9 = 9 \quad \text{Square } 2x - 3.$$
$$5x^2 - 12x = 0 \quad \text{Combine like terms; subtract 9.}$$
$$x(5x - 12) = 0 \quad \text{Factor; GCF is } x.$$
$$x = 0 \quad \text{or} \quad 5x - 12 = 0 \quad \text{Zero-factor property}$$
$$x = \frac{12}{5}$$

**Set both factors equal to 0.**

Let $x = 0$ in equation (3) to get $y = -3$. If $x = \frac{12}{5}$, then $y = \frac{9}{5}$. The solution set of the system is $\left\{(0, -3), \left(\frac{12}{5}, \frac{9}{5}\right)\right\}$. The graph in Figure 21 confirms the two points of intersection.

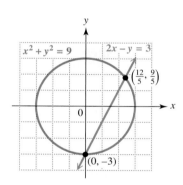

**FIGURE 21**

**NOW TRY** Exercise 19.

---

**EXAMPLE 2** Solving a Nonlinear System by Substitution

Solve the system

$$6x - y = 5 \quad (1)$$
$$xy = 4. \quad (2)$$

The graph of (1) is a line. We have not specifically mentioned equations like (2); however, it can be shown by plotting points that its graph is a hyperbola. Visualizing a line and a hyperbola indicates that there may be zero, one, or two points of intersection. Since neither equation has a squared term, we can solve either equation for one of the variables and then substitute the result into the other equation. Solving $xy = 4$ for $x$ gives $x = \frac{4}{y}$. We substitute $\frac{4}{y}$ for $x$ in equation (1).

$$6\left(\frac{4}{y}\right) - y = 5 \quad \text{Let } x = \frac{4}{y} \text{ in equation (1).}$$

$$\frac{24}{y} - y = 5 \quad \text{Multiply.}$$

$$24 - y^2 = 5y \quad \text{Multiply by } y, y \neq 0.$$

$$y^2 + 5y - 24 = 0 \quad \text{Standard form}$$

$$(y - 3)(y + 8) = 0 \quad \text{Factor.}$$

$$y = 3 \quad \text{or} \quad y = -8 \quad \text{Zero-factor property}$$

We substitute these results into $x = \frac{4}{y}$ to obtain the corresponding values of $x$.

If $y = 3$, then $x = \frac{4}{3}$. If $y = -8$, then $x = -\frac{1}{2}$.

The solution set of the system is $\left\{\left(\frac{4}{3}, 3\right), \left(-\frac{1}{2}, -8\right)\right\}$. The graph in Figure 22 shows these two points of intersection.

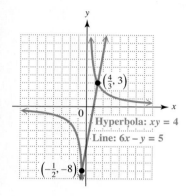

**FIGURE 22**

NOW TRY Exercise 21.

**OBJECTIVE 2** Use the elimination method to solve a system with two second-degree equations. The elimination method **(Section 4.1)** is often used when both equations are second degree.

**EXAMPLE 3** Solving a Nonlinear System by Elimination

Solve the system

$$x^2 + y^2 = 9 \quad (1)$$
$$2x^2 - y^2 = -6. \quad (2)$$

The graph of (1) is a circle, while the graph of (2) is a hyperbola. By analyzing the possibilities we conclude that there may be zero, one, two, three, or four points of intersection. Adding the two equations will eliminate $y$, leaving an equation that can be solved for $x$.

$$\begin{aligned} x^2 + y^2 &= 9 \\ 2x^2 - y^2 &= -6 \\ \hline 3x^2 &= 3 \quad \text{Add.} \\ x^2 &= 1 \quad \text{Divide by 3.} \\ x = 1 \quad &\text{or} \quad x = -1 \quad \text{Square root property} \end{aligned}$$

Each value of $x$ gives corresponding values for $y$ when substituted into one of the original equations. Using equation (1) gives the following.

If $x = 1$, then

$$1^2 + y^2 = 9$$
$$y^2 = 8$$
$$y = \sqrt{8} \quad \text{or} \quad y = -\sqrt{8}$$
$$y = 2\sqrt{2} \quad \text{or} \quad y = -2\sqrt{2}.$$

If $x = -1$, then

$$(-1)^2 + y^2 = 9$$
$$y^2 = 8$$
$$y = 2\sqrt{2} \quad \text{or} \quad y = -2\sqrt{2}.$$

The solution set is $\{(1, 2\sqrt{2}), (1, -2\sqrt{2}), (-1, 2\sqrt{2}), (-1, -2\sqrt{2})\}$. Figure 23 shows the four points of intersection.

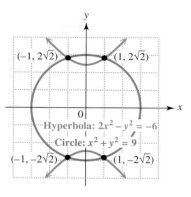

**FIGURE 23**

**NOW TRY** Exercise 35.

**OBJECTIVE 3** Solve a system that requires a combination of methods. Solving a system of second-degree equations may require a combination of methods.

**EXAMPLE 4** Solving a Nonlinear System by a Combination of Methods

Solve the system

$$x^2 + 2xy - y^2 = 7 \quad (1)$$
$$x^2 - y^2 = 3. \quad (2)$$

While we have not graphed equations like (1), its graph is a hyperbola. The graph of (2) is also a hyperbola. Two hyperbolas may have zero, one, two, three, or four points of intersection. We use the elimination method here in combination with the substitution method. We begin by eliminating the squared terms by multiplying each side of equation (2) by $-1$ and then adding the result to equation (1).

$$\begin{array}{rl} x^2 + 2xy - y^2 = & 7 \quad (1) \\ -x^2 \phantom{+2xy} + y^2 = & -3 \\ \hline 2xy \phantom{-y^2} = & 4 \quad \text{Add.} \end{array}$$

Next, we solve $2xy = 4$ for $y$. (Either variable would do.)

$$2xy = 4$$
$$y = \frac{2}{x} \quad \text{Divide by } 2x. \quad (3)$$

Now, we substitute $y = \frac{2}{x}$ into one of the original equations. It is easier to do this with equation (2).

$$x^2 - y^2 = 3 \quad (2)$$
$$x^2 - \left(\frac{2}{x}\right)^2 = 3 \quad \text{Let } y = \frac{2}{x}.$$
$$x^2 - \frac{4}{x^2} = 3 \quad \text{Square .}$$

$$x^4 - 4 = 3x^2 \quad \text{Multiply by } x^2, x \neq 0.$$
$$x^4 - 3x^2 - 4 = 0 \quad \text{Subtract } 3x^2.$$
$$(x^2 - 4)(x^2 + 1) = 0 \quad \text{Factor.}$$
$$x^2 - 4 = 0 \quad \text{or} \quad x^2 + 1 = 0$$
$$x^2 = 4 \quad \text{or} \quad x^2 = -1$$
$$x = 2 \quad \text{or} \quad x = -2 \quad \text{or} \quad x = i \quad \text{or} \quad x = -i$$

Substituting these four values into $y = \frac{2}{x}$ (equation (3)) gives the corresponding values for $y$.

If $x = 2$, then $y = \frac{2}{2} = 1$.

If $x = -2$, then $y = \frac{2}{-2} = -1$.

If $x = i$, then $y = \frac{2}{i} = \frac{2}{i} \cdot \frac{-i}{-i} = -2i$.  *Multiply by the complex conjugate of the denominator.*

If $x = -i$, then $y = \frac{2}{-i} = \frac{2}{-i} \cdot \frac{i}{i} = 2i$.

Note that if we substitute the $x$-values we found into equation (1) or (2) instead of into equation (3), we get extraneous solutions. ***It is always wise to check all solutions in both of the given equations.*** There are four ordered pairs in the solution set, two with real values and two with pure imaginary values. The solution set is

$$\{(2, 1), (-2, -1), (i, -2i), (-i, 2i)\}.$$

The graph of the system, shown in Figure 24, shows only the two real intersection points because the graph is in the real number plane. In general, if solutions contain nonreal complex numbers as components, they do not appear on the graph.

**NOW TRY** Exercise 41.

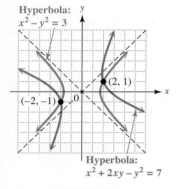

**FIGURE 24**

## 13.3 Exercises

 Exercise

▶ **1.** Write an explanation of the steps you would use to solve the system

$$x^2 + y^2 = 25$$
$$y = x - 1$$

by the substitution method. Why would the elimination method not be appropriate for this system?

▶ **2.** Write an explanation of the steps you would use to solve the system

$$x^2 + y^2 = 12$$
$$x^2 - y^2 = 13$$

by the elimination method.

*Concept Check* Each sketch represents the graphs of a pair of equations in a system. How many points are in each solution set?

**3.** 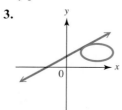    **4.** 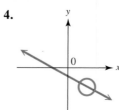    **5.** 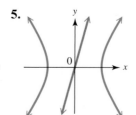    **6.**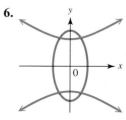

*Concept Check* *Suppose that a nonlinear system is composed of equations whose graphs are those described, and the number of points of intersection of the two graphs is as given. Make a sketch satisfying these conditions. (There may be more than one way to do this.)*

7. A line and a circle; no points
8. A line and a circle; one point
9. A line and a hyperbola; one point
10. A line and an ellipse; no points
11. A circle and an ellipse; four points
12. A parabola and an ellipse; one point
13. A parabola and an ellipse; four points
14. A parabola and a hyperbola; two points

*Solve each system by the substitution method. See Examples 1 and 2.*

15. $y = 4x^2 - x$
    $y = x$

16. $y = x^2 + 6x$
    $3y = 12x$

17. $y = x^2 + 6x + 9$
    $x + y = 3$

18. $y = x^2 + 8x + 16$
    $x - y = -4$

19. $x^2 + y^2 = 2$
    $2x + y = 1$

20. $2x^2 + 4y^2 = 4$
    $x = 4y$

21. $xy = 4$
    $3x + 2y = -10$

22. $xy = -5$
    $2x + y = 3$

23. $xy = -3$
    $x + y = -2$

24. $xy = 12$
    $x + y = 8$

25. $y = 3x^2 + 6x$
    $y = x^2 - x - 6$

26. $y = 2x^2 + 1$
    $y = 5x^2 + 2x - 7$

27. $2x^2 - y^2 = 6$
    $y = x^2 - 3$

28. $x^2 + y^2 = 4$
    $y = x^2 - 2$

29. $x^2 - xy + y^2 = 0$
    $x - 2y = 1$

30. $x^2 - 3x + y^2 = 4$
    $2x - y = 3$

*Solve each system by the elimination method or a combination of the elimination and substitution methods. See Examples 3 and 4.*

31. $3x^2 + 2y^2 = 12$
    $x^2 + 2y^2 = 4$

32. $5x^2 - 2y^2 = -13$
    $3x^2 + 4y^2 = 39$

33. $2x^2 + 3y^2 = 6$
    $x^2 + 3y^2 = 3$

34. $6x^2 + y^2 = 9$
    $3x^2 + 4y^2 = 36$

35. $2x^2 + y^2 = 28$
    $4x^2 - 5y^2 = 28$

36. $x^2 + 6y^2 = 9$
    $4x^2 + 3y^2 = 36$

37. $2x^2 = 8 - 2y^2$
    $3x^2 = 24 - 4y^2$

38. $5x^2 = 20 - 5y^2$
    $2y^2 = 2 - x^2$

39. $x^2 + xy + y^2 = 15$
    $x^2 + y^2 = 10$

40. $2x^2 + 3xy + 2y^2 = 21$
    $x^2 + y^2 = 6$

41. $3x^2 + 2xy - 3y^2 = 5$
    $-x^2 - 3xy + y^2 = 3$

42. $-2x^2 + 7xy - 3y^2 = 4$
    $2x^2 - 3xy + 3y^2 = 4$

*Solve each problem by using a nonlinear system.*

43. The area of a rectangular rug is 84 ft² and its perimeter is 38 ft. Find the length and width of the rug.

44. Find the length and width of a rectangular room whose perimeter is 50 m and whose area is 100 m².

# 13 SUMMARY

## KEY TERMS

**13.1** conic section
circle
center
radius of a circle
center-radius form

ellipse
center of an ellipse
foci of an ellipse

**13.2** hyperbola
asymptotes of a hyperbola
fundamental rectangle

square root function
**13.3** nonlinear equation
nonlinear system of equations

## QUICK REVIEW

| CONCEPTS | EXAMPLES |
|---|---|
| **13.1 The Circle and the Ellipse** | |

**Circle**

The circle with radius $r$ and center at $(h, k)$ has an equation of the form

$$(x - h)^2 + (y - k)^2 = r^2.$$

The circle with equation $(x + 2)^2 + (y - 3)^2 = 25$, which can be written $[x - (-2)]^2 + (y - 3)^2 = 5^2$, has center $(-2, 3)$ and radius 5.

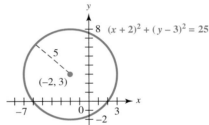

**Ellipse**

The ellipse whose $x$-intercepts are $(a, 0)$ and $(-a, 0)$ and whose $y$-intercepts are $(0, b)$ and $(0, -b)$ has an equation of the form

$$\frac{x^2}{a^2} + \frac{y^2}{b^2} = 1.$$

Graph $\dfrac{x^2}{9} + \dfrac{y^2}{4} = 1$.

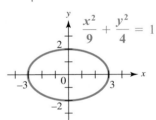

### 13.2 The Hyperbola and Functions Defined by Radicals

**Hyperbola**

A hyperbola with $x$-intercepts $(a, 0)$ and $(-a, 0)$ has an equation of the form

$$\frac{x^2}{a^2} - \frac{y^2}{b^2} = 1,$$

and a hyperbola with $y$-intercepts $(0, b)$ and $(0, -b)$ has an equation of the form

$$\frac{y^2}{b^2} - \frac{x^2}{a^2} = 1.$$

Graph $\dfrac{x^2}{4} - \dfrac{y^2}{4} = 1$.

The graph has $x$-intercepts $(2, 0)$ and $(-2, 0)$.

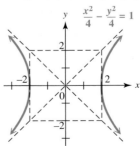

The extended diagonals of the fundamental rectangle with vertices at the points $(a, b)$, $(-a, b)$, $(-a, -b)$, and $(a, -b)$ are the asymptotes of these hyperbolas.

The fundamental rectangle has vertices at $(2, 2)$, $(-2, 2)$, $(-2, -2)$, and $(2, -2)$.

*(continued)*

| CONCEPTS | EXAMPLES |
|---|---|

**Graphing a Square Root Function**

To graph a square root function defined by

$$f(x) = \sqrt{u}$$

for an algebraic expression $u$, with $u \geq 0$, square both sides so that the equation can be easily recognized. Then graph only the part indicated by the original equation.

Graph $y = -\sqrt{4 - x^2}$.

Square both sides and rearrange terms to get

$$x^2 + y^2 = 4.$$

This equation has a circle as its graph. However, graph only the lower half of the circle, since the original equation indicates that $y$ cannot be positive.

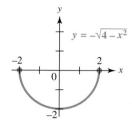

## 13.3 Nonlinear Systems of Equations

**Solving a Nonlinear System**

A nonlinear system can be solved by the substitution method, the elimination method, or a combination of the two.

Solve the system

$$x^2 + 2xy - y^2 = 14 \quad (1)$$
$$x^2 - y^2 = -16. \quad (2)$$

Multiply equation (2) by $-1$ and use elimination.

$$\begin{aligned} x^2 + 2xy - y^2 &= 14 \\ -x^2 \qquad\quad + y^2 &= 16 \\ \hline 2xy \qquad\quad &= 30 \\ xy &= 15 \end{aligned}$$

Solve $xy = 15$ for $y$ to obtain $y = \frac{15}{x}$, and substitute into equation (2).

$$x^2 - y^2 = -16 \quad (2)$$
$$x^2 - \left(\frac{15}{x}\right)^2 = -16$$
$$x^2 - \frac{225}{x^2} = -16$$
$$x^4 + 16x^2 - 225 = 0 \quad \text{Multiply by } x^2; \text{ add } 16x^2.$$
$$(x^2 - 9)(x^2 + 25) = 0 \quad \text{Factor.}$$
$$x = \pm 3 \quad \text{or} \quad x = \pm 5i \quad \text{Zero-factor property}$$

Find corresponding $y$-values to get the solution set

$$\{(3, 5), (-3, -5), (5i, -3i), (-5i, 3i)\}.$$

# 13 REVIEW EXERCISES

*Concept Check* Identify the graph of each equation as a circle or an ellipse.

1. $x^2 + y^2 = 121$
2. $x^2 + 4y^2 = 4$
3. $\dfrac{x^2}{16} + \dfrac{y^2}{25} = 1$
4. $3x^2 + 3y^2 = 300$

*Write an equation for each circle.*

5. Center $(-2, 4)$, $r = 3$
6. Center $(-1, -3)$, $r = 5$
7. Center $(4, 2)$, $r = 6$

*Find the center and radius of each circle.*

8. $x^2 + y^2 + 6x - 4y - 3 = 0$
9. $x^2 + y^2 - 8x - 2y + 13 = 0$
10. $2x^2 + 2y^2 + 4x + 20y = -34$
11. $4x^2 + 4y^2 - 24x + 16y = 48$

*Graph each equation.*

12. $x^2 + y^2 = 16$
13. $\dfrac{x^2}{16} + \dfrac{y^2}{9} = 1$
14. $\dfrac{x^2}{49} + \dfrac{y^2}{25} = 1$
15. $\dfrac{x^2}{16} - \dfrac{y^2}{25} = 1$
16. $\dfrac{y^2}{25} - \dfrac{x^2}{4} = 1$
17. $f(x) = -\sqrt{16 - x^2}$

*Identify the graph of each equation as a parabola, circle, ellipse, or hyperbola.*

18. $x^2 + y^2 = 64$
19. $y = 2x^2 - 3$
20. $y^2 = 2x^2 - 8$
21. $y^2 = 8 - 2x^2$
22. $x = y^2 + 4$

*Solve each system.*

23. $2y = 3x - x^2$
    $x + 2y = -12$
24. $y + 1 = x^2 + 2x$
    $y + 2x = 4$
25. $x^2 + 3y^2 = 28$
    $y - x = -2$
26. $xy = 8$
    $x - 2y = 6$
27. $x^2 + y^2 = 6$
    $x^2 - 2y^2 = -6$
28. $3x^2 - 2y^2 = 12$
    $x^2 + 4y^2 = 18$

29. *Concept Check* How many solutions are possible for a system of two equations whose graphs are a circle and a line?

30. *Concept Check* How many solutions are possible for a system of two equations whose graphs are a parabola and a hyperbola?

# 13 TEST

1. *Multiple Choice* Which one of the following equations is represented by the graph of a circle?
   **A.** $x^2 + y^2 = 0$
   **B.** $x^2 + y^2 = -1$
   **C.** $x^2 + y^2 = x^2 - y^2$
   **D.** $x^2 + y^2 = 1$

2. *Concept Check* For the equation in Exercise 1 that is represented by a circle, what are the coordinates of the center? What is the radius?

3. Find the center and radius of the circle whose equation is $(x - 2)^2 + (y + 3)^2 = 16$. Sketch the graph.

4. Find the center and radius of the circle whose equation is $x^2 + y^2 + 8x - 2y = 8$.

*Graph.*

**5.** $f(x) = \sqrt{9 - x^2}$

**6.** $4x^2 + 9y^2 = 36$

**7.** $16y^2 - 4x^2 = 64$

**8.** $\dfrac{y}{2} = -\sqrt{1 - \dfrac{x^2}{9}}$

*Identify the graph of each equation as a* parabola, hyperbola, ellipse, *or* circle.

**9.** $6x^2 + 4y^2 = 12$

**10.** $16x^2 = 144 + 9y^2$

**11.** $4y^2 + 4x = 9$

*Solve each nonlinear system.*

**12.** $2x - y = 9$
   $xy = 5$

**13.** $x - 4 = 3y$
   $x^2 + y^2 = 8$

**14.** $x^2 + y^2 = 25$
   $x^2 - 2y^2 = 16$

# 14 Trigonometric Functions

Some of the most important concepts in the study of trigonometry deal with the idea of similar right triangles. Informally speaking, two figures in the plane are *similar* if they have the same shape but not necessarily the same size. In the case of two similar right triangles, each has a right angle and two acute angles of corresponding measures. Furthermore, their corresponding sides are proportional. These facts allow us to solve many types of problems in astronomy, navigation, architecture, and other fields.

The depiction of Oliver Wendell Holmes' "Chambered Nautilus" shell can be approximated geometrically by a sequence of similar triangles. It can be found at http://www.maa.org/projectwelcome/calculus%20in%20action/librarya/spiral/spiral5.htm.

In Section 14.2 we investigate *similar triangles* and their properties.

**14.1** Angles

**14.2** Angle Relationships and Similar Triangles

**14.3** Trigonometric Functions

**14.4** Using the Definitions of the Trigonometric Functions

## 14.1 Angles

**OBJECTIVES**

1. Understand the basic terminology of angles.
2. Find the measures of complementary and supplementary angles.
3. Calculate with degrees, minutes, and seconds.
4. Convert between decimal degrees and degrees, minutes, and seconds.
5. Find the measures of coterminal angles.

**OBJECTIVE 1** Understand the basic terminology of angles. Two distinct points $A$ and $B$ determine a line called **line $AB$**. The portion of the line between $A$ and $B$, including points $A$ and $B$ themselves, is **line segment $AB$,** or simply **segment $AB$.** The portion of line $AB$ that starts at $A$ and continues through $B$, and on past $B$, is called **ray $AB$.** Point $A$ is the **endpoint of the ray.** See Figure 1.

In trigonometry, an **angle** consists of two rays in a plane with a common endpoint, or two line segments with a common endpoint. These two rays (or segments) are called the **sides** of the angle, and the common endpoint is called the **vertex** of the angle. Associated with an angle is its measure, generated by a rotation about the vertex. See Figure 2. This measure is determined by rotating a ray starting at one side of the angle, called the **initial side,** to the position of the other side, called the **terminal side.** *A counterclockwise rotation generates a positive measure, while a clockwise rotation generates a negative measure.* The rotation can consist of more than one complete revolution.

Figure 3 shows two angles, one **positive** and one **negative.**

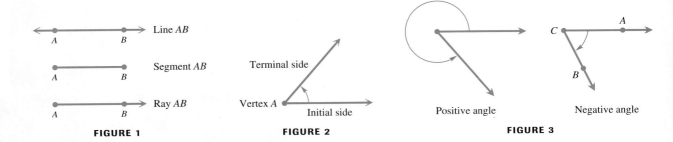

FIGURE 1  FIGURE 2  FIGURE 3

An angle can be named by using the name of its vertex. For example, the angle on the right in Figure 3 can be called angle $C$. Alternatively, an angle can be named using three letters, with the vertex letter in the middle. Thus, the angle on the right also could be named angle $ACB$ or angle $BCA$.

**OBJECTIVE 2** Find the measures of complementary and supplementary angles. The most common unit for measuring angles is the **degree.** Degree measure was developed by the Babylonians, 4000 yr ago. To use degree measure, we assign 360 degrees to a complete rotation of a ray.* In Figure 4, notice that the terminal side of the angle corresponds to its initial side when it makes a complete rotation. One degree, written $1°$, represents $\frac{1}{360}$ of a rotation. Therefore, $90°$ represents $\frac{90}{360} = \frac{1}{4}$ of a complete rotation, and $180°$ represents $\frac{180}{360} = \frac{1}{2}$ of a complete rotation. An angle measuring between $0°$ and $90°$ is called an **acute angle.** An angle measuring exactly $90°$ is a **right angle.** The symbol ⌐ is often used at the vertex of a right angle to denote the $90°$ measure. An angle measuring more than $90°$ but less than $180°$ is an **obtuse angle,** and an angle of exactly $180°$ is a **straight angle.**

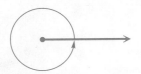

A complete rotation of a ray gives an angle whose measure is $360°$. $\frac{1}{360}$ of a complete rotation gives an angle whose measure is $1°$.

**FIGURE 4**

---

*The Babylonians were the first to subdivide the circumference of a circle into 360 parts. There are various theories as to why the number 360 was chosen. One is that it is approximately the number of days in a year, and it has many divisors, which makes it convenient to work with.

In Figure 5, we use the **Greek letter θ (theta)*** to name each angle.

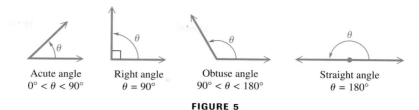

**FIGURE 5**

If the sum of the measures of two positive angles is 90°, the angles are called **complementary** and the angles are **complements** of each other. Two positive angles with measures whose sum is 180° are **supplementary,** and the angles are **supplements.**

**EXAMPLE 1** Finding the Complement and the Supplement of an Angle

For an angle measuring 40°, find the measure of its **(a)** complement and **(b)** supplement.

**Solution**

(a) To find the measure of its complement, subtract the measure of the angle from 90°.

$$90° - 40° = 50° \quad \text{Complement of 40°}$$

(b) To find the measure of its supplement, subtract the measure of the angle from 180°.

$$180° - 40° = 140° \quad \text{Supplement of 40°}$$

**NOW TRY** Exercise 1.

**EXAMPLE 2** Finding Measures of Complementary and Supplementary Angles

Find the measure of each marked angle in Figure 6.

**Solution**

(a) In Figure 6(a), since the two angles form a right angle, they are complementary angles. Thus,

$$6x + 3x = 90$$
$$9x = 90 \quad \text{Combine terms.}$$
**Don't stop here.** $\quad x = 10. \quad \text{Divide by 9.}$

Be sure to determine the measure of each angle by substituting 10 for $x$. The two angles have measures of $6(10) = 60°$ and $3(10) = 30°$.

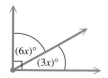

**(a)**

**FIGURE 6**

*In addition to θ (theta), other Greek letters such as α (alpha) and β (beta) are often used.

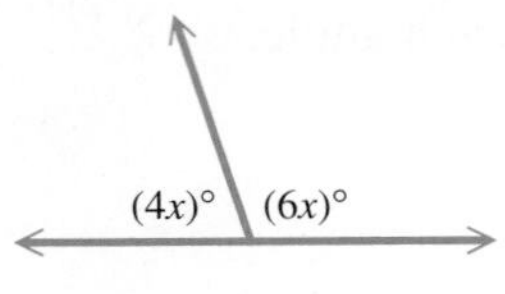

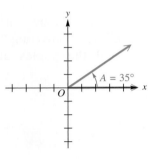

**FIGURE 6**

(b) The angles in Figure 6(b) are supplementary, so their sum must be 180°. Therefore,

$$4x + 6x = 180$$
$$10x = 180$$
$$x = 18.$$

These angle measures are $4(18) = 72°$ and $6(18) = 108°$.

**NOW TRY** Exercises 13 and 15.

**FIGURE 7**

**OBJECTIVE 3** Calculate with degrees, minutes, and seconds. The measure of angle $A$ in Figure 7 is 35°. This measure is often expressed by saying that ***m*(angle *A*)** is 35°, where $m$(angle A) is read **"the measure of angle *A*."** It is convenient, however, to abbreviate the symbolism $m$(angle $A$) = 35° as $A = 35°$.

Traditionally, portions of a degree have been measured with minutes and seconds. One **minute,** written **1′**, is $\frac{1}{60}$ of a degree.

$$1' = \frac{1}{60}° \quad \text{or} \quad 60' = 1°$$

One **second, 1″**, is $\frac{1}{60}$ of a minute.

$$1'' = \frac{1}{60}' = \frac{1}{3600}° \quad \text{or} \quad 60'' = 1'$$

The measure 12° 42′ 38″ represents 12 degrees, 42 minutes, 38 seconds.

**EXAMPLE 3** Calculating with Degrees, Minutes, and Seconds

Perform each calculation.

(a) 51° 29′ + 32° 46′   (b) 90° − 73° 12′

**Solution**

(a)  51° 29′    Add degrees and minutes separately.
    + 32° 46′
    ─────────
     83° 75′

The sum 83° 75′ can be rewritten as

$$83° \; 75' = 83° + 1° \; 15' = 84° \; 15'. \quad 75' = 60' + 15' = 1° \; 15'$$

(b)  89° 60′    Write 90° as 89° 60′.
    − 73° 12′
    ─────────
     16° 48′

**NOW TRY** Exercises 33 and 37.

**OBJECTIVE 4** Convert between decimal degrees and degrees, minutes, and seconds. Because calculators are now so prevalent, angles are commonly measured in decimal degrees. For example, 12.4238° represents

$$12.4238° = 12\frac{4238}{10{,}000}°.$$

### EXAMPLE 4   Converting between Decimal Degrees and Degrees, Minutes, and Seconds

(a) Convert 74° 8′ 14″ to decimal degrees to the nearest thousandth.
(b) Convert 34.817° to degrees, minutes, and seconds.

**Solution**

(a) $74° \, 8' \, 14'' = 74° + \frac{8}{60}° + \frac{14}{3600}°$   $\qquad 1' = \frac{1}{60}°$ and $1'' = \frac{1}{3600}°$

$\approx 74° + 0.1333° + 0.0039°$

$\approx 74.137°$   Add; round to the nearest thousandth.

(b) $34.817° = 34° + 0.817°$

$= 34° + 0.817(60')$   $\qquad 1° = 60'$

$= 34° + 49.02'$

$= 34° + 49' + 0.02'$

$= 34° + 49' + 0.02(60'')$   $\qquad 1' = 60''$

$= 34° + 49' + 1.2''$

$= 34° \, 49' \, 1.2''$

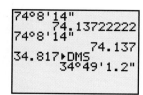

A graphing calculator performs the conversions in Example 4 as shown above. The ▶DMS option is found in the ANGLE Menu of the TI-83/84 Plus calculator.

**NOW TRY** Exercises 53 and 63.

**OBJECTIVE 5** Find the measures of coterminal angles. An angle is in **standard position** if its vertex is at the origin and its initial side lies on the positive *x*-axis. The angles in Figures 8(a) and 8(b) are in standard position. An angle in standard position is said to lie in the quadrant in which its terminal side lies. An acute angle is in quadrant I (Figure 8(a)) and an obtuse angle is in quadrant II (Figure 8(b)). Figure 8(c) shows ranges of angle measures for each quadrant when $0° < \theta < 360°$.

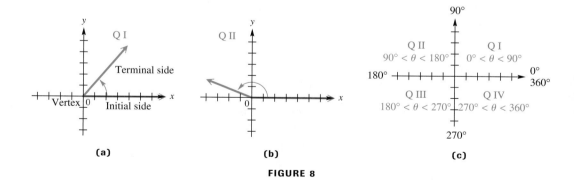

**FIGURE 8**

### Quadrantal Angles

Angles in standard position whose terminal sides lie on the *x*-axis or *y*-axis, such as angles with measures 90°, 180°, 270°, and so on, are called **quadrantal angles**.

A complete rotation of a ray results in an angle measuring 360°. By continuing the rotation, angles of measure larger than 360° can be produced. The angles in Figure 9 (on the next page) with measures 60° and 420° have the same initial side and the same terminal side but different amounts of rotation. Such angles are called **coterminal angles;** *their measures differ by a multiple of* **360°.** As shown in Figure 10, angles with measures 110° and 830° are coterminal.

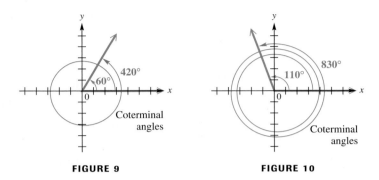

FIGURE 9  FIGURE 10

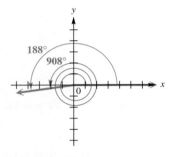

FIGURE 11

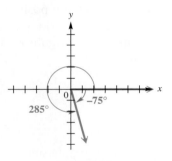

FIGURE 12

### EXAMPLE 5  Finding Measures of Coterminal Angles

Find the angles of least possible positive measure coterminal with each angle.

(a) 908°   (b) −75°   (c) −800°

**Solution**

(a) Add or subtract 360° as many times as needed to obtain an angle with measure greater than 0° but less than 360°. Since

$$908° - 2 \cdot 360° = 188°,$$

an angle of 188° is coterminal with an angle of 908°. See Figure 11.

(b) Use a rotation of 360° + (−75°) = 285°. See Figure 12.

(c) The least integer multiple of 360° greater than 800° is

$$360° \cdot 3 = 1080°.$$

Add 1080° to −800° to obtain

$$1080° + (-800°) = 280°.$$

**NOW TRY** Exercises 73, 83, and 87.

Sometimes it is necessary to find an expression that will generate all angles coterminal with a given angle. For example, we can obtain any angle coterminal with 60° by adding an appropriate integer multiple of 360° to 60°. Let *n* represent any integer; then the expression

$$60° + n \cdot 360° \quad \text{Angles coterminal with } 60°$$

represents all such coterminal angles. The table shows a few possibilities.

| Value of *n* | Angle Coterminal with 60° |
|---|---|
| 2 | 60° + 2 · 360° = 780° |
| 1 | 60° + 1 · 360° = 420° |
| 0 | 60° + 0 · 360° = 60° (the angle itself) |
| −1 | 60° + (−1) · 360° = −300° |

**SECTION 14.1** Angles **665**

**EXAMPLES OF COTERMINAL QUADRANTAL ANGLES**

| Quadrantal Angle θ | Coterminal with θ |
|---|---|
| 0° | ±360°, ±720° |
| 90° | −630°, −270°, 450° |
| 180° | −180°, 540°, 900° |
| 270° | −450°, −90°, 630° |

The table in the margin shows some examples of coterminal quadrantal angles.

**EXAMPLE 6** Analyzing the Revolutions of a CD Player

CAV (Constant Angular Velocity) CD players always spin at the same speed. Suppose a CAV player makes 480 revolutions per min. Through how many degrees will a point on the edge of a CD move in 2 sec?

**Solution** The player revolves 480 times in 1 min or $\frac{480}{60}$ times = 8 times per sec (since 60 sec = 1 min). In 2 sec, the player will revolve $2 \cdot 8 = 16$ times. Each revolution is 360°, so a point on the edge of the CD will revolve $16 \cdot 360° = 5760°$ in 2 sec.

**NOW TRY** Exercise 127.

## 14.1 Exercises

**NOW TRY Exercise**

*Find* **(a)** *the complement and* **(b)** *the supplement of an angle with the given measure. See Examples 1 and 3.*

1. 30°  2. 60°  3. 45°  4. 18°
5. 54°  6. 89°  7. 1°  8. 10°
9. 14° 20′  10. 39° 50′  11. 20° 10′ 30″  12. 50° 40′ 50″

*Find the measure of each unknown angle in Exercises 13–22. See Example 2.*

13.
14.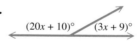
15.

16.
17.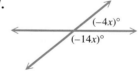
18.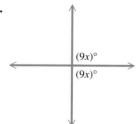

19. supplementary angles with measures $10x + 7$ and $7x + 3$ degrees
20. supplementary angles with measures $6x - 4$ and $8x - 12$ degrees
21. complementary angles with measures $9x + 6$ and $3x$ degrees
22. complementary angles with measures $3x - 5$ and $6x - 40$ degrees
23. *Concept Check* What is the measure of an angle that is its own complement?
24. *Concept Check* What is the measure of an angle that is its own supplement?

*Find the measure of the smaller angle formed by the hands of a clock at the following times.*

**25.**

**26.**

**27.** 3:15

**28.** 9:00

*Concept Check   Answer each question.*

**29.** If an angle measures $x°$, how can we represent its complement?

**30.** If an angle measures $x°$, how can we represent its supplement?

**31.** If a positive angle has measure $x°$ between $0°$ and $60°$, how can we represent the first negative angle coterminal with it?

**32.** If a negative angle has measure $x°$ between $0°$ and $-60°$, how can we represent the first positive angle coterminal with it?

*Perform each calculation. See Example 3.*

**33.** $62° \ 18' + 21° \ 41'$
**34.** $75° \ 15' + 83° \ 32'$
**35.** $71° \ 18' - 47° \ 29'$
**36.** $47° \ 23' - 73° \ 48'$
**37.** $90° - 51° \ 28'$
**38.** $90° - 17° \ 13'$
**39.** $180° - 119° \ 26'$
**40.** $180° - 124° \ 51'$
**41.** $26° \ 20' + 18° \ 17' - 14° \ 10'$
**42.** $55° \ 30' + 12° \ 44' - 8° \ 15'$
**43.** $90° - 72° \ 58' \ 11''$
**44.** $90° - 36° \ 18' \ 47''$

*Convert each angle measure to decimal degrees. If applicable, round to the nearest thousandth of a degree. See Example 4(a).*

**45.** $35° \ 30'$
**46.** $82° \ 30'$
**47.** $112° \ 15'$
**48.** $133° \ 45'$
**49.** $-60° \ 12'$
**50.** $-70° \ 48'$
**51.** $20° \ 54' \ 00''$
**52.** $38° \ 42' \ 00''$
**53.** $91° \ 35' \ 54''$
**54.** $34° \ 51' \ 35''$
**55.** $274° \ 18' \ 59''$
**56.** $165° \ 51' \ 9''$

*Convert each angle measure to degrees, minutes, and seconds. See Example 4(b).*

**57.** $39.25°$
**58.** $46.75°$
**59.** $126.76°$
**60.** $174.255°$
**61.** $-18.515°$
**62.** $-25.485°$
**63.** $31.4296°$
**64.** $59.0854°$
**65.** $89.9004°$
**66.** $102.3771°$
**67.** $178.5994°$
**68.** $122.6853°$

*Find the angle of least positive measure (not equal to the given measure) coterminal with each angle. See Example 5.*

**69.** $32°$
**70.** $86°$
**71.** $26° \ 30'$
**72.** $58° \ 40'$
**73.** $-40°$
**74.** $-98°$
**75.** $-125°$
**76.** $-203°$
**77.** $361°$
**78.** $541°$
**79.** $-361°$
**80.** $-541°$
**81.** $539°$
**82.** $699°$
**83.** $850°$
**84.** $1000°$
**85.** $5280°$
**86.** $8440°$
**87.** $-5280°$
**88.** $-8440°$

*Give two positive and two negative angles that are coterminal with the given quadrantal angle.*

**89.** 90°  **90.** 180°  **91.** 0°  **92.** 270°

*Give an expression that generates all angles coterminal with each angle. Let n represent any integer.*

**93.** 30°  **94.** 45°  **95.** 135°  **96.** 225°

**97.** −90°  **98.** −180°  **99.** 0°  **100.** 360°

**101.** Explain why the answers to Exercises 99 and 100 give the same set of angles.

**102.** *Multiple Choice* Which two of the following are not coterminal with $r°$?

  **A.** $360° + r°$  **B.** $r° − 360°$  **C.** $360° − r°$  **D.** $r° + 180°$

*Concept Check Sketch each angle in standard position. Draw an arrow representing the correct amount of rotation. Find the measure of two other angles, one positive and one negative, that are coterminal with the given angle. Give the quadrant of each angle, if applicable.*

**103.** 75°  **104.** 89°  **105.** 174°  **106.** 234°
**107.** 300°  **108.** 512°  **109.** −61°  **110.** −159°
**111.** 90°  **112.** 180°  **113.** −90°  **114.** −180°

*Concept Check Locate each point in a coordinate system. Draw a ray from the origin through the given point. Indicate with an arrow the angle in standard position having least positive measure. Then find the distance r from the origin to the point, using the distance formula from Section 8.3.*

**115.** $(-3, -3)$  **116.** $(4, -4)$  **117.** $(-3, -5)$  **118.** $(-5, 2)$
**119.** $(\sqrt{2}, -\sqrt{2})$  **120.** $(-2\sqrt{2}, 2\sqrt{2})$  **121.** $(-1, \sqrt{3})$  **122.** $(\sqrt{3}, 1)$
**123.** $(-2, 2\sqrt{3})$  **124.** $(4\sqrt{3}, -4)$  **125.** $(0, -4)$  **126.** $(0, 2)$

*Solve each problem. See Example 6.*

**127.** A turntable in a shop makes 45 revolutions per min. How many revolutions does it make per second?

**128.** A windmill makes 90 revolutions per min. How many revolutions does it make per second?

**129.** A tire is rotating 600 times per min. Through how many degrees does a point on the edge of the tire move in $\frac{1}{2}$ sec?

**130.** An airplane propeller rotates 1000 times per min. Find the number of degrees that a point on the edge of the propeller will rotate in 1 sec.

**131.** A pulley rotates through 75° in 1 min. How many rotations does the pulley make in an hour?

**132.** One student in a surveying class measures an angle as 74.25°, while another student measures the same angle as 74° 20′. Find the difference between these measurements, both to the nearest minute and to the nearest hundredth of a degree.

# 14.2 Angle Relationships and Similar Triangles

**OBJECTIVES**

1. Find the measures of vertical angles and angles that are formed when parallel lines are intersected by a transversal.
2. Apply the angle sum of a triangle property.
3. Classify triangles.
4. Find the unknown angle measures and side lengths in similar triangles.

**OBJECTIVE 1** Find the measures of vertical angles and angles that are formed when parallel lines are intersected by a transversal. In Figure 13, we extended the sides of angle *NMP* to form another angle, *RMQ*. The pair of angles *NMP* and *RMQ* are called **vertical angles.** Another pair of vertical angles, *NMQ* and *PMR*, are also formed. Vertical angles have the following important property.

### Vertical Angles

Vertical angles have equal measures.

**Parallel lines** are lines that lie in the same plane and do not intersect. Figure 14 shows parallel lines *m* and *n*. When a line *q* intersects two parallel lines, *q* is called a **transversal.** In Figure 14, the transversal intersecting the parallel lines forms eight angles, indicated by numbers.

We learn in geometry that the degree measures of angles 1 through 8 in Figure 14 possess some special properties. The following chart gives the names of these angles and rules about their measures.

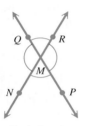

Vertical angles
**FIGURE 13**

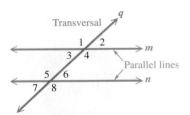

**FIGURE 14**

| Name | Sketch | Rule |
|---|---|---|
| Alternate interior angles | (angles 5 and 4; also 3 and 6) | Angle measures are equal. |
| Alternate exterior angles | (angles 1 and 8; also 2 and 7) | Angle measures are equal. |
| Interior angles on same side of transversal | (angles 6 and 4; also 3 and 5) | Angle measures add to 180°. |
| Corresponding angles | (angles 2 and 6; also 1 and 5, 3 and 7, 4 and 8) | Angle measures are equal. |

### EXAMPLE 1  Finding Angle Measures

Find the measures of angles 1, 2, 3, and 4 in Figure 15, given that lines *m* and *n* are parallel.

**Solution** Angles 1 and 4 are alternate exterior angles, so they are equal.

$$3x + 2 = 5x - 40$$
$$42 = 2x \qquad \text{Subtract } 3x; \text{ add } 40.$$
$$21 = x \qquad \text{Divide by 2.}$$

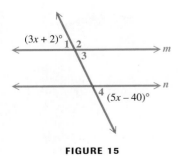

**FIGURE 15**

Angle 1 has measure

$$3x + 2 = 3 \cdot 21 + 2 = 65°, \quad \text{Substitute 21 for } x.$$

and angle 4 has measure

$$5x - 40 = 5 \cdot 21 - 40 = 65°. \quad \text{Substitute 21 for } x.$$

Angle 2 is the supplement of a 65° angle, so it has measure

$$180° - 65° = 115°.$$

Angle 3 is a vertical angle to angle 1, so its measure is 65°. (There are other ways to determine these measures.)

**NOW TRY** Exercises 3 and 11.

**OBJECTIVE 2** Apply the angle sum of a triangle property. An important property of triangles, first proved by Greek geometers, deals with the sum of the measures of the angles of any triangle.

### Angle Sum of a Triangle

*The sum of the measures of the angles of any triangle is **180°**.*

While it is not an actual proof, we give a rather convincing argument for the truth of this statement, using any size triangle cut from a piece of paper. Tear each corner from the triangle, as suggested in Figure 16(a). You should be able to rearrange the pieces so that the three angles form a straight angle, which has measure 180°, as shown in Figure 16(b).

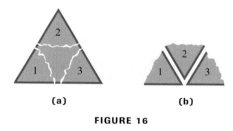

**FIGURE 16**

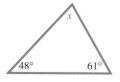

FIGURE 17

**EXAMPLE 2** Applying the Angle Sum of a Triangle Property

The measures of two of the angles of a triangle are 48° and 61°. (See Figure 17.) Find the measure of the third angle, $x$.

**Solution**   $48° + 61° + x = 180°$   The sum of the angles is 180°.
$109° + x = 180°$   Add.
$x = 71°$   Subtract 109°.

The third angle of the triangle measures 71°.

**NOW TRY** Exercises 5 and 15.

**OBJECTIVE 3** Classify triangles. We classify triangles according to angles and sides, as shown on the top of the next page.

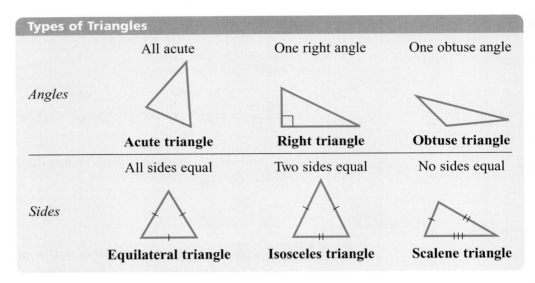

NOW TRY Exercises 25, 27, and 31.

**OBJECTIVE 4** Find the unknown angle measures and side lengths in similar triangles. **Similar triangles** are triangles of exactly the same shape but not necessarily the same size. Figure 18 shows three pairs of similar triangles. The two triangles in Figure 18(c) not only have the same shape but also the same size. Triangles that are both the same size and the same shape are called **congruent triangles.** If two triangles are congruent, then it is possible to pick one of them up and place it on top of the other so that they coincide. *If two triangles are congruent, then they must be similar. However, two similar triangles need not be congruent.*

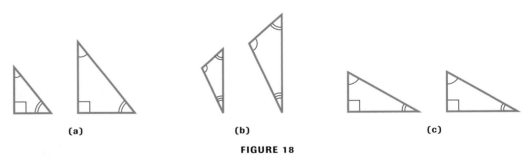

FIGURE 18

As shown in the figure, the triangular supports for a child's swing set are congruent (and thus similar) triangles, machine-produced with exactly the same dimensions each time. These supports are just one example of similar triangles. The supports of a long bridge, all the same shape but decreasing in size toward the center of the bridge, are examples of similar (but not congruent) triangles.

Suppose a correspondence between two triangles $ABC$ and $DEF$ is set up as shown in Figure 19.

Angle $A$ corresponds to angle $D$.
Angle $B$ corresponds to angle $E$.
Angle $C$ corresponds to angle $F$.
Side $AB$ corresponds to side $DE$.
Side $BC$ corresponds to side $EF$.
Side $AC$ corresponds to side $DF$.

The small arcs found at the angles in Figure 19 denote the corresponding angles in the triangles.

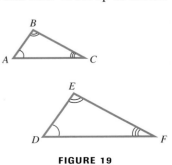

FIGURE 19

### Conditions for Similar Triangles

For triangle *ABC* to be similar to triangle *DEF*, the following conditions must hold.

1. Corresponding angles must have the same measure.
2. Corresponding sides must be proportional. (That is, the ratios of their corresponding sides must be equal.)

**NOW TRY** Exercise 39.

### EXAMPLE 3  Finding Angle Measures in Similar Triangles

In Figure 20, triangles *ABC* and *NMP* are similar. Find the measures of angles *B* and *C*.

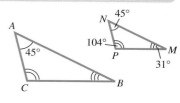

FIGURE 20

**Solution**  Since the triangles are similar, corresponding angles have the same measure. Since *C* corresponds to *P* and *P* measures 104°, angle *C* also measures 104°. Since angles *B* and *M* correspond, *B* measures 31°.

**NOW TRY** Exercise 45.

### EXAMPLE 4  Finding Side Lengths in Similar Triangles

Given that triangle *ABC* and triangle *DFE* in Figure 21 are similar, find the lengths of the unknown sides of triangle *DFE*.

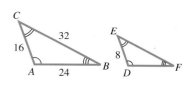

FIGURE 21

**Solution**  Similar triangles have corresponding sides in proportion. Use this fact to find the unknown side lengths in triangle *DFE*.

Side *DF* of triangle *DFE* corresponds to side *AB* of triangle *ABC*, and sides *DE* and *AC* correspond. This leads to the proportion

$$\frac{8}{16} = \frac{DF}{24}.$$

Recall this property of proportions from algebra.

$$\text{If} \quad \frac{a}{b} = \frac{c}{d}, \quad \text{then} \quad ad = bc.$$

We use this property to solve the equation for *DF*.

$$\frac{8}{16} \bowtie \frac{DF}{24}$$

$8 \cdot 24 = 16 \cdot DF$

$192 = 16 \cdot DF$  Multiply.

$12 = DF$  Divide by 16.

Side *DF* has length 12.

Side *EF* corresponds to *CB*. This leads to another proportion.

$$\frac{8}{16} = \frac{EF}{32}$$

$$8 \cdot 32 = 16 \cdot EF$$

$$16 = EF \qquad \text{Solve for } EF.$$

Side *EF* has length 16.

**NOW TRY** Exercise 51.

**EXAMPLE 5** Finding the Height of a Flagpole

Firefighters at the Morganza Fire Station need to measure the height of the station flagpole. They find that at the instant when the shadow of the station is 18 m long, the shadow of the flagpole is 99 ft long. The station is 10 m high. Find the height of the flagpole.

**Solution** Figure 22 shows the information given in the problem. The two triangles are similar, so corresponding sides are in proportion.

$$\frac{MN}{10} = \frac{99}{18}$$

$$\frac{MN}{10} = \frac{11}{2} \qquad \text{Lowest terms}$$

$$MN \cdot 2 = 10 \cdot 11$$

$$MN = 55 \qquad \text{Solve for } MN.$$

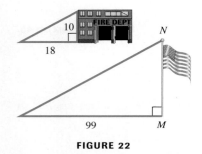

**FIGURE 22**

The flagpole is 55 ft high.

**NOW TRY** Exercise 55.

## 14.2 Exercises

1. *Concept Check* Use the given figure to find the measures of the numbered angles, given that lines *m* and *n* are parallel.

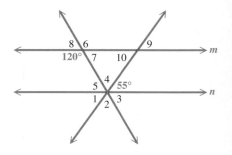

2. Consider Figure 14. If the measure of one of the angles is known, can the measures of the remaining seven angles be determined? Explain.

*Find the measure of each marked angle. In Exercises 11–14, m and n are parallel. See Examples 1 and 2.*

**3.**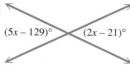

**4.**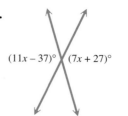

**5.**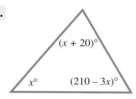

**6.**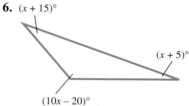

**7.**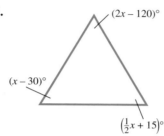

**8.**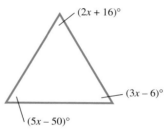

**9.**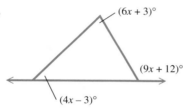

**10.**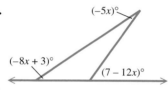

**11.**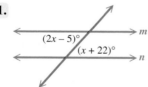

**12.**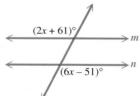

**13.**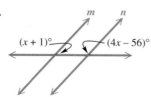

**14.**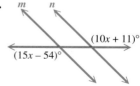

*The measures of two angles of a triangle are given. Find the measure of the third angle. See Example 2.*

**15.** 37°, 52°  **16.** 29°, 104°  **17.** 147° 12′, 30° 19′

**18.** 136° 50′, 41° 38′  **19.** 74.2°, 80.4°  **20.** 29.6°, 49.7°

**21.** 51° 20′ 14″, 106° 10′ 12″  **22.** 17° 41′ 13″, 96° 12′ 10″

**23.** Can a triangle have angles of measures 85° and 100°? Explain.

**24.** Can a triangle have two obtuse angles? Explain.

*Concept Check*  *Classify each triangle in Exercises 25–36 as* acute, right, *or* obtuse. *Also classify each as* equilateral, isosceles, *or* scalene.

**25.**    **26.** 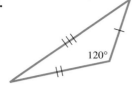   **27.**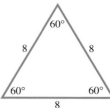

**28.** 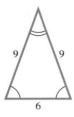   **29.** 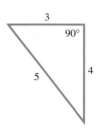   **30.**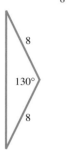

**31.** 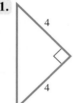   **32.**    **33.**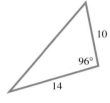

**34.** 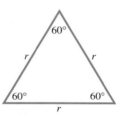   **35.** 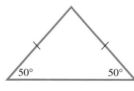   **36.**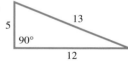

**37.** Write a definition of *isosceles right triangle*.

**38.** Must all equilateral triangles be similar? Explain.

*Concept Check*  *Name the corresponding angles and the corresponding sides of each pair of similar triangles.*

**39.** 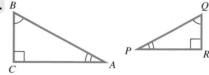   **40.**

**41.** (*EA* is parallel to *CD*.) 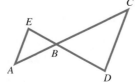   **42.** (*HK* is parallel to *EF*.)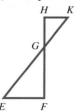

*Find all unknown angle measures in each pair of similar triangles. See Example 3.*

**43.**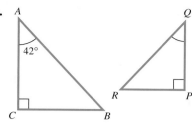

**44.**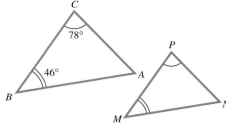

**45.**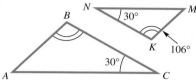

**46.**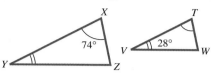

**47.**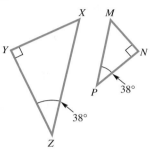

**48.**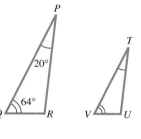

*Find the unknown side lengths labeled with a variable in each pair of similar triangles. See Example 4.*

**49.**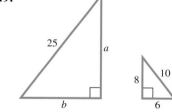

**50.**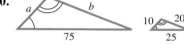

**51.**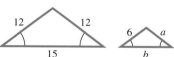

**52.**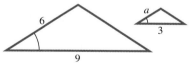

**53.**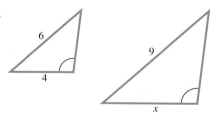

**54.**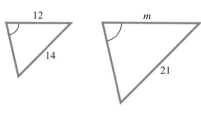

*Solve each problem. See Example 5.*

**55.** A tree casts a shadow 45 m long. At the same time, the shadow cast by a vertical 2-m stick is 3 m long. Find the height of the tree.

**56.** A forest fire lookout tower casts a shadow 180 ft long at the same time that the shadow of a 9-ft truck is 15 ft long. Find the height of the tower.

**57.** On a photograph of a triangular piece of land, the lengths of the three sides are 4 cm, 5 cm, and 7 cm, respectively. The shortest side of the actual piece of land is 400 m long. Find the lengths of the other two sides.

**58.** The Biloxi lighthouse in the figure casts a shadow 28 m long at 7 P.M. At the same time, the shadow of the lighthouse keeper, who is 1.75 m tall, is 3.5 m long. How tall is the lighthouse?

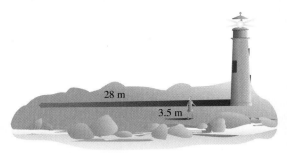

Not to scale

**59.** A house is 15 ft tall. Its shadow is 40 ft long at the same time the shadow of a nearby building is 300 ft long. Find the height of the building.

**60.** Assume that Lincoln was $6\frac{1}{3}$ ft tall and his head $\frac{3}{4}$ ft long. Knowing that the carved head of Lincoln at Mt. Rushmore is 60 ft tall, find how tall his entire body would be if it were carved into the mountain.

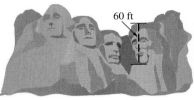

**61.** Two quadrilaterals (four-sided figures) are similar. The lengths of the three shortest sides of the first quadrilateral are 18 cm, 24 cm, and 32 cm. The lengths of the two longest sides of the second quadrilateral are 48 cm and 60 cm. Find the unknown lengths of the sides of these two figures.

**62.** By drawing lines on a map, a triangle can be formed by the cities of Phoenix, Tucson, and Yuma. On the map, the distance between Phoenix and Tucson is 8 cm, the distance between Phoenix and Yuma is 12 cm, and the distance between Tucson and Yuma is 17 cm. The actual straight-line distance from Phoenix to Yuma is 230.0 km. Find the distances between the other pairs of cities to the nearest tenth of a kilometer.

*In each diagram, there are two similar triangles. Find the unknown measurement. (Hint: In the sketch for Exercise 63, the side of length 100 in the small triangle corresponds to the side of the length 100 + 120 = 220 in the large triangle.)*

**63.**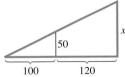

**64.**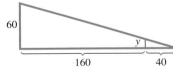

**65.**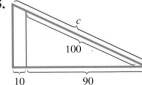

**66.**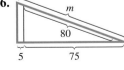

*In each figure, there are two similar triangles. Find the value of each variable.*

**67.**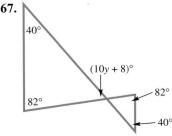

**68.**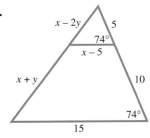

## 14.3 Trigonometric Functions

**OBJECTIVES**
1. Find the values of the six trigonometric functions of an angle.
2. Find the trigonometric function values of a quadrantal angle.

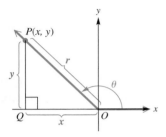

FIGURE 23

**OBJECTIVE 1** Find the values of the six trigonometric functions of an angle. The study of trigonometry covers the six **trigonometric functions** defined here. To define these functions, we start with an angle $\theta$ in standard position and choose any point $P$ having coordinates $(x, y)$ on the terminal side of angle $\theta$. (The point $P$ must not be the vertex of the angle.) See Figure 23. A perpendicular from $P$ to the $x$-axis at point $Q$ determines a right triangle, having vertices at $O$, $P$, and $Q$. We find the distance $r$ from $P(x, y)$ to the origin, $(0, 0)$, using the distance formula.

$$r = \sqrt{(x - 0)^2 + (y - 0)^2} = \sqrt{x^2 + y^2}$$

*Notice that for any angle $\theta$, $r > 0$.*

The six trigonometric functions of angle $\theta$ are **sine, cosine, tangent, cotangent, secant,** and **cosecant.** In the following definitions, we use the customary abbreviations for the names of these functions: **sin, cos, tan, cot, sec,** and **csc.**

### Trigonometric Functions

Let $(x, y)$ be a point other than the origin on the terminal side of an angle $\theta$ in standard position. The distance from the point to the origin is $r = \sqrt{x^2 + y^2}$. The six trigonometric functions of $\theta$ are defined as follows.

$$\sin \theta = \frac{y}{r} \qquad \cos \theta = \frac{x}{r} \qquad \tan \theta = \frac{y}{x} \ (x \neq 0)$$

$$\csc \theta = \frac{r}{y} \ (y \neq 0) \qquad \sec \theta = \frac{r}{x} \ (x \neq 0) \qquad \cot \theta = \frac{x}{y} \ (y \neq 0)$$

### EXAMPLE 1  Finding Function Values of an Angle

The terminal side of an angle $\theta$ in standard position passes through the point $(8, 15)$. Find the values of the six trigonometric functions of angle $\theta$.

**Solution** Figure 24 shows angle $\theta$ and the triangle formed by dropping a perpendicular from the point $(8, 15)$ to the $x$-axis. The point $(8, 15)$ is 8 units to the right of the $y$-axis and 15 units above the $x$-axis, so $x = 8$ and $y = 15$. Since $r = \sqrt{x^2 + y^2}$,

$$r = \sqrt{8^2 + 15^2} = \sqrt{64 + 225} = \sqrt{289} = 17.$$

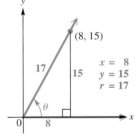

FIGURE 24

We can now find the values of the six trigonometric functions of angle $\theta$.

$$\sin \theta = \frac{y}{r} = \frac{15}{17} \qquad \cos \theta = \frac{x}{r} = \frac{8}{17} \qquad \tan \theta = \frac{y}{x} = \frac{15}{8}$$

$$\csc \theta = \frac{r}{y} = \frac{17}{15} \qquad \sec \theta = \frac{r}{x} = \frac{17}{8} \qquad \cot \theta = \frac{x}{y} = \frac{8}{15}$$

**NOW TRY** Exercise 5.

### EXAMPLE 2  Finding Function Values of an Angle

The terminal side of an angle $\theta$ in standard position passes through the point $(-3, -4)$. Find the values of the six trigonometric functions of angle $\theta$.

**Solution**  As shown in Figure 25, $x = -3$ and $y = -4$. The value of $r$ is

$$r = \sqrt{(-3)^2 + (-4)^2} = \sqrt{25} = 5. \quad \text{Remember that } r > 0.$$

Then use the definitions of the trigonometric functions.

$$\sin \theta = \frac{-4}{5} = -\frac{4}{5} \qquad \cos \theta = \frac{-3}{5} = -\frac{3}{5} \qquad \tan \theta = \frac{-4}{-3} = \frac{4}{3}$$

$$\csc \theta = \frac{5}{-4} = -\frac{5}{4} \qquad \sec \theta = \frac{5}{-3} = -\frac{5}{3} \qquad \cot \theta = \frac{-3}{-4} = \frac{3}{4}$$

**NOW TRY** Exercise 19.

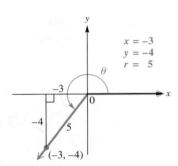

FIGURE 25

We can find the six trigonometric functions using *any* point other than the origin on the terminal side of an angle. To see why any point may be used, refer to Figure 26, which shows an angle $\theta$ and two distinct points on its terminal side. Point $P$ has coordinates $(x, y)$, and point $P'$ (read "*P*-prime") has coordinates $(x', y')$. Let $r$ be the length of the hypotenuse of triangle $OPQ$, and let $r'$ be the length of the hypotenuse of triangle $OP'Q'$. Since corresponding sides of similar triangles are proportional,

$$\frac{y}{r} = \frac{y'}{r'},$$

so $\sin \theta = \frac{y}{r}$ is the same no matter which point is used to find it. A similar result holds for the other five trigonometric functions.

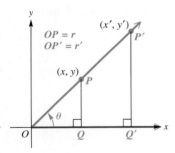

FIGURE 26

We can also find the trigonometric function values of an angle if we know the equation of the line coinciding with the terminal ray. Recall from algebra that the graph of the equation

$$Ax + By = 0$$

is a line that passes through the origin. If we restrict $x$ to have only nonpositive or only nonnegative values, we obtain as the graph a ray with endpoint at the origin. For example, the graph of $x + 2y = 0$, $x \geq 0$, shown in Figure 27, is a ray that can serve as the terminal side of an angle $\theta$ in standard position. By choosing a point on the ray, we can find the trigonometric function values of the angle.

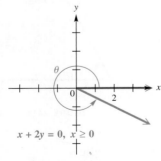

FIGURE 27

### EXAMPLE 3  Finding Function Values of an Angle

Find the six trigonometric function values of the angle $\theta$ in standard position, if the terminal side of $\theta$ is defined by $x + 2y = 0$, $x \geq 0$.

**Solution**  The angle is shown in Figure 28 on the next page. We can use *any* point except $(0, 0)$ on the terminal side of $\theta$ to find the trigonometric function values. We choose $x = 2$ and find the corresponding $y$-value.

$$x + 2y = 0, \, x \geq 0$$
$$2 + 2y = 0 \qquad \text{Let } x = 2.$$
$$2y = -2 \qquad \text{Subtract 2.}$$
$$y = -1 \qquad \text{Divide by 2.}$$

**SECTION 14.3** Trigonometric Functions  **679**

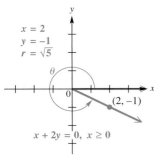

**FIGURE 28**

The point $(2, -1)$ lies on the terminal side, and the corresponding value of $r$ is $r = \sqrt{2^2 + (-1)^2} = \sqrt{5}.$ Now we use the definitions of the trigonometric functions.

$$\sin \theta = \frac{y}{r} = \frac{-1}{\sqrt{5}} = \frac{-1}{\sqrt{5}} \cdot \frac{\sqrt{5}}{\sqrt{5}} = -\frac{\sqrt{5}}{5}$$

Multiply by $\frac{\sqrt{5}}{\sqrt{5}}$, which equals 1, to rationalize the denominators.

$$\cos \theta = \frac{x}{r} = \frac{2}{\sqrt{5}} = \frac{2}{\sqrt{5}} \cdot \frac{\sqrt{5}}{\sqrt{5}} = \frac{2\sqrt{5}}{5}$$

$$\tan \theta = \frac{y}{x} = \frac{-1}{2} = -\frac{1}{2}$$

$$\csc \theta = \frac{r}{y} = \frac{\sqrt{5}}{-1} = -\sqrt{5} \qquad \sec \theta = \frac{r}{x} = \frac{\sqrt{5}}{2} \qquad \cot \theta = \frac{x}{y} = \frac{2}{-1} = -2$$

**NOW TRY** Exercise 45.

Recall that when the equation of a line is written in slope-intercept form $y = mx + b$, the coefficient of $x$ is the slope of the line. In Example 3, the equation $x + 2y = 0$ can be written as $y = -\frac{1}{2}x$, so the slope is $-\frac{1}{2}$. Notice that $\tan \theta = -\frac{1}{2}$. *In general, it is true that* $m = \tan \theta$.

**NOTE** The trigonometric function values we found in Examples 1–3 are *exact*. If we were to use a calculator to approximate these values, the decimal results would not be acceptable if exact values were required.

**OBJECTIVE 2** Find the trigonometric function values of a quadrantal angle. If the terminal side of an angle in standard position lies along the $y$-axis, any point on this terminal side has $x$-coordinate 0. Similarly, an angle with terminal side on the $x$-axis has $y$-coordinate 0 for any point on the terminal side. Since the values of $x$ and $y$ appear in the denominators of some trigonometric functions, and since a fraction is undefined if its denominator is 0, some trigonometric function values of quadrantal angles (i.e., those with terminal side on an axis) are undefined.

When determining trigonometric function values of quadrantal angles, Figure 29 can help find the ratios. Because *any* point on the terminal side can be used, it is convenient to choose the point one unit from the origin, with $r = 1$. (In **Chapter 16** we extend this idea to the *unit circle*.)

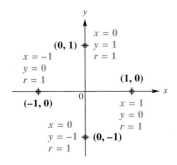

**FIGURE 29**

To find the function values of a quadrantal angle, determine the position of the terminal side, choose the one of these four points that lies on this terminal side, and then use the definitions involving $x$, $y$, and $r$.

**EXAMPLE 4** Finding Function Values of Quadrantal Angles

Find the values of the six trigonometric functions for each angle.

**(a)** an angle of $90°$

**(b)** an angle $\theta$ in standard position with terminal side through $(-3, 0)$

**Solution**

**(a)** The sketch in Figure 30, on the next page, shows that the terminal side passes through $(0, 1)$. So $x = 0, y = 1,$ and $r = 1$. Thus,

$$\sin 90° = \frac{1}{1} = 1 \qquad \cos 90° = \frac{0}{1} = 0 \qquad \tan 90° = \frac{1}{0} \text{ (undefined)}$$

$$\csc 90° = \frac{1}{1} = 1 \qquad \sec 90° = \frac{1}{0} \text{ (undefined)} \qquad \cot 90° = \frac{0}{1} = 0.$$

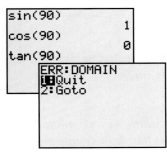

A calculator in degree mode returns the correct values for sin 90° and cos 90°. The second screen shows an ERROR message for tan 90°, because 90° is not in the domain of the tangent function.

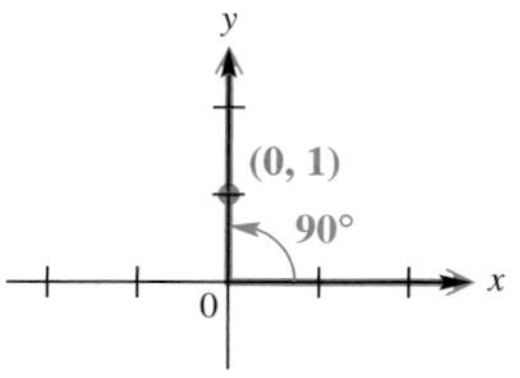

FIGURE 30

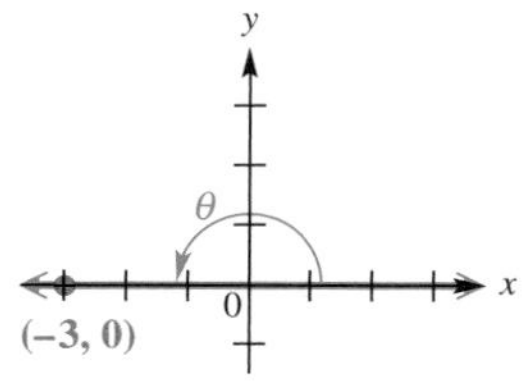

FIGURE 31

(b) Figure 31 shows the angle. Here, $x = -3$, $y = 0$, and $r = 3$, so the trigonometric functions have the following values.

$$\sin \theta = \frac{0}{3} = 0 \qquad \cos \theta = \frac{-3}{3} = -1 \qquad \tan \theta = \frac{0}{-3} = 0$$

$$\csc \theta = \frac{3}{0} \text{ (undefined)} \qquad \sec \theta = \frac{3}{-3} = -1 \qquad \cot \theta = \frac{-3}{0} \text{ (undefined)}$$

(Verify that these values can also be found by using the point $(-1, 0)$.)

**NOW TRY** Exercises 13, 57, 59, 63, and 65.

The conditions under which the trigonometric function values of quadrantal angles are undefined are summarized here.

### Undefined Function Values

If the terminal side of a quadrantal angle lies along the *y*-axis, then the tangent and secant functions are undefined. If it lies along the *x*-axis, then the cotangent and cosecant functions are undefined.

The function values of the most commonly used quadrantal angles, 0°, 90°, 180°, 270°, and 360°, are summarized in the table. They can be determined when needed by using Figure 29 and the method of Example 4(a).

**FUNCTION VALUES OF QUADRANTAL ANGLES**

| $\theta$ | $\sin \theta$ | $\cos \theta$ | $\tan \theta$ | $\cot \theta$ | $\sec \theta$ | $\csc \theta$ |
|---|---|---|---|---|---|---|
| 0° | 0 | 1 | 0 | Undefined | 1 | Undefined |
| 90° | 1 | 0 | Undefined | 0 | Undefined | 1 |
| 180° | 0 | −1 | 0 | Undefined | −1 | Undefined |
| 270° | −1 | 0 | Undefined | 0 | Undefined | −1 |
| 360° | 0 | 1 | 0 | Undefined | 1 | Undefined |

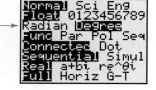

TI-83 Plus

The values given in this table can be found with a calculator that has trigonometric function keys. *Make sure the calculator is set in degree mode.*

TI-84 Plus

FIGURE 32

**CAUTION** *One of the most common errors involving calculators in trigonometry occurs when the calculator is set for radian measure, rather than degree measure.* (Radian measure of angles is discussed in **Chapter 16**.) Be sure you know how to set your calculator in degree mode. See Figure 32, which illustrates degree mode for TI-83/84 Plus calculators.

## 14.3 Exercises

**Concept Check** Sketch an angle $\theta$ in standard position such that $\theta$ has the least possible positive measure, and the given point is on the terminal side of $\theta$. Then find the values of the six trigonometric functions for each angle. Rationalize denominators when applicable. See Examples 1, 2, and 4.

1. $(5, -12)$
2. $(-12, -5)$
3. $(-3, 4)$
4. $(-4, -3)$
5. $(-8, 15)$
6. $(15, -8)$
7. $(7, -24)$
8. $(-24, -7)$
9. $(0, 2)$
10. $(0, 5)$
11. $(-4, 0)$
12. $(-5, 0)$
13. $(0, -4)$
14. $(0, -3)$
15. $(1, \sqrt{3})$
16. $(-1, \sqrt{3})$
17. $(\sqrt{2}, \sqrt{2})$
18. $(-\sqrt{2}, -\sqrt{2})$
19. $(-2\sqrt{3}, -2)$
20. $(-2\sqrt{3}, 2)$

21. For any nonquadrantal angle $\theta$, $\sin \theta$ and $\csc \theta$ will have the same sign. Explain why.

22. *Concept Check* How is the value of $r$ interpreted geometrically in the definitions of the sine, cosine, secant, and cosecant functions?

23. *Concept Check* If $\cot \theta$ is undefined, what is the value of $\tan \theta$?

24. *Concept Check* If the terminal side of an angle $\theta$ is in quadrant III, what is the sign of each of the trigonometric function values of $\theta$?

*Concept Check* Suppose that the point $(x, y)$ is in the indicated quadrant. Decide whether the given ratio is positive or negative. Recall that $r = \sqrt{x^2 + y^2}$. (Hint: Drawing a sketch may help.)

25. II, $\dfrac{x}{r}$
26. III, $\dfrac{y}{r}$
27. IV, $\dfrac{y}{x}$
28. IV, $\dfrac{x}{y}$
29. II, $\dfrac{y}{r}$
30. III, $\dfrac{x}{r}$
31. IV, $\dfrac{x}{r}$
32. IV, $\dfrac{y}{r}$
33. II, $\dfrac{x}{y}$
34. II, $\dfrac{y}{x}$
35. III, $\dfrac{y}{x}$
36. III, $\dfrac{x}{y}$
37. IV, $\dfrac{x}{y}$
38. IV, $\dfrac{y}{x}$
39. I, $\dfrac{x}{y}$
40. I, $\dfrac{y}{x}$
41. I, $\dfrac{y}{r}$
42. I, $\dfrac{x}{r}$
43. I, $\dfrac{r}{x}$
44. I, $\dfrac{r}{y}$

In Exercises 45–54, an equation of the terminal side of an angle $\theta$ in standard position is given with a restriction on $x$. Sketch the least positive such angle $\theta$, and find the values of the six trigonometric functions of $\theta$. See Example 3.

45. $2x + y = 0, x \geq 0$
46. $3x + 5y = 0, x \geq 0$
47. $-6x - y = 0, x \leq 0$
48. $-5x - 3y = 0, x \leq 0$
49. $-4x + 7y = 0, x \leq 0$
50. $6x - 5y = 0, x \geq 0$
51. $x + y = 0, x \geq 0$
52. $x - y = 0, x \geq 0$
53. $-\sqrt{3}x + y = 0, x \leq 0$
54. $\sqrt{3}x + y = 0, x \leq 0$

To work Exercises 55–72, begin by reproducing the graph in Figure 29 on page 679. Keep in mind that for each of the four points labeled in the figure, $r = 1$. For each quadrantal angle, identify the appropriate values of $x$, $y$, and $r$ to find the indicated function value. If it is undefined, say so. See Example 4.

55. $\cos 90°$
56. $\sin 90°$
57. $\tan 180°$
58. $\cot 90°$
59. $\sec 180°$
60. $\csc 270°$
61. $\sin(-270°)$
62. $\cos(-90°)$
63. $\cot 540°$

64. tan 450°  65. csc(−450°)  66. sec(−540°)
67. sin 1800°  68. cos 1800°  69. csc 1800°
70. cot 1800°  71. sec 1800°  72. tan 1800°

*Use the trigonometric function values of quadrantal angles given in this section to evaluate each expression. An expression such as* $\cot^2 90°$ *means* $(\cot 90°)^2$, *which is equal to* $0^2 = 0$.

73. $\cos 90° + 3 \sin 270°$
74. $\tan 0° - 6 \sin 90°$
75. $3 \sec 180° - 5 \tan 360°$
76. $4 \csc 270° + 3 \cos 180°$
77. $\tan 360° + 4 \sin 180° + 5 \cos^2 180°$
78. $2 \sec 0° + 4 \cot^2 90° + \cos 360°$
79. $\sin^2 180° + \cos^2 180°$
80. $\sin^2 360° + \cos^2 360°$
81. $\sec^2 180° - 3 \sin^2 360° + \cos 180°$
82. $5 \sin^2 90° + 2 \cos^2 270° - \tan 360°$
83. $-2 \sin^4 0° + 3 \tan^2 0°$
84. $-3 \sin^4 90° + 4 \cos^3 180°$

*If n is an integer,* $n \cdot 180°$ *represents an integer multiple of* $180°$, $(2n + 1) \cdot 90°$ *represents an odd integer multiple of* $90°$, *and so on. Decide whether each expression is equal to* 0, 1, −1, *or is undefined.*

85. $\cos[(2n + 1) \cdot 90°]$
86. $\sin[n \cdot 180°]$
87. $\tan[n \cdot 180°]$
88. $\tan[(2n + 1) \cdot 90°]$
89. $\sin[270° + n \cdot 360°]$
90. $\cot[n \cdot 180°]$
91. $\cot[(2n + 1) \cdot 90°]$
92. $\cos[n \cdot 360°]$
93. $\sec[(2n + 1) \cdot 90°]$
94. $\csc[n \cdot 180°]$

*Concept Check* In later chapters we will study trigonometric functions of angles other than quadrantal angles, such as 15°, 30°, 60°, 75°, and so on. To prepare for some important concepts, provide conjectures in Exercises 95–98. Be sure that your calculator is in degree mode.

95. The angles 15° and 75° are complementary. With your calculator determine sin 15° and cos 75°. Make a conjecture about the sines and cosines of complementary angles, and test your hypothesis with other pairs of complementary angles. (*Note:* This relationship will be discussed in detail in **Section 15.1**.)

96. The angles 25° and 65° are complementary. With your calculator determine tan 25° and cot 65°. Make a conjecture about the tangents and cotangents of complementary angles, and test your hypothesis with other pairs of complementary angles. (*Note:* This relationship will be discussed in detail in **Section 15.1**.)

97. With your calculator determine sin 10° and sin(−10°). Make a conjecture about the sines of an angle and its negative, and test your hypothesis with other angles. (*Note:* This relationship will be discussed in detail in **Section 18.1**.)

98. With your calculator determine cos 20° and cos(−20°). Make a conjecture about the cosines of an angle and its negative, and test your hypothesis with other angles. (*Note:* This relationship will be discussed in detail in **Section 18.1**.)

## 14.4 Using the Definitions of the Trigonometric Functions

**OBJECTIVES**

1. Use the reciprocal identities to find function values.
2. Determine the signs of the trigonometric functions of nonquadrantal angles.
3. Know the ranges of the trigonometric functions.
4. Find all function values given one value and the quadrant.
5. Use the Pythagorean and quotient identities to find function values.

Identities are equations that are true for all values of the variables for which all expressions are defined. Identities are studied in more detail in **Chapter 18**.

$$(x + y)^2 = x^2 + 2xy + y^2 \qquad 2(x + 3) = 2x + 6 \qquad \text{Identities}$$

**OBJECTIVE 1** Use the reciprocal identities to find function values. Recall the definition of a reciprocal: the **reciprocal** of the nonzero number $x$ is $\frac{1}{x}$. For example, the reciprocal of 2 is $\frac{1}{2}$, and the reciprocal of $\frac{8}{11}$ is $\frac{11}{8}$. There is no reciprocal for 0. Scientific calculators have a reciprocal key, usually labeled $\boxed{1/x}$ or $\boxed{x^{-1}}$. Using this key gives the reciprocal of any nonzero number entered in the display.

The definitions of the trigonometric functions in the previous section on page 677 were written so that functions in the same column are reciprocals of each other. Since $\sin \theta = \frac{y}{r}$ and $\csc \theta = \frac{r}{y}$,

$$\sin \theta = \frac{1}{\csc \theta} \qquad \text{and} \qquad \csc \theta = \frac{1}{\sin \theta}, \quad \text{provided } \sin \theta \neq 0.$$

Also, $\cos \theta$ and $\sec \theta$ are reciprocals, as are $\tan \theta$ and $\cot \theta$. The **reciprocal identities** hold for any angle $\theta$ that does not lead to a 0 denominator.

**Reciprocal Identities**

For all angles $\theta$ for which both functions are defined,

$$\sin \theta = \frac{1}{\csc \theta} \qquad \cos \theta = \frac{1}{\sec \theta} \qquad \tan \theta = \frac{1}{\cot \theta}$$

$$\csc \theta = \frac{1}{\sin \theta} \qquad \sec \theta = \frac{1}{\cos \theta} \qquad \cot \theta = \frac{1}{\tan \theta}.$$

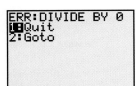

(a)

(b)

**FIGURE 33**

The screen in Figure 33(a) shows how to find $\csc 90°$, $\sec 180°$, and $\csc(-270°)$, using the appropriate reciprocal identities and the reciprocal key of a graphing calculator in degree mode. Attempting to find $\sec 90°$ by entering $\frac{1}{\cos 90°}$ produces an ERROR message, indicating the reciprocal is undefined. See Figure 33(b). Compare these results with the ones found in the table of quadrantal angle function values in **Section 14.3**.

**CAUTION** *Be sure not to use the inverse trigonometric function keys to find reciprocal function values.* For example,

$$\sin^{-1}(90°) \neq \frac{1}{\sin(90°)}$$

Inverse trigonometric functions are covered in **Section 15.3**.

**NOTE** Identities can be written in different forms. For example,

$$\sin \theta = \frac{1}{\csc \theta} \quad \text{can be written} \quad \csc \theta = \frac{1}{\sin \theta}, \quad \text{or} \quad (\sin \theta)(\csc \theta) = 1.$$

### EXAMPLE 1  Using the Reciprocal Identities

Find each function value.

**(a)** $\cos\theta$, given that $\sec\theta = \frac{5}{3}$

**(b)** $\sin\theta$, given that $\csc\theta = -\frac{\sqrt{12}}{2}$

**Solution**

**(a)** Since $\cos\theta$ is the reciprocal of $\sec\theta$,

$$\cos\theta = \frac{1}{\sec\theta} = \frac{1}{\frac{5}{3}} = 1 \div \frac{5}{3} = 1 \cdot \frac{3}{5} = \frac{3}{5}.$$  Simplify the complex fraction.

**(b)** $\sin\theta = \dfrac{1}{-\frac{\sqrt{12}}{2}}$    $\sin\theta = \frac{1}{\csc\theta}$

$= -\dfrac{2}{\sqrt{12}}$    Simplify the complex fraction as in part (a).

$= -\dfrac{2}{2\sqrt{3}}$    $\sqrt{12} = \sqrt{4\cdot 3} = 2\sqrt{3}$

$= -\dfrac{1}{\sqrt{3}}$    We are multiplying by $1 = \frac{\sqrt{3}}{\sqrt{3}}$.

$= -\dfrac{1}{\sqrt{3}}\cdot\dfrac{\sqrt{3}}{\sqrt{3}} = -\dfrac{\sqrt{3}}{3}$    Rationalize the denominator.

**NOW TRY** Exercises 1 and 9.

**OBJECTIVE 2** Determine the signs of the trigonometric functions of nonquadrantal angles. In the definitions of the trigonometric functions, $r$ is the distance from the origin to the point $(x, y)$. This distance is undirected, so $r > 0$. If we choose a point $(x, y)$ in quadrant I, then both $x$ and $y$ will be positive, and the values of all six functions will be positive.

A point $(x, y)$ in quadrant II has $x < 0$ and $y > 0$. This makes the values of sine and cosecant positive for quadrant II angles, while the other four functions take on negative values. Similar results can be obtained for the other quadrants.

***This important information is summarized here.***

**SIGNS OF FUNCTION VALUES**

| $\theta$ in Quadrant | $\sin\theta$ | $\cos\theta$ | $\tan\theta$ | $\cot\theta$ | $\sec\theta$ | $\csc\theta$ |
|---|---|---|---|---|---|---|
| I | + | + | + | + | + | + |
| II | + | − | − | − | − | + |
| III | − | − | + | + | − | − |
| IV | − | + | − | − | + | − |

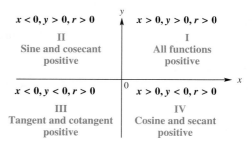

$x < 0, y > 0, r > 0$
II
Sine and cosecant positive

$x > 0, y > 0, r > 0$
I
All functions positive

$x < 0, y < 0, r > 0$
III
Tangent and cotangent positive

$x > 0, y < 0, r > 0$
IV
Cosine and secant positive

### EXAMPLE 2  Determining Signs of Functions of Nonquadrantal Angles

Determine the signs of the trigonometric functions of an angle in standard position with the given measure.

**(a)** 87°    **(b)** 300°    **(c)** −200°

## Solution

(a) An angle of 87° is in the first quadrant, with $x$, $y$, and $r$ all positive, so all of its trigonometric function values are positive.

(b) A 300° angle is in quadrant IV, so the cosine and secant are positive, while the sine, cosecant, tangent, and cotangent are negative.

(c) A $-200°$ angle is in quadrant II. The sine and cosecant are positive, and all other function values are negative.

**NOW TRY** Exercises 19, 21, and 25.

Because numbers that are reciprocals will always have the same sign, knowing the sign of a function value will automatically determine the sign of the reciprocal function value.

### EXAMPLE 3  Identifying the Quadrant of an Angle

Identify the quadrant (or possible quadrants) of an angle $\theta$ that satisfies the given conditions.

(a) $\sin \theta > 0$, $\tan \theta < 0$ 

(b) $\cos \theta < 0$, $\sec \theta < 0$

## Solution

(a) Since $\sin \theta > 0$ in quadrants I and II and $\tan \theta < 0$ in quadrants II and IV, both conditions are met only in quadrant II.

(b) The cosine and secant functions are both negative in quadrants II and III, so in this case $\theta$ could be in either of these two quadrants.

**NOW TRY** Exercises 35 and 41.

**OBJECTIVE 3** Know the ranges of the trigonometric functions. Figure 34 shows an angle $\theta$ as it increases in measure from near 0° toward 90°. In each case, the value of $r$ is the same. As the measure of the angle increases, $y$ increases but never exceeds $r$, so $y \leq r$. Dividing both sides by the positive number $r$ gives $\frac{y}{r} \leq 1$.

In a similar way, angles in quadrant IV suggest that

$$-1 \leq \frac{y}{r},$$

so $$-1 \leq \frac{y}{r} \leq 1$$

and $-1 \leq \sin \theta \leq 1$.    $\frac{y}{r} = \sin \theta$ for any angle $\theta$.

Similarly, $-1 \leq \cos \theta \leq 1$.

The tangent of an angle is defined as $\frac{y}{x}$. It is possible that $x < y$, $x = y$, or $x > y$. Thus, $\frac{y}{x}$ can take any value, so **tan $\theta$ can be any real number, as can cot $\theta$.**

The functions sec $\theta$ and csc $\theta$ are reciprocals of the functions cos $\theta$ and sin $\theta$, respectively, making

$\sec \theta \leq -1$   or   $\sec \theta \geq 1$    and    $\csc \theta \leq -1$   or   $\csc \theta \geq 1$.

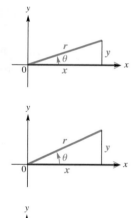

FIGURE 34

In summary, the ranges of the trigonometric functions are as follows.

**Ranges of Trigonometric Functions**

| Trigonometric Function of $\theta$ | Range (Set-Builder Notation) | Range (Interval Notation) |
|---|---|---|
| $\sin \theta$, $\cos \theta$ | $\{y \mid \lvert y \rvert \leq 1\}$ | $[-1, 1]$ |
| $\tan \theta$, $\cot \theta$ | $\{y \mid y \text{ is a real number}\}$ | $(-\infty, \infty)$ |
| $\sec \theta$, $\csc \theta$ | $\{y \mid \lvert y \rvert \geq 1\}$ | $(-\infty, -1] \cup [1, \infty)$ |

**EXAMPLE 4** Deciding Whether a Value is in the Range of a Trigonometric Function

Decide whether each statement is *possible* or *impossible*.

**(a)** $\sin \theta = 2.5$  **(b)** $\tan \theta = 110.47$  **(c)** $\sec \theta = 0.6$

**Solution**

**(a)** For any value of $\theta$, $-1 \leq \sin \theta \leq 1$. Since $2.5 > 1$, it is impossible to find a value of $\theta$ with $\sin \theta = 2.5$.

**(b)** Tangent can take on any real number value. Thus, $\tan \theta = 110.47$ is possible.

**(c)** Since $\lvert \sec \theta \rvert \geq 1$ for all $\theta$ for which the secant is defined, the statement $\sec \theta = 0.6$ is impossible.

**NOW TRY** Exercises 45, 49, and 51.

**OBJECTIVE 4** Find all function values given one value and the quadrant. The six trigonometric functions are defined in terms of $x$, $y$, and $r$, where the Pythagorean theorem shows that $r^2 = x^2 + y^2$ and $r > 0$. With these relationships, knowing the value of only one function and the quadrant in which the angle lies makes it possible to find the values of the other trigonometric functions.

**EXAMPLE 5** Finding All Function Values Given One Value and the Quadrant

Suppose that angle $\theta$ is in quadrant II and $\sin \theta = \frac{2}{3}$. Find the values of the other five trigonometric functions.

**Solution** Choose any point on the terminal side of angle $\theta$. For simplicity, since $\sin \theta = \frac{y}{r}$, choose the point with $r = 3$.

$$\sin \theta = \frac{2}{3} \qquad \text{Given value}$$

$$\frac{y}{r} = \frac{2}{3} \qquad \text{Substitute } \frac{y}{r} \text{ for } \sin \theta.$$

Since $\frac{y}{r} = \frac{2}{3}$ and $r = 3$, then $y = 2$. To find $x$, use the equation $x^2 + y^2 = r^2$.

$$x^2 + y^2 = r^2$$
$$x^2 + 2^2 = 3^2 \qquad \text{Substitute.}$$
$$x^2 + 4 = 9 \qquad \text{Apply exponents.}$$
$$x^2 = 5 \qquad \text{Subtract 4.}$$

**Remember both roots.**

$$x = \sqrt{5} \quad \text{or} \quad x = -\sqrt{5} \qquad \text{Square root property}$$

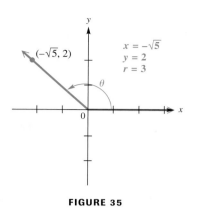

**FIGURE 35**

Since $\theta$ is in quadrant II, $x$ must be negative, as shown in Figure 35, so $x = -\sqrt{5}$, and the point $(-\sqrt{5}, 2)$ is on the terminal side of $\theta$. Now we can find the values of the remaining trigonometric functions.

$$\cos \theta = \frac{x}{r} = \frac{-\sqrt{5}}{3} = -\frac{\sqrt{5}}{3}$$

$$\sec \theta = \frac{r}{x} = \frac{3}{-\sqrt{5}} = -\frac{3}{\sqrt{5}} \cdot \frac{\sqrt{5}}{\sqrt{5}} = -\frac{3\sqrt{5}}{5}$$

$$\tan \theta = \frac{y}{x} = \frac{2}{-\sqrt{5}} = -\frac{2}{\sqrt{5}} \cdot \frac{\sqrt{5}}{\sqrt{5}} = -\frac{2\sqrt{5}}{5}$$

Remember to rationalize denominators.

$$\cot \theta = \frac{x}{y} = \frac{-\sqrt{5}}{2} = -\frac{\sqrt{5}}{2}$$

$$\csc \theta = \frac{r}{y} = \frac{3}{2}$$

**NOW TRY** Exercise 71.

**OBJECTIVE 5** Use the Pythagorean and quotient identities to find function values. We derive three new identities from the relationship $x^2 + y^2 = r^2$.

$$\frac{x^2}{r^2} + \frac{y^2}{r^2} = \frac{r^2}{r^2} \qquad \text{Divide by } r^2.$$

$$\left(\frac{x}{r}\right)^2 + \left(\frac{y}{r}\right)^2 = 1 \qquad \text{Power rule for exponents; } \frac{a^m}{b^m} = \left(\frac{a}{b}\right)^m$$

$$(\cos \theta)^2 + (\sin \theta)^2 = 1 \qquad \theta = \frac{x}{r}, \ \theta = \frac{y}{r}$$

or $\qquad \sin^2 \theta + \cos^2 \theta = 1$

Starting again with $x^2 + y^2 = r^2$ and dividing through by $x^2$ gives

$$\frac{x^2}{x^2} + \frac{y^2}{x^2} = \frac{r^2}{x^2} \qquad \text{Divide by } x^2.$$

$$1 + \left(\frac{y}{x}\right)^2 = \left(\frac{r}{x}\right)^2 \qquad \text{Power rule for exponents}$$

$$1 + (\tan \theta)^2 = (\sec \theta)^2 \qquad \tan \theta = \frac{y}{x}, \sec \theta = \frac{r}{x}$$

or $\qquad \tan^2 \theta + 1 = \sec^2 \theta.$

Similarly, dividing through by $y^2$ leads to

$$1 + \cot^2 \theta = \csc^2 \theta.$$

These three identities are called the **Pythagorean identities** since the original equation that led to them, $x^2 + y^2 = r^2$, comes from the Pythagorean theorem.

**Pythagorean Identities**

For all angles $\theta$ for which the function values are defined,

$$\sin^2 \theta + \cos^2 \theta = 1 \qquad \tan^2 \theta + 1 = \sec^2 \theta \qquad 1 + \cot^2 \theta = \csc^2 \theta.$$

As before, we have given only one form of each identity. However, algebraic transformations produce equivalent identities. For example, by subtracting $\sin^2 \theta$ from both sides of $\sin^2 \theta + \cos^2 \theta = 1$, we get the equivalent identity

$$\cos^2 \theta = 1 - \sin^2 \theta. \qquad \text{Alternative form}$$

*You should be able to transform these identities quickly and recognize their equivalent forms.*

**688** CHAPTER 14 Trigonometric Functions

Consider the quotient of sin $\theta$ and cos $\theta$, for cos $\theta \neq 0$.

$$\frac{\sin\theta}{\cos\theta} = \frac{\frac{y}{r}}{\frac{x}{r}} = \frac{y}{r} \div \frac{x}{r} = \frac{y}{r} \cdot \frac{r}{x} = \frac{y}{x} = \tan\theta$$

Similarly, $\frac{\cos\theta}{\sin\theta} = \cot\theta$, for sin $\theta \neq 0$. Thus, we have the **quotient identities**.

> **Quotient Identities**
>
> For all angles $\theta$ for which the denominators are not zero,
>
> $$\frac{\sin\theta}{\cos\theta} = \tan\theta \qquad \frac{\cos\theta}{\sin\theta} = \cot\theta.$$

**EXAMPLE 6** Finding Other Function Values Given One Value and the Quadrant

Find sin $\theta$ and tan $\theta$, given that cos $\theta = -\frac{\sqrt{3}}{4}$ and sin $\theta > 0$.

**Solution** Start with $\sin^2\theta + \cos^2\theta = 1$.

$$\sin^2\theta + \left(-\frac{\sqrt{3}}{4}\right)^2 = 1 \qquad \text{Replace cos } \theta \text{ with } -\frac{\sqrt{3}}{4}.$$

$$\sin^2\theta + \frac{3}{16} = 1 \qquad \text{Square } -\frac{\sqrt{3}}{4}.$$

$$\sin^2\theta = \frac{13}{16} \qquad \text{Subtract } \frac{3}{16}.$$

$$\sin\theta = \pm\frac{\sqrt{13}}{4} \qquad \text{Take square roots.}$$

*Choose the correct sign here.*

$$\sin\theta = \frac{\sqrt{13}}{4} \qquad \text{Choose the positive square root since sin } \theta \text{ is positive.}$$

To find tan $\theta$, use the quotient identity $\tan\theta = \frac{\sin\theta}{\cos\theta}$.

$$\tan\theta = \frac{\sin\theta}{\cos\theta} = \frac{\frac{\sqrt{13}}{4}}{-\frac{\sqrt{3}}{4}} = \frac{\sqrt{13}}{4}\left(-\frac{4}{\sqrt{3}}\right) = -\frac{\sqrt{13}}{\sqrt{3}}$$

$$= -\frac{\sqrt{13}}{\sqrt{3}} \cdot \frac{\sqrt{3}}{\sqrt{3}} = -\frac{\sqrt{39}}{3} \qquad \text{Rationalize the denominator.}$$

**NOW TRY** Exercise 75.

> **CAUTION** In problems like those in Examples 5 and 6, be careful to choose the correct sign when square roots are taken.

**EXAMPLE 7** Finding Other Function Values Given One Value and the Quadrant

Find sin $\theta$ and cos $\theta$, given that tan $\theta = \frac{4}{3}$ and $\theta$ is in quadrant III.

**Solution** Since $\theta$ is in quadrant III, sin $\theta$ and cos $\theta$ will both be negative. It is tempting to say that since $\tan\theta = \frac{\sin\theta}{\cos\theta}$ and $\tan\theta = \frac{4}{3}$, then sin $\theta = -4$ and cos $\theta = -3$. This is *incorrect*, however, since both sin $\theta$ and cos $\theta$ must be in the interval $[-1, 1]$.

We use the Pythagorean identity $\tan^2 \theta + 1 = \sec^2 \theta$ to find $\sec \theta$, and then the reciprocal identity $\cos \theta = \frac{1}{\sec \theta}$ to find $\cos \theta$.

$$\tan^2 \theta + 1 = \sec^2 \theta$$

$$\left(\frac{4}{3}\right)^2 + 1 = \sec^2 \theta \quad \tan \theta = \frac{4}{3}$$

$$\frac{16}{9} + 1 = \sec^2 \theta$$

$$\frac{25}{9} = \sec^2 \theta$$

**Be careful to choose the correct sign here.**

$$-\frac{5}{3} = \sec \theta \quad \text{Choose the negative square root since } \sec \theta \text{ is negative when } \theta \text{ is in quadrant III.}$$

$$-\frac{3}{5} = \cos \theta \quad \text{Secant and cosine are reciprocals.}$$

Since $\sin^2 \theta = 1 - \cos^2 \theta$,

$$\sin^2 \theta = 1 - \left(-\frac{3}{5}\right)^2 \quad \cos \theta = -\frac{3}{5}$$

$$\sin^2 \theta = 1 - \frac{9}{25}$$

$$\sin^2 \theta = \frac{16}{25}$$

**Again, be careful.**

$$\sin \theta = -\frac{4}{5}. \quad \text{Choose the negative square root.}$$

**NOW TRY** Exercise 73.

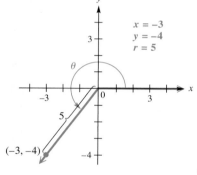

**FIGURE 36**

Example 7 can also be worked by drawing $\theta$ in standard position in quadrant III, finding $r$ to be 5, and then using the definitions of $\sin \theta$ and $\cos \theta$ in terms of $x$, $y$, and $r$. See Figure 36.

When using this method, be sure to choose the correct signs for $x$ and $y$. This is analogous to choosing the correct signs after applying the Pythagorean identities. Always check to be sure that the signs of the functions correspond to those found in the table on page 684.

## 14.4 Exercises

**NOW TRY Exercise**

*Use the appropriate reciprocal identity to find each function value. Rationalize denominators when applicable. See Example 1.*

1. $\sec \theta$, given that $\cos \theta = \frac{2}{3}$
2. $\sec \theta$, given that $\cos \theta = \frac{5}{8}$
3. $\csc \theta$, given that $\sin \theta = -\frac{3}{7}$
4. $\csc \theta$, given that $\sin \theta = -\frac{8}{43}$
5. $\cot \theta$, given that $\tan \theta = 5$
6. $\cot \theta$, given that $\tan \theta = 18$
7. $\cos \theta$, given that $\sec \theta = -\frac{5}{2}$
8. $\cos \theta$, given that $\sec \theta = -\frac{11}{7}$
9. $\sin \theta$, given that $\csc \theta = \frac{\sqrt{8}}{2}$
10. $\sin \theta$, given that $\csc \theta = \frac{\sqrt{24}}{3}$

11. $\tan\theta$, given that $\cot\theta = -2.5$
12. $\tan\theta$, given that $\cot\theta = -0.01$
13. $\sin\theta$, given that $\csc\theta = 1.42716321$
14. $\cos\theta$, given that $\sec\theta = 9.80425133$
15. Can a given angle $\theta$ satisfy both $\sin\theta > 0$ and $\csc\theta < 0$? Explain.
16. Explain what is wrong with the following item that appears on a trigonometry test:

    "*Find* $\sec\theta$, *given that* $\cos\theta = \dfrac{3}{2}$."

17. *Concept Check* What is wrong with the following statement? $\tan 90° = \dfrac{1}{\cot 90°}$.
18. *Concept Check* One form of a particular reciprocal identity is $\tan\theta = \dfrac{1}{\cot\theta}$. Give two other equivalent forms of this identity.

*Determine the signs of the trigonometric functions of an angle in standard position with the given measure. See Example 2.*

19. 74°
20. 84°
21. 218°
22. 195°
23. 178°
24. 125°
25. −80°
26. −15°
27. 845°
28. 1005°
29. −345°
30. −705°

*Identify the quadrant (or possible quadrants) of an angle $\theta$ that satisfies the given conditions. See Example 3.*

31. $\sin\theta > 0$, $\csc\theta > 0$
32. $\cos\theta > 0$, $\sec\theta > 0$
33. $\cos\theta > 0$, $\sin\theta > 0$
34. $\sin\theta > 0$, $\tan\theta > 0$
35. $\tan\theta < 0$, $\cos\theta < 0$
36. $\cos\theta < 0$, $\sin\theta < 0$
37. $\sec\theta > 0$, $\csc\theta > 0$
38. $\csc\theta > 0$, $\cot\theta > 0$
39. $\sec\theta < 0$, $\csc\theta < 0$
40. $\cot\theta < 0$, $\sec\theta < 0$
41. $\sin\theta < 0$, $\csc\theta < 0$
42. $\tan\theta < 0$, $\cot\theta < 0$

43. Explain why the answers to Exercises 33 and 37 are the same.
44. Explain why there is no angle $\theta$ that satisfies $\tan\theta > 0$, $\cot\theta < 0$.

*Decide whether each statement is* possible *or* impossible *for an angle $\theta$. See Example 4.*

45. $\sin\theta = 2$
46. $\sin\theta = 3$
47. $\cos\theta = -0.96$
48. $\cos\theta = -0.56$
49. $\tan\theta = 0.93$
50. $\cot\theta = 0.93$
51. $\sec\theta = -0.3$
52. $\sec\theta = -0.9$
53. $\csc\theta = 100$
54. $\csc\theta = -100$
55. $\cot\theta = -4$
56. $\cot\theta = -6$
57. $\sin\theta = \dfrac{1}{2}$ and $\csc\theta = 2$
58. $\tan\theta = 2$ and $\cot\theta = -2$
59. $\cos\theta = -2$ and $\sec\theta = \dfrac{1}{2}$

60. Explain why there is no angle $\theta$ that satisfies $\cos\theta = \dfrac{1}{2}$ and $\sec\theta = -2$.

*Use identities to solve each of the following. See Examples 5–7.*

61. Find $\cos\theta$, given that $\sin\theta = \dfrac{3}{5}$ and $\theta$ is in quadrant II.
62. Find $\sin\theta$, given that $\cos\theta = \dfrac{4}{5}$ and $\theta$ is in quadrant IV.
63. Find $\csc\theta$, given that $\cot\theta = -\dfrac{1}{2}$ and $\theta$ is in quadrant IV.
64. Find $\sec\theta$, given that $\tan\theta = \dfrac{\sqrt{7}}{3}$ and $\theta$ is in quadrant III.
65. Find $\tan\theta$, given that $\sin\theta = \dfrac{1}{2}$ and $\theta$ is in quadrant II.
66. Find $\cot\theta$, given that $\csc\theta = -2$ and $\theta$ is in quadrant III.
67. Find $\cot\theta$, given that $\csc\theta = -3.5891420$ and $\theta$ is in quadrant III.
68. Find $\tan\theta$, given that $\sin\theta = 0.49268329$ and $\theta$ is in quadrant II.

*Find the five remaining trigonometric function values for each angle θ. See Examples 5–7.*

**69.** $\tan \theta = -\frac{15}{8}$, given that $\theta$ is in quadrant II

**70.** $\cos \theta = -\frac{3}{5}$, given that $\theta$ is in quadrant III

**71.** $\sin \theta = \frac{\sqrt{5}}{7}$, given that $\theta$ is in quadrant I

**72.** $\tan \theta = \sqrt{3}$, given that $\theta$ is in quadrant III

**73.** $\cot \theta = \frac{\sqrt{3}}{8}$, given that $\theta$ is in quadrant I

**74.** $\csc \theta = 2$, given that $\theta$ is in quadrant II

**75.** $\sin \theta = \frac{\sqrt{2}}{6}$, given that $\cos \theta < 0$

**76.** $\cos \theta = \frac{\sqrt{5}}{8}$, given that $\tan \theta < 0$

**77.** $\sec \theta = -4$, given that $\sin \theta > 0$

**78.** $\csc \theta = -3$, given that $\cos \theta > 0$

**79.** $\sin \theta = 0.164215$, given that $\theta$ is in quadrant II

**80.** $\cot \theta = -1.49586$, given that $\theta$ is in quadrant IV

*Work each problem.*

**81.** Derive the identity $1 + \cot^2 \theta = \csc^2 \theta$ by dividing $x^2 + y^2 = r^2$ by $y^2$.

**82.** Using a method similar to the one given in this section showing that $\frac{\sin \theta}{\cos \theta} = \tan \theta$, show that $\frac{\cos \theta}{\sin \theta} = \cot \theta$.

**83.** *True or False* For all angles $\theta$, $\sin \theta + \cos \theta = 1$. If false, give an example showing why it is false.

**84.** *True or False* Since $\cot \theta = \frac{\cos \theta}{\sin \theta}$, if $\cot \theta = \frac{1}{2}$ with $\theta$ in quadrant I, then $\cos \theta = 1$ and $\sin \theta = 2$. If false, explain why.

*Concept Check* Suppose that $90° < \theta < 180°$. Find the sign of each function value.

**85.** $\sin 2\theta$

**86.** $\tan \frac{\theta}{2}$

**87.** $\cot(\theta + 180°)$

**88.** $\cos(-\theta)$

*Concept Check* Suppose that $-90° < \theta < 90°$. Find the sign of each function value.

**89.** $\cos \frac{\theta}{2}$

**90.** $\cos(\theta + 180°)$

**91.** $\sec(-\theta)$

**92.** $\sec(\theta - 180°)$

**93.** *Concept Check* The screen below was obtained with the calculator in degree mode. How can we use it to justify that an angle of 14,879° is a quadrant II angle?

```
cos(14879)
       -.4848096202
sin(14879)
        .8746197071
```

**94.** *Concept Check* The screen below was obtained with the calculator in degree mode. In which quadrant does a 1294° angle lie?

```
tan(1294)
        .6745085168
sin(1294)
       -.5591929035
```

# 14 SUMMARY

## KEY TERMS

**14.1** line
line segment (or segment)
ray
endpoint of a ray
angle
side of an angle
vertex of an angle
initial side
terminal side
positive angle
negative angle
degree
acute angle
right angle
obtuse angle
straight angle
complementary angles (complements)
supplementary angles (supplements)
minute
second
angle in standard position
quadrantal angle
coterminal angles
**14.2** vertical angles
parallel lines
transversal
similar triangles
congruent triangles
**14.3** sine (sin)
cosine (cos)
tangent (tan)
cotangent (cot)
secant (sec)
cosecant (csc)
**14.4** reciprocal

## NEW SYMBOLS

⌐  right angle symbol (for a right triangle)
$\theta$  Greek letter theta
$'$  minute
$''$  second
$P'$  P-prime

## QUICK REVIEW

### CONCEPTS

#### 14.1 Angles

**Types of Angles**

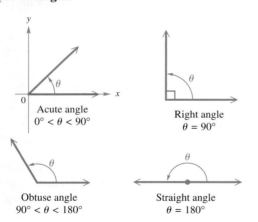

Acute angle $0° < \theta < 90°$

Right angle $\theta = 90°$

Obtuse angle $90° < \theta < 180°$

Straight angle $\theta = 180°$

#### 14.2 Angle Relationships and Similar Triangles

Vertical angles have equal measures.

The sum of the measures of the angles of any triangle is 180°.

### EXAMPLES

If $\theta = 46°$, then angle $\theta$ is an acute angle.

If $\theta = 90°$, then angle $\theta$ is a right angle.

If $\theta = 148°$, then angle $\theta$ is an obtuse angle.

If $\theta = 180°$, then angle $\theta$ is a straight angle.

The acute angle $\theta$ in the figure at the left is in standard position. If $\theta$ measures 46°, find the measure of a negative coterminal angle.

$$46° - 360° = -314°$$

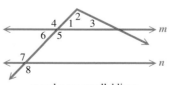

$m$ and $n$ are parallel lines.

Vertical angles 4 and 5 are equal.

The sum of angles 1, 2, and 3 is 180°.

*(continued)*

| CONCEPTS | EXAMPLES |
|---|---|
| When a transversal intersects parallel lines, the following angles formed have equal measure: alternate interior, alternate exterior, and corresponding. Interior angles on the same side of the transversal are supplementary. | Refer to the diagram at the bottom of the previous page. Angles 5 and 7 are alternate interior angles, so they are equal. Angles 4 and 8 are alternate exterior angles, so they are equal. Angles 4 and 7 are corresponding angles, so they are equal. Angles 6 and 7 are interior angles on the same side of the transversal, so they are supplementary. |
| Similar triangles have corresponding angles with the same measures, and corresponding sides proportional. |  Pairs of corresponding angles as marked in triangles $ABC$ and $DEF$ are equal. Also, $\dfrac{AB}{DE} = \dfrac{BC}{EF} = \dfrac{AC}{DF}$. |
| Congruent triangles are the same size and the same shape. | 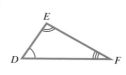 Corresponding angles are equal, and corresponding sides are equal. |

## 14.3 Trigonometric Functions

**Definitions of the Trigonometric Functions**

Let $(x, y)$ be a point other than the origin on the terminal side of an angle $\theta$ in standard position. Let $r = \sqrt{x^2 + y^2}$ represent the distance from the origin to $(x, y)$. Then

$$\sin \theta = \frac{y}{r} \qquad \cos \theta = \frac{x}{r} \qquad \tan \theta = \frac{y}{x} \;(x \neq 0)$$

$$\csc \theta = \frac{r}{y} \;(y \neq 0) \quad \sec \theta = \frac{r}{x} \;(x \neq 0) \quad \cot \theta = \frac{x}{y} \;(y \neq 0).$$

See the summary table of trigonometric function values for quadrantal angles on page 680.

If the point $(-2, 3)$ is on the terminal side of angle $\theta$ in standard position, then $x = -2$, $y = 3$, and $r = \sqrt{(-2)^2 + 3^2} = \sqrt{4 + 9} = \sqrt{13}$. Then

$$\sin \theta = \frac{3\sqrt{13}}{13}, \quad \cos \theta = -\frac{2\sqrt{13}}{13}, \quad \tan \theta = -\frac{3}{2},$$

$$\csc \theta = \frac{\sqrt{13}}{3}, \quad \sec \theta = -\frac{\sqrt{13}}{2}, \quad \cot \theta = -\frac{2}{3}.$$

## 14.4 Using the Definitions of the Trigonometric Functions

**Reciprocal Identities**

$$\sin \theta = \frac{1}{\csc \theta} \qquad \cos \theta = \frac{1}{\sec \theta} \qquad \tan \theta = \frac{1}{\cot \theta}$$

$$\csc \theta = \frac{1}{\sin \theta} \qquad \sec \theta = \frac{1}{\cos \theta} \qquad \cot \theta = \frac{1}{\tan \theta}$$

**Pythagorean Identities**

$$\sin^2 \theta + \cos^2 \theta = 1 \qquad \tan^2 \theta + 1 = \sec^2 \theta$$

$$1 + \cot^2 \theta = \csc^2 \theta$$

If $\cot \theta = -\frac{2}{3}$, find $\tan \theta$.

$$\tan \theta = \frac{1}{\cot \theta} = \frac{1}{-\frac{2}{3}} = -\frac{3}{2}$$

Use the function values for the example from **Section 14.3** to illustrate the Pythagorean identities.

$$\sin^2 \theta + \cos^2 \theta = \left(\frac{3\sqrt{13}}{13}\right)^2 + \left(-\frac{2\sqrt{13}}{13}\right)^2$$

$$= \frac{9}{13} + \frac{4}{13} = 1,$$

$$\tan^2 \theta + 1 = \left(-\frac{3}{2}\right)^2 + 1 = \frac{13}{4} = \left(-\frac{\sqrt{13}}{2}\right)^2 = \sec^2 \theta,$$

$$1 + \cot^2 \theta = 1 + \left(-\frac{2}{3}\right)^2 = \frac{13}{9} = \left(\frac{\sqrt{13}}{3}\right)^2 = \csc^2 \theta.$$

*(continued)*

## CONCEPTS

**Quotient Identities**

$$\frac{\sin\theta}{\cos\theta} = \tan\theta \qquad \frac{\cos\theta}{\sin\theta} = \cot\theta$$

**Signs of the Trigonometric Functions**

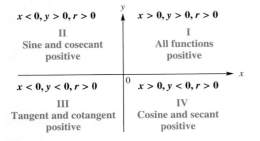

## EXAMPLES

Use the function values for the example from **Section 14.3** to illustrate $\frac{\sin\theta}{\cos\theta} = \tan\theta$.

$$\frac{\sin\theta}{\cos\theta} = \frac{\frac{3\sqrt{13}}{13}}{-\frac{2\sqrt{13}}{13}} = \frac{3\sqrt{13}}{13}\left(-\frac{13}{2\sqrt{13}}\right) = -\frac{3}{2} = \tan\theta$$

Identify the quadrant(s) of any angle $\theta$ that satisfies $\sin\theta < 0$, $\tan\theta > 0$.

Since $\sin\theta < 0$ in quadrants III and IV, while $\tan\theta > 0$ in quadrants I and III, both conditions are met only in quadrant III.

# 14 REVIEW EXERCISES

1. Give the measures of the complement and the supplement of an angle measuring 35°.

*Find the angle of least possible positive measure coterminal with each angle.*

2. −51°      3. −174°     4. 792°

5. Find the measure of each marked angle.

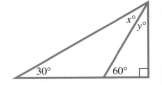

*Work each problem.*

6. A pulley is rotating 320 times per min. Through how many degrees does a point on the edge of the pulley move in $\frac{2}{3}$ sec?

7. The propeller of a speedboat rotates 650 times per min. Through how many degrees will a point on the edge of the propeller rotate in 2.4 sec?

*Convert decimal degrees to degrees, minutes, seconds, and convert degrees, minutes, seconds to decimal degrees. Round to the nearest second or the nearest thousandth of a degree, as appropriate. Use a calculator as necessary.*

8. 47° 25′ 11″     9. 119° 8′ 3″

10. −61.5034°     11. 275.1005°

*Find the measure of each marked angle.*

**12.**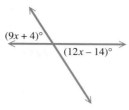

**13.**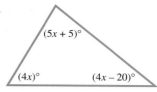

**14.** Express $\theta$ in terms of $\alpha$ and $\beta$.

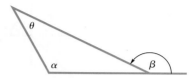

**15.** The flight path $CP$ of a satellite carrying a camera with its lens at $C$ is shown in the figure. Length $PC$ represents the distance from the lens to the film $PQ$, and $BA$ represents a straight road on the ground. Use the measurements given in the figure to find the length of the road. (*Source:* Kastner, B., *Space Mathematics,* NASA, 1985.)

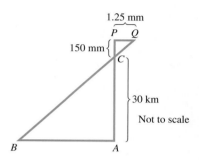

*Find all unknown angle measures in each pair of similar triangles.*

**16.**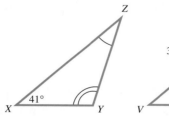

**17.**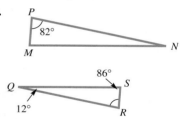

*Find the unknown side lengths in each pair of similar triangles.*

**18.**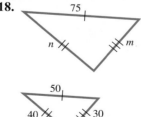

**19.**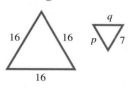

*Find the unknown measurement. There are two similar triangles in each figure.*

**20.**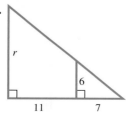

**21.**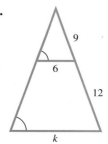

**22.** *Fill in the Blanks* If two triangles are similar, their corresponding sides are _____ and the measures of their corresponding angles are _____.

**23.** If a tree 20 ft tall casts a shadow 8 ft long, how long would the shadow of a 30-ft tree be at the same time and place?

*Find the six trigonometric function values for each angle. If a value is undefined, say so.*

**24.** 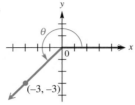     **25.** 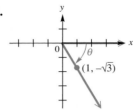     **26.** 180°

*Find the values of the six trigonometric functions for an angle in standard position having each point on its terminal side.*

**27.** $(3, -4)$                            **28.** $(9, -2)$

**29.** $(-8, 15)$                         **30.** $(1, -5)$

**31.** $(6\sqrt{3}, -6)$                  **32.** $(-2\sqrt{2}, 2\sqrt{2})$

**33.** *Concept Check* If the terminal side of a quadrantal angle lies along the *y*-axis, which of its trigonometric functions are undefined?

**34.** Find the values of all six trigonometric functions for an angle in standard position having its terminal side defined by the equation $5x - 3y = 0, x \geq 0$.

*In Exercises 35 and 36, consider an angle $\theta$ in standard position whose terminal side has the equation $y = -5x$, with $x \leq 0$.*

**35.** Sketch $\theta$ and use an arrow to show the rotation if $0° \leq \theta < 360°$.

**36.** Find the exact values of $\sin \theta$, $\cos \theta$, $\tan \theta$, $\cot \theta$, $\sec \theta$, and $\csc \theta$.

*Complete the table with the appropriate function values of the given quadrantal angles. If the value is undefined, say so.*

|     | $\theta$ | $\sin \theta$ | $\cos \theta$ | $\tan \theta$ | $\cot \theta$ | $\sec \theta$ | $\csc \theta$ |
| --- | --- | --- | --- | --- | --- | --- | --- |
| **37.** | 180° | | | | | | |
| **38.** | −90° | | | | | | |

**39.** Decide whether each statement is *possible* or *impossible*.

    **(a)** $\sec \theta = -\dfrac{2}{3}$        **(b)** $\tan \theta = 1.4$        **(c)** $\csc \theta = 5$

*Find all six trigonometric function values for each angle. Rationalize denominators when applicable.*

**40.** $\sin \theta = \dfrac{\sqrt{3}}{5}$, given that $\cos \theta < 0$

**41.** $\cos \theta = -\dfrac{5}{8}$, given that $\theta$ is in quadrant III

42. $\tan\theta = 2$, given that $\theta$ is in quadrant III
43. $\sec\theta = -\sqrt{5}$, given that $\theta$ is in quadrant II
44. $\sin\theta = -\frac{2}{5}$, given that $\theta$ is in quadrant III
45. $\sec\theta = \frac{5}{4}$, given that $\theta$ is in quadrant IV
46. *Concept Check* If, for some particular angle $\theta$, $\sin\theta < 0$ and $\cos\theta > 0$, in what quadrant must $\theta$ lie? What is the sign of $\tan\theta$?

*Solve each problem.*

47. A lifeguard located 20 yd from the water spots a swimmer in distress. The swimmer is 30 yd from shore and 100 yd east of the lifeguard. Suppose the lifeguard runs, then swims to the swimmer in a direct line, as shown in the figure. How far east from his original position will he enter the water? (*Hint:* Find the value of $x$ in the sketch.)

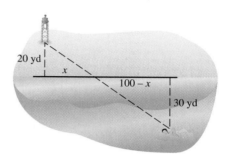

48. At present, the north star Polaris is located very near the celestial north pole. However, because Earth is inclined 23.5°, the Moon's gravitational pull on Earth is uneven. As a result, Earth slowly precesses (moves in) like a spinning top and the direction of the celestial north pole traces out a circular path once every 26,000 yr. See the figure. For example, in approximately A.D. 14,000 the star Vega will be located at the celestial north pole—and not the star Polaris. As viewed from the center $C$ of this circular path, calculate the angle (to the nearest second) that the celestial north pole moves each year. (*Source:* Zeilik, M., S. Gregory, and E. Smith, *Introductory Astronomy and Astrophysics,* Second Edition, Saunders College Publishers, 1998.)

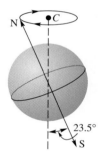

49. The depths of unknown craters on the moon can be approximated by comparing the lengths of their shadows to the shadows of nearby craters with known depths. The crater Aristillus is 11,000 ft deep, and its shadow was measured as 1.5 mm on a photograph. Its companion crater, Autolycus, had a shadow of 1.3 mm on the same photograph. Use similar triangles to determine the depth of the crater Autolycus. (*Source:* Webb, T., *Celestial Objects for Common Telescopes,* Dover Publications, 1962.)

50. The lunar mountain peak Huygens has a height of 21,000 ft. The shadow of Huygens on a photograph was 2.8 mm, while the nearby mountain Bradley had a shadow of 1.8 mm on the same photograph. Calculate the height of Bradley. (*Source:* Webb, T., *Celestial Objects for Common Telescopes,* Dover Publications, 1962.)

# 14 TEST

1. For an angle measuring 67°, give the measure of its **(a)** complement and **(b)** supplement.

*Find the measure of each unknown angle.*

2.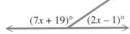

3.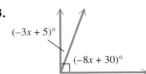

4.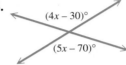

5.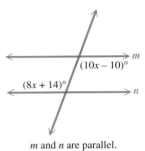

   *m* and *n* are parallel.

6.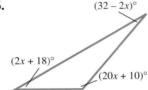

7. Perform the indicated conversion.

   **(a)** 74° 18′ 36″ to decimal degrees   **(b)** 45.2025° to degrees, minutes, seconds

8. Find the least positive measure of an angle coterminal with an angle of the given measure.

   **(a)** 390°   **(b)** −80°   **(c)** 810°

9. A tire rotates 450 times per min. Through how many degrees does a point on the edge of the tire move in 1 sec?

10. If a vertical pole 30 ft tall casts a shadow 8 ft long, how long would the shadow of a 40-ft pole be at the same time and place?

11. Find the unknown side lengths $x$ and $y$ in this pair of similar triangles.

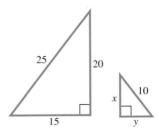

*Draw a sketch of an angle in standard position having the given point on its terminal side. Indicate the angle of least positive measure θ, and give the values of* sin θ, cos θ, tan θ, cot θ, sec θ, *and* csc θ. *If any of these are undefined, say so.*

12. (2, −7)   13. (0, −2)

14. Draw a sketch of an angle in standard position having the equation $3x - 4y = 0, x \leq 0$, as its terminal side. Indicate the angle of least positive measure $\theta$, and give the values of $\sin \theta$, $\cos \theta$, $\tan \theta$, $\cot \theta$, $\sec \theta$, and $\csc \theta$.

15. Complete the table with the appropriate function values of the given quadrantal angles. If the value is undefined, say so.

| $\theta$ | $\sin \theta$ | $\cos \theta$ | $\tan \theta$ | $\cot \theta$ | $\sec \theta$ | $\csc \theta$ |
|---|---|---|---|---|---|---|
| 90° | | | | | | |
| −360° | | | | | | |
| 630° | | | | | | |

16. If the terminal side of a quadrantal angle lies along the negative *x*-axis, which two of its trigonometric function values are undefined?

17. Identify the possible quadrant or quadrants in which $\theta$ must lie under the given conditions.
    (a) $\cos \theta > 0, \tan \theta > 0$  (b) $\sin \theta < 0, \csc \theta < 0$  (c) $\cot \theta > 0, \cos \theta < 0$

18. Decide whether each statement is *possible* or *impossible* for some angle $\theta$.
    (a) $\sin \theta = 1.5$  (b) $\sec \theta = 4$  (c) $\tan \theta = 10{,}000$

19. Find the value of $\sec \theta$ if $\cos \theta = -\frac{7}{12}$.

20. Find the five remaining trigonometric function values of $\theta$ if $\sin \theta = \frac{3}{7}$ and $\theta$ is in quadrant II.

# 15

# Acute Angles and Right Triangles

A right triangle is a basic geometric shape that occurs in many situations and applications. Using trigonometry, we can *solve* any right triangle if we know the lengths of any two sides or the length of one side and the measure of either of the two acute angles. This means that we that can find each unknown side or angle in the triangle without having to measure it, allowing us to find distances and angles that are difficult or impossible to measure directly.

Typical applications include finding the height of a building or a mountain. In Exercise 54 of Section 15.4, we use trigonometry to find the height of Mt. Everest.

**15.1** Trigonometric Functions of Acute Angles

**15.2** Trigonometric Functions of Non-Acute Angles

**15.3** Finding Trigonometric Function Values Using a Calculator

**15.4** Solving Right Triangles

# 15.1 Trigonometric Functions of Acute Angles

**OBJECTIVES**

1. Find the trigonometric function values of acute angles.
2. Use the cofunction identities.
3. Find the trigonometric function values of acute angles.

**OBJECTIVE 1** Find the trigonometric function values of acute angles. In **Section 14.3** we used angles in standard position to define the trigonometric functions. There is another way to approach them: as ratios of the lengths of the sides of right triangles. Figure 1 shows an acute angle $A$ in standard position. The definitions of the trigonometric function values of angle $A$ require $x$, $y$, and $r$. As drawn in Figure 1, $x$ and $y$ are the lengths of the two legs of the right triangle $ABC$, and $r$ is the length of the hypotenuse.

The side of length $y$ is called the **side opposite** angle $A$, and the side of length $x$ is called the **side adjacent** to angle $A$. We use the lengths of these sides to replace $x$ and $y$ in the definitions of the trigonometric functions, and the length of the hypotenuse to replace $r$, to get the following right-triangle-based definitions.

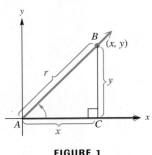

FIGURE 1

### Right-Triangle-Based Definitions of Trigonometric Functions

For any acute angle $A$ in standard position,

$$\sin A = \frac{y}{r} = \frac{\text{side opposite}}{\text{hypotenuse}} \qquad \csc A = \frac{r}{y} = \frac{\text{hypotenuse}}{\text{side opposite}}$$

$$\cos A = \frac{x}{r} = \frac{\text{side adjacent}}{\text{hypotenuse}} \qquad \sec A = \frac{r}{x} = \frac{\text{hypotenuse}}{\text{side adjacent}}$$

$$\tan A = \frac{y}{x} = \frac{\text{side opposite}}{\text{side adjacent}} \qquad \cot A = \frac{x}{y} = \frac{\text{side adjacent}}{\text{side opposite}}.$$

**EXAMPLE 1** Finding Trigonometric Function Values of an Acute Angle

Find the sine, cosine, and tangent values for angles $A$ and $B$ in the right triangle in Figure 2.

**Solution** The length of the side opposite angle $A$ is 7, the length of the side adjacent to angle $A$ is 24, and the length of the hypotenuse is 25. Use the relationships given in the box.

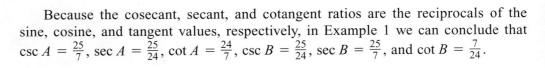

The length of the side opposite angle $B$ is 24, while the length of the side adjacent to $B$ is 7, so

$$\sin B = \frac{24}{25} \qquad \cos B = \frac{7}{25} \qquad \tan B = \frac{24}{7}.$$

**NOW TRY** Exercise 1.

Because the cosecant, secant, and cotangent ratios are the reciprocals of the sine, cosine, and tangent values, respectively, in Example 1 we can conclude that $\csc A = \frac{25}{7}$, $\sec A = \frac{25}{24}$, $\cot A = \frac{24}{7}$, $\csc B = \frac{25}{24}$, $\sec B = \frac{25}{7}$, and $\cot B = \frac{7}{24}$.

**OBJECTIVE 2** Use the cofunction identities. In Example 1, you may have noticed that $\sin A = \cos B$ and $\cos A = \sin B$. Such relationships are always true for the two acute angles of a right triangle. Figure 3 shows a right triangle with acute angles $A$ and $B$ and a right angle at $C$. (Whenever we use $A$, $B$, and $C$ to name angles in a right triangle,

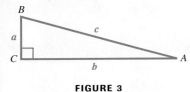

FIGURE 3

$C$ will be the right angle.) The length of the side opposite $A$ is $a$, and the length of the side opposite angle $B$ is $b$. The length of the hypotenuse is $c$.

By the preceding definitions, $\sin A = \frac{a}{c}$. Since $\cos B$ is also equal to $\frac{a}{c}$,

$$\sin A = \frac{a}{c} = \cos B.$$

Similarly, $\tan A = \frac{a}{b} = \cot B$ and $\sec A = \frac{c}{b} = \csc B.$

Since the sum of the three angles in any triangle is 180° and angle $C$ equals 90°, angles $A$ and $B$ must have a sum of 180° − 90° = 90°. As mentioned in **Section 14.1,** angles with a sum of 90° are complementary angles. Since angles $A$ and $B$ are complementary and $\sin A = \cos B$, the functions sine and cosine are called **cofunctions.** Tangent and cotangent are also cofunctions, as are secant and cosecant. And since the angles $A$ and $B$ are complementary, $A + B = 90°$, or $B = 90° − A$, giving

$$\sin A = \cos B = \cos(90° − A).$$

Similar results, called the **cofunction identities,** are true for the other trigonometric functions.

### Cofunction Identities

For any acute angle $A$,

$\sin A = \cos(90° − A)$  $\sec A = \csc(90° − A)$  $\tan A = \cot(90° − A)$
$\cos A = \sin(90° − A)$  $\csc A = \sec(90° − A)$  $\cot A = \tan(90° − A).$

**EXAMPLE 2** Writing Functions in Terms of Cofunctions

Write each function in terms of its cofunction.

(a) $\cos 52°$  (b) $\tan 71°$  (c) $\sec 24°$

**Solution**

(a) Since $\cos A = \sin(90° − A)$,

$$\cos 52° = \sin(90° − 52°) = \sin 38°.$$

(b) $\tan 71° = \cot(90° − 71°) = \cot 19°$  (c) $\sec 24° = \csc 66°$

**NOW TRY** Exercises 19 and 25.

**EXAMPLE 3** Solving Equations Using the Cofunction Identities

Find one solution for each equation. Assume all angles involved are acute angles.

(a) $\cos(\theta + 4°) = \sin(3\theta + 2°)$  (b) $\tan(2\theta − 18°) = \cot(\theta + 18°)$

**Solution**

(a) Since sine and cosine are cofunctions, $\cos(\theta + 4°) = \sin(3\theta + 2°)$ is true if the sum of the angles is 90°.

$(\theta + 4°) + (3\theta + 2°) = 90°$
$4\theta + 6° = 90°$  Combine terms.
$4\theta = 84°$  Subtract 6°.
$\theta = 21°$  Divide by 4.

**(b)** Tangent and cotangent are cofunctions, so

$$(2\theta - 18°) + (\theta + 18°) = 90°$$
$$3\theta = 90°$$
$$\theta = 30°.$$

**NOW TRY** Exercises 27 and 29.

Figure 4 shows three right triangles. From left to right, the length of each hypotenuse is the same, but angle $A$ increases in measure. As angle $A$ increases in measure from 0° to 90°, the length of the side opposite angle $A$ also increases.

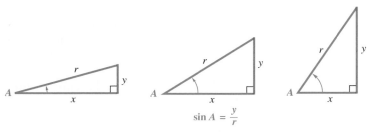

$$\sin A = \frac{y}{r}$$

As $A$ increases, $y$ increases. Since $r$ is fixed, $\sin A$ increases.

**FIGURE 4**

Since
$$\sin A = \frac{\text{side opposite}}{\text{hypotenuse}},$$

as angle $A$ increases, the numerator of this fraction also increases, while the denominator is fixed. This means that $\sin A$ *increases* as $A$ increases from 0° to 90°.

As angle $A$ increases from 0° to 90°, the length of the side adjacent to $A$ decreases. Since $r$ is fixed, the ratio $\frac{x}{r}$ will decrease. This ratio gives $\cos A$, showing that the values of cosine *decrease* as the angle measure changes from 0° to 90°. Finally, increasing $A$ from 0° to 90° causes $y$ to increase and $x$ to decrease, making the values of $\frac{y}{x} = \tan A$ increase.

A similar discussion shows that as $A$ increases from 0° to 90°, the values of $\sec A$ increase, while the values of $\cot A$ and $\csc A$ decrease.

### EXAMPLE 4  Comparing Function Values of Acute Angles

Determine whether each statement is *true* or *false*.

**(a)** $\sin 21° > \sin 18°$ **(b)** $\cos 49° \leq \cos 56°$

**Solution**

**(a)** In the interval from 0° to 90°, as the angle increases, so does the sine of the angle, which makes $\sin 21° > \sin 18°$ a true statement.

**(b)** In the interval from 0° to 90°, as the angle increases, the cosine of the angle decreases. The given statement $\cos 49° \leq \cos 56°$ is false.

**NOW TRY** Exercise 37.

**OBJECTIVE 3** Find the trigonometric function values of acute angles. Certain special angles, such as 30°, 45°, and 60°, occur so often in trigonometry and in more advanced mathematics that they deserve special study. We start with an equilateral triangle, a triangle with all sides of equal length. Each angle of such a triangle measures 60°. While the results we will obtain are independent of the length, for convenience we choose the length of each side to be 2 units. See Figure 5(a).

Bisecting one angle of this equilateral triangle leads to two right triangles, each of which has angles of 30°, 60°, and 90°, as shown in Figure 5(b). Since the hypotenuse of each right triangle has length 2, the shorter leg will have length 1. (Why?) If $x$ represents the length of the longer leg then,

$$2^2 = 1^2 + x^2 \quad \text{Pythagorean theorem}$$
$$4 = 1 + x^2$$
$$3 = x^2 \quad \text{Subtract 1.}$$
$$\sqrt{3} = x. \quad \text{Square root property; choose the positive square root.}$$

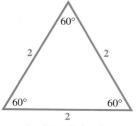

Equilateral triangle

(a)

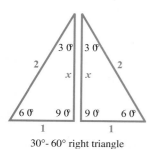

30°- 60° right triangle

(b)

**FIGURE 5**

Figure 6 summarizes our results using a 30°- 60° right triangle. As shown in the figure, the side opposite the 30° angle has length 1; that is, for the 30° angle,

$$\text{hypotenuse} = 2, \quad \text{side opposite} = 1, \quad \text{side adjacent} = \sqrt{3}.$$

Now we use the definitions of the trigonometric functions.

$$\sin 30° = \frac{\text{side opposite}}{\text{hypotenuse}} = \frac{1}{2}$$

$$\cos 30° = \frac{\text{side adjacent}}{\text{hypotenuse}} = \frac{\sqrt{3}}{2}$$

$$\tan 30° = \frac{\text{side opposite}}{\text{side adjacent}} = \frac{1}{\sqrt{3}} = \frac{1}{\sqrt{3}} \cdot \frac{\sqrt{3}}{\sqrt{3}} = \frac{\sqrt{3}}{3}$$

$$\csc 30° = \frac{2}{1} = 2$$

$$\sec 30° = \frac{2}{\sqrt{3}} = \frac{2}{\sqrt{3}} \cdot \frac{\sqrt{3}}{\sqrt{3}} = \frac{2\sqrt{3}}{3}$$

$$\cot 30° = \frac{\sqrt{3}}{1} = \sqrt{3}$$

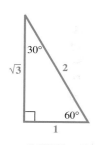

**FIGURE 6**

### EXAMPLE 5  Finding Trigonometric Function Values for 60°

Find the six trigonometric function values for a 60° angle.

**Solution**  Refer to Figure 6 to find the following ratios.

$$\sin 60° = \frac{\sqrt{3}}{2} \qquad \cos 60° = \frac{1}{2} \qquad \tan 60° = \frac{\sqrt{3}}{1} = \sqrt{3}$$

$$\csc 60° = \frac{2}{\sqrt{3}} = \frac{2\sqrt{3}}{3} \qquad \sec 60° = \frac{2}{1} = 2 \qquad \cot 60° = \frac{1}{\sqrt{3}} = \frac{\sqrt{3}}{3}$$

**NOW TRY** Exercises 45, 47, and 49.

The results in Example 5 can also be found using the fact that cofunctions of complementary angles are equal.

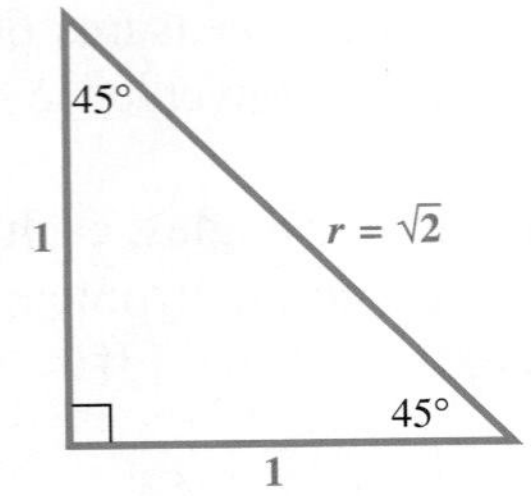

We find the values of the trigonometric functions for 45° by starting with a 45°-45° right triangle, as shown in Figure 7. This triangle is isosceles; we choose the lengths of the equal sides to be 1 unit. (As before, the results are independent of the length of the equal sides.) Since the shorter sides each have length 1, if $r$ represents the length of the hypotenuse, then

$$1^2 + 1^2 = r^2 \quad \text{Pythagorean theorem}$$
$$2 = r^2$$
$$\sqrt{2} = r. \quad \text{Choose the positive square root.}$$

**FIGURE 7** — 45°-45° right triangle

Now we use the measures indicated on the 45°-45° right triangle in Figure 7.

$$\sin 45° = \frac{1}{\sqrt{2}} = \frac{\sqrt{2}}{2} \qquad \cos 45° = \frac{1}{\sqrt{2}} = \frac{\sqrt{2}}{2} \qquad \tan 45° = \frac{1}{1} = 1$$

$$\csc 45° = \frac{\sqrt{2}}{1} = \sqrt{2} \qquad \sec 45° = \frac{\sqrt{2}}{1} = \sqrt{2} \qquad \cot 45° = \frac{1}{1} = 1$$

Function values for 30°, 45°, and 60° are summarized in the table that follows.

**FUNCTION VALUES OF SPECIAL ANGLES**

| $\theta$ | $\sin\theta$ | $\cos\theta$ | $\tan\theta$ | $\cot\theta$ | $\sec\theta$ | $\csc\theta$ |
|---|---|---|---|---|---|---|
| 30° | $\frac{1}{2}$ | $\frac{\sqrt{3}}{2}$ | $\frac{\sqrt{3}}{3}$ | $\sqrt{3}$ | $\frac{2\sqrt{3}}{3}$ | 2 |
| 45° | $\frac{\sqrt{2}}{2}$ | $\frac{\sqrt{2}}{2}$ | 1 | 1 | $\sqrt{2}$ | $\sqrt{2}$ |
| 60° | $\frac{\sqrt{3}}{2}$ | $\frac{1}{2}$ | $\sqrt{3}$ | $\frac{\sqrt{3}}{3}$ | 2 | $\frac{2\sqrt{3}}{3}$ |

**NOW TRY** Exercises 5, 7, and 9.

You will be able to reproduce this table quickly if you learn the values of sin 30°, sin 45°, and sin 60°. Then you can complete the rest of the table using the reciprocal, cofunction, and quotient identities.

Since a calculator finds trigonometric function values at the touch of a key, you may wonder why we spend so much time finding values for special angles. We do this because a calculator gives only *approximate* values in most cases, while we often need *exact* values. For example, tan 30° can be found on a scientific calculator by first setting it in *degree mode,* (see Figure 32 on page 680) then entering 30 and pressing the [tan] key to get

$$\tan 30° \approx 0.57735027. \quad \approx \text{means "is approximately equal to."}$$

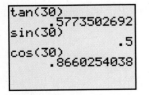

**FIGURE 8**

Earlier, however, we found the exact value:

$$\tan 30° = \frac{\sqrt{3}}{3}.$$

To use a graphing calculator to approximate sine, cosine, or tangent function values, press the appropriate function key *first* and then enter the angle measure. (The calculator must be in degree mode to enter the angle measure in degrees.) See Figure 8.

## 15.1 Exercises

*Find exact values or expressions for* sin A, cos A, *and* tan A. *See Example 1.*

**1.**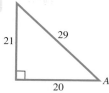

**2.** 

**3.** 

**4.**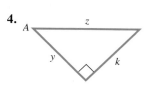

**Matching** *For each trigonometric function in Column I, choose its value from Column II.*

I

II

**5.** sin 30°  **6.** cos 45°   A. $\sqrt{3}$   B. 1   C. $\dfrac{1}{2}$

**7.** tan 45°  **8.** sec 60°   D. $\dfrac{\sqrt{3}}{2}$   E. $\dfrac{2\sqrt{3}}{3}$   F. $\dfrac{\sqrt{3}}{3}$

**9.** csc 60°  **10.** cot 30°   G. 2   H. $\dfrac{\sqrt{2}}{2}$   I. $\sqrt{2}$

*Suppose ABC is a right triangle with sides of lengths a, b, and c and right angle at C. (See Figure 3.) Find the unknown side length using the Pythagorean theorem, and then find the values of the six trigonometric functions for angle B. Rationalize denominators when applicable.*

**11.** $a = 5, b = 12$   **12.** $a = 3, b = 5$   **13.** $a = 6, c = 7$

**14.** $b = 7, c = 12$   **15.** $a = 3, c = 5$   **16.** $b = 8, c = 11$

**17.** *Concept Check* Give a summary of the six cofunction relationships.

*Write each function in terms of its cofunction. Assume that all angles in which an unknown appears are acute angles. See Example 2.*

**18.** cot 73°   **19.** sec 39°   **20.** sin 27°   **21.** sec($\theta + 15°$)

**22.** cos($\alpha + 20°$)   **23.** cot($\theta - 10°$)   **24.** tan 25.4°   **25.** sin 38.7°

**26.** With a calculator, evaluate sin(90° − A) and cos A for various values of A. (Include values greater than 90° and less than 0°.) What do you find?

*Find one solution for each equation. Assume that all angles in which an unknown appears are acute angles. See Example 3.*

**27.** tan $\alpha$ = cot($\alpha + 10°$)   **28.** cos $\theta$ = sin 2$\theta$

**29.** sin(2$\theta$ + 10°) = cos(3$\theta$ − 20°)   **30.** sec($\beta$ + 10°) = csc(2$\beta$ + 20°)

**31.** tan(3B + 4°) = cot(5B − 10°)   **32.** cot(5$\theta$ + 2°) = tan(2$\theta$ + 4°)

**33.** sin($\theta$ − 20°) = cos(2$\theta$ + 5°)   **34.** cos(2$\theta$ + 50°) = sin(2$\theta$ − 20°)

**35.** sec(3$\beta$ + 10°) = csc($\beta$ + 8°)   **36.** csc($\beta$ + 40°) = sec($\beta$ − 20°)

*True or False* See Example 4.

**37.** $\sin 50° > \sin 40°$
**38.** $\tan 28° \leq \tan 40°$
**39.** $\sin 46° < \cos 46°$
(*Hint:* $\cos 46° = \sin 44°$)
**40.** $\cos 28° < \sin 28°$
(*Hint:* $\sin 28° = \cos 62°$)
**41.** $\tan 41° < \cot 41°$
**42.** $\cot 30° < \tan 40°$
**43.** $\sec 60° > \sec 30°$
**44.** $\csc 20° < \csc 30°$

*For each expression, give the exact value. See Example 5.*

**45.** $\tan 30°$
**46.** $\cot 30°$
**47.** $\sin 30°$
**48.** $\cos 30°$
**49.** $\sec 30°$
**50.** $\csc 30°$
**51.** $\csc 45°$
**52.** $\sec 45°$
**53.** $\cos 45°$
**54.** $\cot 45°$
**55.** $\tan 45°$
**56.** $\sin 45°$
**57.** $\sin 60°$
**58.** $\cos 60°$
**59.** $\tan 60°$
**60.** $\csc 60°$

*Concept Check* Work each problem.

**61.** Find the equation of the line passing through the origin and making a 30° angle with the *x*-axis.

**62.** Find the equation of the line passing through the origin and making a 60° angle with the *x*-axis.

**63.** What angle does the line $y = \sqrt{3}x$ make with the positive *x*-axis?

**64.** What angle does the line $y = \frac{\sqrt{3}}{3}x$ make with the positive *x*-axis?

*Find the exact value of each part labeled with a variable in each figure.*

**65.**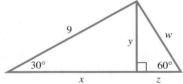

**66.**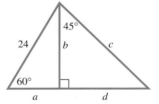

**67.**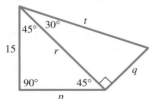

**68.**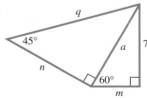

*Find a formula for the area of each figure in terms of s.*

**69.**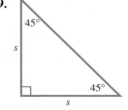

**70.**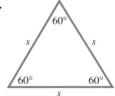

# 15.2 Trigonometric Functions of Non-Acute Angles

**OBJECTIVES**

1. Find the reference angle for a given angle.
2. Find trigonometric function values using special angles as reference angles.
3. Find angle measures given an interval and a function value.

**OBJECTIVE 1** Find the reference angle for a given angle. Associated with every nonquadrantal angle in standard position is a positive acute angle called its *reference angle*. A **reference angle** for an angle $\theta$, written $\theta'$, is the positive acute angle made by the terminal side of angle $\theta$ and the x-axis. Figure 9 shows several angles $\theta$ (each less than one complete counterclockwise revolution) in quadrants II, III, and IV, respectively, with the reference angle $\theta'$ also shown. In quadrant I, $\theta$ and $\theta'$ are the same. If an angle $\theta$ is negative or has measure greater than 360°, its reference angle is found by first finding its coterminal angle that is between 0° and 360°, and then using the diagrams in Figure 9.

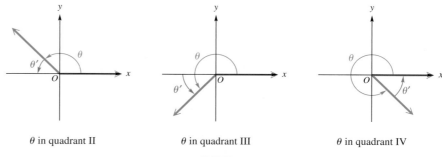

$\theta$ in quadrant II  $\qquad$ $\theta$ in quadrant III  $\qquad$ $\theta$ in quadrant IV

**FIGURE 9**

**CAUTION** A common error is to find the reference angle by using the terminal side of $\theta$ and the y-axis. *The reference angle is always found with reference to the x-axis.*

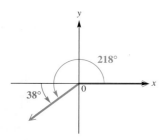

$218° - 180° = 38°$

**FIGURE 10**

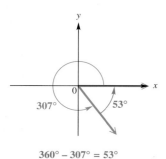

$360° - 307° = 53°$

**FIGURE 11**

**EXAMPLE 1** Finding Reference Angles

Find the reference angle for each angle.

(a) 218°  $\qquad$ (b) 1387°

**Solution**

(a) As shown in Figure 10, the positive acute angle made by the terminal side of this angle and the x-axis is $218° - 180° = 38°$. For $\theta = 218°$, the reference angle $\theta' = 38°$.

(b) First find a coterminal angle between 0° and 360°. Divide 1387° by 360° to get a quotient of about 3.9. Begin by subtracting 360° three times (because of the whole number 3 in 3.9):

$$1387° - 3 \cdot 360° = 1387° - 1080° = 307°.$$

The reference angle for 307° (and thus for 1387°) is $360° - 307° = 53°$. See Figure 11.

**NOW TRY** Exercises 1 and 5.

The preceding example suggests the following table for finding the reference angle $\theta'$ for any angle $\theta$ between 0° and 360°.

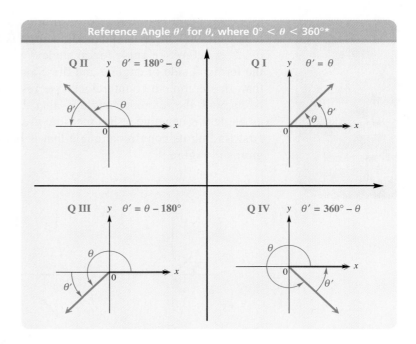

**OBJECTIVE 2** Find trigonometric function values using special angles as reference angles. We can now find exact trigonometric function values of angles with reference angles of 30°, 45°, or 60°.

**EXAMPLE 2** Finding Trigonometric Function Values of a Quadrant III Angle

Find the values of the six trigonometric functions for 210°.

**Solution** An angle of 210° is shown in Figure 12. The reference angle is $210° - 180° = 30°$. To find the trigonometric function values of 210°, choose point $P$ on the terminal side of the angle so that the distance from the origin $O$ to $P$ is 2. By the results from 30°-60° right triangles, the coordinates of point $P$ become $(-\sqrt{3}, -1)$, with $x = -\sqrt{3}$, $y = -1$, and $r = 2$. Then, by the definitions of the trigonometric functions in **Section 15.1**,

$$\sin 210° = \frac{-1}{2} = -\frac{1}{2} \qquad \csc 210° = \frac{2}{-1} = -2$$

$$\cos 210° = \frac{-\sqrt{3}}{2} = -\frac{\sqrt{3}}{2} \qquad \sec 210° = \frac{2}{-\sqrt{3}} = -\frac{2\sqrt{3}}{3}$$

$$\tan 210° = \frac{-1}{-\sqrt{3}} = \frac{\sqrt{3}}{3} \qquad \cot 210° = \frac{-\sqrt{3}}{-1} = \sqrt{3}.$$

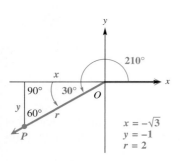

FIGURE 12

**NOW TRY** Exercise 19.

---

*The authors would like to thank Bethany Vaughn and Theresa Matick, of Vincennes Lincoln High School, for their suggestions concerning this table.

Notice in Example 2 that the trigonometric function values of 210° correspond in absolute value to those of its reference angle 30°. The signs are different for the sine, cosine, secant, and cosecant functions because 210° is a quadrant III angle. These results suggest a shortcut for finding the trigonometric function values of a nonacute angle, using the reference angle. In Example 2, the reference angle for 210° is 30°. Using the trigonometric function values of 30°, and choosing the correct signs for a quadrant III angle, we obtain the same results.

We determine the values of the trigonometric functions for any nonquadrantal angle $\theta$ as follows.

### Finding Trigonometric Function Values for Any Nonquadrantal Angle $\theta$

**Step 1** If $\theta > 360°$, or if $\theta < 0°$, then find a coterminal angle by adding or subtracting 360° as many times as needed to get an angle greater than 0° but less than 360°.

**Step 2** Find the reference angle $\theta'$.

**Step 3** Find the trigonometric function values for reference angle $\theta'$.

**Step 4** Determine the correct signs for the values found in Step 3. (Use the table of signs in **Section 14.4**, if necessary.) This gives the values of the trigonometric functions for angle $\theta$.

### EXAMPLE 3  Finding Trigonometric Function Values Using Reference Angles

Find the exact value of each expression.

(a) $\cos(-240°)$      (b) $\tan 675°$

**Solution**

(a) Since an angle of $-240°$ is coterminal with an angle of

$$-240° + 360° = 120°,$$

the reference angle is $180° - 120° = 60°$, as shown in Figure 13(a). Since the cosine is negative in quadrant II,

$$\cos(-240°) = \cos 120° = -\cos 60° = -\frac{1}{2}.$$

↑ Coterminal angle      ↑ Reference angle

(b) Begin by subtracting 360° to get a coterminal angle between 0° and 360°.

$$675° - 360° = 315°$$

As shown in Figure 13(b), the reference angle is $360° - 315° = 45°$. An angle of 315° is in quadrant IV, so the tangent will be negative, and

$$\tan 675° = \tan 315° = -\tan 45° = -1.$$

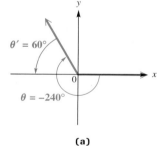

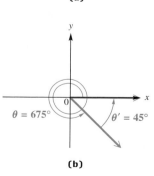

**FIGURE 13**

NOW TRY Exercises 37 and 39.

**EXAMPLE 4** Evaluating an Expression with Function Values of Special Angles

Evaluate $\cos 120° + 2 \sin^2 60° - \tan^2 30°$.

**Solution** Since $\cos 120° = -\frac{1}{2}$, $\sin 60° = \frac{\sqrt{3}}{2}$, and $\tan 30° = \frac{\sqrt{3}}{3}$,

$$\cos 120° + 2\sin^2 60° - \tan^2 30° = -\frac{1}{2} + 2\left(\frac{\sqrt{3}}{2}\right)^2 - \left(\frac{\sqrt{3}}{3}\right)^2$$

$$= -\frac{1}{2} + 2\left(\frac{3}{4}\right) - \frac{3}{9}$$

$$= \frac{2}{3}.$$

**NOW TRY** Exercise 45.

Recall that trigonometric function values of coterminal angles are the same.

**EXAMPLE 5** Using Coterminal Angles to Find Function Values

Evaluate each function by first expressing the function in terms of an angle between 0° and 360°.

**(a)** $\cos 780°$   **(b)** $\tan(-405°)$

**Solution**

**(a)** Add or subtract 360° as many times as necessary to get an angle between 0° and 360°. Subtracting 720°, which is $2 \cdot 360°$, gives

$$\cos 780° = \cos(780° - 2 \cdot 360°)$$

$$= \cos 60°$$

$$= \frac{1}{2}.$$

*Multiply first, and then subtract.*

**(b)** Add 360° twice to get $-405° + 2(360°) = 315°$. This angle is located in quadrant IV and its reference angle is 45°. The tangent function is negative in quadrant IV; thus,

$$\tan(-405°) = \tan 315° = -\tan 45° = -1.$$

**NOW TRY** Exercises 27 and 31.

**OBJECTIVE 3** Find angle measures given an interval and a function value. The ideas discussed in this section can also be used to find the measures of certain angles, given a trigonometric function value and an interval in which the angle must lie. We are most often interested in the interval $[0°, 360°)$.

## EXAMPLE 6 Finding Angle Measures Given an Interval and a Function Value

Find all values of $\theta$, if $\theta$ is in the interval $[0°, 360°)$ and $\cos \theta = -\frac{\sqrt{2}}{2}$.

**Solution** Since $\cos \theta$ is negative, $\theta$ must lie in quadrant II or III. Since the absolute value of $\cos \theta$ is $\frac{\sqrt{2}}{2}$, the reference angle $\theta'$ must be $45°$. The two possible angles $\theta$ are sketched in Figure 14. The quadrant II angle $\theta$ equals

$$180° - 45° = 135°,$$

and the quadrant III angle $\theta$ equals

$$180° + 45° = 225°.$$

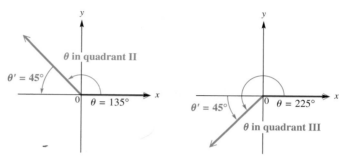

**FIGURE 14**

**NOW TRY** Exercise 67.

# 15.2 Exercises

**NOW TRY Exercise**

*Matching* Match each angle in Column I with its reference angle in Column II. Choices may be used once, more than once, or not at all. See Example 1.

| I | | II | |
|---|---|---|---|
| **1.** 98° | **2.** 212° | **A.** 45° | **B.** 60° |
| **3.** −135° | **4.** −60° | **C.** 82° | **D.** 30° |
| **5.** 750° | **6.** 480° | **E.** 38° | **F.** 32° |

*Give a short explanation in Exercises 7–9.*

7. In Example 2, why was 2 a good choice for $r$? Could any other positive number have been used?

8. Explain how the reference angle is used to find values of the trigonometric functions for an angle in quadrant III.

9. Explain why two coterminal angles have the same values for their trigonometric functions.

Complete the table with exact trigonometric function values. Do not use a calculator. See Examples 2 and 3.

| | $\theta$ | $\sin\theta$ | $\cos\theta$ | $\tan\theta$ | $\cot\theta$ | $\sec\theta$ | $\csc\theta$ |
|---|---|---|---|---|---|---|---|
| 10. | 30° | $\dfrac{1}{2}$ | $\dfrac{\sqrt{3}}{2}$ | | | $\dfrac{2\sqrt{3}}{3}$ | 2 |
| 11. | 45° | | | 1 | 1 | | |
| 12. | 60° | | $\dfrac{1}{2}$ | $\sqrt{3}$ | | 2 | |
| 13. | 120° | $\dfrac{\sqrt{3}}{2}$ | | $-\sqrt{3}$ | | | $\dfrac{2\sqrt{3}}{3}$ |
| 14. | 135° | $\dfrac{\sqrt{2}}{2}$ | $-\dfrac{\sqrt{2}}{2}$ | | | $-\sqrt{2}$ | $\sqrt{2}$ |
| 15. | 150° | | $-\dfrac{\sqrt{3}}{2}$ | $-\dfrac{\sqrt{3}}{3}$ | | | 2 |
| 16. | 210° | $-\dfrac{1}{2}$ | | $\dfrac{\sqrt{3}}{3}$ | $\sqrt{3}$ | | $-2$ |
| 17. | 240° | $-\dfrac{\sqrt{3}}{2}$ | $-\dfrac{1}{2}$ | | | $-2$ | $-\dfrac{2\sqrt{3}}{3}$ |

Find exact values of the six trigonometric functions for each angle. Rationalize denominators when applicable. See Examples 2, 3, and 5.

**18.** 300°   **19.** 315°   **20.** 405°   **21.** −300°   **22.** 420°   **23.** 480°
**24.** 495°   **25.** 570°   **26.** 750°   **27.** 1305°   **28.** 1500°   **29.** 2670°
**30.** −390°   **31.** −510°   **32.** −1020°   **33.** −1290°   **34.** −855°   **35.** −1860°

Find the exact value of each expression. See Example 3.

**36.** sin 1305°   **37.** cos(−510°)   **38.** tan(−1020°)   **39.** sin 1500°
**40.** sec(−495°)   **41.** csc(−855°)   **42.** cot 2280°   **43.** tan 3015°

*True or False* Determine whether each statement is true or false. If false, tell why. See Example 4.

**44.** sin 30° + sin 60° = sin(30° + 60°)
**45.** sin(30° + 60°) = sin 30° · cos 60° + sin 60° · cos 30°
**46.** cos 60° = 2 cos² 30° − 1        **47.** cos 60° = 2 cos 30°
**48.** sin 120° = sin 150° − sin 30°
**49.** sin 120° = sin 180° · cos 60° − sin 60° · cos 180°
**50.** sin(2 · 30°) = 2 sin 30° · cos 30°        **51.** sin² 45° + cos² 45° = 1
**52.** tan² 60° + 1 = sec² 60°

*Concept Check* Find the coordinates of the point P on the circumference of each circle. (Hint: Add x- and y-axes, assuming that the angle is in standard position.)

**53.**

**54.**

**55.** *Concept Check* Does there exist an angle $\theta$ with the function values $\cos\theta = \frac{2}{3}$ and $\sin\theta = \frac{3}{4}$?

**56.** *Concept Check* Does there exist an angle $\theta$ with the function values $\cos\theta = 0.6$ and $\sin\theta = -0.8$?

*Suppose $\theta$ is in the interval $(90°, 180°)$. Find the sign of each of the following.*

**57.** $\cos\dfrac{\theta}{2}$      **58.** $\sin\dfrac{\theta}{2}$      **59.** $\sec(\theta + 180°)$

**60.** $\cot(\theta + 180°)$      **61.** $\sin(-\theta)$      **62.** $\cos(-\theta)$

**63.** Explain why $\cos\theta = \cos(\theta + n \cdot 360°)$ for any angle $\theta$ and any integer $n$.

**64.** Explain why $\sin\theta = \sin(\theta + n \cdot 360°)$ for any angle $\theta$ and any integer $n$.

*Concept Check*

**65.** For what angles $\theta$ between $0°$ and $360°$ does $\cos\theta = \sin\theta$?

**66.** For what angles $\theta$ between $0°$ and $360°$ does $\cos\theta = -\sin\theta$?

*Find all values of $\theta$, if $\theta$ is in the interval $[0°, 360°)$ and has the given function value. See Example 6.*

**67.** $\sin\theta = \dfrac{1}{2}$    **68.** $\cos\theta = \dfrac{\sqrt{3}}{2}$    **69.** $\tan\theta = -\sqrt{3}$    **70.** $\sec\theta = -\sqrt{2}$

**71.** $\cos\theta = \dfrac{\sqrt{2}}{2}$    **72.** $\cot\theta = -\dfrac{\sqrt{3}}{3}$    **73.** $\csc\theta = -2$    **74.** $\sin\theta = -\dfrac{\sqrt{3}}{2}$

**75.** $\tan\theta = \dfrac{\sqrt{3}}{3}$    **76.** $\cos\theta = -\dfrac{1}{2}$    **77.** $\csc\theta = -\sqrt{2}$    **78.** $\cot\theta = -1$

## 15.3 Finding Trigonometric Function Values Using a Calculator

**OBJECTIVES**

1. Find function values using a calculator.
2. Find angle measures using a calculator.

**OBJECTIVE 1** Find function values using a calculator. Calculators are capable of finding trigonometric function values. For example, the values of $\cos(-240°)$ and $\tan 675°$ in Example 3 of **Section 15.2** are found with a calculator as shown in Figure 15.

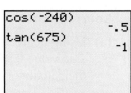

Degree mode

**FIGURE 15**

## CAUTION
*When evaluating trigonometric functions of angles given in degrees, remember that the calculator must be set in degree mode.* Get in the habit of always starting work by entering sin 90. If the displayed answer is 1, then the calculator is set for degree measure. Remember that most calculator values of trigonometric functions are *approximations*.

### EXAMPLE 1  Finding Function Values with a Calculator

Approximate the value of each expression.

(a) sin 49° 12'  (b) sec 97.977°  (c) $\dfrac{1}{\cot 51.4283°}$  (d) sin(−246°)

**Solution**

(a) $49° \, 12' = 49\dfrac{12°}{60} = 49.2°$   Convert 49° 12' to decimal degrees.

$\sin 49° \, 12' = \sin 49.2° \approx 0.75699506$   To eight decimal places

(b) Calculators do not have secant keys. However, $\sec \theta = \dfrac{1}{\cos \theta}$ for all angles $\theta$ where $\cos \theta \neq 0$. First find cos 97.977°, and then take the reciprocal to get

$$\sec 97.977° \approx -7.20587921.$$

(c) Use the reciprocal identity $\tan \theta = \dfrac{1}{\cot \theta}$ from **Section 14.4** to get

$$\dfrac{1}{\cot 51.4283°} = \tan 51.4283° \approx 1.25394815.$$

(d) $\sin(-246°) \approx 0.91354546$

**NOW TRY** Exercises 5, 7, 13, and 17.

**OBJECTIVE 2** Find angle measures using a calculator. Sometimes we need to find the measure of an angle having a certain trigonometric function value. Graphing calculators have three *inverse functions* (denoted **sin⁻¹**, **cos⁻¹**, and **tan⁻¹**) that do just that. If $x$ is an appropriate number, then $\sin^{-1} x$, $\cos^{-1} x$, or $\tan^{-1} x$ gives the measure of an angle whose sine, cosine, or tangent is $x$. For the applications in this chapter, these functions will return values of $x$ in quadrant I.

### EXAMPLE 2  Using Inverse Trigonometric Functions to Find Angles

Use a calculator to find an angle $\theta$ in the interval $[0°, 90°]$ that satisfies each condition.

(a) $\sin \theta \approx 0.9677091705$  (b) $\sec \theta \approx 1.0545829$

**Solution**

(a) Using degree mode and the inverse sine function, we find that an angle $\theta$ having sine value 0.9677091705 is 75.4°. (While there are infinitely many such angles, the calculator gives only this one.) We write this result as

$$\theta \approx \sin^{-1} 0.9677091705 \approx 75.4°.$$

See Figure 16.

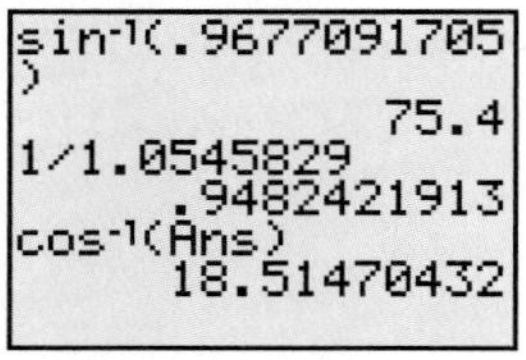

These screens support the results of Example 1. We entered the angle measure in degrees and minutes for part (a). In the fifth line of the first screen, Ans⁻¹ tells the calculator to find the reciprocal of the answer given in the previous line.

Degree mode

**FIGURE 16**

(b) Use the identity $\cos\theta = \frac{1}{\sec\theta}$. Find the reciprocal of 1.0545829 to get $\cos\theta \approx 0.9482421913$. Now find $\theta$ as shown in part (a), using the inverse cosine function. The result is

$$\theta \approx \cos^{-1}(0.9482421913) \approx 18.514704°.$$

See Figure 16.

**NOW TRY** Exercises 25 and 29.

**CAUTION** Compare Examples 1(b) and 2(b). Note that the reciprocal is used *before* the inverse cosine key when finding the angle, but *after* the cosine key when finding the trigonometric function value.

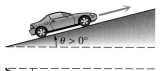
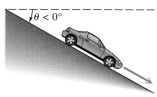

**FIGURE 17**

### EXAMPLE 3 Finding Grade Resistance

When an automobile travels uphill or downhill on a highway, it experiences a force due to gravity. This force $F$ in pounds is called **grade resistance** and is modeled by the equation $F = W \sin\theta$, where $\theta$ is the grade and $W$ is the weight of the automobile. If the automobile is moving uphill, then $\theta > 0°$; if downhill, then $\theta < 0°$. See Figure 17. (*Source:* Mannering, F. and W. Kilareski, *Principles of Highway Engineering and Traffic Analysis,* Second Edition, John Wiley & Sons, 1998.)

(a) Calculate $F$ to the nearest 10 lb for a 2500-lb car traveling an uphill grade with $\theta = 2.5°$.

(b) Calculate $F$ to the nearest 10 lb for a 5000-lb truck traveling a downhill grade with $\theta = -6.1°$.

(c) Calculate $F$ for $\theta = 0°$ and $\theta = 90°$. Do these answers agree with your intuition?

**Solution**

(a) $F = W \sin\theta = 2500 \sin 2.5° \approx 110$ lb

(b) $F = W \sin\theta = 5000 \sin(-6.1°) \approx -530$ lb
$F$ is negative because the truck is moving downhill.

(c) $F = W \sin\theta = W \sin 0° = W(0) = 0$ lb
$F = W \sin\theta = W \sin 90° = W(1) = W$ lb

This agrees with intuition because if $\theta = 0°$, then there is level ground and gravity does not cause the vehicle to roll. If $\theta = 90°$, the road would be vertical and the full weight of the vehicle would be pulled downward by gravity, so $F = W$.

**NOW TRY** Exercises 55 and 57.

## 15.3 Exercises

*Fill in the Blanks* Fill in the blanks to complete each statement.

1. The CAUTION at the beginning of this section suggests verifying that a calculator is in degree mode by finding _____ 90°. If the calculator is in degree mode, the
(sin/cos/tan)
display should be _____.

2. When a scientific or graphing calculator is used to find a trigonometric function value, in most cases the result is an _____ value.
(exact/approximate)

3. To find values of the cotangent, secant, and cosecant functions with a calculator, it is necessary to find the _____ of the _____ function value.

4. The reciprocal is used _____ the inverse function key when finding the angle, but
(before/after)
_____ the function key when finding the trigonometric function value.
(before/after)

*Use a calculator to find a decimal approximation for each value. Give as many digits as your calculator displays. In Exercises 16–22, simplify the expression before using the calculator. See Example 1.*

5. sin 38° 42′        6. cot 41° 24′        7. sec 13° 15′        8. csc 145° 45′

9. cot 183° 48′       10. cos 421° 30′      11. sec 312° 12′      12. tan(−80° 6′)

13. sin(−317° 36′)    14. cot(−512° 20′)    15. cos(−15′)         16. $\dfrac{1}{\sec 14.8°}$

17. $\dfrac{1}{\cot 23.4°}$    18. $\dfrac{\sin 33°}{\cos 33°}$    19. $\dfrac{\cos 77°}{\sin 77°}$

20. cos(90° − 3.69°)    21. cot(90° − 4.72°)    22. $\dfrac{1}{\tan(90° - 22°)}$

*Find a value of θ in the interval [0°, 90°] that satisfies each statement. Leave answers in decimal degrees. See Example 2.*

23. tan θ = 1.4739716    24. tan θ = 6.4358841    25. sin θ = 0.27843196

26. sec θ = 1.1606249    27. cot θ = 1.2575516    28. csc θ = 1.3861147

29. sec θ = 2.7496222    30. sin θ = 0.84802194   31. cos θ = 0.70058013

32. A student, wishing to use a calculator to verify the value of sin 30°, enters the information correctly but gets a display of −.98803162. He knows that the display should be .5, and he also knows that his calculator is in good working order. What do you think is the problem?

33. At one time, a certain make of calculator did not allow the input of angles outside of a particular interval when finding trigonometric function values. For example, trying to find cos 2000° using the methods of this section gave an error message, despite the fact that cos 2000° can be evaluated. Explain how you would find cos 2000° using this calculator.

34. What value of A between 0° and 90° will produce the output in the graphing calculator screen?

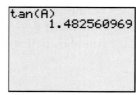

35. What value of A will produce the output (in degrees) in the graphing calculator screen?

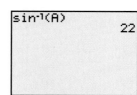

*Use a calculator to evaluate each expression.*

36. sin 35° cos 55° + cos 35° sin 55°        37. cos 100° cos 80° − sin 100° sin 80°

38. cos 75° 29′ cos 14° 31′ − sin 75° 29′ sin 14° 31′

39. sin 28° 14′ cos 61° 46′ + cos 28° 14′ sin 61° 46′

40. sin² 36° + cos² 36°        41. 2 sin 25° 13′ cos 25° 13′ − sin 50° 26′

*True or False  Use a calculator to decide whether each statement is* true *or* false. *It may be that a true statement will lead to results that differ in the last decimal place due to rounding error.*

42. cos 40° = 2 cos 20°        43. sin 10° + sin 10° = sin 20°

**44.** $\cos 70° = 2\cos^2 35° - 1$

**45.** $\sin 50° = 2\sin 25° \cos 25°$

**46.** $2\cos 38° 22' = \cos 76° 44'$

**47.** $\cos 40° = 1 - 2\sin^2 80°$

**48.** $\dfrac{1}{2}\sin 40° = \sin \dfrac{1}{2}(40°)$

**49.** $\sin 39° 48' + \cos 39° 48' = 1$

**50.** $\cos(30° + 20°) = \cos 30° + \cos 20°$

**51.** $\cos(30° + 20°) = \cos 30° \cos 20° - \sin 30° \sin 20°$

**52.** $\tan^2 72° 25' + 1 = \sec^2 72° 25'$

**53.** $1 + \cot^2 42.5° = \csc^2 42.5°$

*See Example 3 to work Exercises 54–58.*

**54.** What is the grade resistance of a 2400-lb car traveling on a $-2.4°$ downhill grade?

**55.** What is the grade resistance of a 2100-lb car traveling on a $1.8°$ uphill grade?

**56.** A car traveling on a $-3°$ downhill grade has a grade resistance of $-145$ lb. What is the weight of the car?

**57.** A 2600-lb car traveling downhill has a grade resistance of $-130$ lb. What is the angle of the grade?

**58.** Which has the greater grade resistance: a 2200-lb car on a $2°$ uphill grade or a 2000-lb car on a $2.2°$ uphill grade?

*Solve each problem.*

When a light ray travels from one medium, such as air, to another medium, such as water or glass, the speed of the light changes, and the direction in which the ray is traveling changes. (This is why a fish under water is in a different position than it appears to be.) These changes are given by **Snell's law**

$$\dfrac{c_1}{c_2} = \dfrac{\sin \theta_1}{\sin \theta_2},$$

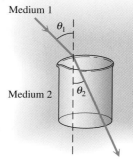

where $c_1$ is the speed of light in the first medium, $c_2$ is the speed of light in the second medium, and $\theta_1$ and $\theta_2$ are the angles shown in the figure. (*Source: The Physics Classroom,* www.glenbrook.k12.il.us) *In Exercises 59 and 60, assume that* $c_1 = 3 \times 10^8$ *m per sec.*

**59.** Find the speed of light in the second medium for each of the following.

   **(a)** $\theta_1 = 46°, \theta_2 = 31°$   **(b)** $\theta_1 = 39°, \theta_2 = 28°$

**60.** Find $\theta_2$ for each of the following values of $\theta_1$ and $c_2$. Round to the nearest degree.

   **(a)** $\theta_1 = 40°, c_2 = 1.5 \times 10^8$ m per sec   **(b)** $\theta_1 = 62°, c_2 = 2.6 \times 10^8$ m per sec

*The figure shows a fish's view of the world above the surface of the water.* (*Source:* Walker, J., "The Amateur Scientist," *Scientific American,* March 1984.) *Suppose that a light ray comes from the horizon, enters the water, and strikes the fish's eye.*

**61.** Assume that this ray gives a value of $90°$ for angle $\theta_1$ in the formula for Snell's law. (In a practical situation, this angle would probably be a little less than $90°$.) The speed of light in water is about $2.254 \times 10^8$ m per sec. Find angle $\theta_2$ to the nearest tenth.

**62.** Refer to Exercise 61. Suppose an object is located at a true angle of $29.6°$ above the horizon. Find the apparent angle above the horizon to a fish.

# 15.4 Solving Right Triangles

**OBJECTIVES**

1. Determine the number of significant digits for measurements.
2. Solve right triangles.
3. Solve applied problems involving the angle of elevation or angle of depression.

**OBJECTIVE 1** Determine the number of significant digits for measurements. A number that represents the result of counting, or a number that results from theoretical work and is not the result of measurement, is an **exact number.** There are 50 states in the United States, so in that statement, 50 is an exact number.

Most values obtained for trigonometric applications are measured values that are *not* exact. Suppose we quickly measure a room as 15 ft by 18 ft. See Figure 18. To calculate the length of a diagonal of the room, we can use the Pythagorean theorem.

$$d^2 = 15^2 + 18^2$$
$$d^2 = 549$$
$$d = \sqrt{549} \approx 23.430749$$

Should this answer be given as the length of the diagonal of the room? Of course not. The number 23.430749 contains six decimal places, while the original data of 15 ft and 18 ft are only accurate to the nearest foot. In practice, the results of a calculation can be no more accurate than the least accurate number in the calculation. Thus, we should indicate that the diagonal of the 15-by-18-ft room is approximately 23 ft.

If a wall measured to the nearest foot is 18 ft long, this actually means that the wall has length between 17.5 ft and 18.5 ft. If the wall is measured more accurately as 18.3 ft long, then its length is really between 18.25 ft and 18.35 ft. The results of physical measurement are only approximately accurate and depend on the precision of the measuring instrument as well as the aptness of the observer. The digits obtained by actual measurement are called **significant digits.** The measurement 18 ft is said to have two significant digits; 18.3 ft has three significant digits.

The following numbers have their significant digits identified in color.

408   21.5   18.00   6.700   0.0025   0.09810   7300

Notice that 18.00 has four significant digits. The zeros in this number represent measured digits and are significant. The number 0.0025 has only two significant digits, 2 and 5, because the zeros here are used only to locate the decimal point. The number 7300 causes some confusion because it is impossible to determine whether the zeros are measured values. The number 7300 may have two, three, or four significant digits. When presented with this situation, we assume that the zeros are not significant, unless the context of the problem indicates otherwise.

To determine the number of significant digits for answers in applications of angle measure, use the following table.

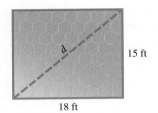

**FIGURE 18**

| Angle Measure to Nearest | Examples | Answer to Number of Significant Digits |
|---|---|---|
| Degree | 62°, 36° | 2 |
| Ten minutes, or nearest tenth of a degree | 52° 30′, 60.4° | 3 |
| Minute, or nearest hundredth of a degree | 81° 48′, 71.25° | 4 |
| Ten seconds, or nearest thousandth of a degree | 10° 52′ 20″, 21.264° | 5 |

To perform calculations with measured numbers, start by identifying the number with the fewest significant digits. Round your final answer to the same number of significant digits as this number. *Remember that your answer is no more accurate than the least accurate number in your calculation.*

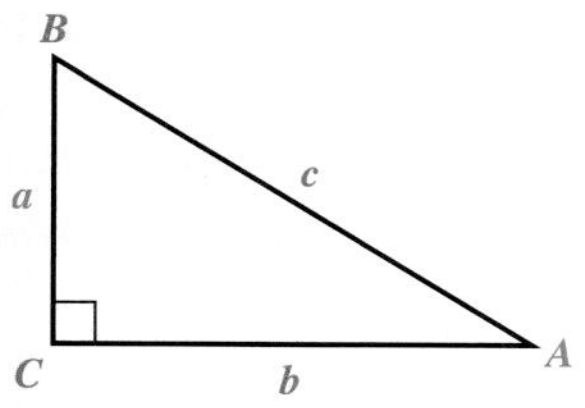

**SECTION 15.4** Solving Right Triangles **721**

**OBJECTIVE 2 Solve right triangles.** To *solve a triangle* means to find the measures of all the angles and sides of the triangle. As shown in Figure 19, we use $a$ to represent the length of the side opposite angle $A$, $b$ for the length of the side opposite angle $B$, and so on. In a right triangle, the letter $c$ is reserved for the hypotenuse.

When solving triangles, a labeled sketch is an important aid.

**FIGURE 19**

**EXAMPLE 1** Solving a Right Triangle Given an Angle and a Side

Solve right triangle $ABC$, if $A = 34° 30'$ and $c = 12.7$ in. See Figure 20.

**Solution** To solve the triangle, find the measures of the remaining sides and angles. To find the value of $a$, use a trigonometric function involving the known values of angle $A$ and side $c$. Since the sine of angle $A$ is given by the quotient of the side opposite $A$ and the hypotenuse, use $\sin A$.

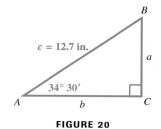

**FIGURE 20**

$$\sin A = \frac{a}{c} \qquad \sin A = \frac{\text{side opposite}}{\text{hypotenuse}}$$

$$\sin 34° 30' = \frac{a}{12.7} \qquad A = 34° 30',\ c = 12.7$$

$$a = 12.7 \sin 34° 30' \qquad \text{Multiply by 12.7; rewrite.}$$

$$a = 12.7 \sin 34.5° \qquad \text{Convert to decimal degrees.}$$

$$a = 12.7(0.56640624) \qquad \text{Use a calculator.}$$

$$a \approx 7.19 \text{ in.} \qquad \text{Three significant digits}$$

Assuming that $34° 30'$ is given to the nearest ten minutes, we rounded the answer to three significant digits.

To find the value of $b$, we could substitute the value of $a$ just calculated and the given value of $c$ in the Pythagorean theorem. It is better, however, to use the information given in the problem rather than a result just calculated. If a mistake is made in finding $a$, then $b$ also would be incorrect. Also, rounding more than once may cause the result to be less accurate. Use $\cos A$.

$$\cos A = \frac{b}{c} \qquad \cos A = \frac{\text{side adjacent}}{\text{hypotenuse}}$$

$$\cos 34° 30' = \frac{b}{12.7}$$

$$b = 12.7 \cos 34° 30'$$

$$b \approx 10.5 \text{ in.}$$

Once $b$ is found, the Pythagorean theorem can be used as a check.

All that remains to solve triangle $ABC$ is to find the measure of angle $B$. Since $A + B = 90°$,

$$B = 90° - A$$
$$B = 89° 60' - 34° 30' \qquad A = 34° 30'$$
$$B = 55° 30'.$$

**NOW TRY** Exercise 21.

In Example 1, we could have found the measure of angle $B$ first and then used the trigonometric function values of $B$ to find the unknown sides. The process of solving a right triangle can usually be done in several ways, each producing the correct answer. ***To maintain accuracy, always use given information as much as possible and avoid rounding off in intermediate steps.***

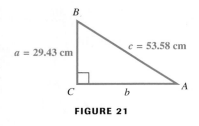

FIGURE 21

**EXAMPLE 2** Solving a Right Triangle Given Two Sides

Solve right triangle $ABC$, if $a = 29.43$ cm and $c = 53.58$ cm.

**Solution** We draw a sketch showing the given information, as in Figure 21. One way to begin is to find angle $A$ by using the sine function.

$$\sin A = \frac{\text{side opposite}}{\text{hypotenuse}} = \frac{29.43}{53.58} \approx 0.5492721165$$

$$A = \sin^{-1} 0.5492721165 \approx 33.32°$$

The measure of $B$ is approximately

$$90° - 33.32° = 56.68°.$$

We now find $b$ from the Pythagorean theorem.

$b^2 = c^2 - a^2$     Pythagorean theorem solved for $b^2$
$b^2 = 53.58^2 - 29.43^2$     $c = 53.58, a = 29.43$
$b = \sqrt{2004.6915}$
$b \approx 44.77$ cm     Keep only the positive square root.

**NOW TRY** Exercise 31.

**OBJECTIVE 3** Solve applied problems involving the angle of elevation or angle of depression. Many applications of right triangles involve angles of elevation or depression. The **angle of elevation** from point $X$ to point $Y$ (above $X$) is the acute angle formed by ray $XY$ and a horizontal ray with endpoint at $X$. See Figure 22(a). The **angle of depression** from point $X$ to point $Y$ (below $X$) is the acute angle formed by ray $XY$ and a horizontal ray with endpoint $X$. See Figure 22(b).

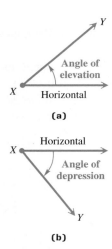

FIGURE 22

**CAUTION** Be careful when interpreting the angle of depression. ***Both the angle of elevation and the angle of depression are measured between the line of sight and a horizontal line.***

To solve applied trigonometry problems, follow the same procedure as solving a triangle.

**Solving an Applied Trigonometry Problem**

*Step 1* Draw a sketch, and label it with the given information. Label the quantity to be found with a variable.

*Step 2* Use the sketch to write an equation relating the given quantities to the variable.

*Step 3* Solve the equation, and check that your answer makes sense.

*Drawing a triangle and labeling it correctly in Step 1 is crucial.*

### EXAMPLE 3  Finding a Length When the Angle of Elevation is Known

Shelly McCarthy knows that when she stands 123 ft from the base of a flagpole, the angle of elevation to the top of the flagpole is 26° 40′. If her eyes are 5.30 ft above the ground, find the height of the flagpole.

**Solution**

*Step 1*  The length of the side adjacent to Shelly is known, and the length of the side opposite her must be found. See Figure 23.

*Step 2*  The tangent ratio involves the given values. Write an equation.

$$\tan A = \frac{\text{side opposite}}{\text{side adjacent}} \qquad \text{Tangent ratio}$$

$$\tan 26° 40' = \frac{a}{123} \qquad A = 26° 40'; \text{ side adjacent} = 123$$

*Step 3*
$$a = 123 \tan 26° 40' \qquad \text{Multiply by 123; rewrite.}$$
$$a = 123 \tan 26.66666667° \qquad \text{Convert to decimal degrees.}$$
$$a = 123(0.50221888) \qquad \text{Use a calculator.}$$
$$a \approx 61.8 \text{ ft} \qquad \text{Three significant digits}$$

Since Shelly's eyes are 5.30 ft above the ground, the height of the flagpole is

$$61.8 + 5.30 = 67.1 \text{ ft.}$$

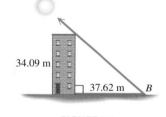

**FIGURE 23**

**NOW TRY** Exercise 51.

### EXAMPLE 4  Finding the Angle of Elevation When Lengths Are Known

The length of the shadow of a building 34.09 m tall is 37.62 m. Find the angle of elevation of the Sun.

**Solution**  As shown in Figure 24, the angle of elevation of the Sun is angle $B$. Since the side opposite $B$ and the side adjacent to $B$ are known, use the tangent ratio to find $B$.

$$\tan B = \frac{34.09}{37.62}, \qquad \text{so} \qquad B = \tan^{-1}\frac{34.09}{37.62} \approx 42.18°.$$

The angle of elevation of the Sun is 42.18°.

**FIGURE 24**

**NOW TRY** Exercise 47.

## 15.4 Exercises

*Concept Check*  Refer to the discussion of accuracy and significant digits in this section to work Exercises 1–8.

1. At the end of the 2004 National Football League season, Jerry Rice was the leading career receiver with 22,895 yd. State the range represented by this number. (*Source:* www.nfl.com)

2. When Mt. Everest was first surveyed, the surveyors obtained a height of 29,000 ft to the nearest foot. State the range represented by this number. (The surveyors thought no one would believe a measurement of 29,000 ft, so they reported it as 29,002.) (*Source:* Dunham, W., *The Mathematical Universe,* John Wiley & Sons, 1994.)

3. The E. Johnson Memorial Tunnel in Colorado, which measures 8959 ft, is one of the longest land vehicular tunnels in the United States. What is the range of this number? (*Source: The World Almanac and Book of Facts.*)

4. Women's National Basketball Association player Lauren Jackson of the Seattle Storm received the 2007 award for most points scored, 604. Is it appropriate to consider this number as between 603.5 and 604.5? Why or why not? (*Source:* www.wnba.com)

5. The formula for the circumference of a circle is $C = 2\pi r$. Suppose you use the $\boxed{\pi}$ key on your calculator to find the circumference of a circle with radius 54.98 cm, getting 345.44953. Since 2 has only one significant digit, the answer should be given as $3 \times 10^2$, or 300 cm. Is this conclusion correct? If not, explain how the answer should be given.

6. Explain the difference between a measurement of 23.0 ft and a measurement of 23.00 ft.

*Fill in the Blanks*

7. If $h$ is the actual height of a building and the height is measured as 58.6 ft, then $|h - 58.6| \leq$ _____.

8. If $w$ is the actual weight of a car and the weight is measured as 1542 lb, then $|w - 1542| \leq$ _____.

*Solve each right triangle. When two sides are given, give angles in degrees and minutes. See Example 1.*

9.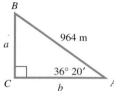

10.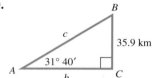

11.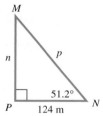

12.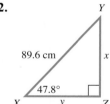

13.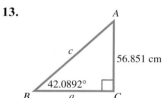

14.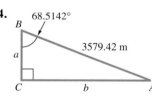

15.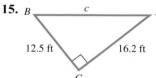

16.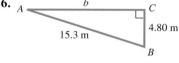

17. Can a right triangle be solved if we are given measures of its two acute angles and no side lengths? Explain.

18. *Concept Check* If we are given an acute angle and a side in a right triangle, what unknown part of the triangle requires the least work to find?

19. Explain why you can always solve a right triangle if you know the measures of one side and one acute angle.

20. Explain why you can always solve a right triangle if you know the lengths of two sides.

*Solve each right triangle. In each case, C = 90°. If angle information is given in degrees and minutes, give answers in the same way. If given in decimal degrees, do likewise in answers. When two sides are given, give angles in degrees and minutes. See Examples 1 and 2.*

21. $A = 28.0°, c = 17.4$ ft
22. $B = 46.0°, c = 29.7$ m
23. $B = 73.0°, b = 128$ in.
24. $A = 61.0°, b = 39.2$ cm
25. $A = 62.5°, a = 12.7$ m
26. $B = 51.7°, a = 28.1$ ft
27. $a = 13$ m, $c = 22$ m
28. $b = 32$ ft, $c = 51$ ft
29. $a = 76.4$ yd, $b = 39.3$ yd
30. $a = 958$ m, $b = 489$ m
31. $a = 18.9$ cm, $c = 46.3$ cm
32. $b = 219$ m, $c = 647$ m
33. $A = 53° 24', c = 387.1$ ft
34. $A = 13° 47', c = 1285$ m
35. $B = 39° 9', c = 0.6231$ m
36. $B = 82° 51', c = 4.825$ cm

37. Explain the meaning of *angle of elevation*.

38. *Concept Check* Can an angle of elevation be more than 90°?

39. Explain why the angle of depression *DAB* has the same measure as the angle of elevation *ABC* in the figure.

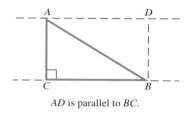

AD is parallel to BC.

40. Why is angle *CAB* not an angle of depression in the figure for Exercise 39?

*Solve each problem involving triangles. See Examples 1–4.*

41. A 13.5-m fire truck ladder is leaning against a wall. Find the distance *d* the ladder goes up the wall (above the top of the fire truck) if the ladder makes an angle of 43° 50′ with the horizontal.

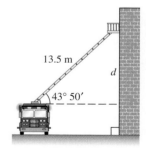

42. To find the distance *RS* across a lake, a surveyor lays off length $RT = 53.1$ m, so that angle $T = 32° 10'$ and angle $S = 57° 50'$. Find length *RS*.

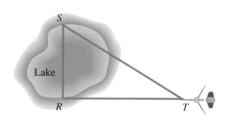

43. From a window 30.0 ft above the street, the angle of elevation to the top of the building across the street is 50.0° and the angle of depression to the base of this building is 20.0°. Find the height of the building across the street.

**44.** To determine the diameter of the Sun, an astronomer might sight with a **transit** (a device used by surveyors for measuring angles) first to one edge of the Sun and then to the other, estimating that the included angle equals 32′. Assuming that the distance $d$ from Earth to the Sun is 92,919,800 mi, approximate the diameter of the Sun.

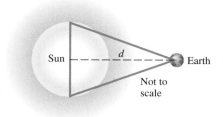

**45.** The length of the base of an isosceles triangle is 42.36 in. Each base angle is 38.12°. Find the length of each of the two equal sides of the triangle. (*Hint:* Divide the triangle into two right triangles.)

**46.** Find the altitude of an isosceles triangle having base 184.2 cm if the angle opposite the base is 68° 44′.

*Solve each problem involving an angle of elevation or depression. See Examples 3 and 4.*

**47.** The Pyramid of the Sun in the ancient Mexican city of Teotihuacan was the largest and most important structure in the city. The base is a square with sides 700 ft long, and the height of the pyramid is 200 ft. Find the angle of elevation of the edge indicated in the figure to two significant digits. (*Hint:* The base of the triangle in the figure is half the diagonal of the square base of the pyramid.)

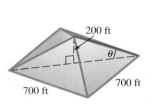

**48.** The U.S. Weather Bureau defines a **cloud ceiling** as the altitude of the lowest clouds that cover more than half the sky. To determine a cloud ceiling, a powerful searchlight projects a circle of light vertically on the bottom of the cloud. An observer sights the circle of light in the crosshairs of a tube called a **clinometer.** A pendant hanging vertically from the tube and resting on a protractor gives the angle of elevation. Find the cloud ceiling if the searchlight is located 1000 ft from the observer and the angle of elevation is 30.0° as measured with a clinometer at eye-height 6 ft. (Assume three significant digits.)

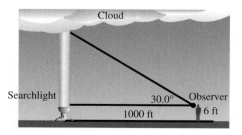

**49.** The shadow of a vertical tower is 40.6 m long when the angle of elevation of the sun is 34.6°. Find the height of the tower.

**50.** The angle of depression from the top of a building to a point on the ground is 32° 30′. How far is the point on the ground from the top of the building if the building is 252 m high?

51. Suppose that the angle of elevation of the Sun is 23.4°. Find the length of the shadow cast by Diane Carr, who is 5.75 ft tall.

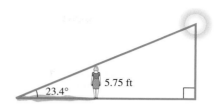

52. An airplane is flying 10,500 ft above the level ground. The angle of depression from the plane to the base of a tree is 13° 50′. How far horizontally must the plane fly to be directly over the tree?

53. The angle of elevation from the top of a small building to the top of a nearby taller building is 46° 40′, while the angle of depression to the bottom is 14°10′. If the shorter building is 28.0 m high, find the height of the taller building.

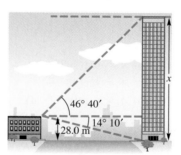

54. The highest mountain peak in the world is Mt. Everest, located in the Himalayas. The height of this enormous mountain was determined in 1856 by surveyors using trigonometry long before it was first climbed in 1953. This difficult measurement had to be done from a great distance. At an altitude of 14,545 ft on a different mountain, the straight line distance to the peak of Mt. Everest is 27.0134 mi and its angle of elevation is $\theta = 5.82°$. (*Source:* Dunham, W., *The Mathematical Universe,* John Wiley & Sons, 1994.)

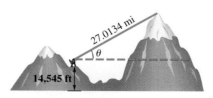

(a) Approximate the height (in feet) of Mt. Everest.
(b) In the actual measurement, Mt. Everest was over 100 mi away and the curvature of Earth had to be taken into account. Would the curvature of Earth make the peak appear taller or shorter than it actually is?

# 15 SUMMARY

## KEY TERMS

**15.1** side opposite
side adjacent
cofunctions

**15.2** reference angle
**15.4** exact number
significant digits

angle of elevation
angle of depression

## QUICK REVIEW

### CONCEPTS — EXAMPLES

### 15.1 Trigonometric Functions of Acute Angles

**Right-Triangle-Based Definitions of the Trigonometric Functions**

For any acute angle $A$ in standard position,

$$\sin A = \frac{y}{r} = \frac{\text{side opposite}}{\text{hypotenuse}} \qquad \csc A = \frac{r}{y} = \frac{\text{hypotenuse}}{\text{side opposite}}$$

$$\cos A = \frac{x}{r} = \frac{\text{side adjacent}}{\text{hypotenuse}} \qquad \sec A = \frac{r}{x} = \frac{\text{hypotenuse}}{\text{side adjacent}}$$

$$\tan A = \frac{y}{x} = \frac{\text{side opposite}}{\text{side adjacent}} \qquad \cot A = \frac{x}{y} = \frac{\text{side adjacent}}{\text{side opposite}}.$$

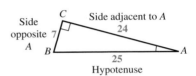

$$\sin A = \frac{7}{25} \qquad \cos A = \frac{24}{25} \qquad \tan A = \frac{7}{24}$$

$$\csc A = \frac{25}{7} \qquad \sec A = \frac{25}{24} \qquad \cot A = \frac{24}{7}$$

**Cofunction Identities**

For any acute angle $A$,

$$\sin A = \cos(90° - A) \qquad \cos A = \sin(90° - A)$$
$$\sec A = \csc(90° - A) \qquad \csc A = \sec(90° - A)$$
$$\tan A = \cot(90° - A) \qquad \cot A = \tan(90° - A).$$

$$\sin 55° = \cos(90° - 55°) = \cos 35°$$
$$\sec 48° = \csc(90° - 48°) = \csc 42°$$
$$\tan 72° = \cot(90° - 72°) = \cot 18°$$

**Function Values of Special Angles**

| $\theta$ | $\sin \theta$ | $\cos \theta$ | $\tan \theta$ | $\cot \theta$ | $\sec \theta$ | $\csc \theta$ |
|---|---|---|---|---|---|---|
| 30° | $\frac{1}{2}$ | $\frac{\sqrt{3}}{2}$ | $\frac{\sqrt{3}}{3}$ | $\sqrt{3}$ | $\frac{2\sqrt{3}}{3}$ | 2 |
| 45° | $\frac{\sqrt{2}}{2}$ | $\frac{\sqrt{2}}{2}$ | 1 | 1 | $\sqrt{2}$ | $\sqrt{2}$ |
| 60° | $\frac{\sqrt{3}}{2}$ | $\frac{1}{2}$ | $\sqrt{3}$ | $\frac{\sqrt{3}}{3}$ | 2 | $\frac{2\sqrt{3}}{3}$ |

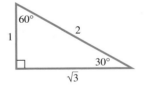 

### 15.2 Trigonometric Functions of Non-Acute Angles

**Reference Angle $\theta'$ for $\theta$ in $(0°, 360°)$**

| $\theta$ in Quadrant | I | II | III | IV |
|---|---|---|---|---|
| $\theta'$ is | $\theta$ | $180° - \theta$ | $\theta - 180°$ | $360° - \theta$ |

See the figure on page 710 for illustrations of reference angles.

Quadrant I:   For $\theta = 25°$, $\theta' = 25°$
Quadrant II:  For $\theta = 152°$, $\theta' = 28°$
Quadrant III: For $\theta = 200°$, $\theta' = 20°$
Quadrant IV:  For $\theta = 320°$, $\theta' = 40°$

*(continued)*

## CONCEPTS

**Finding Trigonometric Function Values for Any Nonquadrantal Angle $\theta$**

**Step 1** Add or subtract 360° as many times as needed to get an angle greater than 0° but less than 360°.

**Step 2** Find the reference angle $\theta'$.

**Step 3** Find the trigonometric function values for $\theta'$.

**Step 4** Determine the correct signs for the values found in Step 3.

## EXAMPLES

Find sin 1050°.

$$1050° - 2(360°) = 330°$$

Thus, $\theta' = 30°$.

$$\sin 1050° = -\sin 30° = -\frac{1}{2}$$

### 15.3 Finding Trigonometric Function Values Using a Calculator

To approximate a trigonometric function value of an angle in degrees, make sure your calculator is in degree mode.

Approximate each value.

$$\cos 50° \ 15' = \cos 50.25° \approx 0.63943900$$

$$\csc 32.5° = \frac{1}{\sin 32.5°} \approx 1.86115900$$

To find the corresponding angle measure given a trigonometric function value, use an appropriate inverse function.

Find an angle $\theta$ in the interval $[0°, 90°]$ that satisfies each condition.

$$\cos \theta \approx 0.73677482$$
$$\theta \approx \cos^{-1}(0.73677482)$$
$$\theta \approx 42.542600°$$

$$\csc \theta \approx 1.04766792$$
$$\sin \theta \approx \frac{1}{1.04766792} \qquad \sin \theta = \frac{1}{\csc \theta}$$
$$\theta \approx \sin^{-1}\left(\frac{1}{1.04766792}\right)$$
$$\theta \approx 72.65°$$

### 15.4 Solving Right Triangles

**Solving an Applied Trigonometry Problem**

**Step 1** Draw a sketch, and label it with the given information. Label the quantity to be found with a variable.

Find the angle of elevation of the Sun if a 48.6-ft flagpole casts a shadow 63.1 ft long.

**Step 1** See the sketch. We must find $\theta$.

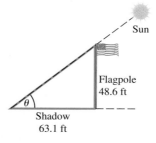

**Step 2** Use the sketch to write an equation relating the given quantities to the variable.

**Step 3** Solve the equation, and check that your answer makes sense.

**Step 2** $\tan \theta = \dfrac{48.6}{63.1} \approx 0.770206$

**Step 3** $\theta = \tan^{-1} 0.770206 \approx 37.6°$

The angle of elevation rounded to three significant digits is 37.6°, or 37° 40'.

# 15 REVIEW EXERCISES

*Find the values of the six trigonometric functions for each angle A.*

**1.**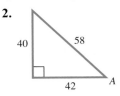

**2.**

*Find one solution for each equation. Assume that all angles are acute angles.*

**3.** $\sin 4\beta = \cos 5\beta$

**4.** $\sec(2\theta + 10°) = \csc(4\theta + 20°)$

**5.** $\tan(5x + 11°) = \cot(6x + 2°)$

**6.** $\cos\left(\dfrac{3\theta}{5} + 11°\right) = \sin\left(\dfrac{7\theta}{10} + 40°\right)$

**True or False** *Tell whether each statement is true or false. If false, tell why.*

**7.** $\sin 46° < \sin 58°$

**8.** $\cos 47° < \cos 58°$

**9.** $\sec 48° \geq \cos 42°$

**10.** $\sin 22° \geq \csc 68°$

**11.** Explain why, in the figure, the cosine of angle $A$ is equal to the sine of angle $B$.

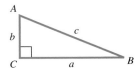

*Find exact values of the six trigonometric functions for each angle. Do not use a calculator. Rationalize denominators when applicable.*

**12.** $120°$

**13.** $1020°$

**14.** $-225°$

**15.** $-1470°$

*Find all values of $\theta$, if $\theta$ is in the interval $[0°, 360°)$ and $\theta$ has the given function value.*

**16.** $\sin \theta = -\dfrac{1}{2}$

**17.** $\cos \theta = -\dfrac{1}{2}$

**18.** $\cot \theta = -1$

**19.** $\sec \theta = -\dfrac{2\sqrt{3}}{3}$

*Evaluate each expression. Give exact values.*

**20.** $\cos 60° + 2 \sin^2 30°$

**21.** $\tan^2 120° - 2 \cot 240°$

**22.** $\sec^2 300° - 2 \cos^2 150° + \tan 45°$

**23.** Find the sine, cosine, and tangent function values for each angle.

(a)

(b)

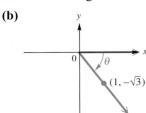

730

*Use a calculator to find each value.*

**24.** sin 72° 30′

**25.** sec 222° 30′

**26.** cot 305.6°

**27.** csc 78° 21′

**28.** sec 58.9041°

**29.** tan 11.7689°

**30.** *Multiple Choice* Which one of the following cannot be *exactly* determined using the methods of this chapter?

    **A.** cos 135°    **B.** cot(−45°)    **C.** sin 300°    **D.** tan 140°

*Use a calculator to find each value of θ, where θ is in the interval [0°, 90°). Give answers in decimal degrees.*

**31.** sin θ = 0.82584121

**32.** cot θ = 1.1249386

**33.** cos θ = 0.97540415

**34.** sec θ = 1.2637891

**35.** tan θ = 1.9633124

**36.** csc θ = 9.5670466

*Find two angles in the interval [0°, 360°) that satisfy each of the following. Leave answers in decimal degrees rounded to the nearest tenth.*

**37.** sin θ = 0.73254290

**38.** tan θ = 1.3865342

*True or False* Tell whether each statement is true *or* false. *If false, tell why. Use a calculator for Exercises 39 and 42.*

**39.** sin 50° + sin 40° = sin 90°

**40.** cos 210° = cos 180° · cos 30° − sin 180° · sin 30°

**41.** sin 240° = 2 sin 120° · cos 120°

**42.** sin 42° + sin 42° = sin 84°

**43.** A student wants to use a calculator to find the value of cot 25°. However, instead of entering $\frac{1}{\tan 25}$, he enters tan⁻¹ 25. Assuming the calculator is in degree mode, will this produce the correct answer? Explain.

*For each angle θ, use a calculator to find* cos θ *and* sin θ. *Use your results to decide in which quadrant the angle lies.*

**44.** θ = 2976°

**45.** θ = 1997°

**46.** θ = 4000°

*Solve each right triangle. In Exercise 48, give angles to the nearest minute. In Exercises 49 and 50, label the triangle ABC as in Exercises 47 and 48.*

**47.**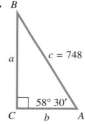

**48.**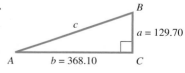

**49.** $A = 39.72°$, $b = 38.97$ m

**50.** $B = 47°\ 53′$, $b = 298.6$ m

Solve each problem. (*Source for Exercises 51 and 52:* Parker, M., Editor, *She Does Math,* Mathematical Association of America, 1995.)

**51.** A civil engineer must determine the height of the tree shown in the figure. The given angle was measured with a **clinometer.** Find the height of the tree to the nearest whole number.

This is a picture of one type of clinometer, called an Abney hand level and clinometer. (Courtesy of Keuffel & Esser Co.)

**52.** To correct mild double vision, a small amount of prism is added to a patient's eyeglasses. The amount of light shift this causes is measured in **prism diopters.** A patient needs 12 prism diopters horizontally and 5 prism diopters vertically. A prism that corrects for both requirements should have length $r$ and be set at angle $\theta$. Find the values of $r$ and $\theta$ in the figure.

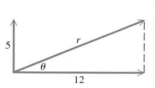

**53.** The angle of elevation from a point 93.2 ft from the base of a tower to the top of the tower is 38° 20′. Find the height of the tower.

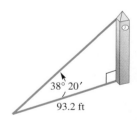

**54.** The angle of depression of a television tower to a point on the ground 36.0 m from the bottom of the tower is 29.5°. Find the height of the tower.

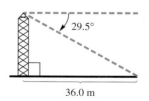

**55.** One side of a rectangle measures 15.24 cm. The angle between the diagonal and that side is 35.65°. Find the length of the diagonal.

**56.** An isosceles triangle has a base of length 49.28 m. The angle opposite the base is 58.746°. Find the length of each of the two equal sides.

# 15 TEST

1. Give the six trigonometric function values of angle $A$.

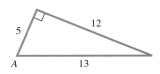

2. Find the exact values of each part labeled with a letter.

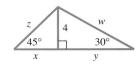

3. Find a solution for $\sin(B + 15°) = \cos(2B + 30°)$.

4. *True or False*  Determine whether each statement is *true* or *false*. If false, tell why.

   (a) $\sin 24° < \sin 48°$   (b) $\cos 24° < \cos 48°$

   (c) $\cos(60° + 30°) = \cos 60° \cdot \cos 30° - \sin 60° \cdot \sin 30°$

*Find the exact values of the six trigonometric functions for each angle. Rationalize denominators when applicable.*

5. $240°$   6. $-135°$   7. $990°$

*Find all values of $\theta$ in the interval $[0°, 360°)$ that have the given function value.*

8. $\cos \theta = -\dfrac{\sqrt{2}}{2}$   9. $\csc \theta = -\dfrac{2\sqrt{3}}{3}$   10. $\tan \theta = 1$

11. Use a calculator to approximate each value.

    (a) $\sin 78° \, 21'$   (b) $\tan 117.689°$   (c) $\sec 58.9041°$

12. Find a value of $\theta$ in the interval $[0°, 90°)$ in decimal degrees, if $\sin \theta = 0.27843196$.

13. Solve the triangle.

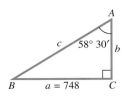

14. A guy wire 77.4 m long is attached to the top of an antenna mast that is 71.3 m high. Find the angle that the wire makes with the ground.

15. To measure the height of a flagpole, Amado Carillo found that the angle of elevation from a point 24.7 ft from the base to the top is $32° \, 10'$. What is the height of the flagpole?

16. The highest point in Texas is Guadalupe Peak. The angle of depression from the top of this peak to a small miner's cabin at an approximate elevation of 2000 ft is $26°$. The cabin is located 14,000 ft horizontally from a point directly under the top of the mountain. Find the altitude of the top of the mountain to the nearest hundred feet.

# Radian Measure and Circular Functions

In August 2003, the planet Mars passed closer to Earth than it had in almost 60,000 years. Like Earth, Mars rotates on its axis and, thus, has days (also called *sols*) and nights. The photo shows a dust cloud/streak in the north polar cap of Mars, taken by the Hubble Space Telescope in 1996. In early 2004, the rovers *Spirit* and *Opportunity* landed on Mars and have provided scientists a wealth of information about the "Red Planet." (*Source:* www.hubblesite.org)

In Exercise 36 of Section 16.4, we examine the length of a Martian *sol* using *radian measure,* an alternative to measuring with degrees.

**16.1** Radian Measure

**16.2** Applications of Radian Measure

**16.3** The Unit Circle and Circular Functions

**16.4** Linear and Angular Speed

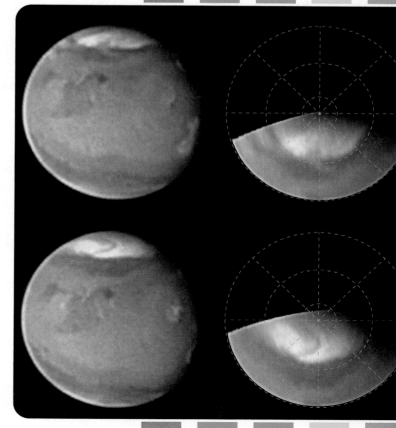

# 16.1 Radian Measure

**OBJECTIVES**

1. Understand the meaning of radian measure.
2. Convert between degrees and radians.
3. Find function values for angles in radians.

**OBJECTIVE 1** Understand the meaning of radian measure. We have seen that angles can be measured in degrees. In more theoretical work in mathematics, *radian measure* of angles is preferred. Radian measure allows us to treat the trigonometric functions as functions with domains of *real numbers,* rather than angles.

Figure 1 shows an angle $\theta$ in standard position along with a circle of radius $r$. The vertex of $\theta$ is at the center of the circle. Because angle $\theta$ intercepts an arc on the circle equal in length to the radius of the circle, we say that angle $\theta$ has a measure of 1 radian.

### Radian

An angle with its vertex at the center of a circle that intercepts an arc on the circle equal in length to the radius of the circle has a measure of **1 radian.**

It follows that an angle of measure 2 radians intercepts an arc equal in length to twice the radius of the circle, an angle of measure $\frac{1}{2}$ radian intercepts an arc equal in length to half the radius of the circle, and so on. In general, if $\theta$ is a central angle of a circle of radius $r$ and $\theta$ intercepts an arc of length $s$, then the radian measure of $\theta$ is $\frac{s}{r}$.

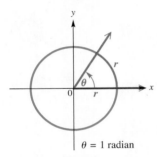

**FIGURE 1**

$\theta = 1$ radian

**OBJECTIVE 2** Convert between degrees and radians. The **circumference** of a circle—the distance around the circle—is given by $C = 2\pi r$, where $r$ is the radius of the circle. The formula $C = 2\pi r$ shows that the radius can be laid off $2\pi$ times around a circle. Therefore, an angle of 360°, which corresponds to a complete circle, intercepts an arc equal in length to $2\pi$ times the radius of the circle. Thus, an angle of 360° has a measure of $2\pi$ radians:

$$360° = 2\pi \text{ radians.}$$

An angle of 180° is half the size of an angle of 360°, so an angle of 180° has half the radian measure of an angle of 360°.

$$180° = \frac{1}{2}(2\pi) \text{ radians} = \pi \text{ radians} \quad \text{Degree/radian relationship}$$

We can use the relationship 180° = $\pi$ radians to develop a method for converting between degrees and radians as follows.

$$180° = \pi \text{ radians}$$

$$1° = \frac{\pi}{180} \text{ radian} \quad \text{Divide by 180.} \quad \text{or} \quad 1 \text{ radian} = \frac{180°}{\pi} \quad \text{Divide by } \pi.$$

### Converting Between Degrees and Radians

1. Multiply a degree measure by $\frac{\pi}{180}$ radian and simplify to convert to radians.
2. Multiply a radian measure by $\frac{180°}{\pi}$ and simplify to convert to degrees.

**EXAMPLE 1** Converting Degrees to Radians

Convert each degree measure to radians.

(a) 45°     (b) −270°     (c) 249.8°

**Solution**

(a) $45° = 45\left(\dfrac{\pi}{180} \text{ radian}\right) = \dfrac{\pi}{4}$ radian    Multiply by $\dfrac{\pi}{180}$ radian.

(b) $-270° = -270\left(\dfrac{\pi}{180} \text{ radian}\right) = -\dfrac{270\pi}{180}$ radians

$\qquad\qquad\qquad\qquad\qquad\qquad = -\dfrac{3\pi}{2}$ radians    Lowest terms

(c) $249.8° = 249.8\left(\dfrac{\pi}{180} \text{ radian}\right) \approx 4.360$ radians    Nearest thousandth

**NOW TRY** Exercises 7, 13, and 43.

### EXAMPLE 2  Converting Radians to Degrees

Convert each radian measure to degrees.

(a) $\dfrac{9\pi}{4}$    (b) $-\dfrac{5\pi}{6}$    (c) 4.25

**Solution**

(a) $\dfrac{9\pi}{4}$ radians $= \dfrac{9\pi}{4}\left(\dfrac{180°}{\pi}\right) = 405°$    Multiply by $\dfrac{180°}{\pi}$.

(b) $-\dfrac{5\pi}{6}$ radians $= -\dfrac{5\pi}{6}\left(\dfrac{180°}{\pi}\right) = -150°$

(c) $4.25$ radians $= 4.25\left(\dfrac{180°}{\pi}\right) \approx 243.5° = 243° \, 30'$    Use a calculator.

**NOW TRY** Exercises 27, 31, and 55.

A TI-83/84 Plus calculator can convert directly between degrees and radians. This radian mode screen shows the conversions for Example 1. Verify that the first two results are *approximations* for the *exact* values of $\dfrac{\pi}{4}$ and $-\dfrac{3\pi}{2}$.

This degree mode screen shows how a TI-83/84 Plus calculator converts the radian measures in Example 2 to degree measures.

Another way to convert a radian measure that is a rational multiple of $\pi$, such as $\dfrac{9\pi}{4}$, to degrees is to just substitute 180° for $\pi$. In Example 2(a), this would be $\dfrac{9(180°)}{4} = 405°$.

One of the most important facts to remember when working with angles and their measures is summarized in the following statement.

### Agreement on Angle Measurement Units

*If no unit of angle measure is specified, then the angle is understood to be measured in radians.*

For example, Figure 2(a) shows an angle of 30°, while Figure 2(b) shows an angle of 30 (which means 30 radians).

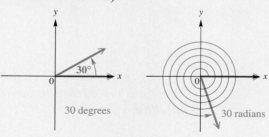

30 degrees

30 radians

Note the difference between an angle of 30 *degrees* and an angle of 30 *radians*.

(a)                    (b)

**FIGURE 2**

The following table and Figure 3 give some equivalent angle measures in degrees and radians. Keep in mind that **180° = π radians.**

| Degrees | Radians | | Degrees | Radians | |
|---|---|---|---|---|---|
| | Exact | Approximate | | Exact | Approximate |
| 0° | 0 | 0 | 90° | $\frac{\pi}{2}$ | 1.57 |
| 30° | $\frac{\pi}{6}$ | 0.52 | 180° | $\pi$ | 3.14 |
| 45° | $\frac{\pi}{4}$ | 0.79 | 270° | $\frac{3\pi}{2}$ | 4.71 |
| 60° | $\frac{\pi}{3}$ | 1.05 | 360° | $2\pi$ | 6.28 |

*These exact values are rational multiples of π.*

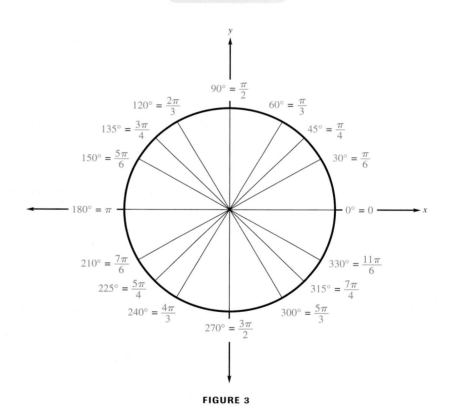

**FIGURE 3**

The angles marked in Figure 3 are extremely important in the study of trigonometry. *You should learn these equivalences, as they will appear often in the chapters to follow.*

**OBJECTIVE 3** Find function values for angles in radians. Trigonometric function values for angles measured in radians can be found by first converting radian measure to degrees. *(You should try to skip this intermediate step as soon as possible, and find the function values directly from radian measure.)*

### EXAMPLE 3  Finding Function Values of Angles in Radian Measure

Find each function value.

**(a)** $\tan \dfrac{2\pi}{3}$    **(b)** $\sin \dfrac{3\pi}{2}$    **(c)** $\cos\left(-\dfrac{4\pi}{3}\right)$

**Solution**

**(a)** First convert $\dfrac{2\pi}{3}$ radians to degrees.

$$\tan \dfrac{2\pi}{3} = \tan\left(\dfrac{2\pi}{3} \cdot \dfrac{180°}{\pi}\right) \quad \text{Multiply by } \dfrac{180°}{\pi}.$$
$$= \tan 120°$$
$$= -\sqrt{3}$$

**(b)** From the table and Figure 3 on the preceding page, $\dfrac{3\pi}{2}$ radians $= 270°$, so

$$\sin \dfrac{3\pi}{2} = \sin 270° = -1.$$

**(c)** $\cos\left(-\dfrac{4\pi}{3}\right) = \cos\left(-\dfrac{4\pi}{3} \cdot \dfrac{180°}{\pi}\right)$
$$= -\cos 60°$$
$$= -\dfrac{1}{2}$$

**NOW TRY** Exercises 65, 75, and 79.

## 16.1 Exercises

**NOW TRY Exercise**

*Concept Check* In Exercises 1–6, each angle θ is an integer when measured in radians. Give the radian measure of the angle.

**1.** 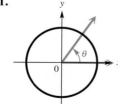   **2.** 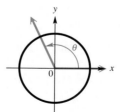   **3.**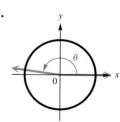

**4.** 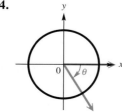   **5.** 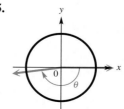   **6.**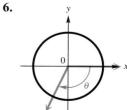

*Convert each degree measure to radians. Leave answers as multiples of $\pi$. See Examples 1(a) and 1(b).*

**7.** 60°  **8.** 30°  **9.** 90°  **10.** 120°

**11.** 150°  **12.** 270°  **13.** −300°  **14.** −315°

**15.** 450°  **16.** 480°  **17.** 1800°  **18.** −3600°

*Give a short explanation in Exercises 19–24.*

**19.** In your own words, explain how to convert degree measure to radian measure.

**20.** In your own words, explain how to convert radian measure to degree measure.

**21.** In your own words, explain the meaning of radian measure.

**22.** Explain the difference between degree measure and radian measure.

**23.** Use an example to show that you can convert from radian measure to degree measure by multiplying by $\frac{180°}{\pi}$.

**24.** Explain why an angle of radian measure $t$ in standard position intercepts an arc of length $t$ on a circle of radius 1.

*Convert each radian measure to degrees. See Examples 2(a) and 2(b).*

**25.** $\frac{\pi}{3}$  **26.** $\frac{8\pi}{3}$  **27.** $\frac{7\pi}{4}$  **28.** $\frac{2\pi}{3}$

**29.** $\frac{11\pi}{6}$  **30.** $\frac{15\pi}{4}$  **31.** $-\frac{\pi}{6}$  **32.** $-\frac{8\pi}{5}$

**33.** $\frac{7\pi}{10}$  **34.** $\frac{11\pi}{15}$  **35.** $-\frac{4\pi}{15}$  **36.** $-\frac{7\pi}{20}$

**37.** $\frac{17\pi}{20}$  **38.** $\frac{11\pi}{30}$  **39.** $-5\pi$  **40.** $15\pi$

*Convert each degree measure to radians. See Example 1(c).*

**41.** 39°  **42.** 74°  **43.** 42.5°  **44.** 264.9°

**45.** 139° 10′  **46.** 174° 50′  **47.** 64.29°  **48.** 85.04°

**49.** 56° 25′  **50.** 122° 37′  **51.** 47.6925°  **52.** 23.0143°

*Convert each radian measure to degrees. Write answers to the nearest minute. See Example 2(c).*

**53.** 2  **54.** 5  **55.** 1.74  **56.** 3.06

**57.** 0.3417  **58.** 9.84763  **59.** −5.01095  **60.** −3.47189

**61.** *Concept Check* The value of sin 30 is not $\frac{1}{2}$. Why is this true?

**62.** Explain in your own words what is meant by an angle of one radian.

*Find the exact value of each expression without using a calculator. See Example 3.*

**63.** $\sin \frac{\pi}{3}$  **64.** $\cos \frac{\pi}{6}$  **65.** $\tan \frac{\pi}{4}$  **66.** $\cot \frac{\pi}{3}$

**67.** $\sec \frac{\pi}{6}$  **68.** $\csc \frac{\pi}{4}$  **69.** $\sin \frac{\pi}{2}$  **70.** $\csc \frac{\pi}{2}$

**71.** $\tan \frac{5\pi}{3}$  **72.** $\cot \frac{2\pi}{3}$  **73.** $\sin \frac{5\pi}{6}$  **74.** $\tan \frac{5\pi}{6}$

**75.** $\cos 3\pi$  **76.** $\sec \pi$  **77.** $\sin\left(-\frac{8\pi}{3}\right)$  **78.** $\cot\left(-\frac{2\pi}{3}\right)$

**79.** $\sin\left(-\frac{7\pi}{6}\right)$  **80.** $\cos\left(-\frac{\pi}{6}\right)$  **81.** $\tan\left(-\frac{14\pi}{3}\right)$  **82.** $\csc\left(-\frac{13\pi}{3}\right)$

**83.** *Concept Check* The figure shows the same angles measured in both degrees and radians. Complete the missing measures.

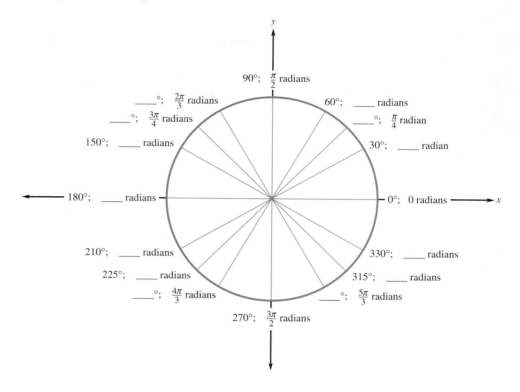

*Solve each problem.*

**84.** The term **grade** has several different meanings in construction work. Some engineers use the term **grade** to represent $\frac{1}{100}$ of a right angle and express grade as a percent. For instance, an angle of 0.9° would be referred to as a 1% grade. (*Source:* Hay, W., *Railroad Engineering*, John Wiley & Sons, 1982.)

(a) By what number should you multiply a grade (disregarding the % symbol) to convert it to radians?

(b) In a rapid-transit rail system, the maximum grade allowed between two stations is 3.5%. Express this angle in degrees and radians.

**85.** Through how many radians will the hour hand on a clock rotate in (a) 24 hr and (b) 4 hr?

**86.** A circular pulley is rotating about its center. Through how many radians would it turn in (a) 8 rotations and (b) 30 rotations?

**87.** A space vehicle is orbiting Earth in a circular orbit. What radian measure corresponds to (a) 2.5 orbits and (b) $\frac{4}{3}$ orbit?

**88.** A stationary horse on a carousel makes 12 complete revolutions. Through what radian measure angle does the horse revolve?

## 16.2 Applications of Radian Measure

**OBJECTIVES**
1. Find arc length on a circle.
2. Find the area of a sector of a circle.

**OBJECTIVE 1** Find arc length on a circle. We use radian measure in the formula to find the length of an arc of a circle. This formula is derived from the fact (proved in geometry) that the length of an arc is proportional to the measure of its central angle. In Figure 4, angle $QOP$ has measure 1 radian and intercepts an arc of length $r$ on the circle. Angle $ROT$ has measure $\theta$ radians and intercepts an arc of length $s$ on the circle. Since the lengths of the arcs are proportional to the measures of their central angles,

$$\frac{s}{r} = \frac{\theta}{1}.$$

Multiplying both sides by $r$ gives the following result.

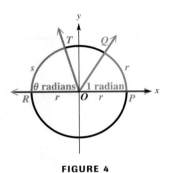

**FIGURE 4**

### Arc Length

The length $s$ of the arc intercepted on a circle of radius $r$ by a central angle of measure $\theta$ radians is given by the product of the radius and the radian measure of the angle, or

$$s = r\theta, \quad \theta \text{ in radians.}$$

**CAUTION** *Avoid the common error of applying this formula with $\theta$ in degree mode. When applying the formula $s = r\theta$, the value of $\theta$ MUST be expressed in radians.*

**EXAMPLE 1** Finding Arc Length Using $s = r\theta$

A circle has radius 18.20 cm. Find the length of the arc intercepted by a central angle having each of the following measures.

(a) $\dfrac{3\pi}{8}$ radians  (b) $144°$

**Solution**

(a) As shown in Figure 5, $r = 18.20$ cm and $\theta = \dfrac{3\pi}{8}$.

$s = r\theta$   Arc length formula

$s = 18.20\left(\dfrac{3\pi}{8}\right)$ cm   Substitute for $r$ and $\theta$.

$s \approx 21.44$ cm

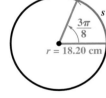

**FIGURE 5**

(b) The formula $s = r\theta$ requires that $\theta$ be measured in radians. First, convert $\theta$ to radians by multiplying $144°$ by $\dfrac{\pi}{180}$ radian.

$$144° = 144\left(\dfrac{\pi}{180}\right) = \dfrac{4\pi}{5} \text{ radians} \quad \text{Convert from degrees to radians.}$$

The length $s$ is given by

$$s = r\theta = 18.20\left(\dfrac{4\pi}{5}\right) \approx 45.74 \text{ cm.}$$

Be sure to use radians for $\theta$ in $s = r\theta$.

**NOW TRY** Exercises 11 and 15.

## EXAMPLE 2 Using Latitudes to Find the Distance between Two Cities

**Latitude** gives the measure of a central angle with vertex at Earth's center whose initial side goes through the equator and whose terminal side goes through the given location. Reno, Nevada, is approximately due north of Los Angeles. The latitude of Reno is 40° N, while that of Los Angeles is 34° N. (The N in 34° N means *north* of the equator.) The radius of Earth is 6400 km. Find the north–south distance between the two cities.

**Solution** As shown in Figure 6, the central angle between Reno and Los Angeles is $40° - 34° = 6°$. The distance between the two cities can be found by the formula $s = r\theta$, after 6° is first converted to radians.

$$6° = 6\left(\frac{\pi}{180}\right) = \frac{\pi}{30} \text{ radian}$$

The distance between the two cities is

$$s = r\theta = 6400\left(\frac{\pi}{30}\right) \approx 670 \text{ km.} \qquad \text{Let } r = 6400 \text{ and } \theta = \frac{\pi}{30}.$$

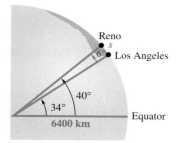

FIGURE 6

**NOW TRY** Exercise 21.

## EXAMPLE 3 Finding a Length Using $s = r\theta$

A rope is being wound around a drum with radius 0.8725 ft. (See Figure 7.) How much rope will be wound around the drum if the drum is rotated through an angle of 39.72°?

**Solution** The length of rope wound around the drum is the arc length for a circle of radius 0.8725 ft and a central angle of 39.72°. Use the formula $s = r\theta$, with the angle converted to radian measure. The length of the rope wound around the drum is approximately

$$s = r\theta = 0.8725\left[39.72\left(\frac{\pi}{180}\right)\right] \approx 0.6049 \text{ ft.}$$

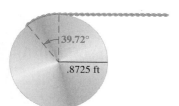

FIGURE 7

**NOW TRY** Exercise 33(a).

## EXAMPLE 4 Finding an Angle Measure Using $s = r\theta$

Two gears are adjusted so that the smaller gear drives the larger one, as shown in Figure 8. If the smaller gear rotates through an angle of 225°, through how many degrees will the larger gear rotate?

**Solution** First find the radian measure of the angle, and then find the arc length on the smaller gear that determines the motion of the larger gear. Since $225° = \frac{5\pi}{4}$ radians, for the smaller gear,

$$s = r\theta = 2.5\left(\frac{5\pi}{4}\right) = \frac{12.5\pi}{4} = \frac{25\pi}{8} \text{ cm.}$$

FIGURE 8

*(continued)*

An arc with this length on the larger gear corresponds to an angle measure $\theta$, in radians, where

$$s = r\theta$$

$$\frac{25\pi}{8} = 4.8\theta \qquad \text{Substitute } \tfrac{25\pi}{8} \text{ for } s \text{ and } 4.8 \text{ for } r.$$

$$\frac{125\pi}{192} = \theta. \qquad 4.8 = \tfrac{48}{10} = \tfrac{24}{5}; \text{ multiply by } \tfrac{5}{24} \text{ to solve for } \theta.$$

Converting $\theta$ back to degrees shows that the larger gear rotates through

$$\frac{125\pi}{192}\left(\frac{180°}{\pi}\right) \approx 117°. \qquad \text{Convert } \theta = \tfrac{125\pi}{192} \text{ to degrees.}$$

**NOW TRY** Exercise 27.

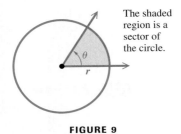

FIGURE 9

The shaded region is a sector of the circle.

**OBJECTIVE 2** Find the area of a sector of a circle. A **sector of a circle** is the portion of the interior of a circle intercepted by a central angle. Think of it as a "piece of pie." See Figure 9. A complete circle can be thought of as an angle with measure $2\pi$ radians. If a central angle for a sector has measure $\theta$ radians, then the sector makes up the fraction $\frac{\theta}{2\pi}$ of a complete circle. The area of a complete circle with radius $r$ is $A = \pi r^2$. Therefore,

$$\text{area of the sector} = \frac{\theta}{2\pi}(\pi r^2) = \frac{1}{2}r^2\theta, \qquad \theta \text{ in radians.}$$

This discussion is summarized as follows.

### Area of a Sector

The area $A$ of a sector of a circle of radius $r$ and central angle $\theta$ is given by

$$A = \frac{1}{2}r^2\theta, \qquad \theta \text{ in radians.}$$

**CAUTION** As in the formula for arc length, *the value of $\theta$ must be in radians when using this formula for the area of a sector.*

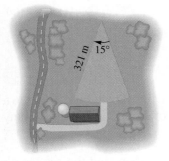

FIGURE 10

**EXAMPLE 5** Finding the Area of a Sector-Shaped Field

Find the area of the sector-shaped field shown in Figure 10.

**Solution** First, convert 15° to radians.

$$15° = 15\left(\frac{\pi}{180}\right) = \frac{\pi}{12} \text{ radian}$$

Now use the formula to find the area of a sector of a circle with radius $r = 321$.

$$A = \frac{1}{2}r^2\theta = \frac{1}{2}(321)^2\left(\frac{\pi}{12}\right) \approx 13{,}500 \text{ m}^2$$

**NOW TRY** Exercise 49.

## 16.2 Exercises

*Concept Check*   Find the exact length of each arc intercepted by the given central angle.

1.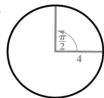
2.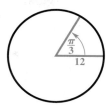
3.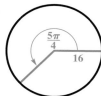

*Concept Check*   Find the radius of each circle.

4.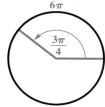
5.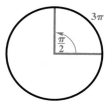
6.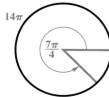

*Concept Check*   Find the measure of each central angle (in radians).

7.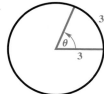
8.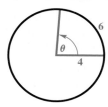
9.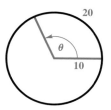

10. Explain in your own words how to find the *degree* measure of a central angle in a circle if both the radius and the length of the intercepted arc are known.

*Unless otherwise directed, give calculator approximations in your answers in the rest of this exercise set.*

Find the length to three significant digits of each arc intercepted by a central angle $\theta$ in a circle of radius $r$. See Example 1.

11. $r = 12.3$ cm, $\theta = \dfrac{2\pi}{3}$ radians
12. $r = 0.892$ cm, $\theta = \dfrac{11\pi}{10}$ radians
13. $r = 1.38$ ft, $\theta = \dfrac{5\pi}{6}$ radians
14. $r = 3.24$ mi, $\theta = \dfrac{7\pi}{6}$ radians
15. $r = 4.82$ m, $\theta = 60°$
16. $r = 71.9$ cm, $\theta = 135°$
17. $r = 15.1$ in., $\theta = 210°$
18. $r = 12.4$ ft, $\theta = 330°$

19. *Concept Check*   If the radius of a circle is doubled, how is the length of the arc intercepted by a fixed central angle changed?

20. *Concept Check*   Radian measure simplifies many formulas, such as the formula for arc length, $s = r\theta$. Give the corresponding formula when $\theta$ is measured in degrees instead of radians.

Find the distance in kilometers between each pair of cities, assuming they lie on the same north–south line. See Example 2.

21. Panama City, Panama, 9° N, and Pittsburgh, Pennsylvania, 40° N
22. Farmersville, California, 36° N, and Penticton, British Columbia, 49° N

23. New York City, New York, 41° N, and Lima, Peru, 12° S

24. Halifax, Nova Scotia, 45° N, and Buenos Aires, Argentina, 34° S

25. Madison, South Dakota, and Dallas, Texas, are 1200 km apart and lie on the same north–south line. The latitude of Dallas is 33° N. What is the latitude of Madison?

26. Charleston, South Carolina, and Toronto, Canada, are 1100 km apart and lie on the same north–south line. The latitude of Charleston is 33° N. What is the latitude of Toronto?

*Work each problem. See Examples 3 and 4.*

27. Two gears are adjusted so that the smaller gear drives the larger one, as shown in the figure. If the smaller gear rotates through an angle of 300°, through how many degrees will the larger gear rotate?

28. Repeat Exercise 27 for gear radii of 4.8 in. and 7.1 in., and for an angle of 315° for the smaller gear.

29. The rotation of the smaller wheel in the figure causes the larger wheel to rotate. Through how many degrees will the larger wheel rotate if the smaller one rotates through 60.0°?

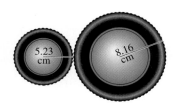

30. Repeat Exercise 29 for wheel radii of 6.84 in. and 12.46 in. and an angle of 150° for the smaller wheel.

31. Find the radius of the larger wheel in the figure if the smaller wheel rotates 80.0° when the larger wheel rotates 50.0°.

32. Repeat Exercise 31 if the smaller wheel of radius 14.6 in. rotates 120° when the larger wheel rotates 60°.

33. (a) How many inches will the weight in the figure rise if the pulley is rotated through an angle of 71° 50′?

    (b) Through what angle, to the nearest minute, must the pulley be rotated to raise the weight 6 in.?

34. Find the radius of the pulley in the figure if a rotation of 51.6° raises the weight 11.4 cm.

35. The figure shows the chain drive of a bicycle. How far will the bicycle move if the pedals are rotated through 180°? Assume the radius of the bicycle wheel is 13.6 in.

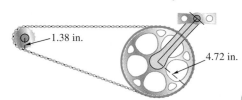

**36.** The speedometer of Terry's Honda CR-V is designed to be accurate with tires of radius 14 in.
  (a) Find the number of rotations of a tire in 1 hr if the car is driven at 55 mph.
  (b) Suppose that oversize tires of radius 16 in. are placed on the car. If the car is now driven for 1 hr with the speedometer reading 55 mph, how far has the car gone? If the speed limit is 55 mph, does Terry deserve a speeding ticket?

*Concept Check* Find the area of each sector.

**37.**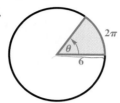

**38.**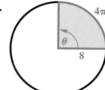

**39.**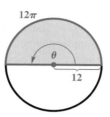

**40.**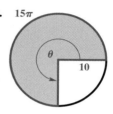

*Concept Check* Find the measure (in degrees) of each central angle. The number inside the sector is the area.

**41.**

**42.**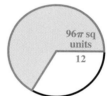

*Concept Check* Find the measure (in radians) of each central angle. The number inside the sector is the area.

**43.**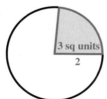

**44.**

*Find the area of a sector of a circle having radius r and central angle $\theta$. Express answers to the nearest tenth. See Example 5.*

**45.** $r = 29.2$ m, $\theta = \dfrac{5\pi}{6}$ radians

**46.** $r = 59.8$ km, $\theta = \dfrac{2\pi}{3}$ radians

**47.** $r = 30.0$ ft, $\theta = \dfrac{\pi}{2}$ radians

**48.** $r = 90.0$ yd, $\theta = \dfrac{5\pi}{6}$ radians

**49.** $r = 12.7$ cm, $\theta = 81°$

**50.** $r = 18.3$ m, $\theta = 125°$

**51.** $r = 40.0$ mi, $\theta = 135°$

**52.** $r = 90.0$ km, $\theta = 270°$

*Work each problem.*

**53.** Find the measure (in radians) of a central angle of a sector of area 16 in.² in a circle of radius 3.0 in.

54. Find the radius of a circle in which a central angle of $\frac{\pi}{6}$ radian determines a sector of area 64 m².

55. The figure shows Medicine Wheel, a Native American structure in northern Wyoming. This circular structure is perhaps 2500 yr old. There are 27 aboriginal spokes in the wheel, all equally spaced.

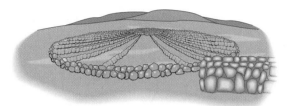

(a) Find the measure of each central angle in degrees and in radians.
(b) If the radius of the wheel is 76.0 ft, find the circumference.
(c) Find the length of each arc intercepted by consecutive pairs of spokes.
(d) Find the area of each sector formed by consecutive spokes.

56. The Ford Model A, built from 1928 to 1931, had a single windshield wiper on the driver's side. The total arm and blade was 10 in. long and rotated back and forth through an angle of 95°. The shaded region in the figure is the portion of the windshield cleaned by the 7-in. wiper blade. What is the area of the region cleaned?

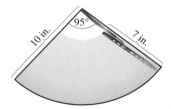

57. In the United States, circular railroad curves are designated by the **degree of curvature,** the central angle subtended by a chord of 100 ft. Suppose a portion of track has curvature 42.0°. (*Source:* Hay, W., *Railroad Engineering,* John Wiley & Sons, 1982.)

(a) What is the radius of the curve?
(b) What is the length of the arc determined by the 100-ft chord?
(c) What is the area of the portion of the circle bounded by the arc and the 100-ft chord?

58. A 300-megawatt solar-power plant requires approximately 950,000 m² of land area in order to collect the required amount of energy from sunlight. If this land area is circular, what is its radius? If this land area is a 35° sector of a circle, what is its radius?

59. A frequent problem in surveying city lots and rural lands adjacent to curves of highways and railways is that of finding the area when one or more of the boundary lines is the arc of a circle. Find the area of the lot (to two significant digits) shown in the figure. (*Source:* Anderson, J. and E. Michael, *Introduction to Surveying,* McGraw-Hill, 1985.)

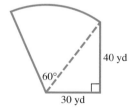

60. **Nautical miles** are used by ships and airplanes. They are different from **statute miles,** which equal 5280 ft. A nautical mile is defined to be the arc length along the equator intercepted by a central angle *AOB* of 1 min, as illustrated in the figure. If the equatorial radius of Earth is 3963 mi, use the arc length formula to approximate the number of statute miles in 1 nautical mile. Round your answer to two decimal places.

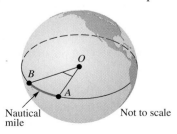

61. The first accurate estimate of the distance around Earth was done by the Greek astronomer Eratosthenes (276–195 B.C.), who noted that the noontime position of the Sun at the summer solstice differed by 7° 12′ from the city of Syene to the city of Alexandria. (See the figure.) The distance between these two cities is 496 mi. Use the arc length formula to estimate the radius of Earth. Then find the circumference of Earth. (*Source:* Zeilik, M., *Introductory Astronomy and Astrophysics,* Third Edition, Saunders College Publishers, 1992.)

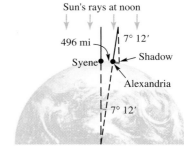

62. The distance to the Moon is approximately 238,900 mi. Use the arc length formula to estimate the diameter $d$ of the Moon if angle $\theta$ in the figure is measured to be 0.5170°.

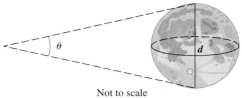

Not to scale

## 16.3 The Unit Circle and Circular Functions

**OBJECTIVES**

1. Learn the definitions and domains of the circular functions.
2. Find exact and approximate circular function values.
3. Find a number with a given circular function value.

**OBJECTIVE 1** Learn the definitions and domains of the circular functions. In Section 14.3, we defined the six trigonometric functions in such a way that the domain of each function was a set of *angles* in standard position. These angles can be measured in degrees or in radians. In advanced courses, such as calculus, it is necessary to modify the trigonometric functions so that their domains consist of *real numbers* rather than angles. We do this by using the relationship between an angle $\theta$ and an arc of length $s$ on a circle.

In Figure 11, we start at the point $(1, 0)$ and measure an arc of length $s$ along the circle. If $s > 0$, then the arc is measured in a counterclockwise direction, and if $s < 0$, then the direction is clockwise. (If $s = 0$, then no arc is measured.) Let the endpoint of this arc be at the point $(x, y)$. The circle in Figure 11 is the **unit circle**—it has center at the origin and radius 1 unit (hence the name *unit circle*). Recall from algebra that the equation of this circle is

$$x^2 + y^2 = 1.$$

Recall that the radian measure of $\theta$ is related to the arc length $s$. For $\theta$ measured in radians, we know that $s = r\theta$. Here, $r = 1$, so $s$, which is measured in linear units such as inches or centimeters, is equal to $\theta$, measured in radians. Thus, the trigonometric functions of angle $\theta$ in radians found by choosing a point $(x, y)$ on the unit circle can be rewritten as functions of the arc length $s$, a real number. When interpreted this way, they are called **circular functions**.

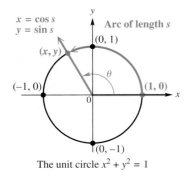

The unit circle $x^2 + y^2 = 1$

**FIGURE 11**

### Circular Functions

For any real number $s$ represented by a directed arc on the unit circle,

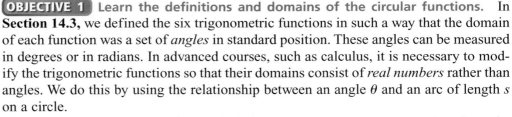

$$\sin s = y \qquad \cos s = x \qquad \tan s = \frac{y}{x} \ (x \neq 0)$$

$$\csc s = \frac{1}{y} \ (y \neq 0) \qquad \sec s = \frac{1}{x} \ (x \neq 0) \qquad \cot s = \frac{x}{y} \ (y \neq 0).$$

Since $x$ represents the cosine of $s$ and $y$ represents the sine of $s$, and because of the discussion in **Section 16.1** on converting between degrees and radians, we can summarize a great deal of information in a concise manner, as seen in Figure 12.*

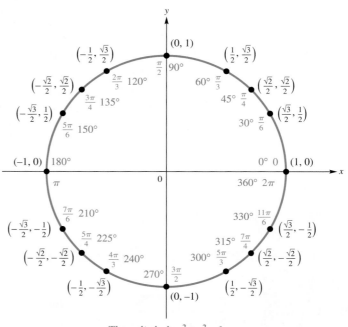

The unit circle $x^2 + y^2 = 1$

**FIGURE 12**

The unit circle is symmetric with respect to the $x$-axis, the $y$-axis, and the origin. (See **Section 10.4**.) Thus, if a point $(a, b)$ lies on the unit circle, so does $(a, -b)$, $(-a, b)$, and $(-a, -b)$. Furthermore, each of these points has a *reference arc* of equal magnitude. For a point on the unit circle, its **reference arc** is the shortest arc from the point itself to the nearest point on the $x$-axis. (This concept is analogous to the reference angle concept introduced in **Chapter 15**.) Using the concept of symmetry makes determining sines and cosines of the real numbers identified in Figure 12 a relatively simple procedure if we know the coordinates of the points labeled in quadrant I.

For example, the quadrant I real number $\frac{\pi}{3}$ is associated with the point $\left(\frac{1}{2}, \frac{\sqrt{3}}{2}\right)$ on the unit circle. Therefore, we can use symmetry to identify the coordinates of the points associated with

$$\pi - \frac{\pi}{3} = \frac{2\pi}{3}, \qquad \pi + \frac{\pi}{3} = \frac{4\pi}{3}, \qquad \text{and} \qquad 2\pi - \frac{\pi}{3} = \frac{5\pi}{3}.$$

↑ Quadrant II    ↑ Quadrant III    ↑ Quadrant IV

---

*The authors thank Professor Marvel Townsend of the University of Florida for her suggestion to include this figure.

The following chart summarizes this information.

| $s$ | Quadrant of $s$ | Symmetry Type and Corresponding Point | $\cos s$ | $\sin s$ |
|---|---|---|---|---|
| $\dfrac{\pi}{3}$ | I | not applicable; $\left(\dfrac{1}{2}, \dfrac{\sqrt{3}}{2}\right)$ | $\dfrac{1}{2}$ | $\dfrac{\sqrt{3}}{2}$ |
| $\pi - \dfrac{\pi}{3} = \dfrac{2\pi}{3}$ | II | y-axis; $\left(-\dfrac{1}{2}, \dfrac{\sqrt{3}}{2}\right)$ | $-\dfrac{1}{2}$ | $\dfrac{\sqrt{3}}{2}$ |
| $\pi + \dfrac{\pi}{3} = \dfrac{4\pi}{3}$ | III | origin; $\left(-\dfrac{1}{2}, -\dfrac{\sqrt{3}}{2}\right)$ | $-\dfrac{1}{2}$ | $-\dfrac{\sqrt{3}}{2}$ |
| $2\pi - \dfrac{\pi}{3} = \dfrac{5\pi}{3}$ | IV | x-axis; $\left(\dfrac{1}{2}, -\dfrac{\sqrt{3}}{2}\right)$ | $\dfrac{1}{2}$ | $-\dfrac{\sqrt{3}}{2}$ |

Since $\sin s = y$ and $\cos s = x$, we can replace $x$ and $y$ in the equation of the unit circle

$$x^2 + y^2 = 1$$

and obtain the Pythagorean identity

$$\cos^2 s + \sin^2 s = 1.$$

The ordered pair $(x, y)$ represents a point on the unit circle, and, therefore,

$$-1 \leq x \leq 1 \quad \text{and} \quad -1 \leq y \leq 1,$$

so
$$-1 \leq \cos s \leq 1 \quad \text{and} \quad -1 \leq \sin s \leq 1.$$

For any value of $s$, both $\sin s$ and $\cos s$ exist, so the domain of these functions is the set of all real numbers.

For $\tan s$, defined as $\dfrac{y}{x}$, $x$ must not equal 0. The only way $x$ can equal 0 is when the arc length $s$ is $\dfrac{\pi}{2}, -\dfrac{\pi}{2}, \dfrac{3\pi}{2}, -\dfrac{3\pi}{2}$, and so on. To avoid a 0 denominator, the domain of the tangent function must be restricted to those values of $s$ satisfying

$$s \neq (2n + 1)\dfrac{\pi}{2}, \quad n \text{ any integer.}$$

The definition of secant also has $x$ in the denominator, so the domain of secant is the same as the domain of tangent. Both cotangent and cosecant are defined with a denominator of $y$. To guarantee that $y \neq 0$, the domain of these functions must be the set of all values of $s$ satisfying

$$s \neq n\pi, \quad n \text{ any integer.}$$

### Domains of the Circular Functions

The domains of the circular functions are as follows:

**Sine and Cosine Functions:** $(-\infty, \infty)$

**Tangent and Secant Functions:** $\left\{s \mid s \neq (2n + 1)\dfrac{\pi}{2}, \text{ where } n \text{ is any integer}\right\}$

**Cotangent and Cosecant Functions:** $\{s \mid s \neq n\pi, \text{ where } n \text{ is any integer}\}$

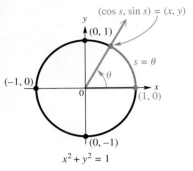

**FIGURE 13**

**OBJECTIVE 2** Find exact and approximate circular function values. The circular functions of real numbers correspond to the trigonometric functions of angles measured in radians. Let us assume that angle $\theta$ is in standard position, superimposed on the unit circle. See Figure 13. Suppose that $\theta$ is the *radian* measure of this angle. Using the arc length formula $s = r\theta$ with $r = 1$, we have $s = \theta$. Thus, the length of the intercepted arc is the real number that corresponds to the radian measure of $\theta$. Using the trigonometric function definitions from **Section 14.3**,

$$\sin\theta = \frac{y}{r} = \frac{y}{1} = y = \sin s, \quad \text{and} \quad \cos\theta = \frac{x}{r} = \frac{x}{1} = x = \cos s,$$

and so on. As shown here, the trigonometric functions and the circular functions lead to the same function values, provided we think of the angles as being in radian measure. This leads to the following important result.

### Evaluating a Circular Function

Circular function values of real numbers are obtained in the same manner as trigonometric function values of angles measured in radians. This applies both to methods of finding exact values (such as reference angle analysis) and to calculator approximations. **Calculators must be in radian mode when finding circular function values.**

**EXAMPLE 1** Finding Exact Circular Function Values

Find the exact values of $\sin\frac{3\pi}{2}$, $\cos\frac{3\pi}{2}$, and $\tan\frac{3\pi}{2}$.

**Solution** Evaluating a circular function at the real number $\frac{3\pi}{2}$ is equivalent to evaluating it at $\frac{3\pi}{2}$ radians. An angle of $\frac{3\pi}{2}$ radians intersects the unit circle at the point $(0, -1)$, as shown in Figure 14. Since

$$\sin s = y, \quad \cos s = x, \quad \text{and} \quad \tan s = \frac{y}{x},$$

it follows that

$$\sin\frac{3\pi}{2} = -1, \cos\frac{3\pi}{2} = 0, \text{ and } \tan\frac{3\pi}{2} \text{ is undefined.}$$

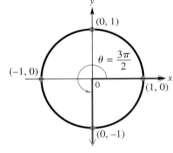

**FIGURE 14**

**NOW TRY** Exercise 1.

**EXAMPLE 2** Finding Exact Circular Function Values

(a) Use Figure 12 to find the exact values of $\cos\frac{7\pi}{4}$ and $\sin\frac{7\pi}{4}$.

(b) Use Figure 12 and the definition of tangent to find the exact value of $\tan\left(-\frac{5\pi}{3}\right)$.

(c) Use reference angles and degree/radian conversion to find the exact value of $\cos\frac{2\pi}{3}$.

## Solution

**(a)** In Figure 12, we see that the real number $\frac{7\pi}{4}$ corresponds to the unit circle point $\left(\frac{\sqrt{2}}{2}, -\frac{\sqrt{2}}{2}\right)$. Thus,

$$\cos\frac{7\pi}{4} = \frac{\sqrt{2}}{2} \quad \text{and} \quad \sin\frac{7\pi}{4} = -\frac{\sqrt{2}}{2}.$$

**(b)** Moving around the unit circle $\frac{5\pi}{3}$ units in the *negative* direction yields the same ending point as moving around $\frac{\pi}{3}$ units in the positive direction. Thus, $-\frac{5\pi}{3}$ corresponds to $\left(\frac{1}{2}, \frac{\sqrt{3}}{2}\right)$, and

$\tan s = \frac{y}{x}$

$$\tan\left(-\frac{5\pi}{3}\right) = \tan\frac{\pi}{3} = \frac{\frac{\sqrt{3}}{2}}{\frac{1}{2}} = \frac{\sqrt{3}}{2} \div \frac{1}{2} = \frac{\sqrt{3}}{2} \cdot \frac{2}{1} = \sqrt{3}.$$

**(c)** An angle of $\frac{2\pi}{3}$ radians corresponds to an angle of 120°. In standard position, 120° lies in quadrant II with a reference angle of 60°, so

Cosine is negative in quadrant II.

$$\cos\frac{2\pi}{3} = \cos 120° = -\cos 60° = -\frac{1}{2}.$$

Reference angle

**NOW TRY** Exercises 7, 17, and 21.

---

**EXAMPLE 3** Approximating Circular Function Values

Find a calculator approximation for each circular function value.

**(a)** cos 1.85  **(b)** cos 0.5149  **(c)** cot 1.3209  **(d)** sec(−2.9234)

## Solution

**(a)** With a calculator in radian mode, we find cos 1.85 ≈ −0.2756.
**(b)** cos 0.5149 ≈ 0.8703    Use a calculator in radian mode.
**(c)** As before, to find cotangent, secant, and cosecant function values, we must use the appropriate reciprocal functions. To find cot 1.3209, first find tan 1.3209 and then find the reciprocal.

$$\cot 1.3209 = \frac{1}{\tan 1.3209} \approx 0.2552$$

**(d)** $\sec(-2.9234) = \dfrac{1}{\cos(-2.9234)} \approx -1.0243$

**NOW TRY** Exercises 23, 29, and 33.

---

```
cos(1.85)
        -.2756
```

Radian mode
This is how the TI-83/84 Plus calculator displays the result of Example 3(a), fixed to four decimal digits.

---

**CAUTION** A common error in trigonometry is using a calculator in degree mode when radian mode should be used. *Remember, if you are finding a circular function value of a real number, the calculator must be in radian mode.*

**OBJECTIVE 3** Find a number with a given circular function value. Recall from Section 15.3 how we used a calculator to determine an angle measure, given a trigonometric function value of the angle. *Remember that the keys marked* **sin⁻¹, cos⁻¹, and tan⁻¹** *do not represent reciprocal functions.*

### EXAMPLE 4 Finding a Number Given its Circular Function Value

(a) Approximate the value of $s$ in the interval $\left[0, \frac{\pi}{2}\right]$, if $\cos s = 0.9685$.

(b) Find the exact value of $s$ in the interval $\left[\pi, \frac{3\pi}{2}\right]$, if $\tan s = 1$.

**Solution**

(a) Since we are given a cosine value and want to determine the real number in $\left[0, \frac{\pi}{2}\right]$ having this cosine value, we use the *inverse cosine* function of a calculator. With the calculator in radian mode, we find
$$\cos^{-1}(0.9685) \approx 0.2517.$$

See Figure 15. (Refer to your owner's manual to determine how to evaluate the $\sin^{-1}$, $\cos^{-1}$, and $\tan^{-1}$ functions with your calculator.)

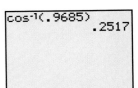

**FIGURE 15**

(b) Recall that $\tan \frac{\pi}{4} = 1$, and in quadrant III tan $s$ is positive. Therefore,
$$\tan\left(\pi + \frac{\pi}{4}\right) = \tan \frac{5\pi}{4} = 1,$$

and $s = \frac{5\pi}{4}$. Figure 16 supports this result.

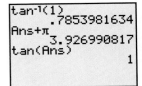

**FIGURE 16**

**NOW TRY** Exercises 55 and 65.

### EXAMPLE 5 Modeling the Angle of Elevation of the Sun

The angle of elevation $\theta$ of the Sun in the sky at any latitude $L$ is calculated with the formula
$$\sin \theta = \cos D \cos L \cos \omega + \sin D \sin L,$$

where $\theta = 0$ corresponds to sunrise and $\theta = \frac{\pi}{2}$ occurs if the Sun is directly overhead. $\omega$ (the Greek letter *omega*) is the number of radians that Earth has rotated through since noon, when $\omega = 0$. $D$ is the declination of the Sun, which varies because Earth is tilted on its axis. (*Source:* Winter, C., R. Sizmann, and L. L. Vant-Hull, Editors, *Solar Power Plants,* Springer-Verlag, 1991.)

Sacramento, California, has latitude $L = 38.5°$ or 0.6720 radian. Find the angle of elevation $\theta$ of the Sun at 3 P.M. on February 29, 2008, where at that time $D \approx -0.1425$ and $\omega \approx 0.7854$.

**Solution** Use the given formula for $\sin \theta$.
$$\sin \theta = \cos D \cos L \cos \omega + \sin D \sin L$$
$$= \cos(-0.1425) \cos(0.6720) \cos(0.7854) + \sin(-0.1425) \sin(0.6720)$$
$$\approx 0.4593426188$$

Thus, $\theta \approx 0.4773$ radian or $27.3°$.   Use inverse sine.

**NOW TRY** Exercise 77.

## 16.3 Exercises

**NOW TRY Exercise**

*For each value of the real number s, find* **(a)** sin s, **(b)** cos s, *and* **(c)** tan s. *See Example 1.*

1. $s = \dfrac{\pi}{2}$
2. $s = \pi$
3. $s = 2\pi$
4. $s = 3\pi$
5. $s = -\pi$
6. $s = -\dfrac{3\pi}{2}$

*Find the exact circular function value for each of the following. See Example 2.*

7. $\sin \dfrac{7\pi}{6}$
8. $\cos \dfrac{5\pi}{3}$
9. $\tan \dfrac{3\pi}{4}$
10. $\sec \dfrac{2\pi}{3}$
11. $\csc \dfrac{11\pi}{6}$
12. $\cot \dfrac{5\pi}{6}$
13. $\cos\left(-\dfrac{4\pi}{3}\right)$
14. $\tan \dfrac{17\pi}{3}$
15. $\cos \dfrac{7\pi}{4}$
16. $\sec \dfrac{5\pi}{4}$
17. $\sin\left(-\dfrac{4\pi}{3}\right)$
18. $\sin\left(-\dfrac{5\pi}{6}\right)$
19. $\sec \dfrac{23\pi}{6}$
20. $\csc \dfrac{13\pi}{3}$
21. $\tan \dfrac{5\pi}{6}$
22. $\cos \dfrac{3\pi}{4}$

*Find a calculator approximation for each circular function value. See Example 3.*

23. sin 0.6109
24. sin 0.8203
25. cos(−1.1519)
26. cos(−5.2825)
27. tan 4.0203
28. tan 6.4752
29. csc(−9.4946)
30. csc 1.3875
31. sec 2.8440
32. sec(−8.3429)
33. cot 6.0301
34. cot 3.8426

*Concept Check* The figure displays a unit circle and an angle of 1 radian. The tick marks on the circle are spaced at every two-tenths radian. Use the figure to estimate each value.

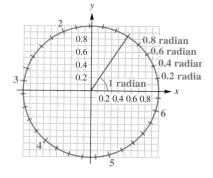

35. cos 0.8
36. cos 0.6
37. sin 2
38. sin 4
39. sin 3.8
40. cos 3.2
41. a positive angle whose cosine is −0.65
42. a positive angle whose sine is −0.95
43. a positive angle whose sine is 0.7
44. a positive angle whose cosine is 0.3

*Concept Check* Without using a calculator, decide whether each function value is positive or negative. (Hint: Consider the radian measures of the quadrantal angles.)

45. cos 2
46. sin(−1)
47. sin 5
48. cos 6
49. tan 6.29
50. tan(−6.29)

*Concept Check* Each figure in Exercises 51–54 shows an angle θ in standard position with its terminal side intersecting the unit circle. Evaluate the six circular function values of θ.

51.

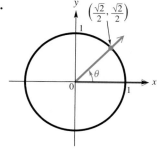

52.

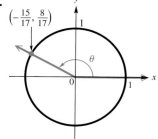

**53.**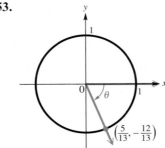

**54.**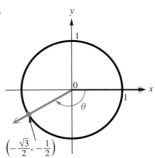

*Find the value of s in the interval $\left[0, \frac{\pi}{2}\right]$ that makes each statement true. See Example 4(a).*

**55.** tan s = 0.2126    **56.** cos s = 0.7826    **57.** sin s = 0.9918

**58.** cot s = 0.2994    **59.** sec s = 1.0806    **60.** csc s = 1.0219

*Find the exact value of s in the given interval that has the given circular function value. Do not use a calculator. See Example 4(b).*

**61.** $\left[\frac{\pi}{2}, \pi\right]$;  sin s = $\frac{1}{2}$    **62.** $\left[\frac{\pi}{2}, \pi\right]$;  cos s = $-\frac{1}{2}$

**63.** $\left[\pi, \frac{3\pi}{2}\right]$;  tan s = $\sqrt{3}$    **64.** $\left[\pi, \frac{3\pi}{2}\right]$;  sin s = $-\frac{1}{2}$

**65.** $\left[\frac{3\pi}{2}, 2\pi\right]$;  tan s = $-1$    **66.** $\left[\frac{3\pi}{2}, 2\pi\right]$;  cos s = $\frac{\sqrt{3}}{2}$

*Suppose an arc of length s lies on the unit circle $x^2 + y^2 = 1$, starting at the point (1, 0) and terminating at the point (x, y). (See Figure 11.) Use a calculator to find the approximate coordinates for (x, y). (Hint: x = cos s and y = sin s.)*

**67.** s = 2.5    **68.** s = 3.4    **69.** s = −7.4    **70.** s = −3.9

*Concept Check    For each value of s, use a calculator to find sin s and cos s and then use the results to decide in which quadrant an angle of s radians lies.*

**71.** s = 51    **72.** s = 49    **73.** s = 65    **74.** s = 79

*Concept Check    In Exercises 75 and 76, each graphing calculator screen shows a point on the unit circle. What is the length of the shortest arc of the circle from (1, 0) to the point?*

**75.**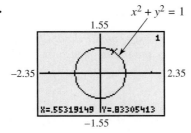

**76.**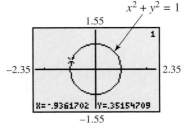

*Solve each problem. See Example 5.*

**77.** Refer to Example 5.

(a) Repeat the example for New Orleans, which has latitude $L = 30°$.

(b) Compare your answers. Do they agree with your intuition?

**78.** The number of daylight hours $H$ at any location can be calculated using the formula

$$\cos(0.1309H) = -\tan D \tan L,$$

where $D$ and $L$ are defined in Example 5. Use this trigonometric equation to calculate the shortest and longest days in Minneapolis, Minnesota, if its latitude $L = 44.88°$, the shortest day occurs when $D = -23.44°$, and the longest day occurs when $D = 23.44°$. Remember to convert degrees to radians. (*Source:* Winter, C., R. Sizmann, and L. L. Vant-Hull, Editors, *Solar Power Plants,* Springer-Verlag, 1991.)

79. Because the values of the circular functions repeat every $2\pi$, they are used to describe things that repeat periodically. For example, the maximum afternoon temperature in a given city might be modeled by

$$t = 60 - 30 \cos \frac{x\pi}{6},$$

where $t$ represents the maximum afternoon temperature in month $x$, with $x = 0$ representing January, $x = 1$ representing February, and so on. Find the maximum afternoon temperature for each of the following months.

(a) January
(b) April
(c) May
(d) June
(e) August
(f) October

80. The temperature in Fairbanks is modeled by

$$T(x) = 37 \sin\left[\frac{2\pi}{365}(x - 101)\right] + 25,$$

where $T(x)$ is the temperature in degrees Fahrenheit on day $x$, with $x = 1$ corresponding to January 1 and $x = 365$ corresponding to December 31. Use a calculator to estimate the temperature on the following days. (*Source:* Lando, B. and C. Lando, "Is the Graph of Temperature Variation a Sine Curve?", *The Mathematics Teacher,* 70, September 1977.)

(a) March 1 (day 60)
(b) April 1 (day 91)
(c) Day 150
(d) June 15
(e) September 1
(f) October 31

## 16.4 Linear and Angular Speed

**OBJECTIVE**

1 Use the linear and angular speed formulas.

**OBJECTIVE 1** Use the linear and angular speed formulas. In many situations we need to know how fast a point on a circular disk is moving or how fast the central angle of such a disk is changing. Some examples occur with machinery involving gears or pulleys or the speed of a car around a curved portion of highway.

Suppose that point $P$ moves at a constant speed along a circle of radius $r$ and center $O$. See Figure 17. The measure of how fast the position of $P$ is changing is called **linear speed.** If $v$ represents linear speed, then

$$\text{speed} = \frac{\text{distance}}{\text{time}}, \quad \text{or} \quad v = \frac{s}{t},$$

where $s$ is the length of the arc traced by point $P$ at time $t$. (This formula is just a restatement of $d = rt$ with $s$ as distance, $v$ as rate (speed), and $t$ as time.)

As point $P$ in Figure 17 moves along the circle, ray $OP$ rotates around the origin. Since ray $OP$ is the terminal side of angle $POB$, the measure of the angle changes as $P$ moves along the circle. The measure of how fast angle $POB$ is changing is called **angular speed.** Angular speed, symbolized $\omega$, is given as

$$\omega = \frac{\theta}{t}, \quad \theta \text{ in radians},$$

where $\theta$ is the measure of angle $POB$ at time $t$. As with earlier formulas in this chapter, $\theta$ must be measured in radians, with $\omega$ expressed as radians per unit of time. Angular speed is used in physics and engineering, among other applications.

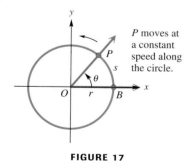

FIGURE 17

In **Section 16.2**, the length $s$ of the arc intercepted on a circle of radius $r$ by a central angle of measure $\theta$ radians was found to be $s = r\theta$. Using this formula, the formula for linear speed, $v = \frac{s}{t}$, becomes

$$v = \frac{s}{t} = \frac{r\theta}{t} = r \cdot \frac{\theta}{t} = r\omega. \qquad s = r\theta;\ \omega = \frac{\theta}{t}$$

The formula $v = r\omega$ relates linear and angular speeds.

As an example of linear and angular speeds, consider the following. The human joint that can be flexed the fastest is the wrist, which can rotate through 90°, or $\frac{\pi}{2}$ radians, in 0.045 sec while holding a tennis racket. The angular speed of a human wrist swinging a tennis racket is

$$\omega = \frac{\theta}{t} = \frac{\frac{\pi}{2}}{0.045} \approx 35 \text{ radians per sec.}$$

If the radius (distance) from the tip of the racket to the wrist joint is 2 ft, then the speed at the tip of the racket is

$$v = r\omega \approx 2(35) = 70 \text{ ft per sec}, \qquad \text{or about 48 mph.}$$

In a tennis serve the arm rotates at the shoulder, so the final speed of the racket is considerably faster. (*Source:* Cooper, J. and R. Glassow, *Kinesiology,* Second Edition, C.V. Mosby, 1968.)

The formulas for angular and linear speed are summarized in the table.

| Angular Speed | Linear Speed |
|---|---|
| $\omega = \dfrac{\theta}{t}$ | $v = \dfrac{s}{t}$ |
| ($\omega$ in radians per unit time, $\theta$ in radians) | $v = \dfrac{r\theta}{t}$ |
| | $v = r\omega$ |

### EXAMPLE 1 Using Linear and Angular Speed Formulas

Suppose that point $P$ is on a circle with radius 10 cm, and ray $OP$ is rotating with angular speed $\frac{\pi}{18}$ radian per sec.

(a) Find the angle generated by $P$ in 6 sec.
(b) Find the distance traveled by $P$ along the circle in 6 sec.
(c) Find the linear speed of $P$ in centimeters per second.

**Solution**

(a) The speed of ray $OP$ is $\omega = \frac{\pi}{18}$ radian per sec. Since $\omega = \frac{\theta}{t}$, then in 6 sec

$$\frac{\pi}{18} = \frac{\theta}{6} \qquad \text{Let } \omega = \frac{\pi}{18} \text{ and } t = 6 \text{ in the angular speed formula.}$$

$$\theta = \frac{6\pi}{18} = \frac{\pi}{3} \text{ radians.} \qquad \text{Solve for } \theta.$$

(b) From part (a), $P$ generates an angle of $\frac{\pi}{3}$ radians in 6 sec. The distance traveled by $P$ along the circle is

$$s = r\theta = 10\left(\frac{\pi}{3}\right) = \frac{10\pi}{3} \text{ cm.}$$

(c) From part (b), $s = \frac{10\pi}{3}$ for 6 sec, so for 1 sec we divide by 6.

$$v = \frac{s}{t} = \frac{\frac{10\pi}{3}}{6} = \frac{10\pi}{3} \div 6 = \frac{10\pi}{3} \cdot \frac{1}{6} = \frac{5\pi}{9} \text{ cm per sec}$$

Be careful simplifying this complex fraction.

**NOW TRY** Exercise 3.

### EXAMPLE 2  Finding Angular Speed of a Pulley and Linear Speed of a Belt

A belt runs a pulley of radius 6 cm at 80 revolutions per min.

(a) Find the angular speed of the pulley in radians per second.
(b) Find the linear speed of the belt in centimeters per second.

**Solution**

(a) In 1 min, the pulley makes 80 revolutions. Each revolution is $2\pi$ radians, so

$$80(2\pi) = 160\pi \text{ radians per min.}$$

Since there are 60 sec in 1 min, we find $\omega$, the angular speed in radians per second, by dividing $160\pi$ by 60.

$$\omega = \frac{160\pi}{60} = \frac{8\pi}{3} \text{ radians per sec}$$

(b) The linear speed of the belt will be the same as that of a point on the circumference of the pulley. Thus,

$$v = r\omega = 6\left(\frac{8\pi}{3}\right) = 16\pi \approx 50 \text{ cm per sec.}$$

**NOW TRY** Exercise 39.

### EXAMPLE 3  Finding Linear Speed and Distance Traveled by a Satellite

A satellite traveling in a circular orbit 1600 km above the surface of Earth takes 2 hr to make an orbit. The radius of Earth is approximately 6400 km. See Figure 18.

(a) Approximate the linear speed of the satellite in kilometers per hour.
(b) Approximate the distance the satellite travels in 4.5 hr.

**Solution**

(a) The distance of the satellite from the center of Earth is approximately

$$r = 1600 + 6400 = 8000 \text{ km.}$$

For one orbit, $\theta = 2\pi$, and

$$s = r\theta = 8000(2\pi) \text{ km.}$$

Since it takes 2 hr to complete an orbit, the linear speed is approximately

$$v = \frac{s}{t} = \frac{8000(2\pi)}{2} = 8000\pi \approx 25,000 \text{ km per hr.}$$

(b) $s = vt = 8000\pi(4.5) = 36,000\pi \approx 110,000 \text{ km}$

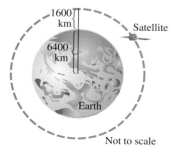

**FIGURE 18**

**NOW TRY** Exercise 37.

# 16.4 Exercises

**NOW TRY Exercise**

1. *Concept Check* If the point $P$ moves around the circumference of the unit circle at an angular velocity of 1 radian per sec, how long will it take for $P$ to move around the entire circle?

2. *Concept Check* If the point $P$ moves around the circumference of the unit circle at a speed of 1 unit per sec, how long will it take for $P$ to move around the entire circle?

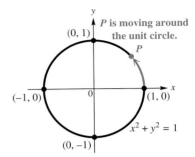

*Suppose that point P is on a circle with radius r, and ray OP is rotating with angular speed $\omega$. For the given values of r, $\omega$, and t, find each of the following.*

**(a)** the angle generated by P in time t
**(b)** the distance traveled by P along the circle in time t
**(c)** the linear speed of P

See Example 1.

3. $r = 20$ cm, $\omega = \dfrac{\pi}{12}$ radian per sec, $t = 6$ sec

4. $r = 30$ cm, $\omega = \dfrac{\pi}{10}$ radian per sec, $t = 4$ sec

*Use the formula $\omega = \dfrac{\theta}{t}$ to find the value of the missing variable.*

5. $\omega = \dfrac{2\pi}{3}$ radians per sec, $t = 3$ sec

6. $\omega = \dfrac{\pi}{4}$ radian per min, $t = 5$ min

7. $\theta = \dfrac{3\pi}{4}$ radians, $t = 8$ sec

8. $\theta = \dfrac{2\pi}{5}$ radians, $t = 10$ sec

9. $\theta = \dfrac{2\pi}{9}$ radians, $\omega = \dfrac{5\pi}{27}$ radian per min

10. $\theta = \dfrac{3\pi}{8}$ radians, $\omega = \dfrac{\pi}{24}$ radian per min

11. $\theta = 3.871142$ radians, $t = 21.4693$ sec

12. $\omega = 0.90674$ radian per min, $t = 11.876$ min

*Use the formula $v = r\omega$ to find the value of the missing variable.*

13. $r = 12$ m, $\omega = \dfrac{2\pi}{3}$ radians per sec

14. $r = 8$ cm, $\omega = \dfrac{9\pi}{5}$ radians per sec

15. $v = 9$ m per sec, $r = 5$ m

16. $v = 18$ ft per sec, $r = 3$ ft

17. $v = 107.692$ m per sec, $r = 58.7413$ m

18. $r = 24.93215$ cm, $\omega = 0.372914$ radian per sec

*The formula $\omega = \dfrac{\theta}{t}$ can be rewritten as $\theta = \omega t$. Using $\omega t$ for $\theta$ changes $s = r\theta$ to $s = r\omega t$. Use the formula $s = r\omega t$ to find the value of the missing variable.*

19. $r = 6$ cm, $\omega = \dfrac{\pi}{3}$ radians per sec, $t = 9$ sec

20. $r = 9$ yd, $\omega = \dfrac{2\pi}{5}$ radians per sec, $t = 12$ sec

21. $s = 6\pi$ cm, $r = 2$ cm, $\omega = \dfrac{\pi}{4}$ radian per sec

22. $s = \dfrac{12\pi}{5}$ m, $r = \dfrac{3}{2}$ m, $\omega = \dfrac{2\pi}{5}$ radians per sec

23. $s = \dfrac{3\pi}{4}$ km, $r = 2$ km, $t = 4$ sec

24. $s = \dfrac{8\pi}{9}$ m, $r = \dfrac{4}{3}$ m, $t = 12$ sec

*Find $\omega$ for each of the following.*

25. the hour hand of a clock
26. a line from the center to the edge of a CD revolving 300 times per min
27. the minute hand of a clock
28. the second hand of a clock

*Find $v$ for each of the following.*

29. the tip of the minute hand of a clock, if the hand is 7 cm long
30. the tip of the second hand of a clock, if the hand is 28 mm long
31. a point on the edge of a flywheel of radius 2 m, rotating 42 times per min
32. a point on the tread of a tire of radius 18 cm, rotating 35 times per min
33. the tip of an airplane propeller 3 m long, rotating 500 times per min (*Hint: $r = 1.5$ m*)
34. a point on the edge of a gyroscope of radius 83 cm, rotating 680 times per min

*Solve each problem. See Examples 1–3.*

35. The tires of a bicycle have radius 13 in. and are turning at the rate of 215 revolutions per min. See the figure. How fast is the bicycle traveling in miles per hour? (*Hint:* 5280 ft $=$ 1 mi)

36. Mars rotates on its axis at the rate of about 0.2552 radian per hr. Approximately how many hours are in a Martian day (or *sol*)? (*Source:* Wright, J. W., General Editor, *The Universal Almanac,* Andrews and McMeel, 1997.)

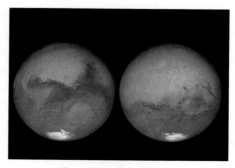

**Opposite sides of Mars**

**37.** Earth travels about the Sun in an orbit that is almost circular. Assume that the orbit is a circle with radius 93,000,000 mi. Its angular and linear speeds are used in designing solar-power facilities.

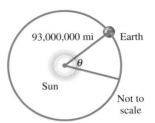

(a) Assume that a year is 365 days, and find the angle formed by Earth's movement in one day.
(b) Give the angular speed in radians per hour.
(c) Find the linear speed of Earth in miles per hour.

**38.** Earth revolves on its axis once every 24 hr. Assuming that Earth's radius is 6400 km, find the following.

(a) angular speed of Earth in radians per day and radians per hour
(b) linear speed at the North Pole or South Pole
(c) linear speed at Quito, Ecuador, a city on the equator
(d) linear speed at Salem, Oregon (halfway from the equator to the North Pole)

**39.** The pulley shown has a radius of 12.96 cm. Suppose it takes 18 sec for 56 cm of belt to go around the pulley.

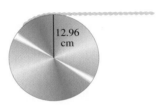

(a) Find the linear speed of the belt in centimeters per second.
(b) Find the angular speed of the pulley in radians per second.

**40.** The two pulleys in the figure have radii of 15 cm and 8 cm, respectively. The larger pulley rotates 25 times in 36 sec. Find the angular speed of each pulley in radians per second.

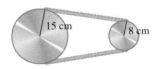

**41.** A thread is being pulled off a spool at the rate of 59.4 cm per sec. Find the radius of the spool if it makes 152 revolutions per min.

**42.** A railroad track is laid along the arc of a circle of radius 1800 ft. The circular part of the track subtends a central angle of $40°$. How long (in seconds) will it take a point on the front of a train traveling 30.0 mph to go around this portion of the track?

**43.** A 90-horsepower outboard motor at full throttle will rotate its propeller at exactly 5000 revolutions per min. Find the angular speed of the propeller in radians per second.

**44.** The shoulder joint can rotate at 25.0 radians per sec. If a golfer's arm is straight and the distance from the shoulder to the club head is 5.00 ft, find the linear speed of the club head from shoulder rotation. (*Source:* Cooper, J. and R. Glasgow, *Kinesiology,* Second Edition, C.V. Mosby, 1968.)

# 16 SUMMARY

## KEY TERMS

**16.1** radian
circumference

**16.2** sector of a circle
**16.3** unit circle

circular functions
reference arc

**16.4** linear speed $v$
angular speed $\omega$

## QUICK REVIEW

| CONCEPTS | EXAMPLES |

### 16.1 Radian Measure

An angle with its vertex at the center of a circle that intercepts an arc on the circle equal in length to the radius of the circle has a measure of **1 radian**.

**Degree/Radian Relationship** $180° = \pi$ **radians**

**Converting between Degrees and Radians**
1. Multiply a degree measure by $\frac{\pi}{180}$ radian and simplify to convert to radians.

2. Multiply a radian measure by $\frac{180°}{\pi}$ and simplify to convert to degrees.

See Figure 1 on page 736.

Convert 135° to radians.

$$135° = 135\left(\frac{\pi}{180} \text{ radian}\right) = \frac{3\pi}{4} \text{ radians}$$

Convert $-\frac{5\pi}{3}$ radians to degrees.

$$-\frac{5\pi}{3} \text{ radians} = -\frac{5\pi}{3}\left(\frac{180°}{\pi}\right) = -300°$$

### 16.2 Applications of Radian Measure

**Arc Length**

The length $s$ of the arc intercepted on a circle of radius $r$ by a central angle of measure $\theta$ radians is given by the product of the radius and the radian measure of the angle, or

$$s = r\theta, \quad \theta \text{ in radians.}$$

In the figure, $s = r\theta$, so

$$\theta = \frac{s}{r} = \frac{3}{4} \text{ radian.}$$

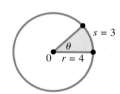

**Area of a Sector**

The area of a sector of a circle of radius $r$ and central angle $\theta$ is given by

$$A = \frac{1}{2}r^2\theta, \quad \theta \text{ in radians.}$$

The area of the sector in the figure above is

$$A = \frac{1}{2}(4)^2\left(\frac{3}{4}\right) = 6 \text{ sq units.}$$

*(continued)*

## CONCEPTS

### 16.3 The Unit Circle and Circular Functions

**Circular Functions**

Start at the point $(1, 0)$ on the unit circle $x^2 + y^2 = 1$ and lay off an arc of length $|s|$ along the circle, going counterclockwise if $s$ is positive and clockwise if $s$ is negative. Let the endpoint of the arc be at the point $(x, y)$. The six circular functions of $s$ are defined as follows. (Assume that no denominators are 0.)

$$\sin s = y \qquad \cos s = x \qquad \tan s = \frac{y}{x}$$

$$\csc s = \frac{1}{y} \qquad \sec s = \frac{1}{x} \qquad \cot s = \frac{x}{y}$$

**The Unit Circle**

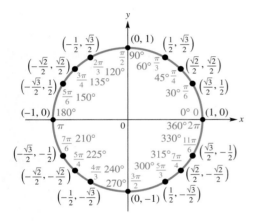

The unit circle $x^2 + y^2 = 1$

## EXAMPLES

Use the unit circle to find each value.

$$\sin \frac{5\pi}{6} = \frac{1}{2}$$

$$\cos \frac{3\pi}{2} = 0$$

$$\tan \frac{\pi}{4} = \frac{\frac{\sqrt{2}}{2}}{\frac{\sqrt{2}}{2}} = 1$$

$$\csc \frac{7\pi}{4} = \frac{1}{-\frac{\sqrt{2}}{2}} = -\sqrt{2}$$

$$\sec \frac{7\pi}{6} = \frac{1}{-\frac{\sqrt{3}}{2}} = -\frac{2\sqrt{3}}{3}$$

$$\cot \frac{\pi}{3} = \frac{\frac{1}{2}}{\frac{\sqrt{3}}{2}} = \frac{\sqrt{3}}{3}$$

$$\sin 0 = 0$$

$$\cos \frac{\pi}{2} = 0$$

$$\tan \pi = \frac{0}{-1} = 0$$

$$\sin\left(-\frac{\pi}{2}\right) = \sin \frac{3\pi}{2} = -1$$

$$\cot\left(-\frac{3\pi}{2}\right) = \cot \frac{\pi}{2} = \frac{0}{1} = 0$$

$$\tan \frac{\pi}{2} \text{ is undefined.}$$

### 16.4 Linear and Angular Speed

**Formulas for Angular and Linear Speed**

| Angular Speed | Linear Speed |
|---|---|
| $\omega = \dfrac{\theta}{t}$ | $v = \dfrac{s}{t}$ |
| ($\omega$ in radians per unit time, $\theta$ in radians) | $v = \dfrac{r\theta}{t}$ |
| | $v = r\omega$ |

A belt runs a pulley of radius 8 in. at 60 revolutions per min. Find
(a) the angular speed $\omega$ in radians per minute, and
(b) the linear speed $v$ of the belt in inches per minute.

(a) $\omega = 60(2\pi) = 120\pi$ radians per min
(b) $v = r\omega = 8(120\pi) = 960\pi$ in. per min

# 16 REVIEW EXERCISES

1. *Concept Check* Which is larger—an angle of 1° or an angle of 1 radian?
2. *Concept Check* Consider each angle in standard position having the given radian measure. In what quadrant does the terminal side lie?
   (a) 3  (b) 4  (c) −2  (d) 7
3. Find three angles coterminal with an angle of 1 radian.
4. Give an expression that generates all angles coterminal with an angle of $\frac{\pi}{6}$ radian. Let $n$ represent any integer.

*Convert each degree measure to radians. Leave answers as multiples of $\pi$.*

5. 45°  6. 120°  7. 175°
8. 330°  9. 800°  10. 1020°

*Convert each radian measure to degrees.*

11. $\frac{5\pi}{4}$  12. $\frac{9\pi}{10}$  13. $\frac{8\pi}{3}$

14. $-\frac{6\pi}{5}$  15. $-\frac{11\pi}{18}$  16. $\frac{21\pi}{5}$

*Suppose the tip of the minute hand of a clock is 2 in. from the center of the clock. For each duration, determine the distance traveled by the tip of the minute hand.*

17. 15 min  18. 20 min
19. 3 hr  20. $10\frac{1}{2}$ hr

*Solve each problem. Use a calculator as necessary.*

21. The radius of a circle is 15.2 cm. Find the length of an arc of the circle intercepted by a central angle of $\frac{3\pi}{4}$ radians.
22. Find the length of an arc intercepted by a central angle of 0.769 radian on a circle with radius 11.4 cm.
23. A circle has radius 8.973 cm. Find the length of an arc on this circle intercepted by a central angle of 49.06°.
24. A central angle of $\frac{7\pi}{4}$ radians forms a sector of a circle. Find the area of the sector if the radius of the circle is 28.69 in.
25. Find the area of a sector of a circle having a central angle of 21° 40′ in a circle of radius 38.0 m.
26. A tree 2000 yd away subtends an angle of 1° 10′. Find the height of the tree to two significant digits.

*Assume that the radius of Earth is 6400 km in Exercises 27 and 28.*

27. Find the distance in kilometers between cities on a north—south line that are on latitudes 28° N and 12° S, respectively.
28. Two cities on the equator have longitudes of 72° E and 35° W, respectively. Find the distance between the cities.

765

*Concept Check* In Exercises 29 and 30, find the measure of the central angle $\theta$ (in radians) and the area of the sector.

29.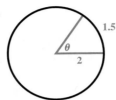

30.

31. *Concept Check* The hour hand of a wall clock measures 6 in. from its tip to the center of the clock.

   (a) Through what angle (in radians) does the hour hand pass between 1 o'clock and 3 o'clock?

   (b) What distance does the tip of the hour hand travel during the time period from 1 o'clock to 3 o'clock?

32. Describe what would happen to the central angle for a given arc length of a circle if the circle's radius were doubled. (Assume everything else is unchanged.)

*Find each exact function value. Do not use a calculator.*

33. $\tan \dfrac{\pi}{3}$

34. $\cos \dfrac{2\pi}{3}$

35. $\sin\left(-\dfrac{5\pi}{6}\right)$

36. $\tan\left(-\dfrac{7\pi}{3}\right)$

37. $\csc\left(-\dfrac{11\pi}{6}\right)$

38. $\cot\left(-\dfrac{17\pi}{3}\right)$

*Without using a calculator, determine which of the following is greater.*

39. $\tan 1$ or $\tan 2$

40. $\sin 1$ or $\tan 1$

41. $\cos 2$ or $\sin 2$

42. $\cos(\sin 0)$ or $\sin(\cos 0)$

*Use a calculator to find an approximation for each circular function value. Be sure your calculator is set in radian mode.*

43. $\sin 1.0472$

44. $\tan 1.2275$

45. $\cos(-0.2443)$

46. $\cot 3.0543$

47. $\sec 7.3159$

48. $\csc 4.8386$

*Find the value of $s$ in the interval $\left[0, \dfrac{\pi}{2}\right]$ that makes each statement true.*

49. $\cos s = 0.9250$

50. $\tan s = 4.0112$

51. $\sin s = 0.4924$

52. $\csc s = 1.2361$

53. $\cot s = 0.5022$

54. $\sec s = 4.5600$

*Find the exact value of $s$ in the given interval that has the given circular function value. Do not use a calculator.*

55. $\left[0, \dfrac{\pi}{2}\right]$; $\cos s = \dfrac{\sqrt{2}}{2}$

56. $\left[\dfrac{\pi}{2}, \pi\right]$; $\tan s = -\sqrt{3}$

57. $\left[\pi, \dfrac{3\pi}{2}\right]$; $\sec s = -\dfrac{2\sqrt{3}}{3}$

58. $\left[\dfrac{3\pi}{2}, 2\pi\right]$; $\sin s = -\dfrac{1}{2}$

*Solve each problem, where $t$, $\omega$, $\theta$, and $s$ are as defined in Section 16.4.*

59. Find $t$ if $\theta = \dfrac{5\pi}{12}$ radians and $\omega = \dfrac{8\pi}{9}$ radians per sec.

60. Find $\theta$ if $t = 12$ sec and $\omega = 9$ radians per sec.

**61.** Find $\omega$ if $t = 8$ sec and $\theta = \frac{2\pi}{5}$ radians.

**62.** Find $\omega$ if $s = \frac{12\pi}{25}$ ft, $r = \frac{3}{5}$ ft, and $t = 15$ sec.

**63.** Find $s$ if $r = 11.46$ cm, $\omega = 4.283$ radians per sec, and $t = 5.813$ sec.

**64.** Find the linear speed of a point on the edge of a flywheel of radius 7 cm if the flywheel is rotating 90 times per sec.

**65.** A Ferris wheel has radius 25 ft. If it takes 30 sec for the wheel to turn $\frac{5\pi}{6}$ radians, what is the angular speed of the wheel?

# 16 TEST

*Convert each degree measure to radians.*

**1.** $120°$

**2.** $-45°$

**3.** $5°$ (to the nearest hundredth)

*Convert each radian measure to degrees.*

**4.** $\frac{3\pi}{4}$

**5.** $-\frac{7\pi}{6}$

**6.** 4 (to the nearest hundredth)

**7.** A central angle of a circle with radius 150 cm intercepts an arc of 200 cm. Find each measure.
   (a) the radian measure of the angle
   (b) the area of a sector with that central angle

**8.** The arrow on a car's gasoline gauge is $\frac{1}{2}$ in. long. See the figure. Through what angle does the arrow rotate when it moves 1 in. on the gauge?

*Find each circular function value.*

**9.** $\sin \frac{3\pi}{4}$

**10.** $\cos\left(-\frac{7\pi}{6}\right)$

**11.** $\tan \frac{3\pi}{2}$

**12.** $\sec \frac{8\pi}{3}$

**13.** $\tan \pi$

**14.** $\cos \frac{3\pi}{2}$

15. Give the sine, cosine, and tangent of s.

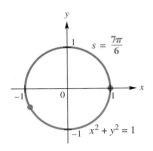

16. Give the domains of the six circular functions.

17. (a) Use a calculator to approximate $s$ in the interval $\left[0, \frac{\pi}{2}\right]$, if $\sin s = 0.8258$.

   (b) Find the exact value of $s$ in the interval $\left[0, \frac{\pi}{2}\right]$, if $\cos s = \frac{1}{2}$.

18. Suppose that point $P$ is on a circle with radius 60 cm, and ray $OP$ is rotating with angular speed $\frac{\pi}{12}$ radian per sec.

   (a) Find the angle generated by $P$ in 8 sec.
   (b) Find the distance traveled by $P$ along the circle in 8 sec.
   (c) Find the linear speed of $P$.

19. It takes Jupiter 11.64 yr to complete one orbit around the Sun. See the figure. If Jupiter's average distance from the Sun is 483,600,000 mi, find its orbital speed (speed along its orbital path) in miles per second. (*Source:* Wright, J. W., General Editor, *The Universal Almanac*, Andrews and McMeel, 1997.)

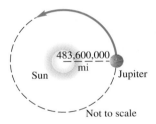

20. A Ferris wheel has radius 50.0 ft. A person takes a seat, and then the wheel turns $\frac{2\pi}{3}$ radians.

   (a) How far is the person above the ground?
   (b) If it takes 30 sec for the wheel to turn $\frac{2\pi}{3}$ radians, what is the angular speed of the wheel?

# Graphs of the Circular Functions

Phenomena that repeat in a regular pattern, such as rotation of a planet on its axis, high and low tides, average monthly temperatures, and biorhythms can be modeled by *periodic functions*. The graph of a periodic function is a picture of how the value of the function (the dependent variable) changes when the independent variable, time, changes. As you go from left to right along the graph, it is as if you were moving forward in time. The origin represents zero time, and each additional second or minute or hour is represented by one unit to the right on the graph. The graph of every periodic function contains a piece that repeats over and over. In Exercise 48 of Section 17.1, you will examine and interpret a graph showing variation in blood pressure, an example of a periodic function.

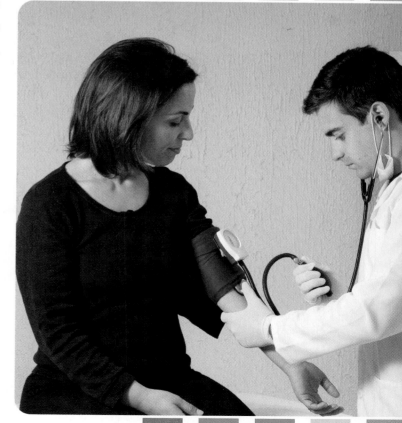

**17.1** Graphs of the Sine and Cosine Functions

**17.2** Translations of the Graphs of the Sine and Cosine Functions

**17.3** Graphs of the Tangent and Cotangent Functions

**17.4** Graphs of the Secant and Cosecant Functions

# 17.1 Graphs of the Sine and Cosine Functions

**OBJECTIVES**

1. Understand the concept of a periodic function.
2. Graph $f(x) = \sin x$.
3. Graph $f(x) = \cos x$.
4. Apply changes of amplitude and period to graphs of the sine and cosine functions.

**OBJECTIVE 1** Understand the concept of a periodic function. Many things in daily life repeat with a predictable pattern, such as weather, tides, and hours of daylight. Because the sine and cosine functions repeat their values in a regular pattern, they are *periodic functions*. Figure 1 shows a periodic graph that represents a normal heartbeat.

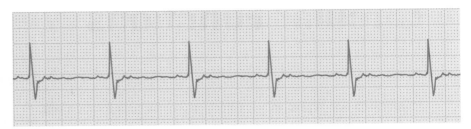

**FIGURE 1**

### Periodic Function

A **periodic function** is a function $f$ such that
$$f(x) = f(x + np),$$
for every real number $x$ in the domain of $f$, every integer $n$, and some positive real number $p$. The least possible positive value of $p$ is the **period** of the function.

The circumference of the unit circle is $2\pi$, so the least value of $p$ for which the sine and cosine functions repeat is $2\pi$. **Therefore, the sine and cosine functions are periodic functions with period $2\pi$.**

**OBJECTIVE 2** Graph $f(x) = \sin x$. In **Section 16.3** we saw that for a real number $s$, the point on the unit circle corresponding to $s$ has coordinates ($\cos s$, $\sin s$). See Figure 2. Trace along the circle to verify the results shown in the table.

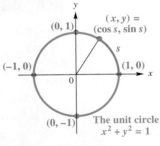

**FIGURE 2**

| As $s$ Increases from | sin $s$ | cos $s$ |
|---|---|---|
| 0 to $\frac{\pi}{2}$ | Increases from 0 to 1 | Decreases from 1 to 0 |
| $\frac{\pi}{2}$ to $\pi$ | Decreases from 1 to 0 | Decreases from 0 to $-1$ |
| $\pi$ to $\frac{3\pi}{2}$ | Decreases from 0 to $-1$ | Increases from $-1$ to 0 |
| $\frac{3\pi}{2}$ to $2\pi$ | Increases from $-1$ to 0 | Increases from 0 to 1 |

To avoid confusion when graphing the sine function, we use $x$ rather than $s$; this corresponds to the letters in the $xy$-coordinate system. Selecting key values of $x$ and finding the corresponding values of $\sin x$ leads to the table in Figure 3.

To obtain the traditional graph in Figure 3, we plot the points from the table, use symmetry, and join them with a smooth curve. Since $y = \sin x$ is periodic with period $2\pi$ and has domain $(-\infty, \infty)$, the graph continues in the same pattern in both directions. This graph is called a **sine wave**, or **sinusoid**.

### Sine Function  $f(x) = \sin x$

Domain: $(-\infty, \infty)$   Range: $[-1, 1]$

| x | y |
|---|---|
| 0 | 0 |
| $\frac{\pi}{6}$ | $\frac{1}{2}$ |
| $\frac{\pi}{4}$ | $\frac{\sqrt{2}}{2}$ |
| $\frac{\pi}{3}$ | $\frac{\sqrt{3}}{2}$ |
| $\frac{\pi}{2}$ | 1 |
| $\pi$ | 0 |
| $\frac{3\pi}{2}$ | $-1$ |
| $2\pi$ | 0 |

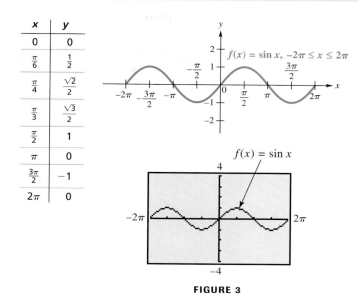

**FIGURE 3**

- The graph is continuous over its entire domain, $(-\infty, \infty)$.
- Its x-intercepts are of the form $(n\pi, 0)$, where $n$ is an integer.
- Its period is $2\pi$.
- The graph is symmetric with respect to the origin, so the function is an odd function. For all $x$ in the domain, $\sin(-x) = -\sin x$.

In algebra, we say that a function $f$ is an **odd function** if for all $x$ in the domain of $f$, $f(-x) = -f(x)$. The graph of an odd function is symmetric with respect to the origin, that is, if $(x, y)$ belongs to the function, then $(-x, -y)$ also belongs to the function. For example, $\left(\frac{\pi}{2}, 1\right)$ and $\left(-\frac{\pi}{2}, -1\right)$ are points on the graph of $y = \sin x$, illustrating the property $\sin(-x) = -\sin x$.

Sine graphs occur in many practical applications. For example, look back at Figure 2 and assume that the line from the origin to some point $(p, q)$ on the circle is part of the pedal of a bicycle, with a foot placed at $(p, q)$. Since $q = \sin x$, the height of the pedal from the horizontal axis in Figure 2 is given by $\sin x$. By choosing various angles for the pedal and calculating $q$ for each angle, the height of the pedal leads to the sine curve shown in Figure 4.

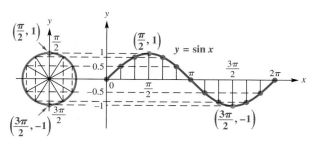

**FIGURE 4**

**OBJECTIVE 3** Graph $f(x) = \cos x$. The graph of $y = \cos x$ in Figure 5 has the same shape as the graph of $y = \sin x$. *The graph of the cosine function is, in fact, the graph of the sine function shifted, or translated, $\frac{\pi}{2}$ units to the left.*

**Cosine Function** $f(x) = \cos x$

Domain: $(-\infty, \infty)$   Range: $[-1, 1]$

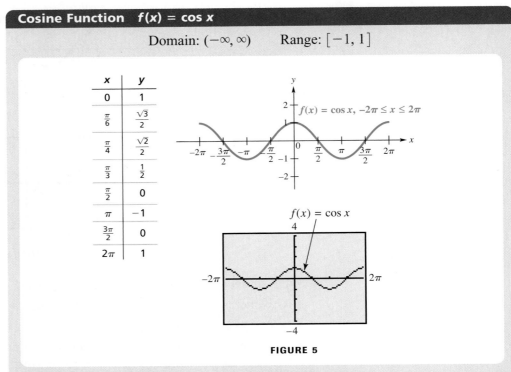

| x | y |
|---|---|
| 0 | 1 |
| $\frac{\pi}{6}$ | $\frac{\sqrt{3}}{2}$ |
| $\frac{\pi}{4}$ | $\frac{\sqrt{2}}{2}$ |
| $\frac{\pi}{3}$ | $\frac{1}{2}$ |
| $\frac{\pi}{2}$ | 0 |
| $\pi$ | $-1$ |
| $\frac{3\pi}{2}$ | 0 |
| $2\pi$ | 1 |

**FIGURE 5**

- The graph is continuous over its entire domain, $(-\infty, \infty)$.
- Its x-intercepts are of the form $\left((2n+1)\frac{\pi}{2}, 0\right)$, where $n$ is an integer.
- Its period is $2\pi$.
- The graph is symmetric with respect to the y-axis, so the function is an even function. For all $x$ in the domain, $\cos(-x) = \cos x$.

A function $f$ is an **even function** if for all $x$ in the domain of $f$, $f(-x) = f(x)$. The graph of an even function is symmetric with respect to the y-axis, that is, if $(x, y)$ belongs to the function, then $(-x, y)$ also belongs to the function. For example, $\left(\frac{\pi}{2}, 0\right)$ and $\left(-\frac{\pi}{2}, 0\right)$ are points on the graph of $y = \cos x$, illustrating the property $\cos(-x) = \cos x$.

The calculator graphs of $f(x) = \sin x$ in Figure 3 and $f(x) = \cos x$ in Figure 5 are graphed in the window $[-2\pi, 2\pi]$ by $[-4, 4]$, with Xscl $= \frac{\pi}{2}$ and Yscl $= 1$. This is called the **trig viewing window**. (Your model may use a different "standard" trigonometric viewing window. Consult your owner's manual.)

**OBJECTIVE 4** Apply changes of amplitude and period to graphs of the sine and cosine functions. The examples that follow show graphs that are "stretched" or "compressed" (shrunk) either vertically, horizontally, or both when compared with the graphs of $y = \sin x$ or $y = \cos x$.

### EXAMPLE 1  Graphing $y = a \sin x$

Graph $y = 2 \sin x$, and compare to the graph of $y = \sin x$.

**Solution** For a given value of $x$, the value of $y$ is twice as large as it would be for $y = \sin x$, as shown in the table of values. The only change in the graph is the range, which becomes $[-2, 2]$. See Figure 6, which includes a graph of $y = \sin x$ for comparison.

| $x$ | 0 | $\frac{\pi}{2}$ | $\pi$ | $\frac{3\pi}{2}$ | $2\pi$ |
|---|---|---|---|---|---|
| $\sin x$ | 0 | 1 | 0 | $-1$ | 0 |
| $2 \sin x$ | 0 | 2 | 0 | $-2$ | 0 |

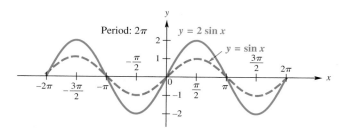

**FIGURE 6**

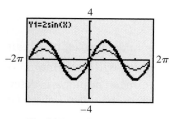

The thick graph style represents the function $y = 2 \sin x$ in Example 1.

The **amplitude** of a periodic function is half the difference between the maximum and minimum values. It describes the height of the graph both above and below a horizontal line passing through the "middle" of the graph. Thus, for both the basic sine and cosine functions, the amplitude is

$$\frac{1}{2}[1 - (-1)] = \frac{1}{2}(2) = 1.$$

Generalizing from Example 1 gives the following.

### Amplitude

The graph of $y = a \sin x$ or $y = a \cos x$, with $a \neq 0$, will have the same shape as the graph of $y = \sin x$ or $y = \cos x$, respectively, except with range $[-|a|, |a|]$. The amplitude is $|a|$.

**NOW TRY** Exercise 15.

No matter what the value of the amplitude, the periods of $y = a \sin x$ and $y = a \cos x$ are still $2\pi$. Consider $y = \sin 2x$. We can complete a table of values for the interval $[0, 2\pi]$.

| $x$ | 0 | $\frac{\pi}{4}$ | $\frac{\pi}{2}$ | $\frac{3\pi}{4}$ | $\pi$ | $\frac{5\pi}{4}$ | $\frac{3\pi}{2}$ | $\frac{7\pi}{4}$ | $2\pi$ |
|---|---|---|---|---|---|---|---|---|---|
| $\sin 2x$ | 0 | 1 | 0 | $-1$ | 0 | 1 | 0 | $-1$ | 0 |

Note that one complete cycle occurs in $\pi$ units, not $2\pi$ units. Therefore, the period here is $\pi$, which equals $\frac{2\pi}{2}$. Now consider $y = \sin 4x$. Look at the next table.

| $x$ | 0 | $\frac{\pi}{8}$ | $\frac{\pi}{4}$ | $\frac{3\pi}{8}$ | $\frac{\pi}{2}$ | $\frac{5\pi}{8}$ | $\frac{3\pi}{4}$ | $\frac{7\pi}{8}$ | $\pi$ |
|---|---|---|---|---|---|---|---|---|---|
| $\sin 4x$ | 0 | 1 | 0 | $-1$ | 0 | 1 | 0 | $-1$ | 0 |

These values suggest that one complete cycle is achieved in $\frac{\pi}{2}$ or $\frac{2\pi}{4}$ units, which is reasonable since

$$\sin\left(4 \cdot \frac{\pi}{2}\right) = \sin 2\pi = 0.$$

***In general, the graph of a function of the form $y = \sin bx$ or $y = \cos bx$, for $b > 0$, will have a period different from $2\pi$ when $b \neq 1$.*** To see why this is so, remember that the values of $\sin bx$ or $\cos bx$ will take on all possible values as $bx$ ranges from 0 to $2\pi$. Therefore, to find the period of either of these functions, we must solve the three-part inequality

$$0 \leq bx \leq 2\pi$$
$$0 \leq x \leq \frac{2\pi}{b}. \quad \text{Divide each part by the positive number } b.$$

***Thus, the period is $\frac{2\pi}{b}$.*** By dividing the interval $\left[0, \frac{2\pi}{b}\right]$ into four equal parts, we obtain the values for which $\sin bx$ or $\cos bx$ is $-1$, 0, or 1. These values will give minimum points, $x$-intercepts, and maximum points on the graph. Once these points are determined, we can sketch the graph by joining the points with a smooth sinusoidal curve. (If a function has $b < 0$, then the identities of the next chapter can be used to rewrite the function so that $b > 0$.)

One method to divide an interval into four equal parts is as follows.

**Step 1** Find the midpoint of the interval by adding the $x$-values of the endpoints and dividing by 2.

**Step 2** Find the midpoints of the two intervals found in Step 1, using the same procedure.

**EXAMPLE 2** Graphing $y = \sin bx$

Graph $y = \sin 2x$, and compare to the graph of $y = \sin x$.

**Solution** In this function the coefficient of $x$ is 2, so $b = 2$ and the period is $\frac{2\pi}{2} = \pi$. Therefore, the graph will complete one period over the interval $[0, \pi]$.

The endpoints are 0 and $\pi$, and the three points between the endpoints are

$$\frac{1}{4}(0 + \pi), \quad \frac{1}{2}(0 + \pi), \quad \text{and} \quad \frac{3}{4}(0 + \pi),$$

which give the following $x$-values:

$$0, \quad \frac{\pi}{4}, \quad \frac{\pi}{2}, \quad \frac{3\pi}{4}, \quad \pi.$$

↑ Left endpoint   ↑ First-quarter point   ↑ Midpoint   ↑ Third-quarter point   ↑ Right endpoint

We plot the points from the table of values given on page 773, and join them with a smooth sinusoidal curve. More of the graph can be sketched by repeating this cycle, as shown in Figure 7. The amplitude is not changed.

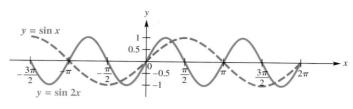

**FIGURE 7**

NOW TRY Exercise 27.

*We can think of the graph of $y = \sin bx$ as a horizontal stretching of the graph of $y = \sin x$ when $0 < b < 1$ and a horizontal shrinking when $b > 1$.*

### Period

For $b > 0$, the graph of $y = \sin bx$ will resemble that of $y = \sin x$, but with period $\frac{2\pi}{b}$. Also, the graph of $y = \cos bx$ will resemble that of $y = \cos x$, but with period $\frac{2\pi}{b}$.

### EXAMPLE 3  Graphing $y = \cos bx$

Graph $y = \cos \frac{2}{3}x$ over one period.

**Solution** The period is $\dfrac{2\pi}{\frac{2}{3}} = 2\pi \div \frac{2}{3} = 2\pi \cdot \frac{3}{2} = 3\pi$. We divide the interval $[0, 3\pi]$ into four equal parts to get the $x$-values $0, \frac{3\pi}{4}, \frac{3\pi}{2}, \frac{9\pi}{4}$, and $3\pi$ that yield minimum points, maximum points, and $x$-intercepts. We use these values to obtain a table of key points for one period.

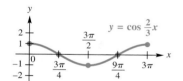

**FIGURE 8**

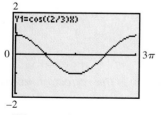

This screen shows a graph of the function in Example 3. By choosing Xscl = $\frac{3\pi}{4}$, the $x$-intercepts, maxima, and minima coincide with tick marks on the $x$-axis.

| $x$ | 0 | $\frac{3\pi}{4}$ | $\frac{3\pi}{2}$ | $\frac{9\pi}{4}$ | $3\pi$ |
|---|---|---|---|---|---|
| $\frac{2}{3}x$ | 0 | $\frac{\pi}{2}$ | $\pi$ | $\frac{3\pi}{2}$ | $2\pi$ |
| $\cos \frac{2}{3}x$ | 1 | 0 | $-1$ | 0 | 1 |

The amplitude is 1 because the maximum value is 1, the minimum value is $-1$, and $\frac{1}{2}(1 - (-1)) = \frac{1}{2}(2) = 1$. We plot these points and join them with a smooth curve. The graph is shown in Figure 8.

NOW TRY Exercise 25.

Look back at the middle row of the table in Example 3. Dividing the interval $\left[0, \frac{2\pi}{b}\right]$ into four equal parts will always give the values $0, \frac{\pi}{2}, \pi, \frac{3\pi}{2}$, and $2\pi$ for this row, in this case resulting in values of $-1$, 0, or 1. These values lead to key points on the graph, which can then be easily sketched.

### Guidelines for Sketching Graphs of Sine and Cosine Functions

To graph $y = a \sin bx$ or $y = a \cos bx$, with $b > 0$, follow these steps.

**Step 1** Find the period, $\frac{2\pi}{b}$. Start at 0 on the $x$-axis, and lay off a distance of $\frac{2\pi}{b}$.

**Step 2** Divide the interval into four equal parts. (See Steps 1 and 2 preceding Example 2 on page 774.)

**Step 3** Evaluate the function for each of the five $x$-values resulting from Step 2. The points will be maximum points, minimum points, and $x$-intercepts.

**Step 4** Plot the points found in Step 3, and join them with a sinusoidal curve having amplitude $|a|$.

**Step 5** Draw the graph over additional periods as needed.

**EXAMPLE 4** Graphing $y = a \sin bx$

Graph $y = -2 \sin 3x$ over one period using the preceding guidelines.

**Solution**

**Step 1** For this function, $b = 3$, so the period is $\frac{2\pi}{3}$. The function will be graphed over the interval $\left[0, \frac{2\pi}{3}\right]$.

**Step 2** Divide the interval $\left[0, \frac{2\pi}{3}\right]$ into four equal parts to get the $x$-values $0, \frac{\pi}{6}, \frac{\pi}{3}, \frac{\pi}{2},$ and $\frac{2\pi}{3}$.

**Step 3** Make a table of values determined by the $x$-values from Step 2.

| $x$ | 0 | $\frac{\pi}{6}$ | $\frac{\pi}{3}$ | $\frac{\pi}{2}$ | $\frac{2\pi}{3}$ |
|---|---|---|---|---|---|
| $3x$ | 0 | $\frac{\pi}{2}$ | $\pi$ | $\frac{3\pi}{2}$ | $2\pi$ |
| $\sin 3x$ | 0 | 1 | 0 | $-1$ | 0 |
| $-2 \sin 3x$ | 0 | $-2$ | 0 | 2 | 0 |

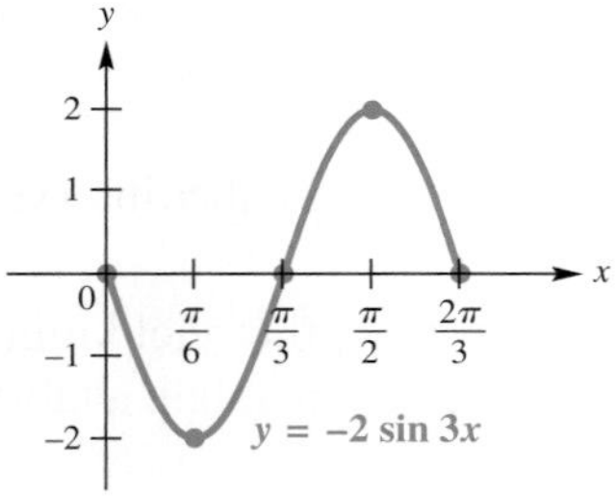

**Step 4** Plot the points $(0, 0)$, $\left(\frac{\pi}{6}, -2\right)$, $\left(\frac{\pi}{3}, 0\right)$, $\left(\frac{\pi}{2}, 2\right)$, and $\left(\frac{2\pi}{3}, 0\right)$, and join them with a sinusoidal curve with amplitude 2. See Figure 9.

**Step 5** The graph can be extended by repeating the cycle.

*Notice that when $a$ is negative, the graph of $y = a \sin bx$ is the reflection across the $x$-axis of the graph of $y = |a| \sin bx$.*

**FIGURE 9**

**NOW TRY** Exercise 29.

**EXAMPLE 5** Graphing $y = a \cos bx$ for $b$ Equal to a Multiple of $\pi$

Graph $y = -3 \cos \pi x$ over one period.

**Solution**

**Step 1** Since $b = \pi$, the period is $\frac{2\pi}{\pi} = 2$, so we will graph the function over the interval $[0, 2]$.

**Step 2** Dividing $[0, 2]$ into four equal parts yields the $x$-values $0, \frac{1}{2}, 1, \frac{3}{2},$ and 2.

*(continued)*

### SECTION 17.1 Graphs of the Sine and Cosine Functions

**Step 3** Make a table using these *x*-values.

| x | 0 | $\frac{1}{2}$ | 1 | $\frac{3}{2}$ | 2 |
|---|---|---|---|---|---|
| πx | 0 | $\frac{\pi}{2}$ | $\pi$ | $\frac{3\pi}{2}$ | $2\pi$ |
| cos πx | 1 | 0 | −1 | 0 | 1 |
| −3 cos πx | −3 | 0 | 3 | 0 | −3 |

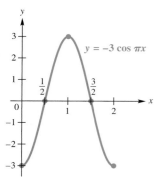

FIGURE 10

**Step 4** Plot the points $(0, -3)$, $\left(\frac{1}{2}, 0\right)$, $(1, 3)$, $\left(\frac{3}{2}, 0\right)$, and $(2, -3)$, and join them with a sinusoidal curve having amplitude $|-3| = 3$. See Figure 10.

**Step 5** The graph can be extended by repeating the cycle.

*Notice that when b is an integer multiple of π, the x-intercepts of the graph are rational numbers.*

**NOW TRY** Exercise 37.

---

### EXAMPLE 6 Interpreting a Sine Function Model

The average temperature (in °F) at Mould Bay, Canada, can be approximated by the function defined by

$$f(x) = 34 \sin\left[\frac{\pi}{6}(x - 4.3)\right],$$

where *x* is the month and $x = 1$ corresponds to January, $x = 2$ to February, and so on.

**(a)** To observe the graph over a two-year interval and to see the maximum and minimum points, graph *f* in the window $[0, 25]$ by $[-45, 45]$.

**(b)** What is the average temperature during the month of May?

**(c)** What would be an approximation for the average *yearly* temperature at Mould Bay?

**Solution**

**(a)** The graph of $f(x) = 34 \sin\left[\frac{\pi}{6}(x - 4.3)\right]$ is shown in Figure 11. Its amplitude is 34, and the period is $\frac{2\pi}{\frac{\pi}{6}} = 2\pi \div \frac{\pi}{6} = 2\pi \cdot \frac{6}{\pi} = 12$. The function *f* has a period of 12 months, or 1 year, which agrees with the changing of the seasons.

**(b)** May is the fifth month, so the average temperature during May is

$$f(5) = 34 \sin\left[\frac{\pi}{6}(5 - 4.3)\right] \approx 12°F. \quad \text{Let } x = 5.$$

See the display at the bottom of the screen in Figure 11.

**(c)** From the graph, it appears that the average yearly temperature is about 0°F since the graph is centered vertically about the line $y = 0$.

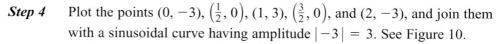

FIGURE 11

**NOW TRY** Exercise 51.

# 17.1 Exercises

*Matching* In Exercises 1–8, match each function with its graph.

1. $y = \sin x$
2. $y = \cos x$
3. $y = -\sin x$
4. $y = -\cos x$
5. $y = \sin 2x$
6. $y = \cos 2x$
7. $y = 2 \sin x$
8. $y = 2 \cos x$

A.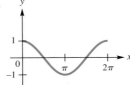
B.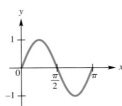
C.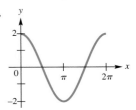

D.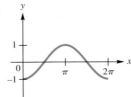
E.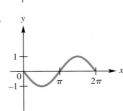
F.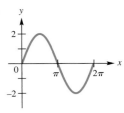

G.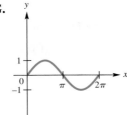
H.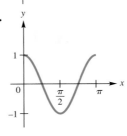

*Matching* In Exercises 9–12, match each function with its calculator graph.

9. $y = \sin 3x$
10. $y = \cos 3x$
11. $y = 3 \cos x$
12. $y = 3 \sin x$

A.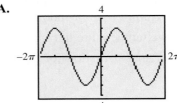
B.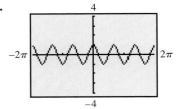

C.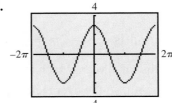
D.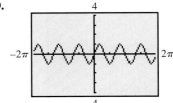

Graph each function over the interval $[-2\pi, 2\pi]$. Give the amplitude. See Example 1.

13. $y = 2 \cos x$
14. $y = 3 \sin x$
15. $y = \dfrac{2}{3} \sin x$

16. $y = \dfrac{3}{4} \cos x$
17. $y = -\cos x$
18. $y = -\sin x$

19. $y = -2 \sin x$
20. $y = -3 \cos x$
21. $y = \sin(-x)$

**22. Concept Check** In Exercise 21, why is the graph the same as that of $y = -\sin x$?

*Graph each function over a two-period interval. Give the period and amplitude. See Examples 2–5.*

**23.** $y = \sin \frac{1}{2}x$  **24.** $y = \sin \frac{2}{3}x$  **25.** $y = \cos \frac{3}{4}x$

**26.** $y = \cos \frac{1}{3}x$  **27.** $y = \sin 3x$  **28.** $y = \cos 2x$

**29.** $y = 2 \sin \frac{1}{4}x$  **30.** $y = 3 \sin 2x$  **31.** $y = -2 \cos 3x$

**32.** $y = -5 \cos 2x$  **33.** $y = \cos \pi x$  **34.** $y = -\sin \pi x$

**35.** $y = -2 \sin 2\pi x$  **36.** $y = 3 \cos 2\pi x$  **37.** $y = \frac{1}{2} \cos \frac{\pi}{2}x$

**38.** $y = -\frac{2}{3} \sin \frac{\pi}{4}x$  **39.** $y = \pi \sin \pi x$  **40.** $y = -\pi \cos \pi x$

*Connecting Graphs with Equations* Each function graphed is of the form $y = a \sin bx$ or $y = a \cos bx$, where $b > 0$. Determine the equation of the graph.

**41.**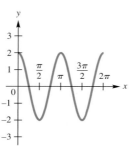

**42.**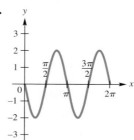

**43.**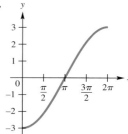

**44.**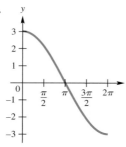

**45.**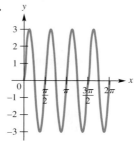

**46.**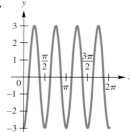

*Solve each problem.*

**47.** Scientists believe that the average annual temperature in a given location is periodic. The average temperature at a given place during a given season fluctuates as time goes on, from colder to warmer, and back to colder. The graph shows an idealized description of the temperature (in °F) for approximately the last 150 thousand years of a location at the same latitude as Anchorage, Alaska.

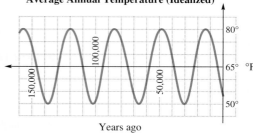

**(a)** Find the highest and lowest temperatures recorded.
**(b)** Use these two numbers to find the amplitude.
**(c)** Find the period of the function.
**(d)** What is the trend of the temperature now?

**48.** The graph gives the variation in blood pressure for a typical person. **Systolic** and **diastolic pressures** are the upper and lower limits of the periodic changes in pressure that produce the pulse. The length of time between peaks is called the period of the pulse.

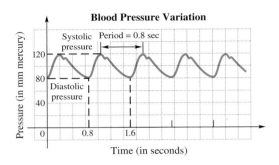

**(a)** Find the amplitude of the graph.
**(b)** Find the pulse rate (the number of pulse beats in 1 min) for this person.

**49.** At Mauna Loa, Hawaii, atmospheric carbon dioxide levels in parts per million (ppm) have been measured regularly since 1958. The function defined by

$$L(x) = 0.022x^2 + 0.55x + 316 + 3.5 \sin 2\pi x$$

can be used to model these levels, where $x$ is in years and $x = 0$ corresponds to 1960. (*Source:* Nilsson, A., *Greenhouse Earth,* John Wiley & Sons, 1992.)

**(a)** Graph $L$ in the window $[15, 35]$ by $[325, 365]$.
**(b)** When do the seasonal maximum and minimum carbon dioxide levels occur?
**(c)** $L$ is the sum of a quadratic function and a sine function. What is the significance of each of these functions? Discuss what physical phenomena may be responsible for each function.

**50.** Refer to Exercise 49. The carbon dioxide content in the atmosphere at Barrow, Alaska, in parts per million (ppm) can be modeled using the function defined by

$$C(x) = 0.04x^2 + 0.6x + 330 + 7.5 \sin 2\pi x,$$

where $x = 0$ corresponds to 1970. (*Source:* Zeilik, M. and S. Gregory, *Introductory Astronomy and Astrophysics,* Brooks/Cole, 1998.)

**(a)** Graph $C$ in the window $[5, 25]$ by $[320, 380]$.
**(b)** Discuss possible reasons why the amplitude of the oscillations in the graph of $C$ is larger than the amplitude of the oscillations in the graph of $L$ in Exercise 49, which models Hawaii.
**(c)** Define a new function $C$ that is valid if $x$ represents the actual year, where $1970 \leq x \leq 1995$. (See horizontal translations in **Section 12.3.**)

**51.** The temperature in a certain city in Alaska is modeled by

$$T(x) = 25 + 37 \sin\left[\frac{2\pi}{365}(x - 101)\right],$$

where $T(x)$ is the temperature in degrees Fahrenheit on day $x$, with $x = 1$ corresponding to January 1 and $x = 365$ corresponding to December 31. Use a calculator to estimate the temperature on the following days. (*Source:* Lando, B. and C. Lando, "Is the Graph of Temperature Variation a Sine Curve?", *The Mathematics Teacher,* 70, September 1977.)

**(a)** March 15 (day 74)    **(b)** April 5 (day 95)    **(c)** Day 200
**(d)** June 25              **(e)** October 1           **(f)** December 31

52. The **solar constant** $S$ is the amount of energy per unit area that reaches Earth's atmosphere from the Sun. It is equal to 1367 watts per sq m but varies slightly throughout the seasons. This fluctuation $\Delta S$ in $S$ can be calculated using the formula

$$\Delta S = 0.034 S \sin\left[\frac{2\pi(82.5 - N)}{365.25}\right].$$

In this formula, $N$ is the day number covering a four-year period, where $N = 1$ corresponds to January 1 of a leap year and $N = 1461$ corresponds to December 31 of the fourth year. (*Source:* Winter, C., R. Sizmann, and L. L. Vant-Hull, Editors, *Solar Power Plants,* Springer-Verlag, 1991.)

(a) Calculate $\Delta S$ for $N = 80$, which is the spring equinox in the first year.
(b) Calculate $\Delta S$ for $N = 1268$, which is the summer solstice in the fourth year.
(c) What is the maximum value of $\Delta S$?
(d) Find a value for $N$ where $\Delta S$ is equal to 0.

## 17.2 Translations of the Graphs of the Sine and Cosine Functions

**OBJECTIVES**

1 Apply horizontal translations to the graphs of sine and cosine functions.
2 Apply vertical translations to the graphs of sine and cosine functions.
3 Sketch graphs of sine and cosine functions using combinations of stretching, shrinking, and translating.

**OBJECTIVE 1** Apply horizontal translations to the graphs of sine and cosine functions. The graph of the function defined by $y = f(x - d)$ is translated *horizontally* compared to the graph of $y = f(x)$. The translation is $d$ units to the right if $d > 0$ and $|d|$ units to the left if $d < 0$. See Figure 12. With circular functions, a horizontal translation is called a **phase shift**. In the function $y = f(x - d)$, the expression $x - d$ is called the **argument**. In Examples 1–3, we give two methods that can be used to sketch the graph of a circular function involving a phase shift.

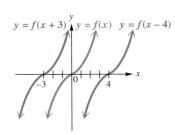

Horizontal translations of $y = f(x)$
**FIGURE 12**

**EXAMPLE 1** Graphing $y = \sin(x - d)$

Graph $y = \sin\left(x - \frac{\pi}{3}\right)$ over one period.

**Solution** *Method 1* For the argument $x - \frac{\pi}{3}$ to result in all possible values throughout one period, it must take on all values between 0 and $2\pi$, inclusive. Therefore, to find an interval of one period, we solve the three-part inequality

$$0 \leq x - \frac{\pi}{3} \leq 2\pi$$

$$\frac{\pi}{3} \leq x \leq \frac{7\pi}{3}. \qquad \text{Add } \tfrac{\pi}{3} \text{ to each part.}$$

*(continued)*

Use the method described right before Example 2 on page 774 to divide the interval $\left[\frac{\pi}{3}, \frac{7\pi}{3}\right]$ into four equal parts, obtaining the following *x*-values.

$$\frac{\pi}{3},\quad \frac{5\pi}{6},\quad \frac{4\pi}{3},\quad \frac{11\pi}{6},\quad \frac{7\pi}{3}$$

These are *key* *x*-values.

A table of values using these *x*-values follows.

| $x$ | $\frac{\pi}{3}$ | $\frac{5\pi}{6}$ | $\frac{4\pi}{3}$ | $\frac{11\pi}{6}$ | $\frac{7\pi}{3}$ |
|---|---|---|---|---|---|
| $x - \frac{\pi}{3}$ | 0 | $\frac{\pi}{2}$ | $\pi$ | $\frac{3\pi}{2}$ | $2\pi$ |
| $\sin\left(x - \frac{\pi}{3}\right)$ | 0 | 1 | 0 | $-1$ | 0 |

We join the corresponding points with a smooth curve to get the solid blue graph shown in Figure 13. The period is $2\pi$, and the amplitude is 1.

**Method 2** We can also graph $y = \sin\left(x - \frac{\pi}{3}\right)$ by using a horizontal translation of the graph of $y = \sin x$. The argument $x - \frac{\pi}{3}$ indicates that the graph will be translated $\frac{\pi}{3}$ units to the *right* (the phase shift) as compared to the graph of $y = \sin x$. See Figure 13.

Therefore, to graph a function using this method, first graph the basic circular function, and then graph the desired function by using the appropriate translation.

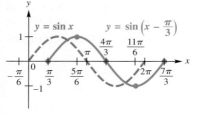

**FIGURE 13**

**NOW TRY** Exercise 35.

The graph in Figure 13 of Example 1 can be extended through additional periods by repeating the given portion of the graph, as necessary.

**EXAMPLE 2**  Graphing $y = a\cos(x - d)$

Graph $y = 3\cos\left(x + \frac{\pi}{4}\right)$ over one period.

**Solution**  **Method 1**  The graph can be sketched over one period by first solving the three-part inequality

$$0 \leq x + \frac{\pi}{4} \leq 2\pi$$

$$-\frac{\pi}{4} \leq x \leq \frac{7\pi}{4}.\qquad \text{Subtract } \frac{\pi}{4} \text{ from each part.}$$

Dividing this interval into four equal parts gives *x*-values of

$$-\frac{\pi}{4},\quad \frac{\pi}{4},\quad \frac{3\pi}{4},\quad \frac{5\pi}{4},\quad \frac{7\pi}{4}.\qquad \text{Key } x\text{-values}$$

A table of points for these *x*-values leads to maximum points, minimum points, and *x*-intercepts.

| $x$ | $-\frac{\pi}{4}$ | $\frac{\pi}{4}$ | $\frac{3\pi}{4}$ | $\frac{5\pi}{4}$ | $\frac{7\pi}{4}$ |
|---|---|---|---|---|---|
| $x + \frac{\pi}{4}$ | 0 | $\frac{\pi}{2}$ | $\pi$ | $\frac{3\pi}{2}$ | $2\pi$ |
| $\cos\left(x + \frac{\pi}{4}\right)$ | 1 | 0 | $-1$ | 0 | 1 |
| $3\cos\left(x + \frac{\pi}{4}\right)$ | 3 | 0 | $-3$ | 0 | 3 |

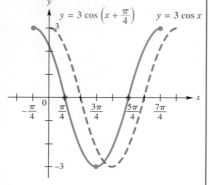

**FIGURE 14**

We join the corresponding points with a smooth curve to get the solid blue graph shown in Figure 14. The period is $2\pi$, and the amplitude is 3.

*(continued)*

### SECTION 17.2 Translations of the Graphs of the Sine and Cosine Functions

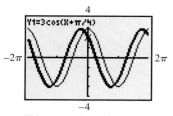

This screen shows the graph of

$$Y_1 = 3 \cos\left(X + \frac{\pi}{4}\right)$$

in Example 2 as a thick line. The graph of

$$Y_2 = 3 \cos X$$

is shown as a thin line for comparison.

**Method 2** Write $3 \cos\left(x + \frac{\pi}{4}\right)$ in the form $a \cos(x - d)$.

$$y = 3 \cos\left(x + \frac{\pi}{4}\right) = 3 \cos\left[x - \left(-\frac{\pi}{4}\right)\right]$$

This result shows that $d = -\frac{\pi}{4}$. Since $-\frac{\pi}{4}$ is negative, the phase shift is $\left|-\frac{\pi}{4}\right| = \frac{\pi}{4}$ unit to the left. The graph is the same as that of $y = 3 \cos x$ (the dashed red graph in Figure 14 on the preceding page), except that it is translated $\frac{\pi}{4}$ unit to the left (the solid blue graph in Figure 14).

**NOW TRY** Exercise 37.

**EXAMPLE 3** Graphing $y = a \cos b(x - d)$

Graph $y = -2 \cos(3x + \pi)$ over two periods.

**Solution**  **Method 1**  The function can be sketched over one period by solving the three-part inequality

$$0 \leq 3x + \pi \leq 2\pi$$

to get the interval $\left[-\frac{\pi}{3}, \frac{\pi}{3}\right]$. Divide this interval into four equal parts to get the points $\left(-\frac{\pi}{3}, -2\right), \left(-\frac{\pi}{6}, 0\right), (0, 2), \left(\frac{\pi}{6}, 0\right),$ and $\left(\frac{\pi}{3}, -2\right)$. Plot these points and join them with a smooth curve. By graphing an additional half period to the left and to the right, we obtain the graph shown in Figure 15.

**Method 2**  First write the expression in the form $a \cos b(x - d)$.

$$y = -2 \cos(3x + \pi) = -2 \cos 3\left(x + \frac{\pi}{3}\right) \quad \text{Rewrite } 3x + \pi \text{ as } 3\left(x + \frac{\pi}{3}\right).$$

Then $a = -2$, $b = 3$, and $d = -\frac{\pi}{3}$. The amplitude is $|-2| = 2$, and the period is $\frac{2\pi}{3}$ (since the value of $b$ is 3). The phase shift is $\left|-\frac{\pi}{3}\right| = \frac{\pi}{3}$ units to the left as compared to the graph of $y = -2 \cos 3x$. Again, see Figure 15.

**NOW TRY** Exercise 43.

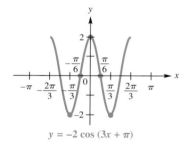

$y = -2 \cos(3x + \pi)$

**FIGURE 15**

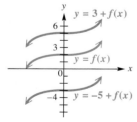

Vertical translations of $y = f(x)$

**FIGURE 16**

**OBJECTIVE 2** Apply vertical translations to the graphs of sine and cosine functions. The graph of a function of the form $y = c + f(x)$ is translated *vertically* as compared with the graph of $y = f(x)$. See Figure 16. The translation is $c$ units up if $c > 0$ and $|c|$ units down if $c < 0$.

**EXAMPLE 4** Graphing $y = c + a \cos bx$

Graph $y = 3 - 2 \cos 3x$ over two periods.

**Solution** The values of $y$ will be 3 greater than the corresponding values of $y$ in $y = -2 \cos 3x$. This means that the graph of $y = 3 - 2 \cos 3x$ is the same as the graph of $y = -2 \cos 3x$, vertically translated 3 units up. Since the period of $y = -2 \cos 3x$ is $\frac{2\pi}{3}$, the key points have $x$-values

$$0, \quad \frac{\pi}{6}, \quad \frac{\pi}{3}, \quad \frac{\pi}{2}, \quad \frac{2\pi}{3}. \qquad \text{Key } x\text{-values}$$

*(continued)*

Use these x-values to make a table of points.

| x | 0 | $\frac{\pi}{6}$ | $\frac{\pi}{3}$ | $\frac{\pi}{2}$ | $\frac{2\pi}{3}$ |
|---|---|---|---|---|---|
| cos 3x | 1 | 0 | −1 | 0 | 1 |
| 2 cos 3x | 2 | 0 | −2 | 0 | 2 |
| 3 − 2 cos 3x | 1 | 3 | 5 | 3 | 1 |

The key points are shown on the graph in Figure 17, along with more of the graph, sketched using the fact that the function is periodic.

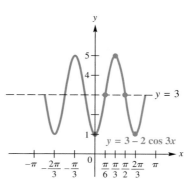

FIGURE 17

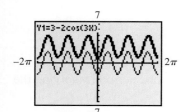

The function in Example 4 is shown using the thick graph style. Notice also the thin graph style for $y = -2 \cos 3x$.

**NOW TRY** Exercise 47.

**OBJECTIVE 3** Sketch graphs of sine and cosine functions using combinations of stretching, shrinking, and translating. A function of the form

$$y = c + a \sin b(x - d) \quad \text{or} \quad y = c + a \cos b(x - d), \quad b > 0,$$

which involves stretching, shrinking, and translating, can be graphed according to the following guidelines.

### Further Guidelines for Sketching Graphs of Sine and Cosine Functions

**Method 1** Follow these steps.

**Step 1** Find an interval whose length is one period $\frac{2\pi}{b}$ by solving the three-part inequality $0 \leq b(x - d) \leq 2\pi$.

**Step 2** Divide the interval into four equal parts. (See page 774.)

**Step 3** Evaluate the function for each of the five x-values resulting from Step 2. The points will be maximum points, minimum points, and points that intersect the line $y = c$ ("middle" points of the wave).

**Step 4** Plot the points found in Step 3, and join them with a sinusoidal curve having amplitude $|a|$.

**Step 5** Draw the graph over additional periods, as needed.

**Method 2** First graph $y = a \sin bx$ or $y = a \cos bx$. The amplitude of the function is $|a|$, and the period is $\frac{2\pi}{b}$. Then use translations to graph the desired function. The vertical translation is c units up if $c > 0$ and $|c|$ units down if $c < 0$. The horizontal translation (phase shift) is d units to the right if $d > 0$ and $|d|$ units to the left if $d < 0$.

## SECTION 17.2 Translations of the Graphs of the Sine and Cosine Functions

**EXAMPLE 5** Graphing $y = c + a \sin b(x - d)$

Graph $y = -1 + 2 \sin(4x + \pi)$ over two periods.

**Solution** We use Method 1. First write the expression on the right side in the form $c + a \sin b(x - d)$.

$$y = -1 + 2 \sin(4x + \pi) = -1 + 2 \sin\left[4\left(x + \frac{\pi}{4}\right)\right]$$

**Step 1** Find an interval whose length is one period.

$$0 \leq 4\left(x + \frac{\pi}{4}\right) \leq 2\pi$$

$$0 \leq x + \frac{\pi}{4} \leq \frac{\pi}{2} \qquad \text{Divide each part by 4.}$$

$$-\frac{\pi}{4} \leq x \leq \frac{\pi}{4} \qquad \text{Subtract } \tfrac{\pi}{4} \text{ from each part.}$$

**Step 2** Divide the interval $\left[-\frac{\pi}{4}, \frac{\pi}{4}\right]$ into four equal parts to get the $x$-values

$$-\frac{\pi}{4}, \quad -\frac{\pi}{8}, \quad 0, \quad \frac{\pi}{8}, \quad \frac{\pi}{4}. \qquad \text{Key } x\text{-values}$$

**Step 3** Make a table of values.

| $x$ | $-\frac{\pi}{4}$ | $-\frac{\pi}{8}$ | $0$ | $\frac{\pi}{8}$ | $\frac{\pi}{4}$ |
|---|---|---|---|---|---|
| $x + \frac{\pi}{4}$ | $0$ | $\frac{\pi}{8}$ | $\frac{\pi}{4}$ | $\frac{3\pi}{8}$ | $\frac{\pi}{2}$ |
| $4\left(x + \frac{\pi}{4}\right)$ | $0$ | $\frac{\pi}{2}$ | $\pi$ | $\frac{3\pi}{2}$ | $2\pi$ |
| $\sin 4\left(x + \frac{\pi}{4}\right)$ | $0$ | $1$ | $0$ | $-1$ | $0$ |
| $2 \sin 4\left(x + \frac{\pi}{4}\right)$ | $0$ | $2$ | $0$ | $-2$ | $0$ |
| $-1 + 2 \sin(4x + \pi)$ | $-1$ | $1$ | $-1$ | $-3$ | $-1$ |

**Steps 4 and 5** Plot the points found in the table and join them with a sinusoidal curve. Figure 18 shows the graph, extended to the right and left to include two full periods.

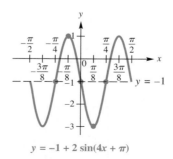

$y = -1 + 2 \sin(4x + \pi)$

**FIGURE 18**

**NOW TRY** Exercise 53.

# 17.2 Exercises

**Matching** In Exercises 1–8, match each function with its graph in A–H.

1. $y = \sin\left(x - \dfrac{\pi}{4}\right)$
2. $y = \sin\left(x + \dfrac{\pi}{4}\right)$
3. $y = \cos\left(x - \dfrac{\pi}{4}\right)$
4. $y = \cos\left(x + \dfrac{\pi}{4}\right)$
5. $y = 1 + \sin x$
6. $y = -1 + \sin x$
7. $y = 1 + \cos x$
8. $y = -1 + \cos x$

A.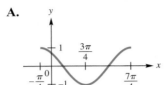

B.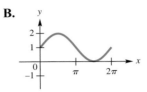

C.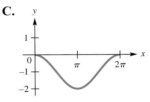

D.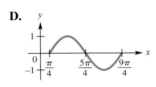

E.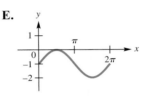

F.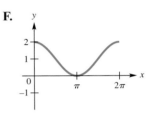

G.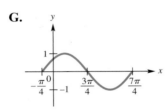

H.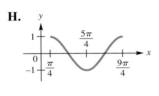

**Matching** In Exercises 9–12, match each function with its calculator graph in the standard trig window.

9. $y = \cos\left(x - \dfrac{\pi}{4}\right)$
10. $y = \sin\left(x - \dfrac{\pi}{4}\right)$
11. $y = 1 + \sin x$
12. $y = -1 + \cos x$

A.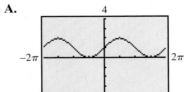

B.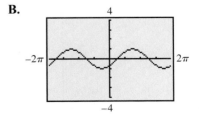

C.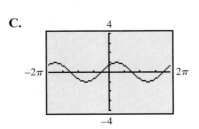

D.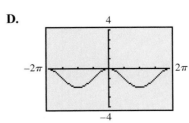

13. The graphs of $y = \sin x + 1$ and $y = \sin(x + 1)$ are **NOT** the same. Explain why this is so.

14. *Concept Check* Refer to Exercise 13. Which one of the two graphs is the same as that of $y = 1 + \sin x$?

*Matching* Match each function in Column I with the appropriate description in Column II.

| I | II |
|---|---|
| 15. $y = 3 \sin(2x - 4)$ | A. amplitude = 2, period = $\frac{\pi}{2}$, phase shift = $\frac{3}{4}$ |
| 16. $y = 2 \sin(3x - 4)$ | B. amplitude = 3, period = $\pi$, phase shift = 2 |
| 17. $y = 4 \sin(3x - 2)$ | C. amplitude = 4, period = $\frac{2\pi}{3}$, phase shift = $\frac{2}{3}$ |
| 18. $y = 2 \sin(4x - 3)$ | D. amplitude = 2, period = $\frac{2\pi}{3}$, phase shift = $\frac{4}{3}$ |

*Fill in the Blanks* In Exercises 19 and 20, fill in the blanks with the word right or the word left.

19. If the graph of $y = \cos x$ is translated $\frac{\pi}{2}$ units horizontally to the _____, it will coincide with the graph of $y = \sin x$.

20. If the graph of $y = \sin x$ is translated $\frac{\pi}{2}$ units horizontally to the _____, it will coincide with the graph of $y = \cos x$.

*Connecting Graphs with Equations* Each function graphed is of the form $y = c + \cos x$, $y = c + \sin x$, $y = \cos(x - d)$, or $y = \sin(x - d)$, where d is the least possible positive value. Determine the equation of the graph.

21.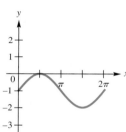

22.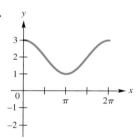

23.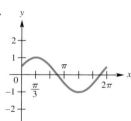

24.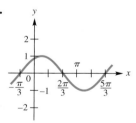

*Find the amplitude, the period, any vertical translation, and any phase shift of the graph of each function. See Examples 1–5.*

25. $y = 2 \sin(x - \pi)$

26. $y = \frac{2}{3} \sin\left(x + \frac{\pi}{2}\right)$

27. $y = 4 \cos\left(\frac{1}{2}x + \frac{\pi}{2}\right)$

28. $y = \frac{1}{2} \sin\left(\frac{1}{2}x + \pi\right)$

29. $y = 3 \cos \frac{\pi}{2}\left(x - \frac{1}{2}\right)$

30. $y = -\cos \pi\left(x - \frac{1}{3}\right)$

31. $y = 2 - \sin\left(3x - \frac{\pi}{5}\right)$

32. $y = -1 + \frac{1}{2} \cos(2x - 3\pi)$

*Graph each function over a two-period interval. See Examples 1 and 2.*

**33.** $y = \cos\left(x - \dfrac{\pi}{2}\right)$  **34.** $y = \sin\left(x - \dfrac{\pi}{4}\right)$

**35.** $y = \sin\left(x + \dfrac{\pi}{4}\right)$  **36.** $y = \cos\left(x - \dfrac{\pi}{3}\right)$

**37.** $y = 2\cos\left(x - \dfrac{\pi}{3}\right)$  **38.** $y = 3\sin\left(x - \dfrac{3\pi}{2}\right)$

*Graph each function over a one-period interval. See Example 3.*

**39.** $y = \dfrac{3}{2}\sin 2\left(x + \dfrac{\pi}{4}\right)$  **40.** $y = -\dfrac{1}{2}\cos 4\left(x + \dfrac{\pi}{2}\right)$

**41.** $y = -4\sin(2x - \pi)$  **42.** $y = 3\cos(4x + \pi)$

**43.** $y = \dfrac{1}{2}\cos\left(\dfrac{1}{2}x - \dfrac{\pi}{4}\right)$  **44.** $y = -\dfrac{1}{4}\sin\left(\dfrac{3}{4}x + \dfrac{\pi}{8}\right)$

*Graph each function over a two-period interval. See Example 4.*

**45.** $y = -3 + 2\sin x$  **46.** $y = 2 - 3\cos x$

**47.** $y = -1 - 2\cos 5x$  **48.** $y = 1 - \dfrac{2}{3}\sin \dfrac{3}{4}x$

**49.** $y = 1 - 2\cos \dfrac{1}{2}x$  **50.** $y = -3 + 3\sin \dfrac{1}{2}x$

**51.** $y = -2 + \dfrac{1}{2}\sin 3x$  **52.** $y = 1 + \dfrac{2}{3}\cos \dfrac{1}{2}x$

*Graph each function over a one-period interval. See Example 5.*

**53.** $y = -3 + 2\sin\left(x + \dfrac{\pi}{2}\right)$  **54.** $y = 4 - 3\cos(x - \pi)$

**55.** $y = \dfrac{1}{2} + \sin 2\left(x + \dfrac{\pi}{4}\right)$  **56.** $y = -\dfrac{5}{2} + \cos 3\left(x - \dfrac{\pi}{6}\right)$

## 17.3 Graphs of the Tangent and Cotangent Functions

**OBJECTIVES**

1. Graph $f(x) = \tan x$.
2. Graph $f(x) = \cot x$.
3. Apply graphing techniques to tangent and cotangent functions.

**OBJECTIVE 1** Graph $f(x) = \tan x$. Consider the table of selected points accompanying the graph of the tangent function in Figure 19 on the next page. These points include special values between $-\dfrac{\pi}{2}$ and $\dfrac{\pi}{2}$. The tangent function is undefined for odd multiples of $\dfrac{\pi}{2}$ and, thus, has *vertical asymptotes* for such values. A **vertical asymptote** is a vertical line that the graph approaches but does not intersect, while function values increase or decrease without bound as $x$-values get closer and closer to the line. Furthermore, since $\tan(-x) = -\tan x$ (see Exercise 43), the graph of the tangent function is symmetric with respect to the origin.

SECTION 17.3 Graphs of the Tangent and Cotangent Functions 789

| x | y = tan x |
|---|---|
| $-\frac{\pi}{3}$ | $-\sqrt{3} \approx -1.7$ |
| $-\frac{\pi}{4}$ | $-1$ |
| $-\frac{\pi}{6}$ | $-\frac{\sqrt{3}}{3} \approx -0.6$ |
| $0$ | $0$ |
| $\frac{\pi}{6}$ | $\frac{\sqrt{3}}{3} \approx 0.6$ |
| $\frac{\pi}{4}$ | $1$ |
| $\frac{\pi}{3}$ | $\sqrt{3} \approx 1.7$ |

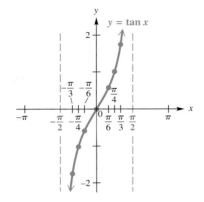

FIGURE 19

The tangent function has period $\pi$. Because $\tan x = \frac{\sin x}{\cos x}$, tangent values are 0 when sine values are 0 and undefined when cosine values are 0. As $x$-values go from $-\frac{\pi}{2}$ to $\frac{\pi}{2}$, tangent values go from $-\infty$ to $\infty$ and increase throughout the interval. Those same values are repeated as $x$ goes from $\frac{\pi}{2}$ to $\frac{3\pi}{2}$, $\frac{3\pi}{2}$ to $\frac{5\pi}{2}$, and so on. The graph of $y = \tan x$ from $-\frac{3\pi}{2}$ to $\frac{3\pi}{2}$ is shown in Figure 20.

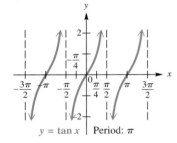

$y = \tan x$ | Period: $\pi$

FIGURE 20

**Tangent Function** $f(x) = \tan x$

Domain: $\{x \mid x \neq (2n + 1)\frac{\pi}{2}, \text{ where } n \text{ is any integer}\}$     Range: $(-\infty, \infty)$

| x | y |
|---|---|
| $-\frac{\pi}{2}$ | undefined |
| $-\frac{\pi}{4}$ | $-1$ |
| $0$ | $0$ |
| $\frac{\pi}{4}$ | $1$ |
| $\frac{\pi}{2}$ | undefined |

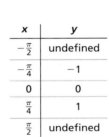
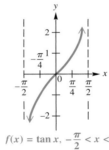

$f(x) = \tan x$, $-\frac{\pi}{2} < x < \frac{\pi}{2}$

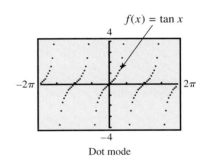

Dot mode

FIGURE 21

- The graph is discontinuous at values of $x$ of the form $x = (2n + 1)\frac{\pi}{2}$ and has vertical asymptotes at these values.
- Its $x$-intercepts are of the form $(n\pi, 0)$.
- Its period is $\pi$.
- Its graph has no amplitude, since there are no minimum or maximum values.
- The graph is symmetric with respect to the origin, so the function is an odd function. For all $x$ in the domain, $\tan(-x) = -\tan x$.

**OBJECTIVE 2** Graph $f(x) = \cot x$.  A similar analysis for selected points between 0 and $\pi$ for the graph of the cotangent function yields the graph in Figure 22 on the next page. Here the vertical asymptotes are at $x$-values that are integer multiples of $\pi$. Because $\cot(-x) = -\cot x$ (see Exercise 44), this graph is also symmetric with respect to the origin. (This can be seen when more of the graph is plotted.)

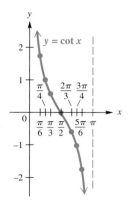

**FIGURE 22**

The cotangent function also has period $\pi$. Cotangent values are 0 when cosine values are 0 and undefined when sine values are 0. As $x$-values go from 0 to $\pi$, cotangent values go from $\infty$ to $-\infty$ and decrease throughout the interval. Those same values are repeated as $x$ goes from $\pi$ to $2\pi$, $2\pi$ to $3\pi$, and so on. The graph of $y = \cot x$ from $-\pi$ to $\pi$ is shown in Figure 23. The graph continues in this pattern.

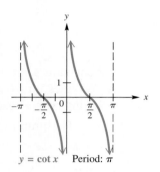

**FIGURE 23**

**Cotangent Function**   $f(x) = \cot x$

Domain: $\{x \mid x \neq n\pi,\ \text{where } n \text{ is any integer}\}$    Range: $(-\infty, \infty)$

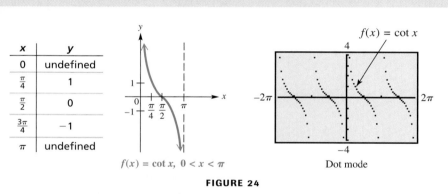

**FIGURE 24**

- The graph is discontinuous at values of $x$ of the form $x = n\pi$ and has vertical asymptotes at these values.
- Its $x$-intercepts are of the form $\left((2n + 1)\frac{\pi}{2}, 0\right)$.
- Its period is $\pi$.
- Its graph has no amplitude, since there are no minimum or maximum values.
- The graph is symmetric with respect to the origin, so the function is an odd function. For all $x$ in the domain, $\cot(-x) = -\cot x$.

The tangent function can be graphed directly with a graphing calculator, using the tangent key. To graph the cotangent function, however, we must use one of the identities $\cot x = \frac{1}{\tan x}$ or $\cot x = \frac{\cos x}{\sin x}$, since graphing calculators generally do not have cotangent keys.

**SECTION 17.3** Graphs of the Tangent and Cotangent Functions

**OBJECTIVE 3** Apply graphing techniques to tangent and cotangent functions.

**Guidelines for Sketching Graphs of Tangent and Cotangent Functions**

To graph $y = a\,\tan bx$ or $y = a\,\cot bx$, with $b > 0$, follow these steps.

**Step 1** Determine the period, $\frac{\pi}{b}$. To locate two adjacent vertical asymptotes, solve the following equations for $x$:

For $y = a\,\tan bx$:  $bx = -\frac{\pi}{2}$  and  $bx = \frac{\pi}{2}$.

For $y = a\,\cot bx$:  $bx = 0$  and  $bx = \pi$.

**Step 2** Sketch the two vertical asymptotes found in Step 1.

**Step 3** Divide the interval formed by the vertical asymptotes into four equal parts. (See page 774.)

**Step 4** Evaluate the function for the first-quarter point, midpoint, and third-quarter point, using the $x$-values found in Step 3.

**Step 5** Join the points with a smooth curve, approaching the vertical asymptotes. Indicate additional asymptotes and periods of the graph as necessary.

**EXAMPLE 1** Graphing $y = \tan bx$

Graph $y = \tan 2x$.

**Solution**

**Step 1** The period of this function is $\frac{\pi}{2}$. To locate two adjacent vertical asymptotes, solve $2x = -\frac{\pi}{2}$ and $2x = \frac{\pi}{2}$ (since this is a tangent function). The two asymptotes have equations $x = -\frac{\pi}{4}$ and $x = \frac{\pi}{4}$.

**Step 2** Sketch the two vertical asymptotes $x = \pm\frac{\pi}{4}$, as shown in Figure 25.

**Step 3** Divide the interval $\left(-\frac{\pi}{4}, \frac{\pi}{4}\right)$ into four equal parts. This gives the following key $x$-values.

first-quarter value: $-\frac{\pi}{8}$,   middle value: $0$,   third-quarter value: $\frac{\pi}{8}$   Key $x$-values

**Step 4** Evaluate the function for the $x$-values found in Step 3.

| $x$ | $-\frac{\pi}{8}$ | $0$ | $\frac{\pi}{8}$ |
|---|---|---|---|
| $2x$ | $-\frac{\pi}{4}$ | $0$ | $\frac{\pi}{4}$ |
| $\tan 2x$ | $-1$ | $0$ | $1$ |

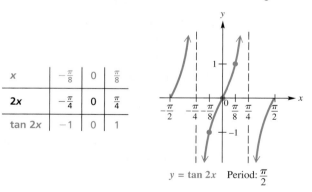

$y = \tan 2x$   Period: $\frac{\pi}{2}$

**FIGURE 25**

**Step 5** Join these points with a smooth curve, approaching the vertical asymptotes. See Figure 25. Another period has been graphed, one half period to the left and one half period to the right.

**NOW TRY** Exercise 7.

### EXAMPLE 2   Graphing y = a tan bx

Graph $y = -3 \tan \frac{1}{2}x$.

**Solution**   The period is $\frac{\pi}{\frac{1}{2}} = \pi \div \frac{1}{2} = \pi \cdot \frac{2}{1} = 2\pi$. Adjacent asymptotes are at $x = -\pi$ and $x = \pi$. Dividing the interval $(-\pi, \pi)$ into four equal parts gives key x-values of $-\frac{\pi}{2}$, 0, and $\frac{\pi}{2}$. Evaluating the function at these x-values gives the following key points.

$$\left(-\frac{\pi}{2}, 3\right), \qquad (0, 0), \qquad \left(\frac{\pi}{2}, -3\right) \qquad \text{Key points}$$

By plotting these points and joining them with a smooth curve, we obtain the graph shown in Figure 26. Because the coefficient $-3$ is negative, the graph is reflected across the x-axis compared to the graph of $y = 3 \tan \frac{1}{2}x$.

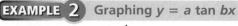

**FIGURE 26**

NOW TRY  Exercise 15.

The function defined by $y = -3 \tan \frac{1}{2}x$ in Example 2, graphed in Figure 26, has a graph that compares to the graph of $y = \tan x$ as follows.

1. The period is larger because $b = \frac{1}{2}$, and $\frac{1}{2} < 1$.
2. The graph is "stretched" because $a = -3$, and $|-3| > 1$.
3. Each branch of the graph goes down from left to right (that is, the function decreases) between each pair of adjacent asymptotes because $a = -3$, and $-3 < 0$. When $a < 0$, the graph is reflected across the x-axis compared to the graph of $y = |a| \tan bx$.

### EXAMPLE 3   Graphing y = a cot bx

Graph $y = \frac{1}{2} \cot 2x$.

**Solution**   Because this function involves the cotangent, we can locate two adjacent asymptotes by solving the equations $2x = 0$ and $2x = \pi$. The lines $x = 0$ (the y-axis) and $x = \frac{\pi}{2}$ are two such asymptotes. Divide the interval $\left(0, \frac{\pi}{2}\right)$ into four equal parts, getting key x-values of $\frac{\pi}{8}, \frac{\pi}{4}$, and $\frac{3\pi}{8}$. Evaluating the function at these x-values gives the key points $\left(\frac{\pi}{8}, \frac{1}{2}\right), \left(\frac{\pi}{4}, 0\right), \left(\frac{3\pi}{8}, -\frac{1}{2}\right)$. Joining these points with a smooth curve approaching the asymptotes gives the graph shown in Figure 27.

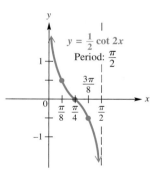

**FIGURE 27**

NOW TRY  Exercise 17.

Like the other circular functions, the graphs of the tangent and cotangent functions may be translated horizontally and vertically.

### EXAMPLE 4  Graphing a Tangent Function with a Vertical Translation

Graph $y = 2 + \tan x$.

**Analytic Solution**

Every value of $y$ for this function will be 2 units more than the corresponding value of $y$ in $y = \tan x$, causing the graph of $y = 2 + \tan x$ to be translated 2 units up compared with the graph of $y = \tan x$. See Figure 28.

**Graphing Calculator Solution**

To see the vertical translation, observe the coordinates displayed at the bottoms of the screens in Figures 29 and 30. For $X = \frac{\pi}{4} \approx .78539816$,

$$Y_1 = \tan X = 1,$$

while for the same X-value,

$$Y_2 = 2 + \tan X = 2 + 1 = 3.$$

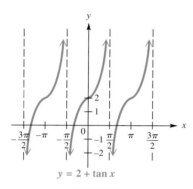

$y = 2 + \tan x$

**FIGURE 28**

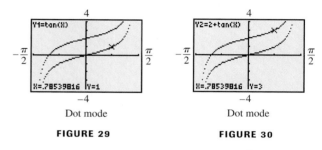

Dot mode

**FIGURE 29**

Dot mode

**FIGURE 30**

**NOW TRY** Exercise 23.

### EXAMPLE 5  Graphing a Cotangent Function with Vertical and Horizontal Translations

Graph $y = -2 - \cot\left(x - \frac{\pi}{4}\right)$.

**Solution**  Here $b = 1$, so the period is $\pi$. The graph will be translated down 2 units (because $c = -2$), reflected across the x-axis (because of the negative sign in front of the cotangent), and will have a phase shift (horizontal translation) $\frac{\pi}{4}$ unit to the right (because of the argument $\left(x - \frac{\pi}{4}\right)$). To locate adjacent asymptotes, since this function involves the cotangent, we solve the following equations:

$$x - \frac{\pi}{4} = 0, \quad \text{so } x = \frac{\pi}{4} \quad \text{and} \quad x - \frac{\pi}{4} = \pi, \quad \text{so } x = \frac{5\pi}{4}.$$

Dividing the interval $\left(\frac{\pi}{4}, \frac{5\pi}{4}\right)$ into four equal parts and evaluating the function at the three key x-values within the interval gives these points.

$$\left(\frac{\pi}{2}, -3\right), \quad \left(\frac{3\pi}{4}, -2\right), \quad (\pi, -1) \quad \text{Key points}$$

Join these points with a smooth curve. This period of the graph, along with the one in the domain interval $\left(-\frac{3\pi}{4}, \frac{\pi}{4}\right)$, is shown in Figure 31.

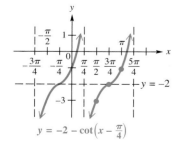

$y = -2 - \cot\left(x - \frac{\pi}{4}\right)$

**FIGURE 31**

**NOW TRY** Exercise 31.

## 17.3 Exercises

**Matching** In Exercises 1–6, match each function with its graph from choices A–F.

1. $y = -\tan x$
2. $y = -\cot x$
3. $y = \tan\left(x - \dfrac{\pi}{4}\right)$
4. $y = \cot\left(x - \dfrac{\pi}{4}\right)$
5. $y = \cot\left(x + \dfrac{\pi}{4}\right)$
6. $y = \tan\left(x + \dfrac{\pi}{4}\right)$

A. 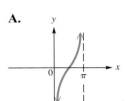   B. 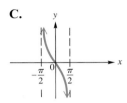   C.

D. 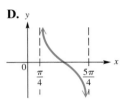   E. 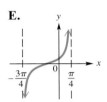   F.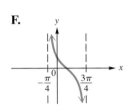

*Graph each function over a one-period interval. See Examples 1–3.*

7. $y = \tan 4x$
8. $y = \tan \dfrac{1}{2}x$
9. $y = 2 \tan x$
10. $y = 2 \cot x$
11. $y = 2 \tan \dfrac{1}{4}x$
12. $y = \dfrac{1}{2} \cot x$
13. $y = \cot 3x$
14. $y = -\cot \dfrac{1}{2}x$
15. $y = -2 \tan \dfrac{1}{4}x$
16. $y = 3 \tan \dfrac{1}{2}x$
17. $y = \dfrac{1}{2} \cot 4x$
18. $y = -\dfrac{1}{2} \cot 2x$

*Graph each function over a two-period interval. See Examples 4 and 5.*

19. $y = \tan(2x - \pi)$
20. $y = \tan\left(\dfrac{x}{2} + \pi\right)$
21. $y = \cot\left(3x + \dfrac{\pi}{4}\right)$
22. $y = \cot\left(2x - \dfrac{3\pi}{2}\right)$
23. $y = 1 + \tan x$
24. $y = 1 - \tan x$
25. $y = 1 - \cot x$
26. $y = -2 - \cot x$
27. $y = -1 + 2 \tan x$
28. $y = 3 + \dfrac{1}{2} \tan x$
29. $y = -1 + \dfrac{1}{2} \cot(2x - 3\pi)$
30. $y = -2 + 3 \tan(4x + \pi)$
31. $y = 1 - 2 \cot 2\left(x + \dfrac{\pi}{2}\right)$
32. $y = \dfrac{2}{3} \tan\left(\dfrac{3}{4}x - \pi\right) - 2$

*Connecting Graphs with Equations* Each function graphed is of the form $y = a \tan bx$ or $y = a \cot bx$, where $b > 0$. Determine the equation of the graph. (Half- and quarter-points are identified by dots.)

**33.**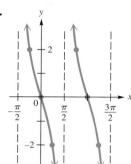

**34.**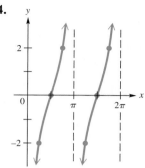

**35.**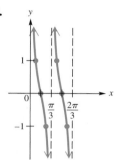

**36.**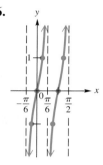

*True or False* In Exercises 37–40, tell whether each statement is true or false. If false, tell why.

**37.** The least positive number $k$ for which $x = k$ is an asymptote for the tangent function is $\frac{\pi}{2}$.

**38.** The least positive number $k$ for which $x = k$ is an asymptote for the cotangent function is $\frac{\pi}{2}$.

**39.** The graph of $y = \tan x$ in Figure 20 suggests that $\tan(-x) = \tan x$ for all $x$ in the domain of $\tan x$.

**40.** The graph of $y = \cot x$ in Figure 23 suggests that $\cot(-x) = -\cot x$ for all $x$ in the domain of $\cot x$.

**41.** *Concept Check* If $c$ is any number, then how many solutions does the equation $c = \tan x$ have in the interval $(-2\pi, 2\pi]$?

**42.** Consider the function defined by $f(x) = -4 \tan(2x + \pi)$. What is the domain of $f$? What is its range?

**43.** Show that $\tan(-x) = -\tan x$ by writing $\tan(-x)$ as $\frac{\sin(-x)}{\cos(-x)}$ and then using the relationships for $\sin(-x)$ and $\cos(-x)$.

**44.** Show that $\cot(-x) = -\cot x$ by writing $\cot(-x)$ as $\frac{\cos(-x)}{\sin(-x)}$ and then using the relationships for $\cos(-x)$ and $\sin(-x)$.

**45.** A rotating beacon is located at point $A$ next to a long wall. (See the figure.) The beacon is 4 m from the wall. The distance $d$ is given by

$$d = 4 \tan 2\pi t,$$

where $t$ is time measured in seconds since the beacon started rotating. (When $t = 0$, the beacon is aimed at point $R$. When the beacon is aimed to the right of $R$, the value of $d$ is positive; $d$ is negative if the beacon is aimed to the left of $R$.) Find $d$ for each time.

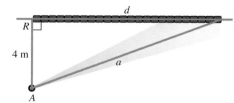

(a) $t = 0$  (b) $t = 0.4$  (c) $t = 0.8$  (d) $t = 1.2$
(e) Why is 0.25 a meaningless value for $t$?

**46.** Simultaneously graph $y = \tan x$ and $y = x$ in the window $[-1, 1]$ by $[-1, 1]$ with a graphing calculator. Write a short description of the relationship between $\tan x$ and $x$ for small $x$-values.

## 17.4 Graphs of the Secant and Cosecant Functions

**OBJECTIVES**

1. Graph $f(x) = \sec x$.
2. Graph $f(x) = \csc x$.
3. Apply graphing techniques to cosecant and secant functions.
4. Determine an equation for a given graph.

**OBJECTIVE 1** Graph $f(x) = \sec x$. Consider the table of selected points accompanying the graph of the secant function in Figure 32. These points include special values between $-\pi$ and $\pi$. The secant function is undefined for odd multiples of $\frac{\pi}{2}$ and, thus, like the tangent function has vertical asymptotes for such values. Furthermore, since $\sec(-x) = \sec x$ (see Exercise 31), the graph of the secant function is symmetric with respect to the $y$-axis.

| $x$ | $y = \sec x$ | $x$ | $y = \sec x$ |
|---|---|---|---|
| $\pm\frac{\pi}{3}$ | 2 | $\pm\frac{2\pi}{3}$ | $-2$ |
| $\pm\frac{\pi}{4}$ | $\sqrt{2} \approx 1.4$ | $\pm\frac{3\pi}{4}$ | $-\sqrt{2} \approx -1.4$ |
| $\pm\frac{\pi}{6}$ | $\frac{2\sqrt{3}}{3} \approx 1.2$ | $\pm\frac{5\pi}{6}$ | $-\frac{2\sqrt{3}}{3} \approx -1.2$ |
| 0 | 1 | $\pm\pi$ | $-1$ |

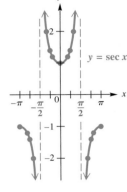

FIGURE 32

Because secant values are reciprocals of corresponding cosine values, the period of the secant function is $2\pi$, the same as for $y = \cos x$. When $\cos x = 1$, the value of $\sec x$ is also 1; likewise, when $\cos x = -1$, $\sec x = -1$ as well. For all $x$, $-1 \leq \cos x \leq 1$, and thus, $|\sec x| \geq 1$ for all $x$ in its domain. Figure 33 shows how the graphs of $y = \cos x$ and $y = \sec x$ are related.

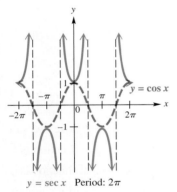

$y = \sec x$  Period: $2\pi$

**FIGURE 33**

**Secant Function**   $f(x) = \sec x$

Domain: $\left\{x \,\middle|\, x \neq (2n + 1)\frac{\pi}{2},\right.$   Range: $(-\infty, -1] \cup [1, \infty)$
$\left.\text{where } n \text{ is any integer}\right\}$

| $x$ | $y$ |
|---|---|
| $-\frac{\pi}{2}$ | undefined |
| $-\frac{\pi}{4}$ | $\sqrt{2}$ |
| 0 | 1 |
| $\frac{\pi}{4}$ | $\sqrt{2}$ |
| $\frac{\pi}{2}$ | undefined |
| $\frac{3\pi}{4}$ | $-\sqrt{2}$ |
| $\pi$ | $-1$ |
| $\frac{3\pi}{2}$ | undefined |

$f(x) = \sec x$

Dot mode

**FIGURE 34**

- The graph is discontinuous at values of $x$ of the form $x = (2n + 1)\frac{\pi}{2}$ and has vertical asymptotes at these values.
- There are no $x$-intercepts.
- Its period is $2\pi$.
- Its graph has no amplitude, since there are no maximum or minimum values.
- The graph is symmetric with respect to the $y$-axis, so the function is an even function. For all $x$ in the domain, $\sec(-x) = \sec x$.

**OBJECTIVE 2** Graph $f(x) = \csc x$.  A similar analysis for selected points between $-\pi$ and $\pi$ for the graph of the cosecant function yields the graph in Figure 35. The vertical asymptotes are at $x$-values that are integer multiples of $\pi$. Because $\csc(-x) = -\csc x$ (see Exercise 32), this graph is symmetric with respect to the origin.

| $x$ | $y = \csc x$ | $x$ | $y = \csc x$ |
|---|---|---|---|
| $\frac{\pi}{6}$ | 2 | $-\frac{\pi}{6}$ | $-2$ |
| $\frac{\pi}{4}$ | $\sqrt{2} \approx 1.4$ | $-\frac{\pi}{4}$ | $-\sqrt{2} \approx -1.4$ |
| $\frac{\pi}{3}$ | $\frac{2\sqrt{3}}{3} \approx 1.2$ | $-\frac{\pi}{3}$ | $-\frac{2\sqrt{3}}{3} \approx -1.2$ |
| $\frac{\pi}{2}$ | 1 | $-\frac{\pi}{2}$ | $-1$ |
| $\frac{2\pi}{3}$ | $\frac{2\sqrt{3}}{3} \approx 1.2$ | $-\frac{2\pi}{3}$ | $-\frac{2\sqrt{3}}{3} \approx -1.2$ |
| $\frac{3\pi}{4}$ | $\sqrt{2} \approx 1.4$ | $-\frac{3\pi}{4}$ | $-\sqrt{2} \approx -1.4$ |
| $\frac{5\pi}{6}$ | 2 | $-\frac{5\pi}{6}$ | $-2$ |

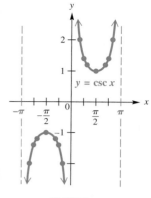

**FIGURE 35**

Because cosecant values are reciprocals of corresponding sine values, the period of the cosecant function is $2\pi$, the same as for $y = \sin x$. When $\sin x = 1$, the value of $\csc x$ is also 1; likewise, when $\sin x = -1$, $\csc x = -1$. For all $x$, $-1 \leq \sin x \leq 1$, and thus $|\csc x| \geq 1$ for all $x$ in its domain. Figure 36 shows how the graphs of $y = \sin x$ and $y = \csc x$ are related.

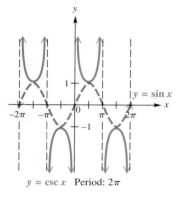

$y = \csc x$  Period: $2\pi$

**FIGURE 36**

### Cosecant Function  $f(x) = \csc x$

Domain: $\{x \mid x \neq n\pi, \text{ where } n \text{ is any integer}\}$    Range: $(-\infty, -1] \cup [1, \infty)$

| $x$ | $y$ |
|---|---|
| 0 | undefined |
| $\frac{\pi}{6}$ | 2 |
| $\frac{\pi}{3}$ | $\frac{2\sqrt{3}}{3}$ |
| $\frac{\pi}{2}$ | 1 |
| $\frac{2\pi}{3}$ | $\frac{2\sqrt{3}}{3}$ |
| $\pi$ | undefined |
| $\frac{3\pi}{2}$ | $-1$ |
| $2\pi$ | undefined |

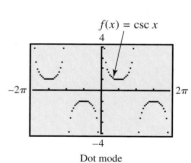

$f(x) = \csc x$

Dot mode

**FIGURE 37**

- The graph is discontinuous at values of $x$ of the form $x = n\pi$ and has vertical asymptotes at these values.
- There are no $x$-intercepts.
- Its period is $2\pi$.
- Its graph has no amplitude, since there are no maximum or minimum values.
- The graph is symmetric with respect to the origin, so the function is an odd function. For all $x$ in the domain, $\csc(-x) = -\csc x$.

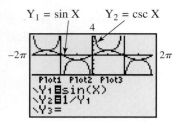

Trig window; connected mode

**FIGURE 38**

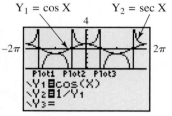

Trig window; connected mode

**FIGURE 39**

🖩 Typically, calculators do not have keys for the cosecant and secant functions. To graph $y = \csc x$ with a graphing calculator, use the fact that

$$\csc x = \frac{1}{\sin x}.$$

The graphs of $Y_1 = \sin X$ and $Y_2 = \csc X$ are shown in Figure 38. The calculator is in split screen and connected modes. Similarly, the secant function is graphed by using the identity

$$\sec x = \frac{1}{\cos x},$$

as shown in Figure 39.

Using dot mode for graphing will eliminate the vertical lines that appear in Figures 38 and 39. While they suggest asymptotes and are sometimes called **pseudo-asymptotes,** they are not actually parts of the graphs.

**OBJECTIVE 3** Apply graphing techniques to cosecant and secant functions. In the previous section, we gave guidelines for sketching graphs of tangent and cotangent functions. We now present similar guidelines for graphing cosecant and secant functions.

### Guidelines for Sketching Graphs of Cosecant and Secant Functions

To graph $y = a \csc bx$ or $y = a \sec bx$, with $b > 0$, follow these steps.

**Step 1** Graph the corresponding reciprocal function as a guide, using a dashed curve.

| To Graph | Use as a Guide |
|---|---|
| $y = a \csc bx$ | $y = a \sin bx$ |
| $y = a \sec bx$ | $y = a \cos bx$ |

**Step 2** Sketch the vertical asymptotes. They will have equations of the form $x = k$, where $k$ is an $x$-intercept of the graph of the guide function.

**Step 3** Sketch the graph of the desired function by drawing the typical U-shaped branches between the adjacent asymptotes. The branches will be above the graph of the guide function when the guide function values are positive and below the graph of the guide function when the guide function values are negative. The graph will resemble those in Figures 34 and 37 in the function boxes on pages 796 and 797.

Like graphs of the sine and cosine functions, graphs of the secant and cosecant functions may be translated vertically and horizontally. The period of both basic functions is $2\pi$.

**EXAMPLE 1** Graphing $y = a \sec bx$

Graph $y = 2 \sec \frac{1}{2}x$.

**Solution**

**Step 1** This function involves the secant, so the corresponding reciprocal function will involve the cosine. The guide function to graph is

$$y = 2 \cos \frac{1}{2}x.$$

*(continued)*

Using the guidelines of **Section 17.1,** we find that this guide function has amplitude 2 and one period of the graph lies along the interval that satisfies the inequality

$$0 \leq \frac{1}{2}x \leq 2\pi, \quad \text{or} \quad [0, 4\pi].$$

Dividing this interval into four equal parts gives the key points

$(0, 2), \quad (\pi, 0), \quad (2\pi, -2), \quad (3\pi, 0), \quad (4\pi, 2), \quad$ Key points

which are joined with a dashed red curve to indicate that this graph is only a guide. An additional period is graphed as seen in Figure 40(a).

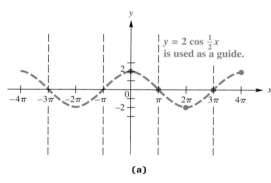
(a)

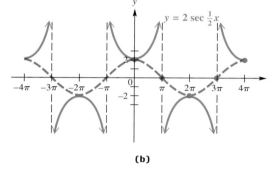
(b)

**FIGURE 40**

**Step 2** Sketch the vertical asymptotes. These occur at *x*-values for which the guide function equals 0, such as

$$x = -3\pi, \quad x = -\pi, \quad x = \pi, \quad x = 3\pi.$$

See Figure 40(a).

**Step 3** Sketch the graph of $y = 2 \sec \frac{1}{2}x$ by drawing the typical U-shaped branches, approaching the asymptotes. See the solid blue graph in Figure 40(b).

**NOW TRY** Exercise 5.

---

**EXAMPLE 2** Graphing $y = a \csc(x - d)$

Graph $y = \frac{3}{2} \csc\left(x - \frac{\pi}{2}\right)$.

**Solution**

**Step 1** Use the guidelines of **Section 17.2** to graph the corresponding reciprocal function defined by

$$y = \frac{3}{2} \sin\left(x - \frac{\pi}{2}\right),$$

shown as a red dashed curve in Figure 41.

**Step 2** Sketch the vertical asymptotes through the *x*-intercepts of the graph of $y = \frac{3}{2} \sin\left(x - \frac{\pi}{2}\right)$. These have the form $x = (2n + 1)\frac{\pi}{2}$, where *n* is any integer. See the black dashed lines in Figure 41.

*(continued)*

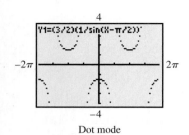

Dot mode
This is a calculator graph of the function in Example 2.

**Step 3** Sketch the graph of $y = \frac{3}{2} \csc\left(x - \frac{\pi}{2}\right)$ by drawing the typical U-shaped branches between adjacent asymptotes. See the solid blue graph in Figure 41.

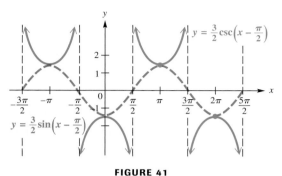

**FIGURE 41**

**NOW TRY** Exercise 7.

**OBJECTIVE 4** Determine an equation for a given graph. Now that the graphs of the six circular functions have been introduced, we can apply the concepts to give an equation of a given graph.

**EXAMPLE 3** Determining an Equation for a Graph

Determine an equation for each graph.

(a)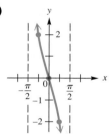

(b)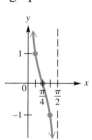

(c)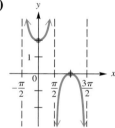

**Solution**

(a) This graph is that of $y = \tan x$ but reflected across the $x$-axis and stretched vertically by a factor of 2. Therefore, an equation for this graph is

$$y = -2 \tan x.$$

↑ $x$-axis reflection ↖ Vertical stretch

(b) This graph is that of $y = \cot x$, but the period is $\frac{\pi}{2}$ rather than $\pi$. Therefore, if $y = \cot bx$, where $b > 0$, we must have $b = 2$. So an equation for this graph is

$$y = \cot 2x.$$

(c) This is the graph of $y = \sec x$, translated one unit upward. An equation is

$$y = 1 + \sec x.$$

**NOW TRY** Exercises 19, 21, and 23.

Because the circular functions are periodic, there are actually infinitely many equations that correspond to each graph in Example 3. We have given only one such equation in each case. For instance, confirm that $y = -\tan\left(2x - \frac{\pi}{2}\right)$ is another equation for the graph in Example 3(b).

## 17.4 Exercises

**Matching** *In Exercises 1–4, match each function with its graph from choices A–D.*

1. $y = -\csc x$
2. $y = -\sec x$
3. $y = \sec\left(x - \dfrac{\pi}{2}\right)$
4. $y = \csc\left(x + \dfrac{\pi}{2}\right)$

A.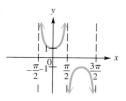

B.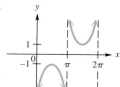

C.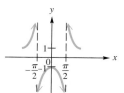

D.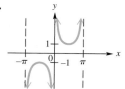

*Graph each function over a one-period interval. See Examples 1 and 2.*

5. $y = 3 \sec \dfrac{1}{4}x$
6. $y = -2 \sec \dfrac{1}{2}x$
7. $y = -\dfrac{1}{2}\csc\left(x + \dfrac{\pi}{2}\right)$
8. $y = \dfrac{1}{2}\csc\left(x - \dfrac{\pi}{2}\right)$
9. $y = \csc\left(x - \dfrac{\pi}{4}\right)$
10. $y = \sec\left(x + \dfrac{3\pi}{4}\right)$
11. $y = \sec\left(x + \dfrac{\pi}{4}\right)$
12. $y = \csc\left(x + \dfrac{\pi}{3}\right)$
13. $y = \sec\left(\dfrac{1}{2}x + \dfrac{\pi}{3}\right)$
14. $y = \csc\left(\dfrac{1}{2}x - \dfrac{\pi}{4}\right)$
15. $y = 2 + 3 \sec(2x - \pi)$
16. $y = 1 - 2 \csc\left(x + \dfrac{\pi}{2}\right)$
17. $y = 1 - \dfrac{1}{2}\csc\left(x - \dfrac{3\pi}{4}\right)$
18. $y = 2 + \dfrac{1}{4}\sec\left(\dfrac{1}{2}x - \pi\right)$

**Connecting Graphs with Equations** *Determine an equation for each graph. See Example 3.*

19.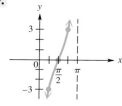

20.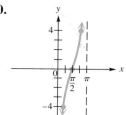

21.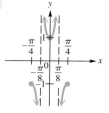

**22.** 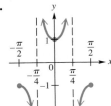 **23.** 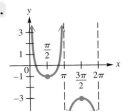 **24.**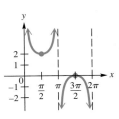

*True or False*   In Exercises 25–28, tell whether each statement is true or false. *If false, tell why.*

25. The tangent and secant functions are undefined for the same values.

26. The secant and cosecant functions are undefined for the same values.

27. The graph of $y = \sec x$ in Figure 34 suggests that $\sec(-x) = \sec x$ for all $x$ in the domain of sec $x$.

28. The graph of $y = \csc x$ in Figure 37 suggests that $\csc(-x) = -\csc x$ for all $x$ in the domain of csc $x$.

29. *Concept Check*   If $c$ is any number such that $-1 < c < 1$, then how many solutions does the equation $c = \sec x$ have over the entire domain of the secant function?

30. Consider the function defined by $g(x) = -2\csc(4x + \pi)$. What is the domain of $g$? What is its range?

31. Show that $\sec(-x) = \sec x$ by writing $\sec(-x)$ as $\frac{1}{\cos(-x)}$ and then using the relationship between $\cos(-x)$ and $\cos x$.

32. Show that $\csc(-x) = -\csc x$ by writing $\csc(-x)$ as $\frac{1}{\sin(-x)}$ and then using the relationship between $\sin(-x)$ and $\sin x$.

33. In the figure for Exercise 45 in **Section 17.3,** the distance $a$ is given by

$$a = 4|\sec 2\pi t|.$$

Find $a$ for each time.

   **(a)** $t = 0$      **(b)** $t = 0.86$      **(c)** $t = 1.24$

34. Between each pair of successive asymptotes, a portion of the graph of $y = \sec x$ or $y = \csc x$ resembles a parabola. Can each of these portions actually be a parabola? Explain.

# 17 SUMMARY

## KEY TERMS

**17.1** periodic function
period
sine wave (sinusoid)

odd function
even function
amplitude

**17.2** phase shift
argument

**17.3** vertical asymptote

## QUICK REVIEW

### CONCEPTS — EXAMPLES

### 17.1 Graphs of the Sine and Cosine Functions

### 17.2 Translations of the Graphs of the Sine and Cosine Functions

**Sine and Cosine Functions**

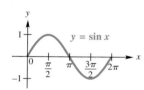

**Domain:** $(-\infty, \infty)$
**Range:** $[-1, 1]$
**Amplitude:** 1
**Period:** $2\pi$

**Domain:** $(-\infty, \infty)$
**Range:** $[-1, 1]$
**Amplitude:** 1
**Period:** $2\pi$

The graph of

$y = c + a \sin b(x - d)$ or $y = c + a \cos b(x - d)$, $b > 0$, has

1. amplitude $|a|$,  2. period $\frac{2\pi}{b}$,
3. vertical translation $c$ units up if $c > 0$ or $|c|$ units down if $c < 0$, and
4. phase shift $d$ units to the right if $d > 0$ or $|d|$ units to the left if $d < 0$.

See pages 776 and 784 for a summary of graphing techniques.

Graph $y = \sin 3x$.

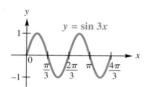

period: $\frac{2\pi}{3}$    amplitude: 1
domain: $(-\infty, \infty)$    range: $[-1, 1]$

Graph $y = -2 \cos x$.

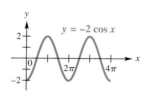

period: $2\pi$    amplitude: 2
domain: $(-\infty, \infty)$    range: $[-2, 2]$

*(continued)*

| CONCEPTS | EXAMPLES |

## 17.3 Graphs of the Tangent and Cotangent Functions

**Tangent and Cotangent Functions**

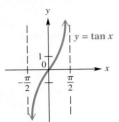 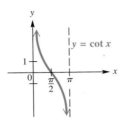

**Domain:** $\{x \mid x \neq (2n+1)\frac{\pi}{2},$ where $n$ is any integer$\}$
**Range:** $(-\infty, \infty)$
**Period:** $\pi$

**Domain:** $\{x \mid x \neq n\pi,$ where $n$ is any integer$\}$
**Range:** $(-\infty, \infty)$
**Period:** $\pi$

See page 791 for a summary of graphing techniques.

Graph one period of $y = 2 \tan x$.

[Graph of $y = 2 \tan x$]

period: $\pi$
domain: $\{x \mid x \neq (2n+1)\frac{\pi}{2},$ where $n$ is any integer$\}$
range: $(-\infty, \infty)$

## 17.4 Graphs of the Secant and Cosecant Functions

**Secant and Cosecant Functions**

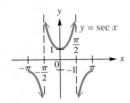 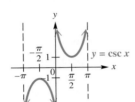

**Domain:** $\{x \mid x \neq (2n+1)\frac{\pi}{2},$ where $n$ is any integer$\}$
**Range:** $(-\infty, -1] \cup [1, \infty)$
**Period:** $2\pi$

**Domain:** $\{x \mid x \neq n\pi,$ where $n$ is any integer$\}$
**Range:** $(-\infty, -1] \cup [1, \infty)$
**Period:** $2\pi$

See page 798 for a summary of graphing techniques.

Graph one period of $y = \sec\left(x + \frac{\pi}{4}\right)$.

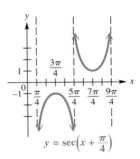

period: $2\pi$
phase shift: $-\frac{\pi}{4}$
domain: $\{x \mid x \neq \frac{\pi}{4} + n\pi,$ where $n$ is any integer$\}$
range: $(-\infty, -1] \cup [1, \infty)$

# 17 REVIEW EXERCISES

1. *Multiple Choice* Which one of the following is true about the graph of $y = 4 \sin 2x$?
   
   **A.** It has amplitude 2 and period $\frac{\pi}{2}$.  **B.** It has amplitude 4 and period $\pi$.
   **C.** Its range is $[0, 4]$.  **D.** Its range is $[-4, 0]$.

2. *Multiple Choice* Which one of the following is false about the graph of $y = -3 \cos \frac{1}{2}x$?
   
   **A.** Its range is $[-3, 3]$.  **B.** Its domain is $(-\infty, \infty)$.
   **C.** Its amplitude is 3, and its period is $4\pi$.  **D.** Its amplitude is 3, and its period is $\pi$.

3. *Concept Check* Which of the basic circular functions can have $y$-value $\frac{1}{2}$?

4. *Concept Check* Which of the basic circular functions can have $y$-value 2?

*For each function, give the amplitude, period, vertical translation, and phase shift, as applicable.*

5. $y = 2 \sin x$
6. $y = \tan 3x$
7. $y = -\frac{1}{2} \cos 3x$

8. $y = 2 \sin 5x$
9. $y = 1 + 2 \sin \frac{1}{4}x$
10. $y = 3 - \frac{1}{4} \cos \frac{2}{3}x$

11. $y = 3 \cos\left(x + \frac{\pi}{2}\right)$
12. $y = -\sin\left(x - \frac{3\pi}{4}\right)$
13. $y = \frac{1}{2} \csc\left(2x - \frac{\pi}{4}\right)$

14. $y = 2 \sec(\pi x - 2\pi)$
15. $y = \frac{1}{3} \tan\left(3x - \frac{\pi}{3}\right)$
16. $y = \cot\left(\frac{x}{2} + \frac{3\pi}{4}\right)$

*Concept Check* *Identify the circular function that satisfies each description.*

17. period is $\pi$, $x$-intercepts are of the form $n\pi$, where $n$ is any integer
18. period is $2\pi$, graph passes through the origin
19. period is $2\pi$, graph passes through the point $\left(\frac{\pi}{2}, 0\right)$
20. period is $2\pi$, domain is $\{x \mid x \neq n\pi,$ where $n$ is any integer$\}$
21. period is $\pi$, function is decreasing on the interval $(0, \pi)$
22. period is $2\pi$, has vertical asymptotes of the form $x = (2n + 1)\frac{\pi}{2}$, where $n$ is any integer
23. Suppose that $f$ is a sine function with period 10 and $f(5) = 2$. Explain why $f(25) = 2$.
24. Suppose that $f$ is a sine function with period $\pi$ and $f\left(\frac{6\pi}{5}\right) = 1$. Explain why $f\left(-\frac{4\pi}{5}\right) = 1$.

*Graph each function over a one-period interval.*

25. $y = 3 \sin x$
26. $y = \frac{1}{2} \sec x$
27. $y = -\tan x$

28. $y = -2 \cos x$
29. $y = 2 + \cot x$
30. $y = -1 + \csc x$

31. $y = \sin 2x$
32. $y = \tan 3x$
33. $y = 3 \cos 2x$

34. $y = \frac{1}{2} \cot 3x$
35. $y = \cos\left(x - \frac{\pi}{4}\right)$
36. $y = \tan\left(x - \frac{\pi}{2}\right)$

37. $y = \sec\left(2x + \frac{\pi}{3}\right)$
38. $y = \sin\left(3x + \frac{\pi}{2}\right)$
39. $y = 1 + 2 \cos 3x$

40. $y = -1 - 3 \sin 2x$
41. $y = 2 \sin \pi x$
42. $y = -\frac{1}{2} \cos(\pi x - \pi)$

43. Explain why a function of the form $f(x) = 2 \sin(bx + c)$ has range $[-2, 2]$.
44. Explain why a function of the form $f(x) = 2 \csc(bx + c)$ has range $(-\infty, -2] \cup [2, \infty)$.

*Solve each problem.*

45. Let a person whose eyes are $h_1$ feet from the ground stand $d$ feet from an object $h_2$ feet tall, where $h_2 > h_1$. Let $\theta$ be the angle of elevation to the top of the object. See the figure.

    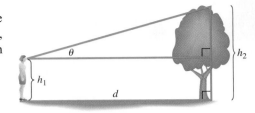

    (a) Show that $d = (h_2 - h_1) \cot \theta$.
    (b) Let $h_2 = 55$ and $h_1 = 5$. Graph $d$ for the interval $0 < \theta \leq \frac{\pi}{2}$.

46. The figure shows a function $f$ that models the tides in feet at Clearwater Beach, Florida, $x$ hours after midnight starting on August 26, 2006. (*Source:* Pentcheff, D., *WWW Tide and Current Predictor.*)

    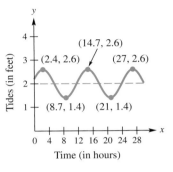

    (a) Find the time between high tides.
    (b) What is the difference in water levels between high tide and low tide?
    (c) The tides can be modeled by
    $$f(x) = 0.6 \cos[0.511(x - 2.4)] + 2.$$
    Estimate the tides when $x = 10$.

47. The maximum afternoon temperature in a given city might be modeled by
    $$t = 60 - 30 \cos \frac{x\pi}{6},$$
    where $t$ represents the maximum afternoon temperature in month $x$, with $x = 0$ representing January, $x = 1$ representing February, and so on. Find the maximum afternoon temperature to the nearest degree for each month.

    (a) January   (b) April   (c) May   (d) June   (e) August   (f) October

48. The figure shows the populations of lynx and hares in Canada for the years 1847–1903. The hares are food for the lynx. An increase in hare population causes an increase in lynx population some time later. The increasing lynx population then causes a decline in hare population. The two graphs have the same period.

    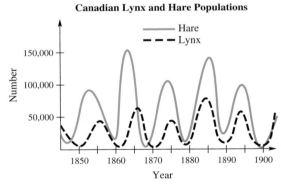

    (a) Estimate the length of one period.
    (b) Estimate maximum and minimum hare populations.

# 17 TEST

1. Identify each of the following basic circular function graphs.

   (a)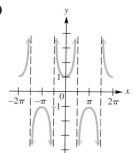
   (b)
   (c)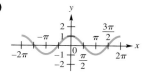
   (d)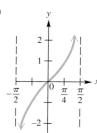
   (e)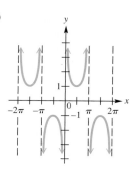
   (f)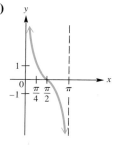

2. One of the following is the graph of $y = \sin 2x$ and the other is the graph of $y = 2 \sin x$. Identify each graph.

   (a)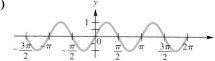

   (b)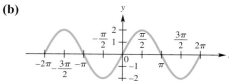

3. Give a short answer to each of the following.
   (a) What is the domain of the cosine function?
   (b) What is the range of the sine function?
   (c) What is the least positive value for which the tangent function is undefined?
   (d) What is the range of the secant function?

4. Consider the function defined by $y = 3 - 6 \sin\left(2x + \frac{\pi}{2}\right)$.
   (a) What is its period?
   (b) What is the amplitude of its graph?
   (c) What is its range?
   (d) What is the y-intercept of its graph?
   (e) What is its phase shift?

*Graph each function over a two-period interval. Identify asymptotes when applicable.*

**5.** $y = \sin(2x + \pi)$  **6.** $y = -\cos 2x$

**7.** $y = 2 + \cos x$  **8.** $y = -1 + 2\sin(x + \pi)$

**9.** $y = \tan\left(x - \dfrac{\pi}{2}\right)$  **10.** $y = -2 - \cot\left(x - \dfrac{\pi}{2}\right)$

**11.** $y = -\csc 2x$  **12.** $y = 3\csc \pi x$

**13.** The average monthly temperature (in °F) in Austin, Texas, can be modeled using the circular function defined by

$$f(x) = 17.5 \sin\left[\dfrac{\pi}{6}(x - 4)\right] + 67.5,$$

where $x$ is the month and $x = 1$ corresponds to January. (*Source:* Miller, A., J. Thompson, and R. Peterson, *Elements of Meteorology, Fourth Edition,* Charles E. Merrill Publishing Co., 1983.)

(a) Graph $f$ in the window $[1, 25]$ by $[45, 90]$.
(b) Determine the amplitude, period, phase shift, and vertical translation of $f$.
(c) What is the average monthly temperature for the month of December?
(d) Determine the maximum and minimum average monthly temperatures and the months when they occur.
(e) What would be an approximation for the average *yearly* temperature in Austin? How is this related to the vertical translation of the sine function in the formula for $f$?

**14.** Explain why the domains of the tangent and secant functions are the same, and then give a similar explanation for the cotangent and cosecant functions.

# Trigonometric Identities

In 1831 Michael Faraday discovered that when a wire passes by a magnet, a small electric current is produced in the wire. Now we generate massive amounts of electricity by simultaneously rotating thousands of wires near large electromagnets. Because electric current alternates its direction on electrical wires, it is modeled accurately by either the sine or the cosine function.

We give many examples of applications of the trigonometric functions to electricity and other phenomena in the examples and exercises in this chapter, including a model of the wattage consumption of a toaster in Section 18.5, Example 6.

**18.1** Fundamental Identities

**18.2** Verifying Trigonometric Identities

**18.3** Sum and Difference Identities for Cosine

**18.4** Sum and Difference Identities for Sine and Tangent

**18.5** Double-Angle Identities

**18.6** Half-Angle Identities

# 18.1 Fundamental Identities

**OBJECTIVES**
1. Learn the fundamental identities.
2. Use the fundamental identities.

Recall that an **identity** is an equation that is satisfied by *every* value in the domain of its variable.

**OBJECTIVE 1** Learn the fundamental identities. As suggested by the circle shown in Figure 1, an angle $\theta$ having the point $(x, y)$ on its terminal side has a corresponding angle $-\theta$ with the point $(x, -y)$ on its terminal side. From the definition of sine,

$$\sin(-\theta) = \frac{-y}{r} \quad \text{and} \quad \sin\theta = \frac{y}{r},$$

so $\sin(-\theta)$ and $\sin\theta$ are negatives of each other, or

$$\sin(-\theta) = -\sin\theta.$$

Figure 1 shows an angle $\theta$ in quadrant II, but the same result holds for $\theta$ in any quadrant. Also, by definition,

$$\cos(-\theta) = \frac{x}{r} \quad \text{and} \quad \cos\theta = \frac{x}{r},$$

so
$$\cos(-\theta) = \cos\theta.$$

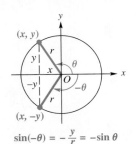

**FIGURE 1**

We use the identities for $\sin(-\theta)$ and $\cos(-\theta)$ to find $\tan(-\theta)$ in terms of $\tan\theta$:

$$\tan(-\theta) = \frac{\sin(-\theta)}{\cos(-\theta)} = \frac{-\sin\theta}{\cos\theta} = -\frac{\sin\theta}{\cos\theta}, \quad \text{or} \quad \tan(-\theta) = -\tan\theta.$$

Similar reasoning gives the remaining three **negative-angle** or **negative-number identities**, which, together with the reciprocal, quotient, and Pythagorean identities from **Chapter 14**, are called the **fundamental identities**.

---

**Fundamental Identities**

**Reciprocal Identities**

$$\cot\theta = \frac{1}{\tan\theta} \qquad \sec\theta = \frac{1}{\cos\theta} \qquad \csc\theta = \frac{1}{\sin\theta}$$

**Quotient Identities**

$$\tan\theta = \frac{\sin\theta}{\cos\theta} \qquad \cot\theta = \frac{\cos\theta}{\sin\theta}$$

**Pythagorean Identities**

$$\sin^2\theta + \cos^2\theta = 1 \qquad \tan^2\theta + 1 = \sec^2\theta \qquad 1 + \cot^2\theta = \csc^2\theta$$

**Negative-Angle Identities**

$$\sin(-\theta) = -\sin\theta \qquad \cos(-\theta) = \cos\theta \qquad \tan(-\theta) = -\tan\theta$$
$$\csc(-\theta) = -\csc\theta \qquad \sec(-\theta) = \sec\theta \qquad \cot(-\theta) = -\cot\theta$$

---

The most commonly recognized forms of the fundamental identities are given in the preceding box. Throughout this chapter you must also recognize alternative forms of these identities. *For example, two other forms of $\sin^2\theta + \cos^2\theta = 1$ are*

$$\sin^2\theta = 1 - \cos^2\theta \quad \text{and} \quad \cos^2\theta = 1 - \sin^2\theta.$$

### SECTION 18.1 Fundamental Identities

**OBJECTIVE 2** Use the fundamental identities. One way we use these identities is to find the values of other trigonometric functions from the value of a given trigonometric function. Although we could find such values using a right triangle, this is a good way to practice using the fundamental identities.

**EXAMPLE 1** Finding Trigonometric Function Values Given One Value and the Quadrant

If $\tan \theta = -\frac{5}{3}$ and $\theta$ is in quadrant II, find each function value.

**(a)** $\sec \theta$  **(b)** $\sin \theta$  **(c)** $\cot(-\theta)$

**Solution**

**(a)** Look for an identity that relates tangent and secant.

$$\tan^2 \theta + 1 = \sec^2 \theta \qquad \text{Pythagorean identity}$$

$$\left(-\frac{5}{3}\right)^2 + 1 = \sec^2 \theta \qquad \tan \theta = -\frac{5}{3}$$

$$\frac{25}{9} + 1 = \sec^2 \theta \qquad \left(-\frac{5}{3}\right)^2 = -\frac{5}{3}\left(-\frac{5}{3}\right) = \frac{25}{9}$$

$$\frac{34}{9} = \sec^2 \theta \qquad \text{Combine terms.}$$

*Choose the correct sign.*

$$-\sqrt{\frac{34}{9}} = \sec \theta \qquad \text{Take the negative square root.}$$

$$-\frac{\sqrt{34}}{3} = \sec \theta \qquad \text{Simplify the radical;} -\sqrt{\frac{34}{9}} = -\frac{\sqrt{34}}{\sqrt{9}} = -\frac{\sqrt{34}}{3}.$$

We chose the negative square root since $\sec \theta$ is negative in quadrant II.

**(b)**
$$\tan \theta = \frac{\sin \theta}{\cos \theta} \qquad \text{Quotient identity}$$

$$\cos \theta \tan \theta = \sin \theta \qquad \text{Multiply each side by } \cos \theta.$$

$$\left(\frac{1}{\sec \theta}\right) \tan \theta = \sin \theta \qquad \text{Reciprocal identity}$$

$$\left(-\frac{3\sqrt{34}}{34}\right)\left(-\frac{5}{3}\right) = \sin \theta \qquad \frac{1}{\sec \theta} = \frac{1}{-\frac{\sqrt{34}}{3}} = -\frac{3}{\sqrt{34}} = -\frac{3}{\sqrt{34}} \cdot \frac{\sqrt{34}}{\sqrt{34}} = -\frac{3\sqrt{34}}{34}; \tan \theta = -\frac{5}{3}$$

$$\sin \theta = \frac{5\sqrt{34}}{34} \qquad \text{Multiply; rewrite.}$$

**(c)**
$$\cot(-\theta) = \frac{1}{\tan(-\theta)} \qquad \text{Reciprocal identity}$$

$$\cot(-\theta) = \frac{1}{-\tan \theta} \qquad \text{Negative-angle identity}$$

$$\cot(-\theta) = \frac{1}{-\left(-\frac{5}{3}\right)} = \frac{3}{5} \qquad \tan \theta = -\frac{5}{3}; \text{ simplify the complex fraction.}$$

**NOW TRY** Exercises 7, 11, and 15.

**CAUTION** *To avoid a common error, when taking the square root, be sure to choose the sign based on the quadrant of θ and the function being evaluated.*

Any trigonometric function of a number or angle can be expressed in terms of any other function.

**EXAMPLE 2** Expressing One Function in Terms of Another

Express $\cos x$ in terms of $\tan x$.

**Solution** Since $\sec x$ is related to both $\cos x$ and $\tan x$ by identities, start with $1 + \tan^2 x = \sec^2 x$.

$$\frac{1}{1 + \tan^2 x} = \frac{1}{\sec^2 x} \qquad \text{Take reciprocals.}$$

$$\frac{1}{1 + \tan^2 x} = \cos^2 x \qquad \text{Reciprocal identity}$$

$$\pm\sqrt{\frac{1}{1 + \tan^2 x}} = \cos x \qquad \text{Take the square root of each side.}$$

Remember both the positive and negative roots.

$$\cos x = \frac{\pm 1}{\sqrt{1 + \tan^2 x}} \qquad \text{Quotient rule for radicals: } \sqrt[n]{\frac{a}{b}} = \frac{\sqrt[n]{a}}{\sqrt[n]{b}}; \text{ rewrite.}$$

$$\cos x = \frac{\pm\sqrt{1 + \tan^2 x}}{1 + \tan^2 x} \qquad \text{Rationalize the denominator.}$$

Choose the $+$ sign or the $-$ sign, depending on the quadrant of $x$.

**NOW TRY** Exercise 33.

We can use a graphing calculator to decide whether two functions are identical. See Figure 2, which supports the identity $\sin^2 x + \cos^2 x = 1$. With an identity, you should see no difference in the two graphs.

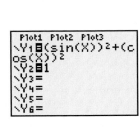
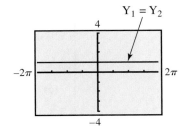

**FIGURE 2**

Each of the functions $\tan \theta$, $\cot \theta$, $\sec \theta$, and $\csc \theta$ can easily be expressed in terms of $\sin \theta$ and/or $\cos \theta$. We often make such substitutions in an expression to simplify it.

### EXAMPLE 3　Rewriting an Expression in Terms of Sine and Cosine

Write $\tan\theta + \cot\theta$ in terms of $\sin\theta$ and $\cos\theta$, and then simplify the expression.

**Solution**

$$\tan\theta + \cot\theta = \frac{\sin\theta}{\cos\theta} + \frac{\cos\theta}{\sin\theta} \qquad \text{Quotient identities}$$

$$= \frac{\sin\theta}{\cos\theta} \cdot \frac{\sin\theta}{\sin\theta} + \frac{\cos\theta}{\sin\theta} \cdot \frac{\cos\theta}{\cos\theta} \qquad \text{Write each fraction with the least common denominator (LCD).}$$

$$= \frac{\sin^2\theta}{\cos\theta \sin\theta} + \frac{\cos^2\theta}{\cos\theta \sin\theta} \qquad \text{Multiply.}$$

$$= \frac{\sin^2\theta + \cos^2\theta}{\cos\theta \sin\theta} \qquad \text{Add fractions; } \frac{a}{c} + \frac{b}{c} = \frac{a+b}{c}.$$

$$\tan\theta + \cot\theta = \frac{1}{\cos\theta \sin\theta} \qquad \text{Pythagorean identity}$$

**NOW TRY** Exercise 45.

The graph supports the result in Example 3. The graphs of $y_1$ and $y_2$ appear to be identical.

$y_1 = \tan x + \cot x$
$y_2 = \dfrac{1}{\cos x \sin x}$

**CAUTION** *When working with trigonometric expressions and identities, be sure to write the argument of the function.* For example, we would *not* write $\sin^2 + \cos^2 = 1$; an argument such as $\theta$ is necessary in this identity.

## 18.1 Exercises

**NOW TRY Exercise**

*Fill in the Blanks*　In Exercises 1–6, use identities to fill in the blanks.

1. If $\tan\theta = 2.6$, then $\tan(-\theta) = $ _____.
2. If $\cos\theta = -0.65$, then $\cos(-\theta) = $ _____.
3. If $\tan\theta = 1.6$, then $\cot\theta = $ _____.
4. If $\cos\theta = 0.8$ and $\sin\theta = 0.6$, then $\tan(-\theta) = $ _____.
5. If $\sin\theta = \frac{2}{3}$, then $-\sin(-\theta) = $ _____.
6. If $\cos\theta = -\frac{1}{5}$, then $-\cos(-\theta) = $ _____.

*Find* $\sin\theta$. *See Example 1.*

7. $\cos\theta = \dfrac{3}{4}$, $\theta$ in quadrant I
8. $\cot\theta = -\dfrac{1}{3}$, $\theta$ in quadrant IV
9. $\cos(-\theta) = \dfrac{\sqrt{5}}{5}$, $\tan\theta < 0$
10. $\tan\theta = -\dfrac{\sqrt{7}}{2}$, $\sec\theta > 0$
11. $\sec\theta = \dfrac{11}{4}$, $\tan\theta < 0$
12. $\csc\theta = -\dfrac{8}{5}$

13. Why is it unnecessary to give the quadrant of $\theta$ in Exercise 12?
14. *Concept Check*　What is **WRONG** with the statement of this problem?
    *Find* $\cos(-\theta)$ *if* $\cos\theta = 3$.

*Find the remaining five trigonometric functions of* $\theta$. *See Example 1.*

15. $\sin\theta = \dfrac{2}{3}$, $\theta$ in quadrant II
16. $\cos\theta = \dfrac{1}{5}$, $\theta$ in quadrant I
17. $\tan\theta = -\dfrac{1}{4}$, $\theta$ in quadrant IV
18. $\csc\theta = -\dfrac{5}{2}$, $\theta$ in quadrant III

**19.** $\cot\theta = \dfrac{4}{3}$, $\sin\theta > 0$   **20.** $\sin\theta = -\dfrac{4}{5}$, $\cos\theta < 0$

**21.** $\sec\theta = \dfrac{4}{3}$, $\sin\theta < 0$   **22.** $\cos\theta = -\dfrac{1}{4}$, $\sin\theta > 0$

*Matching* For each expression in Column I, choose the expression from Column II that completes an identity.

| I | II |
|---|---|
| **23.** $\dfrac{\cos x}{\sin x} = $ \_\_\_\_ | **A.** $\sin^2 x + \cos^2 x$ |
| **24.** $\tan x = $ \_\_\_\_ | **B.** $\cot x$ |
| **25.** $\cos(-x) = $ \_\_\_\_ | **C.** $\sec^2 x$ |
| **26.** $\tan^2 x + 1 = $ \_\_\_\_ | **D.** $\dfrac{\sin x}{\cos x}$ |
| **27.** $1 = $ \_\_\_\_ | **E.** $\cos x$ |

*Matching* For each expression in Column I, choose the expression from Column II that completes an identity. You may have to rewrite one or both expressions.

| I | II |
|---|---|
| **28.** $-\tan x \cos x = $ \_\_\_\_ | **A.** $\dfrac{\sin^2 x}{\cos^2 x}$ |
| **29.** $\sec^2 x - 1 = $ \_\_\_\_ | **B.** $\dfrac{1}{\sec^2 x}$ |
| **30.** $\dfrac{\sec x}{\csc x} = $ \_\_\_\_ | **C.** $\sin(-x)$ |
| **31.** $1 + \sin^2 x = $ \_\_\_\_ | **D.** $\csc^2 x - \cot^2 x + \sin^2 x$ |
| **32.** $\cos^2 x = $ \_\_\_\_ | **E.** $\tan x$ |

*Write the first trigonometric function in terms of the second trigonometric function. See Example 2.*

**33.** $\sin x$;   $\cos x$   **34.** $\cot x$;   $\sin x$   **35.** $\tan x$;   $\sec x$

**36.** $\cot x$;   $\csc x$   **37.** $\csc x$;   $\cos x$   **38.** $\sec x$;   $\sin x$

*Write each expression in terms of sine and cosine, and simplify so that no quotients appear in the final expression. See Example 3.*

**39.** $\cot\theta \sin\theta$   **40.** $\sec\theta \cot\theta \sin\theta$   **41.** $\cos\theta \csc\theta$

**42.** $\cot^2\theta(1 + \tan^2\theta)$   **43.** $\sin^2\theta(\csc^2\theta - 1)$   **44.** $(\sec\theta - 1)(\sec\theta + 1)$

**45.** $(1 - \cos\theta)(1 + \sec\theta)$   **46.** $\dfrac{\cos\theta + \sin\theta}{\sin\theta}$

**47.** $\dfrac{\cos^2\theta - \sin^2\theta}{\sin\theta \cos\theta}$   **48.** $\dfrac{1 - \sin^2\theta}{1 + \cot^2\theta}$

**49.** $\sec\theta - \cos\theta$   **50.** $(\sec\theta + \csc\theta)(\cos\theta - \sin\theta)$

**51.** $\sin\theta(\csc\theta - \sin\theta)$   **52.** $\dfrac{1 + \tan^2\theta}{1 + \cot^2\theta}$

**53.** $\sin^2\theta + \tan^2\theta + \cos^2\theta$   **54.** $\dfrac{\tan(-\theta)}{\sec\theta}$

## 18.2 Verifying Trigonometric Identities

**OBJECTIVES**

1. Verify identities by working with one side.
2. Verify identities by working with both sides.

Recall that an identity is an equation that is satisfied for all meaningful replacements of the variable. One of the skills required for more advanced work in mathematics is the ability to use identities to write expressions in alternative forms. We develop this skill by using the fundamental identities to verify that a trigonometric equation is an identity (for those values of the variable for which it is defined). Here are some hints to help you get started.

### Hints for Verifying Identities

1. **Learn the fundamental identities given in Section 18.1.** Whenever you see either side of a fundamental identity, the other side should come to mind. *Also, be aware of equivalent forms of the fundamental identities.* For example, $\sin^2 \theta = 1 - \cos^2 \theta$ is an alternative form of the identity $\sin^2 \theta + \cos^2 \theta = 1$.

2. **Try to rewrite the more complicated side** of the equation so that it is identical to the simpler side.

3. *It is sometimes helpful to express all trigonometric functions in the equation in terms of sine and cosine* and then simplify the result.

4. *Usually, any factoring or indicated algebraic operations should be performed.* For example, the expression

$$\sin^2 x + 2 \sin x + 1 \quad \text{can be factored as} \quad (\sin x + 1)^2.$$

   The sum or difference of two trigonometric expressions, such as $\frac{1}{\sin \theta} + \frac{1}{\cos \theta}$, can be added or subtracted in the same way as any other rational expression.

$$\frac{1}{\sin \theta} + \frac{1}{\cos \theta} = \frac{\cos \theta}{\sin \theta \cos \theta} + \frac{\sin \theta}{\sin \theta \cos \theta} \quad \text{Write with the LCD.}$$

$$= \frac{\cos \theta + \sin \theta}{\sin \theta \cos \theta} \quad \frac{a}{c} + \frac{b}{c} = \frac{a + b}{c}$$

5. *As you select substitutions, keep in mind the side you are not changing, because it represents your goal.* For example, to verify the identity

$$\tan^2 x + 1 = \frac{1}{\cos^2 x},$$

   try to think of an identity that relates $\tan x$ to $\cos x$. In this case, since $\sec x = \frac{1}{\cos x}$ and $\sec^2 x = \tan^2 x + 1$, the secant function is the best link between the two sides.

6. If an expression contains $1 + \sin x$, **multiplying both numerator and denominator** by $1 - \sin x$ would give $1 - \sin^2 x$, which could be replaced with $\cos^2 x$. Similar results for $1 - \sin x$, $1 + \cos x$, and $1 - \cos x$ may be useful.

**CAUTION** *Verifying identities is not the same as solving equations.* Techniques used in solving equations, such as adding the same terms to both sides, or multiplying both sides by the same term, should not be used when working with identities since you are starting with a statement (to be verified) that may not be true.

**OBJECTIVE 1** **Verify identities by working with one side** To avoid the temptation to use algebraic properties of equations to verify identities, *one strategy is to work with only one side and rewrite it to match the other side,* as shown in Examples 1–4.

**EXAMPLE 1** Verifying an Identity (Working with One Side)

Verify that the following equation is an identity.

$$\cot\theta + 1 = \csc\theta(\cos\theta + \sin\theta)$$

**Solution** We use the fundamental identities from **Section 18.1** to rewrite one side of the equation so that it is identical to the other side. Since the right side is more complicated, we work with it, using the third hint to change all functions to sine or cosine.

| Steps | Reasons |
|---|---|
| $\underbrace{\csc\theta(\cos\theta + \sin\theta)}_{\text{Right side of given equation}} = \dfrac{1}{\sin\theta}(\cos\theta + \sin\theta)$ | $\csc\theta = \dfrac{1}{\sin\theta}$ |
| $= \dfrac{\cos\theta}{\sin\theta} + \dfrac{\sin\theta}{\sin\theta}$ | Distributive property; $a(b + c) = ab + ac$ |
| $= \underbrace{\cot\theta + 1}_{\text{Left side of given equation}}$ | $\dfrac{\cos\theta}{\sin\theta} = \cot\theta;\ \dfrac{\sin\theta}{\sin\theta} = 1$ |

The given equation is an identity. The right side is identical to the left side.

**NOW TRY** Exercise 35.

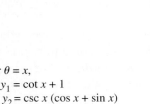

For $\theta = x$,
$y_1 = \cot x + 1$
$y_2 = \csc x\,(\cos x + \sin x)$

The graphs coincide, supporting the conclusion in Example 1.

**EXAMPLE 2** Verifying an Identity (Working with One Side)

Verify that the following equation is an identity.

$$\tan^2 x(1 + \cot^2 x) = \dfrac{1}{1 - \sin^2 x}$$

**Solution** We work with the more complicated left side, as suggested in the second hint. Again, we use the fundamental identities from **Section 18.1.**

| | |
|---|---|
| $\tan^2 x(1 + \cot^2 x) = \tan^2 x + \tan^2 x \cot^2 x$ | Distributive property |
| $= \tan^2 x + \tan^2 x \cdot \dfrac{1}{\tan^2 x}$ | $\cot^2 x = \dfrac{1}{\tan^2 x}$ |
| $= \tan^2 x + 1$ | $\tan^2 x \cdot \dfrac{1}{\tan^2 x} = 1$ |
| $= \sec^2 x$ | $\tan^2 x + 1 = \sec^2 x$ |
| $= \dfrac{1}{\cos^2 x}$ | $\sec^2 x = \dfrac{1}{\cos^2 x}$ |
| $= \dfrac{1}{1 - \sin^2 x}$ | $\cos^2 x = 1 - \sin^2 x$ |

Since the left side is identical to the right side, the given equation is an identity.

**NOW TRY** Exercise 39.

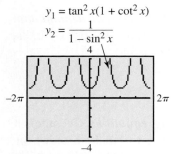

$y_1 = \tan^2 x(1 + \cot^2 x)$
$y_2 = \dfrac{1}{1 - \sin^2 x}$

The screen supports the conclusion in Example 2.

### EXAMPLE 3  Verifying an Identity (Working with One Side)

Verify that the following equation is an identity.

$$\frac{\tan t - \cot t}{\sin t \cos t} = \sec^2 t - \csc^2 t$$

**Solution**  We transform the more complicated left side to match the right side.

$$\frac{\tan t - \cot t}{\sin t \cos t} = \frac{\tan t}{\sin t \cos t} - \frac{\cot t}{\sin t \cos t} \qquad \frac{a-b}{c} = \frac{a}{c} - \frac{b}{c}$$

$$= \tan t \cdot \frac{1}{\sin t \cos t} - \cot t \cdot \frac{1}{\sin t \cos t} \qquad \frac{a}{b} = a \cdot \frac{1}{b}$$

$$= \frac{\sin t}{\cos t} \cdot \frac{1}{\sin t \cos t} - \frac{\cos t}{\sin t} \cdot \frac{1}{\sin t \cos t} \qquad \tan t = \frac{\sin t}{\cos t};\ \cot t = \frac{\cos t}{\sin t}$$

$$= \frac{1}{\cos^2 t} - \frac{1}{\sin^2 t} \qquad \text{Multiply.}$$

$$= \sec^2 t - \csc^2 t \qquad \frac{1}{\cos^2 t} = \sec^2 t;\ \frac{1}{\sin^2 t} = \csc^2 t$$

The third hint about writing all trigonometric functions in terms of sine and cosine was used in the third line of the solution.

**NOW TRY** Exercise 43.

### EXAMPLE 4  Verifying an Identity (Working with One Side)

Verify that the following equation is an identity.

$$\frac{\cos x}{1 - \sin x} = \frac{1 + \sin x}{\cos x}$$

**Solution**  We work on the right side, using the last hint in the list given earlier to multiply numerator and denominator on the right by $1 - \sin x$.

$$\frac{1 + \sin x}{\cos x} = \frac{(1 + \sin x)(1 - \sin x)}{\cos x(1 - \sin x)} \qquad \text{Multiply by 1 in the form } \frac{1 - \sin x}{1 - \sin x}.$$

$$= \frac{1 - \sin^2 x}{\cos x(1 - \sin x)} \qquad (x + y)(x - y) = x^2 - y^2$$

$$= \frac{\cos^2 x}{\cos x(1 - \sin x)} \qquad 1 - \sin^2 x = \cos^2 x$$

$$= \frac{\cos x}{1 - \sin x} \qquad \text{Lowest terms}$$

**NOW TRY** Exercise 49.

**OBJECTIVE 2  Verify identities by working with both sides.**  If both sides of an identity appear to be equally complex, the identity can be verified by working independently on the left side and on the right side, until each side is changed into some common third result. ***Each step, on each side, must be reversible.*** With all steps reversible, the procedure is as shown in the margin. The left side leads to a common third expression, which leads back to the right side. This procedure is just a shortcut for the procedure used in Examples 1–4: one side is changed into the other side, but by going through an intermediate step.

left = right
↘ ↙
common third
expression

**EXAMPLE 5** Verifying an Identity (Working with Both Sides)

Verify that the following equation is an identity.

$$\frac{\sec \alpha + \tan \alpha}{\sec \alpha - \tan \alpha} = \frac{1 + 2 \sin \alpha + \sin^2 \alpha}{\cos^2 \alpha}$$

**Solution** Both sides appear equally complex, so we verify the identity by changing each side into a common third expression. We work first on the left, multiplying numerator and denominator by $\cos \alpha$.

$$\frac{\sec \alpha + \tan \alpha}{\sec \alpha - \tan \alpha} = \frac{(\sec \alpha + \tan \alpha)\cos \alpha}{(\sec \alpha - \tan \alpha)\cos \alpha} \quad \text{Multiply by 1 in the form } \frac{\cos \alpha}{\cos \alpha}.$$

$$= \frac{\sec \alpha \cos \alpha + \tan \alpha \cos \alpha}{\sec \alpha \cos \alpha - \tan \alpha \cos \alpha} \quad \text{Distributive property}$$

$$= \frac{1 + \tan \alpha \cos \alpha}{1 - \tan \alpha \cos \alpha} \quad \sec \alpha \cos \alpha = 1$$

$$= \frac{1 + \frac{\sin \alpha}{\cos \alpha} \cdot \cos \alpha}{1 - \frac{\sin \alpha}{\cos \alpha} \cdot \cos \alpha} \quad \tan \alpha = \frac{\sin \alpha}{\cos \alpha}$$

$$= \frac{1 + \sin \alpha}{1 - \sin \alpha} \quad \text{Simplify.}$$

On the right side of the original equation, begin by factoring.

$$\frac{1 + 2 \sin \alpha + \sin^2 \alpha}{\cos^2 \alpha} = \frac{(1 + \sin \alpha)^2}{\cos^2 \alpha} \quad x^2 + 2xy + y^2 = (x + y)^2$$

$$= \frac{(1 + \sin \alpha)^2}{1 - \sin^2 \alpha} \quad \cos^2 \alpha = 1 - \sin^2 \alpha$$

$$= \frac{(1 + \sin \alpha)^2}{(1 + \sin \alpha)(1 - \sin \alpha)} \quad \text{Factor the denominator;} \; x^2 - y^2 = (x + y)(x - y)$$

$$= \frac{1 + \sin \alpha}{1 - \sin \alpha} \quad \text{Lowest terms}$$

We have shown that

$$\underbrace{\frac{\sec \alpha + \tan \alpha}{\sec \alpha - \tan \alpha}}_{\text{Left side of given equation}} = \underbrace{\frac{1 + \sin \alpha}{1 - \sin \alpha}}_{\text{Common third expression}} = \underbrace{\frac{1 + 2 \sin \alpha + \sin^2 \alpha}{\cos^2 \alpha}}_{\text{Right side of given equation}},$$

verifying that the given equation is an identity.

**NOW TRY** Exercise 65.

**CAUTION** Use the method of Example 5 *only* if the steps are reversible.

There are usually several ways to verify a given identity. For instance, another way to begin verifying the identity in Example 5 is to work on the left as follows.

$$\frac{\sec \alpha + \tan \alpha}{\sec \alpha - \tan \alpha} = \frac{\dfrac{1}{\cos \alpha} + \dfrac{\sin \alpha}{\cos \alpha}}{\dfrac{1}{\cos \alpha} - \dfrac{\sin \alpha}{\cos \alpha}} \quad \text{Fundamental identities (Section 18.1)}$$

Multiply by $\dfrac{\cos \alpha}{\cos \alpha}$ here.

$$= \frac{\dfrac{1 + \sin \alpha}{\cos \alpha}}{\dfrac{1 - \sin \alpha}{\cos \alpha}} \quad \text{Add and subtract fractions.}$$

$$= \frac{1 + \sin \alpha}{1 - \sin \alpha} \quad \text{Simplify the complex fraction.}$$

Compare this with the result shown in Example 5 for the right side to see that the two sides indeed agree.

## 18.2 Exercises

**NOW TRY Exercise**

*Perform each indicated operation and simplify the result.*

1. $\cot \theta + \dfrac{1}{\cot \theta}$

2. $\dfrac{\sec x}{\csc x} + \dfrac{\csc x}{\sec x}$

3. $\tan s (\cot s + \csc s)$

4. $\cos \beta (\sec \beta + \csc \beta)$

5. $\dfrac{1}{\csc^2 \theta} + \dfrac{1}{\sec^2 \theta}$

6. $\dfrac{1}{\sin \alpha - 1} - \dfrac{1}{\sin \alpha + 1}$

7. $\dfrac{\cos x}{\sec x} + \dfrac{\sin x}{\csc x}$

8. $\dfrac{\cos \theta}{\sin \theta} + \dfrac{\sin \theta}{1 + \cos \theta}$

9. $(1 + \sin t)^2 + \cos^2 t$

10. $(1 + \tan s)^2 - 2 \tan s$

11. $\dfrac{1}{1 + \cos x} - \dfrac{1}{1 - \cos x}$

12. $(\sin \alpha - \cos \alpha)^2$

*Factor each trigonometric expression.*

13. $\sin^2 \theta - 1$

14. $\sec^2 \theta - 1$

15. $(\sin x + 1)^2 - (\sin x - 1)^2$

16. $(\tan x + \cot x)^2 - (\tan x - \cot x)^2$

17. $2 \sin^2 x + 3 \sin x + 1$

18. $4 \tan^2 \beta + \tan \beta - 3$

19. $\cos^4 x + 2 \cos^2 x + 1$

20. $\cot^4 x + 3 \cot^2 x + 2$

21. $\sin^3 x - \cos^3 x$

22. $\sin^3 \alpha + \cos^3 \alpha$

*Each expression simplifies to a constant, a single function, or a power of a function. Use fundamental identities to simplify each expression.*

23. $\tan \theta \cos \theta$

24. $\cot \alpha \sin \alpha$

25. $\sec r \cos r$

26. $\cot t \tan t$

27. $\dfrac{\sin \beta \tan \beta}{\cos \beta}$

28. $\dfrac{\csc \theta \sec \theta}{\cot \theta}$

29. $\sec^2 x - 1$

30. $\csc^2 t - 1$

31. $\dfrac{\sin^2 x}{\cos^2 x} + \sin x \csc x$

32. $\dfrac{1}{\tan^2 \alpha} + \cot \alpha \tan \alpha$

33. $1 - \dfrac{1}{\csc^2 x}$

34. $1 - \dfrac{1}{\sec^2 x}$

*In Exercises 35–78, verify that each trigonometric equation is an identity. See Examples 1–5.*

35. $\dfrac{\cot \theta}{\csc \theta} = \cos \theta$

36. $\dfrac{\tan \alpha}{\sec \alpha} = \sin \alpha$

37. $\dfrac{1 - \sin^2 \beta}{\cos \beta} = \cos \beta$

38. $\dfrac{\tan^2 \alpha + 1}{\sec \alpha} = \sec \alpha$

39. $\cos^2 \theta (\tan^2 \theta + 1) = 1$

40. $\sin^2 \beta (1 + \cot^2 \beta) = 1$

41. $\cot s + \tan s = \sec s \csc s$

42. $\sin^2 \alpha + \tan^2 \alpha + \cos^2 \alpha = \sec^2 \alpha$

43. $\dfrac{\cos \alpha}{\sec \alpha} + \dfrac{\sin \alpha}{\csc \alpha} = \sec^2 \alpha - \tan^2 \alpha$

44. $\dfrac{\sin^2 \theta}{\cos \theta} = \sec \theta - \cos \theta$

45. $\sin^4 \theta - \cos^4 \theta = 2 \sin^2 \theta - 1$

46. $\dfrac{\cos \theta}{\sin \theta \cot \theta} = 1$

47. $\dfrac{1 - \cos x}{1 + \cos x} = (\cot x - \csc x)^2$

48. $\sin^2 \theta (1 + \cot^2 \theta) - 1 = 0$

49. $\dfrac{\cos \theta + 1}{\tan^2 \theta} = \dfrac{\cos \theta}{\sec \theta - 1}$

50. $\dfrac{(\sec \theta - \tan \theta)^2 + 1}{\sec \theta \csc \theta - \tan \theta \csc \theta} = 2 \tan \theta$

51. $\dfrac{1}{1 - \sin \theta} + \dfrac{1}{1 + \sin \theta} = 2 \sec^2 \theta$

52. $\dfrac{1}{\sec \alpha - \tan \alpha} = \sec \alpha + \tan \alpha$

53. $\dfrac{\cot \alpha + 1}{\cot \alpha - 1} = \dfrac{1 + \tan \alpha}{1 - \tan \alpha}$

54. $\dfrac{\csc \theta + \cot \theta}{\tan \theta + \sin \theta} = \cot \theta \csc \theta$

55. $\sec^4 x - \sec^2 x = \tan^4 x + \tan^2 x$

56. $(\sec \alpha - \tan \alpha)^2 = \dfrac{1 - \sin \alpha}{1 + \sin \alpha}$

57. $\dfrac{\sec^4 s - \tan^4 s}{\sec^2 s + \tan^2 s} = \sec^2 s - \tan^2 s$

58. $\dfrac{\cot^2 t - 1}{1 + \cot^2 t} = 1 - 2 \sin^2 t$

59. $\dfrac{\tan^2 t - 1}{\sec^2 t} = \dfrac{\tan t - \cot t}{\tan t + \cot t}$

60. $\dfrac{\sin^4 \alpha - \cos^4 \alpha}{\sin^2 \alpha - \cos^2 \alpha} = 1$

61. $(1 - \cos^2 \alpha)(1 + \cos^2 \alpha) = 2 \sin^2 \alpha - \sin^4 \alpha$

62. $\tan^2 \alpha \sin^2 \alpha = \tan^2 \alpha + \cos^2 \alpha - 1$

63. $\sin^2 \alpha \sec^2 \alpha + \sin^2 \alpha \csc^2 \alpha = \sec^2 \alpha$

64. $\dfrac{-1}{\tan \alpha - \sec \alpha} + \dfrac{-1}{\tan \alpha + \sec \alpha} = 2 \tan \alpha$

65. $\dfrac{\tan s}{1 + \cos s} + \dfrac{\sin s}{1 - \cos s} = \cot s + \sec s \csc s$

66. $\dfrac{1 - \cos x}{1 + \cos x} = \csc^2 x - 2 \csc x \cot x + \cot^2 x$

67. $\dfrac{1 - \sin \theta}{1 + \sin \theta} = \sec^2 \theta - 2 \sec \theta \tan \theta + \tan^2 \theta$

68. $\sin \theta + \cos \theta = \dfrac{\sin \theta}{1 - \dfrac{\cos \theta}{\sin \theta}} + \dfrac{\cos \theta}{1 - \dfrac{\sin \theta}{\cos \theta}}$

69. $\dfrac{\sin \theta}{1 - \cos \theta} - \dfrac{\sin \theta \cos \theta}{1 + \cos \theta} = \csc \theta (1 + \cos^2 \theta)$

70. $(1 + \sin x + \cos x)^2 = 2(1 + \sin x)(1 + \cos x)$

71. $\dfrac{1 + \cos x}{1 - \cos x} - \dfrac{1 - \cos x}{1 + \cos x} = 4 \cot x \csc x$

72. $(\sec \alpha + \csc \alpha)(\cos \alpha - \sin \alpha) = \cot \alpha - \tan \alpha$

73. $\dfrac{1 + \sin \theta}{1 - \sin \theta} - \dfrac{1 - \sin \theta}{1 + \sin \theta} = 4 \tan \theta \sec \theta$

74. $\dfrac{1 - \cos \theta}{1 + \cos \theta} = 2 \csc^2 \theta - 2 \csc \theta \cot \theta - 1$

75. $(2 \sin x + \cos x)^2 + (2 \cos x - \sin x)^2 = 5$

76. $\sin^2 x (1 + \cot x) + \cos^2 x (1 - \tan x) + \cot^2 x = \csc^2 x$

77. $\sec x - \cos x + \csc x - \sin x - \sin x \tan x = \cos x \cot x$

78. $\sin^3 \theta + \cos^3 \theta = (\cos \theta + \sin \theta)(1 - \cos \theta \sin \theta)$

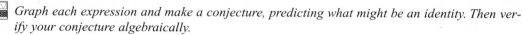
*Graph each expression and make a conjecture, predicting what might be an identity. Then verify your conjecture algebraically.*

79. $(\sec \theta + \tan \theta)(1 - \sin \theta)$

80. $(\csc \theta + \cot \theta)(\sec \theta - 1)$

81. $\dfrac{\cos \theta + 1}{\sin \theta + \tan \theta}$

82. $\tan \theta \sin \theta + \cos \theta$

*By substituting a number for s or t, show that the equation is not an identity.*

83. $\sin(\csc s) = 1$

84. $\sqrt{\cos^2 s} = \cos s$

85. $\csc t = \sqrt{1 + \cot^2 t}$

86. $\cos t = \sqrt{1 - \sin^2 t}$

87. *Concept Check* When is $\sin x = \sqrt{1 - \cos^2 x}$ a true statement?

88. *Concept Check* When is $\cos x = \sqrt{1 - \sin^2 x}$ a true statement?

*Work each problem.*

89. According to Lambert's law, the intensity of light from a single source on a flat surface at point $P$ is given by

$$I = k \cos^2 \theta,$$

where $k$ is a constant. (*Source:* Winter, C., *Solar Power Plants,* Springer-Verlag, 1991.)

(a) Write $I$ in terms of the sine function.
(b) Why does the maximum value of $I$ occur when $\theta = 0$?

90. The distance or displacement $y$ of a weight attached to an oscillating spring from its natural position is modeled by

$$y = 4 \cos(2\pi t),$$

where $t$ is time in seconds. Potential energy is the energy of position and is given by

$$P = ky^2,$$

where $k$ is a constant. The weight has the greatest potential energy when the spring is stretched the most. (*Source:* Weidner, R. and R. Sells, *Elementary Classical Physics,* Vol. 1, Allyn & Bacon, 1973.)

(a) Write an expression for $P$ that involves the cosine function.
(b) Use a fundamental identity to write $P$ in terms of $\sin(2\pi t)$.

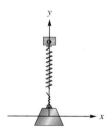

## 18.3 Sum and Difference Identities for Cosine

**OBJECTIVES**

1. Derive the sum and difference identities for cosine.
2. Use the sum and difference identities for cosine to find exact values.
3. Derive and use the cofunction identities.
4. Apply the sum and difference identities for cosine.

**OBJECTIVE 1** Derive the sum and difference identities for cosine. Several examples presented earlier should have convinced you by now that $\cos(A - B)$ does not equal $\cos A - \cos B$. For example, if $A = \frac{\pi}{2}$ and $B = 0$, then

$$\cos(A - B) = \cos\left(\frac{\pi}{2} - 0\right) = \cos\frac{\pi}{2} = 0,$$

while

$$\cos A - \cos B = \cos\frac{\pi}{2} - \cos 0 = 0 - 1 = -1.$$

We can now derive a formula for $\cos(A - B)$. We start by locating angles $A$ and $B$ in standard position on a unit circle, with $B < A$. Let $S$ and $Q$ be the points where the terminal sides of angles $A$ and $B$, respectively, intersect the circle. Let $P$ be the point $(1, 0)$, and locate point $R$ on the unit circle so that angle $POR$ equals the difference $A - B$. See Figure 3.

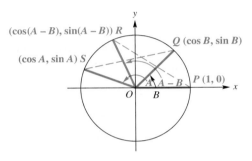

**FIGURE 3**

Point $Q$ is on the unit circle, thus the $x$-coordinate of $Q$ is the cosine of angle $B$, while the $y$-coordinate of $Q$ is the sine of angle $B$.

$Q$ has coordinates $(\cos B, \sin B)$.

In the same way,

$S$ has coordinates $(\cos A, \sin A)$,

and

$R$ has coordinates $(\cos(A - B), \sin(A - B))$.

Angle $SOQ$ also equals $A - B$. Since the central angles $SOQ$ and $POR$ are equal, chords $PR$ and $SQ$ are equal. By the distance formula, since $PR = SQ$,

$$\sqrt{[\cos(A - B) - 1]^2 + [\sin(A - B) - 0]^2}$$
$$= \sqrt{(\cos A - \cos B)^2 + (\sin A - \sin B)^2}.$$

Squaring both sides and clearing parentheses gives

$$\cos^2(A - B) - 2\cos(A - B) + 1 + \sin^2(A - B)$$
$$= \cos^2 A - 2\cos A \cos B + \cos^2 B + \sin^2 A - 2\sin A \sin B + \sin^2 B.$$

Since $\sin^2 x + \cos^2 x = 1$ for any value of $x$, we can rewrite the equation as

$$2 - 2\cos(A - B) = 2 - 2\cos A \cos B - 2\sin A \sin B$$

$$\cos(A - B) = \cos A \cos B + \sin A \sin B. \quad \text{Subtract 2; divide by } -2.$$

This is the identity for $\cos(A - B)$. Although Figure 3 shows angles $A$ and $B$ in the second and first quadrants, respectively, this result is the same for any values of these angles.

To find a similar expression for $\cos(A + B)$, rewrite $A + B$ as $A - (-B)$ and use the identity for $\cos(A - B)$.

$$\begin{aligned}\cos(A + B) &= \cos[A - (-B)] \\ &= \cos A \cos(-B) + \sin A \sin(-B) &&\text{Cosine difference identity} \\ &= \cos A \cos B + \sin A(-\sin B) &&\text{Negative-angle identities}\end{aligned}$$

$$\cos(A + B) = \cos A \cos B - \sin A \sin B$$

### Cosine of a Sum or Difference

$$\cos(A + B) = \cos A \cos B - \sin A \sin B$$
$$\cos(A - B) = \cos A \cos B + \sin A \sin B$$

**OBJECTIVE 2** Use the sum and difference identities for cosine to find exact values. These identities are useful in certain applications. For example, the method shown in Example 1 can be applied to get an exact value for $\cos 15°$, as well as to practice using the sum and difference identities.

**EXAMPLE 1** Finding Exact Cosine Function Values

Find the *exact* value of each expression.

(a) $\cos 15°$  (b) $\cos \dfrac{5\pi}{12}$  (c) $\cos 87° \cos 93° - \sin 87° \sin 93°$

**Solution**

(a) To find $\cos 15°$, we write $15°$ as the sum or difference of two angles with known function values, such as $45°$ and $30°$, since $15° = 45° - 30°$. (We could also use $60° - 45°$.) Then we use the cosine difference identity.

$$\begin{aligned}\cos 15° &= \cos(45° - 30°) &&15° = 45° - 30° \\ &= \cos 45° \cos 30° + \sin 45° \sin 30° &&\text{Cosine difference identity} \\ &= \frac{\sqrt{2}}{2} \cdot \frac{\sqrt{3}}{2} + \frac{\sqrt{2}}{2} \cdot \frac{1}{2} &&\text{Substitute known values.} \\ &= \frac{\sqrt{6} + \sqrt{2}}{4} &&\text{Multiply; add fractions.}\end{aligned}$$

(b) $\begin{aligned}\cos \dfrac{5\pi}{12} &= \cos\left(\dfrac{\pi}{6} + \dfrac{\pi}{4}\right) &&\tfrac{\pi}{6} = \tfrac{2\pi}{12}, \tfrac{\pi}{4} = \tfrac{3\pi}{12} \\ &= \cos \dfrac{\pi}{6} \cos \dfrac{\pi}{4} - \sin \dfrac{\pi}{6} \sin \dfrac{\pi}{4} &&\text{Cosine sum identity} \\ &= \dfrac{\sqrt{3}}{2} \cdot \dfrac{\sqrt{2}}{2} - \dfrac{1}{2} \cdot \dfrac{\sqrt{2}}{2} &&\text{Substitute known values.} \\ &= \dfrac{\sqrt{6} - \sqrt{2}}{4} &&\text{Multiply; subtract fractions.}\end{aligned}$

The screen supports the solution in Example 1(b) by showing that $\cos \dfrac{5\pi}{12} = \dfrac{\sqrt{6} - \sqrt{2}}{4}$.

(c) $\begin{aligned}\cos 87° \cos 93° - \sin 87° \sin 93° &= \cos(87° + 93°) &&\text{Cosine sum identity} \\ &= \cos 180° &&\text{Add.} \\ &= -1\end{aligned}$

**NOW TRY** Exercises 7, 9, and 11.

**OBJECTIVE 3** Derive and use the cofunction identities. We can use the identity for the cosine of the difference of two angles and the fundamental identities to derive the *cofunction identities,* presented in **Section 15.1** for values of $\theta$ in the interval $[0°, 90°]$.

### Cofunction Identities

$$\cos(90° - \theta) = \sin \theta \qquad \cot(90° - \theta) = \tan \theta$$
$$\sin(90° - \theta) = \cos \theta \qquad \sec(90° - \theta) = \csc \theta$$
$$\tan(90° - \theta) = \cot \theta \qquad \csc(90° - \theta) = \sec \theta$$

Similar identities can be obtained for a real number domain by replacing $90°$ with $\frac{\pi}{2}$.

Substituting $90°$ for $A$ and $\theta$ for $B$ in the identity for $\cos(A - B)$ gives

$$\cos(90° - \theta) = \cos 90° \cos \theta + \sin 90° \sin \theta$$
$$= 0 \cdot \cos \theta + 1 \cdot \sin \theta$$
$$= \sin \theta.$$

This result is true for *any* value of $\theta$ since the identity for $\cos(A - B)$ is true for any values of $A$ and $B$.

**EXAMPLE 2** Using Cofunction Identities to Find $\theta$

Find an angle $\theta$ that satisfies each of the following.

**(a)** $\cot \theta = \tan 25°$  **(b)** $\sin \theta = \cos(-30°)$  **(c)** $\csc \dfrac{3\pi}{4} = \sec \theta$

**Solution**

**(a)** Since tangent and cotangent are cofunctions, $\tan(90° - \theta) = \cot \theta$.

$$\cot \theta = \tan 25°$$
$$\tan(90° - \theta) = \tan 25° \qquad \text{Cofunction identity}$$
$$90° - \theta = 25° \qquad \text{Set angle measures equal.}$$
$$\theta = 65° \qquad \text{Solve for } \theta.$$

**(b)**
$$\sin \theta = \cos(-30°)$$
$$\cos(90° - \theta) = \cos(-30°) \qquad \text{Cofunction identity}$$
$$90° - \theta = -30°$$
$$\theta = 120°$$

**(c)**
$$\csc \frac{3\pi}{4} = \sec \theta$$
$$\sec\left(\frac{\pi}{2} - \frac{3\pi}{4}\right) = \sec \theta \qquad \text{Cofunction identity}$$
$$\sec\left(-\frac{\pi}{4}\right) = \sec \theta \qquad \text{Combine terms.}$$
$$-\frac{\pi}{4} = \theta$$

**NOW TRY** Exercises 33 and 37.

Because trigonometric (circular) functions are periodic, the solutions in Example 2 are not unique. We give only one of infinitely many possibilities.

**OBJECTIVE 4** Apply the sum and difference identities for cosine. If one of the angles $A$ or $B$ in the identities for $\cos(A + B)$ and $\cos(A - B)$ is a quadrantal angle, then the identity allows us to write the expression in terms of a single function of $A$ or $B$.

### EXAMPLE 3 Reducing $\cos(A - B)$ to a Function of a Single Variable

Write $\cos(180° - \theta)$ as a trigonometric function of $\theta$ alone.

**Solution**  $\cos(180° - \theta) = \cos 180° \cos \theta + \sin 180° \sin \theta$  Cosine difference identity

$\qquad\qquad\qquad\quad = (-1) \cos \theta + (0) \sin \theta$

$\qquad\qquad\qquad\quad = -\cos \theta$

NOW TRY Exercise 39.

### EXAMPLE 4 Finding $\cos(s + t)$ Given Information about $s$ and $t$

Suppose that $\sin s = \frac{3}{5}$, $\cos t = -\frac{12}{13}$, and both $s$ and $t$ are in quadrant II. Find $\cos(s + t)$.

**Solution**  By the cosine sum identity, $\cos(s + t) = \cos s \cos t - \sin s \sin t$. The values of $\sin s$ and $\cos t$ are given, so we can find $\cos(s + t)$ if we know the values of $\cos s$ and $\sin t$. To find $\cos s$ and $\sin t$, we sketch two angles in the second quadrant, one with $\sin s = \frac{3}{5}$ and the other with $\cos t = -\frac{12}{13}$. See Figure 4.

In Figure 4(a), since $\sin s = \frac{3}{5} = \frac{y}{r}$, we let $y = 3$ and $r = 5$. Substituting in the Pythagorean theorem, we get $x^2 + 3^2 = 5^2$ and solve to find $x = -4$. Thus, $\cos s = -\frac{4}{5}$. In Figure 4(b), $\cos t = -\frac{12}{13} = \frac{x}{r}$, so we let $x = -12$ and $r = 13$. Then $(-12)^2 + y^2 = 13^2$; we solve to get $y = 5$. Thus, $\sin t = \frac{5}{13}$. Now we can find $\cos(s + t)$.

$\cos(s + t) = \cos s \cos t - \sin s \sin t$  Cosine sum identity

$\qquad\qquad = -\frac{4}{5}\left(-\frac{12}{13}\right) - \frac{3}{5} \cdot \frac{5}{13}$  Substitute.

$\qquad\qquad = \frac{48}{65} - \frac{15}{65} = \frac{33}{65}$

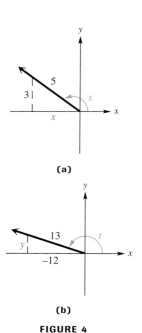

**FIGURE 4**

NOW TRY Exercise 47.

## 18.3 Exercises

NOW TRY Exercise

*Matching*  Match each expression in Column I with the correct expression in Column II to form an identity.

| I | II |
|---|---|
| **1.** $\cos(x + y) = $ _____ | **A.** $\cos x \cos y + \sin x \sin y$ |
| **2.** $\cos(x - y) = $ _____ | **B.** $\cos x$ |
| **3.** $\cos(90° - x) = $ _____ | **C.** $\cos x + \sin x$ |
| **4.** $\sin(90° - x) = $ _____ | **D.** $\cos x - \sin x$ |
| | **E.** $\sin x$ |
| | **F.** $\cos x \cos y - \sin x \sin y$ |

*Use identities to find each exact value. (Do not use a calculator.) See Example 1.*

**5.** $\cos 75°$

**6.** $\cos(-15°)$

**7.** $\cos 105°$
(*Hint:* $105° = 60° + 45°$)

**8.** $\cos(-105°)$
(*Hint:* $-105° = -60° + (-45°)$)

**9.** $\cos \dfrac{7\pi}{12}$

**10.** $\cos\left(-\dfrac{\pi}{12}\right)$

**11.** $\cos 40° \cos 50° - \sin 40° \sin 50°$

**12.** $\cos \dfrac{7\pi}{9} \cos \dfrac{2\pi}{9} - \sin \dfrac{7\pi}{9} \sin \dfrac{2\pi}{9}$

*Use a graphing or scientific calculator to support your answer for each of the following. See Example 1.*

**13.** Exercise 11

**14.** Exercise 12

*Write each function value in terms of the cofunction of a complementary angle. See Example 2.*

**15.** $\tan 87°$

**16.** $\sin 15°$

**17.** $\cos \dfrac{\pi}{12}$

**18.** $\sin \dfrac{2\pi}{5}$

**19.** $\csc(-14° \, 24')$

**20.** $\sin 142° \, 14'$

**21.** $\sin \dfrac{5\pi}{8}$

**22.** $\cot \dfrac{9\pi}{10}$

**23.** $\sec 146° \, 42'$

**24.** $\tan 174° \, 3'$

**25.** $\cot 176.9814°$

**26.** $\sin 98.0142°$

*Fill in the Blanks Use identities to fill in each blank with the appropriate trigonometric function name. See Example 2.*

**27.** $\cot \dfrac{\pi}{3} = $ _____ $\dfrac{\pi}{6}$

**28.** $\sin \dfrac{2\pi}{3} = $ _____ $\left(-\dfrac{\pi}{6}\right)$

**29.** _____ $33° = \sin 57°$

**30.** _____ $72° = \cot 18°$

**31.** $\cos 70° = \dfrac{1}{\underline{\quad} 20°}$

**32.** $\tan 24° = \dfrac{1}{\underline{\quad} 66°}$

*Find an angle θ that makes each statement true. See Example 2.*

**33.** $\tan \theta = \cot(45° + 2\theta)$

**34.** $\sin \theta = \cos(2\theta - 10°)$

**35.** $\sec \theta = \csc\left(\dfrac{\theta}{2} + 20°\right)$

**36.** $\cos \theta = \sin(3\theta + 10°)$

**37.** $\sin(3\theta - 15°) = \cos(\theta + 25°)$

**38.** $\cot(\theta - 10°) = \tan(2\theta + 20°)$

*Use identities to write each expression as a function of θ. See Example 3.*

**39.** $\cos(0° - \theta)$

**40.** $\cos(90° - \theta)$

**41.** $\cos(180° - \theta)$

**42.** $\cos(270° - \theta)$

**43.** $\cos(0° + \theta)$

**44.** $\cos(90° + \theta)$

**45.** $\cos(180° + \theta)$

**46.** $\cos(270° + \theta)$

*Find $\cos(s + t)$ and $\cos(s - t)$. See Example 4.*

**47.** $\cos s = -\dfrac{1}{5}$ and $\sin t = \dfrac{3}{5}$, $s$ and $t$ in quadrant II

**48.** $\sin s = \dfrac{2}{3}$ and $\sin t = -\dfrac{1}{3}$, $s$ in quadrant II and $t$ in quadrant IV

**49.** $\sin s = \dfrac{3}{5}$ and $\sin t = -\dfrac{12}{13}$, $s$ in quadrant I and $t$ in quadrant III

**50.** $\cos s = -\dfrac{8}{17}$ and $\cos t = -\dfrac{3}{5}$, $s$ and $t$ in quadrant III

**51.** $\sin s = \dfrac{\sqrt{5}}{7}$ and $\sin t = \dfrac{\sqrt{6}}{8}$, $s$ and $t$ in quadrant I

**52.** $\cos s = \dfrac{\sqrt{2}}{4}$ and $\sin t = -\dfrac{\sqrt{5}}{6}$, $s$ and $t$ in quadrant IV

*True or False*

53. $\cos 42° = \cos(30° + 12°)$
54. $\cos(-24°) = \cos 16° - \cos 40°$
55. $\cos 74° = \cos 60° \cos 14° + \sin 60° \sin 14°$
56. $\cos 140° = \cos 60° \cos 80° - \sin 60° \sin 80°$
57. $\cos \dfrac{\pi}{3} = \cos \dfrac{\pi}{12} \cos \dfrac{\pi}{4} - \sin \dfrac{\pi}{12} \sin \dfrac{\pi}{4}$
58. $\cos \dfrac{2\pi}{3} = \cos \dfrac{11\pi}{12} \cos \dfrac{\pi}{4} + \sin \dfrac{11\pi}{12} \sin \dfrac{\pi}{4}$
59. $\cos 70° \cos 20° - \sin 70° \sin 20° = 0$
60. $\cos 85° \cos 40° + \sin 85° \sin 40° = \dfrac{\sqrt{2}}{2}$
61. $\tan\left(\theta - \dfrac{\pi}{2}\right) = \cot \theta$
62. $\sin\left(\theta - \dfrac{\pi}{2}\right) = \cos \theta$

*Verify that each equation is an identity.*

63. $\cos\left(\dfrac{\pi}{2} + x\right) = -\sin x$
64. $\sec(\pi - x) = -\sec x$
65. $\cos 2x = \cos^2 x - \sin^2 x$  (*Hint:* $\cos 2x = \cos(x + x)$.)
66. $1 + \cos 2x - \cos^2 x = \cos^2 x$  (*Hint:* Use the result from Exercise 65.)

# 18.4 Sum and Difference Identities for Sine and Tangent

**OBJECTIVES**

1. Derive the sum and difference identities for sine.
2. Derive the sum and difference identities for tangent.
3. Apply the sum and difference identities.

**OBJECTIVE 1** Derive the sum and difference identities for sine. We can use the cosine sum and difference identities to derive similar identities for sine and tangent. Since $\sin \theta = \cos(90° - \theta)$, we replace $\theta$ with $A + B$ to get

$\sin(A + B) = \cos[90° - (A + B)]$      Cofunction identity
$\phantom{\sin(A + B)} = \cos[(90° - A) - B]$
$\phantom{\sin(A + B)} = \cos(90° - A) \cos B + \sin(90° - A) \sin B$      Cosine difference identity

$\sin(A + B) = \sin A \cos B + \cos A \sin B.$      Cofunction identities

Now we write $\sin(A - B)$ as $\sin[A + (-B)]$ and use the identity for $\sin(A + B)$.

$\sin(A - B) = \sin[A + (-B)]$
$\phantom{\sin(A - B)} = \sin A \cos(-B) + \cos A \sin(-B)$      Sine sum identity

$\sin(A - B) = \sin A \cos B - \cos A \sin B$      Negative-angle identities

**Sine of a Sum or Difference**

$$\sin(A + B) = \sin A \cos B + \cos A \sin B$$
$$\sin(A - B) = \sin A \cos B - \cos A \sin B$$

**OBJECTIVE 2** Derive the sum and difference identities for tangent. To derive the identity for $\tan(A + B)$, we start with

$$\tan(A + B) = \frac{\sin(A + B)}{\cos(A + B)} \quad \text{Fundamental identity}$$

$$= \frac{\sin A \cos B + \cos A \sin B}{\cos A \cos B - \sin A \sin B}. \quad \text{Sum identities}$$

We express this result in terms of the tangent function by multiplying both numerator and denominator by $\frac{1}{\cos A \cos B}$.

$$\tan(A + B) = \frac{\dfrac{\sin A \cos B + \cos A \sin B}{1}}{\dfrac{\cos A \cos B - \sin A \sin B}{1}} \cdot \dfrac{\dfrac{1}{\cos A \cos B}}{\dfrac{1}{\cos A \cos B}} \quad \begin{array}{l}\text{Simplify the} \\ \text{complex fraction.}\end{array}$$

$$= \frac{\dfrac{\sin A \cos B}{\cos A \cos B} + \dfrac{\cos A \sin B}{\cos A \cos B}}{\dfrac{\cos A \cos B}{\cos A \cos B} - \dfrac{\sin A \sin B}{\cos A \cos B}} \quad \begin{array}{l}\text{Multiply numerators;} \\ \text{multiply denominators.}\end{array}$$

$$= \frac{\dfrac{\sin A}{\cos A} + \dfrac{\sin B}{\cos B}}{1 - \dfrac{\sin A}{\cos A} \cdot \dfrac{\sin B}{\cos B}} \quad \text{Simplify.}$$

Since $\frac{\sin \theta}{\cos \theta} = \tan \theta$, we have

$$\tan(A + B) = \frac{\tan A + \tan B}{1 - \tan A \tan B}.$$

Replacing $B$ with $-B$ and using the fact that $\tan(-B) = -\tan B$ gives the identity for the tangent of the difference of two angles.

**Tangent of a Sum or Difference**

$$\tan(A + B) = \frac{\tan A + \tan B}{1 - \tan A \tan B} \qquad \tan(A - B) = \frac{\tan A - \tan B}{1 + \tan A \tan B}$$

**OBJECTIVE 3** Apply the sum and difference identities.

**EXAMPLE 1** Finding Exact Sine and Tangent Function Values

Find the *exact* value of each expression.

(a) $\sin 75°$  (b) $\tan \dfrac{7\pi}{12}$  (c) $\sin 40° \cos 160° - \cos 40° \sin 160°$

**Solution**

(a) $\sin 75° = \sin(45° + 30°)$     $75° = 45° + 30°$

$\phantom{\sin 75°} = \sin 45° \cos 30° + \cos 45° \sin 30°$     Sine sum identity

$\phantom{\sin 75°} = \dfrac{\sqrt{2}}{2} \cdot \dfrac{\sqrt{3}}{2} + \dfrac{\sqrt{2}}{2} \cdot \dfrac{1}{2}$     Substitute known values.

$\phantom{\sin 75°} = \dfrac{\sqrt{6} + \sqrt{2}}{4}$     Multiply; add fractions.

*(continued)*

**(b)** $\tan\dfrac{7\pi}{12} = \tan\left(\dfrac{\pi}{3} + \dfrac{\pi}{4}\right)$  $\quad\dfrac{\pi}{3} = \dfrac{4\pi}{12}; \dfrac{\pi}{4} = \dfrac{3\pi}{12}$

$= \dfrac{\tan\frac{\pi}{3} + \tan\frac{\pi}{4}}{1 - \tan\frac{\pi}{3}\tan\frac{\pi}{4}}$  Tangent sum identity

$= \dfrac{\sqrt{3} + 1}{1 - \sqrt{3}\cdot 1}$  Substitute known values.

$= \dfrac{\sqrt{3} + 1}{1 - \sqrt{3}} \cdot \dfrac{1 + \sqrt{3}}{1 + \sqrt{3}}$  Rationalize the denominator.

$= \dfrac{\sqrt{3} + 3 + 1 + \sqrt{3}}{1 - 3}$  Multiply.

$= \dfrac{4 + 2\sqrt{3}}{-2}$  Combine terms.

**Factor first. Then divide out the common factor.**

$= \dfrac{2(2 + \sqrt{3})}{2(-1)}$  Factor out 2.

$= -2 - \sqrt{3}$  Lowest terms

**(c)** $\sin 40° \cos 160° - \cos 40° \sin 160° = \sin(40° - 160°)$  Sine difference identity

$= \sin(-120°)$  Subtract.

$= -\sin 120°$  Negative-angle identity

$= -\dfrac{\sqrt{3}}{2}$

**NOW TRY** Exercises 7, 9, and 13.

---

**EXAMPLE 2** Writing Functions as Expressions Involving Functions of $\theta$

Write each function as an expression involving functions of $\theta$.

**(a)** $\sin(30° + \theta)$  **(b)** $\tan(45° - \theta)$  **(c)** $\sin(180° + \theta)$

**Solution**

**(a)** Using the identity for $\sin(A + B)$,

$\sin(30° + \theta) = \sin 30° \cos\theta + \cos 30° \sin\theta$

$= \dfrac{1}{2}\cos\theta + \dfrac{\sqrt{3}}{2}\sin\theta.$

**(b)** $\tan(45° - \theta) = \dfrac{\tan 45° - \tan\theta}{1 + \tan 45° \tan\theta}$  Tangent difference identity

$= \dfrac{1 - \tan\theta}{1 + \tan\theta}$

**(c)** $\sin(180° + \theta) = \sin 180° \cos\theta + \cos 180° \sin\theta$  Sine sum identity

$= 0 \cdot \cos\theta + (-1)\sin\theta$

$= -\sin\theta$

**NOW TRY** Exercises 27 and 31.

### EXAMPLE 3  Finding Function Values and the Quadrant of A + B

Suppose that $A$ and $B$ are angles in standard position, with $\sin A = \frac{4}{5}$, $\frac{\pi}{2} < A < \pi$, and $\cos B = -\frac{5}{13}$, $\pi < B < \frac{3\pi}{2}$. Find each of the following.

**(a)** $\sin(A + B)$  **(b)** $\tan(A + B)$  **(c)** the quadrant of $A + B$

**Solution**

**(a)** The identity for $\sin(A + B)$ requires $\sin A$, $\cos A$, $\sin B$, and $\cos B$. We are given values of $\sin A$ and $\cos B$. We must find values of $\cos A$ and $\sin B$.

$$\sin^2 A + \cos^2 A = 1 \quad \text{Fundamental identity}$$

$$\frac{16}{25} + \cos^2 A = 1 \quad \sin A = \frac{4}{5}$$

$$\cos^2 A = \frac{9}{25} \quad \text{Subtract } \frac{16}{25}.$$

$$\cos A = -\frac{3}{5} \quad \text{Square root property; since } A \text{ is in quadrant II, } \cos A < 0.$$

In the same way, $\sin B = -\frac{12}{13}$. Now use the formula for $\sin(A + B)$.

$$\sin(A + B) = \frac{4}{5}\left(-\frac{5}{13}\right) + \left(-\frac{3}{5}\right)\left(-\frac{12}{13}\right)$$

$$= -\frac{20}{65} + \frac{36}{65} = \frac{16}{65}$$

**(b)** To find $\tan(A + B)$, first use the values of sine and cosine from part (a), $\sin A = \frac{4}{5}$, $\cos A = -\frac{3}{5}$, $\sin B = -\frac{12}{13}$, and $\cos B = -\frac{5}{13}$, to get $\tan A = -\frac{4}{3}$ and $\tan B = \frac{12}{5}$.

$$\tan(A + B) = \frac{-\frac{4}{3} + \frac{12}{5}}{1 - \left(-\frac{4}{3}\right)\left(\frac{12}{5}\right)} = \frac{\frac{16}{15}}{1 + \frac{48}{15}} = \frac{\frac{16}{15}}{\frac{63}{15}} = \frac{16}{15} \div \frac{63}{15} = \frac{16}{15} \cdot \frac{15}{63} = \frac{16}{63}$$

**(c)** From parts (a) and (b),

$$\sin(A + B) = \frac{16}{65} \quad \text{and} \quad \tan(A + B) = \frac{16}{63},$$

are both positive. Therefore, $A + B$ must be in quadrant I, since it is the only quadrant in which both sine and tangent are positive.

**NOW TRY** Exercise 37.

### EXAMPLE 4  Verifying an Identity Using Sum and Difference Identities

Verify that the equation is an identity.

$$\sin\left(\frac{\pi}{6} + \theta\right) + \cos\left(\frac{\pi}{3} + \theta\right) = \cos\theta$$

**Solution** Work on the left side, using the sum identities for $\sin(A + B)$ and $\cos(A + B)$.

$$\sin\left(\frac{\pi}{6} + \theta\right) + \cos\left(\frac{\pi}{3} + \theta\right)$$

$$= \left(\sin\frac{\pi}{6}\cos\theta + \cos\frac{\pi}{6}\sin\theta\right) \quad \text{Sine sum identity}$$

$$+ \left(\cos\frac{\pi}{3}\cos\theta - \sin\frac{\pi}{3}\sin\theta\right) \quad \text{Cosine sum identity}$$

$$= \left(\frac{1}{2}\cos\theta + \frac{\sqrt{3}}{2}\sin\theta\right) \quad \sin\frac{\pi}{6} = \frac{1}{2};\ \cos\frac{\pi}{6} =$$

$$+ \left(\frac{1}{2}\cos\theta - \frac{\sqrt{3}}{2}\sin\theta\right) \quad \cos\quad\sin$$

$$= \frac{1}{2}\cos\theta + \frac{1}{2}\cos\theta \quad \text{Simplify.}$$

$$= \cos\theta$$

**NOW TRY** Exercise 49.

## 18.4 Exercises

**NOW TRY Exercise**

*Matching* Match each expression in Column I with its value in Column II. *See Example 1.*

| I | II |
|---|---|
| 1. $\sin 15°$ | A. $\dfrac{\sqrt{6} + \sqrt{2}}{4}$ |
| 2. $\sin 105°$ | B. $\dfrac{-\sqrt{6} - \sqrt{2}}{4}$ |
| 3. $\tan 15°$ | C. $\dfrac{\sqrt{6} - \sqrt{2}}{4}$ |
| 4. $\tan 105°$ | D. $2 + \sqrt{3}$ |
| 5. $\sin(-105°)$ | E. $2 - \sqrt{3}$ |
| 6. $\tan(-105°)$ | F. $-2 - \sqrt{3}$ |

*Use identities to find each exact value. See Example 1.*

7. $\sin\dfrac{5\pi}{12}$

8. $\tan\dfrac{5\pi}{12}$

9. $\tan\dfrac{\pi}{12}$

10. $\sin\dfrac{\pi}{12}$

11. $\sin\left(-\dfrac{7\pi}{12}\right)$

12. $\tan\left(-\dfrac{7\pi}{12}\right)$

13. $\sin 76° \cos 31° - \cos 76° \sin 31°$

14. $\sin 40° \cos 50° + \cos 40° \sin 50°$

15. $\dfrac{\tan 80° + \tan 55°}{1 - \tan 80° \tan 55°}$

16. $\dfrac{\tan 80° - \tan(-55°)}{1 + \tan 80° \tan(-55°)}$

17. $\dfrac{\tan 100° + \tan 80°}{1 - \tan 100° \tan 80°}$

18. $\sin 100° \cos 10° - \cos 100° \sin 10°$

19. $\sin\dfrac{\pi}{5}\cos\dfrac{3\pi}{10} + \cos\dfrac{\pi}{5}\sin\dfrac{3\pi}{10}$

20. $\dfrac{\tan\dfrac{5\pi}{12} + \tan\dfrac{\pi}{4}}{1 - \tan\dfrac{5\pi}{12}\tan\dfrac{\pi}{4}}$

*Use identities to write each expression as a single function of x or θ. See Example 2.*

21. $\cos(30° + \theta)$
22. $\cos(45° - \theta)$
23. $\cos(60° + \theta)$
24. $\cos(\theta - 30°)$
25. $\cos\left(\dfrac{3\pi}{4} - x\right)$
26. $\sin(45° + \theta)$
27. $\tan(\theta + 30°)$
28. $\tan\left(\dfrac{\pi}{4} + x\right)$
29. $\sin\left(\dfrac{\pi}{4} + x\right)$
30. $\sin(180° - \theta)$
31. $\sin(270° - \theta)$
32. $\tan(180° + \theta)$
33. $\tan(360° - \theta)$
34. $\sin(\pi + \theta)$
35. $\tan(\pi - \theta)$

36. Show that if $A$, $B$, and $C$ are the angles of a triangle, then $\sin(A + B + C) = 0$.

*Use the given information to find (a) $\sin(s + t)$, (b) $\tan(s + t)$, and (c) the quadrant of $s + t$. See Example 3.*

37. $\cos s = \dfrac{3}{5}$ and $\sin t = \dfrac{5}{13}$, $s$ and $t$ in quadrant I

38. $\cos s = -\dfrac{1}{5}$ and $\sin t = \dfrac{3}{5}$, $s$ and $t$ in quadrant II

39. $\sin s = \dfrac{2}{3}$ and $\sin t = -\dfrac{1}{3}$, $s$ in quadrant II and $t$ in quadrant IV

40. $\sin s = \dfrac{3}{5}$ and $\sin t = -\dfrac{12}{13}$, $s$ in quadrant I and $t$ in quadrant III

41. $\cos s = -\dfrac{8}{17}$ and $\cos t = -\dfrac{3}{5}$, $s$ and $t$ in quadrant III

42. $\cos s = -\dfrac{15}{17}$ and $\sin t = \dfrac{4}{5}$, $s$ in quadrant II and $t$ in quadrant I

*Graph each expression and use the graph to make a conjecture, predicting what might be an identity. Then verify your conjecture algebraically.*

43. $\sin\left(\dfrac{\pi}{2} + \theta\right)$
44. $\sin\left(\dfrac{3\pi}{2} + \theta\right)$
45. $\tan\left(\dfrac{\pi}{2} + \theta\right)$
46. $\dfrac{1 + \tan\theta}{1 - \tan\theta}$

*Verify that each equation is an identity. See Example 4.*

47. $\sin 2x = 2 \sin x \cos x$  (Hint: $\sin 2x = \sin(x + x)$)

48. $\sin(x + y) + \sin(x - y) = 2 \sin x \cos y$

49. $\sin(210° + x) - \cos(120° + x) = 0$

50. $\tan(x - y) - \tan(y - x) = \dfrac{2(\tan x - \tan y)}{1 + \tan x \tan y}$

51. $\dfrac{\cos(\alpha - \beta)}{\cos\alpha \sin\beta} = \tan\alpha + \cot\beta$

52. $\dfrac{\sin(s + t)}{\cos s \cos t} = \tan s + \tan t$

53. $\dfrac{\sin(x - y)}{\sin(x + y)} = \dfrac{\tan x - \tan y}{\tan x + \tan y}$

54. $\dfrac{\sin(x + y)}{\cos(x - y)} = \dfrac{\cot x + \cot y}{1 + \cot x \cot y}$

55. $\dfrac{\sin(s - t)}{\sin t} + \dfrac{\cos(s - t)}{\cos t} = \dfrac{\sin s}{\sin t \cos t}$

56. $\dfrac{\tan(\alpha + \beta) - \tan\beta}{1 + \tan(\alpha + \beta)\tan\beta} = \tan\alpha$

*Solve each problem.*

**57.** If a person bends at the waist with a straight back making an angle of $\theta$ degrees with the horizontal, then the force $F$ exerted on the back muscles can be modeled by the equation

$$F = \frac{0.6W \sin(\theta + 90°)}{\sin 12°},$$

where $W$ is the weight of the person. (*Source:* Metcalf, H., *Topics in Classical Biophysics,* Prentice-Hall, 1980.)

(a) Calculate $F$ when $W = 170$ lb and $\theta = 30°$.
(b) Use an identity to show that $F$ is approximately equal to $2.9W \cos \theta$.
(c) For what value of $\theta$ is $F$ maximum?

**58.** Refer to Exercise 57.

(a) Suppose a 200-lb person bends at the waist so that $\theta = 45°$. Estimate the force exerted on the person's back muscles.
(b) Approximate graphically the value of $\theta$ that results in the back muscles exerting a force of 400 lb.

## 18.5 Double-Angle Identities

**OBJECTIVES**

1. Derive the double-angle identities.
2. Derive and apply multiple-angle identities.

**OBJECTIVE 1** Derive the double-angle identities. When $A = B$ in the identities for the sum of two angles, these identities are called the **double-angle identities.** For example, to derive an expression for $\cos 2A$, we let $B = A$ in the identity $\cos(A + B) = \cos A \cos B - \sin A \sin B$.

$$\cos 2A = \cos(A + A)$$
$$= \cos A \cos A - \sin A \sin A \quad \text{Cosine sum identity}$$
$$\boldsymbol{\cos 2A = \cos^2 A - \sin^2 A}$$

Two other useful forms of this identity can be obtained by substituting either $\cos^2 A = 1 - \sin^2 A$ or $\sin^2 A = 1 - \cos^2 A$. Replace $\cos^2 A$ with the expression $1 - \sin^2 A$ to get

$$\cos 2A = \cos^2 A - \sin^2 A$$
$$= (1 - \sin^2 A) - \sin^2 A \quad \text{Fundamental identity}$$
$$\boldsymbol{\cos 2A = 1 - 2\sin^2 A,}$$

or replace $\sin^2 A$ with $1 - \cos^2 A$ to get

$$\cos 2A = \cos^2 A - \sin^2 A$$
$$= \cos^2 A - (1 - \cos^2 A) \quad \text{Fundamental identity}$$
$$= \cos^2 A - 1 + \cos^2 A$$
$$\boldsymbol{\cos 2A = 2\cos^2 A - 1.}$$

We find sin 2A with the identity $\sin(A + B) = \sin A \cos B + \cos A \sin B$, letting $B = A$.

$$\sin 2A = \sin(A + A)$$
$$= \sin A \cos A + \cos A \sin A \quad \text{Sine sum identity}$$
$$\mathbf{\sin 2A = 2 \sin A \cos A}$$

Using the identity for $\tan(A + B)$, we find $\tan 2A$.

$$\tan 2A = \tan(A + A)$$
$$= \frac{\tan A + \tan A}{1 - \tan A \tan A} \quad \text{Tangent sum identity}$$
$$\mathbf{\tan 2A = \frac{2 \tan A}{1 - \tan^2 A}}$$

**Double-Angle Identities**

$$\mathbf{\cos 2A = \cos^2 A - \sin^2 A} \qquad \mathbf{\cos 2A = 1 - 2 \sin^2 A}$$
$$\mathbf{\cos 2A = 2 \cos^2 A - 1} \qquad \mathbf{\sin 2A = 2 \sin A \cos A}$$
$$\mathbf{\tan 2A = \frac{2 \tan A}{1 - \tan^2 A}}$$

**EXAMPLE 1** Finding Function Values of $2\theta$ Given Information about $\theta$

Given $\cos \theta = \frac{3}{5}$ and $\sin \theta < 0$, find $\sin 2\theta$, $\cos 2\theta$, and $\tan 2\theta$.

**Solution** To find $\sin 2\theta$, we must first find the value of $\sin \theta$.

$$\sin^2 \theta + \left(\frac{3}{5}\right)^2 = 1 \quad \sin^2 \theta + \cos^2 \theta = 1;\ \cos \theta = \tfrac{3}{5}$$

$$\sin^2 \theta = \frac{16}{25} \quad \text{Simplify.}$$

*Pay attention to signs here.*

$$\sin \theta = -\frac{4}{5} \quad \text{Square root property; choose the negative square root since } \sin \theta < 0.$$

Using the double-angle identity for sine,

$$\sin 2\theta = 2 \sin \theta \cos \theta = 2\left(-\frac{4}{5}\right)\left(\frac{3}{5}\right) = -\frac{24}{25}. \quad \sin \theta = -\tfrac{4}{5};\ \cos \theta = \tfrac{3}{5}$$

Now we find $\cos 2\theta$, using the first of the double-angle identities for cosine.

*Any of the three forms may be used.*

$$\cos 2\theta = \cos^2 \theta - \sin^2 \theta = \frac{9}{25} - \frac{16}{25} = -\frac{7}{25}$$

The value of $\tan 2\theta$ can be found in either of two ways. We can use the double-angle identity and the fact that $\tan \theta = \frac{\sin \theta}{\cos \theta} = \frac{-\frac{4}{5}}{\frac{3}{5}} = -\frac{4}{5} \div \frac{3}{5} = -\frac{4}{5} \cdot \frac{5}{3} = -\frac{4}{3}$.

$$\tan 2\theta = \frac{2 \tan \theta}{1 - \tan^2 \theta} = \frac{2\left(-\frac{4}{3}\right)}{1 - \left(-\frac{4}{3}\right)^2} = \frac{-\frac{8}{3}}{-\frac{7}{9}} = \frac{24}{7}$$

*(continued)*

Alternatively, we can find tan $2\theta$ by finding the quotient of sin $2\theta$ and cos $2\theta$.

$$\tan 2\theta = \frac{\sin 2\theta}{\cos 2\theta} = \frac{-\frac{24}{25}}{-\frac{7}{25}} = \frac{24}{7}$$

**NOW TRY** Exercise 15.

**EXAMPLE 2** Finding Function Values of $\theta$ Given Information about $2\theta$

Find the values of the six trigonometric functions of $\theta$ if $\cos 2\theta = \frac{4}{5}$ and $90° < \theta < 180°$.

**Solution** We must obtain a trigonometric function value of $\theta$ alone.

| | |
|---|---|
| $\cos 2\theta = 1 - 2\sin^2 \theta$ | Double-angle identity |
| $\frac{4}{5} = 1 - 2\sin^2 \theta$ | $\cos 2\theta = \frac{4}{5}$ |
| $-\frac{1}{5} = -2\sin^2 \theta$ | Subtract 1. |
| $\frac{1}{10} = \sin^2 \theta$ | Multiply by $-\frac{1}{2}$. |
| $\sin \theta = \sqrt{\frac{1}{10}}$ | Take square roots; choose the positive square root since $\theta$ terminates in quadrant II. |
| $\sin \theta = \frac{\sqrt{1}}{\sqrt{10}} \cdot \frac{\sqrt{10}}{\sqrt{10}} = \frac{\sqrt{10}}{10}$ | Quotient rule; rationalize the denominator. |

Now find values of $\cos \theta$ and $\tan \theta$ by sketching and labeling a right triangle in quadrant II. Since $\sin \theta = \frac{1}{\sqrt{10}}$, the triangle in Figure 5 is labeled accordingly. The Pythagorean theorem is used to find the remaining leg. Now,

$$\cos \theta = \frac{-3}{\sqrt{10}} = -\frac{3\sqrt{10}}{10}, \quad \text{and} \quad \tan \theta = \frac{1}{-3} = -\frac{1}{3}.$$

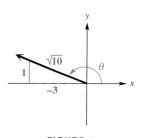

**FIGURE 5**

Find the other three functions using reciprocals.

$$\csc \theta = \frac{1}{\sin \theta} = \sqrt{10}, \quad \sec \theta = \frac{1}{\cos \theta} = -\frac{\sqrt{10}}{3}, \quad \cot \theta = \frac{1}{\tan \theta} = -3$$

**NOW TRY** Exercise 9.

**EXAMPLE 3** Verifying a Double-Angle Identity

Verify that the following equation is an identity.

$$\cot x \sin 2x = 1 + \cos 2x$$

**Solution** We start by working on the left side, using the hint from **Section 18.2** about writing all functions in terms of sine and cosine.

| | |
|---|---|
| $\cot x \sin 2x = \frac{\cos x}{\sin x} \cdot \sin 2x$ | Quotient identity |
| $= \frac{\cos x}{\sin x} (2 \sin x \cos x)$ | Double-angle identity |
| $= 2 \cos^2 x$ | |
| $= 1 + \cos 2x$ | $\cos 2x = 2 \cos^2 x - 1$, so $2 \cos^2 x = 1 + \cos 2x$ |

*Be able to recognize alternative forms of identities.*

**NOW TRY** Exercise 37.

### EXAMPLE 4  Simplifying Expressions Using Double-Angle Identities

Simplify each expression.

**(a)** $\cos^2 7x - \sin^2 7x$ 

**(b)** $\sin 15° \cos 15°$

**Solution**

**(a)** This expression suggests one of the double-angle identities for cosine: $\cos 2A = \cos^2 A - \sin^2 A$. Substituting $7x$ for $A$ gives

$$\cos^2 7x - \sin^2 7x = \cos 2(7x) = \cos 14x.$$

**(b)** If the expression $\sin 15° \cos 15°$ were $2 \sin 15° \cos 15°$, we could apply the identity for $\sin 2A$ directly since

$$\sin 2A = 2 \sin A \cos A.$$

We can still apply the identity with $A = 15°$ by writing the multiplicative identity element 1 as $\frac{1}{2}(2)$.

> This is not an obvious way to begin, but it is indeed valid.

$$\sin 15° \cos 15° = \frac{1}{2}(2) \sin 15° \cos 15° \quad \text{Multiply by 1 in the form } \tfrac{1}{2}(2).$$

$$= \frac{1}{2}(2 \sin 15° \cos 15°) \quad \text{Associative property}$$

$$= \frac{1}{2}\sin(2 \cdot 15°) \quad 2 \sin A \cos A = \sin 2A, \text{ with } A = 15°$$

$$= \frac{1}{2}\sin 30° \quad \text{Multiply.}$$

$$= \frac{1}{2} \cdot \frac{1}{2} \quad \sin 30° = \tfrac{1}{2}$$

$$= \frac{1}{4} \quad \text{Multiply.}$$

**NOW TRY** Exercises 17 and 19.

**OBJECTIVE 2** Derive and apply multiple-angle identities. Identities involving larger multiples of the variable can be derived by repeated use of the double-angle identities and other identities.

### EXAMPLE 5  Deriving a Multiple-Angle Identity

Write $\sin 3x$ in terms of $\sin x$.

**Solution**

> Use the simple fact that $3 = 2 + 1$ here.

$\sin 3x = \sin(2x + x)$

$\quad = \sin 2x \cos x + \cos 2x \sin x$ — Sine sum identity

$\quad = (2 \sin x \cos x)\cos x + (\cos^2 x - \sin^2 x)\sin x$ — Double-angle identities

$\quad = 2 \sin x \cos^2 x + \cos^2 x \sin x - \sin^3 x$ — Multiply.

$\quad = 2 \sin x(1 - \sin^2 x) + (1 - \sin^2 x)\sin x - \sin^3 x$ — $\cos^2 x = 1 - \sin^2 x$

$\quad = 2 \sin x - 2 \sin^3 x + \sin x - \sin^3 x - \sin^3 x$ — Distributive property

$\quad = 3 \sin x - 4 \sin^3 x$ — Combine terms.

**NOW TRY** Exercise 29.

### SECTION 18.5 Double-Angle Identities

#### EXAMPLE 6 Determining Wattage Consumption

If a toaster is plugged into a common household outlet, the wattage consumed is not constant. Instead, it varies at a high frequency according to the model

$$W = \frac{V^2}{R},$$

where $V$ is the voltage and $R$ is a constant that measures the resistance of the toaster in ohms. (*Source:* Bell, D., *Fundamentals of Electric Circuits,* Fourth Edition, Prentice-Hall, 1998.) Graph the wattage $W$ consumed by a typical toaster with $R = 15$ and $V = 163 \sin 120\pi t$ in the window $[0, .05]$ by $[-500, 2000]$. How many oscillations are there?

**Solution** Substituting the given values into the wattage equation gives

$$W = \frac{V^2}{R} = \frac{(163 \sin 120\pi t)^2}{15}.$$

To determine the range of $W$, we note that $\sin 120\pi t$ has maximum value 1, so the expression for $W$ has maximum value $\frac{163^2}{15} \approx 1771$. The minimum value is 0. The graph in Figure 6 shows that there are six oscillations.

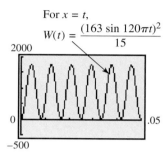

For $x = t$,
$W(t) = \dfrac{(163 \sin 120\pi t)^2}{15}$

**FIGURE 6**

**NOW TRY** Exercise 55.

## 18.5 Exercises

**NOW TRY Exercise**

*Matching* Match each expression in Column I with its value in Column II.

| I | II |
|---|---|
| **1.** $2 \cos^2 15° - 1$ | **A.** $\dfrac{1}{2}$ |
| **2.** $\dfrac{2 \tan 15°}{1 - \tan^2 15°}$ | **B.** $\dfrac{\sqrt{2}}{2}$ |
| **3.** $2 \sin 22.5° \cos 22.5°$ | **C.** $\dfrac{\sqrt{3}}{2}$ |
| **4.** $\cos^2 \dfrac{\pi}{6} - \sin^2 \dfrac{\pi}{6}$ | **D.** $-\sqrt{3}$ |
| **5.** $2 \sin \dfrac{\pi}{3} \cos \dfrac{\pi}{3}$ | **E.** $\dfrac{\sqrt{3}}{3}$ |
| **6.** $\dfrac{2 \tan \dfrac{\pi}{3}}{1 - \tan^2 \dfrac{\pi}{3}}$ | |

*Use identities to find values of the sine and cosine functions for each angle measure. See Examples 1 and 2.*

**7.** $\theta$, given $\cos 2\theta = \dfrac{3}{5}$ and $\theta$ terminates in quadrant I

**8.** $\theta$, given $\cos 2\theta = \dfrac{3}{4}$ and $\theta$ terminates in quadrant III

**9.** $\theta$, given $\cos 2\theta = -\dfrac{5}{12}$ and $\dfrac{\pi}{2} < \theta < \pi$

10. $x$, given $\cos 2x = \dfrac{2}{3}$ and $\dfrac{\pi}{2} < x < \pi$

11. $2\theta$, given $\sin \theta = \dfrac{2}{5}$ and $\cos \theta < 0$

12. $2\theta$, given $\cos \theta = -\dfrac{12}{13}$ and $\sin \theta > 0$

13. $2x$, given $\tan x = 2$ and $\cos x > 0$

14. $2x$, given $\tan x = \dfrac{5}{3}$ and $\sin x < 0$

15. $2\theta$, given $\sin \theta = -\dfrac{\sqrt{5}}{7}$ and $\cos \theta > 0$

16. $2\theta$, given $\cos \theta = \dfrac{\sqrt{3}}{5}$ and $\sin \theta > 0$

*Use an identity to write each expression as a single trigonometric function value or as a single number. See Example 4.*

17. $\cos^2 15° - \sin^2 15°$

18. $\dfrac{2 \tan 15°}{1 - \tan^2 15°}$

19. $1 - 2 \sin^2 15°$

20. $1 - 2 \sin^2 22\dfrac{1}{2}°$

21. $2 \cos^2 67\dfrac{1}{2}° - 1$

22. $\cos^2 \dfrac{\pi}{8} - \dfrac{1}{2}$

23. $\dfrac{\tan 51°}{1 - \tan^2 51°}$

24. $\dfrac{\tan 34°}{2(1 - \tan^2 34°)}$

25. $\dfrac{1}{4} - \dfrac{1}{2} \sin^2 47.1°$

26. $\dfrac{1}{8} \sin 29.5° \cos 29.5°$

27. $\sin^2 \dfrac{2\pi}{5} - \cos^2 \dfrac{2\pi}{5}$

28. $\cos^2 2x - \sin^2 2x$

*Express each function as a trigonometric function of x. See Example 5.*

29. $\sin 4x$

30. $\cos 3x$

31. $\tan 3x$

32. $\cos 4x$

*Graph each expression and use the graph to make a conjecture as to what might be an identity. Then verify your conjecture algebraically.*

33. $\cos^4 x - \sin^4 x$

34. $\dfrac{4 \tan x \cos^2 x - 2 \tan x}{1 - \tan^2 x}$

35. $\dfrac{2 \tan x}{2 - \sec^2 x}$

36. $\dfrac{\cot^2 x - 1}{2 \cot x}$

*Verify that each equation is an identity. See Example 3.*

37. $(\sin x + \cos x)^2 = \sin 2x + 1$

38. $\sec 2x = \dfrac{\sec^2 x + \sec^4 x}{2 + \sec^2 x - \sec^4 x}$

39. $\tan 8\theta - \tan 8\theta \tan^2 4\theta = 2 \tan 4\theta$

40. $\sin 2x = \dfrac{2 \tan x}{1 + \tan^2 x}$

41. $\cos 2\theta = \dfrac{2 - \sec^2 \theta}{\sec^2 \theta}$

42. $-\tan 2\theta = \dfrac{2 \tan \theta}{\sec^2 \theta - 2}$

43. $\sin 4x = 4 \sin x \cos x \cos 2x$

44. $\dfrac{1 + \cos 2x}{\sin 2x} = \cot x$

45. $\dfrac{2 \cos 2\theta}{\sin 2\theta} = \cot \theta - \tan \theta$

46. $\cot 4\theta = \dfrac{1 - \tan^2 2\theta}{2 \tan 2\theta}$

47. $\tan x + \cot x = 2 \csc 2x$

48. $\cos 2x = \dfrac{1 - \tan^2 x}{1 + \tan^2 x}$

49. $1 + \tan x \tan 2x = \sec 2x$

50. $\dfrac{\cot A - \tan A}{\cot A + \tan A} = \cos 2A$

**51.** $\sin 2A \cos 2A = \sin 2A - 4 \sin^3 A \cos A$

**52.** $\sin 4x = 4 \sin x \cos x - 8 \sin^3 x \cos x$

**53.** $\tan(\theta - 45°) + \tan(\theta + 45°) = 2 \tan 2\theta$

**54.** $\cot \theta \tan(\theta + \pi) - \sin(\pi - \theta) \cos\left(\dfrac{\pi}{2} - \theta\right) = \cos^2 \theta$

*Solve each problem.*

**55.** Refer to Example 6. Use an identity to determine values of $a$, $c$, and $\omega$ so that $W = a \cos(\omega t) + c$. Check your answer by graphing both expressions for $W$ on the same coordinate axes.

**56.** Amperage is a measure of the amount of electricity that is moving through a circuit, whereas voltage is a measure of the force pushing the electricity. The wattage $W$ consumed by an electrical device can be determined by calculating the product of the amperage $I$ and voltage $V$. (*Source:* Wilcox, G. and C. Hesselberth, *Electricity for Engineering Technology*, Allyn & Bacon, 1970.)

**(a)** A household circuit has voltage

$$V = 163 \sin 120\pi t$$

when an incandescent lightbulb is turned on with amperage

$$I = 1.23 \sin 120\pi t.$$

Graph the wattage $W = VI$ consumed by the lightbulb in the window $[0, .05]$ by $[-50, 300]$.

**(b)** Determine the maximum and minimum wattages used by the lightbulb.

**(c)** Use identities to determine values for $a$, $c$, and $\omega$ so that $W = a \cos(\omega t) + c$.

**(d)** Check your answer by graphing both expressions for $W$ on the same coordinate axes.

**(e)** Use the graph to estimate the average wattage used by the light. For how many watts (to the nearest integer) do you think this incandescent lightbulb is rated?

# 18.6 Half-Angle Identities

**OBJECTIVES**

1. Derive the half-angle identities.
2. Apply the half-angle identities.

**OBJECTIVE 1** Derive the half-angle identities. From the alternative forms of the identity for $\cos 2A$, we derive three additional identities for $\sin \frac{A}{2}$, $\cos \frac{A}{2}$, and $\tan \frac{A}{2}$. These are known as **half-angle identities.**

To derive the identity for $\sin \frac{A}{2}$, start with the following double-angle identity for cosine and solve for $\sin x$.

$\cos 2x = 1 - 2 \sin^2 x$

$2 \sin^2 x = 1 - \cos 2x$  Add $2 \sin^2 x$; subtract $\cos 2x$.

*Remember both the positive and negative square roots.*

$\sin x = \pm \sqrt{\dfrac{1 - \cos 2x}{2}}$  Divide by 2; square root property.

$\sin \dfrac{A}{2} = \pm \sqrt{\dfrac{1 - \cos A}{2}}$  Let $2x = A$, so $x = \frac{A}{2}$; substitute.

The $\pm$ sign in this identity indicates that the appropriate sign is chosen depending on the quadrant of $\frac{A}{2}$. For example, if $\frac{A}{2}$ is a quadrant III angle, we choose the negative sign since the sine function is negative in quadrant III.

We derive the identity for $\cos \frac{A}{2}$ using the double-angle identity $\cos 2x = 2\cos^2 x - 1$.

$$1 + \cos 2x = 2\cos^2 x \qquad \text{Add 1.}$$

$$\cos^2 x = \frac{1 + \cos 2x}{2} \qquad \text{Rewrite; divide by 2.}$$

$$\cos x = \pm\sqrt{\frac{1 + \cos 2x}{2}} \qquad \text{Take square roots.}$$

$$\cos \frac{A}{2} = \pm\sqrt{\frac{1 + \cos A}{2}} \qquad \text{Replace } x \text{ with } \frac{A}{2}.$$

An identity for $\tan \frac{A}{2}$ comes from the identities for $\sin \frac{A}{2}$ and $\cos \frac{A}{2}$.

$$\tan \frac{A}{2} = \frac{\sin \frac{A}{2}}{\cos \frac{A}{2}} = \frac{\pm\sqrt{\frac{1 - \cos A}{2}}}{\pm\sqrt{\frac{1 + \cos A}{2}}} = \pm\sqrt{\frac{1 - \cos A}{1 + \cos A}}$$

We derive an alternative identity for $\tan \frac{A}{2}$ using double-angle identities.

$$\tan \frac{A}{2} = \frac{\sin \frac{A}{2}}{\cos \frac{A}{2}} = \frac{2 \sin \frac{A}{2} \cos \frac{A}{2}}{2 \cos^2 \frac{A}{2}} \qquad \text{Multiply by } 2 \cos \frac{A}{2} \text{ in numerator and denominator.}$$

$$= \frac{\sin 2\left(\frac{A}{2}\right)}{1 + \cos 2\left(\frac{A}{2}\right)} \qquad \text{Double-angle identities}$$

$$\tan \frac{A}{2} = \frac{\sin A}{1 + \cos A} \qquad \text{Simplify.}$$

From the identity $\tan \frac{A}{2} = \frac{\sin A}{1 + \cos A}$, we can also derive $\tan \frac{A}{2} = \frac{1 - \cos A}{\sin A}$.

### Half-Angle Identities

$$\cos \frac{A}{2} = \pm\sqrt{\frac{1 + \cos A}{2}} \qquad \sin \frac{A}{2} = \pm\sqrt{\frac{1 - \cos A}{2}}$$

$$\tan \frac{A}{2} = \pm\sqrt{\frac{1 - \cos A}{1 + \cos A}} \qquad \tan \frac{A}{2} = \frac{\sin A}{1 + \cos A} \qquad \tan \frac{A}{2} = \frac{1 - \cos A}{\sin A}$$

The final two identities for $\tan \frac{A}{2}$ do not require a sign choice. When using the other half-angle identities, select the plus or minus sign according to the quadrant in which $\frac{A}{2}$ terminates. For example, if an angle $A = 324°$, then $\frac{A}{2} = 162°$, which lies in quadrant II. So when $A = 324°$, $\cos \frac{A}{2}$ and $\tan \frac{A}{2}$ are negative, while $\sin \frac{A}{2}$ is positive.

## SECTION 18.6 Half-Angle Identities

**OBJECTIVE 2** Apply the half-angle identities.

**EXAMPLE 1** Using a Half-Angle Identity to Find an Exact Value

Find the exact value of cos 15° using the half-angle identity for cosine.

**Solution**

$$\cos 15° = \cos \frac{1}{2}(30°) = \sqrt{\frac{1 + \cos 30°}{2}}$$

Choose the positive square root.

$$= \sqrt{\frac{1 + \frac{\sqrt{3}}{2}}{2}} = \sqrt{\frac{\left(1 + \frac{\sqrt{3}}{2}\right) \cdot 2}{2 \cdot 2}} = \frac{\sqrt{2 + \sqrt{3}}}{2}$$

Simplify the radicals.

**NOW TRY** Exercise 11.

**EXAMPLE 2** Using a Half-Angle Identity to Find an Exact Value

Find the exact value of tan 22.5° using the identity $\tan \frac{A}{2} = \frac{\sin A}{1 + \cos A}$.

**Solution** Since $22.5° = \frac{1}{2}(45°)$, replace $A$ with 45°.

$$\tan 22.5° = \tan \frac{45°}{2} = \frac{\sin 45°}{1 + \cos 45°} = \frac{\frac{\sqrt{2}}{2}}{1 + \frac{\sqrt{2}}{2}} = \frac{\frac{\sqrt{2}}{2}}{1 + \frac{\sqrt{2}}{2}} \cdot \frac{2}{2}$$

$$= \frac{\sqrt{2}}{2 + \sqrt{2}} = \frac{\sqrt{2}}{2 + \sqrt{2}} \cdot \frac{2 - \sqrt{2}}{2 - \sqrt{2}} = \frac{2\sqrt{2} - 2}{2} \qquad \text{Rationalize the denominator.}$$

$$= \frac{2(\sqrt{2} - 1)}{2} = \sqrt{2} - 1 \qquad \text{Factor out 2.}$$

Factor first; then divide out the common factor.

**NOW TRY** Exercise 13.

**EXAMPLE 3** Finding Function Values of $\frac{s}{2}$ Given Information about $s$

Given $\cos s = \frac{2}{3}$, with $\frac{3\pi}{2} < s < 2\pi$, find $\cos \frac{s}{2}$, $\sin \frac{s}{2}$, and $\tan \frac{s}{2}$.

**Solution** The angle associated with $\frac{s}{2}$ terminates in quadrant II, since

$$\frac{3\pi}{2} < s < 2\pi$$

and

$$\frac{3\pi}{4} < \frac{s}{2} < \pi. \qquad \text{Divide by 2.}$$

*(continued)*

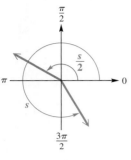

**FIGURE 7**

See Figure 7. In quadrant II, the values of $\cos \frac{s}{2}$ and $\tan \frac{s}{2}$ are negative and the value of $\sin \frac{s}{2}$ is positive. Now use the appropriate half-angle identities and simplify the radicals.

$$\sin \frac{s}{2} = \sqrt{\frac{1 - \frac{2}{3}}{2}} = \sqrt{\frac{1}{6}} = \frac{\sqrt{1}}{\sqrt{6}} \cdot \frac{\sqrt{6}}{\sqrt{6}} = \frac{\sqrt{6}}{6}$$

$$\cos \frac{s}{2} = -\sqrt{\frac{1 + \frac{2}{3}}{2}} = -\sqrt{\frac{5}{6}} = -\frac{\sqrt{5}}{\sqrt{6}} \cdot \frac{\sqrt{6}}{\sqrt{6}} = -\frac{\sqrt{30}}{6}$$

$$\tan \frac{s}{2} = \frac{\sin \frac{s}{2}}{\cos \frac{s}{2}} = \frac{\frac{\sqrt{6}}{6}}{-\frac{\sqrt{30}}{6}} = \frac{\sqrt{6}}{-\sqrt{30}} = -\frac{\sqrt{6}}{\sqrt{30}} \cdot \frac{\sqrt{30}}{\sqrt{30}} = -\frac{\sqrt{180}}{30} = -\frac{6\sqrt{5}}{6 \cdot 5} = -\frac{\sqrt{5}}{5}$$

Notice that it is not necessary to use a half-angle identity for $\tan \frac{s}{2}$ once we find $\sin \frac{s}{2}$ and $\cos \frac{s}{2}$. However, using this identity would provide an excellent check.

**NOW TRY** Exercise 19.

**EXAMPLE 4** Simplifying Expressions Using the Half-Angle Identities

Simplify each expression.

(a) $\pm\sqrt{\dfrac{1 + \cos 12x}{2}}$

(b) $\dfrac{1 - \cos 5\alpha}{\sin 5\alpha}$

**Solution**

(a) This matches part of the identity for $\cos \frac{A}{2}$. Replace $A$ with $12x$ to get

$$\cos \frac{A}{2} = \pm\sqrt{\frac{1 + \cos A}{2}} = \pm\sqrt{\frac{1 + \cos 12x}{2}} = \cos \frac{12x}{2} = \cos 6x.$$

(b) Use the third identity for $\tan \frac{A}{2}$ given earlier with $A = 5\alpha$ to get

$$\frac{1 - \cos 5\alpha}{\sin 5\alpha} = \tan \frac{5\alpha}{2}.$$

**NOW TRY** Exercises 37 and 39.

**EXAMPLE 5** Verifying an Identity

Verify that the following equation is an identity.

$$\left(\sin \frac{x}{2} + \cos \frac{x}{2}\right)^2 = 1 + \sin x$$

**Solution** We work on the more complicated left side.

$$\left(\sin \frac{x}{2} + \cos \frac{x}{2}\right)^2$$

*Remember the term 2ab when squaring a binomial.*

$$= \sin^2 \frac{x}{2} + 2 \sin \frac{x}{2} \cos \frac{x}{2} + \cos^2 \frac{x}{2} \qquad (a + b)^2 = a^2 + 2ab + b^2$$

$$= 1 + 2 \sin \frac{x}{2} \cos \frac{x}{2} \qquad \sin^2 \frac{x}{2} + \cos^2 \frac{x}{2} = 1$$

$$= 1 + \sin 2\left(\frac{x}{2}\right) \qquad 2 \sin \frac{x}{2} \cos \frac{x}{2} = \sin 2\left(\frac{x}{2}\right)$$

$$= 1 + \sin x \qquad \text{Multiply.}$$

**NOW TRY** Exercise 47.

## 18.6 Exercises

**Concept Check** *Determine whether the positive or negative square root should be selected.*

1. $\sin 195° = \pm\sqrt{\dfrac{1 - \cos 390°}{2}}$

2. $\cos 58° = \pm\sqrt{\dfrac{1 + \cos 116°}{2}}$

3. $\tan 225° = \pm\sqrt{\dfrac{1 - \cos 450°}{1 + \cos 450°}}$

4. $\sin(-10°) = \pm\sqrt{\dfrac{1 - \cos(-20°)}{2}}$

*Matching* Match each expression in Column I with its value in Column II. See Examples 1 and 2.

I

5. $\sin 15°$
6. $\tan 15°$
7. $\cos \dfrac{\pi}{8}$
8. $\tan\left(-\dfrac{\pi}{8}\right)$
9. $\tan 67.5°$
10. $\cos 67.5°$

II

A. $2 - \sqrt{3}$

B. $\dfrac{\sqrt{2 - \sqrt{2}}}{2}$

C. $\dfrac{\sqrt{2 - \sqrt{3}}}{2}$

D. $\dfrac{\sqrt{2 + \sqrt{2}}}{2}$

E. $1 - \sqrt{2}$

F. $1 + \sqrt{2}$

*Use a half-angle identity to find each exact value. See Examples 1 and 2.*

11. $\sin 67.5°$
12. $\sin 195°$
13. $\cos 195°$
14. $\tan 195°$
15. $\cos 165°$
16. $\sin 165°$

17. Explain how you could use an identity of this section to find the exact value of $\sin 7.5°$. (*Hint:* $7.5 = \tfrac{1}{2}\left(\tfrac{1}{2}\right)(30)$.)

18. The identity $\tan \dfrac{A}{2} = \pm\sqrt{\dfrac{1 - \cos A}{1 + \cos A}}$ can be used to find $\tan 22.5° = \sqrt{3 - 2\sqrt{2}}$, and the identity $\tan \dfrac{A}{2} = \dfrac{\sin A}{1 + \cos A}$ can be used to find $\tan 22.5° = \sqrt{2} - 1$. Show that these answers are the same without using a calculator. (*Hint:* If $a > 0$ and $b > 0$ and $a^2 = b^2$, then $a = b$.)

*Find each of the following. See Example 3.*

19. $\cos \dfrac{x}{2}$, given $\cos x = \dfrac{1}{4}$, with $0 < x < \dfrac{\pi}{2}$

20. $\sin \dfrac{x}{2}$, given $\cos x = -\dfrac{5}{8}$, with $\dfrac{\pi}{2} < x < \pi$

21. $\tan \dfrac{\theta}{2}$, given $\sin \theta = \dfrac{3}{5}$, with $90° < \theta < 180°$

22. $\cos \dfrac{\theta}{2}$, given $\sin \theta = -\dfrac{4}{5}$, with $180° < \theta < 270°$

23. $\sin \dfrac{x}{2}$, given $\tan x = 2$, with $0 < x < \dfrac{\pi}{2}$

24. $\cos \dfrac{x}{2}$, given $\cot x = -3$, with $\dfrac{\pi}{2} < x < \pi$

25. $\tan \dfrac{\theta}{2}$, given $\tan \theta = \dfrac{\sqrt{7}}{3}$, with $180° < \theta < 270°$

26. $\cot \dfrac{\theta}{2}$, given $\tan \theta = -\dfrac{\sqrt{5}}{2}$, with $90° < \theta < 180°$

27. $\sin \theta$, given $\cos 2\theta = \dfrac{3}{5}$ and $\theta$ terminates in quadrant I

28. $\cos\theta$, given $\cos 2\theta = \dfrac{1}{2}$ and $\theta$ terminates in quadrant II

29. $\cos x$, given $\cos 2x = -\dfrac{5}{12}$, with $\dfrac{\pi}{2} < x < \pi$

30. $\sin x$, given $\cos 2x = \dfrac{2}{3}$, with $\pi < x < \dfrac{3\pi}{2}$

31. *Fill in the Blanks* If $\cos x \approx 0.9682$ and $\sin x = 0.25$, then $\tan \dfrac{x}{2} \approx$ _____.

32. *Fill in the Blanks* If $\cos x = -0.75$ and $\sin x \approx 0.6614$, then $\tan \dfrac{x}{2} \approx$ _____.

*Use an identity to write each expression as a single trigonometric function. See Example 4.*

33. $\sqrt{\dfrac{1 - \cos 40°}{2}}$

34. $\sqrt{\dfrac{1 + \cos 76°}{2}}$

35. $\sqrt{\dfrac{1 - \cos 147°}{1 + \cos 147°}}$

36. $\sqrt{\dfrac{1 + \cos 165°}{1 - \cos 165°}}$

37. $\dfrac{1 - \cos 59.74°}{\sin 59.74°}$

38. $\dfrac{\sin 158.2°}{1 + \cos 158.2°}$

39. $\pm\sqrt{\dfrac{1 + \cos 18x}{2}}$

40. $\pm\sqrt{\dfrac{1 + \cos 20\alpha}{2}}$

41. $\pm\sqrt{\dfrac{1 - \cos 8\theta}{1 + \cos 8\theta}}$

42. $\pm\sqrt{\dfrac{1 - \cos 5A}{1 + \cos 5A}}$

43. $\pm\sqrt{\dfrac{1 + \cos \frac{x}{4}}{2}}$

44. $\pm\sqrt{\dfrac{1 - \cos \frac{3\theta}{5}}{2}}$

*Verify that each equation is an identity. See Example 5.*

45. $\sec^2 \dfrac{x}{2} = \dfrac{2}{1 + \cos x}$

46. $\cot^2 \dfrac{x}{2} = \dfrac{(1 + \cos x)^2}{\sin^2 x}$

47. $\sin^2 \dfrac{x}{2} = \dfrac{\tan x - \sin x}{2 \tan x}$

48. $\dfrac{\sin 2x}{2 \sin x} = \cos^2 \dfrac{x}{2} - \sin^2 \dfrac{x}{2}$

49. $\dfrac{2}{1 + \cos x} - \tan^2 \dfrac{x}{2} = 1$

50. $\tan \dfrac{\theta}{2} = \csc \theta - \cot \theta$

51. $1 - \tan^2 \dfrac{\theta}{2} = \dfrac{2 \cos \theta}{1 + \cos \theta}$

52. $\cos x = \dfrac{1 - \tan^2 \frac{x}{2}}{1 + \tan^2 \frac{x}{2}}$

# 18 SUMMARY

## QUICK REVIEW

| CONCEPTS | EXAMPLES |
|---|---|

### 18.1 Fundamental Identities

**Reciprocal Identities**

$$\cot \theta = \frac{1}{\tan \theta} \qquad \sec \theta = \frac{1}{\cos \theta} \qquad \csc \theta = \frac{1}{\sin \theta}$$

**Quotient Identities**

$$\tan \theta = \frac{\sin \theta}{\cos \theta} \qquad \cot \theta = \frac{\cos \theta}{\sin \theta}$$

**Pythagorean Identities**

$$\sin^2 \theta + \cos^2 \theta = 1 \qquad \tan^2 \theta + 1 = \sec^2 \theta$$
$$1 + \cot^2 \theta = \csc^2 \theta$$

**Negative-Angle Identities**

$$\sin(-\theta) = -\sin \theta \quad \cos(-\theta) = \cos \theta \quad \tan(-\theta) = -\tan \theta$$
$$\csc(-\theta) = -\csc \theta \quad \sec(-\theta) = \sec \theta \quad \cot(-\theta) = -\cot \theta$$

If $\theta$ is in quadrant IV and $\sin \theta = -\frac{3}{5}$, find $\csc \theta$, $\cos \theta$, and $\sin(-\theta)$.

$$\csc \theta = \frac{1}{\sin \theta} = \frac{1}{-\frac{3}{5}} = -\frac{5}{3}$$

$$\sin^2 \theta + \cos^2 \theta = 1$$

$$\left(-\frac{3}{5}\right)^2 + \cos^2 \theta = 1$$

$$\cos^2 \theta = 1 - \frac{9}{25} = \frac{16}{25}$$

$$\cos \theta = +\sqrt{\frac{16}{25}} = \frac{4}{5} \qquad \cos \theta \text{ is positive in quadrant IV.}$$

$$\sin(-\theta) = -\sin \theta = \frac{3}{5}$$

### 18.2 Verifying Trigonometric Identities

See the box titled Hints for Verifying Identities on page 815.

### 18.3 Sum and Difference Identities for Cosine
### 18.4 Sum and Difference Identities for Sine and Tangent

**Cofunction Identities**

$$\cos(90° - \theta) = \sin \theta \qquad \cot(90° - \theta) = \tan \theta$$
$$\sin(90° - \theta) = \cos \theta \qquad \sec(90° - \theta) = \csc \theta$$
$$\tan(90° - \theta) = \cot \theta \qquad \csc(90° - \theta) = \sec \theta$$

**Sum and Difference Identities**

$$\cos(A - B) = \cos A \cos B + \sin A \sin B$$
$$\cos(A + B) = \cos A \cos B - \sin A \sin B$$
$$\sin(A + B) = \sin A \cos B + \cos A \sin B$$
$$\sin(A - B) = \sin A \cos B - \cos A \sin B$$
$$\tan(A + B) = \frac{\tan A + \tan B}{1 - \tan A \tan B}$$
$$\tan(A - B) = \frac{\tan A - \tan B}{1 + \tan A \tan B}$$

Find a value of $\theta$ such that $\tan \theta = \cot 78°$.

$$\tan \theta = \cot 78°$$
$$\cot(90° - \theta) = \cot 78°$$
$$90° - \theta = 78°$$
$$\theta = 12°$$

Find the exact value of $\cos(-15°)$.

$$\cos(-15°) = \cos(30° - 45°)$$
$$= \cos 30° \cos 45° + \sin 30° \sin 45°$$
$$= \frac{\sqrt{3}}{2} \cdot \frac{\sqrt{2}}{2} + \frac{1}{2} \cdot \frac{\sqrt{2}}{2}$$
$$= \frac{\sqrt{6}}{4} + \frac{\sqrt{2}}{4} = \frac{\sqrt{6} + \sqrt{2}}{4}$$

Write $\tan\left(\frac{\pi}{4} + \theta\right)$ in terms of $\tan \theta$.

$$\tan\left(\frac{\pi}{4} + \theta\right) = \frac{\tan \frac{\pi}{4} + \tan \theta}{1 - \tan \frac{\pi}{4} \tan \theta} = \frac{1 + \tan \theta}{1 - \tan \theta} \qquad \tan \frac{\pi}{4} = 1$$

*(continued)*

## CONCEPTS

### 18.5 Double-Angle Identities

**Double-Angle Identities**

$$\cos 2A = \cos^2 A - \sin^2 A \qquad \cos 2A = 1 - 2\sin^2 A$$
$$\cos 2A = 2\cos^2 A - 1 \qquad \sin 2A = 2\sin A \cos A$$
$$\tan 2A = \frac{2\tan A}{1 - \tan^2 A}$$

### 18.6 Half-Angle Identities

**Half-Angle Identities**

$$\cos \frac{A}{2} = \pm\sqrt{\frac{1 + \cos A}{2}} \qquad \sin \frac{A}{2} = \pm\sqrt{\frac{1 - \cos A}{2}}$$

$$\tan \frac{A}{2} = \pm\sqrt{\frac{1 - \cos A}{1 + \cos A}} \qquad \tan \frac{A}{2} = \frac{\sin A}{1 + \cos A}$$

$$\tan \frac{A}{2} = \frac{1 - \cos A}{\sin A}$$

(In the first three identities, the sign is chosen based on the quadrant of $\frac{A}{2}$.)

## EXAMPLES

Given $\cos \theta = -\frac{5}{13}$ and $\sin \theta > 0$, find $\sin 2\theta$.
Sketch a triangle in quadrant II since $\cos \theta < 0$ and $\sin \theta > 0$. Use it to find that $\sin \theta = \frac{12}{13}$.

$$\sin 2\theta = 2 \sin \theta \cos \theta$$
$$= 2\left(\frac{12}{13}\right)\left(-\frac{5}{13}\right) = -\frac{120}{169}$$

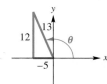

Find the exact value of $\tan 67.5°$.
We choose the last form with $A = 135°$.

$$\tan 67.5° = \tan \frac{135°}{2} = \frac{1 - \cos 135°}{\sin 135°} = \frac{1 - \left(-\frac{\sqrt{2}}{2}\right)}{\frac{\sqrt{2}}{2}}$$

$$= \frac{1 + \frac{\sqrt{2}}{2}}{\frac{\sqrt{2}}{2}} \cdot \frac{2}{2} = \frac{2 + \sqrt{2}}{\sqrt{2}} \quad \text{or} \quad \sqrt{2} + 1$$

Rationalize the denominator; simplify.

# 18 REVIEW EXERCISES

*Matching* For each expression in Column I, choose the expression from Column II that completes an identity.

**I**

1. $\sec x = $ _____  2. $\csc x = $ _____

3. $\tan x = $ _____  4. $\cot x = $ _____

5. $\tan^2 x = $ _____  6. $\sec^2 x = $ _____

**II**

A. $\dfrac{1}{\sin x}$  B. $\dfrac{1}{\cos x}$

C. $\dfrac{\sin x}{\cos x}$  D. $\dfrac{1}{\cot^2 x}$

E. $\dfrac{1}{\cos^2 x}$  F. $\dfrac{\cos x}{\sin x}$

*Use identities to write each expression in terms of $\sin \theta$ and $\cos \theta$, and simplify.*

7. $\sec^2 \theta - \tan^2 \theta$   8. $\dfrac{\cot \theta}{\sec \theta}$   9. $\tan^2 \theta(1 + \cot^2 \theta)$

10. $\csc \theta + \cot \theta$   11. $\tan \theta - \sec \theta \csc \theta$   12. $\csc^2 \theta + \sec^2 \theta$

13. Use the trigonometric identities to find $\sin x$, $\tan x$, and $\cot(-x)$, given $\cos x = \frac{3}{5}$ and $x$ in quadrant IV.

14. Given $\tan x = -\frac{5}{4}$, where $\frac{\pi}{2} < x < \pi$, use the trigonometric identities to find $\cot x$, $\csc x$, and $\sec x$.

15. Find the exact values of the six trigonometric functions of 165°.
16. Find the exact values of sin x, cos x, and tan x, for $x = \frac{\pi}{12}$, using
    (a) difference identities
    (b) half-angle identities.

*Matching* For each expression in Column I, use an identity to choose an expression from Column II with the same value.

| I | | II | |
|---|---|---|---|
| 17. cos 210° | 18. sin 35° | A. sin(−35°) | B. cos 55° |
| 19. tan(−35°) | 20. −sin 35° | C. $\sqrt{\dfrac{1 + \cos 150°}{2}}$ | D. 2 sin 150° cos 150° |
| 21. cos 35° | 22. cos 75° | E. cos 150° cos 60° − sin 150° sin 60° | |
| 23. sin 75° | 24. sin 300° | F. cot(−35°) | G. cos² 150° − sin² 150° |
| 25. cos 300° | 26. cos(−55°) | H. sin 15° cos 60° + cos 15° sin 60° | |
| | | I. cos(−35°) | J. cot 125° |

For each of the following, find sin(x + y), cos(x − y), tan(x + y), and the quadrant of x + y.

27. $\sin x = -\dfrac{3}{5}$, $\cos y = -\dfrac{7}{25}$, x and y in quadrant III

28. $\sin x = \dfrac{3}{5}$, $\cos y = \dfrac{24}{25}$, x in quadrant I, y in quadrant IV

29. $\sin x = -\dfrac{1}{2}$, $\cos y = -\dfrac{2}{5}$, x and y in quadrant III

30. $\sin y = -\dfrac{2}{3}$, $\cos x = -\dfrac{1}{5}$, x in quadrant II, y in quadrant III

31. $\sin x = \dfrac{1}{10}$, $\cos y = \dfrac{4}{5}$, x in quadrant I, y in quadrant IV

32. $\cos x = \dfrac{2}{9}$, $\sin y = -\dfrac{1}{2}$, x in quadrant IV, y in quadrant III

Find sine and cosine of each of the following.

33. θ, given $\cos 2\theta = -\dfrac{3}{4}$, 90° < 2θ < 180°

34. B, given $\cos 2B = \dfrac{1}{8}$, B in quadrant IV

35. 2x, given tan x = 3, sin x < 0

36. 2y, given $\sec y = -\dfrac{5}{3}$, sin y > 0

Find each of the following.

37. $\cos \dfrac{\theta}{2}$, given $\cos \theta = -\dfrac{1}{2}$, with 90° < θ < 180°

38. $\sin \dfrac{A}{2}$, given $\cos A = -\dfrac{3}{4}$, with 90° < A < 180°

39. tan x, given tan 2x = 2, with $\pi < x < \dfrac{3\pi}{2}$

40. sin y, given $\cos 2y = -\dfrac{1}{3}$, with $\dfrac{\pi}{2} < y < \pi$

41. $\tan \dfrac{x}{2}$, given sin x = 0.8, with $0 < x < \dfrac{\pi}{2}$

42. sin 2x, given sin x = 0.6, with $\dfrac{\pi}{2} < x < \pi$

*Verify that each equation is an identity.*

43. $\sin^2 x - \sin^2 y = \cos^2 y - \cos^2 x$

44. $2\cos^3 x - \cos x = \dfrac{\cos^2 x - \sin^2 x}{\sec x}$

45. $\dfrac{\sin^2 x}{2 - 2\cos x} = \cos^2 \dfrac{x}{2}$

46. $\dfrac{\sin 2x}{\sin x} = \dfrac{2}{\sec x}$

47. $2\cos A - \sec A = \cos A - \dfrac{\tan A}{\csc A}$

48. $\dfrac{2\tan B}{\sin 2B} = \sec^2 B$

49. $1 + \tan^2 \alpha = 2\tan \alpha \csc 2\alpha$

50. $\dfrac{2\cot x}{\tan 2x} = \csc^2 x - 2$

51. $\tan \theta \sin 2\theta = 2 - 2\cos^2 \theta$

52. $\csc A \sin 2A - \sec A = \cos 2A \sec A$

53. $2\tan x \csc 2x - \tan^2 x = 1$

54. $2\cos^2 \theta - 1 = \dfrac{1 - \tan^2 \theta}{1 + \tan^2 \theta}$

55. $\tan \theta \cos^2 \theta = \dfrac{2\tan \theta \cos^2 \theta - \tan \theta}{1 - \tan^2 \theta}$

56. $\sec^2 \alpha - 1 = \dfrac{\sec 2\alpha - 1}{\sec 2\alpha + 1}$

57. $\dfrac{\sin^2 x - \cos^2 x}{\csc x} = 2\sin^3 x - \sin x$

58. $\sin^3 \theta = \sin \theta - \cos^2 \theta \sin \theta$

59. $\tan 4\theta = \dfrac{2\tan 2\theta}{2 - \sec^2 2\theta}$

60. $2\cos^2 \dfrac{x}{2} \tan x = \tan x + \sin x$

## 18 TEST

1. If $\cos \theta = \dfrac{24}{25}$ and $\theta$ is in quadrant IV, find the five remaining trigonometric function values of $\theta$.

2. Express $\sec \theta - \sin \theta \tan \theta$ as a single function of $\theta$.

3. Express $\tan^2 x - \sec^2 x$ in terms of $\sin x$ and $\cos x$, and simplify.

4. Find the exact value of $\cos \dfrac{5\pi}{12}$.

5. Express as a function of $x$ alone.
   (a) $\cos(270° - x)$   (b) $\tan(\pi + x)$

6. Use a half-angle identity to find the exact value of $\sin(-22.5°)$.

7. Graph $y = \cot \dfrac{1}{2} x - \cot x$, and make a conjecture as to what might be an identity. Then verify your conjecture algebraically.

8. Given that $\sin A = \dfrac{5}{13}$, $\cos B = -\dfrac{3}{5}$, $A$ is a quadrant I angle, and $B$ is a quadrant II angle, find each of the following.
   (a) $\sin(A + B)$   (b) $\cos(A + B)$   (c) $\tan(A - B)$
   (d) the quadrant of $A + B$

9. Given that $\cos \theta = -\dfrac{3}{5}$ and $\dfrac{\pi}{2} < \theta < \pi$, find each of the following.
   (a) $\cos 2\theta$   (b) $\sin 2\theta$   (c) $\tan 2\theta$   (d) $\cos \dfrac{1}{2}\theta$   (e) $\tan \dfrac{1}{2}\theta$

*Verify each identity.*

10. $\sec^2 B = \dfrac{1}{1 - \sin^2 B}$

11. $\tan^2 x - \sin^2 x = (\tan x \sin x)^2$

12. $\dfrac{\tan x - \cot x}{\tan x + \cot x} = 2\sin^2 x - 1$

13. $\cos 2A = \dfrac{\cot A - \tan A}{\csc A \sec A}$

14. $\dfrac{\sin 2x}{\cos 2x + 1} = \tan x$

# 19

# Inverse Circular Functions and Trigonometric Equations

On the night of October 20, 1955, a new sound exploded from a stage at Brooklyn High School in Cleveland, Ohio. The concert showcased rising stars in the music industry, including Elvis Presley and Bill Haley, and ushered in a new era of popular music—*rock and roll*. The distinctive melodic phrases, or *riffs,* generated by the musicians' electric guitars defined the new genre. (*Source:* Rock and Roll Hall of Fame.)

Sound waves, such as those initiated by musical instruments, travel in sinusoidal patterns that can be graphed as sine or cosine functions and described by trigonometric equations. When sound waves are combined and organized to have rhythm, melody, harmony, and dynamics, the brain interprets them as music.

In Section 19.2, Example 6, and Section 19.3, Example 5, various aspects of music are analyzed using trigonometric equations and graphs.

**19.1** Inverse Circular Functions

**19.2** Trigonometric Equations I

**19.3** Trigonometric Equations II

# 19.1 Inverse Circular Functions

**OBJECTIVES**

1. Understand the concepts of one-to-one and inverse functions.
2. Define and graph the inverse sine function.
3. Define and graph the inverse cosine function.
4. Define and graph the inverse tangent function.
5. Define and graph the remaining inverse circular functions.
6. Find inverse function values.

**OBJECTIVE 1** Understand the concepts of one-to-one and inverse functions. *Recall that for a function f, every element x in the domain corresponds to one and only one element y, or f(x), in the range.* This means the following:

1. If point $(a, b)$ lies on the graph of $f$, then there is no other point on the graph that has $a$ as first coordinate.
2. Other points may have $b$ as second coordinate, however, since the definition of function allows range elements to be used more than once.

If a function is defined so that **each range element is used only once,** then it is called a **one-to-one function.** For example, the function

$$f(x) = x^3 \text{ is a one-to-one function}$$

because every real number has exactly one real cube root. On the other hand,

$$g(x) = x^2 \text{ is not a one-to-one function}$$

because, for example, $g(2) = 4$ and $g(-2) = 4$. There are two domain elements, 2 and $-2$, that correspond to the range element 4.

The **horizontal line test** helps determine graphically whether a function is one-to-one.

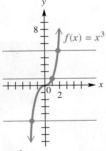

$f(x) = x^3$ is a one-to-one function. It satisfies the conditions of the horizontal line test.

**Horizontal Line Test**

Any horizontal line will intersect the graph of a one-to-one function in at most one point.

This test is applied to the graphs of $f(x) = x^3$ and $g(x) = x^2$ in Figure 1.

By interchanging the components of the ordered pairs of a one-to-one function $f$, we obtain a new set of ordered pairs that satisfies the definition of function. This new function, called the *inverse function,* or *inverse,* of $f$, is symbolized $f^{-1}$.

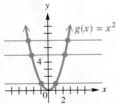

$g(x) = x^2$ is not one-to-one. It does not satisfy the conditions of the horizontal line test.

**FIGURE 1**

**Inverse Function**

The **inverse function** of the one-to-one function $f$ is defined as

$$f^{-1} = \{(y, x) | (x, y) \text{ belongs to } f\}.$$

**CAUTION** *The $-1$ in $f^{-1}(x)$ is not an exponent.* That is,

$$f^{-1}(x) \neq \frac{1}{f(x)}.$$

The following statements summarize our discussion of inverse functions.

### Summary of Inverse Functions

1. In a one-to-one function, each $x$-value corresponds to only one $y$-value and each $y$-value corresponds to only one $x$-value.
2. If a function $f$ is one-to-one, then $f$ has an inverse function $f^{-1}$.
3. The domain of $f$ is the range of $f^{-1}$, and the range of $f$ is the domain of $f^{-1}$.
4. The graphs of $f$ and $f^{-1}$ are reflections of each other across the line $y = x$.
5. To find $f^{-1}(x)$ from $f(x)$, follow these steps.

   **Step 1** Replace $f(x)$ with $y$ and interchange $x$ and $y$.
   **Step 2** Solve for $y$.
   **Step 3** Replace $y$ with $f^{-1}(x)$.

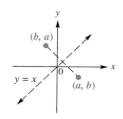

$(b, a)$ is the reflection of $(a, b)$ across the line $y = x$.

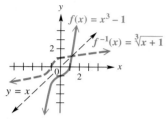

The graph of $f^{-1}$ is the reflection of the graph of $f$ across the line $y = x$.

**FIGURE 2**

Figure 2 illustrates some of these concepts.

We often restrict the domain of a function that is not one-to-one to make it one-to-one, without changing the range. For example, we saw in Figure 1 that $g(x) = x^2$, with its natural domain $(-\infty, \infty)$, is not one-to-one. However, if we restrict its domain to the set of nonnegative numbers $[0, \infty)$, we obtain a new function $f$ that is one-to-one and has the same range as $g$, $[0, \infty)$. See Figure 3.

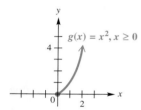

This is a one-to-one function.

**FIGURE 3**

**OBJECTIVE 2** Define and graph the inverse sine function. Refer to the graph of the sine function in Figure 4. Applying the horizontal line test, we see that $y = \sin x$ does not define a one-to-one function. If we restrict the domain to the interval $\left[-\frac{\pi}{2}, \frac{\pi}{2}\right]$, which is the part of the graph in Figure 4 shown in color, this restricted function is one-to-one and has an inverse function. The range of $y = \sin x$ is $[-1, 1]$, so the domain of the inverse function will be $[-1, 1]$, and its range will be $\left[-\frac{\pi}{2}, \frac{\pi}{2}\right]$.

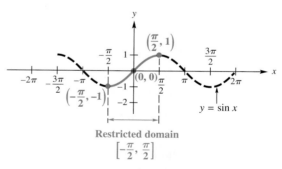

**FIGURE 4**

Reflecting the graph of $y = \sin x$ on the restricted domain, shown in Figure 5(a), across the line $y = x$ gives the graph of the inverse function, shown in Figure 5(b). Some key points are labeled on the graph. The equation of the inverse of $y = \sin x$ is found by interchanging $x$ and $y$ to get

$$x = \sin y.$$

This equation is solved for $y$ by writing

$$y = \sin^{-1} x \quad \text{(read \textbf{``inverse sine of } }x\text{''})}.$$

As Figure 5(b) shows, the domain of $y = \sin^{-1} x$ is $[-1, 1]$, while the restricted domain of $y = \sin x$, $\left[-\frac{\pi}{2}, \frac{\pi}{2}\right]$, is the range of $y = \sin^{-1} x$. An alternative notation for $\sin^{-1} x$ is arcsin $x$.

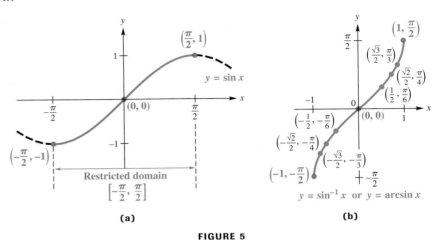

**FIGURE 5**

### Inverse Sine Function

$y = \sin^{-1} x$ or $y = \arcsin x$ means that $x = \sin y$, for $-\frac{\pi}{2} \leq y \leq \frac{\pi}{2}$.

We can think of $y = \sin^{-1} x$ or $y = \arcsin x$ as

*"y is the number in the interval $\left[-\frac{\pi}{2}, \frac{\pi}{2}\right]$ whose sine is x."*

Thus, we can write $y = \sin^{-1} x$ as $\sin y = x$ to evaluate it. We must pay close attention to the domain and range intervals.

---

**EXAMPLE 1** Finding Inverse Sine Values

Find $y$ in each equation.

(a) $y = \arcsin \dfrac{1}{2}$  (b) $y = \sin^{-1}(-1)$  (c) $y = \sin^{-1}(-2)$

**Algebraic Solution**

(a) The graph of the function defined by $y = \arcsin x$ (Figure 5(b)) includes the point $\left(\frac{1}{2}, \frac{\pi}{6}\right)$. Therefore, $\arcsin \frac{1}{2} = \frac{\pi}{6}$.

Alternatively, we can think of $y = \arcsin \frac{1}{2}$ as "y is the number in $\left[-\frac{\pi}{2}, \frac{\pi}{2}\right]$ whose sine is $\frac{1}{2}$." Then we can write the given equation as $\sin y = \frac{1}{2}$. Since $\sin \frac{\pi}{6} = \frac{1}{2}$ and $\frac{\pi}{6}$ is in the range of the arcsine function, $y = \frac{\pi}{6}$.

(b) Writing the equation $y = \sin^{-1}(-1)$ in the form $\sin y = -1$ shows that $y = -\frac{\pi}{2}$. Notice that the point $\left(-1, -\frac{\pi}{2}\right)$ is on the graph of $y = \sin^{-1} x$.

(c) Because $-2$ is not in the domain of the inverse sine function, $\sin^{-1}(-2)$ does not exist.

**Graphing Calculator Solution**

We graph $Y_1 = \sin^{-1} X$ and find the points with X-values $\frac{1}{2} = .5$ and $-1$. For these X-values, Figure 6 shows that $Y = \frac{\pi}{6} \approx .52359878$ and $Y = -\frac{\pi}{2} \approx -1.570796$.

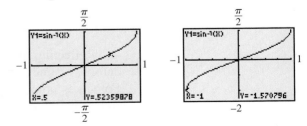

**FIGURE 6**

Since $\sin^{-1}(-2)$ does not exist, a calculator will give an error message for this input.

**NOW TRY** Exercises 13 and 23.

**CAUTION** In Example 1(b), it is tempting to give the value of $\sin^{-1}(-1)$ as $\frac{3\pi}{2}$, since $\sin\frac{3\pi}{2} = -1$. Notice, however, that $\frac{3\pi}{2}$ is not in the range of the inverse sine function. *Be certain that the number given for an inverse function value is in the range of the particular inverse function being considered.*

### Inverse Sine Function  $y = \sin^{-1} x$  or  $y = \arcsin x$

Domain: $[-1, 1]$   Range: $\left[-\frac{\pi}{2}, \frac{\pi}{2}\right]$

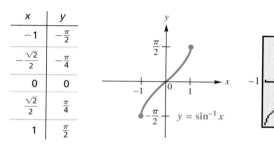

| x | y |
|---|---|
| $-1$ | $-\frac{\pi}{2}$ |
| $-\frac{\sqrt{2}}{2}$ | $-\frac{\pi}{4}$ |
| 0 | 0 |
| $\frac{\sqrt{2}}{2}$ | $\frac{\pi}{4}$ |
| 1 | $\frac{\pi}{2}$ |

**FIGURE 7**

- The inverse sine function is increasing and continuous on its domain $[-1, 1]$.
- Its x-intercept and y-intercept are both (0, 0), the origin.
- Its graph is symmetric with respect to the origin; it is an odd function.

**OBJECTIVE 3** Define and graph the inverse cosine function. The function $y = \cos^{-1} x$ (or $y = \arccos x$) is defined by restricting the domain of the function $y = \cos x$ to the interval $[0, \pi]$ as in Figure 8. This restricted function, which is the part of the graph in Figure 8 shown in color, is one-to-one and has an inverse function. The inverse function, $y = \cos^{-1} x$, is found by interchanging the roles of x and y. Reflecting the graph of $y = \cos x$ across the line $y = x$ gives the graph of the inverse function shown in Figure 9. Again, some key points are shown on the graph.

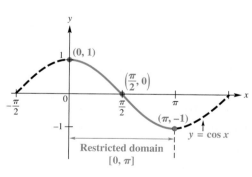

**FIGURE 8**

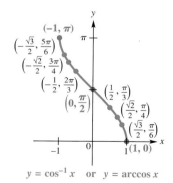

**FIGURE 9**

### Inverse Cosine Function

$y = \cos^{-1} x$ or $y = \arccos x$ means that $x = \cos y$, for $0 \leq y \leq \pi$.

We can think of $y = \cos^{-1} x$ or $y = \arccos x$ as
"*y is the number in the interval* $[0, \pi]$ *whose cosine is x.*"

## EXAMPLE 2 Finding Inverse Cosine Values

Find $y$ in each equation.

**(a)** $y = \arccos 1$ 

**(b)** $y = \cos^{-1}\left(-\dfrac{\sqrt{2}}{2}\right)$

### Solution

**(a)** Since the point $(1, 0)$ lies on the graph of $y = \arccos x$ in Figure 9, the value of $y$ is 0. Alternatively, we can think of $y = \arccos 1$ as

"$y$ is the number in $[0, \pi]$ whose cosine is 1," or $\cos y = 1$.

Thus, $y = 0$, since $\cos 0 = 1$ and 0 is in the range of the arccosine function.

**(b)** We must find the value of $y$ that satisfies $\cos y = -\dfrac{\sqrt{2}}{2}$, where $y$ is in the interval $[0, \pi]$, the range of the function $y = \cos^{-1} x$. The only value for $y$ that satisfies these conditions is $\dfrac{3\pi}{4}$. Again, this can be verified from the graph in Figure 9.

**NOW TRY** Exercises 15 and 25.

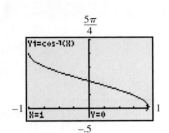

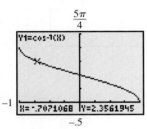

These screens support the results of Example 2, since
$-\dfrac{\sqrt{2}}{2} \approx -0.7071068$ and
$\dfrac{3\pi}{4} \approx 2.3561945$.

Our observations about the inverse cosine function lead to the following generalizations.

---

### Inverse Cosine Function  $y = \cos^{-1} x$  or  $y = \arccos x$

**Domain:** $[-1, 1]$    **Range:** $[0, \pi]$

| x | y |
|---|---|
| $-1$ | $\pi$ |
| $-\dfrac{\sqrt{2}}{2}$ | $\dfrac{3\pi}{4}$ |
| $0$ | $\dfrac{\pi}{2}$ |
| $\dfrac{\sqrt{2}}{2}$ | $\dfrac{\pi}{4}$ |
| $1$ | $0$ |

**FIGURE 10**

- The inverse cosine function is decreasing and continuous on its domain $[-1, 1]$.
- Its $x$-intercept is $(1, 0)$, and its $y$-intercept is $\left(0, \dfrac{\pi}{2}\right)$.
- Its graph is neither symmetric with respect to the $y$-axis nor the origin.

---

**OBJECTIVE 4** Define and graph the inverse tangent function. Restricting the domain of the function $y = \tan x$ to the open interval $\left(-\dfrac{\pi}{2}, \dfrac{\pi}{2}\right)$ yields a one-to-one function. By interchanging the roles of $x$ and $y$, we obtain the inverse tangent function given by $y = \tan^{-1} x$ or $y = \arctan x$. Figure 11 shows the graph of the restricted tangent function. Figure 12 gives the graph of $y = \tan^{-1} x$.

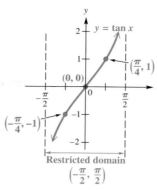

FIGURE 11

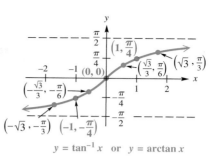

FIGURE 12

### Inverse Tangent Function

$y = \tan^{-1} x$ or $y = \arctan x$ means that $x = \tan y$, for $-\frac{\pi}{2} < y < \frac{\pi}{2}$.

### Inverse Tangent Function $y = \tan^{-1} x$ or $y = \arctan x$

Domain: $(-\infty, \infty)$   Range: $\left(-\frac{\pi}{2}, \frac{\pi}{2}\right)$

| x | y |
|---|---|
| $-1$ | $-\frac{\pi}{4}$ |
| $-\frac{\sqrt{3}}{3}$ | $-\frac{\pi}{6}$ |
| 0 | 0 |
| $\frac{\sqrt{3}}{3}$ | $\frac{\pi}{6}$ |
| 1 | $\frac{\pi}{4}$ |

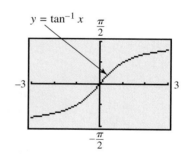

FIGURE 13

- The inverse tangent function is increasing and continuous on its domain $(-\infty, \infty)$.
- Its x-intercept and y-intercept are both $(0, 0)$, the origin.
- Its graph is symmetric with respect to the origin; it is an odd function.
- The lines $y = \frac{\pi}{2}$ and $y = -\frac{\pi}{2}$ are horizontal asymptotes.

**OBJECTIVE 5** Define and graph the remaining inverse circular functions. The remaining three inverse trigonometric functions are defined similarly; their graphs are shown in Figure 14.

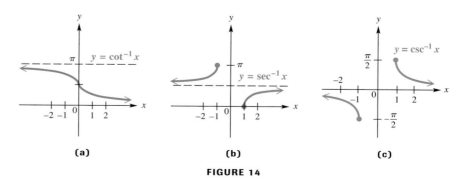

(a)    (b)    (c)

FIGURE 14

> **Inverse Cotangent, Secant, and Cosecant Functions***
>
> $y = \cot^{-1} x$ or $y = \text{arccot } x$ means that $x = \cot y$, for $0 < y < \pi$.
>
> $y = \sec^{-1} x$ or $y = \text{arcsec } x$ means that $x = \sec y$, for $0 \le y \le \pi, y \ne \frac{\pi}{2}$.
>
> $y = \csc^{-1} x$ or $y = \text{arccsc } x$ means that $x = \csc y$, for $-\frac{\pi}{2} \le y \le \frac{\pi}{2}, y \ne 0$.

The table gives all six inverse trigonometric functions with their domains and ranges.

| Inverse Function | Domain | Range Interval | Quadrants of the Unit Circle |
|---|---|---|---|
| $y = \sin^{-1} x$ | $[-1, 1]$ | $\left[-\frac{\pi}{2}, \frac{\pi}{2}\right]$ | I and IV |
| $y = \cos^{-1} x$ | $[-1, 1]$ | $[0, \pi]$ | I and II |
| $y = \tan^{-1} x$ | $(-\infty, \infty)$ | $\left(-\frac{\pi}{2}, \frac{\pi}{2}\right)$ | I and IV |
| $y = \cot^{-1} x$ | $(-\infty, \infty)$ | $(0, \pi)$ | I and II |
| $y = \sec^{-1} x$ | $(-\infty, -1] \cup [1, \infty)$ | $[0, \pi], y \ne \frac{\pi}{2}$ | I and II |
| $y = \csc^{-1} x$ | $(-\infty, -1] \cup [1, \infty)$ | $\left[-\frac{\pi}{2}, \frac{\pi}{2}\right], y \ne 0$ | I and IV |

**OBJECTIVE 6** Find inverse function values. The inverse circular functions are formally defined with real number ranges. However, there are times when it may be convenient to find degree-measured angles equivalent to these real number values. It is also often convenient to think in terms of the unit circle and choose the inverse function values based on the quadrants given in the preceding table.

**EXAMPLE 3** Finding Inverse Function Values (Degree-Measured Angles)

Find the *degree measure* of $\theta$ in the following.

**(a)** $\theta = \arctan 1$  **(b)** $\theta = \sec^{-1} 2$

**Solution**

**(a)** Here $\theta$ must be in $(-90°, 90°)$, but since 1 is positive, $\theta$ must be in quadrant I. The alternative statement, $\tan \theta = 1$, leads to $\theta = 45°$.

**(b)** Write the equation as $\sec \theta = 2$. For $\sec^{-1} x$, $\theta$ is in quadrant I or II. Because 2 is positive, $\theta$ is in quadrant I and $\theta = 60°$, since $\sec 60° = 2$. Note that $60°$ (the degree equivalent of $\frac{\pi}{3}$) is in the range of the inverse secant function.

**NOW TRY** Exercises 35 and 41.

The inverse trigonometric function keys on a calculator give results in the proper quadrant for the inverse sine, inverse cosine, and inverse tangent functions, according to the definitions of these functions. For example, on a calculator, in degrees, $\sin^{-1} 0.5 = 30°$, $\sin^{-1}(-0.5) = -30°$, $\tan^{-1}(-1) = -45°$, and $\cos^{-1}(-0.5) = 120°$.

---

*The inverse secant and inverse cosecant functions are sometimes defined with different ranges. We use intervals that match their reciprocal functions (except for one missing point).

Finding $\cot^{-1} x$, $\sec^{-1} x$, and $\csc^{-1} x$ with a calculator is not as straightforward, because these functions must be expressed in terms of $\tan^{-1} x$, $\cos^{-1} x$, and $\sin^{-1} x$, respectively. If $y = \sec^{-1} x$, for example, then $\sec y = x$, which must be written as a cosine function as follows:

$$\text{If } \sec y = x, \text{ then } \frac{1}{\cos y} = x \quad \text{or} \quad \cos y = \frac{1}{x}, \quad \text{and} \quad y = \cos^{-1}\frac{1}{x}.$$

In summary, to find $\sec^{-1} x$, we find $\cos^{-1}\frac{1}{x}$. Similar statements apply to $\csc^{-1} x$ and $\cot^{-1} x$. There is one additional consideration with $\cot^{-1} x$. Since we take the inverse tangent of the reciprocal to find inverse cotangent, the calculator gives values of inverse cotangent with the same range as inverse tangent, $\left(-\frac{\pi}{2}, \frac{\pi}{2}\right)$, which is not the correct range for inverse cotangent. For inverse cotangent, the proper range must be considered and the results adjusted accordingly.

**EXAMPLE 4** Finding Inverse Function Values with a Calculator

(a) Find $y$ in radians if $y = \csc^{-1}(-3)$.
(b) Find $\theta$ in degrees if $\theta = \text{arccot}(-0.3541)$.

**Solution**

(a) With the calculator in radian mode, enter $\csc^{-1}(-3)$ as $\sin^{-1}\left(-\frac{1}{3}\right)$ to get $y \approx -0.3398369095$. See Figure 15.

(b) Now set the calculator to degree mode. A calculator gives the inverse tangent value of a negative number as a quadrant IV angle. The restriction on the range of arccotangent implies that $\theta$ must be in quadrant II, so enter

$$\text{arccot}(-0.3541) \quad \text{as} \quad \tan^{-1}\left(\frac{1}{-0.3541}\right) + 180°.$$

As shown in Figure 15, $\theta \approx 109.4990544°$.

```
sin⁻¹(1/-3)
        -.3398369095
tan⁻¹(1/-.3541)+1
80
         109.4990544
```

**FIGURE 15**

**NOW TRY** Exercises 51 and 59.

**CAUTION** *Be careful when using your calculator to evaluate the inverse cotangent of a negative quantity.* To do this, we must enter the inverse tangent of the *reciprocal* of the negative quantity, which returns an angle in quadrant IV. Since inverse cotangent is negative in quadrant II, adjust your calculator result by adding 180° or $\pi$ accordingly.

**EXAMPLE 5** Finding Function Values Using Definitions of the Trigonometric Functions

Evaluate each expression without using a calculator.

(a) $\sin\left(\tan^{-1}\frac{3}{2}\right)$ 

(b) $\tan\left(\cos^{-1}\left(-\frac{5}{13}\right)\right)$

*(continued)*

### Solution

(a) Let $\theta = \tan^{-1}\frac{3}{2}$, so $\tan\theta = \frac{3}{2}$. The inverse tangent function yields values only in quadrants I and IV, and since $\frac{3}{2}$ is positive, $\theta$ is in quadrant I. Sketch $\theta$ in quadrant I, and label a triangle, as shown in Figure 16. By the Pythagorean theorem, the hypotenuse is $\sqrt{13}$. The value of sine is the quotient of the side opposite and the hypotenuse, so

$$\sin\left(\tan^{-1}\frac{3}{2}\right) = \sin\theta = \frac{3}{\sqrt{13}} = \frac{3}{\sqrt{13}}\cdot\frac{\sqrt{13}}{\sqrt{13}} = \frac{3\sqrt{13}}{13}.$$

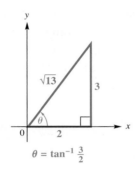

FIGURE 16       FIGURE 17

(b) Let $A = \cos^{-1}\left(-\frac{5}{13}\right)$. Then, $\cos A = -\frac{5}{13}$. Since $\cos^{-1}x$ for a negative value of $x$ is in quadrant II, sketch $A$ in quadrant II, as shown in Figure 17.

$$\tan\left(\cos^{-1}\left(-\frac{5}{13}\right)\right) = \tan A = -\frac{12}{5}$$

**NOW TRY** Exercises 73 and 75.

---

**EXAMPLE 6** Finding Function Values Using Identities

Evaluate each expression without using a calculator.

(a) $\cos\left(\arctan\sqrt{3} + \arcsin\frac{1}{3}\right)$      (b) $\tan\left(2\arcsin\frac{2}{5}\right)$

### Solution

(a) Let $A = \arctan\sqrt{3}$ and $B = \arcsin\frac{1}{3}$, so $\tan A = \sqrt{3}$ and $\sin B = \frac{1}{3}$. Sketch both $A$ and $B$ in quadrant I, as shown in Figure 18. Now, use the cosine sum identity.

$$\cos(A+B) = \cos A \cos B - \sin A \sin B$$

$$\cos\left(\arctan\sqrt{3} + \arcsin\frac{1}{3}\right) = \cos(\arctan\sqrt{3})\cos\left(\arcsin\frac{1}{3}\right)$$

$$- \sin(\arctan\sqrt{3})\sin\left(\arcsin\frac{1}{3}\right) \quad (1)$$

From Figure 18,

$$\cos(\arctan\sqrt{3}) = \cos A = \frac{1}{2}, \quad \cos\left(\arcsin\frac{1}{3}\right) = \cos B = \frac{2\sqrt{2}}{3},$$

$$\sin(\arctan\sqrt{3}) = \sin A = \frac{\sqrt{3}}{2}, \quad \sin\left(\arcsin\frac{1}{3}\right) = \sin B = \frac{1}{3}.$$

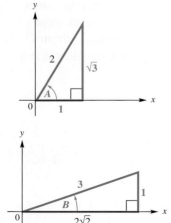

FIGURE 18

*(continued)*

Substitute these values into equation (1) to get

$$\cos\left(\arctan\sqrt{3} + \arcsin\frac{1}{3}\right) = \frac{1}{2} \cdot \frac{2\sqrt{2}}{3} - \frac{\sqrt{3}}{2} \cdot \frac{1}{3} = \frac{2\sqrt{2} - \sqrt{3}}{6}.$$

**(b)** Let $\arcsin\frac{2}{5} = B$. Then, from the double-angle tangent identity,

$$\tan\left(2\arcsin\frac{2}{5}\right) = \tan 2B = \frac{2 \tan B}{1 - \tan^2 B}.$$

Since $\arcsin\frac{2}{5} = B$, $\sin B = \frac{2}{5}$. Sketch a triangle in quadrant I, find the length of the third side, and then find $\tan B$. From the triangle in Figure 19, $\tan B = \frac{2}{\sqrt{21}}$, and

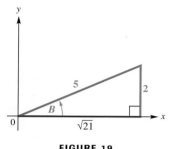

**FIGURE 19**

$$\tan\left(2\arcsin\frac{2}{5}\right) = \frac{2\left(\frac{2}{\sqrt{21}}\right)}{1 - \left(\frac{2}{\sqrt{21}}\right)^2} = \frac{\frac{4}{\sqrt{21}}}{1 - \frac{4}{21}} = \frac{\frac{4}{\sqrt{21}}}{\frac{17}{21}} = \frac{4}{\sqrt{21}} \cdot \frac{\sqrt{21}}{\sqrt{21}} = \frac{\frac{4\sqrt{21}}{21}}{\frac{17}{21}} = \frac{4\sqrt{21}}{17}.$$

> Be careful simplifying the complex fraction.

**NOW TRY** Exercises 77 and 85.

While the work shown in Examples 5 and 6 does not rely on a calculator, we can support our algebraic work with one. By entering $\cos\left(\arctan\sqrt{3} + \arcsin\frac{1}{3}\right)$ from Example 6(a) into a calculator, we get the approximation 0.1827293862, the same approximation as when we enter $\frac{2\sqrt{2} - \sqrt{3}}{6}$ (the exact value we obtained algebraically). Similarly, we obtain the same approximation when we evaluate $\tan\left(2\arcsin\frac{2}{5}\right)$ and $\frac{4\sqrt{21}}{17}$, supporting our answer in Example 6(b).

### EXAMPLE 7 Writing Function Values in Terms of $u$

Write each trigonometric expression as an algebraic expression in $u$.

**(a)** $\sin(\tan^{-1} u)$          **(b)** $\cos(2\sin^{-1} u)$

**Solution**

**(a)** Let $\theta = \tan^{-1} u$, so $\tan \theta = u$. Here, $u$ may be positive or negative. Since $-\frac{\pi}{2} < \tan^{-1} u < \frac{\pi}{2}$, sketch $\theta$ in quadrants I and IV and label two triangles, as shown in Figure 20. Since sine is given by the quotient of the side opposite and the hypotenuse,

$$\sin(\tan^{-1} u) = \sin \theta = \frac{u}{\sqrt{u^2 + 1}} = \frac{u}{\sqrt{u^2 + 1}} \cdot \frac{\sqrt{u^2 + 1}}{\sqrt{u^2 + 1}} = \frac{u\sqrt{u^2 + 1}}{u^2 + 1}.$$

> Rationalize the denominator.

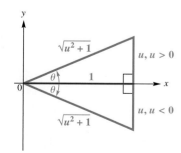

**FIGURE 20**

The result is positive when $u$ is positive and negative when $u$ is negative.

**(b)** Let $\theta = \sin^{-1} u$, so $\sin \theta = u$. To find $\cos 2\theta$, use the double-angle identity $\cos 2\theta = 1 - 2\sin^2 \theta$.

$$\cos(2\sin^{-1} u) = \cos 2\theta = 1 - 2\sin^2 \theta = 1 - 2u^2$$

**NOW TRY** Exercises 93 and 97.

## EXAMPLE 8  Finding the Optimal Angle of Elevation of a Shot Put

The optimal angle of elevation $\theta$ a shot-putter should aim for to throw the greatest distance depends on the velocity $v$ of the throw and the initial height $h$ of the shot. See Figure 21. One model for $\theta$ that achieves this greatest distance is

$$\theta = \arcsin\left(\sqrt{\frac{v^2}{2v^2 + 64h}}\right).$$

(*Source:* Townend, M. S., *Mathematics in Sport,* Chichester, Ellis Horwood Limited, 1984.)

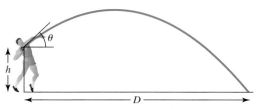

**FIGURE 21**

Suppose a shot-putter can consistently throw the steel ball with $h = 6.6$ ft and $v = 42$ ft per sec. At what angle should he throw the ball to maximize distance?

**Solution**  To find this angle, substitute and use a calculator in degree mode.

$$\theta = \arcsin\left(\sqrt{\frac{42^2}{2(42^2) + 64(6.6)}}\right) \approx 42° \qquad h = 6.6, v = 42$$

**NOW TRY** Exercise 103.

## 19.1 Exercises

**NOW TRY Exercise**

*Fill in the Blanks*  Complete each statement.

1. For a function to have an inverse, it must be _____.
2. The domain of $y = \arcsin x$ equals the _____ of $y = \sin x$.
3. The range of $y = \cos^{-1} x$ equals the _____ of $y = \cos x$.
4. The point $\left(\frac{\pi}{4}, 1\right)$ lies on the graph of $y = \tan x$. Therefore, the point _____ lies on the graph of _____.
5. If a function $f$ has an inverse and $f(\pi) = -1$, then $f^{-1}(-1) = $ _____.
6. How can the graph of $f^{-1}$ be sketched if the graph of $f$ is known?

*Concept Check*  In Exercises 7–10, write short answers.

7. Consider the inverse sine function, defined by $y = \sin^{-1} x$ or $y = \arcsin x$.
   - **(a)** What is its domain?
   - **(b)** What is its range?
   - **(c)** Is this function increasing or decreasing?
   - **(d)** Why is $\arcsin(-2)$ not defined?

8. Consider the inverse cosine function, defined by $y = \cos^{-1} x$ or $y = \arccos x$.
   - **(a)** What is its domain?
   - **(b)** What is its range?
   - **(c)** Is this function increasing or decreasing?
   - **(d)** $\arccos\left(-\frac{1}{2}\right) = \frac{2\pi}{3}$. Why is $\arccos\left(-\frac{1}{2}\right)$ not equal to $-\frac{4\pi}{3}$?

9. Consider the inverse tangent function, defined by $y = \tan^{-1} x$ or $y = \arctan x$.
   (a) What is its domain?  (b) What is its range?
   (c) Is this function increasing or decreasing?
   (d) Is there any real number $x$ for which $\arctan x$ is not defined? If so, what is it (or what are they)?

10. Give the domain and range of the three other inverse trigonometric functions, as defined in this section.
    (a) inverse cosecant function  (b) inverse secant function
    (c) inverse cotangent function

11. *Concept Check* Is $\sec^{-1} a$ calculated as $\cos^{-1} \frac{1}{a}$ or as $\frac{1}{\cos^{-1} a}$?

12. *Concept Check* For positive values of $a$, $\cot^{-1} a$ is calculated as $\tan^{-1} \frac{1}{a}$. How is $\cot^{-1} a$ calculated for negative values of $a$?

*Find the exact value of each real number y. Do not use a calculator. See Examples 1 and 2.*

13. $y = \sin^{-1} 0$
14. $y = \tan^{-1} 1$
15. $y = \cos^{-1}(-1)$
16. $y = \arctan(-1)$
17. $y = \sin^{-1}(-1)$
18. $y = \cos^{-1} \frac{1}{2}$
19. $y = \arctan 0$
20. $y = \arcsin\left(-\frac{\sqrt{3}}{2}\right)$
21. $y = \arccos 0$
22. $y = \tan^{-1}(-1)$
23. $y = \sin^{-1} \frac{\sqrt{2}}{2}$
24. $y = \cos^{-1}\left(-\frac{1}{2}\right)$
25. $y = \arccos\left(-\frac{\sqrt{3}}{2}\right)$
26. $y = \arcsin\left(-\frac{\sqrt{2}}{2}\right)$
27. $y = \cot^{-1}(-1)$
28. $y = \sec^{-1}(-\sqrt{2})$
29. $y = \csc^{-1}(-2)$
30. $y = \text{arccot}(-\sqrt{3})$
31. $y = \text{arcsec} \frac{2\sqrt{3}}{3}$
32. $y = \csc^{-1} \sqrt{2}$
33. $y = \sec^{-1} 1$

34. *Concept Check* Is there a value for $y$ such that $y = \sec^{-1} 0$?

*Give the degree measure of $\theta$. Do not use a calculator. See Example 3.*

35. $\theta = \arctan(-1)$
36. $\theta = \arccos\left(-\frac{1}{2}\right)$
37. $\theta = \arcsin\left(-\frac{\sqrt{3}}{2}\right)$
38. $\theta = \arcsin\left(-\frac{\sqrt{2}}{2}\right)$
39. $\theta = \cot^{-1}\left(-\frac{\sqrt{3}}{3}\right)$
40. $\theta = \csc^{-1}(-2)$
41. $\theta = \sec^{-1}(-2)$
42. $\theta = \csc^{-1}(-1)$
43. $\theta = \tan^{-1} \sqrt{3}$
44. $\theta = \cot^{-1} \frac{\sqrt{3}}{3}$
45. $\theta = \sin^{-1} 2$
46. $\theta = \cos^{-1}(-2)$

*Use a calculator to give each value in decimal degrees. See Example 4.*

47. $\theta = \sin^{-1}(-0.13349122)$
48. $\theta = \cos^{-1}(-0.13348816)$
49. $\theta = \arccos(-0.39876459)$
50. $\theta = \arcsin 0.77900016$
51. $\theta = \csc^{-1} 1.9422833$
52. $\theta = \cot^{-1} 1.7670492$
53. $\theta = \cot^{-1}(-0.60724226)$
54. $\theta = \cot^{-1}(-2.7733744)$
55. $\theta = \tan^{-1}(-7.7828641)$
56. $\theta = \sec^{-1}(-5.1180378)$

Use a calculator to give each real number value. (Be sure the calculator is in radian mode.) See Example 4.

57. $y = \arctan 1.1111111$
58. $y = \arcsin 0.81926439$
59. $y = \cot^{-1}(-0.92170128)$
60. $y = \sec^{-1}(-1.2871684)$
61. $y = \arcsin 0.92837781$
62. $y = \arccos 0.44624593$
63. $y = \cos^{-1}(-0.32647891)$
64. $y = \sec^{-1} 4.7963825$
65. $y = \cot^{-1}(-36.874610)$
66. $y = \cot^{-1}(1.0036571)$

The screen here shows how to define the inverse secant, cosecant, and cotangent functions in order to graph them using a TI-83/84 Plus graphing calculator.

Use this information to graph each inverse circular function and compare your graphs to those in Figure 14.

67. $y = \sec^{-1} x$
68. $y = \csc^{-1} x$
69. $y = \cot^{-1} x$

Graph each inverse circular function by hand.

70. $y = \text{arccsc } 2x$
71. $y = \text{arcsec } \dfrac{1}{2}x$
72. $y = 2 \cot^{-1} x$

Give the exact value of each expression without using a calculator. See Examples 5 and 6.

73. $\tan\left(\arccos \dfrac{3}{4}\right)$
74. $\sin\left(\arccos \dfrac{1}{4}\right)$
75. $\cos(\tan^{-1}(-2))$
76. $\sec\left(\sin^{-1}\left(-\dfrac{1}{5}\right)\right)$
77. $\sin\left(2 \tan^{-1} \dfrac{12}{5}\right)$
78. $\cos\left(2 \sin^{-1} \dfrac{1}{4}\right)$
79. $\cos\left(2 \arctan \dfrac{4}{3}\right)$
80. $\tan\left(2 \cos^{-1} \dfrac{1}{4}\right)$
81. $\sin\left(2 \cos^{-1} \dfrac{1}{5}\right)$
82. $\cos(2 \tan^{-1}(-2))$
83. $\sec(\sec^{-1} 2)$
84. $\csc(\csc^{-1} \sqrt{2})$
85. $\cos\left(\tan^{-1} \dfrac{5}{12} - \tan^{-1} \dfrac{3}{4}\right)$
86. $\cos\left(\sin^{-1} \dfrac{3}{5} + \cos^{-1} \dfrac{5}{13}\right)$
87. $\sin\left(\sin^{-1} \dfrac{1}{2} + \tan^{-1}(-3)\right)$
88. $\tan\left(\cos^{-1} \dfrac{\sqrt{3}}{2} - \sin^{-1}\left(-\dfrac{3}{5}\right)\right)$

Use a calculator to find each value. Give answers as real numbers.

89. $\cos(\tan^{-1} 0.5)$
90. $\sin(\cos^{-1} 0.25)$
91. $\tan(\arcsin 0.12251014)$
92. $\cot(\arccos 0.58236841)$

Write each expression as an algebraic (nontrigonometric) expression in $u$, $u > 0$. See Example 7.

93. $\sin(\arccos u)$
94. $\tan(\arccos u)$
95. $\cos(\arcsin u)$
96. $\cot(\arcsin u)$
97. $\sin\left(2 \sec^{-1} \dfrac{u}{2}\right)$
98. $\cos\left(2 \tan^{-1} \dfrac{3}{u}\right)$
99. $\tan\left(\sin^{-1} \dfrac{u}{\sqrt{u^2 + 2}}\right)$
100. $\sec\left(\cos^{-1} \dfrac{u}{\sqrt{u^2 + 5}}\right)$
101. $\sec\left(\text{arccot} \dfrac{\sqrt{4 - u^2}}{u}\right)$
102. $\csc\left(\arctan \dfrac{\sqrt{9 - u^2}}{u}\right)$

*Solve each problem.*

**103.** Refer to Example 8.

(a) What is the optimal angle when $h = 0$?

(b) Fix $h$ at 6 ft and regard $\theta$ as a function of $v$. As $v$ gets larger and larger, the graph approaches an asymptote. Find the equation of that asymptote.

**104.** The figure shows a stationary communications satellite positioned 20,000 mi above the equator. What percent, to the nearest tenth, of the equator can be seen from the satellite? The diameter of Earth is 7927 mi at the equator.

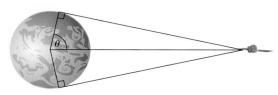

# 19.2 Trigonometric Equations I

**OBJECTIVES**

1. Solve trigonometric equations by linear methods.
2. Solve trigonometric equations by factoring.
3. Solve trigonometric equations by quadratic methods.
4. Solve trigonometric equations by using identities.

In **Chapter 18,** we studied trigonometric equations that were identities. We now consider trigonometric equations that are *conditional;* that is, equations that are satisfied by some values but not others.

**OBJECTIVE 1** Solve trigonometric equations by linear methods. Conditional equations with trigonometric (or circular) functions can usually be solved using algebraic methods and trigonometric identities.

**EXAMPLE 1** Solving a Trigonometric Equation by Linear Methods

Solve $2 \sin \theta + 1 = 0$ over the interval $[0°, 360°)$.

**Solution** Because $\sin \theta$ is the first power of a trigonometric function, we use the same method as we would to solve the linear equation $2x + 1 = 0$.

$$2 \sin \theta + 1 = 0$$
$$2 \sin \theta = -1 \quad \text{Subtract 1.}$$
$$\sin \theta = -\frac{1}{2} \quad \text{Divide by 2.}$$

To find values of $\theta$ that satisfy $\sin \theta = -\frac{1}{2}$, we observe that $\theta$ must be in either quadrant III or IV since the sine function is negative only in these two quadrants. Furthermore, the reference angle must be 30° since $\sin 30° = \frac{1}{2}$. The graphs in Figure 22 show the two possible values of $\theta$, 210° and 330°. The solution set is $\{210°, 330°\}$.

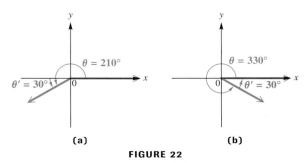

**FIGURE 22**

Alternatively, we could determine the solutions by referring to Figure 12 in **Section 16.3** on page 750.

**NOW TRY** Exercise 11.

**OBJECTIVE 2** Solve trigonometric equations by factoring.

**EXAMPLE 2** Solving a Trigonometric Equation by Factoring

Solve $\sin \theta \tan \theta = \sin \theta$ over the interval $[0°, 360°)$.

**Solution**
$$\sin \theta \tan \theta = \sin \theta$$
$$\sin \theta \tan \theta - \sin \theta = 0 \quad \text{Subtract } \sin \theta.$$
$$\sin \theta (\tan \theta - 1) = 0 \quad \text{Factor out } \sin \theta.$$
$$\sin \theta = 0 \quad \text{or} \quad \tan \theta - 1 = 0 \quad \text{Zero-factor property}$$
$$\tan \theta = 1$$
$$\theta = 0° \quad \text{or} \quad \theta = 180° \quad \theta = 45° \quad \text{or} \quad \theta = 225°$$

The solution set is $\{0°, 45°, 180°, 225°\}$.

**NOW TRY** Exercise 31.

**CAUTION** There are four solutions in Example 2. Trying to solve the equation by dividing each side by $\sin \theta$ would lead to just $\tan \theta = 1$, which would give $\theta = 45°$ or $\theta = 225°$. The other two solutions would not appear. The missing solutions are the ones that make the divisor, $\sin \theta$, equal 0. *For this reason, we avoid dividing by a variable expression.*

**OBJECTIVE 3** Solve trigonometric equations by quadratic methods. An equation in the form $au^2 + bu + c = 0$, where $u$ is an algebraic expression, is solved by quadratic methods. The expression $u$ may be a trigonometric function, as in the equation $\tan^2 x + \tan x - 2 = 0$, which we solve in the next example.

**EXAMPLE 3** Solving a Trigonometric Equation by Factoring

Solve $\tan^2 x + \tan x - 2 = 0$ over the interval $[0, 2\pi)$.

**Solution** This equation is quadratic in form and can be solved by factoring.
$$\tan^2 x + \tan x - 2 = 0$$
$$(\tan x - 1)(\tan x + 2) = 0 \quad \text{Factor.}$$
$$\tan x - 1 = 0 \quad \text{or} \quad \tan x + 2 = 0 \quad \text{Zero-factor property}$$
$$\tan x = 1 \quad \text{or} \quad \tan x = -2 \quad \text{Solve each equation.}$$

The solutions for $\tan x = 1$ over the interval $[0, 2\pi)$ are $x = \frac{\pi}{4}$ and $x = \frac{5\pi}{4}$.

To solve $\tan x = -2$ over that interval, we use a scientific calculator set in *radian* mode. We find that $\tan^{-1}(-2) \approx -1.1071487$. This is a quadrant IV number, based on the range of the inverse tangent function. (Refer to Figure 12 in **Section 16.3** on page 750.) However, since we want solutions over the interval $[0, 2\pi)$, we must first add $\pi$ to $-1.1071487$, and then add $2\pi$.

$$x \approx -1.1071487 + \pi \approx 2.0344439$$
$$x \approx -1.1071487 + 2\pi \approx 5.1760366$$

*(continued)*

The solutions over the required interval form the solution set

$$\left\{ \underbrace{\frac{\pi}{4}, \frac{5\pi}{4},}_{\text{Exact values}} \underbrace{2.0344, \quad 5.1760}_{\substack{\text{Approximate values to} \\ \text{four decimal places}}} \right\}.$$

**NOW TRY** Exercise 21.

### EXAMPLE 4  Solving a Trigonometric Equation Using the Quadratic Formula

Find all solutions of $\cot x(\cot x + 3) = 1$. Write the solution set.

**Solution**  We multiply the factors on the left and subtract 1 to get the equation in standard quadratic form.

$$\cot^2 x + 3 \cot x - 1 = 0$$

Since this equation cannot be solved by factoring, we use the quadratic formula, with $a = 1, b = 3, c = -1$, and $\cot x$ as the variable.

$$\cot x = \frac{-b \pm \sqrt{b^2 - 4ac}}{2a} \qquad \text{Quadratic formula}$$

$$= \frac{-3 \pm \sqrt{3^2 - 4(1)(-1)}}{2(1)} \qquad \text{Be careful with signs.} \qquad a = 1, b = 3, c = -1$$

$$= \frac{-3 \pm \sqrt{9 + 4}}{2} \qquad \text{Simplify}$$

$$= \frac{-3 \pm \sqrt{13}}{2}$$

$$\cot x \approx -3.302775638 \quad \text{or} \quad \cot x \approx 0.3027756377 \qquad \text{Use a calculator.}$$

We cannot find inverse cotangent values directly on a calculator, so we use the fact that $\cot x = \frac{1}{\tan x}$ and take reciprocals to get

$$\tan x \approx \frac{1}{-3.302775638} \quad \text{or} \quad \tan x \approx \frac{1}{0.3027756377}$$

$$\tan x \approx -0.3027756377 \quad \text{or} \quad \tan x \approx 3.302775638$$

$$x \approx -0.2940013018 \quad \text{or} \quad x \approx 1.276795025.$$

To find *all* solutions, we add integer multiples of the period of the tangent function, which is $\pi$, to each solution found previously. Thus, the solution set of the equation is written as

$$\{-0.2940 + n\pi, 1.2768 + n\pi, \text{ where } n \text{ is any integer}\}.$$

**NOW TRY** Exercise 43.

**OBJECTIVE 4** Solve trigonometric equations by using identities. Recall that squaring both sides of an equation, such as $\sqrt{x + 4} = x + 2$, will yield all solutions but may also give extraneous values. (In this equation, 0 is a solution, while $-3$ is extraneous. Verify this.)

### EXAMPLE 5  Solving a Trigonometric Equation by Squaring

Solve $\tan x + \sqrt{3} = \sec x$ over the interval $[0, 2\pi)$.

**Solution**  Since the tangent and secant functions are related by the identity $1 + \tan^2 x = \sec^2 x$, square both sides and express $\sec^2 x$ in terms of $\tan^2 x$.

$$(\tan x + \sqrt{3})^2 = (\sec x)^2$$

*Don't forget the middle term.*

| | |
|---|---|
| $\tan^2 x + 2\sqrt{3} \tan x + 3 = \sec^2 x$ | $(x + y)^2 = x^2 + 2xy + y^2$ |
| $\tan^2 x + 2\sqrt{3} \tan x + 3 = 1 + \tan^2 x$ | Pythagorean identity |
| $2\sqrt{3} \tan x = -2$ | Subtract $3 + \tan^2 x$. |
| $\tan x = -\dfrac{1}{\sqrt{3}} = -\dfrac{\sqrt{3}}{3}$ | Divide by $2\sqrt{3}$; rationalize the denominator. |

The possible solutions are $\dfrac{5\pi}{6}$ and $\dfrac{11\pi}{6}$. Now check them. Try $\dfrac{5\pi}{6}$ first.

Left side: $\tan x + \sqrt{3} = \tan \dfrac{5\pi}{6} + \sqrt{3} = -\dfrac{\sqrt{3}}{3} + \sqrt{3} = \dfrac{2\sqrt{3}}{3}$

Right side: $\sec x = \sec \dfrac{5\pi}{6} = -\dfrac{2\sqrt{3}}{3}$ ← Not equal

The check shows that $\dfrac{5\pi}{6}$ is not a solution. Now check $\dfrac{11\pi}{6}$.

Left side: $\tan \dfrac{11\pi}{6} + \sqrt{3} = -\dfrac{\sqrt{3}}{3} + \sqrt{3} = \dfrac{2\sqrt{3}}{3}$

Right side: $\sec \dfrac{11\pi}{6} = \dfrac{2\sqrt{3}}{3}$ ← Equal

This solution satisfies the equation, so $\left\{\dfrac{11\pi}{6}\right\}$ is the solution set.

**NOW TRY** Exercise 41.

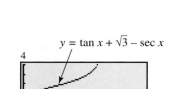

$y = \tan x + \sqrt{3} - \sec x$

Dot mode; radian mode

The graph shows that on the interval $[0, 2\pi)$, the only x-intercept of the graph of $y = \tan x + \sqrt{3} - \sec x$ is 5.7595865, which is an approximation for $\dfrac{11\pi}{6}$, the solution found in Example 5.

Methods for solving trigonometric equations can be summarized as follows.

### Solving a Trigonometric Equation

1. Decide whether the equation is linear or quadratic in form, so you can determine the solution method.
2. If only one trigonometric function is present, solve the equation for that function.
3. If more than one trigonometric function is present, rearrange the equation so that one side equals 0. Then try to factor and set each factor equal to 0 to solve.
4. If the equation is quadratic in form, but not factorable, use the quadratic formula. Check that solutions are in the desired interval.
5. Try using identities to change the form of the equation. It may be helpful to square both sides of the equation first. If this is done, check for extraneous solutions.

## EXAMPLE 6  Describing a Musical Tone from a Graph

A basic component of music is a pure tone. The graph in Figure 23 models the sinusoidal pressure $y = P$ in pounds per square foot from a pure tone at time $x = t$ in seconds.

(a) The frequency of a pure tone is often measured in hertz. One hertz is equal to one cycle per second and is abbreviated Hz. What is the frequency $f$ in hertz of the pure tone shown in the graph?

(b) The time for the tone to produce one complete cycle is called the **period**. Approximate the period $T$ in seconds of the pure tone.

(c) An equation for the graph is $y = 0.004 \sin 300\pi x$. Use a calculator to estimate all solutions to the equation that make $y = 0.004$ over the interval $[0, 0.02]$.

### Solution

(a) From the graph in Figure 23, we see that there are 6 cycles in 0.04 sec. This is equivalent to $\frac{6}{0.04} = 150$ cycles per sec. The pure tone has a frequency of $f = 150$ Hz.

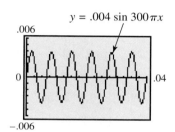

FIGURE 23

(b) Six periods cover a time of 0.04 sec. One period would be equal to $T = \frac{0.04}{6} = \frac{1}{150}$, or $0.00\overline{6}$ sec.

(c) If we reproduce the graph in Figure 23 on a calculator as $Y_1$ and also graph a second function as $Y_2 = 0.004$, we can determine that the approximate values of $x$ at the points of intersection of the graphs over the interval $[0, 0.02]$ are

$$0.0017, \quad 0.0083, \quad \text{and} \quad 0.015.$$

The first value is shown in Figure 24. These values represent time in seconds.

**NOW TRY** Exercise 53.

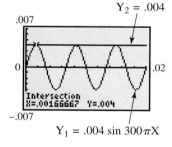

FIGURE 24

## 19.2 Exercises

**NOW TRY Exercise**

*Concept Check* Refer to the summary box on solving a trigonometric equation on page 866. Decide on the appropriate technique to begin the solution of each equation. Do not solve the equation.

1. $2 \cot x + 1 = -1$
2. $\sin x + 2 = 3$
3. $5 \sec^2 x = 6 \sec x$
4. $2 \cos^2 x - \cos x = 1$
5. $9 \sin^2 x - 5 \sin x = 1$
6. $\tan^2 x - 4 \tan x + 2 = 0$
7. $\tan x - \cot x = 0$
8. $\cos^2 x = \sin^2 x + 1$

9. Suppose in solving an equation over the interval $[0°, 360°)$, you reach the step $\sin \theta = -\frac{1}{2}$. Why is $-30°$ not a correct answer?

✎ 10. Lindsay solved the equation $\sin x = 1 - \cos x$ by squaring both sides to get $\sin^2 x = 1 - 2\cos x + \cos^2 x$. Several steps later, using correct algebra, she determined that the solution set for solutions over the interval $[0, 2\pi)$ is $\left\{0, \frac{\pi}{2}, \frac{3\pi}{2}\right\}$. Explain why this is not the correct solution set.

*Solve each equation for exact solutions over the interval $[0, 2\pi)$. See Examples 1–3.*

**11.** $2\cot x + 1 = -1$  
**12.** $\sin x + 2 = 3$  
**13.** $2\sin x + 3 = 4$  
**14.** $2\sec x + 1 = \sec x + 3$  
**15.** $\tan^2 x + 3 = 0$  
**16.** $\sec^2 x + 2 = -1$  
**17.** $(\cot x - 1)(\sqrt{3}\cot x + 1) = 0$  
**18.** $(\csc x + 2)(\csc x - \sqrt{2}) = 0$  
**19.** $\cos^2 x + 2\cos x + 1 = 0$  
**20.** $2\cos^2 x - \sqrt{3}\cos x = 0$  
**21.** $-2\sin^2 x = 3\sin x + 1$  
**22.** $2\cos^2 x - \cos x = 1$

*Solve each equation for solutions over the interval $[0°, 360°)$. Give solutions to the nearest tenth as appropriate. See Examples 2–5.*

**23.** $(\cot\theta - \sqrt{3})(2\sin\theta + \sqrt{3}) = 0$  
**24.** $(\tan\theta - 1)(\cos\theta - 1) = 0$  
**25.** $2\sin\theta - 1 = \csc\theta$  
**26.** $\tan\theta + 1 = \sqrt{3} + \sqrt{3}\cot\theta$  
**27.** $\tan\theta - \cot\theta = 0$  
**28.** $\cos^2\theta = \sin^2\theta + 1$  
**29.** $\csc^2\theta - 2\cot\theta = 0$  
**30.** $\sin^2\theta\cos\theta = \cos\theta$  
**31.** $2\tan^2\theta\sin\theta - \tan^2\theta = 0$  
**32.** $\sin^2\theta\cos^2\theta = 0$  
**33.** $\sec^2\theta\tan\theta = 2\tan\theta$  
**34.** $\cos^2\theta - \sin^2\theta = 0$  
**35.** $9\sin^2\theta - 6\sin\theta = 1$  
**36.** $4\cos^2\theta + 4\cos\theta = 1$  
**37.** $\tan^2\theta + 4\tan\theta + 2 = 0$  
**38.** $3\cot^2\theta - 3\cot\theta - 1 = 0$  
**39.** $\sin^2\theta - 2\sin\theta + 3 = 0$  
**40.** $2\cos^2\theta + 2\cos\theta - 1 = 0$  
**41.** $\cot\theta + 2\csc\theta = 3$  
**42.** $2\sin\theta = 1 - 2\cos\theta$

*Determine the solution set of each equation in radians (for x) to four decimal places or degrees (for $\theta$) to the nearest tenth as appropriate. See Example 4.*

**43.** $3\sin^2 x - \sin x - 1 = 0$  
**44.** $2\cos^2 x + \cos x = 1$  
**45.** $4\cos^2 x - 1 = 0$  
**46.** $2\cos^2 x + 5\cos x + 2 = 0$  
**47.** $5\sec^2\theta = 6\sec\theta$  
**48.** $3\sin^2\theta - \sin\theta = 2$  
**49.** $\dfrac{2\tan\theta}{3 - \tan^2\theta} = 1$  
**50.** $\sec^2\theta = 2\tan\theta + 4$

📈 *The following equations cannot be solved by algebraic methods. Use a graphing calculator to find all solutions over the interval $[0, 2\pi)$. Express solutions to four decimal places.*

**51.** $x^2 + \sin x - x^3 - \cos x = 0$  
**52.** $x^3 - \cos^2 x = \dfrac{1}{2}x - 1$

*Solve each problem.*

**53.** See Example 6. No musical instrument can generate a true pure tone. A pure tone has a unique, constant frequency and amplitude that sounds rather dull and uninteresting. The pressures caused by pure tones on the eardrum are sinusoidal. The change in pressure $P$ in pounds per square foot on a person's eardrum from a pure tone at time $t$ in seconds can be modeled using the equation

$$P = A \sin(2\pi f t + \phi),$$

where $f$ is the frequency in cycles per second, and $\phi$ is the phase angle. When $P$ is positive, there is an increase in pressure and the eardrum is pushed inward; when $P$ is negative, there is a decrease in pressure and the eardrum is pushed outward. (*Source:* Roederer, J., *Introduction to the Physics and Psychophysics of Music,* Second Edition, Springer-Verlag, 1975.) A graph of the tone middle C is shown in the figure.

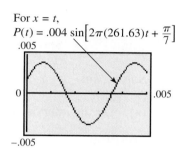

For $x = t$,
$P(t) = .004 \sin\left[2\pi(261.63)t + \dfrac{\pi}{7}\right]$

(a) Determine algebraically the values of $t$ for which $P = 0$ over $[0, .005]$.
(b) From the graph and your answer in part (a), determine the interval for which $P \leq 0$ over $[0, .005]$.
(c) Would an eardrum hearing this tone be vibrating outward or inward when $P < 0$?

**54.** The model

$$0.342D \cos\theta + h \cos^2\theta = \dfrac{16D^2}{V_0^2}$$

is used to reconstruct accidents in which a vehicle vaults into the air after hitting an obstruction. $V_0$ is velocity in feet per second of the vehicle when it hits, $D$ is distance (in feet) from the obstruction to the landing point, and $h$ is the difference in height (in feet) between landing point and takeoff point. Angle $\theta$ is the takeoff angle, the angle between the horizontal and the path of the vehicle. Find $\theta$ to the nearest degree if $V_0 = 60$, $D = 80$, and $h = 2$.

**55.** In an electric circuit, let

$$V = \cos 2\pi t$$

model the electromotive force in volts at $t$ seconds. Find the least value of $t$ where $0 \leq t \leq \frac{1}{2}$ for each value of $V$.

(a) $V = 0$   (b) $V = 0.5$   (c) $V = 0.25$

**56.** A coil of wire rotating in a magnetic field induces a voltage modeled by

$$E = 20 \sin\left(\dfrac{\pi t}{4} - \dfrac{\pi}{2}\right),$$

where $t$ is time in seconds. Find the least positive time to produce each voltage.

(a) 0   (b) $10\sqrt{3}$

# 19.3 Trigonometric Equations II

**OBJECTIVES**

1. Solve trigonometric equations with half-angles.
2. Solve trigonometric equations with multiple angles.

In this section, we discuss trigonometric equations that involve functions of half-angles and multiple angles. Solving these equations often requires adjusting solution intervals to fit given domains.

**OBJECTIVE 1** Solve trigonometric equations with half-angles.

**EXAMPLE 1** Solving an Equation Using a Half-Angle Identity

Solve $2 \sin \dfrac{x}{2} = 1$

**(a)** over the interval $[0, 2\pi)$, and  **(b)** give all solutions.

**Solution**

**(a)** Write the interval $[0, 2\pi)$ as the inequality

$$0 \leq x < 2\pi.$$

The corresponding interval for $\dfrac{x}{2}$ is

$$0 \leq \dfrac{x}{2} < \pi. \quad \text{Divide by 2.}$$

To find all values of $\dfrac{x}{2}$ over the interval $[0, \pi)$ that satisfy the given equation, first solve for $\sin \dfrac{x}{2}$.

$$2 \sin \dfrac{x}{2} = 1$$

$$\sin \dfrac{x}{2} = \dfrac{1}{2} \quad \text{Divide by 2.}$$

The two numbers over the interval $[0, \pi)$ with sine value $\dfrac{1}{2}$ are $\dfrac{\pi}{6}$ and $\dfrac{5\pi}{6}$, so

$$\dfrac{x}{2} = \dfrac{\pi}{6} \quad \text{or} \quad \dfrac{x}{2} = \dfrac{5\pi}{6}$$

$$x = \dfrac{\pi}{3} \quad \text{or} \quad x = \dfrac{5\pi}{3}. \quad \text{Multiply by 2.}$$

The solution set over the given interval is $\left\{\dfrac{\pi}{3}, \dfrac{5\pi}{3}\right\}$.

**(b)** Since this is a sine function with period $4\pi$, the solution set is

$$\left\{\dfrac{\pi}{3} + 4n\pi, \dfrac{5\pi}{3} + 4n\pi, \text{ where } n \text{ is any integer}\right\}.$$

**NOW TRY** Exercise 13.

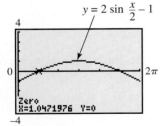

The $x$-intercepts are the solutions found in Example 1. Using Xscl = $\dfrac{\pi}{3}$ makes it possible to support the exact solutions by counting the tick marks from 0 on the graph.

870

**OBJECTIVE 2** Solve trigonometric equations with multiple angles.

**EXAMPLE 2** Solving an Equation with a Double Angle

Solve $\cos 2x = \cos x$ over the interval $[0, 2\pi)$.

**Solution** First change $\cos 2x$ to a trigonometric function of $x$. Use the identity $\cos 2x = 2\cos^2 x - 1$ so the equation involves only $\cos x$. Then factor.

$$\cos 2x = \cos x$$
$$2\cos^2 x - 1 = \cos x \qquad \text{Substitute; double-angle identity}$$
$$2\cos^2 x - \cos x - 1 = 0 \qquad \text{Subtract } \cos x.$$
$$(2\cos x + 1)(\cos x - 1) = 0 \qquad \text{Factor.}$$
$$2\cos x + 1 = 0 \quad \text{or} \quad \cos x - 1 = 0 \qquad \text{Zero-factor property}$$
$$\cos x = -\frac{1}{2} \quad \text{or} \quad \cos x = 1 \qquad \text{Solve each equation.}$$

Cosine is $-\frac{1}{2}$ in quadrants II and III with reference arc $\frac{\pi}{3}$, and has a value of 1 at 0 radians; thus, solutions over the required interval are

$$x = \frac{2\pi}{3} \quad \text{or} \quad x = \frac{4\pi}{3} \quad \text{or} \quad x = 0.$$

The solution set is $\left\{0, \frac{2\pi}{3}, \frac{4\pi}{3}\right\}$.

**NOW TRY** Exercise 15.

**CAUTION** In the solution of Example 2, $\cos 2x$ cannot be changed to $\cos x$ by dividing by 2 since 2 is not a factor of $\cos 2x$, that is, $\frac{\cos 2x}{2} \neq \cos x$. The only way to change $\cos 2x$ to a trigonometric function of $x$ is by using one of the identities for $\cos 2x$.

**EXAMPLE 3** Solving an Equation Using a Multiple-Angle Identity

Solve $4 \sin \theta \cos \theta = \sqrt{3}$ over the interval $[0°, 360°)$.

**Solution** The identity $2 \sin \theta \cos \theta = \sin 2\theta$ is useful here.

$$4 \sin \theta \cos \theta = \sqrt{3}$$
$$2(2 \sin \theta \cos \theta) = \sqrt{3} \qquad 4 = 2 \cdot 2$$
$$2 \sin 2\theta = \sqrt{3} \qquad 2 \sin \theta \cos \theta = \sin 2\theta$$
$$\sin 2\theta = \frac{\sqrt{3}}{2} \qquad \text{Divide by 2.}$$

From the given interval $0° \leq \theta < 360°$, the interval for $2\theta$ is $0° \leq 2\theta < 720°$. Since the sine is positive in quadrants I and II, solutions over this interval are

$$2\theta = 60°, 120°, 420°, 480°,$$

or $$\theta = 30°, 60°, 210°, 240°. \qquad \text{Divide by 2.}$$

The final two solutions for $2\theta$ were found by adding $360°$ to $60°$ and $120°$, respectively, giving the solution set $\{30°, 60°, 210°, 240°\}$.

**NOW TRY** Exercise 35.

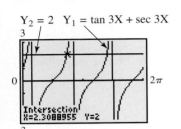

Y₂ = 2   Y₁ = tan 3X + sec 3X

Connected mode; radian mode
The screen shows that one solution is approximately 2.3089. An advantage of using a graphing calculator is that extraneous values do not appear.

**EXAMPLE 4** Solving an Equation with a Multiple Angle

Solve $\tan 3x + \sec 3x = 2$ over the interval $[0, 2\pi)$.

**Solution** Since the tangent and secant functions are related by the identity $1 + \tan^2 \theta = \sec^2 \theta$, one way to begin is to express everything in terms of secant.

$\tan 3x + \sec 3x = 2$   *Don't forget the middle term.*

$\tan 3x = 2 - \sec 3x$   Subtract sec 3x.

$\tan^2 3x = 4 - 4\sec 3x + \sec^2 3x$   Square both sides; $(x-y)^2 = x^2 - 2xy + y^2$.

$\sec^2 3x - 1 = 4 - 4\sec 3x + \sec^2 3x$   Replace with $\sec^2 3x - 1$.

$4 \sec 3x = 5$   Simplify.

$\sec 3x = \dfrac{5}{4}$   Divide by 4.

$\dfrac{1}{\cos 3x} = \dfrac{5}{4}$   $\sec \theta = \dfrac{1}{\cos \theta}$

$\cos 3x = \dfrac{4}{5}$   Use reciprocals.

Multiply each term of the inequality $0 \le x < 2\pi$ by 3 to find the interval for $3x$: $[0, 6\pi)$. Using a calculator and the fact that cosine is positive in quadrants I and IV,

$3x \approx 0.6435, 5.6397, 6.9267, 11.9229, 13.2099, 18.2061$

$x \approx 0.2145, 1.8799, 2.3089, 3.9743, 4.4033, 6.0687.$   Divide by 3.

Since both sides of the equation were squared, each proposed solution must be checked. Verify by substitution in the given equation that the solution set is $\{0.2145, 2.3089, 4.4033\}$.

**NOW TRY** Exercise 31.

A piano string can vibrate at more than one frequency when it is struck. It produces a complex wave that can mathematically be modeled by a sum of several pure tones. If a piano key with a frequency of $f_1$ is played, then the corresponding string will not only vibrate at $f_1$ but it will also vibrate at the higher frequencies of $2f_1, 3f_1, 4f_1, \ldots, nf_1$. $f_1$ is called the **fundamental frequency** of the string, and higher frequencies are called the **upper harmonics**. The human ear will hear the sum of these frequencies as one complex tone. (Source: Roederer, J., *Introduction to the Physics and Psychophysics of Music,* Second Edition, Springer-Verlag, 1975.)

**EXAMPLE 5** Analyzing Pressures of Upper Harmonics

Suppose that the A key above middle C is played. Its fundamental frequency is $f_1 = 440$ Hz, and its associated pressure is expressed as

$$P_1 = 0.002 \sin 880\pi t.$$

The string will also vibrate at

$$f_2 = 880, \quad f_3 = 1320, \quad f_4 = 1760, \quad f_5 = 2200, \ldots \text{ Hz.}$$

*(continued)*

The corresponding pressures of these upper harmonics are

$$P_2 = \frac{0.002}{2} \sin 1760\pi t, \quad P_3 = \frac{0.002}{3} \sin 2640\pi t,$$

$$P_4 = \frac{0.002}{4} \sin 3520\pi t, \quad \text{and} \quad P_5 = \frac{0.002}{5} \sin 4400\pi t.$$

The graph of

$$P = P_1 + P_2 + P_3 + P_4 + P_5,$$

shown in Figure 25, is "saw-toothed."

(a) What is the maximum value of $P$?

(b) At what values of $t = x$ does this maximum occur over the interval $[0, 0.01]$?

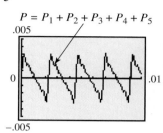

**FIGURE 25**

### Solution

(a) A graphing calculator shows that the maximum value of $P$ is approximately 0.00317. See Figure 26.

(b) The maximum occurs at $t = x \approx 0.000188$, 0.00246, 0.00474, 0.00701, and 0.00928. Figure 26 shows how the second value is found; the others are found similarly.

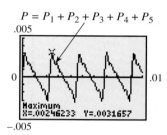

**FIGURE 26**

**NOW TRY** Exercise 41.

## 19.3 Exercises

*Concept Check* Answer each question.

1. Suppose you are solving a trigonometric equation for solutions over the interval $[0, 2\pi)$, and your work leads to $2x = \frac{2\pi}{3}, 2\pi, \frac{8\pi}{3}$. What are the corresponding values of $x$?

2. Suppose you are solving a trigonometric equation for solutions over the interval $[0, 2\pi)$, and your work leads to $\frac{1}{2}x = \frac{\pi}{16}, \frac{5\pi}{12}, \frac{5\pi}{8}$. What are the corresponding values of $x$?

3. Suppose you are solving a trigonometric equation for solutions over the interval $[0°, 360°)$, and your work leads to $3\theta = 180°, 630°, 720°, 930°$. What are the corresponding values of $\theta$?

4. Suppose you are solving a trigonometric equation for solutions over the interval $[0°, 360°)$, and your work leads to $\frac{1}{3}\theta = 45°, 60°, 75°, 90°$. What are the corresponding values of $\theta$?

*Solve each equation for exact solutions over the interval $[0, 2\pi)$. See Examples 1–4.*

5. $\cos 2x = \frac{\sqrt{3}}{2}$      6. $\cos 2x = -\frac{1}{2}$      7. $\sin 3x = -1$

8. $\sin 3x = 0$      9. $3 \tan 3x = \sqrt{3}$      10. $\cot 3x = \sqrt{3}$

11. $\sqrt{2} \cos 2x = -1$      12. $2\sqrt{3} \sin 2x = \sqrt{3}$      13. $\sin \frac{x}{2} = \sqrt{2} - \sin \frac{x}{2}$

14. $\tan 4x = 0$      15. $\sin x = \sin 2x$      16. $\cos 2x - \cos x = 0$

17. $8 \sec^2 \frac{x}{2} = 4$      18. $\sin^2 \frac{x}{2} - 2 = 0$      19. $\sin \frac{x}{2} = \cos \frac{x}{2}$

20. $\sec \frac{x}{2} = \cos \frac{x}{2}$      21. $\cos 2x + \cos x = 0$      22. $\sin x \cos x = \frac{1}{4}$

Solve each equation in Exercises 23–30 for exact solutions over the interval $[0°, 360°)$. In Exercises 31–38, give all solutions. If necessary, express solutions to the nearest tenth of a degree. See Examples 1–4.

23. $\sqrt{2} \sin 3\theta - 1 = 0$
24. $-2 \cos 2\theta = \sqrt{3}$
25. $\cos \dfrac{\theta}{2} = 1$

26. $\sin \dfrac{\theta}{2} = 1$
27. $2\sqrt{3} \sin \dfrac{\theta}{2} = 3$
28. $2\sqrt{3} \cos \dfrac{\theta}{2} = -3$

29. $2 \sin \theta = 2 \cos 2\theta$
30. $\cos \theta - 1 = \cos 2\theta$
31. $1 - \sin \theta = \cos 2\theta$

32. $\sin 2\theta = 2 \cos^2 \theta$
33. $\csc^2 \dfrac{\theta}{2} = 2 \sec \theta$
34. $\cos \theta = \sin^2 \dfrac{\theta}{2}$

35. $2 - \sin 2\theta = 4 \sin 2\theta$
36. $4 \cos 2\theta = 8 \sin \theta \cos \theta$

37. $2 \cos^2 2\theta = 1 - \cos 2\theta$
38. $\sin \theta - \sin 2\theta = 0$

The following equations cannot be solved by algebraic methods. Use a graphing calculator to find all solutions over the interval $[0, 2\pi)$. Express solutions to four decimal places.

39. $2 \sin 2x - x^3 + 1 = 0$
40. $3 \cos \dfrac{x}{2} + \sqrt{x} - 2 = -\dfrac{1}{2}x + 2$

Solve each problem. See Example 5.

41. If a string with a fundamental frequency of 110 Hz is plucked in the middle, it will vibrate at the odd harmonics of 110, 330, 550, . . . Hz but not at the even harmonics of 220, 440, 660, . . . Hz. The resulting pressure $P$ caused by the string can be modeled by the equation

$$P = 0.003 \sin 220\pi t + \frac{0.003}{3} \sin 660\pi t + \frac{0.003}{5} \sin 1100\pi t + \frac{0.003}{7} \sin 1540\pi t.$$

(Source: Benade, A., *Fundamentals of Musical Acoustics*, Dover Publications, 1990; Roederer, J., *Introduction to the Physics and Psychophysics of Music*, Second Edition, Springer-Verlag, 1975.)

(a) Graph $P$ in the window $[0, .03]$ by $[-.005, .005]$.
(b) Use the graph to describe the shape of the sound wave that is produced.
(c) See **Section 19.2**, Exercise 53. At lower frequencies, the inner ear will hear a tone only when the eardrum is moving outward. Determine the times over the interval $[0, .03]$ when this will occur.

42. The seasonal variation in length of daylight can be modeled by a sine function. For example, the daily number of hours of daylight in New Orleans is given by

$$h = \frac{35}{3} + \frac{7}{3} \sin \frac{2\pi x}{365},$$

where $x$ is the number of days after March 21 (disregarding leap year). (Source: Bushaw, D., et al., *A Sourcebook of Applications of School Mathematics*. Copyright © 1980 by The Mathematical Association of America.)

(a) On what date will there be about 14 hr of daylight?
(b) What date has the least number of hours of daylight?
(c) When will there be about 10 hr of daylight?

# 19 SUMMARY

## KEY TERMS

**19.1** one-to-one function
inverse function

## NEW SYMBOLS

$f^{-1}$    inverse of function $f$
$\sin^{-1} x$ (arcsin $x$)    inverse sine of $x$
$\cos^{-1} x$ (arccos $x$)    inverse cosine of $x$
$\tan^{-1} x$ (arctan $x$)    inverse tangent of $x$

$\cot^{-1} x$ (arccot $x$)    inverse cotangent of $x$
$\sec^{-1} x$ (arcsec $x$)    inverse secant of $x$
$\csc^{-1} x$ (arccsc $x$)    inverse cosecant of $x$

## QUICK REVIEW

### CONCEPTS

### 19.1 Inverse Circular Functions

| Inverse Function | Domain | Range Interval | Quadrants of the Unit Circle |
|---|---|---|---|
| $y = \sin^{-1} x$ | $[-1, 1]$ | $\left[-\frac{\pi}{2}, \frac{\pi}{2}\right]$ | I and IV |
| $y = \cos^{-1} x$ | $[-1, 1]$ | $[0, \pi]$ | I and II |
| $y = \tan^{-1} x$ | $(-\infty, \infty)$ | $\left(-\frac{\pi}{2}, \frac{\pi}{2}\right)$ | I and IV |
| $y = \cot^{-1} x$ | $(-\infty, \infty)$ | $(0, \pi)$ | I and II |
| $y = \sec^{-1} x$ | $(-\infty, -1] \cup [1, \infty)$ | $[0, \pi], y \neq \frac{\pi}{2}$ | I and II |
| $y = \csc^{-1} x$ | $(-\infty, -1] \cup [1, \infty)$ | $\left[-\frac{\pi}{2}, \frac{\pi}{2}\right], y \neq 0$ | I and IV |

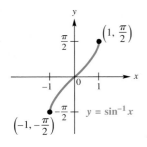

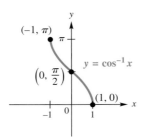

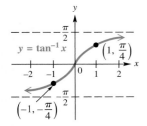

See Figure 14 on page 855 for graphs of the other inverse circular (trigonometric) functions.

### EXAMPLES

Evaluate $y = \cos^{-1} 0$.

Write $y = \cos^{-1} 0$ as $\cos y = 0$. Then $y = \frac{\pi}{2}$, because $\cos \frac{\pi}{2} = 0$ and $\frac{\pi}{2}$ is in the range of $\cos^{-1} x$.

Use a calculator to find $y$ in radians if $y = \sec^{-1}(-3)$.

With the calculator in radian mode, enter $\sec^{-1}(-3)$ as $\cos^{-1}\left(-\frac{1}{3}\right)$ to get $y \approx 1.9106332$.

Evaluate $\sin\left(\tan^{-1}\left(-\frac{3}{4}\right)\right)$.

Let $u = \tan^{-1}\left(-\frac{3}{4}\right)$. Then $\tan u = -\frac{3}{4}$. Since $\tan u$ is negative when $u$ is in quadrant IV, sketch a triangle as shown.

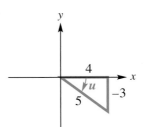

We want $\sin\left(\tan^{-1}\left(-\frac{3}{4}\right)\right) = \sin u$. From the triangle,

$$\sin u = -\frac{3}{5}.$$

*(continued)*

## CONCEPTS

### 19.2 Trigonometric Equations I
### 19.3 Trigonometric Equations II

**Solving a Trigonometric Equation**
1. Decide whether the equation is linear or quadratic in form, so you can determine the solution method.
2. If only one trigonometric function is present, solve the equation for that function.
3. If more than one trigonometric function is present, rearrange the equation so that one side equals 0. Then try to factor and set each factor equal to 0 to solve.
4. If the equation is quadratic in form, but not factorable, use the quadratic formula. Check that solutions are in the desired interval.
5. Try using identities to change the form of the equation. It may be helpful to square both sides of the equation first. If this is done, check for extraneous solutions.

## EXAMPLES

Solve $\tan \theta + \sqrt{3} = 2\sqrt{3}$ over the interval $[0°, 360°)$. Use a linear method.

$$\tan \theta + \sqrt{3} = 2\sqrt{3}$$
$$\tan \theta = \sqrt{3}$$
$$\theta = 60°$$

Another solution over $[0°, 360°)$ is

$$\theta = 60° + 180° = 240°.$$

The solution set is $\{60°, 240°\}$.

Solve $2\cos^2 x = 1$ over the interval $[0, 2\pi)$ using a double-angle identity.

$$2\cos^2 x = 1$$
$$2\cos^2 x - 1 = 0$$
$$\cos 2x = 0 \quad \text{Cosine double-angle identity}$$
$$2x = \frac{\pi}{2}, \frac{3\pi}{2}, \frac{5\pi}{2}, \frac{7\pi}{2} \quad \begin{array}{l} 0 \leq x < 2\pi, \text{ so} \\ 0 \leq 2x < 4\pi. \end{array}$$
$$x = \frac{\pi}{4}, \frac{3\pi}{4}, \frac{5\pi}{4}, \frac{7\pi}{4} \quad \text{Divide by 2.}$$

The solution set is $\left\{\frac{\pi}{4}, \frac{3\pi}{4}, \frac{5\pi}{4}, \frac{7\pi}{4}\right\}$.

# 19 REVIEW EXERCISES

1. Graph the inverse sine, cosine, and tangent functions, indicating three points on each graph. Give the domain and range for each.

*True or False* Tell whether each statement is true or false. If false, tell why.

2. The ranges of the inverse tangent and inverse cotangent functions are the same.
3. It is true that $\sin \frac{11\pi}{6} = -\frac{1}{2}$, and, therefore, $\arcsin\left(-\frac{1}{2}\right) = \frac{11\pi}{6}$.
4. For all $x$, $\tan(\tan^{-1} x) = x$.

*Give the exact real number value of y. Do not use a calculator.*

5. $y = \sin^{-1} \frac{\sqrt{2}}{2}$

6. $y = \arccos\left(-\frac{1}{2}\right)$

7. $y = \tan^{-1}(-\sqrt{3})$

8. $y = \arcsin(-1)$

9. $y = \cos^{-1}\left(-\frac{\sqrt{2}}{2}\right)$

10. $y = \arctan \frac{\sqrt{3}}{3}$

11. $y = \sec^{-1}(-2)$

12. $y = \text{arccsc} \frac{2\sqrt{3}}{3}$

13. $y = \text{arccot}(-1)$

*Give the degree measure of θ. Do not use a calculator.*

**14.** $\theta = \arccos \dfrac{1}{2}$ 
**15.** $\theta = \arcsin\left(-\dfrac{\sqrt{3}}{2}\right)$ 
**16.** $\theta = \tan^{-1} 0$

*Use a calculator to give the degree measure of θ.*

**17.** $\theta = \arctan 1.7804675$ 
**18.** $\theta = \sin^{-1}(-0.66045320)$ 
**19.** $\theta = \cos^{-1} 0.80396577$ 
**20.** $\theta = \cot^{-1} 4.5046388$ 
**21.** $\theta = \operatorname{arcsec} 3.4723155$ 
**22.** $\theta = \csc^{-1} 7.4890096$

*Evaluate the following without using a calculator.*

**23.** $\cos(\arccos(-1))$ 
**24.** $\sin\left(\arcsin\left(-\dfrac{\sqrt{3}}{2}\right)\right)$ 
**25.** $\arccos\left(\cos\dfrac{3\pi}{4}\right)$

**26.** $\operatorname{arcsec}(\sec \pi)$ 
**27.** $\tan^{-1}\left(\tan\dfrac{\pi}{4}\right)$ 
**28.** $\cos^{-1}(\cos 0)$

**29.** $\sin\left(\arccos \dfrac{3}{4}\right)$ 
**30.** $\cos(\arctan 3)$ 
**31.** $\cos(\csc^{-1}(-2))$

**32.** $\sec\left(2\sin^{-1}\left(-\dfrac{1}{3}\right)\right)$ 
**33.** $\tan\left(\arcsin\dfrac{3}{5} + \arccos\dfrac{5}{7}\right)$

*Write each of the following as an algebraic (nontrigonometric) expression in u, u > 0.*

**34.** $\cos\left(\arctan \dfrac{u}{\sqrt{1-u^2}}\right)$ 
**35.** $\tan\left(\operatorname{arcsec} \dfrac{\sqrt{u^2+1}}{u}\right)$

*Solve each equation for solutions over the interval $[0, 2\pi)$. If necessary, express approximate solutions to four decimal places.*

**36.** $\sin^2 x = 1$ 
**37.** $2 \tan x - 1 = 0$ 
**38.** $3 \sin^2 x - 5 \sin x + 2 = 0$ 
**39.** $\tan x = \cot x$ 
**40.** $\sec^4 2x = 4$ 
**41.** $\tan^2 2x - 1 = 0$

*Give all solutions for each equation.*

**42.** $\sec \dfrac{x}{2} = \cos \dfrac{x}{2}$ 
**43.** $\cos 2x + \cos x = 0$ 
**44.** $4 \sin x \cos x = \sqrt{3}$

*Solve each equation for solutions over the interval $[0°, 360°)$. If necessary, express solutions to the nearest tenth of a degree.*

**45.** $\sin^2 \theta + 3 \sin \theta + 2 = 0$ 
**46.** $2 \tan^2 \theta = \tan \theta + 1$ 
**47.** $\sin 2\theta = \cos 2\theta + 1$ 
**48.** $2 \sin 2\theta = 1$ 
**49.** $3 \cos^2 \theta + 2 \cos \theta - 1 = 0$ 
**50.** $5 \cot^2 \theta - \cot \theta - 2 = 0$

*Solve each problem.*

**51.** A 10-ft-wide chalkboard is situated 5 ft from the left wall of a classroom. See the figure. A student sitting next to the wall *x* feet from the front of the classroom has a viewing angle of θ radians.

(a) Show that the value of θ is given by the function defined by
$$f(x) = \arctan\left(\dfrac{15}{x}\right) - \arctan\left(\dfrac{5}{x}\right).$$

(b) Graph $f(x)$ with a graphing calculator to estimate the value of *x* that maximizes the viewing angle.

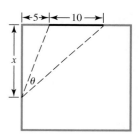

52. Recall Snell's law from Exercises 59 and 60 of **Section 15.3:**

$$\frac{c_1}{c_2} = \frac{\sin \theta_1}{\sin \theta_2},$$

where $c_1$ is the speed of light in one medium, $c_2$ is the speed of light in a second medium, and $\theta_1$ and $\theta_2$ are the angles shown in the figure. Suppose a light is shining up through water into the air as in the figure. As $\theta_1$ increases, $\theta_2$ approaches 90°, at which point no light will emerge from the water. Assume the ratio $\frac{c_1}{c_2}$ in this case is 0.752. For what value of $\theta_1$ does $\theta_2 = 90°$? This value of $\theta_1$ is called the **critical angle** for water.

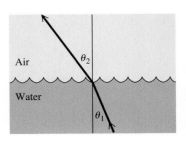

53. Refer to Exercise 52. What happens when $\theta_1$ is greater than the critical angle?

54. The British nautical mile is defined as the length of a minute of arc of a meridian. Since Earth is flat at its poles, the nautical mile, in feet, is given by

$$L = 6077 - 31 \cos 2\theta,$$

where $\theta$ is the latitude in degrees. See the figure. (*Source:* Bushaw, D., et al., *A Sourcebook of Applications of School Mathematics.* Copyright © 1980 by The Mathematical Association of America.)

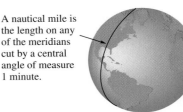

A nautical mile is the length on any of the meridians cut by a central angle of measure 1 minute.

(a) Find the latitude between 0° and 90° at which the nautical mile is 6074 ft.
(b) At what latitude between 0° and 180° is the nautical mile 6108 ft?
(c) In the United States, the nautical mile is defined everywhere as 6080.2 ft. At what latitude between 0° and 90° does this agree with the British nautical mile?

# 19 TEST

1. Graph $y = \sin^{-1} x$, and indicate the coordinates of three points on the graph. Give the domain and range.

2. Find the exact value of $y$ for each equation.

   (a) $y = \arccos\left(-\dfrac{1}{2}\right)$
   
   (b) $y = \sin^{-1}\left(-\dfrac{\sqrt{3}}{2}\right)$
   
   (c) $y = \tan^{-1} 0$
   
   (d) $y = \text{arcsec}(-2)$

3. Give the degree measure of $\theta$.

   (a) $\theta = \arccos \dfrac{\sqrt{3}}{2}$
   
   (b) $\theta = \tan^{-1}(-1)$
   
   (c) $\theta = \cot^{-1}(-1)$
   
   (d) $\theta = \csc^{-1}\left(-\dfrac{2\sqrt{3}}{3}\right)$

4. Use a calculator to give each value in decimal degrees to the nearest hundredth.

   (a) $\sin^{-1} 0.67610476$   (b) $\sec^{-1} 1.0840880$   (c) $\cot^{-1}(-0.7125586)$

5. Find each exact value.

   (a) $\cos\left(\arcsin \dfrac{2}{3}\right)$
   
   (b) $\sin\left(2 \cos^{-1} \dfrac{1}{3}\right)$

6. Explain why $\sin^{-1} 3$ cannot be defined.

7. Explain why $\arcsin\left(\sin \frac{5\pi}{6}\right) \neq \frac{5\pi}{6}$.

8. Write $\tan(\arcsin u)$ as an algebraic (nontrigonometric) expression in $u$, $u > 0$.

*Solve each equation for solutions over the interval $[0°, 360°)$. Express approximate solutions to the nearest tenth of a degree.*

9. $-3 \sec \theta + 2\sqrt{3} = 0$   10. $\sin^2 \theta = \cos^2 \theta + 1$   11. $\csc^2 \theta - 2 \cot \theta = 4$

*Solve each equation for solutions over the interval $[0, 2\pi)$. Express approximate solutions to four decimal places.*

12. $\cos x = \cos 2x$   13. $\sqrt{2} \cos 3x - 1 = 0$   14. $\sin x \cos x = \dfrac{1}{3}$

15. Solve $\sin^2 \theta = -\cos 2\theta$, giving all solutions in degrees.

16. Solve $2\sqrt{3} \sin \dfrac{x}{2} = 3$, giving all solutions in radians.

# 20

# Oblique Triangles and Vectors

*Triangulation* is the process of determining the location of a point by measuring angles to it from known points at either end of a fixed baseline rather than measuring distances to the point directly. Triangulation is used for many purposes, including surveying, navigation, and astronomy.

Surveyors use triangulation to determine distances that may not be otherwise accessible. To begin, surveyors measure a certain length exactly to provide a baseline. From each end of this line, they then measure the angle to a distant point, using a special measuring instrument. They now have a triangle in which they know the length of one side and the two adjacent angles. Using the *law of sines*, they can determine the lengths of the other two sides.

In this chapter, you will use trigonometry to solve problems involving distances that cannot be easily measured. For example, in Exercises 25 and 26 of Section 20.1, you will use the law of sines to find the distance across a river and the distance across a deep canyon.

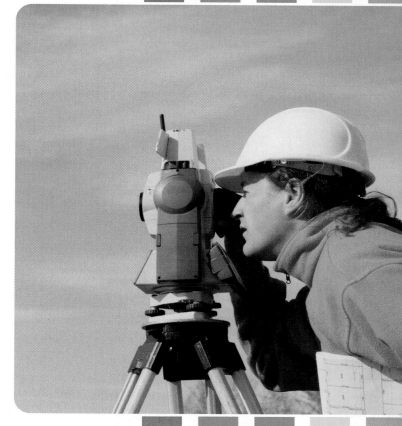

**20.1** Oblique Triangles and the Law of Sines

**20.2** The Ambiguous Case of the Law of Sines

**20.3** The Law of Cosines

**20.4** Vectors

# 20.1 Oblique Triangles and the Law of Sines

**OBJECTIVES**

1. Review the triangle congruence axioms.
2. Derive the law of sines.
3. Solve SSA and ASA triangles (Case 1).
4. Use the SAS formula for the area of a triangle.

The concepts of solving triangles developed in **Chapter 15** can be extended to *all* triangles. If any three of the six side and angle measures of a triangle are known (provided at least one measure is the length of a side), then the other three measures can be found.

**OBJECTIVE 1** Review the triangle congruence axioms. The following axioms from geometry allow us to prove that two triangles are congruent (that is, their corresponding sides and angles are equal).

### Congruence Axioms

**Side-Angle-Side (SAS)**    If two sides and the included angle of one triangle are equal, respectively, to two sides and the included angle of a second triangle, then the triangles are congruent.

**Angle-Side-Angle (ASA)**    If two angles and the included side of one triangle are equal, respectively, to two angles and the included side of a second triangle, then the triangles are congruent.

**Side-Side-Side (SSS)**    If three sides of one triangle are equal, respectively, to three sides of a second triangle, then the triangles are congruent.

If a side and *any* two angles are given (SAA), the third angle is easily determined by the angle sum formula ($A + B + C = 180°$), and then the ASA axiom can be applied. Keep in mind that whenever SAS, ASA, or SSS is given, the triangle is unique. We continue to label triangles as we did earlier with right triangles: side $a$ opposite angle $A$, side $b$ opposite angle $B$, and side $c$ opposite angle $C$.

A triangle that is not a right triangle is called an **oblique triangle.** The measures of the three sides and the three angles of a triangle can be found if at least one side and any other two measures are known. There are four possible cases.

### Data Required for Solving Oblique Triangles

*Case 1*   One side and two angles are known (SAA or ASA).

*Case 2*   Two sides and one angle not included between the two sides are known (SSA). This case may lead to more than one triangle.

*Case 3*   Two sides and the angle included between the two sides are known (SAS).

*Case 4*   Three sides are known (SSS).

**CAUTION** *If we know three angles of a triangle, we cannot find unique side lengths since AAA assures us only of similarity, not congruence.* For example, there are infinitely many triangles $ABC$ with $A = 35°$, $B = 65°$, and $C = 80°$.

Case 1, discussed in this section, and Case 2, discussed in **Section 20.2,** require the *law of sines.* Cases 3 and 4, discussed in **Section 20.3,** require the *law of cosines.*

**OBJECTIVE 2** Derive the law of sines. To derive the law of sines, we start with an oblique triangle, such as the *acute triangle* in Figure 1(a) or the *obtuse triangle* in Figure 1(b). This discussion applies to both triangles. First, construct the perpendicular from $B$ to side $AC$ (or its extension). Let $h$ be the length of this perpendicular. Then $c$ is the hypotenuse of right triangle $ADB$, and $a$ is the hypotenuse of right triangle $BDC$.

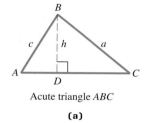

Acute triangle *ABC*

(a)

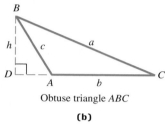

Obtuse triangle *ABC*

(b)

**FIGURE 1**

In triangle $ADB$, $\quad \sin A = \dfrac{h}{c}, \quad$ or $\quad h = c \sin A$.

In triangle $BDC$, $\quad \sin C = \dfrac{h}{a}, \quad$ or $\quad h = a \sin C$.

Since $h = c \sin A$ and $h = a \sin C$,

$$a \sin C = c \sin A$$

$$\dfrac{a}{\sin A} = \dfrac{c}{\sin C}. \quad \text{Divide each side by sin } A \text{ sin } C.$$

In a similar way, by constructing perpendicular lines from the other vertices, it can be shown that

$$\dfrac{a}{\sin A} = \dfrac{b}{\sin B} \quad \text{and} \quad \dfrac{b}{\sin B} = \dfrac{c}{\sin C}.$$

This discussion proves the following theorem.

### Law of Sines

In any triangle $ABC$, with sides $a$, $b$, and $c$,

$$\dfrac{a}{\sin A} = \dfrac{b}{\sin B}, \quad \dfrac{a}{\sin A} = \dfrac{c}{\sin C}, \quad \text{and} \quad \dfrac{b}{\sin B} = \dfrac{c}{\sin C}.$$

This can be written in compact form as

$$\dfrac{a}{\sin A} = \dfrac{b}{\sin B} = \dfrac{c}{\sin C}.$$

That is, according to the law of sines, the lengths of the sides in a triangle are proportional to the sines of the measures of the angles opposite them.

When solving for an angle, we use an alternative form of the law of sines,

$$\dfrac{\sin A}{a} = \dfrac{\sin B}{b} = \dfrac{\sin C}{c}. \quad \text{Alternative form}$$

When using the law of sines, a good strategy is to select an equation so that the unknown variable is in the numerator and all other variables are known.

**OBJECTIVE 3** Solve SSA and ASA triangles (Case 1). If two angles and one side of a triangle are known (Case 1, SAA or ASA), then the law of sines can be used to solve the triangle.

> **Be sure to label a sketch carefully to help set up the correct equation.**

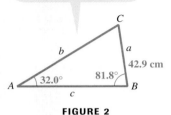

FIGURE 2

**EXAMPLE 1** Using the Law of Sines to Solve a Triangle (SAA)

Solve triangle $ABC$ if $A = 32.0°$, $B = 81.8°$, and $a = 42.9$ cm.

**Solution** Start by drawing a triangle, roughly to scale, and labeling the given parts as in Figure 2. Since the values of $A$, $B$, and $a$ are known, use the form of the law of sines that involves these variables, and then solve for $b$.

*Choose a form that has the unknown variable in the numerator.*

$$\frac{a}{\sin A} = \frac{b}{\sin B} \qquad \text{Law of sines}$$

$$\frac{42.9}{\sin 32.0°} = \frac{b}{\sin 81.8°} \qquad \text{Substitute the given values.}$$

$$b = \frac{42.9 \sin 81.8°}{\sin 32.0°} \qquad \text{Multiply by } \sin 81.8°; \text{ rewrite.}$$

$$b \approx 80.1 \text{ cm} \qquad \text{Approximate with a calculator.}$$

To find $C$, use the fact that the sum of the angles of any triangle is $180°$.

$$A + B + C = 180°$$
$$C = 180° - A - B$$
$$C = 180° - 32.0° - 81.8°$$
$$C = 66.2°$$

Use the law of sines to find $c$. (Why does the Pythagorean theorem not apply?)

$$\frac{a}{\sin A} = \frac{c}{\sin C} \qquad \text{Law of sines}$$

$$\frac{42.9}{\sin 32.0°} = \frac{c}{\sin 66.2°} \qquad \text{Substitute.}$$

$$c = \frac{42.9 \sin 66.2°}{\sin 32.0°} \qquad \text{Multiply by } \sin 66.2°; \text{ rewrite.}$$

$$c \approx 74.1 \text{ cm} \qquad \text{Approximate with a calculator.}$$

**NOW TRY** Exercise 7.

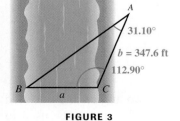

FIGURE 3

**EXAMPLE 2** Using the Law of Sines in an Application (ASA)

Jerry Keefe wishes to measure the distance across the Big Muddy River. See Figure 3. He determines that $C = 112.90°$, $A = 31.10°$, and $b = 347.6$ ft. Find the distance $a$ across the river.

**Solution** To use the law of sines, one side and the angle opposite it must be known. Since $b$ is the only side whose length is given, angle $B$ must be found before the law of sines can be used.

$$B = 180° - A - C$$
$$B = 180° - 31.10° - 112.90° = 36.00°$$

*(continued)*

Now use the form of the law of sines involving $A$, $B$, and $b$ to find $a$.

**Solve for a.**

$$\frac{a}{\sin A} = \frac{b}{\sin B}$$ Law of sines

$$\frac{a}{\sin 31.10°} = \frac{347.6}{\sin 36.00°}$$ Substitute known values.

$$a = \frac{347.6 \sin 31.10°}{\sin 36.00°}$$ Multiply by $\sin 31.10°$.

$$a \approx 305.5 \text{ ft}$$ Use a calculator.

**NOW TRY** Exercise 25.

**OBJECTIVE 4** Use the SAS formula for the area of a triangle. The method used to derive the law of sines can also be used to derive a formula to find the area of a triangle. A familiar formula for the area of a triangle is $A = \frac{1}{2}bh$, where $A$ represents area, $b$ base, and $h$ height. This formula cannot always be used easily since in practice, $h$ is often unknown. To find another formula, refer to acute triangle $ABC$ in Figure 4(a) or obtuse triangle $ABC$ in Figure 4(b).

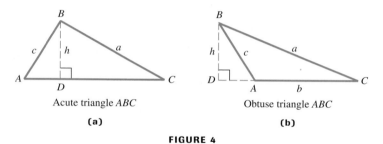

Acute triangle $ABC$
(a)

Obtuse triangle $ABC$
(b)

**FIGURE 4**

A perpendicular has been drawn from $B$ to the base of the triangle (or the extension of the base). This perpendicular forms two right triangles. Using triangle $ABD$,

$$\sin A = \frac{h}{c}, \quad \text{or} \quad h = c \sin A.$$

Substituting into the formula for the area of a triangle,

$$A = \frac{1}{2}bh = \frac{1}{2}bc \sin A.$$

Any other pair of sides and the angle between them could have been used.

### Area of a Triangle (SAS)

In any triangle $ABC$, the area $A$ is given by the following formulas:

$$A = \frac{1}{2}bc \sin A, \quad A = \frac{1}{2}ab \sin C, \quad \text{and} \quad A = \frac{1}{2}ac \sin B.$$

That is, the area is half the product of the lengths of two sides and the sine of the angle included between them. If the included angle measures 90°, its sine is 1, and the formula becomes the familiar $A = \frac{1}{2}bh$.

**EXAMPLE 3** Finding the Area of a Triangle (SAS)

Find the area of triangle $ABC$ in Figure 5.

**Solution** We are given $B = 55° \, 10'$, $a = 34.0$ ft, and $c = 42.0$ ft, so

$$\mathcal{A} = \frac{1}{2}ac \sin B = \frac{1}{2}(34.0)(42.0) \sin 55° \, 10' \approx 586 \text{ ft}^2.$$

**NOW TRY** Exercise 35.

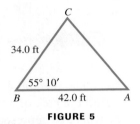

**FIGURE 5**

**EXAMPLE 4** Finding the Area of a Triangle (ASA)

Find the area of triangle $ABC$ if $A = 24° \, 40'$, $b = 27.3$ cm, and $C = 52° \, 40'$.

**Solution** Before the area formula can be used, we must find either $a$ or $c$. Since the sum of the measures of the angles of any triangle is 180°,

$$B = 180° - 24° \, 40' - 52° \, 40' = 102° \, 40'.$$

We can use the law of sines to find $a$.

Solve for $a$.  $\dfrac{a}{\sin A} = \dfrac{b}{\sin B}$  Law of sines

$\dfrac{a}{\sin 24° \, 40'} = \dfrac{27.3}{\sin 102° \, 40'}$  Substitute known values.

$a = \dfrac{27.3 \sin 24° \, 40'}{\sin 102° \, 40'}$  Multiply by $\sin 24° \, 40'$.

$a \approx 11.7$ cm  Use a calculator.

Now, we find the area.

$$\mathcal{A} = \frac{1}{2}ab \sin C = \frac{1}{2}(11.7)(27.3) \sin 52° \, 40' \approx 127 \text{ cm}^2$$

**NOW TRY** Exercise 41.

**CAUTION** Whenever possible, use given values in solving triangles or finding areas rather than values obtained in intermediate steps to avoid possible rounding errors.

## 20.1 Exercises

 Exercise

1. *Multiple Choice* Consider the oblique triangle $ABC$. Which one of the following proportions is *not* valid?

   **A.** $\dfrac{a}{b} = \dfrac{\sin A}{\sin B}$    **B.** $\dfrac{a}{\sin A} = \dfrac{b}{\sin B}$

   **C.** $\dfrac{\sin A}{a} = \dfrac{b}{\sin B}$    **D.** $\dfrac{\sin A}{a} = \dfrac{\sin B}{b}$

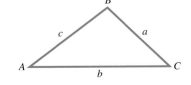

**2.** *Multiple Choice* Which two of the following situations do not provide sufficient information for solving a triangle by the law of sines?

    **A.** You are given two angles and the side included between them.
    **B.** You are given two angles and a side opposite one of them.
    **C.** You are given two sides and the angle included between them.
    **D.** You are given three sides.

*Find the length of each side a. Do not use a calculator.*

**3.**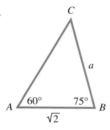

**4.**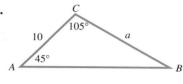

*Determine the remaining sides and angles of each triangle ABC. See Example 1.*

**5.**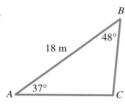

**6.**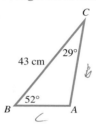

**7.**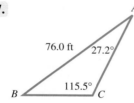

**8.**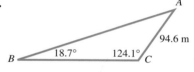

**9.** $A = 68.41°, B = 54.23°, a = 12.75$ ft

**10.** $C = 74.08°, B = 69.38°, c = 45.38$ m

**11.** $A = 87.2°, b = 75.9$ yd, $C = 74.3°$

**12.** $B = 38° \, 40', a = 19.7$ cm, $C = 91° \, 40'$

**13.** $B = 20° \, 50', C = 103° \, 10', AC = 132$ ft

**14.** $A = 35.3°, B = 52.8°, AC = 675$ ft

**15.** $A = 39.70°, C = 30.35°, b = 39.74$ m

**16.** $C = 71.83°, B = 42.57°, a = 2.614$ cm

**17.** $B = 42.88°, C = 102.40°, b = 3974$ ft

**18.** $A = 18.75°, B = 51.53°, c = 2798$ yd

**19.** $A = 39° \, 54', a = 268.7$ m, $B = 42° \, 32'$

**20.** $C = 79° \, 18', c = 39.81$ mm, $A = 32° \, 57'$

**21.** Explain why the law of sines cannot be used to solve a triangle if we are given the lengths of the three sides of the triangle.

**22.** *Concept Check* In Example 1, we asked the question, "Why does the Pythagorean theorem not apply?" Answer this question.

23. Terry Harris, a perceptive trigonometry student, makes the statement, "If we know *any* two angles and one side of a triangle, then the triangle is uniquely determined." Is this a valid statement? Explain, referring to the congruence axioms given in this section.

24. *Concept Check* If $a$ is twice as long as $b$, is $A$ necessarily twice as large as $B$?

*Solve each problem. See Example 2.*

25. To find the distance $AB$ across a river, a surveyor laid off a distance $BC = 354$ m on one side of the river. It is found that $B = 112°\ 10'$ and $C = 15°\ 20'$. Find $AB$. See the figure.

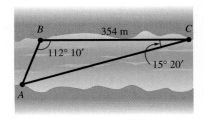

26. To determine the distance $RS$ across a deep canyon, Rhonda lays off a distance $TR = 582$ yd. She then finds that $T = 32°\ 50'$ and $R = 102°\ 20'$. Find $RS$.

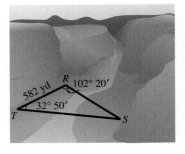

27. A folding chair is to have a seat 12.0 in. deep with angles as shown in the figure. How far down from the seat should the crossing legs be joined? (Find $x$ in the figure.)

28. Three atoms with atomic radii of 2.0, 3.0, and 4.5 are arranged as in the figure. Find the distance between the centers of atoms $A$ and $C$.

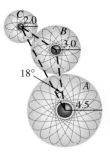

29. Three gears are arranged as shown in the figure. Find angle $\theta$.

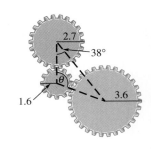

30. A balloonist is directly above a straight road 1.5 mi long that joins two villages. She finds that the town closer to her is at an angle of depression of 35°, and the farther town is at an angle of depression of 31°. How high above the ground is the balloon?

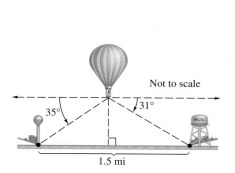

*Find the area of each triangle using the formula $A = \frac{1}{2}bh$, and then verify that the formula $A = \frac{1}{2}ab \sin C$ gives the same result.*

31.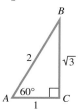

32.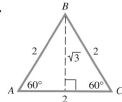

33.

34.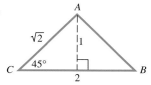

*Find the area of each triangle ABC. See Examples 3 and 4.*

35. $A = 42.5°, b = 13.6$ m, $c = 10.1$ m
36. $C = 72.2°, b = 43.8$ ft, $a = 35.1$ ft
37. $B = 124.5°, a = 30.4$ cm, $c = 28.4$ cm
38. $C = 142.7°, a = 21.9$ km, $b = 24.6$ km
39. $A = 56.80°, b = 32.67$ in., $c = 52.89$ in.
40. $A = 34.97°, b = 35.29$ m, $c = 28.67$ m
41. $A = 30.50°, b = 13.00$ cm, $C = 112.60°$
42. $A = 59.80°, b = 15.00$ m, $C = 53.10°$

*Solve each problem.*

43. A painter is going to apply a special coating to a triangular metal plate on a new building. Two sides measure 16.1 m and 15.2 m. She knows that the angle between these sides is 125°. What is the area of the surface she plans to cover with the coating?

44. A real estate agent wants to find the area of a triangular lot. A surveyor takes measurements and finds that two sides are 52.1 m and 21.3 m, and the angle between them is 42.2°. What is the area of the triangular lot?

# 20.2 The Ambiguous Case of the Law of Sines

**OBJECTIVES**

1. Understand the ambiguous case.
2. Solve SSA triangles (Case 2).
3. Analyze data for the possible number of triangles.

**OBJECTIVE 1** Understand the ambiguous case. We used the law of sines to solve triangles involving Case 1, SAA or ASA, in **Section 20.1**. If we are given the lengths of two sides and the angle opposite one of them (Case 2, SSA), then zero, one, or two such triangles may exist. (There is no SSA axiom.)

Suppose we know the measure of acute angle $A$ of triangle $ABC$, the length of side $a$, and the length of side $b$, as shown in Figure 6. Now we must draw the side of length $a$ opposite angle $A$. The table shows possible outcomes. This situation (SSA) is called the **ambiguous case** of the law of sines.

If angle $A$ is acute, there are four possible outcomes.

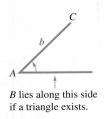

$B$ lies along this side if a triangle exists.

**FIGURE 6**

| Number of Triangles | Sketch | Applying Law of Sines Leads to |
|---|---|---|
| 0 | | $\sin B > 1$, $a < h < b$ |
| 1 | | $\sin B = 1$, $a = h$ and $h < b$ |
| 1 | | $0 < \sin B < 1$, $a \geq b$ |
| 2 | | $0 < \sin B_2 < 1$, $h < a < b$ |

If angle $A$ is obtuse, there are two possible outcomes.

| Number of Triangles | Sketch | Applying Law of Sines Leads to |
|---|---|---|
| 0 | | $\sin B \geq 1$, $a \leq b$ |
| 1 | | $0 < \sin B < 1$, $a > b$ |

### Applying the Law of Sines

1. For any angle $\theta$ of a triangle, $0 < \sin \theta \leq 1$. If $\sin \theta = 1$, then $\theta = 90°$ and the triangle is a right triangle.
2. $\sin \theta = \sin(180° - \theta)$ (Supplementary angles have the same sine value.)
3. The smallest angle is opposite the shortest side, the largest angle is opposite the longest side, and the middle-valued angle is opposite the intermediate side (assuming the triangle has sides that are all of different lengths).

**OBJECTIVE 2** Solve SSA triangles (Case 2).

**EXAMPLE 1** Solving the Ambiguous Case (No Such Triangle)

Solve triangle $ABC$ if $B = 55° \, 40'$, $b = 8.94$ m, and $a = 25.1$ m.

**Solution** Since we are given $B$, $b$, and $a$, we can use the law of sines to find $A$.

*Choose a form that has the unknown variable in the numerator.*

$$\frac{\sin A}{a} = \frac{\sin B}{b} \quad \text{Law of sines (alternative form)}$$

$$\frac{\sin A}{25.1} = \frac{\sin 55° \, 40'}{8.94} \quad \text{Substitute the given values.}$$

$$\sin A = \frac{25.1 \sin 55° \, 40'}{8.94} \quad \text{Multiply by 25.1.}$$

$$\sin A \approx 2.3184379 \quad \text{Use a calculator.}$$

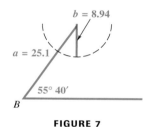

**FIGURE 7**

Since $\sin A$ cannot be greater than 1, there can be no such angle $A$ and, thus, no triangle with the given information. An attempt to sketch such a triangle leads to the situation shown in Figure 7.

**NOW TRY** Exercise 17.

In the ambiguous case, we are given two sides and an angle opposite one of the sides (SSA). For example, suppose $b$, $c$, and angle $C$ are given. This situation represents the ambiguous case because angle $C$ is opposite side $c$.

**EXAMPLE 2** Solving the Ambiguous Case (Two Triangles)

Solve triangle $ABC$ if $A = 55.3°$, $a = 22.8$ ft, and $b = 24.9$ ft.

**Solution** To begin, use the law of sines to find angle $B$.

$$\frac{\sin A}{a} = \frac{\sin B}{b} \quad \text{Solve for sin B.}$$

$$\frac{\sin 55.3°}{22.8} = \frac{\sin B}{24.9} \quad \text{Substitute known values.}$$

$$\sin B = \frac{24.9 \sin 55.3°}{22.8} \quad \text{Multiply by 24.9; rewrite.}$$

$$\sin B \approx 0.8978678 \quad \text{Use a calculator.}$$

There are two angles $B$ between $0°$ and $180°$ that satisfy this condition. Since $\sin B \approx 0.8978678$, to the nearest tenth one value of $B$ is

$$B_1 = 63.9°. \quad \text{Use the inverse sine function.}$$

*(continued)*

Supplementary angles have the same sine value, so another *possible* value of $B$ is

$$B_2 = 180° - 63.9° = 116.1°.$$

To see if $B_2 = 116.1°$ is a valid possibility, add $116.1°$ to the measure of $A$, $55.3°$. Since $116.1° + 55.3° = 171.4°$, and this sum is less than $180°$, it is a valid angle measure for this triangle.

Now separately solve triangles $AB_1C_1$ and $AB_2C_2$ shown in Figure 8. Begin with $AB_1C_1$. Find $C_1$ first.

$$C_1 = 180° - A - B_1$$
$$C_1 = 180° - 55.3° - 63.9°$$
$$C_1 = 60.8°$$

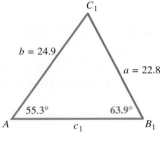

Now, use the law of sines to find $c_1$.

$$\frac{a}{\sin A} = \frac{c_1}{\sin C_1} \quad \text{Solve for } c_1.$$

$$\frac{22.8}{\sin 55.3°} = \frac{c_1}{\sin 60.8°}$$

$$c_1 = \frac{22.8 \sin 60.8°}{\sin 55.3°}$$

$$c_1 \approx 24.2 \text{ ft}$$

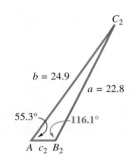

FIGURE 8

To solve triangle $AB_2C_2$, first find $C_2$.

$$C_2 = 180° - A - B_2$$
$$C_2 = 180° - 55.3° - 116.1°$$
$$C_2 = 8.6°$$

Use the law of sines to find $c_2$.

$$\frac{a}{\sin A} = \frac{c_2}{\sin C_2} \quad \text{Solve for } c_2.$$

$$\frac{22.8}{\sin 55.3°} = \frac{c_2}{\sin 8.6°}$$

$$c_2 = \frac{22.8 \sin 8.6°}{\sin 55.3°}$$

$$c_2 \approx 4.15 \text{ ft}$$

**NOW TRY** Exercise 25.

The ambiguous case results in zero, one, or two triangles. The following guidelines can be used to determine how many triangles there are.

### Number of Triangles Satisfying the Ambiguous Case (SSA)

Let sides $a$ and $b$ and angle $A$ be given in triangle $ABC$. (The law of sines can be used to calculate the value of $\sin B$.)

1. If applying the law of sines results in an equation having $\sin B > 1$, then *no triangle* satisfies the given conditions.
2. If $\sin B = 1$, then *one triangle* satisfies the given conditions and $B = 90°$.
3. If $0 < \sin B < 1$, then either *one or two triangles* satisfy the given conditions.
   (a) If $\sin B = k$, then let $B_1 = \sin^{-1} k$ and use $B_1$ for $B$ in the first triangle.
   (b) Let $B_2 = 180° - B_1$. If $A + B_2 < 180°$, then a second triangle exists. In this case, use $B_2$ for $B$ in the second triangle.

### EXAMPLE 3  Solving the Ambiguous Case (One Triangle)

Solve triangle $ABC$, given $A = 43.5°$, $a = 10.7$ in., and $c = 7.2$ in.

**Solution**  To find angle $C$, use an alternative form of the law of sines.

$$\frac{\sin C}{c} = \frac{\sin A}{a}$$

$$\frac{\sin C}{7.2} = \frac{\sin 43.5°}{10.7}$$

$$\sin C = \frac{7.2 \sin 43.5°}{10.7} \approx 0.46319186$$

$$C \approx 27.6° \quad \text{Use the inverse sine function.}$$

There is another angle $C$ that has sine value $0.46319186$; it is $C = 180° - 27.6° = 152.4°$. However, notice in the given information that $c < a$, meaning that in the triangle, angle $C$ must have measure *less than* angle $A$. Notice also that when we add this obtuse value to the given angle $A = 43.5°$, we obtain

$$152.4° + 43.5° = 195.9°,$$

which is greater than $180°$. So either of these approaches shows that there can be only one triangle. See Figure 9. Then

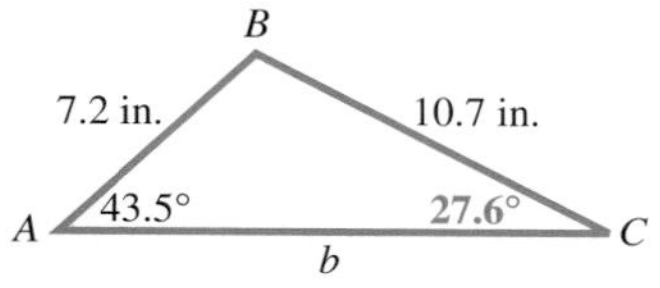

$$B = 180° - 27.6° - 43.5° = 108.9°,$$

and we can find side $b$ with the law of sines.

$$\frac{b}{\sin B} = \frac{a}{\sin A}$$

$$\frac{b}{\sin 108.9°} = \frac{10.7}{\sin 43.5°}$$

$$b = \frac{10.7 \sin 108.9°}{\sin 43.5°}$$

$$b \approx 14.7 \text{ in.}$$

**FIGURE 9**

**NOW TRY** Exercise 21.

**OBJECTIVE 3**  Analyze data for the possible number of triangles.

### EXAMPLE 4  Analyzing Data Involving an Obtuse Angle

Without using the law of sines, explain why $A = 104°$, $a = 26.8$ m, and $b = 31.3$ m cannot be valid for a triangle $ABC$.

**Solution**  Since $A$ is an obtuse angle, it is the largest angle and so the longest side of the triangle must be $a$. However, we are given $b > a$; thus, $B > A$, which is impossible if $A$ is obtuse. Therefore, no such triangle $ABC$ exists.

**NOW TRY** Exercise 33.

## 20.2 Exercises

**1.** *Multiple Choice* Which one of the following sets of data does not determine a unique triangle?
  A. $A = 40°, B = 60°, C = 80°$
  B. $a = 5, b = 12, c = 13$
  C. $a = 3, b = 7, C = 50°$
  D. $a = 2, b = 2, c = 2$

**2.** *Multiple Choice* Which one of the following sets of data determines a unique triangle?
  A. $A = 50°, B = 50°, C = 80°$
  B. $a = 3, b = 5, c = 20$
  C. $A = 40°, B = 20°, C = 30°$
  D. $a = 7, b = 24, c = 25$

*Concept Check* In each figure, a line of length h is to be drawn from the given point to the positive x-axis in order to form a triangle. For what value(s) of h can you draw the following?

  **(a)** two triangles   **(b)** exactly one triangle   **(c)** no triangle

**3.**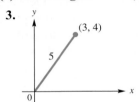

**4.**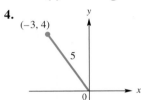

*Determine the number of triangles ABC possible with the given parts. See Examples 1–4.*

**5.** $a = 50, b = 26, A = 95°$
**6.** $b = 60, a = 82, B = 100°$
**7.** $a = 31, b = 26, B = 48°$
**8.** $a = 35, b = 30, A = 40°$
**9.** $a = 50, b = 61, A = 58°$
**10.** $B = 54°, c = 28, b = 23$

*Find each angle B. Do not use a calculator.*

**11.**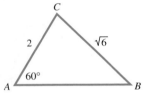

**12.**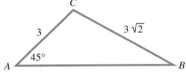

*Find the unknown angles in triangle ABC for each triangle that exists. See Examples 1–3.*

**13.** $A = 29.7°, b = 41.5$ ft, $a = 27.2$ ft
**14.** $B = 48.2°, a = 890$ cm, $b = 697$ cm
**15.** $C = 41° 20', b = 25.9$ m, $c = 38.4$ m
**16.** $B = 48° 50', a = 3850$ in., $b = 4730$ in.
**17.** $B = 74.3°, a = 859$ m, $b = 783$ m
**18.** $C = 82.2°, a = 10.9$ km, $c = 7.62$ km
**19.** $A = 142.13°, b = 5.432$ ft, $a = 7.297$ ft
**20.** $B = 113.72°, a = 189.6$ yd, $b = 243.8$ yd

*Solve each triangle ABC that exists. See Examples 1–3.*

**21.** $A = 42.5°, a = 15.6$ ft, $b = 8.14$ ft
**22.** $C = 52.3°, a = 32.5$ yd, $c = 59.8$ yd
**23.** $B = 72.2°, b = 78.3$ m, $c = 145$ m
**24.** $C = 68.5°, c = 258$ cm, $b = 386$ cm
**25.** $A = 38° 40', a = 9.72$ km, $b = 11.8$ km
**26.** $C = 29° 50', a = 8.61$ m, $c = 5.21$ m
**27.** $A = 96.80°, b = 3.589$ ft, $a = 5.818$ ft
**28.** $C = 88.70°, b = 56.87$ yd, $c = 112.4$ yd
**29.** $B = 39.68°, a = 29.81$ m, $b = 23.76$ m
**30.** $A = 51.20°, c = 7986$ cm, $a = 7208$ cm

31. Apply the law of sines to the following: $a = \sqrt{5}, c = 2\sqrt{5}, A = 30°$. What is the value of $\sin C$? What is the measure of $C$? Based on its angle measures, what kind of triangle is triangle $ABC$?

32. Explain the condition that must exist to determine that there is no triangle satisfying the given values of $a$, $b$, and $B$ once the value of $\sin A$ is found.

33. Without using the law of sines, explain why no triangle $ABC$ exists satisfying $A = 103° \, 20'$, $a = 14.6$ ft, $b = 20.4$ ft.

34. Apply the law of sines to the data given in Example 4. Describe in your own words what happens when you try to find the measure of angle $B$ using a calculator.

*Use the law of sines to solve each problem.*

35. To find the distance between a point $X$ and an inaccessible point $Z$, a line segment $XY$ is constructed. It is found that $XY = 960$ m, angle $XYZ = 43° \, 30'$, and angle $YZX = 95° \, 30'$. Find the distance between $X$ and $Z$ to the nearest meter.

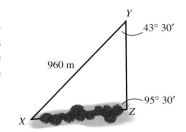

36. The angle of elevation from the top of a building 45.0 ft high to the top of a nearby antenna tower is $15° \, 20'$. From the base of the building, the angle of elevation of the tower is $29° \, 30'$. Find the height of the tower.

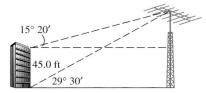

## 20.3 The Law of Cosines

**OBJECTIVES**

1. Derive the law of cosines.
2. Solve SAS and SSS triangles (Cases 3 and 4).
3. Use Heron's formula for the area of a triangle.

As mentioned in **Section 20.1**, if we are given two sides and the included angle (Case 3) or three sides (Case 4) of a triangle, then a unique triangle is determined. These are the SAS and SSS cases, respectively. Both cases require using the *law of cosines*. Remember the following property of triangles when applying the law of cosines to solve a triangle.

**Triangle Side Length Restriction**

In any triangle, the sum of the lengths of any two sides must be greater than the length of the remaining side.

For example, it would be impossible to construct a triangle with sides of lengths 3, 4, and 10. See Figure 10.

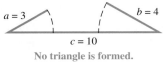

No triangle is formed.

**FIGURE 10**

**OBJECTIVE 1** Derive the law of cosines. To derive the law of cosines, let $ABC$ be any oblique triangle. Choose a coordinate system so that vertex $B$ is at the origin and side $BC$ is along the positive $x$-axis. See Figure 11.

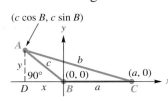

**FIGURE 11**

Let $(x, y)$ be the coordinates of vertex $A$ of the triangle. Verify that for angle $B$, whether obtuse or acute,

$$\sin B = \frac{y}{c} \quad \text{and} \quad \cos B = \frac{x}{c}.$$

$$y = c \sin B \quad \text{and} \quad x = c \cos B. \quad \text{Here } x \text{ is negative when } B \text{ is obtuse.}$$

Thus, the coordinates of point $A$ become $(c \cos B, c \sin B)$.

Point $C$ in Figure 11 has coordinates $(a, 0)$, and $AC$ has length $b$. Since point $A$ has coordinates $(c \cos B, c \sin B)$, by the distance formula,

$$b = \sqrt{(c \cos B - a)^2 + (c \sin B - 0)^2}$$
$$b^2 = (c \cos B - a)^2 + (c \sin B)^2 \quad \text{Square both sides.}$$
$$= (c^2 \cos^2 B - 2ac \cos B + a^2) + c^2 \sin^2 B \quad \text{Multiply; } (x - y)^2 = x^2 - 2xy + y^2$$

**Remember the middle term.**

$$= a^2 + c^2(\cos^2 B + \sin^2 B) - 2ac \cos B \quad \text{Properties of real numbers}$$
$$= a^2 + c^2(1) - 2ac \cos B \quad \text{Fundamental identity}$$
$$b^2 = a^2 + c^2 - 2ac \cos B.$$

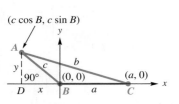

**FIGURE 11 (repeated)**

This result is one form of the law of cosines. In our work, we could just as easily have placed $A$ or $C$ at the origin. This would have given the same result, but with the variables rearranged.

### Law of Cosines

In any triangle $ABC$, with sides $a$, $b$, and $c$,

$$a^2 = b^2 + c^2 - 2bc \cos A,$$
$$b^2 = a^2 + c^2 - 2ac \cos B,$$
$$c^2 = a^2 + b^2 - 2ab \cos C.$$

That is, according to the law of cosines, the square of a side of a triangle is equal to the sum of the squares of the other two sides, minus twice the product of those two sides and the cosine of the angle included between them.

If we let $C = 90°$ in the third form of the law of cosines, then $\cos C = \cos 90° = 0$, and the formula becomes $c^2 = a^2 + b^2$, the Pythagorean theorem. The Pythagorean theorem is a special case of the law of cosines.

**OBJECTIVE 2** Solve SAS and SSS triangles (Cases 3 and 4).

**EXAMPLE 1** Using the Law of Cosines in an Application (SAS)

A surveyor wishes to find the distance between two inaccessible points $A$ and $B$ on opposite sides of a lake. While standing at point $C$, she finds that $AC = 259$ m, $BC = 423$ m, and angle $ACB$ measures $132° 40'$. Find the distance $AB$. See Figure 12.

**Solution** The law of cosines can be used here since we know the lengths of two sides of the triangle and the measure of the included angle.

$$AB^2 = 259^2 + 423^2 - 2(259)(423) \cos 132° 40'$$
$$AB^2 \approx 394{,}510.6 \quad \text{Use a calculator.}$$
$$AB \approx 628 \quad \text{Take the square root of each side.}$$

The distance between the points is approximately 628 m.

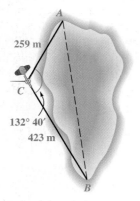

**FIGURE 12**

**NOW TRY** Exercise 35.

## EXAMPLE 2  Using the Law of Cosines to Solve a Triangle (SAS)

Solve triangle $ABC$ if $A = 42.3°$, $b = 12.9$ m, and $c = 15.4$ m.

**Solution**  See Figure 13. We start by finding $a$ with the law of cosines.

$$a^2 = b^2 + c^2 - 2bc \cos A \qquad \text{Law of cosines}$$
$$a^2 = 12.9^2 + 15.4^2 - 2(12.9)(15.4) \cos 42.3° \qquad \text{Substitute.}$$
$$a^2 \approx 109.7 \qquad \text{Use a calculator.}$$
$$a \approx 10.47 \text{ m} \qquad \text{Take square roots.}$$

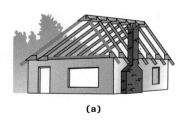

FIGURE 13

Of the two remaining angles $B$ and $C$, $B$ must be the smaller since it is opposite the shorter of the two sides $b$ and $c$. Therefore, $B$ cannot be obtuse.

$$\frac{\sin A}{a} = \frac{\sin B}{b} \qquad \text{Law of sines}$$
$$\frac{\sin 42.3°}{10.47} = \frac{\sin B}{12.9} \qquad \text{Substitute.}$$
$$\sin B = \frac{12.9 \sin 42.3°}{10.47} \qquad \text{Multiply by 12.9; rewrite.}$$
$$B \approx 56.0° \qquad \text{Use the inverse sine function.}$$

The easiest way to find $C$ is to subtract the measures of $A$ and $B$ from $180°$.

$$C = 180° - 42.3° - 56.0° = 81.7°$$

**NOW TRY** Exercise 19.

**CAUTION**  Had we used the law of sines to find $C$ rather than $B$ in Example 2, we would not have known whether $C$ is equal to $81.7°$ or its supplement, $98.3°$.

## EXAMPLE 3  Using the Law of Cosines to Solve a Triangle (SSS)

Solve triangle $ABC$ if $a = 9.47$ ft, $b = 15.9$ ft, and $c = 21.1$ ft.

**Solution**  We can use the law of cosines to solve for any angle of the triangle. We solve for $C$, the largest angle. We will know that $C$ is obtuse if $\cos C < 0$.

$$c^2 = a^2 + b^2 - 2ab \cos C \qquad \text{Law of cosines}$$
$$\cos C = \frac{a^2 + b^2 - c^2}{2ab} \qquad \text{Solve for } \cos C.$$
$$\cos C = \frac{9.47^2 + 15.9^2 - 21.1^2}{2(9.47)(15.9)} \qquad \text{Substitute.}$$
$$\cos C \approx -0.34109402 \qquad \text{Use a calculator.}$$
$$C \approx 109.9° \qquad \text{Use the inverse cosine function.}$$

We can use either the law of sines or the law of cosines to find $B \approx 45.1°$. (Verify this.) Since $A = 180° - B - C$, we obtain $A \approx 25.0°$.

**NOW TRY** Exercise 23.

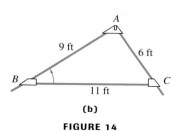

FIGURE 14

Trusses are frequently used to support roofs on buildings, as illustrated in Figure 14(a). The simplest type of roof truss is a triangle, as shown in Figure 14(b). One basic task when constructing a roof truss is to cut the ends of the rafters so that the roof has the correct slope. (*Source:* Riley, W., L. Sturges, and D. Morris, *Statics and Mechanics of Materials,* John Wiley & Sons, 1995.)

### EXAMPLE 4  Designing a Roof Truss (SSS)

Find angle $B$ for the truss shown in Figure 14(b).

**Solution**

$b^2 = a^2 + c^2 - 2ac \cos B$   Law of cosines

$\cos B = \dfrac{a^2 + c^2 - b^2}{2ac}$   Solve for cos $B$.

$\cos B = \dfrac{11^2 + 9^2 - 6^2}{2(11)(9)}$   Let $a = 11$, $b = 6$, and $c = 9$.

$\cos B \approx 0.83838384$   Use a calculator.

$B \approx 33°$   Use the inverse cosine function.

**NOW TRY** Exercise 37.

Four possible cases can occur when solving an oblique triangle, as summarized in the following table. In all four cases, it is assumed that the given information actually produces a triangle.

| Oblique Triangle | Suggested Procedure for Solving |
|---|---|
| **Case 1:** One side and two angles are known. (SAA or ASA) | **Step 1** Find the remaining angle using the angle sum formula ($A + B + C = 180°$). <br> **Step 2** Find the remaining sides using the law of sines. |
| **Case 2:** Two sides and one angle (not included between the two sides) are known. (SSA) | This is the ambiguous case; there may be no triangle, one triangle, or two triangles. <br> **Step 1** Find an angle using the law of sines. <br> **Step 2** Find the remaining angle using the angle sum formula. <br> **Step 3** Find the remaining side using the law of sines. <br> *If two triangles exist, repeat Steps 2 and 3.* |
| **Case 3:** Two sides and the included angle are known. (SAS) | **Step 1** Find the third side using the law of cosines. <br> **Step 2** Find the smaller of the two remaining angles using the law of sines. <br> **Step 3** Find the remaining angle using the angle sum formula. |
| **Case 4:** Three sides are known. (SSS) | **Step 1** Find the largest angle using the law of cosines. <br> **Step 2** Find either remaining angle using the law of sines. <br> **Step 3** Find the remaining angle using the angle sum formula. |

**Heron of Alexandria**

### OBJECTIVE 3  Use Heron's formula for the area of a triangle.

The law of cosines can be used to derive a formula for the area of a triangle given the lengths of the three sides. This formula is known as **Heron's formula,** named after the Greek mathematician Heron of Alexandria, who lived around A.D. 75. It is found in his work *Metrica*. Heron's formula can be used for the case SSS.

### Heron's Area Formula (SSS)

If a triangle has sides of lengths $a$, $b$, and $c$, with **semiperimeter**

$$s = \frac{1}{2}(a + b + c),$$

then the area of the triangle is

$$\mathcal{A} = \sqrt{s(s-a)(s-b)(s-c)}.$$

That is, according to Heron's formula, the area of a triangle is the square root of the product of four factors: (1) the semiperimeter, (2) the semiperimeter minus the first side, (3) the semiperimeter minus the second side, and (4) the semiperimeter minus the third side.

### EXAMPLE 5   Using Heron's Formula to Find an Area (SSS)

The distance "as the crow flies" from Los Angeles to New York is 2451 mi, from New York to Montreal is 331 mi, and from Montreal to Los Angeles is 2427 mi. What is the area of the triangular region having these three cities as vertices? (Ignore the curvature of Earth.)

**Solution**   In Figure 15 we let $a = 2451$, $b = 331$, and $c = 2427$. The semiperimeter $s$ is

$$s = \frac{1}{2}(2451 + 331 + 2427) = 2604.5.$$

Using Heron's formula, the area $\mathcal{A}$ is

$$\mathcal{A} = \sqrt{s(s-a)(s-b)(s-c)}$$

*Don't forget the factor s.*

$$\mathcal{A} = \sqrt{2604.5(2604.5 - 2451)(2604.5 - 331)(2604.5 - 2427)}$$

$$\mathcal{A} \approx 401{,}700 \text{ mi}^2.$$

**NOW TRY** Exercise 59.

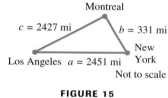

**FIGURE 15**

## 20.3 Exercises

**NOW TRY Exercise**

*Concept Check*   Assume triangle ABC has standard labeling and complete the following.
(a) Determine whether SAA, ASA, SSA, SAS, or SSS is given.
(b) Decide whether the law of sines or the law of cosines should be used to begin solving the triangle.

1. $a$, $b$, and $C$
2. $A$, $C$, and $c$
3. $a$, $b$, and $A$
4. $a$, $b$, and $c$
5. $A$, $B$, and $c$
6. $a$, $c$, and $A$
7. $a$, $B$, and $C$
8. $b$, $c$, and $A$

*Find the length of the remaining side of each triangle. Do not use a calculator.*

9.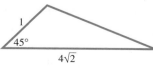

10.

Find the value of θ in each triangle. Do not use a calculator.

11.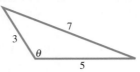

12.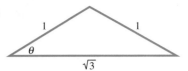

Solve each triangle. Approximate values to the nearest tenth.

13.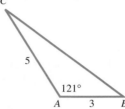

14.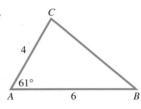

15.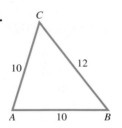

16.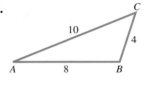

17.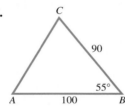

18.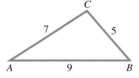

Solve each triangle. See Examples 2 and 3.

19. $A = 41.4°$, $b = 2.78$ yd, $c = 3.92$ yd
20. $C = 28.3°$, $b = 5.71$ in., $a = 4.21$ in.
21. $C = 45.6°$, $b = 8.94$ m, $a = 7.23$ m
22. $A = 67.3°$, $b = 37.9$ km, $c = 40.8$ km
23. $a = 9.3$ cm, $b = 5.7$ cm, $c = 8.2$ cm
24. $a = 28$ ft, $b = 47$ ft, $c = 58$ ft
25. $AB = 1240$ ft, $AC = 876$ ft, $BC = 965$ ft
26. $AB = 298$ m, $AC = 421$ m, $BC = 324$ m
27. $A = 80° 40'$, $b = 143$ cm, $c = 89.6$ cm
28. $C = 72° 40'$, $a = 327$ ft, $b = 251$ ft
29. $B = 74.80°$, $a = 8.919$ in., $c = 6.427$ in.
30. $C = 59.70°$, $a = 3.725$ mi, $b = 4.698$ mi
31. $A = 112.8°$, $b = 6.28$ m, $c = 12.2$ m
32. $B = 168.2°$, $a = 15.1$ cm, $c = 19.2$ cm

33. Refer to Figure 10. If you attempt to find any angle of a triangle with the values $a = 3$, $b = 4$, and $c = 10$ with the law of cosines, what happens?

34. "The shortest distance between two points is a straight line." Explain how this relates to the geometric property that states that the sum of the lengths of any two sides of a triangle must be greater than the length of the remaining side.

*Solve each problem. See Examples 1–4.*

**35.** Points $A$ and $B$ are on opposite sides of Lake Yankee. From a third point, $C$, the angle between the lines of sight to $A$ and $B$ is 46.3°. If $AC$ is 350 m long and $BC$ is 286 m long, find $AB$.

**36.** The sides of a parallelogram are 4.0 cm and 6.0 cm. One angle is 58° while another is 122°. Find the lengths of the diagonals of the parallelogram.

**37.** A triangular truss is shown in the figure. Find angle $\theta$.

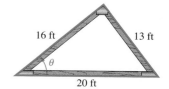

**38.** A crane with a counterweight is shown in the figure. Find the horizontal distance between points $A$ and $B$ to the nearest foot.

**39.** A weight is supported by cables attached to both ends of a balance beam, as shown in the figure. What angles are formed between the beam and the cables?

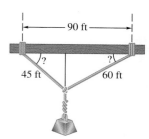

**40.** To measure the distance through a mountain for a proposed tunnel, a point $C$ is chosen that can be reached from each end of the tunnel. (See the figure.) If $AC = 3800$ m, $BC = 2900$ m, and angle $C = 110°$, find the length of the tunnel.

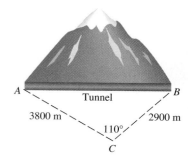

**41.** A baseball diamond is a square, 90.0 ft on a side, with home plate and the three bases as vertices. The pitcher's position is 60.5 ft from home plate. Find the distance from the pitcher's position to each of the bases.

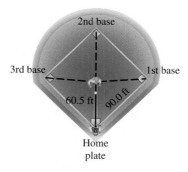

42. The Vietnam Veterans' Memorial in Washington, D.C., is V-shaped with equal sides of length 246.75 ft. The angle between these sides measures 125° 12′. Find the distance between the ends of the two sides. (*Source:* Pamphlet obtained at Vietnam Veterans' Memorial.)

43. The layout for a child's playhouse has the dimensions given in the figure. Find *x*.

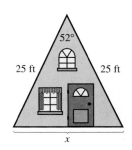

44. To find the distance between two small towns, an electronic distance measuring (EDM) instrument is placed on a hill from which both towns are visible. The distance to each town from the EDM and the angle between the two lines of sight are measured. (See the figure.) Find the distance between the towns.

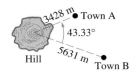

*Find the measure of each angle θ to two decimal places.*

45.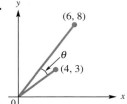

46.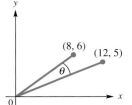

*Find the exact area of each triangle using the formula $A = \frac{1}{2}bh$, and then verify that Heron's formula gives the same result.*

47.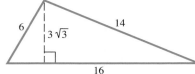

48.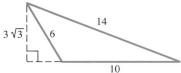

*Find the area of each triangle ABC. See Example 5.*

49. $a = 12$ m, $b = 16$ m, $c = 25$ m
50. $a = 22$ in., $b = 45$ in., $c = 31$ in.
51. $a = 154$ cm, $b = 179$ cm, $c = 183$ cm
52. $a = 25.4$ yd, $b = 38.2$ yd, $c = 19.8$ yd
53. $a = 76.3$ ft, $b = 109$ ft, $c = 98.8$ ft
54. $a = 15.89$ in., $b = 21.74$ in., $c = 10.92$ in.

To help predict eruptions from the volcano Mauna Loa on the island of Hawaii, scientists keep track of the volcano's movement by using a "super triangle" with vertices on the three volcanoes shown on the map at the right. (For example, in one year, Mauna Loa moved 6 in., *a result of increasing internal pressure.*) Refer to the map to work Exercises 55 and 56.

**55.** $AB = 22.47928$ mi, $AC = 28.14276$ mi, $A = 58.56989°$; find $BC$

**56.** $AB = 22.47928$ mi, $BC = 25.24983$ mi, $A = 58.56989°$; find $B$

*Solve each problem. See Example 5.*

**57.** A **perfect triangle** is a triangle whose sides have whole number lengths and whose area is numerically equal to its perimeter. Show that the triangle with sides of length 9, 10, and 17 is perfect.

**58.** A **Heron triangle** is a triangle having integer sides and area. Show that each of the following is a Heron triangle.

(a) $a = 11, b = 13, c = 20$
(b) $a = 13, b = 14, c = 15$
(c) $a = 7, b = 15, c = 20$
(d) $a = 9, b = 10, c = 17$

**59.** Find the area of the Bermuda Triangle if the sides of the triangle have approximate lengths 850 mi, 925 mi, and 1300 mi.

**60.** A painter needs to cover a triangular region 75 m by 68 m by 85 m. A can of paint covers 75 m² of area. How many cans (to the next higher number of cans) will be needed?

## 20.4 Vectors

**OBJECTIVES**

1. Learn the basic terminology of vectors.
2. Understand and apply the algebraic interpretation of vectors.
3. Perform vector operations.

**OBJECTIVE 1** Learn the basic terminology of vectors. Many quantities involve magnitudes, such as 45 lb or 60 mph. These quantities are called **scalars** and can be represented by real numbers. Other quantities, called **vector quantities**, involve both magnitude and direction. Typical vector quantities are velocity, acceleration, and force. For example, traveling 50 mph *east* represents a vector quantity.

A vector quantity is often represented with a directed line segment (a segment that uses an arrowhead to indicate direction) called a **vector**. The length of the vector represents the **magnitude** of the vector quantity. The direction of the vector, indicated by the arrowhead, represents the direction of the quantity. For example, the vector in Figure 16 represents a force of 10 lb applied at an angle of 30° from the horizontal.

The symbol for a vector is often printed in boldface type. When writing vectors by hand, it is customary to use an arrow over the letter or letters. Thus **OP** and $\overrightarrow{OP}$ both represent the vector **OP**. Vectors may be named with either one lowercase or uppercase

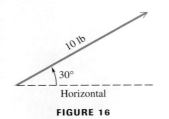

**FIGURE 16**

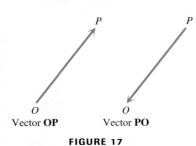

**FIGURE 17**

letter, or two uppercase letters. When two letters are used, the first indicates the **initial point** and the second indicates the **terminal point** of the vector. Knowing these points gives the direction of the vector. For example, vectors **OP** and **PO** in Figure 17 are not the same vector. They have the same magnitude but *opposite* directions. The magnitude of vector **OP** is written $|\mathbf{OP}|$.

Two vectors are equal if and only if they have the same direction and the same magnitude. In Figure 18, vectors **A** and **B** are equal, as are vectors **C** and **D**. As Figure 18 shows, equal vectors need not coincide, but they must be parallel. Vectors **A** and **E** are unequal because they do not have the same direction, while **A** ≠ **F** because they have different magnitudes.

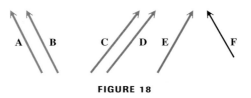

**FIGURE 18**

NOW TRY Exercise 1.

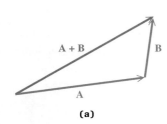

To find the sum of two vectors **A** and **B**, we place the initial point of vector **B** at the terminal point of vector **A**, as shown in Figure 19(a). The vector with the same initial point as **A** and the same terminal point as **B** is the sum **A** + **B**. The sum of two vectors is also a vector.

Another way to find the sum of two vectors is to use the **parallelogram rule.** Place vectors **A** and **B** so that their initial points coincide, as in Figure 19(b). Then, complete a parallelogram that has **A** and **B** as two sides. The diagonal of the parallelogram with the same initial point as **A** and **B** is the sum **A** + **B** found by the definition. Compare Figures 19(a) and (b). Parallelograms can be used to show that vector **B** + **A** is the same as vector **A** + **B**, or that **A** + **B** = **B** + **A**, so vector addition is commutative. The vector sum **A** + **B** is called the **resultant** of vectors **A** and **B**.

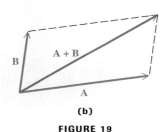

**FIGURE 19**

For every vector **v** there is a vector −**v** that has the same magnitude as **v** but opposite direction. Vector −**v** is called the **opposite** of **v**. (See Figure 20.) The sum of **v** and −**v** has magnitude 0 and is called the **zero vector.** As with real numbers, to subtract vector **B** from vector **A**, find the vector sum **A** + (−**B**). (See Figure 21.)

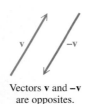

Vectors **v** and −**v** are opposites.

**FIGURE 20**

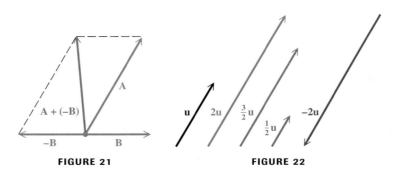

FIGURE 21      FIGURE 22

The **scalar product** of a real number (or scalar) $k$ and a vector **u** is the vector $k \cdot \mathbf{u}$, which has magnitude $|k|$ times the magnitude of **u**. As suggested by Figure 22, the vector $k \cdot \mathbf{u}$ has the same direction as **u** if $k > 0$ and opposite direction if $k < 0$.

NOW TRY Exercises 5, 7, 9, and 11.

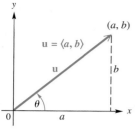

FIGURE 23

**OBJECTIVE 2** Understand and apply the algebraic interpretation of vectors.
We now consider vectors in a rectangular coordinate system. A vector with its initial point at the origin is called a **position vector**. A position vector **u** with its endpoint at the point $(a, b)$ is written $\langle a, b \rangle$, so

$$\mathbf{u} = \langle a, b \rangle.$$

This means that every vector in the real plane corresponds to an ordered pair of real numbers. Thus, geometrically a vector is a directed line segment; algebraically, it is an ordered pair. The numbers $a$ and $b$ are the **horizontal component** and **vertical component**, respectively, of vector **u**. Figure 23 shows the vector $\mathbf{u} = \langle a, b \rangle$. The positive angle between the $x$-axis and a position vector is the **direction angle** for the vector. In Figure 23, $\theta$ is the direction angle for vector **u**.

From Figure 23, we can see that the magnitude and direction of a vector are related to its horizontal and vertical components.

### Magnitude and Direction Angle of a Vector $\langle a, b \rangle$

The magnitude (length) of vector $\mathbf{u} = \langle a, b \rangle$ is given by

$$|\mathbf{u}| = \sqrt{a^2 + b^2}.$$

The direction angle $\theta$ satisfies $\tan \theta = \frac{b}{a}$, where $a \neq 0$.

### EXAMPLE 1 Finding Magnitude and Direction Angle

Find the magnitude and direction angle for $\mathbf{u} = \langle 3, -2 \rangle$.

**Algebraic Solution**

The magnitude is $|\mathbf{u}| = \sqrt{3^2 + (-2)^2} = \sqrt{13}$. To find the direction angle $\theta$, start with $\tan \theta = \frac{b}{a} = \frac{-2}{3} = -\frac{2}{3}$. Vector **u** has a positive horizontal component and a negative vertical component, placing the position vector in quadrant IV. A calculator gives $\tan^{-1}\left(-\frac{2}{3}\right) \approx -33.7°$. Adding 360° yields the direction angle $\theta \approx 326.3°$. See Figure 24.

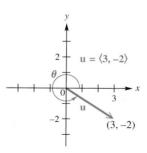

FIGURE 24

**Graphing Calculator Solution**

A calculator returns the magnitude and direction angle, given the horizontal and vertical components. An approximation for $\sqrt{13}$ is given, and the direction angle has a measure with least possible absolute value. We must add 360° to the value of $\theta$ to obtain the positive direction angle. See Figure 25.

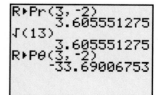

FIGURE 25

For more information, see your owner's manual or the graphing calculator manual that accompanies this text.

**NOW TRY** Exercise 33.

### Horizontal and Vertical Components

The horizontal and vertical components, respectively, of a vector **u** having magnitude $|\mathbf{u}|$ and direction angle $\theta$ are given by

$$a = |\mathbf{u}|\cos\theta \quad \text{and} \quad b = |\mathbf{u}|\sin\theta.$$

That is, $\mathbf{u} = \langle a, b \rangle = \langle |\mathbf{u}|\cos\theta, |\mathbf{u}|\sin\theta \rangle$.

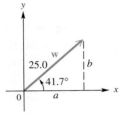

FIGURE 26

**EXAMPLE 2** Finding Horizontal and Vertical Components

Vector **w** in Figure 26 has magnitude 25.0 and direction angle 41.7°. Find the horizontal and vertical components.

**Algebraic Solution**

Use the two formulas in the box, with $|\mathbf{w}| = 25.0$ and $\theta = 41.7°$.

$a = 25.0 \cos 41.7°$ | $b = 25.0 \sin 41.7°$
$a \approx 18.7$ | $b \approx 16.6$

Therefore, $\mathbf{w} = \langle 18.7, 16.6 \rangle$. The horizontal component is 18.7, and the vertical component is 16.6 (rounded to the nearest tenth).

**Graphing Calculator Solution**

See Figure 27. The results support the algebraic solution.

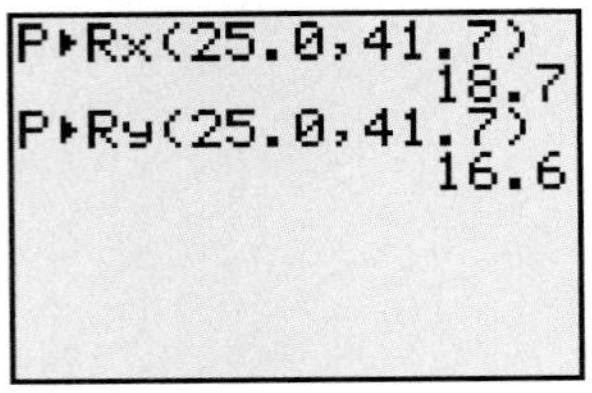

```
P►Rx(25.0,41.7)
            18.7
P►Ry(25.0,41.7)
            16.6
```

FIGURE 27

**NOW TRY** Exercise 37.

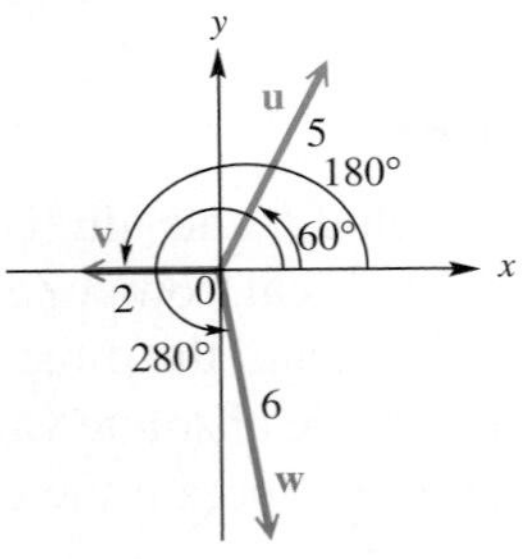

**EXAMPLE 3** Writing Vectors in the Form $\langle a, b \rangle$

Write each vector in Figure 28 in the form $\langle a, b \rangle$.

**Solution**

$$\mathbf{u} = \langle 5\cos 60°, 5\sin 60° \rangle = \left\langle 5\cdot\frac{1}{2}, 5\cdot\frac{\sqrt{3}}{2} \right\rangle = \left\langle \frac{5}{2}, \frac{5\sqrt{3}}{2} \right\rangle$$

$$\mathbf{v} = \langle 2\cos 180°, 2\sin 180° \rangle = \langle 2(-1), 2(0) \rangle = \langle -2, 0 \rangle$$

$$\mathbf{w} = \langle 6\cos 280°, 6\sin 280° \rangle \approx \langle 1.0419, -5.9088 \rangle \quad \text{Use a calculator.}$$

FIGURE 28

**NOW TRY** Exercise 43.

The following properties of parallelograms are helpful when studying vectors.

### Properties of Parallelograms

1. A parallelogram is a quadrilateral whose opposite sides are parallel.
2. The opposite sides and opposite angles of a parallelogram are equal, and adjacent angles of a parallelogram are supplementary.
3. The diagonals of a parallelogram bisect each other but do not necessarily bisect the angles of the parallelogram.

### EXAMPLE 4   Finding the Magnitude of a Resultant

Two forces of 15 and 22 newtons act on a point in the plane. (A **newton** is a unit of force that equals 0.225 lb.) If the angle between the forces is 100°, find the magnitude of the resultant force.

**Solution**   As shown in Figure 29, a parallelogram that has the forces as adjacent sides can be formed. The angles of the parallelogram adjacent to angle $P$ measure 80°, since adjacent angles of a parallelogram are supplementary. Opposite sides of the parallelogram are equal in length. The resultant force divides the parallelogram into two triangles. Use the law of cosines with either triangle.

$|\mathbf{v}|^2 = 15^2 + 22^2 - 2(15)(22) \cos 80°$   Law of cosines
$\approx 225 + 484 - 115$
$|\mathbf{v}|^2 \approx 594$
$|\mathbf{v}| \approx 24$   Take square roots.

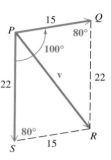

**FIGURE 29**

To the nearest unit, the magnitude of the resultant force is 24 newtons.

**NOW TRY** Exercise 49.

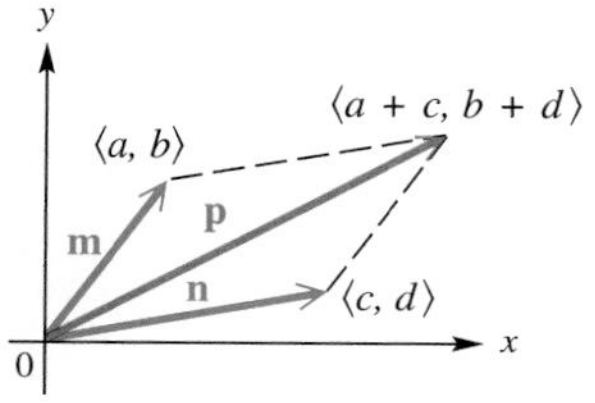

**OBJECTIVE 3**  Perform vector operations.   In Figure 30, $\mathbf{m} = \langle a, b \rangle$, $\mathbf{n} = \langle c, d \rangle$, and $\mathbf{p} = \langle a + c, b + d \rangle$. Using geometry, we can show that the endpoints of the three vectors and the origin form a parallelogram. Since a diagonal of this parallelogram gives the resultant of $\mathbf{m}$ and $\mathbf{n}$, we have $\mathbf{p} = \mathbf{m} + \mathbf{n}$ or

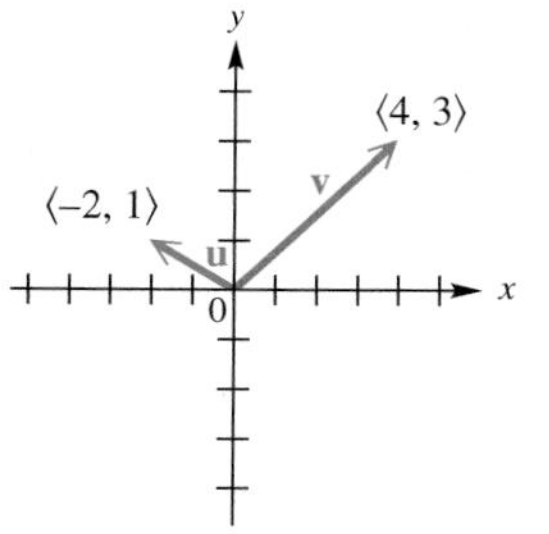

$$\langle a + c, b + d \rangle = \langle a, b \rangle + \langle c, d \rangle.$$

Similarly, we could verify the following vector operations.

---

**Vector Operations**

For any real numbers $a$, $b$, $c$, $d$, and $k$,

$$\langle a, b \rangle + \langle c, d \rangle = \langle a + c, b + d \rangle$$
$$k \cdot \langle a, b \rangle = \langle ka, kb \rangle.$$
If $\mathbf{a} = \langle a_1, a_2 \rangle$, then $-\mathbf{a} = \langle -a_1, -a_2 \rangle$.
$$\langle a, b \rangle - \langle c, d \rangle = \langle a, b \rangle + (-\langle c, d \rangle) = \langle a - c, b - d \rangle$$

---

**FIGURE 30**

**FIGURE 31**

### EXAMPLE 5   Performing Vector Operations

Let $\mathbf{u} = \langle -2, 1 \rangle$ and $\mathbf{v} = \langle 4, 3 \rangle$. (See Figure 31.) Find the following: **(a)** $\mathbf{u} + \mathbf{v}$, **(b)** $-2\mathbf{u}$, **(c)** $4\mathbf{u} - 3\mathbf{v}$.

**Algebraic Solution**

(a) $\mathbf{u} + \mathbf{v} = \langle -2, 1 \rangle + \langle 4, 3 \rangle$
$= \langle -2 + 4, 1 + 3 \rangle$
$= \langle 2, 4 \rangle$

(b) $-2\mathbf{u} = -2 \cdot \langle -2, 1 \rangle$
$= \langle -2(-2), -2(1) \rangle$
$= \langle 4, -2 \rangle$

(c) $4\mathbf{u} - 3\mathbf{v} = 4 \cdot \langle -2, 1 \rangle - 3 \cdot \langle 4, 3 \rangle$
$= \langle -8, 4 \rangle - \langle 12, 9 \rangle$
$= \langle -8 - 12, 4 - 9 \rangle$
$= \langle -20, -5 \rangle$

**Graphing Calculator Solution**

Vector arithmetic can be performed with a graphing calculator, as shown in Figure 32.

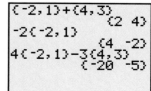

**FIGURE 32**

**NOW TRY** Exercises 59, 61, and 63.

A **unit vector** is a vector that has magnitude 1. Two very useful unit vectors are defined as follows and shown in Figure 33(a).

$$\mathbf{i} = \langle 1, 0 \rangle \qquad \mathbf{j} = \langle 0, 1 \rangle$$

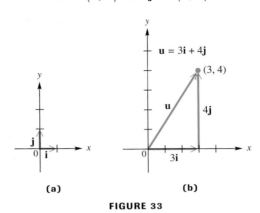

**FIGURE 33**

With the unit vectors $\mathbf{i}$ and $\mathbf{j}$, we can express any other vector $\langle a, b \rangle$ in the form $a\mathbf{i} + b\mathbf{j}$, as shown in Figure 33(b), where $\langle 3, 4 \rangle = 3\mathbf{i} + 4\mathbf{j}$. The vector operations previously given can be restated using $a\mathbf{i} + b\mathbf{j}$ notation.

---

**i, j Form for Vectors**

If $\mathbf{v} = \langle a, b \rangle$, then $\mathbf{v} = a\mathbf{i} + b\mathbf{j}$, where $\mathbf{i} = \langle 1, 0 \rangle$ and $\mathbf{j} = \langle 0, 1 \rangle$.

---

## 20.4 Exercises

*Concept Check* Exercises 1–4 refer to the vectors **m–t** at the right.

1. Name all pairs of vectors that appear to be equal.
2. Name all pairs of vectors that are opposites.
3. Name all pairs of vectors where the first is a scalar multiple of the other with the scalar positive.
4. Name all pairs of vectors where the first is a scalar multiple of the other with the scalar negative.

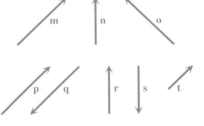

*Concept Check* Refer to vectors **a–h** below. Make a copy or a sketch of each vector, and then draw a sketch to represent each vector in Exercises 5–16. For example, find $\mathbf{a} + \mathbf{e}$ by placing **a** and **e** so that their initial points coincide. Then use the parallelogram rule to find the resultant, as shown in the figure on the right.

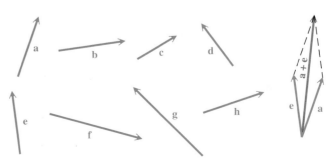

| 5. −b | 6. −g | 7. 3a | 8. 2h |
|---|---|---|---|
| 9. a + b | 10. h + g | 11. a − c | 12. d − e |
| 13. a + (b + c) | 14. (a + b) + c | 15. c + d | 16. d + c |

17. From the results of Exercises 13 and 14, does it appear that vector addition is associative?

18. From the results of Exercises 15 and 16, does it appear that vector addition is commutative?

*In Exercises 19–24, use the figure to find each vector:* **(a)** a + b **(b)** a − b **(c)** −a. *Use $\langle x, y \rangle$ notation as in Example 3.*

19.

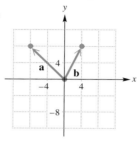

20.

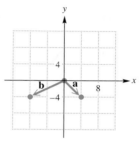

21.

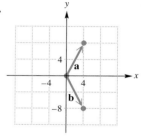

22.

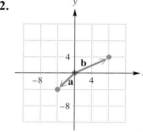

23.

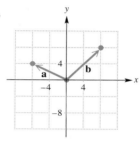

24.

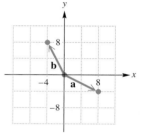

*Given vectors* **a** *and* **b,** *find:* **(a)** 2a **(b)** 2a + 3b **(c)** b − 3a.

25. a = 2i, b = i + j
26. a = −i + 2j, b = i − j
27. a = $\langle -1, 2 \rangle$, b = $\langle 3, 0 \rangle$
28. a = $\langle -2, -1 \rangle$, b = $\langle -3, 2 \rangle$

*For each pair of vectors* **u** *and* **w** *with angle θ between them, sketch the resultant.*

29. $|\mathbf{u}| = 12, |\mathbf{w}| = 20, \theta = 27°$
30. $|\mathbf{u}| = 8, |\mathbf{w}| = 12, \theta = 20°$
31. $|\mathbf{u}| = 20, |\mathbf{w}| = 30, \theta = 30°$
32. $|\mathbf{u}| = 50, |\mathbf{w}| = 70, \theta = 40°$

*Find the magnitude and direction angle for* **u.** *See Example 1.*

33. $\langle 15, -8 \rangle$
34. $\langle -7, 24 \rangle$
35. $\langle -4, 4\sqrt{3} \rangle$
36. $\langle 8\sqrt{2}, -8\sqrt{2} \rangle$

*For each of the following, vector* **v** *has the given magnitude and direction. Find the magnitudes of the horizontal and vertical components of* **v,** *if α is the direction angle of* **v** *from the horizontal. See Example 2.*

37. $\alpha = 20°, |\mathbf{v}| = 50$
38. $\alpha = 50°, |\mathbf{v}| = 26$
39. $\alpha = 35°\,50', |\mathbf{v}| = 47.8$
40. $\alpha = 27°\,30', |\mathbf{v}| = 15.4$
41. $\alpha = 128.5°, |\mathbf{v}| = 198$
42. $\alpha = 146.3°, |\mathbf{v}| = 238$

*Write each vector in the form ⟨a, b⟩. See Example 3.*

**43.** 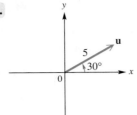    **44.** 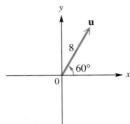    **45.**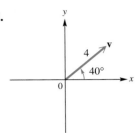

**46.** 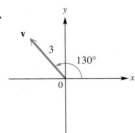    **47.** 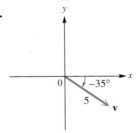    **48.**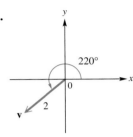

*Two forces act at a point in the plane. The angle between the two forces is given. Find the magnitude of the resultant force. See Example 4.*

**49.** forces of 250 and 450 newtons, forming an angle of 85°

**50.** forces of 19 and 32 newtons, forming an angle of 118°

**51.** forces of 116 and 139 lb, forming an angle of 140° 50′

**52.** forces of 37.8 and 53.7 lb, forming an angle of 68.5°

*Use the parallelogram rule to find the magnitude of the resultant force for the two forces shown in each figure. Round answers to the nearest tenth.*

**53.** 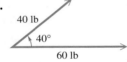    **54.**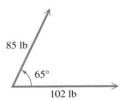

**55.** 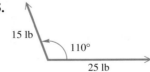    **56.**

**57.** *Concept Check* If $\mathbf{u} = \langle a, b \rangle$ and $\mathbf{v} = \langle c, d \rangle$, what is $\mathbf{u} + \mathbf{v}$?

**58.** Explain how addition of vectors is similar to addition of complex numbers.

*Given $\mathbf{u} = \langle -2, 5 \rangle$ and $\mathbf{v} = \langle 4, 3 \rangle$, find the following. See Example 5.*

**59.** $\mathbf{u} + \mathbf{v}$    **60.** $\mathbf{u} - \mathbf{v}$    **61.** $-4\mathbf{u}$    **62.** $-5\mathbf{v}$

**63.** $3\mathbf{u} - 6\mathbf{v}$    **64.** $-2\mathbf{u} + 4\mathbf{v}$    **65.** $\mathbf{u} + \mathbf{v} - 3\mathbf{u}$    **66.** $2\mathbf{u} + \mathbf{v} - 6\mathbf{v}$

*Write each vector in the form $a\mathbf{i} + b\mathbf{j}$. See Figure 33(b).*

**67.** $\langle -5, 8 \rangle$    **68.** $\langle 6, -3 \rangle$    **69.** $\langle 2, 0 \rangle$    **70.** $\langle 0, -4 \rangle$

# 20 SUMMARY

## KEY TERMS

**20.1** Side-Angle-Side (SAS)
Angle-Side-Angle (ASA)
Side-Side-Side (SSS)
oblique triangle
Side-Angle-Angle (SAA)

**20.2** ambiguous case
**20.3** semiperimeter
**20.4** scalar
vector quantity
vector
magnitude

initial point
terminal point
parallelogram rule
resultant
opposite (of a vector)
zero vector

scalar product
position vector
horizontal component
vertical component
direction angle
unit vector

## NEW SYMBOLS

**OP** or $\overrightarrow{OP}$    vector **OP**
**|OP|**    magnitude of vector **OP**

$\langle a, b \rangle$    position vector
**i, j**    unit vectors

## QUICK REVIEW

### CONCEPTS

### 20.1 Oblique Triangles and the Law of Sines

**Law of Sines**

In any triangle $ABC$, with sides $a$, $b$, and $c$,

$$\frac{a}{\sin A} = \frac{b}{\sin B}, \quad \frac{a}{\sin A} = \frac{c}{\sin C}, \quad \text{and} \quad \frac{b}{\sin B} = \frac{c}{\sin C}.$$

**Area of a Triangle**

In any triangle $ABC$, the area is half the product of the lengths of two sides and the sine of the angle between them.

$$\mathcal{A} = \frac{1}{2} bc \sin A, \quad \mathcal{A} = \frac{1}{2} ab \sin C, \quad \mathcal{A} = \frac{1}{2} ac \sin B$$

### EXAMPLES

In triangle $ABC$, find $c$ if $A = 44°$, $C = 62°$, and $a = 12.00$ units. Then find its area.

$$\frac{a}{\sin A} = \frac{c}{\sin C}$$

$$\frac{12.00}{\sin 44°} = \frac{c}{\sin 62°}$$

$$c = \frac{12.00 \sin 62°}{\sin 44°} \approx 15.25 \text{ units}$$

For triangle $ABC$ above,

$$\mathcal{A} = \frac{1}{2} ac \sin B$$

$$= \frac{1}{2}(12.00)(15.25) \sin 74° \quad B = 180° - 44° - 62°$$

$$\approx 87.96 \text{ sq units.}$$

### 20.2 The Ambiguous Case of the Law of Sines

**Ambiguous Case**

If we are given the lengths of two sides and the angle opposite one of them, for example, $A$, $a$, and $b$ in triangle $ABC$, then it is possible that zero, one, or two such triangles exist. If $A$ is acute, $h$ is the altitude from $C$, and
1. $a < h < b$, then there is no triangle.
2. $a = h$ and $h < b$, then there is one triangle (a right triangle).
3. $a \geq b$, then there is one triangle.
4. $h < a < b$, then there are two triangles.

Solve triangle $ABC$, given $A = 44.5°$, $a = 11.0$ in., and $c = 7.0$ in.

Find angle $C$.

$$\frac{\sin C}{7.0} = \frac{\sin 44.5°}{11.0}$$

$$\sin C \approx 0.4460$$

$$C \approx 26.5°$$

Another angle with this sine value is

$$180° - 26.5° = 153.5°.$$

*(continued)*

## CONCEPTS

If $A$ is obtuse and

1. $a \leq b$, then there is no triangle.
2. $a > b$, then there is one triangle.

See the table on page 890 that illustrates the possible outcomes.

## EXAMPLES

However, $153.5° + 44.5° > 180°$, so there is only one triangle.

$$B = 180° - 44.5° - 26.5°$$
$$B = 109°$$

Using the law of sines again,

$$b \approx 14.8 \text{ in.}$$

### 20.3 The Law of Cosines

**Law of Cosines**

In any triangle $ABC$, with sides $a$, $b$, and $c$,

$$a^2 = b^2 + c^2 - 2bc \cos A$$
$$b^2 = a^2 + c^2 - 2ac \cos B$$
$$c^2 = a^2 + b^2 - 2ab \cos C.$$

In triangle $ABC$, find $C$ if $a = 11$ units, $b = 13$ units, and $c = 20$ units. Then find its area.

$$c^2 = a^2 + b^2 - 2ab \cos C$$
$$20^2 = 11^2 + 13^2 - 2(11)(13) \cos C$$
$$400 = 121 + 169 - 286 \cos C$$
$$\frac{400 - 121 - 169}{-286} = \cos C$$
$$C = \cos^{-1}\left(\frac{400 - 121 - 169}{-286}\right)$$
$$C \approx 112.6°$$

**Heron's Area Formula**

If a triangle has sides of lengths $a$, $b$, and $c$, with semiperimeter

$$s = \frac{1}{2}(a + b + c),$$

then the area of the triangle is

$$\mathcal{A} = \sqrt{s(s - a)(s - b)(s - c)}.$$

The semiperimeter $s$ is

$$s = \frac{1}{2}(11 + 13 + 20) = 22,$$

so

$$\mathcal{A} = \sqrt{22(22 - 11)(22 - 13)(22 - 20)} = 66 \text{ sq units.}$$

### 20.4 Vectors

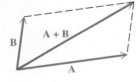 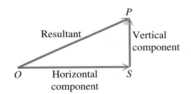 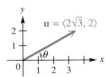

**Magnitude and Direction Angle of a Vector**

The magnitude (length) of vector $\mathbf{u} = \langle a, b \rangle$ is given by

$$|\mathbf{u}| = \sqrt{a^2 + b^2}.$$

$$|\mathbf{u}| = \sqrt{(2\sqrt{3})^2 + 2^2} = \sqrt{16} = 4$$

The direction angle $\theta$ satisfies $\tan \theta = \frac{b}{a}$, where $a \neq 0$.

Since $\tan \theta = \frac{2}{2\sqrt{3}} = \frac{1}{\sqrt{3}} \cdot \frac{\sqrt{3}}{\sqrt{3}} = \frac{\sqrt{3}}{3}$, it follows that $\theta = 30°$.

**Vector Operations**

For any real numbers $a$, $b$, $c$, $d$, and $k$,

$$\langle a, b \rangle + \langle c, d \rangle = \langle a + c, b + d \rangle$$
$$k \cdot \langle a, b \rangle = \langle ka, kb \rangle.$$

If $a = \langle a_1, a_2 \rangle$, then $-a = \langle -a_1, -a_2 \rangle$.

$$\langle a, b \rangle - \langle c, d \rangle = \langle a, b \rangle + (-\langle c, d \rangle) = \langle a - c, b - d \rangle.$$

$$\langle 4, 6 \rangle + \langle -8, 3 \rangle = \langle -4, 9 \rangle$$
$$5\langle -2, 1 \rangle = \langle -10, 5 \rangle$$
$$-\langle -9, 6 \rangle = \langle 9, -6 \rangle$$
$$\langle 4, 6 \rangle - \langle -8, 3 \rangle = \langle 12, 3 \rangle$$

*(continued)*

## CONCEPTS

If $\mathbf{u} = \langle x, y \rangle$ has direction angle $\theta$, then

$$\mathbf{u} = \langle |\mathbf{u}| \cos \theta, |\mathbf{u}| \sin \theta \rangle.$$

### i, j Form for Vectors

If $\mathbf{v} = \langle a, b \rangle$, then $\mathbf{v} = a\mathbf{i} + b\mathbf{j}$, where $\mathbf{i} = \langle 1, 0 \rangle$ and $\mathbf{j} = \langle 0, 1 \rangle$.

## EXAMPLES

For $\mathbf{u}$ defined above,

$$\mathbf{u} = \langle 4 \cos 30°, 4 \sin 30° \rangle$$
$$= \langle 2\sqrt{3}, 2 \rangle$$

and

$$\mathbf{u} = 2\sqrt{3}\mathbf{i} + 2\mathbf{j}.$$

# 20 REVIEW EXERCISES

*Use the law of sines to find the indicated part of each triangle ABC.*

1. Find $b$ if $C = 74.2°$, $c = 96.3$ m, $B = 39.5°$.
2. Find $B$ if $A = 129.7°$, $a = 127$ ft, $b = 69.8$ ft.
3. Find $B$ if $C = 51.3°$, $c = 68.3$ m, $b = 58.2$ m.
4. Find $b$ if $a = 165$ m, $A = 100.2°$, $B = 25.0°$.
5. Find $A$ if $B = 39° 50'$, $b = 268$ m, $a = 340$ m.
6. Find $A$ if $C = 79° 20'$, $c = 97.4$ mm, $a = 75.3$ mm.
7. If we are given $a$, $A$, and $C$ in a triangle $ABC$, does the possibility of the ambiguous case exist? If not, explain why.
8. Can triangle $ABC$ exist if $a = 4.7$, $b = 2.3$, and $c = 7.0$? If not, explain why. Answer this question without using trigonometry.
9. Given $a = 10$ and $B = 30°$, determine the values of $b$ for which $A$ has
   (a) exactly one value  (b) two possible values  (c) no value.
10. Explain why there can be no triangle $ABC$ satisfying $A = 140°$, $a = 5$, and $b = 7$.

*Use the law of cosines to find the indicated part of each triangle ABC.*

11. Find $A$ if $a = 86.14$ in., $b = 253.2$ in., $c = 241.9$ in.
12. Find $b$ if $B = 120.7°$, $a = 127$ ft, $c = 69.8$ ft.
13. Find $a$ if $A = 51° 20'$, $c = 68.3$ m, $b = 58.2$ m.
14. Find $B$ if $a = 14.8$ m, $b = 19.7$ m, $c = 31.8$ m.
15. Find $a$ if $A = 60°$, $b = 5.0$ cm, $c = 21$ cm.
16. Find $A$ if $a = 13$ ft, $b = 17$ ft, $c = 8$ ft.

*Solve each triangle ABC having the given information.*

17. $A = 25.2°$, $a = 6.92$ yd, $b = 4.82$ yd
18. $A = 61.7°$, $a = 78.9$ m, $b = 86.4$ m
19. $a = 27.6$ cm, $b = 19.8$ cm, $C = 42° 30'$
20. $a = 94.6$ yd, $b = 123$ yd, $c = 109$ yd

*Find the area of each triangle ABC with the given information.*

**21.** $b = 840.6$ m, $c = 715.9$ m, $A = 149.3°$

**22.** $a = 6.90$ ft, $b = 10.2$ ft, $C = 35° \, 10'$

**23.** $a = 0.913$ km, $b = 0.816$ km, $c = 0.582$ km

**24.** $a = 43$ m, $b = 32$ m, $c = 51$ m

*Solve each problem.*

**25.** To measure the distance $AB$ across a canyon for a power line, a surveyor measures angles $B$ and $C$ and the distance $BC$, as shown in the figure. What is the distance from $A$ to $B$?

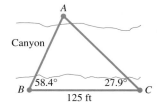

**26.** A banner on an 8.0-ft pole is to be mounted on a building at an angle of 115°, as shown in the figure. Find the length of the brace.

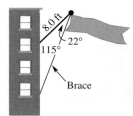

**27.** A tree leans at an angle of 8.0° from the vertical. From a point 7.0 m from the bottom of the tree, the angle of elevation to the top of the tree is 68°. How tall is the tree?

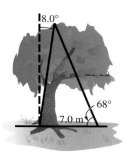

**28.** A hanging sculpture in an art gallery is to be hung with two wires of lengths 15.0 ft and 12.2 ft so that the angle between them is 70.3°. How far apart should the ends of the wire be placed on the ceiling?

**29.** A hill makes an angle of 14.3° with the horizontal. From the base of the hill, the angle of elevation to the top of a tree on top of the hill is 27.2°. The distance along the hill from the base to the tree is 212 ft. Find the height of the tree.

**30.** A pipeline is to run between points $A$ and $B$, which are separated by a protected wetlands area. To avoid the wetlands, the pipe will run from point $A$ to $C$ and then to $B$. The distances involved are $AB = 150$ km, $AC = 102$ km, and $BC = 135$ km. What angle should be used at point $C$?

**31.** Two boats leave a dock together. Each travels in a straight line. The angle between their courses measures 54° 10'. One boat travels 36.2 km per hr, and the other travels 45.6 km per hr. How far apart will they be after 3 hr?

**32.** A ship sailing parallel to shore sights a lighthouse at an angle of 30° from its direction of travel. After the ship travels 2.0 mi farther, the angle has increased to 55°. At that time, how far is the ship from the lighthouse?

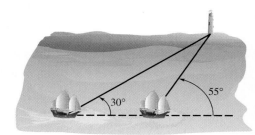

**33.** Find the area of the triangle shown in the figure using Heron's area formula.

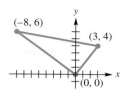

**34.** Show that the triangle in Exercise 33 is a right triangle. Then use the formula $\mathcal{A} = \frac{1}{2}ac \sin B$, with $B = 90°$, to find the area.

*In Exercises 35 and 36, use the given vectors to sketch the following.*

**35.** $\mathbf{a} - \mathbf{b}$

**36.** $\mathbf{a} + 3\mathbf{c}$

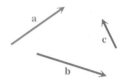

**37.** *True or False*

  (a) Opposite angles of a parallelogram are equal.
  (b) A diagonal of a parallelogram must bisect two angles of the parallelogram.

*Given two forces and the angle between them, find the magnitude of the resultant force.*

**38.** forces of 142 and 215 newtons, forming an angle of 112°

**39.**

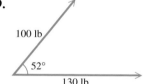

*Vector $\mathbf{v}$ has the given magnitude and direction angle. Find the magnitudes of the horizontal and vertical components of $\mathbf{v}$.*

**40.** $|\mathbf{v}| = 50, \theta = 45°$
(Give exact values.)

**41.** $|\mathbf{v}| = 964, \theta = 154°\ 20'$

*Find the magnitude and direction angle for $\mathbf{u}$ rounded to the nearest tenth.*

**42.** $\mathbf{u} = \langle 21, -20 \rangle$

**43.** $\mathbf{u} = \langle -9, 12 \rangle$

**44.** Let $\mathbf{v} = 2\mathbf{i} - \mathbf{j}$ and $\mathbf{u} = -3\mathbf{i} + 2\mathbf{j}$. Express each in terms of $\mathbf{i}$ and $\mathbf{j}$.

  (a) $2\mathbf{v} + \mathbf{u}$    (b) $2\mathbf{v}$    (c) $\mathbf{v} - 3\mathbf{u}$

*Find the vector of magnitude 1 having the same direction angle as the given vector.*

**45.** $\mathbf{u} = \langle 5, 12 \rangle$    **46.** $\mathbf{u} = \langle -4, 3 \rangle$

# 20 TEST

*Find the indicated part of each triangle ABC.*

1. Find $C$ if $A = 25.2°$, $a = 6.92$ yd, and $b = 4.82$ yd.
2. Find $c$ if $C = 118°$, $a = 75.0$ km, and $b = 131$ km.
3. Find $B$ if $a = 17.3$ ft, $b = 22.6$ ft, $c = 29.8$ ft.
4. Find the area of triangle $ABC$ if $a = 14$, $b = 30$, and $c = 40$.
5. Find the area of triangle $XYZ$ shown here.

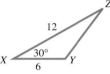

6. Given $a = 10$ and $B = 150°$ in triangle $ABC$, determine the values of $b$ for which $A$ has
   (a) exactly one value  (b) two possible values  (c) no value.

*Solve each triangle ABC.*

7. $A = 60°$, $b = 30$ m, $c = 45$ m
8. $b = 1075$ in., $c = 785$ in., $C = 38° \, 30'$
9. Find the magnitude and the direction angle for the vector shown in the figure.

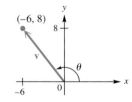

10. Use the given vectors to sketch $\mathbf{a} + \mathbf{b}$.

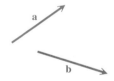

11. For the vectors $\mathbf{u} = \langle -1, 3 \rangle$ and $\mathbf{v} = \langle 2, -6 \rangle$, find each of the following.
    (a) $\mathbf{u} + \mathbf{v}$  (b) $-3\mathbf{v}$  (c) $|\mathbf{u}|$

*Solve each problem.*

12. The angles of elevation of a balloon from two points $A$ and $B$ on level ground are $24° \, 50'$ and $47° \, 20'$, respectively. As shown in the figure, points $A$, $B$, and $C$ are in the same vertical plane and points $A$ and $B$ are 8.4 mi apart. Approximate the height of the balloon above the ground to the nearest tenth of a mile.

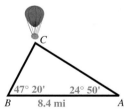

13. Find the horizontal and vertical components of the vector with magnitude 569 that is inclined $127.5°$ from the horizontal. Give your answer in the form $\langle a, b \rangle$.

14. A tree leans at an angle of $8.0°$ from the vertical, as shown in the figure. From a point 8.0 m from the bottom of the tree, the angle of elevation to the top of the tree is $66°$. How tall is the tree?

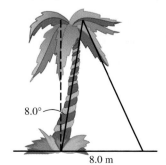

916

# Counting Theory

**OBJECTIVES**
1. Use the fundamental principle of counting.
2. Evaluate expressions involving factorials.
3. Learn the formula $P(n, r)$ for permutations.
4. Use the permutations formula to solve counting problems.
5. Learn the formula $\binom{n}{r}$ for combinations.
6. Use the combinations formula to solve counting problems.
7. Distinguish between permutations and combinations.

If there are 3 roads from Albany to Baker and 2 roads from Baker to Creswich, in how many ways can one travel from Albany to Creswich by way of Baker? For each of the 3 roads from Albany to Baker, there are 2 different roads from Baker to Creswich, so there are $3 \cdot 2 = 6$ different ways to make the trip, as shown in the **tree diagram** in Figure 1.

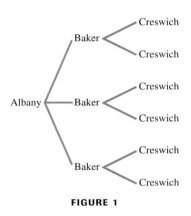

**FIGURE 1**

This example illustrates the following property.

**Fundamental Principle of Counting**

If one event can occur in $m$ ways and a second event can occur in $n$ ways, then both events can occur in $mn$ ways, provided the outcome of the first event does not influence the outcome of the second.

**OBJECTIVE 1** Use the fundamental principle of counting. The fundamental principle of counting can be extended to any number of events, provided the outcome of no one event influences the outcome of another. Such events are called **independent events**.

**EXAMPLE 1** Using the Fundamental Principle of Counting

A restaurant offers a choice of 3 salads, 5 main dishes, and 2 desserts. Use the fundamental principle of counting to find the number of different 3-course meals that can be selected.

Three independent events are involved: selecting a salad, selecting a main dish, and selecting a dessert. The first event can occur in 3 ways, the second event can occur in 5 ways, and the third event can occur in 2 ways; thus there are

$$3 \cdot 5 \cdot 2 = 30 \text{ possible meals.}$$

**NOW TRY** Exercise 21.

917

**EXAMPLE 2** Using the Fundamental Principle of Counting

Eli Maor has 5 different books that he wishes to arrange on his desk. How many different arrangements are possible?

Five events are involved: selecting a book for the first spot, selecting a book for the second spot, and so on. Here the outcome of the first event *does* influence the outcome of the other events (since one book has already been chosen). For the first spot Eli has 5 choices, for the second spot 4 choices, for the third spot 3 choices, and so on. Now use the fundamental principle of counting to find that there are

$$5 \cdot 4 \cdot 3 \cdot 2 \cdot 1 = 120 \text{ different arrangements.}$$

**NOW TRY** Exercise 23.

**OBJECTIVE 2** Evaluate expressions involving factorials. When using the fundamental principle of counting, we often encounter products such as $5 \cdot 4 \cdot 3 \cdot 2 \cdot 1$ from Example 2. For convenience in writing these products, we use the symbol $n!$ (read "*n* factorial"), defined as follows:

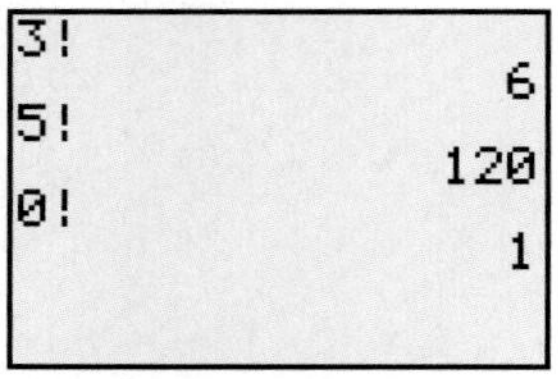

### *n* Factorial (*n*!)

For any positive integer $n$,

$$n(n-1)(n-2)(n-3)\cdots(2)(1) = n!.$$

For example,

$$3! = 3 \cdot 2 \cdot 1 = 6 \quad \text{and} \quad 5! = 5 \cdot 4 \cdot 3 \cdot 2 \cdot 1 = 120.$$

From the definition of $n$ factorial, $n[(n-1)!] = n!$. If $n = 1$, then $1(0!) = 1! = 1$. Because of this, $0!$ is defined as

$$0! = 1.$$

**NOW TRY** Exercise 1.

Scientific and graphing calculators can compute factorials. The three factorial expressions just discussed are shown in Figure 2(a). Figure 2(b) shows some larger factorials.

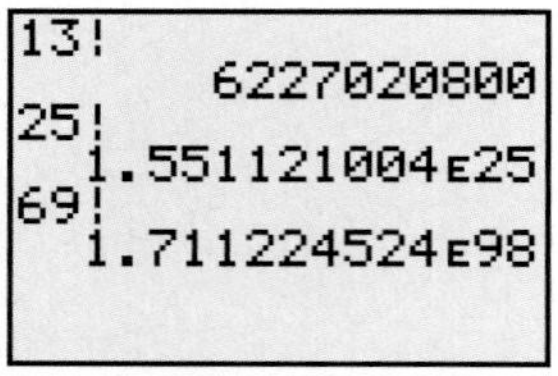

A graphing calculator with a 10-digit display will give the exact value of $n!$ for $n \le 13$ and approximate values of $n!$ for $14 \le n \le 69$.

**(b)**

**FIGURE 2**

**EXAMPLE 3** Evaluating Expressions Involving Factorials

Find the value of each expression.

(a) $\dfrac{5!}{4!1!} = \dfrac{5 \cdot 4 \cdot 3 \cdot 2 \cdot 1}{(4 \cdot 3 \cdot 2 \cdot 1)(1)} = 5$

(b) $\dfrac{5!}{3!2!} = \dfrac{5 \cdot 4 \cdot 3 \cdot 2 \cdot 1}{(3 \cdot 2 \cdot 1)(2 \cdot 1)} = \dfrac{5 \cdot 4}{2 \cdot 1} = 10$

(c) $\dfrac{6!}{3!3!} = \dfrac{6 \cdot 5 \cdot 4 \cdot 3 \cdot 2 \cdot 1}{(3 \cdot 2 \cdot 1)(3 \cdot 2 \cdot 1)} = \dfrac{6 \cdot 5 \cdot 4}{3 \cdot 2 \cdot 1} = 20$

(d) $\dfrac{4!}{4!0!} = \dfrac{4 \cdot 3 \cdot 2 \cdot 1}{(4 \cdot 3 \cdot 2 \cdot 1)(1)} = 1$

**NOW TRY** Exercises 5 and 7.

### EXAMPLE 4  Arranging r of n Items (r < n)

Suppose that Eli (from Example 2) wishes to place only 3 of the 5 books on his desk. How many arrangements of 3 books are possible?

He still has 5 ways to fill the first spot, 4 ways to fill the second spot, and 3 ways to fill the third. Since he wants to use 3 books, there are only 3 spots to be filled (3 events) instead of 5, so there are

$$5 \cdot 4 \cdot 3 = 60 \text{ arrangements.}$$

**NOW TRY** Exercise 29.

**OBJECTIVE 3** Learn the formula P(n, r) for permutations. The number 60 in Example 4 is called the number of *permutations* of 5 things taken 3 at a time, written $P(5, 3) = 60$. Example 2 showed that the number of ways of arranging 5 elements from a set of 5 elements, written $P(5, 5)$, is 120.

A **permutation** of $n$ elements taken $r$ at a time is one of the ways of arranging $r$ elements taken from a set of $n$ elements ($r \leq n$). Generalizing from the examples above, the number of permutations of $n$ elements taken $r$ at a time, denoted by $P(n, r)$, is

$$P(n, r) = n(n-1)(n-2)\cdots(n-r+1)$$
$$= \frac{n(n-1)(n-2)\cdots(n-r+1)(n-r)(n-r-1)\cdots(2)(1)}{(n-r)(n-r-1)\cdots(2)(1)}$$
$$= \frac{n!}{(n-r)!}.$$

This proves the following result.

---

**Permutations of n Elements Taken r at a Time**

If $P(n, r)$ denotes the number of **permutations** of $n$ elements taken $r$ at a time, $r \leq n$, then

$$P(n, r) = \frac{n!}{(n-r)!}.$$

---

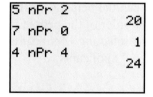

FIGURE 3

Some other symbols used for permutations of $n$ things taken $r$ at a time are $_nP_r$ and $P_r^n$.

Graphing calculators use the notation $_nP_r$ for evaluating permutation expressions. The screen in Figure 3 shows how a calculator evaluates $P(5, 2)$, $P(7, 0)$, and $P(4, 4)$.

**OBJECTIVE 4** Use the permutations formula to solve counting problems.

### EXAMPLE 5  Using the Permutations Formula

Suppose 8 people enter an event in a swim meet. Assuming there are no ties, in how many ways could the gold, silver, and bronze medals be awarded?

Using the fundamental principle of counting, there are 3 choices to be made, giving $8 \cdot 7 \cdot 6 = 336$. However, we can also use the formula for $P(n, r)$ and obtain the same result.

$$P(8, 3) = \frac{8!}{(8-3)!} = \frac{8!}{5!}$$
$$= \frac{8 \cdot 7 \cdot 6 \cdot 5 \cdot 4 \cdot 3 \cdot 2 \cdot 1}{5 \cdot 4 \cdot 3 \cdot 2 \cdot 1}$$
$$= 8 \cdot 7 \cdot 6 = 336$$

**NOW TRY** Exercise 25.

### EXAMPLE 6 Using the Permutations Formula

In how many ways can 6 students be seated in a row of 6 desks?

$$P(6, 6) = \frac{6!}{(6 - 6)!} = \frac{6!}{0!} = 6! = 6 \cdot 5 \cdot 4 \cdot 3 \cdot 2 \cdot 1 = 720$$

**NOW TRY** Exercise 31.

The next example involves using the fundamental counting principle with some restrictions.

### EXAMPLE 7 Using the Fundamental Counting Principle with Restrictions

In how many ways can 3 letters of the alphabet be arranged if a vowel cannot be used in the middle position, and repetitions of the letters are allowed?

We cannot use $P(26, 3)$ here, because of the restriction on the middle position, and because repetition is allowed. In the first and third positions, we can use any of the 26 letters of the alphabet, but in the middle position, we can use only one of $26 - 5 = 21$ letters (since there are 5 vowels). Now, using the fundamental counting principle, there are

$$26 \cdot 21 \cdot 26 = 14{,}196$$

ways to arrange the letters according to the problem.

**NOW TRY** Exercise 27.

**OBJECTIVE 5** Learn the formula $\binom{n}{r}$ for combinations. We have discussed a method for finding the number of ways to arrange $r$ elements taken from a set of $n$ elements. Sometimes, however, the arrangement (or order) of the elements is not important.

For example, suppose three people (Ms. Opelka, Mr. Adams, and Ms. Jacobs) apply for 2 identical jobs. Ignoring all other factors, in how many ways can the personnel officer select 2 people from the 3 applicants? Here the arrangement or order of the people is unimportant. Selecting Ms. Opelka and Mr. Adams is the same as selecting Mr. Adams and Ms. Opelka. Therefore, there are only 3 ways to select 2 of the 3 applicants:

Ms. Opelka and Mr. Adams;

Ms. Opelka and Ms. Jacobs;

Mr. Adams and Ms. Jacobs.

These three choices are called the *combinations* of 3 elements taken 2 at a time. A **combination** of $n$ elements taken $r$ at a time is one of the ways in which $r$ elements can be chosen from $n$ elements.

Each combination of $r$ elements forms $r!$ permutations. Therefore, the number of combinations of $n$ elements taken $r$ at a time is found by dividing the number of permutations, $P(n, r)$, by $r!$ to obtain

$$\frac{P(n, r)}{r!}$$

combinations. This expression can be rewritten as follows.

$$\frac{P(n, r)}{r!} = \frac{\frac{n!}{(n - r)!}}{r!} = \frac{n!}{(n - r)!} \cdot \frac{1}{r!} = \frac{n!}{(n - r)!\, r!}$$

The symbol $\binom{n}{r}$ is used to represent the number of combinations of $n$ things taken $r$ at a time. With this symbol the preceding results are stated as follows.

> **Combinations of $n$ Elements Taken $r$ at a Time**
>
> If $\binom{n}{r}$ represents the number of **combinations** of $n$ elements taken $r$ at a time, for $r \leq n$, then
>
> $$\binom{n}{r} = \frac{n!}{(n-r)!\,r!}.$$

Other symbols used for $\binom{n}{r}$ are $C(n, r)$, $_nC_r$, and $C_r^n$.

The $_nC_r$ notation is used by graphing calculators to find combinations. Figure 4 shows a calculator screen with $_6C_2$ and $_7C_5$.

```
6 nCr 2
            15
7 nCr 5
            21
```

**FIGURE 4**

**OBJECTIVE 6** Use the combinations formula to solve counting problems. In the preceding discussion, it was shown that $\binom{3}{2} = 3$. This can also be found using the formula.

$$\binom{3}{2} = \frac{3!}{(3-2)!\,2!} = \frac{3 \cdot 2 \cdot 1}{1 \cdot 2 \cdot 1} = 3$$

The combinations formula is used in the next examples.

**EXAMPLE 8** Using the Combinations Formula

How many different committees of 3 people can be chosen from a group of 8 people?

Because the order in which the members of the committee are chosen does not affect the result, use combinations.

$$\binom{8}{3} = \frac{8!}{5!\,3!} = \frac{8 \cdot 7 \cdot 6 \cdot 5 \cdot 4 \cdot 3 \cdot 2 \cdot 1}{5 \cdot 4 \cdot 3 \cdot 2 \cdot 1 \cdot 3 \cdot 2 \cdot 1} = 56$$

**NOW TRY** Exercise 33.

**EXAMPLE 9** Using the Combinations Formula

From a group of 30 bank employees, 3 are to be selected to work on a special project.

**(a)** In how many different ways can the employees be selected?

The number of 3-element combinations from a set of 30 elements must be found. (Use combinations because order within the group of 3 does not affect the result.)

$$\binom{30}{3} = \frac{30!}{27!\,3!} = 4060.$$

There are 4060 ways to select the project group.

**(b)** In how many different ways can the group of 3 be selected if it has already been decided that a certain employee must work on the project?

Since one employee has already been selected to work on the project, the problem is reduced to selecting 2 more employees from the 29 employees that are left.

$$\binom{29}{2} = \frac{29!}{27!\,2!} = 406$$

In this case, the project group can be selected in 406 different ways.

**NOW TRY** Exercise 51.

### EXAMPLE 10 Using Combinations to Choose a Delegation

A congressional committee consists of 4 senators and 6 representatives. A delegation of 5 members is to be chosen. In how many ways could this delegation include exactly 3 senators?

"Exactly 3 senators" implies that there must be $5 - 3 = 2$ representatives as well. The 3 senators could be chosen in $\binom{4}{3} = 4$ ways. The 2 representatives could be chosen in $\binom{6}{2} = 15$ ways. Now apply the fundamental principle of counting to find that there are $4 \cdot 15 = 60$ ways to form the delegation.

**NOW TRY** Exercise 45.

**OBJECTIVE 7** Distinguish between permutations and combinations. Students often have difficulty determining whether to use permutations or combinations in solving problems. The following chart lists some of the similarities and differences between these two concepts.

| Permutations | Combinations |
|---|---|
| Number of ways of selecting $r$ items out of $n$ items Repetitions are not allowed. ||
| Order is important. | Order is not important. |
| They are arrangements of $r$ items from a set of $n$ items. | They are subsets of $r$ items from a set of $n$ items. |
| $P(n, r) = \dfrac{n!}{(n - r)!}$ | $\binom{n}{r} = \dfrac{n!}{(n - r)!\,r!}$ |
| Clue words: arrangement, schedule, order | Clue words: group, committee, sample, selection |

### EXAMPLE 11 Distinguishing between Permutations and Combinations

Tell whether permutations or combinations should be used to solve each problem.

**(a)** How many 4-digit codes are possible if no digits are repeated?

Changing the order of the 4 digits results in a different code, so permutations should be used.

**(b)** A sample of 3 light bulbs is randomly selected from a batch of 15 bulbs. How many different samples are possible?

The order in which the 3 light bulbs are selected is not important. The sample is unchanged if the items are rearranged, so combinations should be used.

**(c)** In a basketball tournament with 8 teams, how many games must be played so that each team plays every other team exactly once?

Selection of 2 teams for a game is an *unordered* subset of 2 from the set of 8 teams. Use combinations.

**(d)** In how many ways can 4 stockbrokers be assigned to 6 offices so that each broker has a private office?

The office assignments are an *ordered* selection of 4 offices from the 6 offices. Exchanging the offices of any 2 brokers within a selection of 4 offices gives a different assignment, so permutations should be used.

**NOW TRY** Exercise 41.

# A Exercises

*Evaluate each expression.*

1. 6!
2. 4!
3. $4! \cdot 5$
4. $6! \cdot 7$
5. $\dfrac{6!}{4!2!}$
6. $\dfrac{7!}{3!4!}$
7. $\dfrac{4!}{0!4!}$
8. $\dfrac{5!}{5!0!}$
9. $P(6, 4)$
10. $P(7, 5)$
11. $P(9, 2)$
12. $P(6, 5)$
13. $P(5, 1)$
14. $P(6, 0)$
15. $\dbinom{4}{2}$
16. $\dbinom{9}{3}$
17. $\dbinom{6}{0}$
18. $\dbinom{8}{1}$
19. $_{13}C_{11}$
20. $_{13}C_{2}$

*Use the fundamental counting principle or permutations to solve each problem. See Examples 1, 2, and 4–7.*

21. How many different types of homes are available if a builder offers a choice of 6 basic plans, 4 roof styles, and 2 exterior finishes?

22. A menu offers a choice of 4 salads, 8 main dishes, and 5 desserts. How many different 3-course meals (salad, main dish, dessert) are possible?

23. In an experiment on social interaction, 8 people will sit in 8 seats in a row. In how many different ways can the 8 people be seated?

24. For many years, the state of California used 3 letters followed by 3 digits on its automobile license plates.
    (a) How many different license plates are possible with this arrangement?
    (b) When the state ran out of new plates, the order was reversed to 3 digits followed by 3 letters. How many additional plates were then possible?
    (c) Several years ago, the plates described in (b) were also used up. The state then issued plates with 1 letter followed by 3 digits and then 3 letters. How many plates does this scheme provide?

25. In how many ways can 7 of 10 mice be arranged in a row for a genetics experiment?

26. How many 7-digit telephone numbers are possible if the first digit cannot be 0, and
    (a) only odd digits may be used?
    (b) the telephone number must be a multiple of 10 (that is, it must end in 0)?
    (c) the first three digits must be 456?

27. If your college offers 400 courses, 20 of which are in mathematics, and your counselor arranges your schedule of 4 courses by random selection, how many schedules are possible that do not include a math course?

28. In how many ways can 5 players be assigned to the 5 positions on a basketball team, assuming that any player can play any position? In how many ways can 10 players be assigned to the 5 positions?

29. In a sales force of 35 people, how many ways can 3 salespeople be selected for 3 different leadership jobs?

30. A softball team has 20 players. How many 9-player batting orders are possible?

31. In how many ways can 6 bank tellers be assigned to 6 different stations? In how many ways can 10 tellers be assigned to the 6 stations?

*Use combinations to solve each problem. See Examples 8–10.*

32. How many different samples of 4 light bulbs can be selected from a carton of 2 dozen bulbs?

33. A professional stockbrokers' association has 50 members. If a committee of 6 is to be selected at random, how many different committees are possible?

34. A group of 5 financial planners is to be selected at random from a professional organization with 30 members to participate in a seminar. In how many ways can this be done? In how many ways can the group that will not participate be selected?

35. Harry's Hamburger Heaven sells hamburgers with cheese, relish, lettuce, tomato, onion, mustard, or ketchup. How many different hamburgers can be concocted using any 4 of the extras?

36. How many different 5-card poker hands can be dealt from a deck of 52 playing cards?

37. Seven cards are marked with the numbers 1 through 7 and are shuffled, and then 3 cards are drawn. How many different 3-card combinations are possible?

38. A bag contains 18 marbles. How many samples of 3 can be drawn from it? How many samples of 5 marbles?

39. In Exercise 38, if the bag contains 5 purple, 4 green, and 9 black marbles, how many samples of 3 can be drawn in which all the marbles are black? How many samples of 3 can be drawn in which exactly 2 marbles are black?

40. In Exercise 32, assume it is known that there are 5 defective light bulbs in the carton. How many samples of 4 can be drawn in which all are defective? How many samples of 4 can be drawn in which there are 2 good bulbs and 2 defective bulbs?

41. Determine whether each of the following is a permutation or a combination.

    (a) Your 5-digit postal zip code
    (b) A particular 5-card hand in a game of poker
    (c) A committee of school board members

42. *Concept Check* Padlocks with digit dials are often referred to as "combination locks." According to the mathematical definition of combination, is this an accurate description? Why or why not?

*Solve each problem using any method. See Examples 1, 2, and 4–10.*

43. From a pool of 7 secretaries, 3 are selected to be assigned to 3 managers, with 1 secretary for each manager. In how many ways can this be done?

44. In a game of musical chairs, 12 children, staying in the same order, circle around 11 chairs. Each child who is next to a chair must sit down when the music stops. (One will be left out.) How many seatings are possible?

45. In an office with 8 men and 11 women, how many 5-member groups with the following compositions can be chosen for a training session?

    (a) All men              (b) All women
    (c) 3 men and 2 women    (d) No more than 3 women

46. In an experiment on plant hardiness, a researcher gathers 6 wheat plants, 3 barley plants, and 2 rye plants. Four plants are to be selected at random.

    (a) In how many ways can this be done?
    (b) In how many ways can this be done if exactly 2 wheat plants must be included?

47. From 10 names on a ballot, 4 will be elected to a political party committee. How many different committees are possible? In how many ways can the committee of 4 be formed if each person will have a different responsibility?

48. In how many ways can 5 of 9 plants be arranged in a row on a windowsill?

**49.** Hazel Miller specializes in making different vegetable soups with carrots, celery, onions, beans, peas, tomatoes, and potatoes. How many different soups can she make using any 4 ingredients?

**50.** How many 4-letter radio-station call letters can be made if the first letter must be K or W and no letter may be repeated? How many if repeats are allowed? How many of the call letters with no repeats can end in K?

**51.** A group of 12 workers decides to send a delegation of 3 to their supervisor to discuss their work assignments.
   **(a)** How many delegations of 3 are possible?
   **(b)** How many are possible if one of the 12, the foreman, must be in the delegation?
   **(c)** If there are 5 women and 7 men in the group, how many possible delegations would include exactly 1 woman?

**52.** The Riverdale board of supervisors is composed of 2 liberals and 5 conservatives. Three members are to be selected randomly as delegates to a convention.
   **(a)** How many delegations are possible?
   **(b)** How many delegations could have all liberals?
   **(c)** How many delegations could have 2 conservatives and 1 liberal?
   **(d)** If the supervisor who serves as chairman of the board must be included, how many delegations are possible?

# Basics of Probability

**OBJECTIVES**

1. Learn the terminology of probability theory.
2. Find the probability of an event.
3. Find the probability of the complement of E, given the probability of E.
4. Find the odds in favor of an event.
5. Find the probability of a compound event.

The study of probability has become increasingly popular because it has a wide range of practical applications. The basic ideas of probability are introduced in this section.

**OBJECTIVE 1** Learn the terminology of probability theory. In probability, each repetition of an experiment is called a **trial.** The possible results of each trial are called **outcomes** of the experiment. In this section, we are concerned with outcomes that are equally likely to occur. For example, the experiment of tossing a coin has two equally likely possible outcomes: landing heads up ($H$) or landing tails up ($T$). Also, the experiment of rolling a fair die has 6 equally likely outcomes: landing so the face that is up shows 1, 2, 3, 4, 5, or 6 dots.

The set $S$ of all possible outcomes of a given experiment is called the **sample space** for the experiment. (In this text all sample spaces are finite.) The sample space for the experiment of tossing a coin consists of the outcomes $H$ and $T$. This sample space can be written in set notation as

$$S = \{H, T\}.$$

Similarly, the sample space for the experiment of rolling a single die is

$$S = \{1, 2, 3, 4, 5, 6\}.$$

Any subset of the sample space is called an **event.** In the experiment with the die, for example, "the number showing is a three" is an event, say $E_1$, such that $E_1 = \{3\}$. "The number showing is greater than three" is also an event, say $E_2$, such that $E_2 = \{4, 5, 6\}$. To represent the number of outcomes that belong to event $E$, the notation $n(E)$ is used. Then $n(E_1) = 1$ and $n(E_2) = 3$.

**OBJECTIVE 2** Find the probability of an event. $P(E)$ is used to designate the *probability* of event $E$.

### Probability of Event E

In a sample space with equally likely outcomes, the **probability** of an event $E$, written $P(E)$, is the ratio of the number of outcomes in sample space $S$ that belong to event $E$, $n(E)$, to the total number of outcomes in sample space $S$, $n(S)$. That is

$$P(E) = \frac{n(E)}{n(S)}.$$

This definition is used to find the probability of event $E_1$ given above, by starting with the sample space for the experiment, $S = \{1, 2, 3, 4, 5, 6\}$, and the desired event $E_1 = \{3\}$. Since $n(E_1) = 1$ and since there are 6 outcomes in the sample space,

$$P(E_1) = \frac{n(E_1)}{n(S)} = \frac{1}{6}.$$

### EXAMPLE 1  Finding Probabilities of Events

A single die is rolled. Write the following events in set notation and give the probability for each event.

(a) $E_3$: the number showing is even

Use the definition above. Since $E_3 = \{2, 4, 6\}$, $n(E_3) = 3$. Also shown above, $n(S) = 6$, so

$$P(E_3) = \frac{3}{6} = \frac{1}{2}.$$

(b) $E_4$: the number showing is greater than 4

Again, $n(S) = 6$. Event $E_4 = \{5, 6\}$, with $n(E_4) = 2$. By the definition,

$$P(E_4) = \frac{2}{6} = \frac{1}{3}.$$

(c) $E_5$: the number showing is less than 7

$$E_5 = \{1, 2, 3, 4, 5, 6\} \quad \text{and} \quad P(E_5) = \frac{6}{6} = 1$$

(d) $E_6$: the number showing is 7

$$E_6 = \emptyset \quad \text{and} \quad P(E_6) = \frac{0}{6} = 0$$

**NOW TRY** Exercises 3 and 7.

In Example 1(c), $E_5 = S$. Therefore, the event $E_5$ is certain to occur every time the experiment is performed. An event that is certain to occur, such as $E_5$, always has a probability of 1. On the other hand, in Example 1(d), $E_6 = \emptyset$ and $P(E_6)$ is 0. The probability of an impossible event, such as $E_6$, is always 0, since none of the outcomes in the sample space satisfy the event. For any event $E$, $P(E)$ is between 0 and 1 inclusive.

**OBJECTIVE 3** Find the probability of the complement of $E$, given the probability of $E$. The set of all outcomes in the sample space that do *not* belong to event $E$ is called the **complement** of $E$, written $E'$. For example, in the experiment of drawing a single card from a standard deck of 52 cards, let $E$ be the event "the card is an ace." Then $E'$ is the event "the card is not an ace." From the definition of $E'$, for any event $E$,

$$E \cup E' = S \quad \text{and} \quad E \cap E' = \emptyset.$$

Probability concepts can be illustrated using **Venn diagrams,** as shown in Figure 1. The rectangle in Figure 1 represents the sample space in an experiment. The area inside the circle represents event $E$, while the area inside the rectangle, but outside the circle, represents event $E'$.

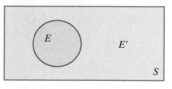

FIGURE 1

A standard deck of 52 cards has four suits: hearts ♥, clubs ♣, diamonds ♦, and spades ♠, with 13 cards of each suit. Each suit has a jack, a queen, and a king (sometimes called the "face cards"), an ace, and cards numbered from 2 to 10. The hearts and diamonds are red, and the spades and clubs are black. We refer to this standard deck of cards in this section.

**EXAMPLE 2** Using the Complement in a Probability Problem

In the experiment of drawing a card from a well-shuffled deck, find the probability of events $E$, the card is an ace, and $E'$.

Because there are four aces in the deck of 52 cards, $n(E) = 4$ and $n(S) = 52$. Therefore, $P(E) = \frac{n(E)}{n(S)} = \frac{4}{52} = \frac{1}{13}$. Of the 52 cards, 48 are not aces, so

$$P(E') = \frac{n(E')}{n(S)} = \frac{48}{52} = \frac{12}{13}.$$

**NOW TRY** Exercise 11.

In Example 2, $P(E) + P(E') = \frac{1}{13} + \frac{12}{13} = 1$. This is always true for any event $E$ and its complement $E'$. That is,

$$P(E) + P(E') = 1.$$

This can be restated as

$$P(E) = 1 - P(E') \quad \text{or} \quad P(E') = 1 - P(E).$$

These two equations suggest an alternative way to compute the probability of an event. For example, if it is known that $P(E) = \frac{1}{10}$, then

$$P(E') = 1 - \frac{1}{10} = \frac{9}{10}.$$

**OBJECTIVE 4** Find the odds in favor of an event. Sometimes probability statements are expressed in terms of *odds,* a comparison of $P(E)$ with $P(E')$. The **odds** in favor of an event $E$ are expressed as the ratio of $P(E)$ to $P(E')$ or as the fraction $\frac{P(E)}{P(E')}$. For example, if the probability of rain can be established as $\frac{1}{3}$, the odds that it will rain are

$$P(\text{rain}) \text{ to } P(\text{no rain}) = \frac{1}{3} \text{ to } \frac{2}{3} = \frac{\frac{1}{3}}{\frac{2}{3}} = \frac{1}{2} \quad \text{or} \quad 1 \text{ to } 2.$$

On the other hand, the odds that it will not rain are 2 to 1 $\left(\text{or } \frac{2}{3} \text{ to } \frac{1}{3}\right)$. If the odds in favor of an event are, say, 3 to 5, then the probability of the event is $\frac{3}{8}$, while the probability of the complement of the event is $\frac{5}{8}$. If the odds favoring event $E$ are $m$ to $n$, then

$$P(E) = \frac{m}{m + n} \quad \text{and} \quad P(E') = \frac{n}{m + n}.$$

> **EXAMPLE 3** Finding Odds in Favor or Against an Event

(a) A manager is to be selected at random from 6 sales managers and 4 office managers. Find the odds in favor of a sales manager being selected.
Let $E$ represent the event "a sales manager is selected." Then

$$P(E) = \frac{6}{10} = \frac{3}{5} \quad \text{and} \quad P(E') = 1 - \frac{3}{5} = \frac{2}{5}.$$

Therefore, the odds in favor of a sales manager being selected are

$$P(E) \text{ to } P(E') = \frac{3}{5} \text{ to } \frac{2}{5} = \frac{\frac{3}{5}}{\frac{2}{5}} = \frac{3}{2} \quad \text{or} \quad 3 \text{ to } 2.$$

(b) In a recent year, the probability that corporate stock was owned by a pension fund was 0.227. (*Source:* Federal Reserve Board and New York Stock Exchange.) Find the odds that year against a corporate stock being owned by a pension fund.
Let $E$ represent the event "corporate stock is owned by a pension fund." Then $P(E) = 0.227$ and $P(E') = 1 - 0.227 = 0.773$. Since

$$\frac{P(E')}{P(E)} = \frac{0.773}{0.227} \approx 3.4,$$

the odds against a corporate stock being owned by a pension fund were about 3.4 to 1.

**NOW TRY** Exercise 13.

> **OBJECTIVE 5** Find the probability of a compound event. A **compound event** involves an *alternative,* such as $E$ or $F$, where $E$ and $F$ are events. For example, in the experiment of rolling a die, suppose $H$ is the event "the result is a 3," and $K$ is the event "the result is an even number." What is the probability of the compound event "the result is a 3 or an even number"? From the information stated above,

$$H = \{3\} \qquad K = \{2, 4, 6\} \qquad H \text{ or } K = \{2, 3, 4, 6\}$$

$$P(H) = \frac{1}{6} \qquad P(K) = \frac{3}{6} = \frac{1}{2} \qquad P(H \text{ or } K) = \frac{4}{6} = \frac{2}{3}.$$

Notice that $P(H) + P(K) = P(H \text{ or } K)$.

Before assuming that this relationship is true in general, consider another event for this experiment, "the result is a 2," event $G$. Now

$$G = \{2\} \qquad K = \{2, 4, 6\} \qquad K \text{ or } G = \{2, 4, 6\}$$

$$P(G) = \frac{1}{6} \qquad P(K) = \frac{3}{6} = \frac{1}{2} \qquad P(K \text{ or } G) = \frac{3}{6} = \frac{1}{2}.$$

In this case $P(K) + P(G) \neq P(K \text{ or } G)$.

As Figure 2 shows, the difference in the two examples above comes from the fact that events $H$ and $K$ cannot occur simultaneously. Such events are called **mutually exclusive events**. In fact, $H \cap K = \emptyset$, which is true for any two mutually exclusive events. Events $K$ and $G$, however, can occur simultaneously. Both are satisfied if the result of the roll is a 2, the element in their intersection ($K \cap G = \{2\}$). This example suggests the following property.

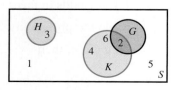

FIGURE 2

### Probability of Alternative Events

For any events $E$ and $F$,
$$P(E \text{ or } F) = P(E \cup F) = P(E) + P(F) - P(E \cap F).$$

**EXAMPLE 4** Finding the Probability of Alternative Events

One card is drawn from a well-shuffled deck of 52 cards. What is the probability of each event?

**(a)** The card is an ace or a spade.

The events "drawing an ace" and "drawing a spade" are not mutually exclusive since it is possible to draw the ace of spades, an outcome satisfying both events. The probability is

$$P(\text{ace or spade}) = P(\text{ace}) + P(\text{spade}) - P(\text{ace and spade})$$
$$= \frac{4}{52} + \frac{13}{52} - \frac{1}{52}$$
$$= \frac{16}{52} = \frac{4}{13}.$$

**(b)** The card is a 3 or a king.

"Drawing a 3" and "drawing a king" are mutually exclusive events because it is impossible to draw one card that is both a 3 and a king. Using the rule given above,

$$P(3 \text{ or } K) = P(3) + P(K) - P(3 \text{ and } K)$$
$$= \frac{4}{52} + \frac{4}{52} - 0$$
$$= \frac{8}{52} = \frac{2}{13}.$$

**NOW TRY** Exercise 17.

**EXAMPLE 5** Finding the Probability of Alternative Events

For the experiment consisting of one roll of a pair of dice, find the probability that the sum of the dots showing is at most 4.

The description "at most 4" can be rewritten as "2 or 3 or 4." (A sum of 1 is meaningless here.) Then

$$P(\text{at most 4}) = P(2 \text{ or } 3 \text{ or } 4)$$
$$= P(2) + P(3) + P(4), \quad (1)$$

since the events represented by "2," "3," and "4" are mutually exclusive.

The sample space for this experiment includes 36 possible pairs of numbers from 1 to 6: (1, 1), (1, 2), (1, 3), (1, 4), (1, 5), (1, 6), (2, 1), (2, 2), and so on. The pair (1, 1) is the only one with a sum of 2, so $P(2) = \frac{1}{36}$. Also $P(3) = \frac{2}{36}$ since both (1, 2) and (2, 1) give a sum of 3. The pairs (1, 3), (2, 2), and (3, 1) have a sum of 4, so $P(4) = \frac{3}{36}$. Substituting into equation (1) above gives

$$P(\text{at most 4}) = \frac{1}{36} + \frac{2}{36} + \frac{3}{36} = \frac{6}{36} = \frac{1}{6}.$$

**NOW TRY** Exercise 21.

### Properties of Probability

For any events $E$ and $F$,
1. $0 \leq P(E) \leq 1$
2. $P(\text{a certain event}) = 1$
3. $P(\text{an impossible event}) = 0$
4. $P(E') = 1 - P(E)$
5. $P(E \text{ or } F) = P(E \cup F) = P(E) + P(F) - P(E \cap F)$.

## B Exercises

*Concept Check* Write a sample space with equally likely outcomes for each experiment.

1. Two ordinary coins are tossed.
2. Three ordinary coins are tossed.
3. Five slips of paper marked with the numbers 1, 2, 3, 4, and 5 are placed in a box. After mixing well, two slips are drawn.
4. A die is rolled and then a coin is tossed.

*Write the events in Exercises 5–8 in set notation and give the probability of each event. See Example 1.*

5. In the experiment from Exercise 1:
   (a) Both coins show the same face.
   (b) At least one coin turns up heads.
6. In Exercise 2:
   (a) The result of the toss is exactly 2 heads and 1 tail.
   (b) At least 2 coins show tails.
7. In Exercise 3:
   (a) Both slips are marked with even numbers.
   (b) Both slips are marked with odd numbers.
   (c) Both slips are marked with the same number.
   (d) One slip is marked with an odd number and the other with an even number.
8. In Exercise 4:
   (a) The die shows an even number.
   (b) The coin shows heads.
   (c) The die shows 6.
   (d) The die shows 2 and the coin shows tails.
9. *Concept Check* A student gives the answer to a probability problem as $\frac{6}{5}$. Tell why this answer must be incorrect.
10. *Concept Check* If the probability of an event is 0.857, what is the probability that the event will not occur?

*Solve each problem. See Examples 2–5.*

11. In the experiment of drawing a card from a well-shuffled deck, find the probability of the events, $E$, the card is a face card (K, Q, J of any suit), and $E'$.
12. A baseball player with a batting average of 0.300 comes to bat. What are the odds in favor of his getting a hit?
13. The probability that a bank with assets greater than or equal to $30 billion will make a loan to a small business is 0.002. (*Source: Wall Street Journal* analysis of *Cal Reports* filed with federal banking authorities.) What are the odds against such a bank making a small business loan?

14. If the odds that it will rain are 4 to 5, what is the probability of rain?

15. Ms. Bezzone invites 10 relatives to a party: her mother, 2 uncles, 3 brothers, and 4 cousins. If the chances of any one guest arriving first are equally likely, find the following probabilities.

    (a) The first guest is an uncle or a cousin.
    (b) The first guest is a brother or a cousin.
    (c) The first guest is an uncle or her mother.

16. A card is drawn from a well-shuffled deck of 52 cards. Find the probability that the card is the following.

    (a) A queen
    (b) Red
    (c) A black 3
    (d) A club or red

17. In Exercise 16, find the probability of the following.

    (a) A face card (K, Q, J of any suit)
    (b) Red or a 3
    (c) Less than a 4 (consider aces as 1s)

18. Two dice are rolled. Find the probability of the following events.

    (a) The sum of the dots is at least 10.
    (b) The sum of the dots is either 7 or at least 10.
    (c) The sum of the dots is 3 or the dice both show the same number.

19. If a marble is drawn from a bag containing 2 yellow, 5 red, and 3 blue marbles, what are the probabilities of the following results?

    (a) The marble is yellow or blue.
    (b) The marble is yellow or red.
    (c) The marble is green.

20. The law firm of Alam, Bartolini, Chinn, Dickinson, and Ellsberg has two senior partners, Alam and Bartolini. Two of the attorneys are to be selected to attend a conference. Assuming that all are equally likely to be selected, find the following probabilities.

    (a) Chinn is selected.
    (b) Alam and Dickinson are selected.
    (c) At least one senior partner is selected.

21. The management of a bank wants to survey its employees, who are classified as follows for the purpose of an interview: 30% have worked for the bank more than 5 yr; 28% are female; 65% contribute to a voluntary retirement plan; half of the female employees contribute to the retirement plan. Find the following probabilities.

    (a) A male employee is selected.
    (b) An employee is selected who has worked for the bank for 5 yr or less.
    (c) An employee is selected who contributes to the retirement plan or is female.

22. The table shows the probabilities of a person accumulating specific amounts of credit card charges over a 12-month period.

    | Charges | Probability |
    | --- | --- |
    | Under $100 | 0.31 |
    | $100–$499 | 0.18 |
    | $500–$999 | 0.18 |
    | $1000–$1999 | 0.13 |
    | $2000–$2999 | 0.08 |
    | $3000–$4999 | 0.05 |
    | $5000–$9999 | 0.06 |
    | $10,000 or more | 0.01 |

    Refer to the table above, and find the probabilities that a person's total charges during the period are the following.

    (a) $500–$999
    (b) $500–$2999
    (c) $5000–$9999
    (d) $3000 or more

# Glossary

## A

**absolute value** The absolute value of a number is the distance between 0 and the number on a number line. (Section 1.1)

**absolute value equation** An absolute value equation is an equation that involves the absolute value of a variable expression. (Section 2.6)

**absolute value function** The function defined by $f(x) = |x|$ with a graph that includes portions of two lines is called the absolute value function. (Section 10.5)

**absolute value inequality** An absolute value inequality is an inequality that involves the absolute value of a variable expression. (Section 2.6)

**acute angle** An acute angle is an angle measuring between 0° and 90°. (Section 14.1)

**addition property of equality** The addition property of equality states that the same number can be added to (or subtracted from) both sides of an equation to obtain an equivalent equation. (Section 2.1)

**addition property of inequality** The addition property of inequality states that the same number can be added to (or subtracted from) both sides of an inequality without changing the solution set. (Section 2.4)

**additive inverse (negative) of a matrix** When two matrices are added and a zero matrix results, the matrices are additive inverses (negatives) of each other. (Section 4.4)

**additive inverse (negative, opposite)** Two numbers that are the same distance from 0 on a number line but on opposite sides of 0 are called additive inverses. (Section 1.1)

**algebraic expression** Any collection of numbers or variables joined by the basic operations of addition, subtraction, multiplication, or division (except by 0), or the operations of raising to powers or taking roots, formed according to the rules of algebra, is called an algebraic expression. (Sections 1.3, 5.2)

**ambiguous case** The situation in which the lengths of two sides of a triangle and the measure of the angle opposite one of them are given (SSA) is called the ambiguous case of the law of sines. Depending on the given measurements, this combination of given parts may result in 0, 1, or 2 possible triangles. (Section 20.2)

**amplitude** The amplitude of a periodic function is half the difference between the maximum and minimum values of the function. (Section 17.1)

**angle** An angle is formed by rotating a ray around its endpoint. (Section 14.1)

**angle in standard position** An angle is in standard position if its vertex is at the origin and its initial side is along the positive $x$-axis. (Section 14.1)

**angle of depression** The angle of depression from point $X$ to point $Y$ (below $X$) is the acute angle formed by ray $XY$ and a horizontal ray with endpoint at $X$. (Section 15.4)

**angle of elevation** The angle of elevation from point $X$ to point $Y$ (above $X$) is the acute angle formed by ray $XY$ and a horizontal ray with endpoint at $X$. (Section 15.4)

**Angle-Side-Angle (ASA)** The Angle-Side-Angle (ASA) congruence axiom states that if two angles and the included side of one triangle are equal, respectively, to two angles and the included side of a second triangle, then the triangles are congruent. (Section 20.1)

**angular speed $\omega$** Angular speed $\omega$ (omega) measures the speed of rotation and is defined by $\omega = \frac{\theta}{t}$, where $\theta$ is the angle of rotation in radians and $t$ is time. (Section 16.4)

**argument of a function** The argument of a function is the expression containing the independent variable of the function. For example, in the function $y = f(x - d)$, the expression $x - d$ is the argument. (Section 17.2)

**array of signs** An array of signs is used when evaluating a determinant using expansion by minors. The signs alternate for each row and column, beginning with $+$ in the first row, first column position. (Section 4.6)

**associative property of addition** The associative property of addition states that the way in which numbers being added are grouped does not change the sum. (Section 1.4)

**associative property of multiplication** The associative property of multiplication states that the way in which numbers being multiplied are grouped does not change the product. (Section 1.4)

**asymptotes of a hyperbola** The two intersecting straight lines that the branches of a hyperbola approach are called asymptotes of the hyperbola. (Section 13.2)

**augmented matrix** An augmented matrix is a matrix that has a vertical bar that separates the columns of the matrix into two groups. (Section 4.3)

**axis (axis of symmetry)** The axis of a parabola is the vertical or horizontal line through the vertex of the parabola. (Sections 10.2, 10.3)

## B

**base** The base is the number that is a repeated factor when written with an exponent. (Sections 1.3, 5.1)

**binomial** A binomial is a polynomial with exactly two terms. (Section 5.2)

**boundary line** In a graph of a linear inequality in two variables, the boundary line is a line that separates the region that satisfies the inequality from the region that does not satisfy the inequality. (Section 3.4)

## C

**center of a circle** The fixed point that is a fixed distance from all the points that form a circle is the center of the circle. (Section 13.1)

**center of an ellipse** The center of an ellipse is the fixed point located exactly halfway between the two foci. (Section 13.1)

**center-radius form of the equation of a circle** The center-radius form of the equation of a circle with center $(h, k)$ and radius $r$ is $(x - h)^2 + (y - k)^2 = r^2$. (Section 13.1)

**circle** A circle is the set of all points in a plane that lie a fixed distance from a fixed point. (Section 13.1)

**circular functions** The trigonometric functions of arc lengths, or real numbers, are called circular functions. (Section 16.3)

**circumference** The circumference of a circle is the distance around the circle. (Section 16.1)

**coefficient (numerical coefficient)** A coefficient is the numerical factor of a term. (Sections 1.4, 5.2)

**cofunctions** The function pairs sine and cosine, tangent and cotangent, and secant and cosecant are called cofunctions. (Section 15.1)

**column matrix** A matrix with just one column is called a column matrix. (Section 4.4)

**column of a matrix** A column of a matrix is a group of elements that are read vertically. (Section 4.3)

**combination** A combination of $n$ elements taken $r$ at a time is one of the ways in which $r$ elements can be chosen from $n$ elements. In combinations, the order of the elements is not important. (Appendix A)

**combined variation** If a problem involves a combination of direct and inverse variation, then it is called a combined variation problem. (Section 7.6)

**combining like terms** Combining like terms is a method of adding or subtracting like terms by using the properties of real numbers. (Section 1.4)

**common logarithm** A common logarithm is a logarithm to base 10. (Section 11.5)

**commutative property of addition** The commutative property of addition states that the order of numbers in an addition problem can be changed without changing the sum. (Section 1.4)

**commutative property of multiplication** The commutative property of multiplication states that the product in a multiplication problem remains the same regardless of the order of the factors. (Section 1.4)

**complement** In probability, the set of all outcomes in a sample space that do *not* belong to an event is called the complement of the event. (Appendix B)

**complementary angles (complements)** Two positive angles are complementary angles (or complements) if the sum of their measures is 90°. (Section 14.1)

**completing the square** The process of adding to a binomial the number that makes it a perfect square trinomial is called completing the square. (Section 9.1)

**complex conjugates** The complex conjugate of $a + bi$ is $a - bi$. (Section 8.7)

**complex fraction** A complex fraction is an expression with one or more fractions in the numerator, denominator, or both. (Section 7.3)

**complex number** A complex number is any number that can be written in the form $a + bi$, where $a$ and $b$ are real numbers. (Section 8.7)

**composite function (composition)** A function in which some quantity depends on a variable that, in turn, depends on another variable is called a composite function. (Sections 5.3, 10.1)

**compound event** In probability, a compound event involves two or more alternative events. (Appendix B)

**compound inequality** A compound inequality consists of two inequalities linked by a connective word such as *and* or *or*. (Section 2.5)

**comprehensive graph** A comprehensive graph of a polynomial function will show the following characteristics: (1) all $x$-intercepts (zeros); (2) the $y$-intercept; (3) all turning points; (4) enough of the domain to show the end behavior. (Section 12.3)

**conditional equation** A conditional equation is true for some replacements of the variable and false for others. (Section 2.1)

**congruent triangles** Triangles that are both the same size and the same shape are called congruent triangles. (Section 14.2)

**conic section** When a plane intersects an infinite cone at different angles, the figures formed by the intersections are called conic sections. (Section 13.1)

**conjugate** The conjugate of $a + b$ is $a - b$. (Section 8.5)

**conjugate zeros theorem** The conjugate zeros theorem states that if $f(x)$ is a polynomial having only real coefficients and if $a + bi$ is a zero of $f(x)$, where $a$ and $b$ are real numbers, then $a - bi$ is also a zero of $f(x)$. (Section 12.2)

**consistent system** A system of equations with a solution is called a consistent system. (Section 4.1)

**constant function** A linear function of the form $f(x) = b$, where $b$ is a constant, is called a constant function. (Section 3.5)

**constant of variation** In the variation equations $y = kx$, or $y = \frac{k}{x}$, or $y = kxz$, the number $k$ is called the constant of variation. (Section 7.6)

**constant on an interval** A function that is neither increasing nor decreasing on an interval is constant on that interval. (Section 10.4)

**contradiction** A contradiction is an equation that is never true. It has no solution. (Section 2.1)

**coordinate on a number line** Each number on a number line is called the coordinate of the point that it labels. (Section 1.1)

**coordinates of a point** The numbers in an ordered pair are called the coordinates of the corresponding point in the plane. (Section 3.1)

**cosecant** Let $P(x, y)$ be a point other than the origin on the terminal side of an angle $\theta$ in standard position. Let $r = \sqrt{x^2 + y^2}$ represent the distance from the origin to $P$. Then the cosecant function is defined by $\csc \theta = \frac{r}{y} \, (y \neq 0)$. (Section 14.3)

**cosine** Let $P(x, y)$ be a point other than the origin on the terminal side of an angle $\theta$ in standard position. Let $r = \sqrt{x^2 + y^2}$ represent the distance from the origin to $P$. Then the cosine function is defined by $\cos \theta = \frac{x}{r}$. (Section 14.3)

**cotangent** Let $P(x, y)$ be a point other than the origin on the terminal side of an angle $\theta$ in standard position. Let $r = \sqrt{x^2 + y^2}$ represent the distance from the origin to $P$. Then the cotangent function is defined by $\cot \theta = \frac{x}{y} \, (y \neq 0)$. (Section 14.3)

**coterminal angles** Two angles that have the same initial side and the same terminal side, but different amounts of rotation, are called coterminal angles. The measures of coterminal angles differ by a multiple of 360°. (Section 14.1)

**Cramer's rule** Cramer's rule uses determinants to solve systems of linear equations. (Section 4.6)

**cube root function** The function defined by $f(x) = \sqrt[3]{x}$ is called the cube root function. (Section 8.1)

**cubing function** The polynomial function defined by $f(x) = x^3$ is called the cubing function. (Section 5.3)

## D

**decreasing function** A function $f$ is a decreasing function on an interval if its graph goes downward from left to right: $f(x_1) > f(x_2)$ whenever $x_1 < x_2$. (Section 10.4)

**degree** The degree is the most common unit of measure for angles. One degree, written $1°$, represents $\frac{1}{360}$ of a complete rotation. (Section 14.1)

**degree of a polynomial** The degree of a polynomial is the greatest degree of any of the terms in the polynomial. (Section 5.2)

**degree of a term** The degree of a term is the sum of the exponents on the variables in the term. (Section 5.2)

**dependent equations** Equations of a system that have the same graph (because they are different forms of the same equation) are called dependent equations. (Section 4.1)

**dependent variable** In an equation relating $x$ and $y$, if the value of the variable $y$ depends on the variable $x$, then $y$ is called the dependent variable. (Section 3.5)

**Descartes's rule of signs** Descartes's rule of signs is a rule that can help determine the number of positive and the number of negative real zeros of a polynomial function. (Section 12.2 Exercises)

**descending powers** A polynomial in one variable is written in descending powers of the variable if the degree of the terms of the polynomial decreases from left to right. (Section 5.2)

**determinant** Associated with every square matrix is a real number called the determinant of the matrix, symbolized by the entries of the matrix placed between two vertical lines. (Section 4.6)

**difference** The answer to a subtraction problem is called the difference. (Section 1.2)

**difference of cubes** The difference of cubes, $x^3 - y^3$, can be factored as $x^3 - y^3 = (x - y)(x^2 + xy + y^2)$. (Section 6.3)

**difference of squares** The difference of squares, $x^2 - y^2$, can be factored as the product of the sum and difference of two terms, or $x^2 - y^2 = (x + y)(x - y)$. (Section 6.3)

**difference quotient** If the coordinates of point $P$ are $(x, f(x))$ and the coordinates of point $Q$ are $(x + h, f(x + h))$, then the expression $\frac{f(x + h) - f(x)}{h}$ is called the difference quotient. (Section 10.1)

**dimensions of a matrix** The number of rows followed by the number of columns gives the dimensions of a matrix. For example, a $3 \times 4$ matrix has 3 rows and 4 columns. (Section 4.3)

**direct variation** $y$ varies directly as $x$ if there exists a real number $k$ such that $y = kx$. (Section 7.6)

**direction angle** The positive angle between the $x$-axis and a position vector is the direction angle for the vector. (Section 20.4)

**directrix** A directrix is a fixed line which, together with the focus (a point not on the line), is used to determine the points that form a parabola. (Section 10.2)

**discontinuous graph** A discontinuous graph is a graph with one or more breaks. (Sections 7.4, 12.4)

**discriminant** The discriminant is the quantity under the radical, $b^2 - 4ac$, in the quadratic formula. (Section 9.2)

**distance** The distance between two points on a number line is the absolute value of the difference between the two numbers. (Section 1.2)

**distributive property** For any real numbers $a$, $b$, and $c$, the distributive property states that $a(b + c) = ab + ac$ and $(b + c)a = ba + ca$. (Section 1.4)

**division algorithm** The division algorithm states that if $f(x)$ and $g(x)$ are polynomials with $g(x)$ of lesser degree than $f(x)$ and $g(x)$ of degree one or more, then there exist unique polynomials $q(x)$ and $r(x)$ such that $f(x) = g(x) \cdot q(x) + r(x)$, where either $r(x) = 0$ or the degree of $r(x)$ is less than the degree of $g(x)$. (Section 12.1)

**domain** The set of all first components ($x$-values) in the ordered pairs of a relation is the domain. (Section 3.5)

**domain of the variable in a rational equation** The domain of the variable in a rational equation is the intersection (overlap) of the domains of the rational expressions in the equation. (Section 7.4)

**dominating term** The dominating term of a polynomial is the term of highest degree. (Section 12.3)

## E

**element of a matrix** The numbers in a matrix are called the elements of the matrix. (Section 4.3)

**elements (members)** Elements are the objects that belong to a set. (Section 1.1)

**elimination method** The elimination method is an algebraic method used to solve a system of equations in which the equations of the system are combined so that one or more variables is eliminated. (Section 4.1)

**ellipse** An ellipse is the set of all points in a plane the sum of whose distances from two fixed points is constant. (Section 13.1)

**empty set (null set)** The empty set, denoted by { } or ∅, is the set containing no elements. (Section 1.1)

**endpoint of a ray** In a given ray $AB$, point $A$ is the endpoint of the ray. (Section 14.1)

**equation** An equation is a statement that two quantities or algebraic expressions are equal. (Sections 1.1, 2.1)

**equivalent equations** Equivalent equations are equations that have the same solution set. (Section 2.1)

**equivalent inequalities** Equivalent inequalities are inequalities that have the same solution set. (Section 2.4)

**even function** A function $f$ is an even function if for all $x$ in the domain of $f$, $f(-x) = f(x)$. (Section 17.1)

**event** In probability, an event is any subset of the sample space. (Appendix B)

**exact number** A number that represents the result of counting, or a number that results from theoretical work and is not the result of a measurement, is an exact number. (Section 15.4)

**expansion by minors** A method of evaluating a $3 \times 3$ or larger determinant is called expansion by minors. (Section 4.6)

**exponent (power)** An exponent is a number that indicates how many times a factor is repeated. (Sections 1.3, 5.1)

**exponential equation** An exponential equation is an equation that has a variable in an exponent. (Section 11.2)

**exponential expression** A number or letter (variable) written with an exponent is an exponential expression. (Section 1.3)

**exponential function** An exponential function is a function defined by an expression of the form $f(x) = a^x$, where $a > 0$ and $a \neq 1$ for all real numbers $x$. (Section 11.2)

**extraneous solution** A solution to a new equation that does not satisfy the original equation is called an extraneous solution. (Section 8.6)

## F

**factor** A factor of a given number is any number that divides evenly (without remainder) into the given number. (Section 1.3)

**factor theorem** The factor theorem states that the polynomial $x - k$ is a factor of the polynomial $f(x)$ if and only if $f(k) = 0$. (Section 12.2)

**factoring** Writing a polynomial as the product of two or more simpler polynomials is called factoring. (Section 6.1)

**factoring by grouping** Factoring by grouping is a method of grouping the terms of a polynomial in such a way that the polynomial can be factored even though the greatest common factor of its individual terms is 1. (Section 6.1)

**factoring out the greatest common factor** Factoring out the greatest common factor is the process of using the distributive property to write a polynomial as a product of the greatest common factor and a simpler polynomial. (Section 6.1)

**finite set** If the number of elements in a set can be counted, the set is called a finite set. (Section 1.1)

**first-degree equation** A first-degree (linear) equation is an equation that has no term with the variable to a power greater than 1. (Sections 2.1, 3.1)

**foci** (singular, **focus**) Foci are fixed points used to determine the points that form a parabola, an ellipse, or a hyperbola. (Sections 10.2, 13.1, 13.2)

**FOIL** FOIL is a method for multiplying two binomials $(A + B)(C + D)$. Multiply **F**irst terms $AC$, **O**uter terms $AD$, **I**nner terms $BC$, and **L**ast terms $BD$. Then combine like terms. (Section 5.4)

**formula** A formula is an equation in which variables are used to describe a relationship. (Section 2.2)

**function** A function is a set of ordered pairs (relation) in which each value of the first component $x$ corresponds to exactly one value of the second component $y$. (Section 3.5)

**function notation** Function notation $f(x)$ represents the value of the function $f$ at $x$, that is, the $y$-value that corresponds to $x$. (Section 3.5)

**fundamental principle of counting** The fundamental principle of counting states that if one event can occur in $m$ ways and a second event can occur in $n$ ways, then both events can occur in $mn$ ways, provided that the outcome of the first event does not influence the outcome of the second event. (Appendix A)

**fundamental rectangle** The asymptotes of a hyperbola are the extended diagonals of its fundamental rectangle, with corners at the points $(a, b)$, $(-a, b)$, $(-a, -b)$, and $(a, -b)$. (Section 13.2)

**fundamental theorem of algebra** The fundamental theorem of algebra states that every polynomial of degree 1 or more has at least one complex zero. (Section 12.2)

## G

**graph of a number** The point on a number line that corresponds to a number is its graph. (Section 1.1)

**graph of a relation** The graph of a relation is the graph of its ordered pairs. (Section 3.5)

**graph of an equation** The graph of an equation is the set of all points that correspond to all of the ordered pairs that satisfy the equation. (Section 3.1)

**greatest common factor (GCF)** The greatest common factor of a list of integers is the largest common factor of those integers. The greatest common factor of a polynomial is the largest term that is a factor of all terms in the polynomial. (Section 6.1)

**greatest integer function** The function defined by $f(x) = [\![x]\!]$, where the symbol $[\![x]\!]$ is used to represent the greatest integer less than or equal to $x$, is called the greatest integer function. (Section 10.5)

## H

**horizontal asymptote** A horizontal line that a graph approaches as $|x|$ gets larger and larger without bound is called a horizontal asymptote. (Section 12.4)

**horizontal component** When a vector **u** is expressed as an ordered pair in the form $\mathbf{u} = \langle a, b \rangle$, the number $a$ is the horizontal component of the vector. (Section 20.4)

**horizontal line test** The horizontal line test states that a function is one-to-one if every horizontal line intersects the graph of the function at most once. (Section 11.1)

**hyperbola** A hyperbola is the set of all points in a plane such that the absolute value of the difference of the distances from two fixed points is constant. (Section 13.2)

**hypotenuse** The hypotenuse is the longest side in a right triangle. It is the side opposite the right angle. (Section 8.3)

## I

**identity** An identity is an equation that is true for all replacements of the variable. It has an infinite number of solutions. (Sections 2.1, 18.1)

**identity element for addition** Since adding 0 to a number does not change the number, 0 is called the identity element for addition. (Section 1.4)

**identity element for multiplication** Since multiplying a number by 1 does not change the number, 1 is called the identity element for multiplication. (Section 1.4)

**identity function** The simplest polynomial function is the identity function, defined by $f(x) = x$. (Section 5.3)

**identity matrix** A multiplicative identity matrix is a matrix $I$ such that $AI = A$ and $IA = A$ for any matrix, where $A$ and $I$ are square matrices of the same size. (Section 4.5)

**identity property** The identity properties state that the sum of 0 and any number equals the number, and the product of 1 and any number equals the number. (Section 1.4)

**imaginary part** The imaginary part of a complex number $a + bi$ is $b$. (Section 8.7)

**imaginary unit** The symbol $i$ is called the imaginary unit. (Section 8.7)

**inconsistent system** An inconsistent system of equations is a system with no solution. (Section 4.1)

**increasing function** A function $f$ is an increasing function on an interval if its graph goes upward from left to right: $f(x_1) < f(x_2)$ whenever $x_1 < x_2$. (Section 10.4)

**independent equations** Equations of a system that have different graphs are called independent equations. (Section 4.1)

**independent events** If the outcome of one event does not influence the outcome of another, then the events are called independent events. (Appendix A)

**independent variable** In an equation relating $x$ and $y$, if the value of the variable $y$ depends on the variable $x$, then $x$ is called the independent variable. (Section 3.5)

**index (order)** In a radical of the form $\sqrt[n]{a}$, $n$ is called the index or order. (Section 8.1)

**inequality** An inequality is a statement that two quantities or expressions are not equal. (Sections 1.1, 2.4)

**infinite set** If the number of elements in a set cannot be counted, the set is called an infinite set. (Section 1.1)

**initial point** When two letters are used to name a vector, the first letter indicates the initial (starting) point of the vector. (Section 20.4)

**initial side** When a ray is rotated around its endpoint to form an angle, the ray in its initial position is called the initial side of the angle. (Section 14.1)

**integers** The set of integers is $\{\ldots, -3, -2, -1, 0, 1, 2, 3, \ldots\}$. (Section 1.1)

**intersection** The intersection of two sets $A$ and $B$, written $A \cap B$, is the set of elements that belong to both $A$ and $B$. (Section 2.5)

**interval** An interval is a portion of a number line. (Section 1.1)

**interval notation** Interval notation is a simplified notation that uses parentheses ( ) and/or brackets [ ] to describe an interval on a number line. (Section 1.1)

**inverse of a function** $f$  If $f$ is a one-to-one function, then the inverse of $f$ is the set of all ordered pairs of the form $(y, x)$, where $(x, y)$ belongs to $f$. (Sections 11.1, 19.1)

**inverse property** The inverse properties state that a number added to its opposite is 0, and a number multiplied by its reciprocal is 1. (Section 1.4)

**inverse variation** $y$ varies inversely as $x$ if there exists a real number $k$ such that $y = \frac{k}{x}$. (Section 7.6)

**irrational numbers** Irrational numbers cannot be written as the quotient of two integers but can be represented by points on the number line. (Section 1.1)

## J

**joint variation** $y$ varies jointly as $x$ and $z$ if there exists a real number $k$ such that $y = kxz$. (Section 7.6)

## L

**least common denominator (LCD)** Given several denominators, the smallest expression that is divisible by all the denominators is called the least common denominator. (Section 7.2)

**legs of a right triangle** The two shorter sides of a right triangle are called the legs. (Section 8.3)

**like terms** Terms with exactly the same variables raised to exactly the same powers are called like terms. (Sections 1.4, 5.2)

**line** Two distinct points $A$ and $B$ determine a line called line $AB$. (Section 14.1)

**line segment (segment)** Line segment $AB$ (or segment $AB$) is the portion of line $AB$ between $A$ and $B$, including $A$ and $B$ themselves. (Section 14.1)

**linear equation in one variable** A linear equation in one variable can be written in the form $Ax + B = C$, where $A$, $B$, and $C$ are real numbers, with $A \neq 0$. (Section 2.1)

**linear equation in two variables** A linear equation in two variables is an equation that can be written in the form $Ax + By = C$, where $A$, $B$, and $C$ are real numbers and $A$ and $B$ are not both 0. (Section 3.1)

**linear function** A function defined by an equation of the form $f(x) = ax + b$, for real numbers $a$ and $b$, is a linear function. The value of $a$ is the slope $m$ of the graph of the function. (Section 3.5)

**linear inequality in one variable** A linear inequality in one variable can be written in the form $Ax + B < C$ or $Ax + B > C$ (or with $\leq$ or $\geq$), where $A$, $B$, and $C$ are real numbers, with $A \neq 0$. (Section 2.4)

**linear inequality in two variables** A linear inequality in two variables can be written in the form $Ax + By < C$ or $Ax + By > C$ (or with $\leq$ or $\geq$), where $A$, $B$, and $C$ are real numbers and $A$ and $B$ are not both 0. (Section 3.4)

**linear speed** $v$  The linear speed $v$ measures the distance traveled per unit of time. (Section 16.4)

**linear system (system of linear equations)** Two or more linear equations form a linear system. (Section 4.1)

**logarithm** A logarithm is an exponent. The expression $\log_a x$ is the exponent on the base $a$ that gives the number $x$. (Section 11.3)

**logarithmic equation** A logarithmic equation is an equation with a logarithm in at least one term. (Section 11.3)

**logarithmic function with base** $a$  If $a$ and $x$ are positive numbers with $a \neq 1$, then $f(x) = \log_a x$ defines the logarithmic function with base $a$. (Section 11.3)

**lowest terms** A fraction is in lowest terms when there are no common factors in the numerator and denominator (except 1). (Section 7.1)

## M

**magnitude** The length of a vector represents the magnitude of the vector quantity. (Section 20.4)

**mathematical model** In a real-world problem, a mathematical model is one or more equations (or inequalities) that describe the situation. (Section 2.2)

**matrix (plural, matrices)** A matrix is a rectangular array of numbers, consisting of horizontal rows and vertical columns. (Section 4.3)

**minor** The minor of an element in a $3 \times 3$ determinant is the $2 \times 2$ determinant remaining when a row and a column of the $3 \times 3$ determinant are eliminated. (Section 4.6)

**minute** One minute, written $1'$, is $\frac{1}{60}$ of a degree. (Section 14.1)

**monomial** A monomial is a polynomial with only one term. (Section 5.2)

**multiplication property of equality** The multiplication property of equality states that the same nonzero number can be multiplied by (or divided into) both sides of an equation to obtain an equivalent equation. (Section 2.1)

**multiplication property of inequality** The multiplication property of inequality states that both sides of an inequality may be multiplied (or divided) by a positive number without changing the direction of the inequality symbol. Multiplying (or dividing) by a negative number reverses the inequality symbol. (Section 2.4)

**multiplication property of 0** The multiplication property of 0 states that the product of any real number and 0 is 0. (Section 1.4)

**multiplicative inverse of a matrix (inverse matrix)** If $A$ is an $n \times n$ matrix, then its multiplicative inverse, written $A^{-1}$, must satisfy both $AA^{-1} = I_n$ and $A^{-1}A = I_n$. (Section 4.5)

**multiplicative inverse (reciprocal)** The multiplicative inverse of a nonzero real number $a$ is $\frac{1}{a}$. (Section 1.2)

**multiplicity of a zero** The multiplicity of a zero $k$ of a polynomial $f(x)$ is the number of factors of $x - k$ that appear when the polynomial is written in factored form. (Section 12.2)

**mutually exclusive events** In probabilty, two events that cannot occur simultaneously are called mutually exclusive events. (Appendix B)

## N

**$n$-factorial ($n!$)** For any positive integer $n$, $n(n-1)(n-2)(n-3)\cdots(2)(1) = n!$. (Appendix A)

**natural logarithm** A natural logarithm is a logarithm to base $e$. (Section 11.5)

**natural numbers (counting numbers)** The set of natural numbers includes the numbers used for counting: $\{1, 2, 3, 4, \ldots\}$. (Section 1.1)

**negative angle** A negative angle is an angle that is formed by clockwise rotation of a ray around its endpoint. (Section 14.1)

**negative of a polynomial** The negative of a polynomial is that polynomial with the sign of every term changed. (Section 5.2)

**nonlinear equation** A nonlinear equation is an equation in which some terms have more than one variable or a variable of degree 2 or higher. (Section 13.3)

**nonlinear system of equations** A nonlinear system of equations is a system that includes at least one nonlinear equation. (Section 13.3)

**number line** A number line is a line with a scale that is used to show how numbers relate to each other. (Section 1.1)

**number of zeros theorem** The number of zeros theorem states that a polynomial of degree $n$ has at most $n$ distinct zeros. (Section 12.2)

**numerical coefficient (coefficient)** The numerical factor in a term is its numerical coefficient. (Sections 1.4, 5.2)

# O

**oblique asymptote** A nonvertical, nonhorizontal line that a graph approaches as $|x|$ gets larger and larger without bound is called an oblique asymptote. (Section 12.4)

**oblique triangle** A triangle that is not a right triangle is called an oblique triangle. (Section 20.1)

**obtuse angle** An obtuse angle is an angle measuring more than 90° but less than 180°. (Section 14.1)

**odd function** A function $f$ is an odd function if for all $x$ in the domain of $f$, $f(-x) = -f(x)$. (Section 17.1)

**odds** The odds in favor of an event is the ratio of the probability of the event to the probability of the complement of the event. (Appendix B)

**one-to-one function** A one-to-one function is a function in which each $x$-value corresponds to only one $y$-value and each $y$-value corresponds to just one $x$-value. (Sections 11.1, 19.1)

**opposite of a vector** The opposite of a vector **v** is a vector $-$**v** that has the same magnitude as **v** but opposite direction. (Section 20.4)

**ordered pair** An ordered pair is a pair of numbers written within parentheses in which the order of the numbers is important. (Section 3.1)

**ordered triple** A solution of an equation in three variables, written $(x, y, z)$, is called an ordered triple. (Section 4.2)

**origin** The point at which the $x$-axis and $y$-axis of a rectangular coordinate system intersect is called the origin. (Section 3.1)

**outcome** In probability, a possible result of each trial in an experiment is called an outcome of the experiment. (Appendix B)

# P

**parabola** The graph of a second-degree (quadratic) equation in two variables is called a parabola. (Sections 5.3, 10.2)

**parallel lines** Parallel lines are two lines in the same plane that never intersect. (Sections 3.2, 14.2)

**parallelogram rule** The parallelogram rule is a way to find the sum of two vectors. If the two vectors are placed so that their initial points coincide and a parallelogram is completed that has these two vectors as two of its sides, then the diagonal vector of the parallelogram that has the same initial point as the two vectors is their sum. (Section 20.4)

**percent** Percent, written with the sign %, means "per one hundred." (Section 2.2)

**perfect square trinomial** A perfect square trinomial is a trinomial that can be factored as the square of a binomial. (Section 6.3)

**period** For a periodic function such that $f(x) = f(x + np)$, the smallest possible positive value of $p$ is the period of the function. (Section 17.1)

**periodic function** A periodic function is a function $f$ such that $f(x) = f(x + np)$, for every real number $x$ in the domain of $f$, every integer $n$, and some positive real number $p$. (Section 17.1)

**permutation** A permutation of $n$ elements taken $r$ at a time is one of the ways of arranging $r$ elements taken from a set of $n$ elements ($r \leq n$). In permutations, the order of the elements is important. (Appendix A)

**perpendicular lines** Perpendicular lines are two lines that intersect to form a right (90°) angle. (Section 3.2)

**phase shift** With trigonometric functions, a horizontal translation is called a phase shift. (Section 17.2)

**piecewise linear function** A function defined with different linear equations for different parts of its domain is called a piecewise linear function. (Section 10.5)

**plot** To plot an ordered pair is to locate it on a rectangular coordinate system. (Section 3.1)

**point of discontinuity** A point of discontinuity is a "hole" in the graph of a function. This appears in the graph of a rational function that has a common variable factor in the numerator and denominator. (Section 12.4)

**point-slope form** A linear equation is written in point-slope form if it is in the form $y - y_1 = m(x - x_1)$, where $m$ is the slope of the line and $(x_1, y_1)$ is a point on the line. (Section 3.3)

**polynomial** A polynomial is a term or a finite sum of terms in which all coefficients are real, all variables have whole number exponents, and no variables appear in denominators. (Section 5.2)

**polynomial function** A function defined by a polynomial in one variable, consisting of one or more terms, is called a polynomial function. (Section 5.3)

**polynomial function of degree $n$** A function defined by $f(x) = a_n x^n + a_{n-1} x^{n-1} + \cdots + a_1 x + a_0$ for complex numbers $a_n, a_{n-1}, \ldots, a_1$, and $a_0$, where $a_n \neq 0$, is called a polynomial function of degree $n$. (Section 12.1)

**polynomial in $x$** A polynomial containing only the variable $x$ is called a polynomial in $x$. (Section 5.2)

**position vector** A vector with its initial point at the origin is called a position vector. (Section 20.4)

**positive angle** A positive angle is an angle that is formed by counterclockwise rotation of a ray around its endpoint. (Section 14.1)

**prime polynomial** A prime polynomial is a polynomial that cannot be factored using only integer coefficients. (Section 6.2)

**principal root (principal $n$th root)** For even indexes, the symbols $\sqrt{\phantom{x}}, \sqrt[4]{\phantom{x}}, \sqrt[6]{\phantom{x}}, \ldots, \sqrt[n]{\phantom{x}}$ are used for nonnegative roots, which are called principal roots. (Section 8.1)

**probability of an event** In a sample space with equally likely outcomes, the probability of an event is the ratio of the number of outcomes in the event to the number of outcomes in the sample space. (Appendix B)

**product** The answer to a multiplication problem is called the product. (Section 1.2)

**product of the sum and difference of two terms** The product of the sum and difference of two terms is the difference of the squares of the terms: $(x + y)(x - y) = x^2 - y^2$. (Section 5.4)

**proportion** A proportion is a statement that two ratios are equal. (Section 7.5)

**proportional** If $y$ varies directly as $x$ and there exists some number (constant) $k$ such that $y = kx$, then $y$ is said to be proportional to $x$. (Section 7.6)

**pure imaginary number** A complex number $a + bi$ with $a = 0$ and $b \neq 0$ is called a pure imaginary number. (Section 8.7)

**Pythagorean theorem** The Pythagorean theorem states that the square of the length of the hypotenuse of a right triangle equals the sum of the squares of the lengths of the two legs. (Section 8.3)

# Q

**quadrant** A quadrant is one of the four regions in the plane determined by a rectangular coordinate system. (Section 3.1)

**quadrantal angle** A quadrantal angle is an angle that, when placed in standard position, has its terminal side along the $x$-axis or the $y$-axis. (Section 14.1)

**quadratic equation** A quadratic equation is an equation that can be written in the form $ax^2 + bx + c = 0$, where $a$, $b$, and $c$ are real numbers, with $a \neq 0$. (Sections 6.5, 9.1)

**quadratic formula** The quadratic formula is a general formula used to solve any quadratic equation. (Section 9.2)

**quadratic function** A function defined by an equation of the form $f(x) = ax^2 + bx + c$, for real numbers $a$, $b$, and $c$, with $a \neq 0$, is a quadratic function. (Section 10.2)

**quadratic in form** A nonquadratic equation that is written in the form $au^2 + bu + c = 0$, for $a \neq 0$ and an algebraic expression $u$, is called quadratic in form. (Section 9.3)

**quadratic inequality** A quadratic inequality can be written in the form $ax^2 + bx + c < 0$ or $ax^2 + bx + c > 0$ (or with $\leq$ or $\geq$), where $a$, $b$, and $c$ are real numbers, with $a \neq 0$. (Section 9.5)

**quotient** The answer to a division problem is called the quotient. (Section 1.2)

## R

**radian** A radian is a unit of measure for angles. An angle with its vertex at the center of a circle that intercepts an arc on the circle equal in length to the radius of the circle has a measure of 1 radian. (Section 16.1)

**radical** A radical sign with a radicand is called a radical. (Section 8.1)

**radical equation** An equation that includes one or more radical expressions with a variable is called a radical equation. (Section 8.6)

**radical expression** A radical expression is an algebraic expression that contains radicals. (Section 8.1)

**radical sign** The symbol $\sqrt{\phantom{x}}$ is called a radical sign. (Section 1.3)

**radicand** The number or expression under a radical sign is called the radicand. (Section 8.1)

**radius** The radius of a circle is the fixed distance between the center and any point on the circle. (Section 13.1)

**range** The set of all second components ($y$-values) in the ordered pairs of a relation is the range. (Section 3.5)

**ratio** A ratio is a comparison of two quantities. (Section 7.5)

**rational equation** A rational equation is an equation that contains at least one rational expression with a variable in the denominator. (Section 7.4)

**rational expression** The quotient of two polynomials with denominator not 0 is called a rational expression, or algebraic fraction. (Section 7.1)

**rational function** A function that is defined by a quotient of polynomials is called a rational function. (Sections 7.1, 12.4)

**rational inequality** An inequality that involves rational expressions called a rational inequality. (Section 9.5)

**rational numbers** Rational numbers can be written as the quotient of two integers, with denominator not 0. (Section 1.1)

**rational zeros theorem** The rational zeros theorem states that if $f(x)$ defines a polynomial function with integer coefficients and $\frac{p}{q}$, a rational number written in lowest terms, is a zero of $f$, then $p$ is a factor of the constant term $a_0$ and $q$ is a factor of the leading coefficient $a_n$. (Section 12.2)

**rationalizing the denominator** The process of removing radicals from a denominator so that the denominator contains only rational numbers is called rationalizing the denominator. (Section 8.5)

**ray** The portion of line $AB$ that starts at $A$ and continues through $B$, and on past $B$, is called ray $AB$. (Section 14.1)

**real numbers** Real numbers include all numbers that can be represented by points on the number line—that is, all rational and irrational numbers. (Section 1.1)

**real part** The real part of a complex number $a + bi$ is $a$. (Section 8.7)

**reciprocal** Pairs of numbers whose product is 1 are called reciprocals of each other. (Sections 1.2, 7.1, 14.4)

**reciprocal function** The reciprocal function is defined by $f(x) = \frac{1}{x}$. (Section 7.1)

**rectangular (Cartesian) coordinate system** The $x$-axis and $y$-axis placed at a right angle at their zero points form a rectangular coordinate system, also called the Cartesian coordinate system. (Section 3.1)

**reduced row echelon form** Reduced row echelon form is an extension of row echelon form that has 0s above and below the diagonal of 1s. (Section 4.3 Exercises)

**reference angle** The reference angle for an angle $\theta$, written $\theta'$, is the positive acute angle made by the terminal side of angle $\theta$ and the $x$-axis. (Section 15.2)

**reference arc** The reference arc for a point on the unit circle is the shortest arc from the point itself to the nearest point on the $x$-axis. (Section 16.3)

**relation** A relation is a set of ordered pairs. (Section 3.5)

**remainder theorem** The remainder theorem states that if the polynomial $f(x)$ is divided by $x - k$, then the remainder is $f(k)$. (Section 12.1)

**resultant** If **A** and **B** are vectors, the vector sum **A** + **B** is called the resultant of vectors **A** and **B**. (Section 20.4)

**right angle** A right angle is an angle measuring exactly 90°. (Section 14.1)

**rise** Rise is the vertical change between two points on a line—that is, the change in $y$-values. (Section 3.2)

**root (or solution)** A root (or solution) of a polynomial equation $f(x) = 0$ is a number $k$ such that $f(k) = 0$. (Section 12.1)

**row echelon form** If a matrix is written with 1s on the diagonal from upper left to lower right and 0s below the 1s, it is said to be in row echelon form. (Section 4.3)

**row matrix** A matrix with just one row is called a row matrix. (Section 4.4)

**row of a matrix** A row of a matrix is a group of elements that are read horizontally. (Section 4.3)

**row operations** Row operations are operations on a matrix that produce equivalent matrices leading to systems that have the same solutions as the original system of equations. (Section 4.3)

**run** Run is the horizontal change between two points on a line—that is, the change in $x$-values. (Section 3.2)

## S

**sample space** In probability, the set of all possible outcomes of a given experiment is called the sample space of the experiment. (Appendix B)

**scalar (with matrices)** In work with matrices, a real number is called a scalar to distinguish it from a matrix. (Section 4.4)

**scalar (with vectors)** A scalar is a quantity that involves a magnitude and can be represented by a real number. (Section 20.4)

**scalar product** The scalar product of a real number (or scalar) $k$ and a vector **u** is the vector $k \cdot \mathbf{u}$, which has magnitude $|k|$ times the magnitude of **u**. (Section 20.4)

**scientific notation** A number is written in scientific notation when it is expressed in the form $a \times 10^n$, where $1 \leq |a| < 10$ and $n$ is an integer. (Section 5.1)

**secant** Let $P(x, y)$ be a point other than the origin on the terminal side of an angle $\theta$ in

**standard position.** Let $r = \sqrt{x^2 + y^2}$ represent the distance from the origin to $P$. Then the secant function is defined by $\sec \theta = \frac{r}{x} (x \neq 0)$. (Section 14.3)

**second** One second, written $1''$, is $\frac{1}{60}$ of a minute. (Section 14.1)

**sector of a circle** A sector of a circle is the portion of the interior of a circle intercepted by a central angle. (Section 16.2)

**semiperimeter** The semiperimeter is half the sum of the lengths of the three sides of a triangle. (Section 20.3)

**set** A set is a collection of objects. (Section 1.1)

**set-builder notation** Set-builder notation is used to describe a set of numbers without actually having to list all of the elements. (Section 1.1)

**Side-Angle-Side (SAS)** The Side-Angle-Side (SAS) congruence axiom states that if two sides and the included angle of one triangle are equal, respectively, to two sides and the included angle of a second triangle, then the triangles are congruent. (Section 20.1)

**side of an angle** One of the two rays (or line segments) with a common endpoint that form an angle is called a side of the angle. (Section 14.1)

**Side-Side-Side (SSS)** The Side-Side-Side (SSS) congruence axiom states that if three sides of one triangle are equal, respectively, to three sides of a second triangle, then the triangles are congruent. (Section 20.1)

**signed numbers** Signed numbers are numbers that can be written with a positive or negative sign. (Section 1.1)

**significant digits** Significant digits are digits obtained by actual measurement. (Section 15.4)

**similar triangles** Triangles that are the same shape, but not necessarily the same size, are called similar triangles. (Section 14.2)

**simplified radical** A simplified radical meets four conditions:
1. The radicand has no factor raised to a power greater than or equal to the index.
2. The radicand has no fractions.
3. No denominator contains a radical.
4. Exponents in the radicand and the index of the radical have no common factor (except 1). (Section 8.3)

**sine** Let $P(x, y)$ be a point other than the origin on the terminal side of an angle $\theta$ in standard position. Let $r = \sqrt{x^2 + y^2}$ represent the distance from the origin to $P$. Then the sine function is defined by $\sin \theta = \frac{y}{r}$. (Section 14.3)

**sine wave (sinusoid)** The graph of a sine function is called a sine wave (or sinusoid). (Section 17.1)

**slope** The ratio of the change in $y$ to the change in $x$ along a line is called the slope of the line. (Section 3.2)

**slope-intercept form** A linear equation is written in slope-intercept form if it is in the form $y = mx + b$, where $m$ is the slope and $(0, b)$ is the $y$-intercept. (Section 3.3)

**solution of an equation** A solution of an equation is any replacement for the variable that makes the equation true. (Section 2.1)

**solution set** The solution set of an equation is the set of all solutions of the equation. (Section 2.1)

**solution set of a linear system** The solution set of a linear system of equations includes all ordered pairs that satisfy all the equations of the system at the same time. (Section 4.1)

**square matrix** A square matrix is a matrix that has the same number of rows as columns. (Section 4.3)

**square matrix of order $n$** An $n \times n$ matrix is called a square matrix of order $n$. (Section 4.4)

**square of a binomial** The square of a binomial is the sum of the square of the first term, twice the product of the two terms, and the square of the last term. That is, $(x + y)^2 = x^2 + 2xy + y^2$ or $(x - y)^2 = x^2 - 2xy + y^2$. (Section 5.4)

**square root** The opposite of squaring a number is called taking its square root; that is, a number $b$ is a square root of $a$ if $b^2 = a$. (Section 1.3)

**square root function** The function defined by $f(x) = \sqrt{x}$, with $x \geq 0$, is called the square root function. (Section 8.1)

**square root function** (extended definition) A function of the form $f(x) = \sqrt{u}$ for an algebraic expression $u$, with $u \geq 0$, is called a square root function. (Section 13.2)

**square root property** The square root property states that if $x^2 = k$, then $x = \sqrt{k}$ or $x = -\sqrt{k}$. (Section 9.1)

**square system** A system of equations that has the same number of equations as variables is called a square system. (Section 4.5)

**squaring function** The polynomial function defined by $f(x) = x^2$ is called the squaring function. (Section 5.3)

**standard form of a complex number** The standard form of a complex number is $a + bi$. (Section 8.7)

**standard form of a linear equation** A linear equation in two variables written in the form $Ax + By = C$, where $A$, $B$, and $C$ are integers with no common factor (except 1) and $A \geq 0$, is in standard form. (Sections 3.1, 3.3)

**standard form of a quadratic equation** A quadratic equation written in the form $ax^2 + bx + c = 0$, where $a$, $b$, and $c$ are real numbers with $a \neq 0$, is in standard form. (Sections 6.5, 9.1)

**step function** A function with a graph that looks like a series of steps is called a step function. (Section 10.5)

**straight angle** A straight angle is an angle measuring exactly $180°$. (Section 14.1)

**substitution method** The substitution method is an algebraic method for solving a system of equations in which one equation is solved for one of the variables and the result is substituted in the other equation. (Section 4.1)

**sum** The answer to an addition problem is called the sum. (Section 1.2)

**sum of cubes** The sum of cubes, $x^3 + y^3$, can be factored as $x^3 + y^3 = (x + y)(x^2 - xy + y^2)$. (Section 6.3)

**supplementary angles (supplements)** Two positive angles are supplementary angles (or supplements) if the sum of their measures is $180°$. (Section 14.1)

**symmetric with respect to the origin** If a graph can be rotated $180°$ about the origin and the result coincides exactly with the original graph, then the graph is symmetric with respect to the origin. (Section 10.4)

**symmetric with respect to the $x$-axis** If a graph can be folded in half along the $x$-axis and the portion of the graph above the $x$-axis exactly matches the portion below the $x$-axis, then the graph is symmetric with respect to the $x$-axis. (Section 10.4)

**symmetric with respect to the $y$-axis** If a graph can be folded in half along the $y$-axis and each half of the graph is the mirror image of the other half, then the graph is symmetric with respect to the $y$-axis. (Section 10.4)

**synthetic division** Synthetic division is a shortcut procedure for dividing a polynomial by a binomial of the form $x - k$. (Section 12.1)

**system of equations** A system of equations consists of two or more equations to be solved at the same time. (Section 4.1)

**system of linear equations**  A system of linear equations is a system of equations in which all of the equations are linear. (Section 4.1)

## T

**tangent**  Let $P(x, y)$ be a point other than the origin on the terminal side of an angle $\theta$ in standard position. Let $r = \sqrt{x^2 + y^2}$ represent the distance from the origin to $P$. Then the tangent function is defined by $\tan \theta = \frac{y}{x}$ ($x \neq 0$). (Section 14.3)

**term**  A term is a number, a variable, or the product or quotient of a number and one or more variables raised to powers. (Sections 1.4, 5.2)

**terminal point**  When two letters are used to name a vector, the second letter indicates the terminal (ending) point of the vector. (Section 20.4)

**terminal side**  When a ray is rotated around its endpoint to form an angle, the ray in its location after rotation is called the terminal side of the angle. (Section 14.1)

**three-part inequality**  An inequality containing two inequality symbols, such as $-2 < x \leq 5$, is called a three-part inequality. (Sections 1.1, 2.4)

**transversal**  A line that intersects two or more other lines, which may be parallel, is called a transversal. (Section 14.2)

**tree diagram**  A tree diagram is a diagram with branches that is used to systematically list all the outcomes of a counting situation or probability experiment. (Appendix A)

**trial**  In probability, each repetition of an experiment is called a trial. (Appendix B)

**trinomial**  A trinomial is a polynomial with exactly three terms. (Section 5.2)

**turning points**  The points on the graph of a function where the function changes from increasing to decreasing or from decreasing to increasing are called turning points. (Section 12.3)

## U

**union**  The union of two sets $A$ and $B$, written $A \cup B$, is the set of elements that belong to either $A$ or $B$ (or both). (Section 2.5)

**unit circle**  The unit circle is the circle with center at the origin and radius 1. (Section 16.3)

**unit vector**  A unit vector is a vector that has magnitude 1. Two useful unit vectors are $\mathbf{i} = \langle 1, 0 \rangle$ and $\mathbf{j} = \langle 0, 1 \rangle$. (Section 20.4)

**universal constant**  A fundamental number in our universe, such as $\pi$ and $e$, is called a universal constant. (Section 11.5)

**unlike terms**  Terms that contain different variables and/or the same variables raised to different powers are called unlike terms. (Section 1.4)

## V

**variable**  A variable is a symbol, usually a letter, used to represent an unknown number. (Section 1.1)

**vary directly (is directly proportional to)**  $y$ varies directly as $x$ if there exists a real number (constant) $k$ such that $y = kx$. (Section 7.6)

**vary inversely**  $y$ varies inversely as $x$ if there exists a real number (constant) $k$ such that $y = \frac{k}{x}$. (Section 7.6)

**vary jointly**  If one variable varies as the product of several other variables (sometimes raised to powers), then the first variable is said to vary jointly as the others. (Section 7.6)

**vector**  A vector is a directed line segment that represents a vector quantity. (Section 20.4)

**vector quantity**  A quantity that involves both magnitude and direction is called a vector quantity. (Section 20.4)

**Venn diagram**  A Venn diagram is a diagram used to illustrate relationships between sets. (Appendix B)

**vertex of a parabola**  For a vertical parabola, the vertex is the lowest point if the parabola opens up and the highest point if the parabola opens down. For a horizontal parabola, the vertex is the point farthest to the left if the parabola opens to the right, and the point farthest to the right if the parabola opens to the left. (Sections 10.2, 10.3)

**vertex of an angle**  The vertex of an angle is the endpoint of the ray that is rotated to form the angle. (Section 14.1)

**vertical angles**  Vertical angles are opposite angles formed by intersecting lines. (Section 14.2)

**vertical asymptote**  A vertical line that a graph approaches, but never touches or intersects, is called a vertical asymptote. The line $x = a$ is a vertical asymptote if $|f(x)|$ gets larger and larger as $x$ approaches $a$. (Sections 7.4, 12.4, 17.3)

**vertical component**  When a vector $\mathbf{u}$ is expressed as an ordered pair in the form $\mathbf{u} = \langle a, b \rangle$, the number $b$ is the vertical component of the vector. (Section 20.4)

**vertical line test**  The vertical line test states that any vertical line drawn through the graph of a function must intersect the graph in at most one point. (Section 3.5)

## W

**whole numbers**  The set of whole numbers is $\{0, 1, 2, 3, 4, \ldots\}$. (Section 1.1)

## X

**$x$-axis**  The horizontal number line in a rectangular coordinate system is called the $x$-axis. (Section 3.1)

**$x$-intercept**  A point where a graph intersects the $x$-axis is called an $x$-intercept. (Section 3.1)

## Y

**$y$-axis**  The vertical number line in a rectangular coordinate system is called the $y$-axis. (Section 3.1)

**$y$-intercept**  A point where a graph intersects the $y$-axis is called a $y$-intercept. (Section 3.1)

## Z

**zero matrix**  A matrix all of whose elements are 0 is a zero matrix. (Section 4.4)

**zero of a polynomial function**  A zero of a a polynomial function $f$ is a value of $k$ such that $f(k) = 0$. (Section 12.1)

**zero-factor property**  The zero-factor property states that if two numbers have a product of 0, then at least one of the numbers must be 0. (Sections 6.5, 9.1)

**zero vector**  The zero vector is the vector with magnitude 0. (Section 20.4)

# Answers to Selected Exercises

In this section, we provide the answers that we think most students will obtain when they work the exercises using the methods explained in the text. If your answer does not look exactly like the one given here, it is not necessarily wrong. In many cases, there are equivalent forms of the answer that are correct. For example, if the answer section shows $\frac{3}{4}$ and your answer is 0.75, you have obtained the right answer but written it in a different (yet equivalent) form. Unless the directions specify otherwise, 0.75 is just as valid an answer as $\frac{3}{4}$.

In general, if your answer does not agree with the one given in the text, see whether it can be transformed into the other form. If it can, then it is the correct answer. If you still have doubts, talk with your teacher.

## 1 Review of the Real Number System

### 1.1 Exercises (pages 12–15)

**1.** $\{1, 2, 3, 4, 5\}$   **3.** $\{5, 6, 7, 8, \ldots\}$
**5.** $\{\ldots, -1, 0, 1, 2, 3, 4\}$   **7.** $\{10, 12, 14, 16, \ldots\}$
**9.** $\emptyset$   **11.** $\{-4, 4\}$
*In Exercises 13 and 15, we give one possible answer.*
**13.** $\{x \mid x \text{ is an even natural number less than or equal to } 8\}$
**15.** $\{x \mid x \text{ is a multiple of 4 greater than } 0\}$   **17.** yes
**19.** [number line: $-6, -4, -2, 0, 2, 4, 6$]   **21.** [number line: $-\frac{6}{5}, -\frac{1}{4}, \frac{5}{6}, \frac{13}{4}, 5.2, \frac{11}{2}$]
**23.** (a) $8, 13, \frac{75}{5}$ (or 15)   (b) $0, 8, 13, \frac{75}{5}$   (c) $-9, 0, 8, 13, \frac{75}{5}$
(d) $-9, -0.7, 0, \frac{6}{7}, 4.\overline{6}, 8, \frac{21}{2}, 13, \frac{75}{5}$   (e) $-\sqrt{6}, \sqrt{7}$   (f) All are real numbers.   **25.** False; some are whole numbers, but negative integers are not.   **27.** False; no irrational number is an integer.
**29.** true   **31.** true   **33.** true   **35.** (a) A   (b) A
(c) B   (d) B   **37.** (a) $-6$   (b) 6   **39.** (a) 12   (b) 12
**41.** (a) $-\frac{6}{5}$   (b) $\frac{6}{5}$   **43.** 8   **45.** $\frac{3}{2}$   **47.** $-5$   **49.** $-2$
**51.** $-4.5$   **53.** 5   **55.** 6   **57.** 0   **59.** (a) Las Vegas; the population increased by 11.5%.   (b) Chicago; the population decreased by 1.2%.   **61.** Pacific Ocean, Indian Ocean, Caribbean Sea, South China Sea, Gulf of California
**63.** true   **65.** true   **67.** false   **69.** true   **71.** true
**73.** $2 < 6$   **75.** $4 > -9$   **77.** $-10 < -5$   **79.** $x > 0$
**81.** $7 > y$   **83.** $5 \geq 5$   **85.** $3t - 4 \leq 10$
**87.** $5x + 3 \neq 0$   **89.** $-3 < t < 5$   **91.** $-3 \leq 3x < 4$
**93.** $-6 < 10$; true   **95.** $10 \geq 10$; true   **97.** $-3 \geq -3$; true
**99.** $-8 > -6$; false
**101.** $(-1, \infty)$ [graph]   **103.** $(-\infty, 6]$ [graph]
**105.** $(0, 3.5)$ [graph]   **107.** $[2, 7]$ [graph]
**109.** $(-4, 3]$ [graph]   **111.** $(0, 3]$ [graph]

### 1.2 Exercises (pages 21–23)

**1.** the numbers are additive inverses; $4 + (-4) = 0$
**3.** negative; $-7 + (-21) = -28$   **5.** the positive number has larger absolute value; $15 + (-2) = 13$   **7.** the number with smaller absolute value is subtracted from the one with larger absolute value; $-15 - (-3) = -12$   **9.** negative; $-5(15) = -75$
**11.** $-19$   **13.** 9   **15.** $-\frac{19}{12}$   **17.** $-1.85$   **19.** $-11$
**21.** 21   **23.** $-13$   **25.** $-10.18$   **27.** $\frac{67}{30}$   **29.** 14
**31.** $-5$   **33.** $-6$   **35.** $-11$   **37.** 16   **39.** $-4$
**41.** 8   **43.** 3.018   **45.** $-\frac{7}{4}$   **47.** $-\frac{7}{8}$   **49.** 1   **51.** 6
**53.** $\frac{13}{2}$, or $6\frac{1}{2}$   **55.** It is true for multiplication (and division). It is false for addition and for subtraction when the number to be subtracted has the smaller absolute value. A more precise statement is, "The product or quotient of two negative numbers is positive."
**57.** $-35$   **59.** 40   **61.** 2   **63.** $-12$   **65.** $\frac{6}{5}$   **67.** 1
**69.** 5.88   **71.** $-10.676$   **73.** $-7$   **75.** 6   **77.** $-4$
**79.** 0   **81.** undefined   **83.** $\frac{25}{102}$   **85.** $-\frac{9}{13}$   **87.** $-2.1$
**89.** 10,000   **91.** 112°F   **93.** $30.13

### 1.3 Exercises (pages 28–30)

**1.** false; $-4^6 = -(4^6)$   **3.** true   **5.** true   **7.** true
**9.** false; the base is 3.   **11.** (a) 64   (b) $-64$   (c) 64
(d) $-64$   **13.** $10^4$   **15.** $\left(\frac{3}{4}\right)^5$   **17.** $(-9)^3$   **19.** $z^7$
**21.** 16   **23.** 0.021952   **25.** $\frac{1}{125}$   **27.** $\frac{256}{625}$   **29.** $-125$
**31.** 256   **33.** $-729$   **35.** $-4096$   **37.** 9   **39.** 13
**41.** $-20$   **43.** $\frac{10}{11}$   **45.** $-0.7$   **47.** not a real number
**49.** (a) B   (b) C   (c) A   **51.** not a real number   **53.** 24
**55.** 15   **57.** 55   **59.** $-91$   **61.** $-8$   **63.** $-48$
**65.** $-2$   **67.** $-79$   **69.** $-10$   **71.** 2   **73.** $-2$
**75.** undefined   **77.** $-7$   **79.** $-1$   **81.** 17   **83.** $-96$
**85.** $-\frac{15}{238}$   **87.** 8   **89.** $-\frac{5}{16}$   **91.** $-2.75$   **93.** $-\frac{3}{16}$

**95.** (a) $22.8 billion  (b) $37.4 billion  (c) $45.7 billion  (d) The amount spent on pets more than doubled from 1998 to 2009.

## 1.4 Exercises (pages 35–36)

**1.** B  **3.** A  **5.** product; 0  **7.** grouping  **9.** like
**11.** $2m + 2p$  **13.** $-12x + 12y$  **15.** $8k$  **17.** $-2r$
**19.** cannot be simplified  **21.** $8a$  **23.** $-2d + f$
**25.** $-6y + 3$  **27.** $p + 11$  **29.** $-2k + 15$
**31.** $-3m + 2$  **33.** $-1$  **35.** $2p + 7$  **37.** $-6z - 39$
**39.** $(5 + 8)x = 13x$  **41.** $(5 \cdot 9)r = 45r$  **43.** $9y + 5x$
**45.** 7  **47.** 0  **49.** $8(-4) + 8x = -32 + 8x$  **51.** 0
**53.** Answers will vary. One example is washing your face and brushing your teeth; one example is putting on your socks and putting on your shoes.  **55.** 1900  **57.** 75  **59.** 431

## 1 Review Exercises (pages 39–40)

**1.** [number line showing $-4, -2, 0, 2, 4$ with points at $\frac{9}{4}$ (or $-2$), $0$, $\frac{12}{3}$ (or $4$)]  **3.** 16  **5.** 5  **7.** $-9, -\sqrt{4}$
**9.** All are real numbers except $\sqrt{-9}$.
**11.** $\{0, 1, 2, 3\}$  **13.** false  **15.** $(-\infty, -5)$ [graph]
**17.** $\frac{41}{24}$  **19.** $-3$  **21.** $-39$  **23.** $\frac{23}{20}$  **25.** $\frac{2}{3}$  **27.** 3.21
**29.** $\frac{5}{7-7}$  **31.** 10,000  **33.** $-125$  **35.** 20  **37.** $-0.9$
**39.** $-4$  **41.** $-30$  **43.** $-\frac{8}{51}$  **45.** $-4z$  **47.** $4p$
**49.** $6r + 18$  **51.** $-p - 3q$  **53.** 0  **55.** $(2 + 3)x = 5x$
**57.** $(2 \cdot 4)x = 8x$  **59.** 0  **61.** 7

## 1 Test (page 41)

**[1.1] 1.** [number line with points at $-2, 0.75, \frac{5}{3}, 2, 4, 6, 6.3$]  **2.** $0, 3, \sqrt{25}$ (or 5), $\frac{24}{2}$ (or 12)
**3.** $-1, 0, 3, \sqrt{25}$ (or 5), $\frac{24}{2}$ (or 12)  **4.** $-1, -0.5, 0, 3, \sqrt{25}$ (or 5), $7.5, \frac{24}{2}$ (or 12)  **5.** All are real numbers except $\sqrt{-4}$.
**6.** $(-\infty, -3)$ [graph]  **7.** $(-4, 2]$ [graph]
**[1.2] 8.** 0  **[1.3] 9.** $-26$  **10.** 19  **11.** 1  **12.** $\frac{16}{7}$
**13.** $\frac{11}{23}$  **[1.2] 14.** 50,395 ft  **15.** 37,486 ft  **16.** 1345 ft
**[1.3] 17.** 14  **18.** $-15$  **19.** not a real number
**20.** (a) $a$ must be positive.  (b) $a$ must be negative.
(c) $a$ must be 0.  **21.** $-\frac{6}{23}$  **[1.4] 22.** $10k - 10$
**23.** It changes the sign of each term. The simplified form is $7r + 2$.  **24.** B  **25.** D  **26.** A  **27.** F
**28.** C  **29.** C  **30.** E

# 2 Linear Equations, Inequalities, and Applications

## 2.1 Exercises (pages 50–52)

**1.** A and C  **3.** Both sides are evaluated as 30, so 6 is a solution.
**5.** (a) equation  (b) expression  (c) equation  (d) expression

**7.** The solution contains a sign error when the distributive property was applied. The left side of the second line of the solution should be $8x - 4x + 6$. The correct solution is 1.  **9.** A conditional equation has one solution, an identity has infinitely many solutions, and a contradiction has no solution.  **11.** $\{-1\}$  **13.** $\{-4\}$
**15.** $\{-7\}$  **17.** $\{0\}$  **19.** $\{4\}$  **21.** $\{-\frac{7}{8}\}$
**23.** $\emptyset$; contradiction  **25.** $\{-\frac{5}{3}\}$  **27.** $\{-\frac{1}{2}\}$  **29.** $\{2\}$
**31.** $\{-2\}$  **33.** {all real numbers}; identity  **35.** $\{-1\}$
**37.** $\{7\}$  **39.** $\{2\}$  **41.** {all real numbers}; identity
**43.** $\{\frac{3}{2}\}$  **45.** 12  **47.** (a) $10^2$, or 100  (b) $10^3$, or 1000
**49.** $\{-\frac{18}{5}\}$  **51.** $\{-\frac{5}{6}\}$  **53.** $\{6\}$  **55.** $\{4\}$
**57.** $\{3\}$  **59.** $\{3\}$  **61.** $\{0\}$  **63.** $\{2000\}$
**65.** $\{25\}$  **67.** $\{40\}$  **69.** $\{3\}$

## 2.2 Exercises (pages 56–60)

**1.** $r = \frac{I}{pt}$  **3.** $L = \frac{P - 2W}{2}$, or $L = \frac{P}{2} - W$  **5.** (a) $W = \frac{V}{LH}$
(b) $H = \frac{V}{LW}$  **7.** $r = \frac{C}{2\pi}$  **9.** (a) $h = \frac{2A}{b + B}$
(b) $B = \frac{2A}{h} - b$, or $B = \frac{2A - hb}{h}$  **11.** $C = \frac{5}{9}(F - 32)$
**13.** D  **15.** $r = \frac{-2k - 3y}{a - 1}$, or $r = \frac{2k + 3y}{1 - a}$  **17.** $y = \frac{-x}{w - 3}$, or $y = \frac{x}{3 - w}$  **19.** 3.699 hr  **21.** 52 mph  **23.** 104°F
**25.** 230 m  **27.** radius: 240 in.; diameter: 480 in.  **29.** 2 in.
**31.** 75% water; 25% alcohol  **33.** 3%
**35.** .520, .500, .480, .385  **37.** $82,304
**39.** $33,890  **41.** 1500  **43.** 18,900

## 2.3 Exercises (pages 67–72)

**1.** (a) $x + 12$  (b) $12 > x$  **3.** (a) $x - 4$  (b) $4 < x$
**5.** D  **7.** $2x - 13$  **9.** $12 + 3x$  **11.** $8(x - 12)$
**13.** $\frac{3x}{7}$  **15.** $x + 6 = -31; -37$  **17.** $x - (-4x) = x + 9; \frac{9}{4}$
**19.** $12 - \frac{2}{3}x = 10; 3$  **21.** expression  **23.** equation
**25.** expression  **27.** Step 1: the number of patents each university secured; Step 2: patents that Stanford secured; Step 3: $x$; $x - 38$; Step 4: 134; Step 5: 134; 96; Step 6: 38; MIT patents: 96; 230  **29.** width: 165 ft; length: 265 ft  **31.** 24.34 in. by 29.88 in.  **33.** 850 mi; 925 mi; 1300 mi  **35.** Exxon Mobil: $443 billion; Wal-Mart: $406 billion  **37.** Eiffel Tower: 984 ft; Leaning Tower: 180 ft  **39.** Obama: 365 votes; McCain: 173 votes  **41.** 45.2%  **43.** $5481  **45.** $35.67
**47.** $225  **49.** $4000 at 3%; $8000 at 4%  **51.** $1800 at 5%; $3200 at 6.5%  **53.** $13,500  **55.** 5 L  **57.** 4 L
**59.** 1 gal  **61.** 150 lb  **63.** We cannot expect the final mixture to be worth more than either of the ingredients.

## 2.4 Exercises (pages 79–83)

**1.** D  **3.** B  **5.** F  **7.** (a) $131 \le s \le 155$
(b) $s > 155$  (c) $9 \le x \le 12$  (d) $x > 18$  **9.** Since $4 > 0$, the student should not have reversed the direction of the inequality symbol when dividing by 4. We reverse the symbol only when multiplying or dividing by a *negative* number. The solution set is $[-16, \infty)$.  **11.** $[16, \infty)$ [graph showing 16]

13. $(7, \infty)$
15. $(-\infty, -4)$
17. $(-\infty, -40]$
19. $(-\infty, 4]$
21. $(-\infty, -10]$
23. $(-\infty, 14)$
25. $\left(-\infty, -\frac{15}{2}\right)$
27. $\left[\frac{1}{2}, \infty\right)$
29. $[2, \infty)$
31. $(3, \infty)$
33. $(-\infty, 4)$
35. $\left(-\infty, \frac{23}{6}\right]$
37. $\left(-\infty, \frac{76}{11}\right)$
39. $(-\infty, \infty)$
41. $\emptyset$
43. $(1, 11)$
45. $[-14, 10]$
47. $[-5, 6]$
49. $\left[-\frac{14}{3}, 2\right]$
51. $\left[-\frac{1}{2}, \frac{35}{2}\right]$
53. $\left(-\frac{1}{3}, \frac{1}{9}\right]$
55. all numbers between $-2$ and $2$—that is, $(-2, 2)$
57. all numbers greater than or equal to 3—that is, $[3, \infty)$
59. all numbers greater than or equal to $-9$—that is, $[-9, \infty)$
61. from about 2:30 P.M. to 6:00 P.M.    63. about 84°F–91°F
65. at least 80    67. 26 months    69. 26 DVDs
71. (a) 140 to 184 lb   (b) Answers will vary.

## 2.5 Exercises (pages 89–90)

1. true    3. false; the union is $(-\infty, 8) \cup (8, \infty)$.    5. false; the intersection is $\emptyset$.    7. $\{1, 3, 5\}$, or $B$    9. $\{4\}$, or $D$
11. $\emptyset$    13. $\{1, 2, 3, 4, 5, 6\}$, or $A$    15.
17.    19. $(-3, 2)$
21. $(-\infty, 2]$    23. $\emptyset$
25. $[5, 9]$    27. $(-3, -1)$
29. $(-\infty, 4]$    31.
33.    35. $(-\infty, 8]$
37. $[-2, \infty)$    39. $(-\infty, \infty)$
41. $(-\infty, -5) \cup (5, \infty)$
43. $(-\infty, -1) \cup (2, \infty)$
45. $(-\infty, \infty)$    47. $[-4, -1]$
49. $[-9, -6]$    51. $(-\infty, 3)$    53. $[3, 9)$
55. intersection; $(-5, -1)$
57. union; $(-\infty, 4)$
59. union; $(-\infty, 0] \cup [2, \infty)$
61. intersection; $[4, 12]$
63. {Tuition and fees}    65. {Tuition and fees, Dormitory charges}

## 2.6 Exercises (pages 96–98)

1. E; C; D; B; A    3. (a) one   (b) two   (c) none
5. $\{-12, 12\}$    7. $\{-5, 5\}$    9. $\{-6, 12\}$
11. $\{-5, 6\}$    13. $\left\{-3, \frac{11}{2}\right\}$    15. $\left\{-\frac{19}{2}, \frac{9}{2}\right\}$
17. $\{-10, -2\}$    19. $\left\{-\frac{32}{3}, 8\right\}$
21. $(-\infty, -3) \cup (3, \infty)$
23. $(-\infty, -4] \cup [4, \infty)$
25. $(-\infty, -25] \cup [15, \infty)$
27. $(-\infty, -12) \cup (8, \infty)$
29. $(-\infty, -2) \cup (8, \infty)$
31. $\left(-\infty, -\frac{9}{5}\right] \cup [3, \infty)$
33. (a)    (b)
35. $[-3, 3]$    37. $(-4, 4)$
39. $[-25, 15]$
41. $[-12, 8]$
43. $[-2, 8]$    45. $\left[-\frac{9}{5}, 3\right]$
47. $(-\infty, -5) \cup (13, \infty)$
49. $(-\infty, -25) \cup (15, \infty)$
51. $\{-6, -1\}$
53. $\left[-\frac{10}{3}, 4\right]$
55. $\left[-\frac{7}{6}, -\frac{5}{6}\right]$
57. $(-\infty, -3] \cup [4, \infty)$

**59.** $\{-5, 1\}$  **61.** $\{3, 9\}$  **63.** $\{-5, 5\}$  **65.** $\{-5, -3\}$
**67.** $(-\infty, -3) \cup (2, \infty)$  **69.** $[-10, 0]$  **71.** $\{-1, 3\}$
**73.** $\{-3, \frac{5}{3}\}$  **75.** $\{-\frac{1}{3}, -\frac{1}{15}\}$  **77.** $\{-\frac{5}{4}\}$  **79.** $(-\infty, \infty)$
**81.** $\emptyset$  **83.** $\{-\frac{1}{4}\}$  **85.** $\emptyset$  **87.** $(-\infty, \infty)$  **89.** $\{-\frac{3}{7}\}$
**91.** $\{\frac{2}{5}\}$  **93.** $(-\infty, \infty)$  **95.** $\emptyset$  **97.** $|x - 1000| \le 100$; $900 \le x \le 1100$

## 2 Review Exercises (pages 102–105)

**1.** $\{-\frac{9}{5}\}$  **3.** $\{-\frac{7}{5}\}$  **5.** $(-\infty, \infty)$; identity
**7.** $\{0\}$; conditional  **9.** $b = \frac{2A - Bh}{h}$, or $b = \frac{2A}{h} - B$
**11.** $x = \frac{4}{3}(P + 12)$, or $x = \frac{4}{3}P + 16$  **13.** 6 ft  **15.** 6.5%
**17.** $525 billion  **19.** $9 - \frac{1}{3}x$  **21.** length: 13 m; width: 8 m
**23.** 12 kg  **25.** 10 L  **27.** $(-9, \infty)$  **29.** $(\frac{3}{2}, \infty)$
**31.** $[3, 5)$  **33.** 38 m or less  **35.** any score greater than or equal to 61  **37.** $\{a, c\}$  **39.** $\{a, c, e, f, g\}$
**41.** $(6, 9)$ 
**43.** $(-\infty, -3] \cup (5, \infty)$ 
**45.** $\emptyset$  **47.** $(-3, 4)$  **49.** $(4, \infty)$  **51.** $\{-7, 7\}$
**53.** $\{-\frac{1}{3}, 5\}$  **55.** $\{0, 7\}$  **57.** $\{-\frac{3}{4}, \frac{1}{2}\}$  **59.** $(-14, 14)$
**61.** $[-3, -2]$

## 2 Test (pages 105–106)

[2.1] **1.** $\{-19\}$  **2.** $\{5\}$  **3.** $(-\infty, \infty)$  **4. (a)** $\emptyset$; contradiction  **(b)** $(-\infty, \infty)$; identity  **(c)** $\{0\}$; conditional equation  [2.2] **5.** $v = \frac{S + 16t^2}{t}$  **6.** $r = \frac{-2 - 6t}{a - 3}$, or $r = \frac{2 + 6t}{3 - a}$  **7.** 3.173 hr  [2.3] **8.** 6.25%  **9.** 73.7%
**10.** $8000 at 3%; $20,000 at 5%
[2.4] **11.** $[1, \infty)$ 
**12.** $(-\infty, 28)$ 
**13.** $[-3, 3]$ 
**14.** C  **15.** 82  **16.** $[500, \infty)$  [2.5] **17. (a)** $\{1, 5\}$
**(b)** $\{1, 2, 5, 7, 9, 12\}$  **18.** $[2, 9)$  **19.** $(-\infty, 3) \cup [6, \infty)$
[2.6] **20.** $[-\frac{5}{2}, 1]$  **21.** $(-\infty, -\frac{7}{6}) \cup (\frac{17}{6}, \infty)$  **22.** $\emptyset$
**23.** $(\frac{1}{3}, \frac{7}{3})$  **24.** $\{-\frac{5}{3}, 3\}$  **25.** $\{-\frac{5}{7}, \frac{11}{3}\}$  **26. (a)** $\emptyset$
**(b)** $(-\infty, \infty)$  **(c)** $\emptyset$

# 3 Graphs, Linear Equations, and Functions

## 3.1 Exercises (pages 114–117)

**1. (a)** $x$ represents the year; $y$ represents the revenue in billions of dollars.  **(b)** about $1850 billion  **(c)** (2002, 1850)
**(d)** In 2000, federal tax revenues were about $2030 billion.
**3.** origin  **5.** $y$; $x$; $x$; $y$  **7.** two  **9. (a)** I  **(b)** III  **(c)** II  **(d)** IV  **(e)** none  **(f)** none  **11. (a)** I or III  **(b)** II or IV  **(c)** II or IV  **(d)** I or III  **13.–21.**

**23. (a)** $-4; -3; -2; -1; 0$  **25. (a)** $-3; 3; 2; -1$
**(b)**   **(b)**

**27. (a)** $\frac{5}{2}; 5; \frac{3}{2}; 1$  **29. (a)** $-4; 5; -\frac{12}{5}; \frac{5}{4}$
**(b)**   **(b)**

**31. (a)** $3; 1; -1; -3$  **33. (a)** 1  **(b)** 2  **(c)** For every increase in $x$ by 1 unit, $y$ increases by 2 units.  **35.** Choose a value *other than* 0 for either $x$ or $y$. For example, if $x = -5$, then $y = 4$.
**(b)**

**37.** $(6, 0); (0, 4)$ 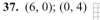  **39.** $(6, 0); (0, -2)$

**41.** $(\frac{21}{2}, 0); (0, -\frac{7}{3})$  **43.** none; $(0, 5)$

**45.** $(2, 0)$; none   **47.** $(-4, 0)$; none

**49.** $(0, 0); (0, 0)$   **51.** $(0, 0); (0, 0)$

**53.** $(0, 0); (0, 0)$   **55.** $(-5, -1)$

**57.** $(\frac{9}{2}, -\frac{3}{2})$  **59.** $(0, \frac{11}{2})$  **61.** $(2.1, 0.9)$  **63.** $(1, 1)$
**65.** $(-\frac{5}{12}, \frac{5}{28})$  **67.** 57,225 thousand (or 57,225,000)

## 3.2 Exercises (pages 125–129)

1. A, B, and D   3. (a) C  (b) A  (c) D  (d) B   5. 0
7. $-\frac{1}{3}$   9. $-4$   11. 2   13. $\frac{5}{2}$   15. 0   17. undefined
19. undefined   21. 8   23. $\frac{5}{6}$   25. 0   27. $-1$   29. 6
31. $-3$   33. $-\frac{5}{2}$   35. undefined   37. B   39. A
41. $-\frac{1}{2}$ 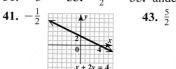   43. $\frac{5}{2}$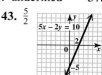
45. 4 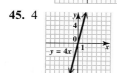   47. undefined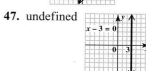
49. 0 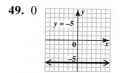   51. 0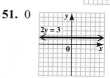
53. 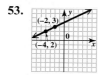   55.    57.
59. 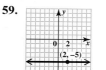   61.
63. $-\frac{4}{9}, \frac{9}{4}$   65. parallel   67. perpendicular   69. neither
71. parallel   73. neither   75. perpendicular   77. $-\$4000$ per yr; The value of the machine is decreasing \$4000 each year during these years.   79. 0% per yr (or no change); The percent of pay raise is not changing—it is 3% each year during these years.
81. 19.5 ft   83. (a) 23,431 per yr; 18,897 per yr; 12,392 per yr; 17,955 per yr; 23,418 per yr   (b) no; no   85. (a) \$18.25 billion per yr   (b) The positive slope means that personal spending on recreation in the United States *increased* by an average of \$18.25 billion each year.   87. $-\$69$ per yr; the price decreased an average of \$69 each year from 1997 to 2002.

## 3.3 Exercises (pages 138–142)

1. A   3. A   5. $3x + y = 10$   7. A   9. C   11. H
13. B   15. $y = 5x + 15$   17. $y = -\frac{2}{3}x + \frac{4}{5}$
19. $y = \frac{2}{5}x + 5$   21. $y = \frac{2}{3}x + 1$   23. (a) $y = x + 4$
(b) 1   (c) $(0, 4)$   (d) 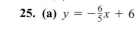   25. (a) $y = -\frac{6}{5}x + 6$
(b) $-\frac{6}{5}$   (c) $(0, 6)$   (d) 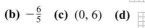   27. (a) $y = \frac{4}{5}x - 4$

(b) $\frac{4}{5}$   (c) $(0, -4)$   (d)    29. (a) $y = -\frac{1}{2}x - 2$

(b) $-\frac{1}{2}$   (c) $(0, -2)$   (d)    31. (a) $y = \frac{4}{3}x + 4$

(b) $\frac{4}{3}$   (c) $(0, 4)$   (d)    33. (a) $2x + y = 18$

(b) $y = -2x + 18$   35. (a) $3x + 4y = 10$   (b) $y = -\frac{3}{4}x + \frac{5}{2}$
37. (a) $x - 2y = -13$   (b) $y = \frac{1}{2}x + \frac{13}{2}$
39. (a) $4x - y = 12$   (b) $y = 4x - 12$
41. (a) $7x - 5y = -20$   (b) $y = 1.4x + 4$   43. $y = 5$
45. $x = 9$   47. $x = 0.5$   49. $y = 8$   51. (a) $2x - y = 2$
(b) $y = 2x - 2$   53. (a) $x + 2y = 8$   (b) $y = -\frac{1}{2}x + 4$
55. (a) $2x - 13y = -6$   (b) $y = \frac{2}{13}x + \frac{6}{13}$   57. (a) $y = 5$
(b) $y = 5$   59. (a) $x = 7$   (b) not possible
61. (a) $y = -3$   (b) $y = -3$   63. (a) $y = 3x - 19$
(b) $3x - y = 19$   65. (a) $y = \frac{1}{2}x - 1$   (b) $x - 2y = 2$
67. (a) $y = -\frac{1}{2}x + 9$   (b) $x + 2y = 18$   69. (a) $y = 7$
(b) $y = 7$   71. $y = 45x$; $(0, 0), (5, 225), (10, 450)$
73. $y = 3.01x$; $(0, 0), (5, 15.05), (10, 30.10)$
75. (a) $y = 39x + 99$   (b) $(5, 294)$; the cost of a 5-month membership is \$294.   (c) \$567   77. (a) $y = 35x + 44.95$
(b) $(5, 219.95)$; the cost of the plan for 5 months is \$219.95.
(c) \$464.95   79. (a) $y = 6x + 30$   (b) $(5, 60)$; it costs \$60 to rent the saw for 5 days.   (c) 18 days
81. (a) $y = -5.6x + 91$; the percent of households accessing the Internet by dial-up is decreasing 5.6% per year.   (b) 57%
83. (a) $y = 906.5x + 17{,}860.5$   (b) about \$26,019; it is slightly lower than the actual value.

## 3.4 Exercises (pages 146–147)

1. solid; below   3. dashed; above   5. The graph of $Ax + By = C$ divides the plane into two regions. In one of the regions, the ordered pairs satisfy $Ax + By < C$; in the other, they satisfy $Ax + By > C$.   7.    9.

11.    13.    15.

**17.**   **19.**   **21.**

**23.**   **25.** $-3 < x < 3$

**27.** $-2 < x + 1 < 2$    **29.**

**31.**   **33.**

## 3.5 Exercises (pages 157–160)

**1.** We give one of many possible answers here. A function is a set of ordered pairs in which each first component corresponds to exactly one second component. For example, $\{(0, 1), (1, 2), (2, 3), (3, 4), \ldots\}$ is a function.   **3.** independent variable
**5.** function   **7.** not a function   **9.** function   **11.** not a function; domain: $\{0, 1, 2\}$; range: $\{-4, -1, 0, 1, 4\}$
**13.** function; domain: $\{2, 3, 5, 11, 17\}$; range: $\{1, 7, 20\}$
**15.** not a function; domain: $\{1\}$; range: $\{5, 2, -1, -4\}$
**17.** function; domain: $(-\infty, \infty)$; range: $(-\infty, \infty)$   **19.** function; domain: $(-\infty, \infty)$; range: $(-\infty, 4]$   **21.** not a function; domain: $[-4, 4]$; range: $[-3, 3]$   **23.** function; domain: $(-\infty, \infty)$
**25.** not a function; domain: $[0, \infty)$   **27.** function; domain: $(-\infty, \infty)$   **29.** not a function; domain: $(-\infty, \infty)$   **31.** function; domain: $[0, \infty)$   **33.** function; domain: $(-\infty, 0) \cup (0, \infty)$
**35.** function; domain: $\left[-\frac{1}{2}, \infty\right)$   **37.** function; domain: $(-\infty, 4) \cup (4, \infty)$   **39.** B   **41.** 4   **43.** $-11$   **45.** 3
**47.** 2.75   **49.** $-3p + 4$   **51.** $3x + 4$   **53.** $-3x - 2$
**55.** $-\pi^2 + 4\pi + 1$   **57.** $-3x - 3h + 4$   **59.** $-9$
**61.** (a) 2   (b) 3   **63.** (a) 15   (b) 10   **65.** (a) 3   (b) $-3$
**67.** (a) $f(x) = -\frac{1}{3}x + 4$   (b) 3   **69.** (a) $f(x) = 3 - 2x^2$
(b) $-15$   **71.** (a) $f(x) = \frac{4}{3}x - \frac{8}{3}$   (b) $\frac{4}{3}$   **73.** line; $-2$; $-2x + 4$; $-2$; $3$; $-2$
**75.** domain: $(-\infty, \infty)$; range: $(-\infty, \infty)$

**77.** domain: $(-\infty, \infty)$; range: $(-\infty, \infty)$

**79.** domain: $(-\infty, \infty)$; range: $(-\infty, \infty)$

**81.** domain: $(-\infty, \infty)$; range: $\{-4\}$

**83.** domain: $(-\infty, \infty)$; range: $\{0\}$

**85.** (a) $8.25   (b) 3 is the value of the independent variable, which represents a package weight of 3 lb; $f(3)$ is the value of the dependent variable, representing the cost to mail a 3-lb package.
(c) $13.75; $f(5) = 13.75$   **87.** (a) 194.53 cm   (b) 177.29 cm
(c) 177.41 cm   (d) 163.65 cm   **89.** (a) $[0, 100]$; $[0, 3000]$
(b) 25 hr; 25 hr   (c) 2000 gal   (d) $f(0) = 0$; the pool is empty at time 0.   (e) $f(25) = 3000$; after 25 hr, there are 3000 gal of water in the pool.

## 3 Review Exercises (pages 163–166)

**1.**

| x | y |
|---|---|
| 0 | 5 |
| $\frac{10}{3}$ | 0 |
| 2 | 2 |
| $\frac{14}{3}$ | $-2$ |

   **3.** $(3, 0); (0, -4)$

**5.** $(10, 0); (0, 4)$

**7.** If both coordinates are positive, the point lies in quadrant I. If the first coordinate is negative and the second is positive, the point lies in quadrant II. To lie in quadrant III, the point must have both coordinates negative. To lie in quadrant IV, the first coordinate must be positive and the second must be negative.
**9.** $-\frac{1}{2}$   **11.** $\frac{3}{4}$   **13.** $\frac{2}{3}$   **15.** undefined   **17.** $-1$
**19.** negative   **21.** 0   **23.** 12 ft   **25.** (a) $y = -\frac{1}{3}x - 1$
(b) $x + 3y = -3$   **27.** (a) $y = -\frac{4}{3}x + \frac{29}{3}$
(b) $4x + 3y = 29$   **29.** (a) not possible   (b) $x = 2$
**31.** (a) $y = \frac{7}{5}x + \frac{16}{5}$   (b) $7x - 5y = -16$
**33.** (a) $y = 4x - 29$   (b) $4x - y = 29$
**35.** (a) $y = 57x + 159$; $843   (b) $y = 47x + 159$; $723
**37.**    **39.**    **41.** D

**43.** domain: {9, 11, 4, 17, 25}; range: {32, 47, 69, 14}; function
**45.** domain: $(-\infty, 0]$; range: $(-\infty, \infty)$; not a function   **47.** not a function; domain: $(-\infty, \infty)$   **49.** function; domain: $\left[-\frac{7}{4}, \infty\right)$
**51.** function; domain: $(-\infty, 6) \cup (6, \infty)$   **53.** $-6$   **55.** $-8$
**57. (a)** yes   **(b)** domain: {1943, 1953, 1963, 1973, 1983, 1993, 2003}; range: {63.3, 68.8, 69.9, 71.4, 74.6, 75.5, 77.6}
**(c)** Answers will vary. Two possible answers are (1943, 63.3) and (1953, 68.8).   **(d)** 71.4; in 1973, life expectancy at birth was 71.4 yr.   **(e)** 1993   **59.** C

### 3 Test (pages 166–168)

[3.1] **1.** $-\frac{10}{3}; -2; 0$

[3.1, 3.3] **2.** $\left(\frac{20}{3}, 0\right); (0, -10)$

**3.** none; (0, 5) 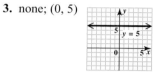   **4.** (2, 0); none

[3.2] **5.** $\frac{1}{2}$   **6.** It is a vertical line.   **7.** perpendicular
**8.** neither   **9.** $-1200$ farms per yr; the number of farms decreased, on the average, by about 1200 each year from 1980 to 2005.   [3.3] **10. (a)** $y = -5x + 19$   **(b)** $5x + y = 19$
**11. (a)** $y = 14$   **(b)** $y = 14$   **12. (a)** $y = -\frac{3}{5}x - \frac{11}{5}$
**(b)** $3x + 5y = -11$   **13. (a)** $y = -\frac{1}{2}x - \frac{3}{2}$
**(b)** $x + 2y = -3$   **14. (a)** $y = -\frac{1}{2}x + 2$   **(b)** $x + 2y = 4$
**15. (a)** not possible   **(b)** $x = 5$   **16.** B
**17. (a)** $y = 1784.17x + 13{,}939.17$   **(b)** \$29,997; it is slightly less than the actual value.
[3.4] **18.**    **19.**

[3.5] **20.** D   **21.** D   **22.** domain: $[0, \infty)$; range: $(-\infty, \infty)$
**23.** domain: $\{0, -2, 4\}$; range: $\{1, 3, 8\}$
**24. (a)** 0   **(b)** $-a^2 + 2a - 1$
**25.** domain: $(-\infty, \infty)$; range: $(-\infty, \infty)$

## 4 Systems and Matrices

### 4.1 Exercises (pages 178–181)

**1.** 3; $-6$   **3.** ∅   **5.** 0   **7.** D; The ordered pair solution must be in quadrant IV, since that is where the graphs of the equations intersect.

**9. (a)** B   **(b)** C   **(c)** A   **(d)** D
**11.** yes   **13.** no   **15.** $\{(-2, -3)\}$

**17.** $\{(1, 2)\}$   **19.** $\{(2, 3)\}$   **21.** $\left\{\left(\frac{22}{9}, \frac{22}{3}\right)\right\}$   **23.** $\{(5, 4)\}$
**25.** $\{(1, 3)\}$   **27.** $\left\{\left(-5, -\frac{10}{3}\right)\right\}$   **29.** $\{(2, 6)\}$
**31.** $\{(x, y) \mid 2x - y = 0\}$; dependent equations   **33.** ∅; inconsistent system   **35.** $\{(2, -4)\}$   **37.** $\{(3, -1)\}$
**39.** $\{(2, -3)\}$   **41.** $\{(x, y) \mid 7x + 2y = 6\}$; dependent equations   **43.** $\left\{\left(\frac{3}{2}, -\frac{3}{2}\right)\right\}$   **45.** ∅; inconsistent system
**47.** $\{(0, 0)\}$   **49.** $\{(0, -4)\}$   **51.** $\{(2, -4)\}$
**53.** $y = -\frac{3}{7}x + \frac{4}{7}$; $y = -\frac{3}{7}x + \frac{3}{14}$; no solution   **55.** Both are $y = -\frac{2}{3}x + \frac{1}{3}$; infinitely many solutions   **57. (a)** Use substitution, since the second equation is solved for $y$.   **(b)** Use elimination, since the coefficients of the $y$-terms are opposites.
**(c)** Use elimination, since the equations are in standard form with no coefficients of 1 or $-1$. Solving by substitution would involve fractions.   **59.** $\{(-3, 2)\}$   **61.** $\{(-4, 6)\}$
**63.** $\{(x, y) \mid 4x - y = -2\}$   **65.** $\left\{\left(1, \frac{1}{2}\right)\right\}$
**67. (a)** $\{(5, 5)\}$   **(b)**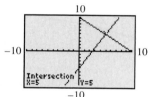

**69. (a)** $\{(0, -2)\}$   **(b)**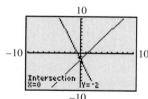

**71. (a)** \$4   **(b)** 300 half-gallons   **(c)** supply: 200 half-gallons; demand: 400 half-gallons   **73.** 2000, 2001, 2002, first half of 2003   **75.** (3.6, 10.4) (Values may vary slightly based on the method of solution used.)

### 4.2 Exercises (pages 187–188)

**1.** B   **3.** $\{(3, 2, 1)\}$   **5.** $\{(1, 4, -3)\}$   **7.** $\{(0, 2, -5)\}$
**9.** $\{(1, 0, 3)\}$   **11.** $\{(-12, 18, 0)\}$   **13.** $\left\{\left(1, \frac{3}{10}, \frac{2}{5}\right)\right\}$
**15.** $\left\{\left(-\frac{7}{3}, \frac{22}{3}, 7\right)\right\}$   **17.** $\{(4, 5, 3)\}$   **19.** $\{(2, 2, 2)\}$
**21.** $\left\{\left(\frac{8}{3}, \frac{2}{3}, 3\right)\right\}$   **23.** $\{(-1, 0, 0)\}$   **25.** $\{(-3, 5, -6)\}$
**27.** ∅; inconsistent system   **29.** $\{(x, y, z) \mid x - y + 4z = 8\}$; dependent equations   **31.** $\{(3, 0, 2)\}$
**33.** $\{(x, y, z) \mid 2x + y - z = 6\}$; dependent equations
**35.** $\{(0, 0, 0)\}$   **37.** ∅; inconsistent system

## 4.3 Exercises (pages 194–196)

**1.** (a) 0, 5, −3  (b) 1, −3, 8  (c) yes; the number of rows is the same as the number of columns (three).  (d) $\begin{bmatrix} 1 & 4 & 8 \\ 0 & 5 & -3 \\ -2 & 3 & 1 \end{bmatrix}$

(e) $\begin{bmatrix} 1 & -\frac{3}{2} & -\frac{1}{2} \\ 0 & 5 & -3 \\ 1 & 4 & 8 \end{bmatrix}$  (f) $\begin{bmatrix} 1 & 15 & 25 \\ 0 & 5 & -3 \\ 1 & 4 & 8 \end{bmatrix}$   **3.** $\left[\begin{array}{cc|c} 1 & 2 & 11 \\ 2 & -1 & -3 \end{array}\right]$;

$\left[\begin{array}{cc|c} 1 & 2 & 11 \\ 0 & -5 & -25 \end{array}\right]$; $\left[\begin{array}{cc|c} 1 & 2 & 11 \\ 0 & 1 & 5 \end{array}\right]$; $x + 2y = 11, y = 5$; $\{(1, 5)\}$

**5.** $\{(4, 1)\}$  **7.** $\{(1, 1)\}$  **9.** $\{(-1, 4)\}$  **11.** ∅

**13.** $\{(x, y) \mid 2x + y = 4\}$  **15.** $\left[\begin{array}{ccc|c} 1 & 1 & -1 & -3 \\ 0 & -1 & 3 & 10 \\ 0 & -6 & 7 & 38 \end{array}\right]$;

$\left[\begin{array}{ccc|c} 1 & 1 & -1 & -3 \\ 0 & 1 & -3 & -10 \\ 0 & -6 & 7 & 38 \end{array}\right]$; $\left[\begin{array}{ccc|c} 1 & 1 & -1 & -3 \\ 0 & 1 & -3 & -10 \\ 0 & 0 & -11 & -22 \end{array}\right]$;

$\left[\begin{array}{ccc|c} 1 & 1 & -1 & -3 \\ 0 & 1 & -3 & -10 \\ 0 & 0 & 1 & 2 \end{array}\right]$; $x + y - z = -3, y - 3z = -10, z = 2$;

$\{(3, -4, 2)\}$  **17.** $\{(4, 0, 1)\}$  **19.** $\{(-1, 23, 16)\}$
**21.** $\{(3, 2, -4)\}$  **23.** $\{(x, y, z) \mid x - 2y + z = 4\}$  **25.** ∅
**27.** $\{(1, 1)\}$  **29.** $\{(-1, 2, 1)\}$  **31.** $\{(1, 7, -4)\}$

## 4.4 Exercises (pages 204–206)

**1.** $w = 3, x = 2, y = -1, z = 4$  **3.** $m = 8, n = -2, z = 2,$ $y = 5, w = 6$  **5.** $a = 2, z = -3, m = 8, k = 1$
**7.** 2 × 2; square  **9.** 3 × 4  **11.** 2 × 1; column
**13.** To add two matrices of the same size, add corresponding elements. For example, $[2 \ \ 1 \ \ -3] + [3 \ \ 4 \ \ -2] = [5 \ \ 5 \ \ -5]$. Matrices of different sizes cannot be added.

**15.** $\begin{bmatrix} -2 & -7 & 7 \\ 10 & -2 & 7 \end{bmatrix}$  **17.** $\begin{bmatrix} -6 & 8 \\ 4 & 2 \end{bmatrix}$

**19.** $\begin{bmatrix} 5x + y & x + y & 7x + y \\ 8x + 2y & x + 3y & 3x + y \end{bmatrix}$  **21.** The matrices cannot be added.  **23.** $\begin{bmatrix} -4 & 8 \\ 0 & 6 \end{bmatrix}$  **25.** $\begin{bmatrix} 2 & 6 \\ -4 & 6 \end{bmatrix}$  **27.** $\begin{bmatrix} -1 & -3 \\ 2 & -3 \end{bmatrix}$

**29.** no  **31.** $\begin{bmatrix} 13 \\ 25 \end{bmatrix}$  **33.** $\begin{bmatrix} -17 \\ -1 \end{bmatrix}$  **35.** $\begin{bmatrix} 17 & -10 \\ 1 & 2 \end{bmatrix}$

**37.** $\begin{bmatrix} -2 & 10 \\ 0 & 8 \end{bmatrix}$  **39.** $\begin{bmatrix} -2 & 5 & 0 \\ 6 & 6 & 1 \\ 12 & 2 & -3 \end{bmatrix}$  **41.** $[2 \ \ 7 \ \ -4]$

**43.** The matrices cannot be multiplied.

**45.** $\begin{bmatrix} 100 & 150 \\ 125 & 50 \\ 175 & 200 \end{bmatrix}$; $\begin{bmatrix} 100 & 125 & 175 \\ 150 & 50 & 200 \end{bmatrix}$

**47.** (a) $\begin{bmatrix} 50 & 100 & 30 \\ 10 & 90 & 50 \\ 60 & 120 & 40 \end{bmatrix}$  (b) $\begin{bmatrix} 12 \\ 10 \\ 15 \end{bmatrix}$  (If the rows and columns are interchanged in part (a), this should be a 1 × 3 matrix.)

(c) $\begin{bmatrix} 2050 \\ 1770 \\ 2520 \end{bmatrix}$  (This may be a 1 × 3 matrix.)  (d) $6340

## 4.5 Exercises (pages 215–217)

**1.** $I = \begin{bmatrix} 1 & 0 \\ 0 & 1 \end{bmatrix}$; $AI = \begin{bmatrix} 4 & -2 \\ 3 & 1 \end{bmatrix}\begin{bmatrix} 1 & 0 \\ 0 & 1 \end{bmatrix} = \begin{bmatrix} 4 & -2 \\ 3 & 1 \end{bmatrix} = A$

**3.** yes  **5.** no  **7.** no  **9.** yes  **11.** $\begin{bmatrix} 2 & 1 \\ -\frac{3}{2} & -\frac{1}{2} \end{bmatrix}$

**13.** The inverse does not exist.  **15.** $\begin{bmatrix} -1 & 1 & 1 \\ 0 & -1 & 0 \\ 2 & -1 & -1 \end{bmatrix}$

**17.** $\begin{bmatrix} 7 & -3 & -3 \\ -1 & 1 & 0 \\ -1 & 0 & 1 \end{bmatrix}$  **19.** $\begin{bmatrix} -\frac{15}{4} & -\frac{1}{4} & -3 \\ \frac{5}{4} & \frac{1}{4} & 1 \\ -\frac{3}{2} & 0 & -1 \end{bmatrix}$

**21.** $\begin{bmatrix} \frac{1}{2} & 0 & \frac{1}{2} & -1 \\ \frac{1}{10} & -\frac{2}{5} & \frac{3}{10} & -\frac{1}{5} \\ -\frac{7}{10} & \frac{4}{5} & -\frac{11}{10} & \frac{12}{5} \\ \frac{1}{5} & \frac{1}{5} & -\frac{2}{5} & \frac{3}{5} \end{bmatrix}$  **23.** $\{(2, 3)\}$  **25.** $\{(-2, 4)\}$

**27.** $\{(4, -6)\}$  **29.** $\{(10, -1, -2)\}$  **31.** $\{(11, -1, 2)\}$
**33.** $\{(1, 0, 2, 1)\}$  **35.** (a) $602.7 = a + 5.543b + 37.14c$
$656.7 = a + 6.933b + 41.30c$
$778.5 = a + 7.638b + 45.62c$
(b) $a \approx -490.547, b = -89, c = 42.71875$
(c) $S = -490.547 - 89A + 42.71875B$  (d) approximately 843.5  (e) $S \approx 1547.5$; Using only three consecutive years to forecast six years into the future is probably not wise.

## 4.6 Exercises (pages 223–225)

**1.** (a) true  (b) true  (c) false; the determinant equals $ad - bc$.
(d) true  **3.** −3  **5.** 14  **7.** 0  **9.** 59  **11.** 14
**13.** Multiply the upper left and lower right entries. Then multiply the upper right and lower left entries. Subtract the second product from the first to obtain the determinant. For example,
$\begin{vmatrix} 4 & 2 \\ 7 & 1 \end{vmatrix} = 4 \cdot 1 - 2 \cdot 7 = 4 - 14 = -10.$  **15.** 16
**17.** −12  **19.** 0  **21.** $\{(1, 0, -1)\}$  **23.** $\{(-3, 6)\}$
**25.** $\left\{\left(\frac{53}{17}, \frac{6}{17}\right)\right\}$  **27.** $\{(-1, 2)\}$  **29.** $\{(4, -3, 2)\}$
**31.** Cramer's rule does not apply.  **33.** $\{(-2, 1, 3)\}$
**35.** $\left\{\left(\frac{49}{9}, -\frac{155}{9}, \frac{136}{9}\right)\right\}$

## 4 Review Exercises (pages 230–232)

1. (a) 1980 and 1985  (b) just less than 500,000   3. D
5. $\{(-\frac{8}{9}, -\frac{4}{3})\}$   7. $\{(2, 4)\}$   9. $\{(0, 1)\}$   11. $\{(-6, 3)\}$
13. $\emptyset$; inconsistent system   15. $\{(1, -5, 3)\}$   17. $\emptyset$; inconsistent system   19. $\{(-1, 5)\}$   21. $\{(1, 2, -1)\}$
23. $k = -8, y = 7, a = 2, m = 3, p = 1, r = -5$

25. $\begin{bmatrix} 9 & -14 & 13 \\ 21 & -4 & 28 \end{bmatrix}$   27. $\begin{bmatrix} 11 & 20 \\ 14 & 40 \end{bmatrix}$   29. $\begin{bmatrix} 2 & 12 \\ 12 & 24 \\ -29 & -38 \end{bmatrix}$

31. $\begin{bmatrix} -3 \\ 10 \end{bmatrix}$   33. $\begin{bmatrix} -\frac{1}{4} & \frac{1}{6} \\ 0 & \frac{1}{3} \end{bmatrix}$   35. The inverse does not exist.
37. $\{(2, 1)\}$   39. $\{(-1, 0, 2)\}$   41. $-25$   43. $-44$
45. $\{(-2, 5)\}$   47. $\{(-4, 6, 2)\}$

## 4 Test (pages 232–233)

[4.1] 1. (a) Houston, Phoenix, Dallas  (b) Philadelphia  (c) Dallas, Phoenix, Philadelphia, Houston   2. (a) 2010; 1.45 million  (b) (2025, 2.8)   3. $\{(6, 1)\}$

4. $\{(6, -4)\}$   5. $\{(-\frac{9}{4}, \frac{5}{4})\}$   6. $\{(x, y) \mid 12x - 5y = 8\}$; dependent equations   7. $\{(3, 3)\}$   8. $\{(0, -2)\}$
9. $\emptyset$; inconsistent system   [4.2] 10. $\{(-\frac{2}{3}, \frac{4}{5}, 0)\}$
11. $\{(3, -2, 1)\}$   [4.3] 12. $\{(\frac{2}{5}, \frac{7}{5})\}$   13. $\{(-1, 2, 3)\}$
[4.4] 14. $x = -1; y = 7; w = -3$   15. $\begin{bmatrix} 8 & 3 \\ 0 & -11 \\ 15 & 19 \end{bmatrix}$
16. The matrices cannot be added.   17. $\begin{bmatrix} -5 & 16 \\ 19 & 2 \end{bmatrix}$
18. The matrices cannot be multiplied.   19. A
[4.5] 20. $\begin{bmatrix} -2 & -5 \\ -3 & -8 \end{bmatrix}$   21. The inverse does not exist.
22. $\begin{bmatrix} -9 & 1 & -4 \\ -2 & 1 & 0 \\ 4 & -1 & 1 \end{bmatrix}$   23. $\{(-7, 8)\}$   24. $\{(0, 5, -9)\}$
[4.6] 25. $-58$   26. $-844$   27. $\{(-6, 7)\}$
28. $\{(1, -2, 3)\}$

## 5 Exponents, Polynomials, and Polynomial Functions

## 5.1 Exercises (pages 246–249)

1. incorrect; $(ab)^2 = a^2 b^2$   3. incorrect; $\left(\frac{4}{a}\right)^3 = \frac{4^3}{a^3}$
5. The product rule says that when exponential expressions with like bases are multiplied, the base stays the same and the exponents are added. For example, $x^5 \cdot x^6 = x^{11}$.   7. $13^{12}$
9. $x^{17}$   11. $-27w^8$   13. $18x^3 y^8$   15. The product rule does not apply.   17. (a) B  (b) C  (c) B  (d) C
19. 1   21. $-1$   23. 1   25. 2   27. 0   29. $-2$
31. (a) B  (b) D  (c) B  (d) D   33. $\frac{1}{5^4}$, or $\frac{1}{625}$   35. $\frac{1}{8}$
37. $\frac{1}{16x^2}$   39. $\frac{4}{x^2}$   41. $-\frac{1}{a^3}$   43. $\frac{1}{a^4}$   45. $\frac{11}{30}$   47. $-\frac{5}{24}$
49. When $n$ is even, the expressions are opposites. When $n$ is odd, they are equal.   51. 16   53. $\frac{27}{4}$   55. $\frac{27}{8}$   57. $\frac{25}{16}$
59. (a) B  (b) D  (c) D  (d) B   61. The quotient rule says that when exponential expressions with like bases are divided, the base stays the same and the exponents are subtracted. For example, $\frac{x^8}{x^5} = x^3$.   63. $4^2$, or 16   65. $x^4$   67. $\frac{1}{r^3}$
69. $6^6$   71. $\frac{1}{6^{10}}$   73. $7^2$, or 49   75. $r^3$   77. The quotient rule does not apply.   79. $x^{18}$   81. $\frac{27}{125}$   83. $64t^3$
85. $-216x^6$   87. $-\frac{64m^6}{t^3}$   89. $\frac{1}{3}$   91. $\frac{1}{a^5}$   93. $\frac{1}{k^2}$
95. $-4r^6$   97. $\frac{625}{a^{10}}$   99. $\frac{z^4}{x^3}$   101. $-14k^3$   103. $\frac{p^4}{5}$
105. $\frac{1}{2pq}$   107. $\frac{4}{a^2}$   109. $\frac{1}{6y^{13}}$   111. $\frac{4k^5}{m^2}$   113. $\frac{4k^{17}}{125}$
115. $\frac{2k^5}{3}$   117. $\frac{8}{3pq^{10}}$   119. $\frac{y^9}{8}$   121. $5.3 \times 10^2$
123. $8.3 \times 10^{-1}$   125. $6.92 \times 10^{-6}$   127. $-3.85 \times 10^4$
129. 72,000   131. 0.00254   133. $-60,000$
135. 0.000012   137. 0.06   139. 0.0000025
141. 200,000   143. 3000   145. (a) $3.059 \times 10^8$
(b) $1 \times 10^{12}$ (or $\$10^{12}$)  (c) $\$3269$   147. 300 sec
149. approximately $5.87 \times 10^{12}$ mi   151. (a) 20,000 hr
(b) 833 days

## 5.2 Exercises (pages 253–254)

1. $2x^3 - 3x^2 + x + 4$   3. $p^7 - 8p^5 + 4p^3$
5. $3m^4 - m^3 + 5m^2 + 10$   7. 7; 1   9. $-15$; 2   11. 1; 4
13. $\frac{1}{6}$; 1   15. $-1$; 6   17. monomial; 0   19. binomial; 1
21. binomial; 8   23. trinomial; 3   25. none of these; 5
27. A   29. $8z^4$   31. $7m^3$   33. $5x$   35. already simplified   37. $-t + 13s$   39. $8k^2 + 2k - 7$
41. $-2n^4 - n^3 + n^2$   43. $-2ab^2 + 20a^2 b$   45. $3m + 11$
47. $-p - 4$   49. A monomial (or term) is a numeral, a variable, or a product of numerals and variables raised to positive integer powers. Some examples of monomials are 6, $x$, and $-4x^2 y^3$. A binomial is a sum or difference of exactly two terms, such as $x^2 + y^2$ and $x^2 - y^2$. A trinomial consists of exactly three terms, such as $x^2 - 3x + 8$. These are all examples of polynomials.
51. $8x^2 + x - 2$   53. $-t^4 + 2t^2 - t + 5$
55. $5y^3 - 3y^2 + 5y + 1$   57. $r + 13$
59. $-2a^2 - 2a - 7$   61. $-3z^5 + z^2 + 7z$   63. $12p - 4$
65. $-9p^2 + 11p - 9$   67. $5a + 18$   69. $14m^2 - 13m + 6$
71. $13z^2 + 10z - 3$   73. $10y^3 - 7y^2 + 5y + 8$
75. $-5a^4 - 6a^3 + 9a^2 - 11$   77. $3y^2 - 4y + 2$
79. $-4m^2 + 4n^2 - 7n$   81. $y^4 - 4y^2 - 4$
83. $10z^2 - 16z$

## 5.3 Exercises (pages 262–264)

1. (a) $-10$  (b) 8   3. (a) 8  (b) 2   5. (a) 8  (b) 74
7. (a) $-11$  (b) 4   9. (a) 15,163  (b) 17,401  (c) 19,656

**11. (a)** 9 million  **(b)** 61 million  **(c)** 118 million
**13. (a)** $8x - 3$  **(b)** $2x - 17$  **15. (a)** $-x^2 + 12x - 12$
**(b)** $9x^2 + 4x + 6$  **17.** $x^2 + 2x - 9$  **19.** 6
**21.** $x^2 - x - 6$  **23.** 6  **25.** $-33$  **27.** 0  **29.** $-\frac{9}{4}$
**31.** $-\frac{9}{2}$  **33.** For example, let $f(x) = 2x^3 + 3x^2 + x + 4$ and
$g(x) = 2x^4 + 3x^3 - 9x^2 + 2x - 4$. For these functions,
$(f - g)(x) = -2x^4 - x^3 + 12x^2 - x + 8$, and
$(g - f)(x) = 2x^4 + x^3 - 12x^2 + x - 8$. Because the two
differences are not equal, subtraction of polynomial functions
is not commutative.  **35.** 16  **37.** 83  **39.** 13
**41.** $(2x + 3)^2 + 4$  **43.** $(x + 5)^2 + 4$  **45.** $2x + 8$
**47.** $\frac{137}{4}$  **49.** 8  **51.** $(f \circ g)(x) = 63,360x$; it computes
the number of inches in $x$ miles.  **53.** $(A \circ r)(t) = 4\pi t^2$; this
is the area of the circular layer as a function of time.

**55.**
domain: $(-\infty, \infty)$;
range: $(-\infty, \infty)$

**57.**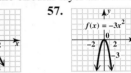
domain: $(-\infty, \infty)$;
range: $(-\infty, 0]$

**59.**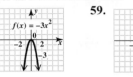
domain: $(-\infty, \infty)$;
range: $(-\infty, \infty)$

## 5.4 Exercises (pages 270–271)

**1.** C  **3.** D  **5.** $-24m^5$  **7.** $-28x^7y^4$  **9.** $-6x^2 + 15x$
**11.** $-2q^3 - 3q^4$  **13.** $18k^4 + 12k^3 + 6k^2$
**15.** $6m^3 + m^2 - 14m - 3$  **17.** $m^3 - 3m^2 - 40m$
**19.** $24z^3 - 20z^2 - 16z$  **21.** $4x^5 - 4x^4 - 24x^3$
**23.** $6y^2 + y - 12$  **25.** $-2b^3 + 2b^2 + 18b + 12$
**27.** $25m^2 - 9n^2$  **29.** $8z^4 - 14z^3 + 17z^2 + 20z - 3$
**31.** $6p^4 + p^3 + 4p^2 - 27p - 6$  **33.** $m^2 - 3m - 40$
**35.** $12k^2 + k - 6$  **37.** $3z^2 + zw - 4w^2$
**39.** $12c^2 + 16cd - 3d^2$  **41.** $0.1x^2 + 0.63x - 0.13$
**43.** $3w^2 - \frac{23}{4}wz - \frac{1}{2}z^2$  **45.** The product of two binomials is
the sum of the product of the first terms, the product of the outer
terms, the product of the inner terms, and the product of the last
terms.  **47.** $4p^2 - 9$  **49.** $25m^2 - 1$  **51.** $9a^2 - 4c^2$
**53.** $16x^2 - \frac{4}{9}$  **55.** $16m^2 - 49n^4$  **57.** $75y^7 - 12y$
**59.** $y^2 - 10y + 25$  **61.** $4p^2 + 28p + 49$
**63.** $16n^2 + 24nm + 9m^2$  **65.** $k^2 - \frac{10}{7}kp + \frac{25}{49}p^2$
**67.** $0.04x^2 - 0.56xy + 1.96y^2$  **69.** Write 101 as $100 + 1$
and 99 as $100 - 1$. Then $101 \cdot 99 = (100 + 1)(100 - 1) = 100^2 - 1^2 = 10,000 - 1 = 9999$.
**71.** $25x^2 + 10x + 1 + 60xy + 12y + 36y^2$
**73.** $4a^2 + 4ab + b^2 - 12a - 6b + 9$
**75.** $4a^2 + 4ab + b^2 - 9$  **77.** $4h^2 - 4hk + k^2 - j^2$
**79.** $y^3 + 6y^2 + 12y + 8$  **81.** $125r^3 - 75r^2s + 15rs^2 - s^3$
**83.** $q^4 - 8q^3 + 24q^2 - 32q + 16$  **85.** $49; 25; 49 \neq 25$
**87.** $2401; 337; 2401 \neq 337$  **89.** $\frac{9}{2}x^2 - 2y^2$
**91.** $15x^2 - 2x - 24$  **93.** $10x^2 - 2x$
**95.** $2x^2 - x - 3$  **97.** $8x^3 - 27$  **99.** $2x^3 - 18x$
**101.** $-20$  **103.** $2x^2 - 6x$  **105.** 36
**107.** $\frac{35}{4}$  **109.** $\frac{1859}{64}$

## 5.5 Exercises (pages 276–278)

**1.** quotient; exponents  **3.** descending  **5.** $3x^3 - 2x^2 + 1$
**7.** $3y + 4 - \frac{5}{y}$  **9.** $3m + 5 + \frac{6}{m}$  **11.** $n - \frac{3n^2}{2m} + 2$
**13.** $\frac{2y}{x} + \frac{3}{4} + \frac{3w}{x}$  **15.** $r^2 - 7r + 6$  **17.** $y - 4$
**19.** $q + 8$  **21.** $t + 5$  **23.** $p - 4 + \frac{44}{p + 6}$
**25.** $m^2 + 2m - 1$  **27.** $m^2 + m + 3$  **29.** $z^2 + 3$
**31.** $x^2 + 2x - 3 + \frac{6}{4x + 1}$  **33.** $2x - 5 + \frac{-4x + 5}{3x^2 - 2x + 4}$
**35.** $x^2 + x + 3$  **37.** $3x^2 + 6x + 11 + \frac{26}{x - 2}$
**39.** $2k^2 + 3k - 1$  **41.** $2y^2 + 2$
**43.** $x^2 - 4x + 2 + \frac{9x - 4}{x^2 + 3}$  **45.** $p^2 + \frac{5}{2}p + 2 + \frac{-1}{2p + 2}$
**47.** $\frac{3}{2}a - 10 + \frac{77}{2a + 6}$  **49.** $p^2 + p + 1$  **51.** $2p + 7$
**53.** $-13; -13$; they are the same, which suggests that when $P(x)$
is divided by $x - r$, the result is $P(r)$. Here, $r = -1$.
**55.** $5x - 1; 0$  **57.** $2x - 3; -1$  **59.** $4x^2 + 6x + 9; \frac{3}{2}$
**61.** $\frac{x^2 - 9}{2x}, x \neq 0$  **63.** $-\frac{5}{4}$  **65.** $\frac{x - 3}{2x}, x \neq 0$  **67.** 0
**69.** $-\frac{35}{4}$  **71.** $\frac{7}{2}$

## 5 Review Exercises (pages 281–283)

**1.** 64  **3.** $-125$  **5.** $\frac{81}{16}$  **7.** $\frac{11}{30}$  **9.** 0  **11.** $x^8$
**13.** $\frac{1}{z^{15}}$  **15.** $\frac{r^{17}}{9}$  **17.** $\frac{1}{96m^7}$  **19.** $-12x^2y^8$  **21.** $\frac{10p^8}{q^7}$
**23.** In $(-6)^0$, the base is $-6$ and the expression simplifies to 1. In
$-6^0$, the base is 6 and the expression simplifies to $-1$.  **25.** yes
**27.** For example, let $x = 2$ and $y = 3$. Then $(x^2 + y^2)^2 = (2^2 + 3^2)^2 = 169$; $x^4 + y^4 = 2^4 + 3^4 = 97 \neq 169$.
**29.** $7.65 \times 10^{-8}$  **31.** 0.0058  **33.** $1.5 \times 10^3$; 1500
**35.** $2.7 \times 10^{-2}$; 0.027  **37. (a)** $5.449 \times 10^3$  **(b)** 63 mi$^2$
**39.** $-1$  **41.** 504  **43. (a)** $9m^7 + 14m^6$  **(b)** binomial
**(c)** 7  **45. (a)** $-7q^5r^3$  **(b)** monomial  **(c)** 8
**47.** $-x^2 - 3x + 1$  **49.** $6a^3 - 4a^2 - 16a + 15$
**51.** $12x^2 + 8x + 5$  **53. (a)** $5x^2 - x + 5$
**(b)** $-5x^2 + 5x + 1$  **(c)** 11  **(d)** $-9$
**55. (a)** 15.9 million  **(b)** 31.6 million  **(c)** 27.244 million
**57.**   **59.** $-12k^3 - 42k$

**61.** $6w^2 - 13wt + 6t^2$  **63.** $3q^3 - 13q^2 - 14q + 20$
**65.** $36r^4 - 1$  **67.** $16m^2 + 24m + 9$  **69.** $y^2 - 3y + \frac{5}{4}$
**71.** $p^2 + 6p + 9 + \frac{54}{2p - 3}$

## 5 Test (pages 283–284)

[5.1] **1. (a)** C  **(b)** A  **(c)** D  **(d)** A  **(e)** E  **(f)** F  **(g)** B
**(h)** G  **(i)** C  **2.** $\frac{4x^7}{9y^{10}}$  **3.** $\frac{6}{r^{14}}$  **4.** $\frac{16}{9p^{10}q^{28}}$  **5.** $\frac{16}{x^6y^{16}}$
**6.** 0.00000091  **7.** $3 \times 10^{-4}$; 0.0003  [5.3] **8. (a)** $-18$
**(b)** $-2x^2 + 12x - 9$  **(c)** $-2x^2 - 2x - 3$  **(d)** $-7$
**9. (a)** 23  **(b)** $3x^2 + 11$  **(c)** $9x^2 + 30x + 27$

**10.** [graph of $f(x) = -2x^2 + 3$]  **11.** [graph of $f(x) = -x^3 + 3$]

**12. (a)** 616 thousand  **(b)** 740 thousand  **(c)** 854 thousand
[5.2] **13.** $x^3 - 2x^2 - 10x - 13$   [5.4] **14.** $10x^2 - x - 3$
**15.** $6m^3 - 7m^2 - 30m + 25$   **16.** $36x^2 - y^2$
**17.** $9k^2 + 6kq + q^2$   **18.** $4y^2 - 9z^2 + 6zx - x^2$
[5.5] **19.** $4p - 8 + \frac{6}{p}$   **20.** $x^2 + 4x + 4$
[5.4] **21. (a)** $x^3 + 4x^2 + 5x + 2$  **(b)** 0
[5.5] **22. (a)** $x + 2, x \neq -1$  **(b)** 0

# 6 Factoring

## 6.1 Exercises (page 290)

**1.** $12(m - 5)$   **3.** $4(1 + 5z)$   **5.** cannot be factored
**7.** $8k(k^2 + 3)$   **9.** $-2p^2q^4(2p + q)$
**11.** $7x^3(3x^2 + 5x - 2)$   **13.** $2t^3(5t^2 - 4t - 2)$
**15.** $5ac(3ac^2 - 5c + 1)$   **17.** $16zn^3(zn^3 + 4n^4 - 2z^2)$
**19.** $7ab(2a^2b + a - 3a^4b^2 + 6b^3)$   **21.** $(m - 4)(2m + 5)$
**23.** $11(2z - 1)$   **25.** $(2 - x)^2(1 + 2x)$
**27.** $(3 - x)(6 + 2x - x^2)$   **29.** $20z(2z + 1)(3z + 4)$
**31.** $5(m + p)^2(m + p - 2 - 3m^2 - 6mp - 3p^2)$
**33.** $r(-r^2 + 3r + 5); -r(r^2 - 3r - 5)$   **35.** $12s^4(-s + 4);$
$-12s^4(s - 4)$   **37.** $2x^2(-x^3 + 3x + 2); -2x^2(x^3 - 3x - 2)$
**39.** $(m + q)(x + y)$   **41.** $(5m + n)(2 + k)$
**43.** $(2 - q)(2 - 3p)$   **45.** $(p + q)(p - 4z)$
**47.** $(2x + 3)(y + 1)$   **49.** $(m + 4)(m^2 - 6)$
**51.** $(a^2 + b^2)(-3a + 2b)$   **53.** $(y - 2)(x - 2)$
**55.** $(3y - 2)(3y^3 - 4)$   **57.** $(1 - a)(1 - b)$   **59.** The
directions said that the student was to factor the polynomial
*completely*. The completely factored form is $4xy^3(xy^2 - 2)$.

## 6.2 Exercises (pages 297–298)

**1.** D   **3.** B   **5.** $(y - 3)(y + 10)$   **7.** $(p + 8)(p + 7)$
**9.** prime   **11.** $(a + 5b)(a - 7b)$   **13.** prime
**15.** $(xy + 9)(xy + 2)$   **17.** $-(6m - 5)(m + 3)$
**19.** $(5x - 6)(2x + 3)$   **21.** $(4k + 3)(5k + 8)$
**23.** $(3a - 2b)(5a - 4b)$   **25.** $(6m - 5)^2$   **27.** prime
**29.** $(2xz - 1)(3xz + 4)$   **31.** $3(4x + 5)(2x + 1)$
**33.** $-5(a + 6)(3a - 4)$   **35.** $-11x(x - 6)(x - 4)$
**37.** $2xy^3(x - 12y)^2$   **39.** $6a(a - 3)(a + 5)$
**41.** $13y(y + 4)(y - 1)$   **43.** $3p(2p - 1)^2$   **45.** She did
not factor the polynomial *completely*. The factor $(4x + 10)$ can
be factored further into $2(2x + 5)$, giving the final form as
$2(2x + 5)(x - 2)$.   **47.** $(6p^3 - r)(2p^3 - 5r)$
**49.** $(5k + 4)(2k + 1)$   **51.** $(3m + 3p + 5)(m + p - 4)$
**53.** $(a + b)^2(a - 3b)(a + 2b)$   **55.** $(p + q)^2(p + 3q)$
**57.** $(z - x)^2(z + 2x)$   **59.** $(p^2 - 8)(p^2 - 2)$
**61.** $(2x^2 + 3)(x^2 - 6)$   **63.** $(4x^2 + 3)(4x^2 + 1)$

## 6.3 Exercises (page 302)

**1.** A, D   **3.** B, C   **5.** The sum of two squares can be factored
only if the binomial has a common factor.   **7.** $(p + 4)(p - 4)$
**9.** $(5x + 2)(5x - 2)$   **11.** $2(3a + 7b)(3a - 7b)$
**13.** $4(4m^2 + y^2)(2m + y)(2m - y)$
**15.** $(y + z + 9)(y + z - 9)$   **17.** $(4 + x + 3y)(4 - x - 3y)$
**19.** $(p^2 + 16)(p + 4)(p - 4)$   **21.** $(k - 3)^2$
**23.** $(2z + w)^2$   **25.** $(4m - 1 + n)(4m - 1 - n)$
**27.** $(2r - 3 + s)(2r - 3 - s)$   **29.** $(x + y - 1)(x - y + 1)$
**31.** $2(7m + 3n)^2$   **33.** $(p + q + 1)^2$   **35.** $(a - b + 4)^2$
**37.** $(x - 3)(x^2 + 3x + 9)$   **39.** $(t - 6)(t^2 + 6t + 36)$
**41.** $(x + 4)(x^2 - 4x + 16)$   **43.** $(10 + y)(100 - 10y + y^2)$
**45.** $(2x + 1)(4x^2 - 2x + 1)$
**47.** $(5x - 6)(25x^2 + 30x + 36)$
**49.** $(x - 2y)(x^2 + 2xy + 4y^2)$
**51.** $(4g - 3h)(16g^2 + 12gh + 9h^2)$
**53.** $(7p + 5q)(49p^2 - 35pq + 25q^2)$
**55.** $3(2n + 3p)(4n^2 - 6np + 9p^2)$
**57.** $(y + z + 4)(y^2 + 2yz + z^2 - 4y - 4z + 16)$
**59.** $(m^2 - 5)(m^4 + 5m^2 + 25)$
**61.** $(10x^3 - 3)(100x^6 + 30x^3 + 9)$
**63.** $(5y^2 + z)(25y^4 - 5y^2z + z^2)$

## 6.4 Exercises (pages 305–306)

**1.** $(10a + 3b)(10a - 3b)$   **3.** $3p^2(p - 6)(p + 5)$
**5.** $3pq(a + 6b)(a - 5b)$   **7.** prime   **9.** $(6b + 1)(b - 3)$
**11.** $(x - 10)(x^2 + 10x + 100)$   **13.** $(p + 2)(4 + m)$
**15.** $9m(m - 5 + 2m^2)$   **17.** $2(3m - 10)(9m^2 + 30m + 100)$
**19.** $(3m - 5n)^2$   **21.** $(k - 9)(q + r)$   **23.** $16z^2x(zx - 2)$
**25.** $(x + 7)(x - 5)$   **27.** $(x - 5)(x + 5)(x^2 + 25)$
**29.** $(p + 1)(p^2 - p + 1)$   **31.** $(8m + 25)(8m - 25)$
**33.** $6z(2z^2 - z + 3)$   **35.** $16(4b + 5c)(4b - 5c)$
**37.** $8(5z + 4)(25z^2 - 20z + 16)$   **39.** $(5r - s)(2r + 5s)$
**41.** $4pq(2p + q)(3p + 5q)$   **43.** $3(4k^2 + 9)(2k + 3)(2k - 3)$
**45.** $(m - n)(m^2 + mn + n^2 + m + n)$
**47.** $(x - 2m - n)(x + 2m + n)$   **49.** $6p^3(3p^2 - 4 + 2p^3)$
**51.** $2(x + 4)(x - 5)$   **53.** $8mn$   **55.** $2(5p + 9)(5p - 9)$
**57.** $4rx(3m^2 + mn + 10n^2)$   **59.** $(7a - 4b)(3a + b)$
**61.** prime   **63.** $(p + 8q - 5)^2$   **65.** $(7m^2 + 1)(3m^2 - 5)$
**67.** $(2r - t)(r^2 - rt + 19t^2)$
**69.** $(x + 3)(x^2 + 1)(x + 1)(x - 1)$
**71.** $(m + n - 5)(m - n + 1)$

## 6.5 Exercises (pages 312–313)

**1.** First rewrite the equation so that one side is 0. Factor the other
side and set each factor equal to 0. The solutions of these linear
equations are solutions of the quadratic equation.   **3.** $\{-10, 5\}$
**5.** $\{-\frac{8}{3}, \frac{5}{2}\}$   **7.** $\{-2, 5\}$   **9.** $\{-6, -3\}$   **11.** $\{-\frac{1}{2}, 4\}$
**13.** $\{-\frac{1}{3}, \frac{4}{5}\}$   **15.** $\{-3, 4\}$   **17.** $\{-5, -\frac{1}{5}\}$   **19.** $\{-4, 0\}$

**21.** $\{0, 6\}$   **23.** $\{-2, 2\}$   **25.** $\{-3, 3\}$   **27.** $\{3\}$
**29.** $\{-\frac{4}{3}\}$   **31.** $\{-4, 2\}$   **33.** $\{-\frac{1}{2}, 6\}$   **35.** $\{1, 6\}$
**37.** $\{-\frac{1}{2}, 0, 5\}$   **39.** $\{-1, 0, 3\}$   **41.** $\{-\frac{4}{3}, 0, \frac{4}{3}\}$
**43.** $\{-\frac{5}{2}, -1, 1\}$   **45.** $\{-3, 3, 6\}$   **47.** By dividing each side by a variable expression, she "lost" the solution 0. The solution set is $\{-\frac{4}{3}, 0, \frac{4}{3}\}$.   **49.** width: 16 ft; length: 20 ft   **51.** base: 12 ft; height: 5 ft   **53.** 50 ft by 100 ft   **55.** length: 15 in.; width: 9 in.   **57.** 5 sec   **59.** $6\frac{1}{4}$ sec

## 6 Review Exercises (pages 315–316)

**1.** $6p(2p - 1)$   **3.** $4qb(3q + 2b - 5q^2b)$
**5.** $(x + 3)(x - 3)$   **7.** $(m + q)(4 + n)$   **9.** $(m + 3)(2 - a)$
**11.** $(3p - 4)(p + 1)$   **13.** $(3r + 1)(4r - 3)$
**15.** $(2k - h)(5k - 3h)$   **17.** $2x(4 + x)(3 - x)$
**19.** $(y^2 + 4)(y^2 - 2)$   **21.** $(p + 2)^2(p + 3)(p - 2)$
**23.** It is not factored because there are two terms: $x^2(y^2 - 6)$ and $5(y^2 - 6)$. The correct answer is $(y^2 - 6)(x^2 + 5)$.
**25.** $(4x + 5)(4x - 5)$   **27.** $(6m - 5n)(6m + 5n)$
**29.** $(3k - 2)^2$   **31.** $(5x - 1)(25x^2 + 5x + 1)$
**33.** $(x^4 + 1)(x^2 + 1)(x + 1)(x - 1)$   **35.** $2b(3a^2 + b^2)$
**37.** $\{4\}$   **39.** $\{2, 3\}$   **41.** $\{-\frac{5}{2}, \frac{10}{3}\}$   **43.** $\{-\frac{3}{2}, -\frac{1}{4}\}$
**45.** $\{-\frac{3}{2}, 0\}$   **47.** $\{4\}$   **49.** $\{-3, -2, 2\}$   **51.** 3 ft
**53.** after 16 sec   **55.** The rock reaches a height of 240 ft once on its way up and once on its way down.

## 6 Test (page 317)

**[6.1–6.4] 1.** $11z(z - 4)$   **2.** $5x^2y^3(2y^2 - 1 - 5x^3)$
**3.** $(x + y)(3 + b)$   **4.** $-(2x + 9)(x - 4)$
**5.** $(3x - 5)(2x + 7)$   **6.** $(4p - q)(p + q)$   **7.** $(4a + 5b)^2$
**8.** $(x + 1 + 2z)(x + 1 - 2z)$   **9.** $(a + b)(a - b)(a + 2)$
**10.** $(3k + 11j)(3k - 11j)$   **11.** $(y - 6)(y^2 + 6y + 36)$
**12.** $(2k^2 - 5)(3k^2 + 7)$   **13.** $(3x^2 + 1)(9x^4 - 3x^2 + 1)$
**[6.1] 14.** It is not in factored form because there are two terms: $(x^2 + 2y)p$ and $3(x^2 + 2y)$. The common factor is $x^2 + 2y$, and the factored form is $(x^2 + 2y)(p + 3)$.   **[6.2] 15.** D
**[6.5] 16.** $\{-2, -\frac{2}{3}\}$   **17.** $\{0, \frac{5}{3}\}$   **18.** $\{-\frac{2}{5}, 1\}$
**19.** length: 8 in.; width: 5 in.   **20.** 2 sec and 4 sec

# 7 Rational Expressions and Functions

## 7.1 Exercises (pages 327–329)

**1.** C   **3.** D   **5.** E   **7.** Replacing $x$ with 2 makes the denominator 0 and the value of the expression undefined. To find the values excluded from the domain, set the denominator equal to 0 and solve the equation. All solutions of the equation are excluded from the domain.   **9.** 7; $\{x \mid x \neq 7\}$   **11.** $-\frac{1}{7}$; $\{x \mid x \neq -\frac{1}{7}\}$
**13.** 0; $\{x \mid x \neq 0\}$   **15.** $-2, \frac{3}{2}$; $\{x \mid x \neq -2, \frac{3}{2}\}$   **17.** none; $(-\infty, \infty)$   **19.** none; $(-\infty, \infty)$   **21. (a)** numerator: $x^2$, $4x$; denominator: $x$, 4   **(b)** First factor the numerator, getting $x(x + 4)$. Then divide the numerator and denominator by the common factor $x + 4$ to get $\frac{x}{1}$, or $x$.   **23.** B   **25.** $x$
**27.** $\frac{x-3}{x+5}$   **29.** $\frac{x+3}{2x(x-3)}$   **31.** It is already in lowest terms.
**33.** $\frac{6}{7}$   **35.** $\frac{t-3}{3}$   **37.** $\frac{2}{t-3}$   **39.** $\frac{x-3}{x+1}$   **41.** $\frac{4x+1}{4x+3}$
**43.** $a^2 - ab + b^2$   **45.** $\frac{c + 6d}{c - d}$   **47.** $\frac{a+b}{a-b}$   **49.** $-1$
In Exercises 51 and 53, there are other acceptable ways to express each answer.   **51.** $-(x + y)$   **53.** $-\frac{x+y}{x-y}$   **55.** $-\frac{1}{2}$
**57.** It is already in lowest terms.   **59.** Multiply the numerators, multiply the denominators, and factor each numerator and denominator. (Factoring can be performed first.) Divide the numerator and denominator by any common factors to write the rational expression in lowest terms. For example,
$$\frac{6r - 5s}{3r + 2s} \cdot \frac{6r + 4s}{5s - 6r} = \frac{(6r - 5s)(6r + 4s)}{(3r + 2s)(5s - 6r)} = \frac{(6r - 5s)2(3r + 2s)}{(3r + 2s)(-1)(6r - 5s)} = \frac{2}{-1} = -2.$$
**61.** $\frac{3y}{x^2}$   **63.** $\frac{3a^3b^2}{4}$   **65.** $\frac{27}{2mn^7}$   **67.** $\frac{x+4}{x-2}$
**69.** $\frac{2x+3}{x+2}$   **71.** $\frac{7x}{6}$   **73.** $-\frac{p+5}{2p}$ (There are other ways.)
**75.** $\frac{35}{4}$   **77.** $-(z + 1)$, or $-z - 1$   **79.** $\frac{14x^2}{5}$   **81.** $-2$
**83.** $\frac{x+4}{x-4}$   **85.** $\frac{a^2 + ab + b^2}{a - b}$   **87.** $\frac{2x-3}{2(x-3)}$
**89.** $\frac{a^2 + 2ab + 4b^2}{a + 2b}$   **91.** $\frac{2x+3}{2x-3}$

## 7.2 Exercises (pages 336–338)

**1.** $\frac{9}{t}$   **3.** $\frac{6x+y}{7}$   **5.** $\frac{2}{x}$   **7.** $-\frac{2}{x^3}$   **9.** 1   **11.** $x - 5$
**13.** $\frac{5}{p+3}$   **15.** $a - b$   **17.** First add or subtract the numerators. Then place the result over the common denominator. Write the answer in lowest terms. We give one example:
$$\frac{5}{x} - \frac{3x+1}{x} = \frac{5 - (3x+1)}{x} = \frac{5 - 3x - 1}{x} = \frac{4 - 3x}{x}.$$
**19.** $72x^4y^5$   **21.** $z(z - 2)$   **23.** $2(y + 4)$
**25.** $(x + 9)^2(x - 9)$   **27.** $(m + n)(m - n)$
**29.** $x(x - 4)(x + 1)$   **31.** $(t + 5)(t - 2)(2t - 3)$
**33.** $2y(y + 3)(y - 3)$   **35.** $2(x + 2)^2(x - 3)$
**37.** The expression $\frac{x - 4x - 1}{x + 2}$ is incorrect. The third term in the numerator should be $+1$, since the $-$ sign should be distributed over both $4x$ and $-1$. The answer should be $\frac{-3x+1}{x+2}$.   **39.** $\frac{31}{3t}$
**41.** $\frac{5 - 22x}{12x^2y}$   **43.** $\frac{16b + 9a^2}{60a^4b^6}$   **45.** $\frac{4pr + 3sq^3}{14p^4q^4}$
**47.** $\frac{a^2b^5 - 2ab^6 + 3}{a^5b^7}$   **49.** $\frac{1}{x(x - 1)}$   **51.** $\frac{5a^2 - 7a}{(a+1)(a-3)}$
**53.** 3   **55.** $\frac{3}{x-4}$, or $\frac{-3}{4-x}$   **57.** $\frac{w+z}{w-z}$, or $\frac{-w-z}{z-w}$
**59.** $\frac{-2}{(x+1)(x-1)}$   **61.** $\frac{2(2x-1)}{x-1}$   **63.** $\frac{7}{y}$   **65.** $\frac{6}{x-2}$
**67.** $\frac{3x-2}{x-1}$   **69.** $\frac{4x-7}{x^2-x+1}$   **71.** $\frac{2x+1}{x}$
**73.** $\frac{4p^2 - 21p + 29}{(p-2)^2}$   **75.** $\frac{x}{(x-2)^2(x-3)}$
**77.** $\frac{2x(x + 12y)}{(x+2y)(x-y)(x+6y)}$   **79.** $\frac{2x^2 + 21xy - 10y^2}{(x+2y)(x-y)(x+6y)}$
**81.** $\frac{3r - 2s}{(2r - s)(3r - s)}$   **83.** $\frac{10x + 23}{(x+2)^2(x+3)}$
**85. (a)** $c(x) = \frac{10x}{49(101 - x)}$   **(b)** approximately 3.23 thousand dollars

## 7.3 Exercises (page 344)

1. *Method 1:* Begin by simplifying the numerator to a single fraction. Then simplify the denominator to a single fraction. Write as a division problem, and multiply by the reciprocal of the denominator. Simplify the result if possible. *Method 2:* Find the LCD of all fractions in the complex fraction. Multiply the numerator and denominator of the complex fraction by this LCD. Simplify the result if possible.   3. $\frac{2x}{x-1}$   5. $\frac{2(k+1)}{3k-1}$
7. $\frac{5x^2}{9z^3}$   9. $\frac{6x+1}{7x-3}$   11. $\frac{y+x}{y-x}$   13. $4x$   15. $x + 4y$
17. $\frac{y+4}{2}$   19. $\frac{a+b}{ab}$   21. $xy$   23. $\frac{3y}{2}$
25. $\frac{x^2+5x+4}{x^2+5x+10}$   27. $\frac{x^2y^2}{y^2+x^2}$   29. $\frac{y^2+x^2}{xy^2+x^2y}$, or $\frac{y^2+x^2}{xy(y+x)}$
31. $\frac{1}{2xy}$

## 7.4 Exercises (pages 348–350)

1. (a) $0$  (b) $\{x \mid x \neq 0\}$   3. (a) $-1, 2$  (b) $\{x \mid x \neq -1, 2\}$
5. (a) $-4, 4$  (b) $\{x \mid x \neq \pm 4\}$   7. (a) $0, 1, -3, 2$
(b) $\{x \mid x \neq 0, 1, -3, 2\}$   9. (a) $-\frac{7}{4}, 0, \frac{13}{6}$
(b) $\{x \mid x \neq -\frac{7}{4}, 0, \frac{13}{6}\}$   11. (a) $4, \frac{7}{2}$  (b) $\{x \mid x \neq 4, \frac{7}{2}\}$
13. No, there is no possibility that the proposed solution will be rejected, because there are no variables in the denominators in the original equation.   15. $\{1\}$   17. $\{-6, 4\}$   19. $\{-\frac{7}{12}\}$
21. $\emptyset$   23. $\{-3\}$   25. $\{0\}$   27. $\{5\}$   29. $\emptyset$
31. $\{\frac{27}{56}\}$   33. $\emptyset$   35. $\{-10\}$   37. $\{-1\}$
39. $\{13\}$   41. $\{x \mid x \neq \pm 3\}$   43. (a) $\{x \mid x \neq -3\}$
(b) $-3$ is not in the domain.   45. 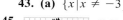 $x = 0$

47. 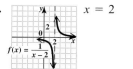   $x = 2$   49. (a) $0$  (b) $1.6$  (c) $4.1$
(d) The waiting time also increases.   51. (a) $500$ ft
(b) It decreases.

## 7.5 Exercises (pages 357–361)

1. A   3. D   5. $24$   7. $\frac{25}{4}$   9. $G = \frac{Fd^2}{Mm}$
11. $a = \frac{bc}{c+b}$   13. $v = \frac{PVt}{pT}$   15. $r = \frac{nE - IR}{In}$, or
$r = \frac{IR - nE}{-In}$   17. $b = \frac{2A}{h} - B$, or $b = \frac{2A - hB}{h}$
19. $r = \frac{eR}{E-e}$   21. Multiply each side by $a - b$.
23. $15$ girls, $5$ boys   25. $\frac{1}{2}$ job per hr   27. $5.351$ in.
29. $7.6$ in.   31. $40$ teachers   33. $210$ deer
35. $25{,}000$ fish   37. $6.6$ more gallons
39. $x = \frac{7}{2}$; $AC = 8$; $DF = 12$   41. $3$ mph   43. $1020$ mi
45. $1750$ mi   47. $190$ mi   49. $6\frac{2}{3}$ min   51. $30$ hr
53. $2\frac{1}{3}$ hr   55. $20$ hr   57. $2\frac{4}{5}$ hr

## 7.6 Exercises (pages 367–370)

1. direct   3. direct   5. inverse   7. inverse
9. inverse   11. direct   13. joint   15. combined
17. increases; decreases   19. $36$   21. $\frac{16}{9}$   23. $0.625$
25. $\frac{16}{5}$   27. $222\frac{2}{9}$   29. $\$2.919$, or $\$2.91\frac{9}{10}$   31. $8$ lb
33. about $450$ in.$^3$   35. $256$ ft   37. $106\frac{2}{3}$ mph
39. $100$ cycles per sec   41. $21\frac{1}{3}$ foot-candles   43. $\$420$
45. $11.8$ lb   47. $448.1$ lb   49. approximately $68{,}600$ calls
51. Answers will vary.   53. If $y$ varies inversely as $x$, then $x$ is in the denominator; however, if $y$ varies directly as $x$, then $x$ is in the numerator. If $k > 0$, then, with inverse variation, as $x$ increases, $y$ decreases. With direct variation, $y$ increases as $x$ increases.

## 7 Review Exercises (pages 375–377)

1. (a) $-6$  (b) $\{x \mid x \neq -6\}$   3. (a) $9$  (b) $\{x \mid x \neq 9\}$
5. $\frac{5m+n}{5m-n}$   7. The reciprocal of a rational expression is another rational expression such that the two rational expressions have a product of $1$.   9. $\frac{-3(w+4)}{w}$   11. $1$   13. $9r^2(3r+1)$
15. $3(x-4)^2(x+2)$   17. $12$   19. $\frac{13r^2 + 5rs}{(5r+s)(2r-s)(r+s)}$
21. $\frac{3+2t}{4-7t}$   23. $\frac{1}{3q+2p}$   25. $\{-3\}$   27. $\{0\}$
29. Although her algebra was correct, $3$ is not a solution because it is not in the domain of the variable. Thus, $\emptyset$ is correct.
31. C; $x = 0$   33. $m = \frac{Fd^2}{GM}$   35. $6000$ passenger-km per day
37. $4\frac{4}{5}$ min   39. C   41. $5.59$ vibrations per sec

## 7 Test (pages 377–378)

[7.1] 1. $-2, \frac{4}{3}$; $\{x \mid x \neq -2, \frac{4}{3}\}$   2. $\frac{2x-5}{x(3x-1)}$   3. $\frac{3(x+3)}{4}$
4. $\frac{y+4}{y-5}$   5. $-2$   6. $\frac{x+5}{x}$   [7.2] 7. $t^2(t+3)(t-2)$
8. $\frac{7-2t}{6t^2}$   9. $\frac{9a+5b}{21a^5b^3}$   10. $\frac{11x+21}{(x-3)^2(x+3)}$   11. $\frac{4}{x+2}$
[7.3] 12. $\frac{72}{11}$   13. $-\frac{1}{a+b}$   14. $\frac{2y^2+x^2}{xy(y-x)}$
[7.4] 15. (a) expression; $\frac{11(x-6)}{12}$  (b) equation; $\{6\}$
16. $\{\frac{1}{2}\}$   17. $\{5\}$   18. A solution cannot make a denominator $0$.   19. $\ell = \frac{2S}{n} - a$, or $\ell = \frac{2S - na}{n}$
20. 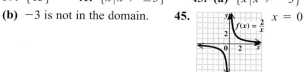   $x = -1$   [7.5] 21. $3\frac{3}{14}$ hr   22. $15$ mph
23. $48{,}000$ fish   24. (a) $3$ units  (b) $0$   [7.6] 25. $200$ amps
26. $0.8$ lb

# 8 Roots, Radicals, and Root Functions

## 8.1 Exercises (pages 384–385)

1. E   3. D   5. A   7. C   9. C   11. (a) It is not a real number.  (b) negative  (c) $0$   13. $-9$   15. $6$

**17.** −4  **19.** −8  **21.** 6  **23.** −2  **25.** It is not a real number.  **27.** 2  **29.** It is not a real number.  **31.** $\frac{8}{9}$
**33.** $\frac{4}{3}$  **35.** $-\frac{1}{2}$  **37.** 0.7  **39.** 0.1

*In Exercises 41–47, we give the domain and then the range.*
**41.** $[-3, \infty); [0, \infty)$  **43.** $[0, \infty); [-2, \infty)$

**45.** $(-\infty, \infty); (-\infty, \infty)$  **47.** $(-\infty, \infty); (-\infty, \infty)$

**49.** 12  **51.** 10  **53.** 2  **55.** −9  **57.** −5  **59.** $|x|$
**61.** $|z|$  **63.** $x$  **65.** $x^5$  **67.** $|x|^5$ (or $|x^5|$)  **69.** 97.381
**71.** 16.863  **73.** −9.055  **75.** 7.507  **77.** 3.162
**79.** 1.885  **81.** 10 mi  **83.** 392,000 mi$^2$

## 8.2 Exercises (pages 391–393)

**1.** C  **3.** A  **5.** H  **7.** B  **9.** D  **11.** 13  **13.** 9
**15.** 2  **17.** $\frac{8}{9}$  **19.** −3  **21.** It is not a real number.
**23.** 1000  **25.** 27  **27.** −1024  **29.** 16  **31.** $\frac{1}{8}$
**33.** $\frac{1}{512}$  **35.** $\frac{9}{25}$  **37.** $\sqrt{10}$  **39.** $(\sqrt[4]{8})^3$
**41.** $(\sqrt[8]{9q})^5 - (\sqrt[3]{2x})^2$  **43.** $\frac{1}{(\sqrt{2m})^3}$  **45.** $(\sqrt[3]{2y+x})^2$
**47.** $\frac{1}{(\sqrt[3]{3m^4+2k^2})^2}$  **49.** $\sqrt{a^2+b^2} = \sqrt{3^2+4^2} = 5$; $a+b = 3+4 = 7; 5 \neq 7$  **51.** 64  **53.** 64  **55.** $x^{10}$
**57.** $\sqrt[6]{x^5}$  **59.** $\sqrt[15]{t^8}$  **61.** 9  **63.** 4  **65.** $y$
**67.** $x^{5/12}$  **69.** $k^{2/3}$  **71.** $x^3y^8$  **73.** $\frac{1}{x^{10/3}}$  **75.** $\frac{1}{m^{1/4}n^{3/4}}$
**77.** $p^2$  **79.** $\frac{c^{11/3}}{b^{11/4}}$  **81.** $\frac{q^{5/3}}{9p^{7/2}}$  **83.** $p + 2p^2$
**85.** $k^{7/4} - k^{3/4}$  **87.** $6 + 18a$  **89.** $x^{17/20}$  **91.** $\frac{1}{x^{3/2}}$
**93.** $y^{5/6}z^{1/3}$  **95.** $m^{1/12}$  **97.** $x^{1/24}$
**99.** 19.0°; the table gives 19°.

## 8.3 Exercises (pages 399–402)

**1.** true; both are equal to $4\sqrt{3}$ and approximately 6.92820323.
**3.** true; both are equal to $6\sqrt{2}$ and approximately 8.485281374.
**5.** D  **7.** $\sqrt{30}$  **9.** $\sqrt{14x}$  **11.** $\sqrt{42pqr}$  **13.** $\sqrt[3]{14xy}$
**15.** $\sqrt[4]{33}$  **17.** $\sqrt[4]{6x^3}$  **19.** This expression cannot be simplified by the product rule.  **21.** To multiply two radical expressions with the same index, multiply the radicands and keep the index. For example, $\sqrt[3]{3} \cdot \sqrt[3]{5} = \sqrt[3]{15}$.  **23.** $\frac{8}{11}$
**25.** $\frac{\sqrt{3}}{5}$  **27.** $\frac{\sqrt{x}}{5}$  **29.** $\frac{p^3}{9}$  **31.** $-\frac{3}{4}$  **33.** $\frac{\sqrt[3]{r^2}}{2}$
**35.** $-\frac{3}{x}$  **37.** $\frac{1}{x^3}$  **39.** $2\sqrt{3}$  **41.** $12\sqrt{2}$  **43.** $-4\sqrt{2}$
**45.** $-2\sqrt{7}$  **47.** This radical cannot be simplified further.
**49.** $4\sqrt[3]{2}$  **51.** $-2\sqrt[3]{2}$  **53.** $2\sqrt[3]{5}$  **55.** $-4\sqrt[4]{2}$

**57.** $2\sqrt[5]{2}$  **59.** His reasoning was incorrect. Here, 8 is a term, not a factor.  **61.** $6k\sqrt{2}$  **63.** $12xy^4\sqrt{xy}$  **65.** $11x^3$
**67.** $-3t^4$  **69.** $-10m^4z^2$  **71.** $5a^2b^3c^4$  **73.** $\frac{1}{2}r^2t^5$
**75.** $5x\sqrt{2x}$  **77.** $-10r^5\sqrt{5r}$  **79.** $x^3y^4\sqrt{13x}$
**81.** $2z^2w^3$  **83.** $-2zt^2\sqrt[3]{2z^2t}$  **85.** $3x^3y^4$
**87.** $-3r^3s^2\sqrt[4]{2r^3s^2}$  **89.** $\frac{y^5\sqrt{y}}{6}$  **91.** $\frac{x^5\sqrt[3]{x}}{3}$  **93.** $4\sqrt{3}$
**95.** $\sqrt{5}$  **97.** $x^2\sqrt{x}$  **99.** $\sqrt[6]{432}$  **101.** $\sqrt[12]{6912}$
**103.** $\sqrt[6]{x^5}$  **105.** 5  **107.** $8\sqrt{2}$  **109.** 13  **111.** $9\sqrt{2}$
**113.** $\sqrt{17}$  **115.** 5  **117.** $6\sqrt{2}$  **119.** $\sqrt{5y^2 - 2xy + x^2}$
**121.** $d = [(x_2 - x_1)^2 + (y_2 - y_1)^2]^{1/2}$  **123.** $2\sqrt{106} + 4\sqrt{2}$
**125.** 15.3 mi  **127.** 27.0 in.

## 8.4 Exercises (pages 404–406)

**1.** B  **3.** 15; Each radical expression simplifies to a whole number.  **5.** −4  **7.** $7\sqrt{3}$  **9.** $14\sqrt[3]{2}$  **11.** $5\sqrt[4]{2}$
**13.** $24\sqrt{2}$  **15.** The expression cannot be simplified further.
**17.** $20\sqrt{5}$  **19.** $4\sqrt{2x}$  **21.** $-11m\sqrt{2}$  **23.** $\sqrt[3]{2}$
**25.** $2\sqrt[3]{x}$  **27.** $-\sqrt[3]{x^2y}$  **29.** $-x\sqrt[3]{xy^2}$  **31.** $19\sqrt[4]{2}$
**33.** $x\sqrt[4]{xy}$  **35.** $9\sqrt[4]{2a^3}$  **37.** $(4+3xy)\sqrt[3]{xy^2}$  **39.** The expression cannot be simplified further.  **41.** $4x\sqrt[3]{x} + 6x\sqrt[4]{x}$
**43.** $2\sqrt{2} - 2$  **45.** $\frac{5\sqrt{5}}{6}$  **47.** $\frac{7\sqrt{2}}{6}$  **49.** $\frac{5\sqrt{2}}{3}$
**51.** $5\sqrt{2} + 4$  **53.** $\frac{5-3x}{x^4}$  **55.** $\frac{m\sqrt[3]{m^2}}{2}$  **57.** $\frac{3x\sqrt[3]{2}-4\sqrt[3]{5}}{x^3}$
**59.** A; 42 m  **61.** $(12\sqrt{5} + 5\sqrt{3})$ in.
**63.** $(24\sqrt{2} + 12\sqrt{3})$ in.

## 8.5 Exercises (pages 411–413)

**1.** E  **3.** A  **5.** D  **7.** $3\sqrt{6} + 2\sqrt{3}$  **9.** $20\sqrt{2}$
**11.** −2  **13.** −1  **15.** 6  **17.** $\sqrt{6} - \sqrt{2} + \sqrt{3} - 1$
**19.** $\sqrt{22} + \sqrt{55} - \sqrt{14} - \sqrt{35}$  **21.** $8 - \sqrt{15}$
**23.** $9 + 4\sqrt{5}$  **25.** $26 - 2\sqrt{105}$  **27.** $4 - \sqrt[3]{36}$
**29.** 10  **31.** $6x + 3\sqrt{x} - 2\sqrt{5x} - \sqrt{5}$  **33.** $9r - s$
**35.** $4\sqrt[3]{4y^2} - 19\sqrt[3]{2y} - 5$  **37.** $3x - 4$  **39.** $4x - y$
**41.** $2\sqrt{6} - 1$  **43.** $\sqrt{7}$  **45.** $5\sqrt{3}$  **47.** $\frac{\sqrt{6}}{2}$
**49.** $\frac{9\sqrt{15}}{5}$  **51.** $-\sqrt{2}$  **53.** $\frac{\sqrt{14}}{2}$  **55.** $-\frac{\sqrt{14}}{10}$  **57.** $\frac{2\sqrt{6x}}{x}$
**59.** $\frac{-8\sqrt{3k}}{k}$  **61.** $\frac{-5m^2\sqrt{6mn}}{n^2}$  **63.** $\frac{12x^3\sqrt{2xy}}{y^5}$  **65.** $\frac{5\sqrt{2my}}{y^2}$
**67.** $-\frac{4k\sqrt{3z}}{z}$  **69.** $\frac{\sqrt[3]{18}}{3}$  **71.** $\frac{\sqrt[3]{12}}{3}$  **73.** $\frac{\sqrt[3]{18}}{4}$
**75.** $-\frac{\sqrt[3]{2pr}}{r}$  **77.** $\frac{x^2\sqrt[3]{y^2}}{y}$  **79.** $\frac{2\sqrt[4]{x^3}}{x}$  **81.** $\frac{\sqrt[4]{2yz^3}}{z}$
**83.** $\frac{3(4-\sqrt{5})}{11}$  **85.** $\frac{6\sqrt{2}+4}{7}$  **87.** $\frac{2(3\sqrt{5}-2\sqrt{3})}{33}$
**89.** $2\sqrt{3} + \sqrt{10} - 3\sqrt{2} - \sqrt{15}$  **91.** $\sqrt{m} - 2$
**93.** $\frac{4(\sqrt{x}+2\sqrt{y})}{x-4y}$  **95.** $\frac{x-2\sqrt{xy}+y}{x-y}$  **97.** $\frac{5\sqrt{k}(2\sqrt{k}-\sqrt{q})}{4k-q}$
**99.** $3 - 2\sqrt{6}$  **101.** $1 - \sqrt{5}$  **103.** $\frac{4-2\sqrt{2}}{3}$
**105.** $\frac{6+2\sqrt{6p}}{3}$  **107.** $\frac{\sqrt{x+y}}{x+y}$  **109.** $\frac{p\sqrt{p+2}}{p+2}$

## 8.6 Exercises (pages 418–419)

1. (a) yes  (b) no   3. (a) yes  (b) no   5. no; there is no solution. The radical expression, which is positive, cannot equal a negative number.   7. $\{11\}$   9. $\{\frac{1}{3}\}$   11. $\emptyset$
13. $\{5\}$   15. $\{18\}$   17. $\{5\}$   19. $\{4\}$   21. $\{17\}$
23. $\{5\}$   25. $\emptyset$   27. $\{0\}$   29. $\{0\}$   31. $\{-\frac{1}{3}\}$
33. $\emptyset$   35. You cannot just square each term. The right side should be $(8 - x)^2 = 64 - 16x + x^2$. The correct first step is $3x + 4 = 64 - 16x + x^2$, and the solution set is $\{4\}$.
37. $\{1\}$   39. $\{-1\}$   41. $\{14\}$   43. $\{8\}$   45. $\{0\}$
47. $\emptyset$   49. $\{7\}$   51. $\{7\}$   53. $\{4, 20\}$   55. $\emptyset$
57. $\{\frac{5}{4}\}$   59. $\emptyset$; domain: $[-\frac{2}{3}, 1]$   61. $K = \frac{V^2 m}{2}$
63. $L = \frac{1}{4\pi^2 f^2 C}$

## 8.7 Exercises (pages 425–426)

1. $i$   3. $-1$   5. $-i$   7. $13i$   9. $-12i$   11. $i\sqrt{5}$
13. $4i\sqrt{3}$   15. $-\sqrt{105}$   17. $-10$   19. $i\sqrt{33}$
21. $\sqrt{3}$   23. $5i$   25. $-1 + 7i$   27. $0$   29. $7 + 3i$
31. $-2$   33. $1 + 13i$   35. $6 + 6i$   37. $4 + 2i$
39. $-81$   41. $-16$   43. $-10 - 30i$   45. $10 - 5i$
47. $-9 + 40i$   49. $-16 + 30i$   51. $153$   53. $97$
55. $a - bi$   57. $1 + i$   59. $-1 + 2i$   61. $2 + 2i$
63. $-\frac{5}{13} - \frac{12}{13}i$   65. $1 - 3i$   67. $-1$   69. $i$   71. $-1$
73. $-i$   75. $-i$   77. Since $i^{20} = (i^4)^5 = 1^5 = 1$, the student multiplied by 1, which is justified by the identity property for multiplication.

## 8 Review Exercises (pages 430–433)

1. $42$   3. $6$   5. $-3$   7. $\sqrt[n]{a}$ is not a real number if $n$ is even and $a$ is negative.   9. $-6.856$   11. $4.960$
13. $-3968.503$   15.   domain: $[1, \infty)$;
range: $[0, \infty)$   17. B   19. A   21. It is not a real number.
23. $-11$   25. $-4$   27. $-32$   29. It is not a real number.
31. The radical $\sqrt[n]{a^m}$ is equivalent to $a^{m/n}$. For example, $\sqrt[3]{8^2} = \sqrt[3]{64} = 4$, and $8^{2/3} = (8^{1/3})^2 = 2^2 = 4$.
33. $\frac{1}{(\sqrt[3]{3a+b})^5}$, or $\frac{1}{\sqrt[3]{(3a+b)^5}}$   35. $p^{4/5}$   37. $96$   39. $\frac{1}{y^{1/2}}$
41. $r^{1/2} + r$   43. $r^{3/2}$   45. $k^{9/4}$   47. $z^{1/12}$   49. $x^{1/15}$
51. The product rule for exponents applies only if the bases are the same.   53. $\sqrt{5r}$   55. $\sqrt[4]{21}$   57. $5\sqrt{3}$
59. $-3\sqrt[3]{4}$   61. $4pq^2\sqrt[3]{p}$   63. $2r^2 t\sqrt[3]{79r^2 t}$   65. $\frac{m^5}{3}$
67. $\frac{a^2 \sqrt[4]{a}}{3}$   69. $p\sqrt{p}$   71. $\sqrt[10]{x^7}$   73. $\sqrt{197}$
75. $23\sqrt{5}$   77. $26m\sqrt{6m}$   79. $-8\sqrt[4]{2}$   81. $\frac{16 + 5\sqrt{5}}{20}$
83. $(12\sqrt{3} + 5\sqrt{2})$ ft   85. $2$   87. $15 - 2\sqrt{26}$
89. $2\sqrt[3]{2y^2} + 2\sqrt[3]{4y} - 3$   91. The denominator would become $\sqrt[3]{6^2} = \sqrt[3]{36}$, which is not rational.   93. $-3\sqrt{6}$   95. $\frac{\sqrt{22}}{4}$
97. $\frac{3m\sqrt[3]{4n}}{n^2}$   99. $\frac{5(\sqrt{6}+3)}{3}$   101. $\frac{1-4\sqrt{2}}{3}$   103. $\{2\}$
105. $\emptyset$   107. $\{9\}$   109. $\{7\}$   111. $\{-13\}$
113. $\{14\}$   115. $\emptyset$   117. $\{7\}$
119. (a) $H = \sqrt{L^2 - W^2}$   (b) 7.9 ft   121. $10i\sqrt{2}$
123. $-10 - 2i$   125. $-\sqrt{35}$   127. $3$   129. $32 - 24i$
131. $4 + i$   133. $1$   135. $1$

## 8 Test (pages 433–434)

[8.1] 1. $-29$   2. $-8$   [8.2] 3. $5$   [8.1] 4. C
5. $21.863$   6. $-9.405$   7. domain: $[-6, \infty)$; range: $[0, \infty)$

[8.2] 8. $\frac{125}{64}$   9. $\frac{1}{256}$   10. $\frac{9y^{3/10}}{x^2}$
11. $x^{4/3} y^6$   12. $7^{1/2}$, or $\sqrt{7}$   [8.3] 13. $a^3 \sqrt[3]{a^2}$, or $a^{11/3}$
14. $\sqrt{145}$   15. $10$   16. $3x^2 y^3 \sqrt{6x}$   17. $2ab^3 \sqrt[4]{2a^3 b}$
18. $\sqrt[6]{200}$   [8.4] 19. $26\sqrt{5}$   20. $(2ts - 3t^2)\sqrt[3]{2s^2}$
[8.5] 21. $66 + \sqrt{5}$   22. $23 - 4\sqrt{15}$   23. $-\frac{\sqrt{10}}{4}$
24. $\frac{2\sqrt[3]{25}}{5}$   25. $-2(\sqrt{7} - \sqrt{5})$   26. $3 + \sqrt{6}$
[8.6] 27. (a) $59.8$   (b) $T = \frac{V_0^2 - V^2}{-V^2 k}$, or $T = \frac{V^2 - V_0^2}{V^2 k}$
28. $\{-1\}$   29. $\{3\}$   30. $\{-3\}$   [8.7] 31. $-5 - 8i$
32. $-2 + 16i$   33. $3 + 4i$   34. $i$   35. (a) true
(b) true   (c) false   (d) true

## 9 Quadratic Equations and Inequalities

### 9.1 Exercises (pages 442–444)

1. The equation is also true for $x = -4$. The solution set is $\{-4, 4\}$.   3. (a) A quadratic equation in standard form has a second-degree polynomial in decreasing powers equal to 0.
(b) The zero-factor property states that if a product equals 0, then at least one of the factors equals 0.   (c) The square root property states that if the square of a quantity equals a number, then the quantity equals the positive or negative square root of the number.
5. $\{9, -9\}$   7. $\{\sqrt{17}, -\sqrt{17}\}$   9. $\{4\sqrt{2}, -4\sqrt{2}\}$
11. $\{2\sqrt{5}, -2\sqrt{5}\}$   13. $\{2\sqrt{6}, -2\sqrt{6}\}$   15. $\{-7, 3\}$
17. $\{4 + \sqrt{3}, 4 - \sqrt{3}\}$   19. $\{-5 + 4\sqrt{3}, -5 - 4\sqrt{3}\}$
21. $\{\frac{1+\sqrt{7}}{3}, \frac{1-\sqrt{7}}{3}\}$   23. $\{\frac{-1+2\sqrt{6}}{4}, \frac{-1-2\sqrt{6}}{4}\}$
25. $\{\frac{2+2\sqrt{3}}{5}, \frac{2-2\sqrt{3}}{5}\}$   27. $5.6$ sec   29. $(2x + 1)^2 = 5$ is more suitable for solving by the square root property, while $x^2 + 4x = 12$ is more suitable for solving by completing the square.   31. $9; (x + 3)^2$   33. $36; (p - 6)^2$

**35.** $\frac{81}{4}$; $\left(q + \frac{9}{2}\right)^2$   **37.** $\frac{1}{64}$; $\left(x + \frac{1}{8}\right)^2$   **39.** 0.16; $(x - 0.4)^2$
**41.** 4   **43.** 25   **45.** $\frac{1}{36}$   **47.** $\{-4, 6\}$
**49.** $\{-2 + \sqrt{6}, -2 - \sqrt{6}\}$   **51.** $\left\{\frac{-7 + \sqrt{53}}{2}, \frac{-7 - \sqrt{53}}{2}\right\}$
**53.** $\left\{-\frac{8}{3}, 3\right\}$   **55.** $\left\{\frac{-5 + \sqrt{41}}{4}, \frac{-5 - \sqrt{41}}{4}\right\}$
**57.** $\left\{\frac{5 + \sqrt{15}}{5}, \frac{5 - \sqrt{15}}{5}\right\}$   **59.** $\left\{\frac{4 + \sqrt{3}}{3}, \frac{4 - \sqrt{3}}{3}\right\}$
**61.** $\left\{\frac{2 + \sqrt{3}}{3}, \frac{2 - \sqrt{3}}{3}\right\}$   **63.** $\{1 + \sqrt{2}, 1 - \sqrt{2}\}$
**65.** $\{2i\sqrt{3}, -2i\sqrt{3}\}$   **67.** $\{5 + 2i, 5 - 2i\}$
**69.** $\left\{\frac{1}{6} + \frac{\sqrt{2}}{3}i, \frac{1}{6} - \frac{\sqrt{2}}{3}i\right\}$   **71.** $\{-2 + 3i, -2 - 3i\}$
**73.** $\left\{-\frac{2}{3} + \frac{2\sqrt{2}}{3}i, -\frac{2}{3} - \frac{2\sqrt{2}}{3}i\right\}$   **75.** $\{-3 + i\sqrt{3}, -3 - i\sqrt{3}\}$

### 9.2 Exercises (pages 449–450)

**1.** The patron forgot the $\pm$ sign in the numerator. The correct formula is $x = \frac{-b \pm \sqrt{b^2 - 4ac}}{2a}$.   **3.** No, the quadratic formula can be used to solve *any* quadratic equation. Here, the quadratic formula can be used with $a = 2$, $b = 0$, and $c = -5$.
**5.** $\{3, 5\}$   **7.** $\left\{\frac{-2 + \sqrt{2}}{2}, \frac{-2 - \sqrt{2}}{2}\right\}$   **9.** $\left\{\frac{1 + \sqrt{3}}{2}, \frac{1 - \sqrt{3}}{2}\right\}$
**11.** $\{5 + \sqrt{7}, 5 - \sqrt{7}\}$   **13.** $\left\{\frac{-1 + \sqrt{2}}{2}, \frac{-1 - \sqrt{2}}{2}\right\}$
**15.** $\left\{\frac{-1 + \sqrt{7}}{3}, \frac{-1 - \sqrt{7}}{3}\right\}$   **17.** $\{1 + \sqrt{5}, 1 - \sqrt{5}\}$
**19.** $\left\{\frac{-2 + \sqrt{10}}{2}, \frac{-2 - \sqrt{10}}{2}\right\}$   **21.** $\{-1 + 3\sqrt{2}, -1 - 3\sqrt{2}\}$
**23.** $\left\{\frac{1 + \sqrt{29}}{2}, \frac{1 - \sqrt{29}}{2}\right\}$   **25.** $\left\{\frac{-4 + \sqrt{91}}{3}, \frac{-4 - \sqrt{91}}{3}\right\}$
**27.** $\left\{\frac{-3 + \sqrt{57}}{8}, \frac{-3 - \sqrt{57}}{8}\right\}$   **29.** $\left\{\frac{3}{2} + \frac{\sqrt{15}}{2}i, \frac{3}{2} - \frac{\sqrt{15}}{2}i\right\}$
**31.** $\{3 + i\sqrt{5}, 3 - i\sqrt{5}\}$   **33.** $\left\{\frac{1}{2} + \frac{\sqrt{6}}{2}i, \frac{1}{2} - \frac{\sqrt{6}}{2}i\right\}$
**35.** $\left\{-\frac{2}{3} + \frac{\sqrt{2}}{3}i, -\frac{2}{3} - \frac{\sqrt{2}}{3}i\right\}$   **37.** $\left\{\frac{1}{2} + \frac{1}{4}i, \frac{1}{2} - \frac{1}{4}i\right\}$
**39.** B   **41.** C   **43.** A   **45.** D   **47.** B   **49.** $\left\{-\frac{7}{5}\right\}$
**51.** $\left\{-\frac{1}{3}, 2\right\}$   **53. (a)** Discriminant is 25, or $5^2$; solve by factoring; $\left\{-3, -\frac{4}{3}\right\}$   **(b)** Discriminant is 44; use the quadratic formula; $\left\{\frac{7 + \sqrt{11}}{2}, \frac{7 - \sqrt{11}}{2}\right\}$   **55.** $-10$ or $10$   **57.** 16
**59.** 25   **61.** $b = \frac{44}{5}; \frac{3}{10}$

### 9.3 Exercises (pages 457–460)

**1.** square root property   **3.** quadratic formula   **5.** factoring
**7.** Multiply by the LCD, $x$.   **9.** Substitute a variable for $x^2 + x$.
**11.** The proposed solution $-1$ does not check. The solution set is $\{4\}$.   **13.** $\{-2, 7\}$   **15.** $\{-4, 7\}$   **17.** $\left\{-\frac{2}{3}, 1\right\}$
**19.** $\left\{-\frac{14}{17}, 5\right\}$   **21.** $\left\{-\frac{11}{7}, 0\right\}$   **23.** $\left\{\frac{-1 + \sqrt{13}}{2}, \frac{-1 - \sqrt{13}}{2}\right\}$
**25.** $\left\{-\frac{8}{3}, -1\right\}$   **27.** $\left\{\frac{2 + \sqrt{22}}{3}, \frac{2 - \sqrt{22}}{3}\right\}$
**29.** $\left\{\frac{-1 + \sqrt{5}}{4}, \frac{-1 - \sqrt{5}}{4}\right\}$   **31. (a)** $(20 - t)$ mph   **(b)** $(20 + t)$ mph   **33.** 25 mph   **35.** 50 mph   **37.** 3.6 hr

**39.** Rusty: 25.0 hr; Nancy: 23.0 hr   **41.** 3 hr; 6 hr   **43.** $\{2, 5\}$
**45.** $\{3\}$   **47.** $\left\{\frac{8}{9}\right\}$   **49.** $\{9\}$   **51.** $\left\{\frac{2}{5}\right\}$   **53.** $\{-2\}$
**55.** $\{-5, -2, 2, 5\}$   **57.** $\left\{-\frac{3}{2}, -1, 1, \frac{3}{2}\right\}$
**59.** $\{-2\sqrt{3}, -2, 2, 2\sqrt{3}\}$   **61.** $\{-6, -5\}$
**63.** $\left\{-\frac{16}{3}, -2\right\}$   **65.** $\left\{-\frac{1}{3}, \frac{1}{6}\right\}$   **67.** $\left\{-\frac{1}{2}, 3\right\}$
**69.** $\{-8, 1\}$   **71.** $\{-64, 27\}$   **73.** $\left\{-\frac{27}{8}, -1, 1, \frac{27}{8}\right\}$
**75.** $\{25\}$   **77.** $\left\{-1, 1, -\frac{\sqrt{6}}{2}i, \frac{\sqrt{6}}{2}i\right\}$   **79.** $\left\{-\frac{\sqrt{6}}{3}, -\frac{1}{2}, \frac{1}{2}, \frac{\sqrt{6}}{3}\right\}$
**81.** $\{3, 11\}$   **83.** $\left\{-\sqrt[3]{5}, -\frac{\sqrt[3]{4}}{2}\right\}$   **85.** $\left\{\frac{4}{3}, \frac{9}{4}\right\}$
**87.** $\left\{\frac{\sqrt{9 + \sqrt{65}}}{2}, -\frac{\sqrt{9 + \sqrt{65}}}{2}, \frac{\sqrt{9 - \sqrt{65}}}{2}, \frac{\sqrt{9 - \sqrt{65}}}{2}\right\}$

### 9.4 Exercises (pages 465–469)

**1.** Find a common denominator, and then multiply both sides by the common denominator.   **3.** Write it in standard form (with 0 on one side, in decreasing powers of $w$).   **5.** $m = \sqrt{p^2 - n^2}$
**7.** $t = \frac{\pm\sqrt{dk}}{k}$   **9.** $d = \frac{\pm\sqrt{skl}}{l}$   **11.** $v = \frac{\pm\sqrt{kAF}}{F}$
**13.** $r = \frac{\pm\sqrt{3\pi Vh}}{\pi h}$   **15.** $t = \frac{-B\sqrt{B^2 - 4AC}}{2A}$   **17.** $h = \frac{D^2}{k}$
**19.** $\ell = \frac{p^2 g}{k}$   **21.** $r = \frac{\pm\sqrt{S\pi}}{2\pi}$   **23.** $R = \frac{E^2 - 2prE\sqrt{E^2 - 4pr}}{2p}$
**25.** $r = \frac{5pc}{4}$ or $r = -\frac{2pc}{3}$   **27.** $I = \frac{-cR\sqrt{c^2R^2 - 4cL}}{2cL}$   **29.** 7.9, 8.9, 11.9   **31.** eastbound ship: 80 mi; southbound ship: 150 mi
**33.** 8 in., 15 in., 17 in.   **35.** length: 24 ft; width: 10 ft
**37.** 2 ft   **39.** 7 m by 12 m   **41.** 20 in. by 12 in.
**43.** 1 sec and 8 sec   **45.** 2.4 sec and 5.6 sec   **47.** 9.2 sec
**49.** It reaches its *maximum* height at 5 sec because this is the only time it reaches 400 ft.   **51.** 5 or 14   **53. (a)** 2750 billion
**(b)** 2750 billion; They are the same.   **55.** 2001; The graph indicates that vehicle-miles reached 2800 in 2001.

### 9.5 Exercises (page 474)

**1.** false   **3.** true   **5.** Include the endpoints if the symbol is $\geq$ or $\leq$. Exclude the endpoints if the symbol is $>$ or $<$.
**7.** $(-\infty, -1) \cup (5, \infty)$
**9.** $(-4, 6)$
**11.** $(-\infty, 1] \cup [3, \infty)$
**13.** $\left(-\infty, -\frac{3}{2}\right] \cup \left[\frac{3}{5}, \infty\right)$
**15.** $\left[-\frac{3}{2}, \frac{3}{2}\right]$
**17.** $\left(-\infty, -\frac{1}{2}\right] \cup \left[\frac{1}{3}, \infty\right)$
**19.** $(-\infty, 0] \cup [4, \infty)$
**21.** $\left[0, \frac{5}{3}\right]$

23. $(-\infty, 3 - \sqrt{3}] \cup [3 + \sqrt{3}, \infty)$

25. $(-\infty, \infty)$   27. $\emptyset$

29. $(-\infty, 1) \cup (2, 4)$

31. $\left[-\frac{3}{2}, \frac{1}{3}\right] \cup [4, \infty)$

33. $(-\infty, 1) \cup (4, \infty)$

35. $\left[-\frac{3}{2}, 5\right)$   37. $(2, 6]$

39. $\left(-\infty, \frac{1}{2}\right) \cup \left(\frac{5}{4}, \infty\right)$

41. $[-7, -2)$

43. $(-\infty, 2) \cup (4, \infty)$

45. $\left(0, \frac{1}{2}\right) \cup \left(\frac{5}{2}, \infty\right)$

47. $\left[\frac{3}{2}, \infty\right)$

49. $\left(-2, \frac{5}{3}\right) \cup \left(\frac{5}{3}, \infty\right)$

## 9 Review Exercises (pages 477–480)

1. $\{-11, 11\}$   3. $\left\{-\frac{15}{2}, \frac{5}{2}\right\}$   5. $\{-2 + \sqrt{19}, -2 - \sqrt{19}\}$
7. By the square root property, the first step should be $x = \sqrt{12}$ or $x = -\sqrt{12}$. The solution set is $\{-2\sqrt{3}, 2\sqrt{3}\}$.   9. C
11. D   13. $\left\{-\frac{7}{3}, 3\right\}$   15. $\left\{\frac{1 + \sqrt{41}}{2}, \frac{1 - \sqrt{41}}{2}\right\}$
17. $\left\{\frac{2}{3} + \frac{\sqrt{2}}{3}i, \frac{2}{3} - \frac{\sqrt{2}}{3}i\right\}$   19. $\left\{-\frac{5}{2}, 3\right\}$   21. $\{-4\}$
23. $\left\{-\frac{343}{8}, 64\right\}$   25. 7 mph   27. 4.6 hr   29. $v = \frac{\pm\sqrt{rFkw}}{kw}$
31. $t = \frac{3m \pm \sqrt{9m^2 + 24m}}{2m}$   33. 12 cm by 20 cm   35. 3 min
37. 0.7 sec and 4.0 sec   39. 4.5%
41. $[-4, 3]$   43. $\emptyset$
45. $[-3, 2)$

## 9 Test (pages 480–481)

[9.1] 1. $\{3\sqrt{6}, -3\sqrt{6}\}$   2. $\left\{-\frac{8}{7}, \frac{2}{7}\right\}$
3. $\{-1 + \sqrt{5}, -1 - \sqrt{5}\}$   [9.2] 4. $\left\{\frac{3 + \sqrt{17}}{4}, \frac{3 - \sqrt{17}}{4}\right\}$
5. $\left\{\frac{2}{3} + \frac{\sqrt{11}}{3}i, \frac{2}{3} - \frac{\sqrt{11}}{3}i\right\}$   6. $\left\{\frac{-5 + \sqrt{97}}{12}, \frac{-5 - \sqrt{97}}{12}\right\}$
[9.1] 7. A   [9.2] 8. discriminant: 88; There are two irrational solutions.   [9.1–9.3] 9. $\left\{-\frac{2}{3}, 6\right\}$   10. $\left\{\frac{-7 + \sqrt{97}}{8}, \frac{-7 - \sqrt{97}}{8}\right\}$
11. $\left\{\frac{2}{3}\right\}$   12. $\left\{-2, -\frac{1}{3}, \frac{1}{3}, 2\right\}$   13. $\left\{-\frac{5}{2}, 1\right\}$
[9.4] 14. $r = \frac{\pm\sqrt{\pi S}}{2\pi}$   [9.3] 15. Andrew: 11.1 hr; Kent: 9.1 hr

16. 7 mph   [9.4] 17. 2 ft   18. 16 m
[9.5] 19. $(-\infty, -5) \cup \left(\frac{3}{2}, \infty\right)$
20. $(-\infty, 4) \cup [9, \infty)$

## 10 Additional Graphs of Functions and Relations

### 10.1 Exercises (pages 488–490)

1. 55   3. 1848   5. $-\frac{7}{6}$   7. 1122   9. 97   11. 930
13. (a) $10x + 2$   (b) $-2x - 4$   (c) $24x^2 + 6x - 3$
(d) $\frac{4x - 1}{6x + 3}$; All domains are $(-\infty, \infty)$, except for $\frac{f}{g}$, which is
$\left(-\infty, -\frac{1}{2}\right) \cup \left(-\frac{1}{2}, \infty\right)$.   15. (a) $4x^2 - 4x + 1$   (b) $2x^2 - 1$
(c) $(3x^2 - 2x)(x^2 - 2x + 1)$   (d) $\frac{3x^2 - 2x}{x^2 - 2x + 1}$; All domains
are $(-\infty, \infty)$, except for $\frac{f}{g}$, which is $(-\infty, 1) \cup (1, \infty)$.
17. (a) $\sqrt{2x + 5} + \sqrt{4x + 9}$   (b) $\sqrt{2x + 5} - \sqrt{4x + 9}$
(c) $\sqrt{(2x + 5)(4x + 9)}$   (d) $\sqrt{\frac{2x + 5}{4x + 9}}$   19. The function
values for $f + g$ are found by adding $f(x) + g(x)$. For example, if
$f(x) = 2x + 3$ and $g(x) = x^2$, then $(f + g)(x) = x^2 + 2x + 3$.
21. (a) $4xh + 2h^2$   (b) $4x + 2h$   23. (a) $2xh + h^2 + 4h$
(b) $2x + h + 4$   In Exercises 25–31, we give $(f \circ g)(x)$,
$(g \circ f)(x)$, and the domains.   25. $-5x^2 + 20x + 18$;
$-25x^2 - 10x + 6$; both domains are $(-\infty, \infty)$.   27. $\frac{1}{x^2}; \frac{1}{x^2}$;
both domains are $(-\infty, 0) \cup (0, \infty)$.   29. $2\sqrt{2x - 1}$;
$8\sqrt{x + 2} - 6$; domain of $f \circ g$: $\left[\frac{1}{2}, \infty\right)$; domain of $g \circ f$: $[-2, \infty)$
31. $\frac{x}{2 - 5x}$; $2(x - 5)$; domain of $f \circ g$: $(-\infty, 0) \cup \left(0, \frac{2}{5}\right) \cup \left(\frac{2}{5}, \infty\right)$;
domain of $g \circ f$: $(-\infty, 5) \cup (5, \infty)$   33. To find values of
$(f \circ g)(x)$, replace $x$ in $f$ with $g(x)$. For example, if $f(x) = 2x - 5$
and $g(x) = x^2 + 3$, then $(f \circ g)(x) = 2(x^2 + 3) - 5 =$
$2x^2 + 6 - 5 = 2x^2 + 1$.   35. 4   37. 0   39. 1
41. 2   43. 1   45. 9   47. 1   49. $g(1) = 9$, and $f(9)$
cannot be determined from the table.   Other correct answers
are possible in Exercises 51 and 53.   51. $f(x) = x^2$;
$g(x) = 6x - 2$   53. $f(x) = \frac{1}{x + 2}$; $g(x) = x^2$
55. $(f \circ g)(x) = 5280x$. It computes the number of feet in $x$ miles.
57. $D(c) = \frac{-c^2 + 10c - 25}{25} + 500$

### 10.2 Exercises (pages 497–500)

1. (a) B   (b) C   (c) A   (d) D   3. $(0, 0)$   5. $(0, 4)$
7. $(1, 0)$   9. $(-3, -4)$   11. $(5, 6)$   13. down; wider
15. up; narrower   17. (a) I   (b) IV   (c) II   (d) III
19. (a) D   (b) B   (c) C   (d) A
21.   23.   25.

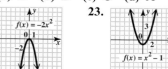

**27.** vertex: $(4, 0)$; axis: $x = 4$; domain: $(-\infty, \infty)$; range: $[0, \infty)$

**29.** vertex: $(-2, -1)$; axis: $x = -2$; domain: $(-\infty, \infty)$; range: $[-1, \infty)$

**31.** vertex: $(2, -4)$; axis: $x = 2$; domain: $(-\infty, \infty)$; range: $[-4, \infty)$

**29.** vertex: $(1, 5)$; axis: $y = 5$; domain: $(-\infty, 1]$; range: $(-\infty, \infty)$

**31.** vertex: $(-7, -2)$; axis: $y = -2$; domain: $[-7, \infty)$; range: $(-\infty, \infty)$

**33.** vertex: $(-1, 2)$; axis: $x = -1$; domain: $(-\infty, \infty)$; range: $(-\infty, 2]$

**35.** vertex: $(2, -3)$; axis: $x = 2$; domain: $(-\infty, \infty)$; range: $[-3, \infty)$

**33.** 140 ft by 70 ft; 9800 ft² **35.** 16 ft; 2 sec
**37.** (a) maximum (b) 1993; 13.2%
**39.** (a) $R(x) = (100 - x)(200 + 4x) = 20{,}000 + 200x - 4x^2$
(b) 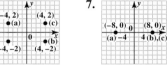 (c) 25 (d) $22,500

**37.** linear; positive **39.** quadratic; positive
**41.** (a)
(b) quadratic; positive
(c) $y = 2.969x^2 - 23.125x + 115$
(d) 265 (e) No. About 16 companies filed for bankruptcy each month, so at this rate, filings for 2002 would be about 192. The approximation from the model seems high.

**43.** (a) 222.7 (per 100,000) (b) The approximation using the model is high. **45.** $y = \frac{1}{8}x^2$ **47.** $y = -\frac{1}{4}x^2$
**49.** $y = \frac{1}{4}x^2 + 3$

## 10.3 Exercises (pages 508–510)

**1.** If $x$ is squared, it has a vertical axis; if $y$ is squared, it has a horizontal axis. **3.** Use the discriminant of the function. If it is positive, there are two $x$-intercepts. If it is 0, there is one $x$-intercept (at the vertex), and if it is negative, there is no $x$-intercept. **5.** $(-4, -6)$ **7.** $(1, -3)$ **9.** $\left(-\frac{1}{2}, -\frac{29}{4}\right)$
**11.** $(-1, 3)$; up; narrower; no $x$-intercepts **13.** $\left(\frac{5}{2}, \frac{37}{4}\right)$; down; same; two $x$-intercepts **15.** $(-3, -9)$; to the right; wider
**17.** F **19.** C **21.** D
**23.** vertex: $(-4, -6)$; axis: $x = -4$; domain: $(-\infty, \infty)$; range: $[-6, \infty)$

**25.** vertex: $(1, -3)$; axis: $x = 1$; domain: $(-\infty, \infty)$; range: $(-\infty, -3]$

**27.** vertex: $(1, -2)$; axis: $y = -2$; domain: $[1, \infty)$; range: $(-\infty, \infty)$

## 10.4 Exercises (pages 515–516)

**1.** (a)  The graph of $f(x)$ is reflected about the $x$-axis.

(b) 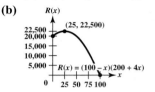 The graph is the same shape as that of $f(x)$, but stretched vertically by a factor of 2.

**3.** (a) Replace $y$ with $-y$. If the equation is equivalent to the given equation, its graph is symmetric with respect to the $x$-axis.
(b) Replace $x$ with $-x$. If the equation is equivalent to the given equation, its graph is symmetric with respect to the $y$-axis.
(c) If the equation is equivalent to the given equation when both $-x$ replaces $x$ and $-y$ replaces $y$, its graph is symmetric with respect to the origin. **5.** **7.**

**9.** $x$-axis, $y$-axis, origin **11.** None of the symmetries apply.
**13.** $y$-axis **15.** origin **17.** $y$-axis **19.** origin
**21.** $f$ is increasing on $(-\infty, -3]$; $f$ is decreasing on $[0, \infty)$.
**23.** $f$ is increasing on $(-\infty, -2]$ and $[1, \infty)$; $f$ is decreasing on $[-2, 1]$. **25.** $f$ is increasing on $[0, \infty)$; $f$ is decreasing on $(-\infty, 0]$. **27.** (a) symmetric (b) symmetric
**29.** (a) not symmetric (b) symmetric **31.** $f(-2) = -3$
**33.** $f(4) = 3$ **35.**

## 10.5 Exercises (pages 521–523)

**1.** B  **3.** A  **5.**   **7.**

**9.**   **11.**   **13.** (a) $-10$

(b) $-2$  (c) $-1$  (d) $2$  (e) $4$  **15.**

**17.**   **19.**

**21.**   **23.**

**25.**   **27.**  **29.**

**31.** for $[3, 6]$: $y = -\frac{1}{3}x + 74$; for $(6, 9]$: $y = -x + 78$;

$f(x) = \begin{cases} -\frac{1}{3}x + 74 & \text{if } 3 \leq x \leq 6 \\ -x + 78 & \text{if } 6 < x \leq 9 \end{cases}$  **33.** (a) $\$11$  (b) $\$18$

(c) $\$32$  (d)   (e) domain: $(0, \infty)$; range: $\{11, 18, 25, \ldots\}$

## 10 Review Exercises (pages 527–528)

**1.** $x^2 + 3x + 3$; $(-\infty, \infty)$  **3.** $(x^2 - 2x)(5x + 3)$; $(-\infty, \infty)$
**5.** $5x^2 - 10x + 3$; $(-\infty, \infty)$  **7.** $3$  **9.** $7\sqrt{5}$
**11.** $(4b - 3)(\sqrt{2b})$, $b \geq 0$  **13.** $(f \circ g)(2) = 2\sqrt{2} - 3$; The answers are not equal, so composition of functions is not commutative.  **15.** $(0, 6)$  **17.** $(3, 7)$  **19.** $(-4, 3)$

**21.** vertex: $(0, -2)$; axis: $x = 0$; domain: $(-\infty, \infty)$; range: $[-2, \infty)$

**23.** vertex: $(2, -3)$; axis: $x = 2$; domain: $(-\infty, \infty)$; range: $[-3, \infty)$

**25.** vertex: $\left(-\frac{3}{2}, -\frac{1}{4}\right)$; axis: $x = -\frac{3}{2}$; domain: $(-\infty, \infty)$; range: $\left[-\frac{1}{4}, \infty\right)$

**27.** vertex: $(-4, -3)$; axis: $y = -3$; domain: $[-4, \infty)$; range: $(-\infty, \infty)$

**29.** length: 50 m; width: 50 m; maximum area: 2500 m²
**31.** $x$-axis, $y$-axis, origin  **33.** $x$-axis  **35.** no symmetries
**37.** The vertical line test shows that a circle does not represent a function.  **39.** decreasing  **41.** increasing on $[0, \infty)$
**43.** increasing on $[1, \infty)$; decreasing on $(-\infty, 1]$

**45.**   **47.**   **49.**

**51.** The graph is narrower than the graph of $y = |x|$, and it is shifted (translated) 4 units to the left and 3 units down.

## 10 Test (pages 528–530)

[10.1] **1.** $2$  **2.** $-7$  **3.** $-\frac{7}{3}$  **4.** $-2$
**5.** $x^2 + 4x - 1$; $(-\infty, \infty)$  [10.2] **6.** A
**7.** vertex: $(0, -2)$; axis: $x = 0$; domain: $(-\infty, \infty)$; range: $[-2, \infty)$

[10.3] **8.** vertex: $(2, 3)$; axis: $x = 2$; domain: $(-\infty, \infty)$; range: $(-\infty, 3]$

9. vertex: (2, 2); axis: $y = 2$; domain: $(-\infty, 2]$; range: $(-\infty, \infty)$

10. (a) 6.5%  (b) 1996; 3.5%

11. 160 ft by 320 ft; 51,200 ft² [10.4] 12. $y$-axis
13. $x$-axis  14. $x$-axis, $y$-axis, origin  [10.5] 15. (a) C
(b) A  (c) D  (d) B  [10.4] 16. increasing: $[-2, 1]$; decreasing: $(-\infty, -2]$; constant: $[1, \infty)$
[10.5] 17. 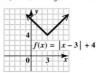  18.

19.   20.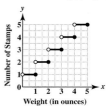

## 11 Inverse, Exponential, and Logarithmic Functions

### 11.1 Exercises (pages 537–539)

1. This function is not one-to-one because both France and the United States are paired with the same trans fat percentage, 11. Also both Hungary and Poland are paired with the same trans fat percentage, 8.  3. Yes. By adding 1 to 1058 two distances would be the same, so the function would not be one-to-one.
5. B  7. A  9. $\{(6, 3), (10, 2), (12, 5)\}$  11. not one-to-one  13. $f^{-1}(x) = \frac{x-4}{2}$, or $f^{-1}(x) = \frac{1}{2}x - 2$
15. $g^{-1}(x) = x^2 + 3, x \geq 0$  17. not one-to-one
19. $f^{-1}(x) = \sqrt[3]{x+4}$  21. (a) 8  (b) 3  23. (a) 1  (b) 0  25. (a) one-to-one  (b)

27. (a) not one-to-one  (b)  29. (a) one-to-one  31.  33.

35. | $x$ | $f(x)$ |
|---|---|
| 0 | 0 |
| 1 | 1 |
| 4 | 2 |

37. | $x$ | $f(x)$ |
|---|---|
| $-1$ | $-3$ |
| 0 | $-2$ |
| 1 | $-1$ |
| 2 | 6 |

### 11.2 Exercises (pages 545–547)

1. C  3. A  5.   7.

9.   11.   13. (a) rises; falls
(b) It is one-to-one and thus has an inverse.  15. $\{2\}$
17. $\{\frac{3}{2}\}$  19. $\{7\}$  21. $\{-3\}$  23. $\{-1\}$  25. $\{-3\}$
27. 639.545  29. 0.066  31. 12.179  33. (a) 0.5°C
(b) 0.35°C  35. (a) 1.6°C  (b) 0.5°C  37. (a) 220,717 thousand tons  (b) 129,048 thousand tons  (c) It is slightly greater than what the model provides (111,042 thousand tons).
39. (a) $5000  (b) $2973  (c) $1768
(d) 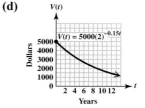  41. 6.67 yr after it was purchased

### 11.3 Exercises (pages 552–554)

1. (a) B  (b) E  (c) D  (d) F  (e) A  (f) C
3. $\log_4 1024 = 5$  5. $\log_{1/2} 8 = -3$  7. $\log_{10} 0.001 = -3$
9. $\log_{625} 5 = \frac{1}{4}$  11. $4^3 = 64$  13. $10^{-4} = \frac{1}{10,000}$
15. $6^0 = 1$  17. $9^{1/2} = 3$  19. (a) C  (b) B  (c) B
(d) C  21. $\{\frac{1}{3}\}$  23. $\{81\}$  25. $\{\frac{1}{5}\}$  27. $\{1\}$
29. $\{x \mid x > 0, x \neq 1\}$  31. $\{5\}$  33. $\{\frac{5}{3}\}$  35. $\{4\}$
37. $\{\frac{3}{2}\}$  39. $\{30\}$  41.  43.

45. Every power of 1 is equal to 1, and thus it cannot be used as a base.  47. $(0, \infty); (-\infty, \infty)$  49. 8  51. 24
53. (a) 4385 billion ft³  (b) 5555 billion ft³  (c) 6140 billion ft³
55. (a) 130 thousand units  (b) 190 thousand units
(c)   57. about 4 times as powerful

### 11.4 Exercises (pages 560–561)

1. $\log_{10} 3 + \log_{10} 4$  3. 4  5. 4  7. $\log_7 4 + \log_7 5$
9. $\log_5 8 - \log_5 3$  11. $2 \log_4 6$
13. $\frac{1}{3} \log_3 4 - 2 \log_3 x - \log_3 y$

**15.** $\frac{1}{2}\log_3 x + \frac{1}{2}\log_3 y - \frac{1}{2}\log_3 5$
**17.** $\frac{1}{3}\log_2 x + \frac{1}{5}\log_2 y - 2\log_2 r$
**19.** The distributive property tells us that the *product* $a(x + y)$ equals the sum $ax + ay$. In the notation $\log_a(x + y)$, the parentheses do not indicate multiplication. They indicate that $x + y$ is the result of raising $a$ to some power.
**21.** $\log_b xy$  **23.** $\log_a \frac{m}{n}$  **25.** $\log_a \frac{rt^3}{s}$  **27.** $\log_a \frac{125}{81}$
**29.** $\log_{10}(x^2 - 9)$  **31.** $\log_p \frac{x^3 y^{1/2}}{z^{3/2} a^3}$  **33.** 1.2552
**35.** $-0.6532$  **37.** 1.5562  **39.** 0.4771  **41.** 0.2386
**43.** 4.7710  **45.** false
**47.** true  **49.** true  **51.** false

## 11.5 Exercises (pages 566–569)

**1.** C  **3.** C  **5.** 19.2  **7.** 1.6335  **9.** 2.5164
**11.** $-1.4868$  **13.** 9.6776  **15.** 2.0592  **17.** $-2.8896$
**19.** 5.9613  **21.** 4.1506  **23.** 2.3026
**25.** (a) 2.552424846  (b) 1.552424846  (c) 0.552424846
(d) The whole number parts will vary but the decimal parts are the same.  **27.** An error message appears, because we cannot find the common logarithm of a negative number.  **29.** bog
**31.** 11.6  **33.** 4.3  **35.** $4.0 \times 10^{-8}$  **37.** $4.0 \times 10^{-6}$
**39.** (a) 107 dB  (b) 100 dB  (c) 98 dB  **41.** (a) 800 yr
(b) 5200 yr  (c) 11,500 yr  **43.** (a) 77%  (b) 1989
**45.** (a) $54 per ton  (b) If $p = 0$, then $\ln(1 - p) = \ln 1 = 0$, so $T$ would be negative. If $p = 1$, then $\ln(1 - p) = \ln 0$, but the domain of $\ln x$ is $(0, \infty)$.  **47.** 2.2619  **49.** 0.6826
**51.** 0.3155  **53.** 0.8736  **55.** 2.4849  **57.** 6446 billion ft³

## 11.6 Exercises (pages 576–578)

**1.** {0.827}  **3.** {0.833}  **5.** {1.201}  **7.** {2.269}
**9.** {15.967}  **11.** {261.291}  **13.** {$-10.718$}  **15.** {3}
**17.** {5.879}  **19.** Natural logarithms are a better choice because $e$ is the base.  **21.** $\{\frac{2}{3}\}$  **23.** $\{\frac{33}{2}\}$  **25.** $\{-1 + \sqrt[3]{49}\}$
**27.** 2 cannot be a solution because $\log(2 - 3) = \log(-1)$, and $-1$ is not in the domain of $\log x$.  **29.** $\{\frac{1}{3}\}$  **31.** {2}
**33.** ∅  **35.** {8}  **37.** $\{\frac{4}{3}\}$  **39.** {8}  **41.** (a) $2539.47
(b) 10.2 yr  **43.** (a) $4934.71  (b) 19.8 yr
**45.** (a) $11,260.96  (b) $11,416.64  (c) $11,497.99
(d) $11,580.90  (e) $11,581.83  **47.** $137.41
**49.** (a) 15.8 million tons  (b) 22.5 million tons  (c) 45.7 million tons  (d) 80.5 million tons  **51.** $40,693 million
**53.** (a) 1.62 g  (b) 1.18 g  (c) 0.69 g  (d) 2.00 g
**55.** (a) 179.73 g  (b) 21.66 yr  **57.** 2006

## 11 Review Exercises (pages 582–585)

**1.** not one-to-one  **3.** This function is not one-to-one because two sodas in the list have 41 mg of caffeine.  **5.** $f^{-1}(x) = \frac{x^3 + 4}{6}$
**7.**   **9.**   **11.**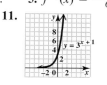

**13.** {4}  **15.** (a) 10.3 million tons  (b) 18.4 million tons
(c) 24.5 million tons  **17.**   **19.** $\{\frac{3}{2}\}$

**21.** {8}  **23.** $\{b \mid b > 0, b \neq 1\}$  **25.** $a$
**27.** $\log_2 3 + \log_2 x + 2 \log_2 y$  **29.** $\log_b \frac{3x}{y^2}$  **31.** 1.4609
**33.** 3.3638  **35.** 0.9251  **37.** 6.4  **39.** $2.5 \times 10^{-5}$
**41.** (a) 18 yr  (b) 12 yr  (c) 7 yr  (d) 6 yr  (e) Each comparison shows approximately the same number. For example, in part (a) the doubling time is 18 yr (rounded) and $\frac{72}{4} = 18$. Thus, the formula $t = \frac{72}{100r}$ (called the *rule of 72*) is an excellent approximation of the doubling time formula. (It is used by bankers for that purpose.)  **43.** {4.907}  **45.** $\{\frac{1}{9}\}$  **47.** {2}
**49.** {4}  **51.** When the power rule was applied in the second step, the domain was changed from $\{x \mid x \neq 0\}$ to $\{x \mid x > 0\}$. The valid solution $-10$ was "lost." The solution set is $\{\pm 10\}$.
**53.** $11,972.17  **55.** about 13.9 days  **57.** about 67%

## 11 Test (pages 585–586)

[11.1] **1.** (a) not one-to-one  (b) one-to-one
**2.** $f^{-1}(x) = x^3 - 7$  **3.** 

[11.2] **4.**   [11.3] **5.** 

[11.1–11.3] **6.** Once the graph of $f(x) = 6^x$ is sketched, interchange the x- and y-values of its ordered pairs. The resulting points will be on the graph of $g(x) = \log_6 x$ since $f$ and $g$ are inverses.

[11.2] **7.** $\{-4\}$  **8.** $\{-\frac{13}{3}\}$  [11.5] **9.** (a) 37.7 million
(b) 42.3 million  [11.3] **10.** $\log_4 0.0625 = -2$
**11.** $7^2 = 49$  **12.** {32}  **13.** $\{\frac{1}{2}\}$  **14.** {2}
**15.** 5; 2; 5th; 32  [11.4] **16.** $2 \log_3 x + \log_3 y$
**17.** $\frac{1}{2}\log_5 x - \log_5 y - \log_5 z$  **18.** $\log_b \frac{s^3}{t}$  **19.** $\log_b \frac{r^{1/4} s^2}{t^{2/3}}$
[11.5] **20.** (a) 1.3636  (b) $-0.1985$  **21.** (a) $\frac{\log 19}{\log 3}$  (b) $\frac{\ln 19}{\ln 3}$
(c) 2.6801  [11.6] **22.** {3.9656}  **23.** {3}
**24.** $12,507.51  **25.** (a) $19,260.38
(b) approximately 13.9 yr

## 12 Polynomial and Rational Functions

### 12.1 Exercises (pages 592–593)

**1.** Synthetic division provides a quick, easy way to divide a polynomial by a binomial of the form $x - k$.  **3.** $x - 5$

**5.** $4m - 1$   **7.** $2a + 4 + \frac{5}{a+2}$   **9.** $p - 4 + \frac{9}{p+1}$
**11.** $4a^2 + a + 3$   **13.** $x^4 + 2x^3 + 2x^2 + 7x + 10 + \frac{18}{x-2}$
**15.** $-4r^5 - 7r^4 - 10r^3 - 5r^2 - 11r - 8 + \frac{-5}{r-1}$
**17.** $-3y^4 + 8y^3 - 21y^2 + 36y - 72 + \frac{143}{y+2}$
**19.** $y^2 + y + 1 + \frac{2}{y-1}$
**21.** $f(x) = (x+1)(2x^2 - x + 2) + (-10)$
**23.** $f(x) = (x+2)(-x^2 + 4x - 8) + 20$
**25.** $f(x) = (x-3)(4x^3 + 9x^2 + 7x + 20) + 60$   **27.** 2
**29.** $-1$   **31.** $-6$   **33.** 0   **35.** 11   **37.** $-6 - i$
**39.** By the remainder theorem, a 0 remainder means that $f(k) = 0$; that is, $k$ is a number that makes $f(x) = 0$.   **41.** no
**43.** yes   **45.** no   **47.** no

## 12.2 Exercises *(pages 599–600)*

**1.** true   **3.** false   **5.** no   **7.** yes   **9.** yes
**11.** $f(x) = (x-2)(2x-5)(x+3)$
**13.** $f(x) = (x+3)(3x-1)(2x-1)$   **15.** $-1 \pm i$
**17.** $3, 2 + i$   **19.** $i, \pm 2i$   **21. (a)** $\pm 1, \pm 2, \pm 5, \pm 10$
**(b)** $-1, -2, 5$   **(c)** $f(x) = (x+1)(x+2)(x-5)$
**23. (a)** $\pm 1, \pm 2, \pm 3, \pm 5, \pm 6, \pm 10, \pm 15, \pm 30$
**(b)** $-5, -3, 2$   **(c)** $f(x) = (x+5)(x+3)(x-2)$
**25. (a)** $\pm 1, \pm 2, \pm 3, \pm 4, \pm 6, \pm 12, \pm\frac{1}{2}, \pm\frac{3}{2}, \pm\frac{1}{3}, \pm\frac{2}{3}, \pm\frac{4}{3}, \pm\frac{1}{6}$
**(b)** $-4, -\frac{1}{3}, \frac{3}{2}$   **(c)** $f(x) = (x+4)(3x+1)(2x-3)$
**27. (a)** $\pm 1, \pm 2, \pm 3, \pm 6, \pm\frac{1}{2}, \pm\frac{3}{2}, \pm\frac{1}{3}, \pm\frac{2}{3}, \pm\frac{1}{6}, \pm\frac{1}{12}, \pm\frac{1}{4}, \pm\frac{3}{4}$
**(b)** $-\frac{3}{2}, -\frac{2}{3}, \frac{1}{2}$   **(c)** $f(x) = (3x+2)(2x+3)(2x-1)$
**29.** $-4$ (mult. 2), $\pm\sqrt{7}, -1$   **31.** $2, -3, 1, -1$
**33.** $-2$ (mult. 5), 1 (mult. 5), $1 - \sqrt{3}$ (mult. 2)
**35.** $f(x) = x^2 - 6x + 10$   **37.** $f(x) = x^3 - 5x^2 + 5x + 3$
**39.** $f(x) = x^4 + 4x^3 - 4x^2 - 36x - 45$
**41.** $f(x) = x^3 - 2x^2 + 9x - 18$
**43.** $f(x) = x^4 - 6x^3 + 17x^2 - 28x + 20$
**45.** $f(x) = -3x^3 + 6x^2 + 33x - 36$
**47.** $f(x) = -\frac{1}{2}x^3 - \frac{1}{2}x^2 + x$   **49.** $f(x) = -\frac{1}{3}x^3 + \frac{5}{3}x^2 - \frac{1}{3}x + \frac{5}{3}$
**51.** $-1; 3; f(x) = (x+2)^2(x+1)(x-3)$
**53.** 2 or 0 positive; 1 negative   **55.** 1 positive; 1 negative
**57.** 2 or 0 positive; 3 or 1 negative

## 12.3 Exercises *(pages 610–613)*

**1.**    **3.**    **5.**

**7.**    **9.**    **11.** 2   **13.** 3

**15.** 3   **17.** A   **19.** one   **21.** B and D   **23.** one
**25.** 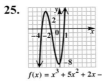   **27.**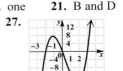

**29.**    **31.**

**33.**    **35.**

**37.**    **39.** $-1, \frac{3}{2}, 2$; $f(x) = (x+1)(2x-3)(x-2)$

**41.** $-2$ (multiplicity 2), 3; $f(x) = (x+2)^2(x-3)$   **43.** $-2$ (multiplicity 2), 3; $f(x) = (x+2)^2(-x+3)$

**45.** 3 (multiplicity 2), $-3$ (multiplicity 2); $f(x) = (x-3)^2(x+3)^2$   **47. (a)** $f(-2) = 8 > 0$ and $f(-1) = -2 < 0$   **(b)** $-1.236, 3.236$

**49. (a)** $f(-4) = 76 > 0$ and $f(-3) = -75 < 0$   **(b)** $-3.646, -0.317, 1.646, 6.317$   **57.** $f(x) = 0.5(x+6)(x-2)(x-5) = 0.5x^3 - 0.5x^2 - 16x + 30$   **59.** $-0.88, 2.12, 4.86$
**61.** $-1.52$   **63.** $-0.40, 2.02$   **65.** $(-3.44, 26.15)$
**67.** $(-0.09, 1.05)$   **69.** $(-0.20, -28.62)$   **71. (a)** See part (b).
**(b)** $g(x) = 1818(x-2)^2 + 620$

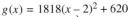
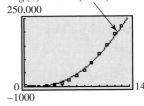

(c) 46,070; This figure is a bit higher than the figure 40,820 given in the table.  **73. (a)** $0 < x < 10$  **(b)** $A(x) = x(20 - 2x)$ or $A(x) = -2x^2 + 20x$  **(c)** $x = 5$; maximum cross section area: 50 in.²  **(d)** between 0 and 2.76 in. or between 7.24 and 10 in.
**75.** 1.732

## 12.4 Exercises (pages 623–625)

**1.** A, B, C  **3.** A  **5.** A, C, D  **7.**

**9.**   **11.**

**13. (a)** 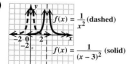  **(b)**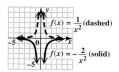

**(c)**

**15.** vertical asymptote: $x = \frac{7}{3}$; horizontal asymptote: $y = 0$
**17.** vertical asymptotes: $x = \frac{5}{3}, x = -2$; horizontal asymptote: $y = 0$  **19.** vertical asymptote: $x = -2$; horizontal asymptote: $y = -1$
**21.** vertical asymptote: $x = -\frac{9}{2}$; horizontal asymptote: $y = \frac{3}{2}$
**23.** vertical asymptotes: $x = 3, x = 1$; horizontal asymptote: $y = 0$  **25.** vertical asymptote: $x = -3$; oblique asymptote: $y = x - 3$  **27.** vertical asymptotes: $x = -2, x = \frac{5}{2}$; horizontal asymptote: $y = \frac{1}{2}$  **29.** A  **31.**

**33.**   **35.**   **37.**

**39.**   **41.**   **43.**

**45.**   **47.**   **49.**

**51.**   **53.**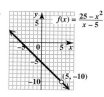

**55.** It is not correct. There is a hole in the graph at $(-5, -10)$.
**57. (a)** 26 per min   For $r = x$,   **(b)** 5

$$y = \frac{2x - 25}{2x^2 - 50x} \quad y = 0.5$$

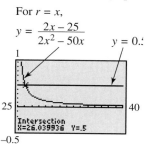

## 12 Review Exercises (pages 629–630)

**1.** $3x + 2$  **3.** $2x^2 + x + 3 + \frac{21}{x - 3}$  **5.** yes  **7.** $-13$
**9.** $2x^3 + x - 6 = (x + 2) \cdot (2x^2 - 4x + 9) + (-24)$
**11.** $\frac{1}{2}, -1, 5$  **13.** $4, -\frac{1}{2}, -\frac{2}{3}$  **15.** no  **17.** no
**19.** $f(x) = -2x^3 + 6x^2 + 12x - 16$
**21.** $f(x) = x^4 - 3x^2 - 4$ (There are others.)
**23.** $f(x) = x^3 + x^2 - 4x + 6$ (There are others.)
**25.** No. Because zeros that are not real come in conjugate *pairs*, there must be an even number of them. 3 is odd.
**27.** zero; solution; $x$-intercept  **29.** two
**31.**   **33.**   **35.**

**37.** $f(-1) = -10$ and $f(0) = 2$; $f(2) = -4$; and $f(3) = 14$
**39.** $f(-1) = 15$ and $f(0) = -8$; $f(-6) = -20$ and $f(-5) = 27$
**43.**   **45.**

**47.**   **49.**   **51.**

**53.**   $f(x) = \frac{x^2-1}{x}$

**55.** 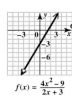  $f(x) = \frac{4x^2-9}{2x+3}$

## 12 Test (pages 631–632)

**[12.1] 1.** $2x^2 + 4x + 5$   **2.** $x^4 + 2x^3 - x^2 + 3x - 5 = (x+1) \cdot (x^3 + x^2 - 2x + 5) + (-10)$   **3.** yes
**4.** $-227$   **5.** Yes, 3 is a zero because the last term in the bottom row of the synthetic division is 0.
**[12.2] 6.** $f(x) = 2x^4 - 2x^3 - 2x^2 - 2x - 4$
**7. (a)** $\pm 1, \pm\frac{1}{2}, \pm\frac{1}{3}, \pm\frac{1}{6}, \pm 7, \pm\frac{7}{2}, \pm\frac{7}{3}, \pm\frac{7}{6}$   **(b)** $-\frac{1}{3}, 1, \frac{7}{2}$
**[12.3] 9. (a)** $f(-2) = -11 < 0$ and $f(-1) = 2 > 0$
**(b)** $-1.290$   **10. (a)** 3   **(b)** 2   **11.**  $f(x) = (x-1)^4$

**12.**  $f(x) = x(x+1)(x-2)$   **13.**  $f(x) = 2x^3 - 7x^2 + 2x + 3$

**14.**  $f(x) = x^4 - 5x^2 + 6$   **15.** 49°F   **[12.4] 16.**  $f(x) = \frac{-2}{x+3}$

**17.**  $f(x) = \frac{3x-1}{x-2}$   **18.**  $f(x) = \frac{x^2-1}{x^2-9}$

**19.** $y = 2x + 3$   **20.** D

## 13 Conic Sections and Nonlinear Systems

### 13.1 Exercises (pages 638–640)

**1. (a)** $(0, 0)$   **(b)** 5   **(c)**  $x^2+y^2=25$   **3.** B   **5.** D
**7.** $(x+4)^2 + (y-3)^2 = 4$   **9.** $(x+8)^2 + (y+5)^2 = 5$
**11.** center: $(-2, -3)$; $r = 2$   **13.** center: $(-5, 7)$; $r = 9$

**15.** center: $(2, 4)$; $r = 4$   **17.** 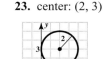 $x^2+y^2=9$

**19.**  $2y^2 = 10 - 2x^2$   **21.** center: $(-3, 2)$   **23.** center: $(2, 3)$

$(x+3)^2 + (y-2)^2 = 9$   $x^2+y^2-4x-6y+9=0$

**25.** center: $(-3, 3)$    **27.** The thumbtack acts as the center, and the length of the string acts as the radius.

$x^2+y^2+6x-6y+9=0$

**29.**    **31.**    **33.**

$\frac{x^2}{9}+\frac{y^2}{25}=1$   $\frac{x^2}{36}=1-\frac{y^2}{16}$   $\frac{y^2}{25}=1-\frac{x^2}{49}$

**35.**   **37.**   **39.**

$\frac{x^2}{16}+\frac{y^2}{4}=1$   $\frac{(x+1)^2}{64}+\frac{(y-2)^2}{49}=1$   $\frac{(x-2)^2}{16}+\frac{(y-1)^2}{9}=1$

**41.** By the vertical line test the set is not a function, because a vertical line may intersect the graph of an ellipse in two points.
**43. (a)** 10 m   **(b)** 36 m   **45. (a)** 154.7 million mi
**(b)** 128.7 million mi (Answers are rounded.)

### 13.2 Exercises (pages 646–647)

**1.** C   **3.** D   **5.** When written in one of the forms given in the box titled "Equations of Hyperbolas" in this section, it will open up and down if the $-$ sign precedes the $x^2$-term; it will open left and right if the $-$ sign precedes the $y^2$-term.

**7.**   **9.**   **11.**

$\frac{x^2}{16}-\frac{y^2}{9}=1$   $\frac{y^2}{4}-\frac{x^2}{25}=1$   $\frac{x^2}{25}-\frac{y^2}{36}=1$

**13.**    **15.** hyperbola

$\frac{y^2}{16}-\frac{x^2}{16}=1$   $x^2-y^2=16$

**17.** ellipse   **19.** parabola

**21.** hyperbola   **23.** hyperbola

**25.** domain: $[-4, 4]$; range: $[0, 4]$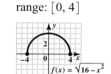

**27.** domain: $[-6, 6]$; range: $[-6, 0]$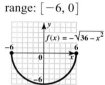

**29.** domain: $(-\infty, \infty)$; range: $[3, \infty)$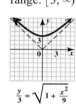

**31.** $\frac{(x-2)^2}{4} - \frac{(y+1)^2}{9} = 1$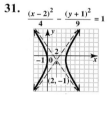

**33.** $\frac{y^2}{36} - \frac{(x-2)^2}{49} = 1$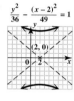

**35.** (a) 50 m  (b) 69.3 m

## 13.3 Exercises (pages 652–653)

**1.** Substitute $x - 1$ for $y$ in the first equation. Then solve for $x$. Find the corresponding $y$-values by substituting back into $y = x - 1$. In the first equation, both variables are squared and in the second, both variables are to the first power, so the elimination method is not appropriate.   **3.** one   **5.** none   **7.**

**9.**    **11.**    **13.**

**15.** $\{(0, 0), (\frac{1}{2}, \frac{1}{2})\}$   **17.** $\{(-6, 9), (-1, 4)\}$
**19.** $\{(-\frac{1}{5}, \frac{7}{5}), (1, -1)\}$   **21.** $\{(-2, -2), (-\frac{4}{3}, -3)\}$
**23.** $\{(-3, 1), (1, -3)\}$   **25.** $\{(-\frac{3}{2}, -\frac{9}{4}), (-2, 0)\}$
**27.** $\{(-\sqrt{3}, 0), (\sqrt{3}, 0), (-\sqrt{5}, 2), (\sqrt{5}, 2)\}$
**29.** $\{(\frac{\sqrt{3}}{3}i, -\frac{1}{2} + \frac{\sqrt{3}}{6}i), (-\frac{\sqrt{3}}{3}i, -\frac{1}{2} - \frac{\sqrt{3}}{6}i)\}$
**31.** $\{(-2, 0), (2, 0)\}$   **33.** $\{(\sqrt{3}, 0), (-\sqrt{3}, 0)\}$
**35.** $\{(-2\sqrt{3}, -2), (-2\sqrt{3}, 2), (2\sqrt{3}, -2), (2\sqrt{3}, 2)\}$

**37.** $\{(-2i\sqrt{2}, -2\sqrt{3}), (-2i\sqrt{2}, 2\sqrt{3}), (2i\sqrt{2}, -2\sqrt{3}), (2i\sqrt{2}, 2\sqrt{3})\}$   **39.** $\{(-\sqrt{5}, -\sqrt{5}), (\sqrt{5}, \sqrt{5})\}$
**41.** $\{(i, 2i), (-i, -2i), (2, -1), (-2, 1)\}$
**43.** length: 12 ft; width: 7 ft

## 13 Review Exercises (page 656)

**1.** circle   **3.** ellipse   **5.** $(x + 2)^2 + (y - 4)^2 = 9$
**7.** $(x - 4)^2 + (y - 2)^2 = 36$   **9.** center: $(4, 1)$; $r = 2$
**11.** center: $(3, -2)$; $r = 5$   **13.**

**15.**    **17.**    **19.** parabola

**21.** ellipse   **23.** $\{(6, -9), (-2, -5)\}$
**25.** $\{(4, 2), (-1, -3)\}$   **27.** $\{(-\sqrt{2}, 2), (-\sqrt{2}, -2), (\sqrt{2}, -2), (\sqrt{2}, 2)\}$   **29.** 0, 1, or 2

## 13 Test (pages 656–657)

[13.1] **1.** D   **2.** $(0, 0)$; 1   **3.** center: $(2, -3)$; radius: 4

**4.** center: $(-4, 1)$; radius: 5   [13.2] **5.**

[13.1] **6.**    [13.2] **7.**

**8.**    **9.** ellipse   **10.** hyperbola

**11.** parabola   [13.3] **12.** $\{(-\frac{1}{2}, -10), (5, 1)\}$
**13.** $\{(-2, -2), (\frac{14}{5}, -\frac{2}{5})\}$   **14.** $\{(-\sqrt{22}, -\sqrt{3}), (-\sqrt{22}, \sqrt{3}), (\sqrt{22}, -\sqrt{3}), (\sqrt{22}, \sqrt{3})\}$

# 14 Trigonometric Functions

## 14.1 Exercises (pages 665–667)

**1.** (a) 60°  (b) 150°   **3.** (a) 45°  (b) 135°   **5.** (a) 36°
(b) 126°   **7.** (a) 89°  (b) 179°   **9.** (a) 75° 40′
(b) 165° 40′   **11.** (a) 69° 49′ 30″  (b) 159° 49′ 30″
**13.** 70°; 110°   **15.** 30°; 60°   **17.** 40°; 140°
**19.** 107°; 73°   **21.** 69°; 21°   **23.** 45°   **25.** 150°
**27.** 7° 30′   **29.** $(90 - x)°$   **31.** $(x - 360)°$
**33.** 83° 59′   **35.** 23° 49′   **37.** 38° 32′   **39.** 60° 34′
**41.** 30° 27′   **43.** 17° 1′ 49″   **45.** 35.5°   **47.** 112.25°
**49.** −60.2°   **51.** 20.9°   **53.** 91.598°   **55.** 274.316°
**57.** 39° 15′ 00″   **59.** 126° 45′ 36″   **61.** −18° 30′ 54″
**63.** 31° 25′ 47″   **65.** 89° 54′ 1″   **67.** 178° 35′ 58″
**69.** 392°   **71.** 386° 30′   **73.** 320°   **75.** 235°   **77.** 1°
**79.** 359°   **81.** 179°   **83.** 130°   **85.** 240°   **87.** 120°

*In Exercises 89 and 91, answers may vary.*
**89.** 450°, 810°; −270°, −630°   **91.** 360°, 720°; −360°, −720°
**93.** $30° + n \cdot 360°$   **95.** $135° + n \cdot 360°$
**97.** $-90° + n \cdot 360°$   **99.** $0° + n \cdot 360°$, or $n \cdot 360°$
**101.** Angles with measures 0° and 360° are coterminal, so they will have all of the same coterminal angles.

*Angles other than those given are possible in Exercises 103–113.*

**103.**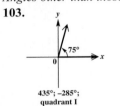
435°; −285°; quadrant I

**105.**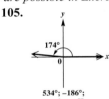
534°; −186°; quadrant II

**107.**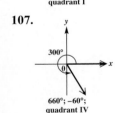
660°; −60°; quadrant IV

**109.**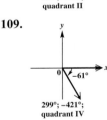
299°; −421°; quadrant IV

**111.**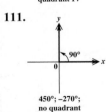
450°; −270°; no quadrant

**113.**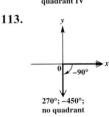
270°; −450°; no quadrant

**115.** $3\sqrt{2}$   **117.** $\sqrt{34}$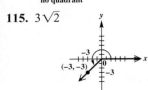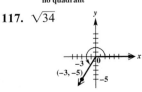

**119.** 2   **121.** 2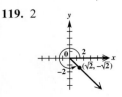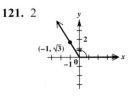

**123.** 4 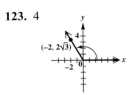   **125.** 4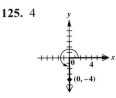

**127.** $\frac{3}{4}$   **129.** 1800°   **131.** 12.5 rotations per hr

## 14.2 Exercises (pages 672–676)

**1.** Answers are given in numerical order: 55°; 65°; 60°; 65°; 60°; 120°; 60°; 60°; 55°; 55°   **3.** 51°; 51°   **5.** 50°; 60°; 70°
**7.** 60°; 60°; 60°   **9.** 45°; 75°; 120°   **11.** 49°; 49°
**13.** 48°; 132°   **15.** 91°   **17.** 2° 29′   **19.** 25.4°
**21.** 22° 29′ 34″   **23.** No, a triangle cannot have angles of measures 85° and 100°. The sum of the measures of these two angles is 185°, which is greater than 180°, the sum of the measures of all three angles in any triangle.   **25.** right; scalene
**27.** acute; equilateral   **29.** right; scalene   **31.** right; isosceles
**33.** obtuse; scalene   **35.** acute; isosceles   **37.** An isosceles right triangle is a triangle with one right angle and two sides of equal length.   **39.** $A$ and $P$; $B$ and $Q$; $C$ and $R$; $AC$ and $PR$; $BC$ and $QR$; $AB$ and $PQ$   **41.** $A$ and $C$; $E$ and $D$; $ABE$ and $CBD$; $EB$ and $DB$; $AB$ and $CB$; $AE$ and $CD$   **43.** $Q = 42°$; $B = R = 48°$
**45.** $B = 106°$; $A = M = 44°$   **47.** $X = M = 52°$
**49.** $a = 20$; $b = 15$   **51.** $a = 6$; $b = 7\frac{1}{2}$   **53.** $x = 6$
**55.** 30 m   **57.** 500 m; 700 m   **59.** 112.5 ft
**61.** The unknown side in the first quadrilateral is 40 cm. The unknown sides in the second quadrilateral are 27 cm and 36 cm.
**63.** $x = 110$   **65.** $c \approx 111.1$   **67.** $y = 5$

## 14.3 Exercises (pages 681–682)

*In Exercises 1–19 and 45–53, we give, in order, sine, cosine, tangent, cotangent, secant, and cosecant.*

**1.** 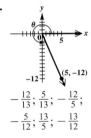   **3.**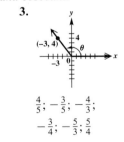

$-\frac{12}{13}$; $\frac{5}{13}$; $-\frac{12}{5}$;
$-\frac{5}{12}$; $\frac{13}{5}$; $-\frac{13}{12}$

$\frac{4}{5}$; $-\frac{3}{5}$; $-\frac{4}{3}$;
$-\frac{3}{4}$; $-\frac{5}{3}$; $\frac{5}{4}$

**5.** 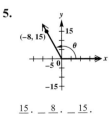   **7.**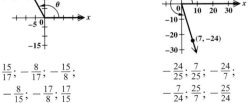

$\frac{15}{17}$; $-\frac{8}{17}$; $-\frac{15}{8}$;
$-\frac{8}{15}$; $-\frac{17}{8}$; $\frac{17}{15}$

$-\frac{24}{25}$; $\frac{7}{25}$; $-\frac{24}{7}$;
$-\frac{7}{24}$; $\frac{25}{7}$; $-\frac{25}{24}$

**9.**
1; 0; undefined;
0; undefined; 1

**11.**
0; −1; 0; undefined;
−1; undefined

**13.**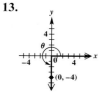
−1; 0; undefined;
0; undefined; −1

**15.**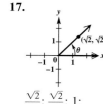
$\frac{\sqrt{3}}{2}; \frac{1}{2}; \sqrt{3};$
$\frac{\sqrt{3}}{3}; 2; \frac{2\sqrt{3}}{3}$

**17.**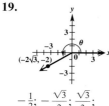
$\frac{\sqrt{2}}{2}; \frac{\sqrt{2}}{2}; 1;$
$1; \sqrt{2}; \sqrt{2}$

**19.** 
$-\frac{1}{2}; -\frac{\sqrt{3}}{2}; \frac{\sqrt{3}}{3};$
$\sqrt{3}; -\frac{2\sqrt{3}}{3}; -2$

**21.** For any nonquadrantal angle $\theta$, $\sin\theta$ and $\csc\theta$ are reciprocals, and reciprocals always have the same sign because their product is 1.   **23.** 0   **25.** negative   **27.** negative
**29.** positive   **31.** positive   **33.** negative   **35.** positive
**37.** negative   **39.** positive   **41.** positive   **43.** positive

**45.**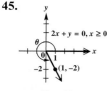
$-\frac{2\sqrt{5}}{5}; \frac{\sqrt{5}}{5}; -2;$
$-\frac{1}{2}; \sqrt{5}; -\frac{\sqrt{5}}{2}$

**47.**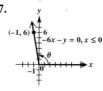
$\frac{6\sqrt{37}}{37}; -\frac{\sqrt{37}}{37}; -6;$
$-\frac{1}{6}; -\sqrt{37}; \frac{\sqrt{37}}{6}$

**49.**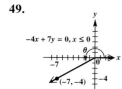
$-\frac{4\sqrt{65}}{65}; -\frac{7\sqrt{65}}{65}; \frac{4}{7};$
$\frac{7}{4}; -\frac{\sqrt{65}}{7}; -\frac{\sqrt{65}}{4}$

**51.**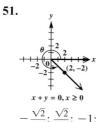
$-\frac{\sqrt{2}}{2}; \frac{\sqrt{2}}{2}; -1;$
$-1; \sqrt{2}; -\sqrt{2}$

**53.**
$-\frac{\sqrt{3}}{2}; -\frac{1}{2}; \sqrt{3};$
$\frac{\sqrt{3}}{3}; -2; -\frac{2\sqrt{3}}{3}$

**55.** 0   **57.** 0   **59.** −1   **61.** 1   **63.** undefined
**65.** −1   **67.** 0   **69.** undefined   **71.** 1   **73.** −3
**75.** −3   **77.** 5   **79.** 1   **81.** 0   **83.** 0   **85.** 0
**87.** 0   **89.** −1   **91.** 0   **93.** undefined
**95.** They are equal.   **97.** They are negatives of each other.

## 14.4 Exercises (pages 689–691)

**1.** $\frac{3}{2}$   **3.** $-\frac{7}{3}$   **5.** $\frac{1}{5}$   **7.** $-\frac{2}{5}$   **9.** $\frac{\sqrt{2}}{2}$   **11.** −0.4
**13.** 0.70069071   **15.** No, the sine and cosecant of an angle cannot have opposite signs. The sine and cosecant are reciprocal functions, so, for all nonquadrantal angles $\theta$, the values of $\sin\theta$ and $\csc\theta$ are reciprocals and must have the same sign.
**17.** Because $\cot 90° = 0$, $\frac{1}{\cot 90°}$ and consequently $\tan 90°$ are undefined.   **19.** All are positive.   **21.** Tangent and cotangent are positive; all others are negative.   **23.** Sine and cosecant are positive; all others are negative.   **25.** Cosine and secant are positive; all others are negative.   **27.** Sine and cosecant are positive; all others are negative.   **29.** All are positive.   **31.** I, II   **33.** I   **35.** II   **37.** I   **39.** III
**41.** III, IV   **43.** Because the cosine and secant are reciprocal functions, their values are positive in the same quadrants. Likewise, because the sine and cosecant functions are reciprocal functions, their values are positive in the same quadrants.   **45.** impossible
**47.** possible   **49.** possible   **51.** impossible   **53.** possible
**55.** possible   **57.** possible   **59.** impossible
**61.** $-\frac{4}{5}$   **63.** $-\frac{\sqrt{5}}{2}$   **65.** $-\frac{\sqrt{3}}{3}$   **67.** 3.44701905

*In Exercises 69–79, we give, in order, sine, cosine, tangent, cotangent, secant, and cosecant.*

**69.** $\frac{15}{17}; -\frac{8}{17}; -\frac{15}{8}; -\frac{8}{15}; -\frac{17}{8}; \frac{17}{15}$   **71.** $\frac{\sqrt{5}}{7}; \frac{2\sqrt{11}}{7}; \frac{\sqrt{55}}{22}; \frac{2\sqrt{55}}{5}; \frac{7\sqrt{11}}{22}; \frac{7\sqrt{5}}{5}$   **73.** $\frac{8\sqrt{67}}{67}; \frac{\sqrt{201}}{67}; \frac{8\sqrt{3}}{3}; \frac{\sqrt{3}}{8}; \frac{\sqrt{201}}{3}; \frac{\sqrt{67}}{8}$   **75.** $\frac{\sqrt{2}}{6}; -\frac{\sqrt{34}}{6}; -\frac{\sqrt{17}}{17}; -\sqrt{17}; -\frac{3\sqrt{34}}{17}; 3\sqrt{2}$   **77.** $\frac{\sqrt{15}}{4}; -\frac{1}{4}; -\sqrt{15}; -\frac{\sqrt{15}}{15}; -4; \frac{4\sqrt{15}}{15}$   **79.** 0.164215; −0.986425; −0.166475; −6.00691; −1.01376; 6.08958   **83.** This statement is false. For example, $\sin 180° + \cos 180° = 0 + (-1) = -1 \neq 1$.
**85.** negative   **87.** negative   **89.** positive   **91.** positive
**93.** Quadrant II is the only quadrant in which the cosine is negative and the sine is positive.

## 14 Review Exercises (pages 694–697)

**1.** complement: 55°; supplement: 145°   **3.** 186°
**5.** $x = 30; y = 30$   **7.** 9360°   **9.** 119.134°
**11.** 275° 6′ 2″   **13.** 40°; 60°; 80°   **15.** 0.25 km
**17.** $N = 12°; R = 82°; M = 86°$   **19.** $p = 7; q = 7$
**21.** $k = 14$   **23.** 12 ft

*In Exercises 25–31, we give, in order, sine, cosine, tangent, cotangent, secant, and cosecant.*

**25.** $-\frac{\sqrt{3}}{2}; \frac{1}{2}; -\sqrt{3}; -\frac{\sqrt{3}}{3}; 2; -\frac{2\sqrt{3}}{3}$   **27.** $-\frac{4}{5}; \frac{3}{5}; -\frac{4}{3}; -\frac{3}{4}; \frac{5}{3}; -\frac{5}{4}$   **29.** $\frac{15}{17}; -\frac{8}{17}; -\frac{15}{8}; -\frac{8}{15}; -\frac{17}{8}; \frac{17}{15}$   **31.** $-\frac{1}{2}; \frac{\sqrt{3}}{2}; -\frac{\sqrt{3}}{3}; -\sqrt{3}; \frac{2\sqrt{3}}{3}; -2$   **33.** tangent and secant

**35.**    **37.** 0; −1; 0; undefined; −1; undefined

**39. (a)** impossible  **(b)** possible  **(c)** possible

*In Exercises 41–45, we give, in order, sine, cosine, tangent, cotangent, secant, and cosecant.*

**41.** $-\frac{\sqrt{39}}{8}; -\frac{5}{8}; \frac{\sqrt{39}}{5}; \frac{5\sqrt{39}}{39}; -\frac{8}{5}; -\frac{8\sqrt{39}}{39}$  **43.** $\frac{2\sqrt{5}}{5}; -\frac{\sqrt{5}}{5}; -2; -\frac{1}{2}; -\sqrt{5}; \frac{\sqrt{5}}{2}$  **45.** $-\frac{3}{5}; \frac{4}{5}; -\frac{3}{4}; -\frac{4}{3}; \frac{5}{4}; -\frac{5}{3}$  **47.** 40 yd

**49.** approximately 9500 ft

## 14 Test *(pages 698–699)*

[14.1] **1. (a)** 23°  **(b)** 113°   **2.** 145°; 35°   **3.** 20°; 70°
[14.2] **4.** 130°; 130°   **5.** 110°; 110°   **6.** 20°; 30°; 130°
[14.1] **7. (a)** 74.31°  **(b)** 45° 12′ 9″   **8. (a)** 30°  **(b)** 280°
**(c)** 90°   **9.** 2700°   [14.2] **10.** $10\frac{2}{3}$ ft, or 10 ft, 8 in.
**11.** $x = 8$; $y = 6$
[14.3] **12.**  $\sin\theta = -\frac{7\sqrt{53}}{53}$; $\cos\theta = \frac{2\sqrt{53}}{53}$; $\tan\theta = -\frac{7}{2}$; $\cot\theta = -\frac{2}{7}$; $\sec\theta = \frac{\sqrt{53}}{2}$; $\csc\theta = -\frac{\sqrt{53}}{7}$

**13.** $\sin\theta = -1$; $\cos\theta = 0$; $\tan\theta$ is undefined; $\cot\theta = 0$; $\sec\theta$ is undefined; $\csc\theta = -1$

**14.** $\sin\theta = -\frac{3}{5}$; $\cos\theta = -\frac{4}{5}$; $\tan\theta = \frac{3}{4}$; $\cot\theta = \frac{4}{3}$; $\sec\theta = -\frac{5}{4}$; $\csc\theta = -\frac{5}{3}$

**15.** row 1: 1, 0, undefined, 0, undefined, 1;
row 2: 0, 1, 0, undefined, 1, undefined;
row 3: −1, 0, undefined, 0, undefined, −1
**16.** cosecant and cotangent   [14.4] **17. (a)** I  **(b)** III, IV
**(c)** III   **18. (a)** impossible  **(b)** possible  **(c)** possible
**19.** $\sec\theta = -\frac{12}{7}$   **20.** $\cos\theta = -\frac{2\sqrt{10}}{7}$; $\tan\theta = -\frac{3\sqrt{10}}{20}$; $\cot\theta = -\frac{2\sqrt{10}}{3}$; $\sec\theta = -\frac{7\sqrt{10}}{20}$; $\csc\theta = \frac{7}{3}$

# 15 Acute Angles and Right Triangles

## 15.1 Exercises *(pages 707–708)*

*In Exercises 1 and 3, we give, in order, sine, cosine, and tangent.*
**1.** $\frac{21}{29}; \frac{20}{29}; \frac{21}{20}$   **3.** $\frac{n}{p}; \frac{m}{p}; \frac{n}{m}$   **5.** C   **7.** B   **9.** E

*In Exercises 11–15, we give, in order, the unknown side, sine, cosine, tangent, cotangent, secant, and cosecant.*
**11.** $c = 13$; $\frac{12}{13}; \frac{5}{13}; \frac{12}{5}; \frac{5}{12}; \frac{13}{5}; \frac{13}{12}$   **13.** $b = \sqrt{13}$; $\frac{\sqrt{13}}{7}; \frac{6}{7}; \frac{\sqrt{13}}{6}; \frac{6\sqrt{13}}{13}; \frac{7}{6}; \frac{7\sqrt{13}}{13}$   **15.** $b = 4$; $\frac{4}{5}; \frac{3}{5}; \frac{4}{3}; \frac{3}{4}; \frac{5}{3}; \frac{5}{4}$
**17.** $\sin A = \cos(90° - A)$; $\cos A = \sin(90° - A)$; $\tan A = \cot(90° - A)$; $\cot A = \tan(90° - A)$; $\sec A = \csc(90° - A)$; $\csc A = \sec(90° - A)$
**19.** $\csc 51°$   **21.** $\csc(75° - \theta)$   **23.** $\tan(100° - \theta)$
**25.** $\cos 51.3°$   **27.** 40°   **29.** 20°   **31.** 12°   **33.** 35°
**35.** 18°   **37.** true   **39.** false   **41.** true   **43.** true
**45.** $\frac{\sqrt{3}}{3}$   **47.** $\frac{1}{2}$   **49.** $\frac{2\sqrt{3}}{3}$   **51.** $\sqrt{2}$   **53.** $\frac{\sqrt{2}}{2}$   **55.** 1
**57.** $\frac{\sqrt{3}}{2}$   **59.** $\sqrt{3}$   **61.** $y = \frac{\sqrt{3}}{3}x$   **63.** 60°
**65.** $x = \frac{9\sqrt{3}}{2}$; $y = \frac{9}{2}$; $z = \frac{3\sqrt{3}}{2}$; $w = 3\sqrt{3}$   **67.** $p = 15$; $r = 15\sqrt{2}$; $q = 5\sqrt{6}$; $t = 10\sqrt{6}$   **69.** $A = \frac{s^2}{2}$

## 15.2 Exercises *(pages 713–715)*

**1.** C   **3.** A   **5.** D   **7.** In a 30°-60° right triangle, the length of the shorter leg is half the length of the hypotenuse, so if the length of the hypotenuse is 2, the length of the shorter leg (opposite the 30° angle) will be 1, a convenient length. Because the angle in Example 2 is in quadrant III, choosing $r = 2$ gives $y = -1$. Any other positive number could have been used for $r$, but $r = 2$ was chosen to make the computations as easy as possible.
**9.** The trigonometric function values of an angle in standard position are determined only by the position of its terminal side. Because coterminal angles by definition have the same terminal side, they must have all the same values for their trigonometric functions.   **11.** $\frac{\sqrt{2}}{2}; \frac{\sqrt{2}}{2}; \sqrt{2}; \sqrt{2}$   **13.** $-\frac{1}{2}; -\frac{\sqrt{3}}{3}; -2$
**15.** $\frac{1}{2}; -\sqrt{3}; -\frac{2\sqrt{3}}{3}$   **17.** $\sqrt{3}; \frac{\sqrt{3}}{3}$

*In Exercises 19–35, we give, in order, sine, cosine, tangent, cotangent, secant, and cosecant.*
**19.** $-\frac{\sqrt{2}}{2}; \frac{\sqrt{2}}{2}; -1, -1; \sqrt{2}; -\sqrt{2}$   **21.** $\frac{\sqrt{3}}{2}; \frac{1}{2}; \sqrt{3}; \frac{\sqrt{3}}{3}; 2; \frac{2\sqrt{3}}{3}$   **23.** $\frac{\sqrt{3}}{2}; -\frac{1}{2}; -\sqrt{3}; -\frac{\sqrt{3}}{3}; -2; \frac{2\sqrt{3}}{3}$   **25.** $-\frac{1}{2}; -\frac{\sqrt{3}}{2}; \frac{\sqrt{3}}{3}; \sqrt{3}; -\frac{2\sqrt{3}}{3}; -2$   **27.** $-\frac{\sqrt{2}}{2}; -\frac{\sqrt{2}}{2}; 1; 1; -\sqrt{2}; -\sqrt{2}$
**29.** $\frac{1}{2}; -\frac{\sqrt{3}}{2}; -\frac{\sqrt{3}}{3}; -\sqrt{3}; -\frac{2\sqrt{3}}{3}; 2$   **31.** $-\frac{1}{2}; -\frac{\sqrt{3}}{2}; \frac{\sqrt{3}}{3}; \sqrt{3}; -\frac{2\sqrt{3}}{3}; -2$   **33.** $\frac{1}{2}; -\frac{\sqrt{3}}{2}; -\frac{\sqrt{3}}{3}; -\sqrt{3}; -\frac{2\sqrt{3}}{3}; 2$   **35.** $-\frac{\sqrt{3}}{2}; \frac{1}{2}; -\sqrt{3}; -\frac{\sqrt{3}}{3}; 2; -\frac{2\sqrt{3}}{3}$   **37.** $-\frac{\sqrt{3}}{2}$   **39.** $\frac{\sqrt{3}}{2}$   **41.** $-\sqrt{2}$

43. $-1$   45. true   47. false; $\frac{1}{2} \neq \sqrt{3}$   49. true
51. true   53. $(-3\sqrt{3}, 3)$   55. no   57. positive
59. positive   61. negative   63. For any angle $\theta$ and any integer $n$, $\theta$ and $\theta + n \cdot 360°$ are coterminal angles, so they will have all of the same trigonometric function values.
65. 45°; 225°   67. 30°; 150°   69. 120°; 300°
71. 45°; 315°   73. 210°; 330°   75. 30°; 210°
77. 225°; 315°

## 15.3 Exercises (pages 717–719)

1. sin; 1   3. reciprocal; reciprocal

*In Exercises 5–21, the number of decimal places may vary depending on the calculator used.*

5. 0.6252427   7. 1.0273488   9. 15.055723
11. 1.4887142   13. 0.6743024   15. 0.9999905
17. tan 23.4° ≈ 0.4327386   19. cot 77° ≈ 0.2308682
21. tan 4.72° ≈ 0.0825664   23. 55.845496°
25. 16.166641°   27. 38.491580°   29. 68.673241°
31. 45.526434°   33. Subtract enough multiples of 360° from 2000° to get a coterminal angle between 0° and 360° and then find the cosine of that angle. In this case, the coterminal angle to use would be $2000° - 5 \cdot 360° = 200°$.   35. 0.3746065934
37. $-1$   39. 1   41. 0   43. false   45. true
47. false   49. false   51. true   53. true   55. 65.96 lb
57. $-2.87°$   59. (a) $2 \times 10^8$ m per sec
(b) $2 \times 10^8$ m per sec   61. 48.7°

## 15.4 Exercises (pages 723–727)

1. 22,894.5 to 22,895.5   3. 8958.5 to 8959.5
5. No, the conclusion is not correct. The constant 2 in the formula is an exact number, not the result of measurement, so it does not affect the number of significant digits in the answer. The only measured quantity is the radius, which has 4 significant digits, so the circumference should be given with 4 significant digits as 345.4 cm.   7. 0.05

*Note to student: While most of the measures resulting from solving triangles in this chapter are approximations, for convenience we use = rather than ≈.*

9. $B = 53° 40'$; $a = 571$ m; $b = 777$ m   11. $M = 38.8°$; $n = 154$ m; $p = 198$ m   13. $A = 47.9108°$; $c = 84.816$ cm; $a = 62.942$ cm   15. $A = 37° 40'$; $B = 52° 20'$; $c = 20.5$ ft
17. No; the three angle measures of a triangle do not determine a unique triangle. Two triangles with the same angle measures are similar, but not necessarily congruent. Therefore, it is not possible to find the side lengths of a triangle if only the angle measures are known.   19. You can find the lengths of the other two sides by using trigonometric ratios that involve the given side, the unknown side, and the given acute angle. You can find the measure of the other acute angle by subtracting the measure of the given acute angle from 90°.   21. $B = 62.0°$; $a = 8.17$ ft; $b = 15.4$ ft   23. $A = 17.0°$; $a = 39.1$ in.; $c = 134$ in.
25. $B = 27.5°$; $b = 6.61$ m; $c = 14.3$ m   27. $A = 36°$; $B = 54°$; $b = 18$ m   29. $c = 85.9$ yd; $A = 62° 50'$; $B = 27° 10'$   31. $b = 42.3$ cm; $A = 24° 10'$; $B = 65° 50'$
33. $B = 36° 36'$; $a = 310.8$ ft; $b = 230.8$ ft

35. $A = 50° 51'$; $a = 0.4832$ m; $b = 0.3934$ m
37. The angle of elevation from a point $X$ to a point $Y$ above $X$ is the acute angle formed by the horizontal ray with endpoint $X$ and ray $XY$.   39. Angles $DAB$ and $ABC$ are alternate interior angles formed when transversal $AB$ intersects parallel lines $AD$ and $BC$, so these two angles have the same measure.
41. 9.35 m   43. 128 ft   45. 26.92 in.
47. 22°   49. 28.0 m   51. 13.3 ft   53. 146 m

## 15 Review Exercises (pages 730–732)

*In Exercises 1, 13, and 15, we give, in order, sine, cosine, tangent, cotangent, secant, and cosecant.*

1. $\frac{60}{61}$; $\frac{11}{61}$; $\frac{60}{11}$; $\frac{11}{60}$; $\frac{61}{11}$; $\frac{61}{60}$   3. 10°   5. 7°   7. true
9. true   11. Angle $A$ is the complement of angle $B$, so $\cos A = \sin B = \frac{b}{c}$.   13. $-\frac{\sqrt{3}}{2}$; $\frac{1}{2}$; $-\sqrt{3}$; $-\frac{\sqrt{3}}{3}$; 2; $-\frac{2\sqrt{3}}{3}$
15. $-\frac{1}{2}$; $\frac{\sqrt{3}}{2}$; $-\frac{\sqrt{3}}{3}$; $-\sqrt{3}$; $\frac{2\sqrt{3}}{3}$; $-2$   17. 120°; 240°
19. 150°; 210°   21. $3 - \frac{2\sqrt{3}}{3}$   23. (a) $-\frac{\sqrt{2}}{2}$; $-\frac{\sqrt{2}}{2}$; 1
(b) $-\frac{\sqrt{3}}{2}$; $\frac{1}{2}$; $-\sqrt{3}$   25. $-1.3563417$   27. 1.0210339
29. 0.20834446   31. 55.673870°   33. 12.733938°
35. 63.008286°   37. 47.1°; 132.9°   39. false; $1.4088321 \neq 1$   41. true   43. No, this will not produce the correct answer. Entering $\tan^{-1} 25$ will give an angle in degrees whose tangent is 25. On a calculator, $\sin^{-1}$, $\cos^{-1}$, and $\tan^{-1}$ are used for inverse trigonometric functions, not reciprocal functions.
45. III   47. $B = 31° 30'$; $a = 638$; $b = 391$
49. $B = 50.28°$; $a = 32.38$ m; $c = 50.66$ m   51. 137 ft
53. 73.7 ft   55. 18.75 cm

## 15 Test (page 733)

[15.1] 1. $\sin A = \frac{12}{13}$; $\cos A = \frac{5}{13}$; $\tan A = \frac{12}{5}$; $\cot A = \frac{5}{12}$; $\sec A = \frac{13}{5}$; $\csc A = \frac{13}{12}$   2. $x = 4$; $y = 4\sqrt{3}$; $z = 4\sqrt{2}$; $w = 8$   3. 15°   [15.1, 15.2] 4. (a) true   (b) false; For $0° \leq \theta \leq 90°$, cosine is decreasing.   (c) true

*In Exercises 5–7, we give, in order, sine, cosine, tangent, cotangent, secant, and cosecant.*

[15.2] 5. $-\frac{\sqrt{3}}{2}$; $-\frac{1}{2}$; $\sqrt{3}$; $\frac{\sqrt{3}}{3}$; $-2$; $-\frac{2\sqrt{3}}{3}$   6. $-\frac{\sqrt{2}}{2}$; $-\frac{\sqrt{2}}{2}$; 1; 1; $-\sqrt{2}$; $-\sqrt{2}$   7. $-1$; 0; undefined; 0; undefined; $-1$
8. 135°; 225°   9. 240°; 300°   10. 45°; 225°
[15.3] 11. (a) 0.97939940   (b) $-1.9056082$   (c) 1.9362132
12. 16.16664145°   [15.4] 13. $B = 31° 30'$; $c = 877$; $b = 458$
14. 67.1°, or 67° 10'   15. 15.5 ft   16. 8800 ft

# 16 Radian Measure and Circular Functions

## 16.1 Exercises (pages 739–741)

1. 1   3. 3   5. $-3$   7. $\frac{\pi}{3}$   9. $\frac{\pi}{2}$   11. $\frac{5\pi}{6}$
13. $-\frac{5\pi}{3}$   15. $\frac{5\pi}{2}$   17. $10\pi$   19. Answers will vary.
21. Answers will vary.   23. Answers will vary.   25. 60°
27. 315°   29. 330°   31. $-30°$   33. 126°   35. $-48°$
37. 153°   39. $-900°$   41. 0.68   43. 0.742   45. 2.43

**47.** 1.122   **49.** 0.9847   **51.** 0.832391   **53.** 114° 35′
**55.** 99° 42′   **57.** 19° 35′   **59.** −287° 6′   **61.** In the expression "sin 30," 30 means 30 radians; sin 30° = $\frac{1}{2}$, while sin 30 ≈ −0.9880.   **63.** $\frac{\sqrt{3}}{2}$   **65.** 1   **67.** $\frac{2\sqrt{3}}{3}$   **69.** 1
**71.** $-\sqrt{3}$   **73.** $\frac{1}{2}$   **75.** −1   **77.** $-\frac{\sqrt{3}}{2}$   **79.** $\frac{1}{2}$
**81.** $\sqrt{3}$   **83.** We begin the answers with the blank next to 30°, and then proceed counterclockwise from there: $\frac{\pi}{6}$; 45; $\frac{\pi}{3}$; 120; 135; $\frac{5\pi}{6}$; $\pi$; $\frac{7\pi}{6}$, $\frac{5\pi}{4}$; 240; 300; $\frac{7\pi}{4}$; $\frac{11\pi}{6}$.   **85. (a)** $4\pi$   **(b)** $\frac{2\pi}{3}$
**87. (a)** $5\pi$   **(b)** $\frac{8\pi}{3}$

## 16.2 Exercises (pages 745–749)

**1.** $2\pi$   **3.** $20\pi$   **5.** 6   **7.** 1   **9.** 2   **11.** 25.8 cm
**13.** 3.61 ft   **15.** 5.05 m   **17.** 55.3 in.   **19.** The length is doubled.   **21.** 3500 km   **23.** 5900 km   **25.** 44° N
**27.** 156°   **29.** 38.5°   **31.** 18.7 cm   **33. (a)** 11.6 in.
**(b)** 37° 5′   **35.** 146 in.   **37.** $6\pi$   **39.** $72\pi$   **41.** 60°
**43.** 1.5   **45.** 1116.1 m²   **47.** 706.9 ft²   **49.** 114.0 cm²
**51.** 1885.0 mi²   **53.** 3.6   **55. (a)** $13\frac{1}{3}°$; $\frac{2\pi}{27}$   **(b)** 478 ft
**(c)** 17.7 ft   **(d)** approximately 672 ft²   **57. (a)** 140 ft
**(b)** 102 ft   **(c)** 622 ft²   **59.** 1900 yd²   **61.** radius: 3950 mi; circumference: 24,800 mi

## 16.3 Exercises (pages 755–757)

**1. (a)** 1   **(b)** 0   **(c)** undefined   **3. (a)** 0   **(b)** 1   **(c)** 0
**5. (a)** 0   **(b)** −1   **(c)** 0   **7.** $-\frac{1}{2}$   **9.** −1   **11.** −2
**13.** $-\frac{1}{2}$   **15.** $\frac{\sqrt{2}}{2}$   **17.** $\frac{\sqrt{3}}{2}$   **19.** $\frac{2\sqrt{3}}{3}$   **21.** $-\frac{\sqrt{3}}{3}$
**23.** 0.5736   **25.** 0.4068   **27.** 1.2065   **29.** 14.3338
**31.** −1.0460   **33.** −3.8665   **35.** 0.7   **37.** 0.9
**39.** −0.6   **41.** 2.3 or 4.0   **43.** 0.8 or 2.4   **45.** negative
**47.** negative   **49.** positive   **51.** sin θ = $\frac{\sqrt{2}}{2}$; cos θ = $\frac{\sqrt{2}}{2}$; tan θ = 1; cot θ = 1; sec θ = $\sqrt{2}$; csc θ = $\sqrt{2}$
**53.** sin θ = $-\frac{12}{13}$; cos θ = $\frac{5}{13}$; tan θ = $-\frac{12}{5}$; cot θ = $-\frac{5}{12}$; sec θ = $\frac{13}{5}$; csc θ = $-\frac{13}{12}$   **55.** 0.2095   **57.** 1.4426
**59.** 0.3887   **61.** $\frac{5\pi}{6}$   **63.** $\frac{4\pi}{3}$   **65.** $\frac{7\pi}{4}$
**67.** (−0.8011, 0.5985)   **69.** (0.4385, −0.8987)   **71.** I
**73.** II   **75.** 0.9846   **77. (a)** 32.4°   **(b)** Answers will vary.
**79. (a)** 30°   **(b)** 60°   **(c)** 75°   **(d)** 86°   **(e)** 86°   **(f)** 60°

## 16.4 Exercises (pages 760–762)

**1.** $2\pi$ sec   **3. (a)** $\frac{\pi}{2}$ radians   **(b)** $10\pi$ cm   **(c)** $\frac{5\pi}{3}$ cm per sec
**5.** $2\pi$ radians   **7.** $\frac{3\pi}{32}$ radian per sec   **9.** $\frac{6}{5}$ min
**11.** 0.180311 radian per sec   **13.** $8\pi$ m per sec
**15.** $\frac{9}{5}$ radians per sec   **17.** 1.83333 radians per sec
**19.** $18\pi$ cm   **21.** 12 sec   **23.** $\frac{3\pi}{32}$ radian per sec
**25.** $\frac{\pi}{6}$ radian per hr   **27.** $\frac{\pi}{30}$ radian per min   **29.** $\frac{7\pi}{30}$ cm per min   **31.** $168\pi$ m per min   **33.** $1500\pi$ m per min

**35.** 16.6 mph   **37. (a)** $\frac{2\pi}{365}$ radian   **(b)** $\frac{\pi}{4380}$ radian per hr
**(c)** about 67,000 mph   **39. (a)** 3.1 cm per sec
**(b)** 0.24 radian per sec   **41.** 3.73 cm
**43.** 523.6 radians per sec

## 16 Review Exercises (pages 765–767)

**1.** An angle of 1 radian is larger.   **3.** Three of many possible answers are 1 + 2π, 1 + 4π, and 1 + 6π.   **5.** $\frac{\pi}{4}$   **7.** $\frac{35\pi}{36}$
**9.** $\frac{40\pi}{9}$   **11.** 225°   **13.** 480°   **15.** −110°   **17.** π in.
**19.** $12\pi$ in.   **21.** 35.8 cm   **23.** 7.683 cm   **25.** 273 m²
**27.** 4500 km   **29.** $\frac{3}{4}$; 1.5 sq units   **31. (a)** $\frac{\pi}{3}$ radians
**(b)** $2\pi$ in.   **33.** $\sqrt{3}$   **35.** $-\frac{1}{2}$   **37.** 2   **39.** tan 1
**41.** sin 2   **43.** 0.8660   **45.** 0.9703   **47.** 1.9513
**49.** 0.3898   **51.** 0.5148   **53.** 1.1054   **55.** $\frac{\pi}{4}$   **57.** $\frac{7\pi}{6}$
**59.** $\frac{15}{32}$ sec   **61.** $\frac{\pi}{20}$ radian per sec   **63.** 285.3 cm
**65.** $\frac{\pi}{36}$ radian per sec

## 16 Test (pages 767–768)

[16.1] **1.** $\frac{2\pi}{3}$   **2.** $-\frac{\pi}{4}$   **3.** 0.09   **4.** 135°   **5.** −210°
**6.** 229.18°   [16.2] **7. (a)** $\frac{4}{3}$   **(b)** 15,000 cm²   **8.** 2 radians
[16.3] **9.** $\frac{\sqrt{2}}{2}$   **10.** $-\frac{\sqrt{3}}{2}$   **11.** undefined   **12.** −2
**13.** 0   **14.** 0   **15.** sin $\frac{7\pi}{6}$ = $-\frac{1}{2}$; cos $\frac{7\pi}{6}$ = $-\frac{\sqrt{3}}{2}$; tan $\frac{7\pi}{6}$ = $\frac{\sqrt{3}}{3}$   **16.** sine and cosine: (−∞, ∞); tangent and secant: $\{s \mid s \neq (2n+1)\frac{\pi}{2}$, where $n$ is any integer$\}$; cotangent and cosecant: $\{s \mid s \neq n\pi$, where $n$ is any integer$\}$   **17. (a)** 0.9716
**(b)** $\frac{\pi}{3}$   [16.4] **18. (a)** $\frac{2\pi}{3}$ radians   **(b)** $40\pi$ cm   **(c)** $5\pi$ cm per sec   **19.** approximately 8.278 mi per sec   **20. (a)** 75 ft
**(b)** $\frac{\pi}{45}$ radian per sec

# 17 Graphs of the Circular Functions

## 17.1 Exercises (pages 778–781)

**1.** G   **3.** E   **5.** B   **7.** F   **9.** D   **11.** C
**13.** 2   **15.** $\frac{2}{3}$   **17.** 1

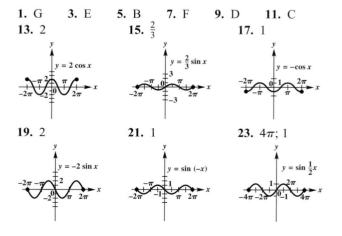

**19.** 2   **21.** 1   **23.** $4\pi$; 1

**25.** $\frac{8\pi}{3}$; 1

**27.** $\frac{2\pi}{3}$; 1

**29.** $8\pi$; 2

**31.** $\frac{2\pi}{3}$; 2

**33.** 2; 1

**35.** 1; 2

**37.** 4; $\frac{1}{2}$

**39.** 2; $\pi$

**41.** $y = 2\cos 2x$

**43.** $y = -3\cos\frac{1}{2}x$  **45.** $y = 3\sin 4x$  **47.** (a) 80°; 50°
(b) 15  (c) about 35,000 yr  (d) downward

**49.** (a) $L(x) = 0.022x^2 + 0.55x + 316 + 3.5\sin 2\pi x$

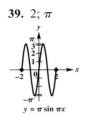

(b) maximums: $x = \frac{1}{4}, \frac{5}{4}, \frac{9}{4}, \ldots$; minimums: $x = \frac{3}{4}, \frac{7}{4}, \frac{11}{4}, \ldots$
(c) Answers will vary.

**51.** (a) 8°  (b) 21°  (c) 62°  (d) 61°  (e) 31°  (f) −11°

## 17.2 Exercises (pages 786–788)

**1.** D  **3.** H  **5.** B  **7.** F  **9.** C  **11.** A
**13.** The graph of $y = \sin x + 1$ is a vertical translation of the graph of $y = \sin x$ 1 unit up, while the graph of $y = \sin(x + 1)$ is a horizontal translation of the graph of $y = \sin x$ 1 unit to the left.
**15.** B  **17.** C  **19.** right  **21.** $y = -1 + \sin x$
**23.** $y = \cos\left(x - \frac{\pi}{3}\right)$  **25.** 2; $2\pi$; none; $\pi$ to the right
**27.** 4; $4\pi$; none; $\pi$ to the left  **29.** 3; 4; none; $\frac{1}{2}$ to the right
**31.** 1; $\frac{2\pi}{3}$; up 2; $\frac{\pi}{15}$ to the right  **33.**

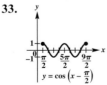

**35.**

**37.**

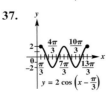

**39.**

**41.**

**43.**

**45.**

**47.**

**49.**

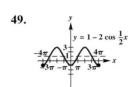

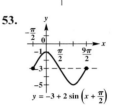

**51.**

**53.**

**55.**

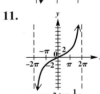

## 17.3 Exercises (pages 794–795)

**1.** C  **3.** B  **5.** F  **7.**

**9.**

**11.**

**13.**

**15.**

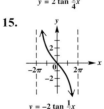

**17.**

**19.**

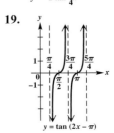

**21.** 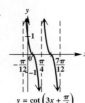  **23.**

**25.**   **27.**

**29.**   **31.**

**33.** $y = -2 \tan x$  **35.** $y = \cot 3x$  **37.** true
**39.** false; $\tan(-x) = -\tan x$ for all $x$ in the domain.
**41.** four  **45. (a)** 0 m  **(b)** $-2.9$ m  **(c)** $-12.3$ m
**(d)** 12.3 m  **(e)** It leads to $\tan \frac{\pi}{2}$, which is undefined.

### 17.4 Exercises (pages 801–802)

**1.** B  **3.** D  **5.**   **7.**

**9.**   **11.**

**13.**   **15.**

**17.** 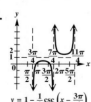  **19.** $y = -3 \cot x$  **21.** $y = \sec 4x$
**23.** $y = -2 + \csc x$  **25.** true
**27.** true  **29.** none  **33. (a)** 4 m
**(b)** 6.3 m  **(c)** 63.7 m

### 17 Review Exercises (pages 805–806)

**1.** B  **3.** sine, cosine, tangent, cotangent  **5.** 2; $2\pi$; none;
none  **7.** $\frac{1}{2}$; $\frac{2\pi}{3}$; none; none  **9.** 2; $8\pi$; 1 up; none

**11.** 3; $2\pi$; none; $\frac{\pi}{2}$ to the left  **13.** not applicable; $\pi$; none; $\frac{\pi}{8}$
to the right  **15.** not applicable; $\frac{\pi}{3}$; none; $\frac{\pi}{9}$ to the right
**17.** tangent  **19.** cosine  **21.** cotangent  **23.** Every sine
function is a periodic function, so the definition of a periodic
function applies to this function $f$: $f(x) = f(x + np)$. Substituting
5 for $x$, 2 for $n$, and 10 for $p$ in this equation gives $f(25) = f(5 + 2 \cdot 10) = f(5) = 2$.  **25.**

**27.**   **29.**

**31.**   **33.**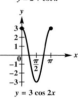

**35.**   **37.**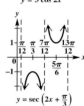

**39.**   **41.**

**43.** The graph of $f(x) = 2 \sin(bx + c)$ has amplitude 2 and
there is no vertical translation, so the minimum $y$-value is $-2$
and the maximum $y$-value is 2, that is, $-2 \leq y \leq 2$. This is
equivalent to saying that the range of the function is $[-2, 2]$.
**45. (b)**   **47. (a)** 30°  **(b)** 60°  **(c)** 75°
**(d)** 86°  **(e)** 86°  **(f)** 60°

### 17 Test (pages 807–808)

[17.1–17.4] **1. (a)** $y = \sec x$  **(b)** $y = \sin x$  **(c)** $y = \cos x$
**(d)** $y = \tan x$  **(e)** $y = \csc x$  **(f)** $y = \cot x$
[17.1] **2. (a)** $y = \sin 2x$  **(b)** $y = 2 \sin x$
[17.1, 17.3, 17.4] **3. (a)** $(-\infty, \infty)$  **(b)** $[-1, 1]$  **(c)** $\frac{\pi}{2}$
**(d)** $(-\infty, -1] \cup [1, \infty)$  [17.2] **4. (a)** $\pi$  **(b)** 6  **(c)** $[-3, 9]$

**(d)** $-3$  **(e)** $\frac{\pi}{4}$ to the left (that is, $-\frac{\pi}{4}$)   **5.**

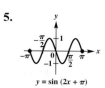
$y = \sin(2x + \pi)$

[17.1] **6.**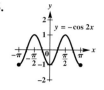
$y = -\cos 2x$

[17.2] **7.**
$y = 2 + \cos x$

**8.**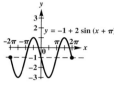
$y = -1 + 2\sin(x + \pi)$

[17.3] **9.**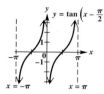
$y = \tan\left(x - \frac{\pi}{2}\right)$

**10.**
$y = -2 - \cot\left(x - \frac{\pi}{2}\right)$

[17.4] **11.**
$y = -\csc 2x$

**12.**
$y = 3\csc \pi x$

[17.1, 17.2] **13. (a)** $f(x) = 17.5 \sin\left[\frac{\pi}{6}(x-4)\right] + 67.5$

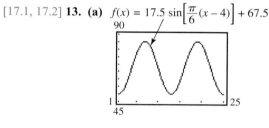

**(b)** 17.5; 12; 4 to the right; 67.5 up  **(c)** approximately 52°F
**(d)** 50°F in January; 85°F in July  **(e)** approximately 67.5°; This is the vertical translation.  [17.3, 17.4] **14.** The tangent and secant functions are both defined by ratios that have $x$ in the denominator, so these functions are both undefined whenever $x = 0$, where $(x, y)$ is a point on the unit circle. Therefore, these two functions have the same domain, $\{x \mid x \neq (2n + 1)\frac{\pi}{2}, \text{ where } n \text{ is any integer}\}$. The cotangent and cosecant functions are both defined by ratios that have $y$ in the denominator, so the cotangent and cosecant functions are both undefined whenever $y = 0$, where $(x, y)$ is a point on the unit circle. Therefore, these two functions have the same domain, $\{x \mid x \neq n\pi, \text{ where } n \text{ is any integer}\}$.

## 18 Trigonometric Identities

### 18.1 Exercises (pages 813–814)

**1.** $-2.6$  **3.** $0.625$  **5.** $\frac{2}{3}$  **7.** $\frac{\sqrt{7}}{4}$  **9.** $-\frac{2\sqrt{5}}{5}$
**11.** $-\frac{\sqrt{105}}{11}$  **13.** The sine and cosecant are reciprocal functions, so they have the same sign for any angle $\theta$. In Exercise 12, $\sin \theta = -\frac{5}{8}$ whether $\theta$ is in quadrant III or quadrant IV, so the quadrant does not need to be given.  **15.** $\cos \theta = -\frac{\sqrt{5}}{3}$; $\tan \theta = -\frac{2\sqrt{5}}{5}$; $\cot \theta = -\frac{\sqrt{5}}{2}$; $\sec \theta = -\frac{3\sqrt{5}}{5}$; $\csc \theta = \frac{3}{2}$
**17.** $\sin \theta = -\frac{\sqrt{17}}{17}$; $\cos \theta = \frac{4\sqrt{17}}{17}$; $\cot \theta = -4$; $\sec \theta = \frac{\sqrt{17}}{4}$; $\csc \theta = -\sqrt{17}$   **19.** $\sin \theta = \frac{3}{5}$; $\cos \theta = \frac{4}{5}$; $\tan \theta = \frac{3}{4}$; $\sec \theta = \frac{5}{4}$; $\csc \theta = \frac{5}{3}$   **21.** $\sin \theta = -\frac{\sqrt{7}}{4}$; $\cos \theta = \frac{3}{4}$; $\tan \theta = -\frac{\sqrt{7}}{3}$; $\cot \theta = -\frac{3\sqrt{7}}{7}$; $\csc \theta = -\frac{4\sqrt{7}}{7}$   **23.** B   **25.** E
**27.** A   **29.** A   **31.** D   **33.** $\sin x = \pm\sqrt{1 - \cos^2 x}$
**35.** $\tan x = \pm\sqrt{\sec^2 x - 1}$   **37.** $\csc x = \frac{\pm\sqrt{1 - \cos^2 x}}{1 - \cos^2 x}$
**39.** $\cos \theta$   **41.** $\cot \theta$   **43.** $\cos^2 \theta$   **45.** $\sec \theta - \cos \theta$
**47.** $\cot \theta - \tan \theta$   **49.** $\tan \theta \sin \theta$   **51.** $\cos^2 \theta$
**53.** $\sec^2 \theta$

### 18.2 Exercises (pages 819–821)

**1.** $\csc \theta \sec \theta$, or $\frac{1}{\sin \theta \cos \theta}$   **3.** $1 + \sec s$   **5.** 1   **7.** 1
**9.** $2 + 2\sin t$   **11.** $-\frac{2\cos x}{\sin^2 x}$, or $-2\cot x \csc x$
**13.** $(\sin \theta + 1)(\sin \theta - 1)$   **15.** $4\sin x$
**17.** $(2\sin x + 1)(\sin x + 1)$   **19.** $(\cos^2 x + 1)^2$
**21.** $(\sin x - \cos x)(1 + \sin x \cos x)$   **23.** $\sin \theta$   **25.** 1
**27.** $\tan^2 \beta$   **29.** $\tan^2 x$   **31.** $\sec^2 x$   **33.** $\cos^2 x$
**79.** $(\sec \theta + \tan \theta)(1 - \sin \theta) = \cos \theta$
**81.** $\frac{\cos \theta + 1}{\sin \theta + \tan \theta} = \cot \theta$   **87.** It is true when $\sin x \geq 0$.
**89. (a)** $I = k(1 - \sin^2 \theta)$  **(b)** For $\theta = 2\pi n$ and all integers $n$, $\cos^2 \theta = 1$, its maximum value, and $I$ attains a maximum value of $k$.

### 18.3 Exercises (pages 825–827)

**1.** F   **3.** E   **5.** $\frac{\sqrt{6} - \sqrt{2}}{4}$   **7.** $\frac{\sqrt{2} - \sqrt{6}}{4}$   **9.** $\frac{\sqrt{2} - \sqrt{6}}{4}$
**11.** 0   **13.** The calculator gives a value of 0 for the expression.
**15.** $\cot 3°$   **17.** $\sin \frac{5\pi}{12}$   **19.** $\sec 104° 24'$   **21.** $\cos\left(-\frac{\pi}{8}\right)$
**23.** $\csc(-56° 42')$   **25.** $\tan(-86.9814°)$   **27.** $\tan$
**29.** $\cos$   **31.** $\csc$

*For Exercises 33–37, other answers are possible.*
**33.** $15°$   **35.** $\frac{140°}{3}$   **37.** $20°$   **39.** $\cos \theta$   **41.** $-\cos \theta$
**43.** $\cos \theta$   **45.** $-\cos \theta$   **47.** $\frac{4 - 6\sqrt{6}}{25}$; $\frac{4 + 6\sqrt{6}}{25}$   **49.** $\frac{16}{65}$; $-\frac{56}{65}$   **51.** $\frac{2\sqrt{638} - \sqrt{30}}{56}$; $\frac{2\sqrt{638} + \sqrt{30}}{56}$   **53.** true   **55.** false
**57.** true   **59.** true   **61.** false

## 18.4 Exercises (pages 831–833)

**1.** C  **3.** E  **5.** B  **7.** $\frac{\sqrt{6}+\sqrt{2}}{4}$  **9.** $2-\sqrt{3}$
**11.** $\frac{-\sqrt{6}-\sqrt{2}}{4}$  **13.** $\frac{\sqrt{2}}{2}$  **15.** $-1$  **17.** 0  **19.** 1
**21.** $\frac{\sqrt{3}\cos\theta - \sin\theta}{2}$  **23.** $\frac{\cos\theta - \sqrt{3}\sin\theta}{2}$  **25.** $\frac{\sqrt{2}(\sin x - \cos x)}{2}$
**27.** $\frac{\sqrt{3}\tan\theta + 1}{\sqrt{3} - \tan\theta}$  **29.** $\frac{\sqrt{2}(\cos x + \sin x)}{2}$  **31.** $-\cos\theta$
**33.** $-\tan\theta$  **35.** $-\tan\theta$  **37.** (a) $\frac{63}{65}$  (b) $\frac{63}{16}$  (c) I
**39.** (a) $\frac{4\sqrt{2}+\sqrt{5}}{9}$  (b) $\frac{-8\sqrt{5}-5\sqrt{2}}{20-2\sqrt{10}}$ (Other forms are possible.)  (c) II  **41.** (a) $\frac{77}{85}$  (b) $-\frac{77}{36}$  (c) II
**43.** $\sin\left(\frac{\pi}{2}+\theta\right) = \cos\theta$  **45.** $\tan\left(\frac{\pi}{2}+\theta\right) = -\cot\theta$
**57.** (a) 425 lb  (c) 0°

## 18.5 Exercises (pages 837–839)

**1.** C  **3.** B  **5.** C  **7.** $\cos\theta = \frac{2\sqrt{5}}{5}$; $\sin\theta = \frac{\sqrt{5}}{5}$
**9.** $\cos\theta = -\frac{\sqrt{42}}{12}$; $\sin\theta = \frac{\sqrt{102}}{12}$  **11.** $\cos 2\theta = \frac{17}{25}$; $\sin 2\theta = -\frac{4\sqrt{21}}{25}$  **13.** $\cos 2x = -\frac{3}{5}$; $\sin 2x = \frac{4}{5}$  **15.** $\cos 2\theta = \frac{39}{49}$; $\sin 2\theta = -\frac{4\sqrt{55}}{49}$  **17.** $\frac{\sqrt{3}}{2}$  **19.** $\frac{\sqrt{3}}{2}$  **21.** $-\frac{\sqrt{2}}{2}$
**23.** $\frac{1}{2}\tan 102°$  **25.** $\frac{1}{4}\cos 94.2°$  **27.** $-\cos\frac{4\pi}{5}$
**29.** $\sin 4x = 4\sin x \cos^3 x - 4\sin^3 x \cos x$
**31.** $\tan 3x = \frac{3\tan x - \tan^3 x}{1 - 3\tan^2 x}$  **33.** $\cos^4 x - \sin^4 x = \cos 2x$
**35.** $\frac{2\tan x}{2 - \sec^2 x} = \tan 2x$  **55.** $a = -885.6$; $c = 885.6$; $\omega = 240\pi$

## 18.6 Exercises (pages 843–844)

**1.** $-$  **3.** $+$  **5.** C  **7.** D  **9.** F  **11.** $\frac{\sqrt{2+\sqrt{2}}}{2}$
**13.** $-\frac{\sqrt{2+\sqrt{3}}}{2}$  **15.** $-\frac{\sqrt{2+\sqrt{3}}}{2}$  **17.** First find the exact value for $\cos 15°$ by using the cosine half-angle identity: $\cos 15° = \cos\frac{30°}{2} = \sqrt{\frac{1+\cos 30°}{2}}$. Substitute $\frac{\sqrt{3}}{2}$ for $\cos 30°$ and simplify the result. Then find the exact value for $\sin 7.5°$ by using the sine half-angle identity: $\sin 7.5° = \sin\frac{15°}{2} = \sqrt{\frac{1-\cos 15°}{2}}$. Substitute the exact value found for $\cos 15°$ and simplify the result. (Choose the positive square roots because 15° and 7.5° are both in quadrant I.)  **19.** $\frac{\sqrt{10}}{4}$  **21.** 3  **23.** $\frac{\sqrt{50-10\sqrt{5}}}{10}$
**25.** $-\sqrt{7}$  **27.** $\frac{\sqrt{5}}{5}$  **29.** $-\frac{\sqrt{42}}{12}$  **31.** 0.127
**33.** $\sin 20°$  **35.** $\tan 73.5°$  **37.** $\tan 29.87°$  **39.** $\cos 9x$
**41.** $\tan 4\theta$  **43.** $\cos\frac{x}{8}$

## 18 Review Exercises (pages 846–848)

**1.** B  **3.** C  **5.** D  **7.** 1  **9.** $\frac{1}{\cos^2\theta}$  **11.** $-\frac{\cos\theta}{\sin\theta}$
**13.** $\sin x = -\frac{4}{5}$; $\tan x = -\frac{4}{3}$; $\cot(-x) = \frac{3}{4}$  **15.** $\sin 165° = \frac{\sqrt{6}-\sqrt{2}}{4}$; $\cos 165° = \frac{-\sqrt{6}-\sqrt{2}}{4}$; $\tan 165° = -2+\sqrt{3}$; $\csc 165° = \sqrt{6}+\sqrt{2}$; $\sec 165° = -\sqrt{6}+\sqrt{2}$; $\cot 165° = -2-\sqrt{3}$  **17.** E  **19.** J  **21.** I
**23.** H  **25.** G  **27.** $\frac{117}{125}$; $\frac{4}{5}$; $-\frac{117}{44}$; II

In Exercises 29 and 31, other forms are possible for $\tan(x+y)$.
**29.** $\frac{2+3\sqrt{7}}{10}$; $\frac{2\sqrt{3}+\sqrt{21}}{10}$; $\frac{2+3\sqrt{7}}{2\sqrt{3}-\sqrt{21}}$; II  **31.** $\frac{4-9\sqrt{11}}{50}$; $\frac{12\sqrt{11}-3}{50}$; $\frac{4-9\sqrt{11}}{12\sqrt{11}+3}$; IV  **33.** $\sin\theta = \frac{\sqrt{14}}{4}$; $\cos\theta = \frac{\sqrt{2}}{4}$
**35.** $\sin 2x = \frac{3}{5}$; $\cos 2x = -\frac{4}{5}$  **37.** $\frac{1}{2}$  **39.** $\frac{\sqrt{5}-1}{2}$
**41.** 0.5

## 18 Test (page 848)

[18.1] **1.** $\sin\theta = -\frac{7}{25}$; $\tan\theta = -\frac{7}{24}$; $\cot\theta = -\frac{24}{7}$; $\sec\theta = \frac{25}{24}$; $\csc\theta = -\frac{25}{7}$  **2.** $\cos\theta$  **3.** $-1$  [18.3] **4.** $\frac{\sqrt{6}-\sqrt{2}}{4}$
[18.3, 18.4] **5.** (a) $-\sin x$  (b) $\tan x$  [18.6] **6.** $-\frac{\sqrt{2-\sqrt{2}}}{2}$
**7.** $\cot\frac{1}{2}x - \cot x = \csc x$  [18.3, 18.4] **8.** (a) $\frac{33}{65}$  (b) $-\frac{56}{65}$
(c) $\frac{63}{16}$  (d) II  [18.5, 18.6] **9.** (a) $-\frac{7}{25}$  (b) $-\frac{24}{25}$  (c) $\frac{24}{7}$
(d) $\frac{\sqrt{5}}{5}$  (e) 2

# 19 Inverse Circular Functions and Trigonometric Equations

## 19.1 Exercises (pages 860–863)

**1.** one-to-one  **3.** domain  **5.** $\pi$  **7.** (a) $[-1, 1]$
(b) $\left[-\frac{\pi}{2}, \frac{\pi}{2}\right]$  (c) increasing  (d) $-2$ is not in the domain.
**9.** (a) $(-\infty, \infty)$  (b) $\left(-\frac{\pi}{2}, \frac{\pi}{2}\right)$  (c) increasing  (d) no
**11.** $\cos^{-1}\frac{1}{a}$  **13.** 0  **15.** $\pi$  **17.** $-\frac{\pi}{2}$  **19.** 0  **21.** $\frac{\pi}{2}$
**23.** $\frac{\pi}{4}$  **25.** $\frac{5\pi}{6}$  **27.** $\frac{3\pi}{4}$  **29.** $-\frac{\pi}{6}$  **31.** $\frac{\pi}{6}$  **33.** 0
**35.** $-45°$  **37.** $-60°$  **39.** 120°  **41.** 120°  **43.** 60°
**45.** $\sin^{-1} 2$ does not exist.  **47.** $-7.6713835°$
**49.** 113.500970°  **51.** 30.987961°  **53.** 121.267893°
**55.** $-82.678329°$  **57.** 0.83798122  **59.** 2.3154725
**61.** 1.1900238  **63.** 1.9033723  **65.** 3.1144804
**67.** 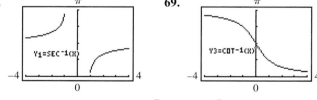  **69.**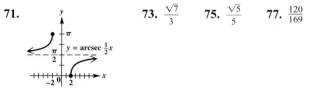
**71.** 

$y = \text{arcsec}\frac{1}{2}x$

  **73.** $\frac{\sqrt{7}}{3}$  **75.** $\frac{\sqrt{5}}{5}$  **77.** $\frac{120}{169}$
**79.** $-\frac{7}{25}$  **81.** $\frac{4\sqrt{6}}{25}$  **83.** 2  **85.** $\frac{63}{65}$  **87.** $\frac{\sqrt{10}-3\sqrt{30}}{20}$
**89.** 0.894427191  **91.** 0.1234399811  **93.** $\sqrt{1-u^2}$
**95.** $\sqrt{1-u^2}$  **97.** $\frac{4\sqrt{u^2-4}}{u^2}$  **99.** $\frac{u\sqrt{2}}{2}$  **101.** $\frac{2\sqrt{4-u^2}}{4-u^2}$
**103.** (a) 45°  (b) $\theta = 45°$

## 19.2 Exercises (pages 867–869)

1. Solve the linear equation for cot $x$.  3. Solve the quadratic equation for sec $x$ by factoring.  5. Solve the quadratic equation for sin $x$ using the quadratic formula.  7. Use an identity to rewrite as an equation with one trigonometric function.  9. $-30°$ is not a correct answer because it is not in the interval $[0°, 360°)$.  11. $\left\{\frac{3\pi}{4}, \frac{7\pi}{4}\right\}$  13. $\left\{\frac{\pi}{6}, \frac{5\pi}{6}\right\}$  15. $\emptyset$
17. $\left\{\frac{\pi}{4}, \frac{2\pi}{3}, \frac{5\pi}{4}, \frac{5\pi}{3}\right\}$  19. $\{\pi\}$  21. $\left\{\frac{7\pi}{6}, \frac{3\pi}{2}, \frac{11\pi}{6}\right\}$
23. $\{30°, 210°, 240°, 300°\}$  25. $\{90°, 210°, 330°\}$
27. $\{45°, 135°, 225°, 315°\}$  29. $\{45°, 225°\}$
31. $\{0°, 30°, 150°, 180°\}$
33. $\{0°, 45°, 135°, 180°, 225°, 315°\}$
35. $\{53.6°, 126.4°, 187.9°, 352.1°\}$
37. $\{149.6°, 329.6°, 106.3°, 286.3°\}$  39. $\emptyset$
41. $\{57.7°, 159.2°\}$  43. $\{0.8751 + 2n\pi, 2.2665 + 2n\pi, 3.5908 + 2n\pi, 5.8340 + 2n\pi$, where $n$ is any integer$\}$
45. $\left\{\frac{\pi}{3} + n\pi, \frac{2\pi}{3} + n\pi\right.$, where $n$ is any integer$\}$
47. $\{33.6° + 360°n, 326.4° + 360°n$, where $n$ is any integer$\}$
49. $\{45° + 180°n, 108.4° + 180°n$, where $n$ is any integer$\}$
51. $\{0.6806, 1.4159\}$  53. (a) 0.00164 and 0.00355
(b) $[0.00164, 0.00355]$  (c) outward  55. (a) $\frac{1}{4}$ sec
(b) $\frac{1}{6}$ sec  (c) 0.21 sec

## 19.3 Exercises (pages 873–874)

1. $\left\{\frac{\pi}{3}, \pi, \frac{4\pi}{3}\right\}$  3. $\{60°, 210°, 240°, 310°\}$
5. $\left\{\frac{\pi}{12}, \frac{11\pi}{12}, \frac{13\pi}{12}, \frac{23\pi}{12}\right\}$  7. $\left\{\frac{\pi}{2}, \frac{7\pi}{6}, \frac{11\pi}{6}\right\}$
9. $\left\{\frac{\pi}{18}, \frac{7\pi}{18}, \frac{13\pi}{18}, \frac{19\pi}{18}, \frac{25\pi}{18}, \frac{31\pi}{18}\right\}$  11. $\left\{\frac{3\pi}{8}, \frac{5\pi}{8}, \frac{11\pi}{8}, \frac{13\pi}{8}\right\}$
13. $\left\{\frac{\pi}{2}, \frac{3\pi}{2}\right\}$  15. $\left\{0, \frac{\pi}{3}, \pi, \frac{5\pi}{3}\right\}$  17. $\emptyset$  19. $\left\{\frac{\pi}{2}\right\}$
21. $\left\{\frac{\pi}{3}, \pi, \frac{5\pi}{3}\right\}$  23. $\{15°, 45°, 135°, 165°, 255°, 285°\}$
25. $\{0°\}$  27. $\{120°, 240°\}$  29. $\{30°, 150°, 270°\}$
31. $\{180°n, 30° + 360°n, 150° + 360°n$, where $n$ is any integer$\}$
33. $\{60° + 360°n, 300° + 360°n$, where $n$ is any integer$\}$
35. $\{11.8° + 180°n, 78.2° + 180°n$, where $n$ is any integer$\}$
37. $\{30° + 180°n, 90° + 180°n, 150° + 180°n$, where $n$ is any integer$\}$  39. $\{1.2802\}$
41. (a) For $x = t$,
$P(t) = 0.003 \sin 220\pi t +$
$\frac{0.003}{3} \sin 660\pi t +$
$\frac{0.003}{5} \sin 1100\pi t +$
$\frac{0.003}{7} \sin 1540\pi t$
(b) The graph is periodic, and the wave has "jagged square" tops and bottoms.
(c) This will occur when $t$ is in one of these intervals: (0.0045, 0.0091), (0.0136, 0.0182), (0.0227, 0.0273).

## 19 Review Exercises (pages 876–878)

1. 

$[-1, 1]; \left[-\frac{\pi}{2}, \frac{\pi}{2}\right]$  $[-1, 1]; [0, \pi]$  $(-\infty, \infty); \left(-\frac{\pi}{2}, \frac{\pi}{2}\right)$

3. false; $\arcsin\left(-\frac{1}{2}\right) = -\frac{\pi}{6}$, not $\frac{11\pi}{6}$.  5. $\frac{\pi}{4}$
7. $-\frac{\pi}{3}$  9. $\frac{3\pi}{4}$  11. $\frac{2\pi}{3}$  13. $\frac{3\pi}{4}$  15. $-60°$
17. $60.67924514°$  19. $36.4895081°$  21. $73.26220613°$
23. $-1$  25. $\frac{3\pi}{4}$  27. $\frac{\pi}{4}$  29. $\frac{\sqrt{7}}{4}$  31. $\frac{\sqrt{3}}{2}$
33. $\frac{294 + 125\sqrt{6}}{92}$  35. $\frac{1}{u}$  37. $\{0.4636, 3.6052\}$
39. $\left\{\frac{\pi}{4}, \frac{3\pi}{4}, \frac{5\pi}{4}, \frac{7\pi}{4}\right\}$  41. $\left\{\frac{\pi}{8}, \frac{3\pi}{8}, \frac{5\pi}{8}, \frac{7\pi}{8}, \frac{9\pi}{8}, \frac{11\pi}{8}, \frac{13\pi}{8}, \frac{15\pi}{8}\right\}$
43. $\left\{\frac{\pi}{3} + 2n\pi, \pi + 2n\pi, \frac{5\pi}{3} + 2n\pi\right.$, where $n$ is any integer$\}$
45. $\{270°\}$  47. $\{45°, 90°, 225°, 270°\}$
49. $\{70.5°, 180°, 289.5°\}$
51. (b) 8.6602567 ft; There may be a discrepancy in the final digits.

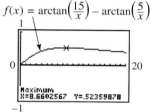

$f(x) = \arctan\left(\frac{15}{x}\right) - \arctan\left(\frac{5}{x}\right)$

53. The light beam is completely underwater.

## 19 Test (pages 878–879)

[19.1] 1. $[-1, 1]; \left[-\frac{\pi}{2}, \frac{\pi}{2}\right]$

2. (a) $\frac{2\pi}{3}$  (b) $-\frac{\pi}{3}$  (c) 0  (d) $\frac{2\pi}{3}$  3. (a) 30°  (b) $-45°$
(c) 135°  (d) $-60°$  4. (a) 42.54°  (b) 22.72°  (c) 125.47°
5. (a) $\frac{\sqrt{5}}{3}$  (b) $\frac{4\sqrt{2}}{9}$  6. The domain of the inverse sine function, $y = \sin^{-1} x$, is $[-1, 1]$. Because 3 is not in this interval, $\sin^{-1} 3$ cannot be defined.  7. $\sin \frac{5\pi}{6} = \frac{1}{2}$, but $\arcsin \frac{1}{2} = \frac{\pi}{6}$, so $\arcsin\left(\sin \frac{5\pi}{6}\right) = \arcsin \frac{1}{2} = \frac{\pi}{6} \neq \frac{5\pi}{6}$. This happens because $\frac{5\pi}{6}$ is not in the range of the inverse sine function.  8. $\frac{u\sqrt{1-u^2}}{1-u^2}$
[19.2, 19.3] 9. $\{30°, 330°\}$  10. $\{90°, 270°\}$
11. $\{18.4°, 135°, 198.4°, 315°\}$  12. $\left\{0, \frac{2\pi}{3}, \frac{4\pi}{3}\right\}$
13. $\left\{\frac{\pi}{12}, \frac{7\pi}{12}, \frac{3\pi}{4}, \frac{5\pi}{4}, \frac{17\pi}{12}, \frac{23\pi}{12}\right\}$
14. $\{0.3649, 1.2059, 3.5065, 4.3475\}$

**15.** $\{90° + 180°n,$ where $n$ is any integer$\}$
**16.** $\{\frac{2\pi}{3} + 4n\pi, \frac{4\pi}{3} + 4n\pi,$ where $n$ is any integer$\}$

## 20 Oblique Triangles and Vectors

**Note to student:** While most of the measures resulting from solving triangles in this chapter are approximations, for convenience we use = rather than ≈ in the answers.

### 20.1 Exercises (pages 886–889)

**1.** C **3.** $\sqrt{3}$ **5.** $C = 95°, b = 13$ m, $a = 11$ m
**7.** $B = 37.3°, a = 38.5$ ft, $b = 51.0$ ft **9.** $C = 57.36°,$ $b = 11.13$ ft, $c = 11.55$ ft **11.** $B = 18.5°, a = 239$ yd, $c = 230$ yd **13.** $A = 56°\,00', AB = 361$ ft, $BC = 308$ ft
**15.** $B = 110.0°, a = 27.01$ m, $c = 21.37$ m **17.** $A = 34.72°,$ $a = 3326$ ft, $c = 5704$ ft **19.** $C = 97°\,34', b = 283.2$ m, $c = 415.2$ m **21.** In order to use the law of sines, at least one angle measure must be given. **23.** yes; If the given side is included between the two given angles, we have ASA. If not, we can use the angle sum formula to find the measure of the third angle, giving ASA. Thus, by the ASA congruence axiom, the triangle is uniquely determined. **25.** 118 m **27.** 10.4 in.
**29.** 111° **31.** $\frac{\sqrt{3}}{2}$ sq unit **33.** $\frac{\sqrt{2}}{2}$ sq unit **35.** 46.4 m²
**37.** 356 cm² **39.** 722.9 in.² **41.** 65.94 cm² **43.** 100 m²

### 20.2 Exercises (pages 894–895)

**1.** A **3. (a)** $4 < h < 5$ **(b)** $h = 4$ or $h > 5$ **(c)** $h < 4$
**5.** 1 **7.** 2 **9.** 0 **11.** 45° **13.** $B_1 = 49.1°,$ $C_1 = 101.2°, B_2 = 130.9°, C_2 = 19.4°$ **15.** $B = 26°\,30',$ $A = 112°\,10'$ **17.** no such triangle **19.** $B = 27.19°,$ $C = 10.68°$ **21.** $B = 20.6°, C = 116.9°, c = 20.6$ ft
**23.** no such triangle **25.** $B_1 = 49°\,20', C_1 = 92°\,00',$ $c_1 = 15.5$ km; $B_2 = 130°\,40', C_2 = 10°\,40', c_2 = 2.88$ km
**27.** $B = 37.77°, C = 45.43°, c = 4.174$ ft **29.** $A_1 = 53.23°,$ $C_1 = 87.09°, c_1 = 37.16$ m; $A_2 = 126.77°, C_2 = 13.55°,$ $c_2 = 8.719$ m **31.** 1; 90°; a right triangle **33.** Because $A$ is an obtuse angle, it must be the largest angle in the triangle; thus, $a$ must be the longest side. However, we are given that $b > a$, so no such triangle exists. **35.** 664 m

### 20.3 Exercises (pages 899–903)

**1. (a)** SAS **(b)** law of cosines **3. (a)** SSA **(b)** law of sines
**5. (a)** ASA **(b)** law of sines **7. (a)** ASA **(b)** law of sines
**9.** 5 **11.** 120° **13.** $a = 7.0, B = 37.6°, C = 21.4°$
**15.** $A = 73.7°, B = 53.1°, C = 53.1°$ (The angles do not sum to 180° due to rounding.) **17.** $b = 88.2, A = 56.7°, C = 68.3°$
**19.** $a = 2.60$ yd, $B = 45.1°, C = 93.5°$ **21.** $c = 6.46$ m, $A = 53.1°, B = 81.3°$ **23.** $A = 82°, B = 37°, C = 61°$
**25.** $C = 84°\,30', B = 44°\,40', A = 50°\,50'$ **27.** $a = 156$ cm, $B = 64°\,50', C = 34°\,30'$ **29.** $b = 9.529$ in., $A = 64.59°,$ $C = 40.61°$ **31.** $a = 15.7$ m, $B = 21.6°, C = 45.6°$

**33.** The value of cos $\theta$ will be greater than 1; your calculator will give you an error message when using the inverse cosine function.
**35.** 257 m **37.** 40° **39.** 26° and 36° **41.** second base: 66.8 ft; first and third bases: 63.7 ft **43.** 22 ft **45.** 16.26°
**47.** $24\sqrt{3}$ sq units **49.** 78 m² **51.** 12,600 cm²
**53.** 3650 ft² **55.** 25.24983 mi **57.** Area and perimeter are both 36. **59.** 390,000 mi²

### 20.4 Exercises (pages 908–910)

**1.** **m** and **p**; **n** and **r** **3.** **m** and **p** equal 2**t**, or **t** is one-half **m** or **p**; also **m** = 1**p** and **n** = 1**r** **5.**

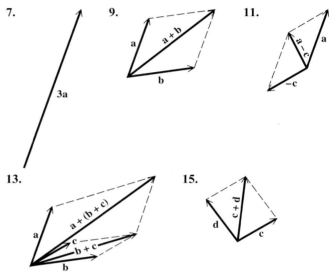

**17.** Yes, it appears that vector addition is associative (and this is true, in general). **19. (a)** $\langle -4, 16 \rangle$ **(b)** $\langle -12, 0 \rangle$
**(c)** $\langle 8, -8 \rangle$ **21. (a)** $\langle 8, 0 \rangle$ **(b)** $\langle 0, 16 \rangle$ **(c)** $\langle -4, -8 \rangle$
**23. (a)** $\langle 0, 12 \rangle$ **(b)** $\langle -16, -4 \rangle$ **(c)** $\langle 8, -4 \rangle$
**25. (a)** $4\mathbf{i}$ **(b)** $7\mathbf{i} + 3\mathbf{j}$ **(c)** $-5\mathbf{i} + \mathbf{j}$ **27. (a)** $\langle -2, 4 \rangle$
**(b)** $\langle 7, 4 \rangle$ **(c)** $\langle 6, -6 \rangle$ **29.**

**31.** **33.** 17; 331.9° **35.** 8; 120°

**37.** 47, 17 **39.** 38.8, 28.0 **41.** 123, 155 **43.** $\langle \frac{5\sqrt{3}}{2}, \frac{5}{2} \rangle$
**45.** $\langle 3.0642, 2.5712 \rangle$ **47.** $\langle 4.0958, -2.8679 \rangle$
**49.** 530 newtons **51.** 88.2 lb **53.** 94.2 lb **55.** 24.4 lb
**57.** $\langle a + c, b + d \rangle$ **59.** $\langle 2, 8 \rangle$ **61.** $\langle 8, -20 \rangle$
**63.** $\langle -30, -3 \rangle$ **65.** $\langle 8, -7 \rangle$ **67.** $-5\mathbf{i} + 8\mathbf{j}$
**69.** $2\mathbf{i}$, or $2\mathbf{i} + 0\mathbf{j}$

### 20 Review Exercises (pages 913–915)

**1.** 63.7 m **3.** 41.7° **5.** 54° 20′ or 125° 40′ **7.** no; We are not given SSA, the ambiguous case. Instead, we are given SAA, which determines a unique triangle. **9. (a)** $b = 5, b \geq 10$
**(b)** $5 < b < 10$ **(c)** $b < 5$ **11.** 19.87°, or 19° 52′
**13.** 55.5 m **15.** 19 cm **17.** $B = 17.3°, C = 137.5°,$

$c = 11.0$ yd   **19.** $c = 18.7$ cm, $A = 91°\,40'$, $B = 45°\,50'$
**21.** 153,600 m$^2$   **23.** 0.234 km$^2$   **25.** 58.6 ft   **27.** 13 m
**29.** 53.2 ft   **31.** 115 km   **33.** 25 sq units
**35.**

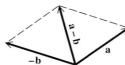

**37. (a)** true   **(b)** false
**39.** 207 lb
**41.** 869; 418   **43.** 15; 126.9°   **45.** $\langle \frac{5}{13}, \frac{12}{13} \rangle$

## 20 Test *(page 916)*

[20.1] **1.** 137.5°   [20.3] **2.** 179 km   **3.** 49.0°
**4.** 168 sq units   [20.1] **5.** 18 sq units
[20.2] **6. (a)** $b > 10$   **(b)** none   **(c)** $b \leq 10$
[20.1–20.3] **7.** $a = 40$ m, $B = 41°$, $C = 79°$
**8.** $B_1 = 58°\,30'$, $A_1 = 83°\,00'$, $a_1 = 1250$ in.; $B_2 = 121°\,30'$, $A_2 = 20°\,00'$, $a_2 = 431$ in.   [20.4] **9.** $|\mathbf{v}| = 10$; $\theta = 126.9°$
**10.**

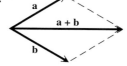

**11. (a)** $\langle 1, -3 \rangle$   **(b)** $\langle -6, 18 \rangle$
**(c)** $\sqrt{10}$

[20.1] **12.** 2.7 mi   [20.4] **13.** $\langle -346, 451 \rangle$
[20.1] **14.** 14 m

# Appendixes

## Appendix A *(pages 923–925)*

**1.** 720   **3.** 120   **5.** 15   **7.** 1   **9.** 360   **11.** 72
**13.** 5   **15.** 6   **17.** 1   **19.** 78   **21.** 48   **23.** 40,320
**25.** 604,800   **27.** $2.052371412 \times 10^{10}$   **29.** 39,270
**31.** 720; 151,200   **33.** 15,890,700   **35.** 35   **37.** 35
**39.** 84; 324   **41. (a)** permutation   **(b)** combination
**(c)** combination   **43.** 210   **45. (a)** 56   **(b)** 462
**(c)** 3080   **(d)** 8526   **47.** 210; 5040   **49.** 35
**51. (a)** 220   **(b)** 55   **(c)** 105

## Appendix B *(pages 932–933)*

**1.** $S = \{HH, HT, TH, TT\}$   **3.** $S = \{(1, 2), (1, 3), (1, 4),$
$(1, 5), (2, 3), (2, 4), (2, 5), (3, 4), (3, 5), (4, 5)\}$
**5. (a)** $\{HH, TT\}, \frac{1}{2}$   **(b)** $\{HH, HT, TH\}, \frac{3}{4}$
**7. (a)** $\{(2, 4)\}, \frac{1}{10}$   **(b)** $\{(1, 3), (1, 5), (3, 5)\}, \frac{3}{10}$   **(c)** $\emptyset, 0$
**(d)** $\{(1, 2), (1, 4), (2, 3), (2, 5), (3, 4), (4, 5)\}, \frac{3}{5}$
**9.** A probability cannot be greater than 1.   **11.** $\frac{3}{13}; \frac{10}{13}$
**13.** 499 to 1   **15. (a)** $\frac{3}{5}$   **(b)** $\frac{7}{10}$   **(c)** $\frac{3}{10}$
**17. (a)** $\frac{3}{13}$   **(b)** $\frac{7}{13}$   **(c)** $\frac{3}{13}$   **19. (a)** $\frac{1}{2}$   **(b)** $\frac{7}{10}$   **(c)** 0
**21. (a)** 0.72   **(b)** 0.70   **(c)** 0.79

# Index

## A

Absolute value
   definition, 7
   distance definition, 91
   as negative number, 382
   simplifying square roots, 383
Absolute value equations
   definition, 91
   solution of, 92
   steps to solve, 92
Absolute value function
   definition, 516
   graph of, 517
Absolute value inequalities
   definition, 91
   solution of, 92
   steps to solve, 92
Acute angles
   definition, 660
   reference angles as, 710
   trigonometric function values of, 704–705
   trigonometric functions of, 702
Acute triangle, 670, 883
Addition
   associative property of, 33
   commutative property of, 33
   of complex numbers, 422
   of functions, 256, 484
   identity element for, 32
   identity property of, 32
   inverse property of, 31–32
   of matrices, 198
   of polynomial functions, 256
   of polynomials, 252
   of radical expressions, 402
   of radicals, 402
   of rational expressions, 330
   of real numbers, 15
Addition property
   of equality, 45
   of inequality, 73
Additive identity, 32
Additive inverse
   of a matrix, 198
   of a real number, 6, 31–32
Adjacent side to an angle, 702
Agreement on domain, 151
Algebraic expressions
   definition, 27, 250
   evaluating, 27
   examples of, 44
Algebraic fraction, 320
Alternate exterior angles, 668
Alternate interior angles, 668

Alternative events
   definition, 930
   probability of, 931
Ambiguous case of the law of sines, 890
Amplitude
   of cosine function, 773
   definition, 773
   of sine function, 773
Angle(s)
   acute, 660
   alternate exterior, 668
   alternate interior, 668
   complementary, 661
   corresponding, 668
   coterminal, 664
   definition, 660
   of depression, 722
   of elevation, 722
   initial side of, 660
   measure of, 660, 712, 716
   negative, 660
   obtuse, 660
   positive, 660
   quadrantal, 663
   reference, 709
   right, 660
   side adjacent to, 702
   side opposite, 702
   significant digits for, 720
   standard position of, 663
   straight, 660
   supplementary, 661
   terminal side of, 660
   types of, 660
   vertex of, 660
   vertical, 668
Angle-Side-Angle (ASA), 882
Angle sum of a triangle, 669
Angular speed
   applications of, 758–759
   definition, 757
Apogee
   definition, 365, 640
   of an ellipse, 640
Applied problems, steps for solving, 62
Applied trigonometry problems, steps to solve, 722
Approximately equal to
   calculator values, 706
   symbol for, 383, 706
Arc length, 742
arccos $x$, 853–854
arccot $x$, 856
arccsc $x$, 856
arcsec $x$, 856

arcsin $x$, 852–853
arctan $x$, 854–855
Area of a sector, 744
Area of a triangle
   formulas for, 885
   Heron's formula for, 898–899
Area problem, 463
Argument
   definition, 781
   of a function, 781, 813
Array of signs for a determinant, 219
Associative properties, 33
Asymptote(s)
   definition, 320
   graph of, 320
   horizontal, 614, 615
   of a hyperbola, 641
   oblique, 615, 616
   procedure to determine, 617
   pseudo-, 798
   vertical, 348, 614, 615, 788
Augmented matrix
   definition, 189, 209
   reduced row echelon form of, 196
Average rate of change, 107, 123
Average speed, 54
Axis
   of a coordinate system, 108
   of a parabola, 491, 493

## B

Base of an exponential expression, 23, 236
Binomials
   conjugate of, 409
   definition, 251
   factoring of, 303
   multiplication of, 266, 406
   squares of, 268
Boundary line, 143
Boundedness theorem, 607
Braces, 2

## C

Calculator graphing
   of a hyperbola, 642
   of rational functions, 621
   of a root function, 645
   for solving exponential equations, 575
   for solving logarithmic equations, 575
   for solving radical equations, 417
Cartesian coordinate system, 108
Celsius-Fahrenheit relationship, 57
Center
   of a circle, 634
   of an ellipse, 636

I-1

Center-radius form of equation of a circle, 635
Change-of-base rule for logarithms, 565
Circle(s)
 arc length of, 742
 center of, 634
 center-radius form of equation of, 635
 circumference of, 736
 definition, 634
 equation of, 634
 graph of, 634
 radius of, 634
 sector of, 744
 unit, 749
Circular functions
 applications of, 754
 definition, 749
 domains of, 751
 evaluating, 752
Circumference of a circle, 736
Clinometer, 726, 732
Cloud ceiling, 726
Coefficient, 33, 249
Cofunction identities, 703, 824
Cofunctions of trigonometric functions, 703
Column matrix, 197
Columns of a matrix, 189, 200–201
Combinations
 definition, 921
 distinguishing from permutations, 922
 formula for, 921
Combined functions, graph of, 793
Combined variation, 366
Combining like terms, 33, 252
Common logarithms
 applications of, 562
 definition, 562
 evaluating, 562
Commutative properties, 33
Complement of an event, 928
Complementary angles, 661
Completing the square
 to find the vertex, 500
 solving quadratic equations by, 439
Complex conjugates, 423
Complex fraction
 definition, 20, 339
 steps to simplify, 339
Complex numbers
 addition of, 422
 conjugate of, 423
 definition, 421
 division of, 423
 imaginary part of, 421
 multiplication of, 423
 nonreal, 422
 real part of, 421
 standard form of, 422
 subtraction of, 422
Components of a vector
 definition, 904
 horizontal, 905
 vertical, 905
Composite function
 definition, 258
 domain of, 487
 evaluating, 259
Composition of functions, 258, 486
Compound event, 930
Compound inequalities
 with *and*, 84
 definition, 84
 with *or*, 86
 solution of, 92
Compound interest
 continuous, 573
 definition, 65, 572
 formula for, 572
Comprehensive graph, 604
Conditional equation, 49
Conditional trigonometric equations
 definition, 863
 factoring method for solving, 864
 with half-angles, 870
 linear methods for solving, 863
 with multiple angles, 871
 solving with the quadratic formula, 865
 steps to solve algebraically, 866
 using trigonometric identities to solve, 865
Congruence axioms, 882
Congruent triangles, 670, 882
Conic sections
 definition, 633, 634
 geometric interpretation of, 634
 identifying by equation, 644
 summary of, 643
Conjugate(s)
 of a binomial, 409
 of a complex number, 423
Conjugate zeros theorem, 597
Consistent system, 172
Constant function, 156
Constant of variation, 362
Continuous compounding
 definition, 573
 formula for, 573
Contradiction, 49
Coordinate(s)
 on a line, 4
 in a plane, 108
 of a point, 4
Coordinate system
 Cartesian, 108
 rectangular, 108
Corresponding angles, 668
Cosecant function
 characteristics of, 797
 definition, 677, 797
 domain of, 797
 graph of, 797
 inverse of, 856
 period of, 797
 range of, 686, 797
 steps to graph, 798
Cosine function
 amplitude of, 773
 characteristics of, 772
 definition, 677, 772
 difference identity for, 822–823
 domain of, 772
 double-angle identity for, 833–834
 graph of, 772
 half-angle identity for, 839–840
 horizontal translation of, 781
 inverse of, 853–854
 period of, 772, 775
 range of, 686, 772
 steps to graph, 776, 784
 sum identity for, 822–823
 translating graphs of, 781, 784
 vertical translation of, 783
Cosine wave, 772

Cosines, law of, 895
Cost-benefit equation, 568
Cost-benefit model, 338
Cotangent function
 characteristics of, 790
 definition, 677, 790
 domain of, 790
 graph of, 790
 horizontal translation of, 793
 period of, 790
 range of, 686, 790
 steps to graph, 791
 vertical translation of, 793
Coterminal angles, 664
Counting numbers, 2, 5
Cramer's rule
 applying to solve linear systems, 221–223
 derivation of, 220
 summary of, 221
Cube(s)
 difference of, 300, 303
 of a number, 24
 sum of, 300–301, 303
Cube root function
 definition, 381
 graph of, 382
Cubing function, 261
Curvature, degree of, 748

D

Data modeling, 135
Decay
 applications of, 544
 exponential, 544, 574
Decibel, 563
Decimal degrees, 662–663
Decimals, linear equations with, 48
Decreasing function
 definition, 514
 on an interval, 513–514
Degree
 of a polynomial, 251
 of a term, 251
Degree measure
 converting to radian measure, 736
 definition, 660
Degree mode, calculator in, 679, 680
Degree of curvature, 748
Degree/radian relationship
 definition, 736
 table of, 738
Delta, 118
Denominator
 least common, 331
 rationalizing, 407
Dependent equations
 definition, 172
 solving a system of, 177
Dependent variable, 147–148
Depression, angle of, 722
Descartes, René, 108
Descartes' rule of signs, 600
Descending powers, 250
Determinant of a square matrix
 array of signs for, 219
 definition, 217
 evaluating, 217
 expansion by minors, 218
 minor of, 218
Diastolic pressure, 780

Difference, 16
Difference identity
  application of, 825, 828
  for cosine, 822–823
  for sine, 827
  for tangent, 828
Difference of cubes
  definition, 300, 303
  factoring of, 300
Difference of squares
  definition, 303
  factoring of, 298
Difference quotient, 485
Digits, significant, 720
Direct variation
  definition, 362
  as a power, 363
Direction angle for a vector, 905
Directrix of a parabola, 496
Discontinuity, point of, 620
Discontinuous graph, 613
Discriminant, 447–448, 503
Distance, rate, time, relationship, 353, 364–365, 452
Distance between points
  definition, 18
  formula for, 398
Distance to the horizon formula, 379, 401
Distributive property, 30, 46
Division
  of complex numbers, 423
  of functions, 258, 484
  of polynomial functions, 276
  of polynomials, 272
  of rational expressions, 326
  of real numbers, 19
  synthetic, 588
  by zero, 19
Division algorithm, 588, 590
Domain(s)
  agreement on, 151
  of circular functions, 751
  of composite functions, 487
  of a function, 149
  of inverse circular functions, 856
  of polynomial functions, 260
  of rational equations, 345
  of a rational function, 320, 348
  of a relation, 149
Dominating term, 604
Double-angle identities
  application of, 836
  definition, 833–834
  simplifying expressions using, 836
  verifying, 835
Double negative property, 7
Doubling time, 567
Downward opening parabola, 493

E

$e$, 564
Elements of a matrix, 189
Elements of a set
  definition, 2
  symbol for, 2
Elevation, angle of, 722
Elimination method for solving systems, 183, 650
Ellipse
  apogee of, 640
  center of, 636

definition, 633, 634, 636
equation of, 637
foci of, 636
graph of, 637
intercepts of, 636
perigee of, 640
Empty set
  definition, 2
  notation for, 2
Endpoint of a ray, 660
Equal matrices, 197
Equality
  addition property of, 45
  multiplication property of, 45
Equation(s)
  absolute value, 91
  of a circle, 634
  conditional, 49
  conditional trigonometric, 863
  contradiction, 49
  definition, 8, 44
  dependent, 172
  distinguishing from expressions, 62
  of ellipses, 637
  equivalent, 44
  exponential, 542, 569
  first-degree, 44, 110
  graph of, 110
  of horizontal asymptotes, 615
  of a horizontal line, 112, 133
  of hyperbolas, 641
  identity, 49
  independent, 172
  of the inverse of a function, 535
  linear in one variable, 44
  linear in three variables, 182
  linear in two variables, 110
  linear system of, 170, 182
  logarithmic, 548
  matrix, 213
  with no solution, 346
  nonlinear, 44, 648
  power rule for, 413
  quadratic, 307, 436
  quadratic in form, 451
  with radicals, 413
  with rational expressions, 345
  second-degree, 436
  solution of, 44
  square root property for, 436
  translating words into, 61
  trigonometric, 863
  of vertical asymptotes, 615
  of a vertical line, 112, 133
Equilateral triangle, 670
Equivalent equations, 44
Equivalent forms of fractions, 21
Equivalent inequalities, 73
Euler, Leonhard, 564
Even function, 772
Event(s)
  alternative, 930
  complement of, 928
  compound, 930
  definition, 927
  independent, 917
  mutually exclusive, 930
  odds in favor of, 929
  probability of, 927
Exact number, 720
Expansion of a determinant by minors, 218

Exponential decay, 544, 574
Exponential equations
  applications of, 572
  definition, 542, 569
  general method for solving, 570
  graphing calculator method for solving, 575
  properties for solving, 542, 569
  steps to solve, 542
Exponential expressions
  base of, 23
  definition, 23
  evaluating, 24
  simplifying, 242
Exponential functions
  applications of, 543–544
  converting to logarithmic form, 548
  definition, 539
  graph of, 540
  properties of, 557
Exponential growth, 544, 574
Exponential notation, 24, 386
Exponents
  base of, 23, 236
  definition, 23, 236
  fractional, 387
  integer, 236
  negative, 237
  power rules for, 240, 242
  product rule for, 236, 241
  quotient rule for, 239, 241
  rational, 386
  summary of rules for, 241–242
  zero, 237
Expressions
  algebraic, 27, 44, 250
  distinguishing from equations, 62
  exponential, 23
  radical, 381, 402
  rational, 320
Extraneous solutions, 414

F

Factor(s)
  greatest common, 286
  of numbers, 23
Factor theorem for polynomial functions, 593
Factorial notation, 918
Factoring
  binomials, 303
  definition, 52
  difference of cubes, 300
  difference of squares, 298
  by grouping, 288
  perfect square trinomials, 299
  polynomials, 286, 304
  solving quadratic equations by, 307
  substitution method for, 296
  sum of cubes, 300–301
  summary of special types of, 301
  trinomials, 291, 304
  using FOIL, 291
Fahrenheit-Celsius relationship, 57
Faraday, Michael, 809
Finite set, 2
First-degree equations, 44, 110
Focus (foci)
  of an ellipse, 636
  of a hyperbola, 640
  of a parabola, 496
FOIL method, 266, 406

Formula(s)
  compound interest, 572
  definition, 52, 461
  distance, 379, 398
  Galileo's, 437
  Heron's, 385
  midpoint, 113–114
  Pythagorean, 397, 462
  quadratic, 445
  with rational expressions, 351
  simple interest, 572
  solving for a specified variable of, 52, 351, 461
  with square roots, 461
  vertex, 502
Fractional exponents
  definition, 387
  radical form of, 389
Fractions
  algebraic, 320
  complex, 20, 339
  equivalent forms of, 21
  linear equations with, 47
  linear inequalities with, 76
Frequency, fundamental, 872
Function(s). *See also* Trigonometric functions
  absolute value, 516
  argument of, 781, 813
  circular, 749
  composite, 258, 486
  composition of, 258, 486
  constant, 156
  cosecant, 798
  cosine, 772
  cotangent, 790
  cube root, 381
  cubing, 261
  decreasing, 513–514
  definition, 148
  definitions of, 148, 153
  division of, 276
  domain of, 149
  equation of the inverse of, 535
  even, 772
  exponential, 539
  greatest integer, 519–520
  identity, 260
  increasing, 513–514
  inverse, 850
  inverse cosecant, 856
  inverse cosine, 853–854
  inverse cotangent, 856
  inverse of, 532
  inverse secant, 856
  inverse sine, 851–853
  inverse tangent, 854–855
  linear, 156, 260
  logarithmic, 548
  notation, 153, 255
  odd, 771
  one-to-one, 532, 850
  operations on, 256, 269, 276, 484
  periodic, 770
  piecewise linear, 516, 518
  polynomial, 255, 591
  quadratic, 490
  radical, 382
  range of, 149
  rational, 320, 347, 364, 613
  root, 381
  secant, 796
  sine, 770–771
  square root, 381, 644–645
  squaring, 261
  step, 519–520
  tangent, 789
  trigonometric, 677, 702
  vertical line test for, 151
Fundamental frequency, 872
Fundamental identities, 810
Fundamental principle of counting, 917
Fundamental property of rational numbers, 321
Fundamental rectangle of a hyperbola, 641
Fundamental theorem of algebra, 596

## G

Galilei, Galileo, 437
Galileo's formula, 437
Gauss, Carl Friederich, 596
Grade, 117, 741
Grade resistance, 717
Graph(s)
  of absolute value functions, 517
  of an asymptote, 320
  of circles, 634
  of combined functions, 793
  comprehensive, 604
  of cosecant function, 798
  of cosine function, 772
  of cotangent function, 790
  of cube root functions, 382
  definition, 4
  discontinuous, 613
  of ellipses, 637
  of equations, 110
  of exponential functions, 540
  of first-degree equations, 110
  of a greatest integer function, 520
  of horizontal lines, 112
  of a horizontal parabola, 506
  of hyperbolas, 640
  of inverse cosecant function, 855
  of inverse cosine function, 855
  of inverse cotangent function, 855
  of inverse functions, 853–855
  of inverse secant function, 855
  of inverse sine function, 853
  of inverse tangent function, 855
  of inverses, 536
  of linear equations, 110
  of linear inequalities, 143
  of linear systems, 172, 183
  of logarithmic functions, 550
  of numbers, 4
  of ordered pairs of numbers, 109
  of parabolas, 490–492, 506
  of a piecewise linear function, 518
  of polynomial functions, 260, 601
  of quadratic functions, 490–492
  of radical expressions, 381
  of radical functions, 382
  of rational functions, 348, 614
  reflection of, 261, 510
  of secant function, 796
  of sets of numbers, 10
  shrinking, 510
  of sine function, 771
  of square root functions, 381, 644
  of a step function, 520
  stretching, 510
  symmetric with respect to the origin, 512
  symmetric with respect to the $x$-axis, 511
  symmetric with respect to the $y$-axis, 511
  of tangent function, 789
  of three-part inequalities, 11
  translations of, 602
  turning points of, 604
  of vertical lines, 112
  of a vertical parabola, 490, 492
Graphing calculator. *See* Calculator graphing
Graphing method for solving systems, 171
Greater than, 9
Greatest common factor (GCF)
  definition, 286
  factoring out, 286
Greatest integer function
  application of, 520
  definition, 519
  graph of, 520
Grouping
  factoring by, 288
  steps to factor by, 289
Growth
  applications of, 543–544
  exponential, 544, 574

## H

Half-angle identities
  application of, 841
  definition, 839–840
  simplifying expressions using, 842
  trigonometric equations with, 870
  verifying, 842
Half-life, 575
Height of a propelled object, 505
Heron of Alexandria, 898
Heron triangles, 903
Heron's formula
  application of, 899
  definition, 385, 898–899
Horizontal asymptotes
  definition, 614, 615
  equation of, 615
Horizontal component of a vector, 905
Horizontal line
  definition, 112
  equation of, 112, 133
  graph of, 112
  slope of, 119–120
Horizontal line test
  for a one-to-one function, 534
  for one-to-one functions, 850
Horizontal parabola
  definition, 506
  graph of, 506
Horizontal shift of a parabola, 493
Horizontal translations, 781
Hyperbola
  asymptotes of, 641
  definition, 634, 640
  equation of, 641
  foci of, 640
  fundamental rectangle of, 641
  graph of, 640
  intercepts of, 641
  steps to graph, 642
Hypotenuse of a right triangle, 397

## I

$i$
  definition, 420
  powers of, 424

**i, j** unit vectors, 908
Identities, 683. *See also* Trigonometric identities
Identity elements, 32
Identity equations, 49
Identity function, 260
Identity matrix, 207
Identity properties, 32
Imaginary part of a complex number, 421
Imaginary unit, 420
Incidence rate, 350
Inconsistent system
   definition, 172
   recognizing, 193
   solving, 177
   solving with three variables, 186
Increasing function
   definition, 514
   on an interval, 513–514
Independent equations, 172
Independent events, 917
Independent variable, 148
Index of a radical, 380
Inequalities
   absolute value, 91
   addition property of, 73
   compound, 84
   definition, 8
   equivalent, 73
   interval notation for, 10
   linear in one variable, 73
   linear in two variables, 142
   multiplication property of, 75
   nonlinear, 469
   polynomial, 471
   quadratic, 469
   rational, 472
   summary of symbols for, 8
   third-degree, 471
   three-part, 11, 77
Infinite set, 2
Infinity symbol, 10
Initial point of a vector, 904
Initial side of an angle, 660
Integer(s), 4, 5
Integer exponents, 236
Intercepts
   of an ellipse, 636
   of a hyperbola, 641
   of a parabola, 503
   $x$, 110
   $y$, 110
Interest
   compound, 65, 572
   simple, 65, 572
Intermediate value theorem, 606
Intersection
   of linear inequalities, 145
   of sets, 83
Interval notation, 10, 72
Interval on a number line, 10
Inverse
   additive, 6, 31–32
   multiplicative, 19, 31–32
   of a one-to-one function, 532
Inverse cosecant function
   definition, 856
   graph of, 855
Inverse cosine function
   definition, 853–854
   graph of, 855

Inverse cotangent function
   definition, 856
   graph of, 855
Inverse functions
   general statement about, 850
   graphs of, 851
   notation for, 850
   summary of concepts, 851
Inverse matrix method for solving systems, 213
Inverse of a function
   definition, 532, 533
   equation of, 535
   graph of, 536
   steps to find the equation of, 535
   symbol for, 532
Inverse properties, 31–32
Inverse secant function
   definition, 856
   graph of, 855
Inverse sine function
   definition, 852–853
   graph of, 853
Inverse tangent function
   definition, 854–855
   graph of, 855
Inverse trigonometric functions
   domains and ranges of, 856
   graphs of, 853–855
   notation for, 850
   summary of, 856
Inverse variation
   definition, 364
   as a power of, 364
Irrational numbers, 4, 5
Isosceles triangle, 670

## J

Joint variation, 366

## L

Lambert's law, 821
Latitude, 743
Law of cosines
   applications of, 896
   definition, 895
   derivation of, 895
Law of sines
   ambiguous case of, 890
   applications of, 884
   definition, 883
   derivation of, 883
Least common denominator (LCD)
   definition, 331
   steps to find, 331
Legs of a right triangle, 397
Less than, 8
Light-year, 249
Like terms
   combining, 33, 252
   definition, 33, 251
Limit notation, 614
Line(s)
   definition, 660
   horizontal, 112
   number, 3
   parallel, 668
   segment, 660
   slope of, 117–118
   vertical, 112
Line graph, 108

Line segment
   definition, 660
   midpoint of, 113–114
Linear equations in one variable
   with decimal coefficients, 48
   definition, 44
   with fractional coefficients, 47
   solution of, 44
   solution set of, 44, 78
   solving, 44
   steps to solve, 46
   types of, 49
Linear equations in three variables
   definition, 182
   graphs of, 183
Linear equations in two variables
   definition, 110
   graph of, 110
   point-slope form of, 132
   slope-intercept form of, 130
   standard form of, 110, 132
   summary of forms of, 135
   system of, 170, 182
   $x$-intercept of, 110
   $y$-intercept of, 110
Linear functions
   definition, 156, 260
   piecewise, 516, 518
Linear inequalities in one variable
   application of, 78
   definition, 73
   with fractional coefficients, 76
   solution set of, 78
   steps to solve, 76
   three-part, 77
Linear inequalities in two variables
   boundary line of graph of, 143
   definition, 142
   graph of, 143
   intersection of, 145
   region of solution of, 143
   union of, 145–146
Linear methods for solving trigonometric equations, 863
Linear speed, 757
Logarithm(s)
   alternative forms of, 558
   change-of-base rule for, 565
   common, 562
   definition, 548
   evaluating, 562
   exponential form of, 548
   multiplying using, 555
   natural, 564
   power rule for, 557
   product rule for, 555
   properties of, 549, 555
   quotient rule for, 556
   summary of properties of, 558
Logarithmic equations
   definition, 548
   graphing calculator method for solving, 575
   properties for solving, 569
   solving, 570
   steps to solve, 572
Logarithmic functions
   applications of, 550
   with base $a$, 550
   converting to exponential form, 548
   definition, 548

Logarithmic functions (*continued*)
   graph of, 550
   properties of, 557
Lowest terms of a rational expression, 321

## M

Magnitude of a vector, 903
Mapping of sets, 149
Mathematical expressions from word expressions, 60–61
Mathematical models, 52
Matrix (matrices)
   addition of, 198
   additive inverse of, 198
   augmented, 189, 209
   calculator display of, 189
   column, 197
   columns of, 189, 200–201
   definition, 189, 197, 217
   elements of, 189
   equal, 197
   identity, 207
   multiplication by a scalar, 199
   multiplication of, 201
   multiplicative inverse of, 208
   negative of, 198
   properties of, 197
   reduced row echelon form of, 196
   row, 197
   row echelon form of, 190
   row operations on, 190
   rows of, 189, 200–201
   size of, 197
   square, 189, 197, 217
   steps to find inverse of, 211
   subtraction of, 199
   zero, 198
Matrix equation, 213
Matrix method for solving systems, 191, 192
Matrix multiplication
   application of, 203
   properties of, 202
Maximum value of a quadratic function, 504
Measure of an angle, 660
Members of a set, 2
Midpoint of a line segment
   definition, 113–114
   formula for, 114
Minimum value of a quadratic function, 504
Minor of a determinant, 218
Minute, 662
Mixture problems, 65–67
Model(s)
   to approximate data, 256
   definition, 135
   mathematical, 52
   polynomial, 609
   quadratic, 494–495
Modeling multiple births, 495
Monomials
   definition, 251
   multiplication of, 264
Motion problems, 354, 452
Multiple-angle identity, deriving, 836
Multiple angles, trigonometric equations with, 871
Multiplication
   associative property of, 33
   of binomials, 266, 406
   commutative property of, 33
   of complex numbers, 423

   FOIL method of, 266, 406
   of functions, 269, 484
   identity element for, 32
   identity property of, 32
   inverse property of, 31–32
   of matrices, 201
   of a matrix by a scalar, 199
   of monomials, 264
   of polynomial functions, 269
   of polynomials, 265
   of radical expressions, 406
   of radicals, 393
   of radicals with different indexes, 397
   of rational expressions, 324
   of real numbers, 18
   of sum and difference of two terms, 267
   using logarithms, 555
   by zero, 19, 35
Multiplication property
   of equality, 45
   of inequality, 75
   of zero, 35
Multiplicative identity, 32
Multiplicative inverse
   of matrices, 208
   of a real number, 19, 31–32
Mutually exclusive events, 930

## N

$n$ factorial, 918
Natural logarithms
   applications of, 565
   definition, 564
   evaluating, 564
Natural numbers, 2
Nautical mile, 748
Negative
   of a matrix, 198
   of a number, 6
   of a polynomial, 252
Negative angle, 660
Negative-angle identities, 810
Negative exponents
   definition, 237
   in rational expressions, 343
   rules for, 239, 241
Negative-number identities, 810
Negative root, 380
Negative slope, 121
Newton, 362, 907
Nonlinear equations, 44, 648
Nonlinear inequalities, 469
Nonlinear system of equations
   definition, 648
   elimination method for solving, 650
   substitution method for solving, 648
Not equal to
   definition, 2
   symbol for, 2
Notation
   exponential, 24, 386
   factorial, 918
   function, 153, 255
   interval, 10, 72
   inverse function, 850
   inverse trigonometric function, 850
   limit, 614
   scientific, 243
   set, 2
   set-builder, 3
   square root, 380
   subscript, 114, 197

$n$th root
   definition, 380
   exponential notation for, 386
   radical notation for, 383
Null set
   definition, 2
   notation for, 2
Number(s)
   absolute value of, 6, 382, 383
   additive inverse of, 6, 31–32
   complex, 421
   counting, 2, 5
   cubes of, 24
   exact, 720
   factors of, 23
   graph of, 4
   imaginary, 422
   integers, 4, 5
   irrational, 4, 5
   natural, 2, 5
   negative of, 6
   opposite of, 6
   ordered pair of, 108
   rational, 4, 5
   real, 4, 5
   reciprocal of, 19
   roots of, 380
   sets of, 5
   signed, 6
   square of, 24
   square roots of, 25, 380, 420
   whole, 2, 5
Number line
   coordinate of a point on, 4
   definition, 3
   distance between points on, 18
   graph of a point, 4
   intervals on, 10
Number of zeros theorem for polynomials, 596
Numerator, rationalizing, 411
Numerical coefficient, 33, 249

## O

Oblique asymptotes, 615, 616
Oblique triangle
   data required for solving, 882
   definition, 882
   solving procedures for, 883
Obtuse angle, 660
Obtuse triangle, 883
Odd function, 771
Odds in favor of an event, 929
One-to-one function
   definition, 532, 850
   horizontal line test for, 534, 850
   inverse of, 532
Operations
   on functions, 256, 484
   order of, 25–26
   on real numbers, 15–21
   on sets, 83, 84, 86
Opposite of a number, 6
Opposite of a vector, 904
Opposite side to an angle, 702
Order of a radical, 380
Order of operations, 25–26
Ordered pairs
   definition, 108
   graph of, 109
   table of, 109
Ordered triple, 182

Origin
  definition, 108
  graphing line passing through, 113
  symmetry with respect to, 512
Outcome of an experiment, 927

## P

Pairs, ordered, 108
Parabola
  axis of, 491, 493
  definition, 261, 491, 634
  directrix of, 496
  downward opening, 493
  focus of, 491, 496
  geometric definition, 496
  graph of, 490–492, 506
  horizontal, 506
  horizontal shift of, 493
  intercepts of, 503
  summary of graphs of, 507
  symmetry of, 491
  vertex formula for, 502
  vertex of, 491, 493, 500
  vertical, 491–492
  vertical shift of, 491–492
Parallel lines
  definition, 668
  transversal intersecting, 668
Parallel lines, slope of, 122
Parallelogram rule for vectors, 904
Parallelograms, properties of, 906
Parentheses, solving a formula with, 53
Percent problems
  definition, 43
  interpreting from graphs, 56
Perfect square trinomial
  definition, 299
  factoring of, 299
Perfect triangle, 903
Perigee
  definition, 365, 640
  of an ellipse, 640
Periodic functions
  definition, 770
  period of, 770
Permutations
  definition, 919
  distinguishing from combinations, 922
  formula for, 919
Perpendicular lines, slope of, 122
pH
  application of, 563
  definition, 562
Phase shift, 781
Pi ($\pi$), 4, 736
Piecewise linear function
  application of, 518
  definition, 516, 518
  graph of, 518
Plane
  coordinates of points in, 108
  definition, 182
  plotting points in, 108
Point(s)
  coordinate on a line, 4
  coordinates in a plane, 108
  of discontinuity, 620
Point-slope form, 132
Polynomial(s)
  addition of, 252
  definition, 249, 286
  degree of, 251
  in descending powers, 250
  division by a monomial, 272
  division of, 272
  evaluating by remainder theorem, 590–591
  factoring, 286, 304
  factoring by substitution, 296
  greatest common factor of, 286
  multiplication of, 265
  negative of, 252
  prime, 292
  steps to factor, 303
  subtraction of, 252
  term of, 250
  in $x$, 250
Polynomial function(s)
  addition of, 256
  approximating real zeros by calculator, 608
  boundedness theorem for, 607
  comprehensive graph of, 604
  conjugate zeros theorem for, 597
  definition, 255, 591
  of degree $n$, 255, 591
  division of, 276
  domain of, 260
  evaluating, 256
  factor theorem for, 593
  graphs of, 260, 601
  identity, 260
  intermediate value theorem for, 606
  modeling data using, 256
  multiplication of, 269
  number of zeros theorem for, 596
  range of, 260
  squaring, 261
  subtraction of, 256
  zero of, 591
Polynomial inequalities
  definition, 471
  third-degree, 471
Polynomial model, 609
Position vector, 905
Positive angle, 660
Positive root, 380
Positive slope, 121
Power rules
  for exponents, 240, 242
  for logarithms, 557
  for radical equations, 413
Powers, descending, 250
Powers of $i$
  definition, 424
  simplifying, 424
Prime polynomial, 292
Principal $n$th root, 380
Principal square root, 25, 380
Principle of counting, 917
Prism diopters, 732
Probability of alternative events, 931
Probability of an event, 927
Product
  definition, 18
  of sum and difference of two terms, 267
Product rule
  for exponents, 236, 241
  for logarithms, 555
  for radicals, 393
Properties of parallelograms, 906
Properties of probability, summary of, 932
Proportion
  definition, 352
  solving, 352
Proportional, 362
Pseudo-asymptotes, 798
Pythagorean formula, 397, 462
Pythagorean identities, 687, 810

## Q

Quadrantal angles, 663
Quadrants, 109
Quadratic equations
  applications of, 310, 462
  completing the square method for solving, 439
  definition, 307, 436
  discriminant of, 447–448, 503
  factoring method for solving, 307
  with nonreal complex solutions, 442
  quadratic formula for solving, 445
  solving with substitution method, 455
  square root property for solving, 436
  standard form of, 307, 436
  steps to solve by completing the square, 440
  steps to solve by factoring, 307
  summary of methods for solving, 451
  types of solutions of, 448
  zero-factor property for solving, 436
Quadratic formula
  definition, 445
  derivation of, 444
  solving quadratic equations using, 445
Quadratic functions
  application using, 310
  definition, 311, 464, 490–491
  general principles for graphing, 494
  graphs of, 490–492
  maximum value of, 504
  minimum value of, 504
  modeling multiple births by, 495
  steps to graph, 602
Quadratic in form equations
  definition, 454
  substitution method for solving, 455
Quadratic inequalities
  definition, 469
  graph of solutions of, 470
  steps to solve, 470
Quotient, 19
Quotient identities, 687, 810
Quotient rule
  for exponents, 239, 241
  for logarithms, 556
  of opposites, 324
  for radicals, 394

## R

Radian, 736
Radian/degree relationship
  definition, 736
  table of, 738
Radian measure
  applications of, 742
  converting to degree measure, 737
  definition, 736
Radical(s)
  addition of, 402
  conditions for simplified form, 395
  definition, 380
  equations with, 413
  index of, 380

Radical(s) (*continued*)
   multiplication of, 393
   order of, 380
   product rule for, 393
   quotient rule for, 394
   simplified form of, 395
   steps to simplify, 395
   subtraction of, 402
Radical equations
   definition, 413
   extraneous solutions of, 414
   graphing calculator method for solving, 417
   power rule for solving, 413
   steps to solve, 414
Radical expressions
   addition of, 402
   definition, 381, 402
   graphs of, 382
   multiplication of, 406
   rationalizing the denominator of, 407
   rationalizing the numerator of, 411
   simplifying, 394
   subtraction of, 402
Radical function, 382
Radical symbol, 25
Radicand, 380
Radius of a circle, 634
Range(s)
   of a function, 149
   of inverse circular functions, 856
   of a polynomial function, 260
   of a relation, 149
   of trigonometric functions, 686
Rate of change
   average, 107, 123
   definition, 7, 123
Rate of work, 356
Ratio, 352
Rational equations
   definition, 345
   domain of, 345
   with no solutions, 346
Rational exponents
   definition, 386
   evaluating terms with, 387
   radical form of, 389
   rules for, 390
Rational expressions
   addition of, 330
   addition with different denominators, 332
   applications of, 351
   definition, 320
   division of, 326
   equations with, 345
   formulas with, 351
   lowest terms of, 321
   multiplication of, 324
   with negative exponents, 343
   reciprocals of, 326
   steps to multiply, 324
   subtraction of, 330
   subtraction with different denominators, 332
Rational functions
   calculator graphing of, 621
   definition, 320, 347, 364, 613
   discontinuous, 347
   domains of, 320, 348
   graphs of, 348, 614
   with point of discontinuity, 620
   steps to graph, 617

Rational inequalities
   definition, 472
   steps to solve, 472
Rational numbers
   definition, 4, 5
   as exponents, 386
   fundamental property of, 321
Rational zeros theorem, 594
Rationalizing a binomial denominator, 410
Rationalizing the denominator, 407
Rationalizing the numerator, 411
Ray
   definition, 660
   endpoint of, 660
Real numbers
   additive inverse of, 6, 31–32
   definition, 4, 5
   graphing sets of, 10
   multiplicative inverse of, 19, 31–32
   operations on, 15–21
   properties of, 30
   reciprocals of, 19
Real part of a complex number, 421
Reciprocal
   definition, 683
   of a rational expression, 326
   of a real number, 19
Reciprocal identities, 683, 810
Rectangular coordinate system
   definition, 108
   plotting points in, 108
   quadrants of, 109
Reduced row echelon form, 196
Reference angles
   definition, 709
   table of, 709
Reference arc, 750
Reflection of a graph, 261, 510
Regions in the real number plane, 143
Relation
   definition, 148
   domain of, 149
   range of, 149
Remainder theorem, 591
Resultant vectors, 904
Richter scale, 554
Right angle, 660
Right triangle(s)
   definition, 397, 670
   hypotenuse of, 397
   legs of, 397
   solving, 721
Right-triangle-based definitions of trigonometric functions, 702
Rise, 118
Root(s)
   approximating by calculator, 383
   negative, 380
   $n$th, 380
   of numbers, 380
   positive, 380
   principal, 380
   simplifying, 380
   square, 25, 380
Root functions
   calculator graphing of, 645
   definition, 381
Row echelon form, 191
Row matrix, 197
Row operations on a matrix, 190
Rows of a matrix, 189, 200–201
Run, 118

## S

Sample space, 927
Scalar, 199
Scalar multiplication, properties of, 199
Scalar product of a vector, 904
Scalars, 903
Scalene triangle, 670
Scatter diagram, 136
Scientific notation
   application of, 245
   converting from scientific notation, 244
   converting to standard notation, 244
   definition, 243
Secant function
   characteristics of, 796
   definition, 677, 796
   domain of, 796
   graph of, 796
   period of, 796
   range of, 686, 796
   steps to graph, 798
Second, 662
Second-degree equation, 436
Sector of a circle
   area of, 744
   definition, 744
Segment of a line, 660
Semiperimeter, 385, 899
Set(s)
   definition, 2
   elements of, 2
   empty, 2
   finite, 2
   graph of, 10
   graph of real numbers of, 10
   infinite, 2
   intersection of, 83
   mapping of, 149
   members of, 2
   null, 2
   of numbers, 5
   operations on, 83, 84, 86
   solution, 44
   union of, 86
Set braces, 2
Set-builder notation, 3
Set operations, 83, 84, 86
Shrinking of a graph, 510
Side adjacent to an angle, 702
Side-Angle-Angle (SAA), 882
Side-Angle-Side (SAS), 882
Side opposite an angle, 702
Side-Side-Side (SSS), 882
Signed numbers, 6
Significant digits, 720
Signs of trigonometric function values, 684
Similar triangles
   conditions of, 671
   definition, 359, 670
Simple interest, 65, 572
Simplified form of a radical, 395
Sine function
   amplitude of, 773
   characteristics of, 771
   definition, 677, 771
   difference identity for, 827
   domain of, 771
   double-angle identity for, 833–834
   graph of, 771
   half-angle identity for, 839–840
   horizontal translation of, 781
   inverse of, 852–853